# Two-Dimensional (2D) NMR Methods

# Two-Dimensional (2D) NMR Methods

*Edited by*

*K. Ivanov*‡
*International Tomography Center, Novosibirsk, Russia*

*P.K. Madhu*
*Tata Institute of Fundamental Research, Hyderabad, India*

*G. Rajalakshmi*
*Tata Institute of Fundamental Research, Hyderabad, India*

WILEY

This edition first published 2023.


*Registered Office(s)*
John Wiley & Sons, Inc., 111 River Street, Hoboken, NJ 07030, USA
John Wiley & Sons Ltd, The Atrium, Southern Gate, Chichester, West Sussex, PO19 8SQ, UK

For details of our global editorial offices, customer services, and more information about Wiley products visit us at www.wiley.com.

Wiley also publishes its books in a variety of electronic formats and by print-on-demand. Some content that appears in standard print versions of this book may not be available in other formats.



A catalogue record for this book is available from the Library of Congress

Hardback ISBN: 9781119806691; ePDF ISBN: 9781119806707; epub ISBN: 9781119806714; oBook ISBN: 9781119806721

Cover Images: *Centre and bottom right-hand image courtesy of Nathaniel J. Traaseth. Bottom right-hand image adapted from Figure 2 from dx.doi.org/10.1021/jp303269m | J. Phys. Chem. B 2012, 116, 7138–7144; Bottom left-hand images courtesy of Pramodh Vallurupalli. Bottom left-hand image adapted from Figure 1 from DOI: 10.1039/c5sc03886c, Chem. Sci., 2016, 7, 3602*
Cover design by Wiley

Set in 9.5/12.5pt STIXTwoText by Integra Software Services Pvt. Ltd, Pondicherry, India

# Dedication

*Dedication and tribute to Kostya*

*K. Ivanov at the meeting of the Alexander von Humboldt foundation scholarship holders in Novosibirsk, 2012.*

*Our esteemed colleague and good friend Konstantin L'vovich ("Kostya") Ivanov became one of the first victims of the Covid-19 pandemic in the NMR community. He passed away at a hospital in Novosibirsk on March 5, 2021. He was not only a great scientist, but also a good human being, always sincere, honest, joyous, and considerate. In addition, he was a great citizen of the scientific community. Aside from his demanding job as the Director of the International Tomography Center (ITC), Novosibirsk, he kept his research at a very high level and organized a multitude of meetings, seminars, and webinars.*

*Alexandra Yurkovskaya met Kostya for the first time in 1998 at the ITC when he was a master student in the theoretical chemistry group headed by Nikita Lukzen. Having defended his Ph.D. thesis in 2002 at the ITC, Kostya teamed up with the experimental group of Alexandra Yurkovskaya in solving theoretical problems related to chemically induced dynamic nuclear polarization. Since 2005 both of them worked in Hans-Martin Vieth group at the Free University of Berlin as part of the large EU project "Bio-DNP" under the leadership of Thomas Prisner. In 2007, Kostya became a fellow of the Alexander von Humboldt Foundation at the Free University of Berlin and*

*Alexandra had a Marie Curie International fellowship with Hans-Martin Vieth as the host professor. The scientific collaboration had since continued for almost 20 years and led to over 100 joint publications on nuclear hyperpolarization and polarization transfer processes, combining modern theoretical and sophisticated experimental techniques. In particular, over the last decade, Alexey Kiryutin from the Yurkovskyaya group at the ITC pushed the limits by developing unique high-resolution fast field-cycling NMR apparatus over nine orders of magnitude of magnetic field. Kostya proposed multiple smart applications of the technique to hyperpolarization and relaxation, establishing Kostya as one of the leaders of a new scientific direction. In 2007, Kostya defended his second doctoral thesis, which is the second scientific degree in Russia, awarded for a significant and sustainable contribution to scientific knowledge, which is analogous to the habilitation degree in Germany. The thesis title was: "Kinetics of multi-stage liquid-phase processes involving particles with spin degrees of freedom". He completed his degree at the Institute of Chemical Physics in Moscow, which is the leading institute of the Russian Academy in the field of chemical physics. His scientific work was honored with several important prizes and awards, including the Medal of the European Academy of Science in 2010, Voevodsky prize in 2012, a Fellowship of the Japanese Society for Promotion of Science in 2016, and the Laukien Prize in 2020 for his contribution to the SABRE research field.*

*The cooperation between French (ENS) and Russian (ITC) groups doing fast field-cycling NMR started in 2017 with a visit of Kostya to École Normale Supérieure (ENS, Paris) organized by Geoffrey Bodenhausen. This led to several collaborations with Bodenhausen, Daniel Abergel, and Fabien Ferrage on topics as diverse as long-lived states, mechanisms of dynamic nuclear polarization, and coherence effects in field-cycling experiments. A comparative study of ZULF spectroscopy with NMR detection using an atomic magnetometer and inductive detection at high magnetic field was inspired by Kostya's tutorial talk at the ZULF seminar organized by the Marie Curie ZULF-NMR Innovative Training Network in Mainz in the fall of 2019. During those several unforgettable days, followed by fantastic musical evenings including Kostya and Dima Budker singing together with a band formed by Dima's research group members, the atmosphere of warm mutual cooperation was created and the fruitful although dramatically short joint work started.*

*Dima Budker's acquaintance and collaboration with Kostya were both relatively short, some two-three years, depending on where one marks the beginning. Nevertheless, this collaboration, consisting of several joint projects that have resulted in at least five joint papers and book chapters, has had a profound effect on Dima's research interests. In most cases, the collaboration was initiated by Kostya and was centered around an idea formulated by him with such clarity and enthusiasm that it was impossible not to embrace the project wholeheartedly.*

*This was also the case with the chapter for this 2D-NMR book. On the February 3, 2021, Dima received an e-mail from Kostya that said: "Madhu and myself have got an offer from Wiley to edit a book on 2D-NMR. As a part of this initiative, I want to cover some aspects of field-cycling NMR and zero-field NMR. Although 2D methods are not that widespread in these branches of NMR, they are used as well and can be of interest to NMR people. Would you like to write such a chapter together with me and Fabien Ferrage, with whom I cooperate on field-cycling NMR?" This was followed by a detailed outline of the chapter proposed by Kostya. Kostya and Dima then met on Skype to discuss this suggestion, including the scope and whom to invite as co-authors. Their last communication was on the February 11, 2021, and they agreed to talk again very soon... That conversation was not destined to occur.*

*Fabien Ferrage's first acquaintance with Kostya was by reading his work, particularly the series of articles he published with Hans-Martin Vieth, Rob Kaptein, and Alexandra Yurkovskaya on level anti-crossings and coherent effects in field-cycling experiments. As Fabien was making his first steps in relaxometry, these were enlightening contributions that defined very clearly what he should not do if he wanted to measure pure relaxation rates at low fields. Geoffrey Bodenhausen and Fabien discussed several of Kostya's articles before meeting him, often with praise, in particular the introduction of SABRE at high field. Fabien met Kostya in person in May 2015, when Kostya came to Paris for a COST meeting on hyperpolarization. Never wasting any time, Kostya came to ENS to visit the lab and give a seminar during the lunch break of the meeting. Over the years, several meetings and an invited professorship at ENS (an honour rarely given to scientists of such a young age), Kostya and Fabien discussed many topics on field cycling, relaxometry, and hyperpolarization. If Kostya was always asking for chemical insight from Fabien side, Fabien was always impressed by Kostya's command of theory, in quantum mechanics in general as well as, of course, in both nuclear and electron magnetic resonance, the latter subject being invariably challenging for many NMR specialists. The collaboration between Fabien and Kostya really took off when Ivan Zhukov, a talented graduate student supervised by Alexandra Yurkovskaya, one of Kostya's mentors, came to work on field cycling with the Ferrage group in 2018. As often happens, Ivan's work with at ENS ended up being guided by serendipity and led us to an investigation of interesting effects of scalar couplings at low fields discussed in this book. This work inspired Kostya to propose the ZULF-TOCSY experiment, such a beautiful idea, that they have just started exploring.*

*Fabien Ferrage recalls: Kostya was a true force, initiating and leading many projects with intelligence, ambition and intensity while, at the same time, being the nicest, kindest and most generous colleague. Such a rare combination. Kostya and I were born the same year. His daughter and my son are about the same age. I considered him as a compass, setting a path to inspire others. In February 2021, we exchanged about the scope of this book chapter. One of the last emails I received from Kostya, on February 15, 2021, was about a conversation he had had with Dmitry Budker to write this book chapter together, which I am glad we accomplished. On March 5, my colleague Daniel Abergel called me to inform me that Kostya had died. The compass was broken, but the inspiration lives on.*

*Madhu recollects his association with Kostya which started in 2016 at the Windischleuba School on NMR, a series being organised by Joerg Matysik: We had been regular teachers in this workshop series in 2017, 2018, and 2020. Somewhere in 2018, we started discussing the prospects of writing a book on solid-state NMR. We met up in Leipzig in June 2019 for a discussion regarding this, having done some amount of writing already. During this meeting Kostya had this wonderful spark and told me that why do not we write a review on Floquet theory with new ideas embedded and with more examples than existing reviews. This review kept us busy till he passed away when we were refining it after comments from the editors. In fact, we in between took up the assignment with Wiley to bring out this book and Kostya was, as usual, very enthusiastic and full of ideas about what to be included and how they should be. Writing the review with him, in particular, was very rewarding. I remember the large number of extensive discussions we had from which I learned quite a bit. We actually had the Skype window always active in our computers, and either of us would call the other informally in case of any questions. We were also discussing some of the solid-state NMR experiments that my group was carrying out which were the topics of discussion just before he got hospitalized. Needless to say, the other major thing Kostya did for the magnetic resonance community was starting the Friday Intercontinental Seminar Series on April 8, 2020 (which was a Wednesday). A week before that as we were talking about the review, he as usual wanted to do something for the community. We got in touch with Daniel Abergel and Gerd Buntkowsky and the seminar series has been running since then smoothly with great talks. As Kostya wanted, it combines talks by senior and younger researchers. He also started the ICONS conference in 2020 and just before he passed away, we had the second of that meeting. ICONS conferences have been continuing since then on a regular basis with the third held in September 2021, the fourth in February 2022, the fifth in August-September 2022, and the sixth in January 2023. All these clearly reveal the vitality in Kostya with a sharp eye for details. Besides science, we had fun discussing politics, literature, music, and many other things.*

*Muslim Dvoyashkin recalls the first meeting with Kostya on the way to Windischleuba NMR school in 2020, picking up Kostya at the train station in Leipzig. Kostya made the first impression of a very intelligent and very modest person, which later turned out to be quite true. As his countryman, it was very interesting for Muslim to find out how he and his colleagues manage to remain at the forefront of fundamental science given its very limited financial support in Russia. Therefore, Kostya will remain in Muslim's memory as a real hero, and thanks to whom, many students were able to find their way to the academia later on.*

*Malcolm Levitt recalls that his close association with Kostya started around 2017 when Malcolm expressed his appreciation of Kostya's conference talk and his insightful theoretical approach. This led to extensive discussions and the identification of various possible themes for collaboration. Vigorous discussions led to several collaborative papers on the theme of singlet NMR. In 2019 Malcolm visited Novosibirsk and had the pleasure of getting to know*

*Kostya and his colleagues, in their "natural habitat". It was a very pleasant and instructive visit. Malcolm's most striking impression of Kostya was how he managed this large and prestigious institute, full of competing egos, with the minimum of perceptible friction, and always with remarkable good humour. He was basically an extremely rare person who combined refined political skills with a sharp and creative scientific mind as well as possessing remarkable patience and energy, a great sense of humour, and legendary powers of concentration. His loss is extraordinarily tragic.*

*We shall deeply miss Kostya as an exceptional human being. He was a creative and rigorous scientist, a generous and attentive friend, and a considerate and eminently civilized colleague. We dedicate our contributions to him and the whole book itself, which was initiated by Kostya, that will be a brilliant scientific testament to his memory.*

# Contents

# List of Contributors

***John W. Blanchard***
Quantum Technology Center, University of Maryland, Maryland, USA

***Rolf Boelens***
Department of Chemistry, Utrecht University, The Netherlands

***Dmitry Budker***
Helmholtz Institut Mainz, Johannes Gutenberg-Universität Mainz, Germany

Department of Physics, University of California, Berkeley, California, USA

***Muslim Dvoyashkin***
Institute of Chemical Technology, Leipzig University, Leipzig, Germany

***Matthias Ernst***
Physical Chemistry, ETH Zürich, Switzerland

***Fabien Ferrage***
Laboratoire des Biomolécules, LBM, Département de chimie, École normale supérieure, PSL University, Sorbonne Université, Paris, France

***Roberto R. Gil***
Department of Chemistry, Carnegie Mellon University, Pittsburgh, USA

***Boris Gouilleux***
Université Paris-Saclay, laboratoire ICMMO, Orsay, France

***K. Ivanov***
International Tomography Center SB RAS, Novosibirsk, Russia

***Alexej Jerschow***
Department of Chemistry, New York University, New York, USA

***Harindranath Kadavath***
Laboratory of Physical Chemistry, ETH Zurich, Switzerland

***Robert Kaptein***
Department of Chemistry, Utrecht University, The Netherlands

***Pawel Kasprzak***
Faculty of Physics, University of Warsaw, Warsaw, Pasteura, Poland

Centre of New Technologies, University of Warsaw, Warsaw, Poland

***Krzysztof Kazimierczuk***
Faculty of Physics, University of Warsaw, Warsaw, Poland

***Alexey Kiryutin***
International Tomography Center SB RAS, Novosibirsk, Russia

***David Korenchan***
Athinoula A. Martinos Center for Biomedical Imaging, Massachusetts General Hospital Boston, Massachusetts, USA

***Michael Kovermann***
Department of Chemistry, Universität Konstanz, Konstanz, Germany

Graduate School Chemical Biology KoRS-CB, Universität Konstanz, Konstanz, Germany

***Philippe Lesot***
Université Paris-Saclay, RMN en Milieu Orienté, France

***Malcolm H. Levitt***
Department of Chemistry, University of Southampton, Southampton, UK

***P.K. Madhu***
Department of Chemical Sciences, Tata Institute of Fundamental Research Hyderabad, India

***Vladislav Orekhov***
Department of Chemistry and Molecular Biology, University of Gothenburg, Gothenburg, Sweden

***Andrew J. Pell***
CRMN, Centre de RMN à Très Hauts Champs de Lyon Université de Lyon, Villeurbanne, France

***Guido Pintacuda***
CRMN, Centre de RMN à Très Hauts Champs de Lyon Université de Lyon, Villeurbanne, France

***Roland Riek***
Laboratory of Physical Chemistry, ETH Zürich, Switzerland

***Thomas Robinson***
CRMN, Centre de RMN à Très Hauts Champs de Lyon Université de Lyon, Villeurbanne, France

***Kevin J. Sanders***
CRMN, Centre de RMN à Très Hauts Champs de Lyon Université de Lyon, Villeurbanne, France

***Tobias Schneider***
Department of Chemistry, Universität Konstanz, Konstanz, Germany

Graduate School Chemical Biology KoRS-CB, Universität Konstanz, Konstanz, Germany

***Monika Schönhoff***
Institute of Physical Chemistry, University of Münster, Münster, Germany

***Ashok Sekhar***
Molecular Biophysics Unit, Indian Institute of Science, Bengaluru, Karnataka, India

***Tobias Sjolander***
Department of Physics, University of Basel, Klingelbergstrasse 82, Basel, Switzerland

***Ville-Veikko Telkki***
NMR Research Unit, University of Oulu, Oulu, Finland

***Nathaniel J. Traaseth***
Department of Chemistry, New York University, New York, USA

***Pramodh Vallurupalli***
Tata Institute of Fundamental Research Hyderabad, Hyderabad, India

***Alexandra Yurkovskaya***
International Tomography Center SB RAS, Novosibirsk, Russia

***Yvan Zhukov***
International Tomography Center SB RAS, Novosibirsk, Russia

# Preface

NMR spectroscopy has grown tremendously since its inception in the 1940's when NMR signals were first recorded by Felix Bloch and Edward Purcell. The developments in the field have been fuelled by advancements in various fields, such as, superconducting magnets, spectrometer hardware and software, isotope labelling, and the theory behind understanding the various NMR experiments. The applications have been vast from small molecules, materials, structure and dynamics of biomolecular complexes to medical applications of NMR and MRI. NMR has become a standard characterisation tool in many Chemistry laboratories and widely used in various industrial applications, for example, in the pharmaceutical industry.

One of the most important milestones in the time line of NMR is the development of 2D NMR methods, which provide an elegant way to get around the problem of "spectral crowding" and provide versatile and highly detailed information about molecular structure and dynamics through correlation spectroscopy. This book attempts to give a rigorous account of the basic concepts in 2D NMR methods, both in terms of theory and applications. Whilst the list of content covered here is not exhaustive, we believe it certainly gives the reader a good flair and comprehension of the field and necessary background to grasp more advanced topics, including higher-dimensional NMR methods.

The book has Chapters on the basics of NMR, analysis of pulse sequences, zero-to ultralow-field NMR, NMR methods in biomolecules in solution state, methods on data processing, pushing the frontiers of resolution, and speeding up of 2D data acquisition, diffusion spectroscopy, NMR of anisotropic and paramagnetic systems, and chemical exchange. We have also included a Chapter as Appendix that we hope will complement many of the other Chapters, and in particular Chapters 11 and 12. The Chapters are all written by leading researchers in the field who have also contributed significantly to the topic of their respective Chapter. We hope that this book will be useful to established as well as beginning researchers who want to explore the benefits of NMR spectroscopy and new areas of its multidimensional facets.

We thank all the authors who have contributed to this book and also in a timely fashion. The significant amount of care that has gone into each Chapter merits attention and deep acknowledgement. We also thank the people at Wiley for putting up this book very nicely, some of them being, Jenny Cossham, Sarah Higginbotham, Elke Morice-Atkinson, and Richa John.

Finally, this book also has a tribute section written for Kostya (K. Ivanov) by some of the authors. The concept of this book was Kostya's idea, however, unfortunately he passed away on March 5, 2021, at the young age of 44. A great scientist, a fantastic human being, we dedicate this book to his memories.

# 1

# Basics of Two-dimensional NMR

*Malcolm H. Levitt*

*Department of Chemistry, University of Southampton, SO17 1BJ, Southampton, UK*

## 1.1 Introduction

### 1.1.1 Time-domain NMR

The introduction of pulse-Fourier transform (pulse-FT) NMR in 1966 by Ernst and Anderson [1] represented a paradigm shift, not only in nuclear magnetic resonance but also in many other forms of spectroscopy. Prior to this seminal experiment, there were two forms of NMR, which were generally viewed as being quite distinct and practiced mainly by chemists on the one hand, and physicists on the other.

Chemical applications of NMR spectroscopy used a "continuous-wave" (cw) method, in which chemical shifts and spin-spin couplings were probed, either by (i) varying the frequency of applied radiofrequency irradiation and detecting a change in the nuclear magnetic response when the applied frequency matches a nuclear energy level spacing, or (ii) by applying radiofrequency irradiation of fixed frequency and varying the applied magnetic field while monitoring the response. For technical reasons, the latter method (fixed frequency, variable field) was easier to perform and more common. A residue of this historical method still persists in the common nomenclature of modern NMR, where the terms "high field" and "low field" are still often used (despite a jarring logical inconsistency) to characterize nuclei in electron environments, which are relatively weakly shielded from the external magnetic field (low field!), and for those that are relatively strongly shielded from the external magnetic field (high field!).

In 1966, many important time-domain NMR experiments and phenonema also existed, but they were largely developed and used by physicists. These include the spin echo [2], the modulation of spin echoes by spin-spin couplings [3], the use of spin echoes to probe molecular diffusion in a field gradient [3], the development of multiple spin echoes [3, 4], and Hartmann-Hahn cross-polarization between different nuclear species in solids [5].

The introduction of pulse-FT NMR established a permanent link between the time-domain "physics" phenomena and the continuous-wave "chemistry" procedures.

The perturbation of nuclear spins by a strong, short, resonant radiofrequency pulse elicits an extended time-domain electrical response termed a *free-induction decay*, following Bloch [6]. Ernst realized that Fourier transformation of the time-domain NMR signal $s(t)$ yields a frequency-domain function $S(\omega)$, which under certain assumptions, is identical to the NMR spectrum generated far more slowly by a continuous-wave frequency-domain NMR experiment:

*Two-Dimensional (2D) NMR Methods*, First Edition. Edited by K. Ivanov, P.K. Madhu and G. Rajalakshmi.

$$S(\omega) = \int_0^\infty s(t) \exp -i\omega t \mathrm{d}t. \tag{1.1}$$

In one stroke, Ernst unified time-domain NMR, with its central object the free-induction decay $s(t)$, and frequency-domain NMR, with its central object the NMR spectrum at fixed field $S(\omega)$. Furthermore, this ground-breaking theoretical unification was accompanied by a large increase in signal-to-noise ratio, of great practical importance. The timing of this advance was exceptionally favorable. Practical application of the Fourier transform requires numerical computation. Sufficiently powerful computational hardware was becoming widely available, and furthermore, a fast algorithm for numerical Fourier transformation had just been developed [7]. As they say, the rest is history. NMR was revolutionized, with a huge impact on many other sciences, as recognized by Ernst's Nobel prize in 1991. It was probably one of the least contentious Nobel prizes in history.

### 1.1.2 Hans Primas and the "Correlation Function of the Spectrum"

The seminal contribution of Ernst and Anderson is well known to most NMR spectroscopists. Less well known is that an Equation very similar to Equation 1.1 had been published by a close colleague of Ernst at ETH-Zürich a few years earlier. This colleague was Hans Primas, who enjoyed a particularly remarkable career. Primas was a self-taught genius who rose to become a Professor of Physical and Theoretical Chemistry at the ETH-Zürich and an authority in quantum theory and the philosophy of science, despite the absence of secondary school education or a university degree of any kind [8]. The 1963 paper by Banwell and Primas [9] introduces a quantity called the "correlation function of the spectrum" and given by

$$K(t) = \int_{-\infty}^{\infty} S(\omega) \exp +i\omega t \mathrm{d}\omega \tag{1.2}$$

(omitting an unimportant normalization factor). Since the Fourier transform is reversible, Equations 1.1 and 1.2 are entirely equivalent (overlooking a technical difference in integration limits). Banwell and Primas introduced the term $K(t)$ as an object of theoretical interest and presumably missed its significance as being identical to the free-induction decay $s(t)$ generated by a nuclear spin system subjected to pulse excitation – and was certainly not aware of its practical significance.

I mention Primas' work not to disparage Ernst's achievement – on the contrary, I feel that the work by Primas and colleagues helps us understand better the historical provenance of Ernst's insight. Ernst's skill was in marrying the contemporary thinking at the ETH-Zürich with the highly practical motivation of Anderson at the Varian Corporation, with whom Ernst developed the concept and application of FT-NMR [1].

Remarkably, Equation 1.2 is far from being the only important insight in reference [9]. This paper also introduces the concepts of superoperators and Liouville space and also anticipates Cartesian product operator techniques, brought to fruition by Sørensen et al. [10] and other groups nearly 20 years later (see Chapter 3).

## 1.2 Spin Dynamics

### 1.2.1 Density Operator

Time-domain Fourier NMR is most conveniently analyzed by the modern operator description of NMR, as expounded in the textbook by Ernst, Bodenhausen, and Wokaun [11].

The quantum state of the nuclear spins in the sample is described by a spin density operator $\rho$, defined:

$$\rho = \overline{|\psi\rangle\langle\psi|} \tag{1.3}$$

where $|\psi\rangle$ is the spin state of an individual spin system and the overbar indicates an ensemble average.

### 1.2.2 Spin Hamiltonian

In many cases, the spin Hamiltonian $\mathcal{H}$ may be written as the sum of a *coherent* term, which is identical for all members of the spin ensemble, and a *fluctuating term*, which is strongly time-independent, has a different value for different ensemble members, and a zero average over the ensemble:

$$\mathcal{H} = \mathcal{H}_{\text{coh}} + \mathcal{H}_{\text{fluc}}(t). \tag{1.4}$$

The coherent part of the spin Hamiltonian $\mathcal{H}_{\text{coh}}$ contains terms, which are responsible for major features of the NMR spectrum such as chemical shifts, spin-spin couplings, and residual dipolar couplings in anisotropic phase [12].

In general, the coherent spin Hamiltonian $\mathcal{H}_{\text{coh}}$ possesses a set of $N_H$ eigenstates $|r\rangle$ and eigenvalues $\omega_r$, satisfying the following eigenequation:

$$\mathcal{H}_{\text{coh}}\,|r\rangle = \omega_r\,|r\rangle\,. \tag{1.5}$$

When multiplied by $\hbar$, the eigenvalues $\omega_r$ correspond to the quantum energy levels of the spin system.

The fluctuating part of the spin Hamiltonian $\mathcal{H}_{\text{fluc}}(t)$ is responsible for *dissipative* phenomena, including relaxation processes, which bring the spin system back to equilibrium with the molecular environment as well as important dissipative effects such as the nuclear Overhauser effect, which underpins many applications of NMR to structural biology [13].

### 1.2.3 Liouville Space

The principles of time-domain NMR are expressed most compactly in *Liouville space*, meaning the space of all orthogonal operators for the spin system [11, 14, 15]. The dimension of this space is $N_L = N_H^2$.

A convenient basis for Liouville space is given by all possible products of the kets and bras of the coherent Hamiltonian eigenstates, each operator having the form

$$C_{rs} = |r\rangle\,\langle s|\,. \tag{1.6}$$

The $N_H$ operators with $r = s$ are known as *population operators*:

$$P_r = C_{rr} = |r\rangle\,\langle r|\,. \tag{1.7}$$

The $N_H(N_H - 1)$ operators $C_{rs}$ with $r \neq s$ are known as *coherence operators*.

In this representation, the density operator may be written as a $N_L$-dimensional vector, termed a *Liouville ket* $|\rho)$, as follows:

$$|\rho) = \begin{pmatrix} \rho_{11} \\ \rho_{12} \\ \rho_{13} \\ \vdots \\ \rho_{1N_H} \\ \rho_{21} \\ \rho_{22} \\ \rho_{23} \\ \vdots \\ \rho_{2N_H} \\ \vdots \\ \rho_{N_H N_H} \end{pmatrix}. \tag{1.8}$$

The $N_H$ diagonal elements of the density operator, called *populations*, are defined by

$$\rho_{rr} = \langle r|\,\rho\,|r\rangle = \left(P_r\middle|\rho\right) \tag{1.9}$$

where the *Liouville bracket* is defined [11, 14]:

$$\left(A\middle|B\right) = \mathrm{Tr}\{A^{\dagger}B\}. \tag{1.10}$$

The $N_H(N_H - 1)$ off-diagonal elements of the density operator, called *coherences*, are defined by

$$\rho_{rs} = \langle r|\,\rho\,|s\rangle = \left(C_{rs}\middle|\rho\right) \tag{1.11}$$

where, by symmetry, $\rho_{sr} = \rho^*_{rs}$.

In a high magnetic field and in the absence of additional applied fields, the *secular approximation* may be made. This implies that all components of the spin Hamiltonian are neglected if they do not commute with the total angular momentum operator along the static magnetic field. In these circumstances the coherence operators obey the following eigenequation:

$$\hat{I}_z\,|C_{rs}\rangle = p_{rs}\,|C_{rs}\rangle \tag{1.12}$$

where $\hat{I}_z$ is the *commutation superoperator* of the total angular momentum operator $I_z$ in the field direction (by convention the z-axis), defined by the following property:

$$\hat{I}_z\,|A\rangle = |\,[I_z,\,A]\,\rangle = |I_zA - AI_z\rangle\,. \tag{1.13}$$

The eigenvalues $p_{rs}$ in Equation 1.12 are real integers called *coherence orders*. Coherence orders play a prominent role in the principles of 2D spectroscopy (see Chapter 5).

### 1.2.4 Liouvillian

The equation of motion of the density operator is a first-order differential equation

$$\frac{\mathrm{d}}{\mathrm{d}t}\,|\rho(t)\rangle = \hat{L}\,|\rho(t)\rangle \tag{1.14}$$

where the superoperator $\hat{L}$ is called the *Liouvillian*. Under suitable conditions the Liouvillian may be written as the sum of two terms:

$$\hat{L} = -i\hat{\mathcal{H}}_{\mathrm{coh}} + \hat{\Gamma} \tag{1.15}$$

where $\hat{\mathcal{H}}_{\mathrm{coh}}$ is the commutation superoperator of the coherent Hamiltonian, defined as follows:

$$\hat{\mathcal{H}}_{\mathrm{coh}}\,|A\rangle = |\,[\mathcal{H}_{\mathrm{coh}},\,A]\,\rangle = |\mathcal{H}_{\mathrm{coh}}A - A\mathcal{H}_{\mathrm{coh}}\rangle \tag{1.16}$$

and $\hat{\Gamma}$ is the relaxation superoperator, representing the dissipative behavior of the spin system, generated by the fluctuating Hamiltonian $\mathcal{H}_{\mathrm{fluc}}$. If necessary, the effects of chemical exchange, diffusion, molecular transport, and mechanical motion may also be included in the Liouvillian by using an extended Liouville space (see Chapters 7 and 15) [11, 16].

The construction of the relaxation superoperator $\hat{\Gamma}$ from the correlation function of the fluctuating Hamiltonian $\mathcal{H}_{\mathrm{fluc}}$ occupies a large body of theory by itself, which is beyond the scope of this chapter (see Chapter 4). Lindbladian techniques may be used to construct $\hat{\Gamma}$ in such a way that the approach to thermal equilibrium is treated correctly, even for spin systems, which are far from equilibrium [15, 17].

In those cases that do not require an extended Liouville space, the Liouvillian superoperator may be represented by a $N_L \times N_L$ matrix.

In general, the Liouvillian has $N_L$ *right eigenkets* or *right eigenoperators* denoted $\left|{}^R Q_\ell\right\rangle$, and $N_L$ *left eigenbras* or *left eigenoperators* denoted $\left\langle{}^L Q_\ell\right|$, with the following properties:

$$\begin{aligned}\hat{L}\left|{}^R Q_\ell\right\rangle &= \Lambda_\ell \left|{}^R Q_\ell\right\rangle \\ \left\langle{}^L Q_\ell\right| \hat{L} &= \Lambda_\ell^* \left\langle{}^L Q_\ell\right|\end{aligned} \tag{1.17}$$

where the index $\ell$ takes values $\{1, 2 \ldots N_L\}$ and $\Lambda_\ell$ are the Liouvillian eigenvalues. By convention, the left and right eigenoperators are normalized:

$$\left({}^L Q_\ell\middle|{}^L Q_\ell\right) = \left({}^R Q_\ell\middle|{}^R Q_\ell\right) = 1. \tag{1.18}$$

where

$$\begin{aligned}\left\langle{}^R Q_\ell\right| &= \left|{}^R Q_\ell\right\rangle^\dagger . \\ \left|{}^L Q_\ell\right\rangle &= \left\langle{}^L Q_\ell\right|^\dagger\end{aligned} \tag{1.19}$$

The left and right eigenkets are orthogonal to each other:

$$\left({}^L Q_\ell\middle|{}^R Q_{\ell'}\right) = \delta_{\ell\ell'} \tag{1.20}$$

where the Kronecker delta $\delta_{ab}$ is equal to 1 for $a = b$, and 0 otherwise. However, the left and right eigenoperators are not, in general, mutually adjoint:

$$\left\langle{}^L Q_\ell\right| \neq \left|{}^R Q_\ell\right\rangle^\dagger . \tag{1.21}$$

The inequality in Equation 1.21 may only be replaced by an equality in the case that relaxation is ignored, or the environmental temperature is assumed to be infinite. In order to maintain generality, the left and right eigenoperators are kept distinct in this chapter.

The Liouvillian eigenvalues are complex in general, and may be written

$$\Lambda_\ell = -\lambda_\ell + i\omega_\ell \tag{1.22}$$

where $\lambda_\ell$ and $\omega_\ell$ are real, and $\lambda_\ell \geq 0$. For each eigenoperator $\left|{}^R Q_\ell\right\rangle$, the quantity $\omega_\ell$ represents its oscillation frequency, and $\lambda_\ell$ its decay rate constant.

Eigenoperators $\left|{}^R Q_\ell\right\rangle$ with real eigenvalues represent particular configurations of the spin state populations that decay monotonically and exponentially, without oscillating, as the evolution proceeds. Eigenoperators $\left|{}^R Q_\ell\right\rangle$ with complex eigenvalues, on the other hand, oscillate while decaying. The oscillation frequency of an eigenoperator $\left|{}^R Q_\ell\right\rangle$ is given by the imaginary part $\omega_\ell$ of the eigenvalue $\Lambda_\ell$. Their decay rate constant is given by the quantity $\lambda_\ell$, which is minus the real part of the eigenvalue $\Lambda_\ell$.

In most cases, an eigenoperator $\left|{}^R Q_\ell\right\rangle$ with a complex eigenvalue is very close to a coherence operator $|C_{rs}\rangle$, with $r \neq s$. The eigenfrequency $\omega_\ell$ of such an operator is (minus) the difference in eigenvalues between the relevant Hamiltonian eigenstates [9]:

$$\begin{aligned}\left|{}^R Q_\ell\right\rangle &\simeq |C_{rs}\rangle \\ \omega_\ell &\simeq -(\omega_r - \omega_s).\end{aligned} \tag{1.23}$$

The correspondence in Equation 1.23 is not exact in general. This is because the superoperators $\hat{\mathcal{H}}_{\text{coh}}$ and $\hat{\Gamma}$ do not always commute. This implies that new properties may emerge when the two terms are combined. For example, the relaxation properties of the spin ensemble may be profoundly modified by changing the coherent Hamiltonian, even when the relaxation superoperator $\hat{\Gamma}$ is left untouched. One example is the emergence of slowly relaxing long-lived states under radiofrequency irradiation [18, 19]. It is also possible for the relaxation superoperator to effectively modify the coherent spin Hamiltonian. This happens, for example, when rapidly relaxing spins effectively "self-decouple" from the rest of the spin system. Numerous intermediate cases are known.

### 1.2.5 Propagation Superoperator

Consider two different time points $t_a$ and $t_b$, where $t_b \geq t_a$. The density operator at the later time point $t_b$ may be deduced from its value at an earlier time point $t_a$ by applying the *propagation superoperator*, denoted $\hat{V}$:

$$|\rho(t_b)\rangle = \hat{V}(t_b, t_a)\,|\rho(t_a)\rangle\,. \tag{1.24}$$

If the Liouvillian is time-independent, the propagation superoperator is given by

$$\hat{V}(t_b, t_a) = \exp \hat{L}\tau_{ba} \tag{1.25}$$

where the time interval is $\tau_{ba} = t_b - t_a$. The propagator $\hat{V}$ may be written in terms of the left and right eigenoperators of the Liouvillian as follows:

$$\hat{V}(t_b, t_a) = \sum_{\ell=1}^{N_L} |{}^R Q_\ell\rangle\langle{}^L Q_\ell| \exp \Lambda_\ell \tau_{ba}. \tag{1.26}$$

## 1.3 One-dimensional Fourier NMR

### 1.3.1 The One-dimensional NMR Experiment

A general 1D NMR experiment consists of two main sections (see Figure 1.1): (i) a *preparation sequence*, in which observable spin coherences are prepared by applying a sequence of resonant radiofrequency fields to a spin system in some initial state and (ii) a *detection* interval, during which the NMR signal is observed. In the simplest cases, the detection of the NMR signal is carried out while the spin system evolves under a time-independent Hamiltonian $\mathcal{H}_{\text{det}}$.

Figure 1.1 indicates the timeline of a 1D NMR experiment. Although it is common to define the time origin $t = 0$ as the start of the preparation sequence, a more consistent theoretical description is achieved by setting the time origin $t = 0$ to the start of the *detection* interval, since this conforms to the definition of the Fourier transform, Equation 1.1. If the preparation sequence has duration $\tau_{\text{prep}}$, this implies that the preparation sequence extends from the initial time point $t = -\tau_{\text{prep}}$ to the end of preparation and start of detection at time point $t = 0$, as shown in Figure 1.1.

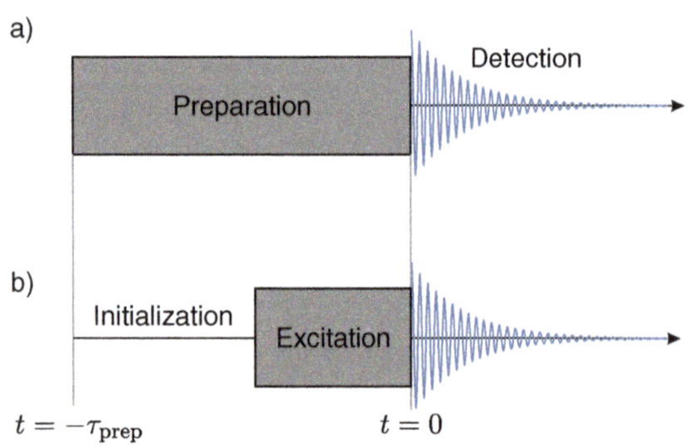


**Figure 1.1** (a) A general 1D NMR procedure, consisting of a *preparation* sequence and a *detection* interval. (b) In many 1D experiments, the preparation sequence consists of *initialization* followed by *excitation*.

Although the precise definition of the time origin is often unimportant for NMR experiments on static samples, a higher level of rigor is often required when treating experiments on moving samples, such as in magic-angle-spinning solid-state NMR [20].

#### 1.3.1.1 Preparation

In many cases, the preparation sequence itself consists of two parts. First, a reproducible initial condition is established by an *initialization* sequence. This establishes an initial density operator $|\rho_{\mathrm{ini}}\rangle$. A sequence of radiofrequency fields, called an *excitation sequence* is then applied, which generates the spin coherences, which induce an NMR signal in the subsequent detection interval.

The density operator at the end of the preparation sequence and start of detection (time point $t = 0$) is therefore given by

$$|\rho(0)\rangle = \hat{V}_{\mathrm{exc}}\,|\rho_{\mathrm{ini}}\rangle \tag{1.27}$$

where $\hat{V}_{\mathrm{exc}}$ is the propagation superoperator for the excitation sequence.

##### *Initialization*

**Thermal equilibrium** Most NMR experiments employ an initial condition $|\rho_{\mathrm{ini}}\rangle$ corresponding to thermal equilibrium of the spin system at the sample temperature $T$:

$$|\rho_{\mathrm{ini}}\rangle = |\rho_{\mathrm{eq}}(T)\rangle \tag{1.28}$$

where the thermal equilibrium state is given by the Boltzmann distribution:

$$|\rho_{\mathrm{eq}}(T)\rangle = \frac{|\exp -\hbar\mathcal{H}_{\mathrm{lab}}/(k_B T)\rangle}{\mathrm{Tr}\{\exp -\hbar\mathcal{H}_{\mathrm{lab}}/(k_B T)\}}. \tag{1.29}$$

Here $k_B$ is the Boltzmann constant, $T$ is the temperature of the sample, and $\mathcal{H}_{\mathrm{lab}}$ is the laboratory-frame spin Hamiltonian, which is often dominated by the interaction of the nuclear spins with the strong applied magnetic field.

The thermal equilibrium state in Equation 1.29 may be established by simply allowing the sample to rest in the magnetic field for a sufficiently long time.

In most NMR experiments, the *high-temperature approximation* may be made, allowing the exponential factors in Equation 1.29 to be cut off after the first two terms:

$$|\rho_{\mathrm{eq}}(T)\rangle \simeq N_H^{-1}\,|1 - \hbar\mathcal{H}_{\mathrm{lab}}/(k_B T)\rangle. \tag{1.30}$$

It is also common to apply the *high-field approximation* in which it is assumed that the Zeeman interaction with the applied magnetic field dominates the laboratory-frame Hamiltonian, allowing this equation to be simplified further:

$$|\rho_{\mathrm{eq}}(T)\rangle \simeq N_H^{-1}\left|1 - \sum_I \frac{\hbar\omega_I^0}{k_B T} I_z\right\rangle \tag{1.31}$$

where the sum is taken over all isotopic species $I$ in the spin system, and the Larmor frequency of each species is given by

$$\omega_I^0 = -\gamma_I B^0 \tag{1.32}$$

where $\gamma_I$ is the magnetogyric ratio of species $I$ and $B^0$ is the static magnetic field.

Equation 1.31 is often stripped down to its barest bones by ignoring the numerical factors and the unity operator and concentrating on the thermal equilibrium polarization of only a single spin species, allowing the use of a highly simplified form of the thermal equilibrium density operator:

$$|\rho_{\text{eq}}\rangle \sim |I_z\rangle . \tag{1.33}$$

Although commonly used in the description of NMR experiments, the use of Equation 1.33 is hazardous, since the operator $I_z$ does not fulfil the conditions of a valid density operator (trace of unity), and the series of approximations break down for highly polarized spin systems, or systems at low spin temperature. Nevertheless, the great convenience of Equation 1.33 often trumps most reservations about its validity.

**Hyperpolarization** In hyperpolarized NMR experiments, the initial state $|\rho_{\text{ini}}\rangle$ does not correspond to thermal equilibrium, but to a non-equilibrium state, which contains spin order terms greatly exceeding the thermal equilibrium Zeeman polarization. Numerous techniques exist for generating such far-from-equilibrium states, including dynamic nuclear polarization of systems containing unpaired electrons [21, 22], optical pumping of noble gases [23], quantum-rotor-induced polarization of freely rotating molecular groups [24, 25], chemically-induced dynamic nuclear polarization [26], and many variants of parahydrogen-induced polarization [27, 28].

A definition of the term *hyperpolarization* in terms of von Neumann entropy has been given [17].

### *Excitation*

A great variety of excitation sequences exists for converting the initial state $|\rho_{\text{ini}}\rangle$, which usually does not contain observable terms, to a state $|\rho(0)\rangle$, which induces an observable NMR signal in the following detection interval, as described by Equation 1.27.

**Single-pulse excitation** In the simplest case, the excitation sequence consists of a single strong, short, radiofrequency pulse, which induces a clean rotation of the spin states:

$$\hat{V}_{\text{exc}} = \hat{R}_\phi(\beta) \tag{1.34}$$

where $\phi$ is the pulse phase and $\beta$ is the pulse flip angle. The rotation superoperator is given by

$$\hat{R}_\phi(\beta) = \exp -i\beta(\hat{I}_x \cos\phi + \hat{I}_y \sin\phi). \tag{1.35}$$

If the pulse phase is $\phi = 0$ and the flip angle is $\beta = \pi/2$ ($90^\circ_x$ pulse) and the initial condition is represented by the greatly simplified thermal equilibrium density operator in Equation 1.33, the density operator at the end of the preparation sequence has a very simple form:

$$|\rho(0)\rangle = \hat{R}_0(\pi/2)\,|I_z\rangle = -\,|I_y\rangle . \tag{1.36}$$

This corresponds to the conversion of longitudinal nuclear polarization $|I_z\rangle$ to transverse nuclear polarization $|I_y\rangle$ by the strong radiofrequency pulse.

**Complex excitation sequences** Excitation sequences may be complex affairs in Fourier NMR, containing many elements. In general, the excitation superoperator is given by a product of the individual superoperators, i.e.

$$\hat{V}_{\text{exc}} = \hat{V}^X_{\text{exc}} \dots \hat{V}^B_{\text{exc}} \hat{V}^A_{\text{exc}} \tag{1.37}$$

where the excitation sequence is composed of elements $\{A, B \dots X\}$ in left-to-right chronological order.

#### 1.3.1.2 Detection

Assume that the coherent Hamiltonian and relaxation superoperator during the detection interval are given by $\mathcal{H}_{\text{det}}$ and $\hat{\Gamma}_{\text{det}}$, respectively. The detection Liouvillian is Liouvillian is given by $\hat{L}_{\text{det}} = -i\hat{\mathcal{H}}_{\text{det}} + \hat{\Gamma}_{\text{det}}$. The left eigenoperators $\langle {}^{L}Q_{\ell}^{\text{det}}|$, right eigenoperators $|{}^{R}Q_{\ell}^{\text{det}}\rangle$, and eigenvalues $\Lambda_{\ell}^{\text{det}}$ of the detection Liouvillian have the following properties:

$$\begin{aligned} \langle {}^{L}Q_{\ell}^{\text{det}}| \hat{L}_{\text{det}} &= \Lambda_{\ell}^{\text{det}} \langle {}^{L}Q_{\ell}^{\text{det}}| \\ \hat{L}_{\text{det}} |{}^{R}Q_{\ell}^{\text{det}}\rangle &= \Lambda_{\ell}^{\text{det}} |{}^{R}Q_{\ell}^{\text{det}}\rangle \end{aligned} \tag{1.38}$$

where the complex eigenvalues $\Lambda_{\ell}^{\text{det}}$ may be written:

$$\Lambda_{\ell}^{\text{det}} = \lambda_{\ell}^{\text{det}} + i\omega_{\ell}^{\text{det}}. \tag{1.39}$$

The density operator at the start of signal detection, Equation 1.27, may be expressed in terms of the Liouvillian eigenoperators as follows:

$$|\rho(0)\rangle = \sum_{\ell=1}^{N_L} |{}^{R}Q_{\ell}^{\text{det}}\rangle \langle {}^{L}Q_{\ell}^{\text{det}}|\rho(0)\rangle = \sum_{\ell=1}^{N_L} |{}^{R}Q_{\ell}^{\text{det}}\rangle \langle {}^{L}Q_{\ell}^{\text{det}}| \hat{V}_{\text{exc}} |\rho_{\text{ini}}\rangle. \tag{1.40}$$

The density operator is allowed to evolve freely during the detection interval, in the presence of the Hamiltonian $\mathcal{H}_{\text{det}}$. If the detection interval has duration $\tau_{\text{det}}$, the density operator at time $0 \leq t \leq \tau_{\text{det}}$ is given by

$$|\rho(t)\rangle = \sum_{\ell=1}^{N_L} |{}^{R}Q_{\ell}^{\text{det}}\rangle \langle {}^{L}Q_{\ell}^{\text{det}}| \hat{V}_{\text{exc}} |\rho_{\text{ini}}\rangle \exp \Lambda_{\ell}^{\text{det}} t. \tag{1.41}$$

The detection of the NMR signal may be represented by an observable operator $|Q_{\text{obs}}\rangle$. The NMR signal at time $0 \leq t \leq \tau_{\text{det}}$ is given by

$$s(t) = \sum_{\ell=1}^{N_L} s_{\ell}(t) \tag{1.42}$$

where each signal component $s_{\ell}(t)$ is given by

$$s_{\ell}(t) = a_{\ell} \exp(i\omega_{\ell}^{\text{det}} - \lambda_{\ell}^{\text{det}})t. \tag{1.43}$$

The complex amplitude $a_{\ell}$ of the signal component $s_{\ell}(t)$ is given by:

$$a_{\ell} = (Q_{\text{obs}}|{}^{R}Q_{\ell}^{\text{det}}\rangle \langle {}^{L}Q_{\ell}^{\text{det}}| \hat{V}_{\text{exc}} |\rho_{\text{ini}}\rangle. \tag{1.44}$$

The complex amplitude $a_{\ell}$ in Equation 1.44 has a straightforward physical interpretation. In order to generate an observable signal, an eigenoperator $|{}^{R}Q_{\ell}^{\text{det}}\rangle$ must be excited and must also induce an observable signal. Hence the amplitude of a signal component is proportional to the product of two factors, one representing the excitation amplitude of a particular eigenoperator, and one representing the amplitude of the signal induced by that eigenoperator. The first factor is given by the term $\langle {}^{L}Q_{\ell}^{\text{det}}| \hat{V}_{\text{exc}} |\rho_{\text{ini}}\rangle$, which indicates the amplitude for conversion of the initial density operator $|\rho_{\text{ini}}\rangle$ into the eigenoperator $|{}^{R}Q_{\ell}^{\text{det}}\rangle$ by the excitation sequence. The second factor $(Q_{\text{obs}}|{}^{R}Q_{\ell}^{\text{det}}\rangle$ indicates the signal amplitude for coupling the eigenoperator $|{}^{R}Q_{\ell}^{\text{det}}\rangle$ to the observable operator $|Q_{\text{obs}}\rangle$.

The observable operator $|Q_{\text{obs}}\rangle$ depends on the type of NMR experiment.

##### *Quadrature Detection*

In the majority of NMR experiments, the nuclear magnetization is detected by the Faraday induction of an electrical signal in a coil of wire close to the NMR sample, induced by precessing transverse magnetization. The electrical signal is converted into complex form by a device known as a quadrature receiver (or its modern equivalent). This process is described in some detail in ref. [29] and corresponds to an observable operator of the form:

$$\begin{aligned}|Q_{\text{obs}}\rangle &\sim -2i \exp i\phi_{\text{rec}} \,|\mathfrak{I}\rangle \\ &= -2i \exp i\phi_{\text{rec}} \left|I_x - iI_y\right\rangle \quad \text{(quadrature detection)}\end{aligned} \tag{1.45}$$

where $\phi_{\text{rec}}$ is the receiver reference phase.

Within the high-field approximation, the observable operator in Equation 1.45, and the operator $\left|Q_\ell^{\text{det}}\right\rangle$ are both eigenoperators of the commutation superoperator $\hat{I}_z$, with eigenvalues $-1$ and $p_\ell$, where $p_\ell$ is the coherence order (Equation 1.12):

$$\begin{aligned}\hat{I}_z \,|Q_{\text{obs}}\rangle &= (-1)\,|Q_{\text{obs}}\rangle \\ \hat{I}_z \left|Q_\ell^{\text{det}}\right\rangle &= p_\ell \left|Q_\ell^{\text{det}}\right\rangle.\end{aligned} \tag{1.46}$$

It follows that the matrix element $\left(Q_{\text{obs}}\middle|Q_\ell^{\text{det}}\right)$ is identically zero unless $p_\ell = -1$. Hence, only $(-1)$-quantum coherences are observable by quadrature detection.

#### *Magnetometer Detection*

In a very low magnetic field, it is possible to detect the nuclear magnetization directly using a sensitive magnetometer, rather than exploiting Faraday induction (see Chapter 13) [30, 31]. In these cases, the nuclear magnetization is detected directly in a particular direction (for example, the z-axis), leading to an observable operator of the form

$$|Q_{\text{obs}}\rangle \sim \sum_I \gamma_I \,|I_z\rangle \quad \text{(z-magnetometry)} \tag{1.47}$$

where the sum is over the nuclear species. The selection rule $p_\ell = -1$ does not apply in this case.

### 1.3.2 One-dimensional NMR Spectrum

In 1D Fourier NMR, the complex signal $s(t)$ is digitized by the spectrometer electronics, and subjected to a Fourier transform according to Equation 1.1, in order to generate the NMR spectrum $S(\omega)$, which has the form:

$$S(\omega) = \sum_{\ell=1}^{N_L} S_\ell(\omega) \tag{1.48}$$

where each spectral component $S_\ell(\omega)$ is given by

$$S_\ell(\omega) = a_\ell \,\mathcal{L}(\omega; \omega_\ell^{\text{det}}, \lambda_\ell^{\text{det}}). \tag{1.49}$$

The complex Lorentzian lineshape is given by

$$\mathcal{L}(\omega, \omega_\ell, \lambda_\ell) = \mathcal{A}(\omega; \omega_\ell, \lambda_\ell) + i\mathcal{D}(\omega; \omega_\ell, \lambda_\ell). \tag{1.50}$$

The center frequency of the Lorentzian is given by $\omega_\ell$, and its linewidth is given by $\lambda_\ell$. The absorption and dispersion-mode components are given by

$$\begin{aligned}\mathcal{A}(\omega; \omega_\ell, \lambda_\ell) &= \frac{\lambda_\ell}{\lambda_\ell^2 + (\omega - \omega_\ell)^2} \\ \mathcal{D}(\omega; \omega_\ell, \lambda_\ell) &= -\frac{\omega - \omega_\ell}{\lambda_\ell^2 + (\omega - \omega_\ell)^2}.\end{aligned} \tag{1.51}$$

In the case that all amplitudes $a_\ell$ are real, the real part of the spectrum $S(\omega)$ is a superposition of absorption-mode Lorentzians, each one centered at the frequency $\omega_\ell$ of a particular eigenoperator, with a half-linewidth-at-half-height given by $\lambda_\ell$.

This is the essence of 1D Fourier NMR: An excitation sequence generates a set of Liouvillian eigenoperators some of which induce a complex NMR signal. The signal is detected and Fourier transformed to give a NMR spectrum with peaks at the Liouvillian eigenfrequencies $\omega_\ell$, which contain information on chemical shifts, paramagnetic shifts, and scalar spin-spin couplings in isotropic solution and also dipole-dipole couplings, chemical shift anisotropies, and quadrupolar couplings in anisotropic phases. The widths of the peaks are proportional to the decay rate constants $\lambda_\ell$ and contain information on the dissipative behavior of the nuclear spin system. In solution phase, this dissipative information can be a sensitive marker of molecular mobility and anisotropic nuclear spin interactions.

## 1.4 Two-dimensional NMR

Jean Jeener realized that 1D Fourier NMR could readily be extended by introducing two new elements: (i) a variable time interval in the pulse sequence, allowing the NMR signal to be described as a function of two time variables instead of just one and (ii) an additional Fourier transform with respect to the new time variable. Jeener's idea was first communicated at a 1971 workshop in the Croatian resort of Basko Polje [32]. Ernst and co-workers soon extended Jeener's proposal into a massive edifice of novel concepts, theory, and experimental techniques. The landmark 1976 paper by Aue et al. still underpins much of modern NMR [33].

### 1.4.1 The Two-dimensional NMR Experiment

A general 2D NMR pulse sequence is constructed by inserting an *evolution interval* and a *mixing sequence* into a 1D pulse sequence (see Figure 1.2).

In the simplest version of 2D spectroscopy, the evolution element consists of a variable interval of duration $t_1$, during which the spin system evolves freely under a time-independent Hamiltonian $\mathcal{H}_{\text{ev}}$. The NMR signal is acquired during the detection interval as a function of the time variable $t_2$. In the original procedure for performing 2D spectroscopy, the entire pulse sequence is repeated many times, for different values of the evolution interval $t_1$. In this way a 2D time-domain signal surface $s(t_1, t_2)$ is compiled. The 2D time-domain signal is subjected to a 2D Fourier transform to generate the 2D spectrum $S(\omega_1, \omega_2)$, which is a function of two frequency variables $\omega_1$ and

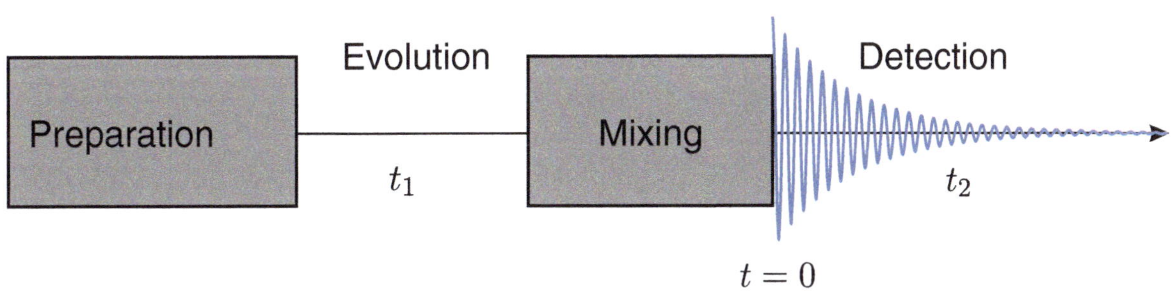


Figure 1.2 A general 2D NMR sequence, consisting of a *preparation* sequence, an *evolution* interval, a *mixing sequence*, and a *detection* interval. The preparation sequence often consists of *initialization* followed by *excitation*, as in 1D Fourier spectroscopy (Figure 1.1). In a conventional 2D acquisition scheme, the sequence is repeated many times for different values of the evolution interval $t_1$, thereby compiling a 2D signal surface $s(t_1, t_2)$.

$\omega_2$. The 2D Fourier transform is defined as follows:

$$S(\omega_1, \omega_2) = \int_0^{\infty} \int_0^{\infty} s(t_1, t_2) \exp -i(\omega_1 t_1 + \omega_2 t_2)\, \mathrm{d}t_1 \mathrm{d}t_2. \tag{1.52}$$

The digitization of the surface $s(t_1, t_2)$ by repetition of a 1D experiment for many equally spaced values of the $t_1$ interval can be slow. A variety of methods have been developed for the accelerated acquisition and processing of 2D NMR data (see Chapter 2 and Chapter 10).

A different approach is to use magnetic field gradients to encode the $t_1$-dependence over the spatial extent of the NMR sample. In favorable cases, this allows acquisition of 2D spectroscopic information in a single free-induction decay, albeit with a signal-to-noise penalty [34, 35].

### 1.4.2 Two-dimensional NMR Signal

The evolution Liouvillian is given by $\hat{L}_{\text{ev}} = -i\hat{\mathcal{H}}_{\text{ev}} + \hat{\Gamma}_{\text{ev}}$. The left eigenoperators $\left\langle {}^{L}Q_{\ell}^{\text{ev}}\right|$, right eigenoperators $\left|{}^{R}Q_{\ell}^{\text{ev}}\right\rangle$, and eigenvalues $\Lambda_{\ell}^{\text{ev}}$ of the evolution Liouvillian have the following properties:

$$\begin{aligned} \left\langle {}^{L}Q_{\ell}^{\text{ev}}\right| \hat{L}_{\text{ev}} &= \Lambda_{\ell}^{\text{ev}} \left\langle {}^{L}Q_{\ell}^{\text{ev}}\right| \\ \hat{L}_{\text{ev}} \left|{}^{R}Q_{\ell}^{\text{ev}}\right\rangle &= \Lambda_{\ell}^{\text{ev}} \left|{}^{R}Q_{\ell}^{\text{ev}}\right\rangle \end{aligned} \tag{1.53}$$

where the complex eigenvalues $\Lambda_{\ell}^{\text{ev}}$ may be written:

$$\Lambda_{\ell}^{\text{ev}} = \lambda_{\ell}^{\text{ev}} + i\omega_{\ell}^{\text{ev}}. \tag{1.54}$$

The 2D NMR signal is given in general by a sum over $N_L^2$ signal components:

$$s(t_1, t_2) = \sum_{\ell=1}^{N_L} \sum_{\ell'=1}^{N_L} s_{\ell\ell'}(t_1, t_2) \tag{1.55}$$

where each component is given by

$$s_{\ell\ell'}(t_1, t_2) = a_{\ell\ell'} \exp \Lambda_{\ell}^{\text{ev}} t_1 + \Lambda_{\ell'}^{\text{det}} t_2 \tag{1.56}$$

$\Lambda_{\ell}^{\text{ev}}$ and $\Lambda_{\ell'}^{\text{det}}$ are the eigenvalues of the Liouvillian during the evolution and detection intervals, respectively.

A straightforward extension of the treatment of 1D Fourier spectroscopy gives the following expression for the complex amplitude of a 2D signal component:

$$a_{\ell\ell'} = \left(Q_{\text{obs}}\middle|{}^{R}Q_{\ell'}^{\text{det}}\right) \left\langle {}^{L}Q_{\ell'}^{\text{det}}\right| \hat{V}_{\text{mix}} \left|{}^{R}Q_{\ell}^{\text{ev}}\right\rangle \left\langle {}^{L}Q_{\ell}^{\text{ev}}\right| \hat{V}_{\text{exc}} \left|\rho_{\text{ini}}\right\rangle. \tag{1.57}$$

The physical interpretation of this expression is as follows: the term $\left\langle {}^{L}Q_{\ell}^{\text{ev}}\right| \hat{V}_{\text{exc}} \left|\rho_{\text{ini}}\right\rangle$ represents the amplitude for conversion of the initial density operator $\left|\rho_{\text{ini}}\right\rangle$ into the eigenoperator $\left|{}^{R}Q_{\ell}^{\text{ev}}\right\rangle$ by the excitation sequence; the term $\left\langle {}^{L}Q_{\ell'}^{\text{det}}\right| \hat{V}_{\text{mix}} \left|{}^{R}Q_{\ell}^{\text{ev}}\right\rangle$ represents the amplitude for the conversion of the evolution eigenoperator $\left|{}^{R}Q_{\ell}^{\text{ev}}\right\rangle$ into the detection eigenoperator $\left|{}^{R}Q_{\ell'}^{\text{det}}\right\rangle$ by the mixing sequence; the term $\left(Q_{\text{obs}}\middle|{}^{R}Q_{\ell'}^{\text{det}}\right)$ represents the detection amplitude for the eigenoperator $\left|{}^{R}Q_{\ell'}^{\text{det}}\right\rangle$.

### 1.4.3 Two-dimensional NMR Spectrum

After 2D Fourier transformation (Equation 1.52), the 2D is a sum of signal components, each representing the correlation between the evolution and the detection eigenoperators:

$$S(\omega_1, \omega_2) = \sum_{\ell=1}^{N_L} \sum_{\ell'=1}^{N_L} S_{\ell\ell'}(\omega_1, \omega_2) \tag{1.58}$$

where each component is given by

$$S_{\ell\ell'}(\omega_1, \omega_2) = a_{\ell\ell'} \mathcal{L}(\omega_1; \omega_\ell^{\text{ev}}, \lambda_\ell^{\text{ev}}) \mathcal{L}(\omega_2; \omega_{\ell'}^{\text{det}}, \lambda_{\ell'}^{\text{det}}). \tag{1.59}$$

Here $\omega_\ell^{\text{ev}}$ and $\lambda_\ell^{\text{ev}}$ are the imaginary and (minus) the real parts of the Liouvillian eigenvalues during the evolution interval, respectively:

$$\Lambda_\ell^{\text{ev}} = -\lambda_\ell^{\text{ev}} + i\omega_\ell^{\text{ev}}. \tag{1.60}$$

Equation 1.59 represents a 2D spectral peak, centered at frequency coordinates $(\omega_1, \omega_2) = (\omega_\ell^{\text{ev}}, \omega_{\ell'}^{\text{det}})$, and with linewidth parameters $(\lambda_\ell^{\text{ev}}, \lambda_{\ell'}^{\text{det}})$ in the two dimensions. The 2D NMR spectrum $S(\omega_1, \omega_2)$ consists of a superposition of many such peaks.

The 2D peakshape described by Equation 1.59 has an undesirable feature, which has preoccupied NMR spectroscopists since the earliest days [33]. Unlike one-dimensional NMR, the real part of the 2D spectral function mixes together the absorption and dispersion Lorentzian functions, even in the case that the amplitude $a_{\ell\ell'}$ is a real number:

$$\text{Re}\{\mathcal{L}_1 \mathcal{L}_2\} = \mathcal{A}_1 \mathcal{A}_2 - \mathcal{D}_1 \mathcal{D}_2. \tag{1.61}$$

The resulting function is called a *phase-twist* lineshape and is quite unsuitable for high-resolution spectroscopy. So-called "pure absorption" or "pure phase" procedures use a combination of data acquisition and data processing schemes to cancel out the undesirable dispersion components [13, 36]. It has been shown that these methods are equivalent to the combination of the signals from *conjugate pairs* of spectral peaks [11, 29]. The peaks belonging to a conjugate pair have the same amplitude $a_{\ell\ell'}$ and the same frequency coordinate $\omega_{\ell'}^{\text{det}}$ in the $\omega_2$ dimension, but with frequency coordinates $\omega_\ell^{\text{ev}}$ of opposite sign in the $\omega_1$-dimension [11, 29]. "Pure absorption" methods are only applicable to those 2D experiments that generate peaks in conjugate pairs; most popular 2D experiments are members of that class.

### 1.4.4 Two-dimensional Experiments

The number and variety of 2D NMR experiments is enormous. The remaining chapters in this book cover some of the ground.

Two-dimensional NMR is used for many and various purposes. Some of these may be summarized as follows:

***Assignment***

Suppose that the 1D NMR spectrum contains a set of peaks at frequencies $\{\omega_A, \omega_B, \omega_C \ldots\}$. In 1D NMR, it is hard to establish whether a particular pair of peaks, or group of peaks, are generated by the same spin system, or by different spin systems. In 2D NMR, on the other hand, the existence of a peak at frequency coordinates $\{\omega_1, \omega_2\} = \{\omega_A, \omega_B\}$ is an unambiguous proof that the frequencies $\omega_A$ and $\omega_B$ are generated by the *same* spin system. In the case of complex NMR spectra containing hundreds or thousands of peaks, this is an invaluable aid to spectral assignment [13].

***Observation of "Forbidden" Transitions***

In 1D NMR, the amplitude of a spectral peak at frequency $\omega = \omega_\ell$ is given by Equation 1.44. This amplitude vanishes except for those Liouvillian eigenoperators for which $(Q_{\text{obs}}|^R Q_\ell^{\text{det}}) \neq 0$. Such "observable" eigenoperators have coherence order $p_\ell = -1$. Eigenoperators with coherence order $p_\ell \neq -1$ have an observable matrix element $(Q_{\text{obs}}|^R Q_\ell^{\text{det}}) = 0$ and are not directly observable by quadrature detection in high magnetic field.

In 2D NMR, the amplitude of a 2D peak is given by Equation 1.57. This expression does not contain the matrix element $(Q_{\text{obs}}|^R Q_\ell^{\text{ev}})$, and hence, there are no constraints on the coherence order of the eigenoperator $|^R Q_\ell^{\text{ev}})$ in the evolution interval. Eigenoperators of arbitrary coherence order are observable in 2D NMR, providing that the excitation and mixing matrix elements $(^L Q_\ell^{\text{ev}}| \hat{V}_{\text{exc}} |\rho_{\text{ini}})$ and $(^L Q_{\ell'}^{\text{det}}| \hat{V}_{\text{mix}} |^R Q_\ell^{\text{ev}})$ are sufficiently large. It is possible to ensure that both matrix elements are finite by sufficiently ingenious experimental design.

Two-dimensional experiments may therefore be used to probe the properties of "forbidden" transitions, such as zero-quantum and multiple-quantum transitions. Such transitions sometimes exhibit favorable relaxation and/or fine-structure properties [11, 37].

Paradoxically, multiple-quantum coherences are sometimes easier to exploit than single-quantum coherences. This is the case in *heteronuclear multiple-quantum coherence spectroscopy* (HMQC), which is a popular method for establishing correlations between nuclei of different isotopic type in solution NMR (see Chapter 11) [13, 38].

***Correlation***

One of the most powerful and versatile principles of 2D NMR is that it allows the experimental determination of correlations in the NMR behavior of the nuclear spin system at two different points in time. This is a very powerful principle since the behavior of nuclear spins depends in turn on numerous factors including chemical identity, material phase, spatial position (in the presence of magnetic field gradients), molecular mobility, molecular ordering and orientation, temperature, direction, and speed of motion (again in the presence of magnetic field gradients), etc. Hence the correlation of NMR behavior at two different time points allows the experimental correlation of different chemical and physical states of the sample at two different time points. The possible combinations are almost endless.

One very early example is 2D Fourier imaging, which involves the application of magnetic field gradients in orthogonal directions during the $t_1$ and $t_2$ intervals of a 2D NMR experiment. The resulting 2D spectrum is an image of an object in two dimensions [39]. This method underpins most modern magnetic resonance imaging (MRI) techniques.

Another early application is chemical exchange spectroscopy (EXSY). Here the observed molecules are in two different chemical states during the $t_1$ and $t_2$ intervals, with a chemical process allowing interconversion of the chemical forms during the mixing interval (see Chapter 15) [40].

In nuclear Overhauser spectroscopy (NOESY) the correlated species are nuclear sites at separate locations in the same molecule. Nuclear spin order is transmitted from one site to another during an extended mixing interval. The dynamics of the spin order transfer depend on the spatial separation of the two nuclear sites, as well as the molecular dynamics, through well-understood processes (see Chapter 6) [41]. NOESY allows the simultaneous estimation of the spatial proximity between a large number of pairs of nuclear sites, allowing strong constraints to be placed on the 3D molecular structure [42]. This is one of the few methods that are capable of estimating 3D molecular structures in solution.

## 1.5 Summary

In summary, 2D Fourier spectroscopy is an enormously versatile and powerful principle that has profoundly influenced all varieties of magnetic resonance. The remaining chapters of this book explore some of the varieties of 2D NMR in more detail.

Extensions to more than two dimensions are feasible and are particularly useful in biomolecular [43] and imaging contexts.

Two-dimensional methods are increasingly applied to other forms of spectroscopy, such as ion cyclotron resonance [44], infrared spectroscopy [45, 46], terahertz spectroscopy [47], optical spectroscopy [48], and combinations of spectroscopies at different wavelengths [49].

## Acknowledgments

This article benefitted from discussions with Geoffrey Bodenhausen, Christian Bengs, and James Whipham.

## References


1 Ernst, R.R. and Anderson, W.A. (1966 January). Application of fourier transform spectroscopy to magnetic resonance. *Rev. Sci. Instrum.* 37 (1): 93–102. ISSN 0034-6748.

2 Hahn, E.L. (1950). Spin Echoes. *Phys. Rev.* 80: 580–594.

3 Hahn, E.L. and Maxwell, D.E. (1952). Spin echo measurements of nuclear spin coupling in molecules. *Phys. Rev.* 88 (5): 1070–1084.

4 Meiboom, S. and Gill, D. (1958 August). Modified spin-echo method for measuring nuclear relaxation times. *Rev. Sci. Instrum.* 29 (8): 688–691. ISSN 0034-6748.

5 Hartmann, S.R. and Hahn, E.L. (1962). Nuclear double resonance in the rotating frame. *Phys. Rev.* 128: 2042–2053.

6 Bloch, F. (1946 October). Nuclear Induction. *Phys. Rev.* 70 (7-8): 460–474.

7 Cooley, J.W. and Tukey, J.W. (1965). An algorithm for the machine calculation of complex Fourier series. *Math. Comp.* 19 (90): 297–301. ISSN 0025-5718, 1088-6842.

8 Atmanspacher, H. Remembering Hans Primas (1928–2014). (2014 January). *Mind Matter.*

9 Banwell, C.N. and Primas, H. (1963 April). On the analysis of high-resolution nuclear magnetic resonance spectra. *Mol. Phys.* 6 (3): 225–256 ISSN 0026-8976.

10 Sørensen, O.W., Eich, G.W., Levitt, M.H., Bodenhausen, G. and Ernst, R.R. (1984). Product operator formalism for the description of NMR pulse experiments. *Prog. Nucl. Magn. Reson. Spectrosc.* 16: 163–192. ISSN 0079-6565.

11 Ernst, R.R., Bodenhausen, G. and Wokaun, A. (1990 September). *Principles of Nuclear Magnetic Resonance in One and Two Dimensions*, volume 1. Clarendon Press, Oxford.

12 Tjandra, N. and Bax, A. (1997 November). Direct Measurement of distances and angles in biomolecules by NMR in a dilute liquid crystalline medium. *Science.*

13 Cavanagh, J., Fairbrother, W.J., Palmer III, A.G., Rance, M. and Skelton, N.J. (2007). *Protein NMR Spectroscopy. Principles and Practice*, 2e. Amsterdam: Elsevier Academic Press.

14 Jeener, J. (1982 January). Superoperators in Magnetic Resonance. In: *Advances in Magnetic and Optical Resonance*, (ed. Waugh, J.S.), volume 10, 1–51. Academic Press.

15 Levitt, M.H. and Bengs, C. (2021 June). Hyperpolarization and the physical boundary of Liouville space. *Magn. Reson.* 2 (1): 395–407. ISSN 2699-0016.

16 Kuprov, I. (2016 September). Fokker-Planck formalism in magnetic resonance simulations. *J. Magn. Reson.* 270: 124–135. ISSN 1090-7807.

17 Bengs, C. and Levitt, M.H. (2020 January). A master equation for spin systems far from equilibrium. *J. Magn. Reson.* 310: 106645. ISSN 1090-7807.

18 Levitt, M.H. (2012). Singlet Nuclear Magnetic Resonance. *Annu. Rev. Phys. Chem.* 63 (1): 89–105.

19 Carravetta, M. and Levitt, M.H. (2004 May). Long-lived nuclear spin states in high-field solution NMR. *J. Am. Chem. Soc.*, 126 (20): 6228–6229. ISSN 0002-7863.

**20** Maricq, M.M. and Waugh, J.S. (1979 April). NMR in rotating solids. *J. Chem. Phys.* 70 (7): 3300–3316. ISSN 0021-9606.

**21** Hall, D.A., Maus, D.C., Gerfen, G.J., Inati, S.J., Becerra, L.R., Dahlquist, F.W. and Griffin, R.G. (1997 May). Polarization-enhanced NMR spectroscopy of biomolecules in frozen solution. *Science*, 276 (5314): 930–932. ISSN 0036-8075, 1095-9203.

**22** Ardenkjaer-Larsen, J.H., Fridlund, B., Gram, A., Hansson, G., Hansson, L., Lerche, M.H., Servin, R., Thaning, M. and Golman, K. (2003 September). Increase in signal-to-noise ratio of > 10,000 times in liquid-state NMR. *Proc. Natl. Acad. Sci. USA* 100 (18): 10158–10163.

**23** Navon, G., Song, Y.Q., Room, T., Appelt, S.T.R.E., Taylor, R.E. and Pines, A. Enhancement of solution NMR and MRI with laser-polarized xenon. *Science* 271 (5257): 1848–1851. ISSN 0036-8075, 1095-9203.

**24** Icker, M. and Berger, S. (2012). Unexpected multiplet patterns induced by the Haupt-effect. *J. Magn. Reson.* 219 (0): 1–3. ISSN 1090-7807.

**25** Dumez, J.N., Håkansson, P., Mamone, S., Meier, B., Stevanato, G., Hill-Cousins, J.T., Roy, S.S., Brown, R.C., Pileio, G. and Levitt, M.H. (2015). Theory of long-lived nuclear spin states in methyl groups and quantum-rotor induced polarisation. *J. Chem. Phys.* 142 (4): 044506.

**26** Kaptein, R. (1972, September). Chemically induced dynamic nuclear polarization. VIII. Spin dynamics and diffusion of radical pairs. *J. Am. Chem. Soc.* 94 (18): 6251–6262. ISSN 0002-7863.

**27** Bowers, C.R. and Weitekamp, D.P. (1987 September). Parahydrogen and synthesis allow dramatically enhanced nuclear alignment. *J. Am. Chem. Soc.* 109 (18): 5541–5542. ISSN 0002-7863.

**28** Adams, R.W., Aguilar, J.A., Atkinson, K.D., Cowley, M.J., Elliott, P.I., Duckett, S.B., Green, G.G., Khazal, I.G., López-Serrano, J. and Williamson, D.C (2009 March). Reversible interactions with para-hydrogen enhance NMR sensitivity by polarization transfer. *Science* 323 (5922): 1708–1711.

**29** Levitt, M.H. (2007). *Spin Dynamics. Basics of Nuclear Magnetic Resonance*. Wiley, Chichester, second edition. ISBN 0-471-48921-2.

**30** Tayler, M.C., Theis, T., Sjolander, T.F., Blanchard, J.W., Kentner, A., Pustelny, S., Pines, A. and Budker, D. (2017 September). Invited review article: instrumentation for nuclear magnetic resonance in zero and ultralow magnetic field. *Rev. Sci. Instrum.* 88 (9): 091101. ISSN 0034-6748.

**31** Jiang, M., Bian, J., Li, Q., Wu, Z., Su, H., Xu, M., Wang, Y., Wang, X. and Peng, X. (2021 January). Zero- to ultralow-field nuclear magnetic resonance and its applications. *Fundam. Res.* 1 (1): 68–84. ISSN 2667-3258.

**32** Jeener, J. and Alewaeters, G. (2016 May). "Pulse pair technique in high resolution NMR" a reprint of the historical 1971 lecture notes on two-dimensional spectroscopy. *Prog. Nucl. Magn. Reson. Spectrosc.* 94–95: 75–80. ISSN 0079-6565.

**33** Aue, W.P., Bartholdi, E. and Ernst, R.R. (1976). Two-dimensional spectroscopy. Application to nuclear magnetic resonance. *J. Chem. Phys.* 64: 2229.

**34** Frydman, L., Scherf, T. and Lupulescu, A. (2002). The acquisition of multidimensional NMR spectra within a single scan. *Proc. Natl. Acad. Sci. USA* 99 (25): 15858–15862.

**35** Dumez, J.N. (2018 December). Spatial encoding and spatial selection methods in high-resolution NMR spectroscopy. *Prog. Nucl. Magn. Reson. Spectrosc.* 109: 101–134. ISSN 0079-6565.

**36** States, D.J., Haberkorn, R.A. and Ruben, D.J. (1982 June). A two-dimensional nuclear overhauser experiment with pure absorption phase in four quadrants. *J. Magn. Reson. (1969)*, 48 (2): 286–292. ISSN 0022-2364.

**37** Wokaun, A. and Ernst, R.R. (1978 August). The use of multiple quantum transitions for relaxation studies in coupled spin systems. *Mol. Phys.* 36 (2): 317–341. ISSN 0026-8976.

**38** Mueller, L. (1979 August). Sensitivity enhanced detection of weak nuclei using heteronuclear multiple quantum coherence. *J. Am. Chem. Soc.* 101 (16): 4481–4484. ISSN 0002-7863.

**39** Kumar, A., Welti, D. and Ernst, R.R. (1975 April). NMR Fourier zeugmatography. *J. Magn. Reson. (1969)*, 18 (1): 69–83. ISSN 0022-2364.

**40** Jeener, J.M.B.H., Meier, B.H., Bachmann, P. and Ernst, R.R. (1979). Investigation of exchange processes by 2-dimensional NMR spectroscopy. *J. Chem. Phys.* 71 (11): 4546–4553. ISSN 0021-9606.

**41** Solomon, I. (1955 July). Relaxation processes in a system of two spins. *Phys. Rev.* 99 (2): 559–565.

**42** Kumar, A., Ernst, R.R. and Wthrich, K. (1980 July). A two-dimensional nuclear Overhauser enhancement (2D NOE) experiment for the elucidation of complete proton-proton cross-relaxation networks in biological macromolecules. *Biochem. Biophys. Res. Commun.* 95 (1): 1–6. ISSN 0006-291X.

**43** Oschkinat, H., Griesinger, C., Kraulis, P.J., Srensen, O.W., Ernst, R.R., Gronenborn, A.M. and Clore, G.M. Three-dimensional NMR spectroscopy of a protein in solution. *Nature*, 332 (6162): 374–376. ISSN 1476-4687.

**44** Bodenhausen, G. (2021 January). Early days of two-dimensional ion cyclotron resonance. *Molecule*, 26 (11): 3381.

**45** Zanni, M.T. and Hochstrasser, R.M. (2001 September). Two-dimensional infrared spectroscopy: a promising new method for the time resolution of structures. *Curr. Opin. Struct. Biol.* 11 (5): 516–522. ISSN 0959-440X.

**46** Cahoon, J.F., Sawyer, K.R., Schlegel, J.P. and Harris, C.B. (2008 March). Determining transition-state geometries in liquids using 2D-IR. *Science*.

**47** Finneran, I.A., Welsch, R., Allodi, M.A., Miller III, T.F. and Blake, G.A. (2016 June). Coherent two-dimensional terahertz-terahertz-Raman spectroscopy. *PNAS* 113 (25): 6857–6861. ISSN 0027-8424, 1091-6490.

**48** Cho, M. (2008 April). Coherent two-dimensional optical spectroscopy. *Chem. Rev.* 108 (4): 1331–1418. ISSN 0009-2665.

**49** Mukamel, S. (2000 October). Multidimensional femtosecond correlation spectroscopies of electronic and vibrational excitations. *Annu. Rev. Phys. Chem.* 51 (1): 691–729. ISSN 0066-426X.

# 2

# Data Processing Methods: Fourier and Beyond

*Vladislav Orekhov*[1,*], *Paweł Kasprzak*[2,3], *and Krzysztof Kazimierczuk*[3]

[1] Department of Chemistry and Molecular Biology, Swedish NMR Centre, University of Gothenburg, 40530, Gothenburg, Box 465, Sweden
[2] Faculty of Physics, University of Warsaw, 02-093, Warsaw, Pasteura 5, Poland
[3] Centre of New Technologies, University of Warsaw, 02-097, Warsaw, Banacha 2C, Poland
* Corresponding Author

## 2.1 Introduction

The modern NMR signal processing, which emerged with the introduction of Fourier NMR at the beginning of the 1970s is still a rapidly developing field due to demands posed by the introduction of evermore complex experiments and analyses as well as because of tremendous advances in mathematics, algorithms, and computer hardware. In this chapter, we attempt to present a compact overview of both practical and rigorously mathematical aspects of modern NMR signal processing.

## 2.2 Time-domain NMR Signal

As explained in previous chapter, in an NMR experiment, the radio-frequency (RF) pulse generates a coherent precession of the affected, polarized nuclear magnetic moments. The precession, in turn, induces an oscillating voltage in the receiver coil. The signal measured this way decays with time as the spin system returns to its thermal equilibrium and is called free induction decay (FID). The analytical shape of an NMR signal stems, to a good approximation, from the equation of motion of the magnetic moment exposed to the combination of constant vertical and oscillating transverse magnetic fields. It is a sum of damped oscillations, each corresponding to one among $M$ groups of magnetically equivalent nuclei characterized by specific resonance frequencies $(\Omega_j)_{j=1}^{M}$, amplitudes $(A_j)_{j=1}^{M}$, and transverse relaxation rates $(R_j)_{j=1}^{M}$. Thus, the signal can be described as a real function of time $t \geq 0$.

$$s(t) = \sum_{j=1}^{M} A_j \cos(\Omega_j t)\, e^{-R_j t}. \tag{2.1}$$

As explained later in Section 2.10 determination of both the absolute value and the sign of a resonance frequency $\Omega_j$ requires generation of two signals that may be interpreted as real and imaginary parts of one complex signal:

$$s(t) = \sum_{j=1}^{M} A_j e^{-R_j t} \left(\cos(\Omega_j t) + \imath \sin(\Omega_j t)\right) = \sum_{j=1}^{M} A_j e^{\imath \Omega_j t - R_j t} \tag{2.2}$$

*Two-Dimensional (2D) NMR Methods*, First Edition. Edited by K. Ivanov, P.K. Madhu and G. Rajalakshmi.

where throughout this chapter $\imath$ denotes the imaginary unit. Furthermore, viewing $\Omega_j$ and $R_j$ as real and imaginary parts of one, complex frequency $\zeta_j = \Omega_j + \imath R_j \in \mathbb{C}$, the signal simplifies to:

$$s(t) = \sum_{j=1}^{M} A_j e^{\imath \zeta_j t}. \tag{2.3}$$

In general, the amplitude $A_j$ can also be complex, which represents the phase distortion (see Section 2.5).

## 2.3 NMR Spectrum

Parameters $A_j$ and $\zeta_j$, which are of interest for the researcher, are difficult to determine directly from an FID signal. However, the operation known as Fourier transform (FT) converts FID into an easily interpretable frequency-domain representation called *spectrum* and denoted $S(\omega)$ (see Figure 2.1). The spectrum of an FID signal is obtained by calculating an FT integral:

$$S(\omega) = \int_{t\in\mathbb{R}} s(t) e^{-\imath\omega t} dt. \tag{2.4}$$

Each of $M$ components in an FID signal corresponds to one "peak" $S_j(\omega)$ in its spectrum. The peak is described by the Lorentzian function, which is a Fourier transform of a single decaying oscillation $s_j(t) = A_j e^{\imath\zeta_j t}$ (see Figure 2.2). Remembering that $s_j(t) = 0$ for $t < 0$, the integral (2.4) is restricted to $t \in [0, \infty[$ and:

$$\int_{t\geq 0} A_j e^{\imath(\zeta_j - \omega)t} dt = \frac{A_j}{\imath(\omega - \zeta_j)} = \frac{A_j}{R_j + \imath(\omega - \Omega_j)}$$

$$= A_j \left( \frac{R_j}{R_j^2 + (\omega - \Omega_j)^2} + \imath \frac{\Omega_j - \omega}{R_j^2 + (\omega - \Omega_j)^2} \right)$$

The real part of the Lorentzian function $\frac{1}{R_j + \imath(\omega - \Omega_j)}$ is referred to as the absorptive lineshape and the imaginary part is referred to as the dispersive lineshape. They are of the form:

$$\Re\left(\frac{1}{R_j + \imath(\omega - \Omega_j)}\right) = \frac{R_j}{R_j^2 + (\omega - \Omega_j)^2} \tag{2.5}$$

$$\Im\left(\frac{1}{R_j + \imath(\omega - \Omega_j)}\right) = \frac{\Omega_j - \omega}{R_j^2 + (\omega - \Omega_j)^2}. \tag{2.6}$$

## 2.4 The Most Important Features of FT

In this section, we will discuss the properties of the Fourier transform, which will be later useful to explain the effects of the experimental imperfections and signal processing procedures. FT is given by the integral formula (Equation 2.4) and binds the spectrum $S(\omega)$ with the signal $s(t)$. In order for the integral in Equation 2.4 to be well defined, the signal must be sufficiently regular and vanishing rapidly at infinity.

For the purpose of further discussion, we should know, however, that the Fourier transform can be also computed for a wider class of "functions" known as tempered distributions. A useful example is a famous Dirac delta $\delta_p(t)$ at a point $p \in \mathbb{R}$, which is a distribution such that:

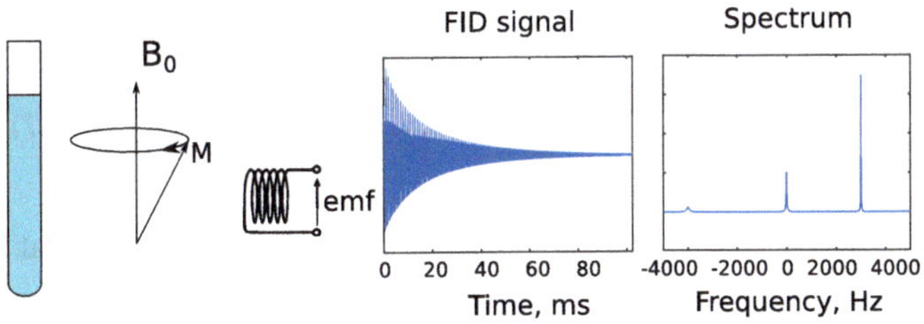


**Figure 2.1** Following the radio-frequency pulse, the macroscopic magnetization of the sample placed in a constant vertical magnetic field ($B_0$) undergoes precession while returning to the equilibrium state (left). The precession induces an electromotive force (emf) in the receiver coil. The resulting voltage signal (FID, in the middle) is a sum of damped oscillations (see Equation 2.2). Its spectrum (right) is a sum of Lorentzian peaks, each corresponding to one frequency component, with linewidth proportional to the decay rate.

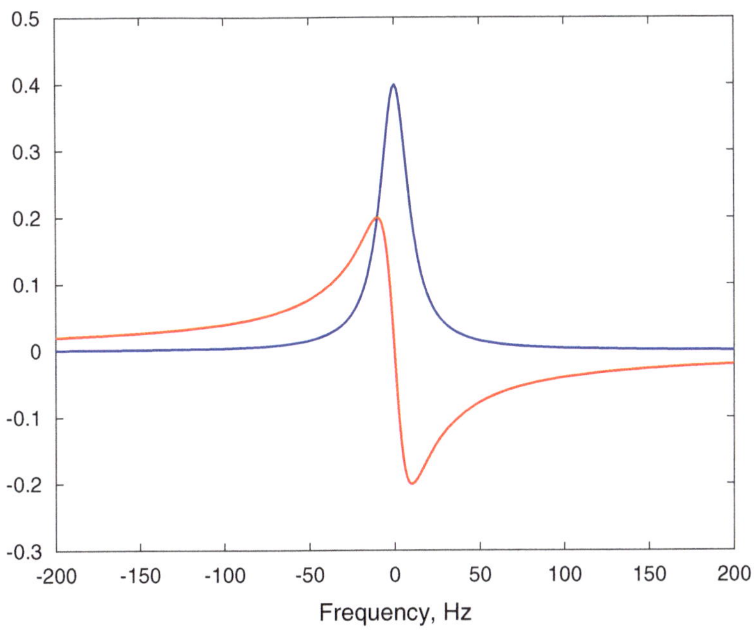


**Figure 2.2** A perfect peak in an NMR spectrum – the Lorentzian function whose real (blue) and imaginary (red) parts are described by Equations 2.5 and 2.6, respectively.

$$\int_{t\in\mathbb{R}} \delta_p(t) f(t) dt = f(p)$$

for every test function $f : \mathbb{R} \to \mathbb{C}$. A slightly more complicated example is so-called Dirac comb, which is an infinite sum of equally spaced Dirac delta functions, $\sum_{n\in\mathbb{Z}} \delta_{n\Delta T}(t)$, where $\Delta T > 0$ is a distance between consecutive "peaks" in the comb. In Section 2.9 we will employ the concept of Dirac comb to discuss the theory of sampling of NMR signals.

The Fourier transform has a number of general properties that are important in the context of NMR. The most significant ones include:

- Fourier transform of a linear combination $\alpha_1 s_1 + \alpha_2 s_2$, $\alpha_1, \alpha_2 \in \mathbb{C}$ of signals $s_1, s_2$ is the linear combination $\alpha_1 S_1 + \alpha_2 S_2$ of the corresponding spectra $S_1, S_2$.

- Fourier transform is invertible and for a spectrum $S$, the signal $s$ is given by:

$$s(t) = \frac{1}{2\pi} \int_{\omega\in\mathbb{R}} S(\omega)e^{\imath\omega t}d\omega. \tag{2.7}$$

In other words, the transformation is faithful, and thus $S$ and $s$ are equivalent representations of an NMR signal. Let us also note incidentally that the integral of the spectrum $\frac{1}{2\pi}\int_{\omega\in\mathbb{R}} S(\omega)d\omega$ is nothing but $s(0)$, which in turn is a sum of the amplitudes $\sum_{j=1}^{M} A_j$ - c.f. Equation 2.2.

- Fourier transform of the product $s_1(t)s_2(t)$ is equal to $\frac{1}{2\pi}(S_1 * S_2)(\omega)$ where the convolution $S_1 * S_2$ is given by:

$$(S_1 * S_2)(\omega) = \int_{u\in\mathbb{R}} S_1(u)S_2(\omega - u)du. \tag{2.8}$$

- Fourier transform of delta function $\delta_p(t)$ is equal to:

$$\int_{t\in\mathbb{R}} \delta_p(t)e^{-\imath\omega t}dt = e^{-\imath\omega p}.$$

- The convolution with $\delta_p$ is the shift of a convolved function:

$$(\delta_p * f)(t) = \int_{u\in\mathbb{R}} \delta_p(u)f(t-u)du = f_p(t) \tag{2.9}$$

where we write $f_p(t)$ to denote the function $f$ shifted by $p$: $f_p(t) = f(t-p)$.

- Fourier transform of Dirac comb is the multiple of a Dirac comb in the frequency domain:

$$\int_{t\in\mathbb{R}} \sum_{n\in\mathbb{Z}} \delta_{n\Delta T}(t)e^{-\imath\omega t}dt = F\sum_{n\in\mathbb{Z}} \delta_{nF}(\omega)$$

with $F = \frac{2\pi}{\Delta T}$. In other words, the "denser" the comb in the primary domain the "sparser" it becomes after FT.

- Finally, it is worth noting the Fourier uncertainty principle (FUP), which is formulated in terms of the variance $\sigma_t^2$ of time variable and the variance $\sigma_\omega^2$ of frequency variable. The distribution $\hat{\rho}$ of the frequency variable $\omega$ is proportional to $|S(\omega)|^2$, $\hat{\rho}(\omega) = \frac{|S(\omega)|^2}{\|S\|_2^2}$, and hence, the expected value and the variance of $\omega$ are given by:

$$\langle\omega\rangle = \int_{\omega\in\mathbb{R}} \omega\hat{\rho}(\omega)d\omega, \quad \sigma_\omega^2 = \langle(\omega - \langle\omega\rangle)^2\rangle = \int_{\omega\in\mathbb{R}} (\omega - \langle\omega\rangle)^2\,\hat{\rho}(\omega)d\omega,$$

Similarly, using the distribution $\rho(t) = \frac{|s(t)|^2}{\|s\|_2^2}$ of the time variable we can define $\langle t\rangle$ and $\sigma_t^2 = \langle(t - \langle t\rangle)^2\rangle$. FUP is stated as the inequality:

$$\sigma_t^2\sigma_\omega^2 \geq \frac{1}{2}, \tag{2.10}$$

which is a quantitative expression of the rule that the more "concentrated" a signal is in the time domain, the "broader" it is in the frequency domain. For more explanations see works of Szantay [1] and next papers in the series.

FUP helps to understand many effects in NMR spectroscopy. Everything that causes signal's "concentration" in the time domain: truncation, relaxation, weighting, etc., results in peak broadening and thus a drop of resolution.

## 2.5 Distortion: Phase

NMR spectrum $S(\omega)$, which is a Fourier transform of an FID signal $s(t)$ (Equation 2.3) is a superposition of Lorentzian functions (peaks) described by Equations 2.5 and 2.6. For the optimal resolution of peaks and thus convenient analysis, we want to work with *phased* spectra, which refers to the situation when all complex amplitudes $A_j$ are actually real. In this case, the real part $\Re(S(\omega))$ of the spectrum consists only of the absorptive lineshapes:

$$\Re(S(\omega)) = \sum_{j=1}^{M} \frac{A_j R_j}{R_j^2 + (\omega - \Omega_j)^2}.$$

The case of non-zero imaginary part of amplitude, $A_j \notin \mathbb{R}$, is referred to as the *phase error*, and it is manifested in the real part of the spectrum $S(\omega)$, which now contains the combination of absorptive and dispersive lineshapes:

$$\Re(S(\omega)) = \sum_{j=1}^{M} \left( \frac{\Re(A_j) R_j}{R_j^2 + (\omega - \Omega_j)^2} + \frac{\Im(A_j)(\omega - \Omega_j)}{R_j^2 + (\omega - \Omega_j)^2} \right).$$

The real part of the spectrum in the presence of the phase error has degraded resolution and distorted positions of peak's maxima (see Figure 2.3).

The phase errors in the NMR signal are usually of two types:

- the zero-order error is represented by the global phase factor $e^{\imath\alpha}$ that modifies every amplitude $A_j$ independently of $j$ and its effect on the spectrum is illustrated in Figure 2.3;
- the first-order error is represented by a local phase factor $e^{\imath\alpha_j}$ that depends linearly on resonance frequency $\Omega_j$ and its effect on the spectrum is illustrated in Figure 2.4.

## 2.6 Kramers-Kronig Relations and Hilbert Transform

As a consequence of the causality principle (see [2, 3]), the real and imaginary parts of a spectrum $S(\omega) = \Re(S(\omega)) + \imath\Im(S(\omega))$ $s(t)$ can be recovered from each other using the Kramers-Kronig relations:

$$\Re(S(\omega)) = \frac{1}{\pi} \int\limits_{\omega' \in \mathbb{R}} \frac{\Im(S(\omega'))}{\omega' - \omega} d\omega'$$

$$\Im(S(\omega)) = -\frac{1}{\pi} \int_{\omega' \in \mathbb{R}} \frac{\Re(S(\omega'))}{\omega' - \omega} d\omega'. \tag{2.11}$$

The integral operation (Equation 2.11) is also known as the Hilbert transform. The Kramers-Kronig relations are explained in Figure 2.5. Signal $s(t)$ (Figure 2.5a) and its spectrum $S(\omega)$ (Figure 2.5b) can be obtained from each other via the Fourier transform. The spectrum shown in Figure 2.5d is obtained from that in Figure 2.5b) by setting the imaginary part to zero. The inverse Fourier transform of such a real spectrum results in a complex signal in

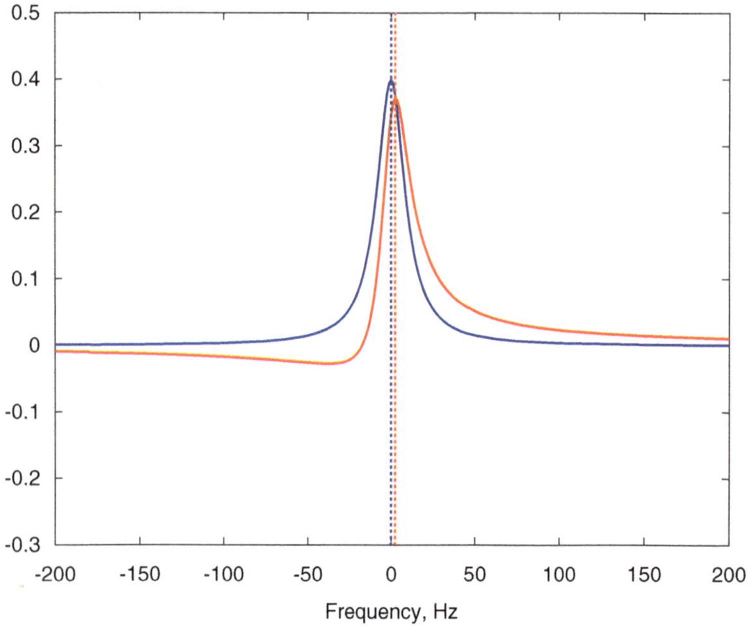


**Figure 2.3** The NMR peak disturbed by the phase error. As can be seen, both real (blue) and imaginary (red) parts of the function are mixtures of absorptive and dispersive lineshapes. Notably, the peak maximum (marked with a dashed line) is shifted due to phase error.

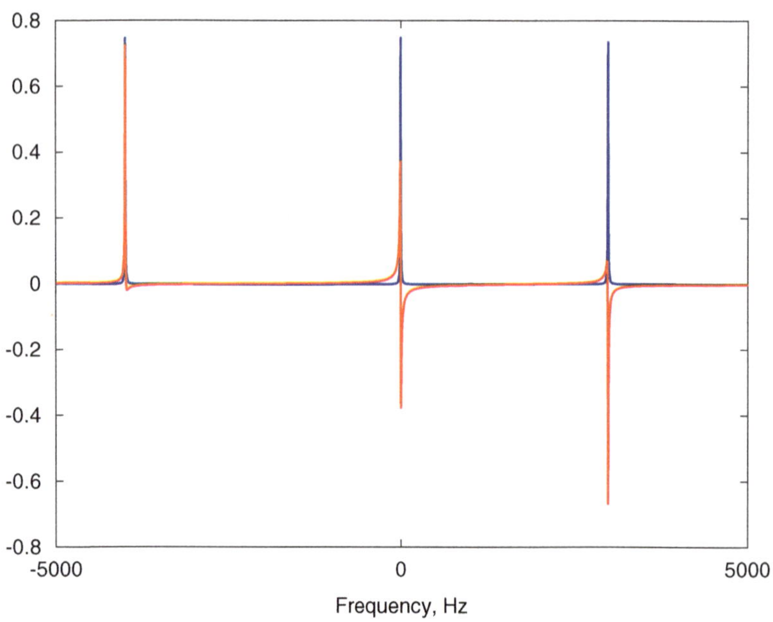


**Figure 2.4** The phase error of the first order, i.e. linearly dependent on the resonance frequency. Spectrum without phase error (with absorptive lineshapes) is marked with blue, spectrum with linear phase error is marked with red.

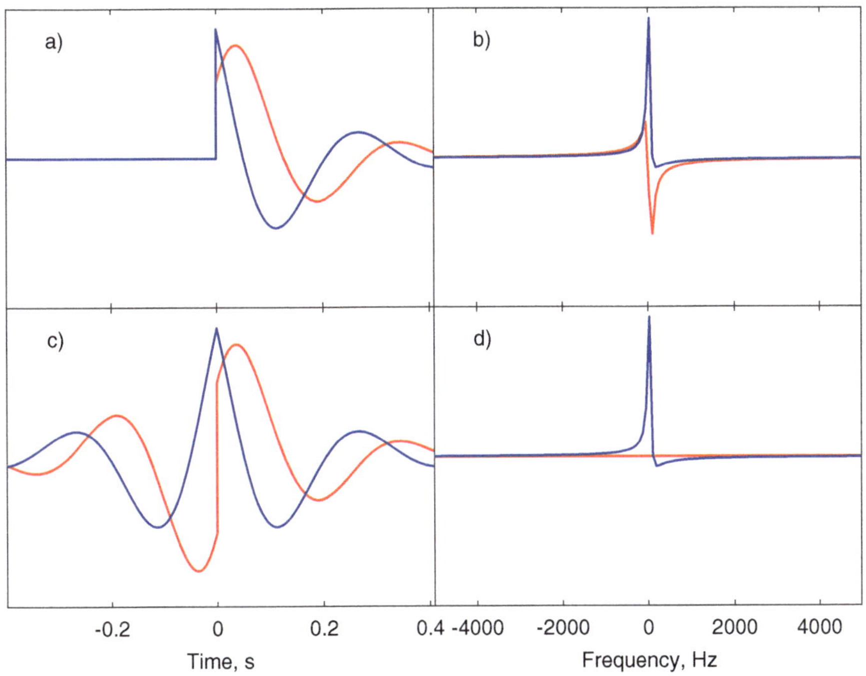


**Figure 2.5** The visualization of Kramers-Kronig relations. Two representations of the NMR time-domain signals: (a) FID and (c) virtual-echo and the corresponding spectra (b) and (d). Real and imaginary parts are marked in blue and red, respectively. Importantly, the spectrum shown in panel (d) has an imaginary part equal to zero. The effect of the non-zero phase on the signal in the time and frequency domains is also shown.

the time domain (Figure 2.5c). Its real and imaginary parts are just even and odd parts of the real and imaginary components of $s(t)$ (Figure 2.5a), accordingly. In this way, the signal in Figure 2.5c) can also be obtained by the time-reversal and a complex conjugate of $s(t)$, which can be written as:

$$s_{\mathrm{VE}}(t) = \begin{cases} s(t) & t \geq 0 \\ \bar{s}(-t) & t < 0. \end{cases} \tag{2.12}$$

We call the $s_{\mathrm{VE}}(t)$ signal in (Equation 2.12) the virtual echo (VE). The original signal $s(t)$ can be obtained from $s_{\mathrm{VE}}(t)$ by zeroing the signal for $t < 0$. Hilbert transform corresponds to a direct transition from panel (d) to panel (b) in Figure 2.5. In practice, it follows the way of $d) \rightarrow c) \rightarrow a) \rightarrow b)$, which allows using the efficient fast Fourier transform algorithm. For extended discussion of the VE theory and practice see [3].

## 2.7 Distortion: Truncation

While the Fourier transform involves the integration over an infinite time domain (see Equation 2.4), in practice, the signal measurement stops well before it completely vanish in noise. As explained later (Section 2.9), in the indirect dimensions the collection of data points is so time-consuming that we rarely sample evolution times so long that the signal fully decays well below the noise level. Therefore, the signal is often truncated, i.e. measured shorter than justified by the relaxation rate.

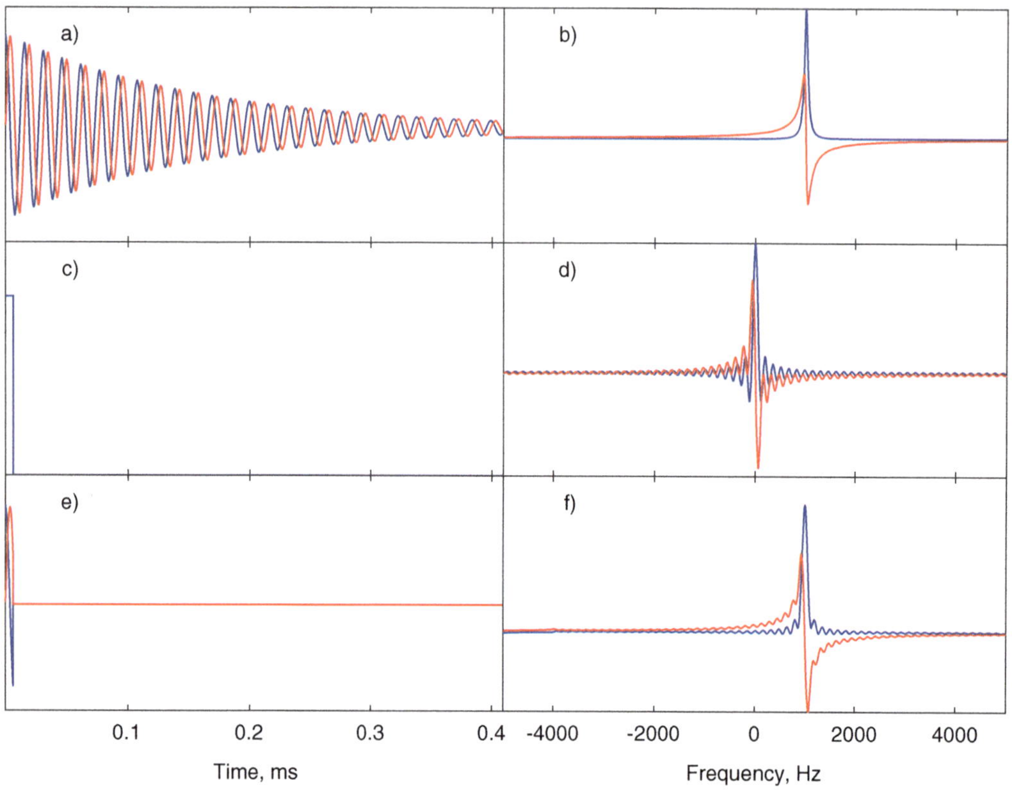


**Figure 2.6** The effect of NMR signal truncation is explained by the convolution theorem. The panels show: perfect (non-truncated) signal (a) and its spectrum (b); step function representing truncation (c) and its Fourier transform (d); the product of (a) and (c) i.e. a truncated signal and its Fourier transform, i.e. the convolution of (b) and (d).

As shown in Figure 2.6, the truncation can be described by replacing an FID component $\exp(\imath\zeta t)$ by its product with the step function:

$$\chi_T(t) = \begin{cases} 1 & 0 \le t \le T \\ 0 & \text{otherwise.} \end{cases}$$

where $T$ stands for the maximum evolution time. Exploiting the convolution property of the Fourier transform (Section 2.4) we can write Fourier transform of the product $\chi_T(t)e^{\imath\zeta t}$ as the convolution $\hat{\chi}_T(\omega) * \frac{\imath}{(\zeta-\omega)}$ where the Fourier transform $\hat{\chi}_T(\omega)$ of the step function $\chi_T(t)$ is the sinc (*sinus cardinalis*) function given by:

$$\hat{\chi}_T(\omega) = -2\frac{e^{\frac{\imath\omega T}{2}}\sin\left(\frac{\omega T}{2}\right)}{\omega}.$$

The convolution with sinc affects spectral lineshapes in two unfavorable ways. First, it broadens the peaks inversely proportionally to $T$. Second, when combined with zero-filling (Section 2.12), it generates "sinc wiggles." These artifacts may heavily distort the spectrum. They can be avoided by using weighting, as described in Section 2.11, but at the cost of further line broadening.

Notably, the fact that truncation leads to line broadening is the manifestation of the FUP mentioned in Section 2.4. The shorter is the signal's component in the time domain, the broader is its representation, that is a peak, in the frequency domain.

## 2.8 Distortion: Noise and Multiple Scans

In a real experiment, the ideal NMR signal $s(t)$ gets contaminated with noise $\varepsilon$ and takes the form of:

$$(s(0) + \varepsilon_0, s(\Delta T) + \varepsilon_1, \dots, s(N\Delta T) + \varepsilon_N).$$

The noise content in the signal is characterized by the signal-to-noise ratio (SNR), which can be increased by adding the FIDs from repeated experiments. In order to explain the mathematics behind this procedure, let us focus on an arbitrary (e.g. $s(0)$) point of the multi-scan FID. Clearly by adding the scans, the resulting signal gets multiplied by the number $K$ of measurements, $Ks(0)$. However, in this summation process, the noise level is multiplied only by the square root $\sqrt{K}$. Indeed, assuming that the noise in the $j$th measurement of $\varepsilon(0)_j$ is described by a Gaussian distribution with a mean value 0 and standard deviation $\sigma$, and the noise values are mutually independent between scans, the sum $\varepsilon(0)_1+\varepsilon(0)_2+\dots+\varepsilon(0)_j+\dots+\varepsilon(0)_K$ is a random variable with a mean value 0 and standard deviation $\sqrt{K}\sigma$. In particular the resulting SNR grows by the factor $K/\sqrt{K} = \sqrt{K}$, which is illustrated in Figure 2.7.

## 2.9 Distortion: Sampling and DFT

The complex function $s(t)$ of real parameter $t \in \mathbb{R}$ (time) described by Equation 2.3 is a relevant mathematical representation of an FID signal. In practice, however, the NMR spectrometer cannot measure continuously $s(t)$ (i.e. for all reals) but samples it at certain discrete values $n\Delta T$. Indeed, the sampling in the direct dimension of an NMR signal is performed by an analog-to-digital converter, while the indirect dimensions are sampled by setting appropriate lengths of the delays in the pulse sequence.

The regular sampling can conveniently be represented by the Dirac comb $\sum_{n\in\mathbb{Z}} \delta_{n\Delta T}(t)$ discussed in Section 2.4. Its product with the signal $s(t)$ gives a representation of sampled signal (see Figure 2.8):

$$\left(\sum_{n\in\mathbb{Z}} \delta_{n\Delta T}(t)\right) \cdot s(t) = \sum_{n\in\mathbb{Z}} s(n\Delta T)\delta_{n\Delta T}(t).$$

Fourier transform of the sampled signal $\left(\sum_{n\in\mathbb{Z}} \delta_{n\Delta T}(t)\right) \cdot s(t)$ can in turn be computed using Equation 2.8 and it is equal $\frac{F}{2\pi}\sum_{n\in\mathbb{Z}} (\delta_{nF} * S)(\omega)$. Using Equation 2.9 we see that this is the sum of the shifts of $S$:

$$\frac{F}{2\pi}\sum_{n\in\mathbb{Z}} (\delta_{nF} * S)(\omega) = \frac{F}{2\pi}\sum_{n\in\mathbb{Z}} S_{nF}(\omega)$$

where $S_{nF}$ is the copy of $S$ shifted by $nF$ where $F = \frac{2\pi}{\Delta T}$.

Note that if the spectrum $S$ is empty outside the interval $[F_1, F_2] \subset \mathbb{R}$ of length $F_2 - F_1$ smaller than $F$ then the sampled signal can be used for the perfect recovery of $S$. Moreover, due to the invertibility of the Fourier transform the sampled signal encodes the full, continuous signal $s(t)$. The observation that the discrete sampling

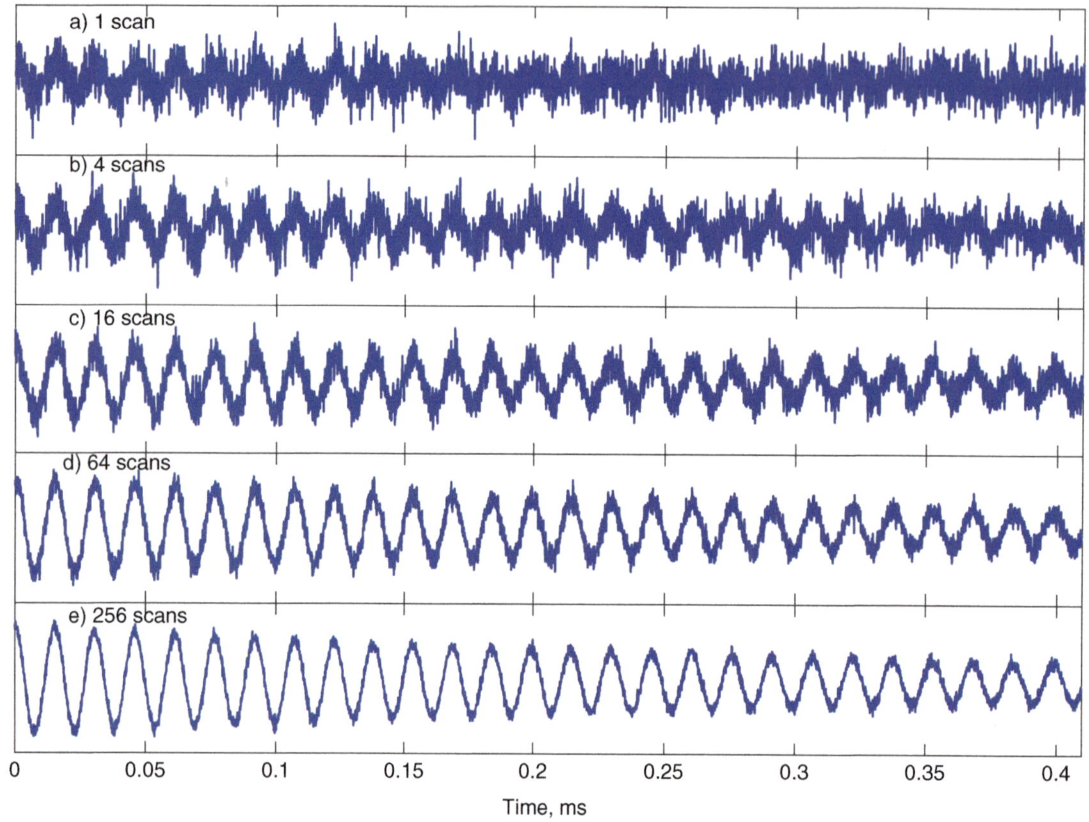


**Figure 2.7** The average FID in a multi-scan experiment. It can be seen, that the signal-to-noise ratio grows as a square root of the number of accumulated scans. Note, that for a given FID it also decreases in time, since signal decays and noise level is constant.

at the rate equal or higher than the spectral bandwidth of a signal completely describes the signal, is known as the Kotelnikov-Shannon-Nyquist sampling theorem. (see [4])

In practical NMR experiments, we assume certain bandwidth of a measured FID signal and based on this assumption, the spectrometer sets the sampling rate. If the assumption is wrong, i.e. there are peaks outside the assumed band, then they appear at wrong positions or are *folded* (see Figure 2.9). This phenomenon, known also as *aliasing*, is sometimes triggered intentionally, especially in the wide-band spectral indirect dimensions like $^{13}$C. During analysis, the true positions of the folded peaks can be recovered by simple calculations. One has to be careful, however, since sometimes the folded peaks may overlap with the peaks of properly sampled components.

The infinite Dirac comb $\sum_{n\in\mathbb{Z}} \delta_{n\Delta T}(t)$ represents an impossible situation of sampling carried out forever. To describe the real experiment, it should be replaced by the finite Dirac comb $\sum_{n=0}^{K-1} \delta_{n\Delta T}(t)$ representing the finite sampling. The Fourier transform $S_{\text{fin}}(\omega)$ of the finitely sampled signal $\sum_{n=0}^{K-1} s(n\Delta T)\delta_{n\Delta T}(t)$ is equal:

$$S_{\text{fin}}(\omega) = \sum_{n=0}^{N} s(n\Delta T)e^{-\imath n\Delta T\omega}. \tag{2.13}$$

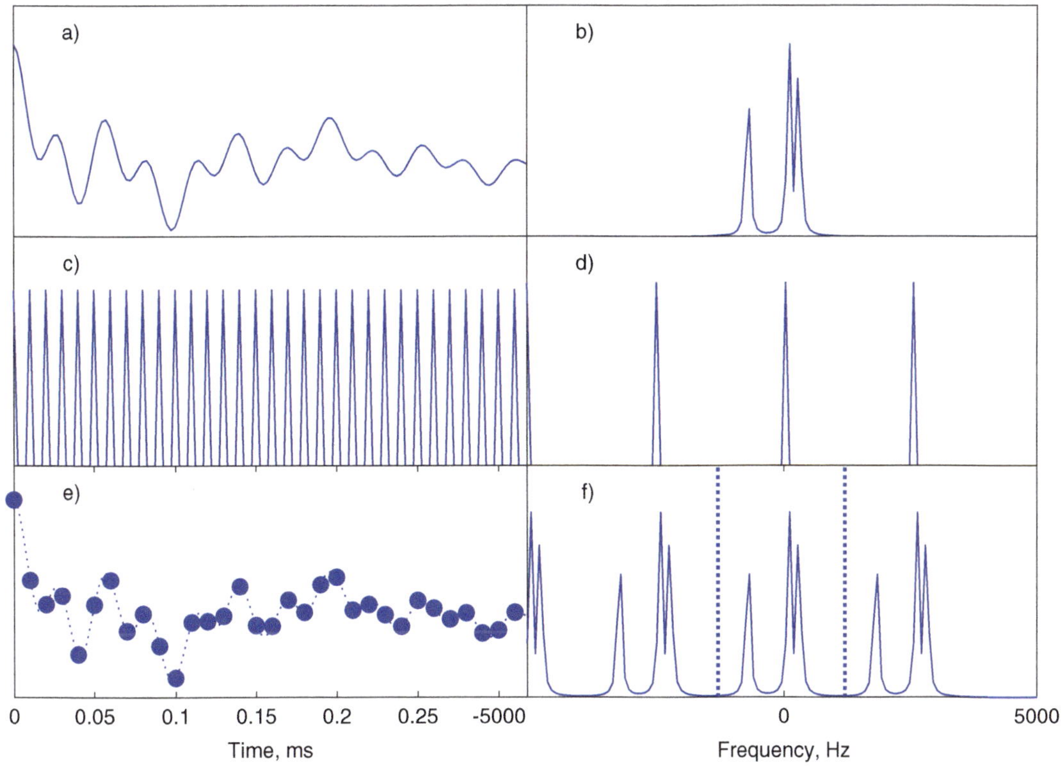


**Figure 2.8** The sampling of an NMR signal at the proper rate. (a) Continuous (perfect) FID signal and its spectrum (b); (c) the Dirac comb representing the regular sampling and its Fourier transform (d); (e) the sampled FID signal, i.e. the product of (a) and (c) and its Fourier transform (f). According to the convolution theorem, the latter is a convolution of (b) and (d) resulting in an infinite number of "copies" of (b). The central "copy", marked with dashed lines in panel (f), is a proper NMR spectrum obtained using discrete sampling.

Remarkably the sampled signal:

$$(s(0), s(\Delta T), \dots, s((K-1)\Delta T)) \tag{2.14}$$

can be recovered from the finite number (of $K$) samples of $S_{\text{fin}}$. Indeed, the samples of $S_{\text{fin}}$ at $\omega_l = \dfrac{2\pi l}{K\Delta T}$ equal:

$$S_{\text{fin}}(\omega_l) = \sum_{n=0}^{K-1} s(n\Delta T) e^{-2\pi\imath\frac{ln}{K}} \tag{2.15}$$

or in other words the finite sequence:

$$\left(S_{\text{fin}}(0), S_{\text{fin}}\left(\frac{2\pi}{K\Delta T}\right), \dots, S_{\text{fin}}\left(\frac{2\pi(K-1)}{K\Delta T}\right)\right) \tag{2.16}$$

is the discrete fourier transform (DFT) of the sampled signal (2.14). Using the inverse formula for DFT we get:

$$s(n\Delta T) = \frac{1}{K}\sum_{l=0}^{K-1} S_{\text{fin}}\left(\frac{2\pi l}{K\Delta T}\right) e^{2\pi\imath\frac{ln}{K}}, \tag{2.17}$$

where $n = 0, 1, \dots, N-1$.

Notably, the DFT can be computed using an efficient *Fast Fourier Transform* algorithm (see [5]), which in its fastest version, requires the number of sampling points to be the power of 2.

## 2.10 Quadrature Detection

The resonance frequencies $\Omega_j$ usually differ only slightly, which is why the ppm units are used to describe the horizontal spectral axis. However, their absolute values can be high (up to 1.2 GHz for $^1$H for the largest magnets commercially available in 2021). As explained in Section 2.9, the sampling rate has to be at least equal to the bandwidth of the signal. A high sampling rate of analog-to-digital converters always comes at the cost of resolution. To overcome this problem, the frequencies present in the signal are usually shifted by mixing the detected FID with the carrier frequency of the pulse followed by the low-pass filtering. This corresponds to the conversion of the coordinates to the rotating frame in the vector model (see Chapter 1). The conversion is done prior to sampling and reduces the detected frequencies by $3-4$ orders of magnitude, typically from MHz to kHz. Then, it is very easy to sample at the rate required by the Kotelnikov-Shannon-Nyquist theorem.

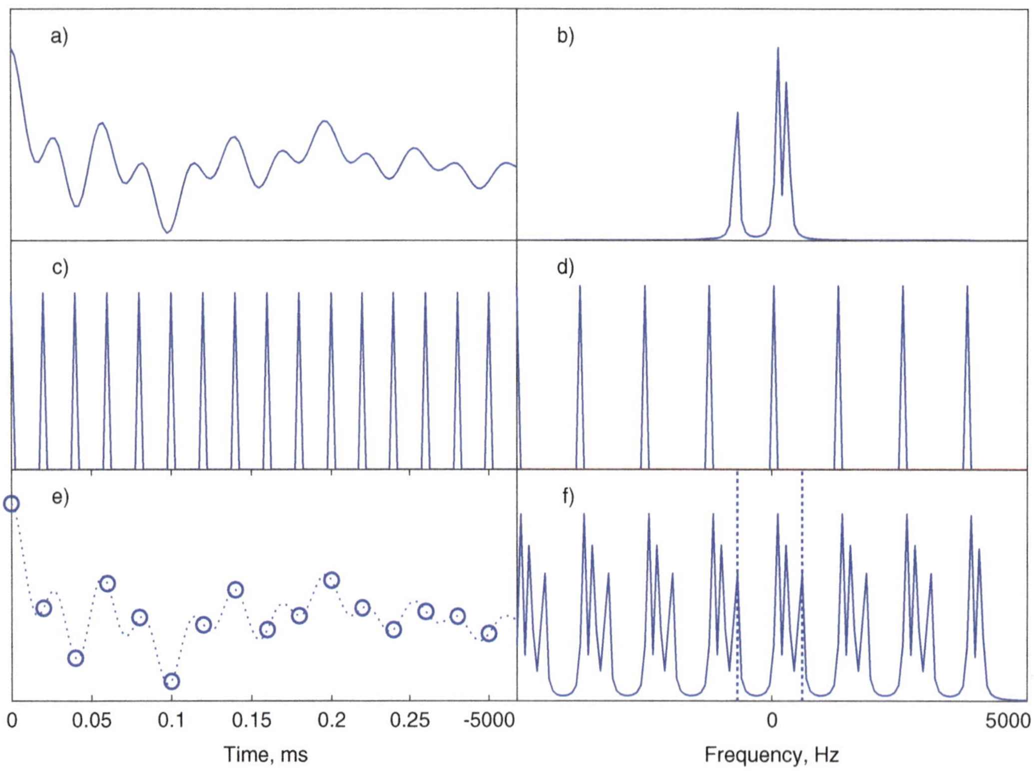


**Figure 2.9** The sampling of an NMR signal at a too low rate. (a) Continuous (perfect ) FID signal and its spectrum (b); (c) the Dirac comb representing the regular sampling below Nyquist rate and its Fourier transform (d); (e) the sampled FID signal, i.e. the product of (a) and (c) and its Fourier transform (f). According to the convolution theorem, the latter is a convolution of (b) and (d) resulting in an infinite number of "copies" of (b). The region marked with dashed lines in panel (f), is assumed spectral width of a signal corresponding to a sampling rate in (c). A too low sampling rate results in a too "dense" train in (d) and thus overlap of "copies" in (f) i.e. *aliasing*.

However, the frequencies in the spectrum can be either lower or higher than the frequency of the rotating frame and thus appearing negative and positive after subtraction of the carrier frequency. Therefore, the Fourier transform has to determine the sign of the peak frequency. As shown in Figure 2.10, only the complex Fourier transform can do this. Thus, two phase-shifted FIDs have to be acquired in an NMR experiment.

The procedure of acquiring complex NMR signal is referred to as *the quadrature detection*. While it is quite straightforward to mathematically describe both acquisition and processing of the complex signal in one dimension, the situation gets complicated in the case of multidimensional signals (see Section 2.14).

## 2.11 Processing: Weighting

The FID signal $s(t)$ is expected to be a combination of decaying periodic components $e^{\imath\zeta t}$ where $\zeta = \Omega + \imath R$ and $\Omega \in \mathbb{R}$ and $R > 0$. In particular, since the signal decays in time, for large times the noise prevails in the measured signal (as was demonstrated in Figure 2.7).

During measurement, we always try to find a compromise between two undesired effects: signal truncation leading to line broadening (and *sinc* wiggles, see Section 2.7) and too long acquisition resulting in an increased noise.

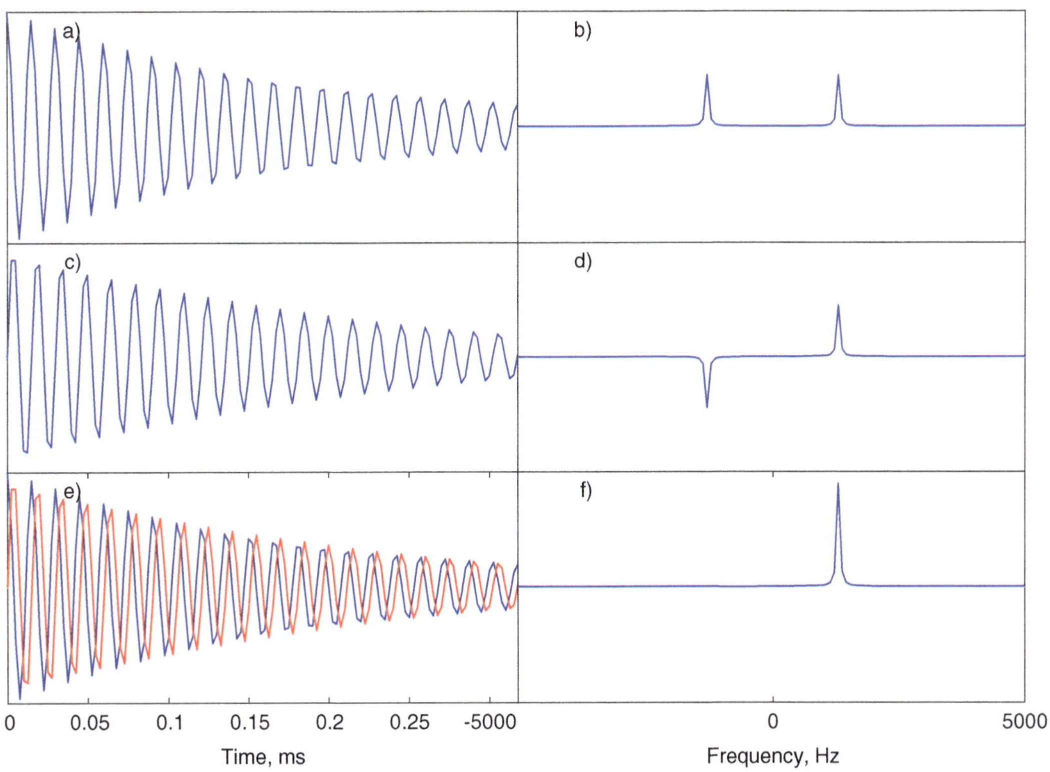


**Figure 2.10** The concept of quadrature detection. (a) The real FID signal with cosine modulation and a real part of its spectrum (b); (c) The real FID with sine modulation and a real part of its spectrum (d); (e) The complex FID whose real (blue) and imaginary (red) parts correspond to (a) and (c), respectively and a real part of its spectrum. Note that spectra of real signals (b) and (d) allow to determine the value of the frequency, but not its sign. Both sign and frequency are determined by their sum, (f), corresponding to the complex FT.

Luckily, to a certain extent, the non-optimally measured signal can be improved in a post-acquisition procedure known as *weighting* or *apodization*.

The weighting is performed by multiplying the measured signal $s(t)$ by an arbitrarily chosen function $W(t)$, which is usually equal to 1 for the initial point of $s(t)$ and decays toward zero for large $t$. As shown in Figure 2.12, this results in an improved signal-to-noise ratio, since we assign higher weights to points with higher SNR. However, it also leads to line broadening, according to the FUP introduced in Section 2.4. The multiplication by the decaying weighting function, effectively "concentrates" the signal in time domain, broadening its frequency-domain representation. The same compromise between desired and undesired effects of weighting is shown in Figure 2.13 depicting the elimination of the truncation effects.

The popular examples of such peak-broadening (sensitivity-boosting) weighting functions are: cosine-bell (Figure 2.11a) and squared cosine-bell (Figure 2.11c). The exponentially decaying weighting functions are also commonly used (Figure 2.11e). All of them improve the signal-to-noise ratio while decreasing the resolution. The spectrum of a weighted FID signal is a convolution of the original spectrum and the FT of the weighting function (Figures 2.11b and 2.11d).

The weighting can be also used in an opposite manner, i.e. improve resolution (narrow linewidths) by sacrificing sensitivity. An example of such weighting function is a shifted sine-bell, whose time- and frequency-domain representations are shown in Figures 2.11g and 2.11h.

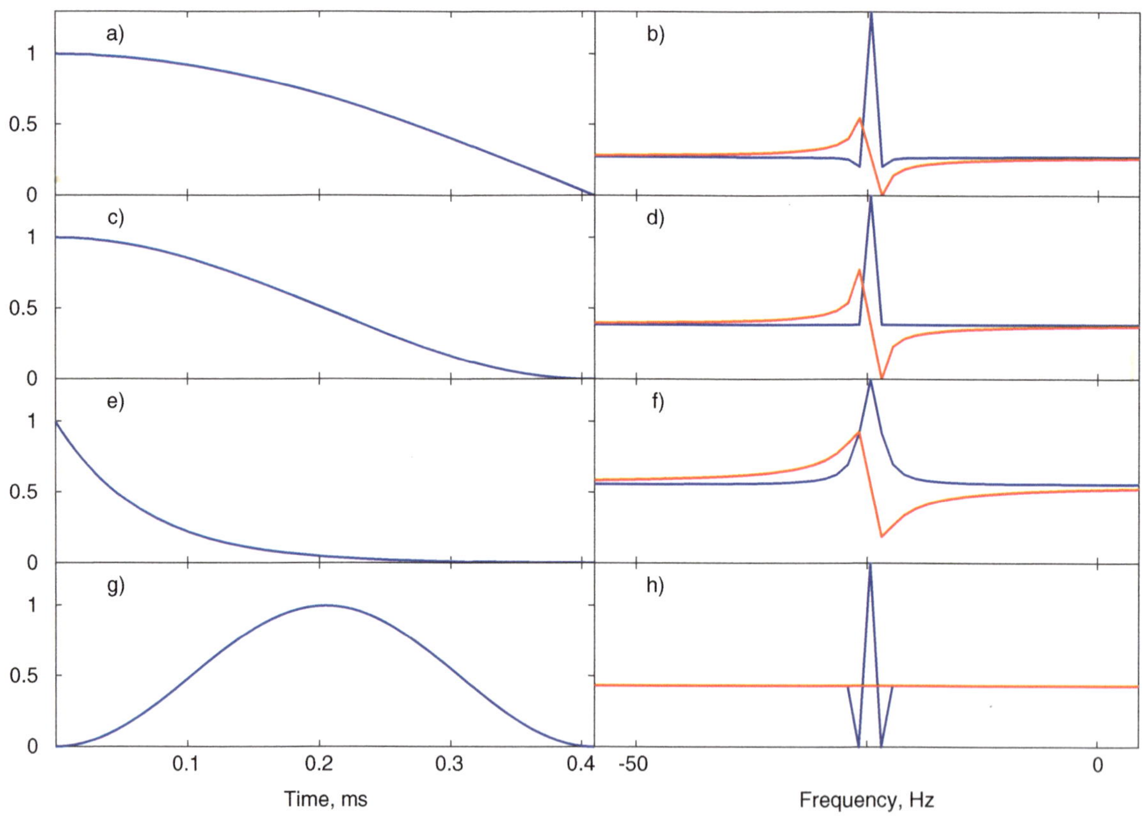


**Figure 2.11** Four examples of commonly used weighting functions: (a) cosine-bell, (c) squared cosine-bell, (e) exponential (decay 15 Hz), (g) shifted squared cosine-bell and their Fourier transforms (lineshape effects) (b),(d),(f), and (h), respectively.

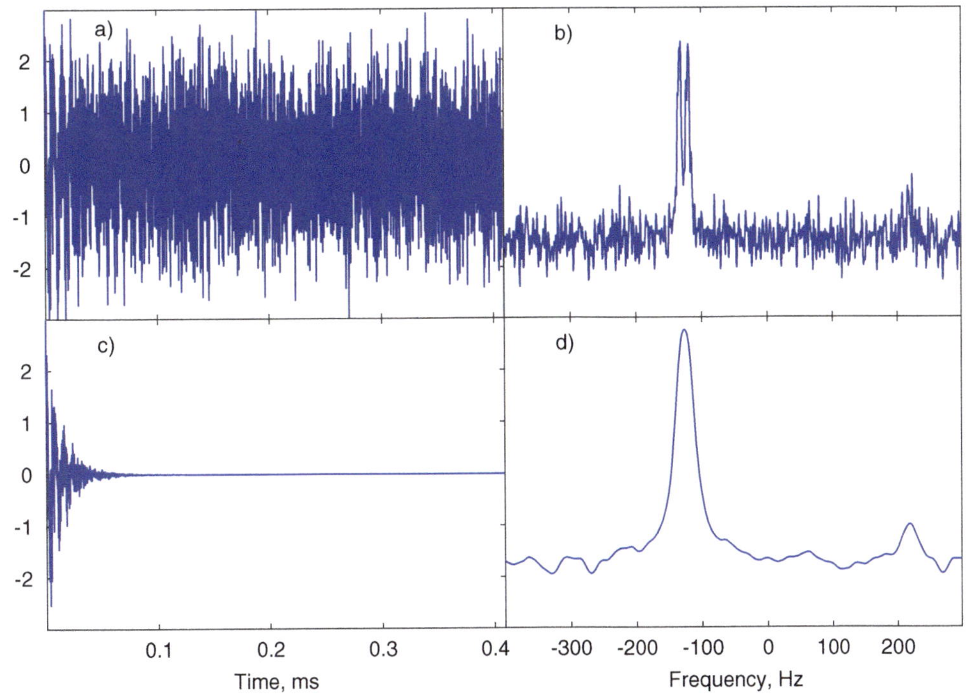


**Figure 2.12** How weighting allows to balance between resolution and sensitivity. (a) non-weighted signal with 3 decaying components and its spectrum (b). The peak at 220 Hz is hardly visible, but the doublet at -125 Hz is well resolved. (c) The situation changes when FID is multiplied (weighted) by an exponentially decaying function (decay rate of 64 Hz). In the spectrum of a weighted signal (d) the peak at 220 Hz is well visible, but the doublet is not resolved anymore.

## 2.12 Processing: Zero Filling

The fast Fourier transform algorithm, used to calculate the discrete FT (see Section 2.9) requires the same number of points in the input (time domain) and output (frequency domain). However, one can increase the number of spectral points to any desired value by *zero filling*, i.e. extending the FID by adding artificial data points equal to zero at the end. As shown in Figure 2.14, the resulting digital resolution in the frequency domain is increased, although no new information is added, i.e. the spectrum from Figure 2.14d is effectively just an interpolation of that from Figure 2.14b.

Besides, the "cosmetic" effect, i.e. smooth-looking spectrum, zero filling has several other advantages. First, as shown in Figure 2.15, it improves sensitivity, at least up to a certain level, corresponding to the moment when the top of the peak is perfectly digitized (there is an actual spectral point on the peak top). Second, the Hilbert transform (see Section 2.6) can be applied to recover both the real and imaginary parts of the complex FID having only the real part of the spectrum, assuming that it was obtained from a signal extended with zeros at least 2×.

## 2.13 Fourier Transform in Multiple Dimensions

The definition of Fourier transform of a function of a real variable $t \in \mathbb{R}$ given by (2.4) can be extended to the multivariable case $\boldsymbol{t} \in \mathbb{R}^n$. The resulting $n$-dimensional version of Fourier analysis proved to be particularly

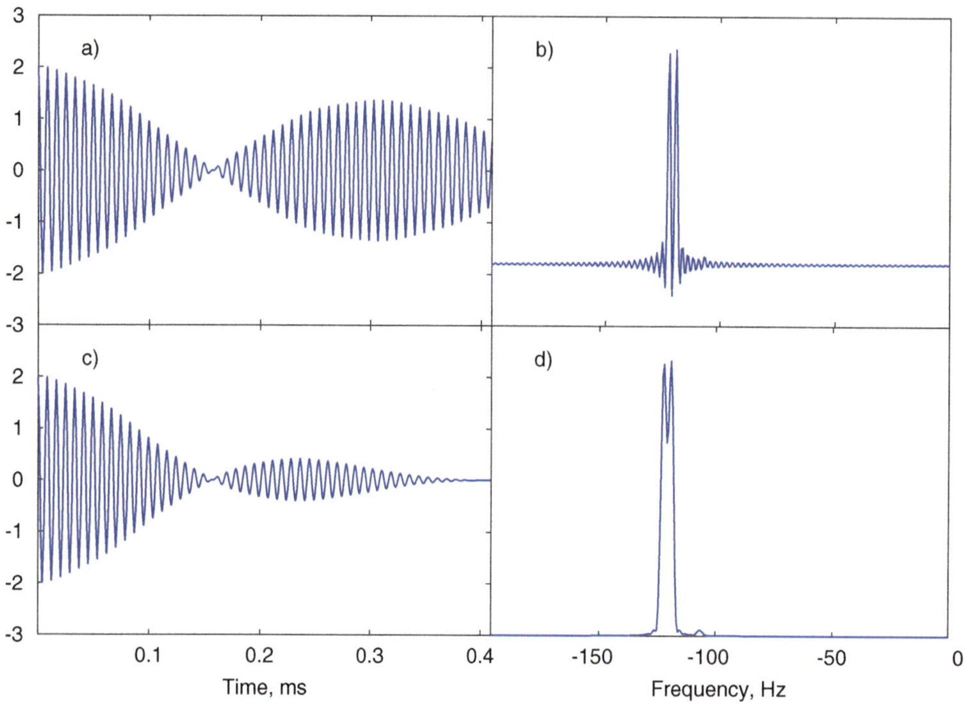


**Figure 2.13** How weighting eliminates the effects of truncation. (a) The original truncated signal with three components and its spectrum (b). The small peak at -106 Hz is heavily disturbed by *sinc* wiggles. Panel (c) shows the same signal but weighted using the squared cosine-bell function. Its spectrum (d) is free of *sinc* wiggles. However, the increased linewidth affects the resolution of the doublet peaks.

useful for the processing of a signal $s(t_1, t_2, \dots, t_n)$ acquired in the $n$-dimensional NMR experiment. The definition of $n$-dimensional Fourier transform looks particularly neat when written in the vectorial notation: denoting $\boldsymbol{t} = (t_1, t_2, \dots, t_n)$ and $\boldsymbol{\omega} = (\omega_1, \omega_2, \dots, \omega_n)$, the Fourier transform $S(\boldsymbol{\omega})$ of a function $s(\boldsymbol{t})$ is defined by the formula:

$$S(\boldsymbol{\omega}) = \int_{\boldsymbol{t} \in \mathbb{R}^n} s(\boldsymbol{t})\, e^{-\imath \boldsymbol{\omega} \cdot \boldsymbol{t}} d^n \boldsymbol{t} \tag{2.18}$$

where $\boldsymbol{\omega} \cdot \boldsymbol{t}$ denotes the scalar product of $\boldsymbol{\omega}$ and $\boldsymbol{t}$:

$$\boldsymbol{\omega} \cdot \boldsymbol{t} = t_1\omega_1 + t_2\omega_2 + \dots t_n\omega_n.$$

The inversion formula has the form (cf. Equation 2.7):

$$s(\boldsymbol{t}) = \frac{1}{(2\pi)^n} \int_{\boldsymbol{\omega} \in \mathbb{R}^n} S(\boldsymbol{\omega})\, e^{\imath \boldsymbol{t} \cdot \boldsymbol{\omega}} d^n \boldsymbol{\omega}.$$

As explained in Section 2.9, the continuous Fourier transform in practice needs to be replaced by its finite discrete version for which the following notation will be needed: if $K \in \mathbb{N}$ then we denote $\{0, 1, \dots, K-1\} = X_K$ and more generally given an $n$-tuple of natural numbers $\boldsymbol{K} = (K_1, K_2, \dots, K_n)$ we denote $X_{K_1} \times X_{K_2} \times \dots X_{K_n}$ by $\boldsymbol{X}_{\boldsymbol{K}}$; generic

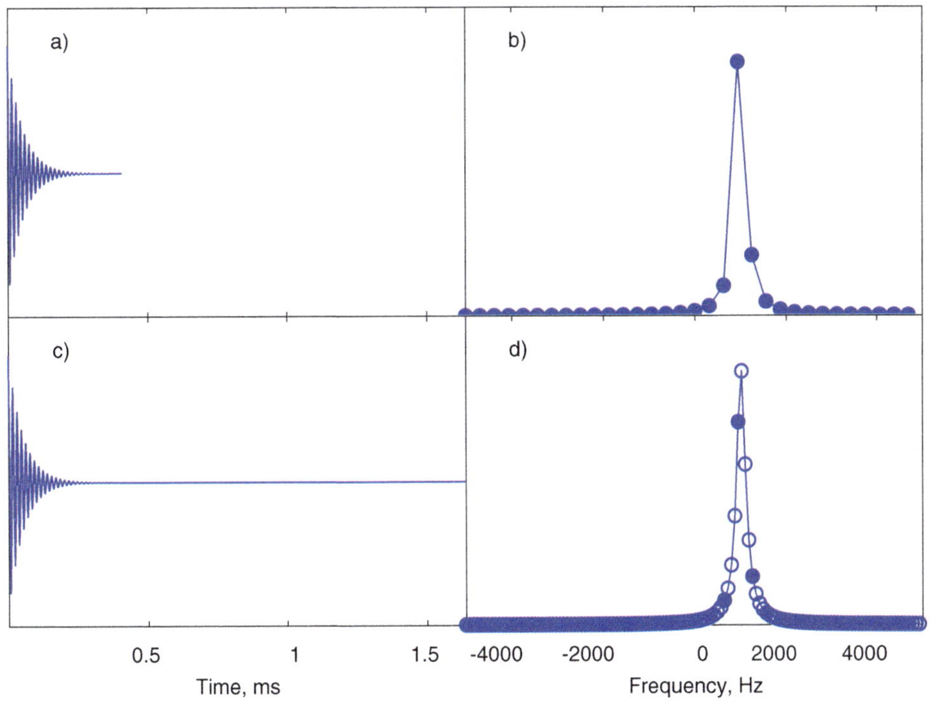


**Figure 2.14** The effect of zero filling on digital resolution. (a) The FID signal without zero filling and its spectrum (b); (c) The same FID signal extended 4× by adding zeros and its spectrum (d). Blue and red lines correspond to real and imaginary parts, respectively.

elements of $\boldsymbol{X_K}$ will be denoted by letters $\boldsymbol{a} = (a_1, a_2, \dots, a_n), \boldsymbol{b} = (b_1, b_2, \dots, b_n) \in \boldsymbol{X_K}$. The DFT of a function $f : \boldsymbol{X_k} \to \mathbb{C}$ is a function $g : \boldsymbol{X_k} \to \mathbb{C}$ given by (cf. Equation 2.15):

$$g(\boldsymbol{b}) = \sum_{\boldsymbol{a} \in \boldsymbol{X_K}} f(\boldsymbol{a}) e^{-2\pi\imath \sum_{i=1}^{n} \frac{a_i b_i}{K_i}}$$

and the inverse of the DFT is of the form (cf. Equation 2.17):

$$f(\boldsymbol{a}) = \frac{1}{K_1 \cdot K_2 \cdot \dots \cdot K_n} \sum_{\boldsymbol{b} \in \boldsymbol{X_K}} g(\boldsymbol{b}) e^{2\pi\imath \sum_{i=1}^{n} \frac{b_i a_i}{K_i}}.$$

In order to link discrete version of $n$-dimensional Fourier transform with the NMR signal processing let us consider an $n$-dimensional NMR signal $s(\boldsymbol{t})$. Each time dimension $t_j$ of $\boldsymbol{t}$ is sampled according to the sampling schedule fixed by $\Delta T_j$ and $K_j$. The regular sampling defines a function:

$$f : \boldsymbol{X_K} \ni \boldsymbol{a} \mapsto s(a_1 \Delta T_1, a_2 \Delta T_2, \dots, a_n \Delta T_n) \in \mathbb{C}$$

that represents the signal that is discretely sampled on a *Nyquist grid*. The DFT $g$ of $f$ corresponds to the discretized version of the spectrum $S_{\text{fin}}$ by the formula (cf. Equation 2.16):

$$S_{\text{fin}}\left(b_1 \frac{2\pi}{K_1 \Delta T_1}, b_2 \frac{2\pi}{K_2 \Delta T_2}, \dots, b_n \frac{2\pi}{K_n \Delta T_n}\right) = g(\boldsymbol{b}).$$

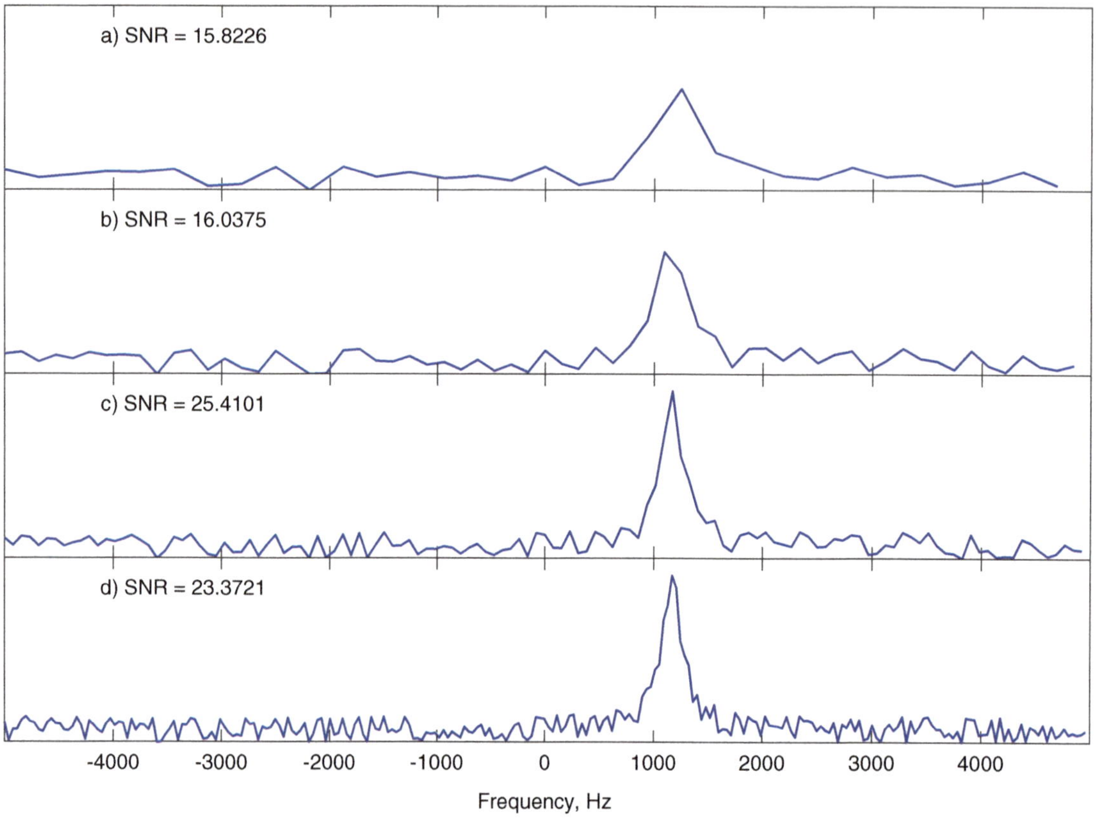


**Figure 2.15** The digital resolution and signal-to-noise ratio (SNR). Panels (a)–(d) show the same spectrum using zero-filling up to: (a) 64; (b) 128; (c) 256; (d) 512 points. Interestingly, the SNR is maximum in (c), when the spectral point appears exactly at the top of the peak. The SNR does not increase with further zerofilling.

In particular, if in the $j$th time dimension we measure $K_j$ samples at the multiples of $\Delta T_j$ then the resolution in the $j$th spectral dimension equals $\frac{2\pi}{K_j \Delta T_j}$.

## 2.14 Quadrature Detection in Multiple Dimensions

The quadrature detection in 1D spectra is realized through the acquisition of the two modulations, interpreted as real and imaginary parts of a complex NMR signal (see Section 2.10). Similarly, in multiple ($N$) dimensions one has to acquire $2^N$ modulations to obtain absorptive Lorentzian line shapes in all dimensions. For example, in a 2D experiment acquired using States quadrature method (see [6]), we acquire two modulations $s_{\cos}(t_1, t_2)$ and $s_{\sin}(t_1, t_2)$, whose shape for a single peak $(\Omega_1, \Omega_2)$ is as follows:

$$s_{\cos}(t_1, t_2) = e^{\iota\Omega_2 t_2} \cos(\Omega_1 t_1) \tag{2.19}$$

and:

$$s_{\sin}(t_1, t_2) = e^{\iota\Omega_2 t_2} \sin(\Omega_1 t_1). \tag{2.20}$$

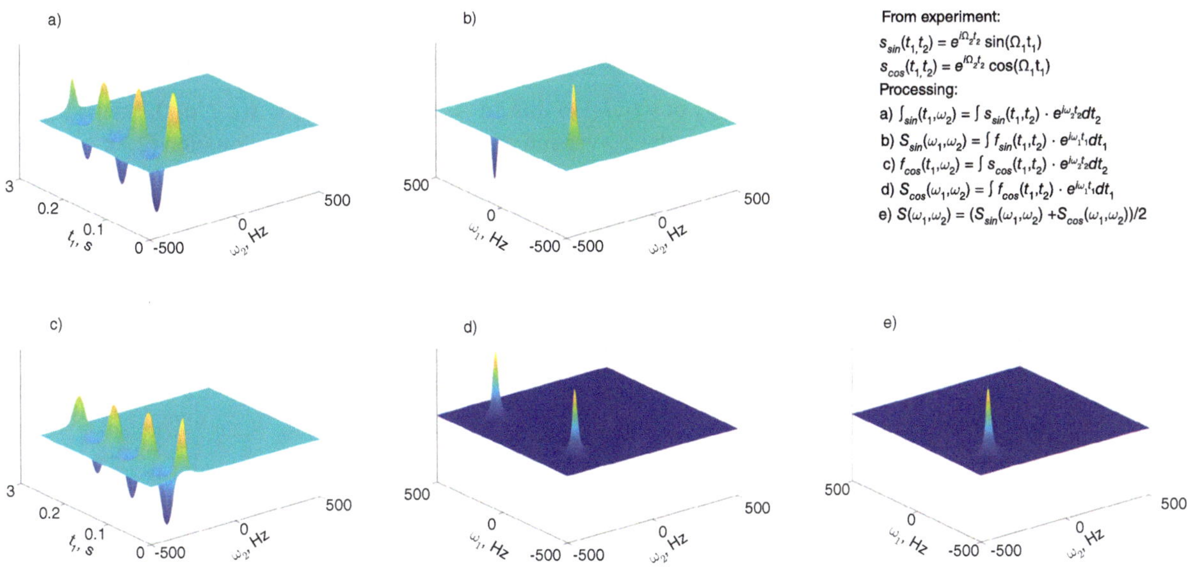


**Figure 2.16** The quadrature detection (States method) of a multidimensional signal. The two modulations $s_{\sin}(t_1, t_2)$ and $s_{\cos}(t_1, t_2)$ acquired in the experiment are processed according to the scheme described in the top right panel.

As shown in Figure 2.16, these two modulations can be processed separately using 2D complex FT and the results can be added to obtain unique sign and value of both $\Omega_1$ and $\Omega_2$.

There are many ways of acquiring a complete multidimensional signal providing absorptive line shapes. An alternative to States method is, e.g. the echo-antiecho (or Rance-Kay, see [7]) method, where the phase modulation is used in the indirect dimensions:

$$s_{\text{echo}}(t_1, t_2) = e^{\iota\Omega_2 t_2} e^{-\iota\Omega_2 t_2} \tag{2.21}$$

and

$$s_{\text{antiecho}}(t_1, t_2) = e^{\iota\Omega_2 t_2} e^{\iota\Omega_2 t_2} \tag{2.22}$$

The older approach, called time-proportional phase incrementation (TPPI, see [8]) is less commonly used. In this method, only one modulation is acquired but with a double-sampling rate. The frequency offset is shifted by a half spectral width by incrementing the phase of the pulse before the $t_1$ evolution delay by $\pi/2$ on each $t_1$ increment.

However, the combination of TPPI with the States method is quite popular. In the States-TPPI approach (see [9]), the frequency sign discrimination is achieved in the same way as in the States method, but additionally, the zero-frequency artifact (axial peak) is shifted to the edge of the spectrum thanks to TPPI-like modulation.

## 2.15 Projection Theorem

The Projection Theorem is a powerful tool, useful in accelerating NMR experiments of dimensionality three and more. The mathematics of projection theorem will be explained in the case of 3D signal $s(t_1, t_2, t_3)$, with straightforward extensions to higher dimensions.

The complete Nyquist sampling of the indirect time dimensions $t_1, t_2$ is very time-consuming and one way of overcoming this difficulty is to sub-sample these dimensions. Since the full sampling of the direct time dimension is not an issue, $t_3$ variable will be ignored in what follows and we shall consider the signal $s(t_1, t_2)$.

In a popular sampling strategy, known as reduced dimensionality, one samples $s(t_1, t_2)$ on a small number of linear subspaces of $(t_1, t_2)$ evolution time domain (for review see [10]). Given a normalized vector $\boldsymbol{a} = \begin{bmatrix} a_1 \\ a_2 \end{bmatrix}$ generating the line $\{t\boldsymbol{a} \,:\, t \in \mathbb{R}\}$, the restriction $s_{\boldsymbol{a}}$ of $s(t_1, t_2)$ to this line is given by $s_{\boldsymbol{a}}(t) = s(a_1 t, a_2 t)$. Interestingly, the Fourier transform of $s_{\boldsymbol{a}}$ can be derived using the 2D version of the inversion formula Equation (2.7):

$$s_{\boldsymbol{a}}(t) = \frac{1}{4\pi^2} \int\limits_{\omega \in \mathbb{R}^2} e^{\imath\omega_1 a_1 t + \omega_2 a_2 t} S(\omega_1, \omega_2) d\omega_1 d\omega_2. \tag{2.23}$$

In order to derive FT of 1D signal $s_{\boldsymbol{a}}$, let us note that the change of variable given by the rotation:

$$\begin{bmatrix} \tilde{\omega}_1 \\ \tilde{\omega}_2 \end{bmatrix} = \begin{bmatrix} a_1 & a_2 \\ -a_2 & a_1 \end{bmatrix} \begin{bmatrix} \omega_1 \\ \omega_2 \end{bmatrix}, \quad \begin{bmatrix} \omega_1 \\ \omega_2 \end{bmatrix} = \begin{bmatrix} a_1 & -a_2 \\ a_2 & a_1 \end{bmatrix} \begin{bmatrix} \tilde{\omega}_1 \\ \tilde{\omega}_2 \end{bmatrix}$$

allows to replace Equation 2.23 with:

$$\begin{aligned} s_{\boldsymbol{a}}(t) &= \frac{1}{4\pi^2} \int\limits_{\tilde{\omega} \in \mathbb{R}^2} e^{\imath\tilde{\omega}_1 t} \tilde{S}(\tilde{\omega}_1, \tilde{\omega}_2) d\tilde{\omega}_1 d\tilde{\omega}_2 \\ &= \frac{1}{2\pi} \int\limits_{\tilde{\omega}_1 \in \mathbb{R}} e^{\imath\tilde{\omega}_1 t} \left( \frac{1}{2\pi} \int\limits_{\tilde{\omega}_2 \in \mathbb{R}} \tilde{S}(\tilde{\omega}_1, \tilde{\omega}_2) d\tilde{\omega}_2 \right) d\tilde{\omega}_1 \end{aligned}$$

where $\tilde{S}(\tilde{\omega}_1, \tilde{\omega}_2) = S(\omega_1, \omega_2)$. In particular, Fourier transform $S_{\boldsymbol{a}}(\omega)$ of $s_{\boldsymbol{a}}$ is given by the expression in the bracket above that is:

$$S_{\boldsymbol{a}}(\omega) = \left( \frac{1}{2\pi} \int\limits_{\tilde{\omega} \in \mathbb{R}} \tilde{S}(\omega, \tilde{\omega}) d\tilde{\omega} \right) = \frac{1}{2\pi} \int\limits_{\tilde{\omega} \in \mathbb{R}} S(\omega a_1 - \tilde{\omega} a_2, \omega a_2 + \tilde{\omega} a_1) d\tilde{\omega}.$$

The latter, can be written as:

$$S_{\boldsymbol{a}}(\omega) = \frac{1}{2\pi} \int\limits_{\tilde{\omega}_2 \in \mathbb{R}} S(\omega\boldsymbol{a} + \tilde{\omega}\boldsymbol{a}^{\perp}) d\tilde{\omega} \tag{2.24}$$

where we denote $\boldsymbol{a}^{\perp} = \begin{bmatrix} -a_2 \\ a_1 \end{bmatrix}$. Equation 2.24 allows to view $S_{\boldsymbol{a}}$ as the orthogonal projection of $S$ along the vector $\boldsymbol{a}$, which is known as the Projection Theorem.

Since multidimensional spectra are reconstructed on the Nyquist grid, we introduce also a discrete version of the Projection Theorem (cf. [11], [12]) exemplified in 2D case in Fig 2.17. Consider the discrete time signal $s(i, j)$, $i, j \in \mathbb{Z}$, which is $K_1$ periodic in $i$ variable $K_2$ periodic in $j$ variable:

$$s(i + K_1, j) = s(i, j + K_2) = s(i, j).$$

The DFT expansion of $s$ has the form:

$$s(i, j) = \frac{1}{K_1 K_2} \sum_{m,n} S_{m,n} e^{2\pi\imath\left(\frac{im}{K_1} + \frac{jn}{K_2}\right)}$$

where $0 \leq m \leq K_1 - 1$ and $0 \leq n \leq K_2 - 1$ and $S_{m,n}$ are the Fourier coefficients of $s$.

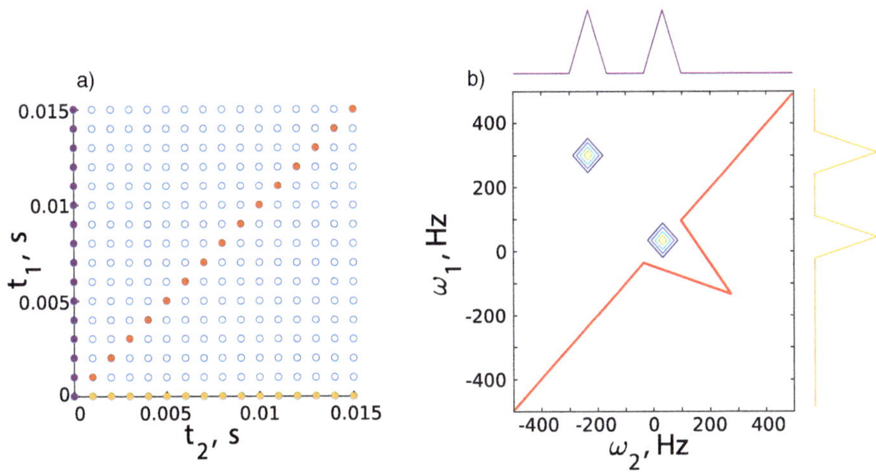


**Figure 2.17** The projection theorem. (a) Sampling schemes in two indirect dimensions: full (blue circles) and corresponding to 1D projections at angles 90°, 45°, and 0° (purple, red and yellow). (b) Example 2D spectrum and projections obtained by Fourier transform of the data sampled according to schemes in (a).

In analogy with the continuous case considered above, let us fix a pair of integers $\boldsymbol{a} = (a_1, a_2) \in \mathbb{Z}^2$ and define the signal:

$$s_{\boldsymbol{a}}(l) = s(a_1 l, a_2 l) \text{ where } l \in \mathbb{Z}.$$

In order to establish the range of periodicity of $s_{\boldsymbol{a}}$ and later write the Fourier expansion we observe that:

$$s_{\boldsymbol{a}}(l + K) = s(a_1 l + a_1 K, a_2 l + a_2 K).$$

In particular, the periodicity range of $s_{\boldsymbol{a}}$ is the smallest natural number $K$ such that $K_1$ divides $a_1 K$ and $K_2$ divides $a_2 K$. Choosing $b_1$ and $b_2$ such that $a_1 K = b_1 K_1$ and $a_2 K = b_2 K_2$ we have:

$$s_{\boldsymbol{a}}(l) = \frac{1}{K_1 K_2} \sum_{m,n} S_{m,n} e^{2\pi\imath\left(\frac{a_1 lm}{K_1} + \frac{a_2 ln}{K_2}\right)} = \frac{1}{K_1 K_2} \sum_{m,n} S_{m,n} e^{\frac{2\pi\imath(b_1 m + b_2 n)l}{K}}. \tag{2.25}$$

In order to derive DFT of $s_{\boldsymbol{a}}$ we write Equation 2.25 as:

$$s_{\boldsymbol{a}}(l) = \frac{1}{K_1 K_2} \sum_k \left( \sum_{b_1 m + b_2 n = k \bmod K} S_{m,n} \right) e^{\frac{2\pi\imath kl}{K}} \tag{2.26}$$

where $0 \leq k \leq K - 1$. Equation 2.26 allows us to state Projection Theorem in the discrete case: Fourier components $S_{\boldsymbol{a}}(k)$ of $s_{\boldsymbol{a}}$ are given by:

$$S_{\boldsymbol{a}}(k) = \frac{K}{K_1 K_2} \left( \sum_{\{(m,n):\, b_1 m + b_2 n = k \bmod K\}} S_{m,n} \right) \tag{2.27}$$

where $b_1 = \frac{a_1 K}{K_1}$ and $b_2 = \frac{a_2 K}{K_2}$, where $K$ is the smallest integer such that $b_1$ and $b_2$ are both integers and $0 \leq k \leq K - 1$.

Summarizing, projection theorem states that signal sampled over a linear (e.g. 1D) sub-space of an ND time domain data leads to the spectrum with reduced dimensionality (1D) corresponding to orthogonal projection from the ND spectrum.

## 2.16 ND Sampling Aspects and Sparse Sampling

The number of measured points rapidly rises with the increase of a spectrum dimensionality, frequency bandwidth, and resolution. Accordingly, it increases the total experiment time, because we need to have a time delay for the spin system to recover to the thermal equilibrium between the scans. Thus, acquiring a single point in the indirect dimensions typically costs several seconds of experiment time

The systematic sampling of the entire time domain $\boldsymbol{X}_K$ (we use the notation of Section 2.13) may be impractical or even impossible, when it takes months or years. Fortunately, the information contained in ND spectrum, i.e. peak positions, can often be obtained from a few spectra projections (Figure 2.17). This approach, which is based on the projection theorem (see Section 2.15), is called reduced dimensionality ([13–15]). Alternatively, we can measure a small subset of time-domain $\boldsymbol{X}_K$ selected (pseudo)randomly from the full grid. Using specific probability distribution in the random sampling generator may result with a better SNR and/or the artifacts suppression. This approach of data acquisition is called non-uniform sampling (NUS) [16–18].

Application of the Fourier transform to NUS time-domain data leads to a spectrum with significant noise-looking artifacts (Figure 2.18). These artifacts are easily understood if we present the NUS time domain signal as a product of the completely sampled signal and the NUS mask, where NUS mask assigns 1 and 0 to the point on the time grid, 1 for the measured and 0 for skipped points, respectively. Spectrum with the artifacts is the

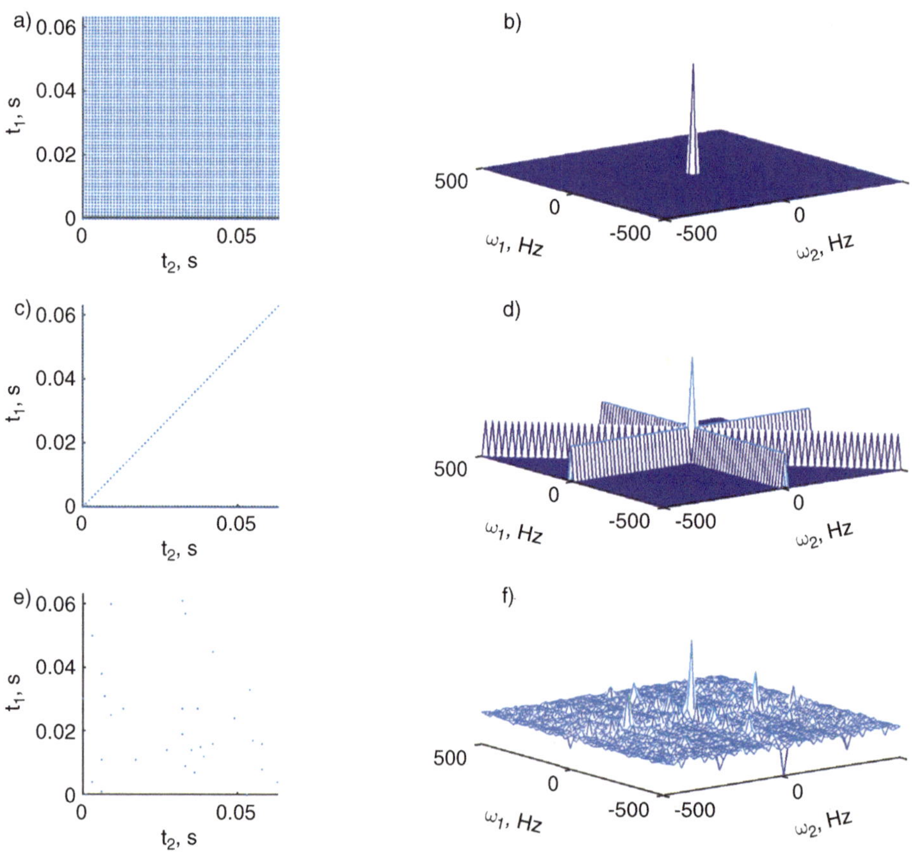


**Figure 2.18** The sampling schemes in two indirect dimensions (3D experiments) and the corresponding point spread functions. (a) full sampling (256 × 256 grid); (b) corresponding PSF; (c) radial sampling along three lines ($t_1 = 0$, $t_2 = 0$, and $t_1 = t_2$ from 256 × 256 grid); (d) corresponding PSF; (e) random non-uniform sampling (32 points from 256 × 256 grid); (f) corresponding PSF.

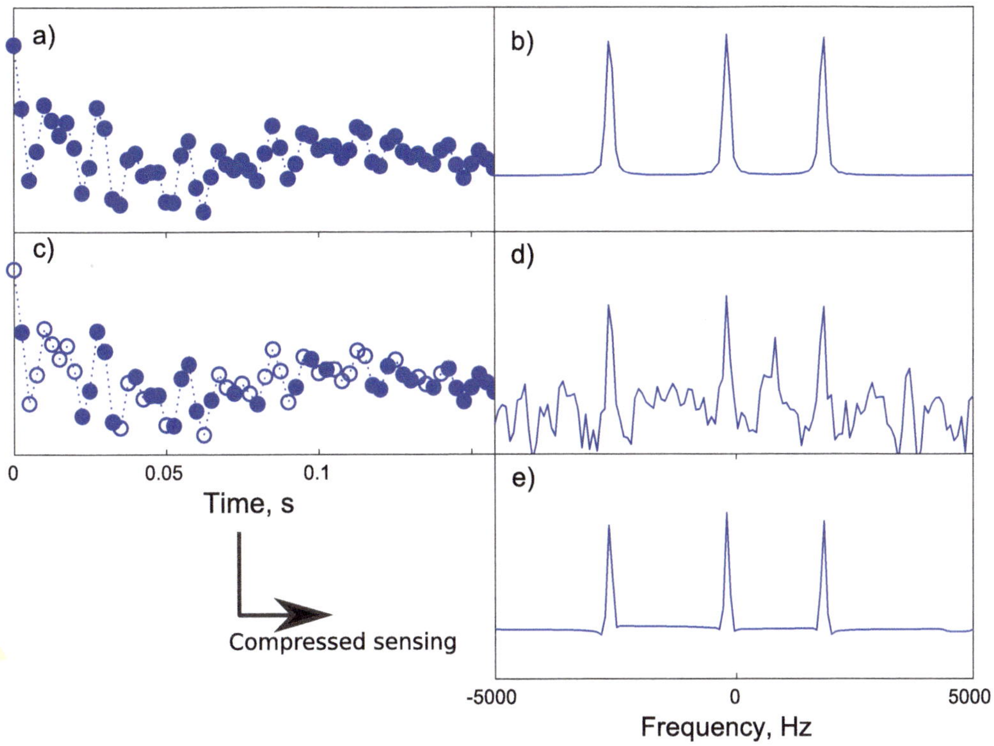


**Figure 2.19** The spectral reconstruction using compressed sensing (iteratively reweighted least squares algorithm[19]). (a) Fully sampled FID; (b) Its spectrum; (c) Undersampled FID; (d) Fourier transform of the undersampled FID (peaks convolved with PSF; (e) Compressed sensing reconstruction of (c). Cosine-bell weighting has been used in all cases.

convolution of the true spectrum and the so-called spectrum density function (SDF), which is the spectrum of the NUS mask. The artifacts are largely suppressed and much cleaner spectra are reconstructed by using one of the nonlinear methods described in the next section (Figure 2.19).

## 2.17 Reconstructing Sparsely Sampled Data Sets

Among the numerous NUS data processing methods that appeared over the last three decades, the compressed sensing (CS) algorithms gained particular attention and is often a default approach in the most of NMR software. Thus, we will focus below on this type of method.

Although our description of the CS framework in this section is limited to the 1D case, it extends to more dimensions in a straightforward way through replacing FT in one dimension with its higher dimensional counterpart (also denoted by FT, see Section 2.13).

The 1D spectrum $S$ is obtained from the fully sampled time domain signal $s$ as the unique solution of a system of linear equations, which in matrix notation can be written as:

$$s = FS \tag{2.28}$$

where $F$ is the matrix of inverse Fourier transform. The solution corresponds to the Fourier transform $S = F^{-1}s$, described equivalently by Equation 2.17.

The NUS signal is measured at time points $\{t_1, \dots, t_k\}$, which is a fixed subset of the full sampling grid. For such a signal, we can still relate subsampled $s \in \mathbb{C}^k$ with the actual $S \in \mathbb{C}^n$ $(k < n)$ using Equation 2.28. In this case, however, $F \in \mathbb{C}^{k \times n}$ consists of $k$ rows from the $n \times n$ inverse Fourier matrix that correspond to the sampling schedule. Unlike in the fully sampled case, Equation 2.28 has infinitely many solutions.

To obtain a faithful reconstruction of the missing data points we need to constrain the solution with additional assumptions about the spectrum, the FID, or both. The CS methodology is based on the assumption that $S$ is *sparse*, i.e. "compressible" or "almost empty." Indeed, the NMR spectrum usually consists of a small number of relatively sharp peaks sticking out from the baseline noise.

The CS theory specifies the NMR spectrum $S$ as the unique solution of the convex optimization problem

$$S = \arg\min_{x \in \mathbb{C}^n} \left( \|Fx - s\|_2^2 + \lambda \|x\|_1 \right). \tag{2.29}$$

The first term in this sum promotes the consistency of tested $x$ with the measured data, cf. Equation 2.28, whereas, the second term promotes the spectrum sparseness by adding penalty to the $\ell_1$ norm of the tested spectrum $x$. Finally, $\lambda > 0$ fixes the balance between the two.

The $\ell_1$ and $\ell_2$ norms used in Equation 2.29 are the special cases of $\ell_p$ norms, where the $\ell_p$-norm of a vector $y \in \mathbb{C}^m$ is given by:

$$\|y\|_p = \begin{cases} \sqrt[p]{|y_1|^p + |y_2|^p + \ldots + |y_m|^p} & p \geq 1 \\ |y_1|^p + |y_2|^p + \ldots + |y_m|^p & 0 \leq p \leq 1 \end{cases}.$$

Note incidentally that $\|y\|_0$ returns a number of non-zero components $y_j$ of a vector $y$, and it quantifies the sparseness of $y$. The CS theorem states that for sufficiently sparse $S$ and $p < 1$ the solution of the non-convex CS problem ($\|\cdot\|_p$- is a non-convex function):

$$\arg\min_{x \in \mathbb{C}^n} \left( \|Fx - s\|_2^2 + \lambda \|x\|_p \right) \tag{2.30}$$

coincides (with very high probability) with the solution of Equation 2.29. This is indeed a remarkable result since the non-convex CS problem is NP-hard while its convexitization is solvable in polynomial time by the linear programming methods. Among them, the iterative soft thresholding is particularly popular (see [20–23]). Alternatively, one may employ iteratively reweighted least squares that approximates $\ell_p$ minimization for $p < 1$ [24].

Apart from promoting sparseness by controlling the $\ell_1$ norm, many other approaches have been proposed. These include: maximum entropy (see [25]), presence of dark regions (see [26]), multidimensional decomposition (see [27]), low-rank reconstruction (see [28]), machine learning (see [29–31]), and others (see [32, 33]). Special algorithms were proposed for spectra with five and more dimensions [12, 34–36]. These spectra are so large that they not only cannot be sampled in full in the time domain but also cannot be fully reconstructed and handled using modern computers in the frequency domain.

## 2.18 Deconvolution

Shape of a peak in the spectrum can be seen as a result of convolution of a "pure," i.e. sharp, peak, and a line distortion. The later may be due to instrumental artifacts such as inhomogeneity of the $B_0$ field or a natural physical property of the spin system exemplified by relaxation or scalar coupling. Correspondingly in time domain, the signal is a product of the pure signal and the modulation function causing the distortion. It is tempting to remove the distortions and obtain the pure spectrum by simply dividing the time domain signal by the distorting modulation, and this works in many practical cases (see [37]). However, when the modulating function has close to zero values, this naive approach to the distortions' deconvolution significantly amplifies the noise existing in the time domain signal.

As a practical example, below we consider deconvolution of peak splitting caused by the scalar coupling. A doublet peak in the spectrum corresponds to a time domain signal modulated by $\cos(\pi J t)$ function. Since this modulation crosses zero at time points $t_k = (0.5 + k)/J$ with $k = 0, 1$ it is particularly problematic for the above-mentioned division approach.

We present a modification of the CS methods designed to avoid the division problem by addressing the modulation of the noise in the process of deconvolution.

Usually for the time domain signal $s$, we can safely assume that the noise in different time points is statistically independent and has the same amplitude. Thus, the noise covariance matrix $\Sigma$ is isotropic, $\Sigma = \sigma^2 \mathbb{1} \in \mathbb{C}^{k \times k}$ where $\sigma$ is the standard deviation of the noise. In order to take into account the level of noise in Equation 2.29 it is sufficient to set $\lambda$ proportional to $\sigma^2$.

In the case of deconvolution described below, the covariance noise matrix $\Sigma$ is not necessarily isotropic and the $\ell_2$ norm used in the data consistency term $||Fx - s||_2^2$ in Equation 2.29 requires a modification of the form $||Fx - s||_Q^2$, $Q = \Sigma^{-1}$ where for a given complex positive definite matrix $G$ and a complex vector $y$ we define $||y||_G^2 = y^\dagger G y$ ($y^\dagger$ is the conjugated transpose of $y$).

Let's illustrate that this modification takes a correct care of the noise. For that matter, let us consider two independent measurements $s_1 = s(t_1)$ and $s_2 = s(t_2)$ of the signal in which the noise level of $s(t_1)$ is two times smaller then that of $s(t_2)$. The corresponding covariance matrix $\Sigma$ is of the form:

$$\Sigma = \begin{bmatrix} \sigma^2 & 0 \\ 0 & 4\sigma^2 \end{bmatrix}$$

where $\sigma$ is the noise level entering $s(t_1)$. Let us denote the signal corresponding to the spectrum $x$ by $s^\# = Fx$. The "data consistency term:"

$$||Fx - s||_2^2 = |s_1^\# - s_1|^2 + |s_2^\# - s_2|^2$$

entering the standard formulation of the CS problem, Equation 2.29, is replaced by:

$$||Fx - s||_Q^2 = (s^{\#\dagger} - s^\dagger)\Sigma^{-1}(s^\# - s) = \frac{1}{\sigma^2}|s_1^\# - s_1|^2 + \frac{1}{4\sigma^2}|s_2^\# - s_2|^2.$$

Thus our modification introduces the weights in the data consistency term that correctly reflect the noise level of the corresponding measurements. Points with larger noise enter the sum with smaller weights. The above discussion and conclusion easily generalizes to a larger number of independent measurements. Let us note that for the unstructured noise ($\Sigma = \sigma^2 \mathbb{1}$) we have:

$$||Fx - s||_Q^2 = \sigma^{-2}||Fx - s||_2^2 \tag{2.31}$$

and thus the cases of structured and unstructured noise are consistent.

Let us apply our framework to the scalar coupling J-modulation. In this case J-modulation is represented by a vector $M \in \mathbb{R}^k$ where for example, $M$ corresponding to a doublet signal with splitting of $J$ Hz in the spectrum has the form $M(t_j) = \cos(\pi J t_j)$. The unmodulated version $\tilde{s}$ of $s$ is defined by the equality $s = M\tilde{s}$, or more precisely $s(t_i) = M(t_i)\tilde{s}(t_i)$ where $i \in \{1, \dots, k\}$. Let $C \in M_{k \times k}(\mathbb{C})$ denote the modulation matrix:

$$C = \mathrm{diag}(M(t_1), \dots, M(t_k)).$$

The relation between $\tilde{s}$, $s$, and the spectrum $\tilde{S}$ corresponding to unmodulated signal $\tilde{s}$ is of the form:

$$s = C\tilde{s} = CF\tilde{S}.$$

Assuming that the J-modulated (measured) signal $s$ is corrupted by the unstructured noise with covariance matrix $\Sigma$, the de-modulated signal $\tilde{s} = C^{-1}s$ is corrupted by the structured noise with the covariance matrix $\tilde{\Sigma} = C^{-1}\Sigma C^{\dagger -1}$. In particular, the weighting matrix $\tilde{Q} = \tilde{\Sigma}^{-1}$ is equal to $C^\dagger Q C$ where $Q = \sigma^{-2}\mathbb{1}$ and we get:

$$\begin{aligned}
\|Fx - \tilde{s}\|_{\tilde{Q}}^2 &= (Fx - \tilde{s})^\dagger C^\dagger QC(Fx - s) \\
&= (CFx - C\tilde{s})^\dagger Q(CFx - C\tilde{s}) \\
&= (CFx - s)^\dagger Q(CFx - s) \\
&= \|CFx - s\|_Q^2 \\
&= \sigma^{-2}\|CFx - s\|_2^2.
\end{aligned}$$

This computation shows that the solution of the CS minimization problem, which returns deconvoluted spectrum $\tilde{S}$:

$$\tilde{S} = \arg\min_{x\in\mathbb{C}^n} \left(\|Fx - \tilde{s}\|_{\tilde{Q}}^2 + \|x\|_1\right)$$

is equivalent to Equation 2.29 with matrix $F$ substituted by $CF$:

$$\tilde{S} = \arg\min_{x\in\mathbb{C}^n} \left(\|CFx - s\|_2^2 + \lambda\|x\|_1\right).$$

## References


1 Szántay, C. (2007). NMR and the uncertainty principle: how to and how not to interpret homogeneous line broadening and pulse nonselectivity. I. The fundamentals. *Concepts. Magn. Reson. Part A Bridg. Educ. Res.* doi: 10.1002/cmr.a.20098. ISSN 15466086.

2 Bartholdi, E. and Ernst, R.R. (1973). Fourier spectroscopy and the causality principle. *J. Magn. Reson. (1969)*, 11 (1): 9–19. ISSN 00222364.

3 Mayzel, M., Kazimierczuk, K. and Orekhov, V.Y. (2014). The causality principle in the reconstruction of sparse NMR spectra. *Chem. Comm.* 50 (64): 8947–8950. doi: 10.1039/c4cc03047h. ISSN 1364548X.

4 Nyquist, H. Certain topics in telegraph transmission theory. *Trans. Electr. Electron. Eng.* 47 (2): 617–644. doi: 10.1109/5.989875. ISSN 00189219.

5 Cooley, J.W. and Tukey, J.W. 1965. An algorithm for the machine calculation of complex fourier series. *Math. Comput.* 19 (90): 297. doi: 10.2307/2003354. ISSN 00255718.

6 States, D.J., Haberkorn, R.A. and Ruben, D.J. (1982). A 2D NMR experiment with pure absorption phase in four quadrants. *J. Magn. Res.*, 48: 286–292.

7 Palmer III, A.G., Cavanagh, J., Wright, P.E. and Rance, M. (1991). Sensitivity improvement in proton-detected two-dimensional heteronuclear correlation nmr spectroscopy. *J. Magn. Reson. (1969)*, 93 (1): 151–170. ISSN 0022-2364.

8 Marion, D. and Wthrich, K. (1983). Application of phase sensitive two-dimensional correlated spectroscopy (cosy) for measurements of 1h-1h spin-spin coupling constants in proteins. *Biochem. Biophys. Res. Commun.* 113 (3): 967–74.

9 Marion, D., Ikura, M., Tschudin, R. and Bax, A.D. (1989). Rapid recording of 2d nmr spectra without phase cycling. application to the study of hydrogen exchange in proteins. *J. Magn. Reson.* 85 (2): 393–399.

10 Coggins, B.E., Venters, R.A. and Zhou, P. (2010). Radial sampling for fast NMR: Concepts and practices over three decades. *Prog. Nucl. Magn. Reson. Spectrosc.* 57 (4): 381–419. doi: 10.1016/j.pnmrs.2010.07.001. ISSN 1873-3301.

11 Hassanieh, H., Mayzel, M., Shi, L., Katabi, D. and Orekhov, V.Y. (2015). Fast multi-dimensional NMR acquisition and processing using the sparse FFT. *J. Biomol. NMR*, 63 (1): 9–19. doi: 10.1007/s10858-015-9952-5. ISSN 15735001.

12 Pustovalova, Y., Mayzel, M. and Orekhov, V.Y. (2018). XLSY: Extra-Large NMR Spectroscopy. *Angew. Chem.* 130 (43): 14239–14241. doi: 10.1002/ange.201806144. ISSN 1521-3757.

**13** Nagayama, K., Bachmann, P., Wuthrich, K. and Ernst, R.R. (1978). The use of cross-sections and of projections in two-dimensional NMR spectroscopy. *J. Magn. Reson. (1969)* 31 (1): 133–148. doi: 10.1016/0022-2364(78)90176-2. ISSN 00222364.

**14** Szyperski, T., Wider, G., Bushweller, J.H. and Wuethrich, K. (1993). Reduced Dimensionality in Triple-Resonance NMR Experiments. *J. Am. Chem. Soc.* 115 (20): 9307–9308. doi: 10.1021/ja00073a064. ISSN 15205126.

**15** Hiller, S., Fiorito, F., Wüthrich, K. and Wider, G. (2005). Automated projection spectroscopy (APSY). *Proc. Natl. Acad. Sci. U.S.A.* 102 (31): 10876–10881. doi: 10.1073/pnas.0504818102. ISSN 00278424.

**16** Barna, J.C.J., Laue, E.D., Mayger, M.R., Skilling, J. and Worrall, S.J.P. (1987). Exponential sampling, an alternative method for sampling in two-dimensional NMR experiments. *J. Magn. Reson. (1969)* 73 (1): 69–77. doi: 10.1016/0022-2364(87)90225-3. ISSN 00222364.

**17** Hoch, J.C. and Stern, A. (1996). *NMR Data Processing*. Wiley-Liss. ISBN 0471039004.

**18** Kazimierczuk, K. and Orekhov, V. (2015). Non-uniform sampling: Post-Fourier era of NMR data collection and processing. *Magn. Reson. Chem.* 53 (11): 921–926. doi: 10.1002/mrc.4284. URL http://www.ncbi.nlm.nih.gov/pubmed/26290057. ISSN 1097458X.

**19** Kazimierczuk, K. and Orekhov, V.Y. (2012). A comparison of convex and non-convex compressed sensing applied to multidimensional NMR. *J. Magn. Reson.* 223: 1–10. doi: 10.1016/j.jmr.2012.08.001. ISSN 10907807.

**20** Kazimierczuk, K. and Orekhov, V.Y. (2011). Accelerated NMR spectroscopy by using compressed sensing. *Angew. Chem. Int. Ed. Engl.* 50 (24): 556–5559. ISSN 1521-3773.

**21** Holland, D.J., Bostock, M.J., Gladden, L.F. and Nietlispach, D. (2011). Fast multidimensional NMR spectroscopy using compressed sensing. *Angew. Chem. Int. Ed. Engl.* 50 (29): 6548–6551. ISSN 1521-3773.

**22** Hyberts, S.G., Milbradt, A.G., Wagner, A.B., Arthanari, H. and Wagner, G. (2012). Application of iterative soft thresholding for fast reconstruction of NMR data non-uniformly sampled with multidimensional Poisson Gap scheduling. *J. Biomol. NMR* 52: 1–13, 2012. ISSN 0925-2738.

**23** Sun, S., Gill, M., Li, Y., Huang, M. and Byrd, R.A. (2015). Efficient and generalized processing of multidimensional NUS NMR data: the NESTA algorithm and comparison of regularization terms. *J. Biomol. NMR* 62 (1): 105–17. doi: 10.1007/s10858-015-9923-x. ISSN 1573-5001.

**24** Kazimierczuk, K. and Orekhov, V.Y. (2012b). A comparison of convex and non-convex compressed sensing applied to multidimensional NMR. *J. Magn. Reson.* 223 (0): 1–10. doi: 10.1016/j.jmr.2012.08.001. ISSN 10907807.

**25** Mobli, M., Maciejewski, M.W., Gryk, M.R. and Hoch, J.C. (2007). Automatic maximum entropy spectral reconstruction in NMR. *J. Biomol. NMR*, 39 (2): 133–139. doi: 10.1007/s10858-007-9180-8. ISSN 09252738.

**26** Matsuki, Y., Eddy, M.T. and Herzfeld, J. (2009 April). Spectroscopy by Integration of Frequency and Time Domain Information for Fast Acquisition of High-Resolution Dark Spectra. *J. Am. Chem. Soc.* 131 (13): 4648–4656. doi: 10.1021/ja807893k. ISSN 1520-5126.

**27** Orekhov, V.Y. and Jaravine, V.A. (2011). Analysis of non-uniformly sampled spectra with multi-dimensional decomposition. *Prog. Nucl. Magn. Reson. Spectrosc.* 59 (3): 271–292. doi: 10.1016/j.pnmrs.2011.02.002. ISSN 00796565.

**28** Qu, X., Mayzel, M., Cai, J.F., Chen, Z. and Orekhov, V. (2015). Accelerated NMR spectroscopy with low-rank reconstruction. *Angew. Chem. Int. Ed.* 54 (3): 852–854. doi: 10.1002/anie.201409291. ISSN 15213773.

**29** Hansen, D.F. (2019). Using Deep Neural Networks to Reconstruct Non-uniformly Sampled NMR Spectra. *J. Biomol. NMR* 73 (10-11): 577–585. doi: 10.1007/s10858-019-00265-1. ISSN 15735001.

**30** Qu, X., Huang, Y., Lu, H., Qiu, T., Guo, D., Agback, T., Orekhov, V. and Chen, Z. (2019). Accelerated nuclear magnetic resonance spectroscopy with deep learning. *Angew. Chem.* 59 (26): 10297–10300. doi: 10.1002/anie.201908162. ISSN 0044-8249.

**31** Karunanithy, G. and Hansen, D.F. (2020). FID-net: a versatile deep neural network architecture for NMR spectral reconstruction and virtual decoupling. *J. Biomol. NMR* 75: 1–19. doi: 10.1007/s10858-021-00366-w. ISSN 1573-5001.

**32** Stanek, J. and Koźmiński, W. (2010). Iterative algorithm of discrete Fourier transform for processing randomly sampled NMR data sets. *J. Biomol. NMR*, 47 (1): 65–77. doi: 10.1007/s10858-010-9411-2. ISSN 09252738.

**33** Ying, J., Delaglio, F., Torchia, D.A. and Bax, A. (2017). Sparse multidimensional iterative lineshape-enhanced (SMILE) reconstruction of both non-uniformly sampled and conventional NMR data. *J. Biomol. NMR* 68 (2): 101–118. doi: 10.1007/s10858-016-0072-7. ISSN 15735001.

**34** Kazimierczuk, K., Zawadzka, A. and Koźmiński, W. (2009). Narrow peaks and high dimensionalities: exploiting the advantages of random sampling. *J. Magn. Reson.* 197 (2): 219–228. doi: 10.1016/j.jmr.2009.01.003.

**35** Motáčková, V., Nováček, J., Zawadzka-Kazimierczuk, A., Kazimierczuk, K., Žídek, L., Šanderová, H., Krásný, L., Koźmiński, W. and Sklenář, V. (2010 November). Strategy for complete NMR assignment of disordered proteins with highly repetitive sequences based on resolution-enhanced 5D experiments. *J. Biomol. NMR* 48 (3): 169–177.

**36** Kosiński, K., Stanek, J., Górka, M.J., Żerko, S. and Koźmiński, W. (2017). Reconstruction of non-uniformly sampled five-dimensional NMR spectra by signal separation algorithm. *J. Biomol. NMR* 68 (2): 129–138.

**37** Morris, G.A., Barjat, H. and Home, T.J. (1997). Reference deconvolution methods. *Prog. Nucl. Magn. Reson. Spectr.* 31 (2-3): 197–257. doi: 10.1016/S0079-6565(97)00011-3. ISSN 00796565.

# 3

# Product Operator Formalism

*Rolf Boelens* and Robert Kaptein*

*Department of Chemistry, Utrecht University, Padualaan 8, 3584 CH, Utrecht, The Netherlands*
** Corresponding Author*

## 3.1 Introduction

In quantum mechanics, measurable quantities are described by operators (see Appendix 3.A for an introduction to quantum mechanics and operators). A well-known example of an operator is the Hamiltonian operator ($\mathcal{H}$), which, when applied to a wave function describing a stationary state (an eigenfunction of $\mathcal{H}$), yields the energy of that state:

$$\mathcal{H}\,\phi = E\,\phi. \tag{3.1}$$

When states are not stationary, for example in NMR due to irradiation with a radio-frequency (RF) source, a system can be described by the time-dependent Schrödinger equation:

$$\mathrm{i}\,\hbar\,\frac{\mathrm{d}\psi}{\mathrm{dt}} = \mathcal{H}\,\psi. \tag{3.2}$$

The equation describes how $\psi$ evolves in time. Assume we have two stationary states $\phi_1$ and $\phi_2$ with energies $E_1$ and $E_2$, then a solution for the wave function describing the system under irradiation will be given by:

$$\psi(\mathrm{t}) = c_1(t)\,\phi_1 + c_2(t)\,\phi_2. \tag{3.3}$$

The time dependence of the system is now represented by the coefficients and the system will oscillate between the two stationary states. The probabilities to find the system in one of the two states will then be:

$$\begin{aligned} P_1(t) &= c_1(t)\,c_1^*(t) \\ P_2(t) &= c_2(t)\,c_2^*(t). \end{aligned} \tag{3.4}$$

When there are multiple states, the wave function can be described by:

$$\psi(\mathrm{t}) = \sum_n c_n(t)\,\phi_\mathrm{n}. \tag{3.5}$$

*Two-Dimensional (2D) NMR Methods*, First Edition. Edited by K. Ivanov, P.K. Madhu and G. Rajalakshmi.

With the wave function we can calculate the expectation values of measurable quantities as the trace of the product of the population matrix **P** and the matrix representation **A** of the operator A:

$$\langle A\rangle = \int \psi^* \, A \, \psi \, d\tau = \sum_{i,\,j} c_i^*(t)\, c_j(t) \int \phi_{\mathrm{i}}^* \, A \, \phi_{\mathrm{j}} \, d\tau$$
$$= \sum_{i,\,j} P_{ji}\, A_{ij} = \mathrm{Tr}\,(\mathbf{P}\,\mathbf{A}). \tag{3.6}$$

To describe an ensemble of spins we use the density matrix, $\boldsymbol{\rho}$, which is the matrix representation of the density operator (see Appendix 3.A.7) and the ensemble average of the population matrix:

$$\rho_{nm} = \overline{P_{nm}} = \overline{c_n\, c_m^*}. \tag{3.7}$$

The terms density matrix and density operator are often used interchangeably.

The expectation value $\langle \overline{A} \rangle$ of an ensemble will be the trace of the product of the density matrix and matrix representation of the operator **A**:

$$\langle \overline{A} \rangle = \sum_{i,\,j} \overline{P_{ji}}\, A_{ij} = \sum_{i,\,j} \rho_{ji}\, A_{ij} = \mathrm{Tr}\,(\boldsymbol{\rho}\,\mathbf{A}) \tag{3.8}$$

where **A** is time independent and the same for all identical spins in the ensemble.

For further reading on the density matrix we refer to [1, 2].

The time dependence of the expectation value $\langle \overline{A} \rangle$ is given by the coefficients $c_n(t)$ and therefore in the density matrix $\boldsymbol{\rho}$. Thus, calculating the time dependence of measurable quantities comes down to calculating the time evolution of the density matrix $\boldsymbol{\rho}$. This can be tedious, especially for large spin systems. By expressing the density matrix as a linear combination of product operators, as introduced in [3] and [4], the complexity of the calculations can be considerably reduced.

In this chapter, we will introduce these product operators, show their matrix representations, and show how these operators will evolve in time or be changed upon RF irradiation. We will present several examples of the use of product operators in describing NMR experiments. A full description of product operators can be found in the review [5]. For additional introductory reading, we also refer the reader to the excellent textbooks [6], [7] and [8]. More detailed descriptions including many topics not addressed in this chapter can be found in the monograph [9].

## 3.2 Product Operators and Time Evolution

By rewriting the time-dependent Schrödinger equation, the time evolution of the density matrix $\rho$ can be described by:

$$\frac{d\rho}{dt} = \frac{i}{\hbar}\,[\rho,\,\mathcal{H}] \tag{3.9}$$

where $\hbar = \frac{h}{2\pi}$ and $h$ is Planck's constant. Or if the Hamiltonian is expressed in $\hbar$ units:

$$\frac{d\rho}{dt} = i\,[\rho,\,\mathcal{H}]. \tag{3.10}$$

The spin Hamiltonian used in solution NMR can include chemical shifts, J-couplings, and time-dependent interactions with RF fields. For example, the homonuclear spin Hamiltonian for two weakly J-coupled spins is:

$$\mathcal{H} = -\omega_I\, I_z - \omega_S\, S_z + 2\pi J\, I_z\, S_z - \omega_1\,(I_x + S_x)\,\cos(\omega_0\, t) \quad (\text{in } \hbar \text{ units}) \tag{3.11}$$

where $\omega_I$ and $\omega_S$ are the (angular) NMR frequencies of spins I and S, J the J-coupling in Hz between both spins, and $\omega_0$ and $\omega_1(=\gamma B_1)$ the frequency and strength of the applied RF field ($B_1$).

In a suitable rotating frame, the spin Hamiltonian can be made time independent. The time evolution can then be expressed by a sequence of unitary transformations like:

$$\rho\,(t+\tau_1+\tau_2) = e^{-i\mathcal{H}\tau_2}\, e^{-i\mathcal{H}\tau_1}\, \rho(t)\, e^{\,i\mathcal{H}\tau_1}\, e^{\,i\mathcal{H}\tau_2}. \tag{3.12}$$

After calculating the time evolution of the density matrix the observable magnetization can subsequently be calculated from the trace:

$$\begin{aligned} &< M_x >= N\gamma\hbar\; Tr\{F_x\, \rho(t)\} \\ &< M_y >= N\gamma\hbar\; Tr\{F_y\, \rho(t)\} \end{aligned} \tag{3.13}$$

with the observable operators

$$\begin{aligned} F_x &= \sum_k I_{kx} \\ F_y &= \sum_k I_{ky}. \end{aligned} \tag{3.14}$$

For the description of NMR pulse experiments with more than one spin, the density matrixes tend to become quite complex. For example, for two spins $\frac{1}{2}$ we have a $4\times4$ matrix with 16 elements, for three spins $\frac{1}{2}$ a $8\times8$ matrix with 64 elements, etc.

| number of spins | $\rho$ | number of elements |
|---|---|---|
| 1 | $2\times2$ | 4 |
| 2 | $4\times4$ | 16 |
| 3 | $8\times8$ | 64 |
| $N$ | $2^N\times2^N$ | $4^N$ |

The density operator of spin systems can be written as an expansion of base operators $B_s$ [5]:

$$\rho\,(t) = \sum_s c_s(t)\, B_s(t). \tag{3.15}$$

The complexity of the practical calculations depends mainly on the choice of $B_s$. [5] and [4] proposed to express the density matrix in terms of products of single spin operators. For these base operators we can derive simple relationships.

For a single spin $\frac{1}{2}$ the density matrix can be expressed as a $2\times2$ matrix. The natural base set of a $2\times2$ matrix contains the following four matrixes:

$$\begin{pmatrix}1 & 0\\ 0 & 0\end{pmatrix} \quad \begin{pmatrix}0 & 1\\ 0 & 0\end{pmatrix} \quad \begin{pmatrix}0 & 0\\ 1 & 0\end{pmatrix} \quad \begin{pmatrix}0 & 0\\ 0 & 1\end{pmatrix}. \tag{3.16}$$

The density matrix of a single spin can be described as a linear combination of these base matrixes.

$$\rho = \begin{pmatrix}a & b\\ c & d\end{pmatrix} = a\begin{pmatrix}1 & 0\\ 0 & 0\end{pmatrix} + b\begin{pmatrix}0 & 1\\ 0 & 0\end{pmatrix} + c\begin{pmatrix}0 & 0\\ 1 & 0\end{pmatrix} + d\begin{pmatrix}0 & 0\\ 0 & 1\end{pmatrix}. \tag{3.17}$$

In NMR, it is more practical to choose another base set, the set of matrix representations of the spin operators $I_x, I_y, I_z$ and $\frac{1}{2}E$, where $E$ is the identity operator:

$$I_x = \frac{1}{2}\begin{pmatrix} 0 & 1 \\ 1 & 0 \end{pmatrix} \quad I_y = \frac{1}{2i}\begin{pmatrix} 0 & 1 \\ -1 & 0 \end{pmatrix}$$
$$I_z = \frac{1}{2}\begin{pmatrix} 1 & 0 \\ 0 & -1 \end{pmatrix} \quad \frac{1}{2}E = \frac{1}{2}\begin{pmatrix} 1 & 0 \\ 0 & 1 \end{pmatrix}. \tag{3.18}$$

The density matrix for a single spin $\frac{1}{2}$ then becomes:

$$\rho = p\, I_x + q\, I_y + r\, I_z + s\, E. \tag{3.19}$$

For instance, at thermal equilibrium the density matrix will be:

$$\rho^{eq} = \frac{e^{-\mathcal{H}/kT}}{Z} \approx \frac{1}{2}(E - \frac{\mathcal{H}}{kT}) = \frac{1}{2}E + \frac{\gamma\hbar B_z}{2kT}\, I_z \tag{3.20}$$

where $k$ is Boltzmann's constant, $T$ the temperature (in K), and $B_z$ the magnetic field strength. In this expression we can remove $E$, since it does not transform under any operator and will not contribute to the trace of the NMR observables, and we ignore the constant $\frac{\gamma\hbar B_z}{2kT}$ at high temperature. Therefore, for the start of experiments at thermal equilibrium we will generally use:

$$\rho^{eq} = I_z. \tag{3.21}$$

To describe the motion of the density matrix under a time-dependent Hamiltonian, we use the so-called interaction representation, in which we split off the time-dependent part of the Hamiltonian:

$$\mathcal{H}(t) = \mathcal{H}^0 + \mathcal{H}^1(t). \tag{3.22}$$

We define the interaction representations of $\rho$ and $\mathcal{H}$ as:

$$\rho^r = e^{\frac{i}{\hbar}\mathcal{H}^0 t}\, \rho\, e^{-\frac{i}{\hbar}\mathcal{H}^0 t}$$
$$\mathcal{H}^r = e^{\frac{i}{\hbar}\mathcal{H}^0 t}\, \mathcal{H}\, e^{-\frac{i}{\hbar}\mathcal{H}^0 t}. \tag{3.23}$$

In the interaction representation, which is equivalent to a rotating frame, the motional equation for $\rho^r$ becomes independent of $\mathcal{H}^0$:

$$\frac{d\rho^r}{dt} = \frac{i}{\hbar}\,[\rho^r,\, \mathcal{H}^{1r}]. \tag{3.24}$$

The time evolution (transformation) of the density matrix $\rho^r$ becomes:

$$\rho^r(t) = e^{-\frac{i}{\hbar}\mathcal{H}^{1r} t}\, \rho^r(0)\, e^{-\frac{i}{\hbar}\mathcal{H}^{1r} t}. \tag{3.25}$$

The spin Hamiltonian $\mathcal{H}^{1r}$ that describes the effect of an RF field in the rotating frame is:

$$\mathcal{H}^{1r} = -\gamma\, \hbar\, B_1 I_x = -\hbar\, \omega_1 I_x. \tag{3.26}$$

Thus, the density matrix calculations of evolution under a Hamiltonian become transformations of spin operators by exponential spin operators:

$$e^{\,i\,\omega_1 t\, I_x}\, I_z\, e^{-i\,\omega_1 t\, I_x} \tag{3.27}$$

or

$$e^{i\,\alpha\,I_x}\,I_z\,e^{-i\,\alpha\,I_x}\text{ , where }\alpha = \omega_1 t. \tag{3.28}$$

For systems with two and more spins, the base set will be products of single spin operators. For N spins $\frac{1}{2}$ the base set will have $4^N$ elements. They can be derived from:

$$B_s = 2^{q-1}\prod_{k=1}^{N}(I_{k\nu})^{a_{sk}} \tag{3.29}$$

where

$$\nu = x, y\ or\ z$$
$$q = \text{number of operators in product}$$
$$a^{sk} = \begin{cases} 1 & \text{for}\quad q\quad \text{nuclei} \\ 0 & \quad N-q \end{cases}.$$

The set of product operators $\{B_s\}$ for spin $\frac{1}{2}$ nuclei is orthogonal (with respect to the trace) but not normalized:

$$Tr\,(B_r\,B_s) = \delta_{rs}\,2^{N-2}. \tag{3.30}$$

As an example, the 16 operators $B_s$ for a two-spin system become:

$$\begin{array}{llll} q=0 & \frac{1}{2}E & & \\ q=1 & I_x, I_y, I_z & S_x, S_y, S_z & \\ & 2I_xS_x & 2I_xS_y & 2I_xS_z \\ q=2 & 2I_yS_x & 2I_yS_y & 2I_yS_z \\ & 2I_zS_x & 2I_zS_y & 2I_zS_z \end{array}. \tag{3.31}$$

Each two-spin density matrix (a 4×4 matrix) can be expressed as a linear combination of these 16 product operators:

$$\rho\,(t) = \sum_{s=1}^{16} c_s(t)\,B_s. \tag{3.32}$$

The effects of free precession and of RF pulses will then all become transformations of the type:

$$e^{i\,\theta\,B_r}\,B_s\,e^{-i\,\theta\,B_r}. \tag{3.33}$$

### 3.2.1 Advantages of Product Operators

The description of the density matrix in terms of product operators has many advantages:

- Observable quantities can be easily calculated, e.g. the magnetization $< I_y >$

$$< I_y >= Tr\left(\rho\,I_y\right). \tag{3.34}$$

The density matrix is $\rho = \sum_{s=1}^{16} c_s(t)\,B_s = \cdots c_y(t)\,I_y \cdots$ and therefore:

$$< I_y >= Tr\left(\rho\,I_y\right) = \sum_{s=1}^{16} c_s(t)\,Tr\left(B_s\,I_y\right) = \sum_{s=1}^{16} c_s(t)\,\delta_{sy} = c_y(t). \tag{3.35}$$

- All operators $B_s$ have a physical meaning.
- The transformation of the operators $B_s$ by RF pulses or the Hamiltonian $\mathcal{H}$ into other operators can be easily followed through pulse sequences.

#### 3.2.1.1 Shape and Physical Meaning of the Product Operators

As an example we take two weakly J-coupled spins $I$ and $S$. The time-independent Hamiltonian for this spin system is

$$\mathcal{H} = -\omega_I\, I_z - \omega_S\, S_z + 2\pi J\, I_z\, S_z \quad (\text{in } \hbar \text{ units}) \tag{3.36}$$

and has the spin eigenfunctions $\alpha\alpha, \alpha\beta, \beta\alpha, \beta\beta$ (see Appendix 3.A.6). Below we derive the matrix representation of the product operators using these spin functions (see Appendix 3.A.5).

The diagonal terms of the matrixes are populations and can contribute to the probability to find the spin system in state n. The off-diagonal terms $\rho_{ij}$ oscillate during evolution under the Hamiltonian as $e^{\frac{i}{\hbar}(E_j - E_i)t}$ and represent coherences.

- The unity matrix:

$$\frac{1}{2}E \qquad \frac{1}{2}\begin{pmatrix} 1 & & & \\ & 1 & & \\ & & 1 & \\ & & & 1 \end{pmatrix}. \tag{3.37}$$

The unity matrix will not contribute to population differences.
- The z-magnetization of spins $I$ and $S$:

$$I_z \qquad \frac{1}{2}\begin{pmatrix} 1 & & & \\ & 1 & & \\ & & -1 & \\ & & & -1 \end{pmatrix} \qquad S_z \qquad \frac{1}{2}\begin{pmatrix} 1 & & & \\ & -1 & & \\ & & 1 & \\ & & & -1 \end{pmatrix}. \tag{3.38}$$

The z-magnetization of spin $I$ is due to the population differences between states $\alpha\alpha$ and $\beta\alpha$ plus that between $\alpha\beta$ and $\beta\beta$, and the z-magnetization of spin $S$ is due to the population differences between states $\alpha\alpha$ and $\alpha\beta$ plus that between $\beta\alpha$ and $\beta\beta$. The coefficients $c_{I_z}$ and $c_{S_z}$ in $\rho$ represent the expectation values of the z-magnetization of spins $I$ and $S$.
- The x-magnetization of spins $I$ and $S$:

$$I_x \qquad \frac{1}{2}\begin{pmatrix} & & 1 & 0 \\ & & 0 & 1 \\ 1 & 0 & & \\ 0 & 1 & & \end{pmatrix} \qquad S_x \qquad \frac{1}{2}\begin{pmatrix} 0 & 1 & & \\ 1 & 0 & & \\ & & 0 & 1 \\ & & 1 & 0 \end{pmatrix} \tag{3.39}$$

These terms oscillate with frequencies $\nu_I \pm \frac{1}{2}J$ and $\nu_S \pm \frac{1}{2}J$.

- Longitudinal spin-order $2I_zS_z$:

$$\frac{1}{2}\begin{pmatrix} 1 & & & \\ & -1 & & \\ & & -1 & \\ & & & 1 \end{pmatrix} \quad (2I_zS_z) \tag{3.40}$$

The $\alpha\alpha$ and $\beta\beta$ states are highly populated with respect to the $\alpha\beta$ and $\beta\alpha$ states (can be detected using a small RF pulse).

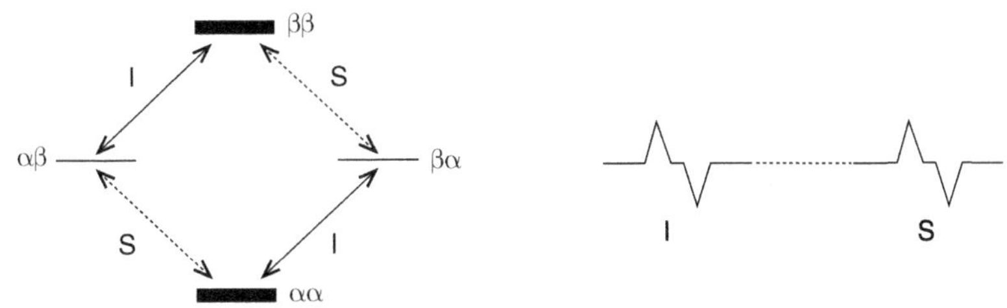


- Antiphase magnetization:

$$\underset{2I_xS_z}{\frac{1}{2}\begin{pmatrix} 0 & 0 & 1 & 0 \\ 0 & 0 & 0 & -1 \\ 1 & 0 & 0 & 0 \\ 0 & -1 & 0 & 0 \end{pmatrix}} \quad \underset{2I_zS_x}{\frac{1}{2}\begin{pmatrix} 0 & 1 & 0 & 0 \\ 1 & 0 & 0 & 0 \\ 0 & 0 & 0 & -1 \\ 0 & 0 & -1 & 0 \end{pmatrix}}. \tag{3.41}$$

These terms oscillate with frequencies $\nu_I \pm \frac{1}{2}J$ and $\nu_S \pm \frac{1}{2}J$ as in 3.39, but the terms in $2I_xS_z$ with frequencies $\nu_I + \frac{1}{2}J$ and $\nu_I - \frac{1}{2}J$ have opposite sign, as do the terms in $2I_zS_x$ with frequencies $\nu_S + \frac{1}{2}J$ and $\nu_S - \frac{1}{2}J$.

Analogously, one can express the matrix representations of the product operators $I_y$, $S_y$ $2I_yS_z$, and $2I_zS_y$, which we leave for the reader as an exercise.

- Multiple-quantum coherence. The product operators $2I_xS_x$ and $2I_yS_y$ represent combinations of zero- and double-quantum coherences and by adding and subtracting, the (real) 'x'-terms of the zero-quantum (ZQ) and double-quantum (DQ) terms can be obtained.

$$2I_xS_x \quad \frac{1}{2}\begin{pmatrix} 0 & 0 & 0 & 1 \\ 0 & 0 & 1 & 0 \\ 0 & 1 & 0 & 0 \\ 1 & 0 & 0 & 0 \end{pmatrix} \qquad 2I_yS_y \quad \frac{1}{2}\begin{pmatrix} 0 & 0 & 0 & -1 \\ 0 & 0 & 1 & 0 \\ 0 & 1 & 0 & 0 \\ -1 & 0 & 0 & 0 \end{pmatrix} \tag{3.42}$$

$$(ZQ)_x = 2I_xS_x + 2I_yS_y \quad \begin{pmatrix} 0 & 0 & 0 & 0 \\ 0 & 0 & 1 & 0 \\ 0 & 1 & 0 & 0 \\ 0 & 0 & 0 & 0 \end{pmatrix} \qquad (DQ)_x = 2I_xS_x - 2I_yS_y \quad \begin{pmatrix} 0 & 0 & 0 & 1 \\ 0 & 0 & 0 & 0 \\ 0 & 0 & 0 & 0 \\ 1 & 0 & 0 & 0 \end{pmatrix} \tag{3.43}$$

The ZQ terms oscillate with frequencies $\nu_I - \nu_S$ and DQ terms with frequencies $\nu_I + \nu_S$.

The presence of ZQ and DQ coherences in these operators can be easily demonstrated by using the raising and lowering operators (see Appendix 3.A.4):

$$\begin{aligned} I_+ &= I_x + iI_y \\ I_- &= I_x - iI_y \end{aligned} \tag{3.44}$$

or equivalently

$$\begin{aligned} I_x &= \frac{1}{2}(I_+ + I_-) \\ I_y &= \frac{1}{2i}(I_+ - I_-). \end{aligned} \tag{3.45}$$

Thus, the terms $2I_xS_x$ and $2I_yS_y$ will become:

$$\begin{aligned} 2I_xS_x &= \frac{1}{2}(I_+S_+ + I_-S_+ + I_+S_- + I_-S_-) \\ 2I_yS_y &= -\frac{1}{2}(I_+S_+ - I_-S_+ - I_+S_- + I_-S_-) \end{aligned} \tag{3.46}$$

and their linear combinations:

$$\begin{aligned} (ZQ)_x &= 2I_xS_x + 2I_yS_y = I_+S_- + I_-S_+ \\ (DQ)_x &= 2I_xS_x - 2I_yS_y = I_+S_+ + I_-S_-. \end{aligned} \tag{3.47}$$

Analogously one can express the linear combinations of $2I_xS_y$ and $2I_yS_x$, from which the 'y'-components of the ZQ and DQ coherences can be obtained:

$$\begin{aligned} (ZQ)_y &= 2I_yS_x - 2I_xS_y = \frac{1}{i}(I_+S_- - I_-S_+) \\ (DQ)_y &= 2I_yS_x + 2I_xS_y = \frac{1}{i}(I_+S_+ - I_-S_-). \end{aligned} \tag{3.48}$$

The ZQ terms in the density matrix connect the $\alpha\beta$ and $\beta\alpha$ states, whereas the DQ terms connect the $\alpha\alpha$ and $\beta\beta$ states.

## 3.3 Time Evolution of the Product Operators

Assume that we apply the following pulse sequence on the density matrix $\rho$

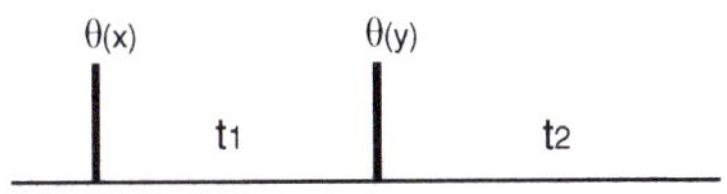


where $\theta(x)$ is a $\theta$-pulse along the x'-axis in the rotating frame.
A shorthand notation for this experiment would be:

$$\rho(-0) \xrightarrow{\theta(I_x+S_x)} \rho(+0) \xrightarrow{Ht_1} \rho(-t_1) \xrightarrow{\theta(I_y+S_y)} \rho(+t_1) \xrightarrow{Ht_2} \rho(t_1+t_2) \tag{3.49}$$

where:

$\rho(-t)$ = density matrix before a pulse at time t
$\rho(+t)$ = density matrix after the pulse at time t
$\theta(I_x + S_x)$ = RF pulse along x on both spins I and S
$Ht$ = evolution during time t under the Hamiltonian $\mathcal{H}$.

We can discriminate the evolution (transformation) of $\rho$ under a pulse and the Hamiltonian:

a. **pulses**
The transformation of $\rho$ by a $\theta$ pulse becomes:

$$e^{i\theta I_v}\,\rho\, e^{-i\theta I_v} \tag{3.50}$$

b. **evolution** under the Hamiltonian (in the rotating frame):

$$\mathcal{H}^r = -\Omega_I\, I_z - \Omega_S\, S_z + 2\pi J\, I_z S_z \quad (\text{in } \hbar \text{ units}) \tag{3.51}$$

where $\Omega_I = \omega_I - \omega_r$ and $\omega_r$ = rotating frame frequency ('carrier').

The transformation of $\rho$ due to evolution under the Hamiltonian becomes:

$$e^{-i\mathcal{H}^r t}\,\rho\, e^{i\mathcal{H}^r t}. \tag{3.52}$$

Note that $\mathcal{H}$ also contains the operators $B_s$ (after rearrangement):

$$\mathcal{H}^r = -\Omega_I\, I_z - \Omega_S\, S_z + \pi J\,(2I_z S_z). \tag{3.53}$$

Thus, all evolutions and transformations by pulses have the following general form:

$$e^{i\theta B_s}\, B_r\, e^{-i\theta B_s}. \tag{3.54}$$

The exponential product operator in this equation (see Appendix 3.A.8) can be expressed as:

$$e^{i\theta B_s} = E\, \cos\frac{\theta}{2} + 2iB_s\, \sin\frac{\theta}{2}. \tag{3.55}$$

From this result one can derive how a product operator transforms under another product operator:

$$\boxed{e^{i\theta B_s}\, B_r\, e^{-i\theta B_s} = B_r\, \cos\theta + i\,[B_s, B_r]\, \sin\theta \qquad \text{if } [B_s, B_r] \neq 0}\,. \tag{3.56}$$

A shorthand notation for this last transformation is:

$$B_r \xrightarrow{\theta B_s} B_r\, \cos\theta + i\,[B_s, B_r]\, \sin\theta. \tag{3.57}$$

The operator remains unchanged if $B_r$ and $B_s$ commute:

$$e^{i\theta\, B_s}\, B_r\, e^{-i\theta B_s} = B_r \qquad \text{if } [B_s, B_r] = 0. \tag{3.58}$$

### 3.3.1 Effect of Pulses

Using the Equation 3.56 we can simply derive the effect of pulses on the product operators. For example, when we apply an x-pulse on z-magnetization, we need to calculate:

$$e^{i\theta I_x}\, I_z\, e^{-i\theta I_x} = I_z\, \cos\theta + i\,[I_x, I_z]\, \sin\theta. \tag{3.59}$$

For this we need to know the commutation relations of the operators $I_x$, $I_y$ and $I_z$. These are (see Appendix 3.A.4):

$$\begin{aligned}[I_x, I_y] &= iI_z\\ [I_y, I_z] &= iI_x\\ [I_z, I_x] &= iI_y.\end{aligned} \tag{3.60}$$

Thus, the effect of an x-pulse on z-magnetization will become:

$$e^{i\theta I_x}\, I_z\, e^{-i\theta I_x} = I_z\, \cos\theta + I_y\, \sin\theta. \tag{3.61}$$

Or in shorthand notation:

$$I_z \xrightarrow{\theta I_x} I_z \cos\theta + I_y \sin\theta. \tag{3.62}$$

Table 3.1 summarizes the effect of RF pulses and $\mathcal{H}$-evolution on the product operators. A pictorial representation of this is given in Figure 3.1. Please note that in our description the definition of + and - RF pulses is opposite to what [5] used in their review. In our notation, the effect of a + RF pulse leads to a right-handed rotation.

A non-selective $\theta(x)$ pulse in homonuclear experiments works on both spins. Therefore, the transformation becomes:

$$e^{i\theta(I_x+S_x)} \tag{3.63}$$

and since $[I_x, S_x] = 0$ we can replace this by two separate transformations, which can be applied in any order:

$$e^{i\theta(I_x+S_x)} = e^{i\theta I_x}\, e^{i\theta S_x} = e^{i\theta S_x}\, e^{i\theta I_x}. \tag{3.64}$$

**Table 3.1** Effect of RF pulses and $\mathcal{H}$-evolution on product operators.

| | $I_x$ | $I_y$ | $I_z$ | |
|---|---|---|---|---|
| x-pulse | $I_x$ | $I_y \cos\theta - I_z \sin\theta$ | $I_z \cos\theta + I_y \sin\theta$ | |
| y-pulse | $I_x \cos\theta + I_z \sin\theta$ | $I_y$ | $I_z \cos\theta - I_x \sin\theta$ | |
| z-pulse | $I_x \cos\theta - I_y \sin\theta$ | $I_y \cos\theta + I_x \sin\theta$ | $I_z$ | |
| $\Omega_I t$ | $I_x \cos\Omega_I t - I_y \sin\Omega_I t$ | $I_y \cos\Omega_I t + I_x \sin\Omega_I t$ | $I_z$ | |
| $\pi Jt(2I_zS_z)$ | $I_x \cos\pi Jt + 2I_yS_z \sin\pi Jt$ | $I_y \cos\pi Jt - 2I_xS_z \sin\pi Jt$ | $I_z$ | |
| | $2I_xS_z$ | $2I_yS_z$ | $2I_xS_x$ | $2I_xS_y$ |
| $\pi Jt(2I_zS_z)$ | $2I_xS_z \cos\pi Jt + I_y \sin\pi Jt$ | $2I_yS_z \cos\pi Jt - I_x \sin\pi Jt$ | $2I_xS_x$ | $2I_xS_y$ |

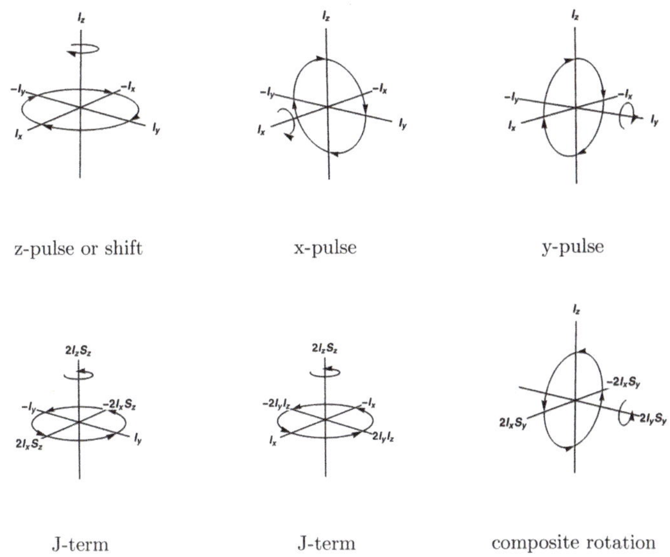


**Figure 3.1** Effect of RF pulses and $\mathcal{H}$-evolution on product operators.

The transform of the density operator therefore becomes two successive transforms of $\rho$. When we do this in steps from the inside out, first a transform by the operator $S_x$ and thereafter a transform by the operator $I_x$:

$$e^{i\theta I_x}\, e^{i\theta S_x}\, \rho\, e^{-i\theta S_x}\, e^{-i\theta I_x}. \tag{3.65}$$

Since $[I_x, S_x] = 0$, the opposite order, first a transform by the operator $I_x$ and thereafter a transform by the operator $S_x$ would give the same result:

$$e^{i\theta S_x}\, e^{i\theta I_x}\, \rho\, e^{-i\theta I_x}\, e^{-i\theta S_x}. \tag{3.66}$$

The effect of RF pulses on other product operators can be calculated as well, e.g. the application of a $\theta(y)$ pulse on antiphase magnetization $2I_xS_z$ would be:

$$\begin{aligned} e^{i\theta I_y}\, e^{i\theta S_y}\, 2I_xS_z\, e^{-i\theta S_y}\, e^{-i\theta I_y} &= 2\, e^{i\theta I_y}\, I_x\, e^{-i\theta I_y}\, e^{i\theta S_y}\, S_z\, e^{-i\theta S_y} \\ &= 2\,(I_x \cos\theta + I_z \sin\theta)\,(S_z \cos\theta - S_x \sin\theta). \end{aligned} \tag{3.67}$$

Thus, a $\frac{\pi}{2}(y)$ pulse will convert antiphase magnetization on spin I, $2I_xS_z$ to antiphase magnetization on spin S, $-2I_zS_x$. In shorthand notation,

$$2I_xS_z \xrightarrow{\frac{\pi}{2}(I_y+S_y)} -2I_zS_x \tag{3.68}$$

This type of transfer is the basis of most coherence transfer experiments.

Depending on the phase of the RF pulse or the starting situation, the pulses would create different coherences:

- a $\frac{\pi}{2}(x)$ pulse on antiphase magnetization, $2I_xS_z$, will create ZQ and DQ coherences:

$$2I_xS_z \xrightarrow{\frac{\pi}{2}(I_x+S_x)} 2I_xS_y \tag{3.69}$$

  This transfer is used for the creation of multiple-quantum (MQ) coherences in MQ spectroscopy.
- a $\frac{\pi}{2}(x)$ pulse on DQ coherence will create antiphase doublets:

$$(2I_xS_y + 2I_yS_x) \xrightarrow{\frac{\pi}{2}(I_x+S_x)} (-2I_xS_z - 2I_zS_x). \tag{3.70}$$

  In a spectrum this will be observed as

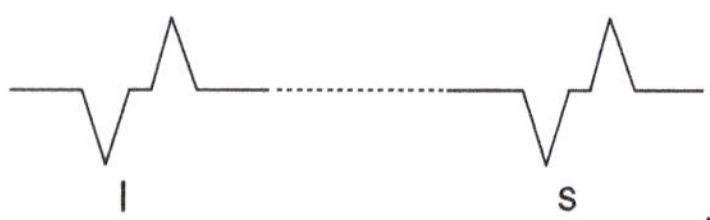


  This transfer is used for detecting DQ coherences.

### 3.3.2 Effect of Evolution Under the Hamiltonian

The Hamiltonian of a weakly coupled two-spin system is:

$$\mathcal{H} = -\Omega_I I_z - \Omega_S S_z + 2\pi J I_z S_z. \tag{3.71}$$

Since all terms commute, we can calculate the effect of each term separately and in any order:

$$\rho(t) = e^{i\Omega_I I_z t}\, e^{i\Omega_S S_z t}\, e^{-i\pi J 2I_zS_z t}\rho(0)\, e^{i\pi J 2I_zS_z t}\, e^{-i\Omega_S S_z t}\, e^{-i\Omega_I I_z t}. \tag{3.72}$$

- Chemical shift evolution (free precession) behaves like a $\theta(z)$ pulse where $\theta = \Omega_I t$. Thus, when $\rho(0) = I_x$ the coherence will evolve as:

$$e^{i\theta I_z}\, I_x\, e^{-i\theta I_z} = I_x\, \cos\theta - I_y\, \sin\theta \tag{3.73}$$

  or in a shorthand notation:

$$I_x \xrightarrow{\theta(I_z)} I_x\, \cos\theta - I_y\, \sin\theta. \tag{3.74}$$

- Coupling terms evolve through the effect of the $2I_zS_z$ product operator. Since it can be shown that $[2I_zS_z, I_x] = i\,2I_yS_z$, $I_x$ will evolve as:

$$I_x \xrightarrow{\pi Jt\; 2I_zS_z} I_x\, \cos\pi Jt + 2I_yS_z\, \sin\pi Jt. \tag{3.75}$$

  Similarly, since $[2I_zS_z, I_y] = -i2I_xS_z$, $I_y$ will evolve due to the J-coupling as:

$$I_y \xrightarrow{\pi Jt 2I_zS_z} I_y\cos\pi Jt - 2I_xS_z\sin\pi Jt. \tag{3.76}$$

  The J-coupling term creates pure antiphase magnetization from $I_x$ or $I_y$ at time $t = \frac{1}{2J}$.

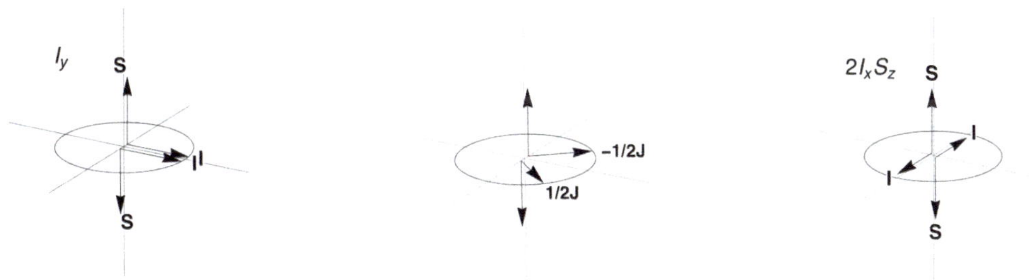


It can also be shown that the following commutation relations hold:

$$\begin{aligned}&[2I_zS_z, 2I_xS_z] = iI_y\\&[2I_zS_z, 2I_xS_x] = 0\\&\left[2I_zS_z, 2I_xS_y\right] = 0.\end{aligned} \tag{3.77}$$

From now on we will shorten the transformation $\xrightarrow{\pi Jt\,(2I_zS_z)}$ to $\xrightarrow{\text{Jt}}$.

Using the commutators above (3.77) it can be shown how antiphase magnetization and MQ terms evolve due to the J-coupling:

$$\begin{aligned}&2I_xS_z \xrightarrow{\text{Jt}} 2I_xS_z \cos \pi Jt + I_y \sin \pi Jt\\&2I_xS_x \xrightarrow{\text{Jt}} 2I_xS_x\\&2I_xS_y \xrightarrow{\text{Jt}} 2I_xS_y.\end{aligned} \tag{3.78}$$

Thus, antiphase magnetization $2I_xS_z$ will evolve to observable magnetization $I_y$ by the J-term, but the J-term has no effect on the evolution of ZQ and DQ coherences.

The combined evolution by chemical shift and J becomes:

$$\begin{aligned}&I_y \xrightarrow{\Omega_I t} I_y \cos \Omega_I t + I_x \sin \Omega_I t \xrightarrow{Jt}\\&\qquad (I_y \cos \pi Jt - 2I_xS_z \sin \pi Jt) \cos \Omega_I t + (I_x \cos \pi Jt + 2I_yS_z \sin \pi Jt) \sin \Omega_I t.\end{aligned} \tag{3.79}$$

The same result is obtained if we first apply the J-term and thereafter the shift-term.

## 3.4 Applications

We will now describe a few basic NMR experiments using the product operators

- spin-echo experiments
- multiple-quantum spectroscopy
- composite pulses

In the next Section (3.5), we will show the use of product operators in several 2D experiments.

### 3.4.1 Spin-echo Experiments

We will describe the experiment 90($x$)-$\tau$-180($x$)-$\tau$ with product operators for a few cases. First, we apply only pulses on spin I.

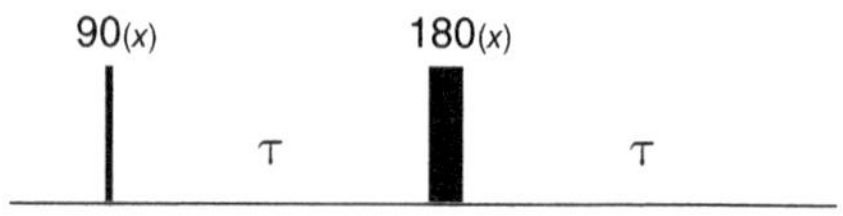


The initial condition for $\rho(0)$ is thermal equilibrium. Thus, for one spin I we start with:

$$\rho(0) = I_z. \tag{3.80}$$

Subsequently, we apply all transformations by the pulses and the evolution of the Hamiltonian.

- Spin-echo experiment with one spin I:

$$\begin{aligned}
I_z &\xrightarrow{\frac{\pi}{2}(x)} I_y \xrightarrow{\Omega\tau} I_y \cos\Omega\tau + I_x \sin\Omega\tau \longrightarrow \\
&\xrightarrow{\pi(x)} -I_y \cos\Omega\tau + I_x \sin\Omega\tau \longrightarrow \\
&\xrightarrow{\Omega\tau} -(I_y \cos\Omega\tau + I_x \sin\Omega\tau)\cos\Omega\tau \\
&\qquad + (I_x \cos\Omega\tau - I_y \sin\Omega\tau)\sin\Omega\tau \\
&\quad = -I_y(c^2 + s^2) + I_x(cs - sc) = -I_y.
\end{aligned} \tag{3.81}$$

Thus, in this case there is a refocusing of the chemical shift. The echo intensity is independent from $\Omega$ and independent from $\tau$ (if we neglect relaxation). This is always true: the sequence $\tau-180-\tau$ will always refocus the chemical shift (linear term in $I_z$ only), also when the Hamiltonian contains terms of other spins and couplings with the other spins.

- Heteronuclear spin-echo experiment with two coupled spins I and S, where we will only apply pulses to spin I (e.g. $I = {}^{13}C$ and $S = {}^{1}H$).

  In this case, we only consider the effect of the J-coupling, since we have already seen before that the shift terms will refocus.

$$\begin{aligned}
I_z &\xrightarrow{\frac{\pi}{2}(x)} I_y \xrightarrow{J\tau} I_y \cos\pi J\tau - 2I_xS_z \sin\pi J\tau \longrightarrow \\
&\xrightarrow{\pi(x)} -I_y \cos\pi J\tau - 2I_xS_z \sin\pi J\tau \longrightarrow \\
&\xrightarrow{J\tau} -(I_y \cos\pi J\tau - 2I_xS_z \sin\pi J\tau)\cos\pi J\tau \\
&\qquad - (2I_xS_z \cos\pi J\tau + I_y \sin\pi J\tau)\sin\pi J\tau \\
&\quad = -I_y(c^2 + s^2) - 2I_xS_z(cs - sc) = -I_y.
\end{aligned} \tag{3.82}$$

  Thus, we will have again complete refocusing of both ($+\frac{1}{2J}$ and $-\frac{1}{2J}$) components of $I_y$, independent on the values of $J$ and $\tau$.

- Spin-echo experiment with two coupled spins I and S, where we apply $180(x)$ pulses to both spin I and S.

  This is the case of a homonuclear spin-echo experiment (e.g. two protons) or a heteronuclear spin-echo experiment where we apply $180(x)$ pulses to both spins.

$$
\begin{aligned}
I_z &\xrightarrow{\frac{\pi}{2}(x)} I_y \xrightarrow{J\tau} I_y \cos \pi J\tau - 2I_xS_z \sin \pi J\tau \longrightarrow \\
&\xrightarrow{\pi(I_x)} -I_y \cos \pi J\tau - 2I_xS_z \sin \pi J\tau \longrightarrow \\
&\xrightarrow{\pi(S_x)} -I_y \cos \pi J\tau + 2I_xS_z \sin \pi J\tau \longrightarrow \\
&\xrightarrow{J\tau} -(I_y \cos \pi J\tau - 2I_xS_z \sin \pi J\tau) \cos \pi J\tau \\
&\qquad + (2I_xS_z \cos \pi J\tau + I_y \sin \pi J\tau) \sin \pi J\tau \\
&= -I_y(c^2 - s^2) + 2I_xS_z(cs + sc) \\
&= -I_y \cos 2\pi J\tau + 2I_xS_z \sin 2\pi J\tau.
\end{aligned} \tag{3.83}
$$

The signal intensity is now dependent on $J$ and $\tau$, it is 'J-modulated'. The observable echo intensity $< I_y >$ at $2\tau$ disappears completely when $\tau = \frac{1}{4J}$. At that moment, the antiphase term $2I_xS_z$ will be at its maximum.

Using $e^{i\alpha I_z} = E \cos \frac{\alpha}{2} + 2iI_z \sin \frac{\alpha}{2}$ (see Appendix 3.A.8), we can show that $e^{-i\pi I_x}\, e^{i\alpha I_z}\, e^{i\pi I_x} = e^{-i\alpha I_z}$. From this it follows that $e^{i\alpha I_z}\, e^{i\pi I_x} = e^{i\pi I_x}\, e^{-i\alpha I_z}$, and therefore that:

$$
e^{i\alpha I_z}\, e^{i\pi I_x}\, e^{i\alpha I_z} = e^{i\pi I_x}\, e^{-i\alpha I_z}\, e^{i\alpha I_z} = e^{i\pi I_x}. \tag{3.84}
$$

Thus, a $\tau - 180(I_x) - \tau$ element removes all terms linear in $I_z$, and the net effect is a 180° pulse, or in shorthand notation:

$$
\begin{aligned}
&\rho(z) \xrightarrow{\alpha I_z} \xrightarrow{\pi I_x} \xrightarrow{\alpha I_z} \rho(2\tau) \\
&\rho(z) \xrightarrow{\pi I_x} \xrightarrow{-\alpha I_z} \xrightarrow{\alpha I_z} \rho(2\tau) \\
&\rho(z) \xrightarrow{\pi I_x} \rho(2\tau).
\end{aligned} \tag{3.85}
$$

If we include the effect of J-coupling, the density matrix in a spin-echo experiment with two coupled spins I and S will evolve as:

$$
\rho(0) \xrightarrow{Jt} \xrightarrow{\Omega_I t} \xrightarrow{\Omega_S t} \xrightarrow{\pi(I_x+S_x)} \xrightarrow{\Omega_I t} \xrightarrow{\Omega_S t} \xrightarrow{Jt} \rho(2\tau). \tag{3.86}
$$

We can reduce the total number of different transformations. First, since $[S_z, I_x] = 0$, we can swap the order of two arrows that contain commutating operators, such as $\xrightarrow{\Omega_S t} \xrightarrow{\pi(I_x)}$, and replace it by $\xrightarrow{\pi(I_x)} \xrightarrow{\Omega_S t}$, and second we can swap the order of the arrows $\xrightarrow{Jt} \xrightarrow{\pi(I_x)} \xrightarrow{\pi(I_x)}$ and replace it by $\xrightarrow{\pi(I_x)} \xrightarrow{\pi(I_x)} \xrightarrow{Jt}$, since:

$$
\begin{aligned}
&e^{i\alpha I_zS_z} e^{i\pi I_x} = e^{i\pi I_x} e^{-i\alpha I_zS_z} \\
&e^{i\alpha I_zS_z} e^{i\pi I_x} e^{i\pi S_x} = e^{i\pi I_x} e^{-i\alpha I_zS_z} e^{i\pi S_x} = e^{i\pi I_x} e^{i\pi S_x} e^{i\alpha I_zS_z}.
\end{aligned} \tag{3.87}
$$

When we step-by-step include the various arrow swaps, we note that only the effect of the J-modulation and the two 180° pulses will remain:

$$
\begin{aligned}
&\rho(0) \xrightarrow{Jt} \xrightarrow{\Omega_I t} \xrightarrow{\Omega_S t} \xrightarrow{\pi(I_x+S_x)} \xrightarrow{\Omega_I t} \xrightarrow{\Omega_S t} \xrightarrow{Jt} \rho(2t)\\
&\rho(0) \xrightarrow{Jt} \xrightarrow{\Omega_I t} \xrightarrow{\Omega_S t} \xrightarrow{\pi I_x} \xrightarrow{\pi S_x} \xrightarrow{\Omega_I t} \xrightarrow{\Omega_S t} \xrightarrow{Jt} \rho(2t)\\
&\rho(0) \xrightarrow{Jt} \xrightarrow{\Omega_I t} \xrightarrow{\pi I_x} \xrightarrow{\Omega_S t} \xrightarrow{\pi S_x} \xrightarrow{\Omega_I t} \xrightarrow{\Omega_S t} \xrightarrow{Jt} \rho(2t)\\
&\rho(0) \xrightarrow{Jt} \xrightarrow{\Omega_I t} \xrightarrow{\pi I_x} \xrightarrow{\Omega_I t} \xrightarrow{\Omega_S t} \xrightarrow{\pi S_x} \xrightarrow{\Omega_S t} \xrightarrow{Jt} \rho(2t)\\
&\rho(0) \xrightarrow{Jt} \xrightarrow{\pi I_x} \xrightarrow{-\Omega_I t} \xrightarrow{\Omega_I t} \xrightarrow{\pi S_x} \xrightarrow{-\Omega_S t} \xrightarrow{\Omega_S t} \xrightarrow{Jt} \rho(2t)\\
&\rho(0) \xrightarrow{Jt} \xrightarrow{\pi I_x} \xrightarrow{\pi S_x} \xrightarrow{Jt} \rho(2t)\\
&\rho(0) \xrightarrow{\pi I_x} \xrightarrow{\pi S_x} \xrightarrow{Jt} \xrightarrow{Jt} \rho(2t)\\
&\rho(0) \xrightarrow{\pi(I_x+S_x)} \xrightarrow{2Jt} \rho(2t).
\end{aligned}
\tag{3.88}
$$

In this way, the total number of transformations can be reduced considerably, which means less math and not unimportant either, fewer mistakes. Therefore, when we start with $\rho(0) = I_y$ only two transformations need to be calculated:

$$I_y \xrightarrow{\pi I_x} -I_y \xrightarrow{2Jt} -(I_y \cos \pi J\,2t - 2I_xS_z \sin \pi J\,2t). \tag{3.89}$$

### 3.4.2 Multiple-quantum Coherence

Multiple-quantum coherence can be created in a number of ways. The most common method for creating DQ coherences, however, is the $90(x) - \tau - 180(x) - \tau - 90(x)$ sequence, with $\tau = \frac{1}{4J}$. This has the advantage that the excitation is independent of the chemical shift of I and S.

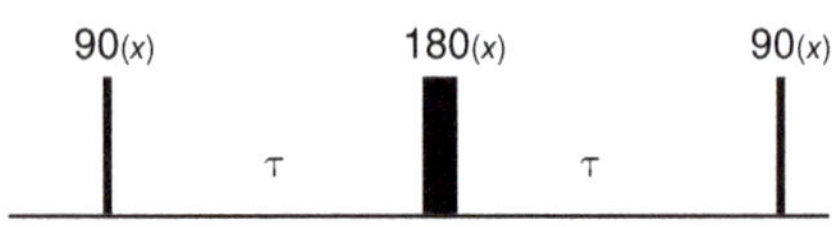


The first part of the sequence is the same as the homonuclear spin-echo experiment discussed above (in Section 3.4.1):

$$I_z \rightarrow \ldots \rightarrow -I_y \cos 2\pi J\tau + 2I_xS_z \sin 2\pi J\tau \tag{3.90}$$

and since we start with both $I_z$ and $S_z$:

$$S_z \rightarrow \ldots \rightarrow -S_y \cos 2\pi J\tau + 2I_zS_x \sin 2\pi J\tau. \tag{3.91}$$

At $\tau = \frac{1}{4J}$ the result becomes $2I_xS_z + 2I_zS_x$. The effect of the last 90(x) pulse therefore is:

$$2I_xS_z + 2I_zS_x \xrightarrow{\frac{\pi}{2}(I_x+S_x)} 2I_xS_y + 2I_yS_x = (DQ)_y. \tag{3.92}$$

In this way, from antiphase magnetization pure DQ coherence (see 3.48) will be created.
We will now follow the evolution of $(DQ)_y$ using the pulse sequence:

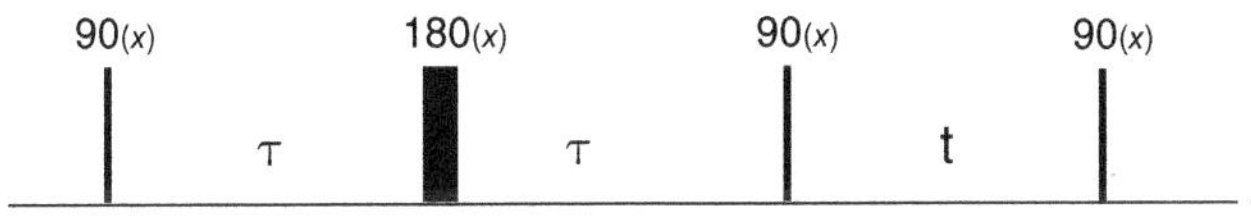


During the period t the MQ coherence will evolve. Since we know that $[I_xS_y + I_yS_x, I_zS_z] = 0$, the J-term will have no effect on the DQ term and we only need to consider the effect of the chemical shift evolution during the period t:

$$
\begin{aligned}
(DQ)_y & \xrightarrow{\Omega_I t} \xrightarrow{\Omega_S t} \\
& \to 2(I_x \cos \Omega_I t - I_y \sin \Omega_I t)(S_y \cos \Omega_S t + S_x \sin \Omega_S t) \\
& \quad + 2(I_y \cos \Omega_I t + I_x \sin \Omega_I t)(S_x \cos \Omega_S t - S_y \sin \Omega_S t) \\
& = [2I_xS_y + 2I_yS_x](\cos \Omega_I t \cos \Omega_S t - \sin \Omega_I t \sin \Omega_S t) \\
& \quad + [2I_xS_x - 2I_yS_y](\cos \Omega_I t \sin \Omega_S t + \sin \Omega_I t \cos \Omega_S t) \\
& = (DQ)_y \cos(\Omega_I + \Omega_S)t + (DQ)_x \sin(\Omega_I + \Omega_S)t \\
& = (DQ)_y \cos(\omega_I + \omega_S - 2\omega_R)t + (DQ)_x \sin(\omega_I + \omega_S - 2\omega_R)t.
\end{aligned}
\tag{3.93}
$$

The DQ term modulates during time t with a frequency $\Omega(DQ) = \Omega_I + \Omega_S = \omega_I + \omega_S - 2\omega_R$ (see Appendix 3.A.8 for the trigonometric equations used). The DQ oscillations are thus strongly dependent on the carrier offset.

ZQ coherences can be created by other pulse sequences. For instance, if the last $90(x)$ pulses on I and S of the $90(x) - \tau - 180(x) - \tau - 90(x)$ sequence would have opposite sign (compare with Equation 3.92), we could create pure $2I_yS_x - 2I_xS_y = (ZQ)_y$ (see Equation 3.48). Another method would be to take out the $180(x)$ pulse, in which case we would create a mixture of coherences. We will now follow the evolution of this ZQ term during time t:

$$
\begin{aligned}
(ZQ)_y & \xrightarrow{\Omega_I t} \xrightarrow{\Omega_S t} \\
& = (ZQ)_y \cos(\Omega_I - \Omega_S)t + (ZQ)_x \sin(\Omega_I - \Omega_S)t.
\end{aligned}
\tag{3.94}
$$

The frequency of the ZQ term is $\Omega(ZQ) = \Omega_I - \Omega_S = \omega_I - \omega_S$. It is independent of the carrier offset.

Finally, we will describe the effect of the last $90(x)$ pulse on the DQ coherence that is created at time t in the pulse sequence described above.

The last $90(x)$ pulse on both spins will convert the $(DQ)_y$ term in Equation 3.93 into two antiphase magnetization terms:

$$2I_xS_y + 2I_yS_x \xrightarrow{\frac{\pi}{2}(I_x+S_x)} -2I_xS_z - 2I_zS_x. \tag{3.95}$$

The $(DQ)_x$ term in scheme 3.93 becomes a mixture of ZQ, DQ, and longitudinal spin-order $2I_zS_z$ after the last 90(x) pulse:

$$2I_xS_x - 2I_yS_y \xrightarrow{\frac{\pi}{2}(I_x+S_x)} 2I_xS_x - 2I_zS_z. \tag{3.96}$$

Thus, only the $(DQ)_y$ term in Equation 3.93 will become observable during acquisition, as antiphase doublets, since it can evolve to observable $I_x$, $I_y$, $S_x$, and $S_y$ magnetization terms due to the J-term.

A quick way to calculate the $90(x) - \tau - 180(x) - \tau - 90(x)$ sequence uses the following transformation:

$$\begin{aligned} e^{-i\frac{\pi}{2}I_x}\, e^{i\alpha I_z}\, e^{i\frac{\pi}{2}I_x} &= E\cos\frac{\alpha}{2} + 2i\, e^{-i\frac{\pi}{2}I_x} I_z e^{i\frac{\pi}{2}I_x} \sin\frac{\alpha}{2} \\ &= E\cos\frac{\alpha}{2} - 2i\, I_y \sin\frac{\alpha}{2} = e^{-i\alpha I_y}. \end{aligned} \tag{3.97}$$

From this we can derive:

$$e^{-i\frac{\pi}{2}(I_x+S_x)}\, e^{i\alpha I_z S_z e}\, e^{i\frac{\pi}{2}(I_x+S_x)} = e^{i\alpha I_y S_y}. \tag{3.98}$$

Thus, the sequence simplifies as follows

$$\begin{aligned} &\rho(0) \xrightarrow{\frac{\pi}{2}(x)} \xrightarrow{\pi J\,\tau\, 2I_zS_z} \xrightarrow{\pi(x)} \xrightarrow{\pi J\,\tau\, 2I_zS_z} \xrightarrow{\frac{\pi}{2}(x)} \rho(2\tau) \\ &\rho(0) \xrightarrow{\frac{3\pi}{2}(x)} \xrightarrow{\pi J\, 2\tau\, 2I_zS_z} \xrightarrow{\frac{\pi}{2}(x)} \rho(2\tau) \\ &\rho(0) \xrightarrow{2\pi(x)} \xrightarrow{-\frac{\pi}{2}(x)} \xrightarrow{\pi J\, 2\tau\, 2I_zS_z} \xrightarrow{\frac{\pi}{2}(x)} \rho(2\tau) \\ &\rho(0) \xrightarrow{2\pi(x)} \xrightarrow{\pi J\, 2\tau\, 2I_yS_y} \rho(2\tau) \\ &\rho(0) \xrightarrow{\pi J\, 2\tau\, 2I_yS_y} \rho(2\tau). \end{aligned} \tag{3.99}$$

The effect of the pulse train on the initial state $\rho(0) = I_z$ can be calculated as a 'composite rotation' by a single transformation using the commutation relation $[2I_yS_y, I_z] = 2iI_xS_y$:

$$I_z \xrightarrow{\pi J\, 2\tau\, 2I_yS_y} I_z \cos 2\pi J\tau + 2I_xS_y \sin 2\pi J\tau. \tag{3.100}$$

An application of DQ coherence is the so-called INADEQUATE experiment [10], which can be used for detecting $^{13}C$ - $^{13}C$ J-couplings in natural abundance. The aim is to filter away all magnetization which is not due to DQ coherence.

The above sequence has to be modified since it will also transmit magnetization of uncoupled spins. If applied on an uncoupled spin I, starting with $\rho(0) = I_z$, the pulse train $90(x) - \tau - 180(x) - \tau - 90(x)$ would create $I_z$ and the last $90(x)$ of the MQ sequence would create observable magnetization $I_y$. Thus, the last detection pulse in this first experiment, $90(x) - \tau - 180(x) - \tau - 90(x) - 90(x)$, would create these coherences:

$$\begin{aligned} 2I_xS_y + 2I_yS_x &\xrightarrow{\frac{\pi}{2}(I_x+S_x)} -2I_xS_z - 2I_zS_x \\ I_z &\xrightarrow{\frac{\pi}{2}(I_y+S_y)} +I_y. \end{aligned} \tag{3.101}$$

We can remove this single-quantum (SQ) coherence by phase cycling the last detection pulse together with the receiver phase. Assume that the last 90° pulse was a $90(y)$ pulse. We would create now:

$$\begin{aligned} 2I_xS_y + 2I_yS_x &\xrightarrow{\frac{\pi}{2}(I_y+S_y)} 2I_zS_y + 2I_yS_z \\ I_z &\xrightarrow{\frac{\pi}{2}(I_y+S_y)} -I_x. \end{aligned} \tag{3.102}$$

We can now shift the phase of all pulses in this last experiment by 90°:

$$\begin{array}{ccc} x & \longrightarrow & y \\ \uparrow & & \downarrow \\ -y & \longleftarrow & -x \end{array} \tag{3.103}$$

The last detection pulse in this second experiment, $90(y) - \tau - 180(y) - \tau - 90(y) - 90(-x)$, would then create these coherences:

$$\begin{aligned} -2I_yS_x - 2I_xS_y &\xrightarrow{-\frac{\pi}{2}(I_x+S_x)} -2I_zS_x - 2I_xS_z \\ I_z &\xrightarrow{-\frac{\pi}{2}(I_x+S_x)} -I_y. \end{aligned} \tag{3.104}$$

Adding this experiment to the first would create pure antiphase magnetization. Thus, only the pathway that went through DQ coherence remains, and we filter away the SQ resonances of uncoupled spins or spins with $J \neq \frac{1}{4\tau}$. In this way, pure DQ spectra can be obtained.

In practice, the experiment is done with a more complete and slightly different phase cycle in order to compensate for relaxation effects and instrument artifacts.

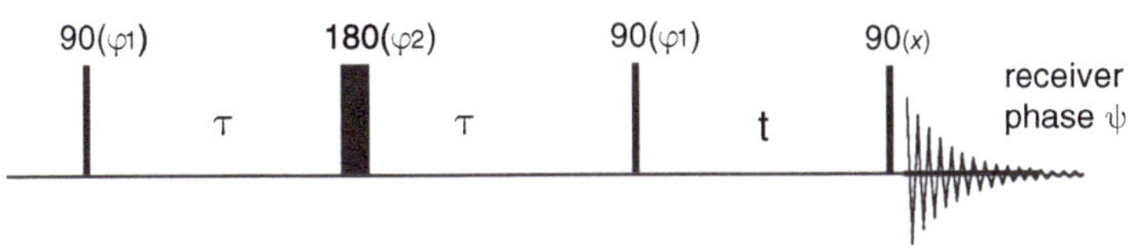


| $\phi_1$ | $\phi_2$ | $\psi$ | SQ | DQ |
|---|---|---|---|---|
| x | y | x | -x | x |
| y | -x | -x | -y | x |
| -x | -y | x | x | x |
| -y | x | -x | y | x |

By adding all experiments the DQ coherence adds up, and SQ coherences and artifacts are canceled.

### 3.4.3 Composite Pulses

In the previous paragraph (in Equation 3.97) we derived the relation:

$$e^{-i\frac{\pi}{2}I_x} e^{i\alpha I_z} e^{i\frac{\pi}{2}I_x} = e^{-i\alpha I_y}. \tag{3.105}$$

And similarly:

$$e^{i\frac{\pi}{2}I_x} e^{-i\alpha I_y} e^{-i\frac{\pi}{2}I_x} = e^{i\alpha I_z}. \tag{3.106}$$

This shows that the effect of a sequence $90(x) - \alpha(-y) - 90(-x)$ is the same as an RF pulse along z ('z-pulse'). Since normal NMR specrometers cannot pulse along the z-axis, we can use this composite RF pulse to mimic arbitrary RF phase shifts ('the poor mans phase shifter'). An application of (z) pulses is the separation of various coherences.

$$\begin{aligned}
I_z &\xrightarrow{\alpha I_z} I_z \\
I_y &\xrightarrow{\alpha I_z} I_y \cos\alpha + I_x \sin\alpha \\
(DQ)_y = 2I_xS_y + 2I_yS_x &\xrightarrow{\alpha(I_z+S_z)} (I_x\cos\alpha - I_y\sin\alpha)(S_y\cos\alpha + S_x\sin\alpha) \\
&\qquad + (I_y\cos\alpha + I_x\sin\alpha)(S_x\cos\alpha - S_y\sin\alpha) \\
&\qquad = (DQ)_y\cos 2\alpha + (DQ)_x \sin 2\alpha.
\end{aligned} \tag{3.107}$$

Another composite pulse, used for magnetization inversion, is the sequence $90(x)$–$180(y)$–$90(x)$. This sequence will still give good inversion, even when the $90(x)$ pulses are not perfect, as can be shown as follows:

$$\begin{aligned}
e^{-i\pi I_y} e^{i\alpha I_x} e^{i\pi I_y} &= e^{-i\alpha I_x} \\
e^{i\alpha I_x} e^{i\pi I_y} &= e^{i\pi I_y} e^{-i\alpha I_x} \\
e^{i\alpha I_x} e^{i\pi I_y} e^{i\alpha I_x} &= e^{i\pi I_y}.
\end{aligned} \tag{3.108}$$

Independent of the value of $\alpha$ the composite pulse will act as a $180(y)$ pulse.

The exact behavior of composite pulses with respect to resonances at an offset to the carrier cannot easily be derived from the product operators. The length of the pulses should be explicitly taken into account, i.e. one should calculate the motion of the magnetization under the influence of both $B_0$ and $B_1$ either using the complete density matrixes or in the case of uncoupled spins by solving the Bloch equations. Such calculations have led to a large number of composite pulses with special properties. Applications are creation of 'ideal' (i.e. off-resonance independent and $B_1$ strength independent) 90 or 180 pulses and that have a uniform phase at each frequency.

Composite pulses are used routinely in decoupling sequences, where a low power is required, which can decouple the J-couplings for spins over a large spectral width (e.g. $^{13}C$) with another heteronuclei (e.g. $^{1}H$). Well-known composite pulse trains for this purpose are the WALTZ, MLEV, and GARP sequences. The net properties of those decoupling sequences are often described with a so-called averaged Hamiltonian (see [9] for further reading).

## 3.5 Two-dimensional Experiments

In this section we will describe several 2D NMR experiments using the product operators.

- 2D J-resolved
- COSY
- 2D NOE (or NOESY)
- DQF-COSY
- 2D DQ (or 2D INADEQUATE)
- Relayed-COSY
- TOCSY
- INEPT
- HMQC, HMBC

The general scheme of all 2D experiments is:

$$\text{preparation} \mid \underset{t_1}{\text{evolution}} \mid \text{mixing} \mid \underset{t_2}{\text{detection}}\,. \tag{3.109}$$

The different 2D experiments vary mainly in the type of the mixing period, but different preparation periods are possible as well. The magnetization transfer mechanisms are either incoherent (cross-relaxation or chemical exchange) or coherent ( J-coupling). The first can be measured by the NOESY pulse scheme, the second by the COSY pulse scheme. Most other 2D experiments are variations on a theme.

### 3.5.1 Two-dimensional J-Resolved

The experiment is similar to the spin-echo experiments (Section 3.4.1), but now we increase the time $t_1$ for each newly acquired data-set.

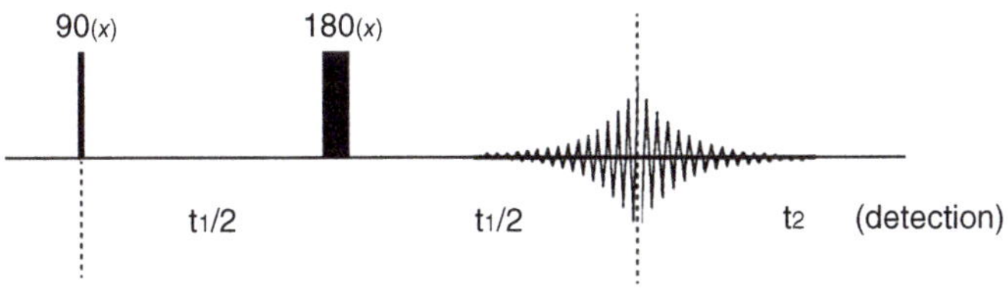


We will obtain a data-set, which is a function of two time variables, $t_1$ and $t_2$. We start just before the acquisition:

$$I_z \xrightarrow{\frac{\pi}{2}(I_x)} \xrightarrow{\pi I_x} \xrightarrow{Jt_1} -I_y \cos \pi J t_1 + 2I_x S_z \sin \pi J t_1. \tag{3.110}$$

The evolution in the detection period $t_2$ becomes:

$$\begin{aligned}
&\xrightarrow{\Omega_I t_2} -[I_y \cos \Omega_I t_2 + I_x \sin \Omega_I t_2] \cos \pi J t_1 \\
&\qquad\quad + [2I_x S_z \cos \Omega_I t_2 - 2I_y S_z \sin \Omega_I t_2] \sin \pi J t_1 \\
&\xrightarrow{Jt_2} -[I_y \cos \pi J t_2 - 2I_x S_z \sin \pi J t_2] \cos \Omega_I t_2 \cos \pi J t_1 \\
&\quad - [I_x \cos \pi J t_2 + 2I_y S_z \sin \pi J t_2] \sin \Omega_I t_2 \cos \pi J t_1 \\
&\quad + [2I_x S_z \cos \pi J t_2 + I_y \sin \pi J t_2] \cos \Omega_I t_2 \sin \pi J t_1 \\
&\quad - [2I_y S_z \cos \pi J t_2 - I_x \sin \pi J t_2] \sin \Omega_I t_2 \sin \pi J t_1.
\end{aligned} \tag{3.111}$$

Assume we detect $I_y$ during $t_2$. Then the observable components ($I_y$) are:

$$\begin{aligned}
&-I_y \cos \pi J t_2 \ \cos \Omega_I t_2 \ \cos \pi J t_1 \\
&\qquad = -\frac{I_y}{2}[\cos(\Omega_I - \pi J)t_2 + \cos(\Omega_I + \pi J)t_2] \cos \pi J t_1
\end{aligned} \tag{3.112}$$

and

$$\begin{aligned}
&-I_y \sin \pi J t_2 \ \cos \Omega_I t_2 \ \sin \pi J t_1 \\
&\qquad = +\frac{I_y}{2}[\sin(\Omega_I - \pi J)t_2 + \sin(\Omega_I + \pi J)t_2] \sin \pi J t_1.
\end{aligned} \tag{3.113}$$

For the other spin S we can derive the same expressions. This will lead to the following 2D spectrum after a 2D Fourier transformation

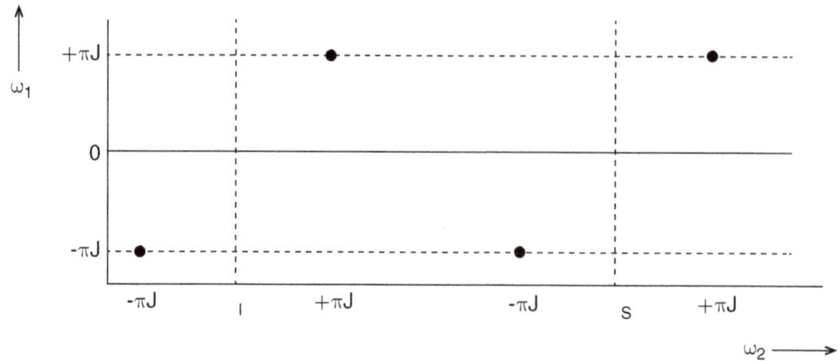


Note that the lineshape of the crosspeaks is a mixture of absorptive and dispersive components. The multiplet structure is observed under an angle of 45°. By a geometrical transformation the spectrum can be rearranged to a spectrum where pure J-coupling is in $\omega_1$ and pure chemical shift is in $\omega_2$.

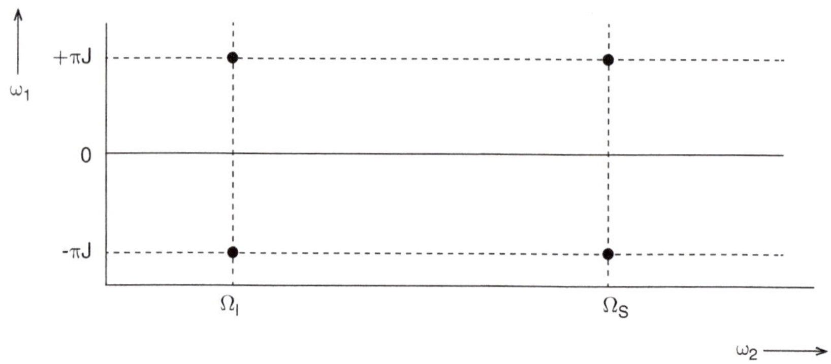


Conceptually the pure shift in one dimension and multiple pattern in the other is elegant. However, the mixed lineshape and appearance of additional crosspeaks in the case of strong coupling of the original 2D J-resolved experiment can be problematic. A number of improved versions of the 2D J-resolved experiment exist now as summarized in [11].

### 3.5.2 COSY

This was the first 2D experiment. The concept of the COSY experiment was originally proposed by Jeener (Brussels, Belgium) at a summer school (Ampere Summer School, Basko Polje, Yugoslavia, 1971; see description in [12]). The sequence was worked out further by the group of Richard Ernst (ETH Zürich, Switzerland) and described in detail in the groundbreaking paper [13].

The sequence contains only two RF pulses.

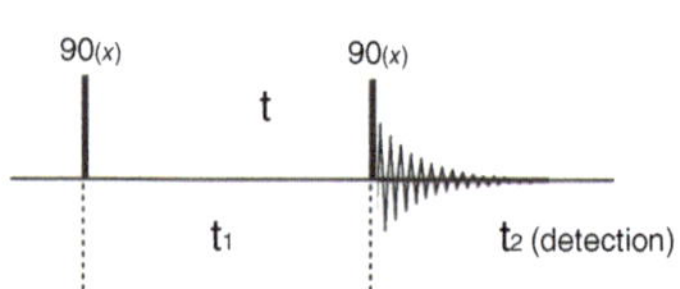

The sequence can be described as:

$$\begin{aligned}
I_z &\xrightarrow{\frac{\pi}{2}(x)} I_y \xrightarrow{\Omega_I t_1} I_y \cos\Omega_I t_1\, t_1 + I_x \sin\Omega_I t_1 \\
&\xrightarrow{Jt_1} [I_y \cos\pi J t_1 - 2I_x S_z \sin\pi J t_1]\cos\Omega_I t_1 \\
&\qquad + [I_x \cos\pi J t_1 + 2I_y S_z \sin\pi J t_1]\sin\Omega_I t_1 \\
&\xrightarrow{\frac{\pi}{2}(I_x+S_x)} -I_z\, cc - 2I_x S_y\, sc + \underline{I_x}\, cs - \underline{2I_z S_y}\, ss.
\end{aligned} \tag{3.114}$$

Only the underlined terms $I_x$ and $2I_zS_y$ can become observable during the detection period:

$$\begin{aligned}
I_x \cos\pi J t_1 \sin\Omega_I t_1 &= \frac{I_x}{2}[\sin(\Omega_I - \pi J)t_1 + \sin(\Omega_I + \pi J)t_1] \\
-2I_z S_y \sin\pi J t_1 \sin\Omega_I t_1 &= -I_z S_y\, [\cos(\Omega_I - \pi J)t_1 - \cos(\Omega_I + \pi J)t_1]
\end{aligned} \tag{3.115}$$

and the same for $S_z$.

Now the detection period: the $I_x$ term will precess during $t_2$ with frequencies $\Omega_I \pm \pi J$, which will give rise to diagonal peaks (same chemical shifts in $t_1$ and $t_2$) and the antiphase $2I_zS_y$ term will precess during $t_2$ with frequencies $\Omega_S \pm \pi J$, which will give rise to crosspeaks, linking the chemical shifts of I ($\Omega_I$) and S ($\Omega_S$).

The COSY spectrum has the following shape

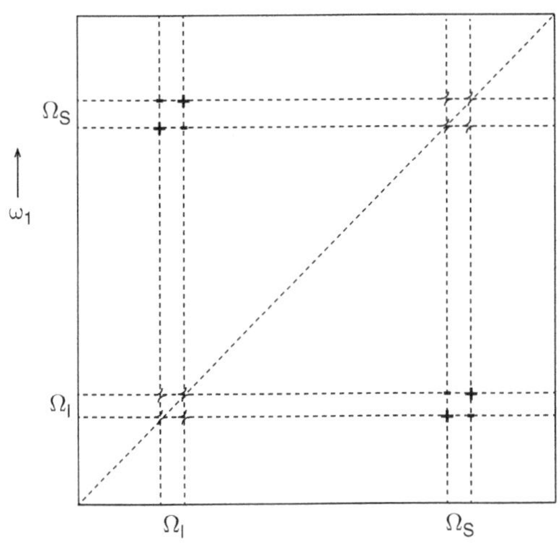


The phase of the diagonal peaks and cross peaks is different (if one is absorptive, the other is dispersive) and there is a multiplet fine structure. On the diagonal all multiplet lines have the same phase and there is net intensity, in the crosspeak there is an up-down multiplet pattern and no net intensity. If the linewidth is much larger than J, the crosspeak will therefore disappear. Generally, COSY spectra are processed in absolute value mode, which leads to loss in resolution, and the multiple fine structure is hardly visible.

The crosspeaks derive from the antiphase $2I_yS_z$ term during $t_1$. Therefore $t_1$ should be incremented to at least $\frac{1}{2J}$ to lead to observable crosspeaks.

### 3.5.3 Two-dimensional NOE

The pulse sequences for the 2D NOE (or NOESY) experiment and the 2D exchange (or EXSY) experiment are the same [14].

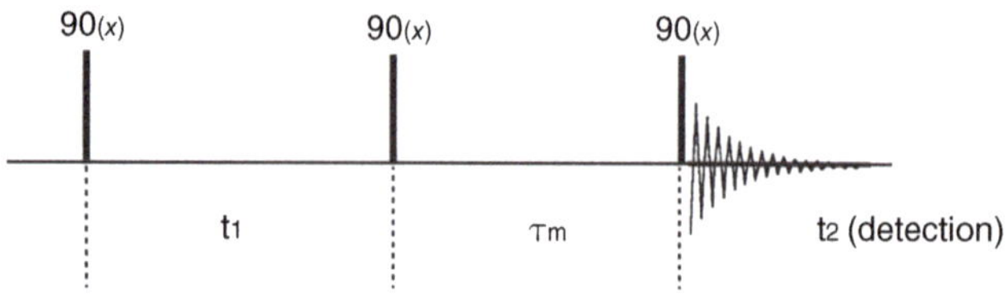


During the mixing time, $\tau_m$, z-magnetization can exchange due to cross-relaxation in the 2D NOE experiment or due to chemical exchange in a 2D exchange experiment [14].

For two weakly coupled spins the sequence would create various coherences during the mixing time:

$$
\begin{aligned}
I_z \xrightarrow{\frac{\pi}{2}(x)} \xrightarrow{\Omega_I t_1} \xrightarrow{J t_1} \xrightarrow{\frac{\pi}{2}(x)} & -I_z \cos\Omega_I t_1 \cos\pi J t_1 - 2I_x S_y \cos\Omega_I t_1 \sin\pi J t_1 \\
& \quad \text{(Z)} \qquad\qquad\qquad\qquad \text{(ZQ+DQ)} \\
& +I_x \sin\Omega_I t_1 \cos\pi J t_1 - 2I_z S_y \sin\Omega_I t_1 \sin\pi J t_1. \\
& \quad \text{(SQ)} \qquad\qquad\qquad\qquad \text{(antiphase)}
\end{aligned}
\tag{3.116}
$$

There are two ways to suppress the SQ, DQ and higher coherences: either the phases of the pulses are cycled and experiments are added, or a field-gradient pulse (short duration of an inhomogeneous magnetic field) is applied during the mixing period. Since ZQ coherences behave like z-magnetization, phase cycling cannot suppress the ZQ coherence.

$$
\begin{aligned}
[ZQ_x, I_z + S_z] &= [2I_x S_x + 2I_y S_y, I_z + S_z] = 0 \\
[ZQ_y, I_z + S_z] &= [2I_y S_x - 2I_x S_y, I_z + S_z] = 0
\end{aligned}
\tag{3.117}
$$

Also a field-gradient pulse cannot suppress the ZQ coherence, since the field gradient is much too weak. Since the ZQ coherence modulates during the mixing period, some suppression can be obtained by a random variation of the mixing time in successive $t_1$ experiments. However, for large molecules, where the relaxation times $T_2 < T_1$, all coherences, including the ZQ coherence, will relax rapidly during the mixing time and their intensities will be weak with respect to the z-magnetization term.

We will now follow only the magnetization of $I_z$ and $S_z$ during the experiment (no J).

$$
\begin{aligned}
I_z \ldots\ldots\ldots -I_z \cos\Omega_I t_1 \quad & \xrightarrow{\tau_m} \quad [-I_z\, a_{II} - S_z\, a_{IS}] \cos\Omega_I t_1 \\
& \xrightarrow{\frac{\pi}{2}(I_x + S_x)} -I_y\, a_{II} \cos\Omega_I t_1 - S_y\, a_{IS} \cos\Omega_I t_1 \\
S_z \ldots\ldots\ldots \qquad\qquad\qquad & \longrightarrow\!\!\!\rightarrow\!\!\!\rightarrow \quad [-S_y\, a_{SS} \cos\Omega_S t_1 - I_y\, a_{SI}] \cos\Omega_S t_1. \\
& \qquad\qquad \text{diagonal peaks} \qquad \text{crosspeaks}
\end{aligned}
\tag{3.118}
$$

Assuming that spins I and S exchange magnetization with a rate constant k (either chemical exchange or cross-relaxation):

$$
I \underset{k}{\overset{k}{\rightleftarrows}} S, \tag{3.119}
$$

the mixing coefficients will be:

$$a_{II} = a_{SS} = \frac{1}{2}e^{-\tau_m/T_1}[1 + e^{-2k\tau_m}]$$
$$a_{IS} = a_{SI} = \frac{1}{2}e^{-\tau_m/T_1}[1 - e^{-2k\tau_m}] \tag{3.120}$$

and the intensities of the diagonal peaks and crosspeaks will develop in the following way during the mixing time as:

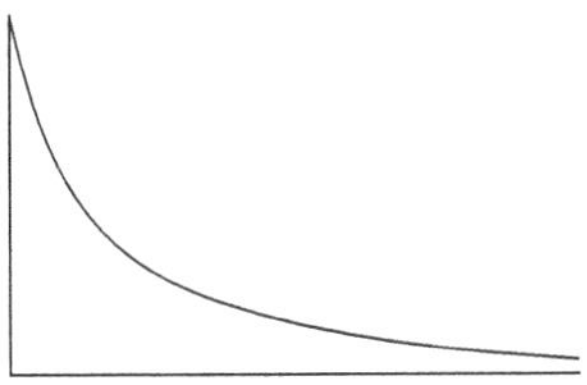

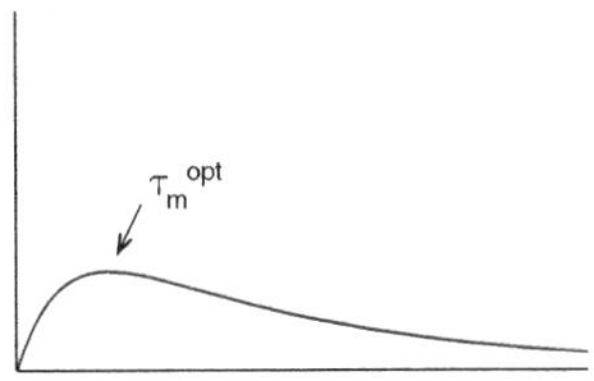


For a two-spin system the energy-level diagram used to describe dipolar cross-relaxation is:

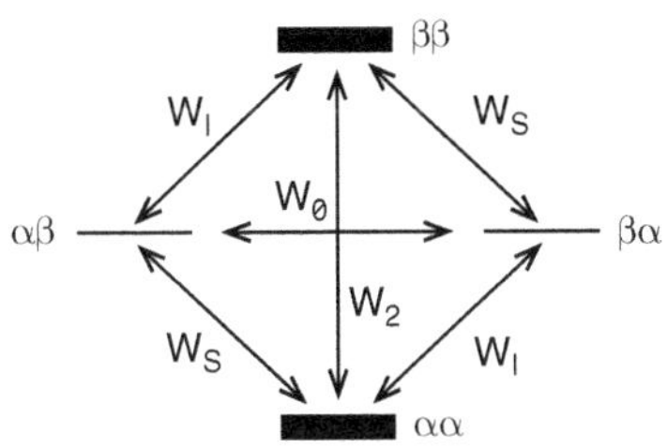


In this diagram the relaxation transition probabities between the energy levels are represented by $W_1$, $W_2$ and $W_0$. The expressions for $T_1$-relaxation and cross-relaxation ($\sigma$) rates in terms of the transition probabilities will be:

$$\frac{1}{T_1} = \frac{1}{T_{1\ leak}} + W_0 + 2W_1 + W_2 - |W_2 - W_0|$$
$$\sigma = W_2 - W_0 \tag{3.121}$$
$$k = |\sigma|$$

and the time-evolutions of the diagonal peaks and crosspeaks are:

$$a_{II} = a_{SS} = \frac{1}{2}e^{-\tau_m/T_1}[1 + e^{-2|\sigma|\tau_m}]$$
$$a_{IS} = a_{SI} = -\frac{1}{2}\frac{W_2 - W_0}{|W_2 - W_0|}e^{-\tau_m/T_1}[1 - e^{-2|\sigma|\tau_m}]. \tag{3.122}$$

For small molecules $W_2 > W_0$ and the crosspeaks and diagonal peaks have opposite sign; for large molecules $W_0 > W_2$ and they have the same sign.

In the initial rate regime (or for short mixing times) the crosspeak intensity is:

$$a_{IS} = -\sigma\ \tau_m\ (1 - \tau_m/T_1). \tag{3.123}$$

Then, if the $T_1$ relaxation times are the same, the ratio of crosspeak intensities just provides the ratio of the cross-relaxation rates $\sigma$.

The 2D NOE spectrum has the following shape:

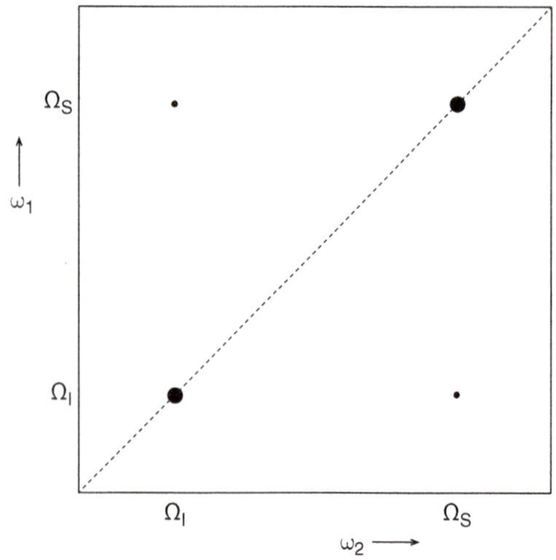


For large molecules diagonal peaks and crosspeaks are both positive and have the same phase (both absorptive in a properly phased spectrum).

### 3.5.4 Double-quantum Filtered COSY

The sequence for a DQ-filtered COSY [15] is:

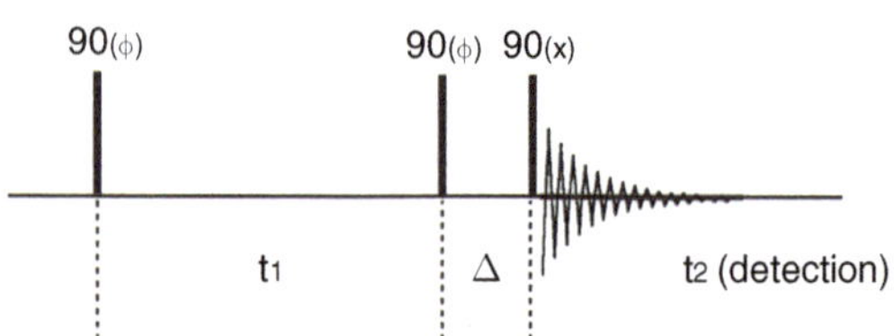


The aim is the observation of crosspeaks between coupled spins only, with suppression of all SQ coherences. The diagonal will obtain the same intensity and phase, as that of the crosspeaks. The delay $\Delta$ is very small ($\sim \mu$s).

The terms created after the first two RF pulses are the same as with the COSY sequence:

$$
\begin{aligned}
I_z \xrightarrow{\frac{\pi}{2}(x)} \xrightarrow{\Omega_I t_1} \xrightarrow{Jt_1} \xrightarrow{\frac{\pi}{2}(x)} & -I_z \cos \Omega_I t_1 \cos \pi J t_1 - 2I_x S_y \cos \Omega_I t_1 \sin \pi J t_1 \\
& + I_x \sin \Omega_I t_1 \cos \pi J t_1 - 2I_z S_y \sin \Omega_I t_1 \sin \pi J t_1 \\
= & -I_z \ldots \quad -(I_x S_y - I_y S_x) \ldots \quad -(I_x S_y + I_y S_x) \\
& \qquad\qquad\quad (ZQ)_y \qquad\qquad\qquad (DQ)_y \\
& + I_x \ldots \qquad\qquad\qquad\qquad -2I_z S_y .
\end{aligned}
\tag{3.124}
$$

Thus, we created all types of coherences. We will now shift the RF phase of the first two pulses by 90° and shift the created coherences accordingly:

$$\begin{array}{ccc} x & \longrightarrow & y \\ \uparrow & & \downarrow \\ -y & \longleftarrow & -x \end{array} \tag{3.125}$$

The terms created in the second experiment become (the time development is unchanged):

$$\begin{array}{lll} -I_z \ldots & -(-I_yS_x + I_xS_y)\ldots & -(-I_yS_x - I_xS_y) \\ & (ZQ)_y & (DQ)_y \\ +I_y \ldots & & +2I_zS_x. \end{array} \tag{3.126}$$

We see that a 90° phase-shift does not change z-magnetization or the ZQ coherences, but shifts the SQ coherence by 90° and shifts the DQ coherence by 2× 90°.

This rule is general [15]:

- A $\phi$ phase-shift of the first two RF pulses shifts the created nQ coherence by a phase $n\times\phi$.
- The selection of an n-quantum coherence (n>0) requires a 2n-step phase cycle.

Thus, another shift by 90° (net 180°) for the third experiment will give:

$$\begin{array}{lll} -I_z \ldots & -(I_xS_y - I_yS_x)\ldots & -(I_xS_y + I_yS_x) \\ & (ZQ)_y & (DQ)_y \\ -I_x \ldots & & +2I_zS_y. \end{array} \tag{3.127}$$

Adding this third experiment ($\phi = 180°$) with the very first one ($\phi = 0°$) will cancel the SQ coherence ($I_x$ and $2I_zS_y$) and select the DQ coherence (plus the ZQ and the z-term).

In a fourth experiment we shift the phases another 90° (net 270°). In the sum of experiment 2 ($\phi = 90°$) and experiment 4 ($\phi = 270°$), the ZQ coherence (and $I_z$) and DQ will have opposite sign.

Thus, when we add all four experiments as 1 + 3 − (2 + 4), we will get the desired DQ coherence. The complete phase for a DQF-COSY therefore becomes:

| 1 | 2 | 3 | receiver |
|---|---|---|---|
| x | x | x | + |
| y | y | x | - |
| -x | -x | x | + |
| -y | -y | x | - |

For a triple-quantum filtered COSY:

| 1 | 2 | 3 | receiver |
|---|---|---|---|
| x | x | x | + |
| $x+\frac{\pi}{3}$ | $x+\frac{\pi}{3}$ | x | - |
| $x+\frac{2\pi}{3}$ | $x+\frac{2\pi}{3}$ | x | + |
| y | y | x | - |
| $y+\frac{\pi}{3}$ | $y+\frac{\pi}{3}$ | x | + |
| $y+\frac{2\pi}{3}$ | $y+\frac{2\pi}{3}$ | x | + |

In order to suppress instrument artifacts and artifacts due to incomplete relaxation between different experiments, the real phase cycles are normally longer and they may have a different order.

The DQF-COSY spectrum of a two-spin system has the following shape:

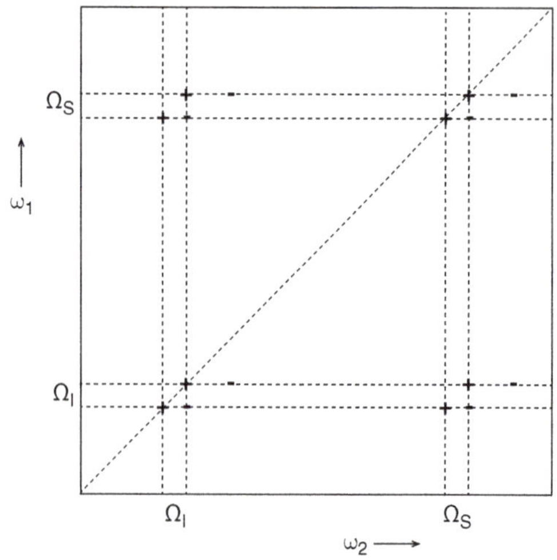


The diagonal peaks and crosspeaks have the same intensity, phase and multiplet structure.

### 3.5.5 Two-dimensional Double-quantum Spectroscopy

The possibility to observe the 'invisible' MQ coherences via the indirect time domain of a 2D experiment was already noted by [12]. Multiple-quantum NMR has been reviewed in [16]. A well-known 2D, DQ experiment is the 2D INADEQUATE experiment [17], which has the same phase cycle as the DQF- and relayed-COSY but a slightly different order in the sequence:

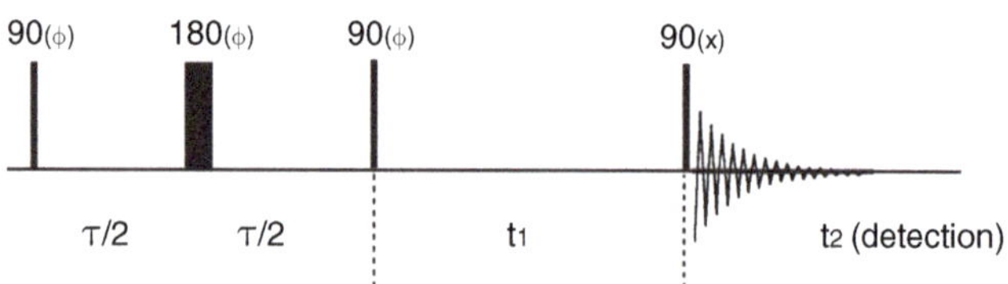


The phases of the first pulses can be cycled and the experiments added as in a DQF-COSY. In addition, the creation of DQ coherence can be tuned by the choice of $\tau = 1/2J$. This can be used for a further spectral simplification: observation of DQ coherences of coupled spins only within a narrow range of J-values. During the evolution time $t_1$, these DQ coherences evolve with the frequencies $\Omega_I + \Omega_S = \omega_I + \omega_S - 2\,\omega_r$. The third 90° pulse of the pulse sequence will convert DQ coherence into antiphase magnetization, which becomes observable in the detection period.

$$
\begin{aligned}
I_z &\xrightarrow{\pi J(2I_yS_y)} I_z \cos \pi J\tau + 2I_xS_y \sin \pi J\tau \\
&\qquad\qquad\qquad \underline{(DQ)_y} + (ZQ)_y \\
\underline{(DQ)_y} &\xrightarrow{\Omega_I t_1}\xrightarrow{\Omega_S t_1} \underline{(DQ)_y} \cos(\Omega_I + \Omega_S)t_1 + (DQ)_x \sin(\Omega_I + \Omega_S)t_1 \\
\underline{(DQ)_y} &\xrightarrow{\frac{\pi}{2}(x)} \quad -\underline{2I_zS_x} - \underline{2I_xS_z}
\end{aligned}
\tag{3.128}
$$

If the $t_1$ time is sufficiently long, for larger spin systems "relay" peaks can occur (see relayed-COSY), which have a dispersive phase. Two-dimensional, DQ spectra of a two-spin and a three-spin system will have the following shape:

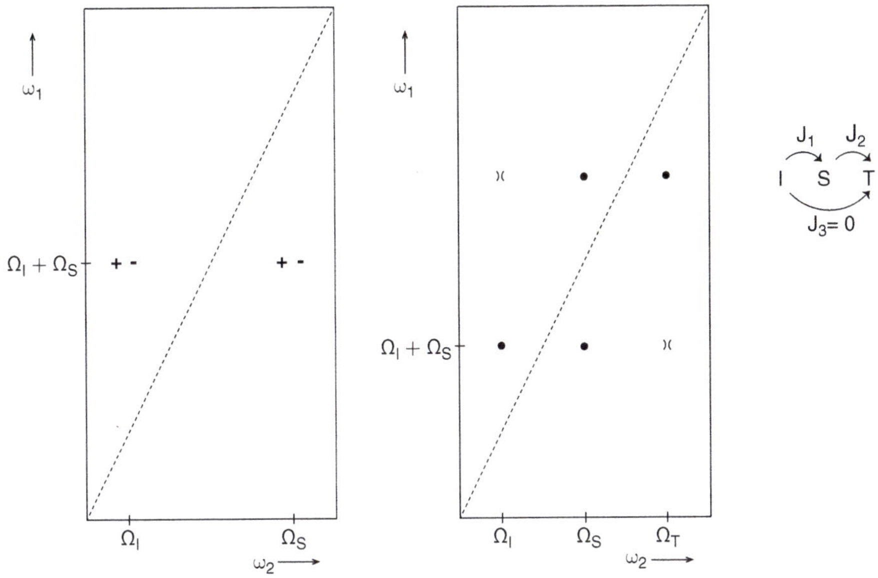


## 3.5.6 Relayed-COSY

The relayed-COSY sequence [18] is similar to the previous 2D, DQ experiment. The phase cycle is different, however, so that SQ coherences ($I_x$, $I_y$, $S_x$, $S_y$, and antiphase terms) will be selected.

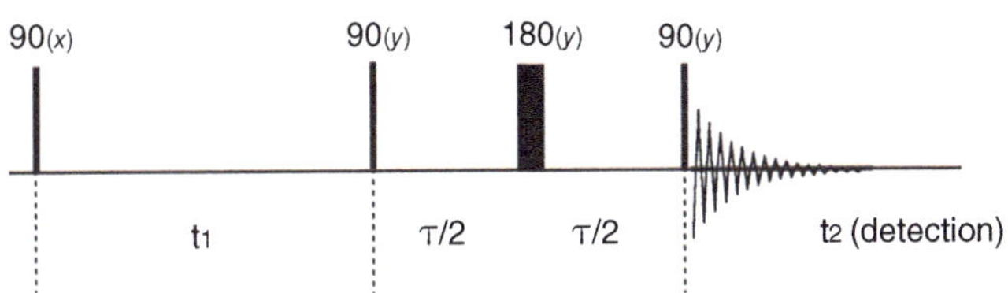


In the evolution time between the second and third 90° RF pulse the magnetization on S can become antiphase with respect to a third spin T and after a third pulse there can be polarization transfer from I to T, even in the absence of a direct J-coupling between I and T.

$$I_z \xrightarrow{\frac{\pi}{2}(x)} I_y \xrightarrow{\Omega_I t_1} \xrightarrow{\pi J_1 t_1} -2I_xS_z \xrightarrow{\frac{\pi}{2}(y)} 2I_zS_x \xrightarrow{\pi(y)} -2I_zS_x$$
$$\xrightarrow{\pi J_1 \tau} -S_y \xrightarrow{\pi J_2 \tau} 2S_xT_z \xrightarrow{\frac{\pi}{2}(y)} -2S_zT_x \xrightarrow{\Omega_T t_2} \xrightarrow{\pi J_2 t_2} T_y \tag{3.129}$$

The spectrum will show extra (relay) crosspeaks at $(\Omega_I, \Omega_T)$ with respect to a regular COSY spectrum. These relay peaks can be helpful for spectral interpretation in the case of resonance overlap of spin S with another spin.

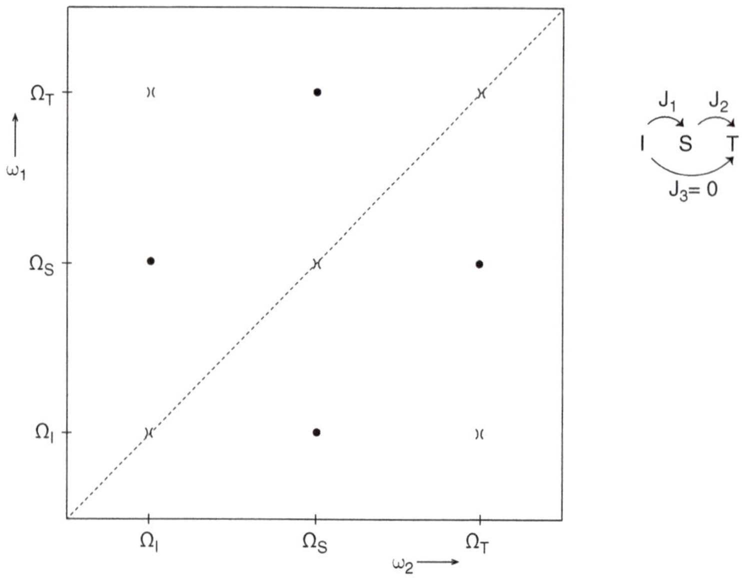


For most applications the relayed-COSY has been superseded by the TOCSY experiment. Conceptually, however, the multiple coherence transfer in the relayed-COSY has been at the basis of developing many 3D NMR experiments [19, 20].

### 3.5.7 TOCSY or Homonuclear Hartmann-Hahn Transfer

The sequence for the TOCSY experiment [21] or its improved version, the HOHAHA experiment (reviewed in [22]), is:

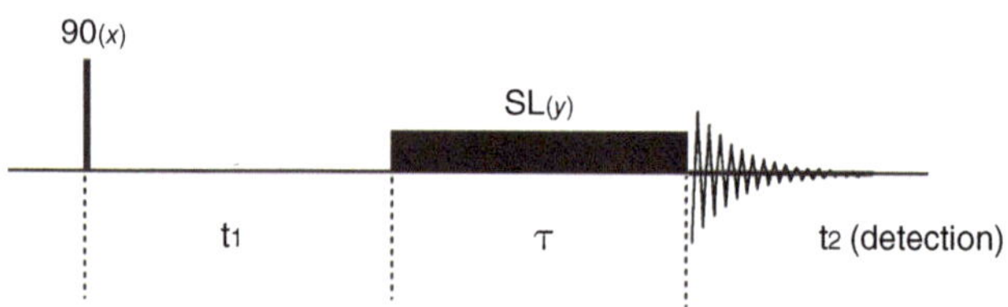


In the strong spin-lock field along y, $SL_y$, the motion of the spins can be described by a transformation to the rotating frame. The spins feel an effective field:

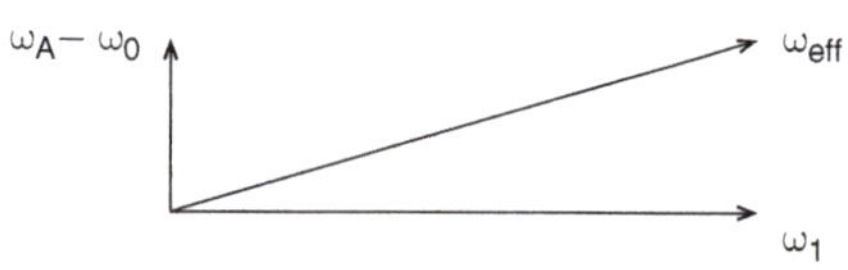


where

$\omega_1 = \gamma B_1$ is the spin-lock field strength
$\omega_0 = \gamma B_0$ is the RF frequency
$\omega_A = \gamma(1 - \sigma_A)B_0$ is the Larmor frequency of spin A
$\sigma_A =$ the chemical shift of spin A.

Thus, on-resonance the spins will only observe a field along the spin-lock axis.

If the frequency of the spin is different from the RF frequency:

$$\omega_{eff}^2 = \omega_1^2 + (\omega_A - \omega_0)^2. \tag{3.130}$$

For very strong RF fields, $\omega_1 \gg (\omega_A - \omega_0)$, the frequency becomes independent from the chemical shift:

$$\omega_{eff} \approx \omega_1 (1 + \frac{|\omega_A - \omega_0|}{2\omega_1}) \approx \omega_1, \tag{3.131}$$

and the Hamiltonian in the rotating frame is:

$$\mathcal{H}^r = -\omega_1 I_z - \omega_1 S_z + 2\pi J(I_x S_x + I_y S_y + I_z S_z). \tag{3.132}$$

Since there is no difference in the Zeeman energies of the spins, we can neglect the Zeeman terms for the motion of the spins:

$$\mathcal{H}^r = \pi J\ (2I_x S_x + 2I_y S_y + 2I_z S_z). \tag{3.133}$$

Thus, the density matrix in the spin-lock field changes as:

$$\rho(-\tau) \xrightarrow{2I_xS_x} \xrightarrow{2I_yS_y} \xrightarrow{2I_zS_z} \rho(+\tau). \tag{3.134}$$

Using $[2I_xS_x, I_y] = i2I_zS_x$, $[2I_zS_x, 2I_yS_y] = 0$ and $[2I_zS_z, 2I_zS_x] = iS_y$ it can be easily shown that the pulse sequence yields:

$$I_y \rightarrow I_y \cos^2 \pi J\tau + S_y \sin^2 \pi J\tau + (2I_zS_x - 2I_xS_z) \cos \pi J\tau \sin \pi J\tau. \tag{3.135}$$

In contrast to the COSY there will be a net transfer $I_y \rightarrow S_y$, which is optimal at $\tau = \frac{1}{2J}$. Even when $\tau$ is suboptimal, for large molecules this net transfer will dominate, whereas the antiphase terms will largely cancel. For large values of $\tau$ the coherence transfer can efficiently relay through the whole spin system in large spin systems. Because of the high transfer efficiency and the same phase of the direct and relayed crosspeaks, the TOCSY experiment is generally preferred over the relayed-COSY.

### 3.5.8 INEPT and HSQC

For heteronuclear polarization transfer in [23] the INEPT (insensitive nuclei enhanced by polarization transfer) pulse sequence was introduced, which is used as a building block in many heteronuclear multidimensional NMR experiments. The pulse sequence for INEPT is:

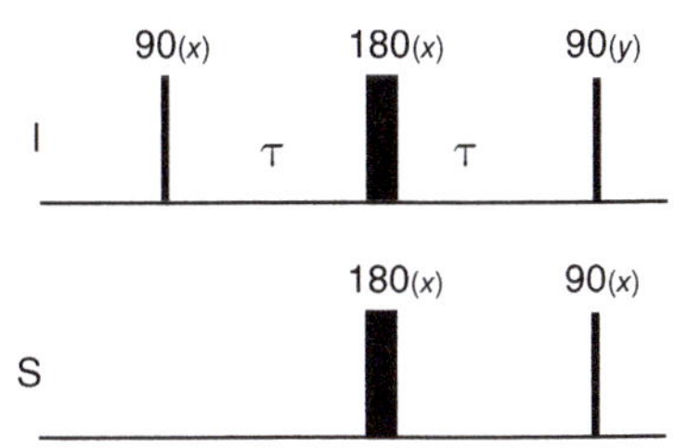


The first part of INEPT is like the heteronuclear spin-echo experiment (Section 3.4.1). The final 90° pulses on spin I and S transfer the magnetization from I to S as in the COSY sequence (Section 3.5.2). The second 90° pulse on **I** must be 90° phase shifted with respect to the first pulse.

$$
\begin{aligned}
I_z &\xrightarrow{\frac{\pi}{2}(I_x)} I_y \xrightarrow{\pi(I_x+S_x)}\xrightarrow{2\pi J\tau} -I_y\cos 2\pi J\tau + 2I_xS_z\sin 2\pi J\tau \\
&\xrightarrow{\frac{\pi}{2}(I_y)}\xrightarrow{\frac{\pi}{2}(S_x)} -I_y\cos 2\pi J\tau + 2I_zS_y\sin 2\pi J\tau.
\end{aligned}
\tag{3.136}
$$

INEPT can be used for sensitivity enhancement. For example when **I** $={}^1H$ and **S** $= {}^{13}C$ (or ${}^{15}N$) we create magnetization on ${}^{13}C$ (or ${}^{15}N$) but its intensity derives from the Boltzman distribution of the ${}^1H$ nuclei. The sensitivity gain is $\gamma_I/\gamma_S$.

Optimal transfer from spin I to S occurs when $\tau = \frac{1}{4J}$. We could directly detect the antiphase magnetization $2I_zS_y$. However, it may be convenient to add a spin-echo and transform the antiphase to $S_x$, which allows decoupling doubling the intensities and giving fewer signals.

We can also use INEPT to create ${}^{13}C$ (or ${}^{15}N$) magnetization for the $t_1$ evolution period of a 2D experiment. In that case we bring the magnetization back to the protons by a reversed INEPT, as proposed by [24], in the HSQC (heteronuclear single-quantum coherence) experiment. Since in HSQC we will detect the large proton magnetic moments (leading to a stronger signal in the receiver coil), this will lead to another gain of $(\gamma_I/\gamma_S)^{1/2}$. The HSQC pulse sequence is:

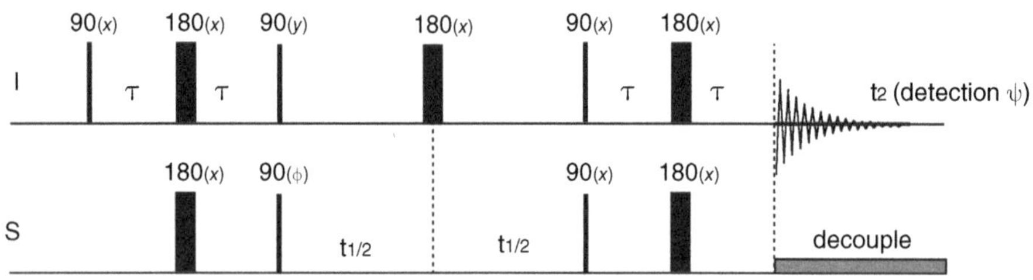


With $\tau = \frac{1}{4J}$ the first INEPT creates $2I_zS_y$. During the $t_1$ evolution period this will evolve to:

$$
2I_zS_y \xrightarrow{\pi(I_x)}\xrightarrow{\Omega_S t_1} -2I_zS_y\cos\Omega_S t_1 - 2I_zS_x\sin\Omega_S t_1.
\tag{3.137}
$$

The reversed INEPT in the HSQC will make the $2I_zS_y$ term observable:

$$
\begin{aligned}
&\xrightarrow{\frac{\pi}{2}(I_x)}\xrightarrow{\frac{\pi}{2}(S_x)} \underline{2I_yS_z}\cos\Omega_S t_1 - 2I_yS_x\sin\Omega_S t_1 \\
&\xrightarrow{\pi(I_x+S_x)}\xrightarrow{2\pi J\tau} -\underline{I_x}\cos\Omega_S t_1 + 2I_yS_x\sin\Omega_S t_1.
\end{aligned}
\tag{3.138}
$$

The $I_x$ magnetization can be decoupled during the detection period and will evolve at frequency $\Omega_I$. The MQ term $2I_yS_x$ remains unobservable. Since there is decoupling during $t_2$ and a 180° pulse in the middle of $t_1$, splitting by the heteronuclear J-coupling is removed and the HSQC spectrum will only correlate the frequency $\Omega_I$ of spin I with the frequency $\Omega_S$ of spin S and will have the following shape:

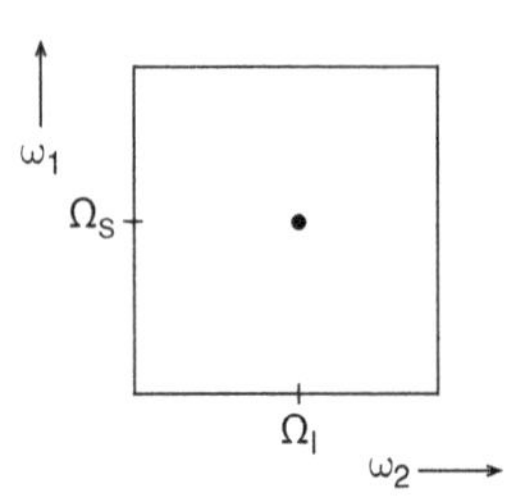

The HSQC can be found as a building block in many heteronuclear 2D and 3D NMR experiments (for an overview see [20]).

### 3.5.9 HMQC and HMBC

The sequence for the HMQC (heteronuclear multiple-quantum coherence) experiment [25, 26] is:

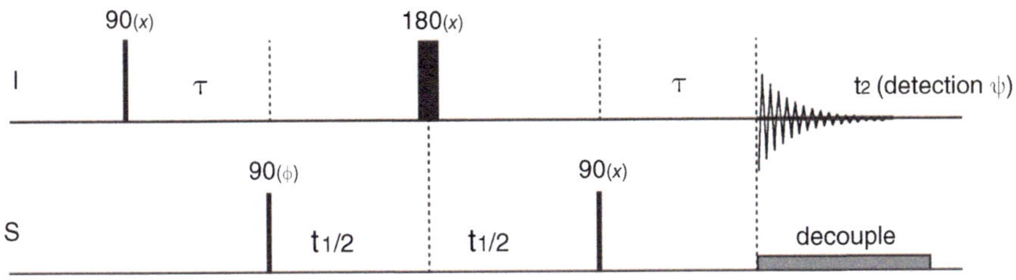


The sequence simplifies as:

$$
\begin{aligned}
&\rho(0) \xrightarrow{\frac{\pi}{2}(I_x)} \xrightarrow{\pi J\tau} \xrightarrow{\frac{\pi}{2}(S_x)} \xrightarrow{\Omega_I t_1/2} \xrightarrow{\Omega_S t_1/2} \xrightarrow{\pi J t_1/2} \xrightarrow{\pi(I_x)} \xrightarrow{\Omega_I t_1/2} \xrightarrow{\Omega_S t_1/2} \xrightarrow{\pi J t_1/2} \xrightarrow{\frac{\pi}{2}(S_x)} \xrightarrow{\pi J\tau} \rho(2\tau)\\
&\rho(0) \xrightarrow{\frac{\pi}{2}(I_x)} \xrightarrow{\pi J\tau} \xrightarrow{\frac{\pi}{2}(S_x)} \xrightarrow{\Omega_I t_1/2} \xrightarrow{\Omega_S t_1/2} \xrightarrow{\pi J t_1/2} \xrightarrow{-\Omega_I t_1/2} \xrightarrow{\pi(I_x)} \xrightarrow{\Omega_S t_1/2} \xrightarrow{\pi J t_1/2} \xrightarrow{\frac{\pi}{2}(S_x)} \xrightarrow{\pi J\tau} \rho(2\tau)\\
&\rho(0) \xrightarrow{\frac{\pi}{2}(I_x)} \xrightarrow{\pi J\tau} \xrightarrow{\frac{\pi}{2}(S_x)} \xrightarrow{\pi J t_1/2} \xrightarrow{\Omega_S t_1} \xrightarrow{\pi(I_x)} \xrightarrow{\pi J t_1/2} \xrightarrow{\frac{\pi}{2}(S_x)} \xrightarrow{\pi J\tau} \rho(2\tau)\\
&\rho(0) \xrightarrow{\frac{\pi}{2}(I_x)} \xrightarrow{\pi J\tau} \xrightarrow{\frac{\pi}{2}(S_x)} \xrightarrow{\pi J t_1/2} \xrightarrow{\Omega_S t_1} \xrightarrow{-\pi J t_1/2} \xrightarrow{\pi(I_x)} \xrightarrow{\frac{\pi}{2}(S_x)} \xrightarrow{\pi J\tau} \rho(2\tau)\\
&\rho(0) \xrightarrow{\frac{\pi}{2}(I_x)} \xrightarrow{\pi J\tau} \xrightarrow{\frac{\pi}{2}(S_x)} \xrightarrow{\Omega_S t_1} \xrightarrow{\pi(I_x)} \xrightarrow{\frac{\pi}{2}(S_x)} \xrightarrow{\pi J\tau} \rho(2\tau)\\
&\rho(0) \xrightarrow{\frac{3\pi}{2}(I_x)} \xrightarrow{-\pi J\tau} \xrightarrow{\frac{\pi}{2}(S_x)} \xrightarrow{\Omega_S t_1} \xrightarrow{\frac{\pi}{2}(S_x)} \xrightarrow{\pi J\tau} \rho(2\tau).
\end{aligned}
\tag{3.139}
$$

When $\tau = \frac{1}{2J}$ the sequence creates DQ and ZQ coherences during the $t_1$ evolution period (see Equation 3.48):

$$
I_z \xrightarrow{\frac{3\pi}{2}(I_x)} -I_y \xrightarrow{-\pi J\tau} -2I_xS_z \xrightarrow{\frac{\pi}{2}(S_x)} -2I_xS_y = -\frac{1}{2}\,((DQ)_y - ZQ)_y). \tag{3.140}
$$

These DQ and ZQ terms only develop during $t_1$ due to the shift terms of spin S in the Hamiltonian.

$$
-2I_xS_y \xrightarrow{\Omega_S t_1} -2I_xS_y \cos\Omega_S t_1 - 2I_xS_x \sin\Omega_S t_1 \tag{3.141}
$$

The magnetization oscillates during $t_1$ only with the $\Omega_S$ frequency. The final $90(S_x)$ pulse converts the MQ term to antiphase magnetization and after another delay $\tau$ we have in-phase magnetization $2I_y$, which can be decoupled:

$$
\begin{aligned}
&\xrightarrow{\frac{\pi}{2}(S_x)} \underline{2I_xS_z} \cos\Omega_S t_1 - 2I_xS_x \sin\Omega_S t_1\\
&\xrightarrow{\pi J\tau} \underline{2I_y} \cos\Omega_S t_1 - 2I_xS_x \sin\Omega_S t_1.
\end{aligned}
\tag{3.142}
$$

The MQ term $2I_xS_x$ remains unobservable.

The HMQC spectrum has the same appearance as the HSQC spectrum:

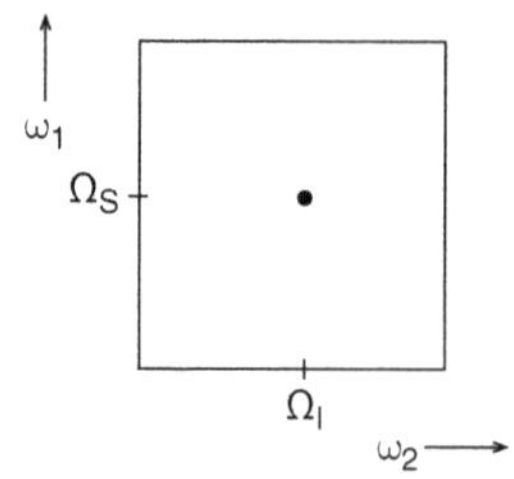


The sequence will also observe uncoupled SQ coherences of spin **I**, which can suppressed by alternating the phase of the $^{13}C$ pulse and switching the receiver. The phase cycle for HMQC will be:

| $\phi$ | $\psi$ |
|---|---|
| x | + |
| -x | - |

Like the HSQC, the HMQC experiment is a common building block for 2D and 3D NMR.

A variant HMQC that suppresses MQ coherences of strongly coupled nuclei is the HMBC (heteronuclear multiple-bond connectivity) experiment [27]. The sequence for HMBC is

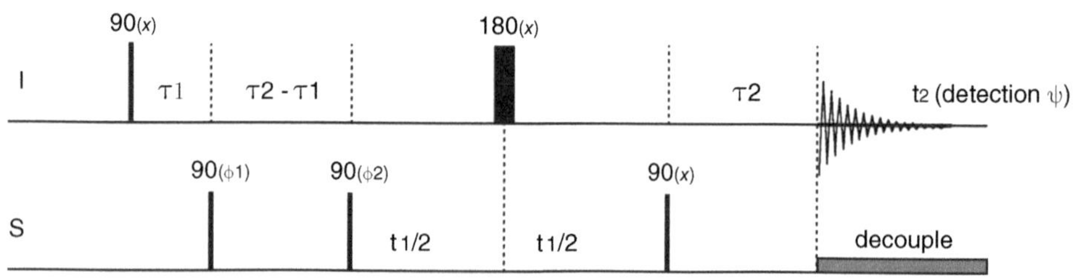


The first delay ($\tau_1$) develops $^{13}C - {}^1H$ MQ coherences of directly attached $^{13}C$ nuclei, which are suppressed by alternating the phase of the first $^{13}C$ pulse without switching the receiver. The next $^{13}C$ pulse is given at time $\tau_2$ when the $^1H_y$ magnetization is antiphase to the weakly coupled $^{13}C$ nuclei. The rest of the pulse scheme is similar to the HMQC. The phase cycle for HMBC will be:

| $\phi_1$ | $\phi_2$ | $\psi$ |
|---|---|---|
| x | x | + |
| -x | x | + |
| x | -x | - |
| -x | -x | - |

Typical values for this 'low-pass filter' J-filter, that suppresses $^1$J = 140 Hz would be $\tau_1 = \frac{1}{2\,{}^1J} = 3.57$ ms, whereas the correlations corresponding to a weak coupling of $^1$J = 4.5 Hz would be well visible with $\tau_2 = \frac{1}{2\,{}^2J} \approx 110$ ms.

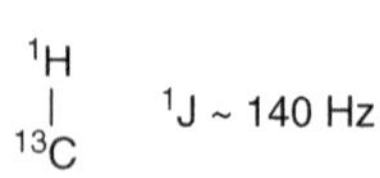


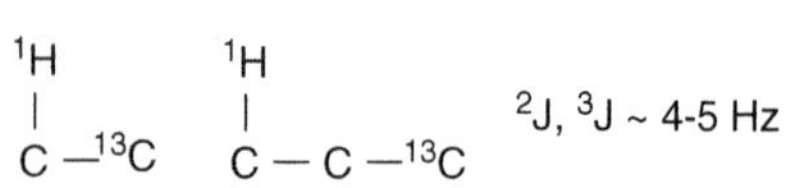

## References


1 Fano, U. Description of states in quantum mechanics by density matrix and operator techniques. *Rev. Mod. Phys.* 29: 74–93. doi: 10.1103/RevModPhys.29.74.

2 Slichter, C.P. (1990). *Principles of Magnetic Resonance*, 3e. Berlin, Heidelberg: Springer. ISBN 978-3-540-50157-2. doi: 10.1007/978-3-662-09441-9.

3 Sørensen, O.W. and Ernst, R.R. (1983). Elimination of spectral distortion in polarization transfer experiments. Improvements and comparison of techniques. *J. Mag. Res.* (1969) 51 (3): 477–489. doi: 10.1016/0022-2364(83)90300-1.

4 van de Ven, F.J.M. and Hilbers, C.W. (1983). A simple formalism for the description of multiple-pulse experiments. Application to a weakly coupled two-spin (I= 1/2) system. *J. Mag. Res.* (1969) 54 (3): 512–520. doi: 10.1016/0022-2364(83)90331-1.

5 Sørensen, O.W., Eich, G.W., Levitt, M.H., Bodenhausen, G., and Ernst, R.R. (1983). Product operator formalism for the description of NMR pulse experiments. *Prog. Nucl. Mag. Res. Spectrosc.* 16: 163–192. doi: 10.1016/0079-6565(84)80005-9.

6 Keeler, J. (2010). *Understanding NMR Spectroscopy*, 2e. Chichester, U.K.: Wiley. ISBN 978-0-470-74608-0.

7 Hore, P.J., Jones, J.R., and Wimperis, S. (2015). *NMR: The Toolkit. How Pulse Sequences Work*. Oxford, U.K: Oxford University Press. ISBN 9780198703426.

8 Levitt, M. (2008). *Spin Dynamics: Basics of Nuclear Magnetic Resonance*, 2e. Chichester, England: JohnWiley & Sons Ltd. ISBN 978-0-470- 51117-6.

9 Ernst, R.R., Bodenhausen, G., and Wokaun, A. (1987). *Principles of Nuclear Magnetic Resonance in One and Two Dimensions*. Oxford: Clarendon Press. ISBN 0198556292.

10 Bax, A., Freeman, R., and Kempsell, S.P. (1980). Natural abundance carbon-13-carbon-13 coupling observed via double-quantum coherence. *J. Am. Chem.Soc.* 102 (14): 4849–4851. doi: 10.1021/ja00534a056.

11 Parella, T. (2018). Current developments in homonuclear and heteronuclear j-resolved NMR experiments. *Magn. Reson. Chem.* 56 (4): 230–250. doi: 10.1002/mrc.4706.

12 Jeener, J. and Alewaeters, G. (2016). “Pulse pair technique in high resolution NMR” a reprint of the historical 1971 lecture notes on two-dimensional spectroscopy. *Prog. Nucl. Magn. Reson. Spectrosc.* 94–95: 75–80. doi: 10.1016/j.pnmrs.2016.03.002.

13 Aue, W.P., Bartholdi, E., and Ernst, R.R. (1976). Two-dimensional spectroscopy. Application to nuclear magnetic resonance. *J. Chem. Phys.* 64 (5): 2229–2246. doi: 10.1063/1.432450.

14 Jeener, J., Meier, B.H., Bachmann, P., and Ernst, R.R. Investigation of exchange processes by two-dimensional NMR spectroscopy. *J. Chem. Phys.* 71 (11): 4546–4553. doi: 10.1063/1.438208.

15 Piantini, U., Sørensen, O.W., and Ernst, R.R. (1982). Multiple quantum filters for elucidating NMR coupling networks. *J. Am. Chem. Soc.* 104 (24): 6800–6801. doi: 10.1021/ja00388a062.

16 Norwood, T.J. (1992). Multiple-quantum nmr methods. Prog. Nucl. Magn. Res. Spectrosc. 24 (4): 295–375. doi: https://doi.org/10.1016/0079-6565(92)80005-Z. ISSN 0079-6565.

17 Bax, A., Freeman, R., Frenkiel, T.A., and Levitt, M.H. (1981). Assignment of carbon-13 NMR spectra via double-quantum coherence. *J. Mag. Reson.* 43 (3): 478–483. doi: https://doi.org/10.1016/0022-2364(81)90060-3.

18 Eich, G., Bodenhausen, G., and Ernst, R.R. (1982). Exploring nuclear spin systems by relayed magnetization transfer. *J. Am. Chem. Soc.* 104 (13): 3731–3732. doi: 10.1021/ja00377a036.

19 Griesinger, C., Sørensen, O.W., and Ernst, R.R. (1969). Three-dimensional Fourier spectroscopy. Application to high-resolution NMR. *J. Mag. Reson.* 84 (1): 14–63. doi: 10.1016/0022-2364(89)90004-8.

20 Cavanagh, J., Fairbrother, W.J., Palmer III, A.G., Rance, M., and Skelton, N.J. (2006). *Protein NMR Spectroscopy: Principles and Practice*, 2e. Academic Press. ISBN 9780121644918.

21 Braunschweiler, L. and Ernst, R.R. (1969). Coherence transfer by isotropic mixing: application to proton correlation spectroscopy. *J. Mag. Reson.* 53 (3): 521–528. doi: 10.1016/0022-2364(83)90226-3.

**22** Bax, A. (1989). [8] Homonuclear Hartmann-Hahn experiments. In: *Nuclear Magnetic Resonance Part A: Spectral Techniques and Dynamics*, volume 176 of Methods in Enzymology, (eds. Norman J. Oppenheimer and Thomas L. James) 151Ű-168. Academic Press. doi: 10.1016/0076-6879(89)76010-9.

**23** Morris, G.A. and Freeman, R. (1979). Enhancement of nuclear magnetic-resonance signals by polarization transfer. *J. Am. Chem. Soc.* 101 (3): 760–762. doi: 10.1021/ja00497a058.

**24** Bodenhausen, G. and Ruben, D.J. Natural abundance nitrogen-15 NMR by enhanced heteronuclear spectroscopy. *Chem. Phys Lett.* 69 (1): 185–189. doi: 10.1016/0009-2614(80)80041-8.

**25** Muller, L. (1979). Sensitivity enhanced detection of weak nuclei using heteronuclear multiple quantum coherence. *J. Am. Chem. Soci.* 101 (16): 4481–4484. doi: 10.1021/ja00510a007.

**26** Bax, A., Griffey, R.H., and Hawkins, B.L. (1983). Correlation of proton and N-15 chemical-shifts by multiple quantum NMR. *J. Mag. Reson.* 55 (2): 301–315. doi: 10.1016/0022-2364(83)90241-x.

**27** Bax, A. and Summers, M.F. (1986). H-1 and C-13 assignments from sensitivityenhanced detection of heteronuclear multiple-bond connectivity by 2D multiple quantum NMR. *J. Am. Chem. Soc.* 108 (8): 2093–2094. doi: 10.1021/ja00268a061.

# Appendix 3.A

# Quantum Mechanics Dictionary

## 3.A.1 Operators

In classical mechanics we use coordinates, momentum, angular momentum, energy, etc. In quantum mechanics these observables are replaced by operators (Table 3.A.1).

A function works on a number and transforms it into another number. Similarly, an operator acts on a function and transforms it into another function:

$$\begin{aligned} &f(x) = x^2 \quad &&f(3) = 9 \\ &A = 3+ \quad &&Af(x) = 3 + x^2 \\ &B = \frac{d}{dx} \quad &&Bf(x) = 2x \end{aligned} \tag{3.A.1}$$

If we apply two operators on a function, the order is important. When we write $A\,B\,f(x)$, we will first apply the operator B on $f(x)$ and successively we will apply the operator A on the resulting function. In general, the result of $A\,B\,f(x)$ is different from the result $B\,A\,f(x)$. The commutator of two operators is defined as $[A,\ B] = A\,B - B\,A$. With the examples above:

$$\begin{aligned} &f(x) = x^2 \\ &[A, B] = 3 + 2x - 2x = 3 \qquad \neq 0 \end{aligned} \tag{3.A.2}$$

Two operators commute if $[A,\ B] = 0$.

**Table 3.A.1** Description of observables.

| Classical mechanics | | Quantum mechanics |
|---|---|---|
| $x, y, z$ | $\longrightarrow$ | $x, y, z$ |
| $m\vec{v}$ | $\longrightarrow$ | $\frac{\hbar}{i}\frac{\partial}{\partial x}$ |
| $\frac{1}{2}mv^2$ | $\longrightarrow$ | $\mathcal{H}$ |

*Two-Dimensional (2D) NMR Methods*, First Edition. Edited by K. Ivanov, P.K. Madhu and G. Rajalakshmi.

Some functions are special. When we apply the operator on a so-called eigenfunction the result is the same function, with a proportionality constant (eigenvalue):

$$A\,f(x) = a\,f(x) \tag{3.A.3}$$

where $a$ = eigenvalue

$f(x)$ = eigenfunction of the operator A

Operators with real eigenvalues (such as the operators that represent observables in physics) are Hermitian

$$\langle f|A\,g\rangle = \langle A\,g|f\rangle \tag{3.A.4}$$

Proof:

$$A\,f = a\,f$$

$$A^*\,f^* = a^*\,f^*$$

$$\langle f|Af\rangle = a\langle f|f\rangle$$

$$\langle Af|f\rangle = a^*\,\langle f|f\rangle$$

thus $a = a^*$, which is true for a real number.

Commuting operators have common eigenfunctions

$$\begin{aligned} A\,f &= a\,f \\ B\,f &= b\,f \\ [\mathrm{A},\ \mathrm{B}]f &= (A\,B\ - B\,A)f = (ab - ba)\,f = 0 \\ \text{thus } [\mathrm{A},\ \mathrm{B}] &= 0 \end{aligned} \tag{3.A.5}$$

Thus, commuting operators belong to the same state and their associated physical quantities can be observed simultaneously. When two operators do not commute, the system is not stationary under both operators and the simultaneous application of both operators will transform the system.

## 3.A.2 Schrödinger Equation

The Schrödinger equation describes how a wavefunction $\psi$ that describes a state evolves in time under the Hamiltonian (energy) operator $\mathcal{H}$

$$i\,\hbar\,\frac{\partial\psi}{\partial t} = \mathcal{H}\,\psi \tag{3.A.6}$$

If the Hamiltonian is time independent, the wavefunction $\psi$ can be written as a product of a position dependent and a time dependent part:

$$\psi = \phi(x)\,T(t) \tag{3.A.7}$$

When we substitute this in the Schrödinger equation, we will see that the equation reduces to

$$T(t) = e^{-i\,(E/\hbar)\,t} \tag{3.A.8}$$

and

$$\mathcal{H}\,\phi = E\,\phi \tag{3.A.9}$$

This last equation, the time-independent Schrödinger equation, is the eigenvalue equation from which energies of stationary states can be calculated.

## 3.A.3 Energies of Magnetic Spin States

As we can read in many textbooks on magnetic resonance, the energy of a magnetic moment in a magnetic field B can be calculated from the spin Hamiltonian.

$$\mathcal{H} = -\vec{\mu} \cdot \vec{B} \tag{3.A.10}$$

where

$$\vec{\mu} = \mu_x \vec{i} + \mu_y \vec{j} + \mu_z \vec{k}$$

$$\vec{B} = B_x \vec{i} + B_y \vec{j} + B_z \vec{k}$$

The magnetic moment along the z-axis for spin I is represented by the $I_z$ operator

$$\begin{aligned} &\mu_z = \gamma \hbar I_z \\ \text{where} \quad &\gamma = \text{gyromagnetic ratio of a particle} \\ &\hbar = \frac{h}{2\pi} \\ &h = \text{Planck's constant} \end{aligned} \tag{3.A.11}$$

With the magnetic field along the z-axis, $\vec{B} = B_z \vec{k}$, the spin Hamiltonian becomes

$$\mathcal{H} = -\mu_z B_z = -\gamma \hbar B_z I_z \tag{3.A.12}$$

The operator $I_z$ has as eigenfunctions the (normalized) spin functions $|m_i\rangle$

$$I_z |m_i\rangle = m_i |m_i\rangle \tag{3.A.13}$$

where the eigenvalues are $m_i$, the spin quantum numbers, which can take the values

$$I, I-1, I-2 \ldots\ldots, -I$$

for the different spin states. The spin functions are normalized, $\langle m_i|m_i\rangle = 1$, and orthogonal for different spin functions, $\langle m_i|m_j\rangle = 0$.
We can calculate the energy $E$ for the state $|m_i\rangle$ as

$$\begin{aligned} \mathcal{H} |m_i\rangle &= -\gamma \hbar B_z I_z |m_i\rangle = -m_i \gamma \hbar B_z |m_i\rangle \\ E &= \langle m_i| \mathcal{H} |m_i\rangle = -m_i \gamma \hbar B_z \langle m_i|m_i\rangle = -m_i \gamma \hbar B_z \end{aligned} \tag{3.A.14}$$

For a spin $I = \frac{1}{2}$ system there are two states and the spin functions are $|\alpha\rangle$ and $|\beta\rangle$ with eigenvalues $\frac{1}{2}$ and $-\frac{1}{2}$

$$\begin{aligned} I_z |\alpha\rangle &= \frac{1}{2} |\alpha\rangle \\ I_z |\beta\rangle &= -\frac{1}{2} |\beta\rangle \end{aligned} \tag{3.A.15}$$

The spin functions are normalized and orthogonal, $\langle\alpha|\alpha\rangle = 1$, $\langle\beta|\alpha\rangle = 0$, $\langle\alpha|\beta\rangle = 0$, $\langle\beta|\beta\rangle = 1$.

The energies of the two states are

$$\begin{aligned} &\text{state } \alpha \qquad E = -\gamma\hbar B_z \langle\alpha| I_z |\alpha\rangle = -\tfrac{1}{2}\gamma\hbar B_z \\ &\text{state } \beta \qquad E = -\gamma\hbar B_z \langle\beta| I_z |\beta\rangle = \tfrac{1}{2}\gamma\hbar B_z \end{aligned} \tag{3.A.16}$$

Thus, the energy difference will be $\Delta E = \gamma\hbar B_z$. Using $\Delta E = h\nu$ the magnetic resonance transition frequency can be calculated as

$$\nu = \frac{\gamma}{2\pi} B_z \tag{3.A.17}$$

### 3.A.4 Angular Momentum

The angular momentum is defined in classic mechanics as $\vec{L} = \vec{r} \times \vec{p}$. In quantum mechanics this becomes (see Table 3.A.1) $\vec{L} = -i(\vec{r} \times \vec{\nabla})$ or

$$\vec{L} = -i \begin{vmatrix} \vec{i} & \vec{j} & \vec{k} \\ x & y & z \\ \frac{\partial}{\partial x} & \frac{\partial}{\partial y} & \frac{\partial}{\partial z} \end{vmatrix} \tag{3.A.18}$$

thus

$$\begin{aligned} L_x &= -i\left(y\frac{\partial}{\partial z} - z\frac{\partial}{\partial y}\right) \\ L_y &= -i\left(z\frac{\partial}{\partial x} - x\frac{\partial}{\partial z}\right) \\ L_z &= -i\left(x\frac{\partial}{\partial y} - y\frac{\partial}{\partial x}\right) \end{aligned} \tag{3.A.19}$$

The following commutator relations exist between the $L_x$, $L_y$ and $L_z$ operators

$$\begin{aligned} [L_x, L_y] &= i\,L_z \\ [L_y, L_z] &= i\,L_x \\ [L_z, L_x] &= i\,L_y \end{aligned} \tag{3.A.20}$$

From these three commutators many properties can be derived.

The properties of nuclear and electron spin operators can be calculated with spin angular-momentum operators. The spin angular-momentum behaves as the classic angular-momentum

$$\begin{aligned} [S_x, S_y] &= i\,S_z \\ [S_y, S_x] &= -i\,S_z \end{aligned} \tag{3.A.21}$$

Other operators that are often used, are the raising and lowering operators

$$\begin{aligned} S_+ &= S_x + i\,S_y \\ S_- &= S_x - i\,S_y \end{aligned} \tag{3.A.22}$$

and

$$S_x = \frac{1}{2}(S_+ + S_-)$$
$$S_y = \frac{1}{2i}(S_+ - S_-) \tag{3.A.23}$$

Other commutators that can be derived for the spin angular momentum are

$$[S^2, S_z] = 0 \tag{3.A.24}$$

$$[S_z, S_\pm] = [S_z, S_x] \pm i[S_z, S_y] = iS_y \pm S_x = \pm S_\pm \tag{3.A.25}$$

How do we use these operators in NMR? In a magnetic field B along the z-axis the spins are in a spin up ($\alpha$) or spin down ($\beta$) state. The spin functions $|\alpha\rangle$ and $|\beta\rangle$ belong to these states

$$S_z\,|\alpha\rangle = \frac{1}{2}\,|\alpha\rangle$$
$$S_z\,|\beta\rangle = -\frac{1}{2}\,|\beta\rangle \tag{3.A.26}$$

The commutation relation $[S_z, S_+] = S_+$ tells us that the operation $S_z S_+ = S_+S_z + S_+$, thus

$$S_z S_+\,|\beta\rangle = S_+(S_z + 1)\,|\beta\rangle = S_+\frac{1}{2}\,|\beta\rangle = \frac{1}{2}\,S_+\,|\beta\rangle \tag{3.A.27}$$

Since we also know that $S_z\,|\alpha\rangle = \frac{1}{2}\,|\alpha\rangle$, the transform of $S_+$ on the function $|\beta\rangle$ must give $|\alpha\rangle$ (within a proportionality constant)

$$S_+\,|\beta\rangle = C\,|\alpha\rangle \tag{3.A.28}$$

It can be shown that for spin $\frac{1}{2}$ systems C = 1 for transitions. Thus, the effect of the $S_+$ operator (raising operator) is a conversion from spin down to spin up. Similarly, there exists a lowering operator $S_-$ :

$$S_+\,|\beta\rangle = |\alpha\rangle$$
$$S_+\,|\alpha\rangle = 0$$
$$S_-\,|\beta\rangle = 0$$
$$S_-\,|\alpha\rangle = |\beta\rangle \tag{3.A.29}$$

The effect of the $S_x$ and $S_y$ operators on a spin system will be transitions between the $\alpha$ and $\beta$ states

$$S_x\,|\alpha\rangle = \frac{1}{2}(S_+ + S_-)\,|\alpha\rangle = \frac{1}{2}\,|\beta\rangle$$
$$S_x\,|\beta\rangle = \frac{1}{2}(S_+ + S_-)\,|\beta\rangle = \frac{1}{2}\,|\alpha\rangle \tag{3.A.30}$$

## 3.A.5 Matrix Representation of the Angular Momentum

The spin angular momentum operators can be written in a matrix representation. For a spin $\frac{1}{2}$ system the chosen spin functions are $|\alpha\rangle$ and $|\beta\rangle$, the eigenfunctions of the operator $S_z$ :

$$S_z\,|\alpha\rangle = \frac{1}{2}\,|\alpha\rangle$$

and thus

$$\langle\alpha|\,S_z\,|\alpha\rangle = \frac{1}{2}$$

$$\langle\beta|\,S_z\,|\alpha\rangle = 0$$

similarly

$$S_z\ |\beta\rangle = -\tfrac{1}{2}\ |\beta\rangle$$

and thus

$$\langle\alpha|\ S_z\ |\beta\rangle = 0$$

$$\langle\beta|\ S_z\ |\beta\rangle = -\tfrac{1}{2}$$

The matrix representation of the operator $S_z$ will be

$$S_z = \begin{pmatrix} \langle\alpha|\ S_z\ |\alpha\rangle & \langle\alpha|\ S_z\ |\beta\rangle \\ \langle\beta|\ S_z\ |\alpha\rangle & \langle\beta|\ S_z\ |\beta\rangle \end{pmatrix} = \frac{1}{2}\begin{pmatrix} 1 & 0 \\ 0 & -1 \end{pmatrix} \tag{3.A.31}$$

The raising and lowering operators will have the following shape

$$S_+ = \begin{pmatrix} \langle\alpha|\ S_+\ |\alpha\rangle & \langle\alpha|\ S_+\ |\beta\rangle \\ \langle\beta|\ S_+\ |\alpha\rangle & \langle\beta|\ S_+\ |\beta\rangle \end{pmatrix} = \begin{pmatrix} 0 & 1 \\ 0 & 0 \end{pmatrix} \tag{3.A.32}$$

$$S_- = \begin{pmatrix} \langle\alpha|\ S_-\ |\alpha\rangle & \langle\alpha|\ S_-\ |\beta\rangle \\ \langle\beta|\ S_-\ |\alpha\rangle & \langle\beta|\ S_-\ |\beta\rangle \end{pmatrix} = \begin{pmatrix} 0 & 0 \\ 1 & 0 \end{pmatrix} \tag{3.A.33}$$

Using equation 3.A.23 the matrix representations of the $S_x$ and $S_y$ operators become

$$S_x = \frac{1}{2}\,(S_+ + S_-) = \frac{1}{2}\begin{pmatrix} 0 & 1 \\ 1 & 0 \end{pmatrix} \tag{3.A.34}$$

$$S_y = \frac{1}{2i}\,(S_+ - S_-) = \frac{1}{2i}\begin{pmatrix} 0 & 1 \\ -1 & 0 \end{pmatrix} \tag{3.A.35}$$

Of course for these matrixes the commutator relations should still hold. For example

$$[\,S_x,\ S_y] = i\,S_z$$

which is in matrix notation

$$S_x\,S_y - S_y\,S_x = \frac{1}{4i}\begin{pmatrix} 0 & 1 \\ 1 & 0 \end{pmatrix}\begin{pmatrix} 0 & 1 \\ -1 & 0 \end{pmatrix} - \frac{1}{4i}\begin{pmatrix} 0 & 1 \\ -1 & 0 \end{pmatrix}\begin{pmatrix} 0 & 1 \\ 1 & 0 \end{pmatrix} = \frac{i}{2}\begin{pmatrix} 1 & 0 \\ 0 & -1 \end{pmatrix} = i\,S_z$$

The matrix reprensentation of an operator is only defined for a chosen set of spin functions. For two spins the chosen basis set will be $|\alpha\alpha\rangle$, $|\alpha\beta\rangle$, $|\beta\alpha\rangle$, and $|\beta\beta\rangle$. The matrix representation of spin operators of larger spin systems can be calculated by combining the single spin operators.

For example:

$$S_x = S_{Ax} + S_{Bx} = \frac{1}{2}\begin{array}{c} \\ \langle\alpha\alpha| \\ \langle\alpha\beta| \\ \langle\beta\alpha| \\ \langle\beta\beta| \end{array}\begin{array}{cccc} |\alpha\alpha\rangle & |\alpha\beta\rangle & |\beta\alpha\rangle & |\beta\beta\rangle \\ \end{array}\begin{pmatrix} 0 & 1 & 1 & 0 \\ 1 & 0 & 0 & 1 \\ 1 & 0 & 0 & 1 \\ 0 & 1 & 1 & 0 \end{pmatrix} \tag{3.A.36}$$

and

$$S_A\, S_B = \frac{1}{2}\,(S_{A+}\, S_{B-} + S_{A-}\, S_{B+}) + \, S_{Az}\, S_{Bz}$$
$$S_A\, S_B\, |\beta\alpha\rangle = \frac{1}{2}\ |\alpha\beta\rangle - \frac{1}{4}\ |\beta\alpha\rangle \tag{3.A.37}$$

and thus

$$\langle\alpha\beta|\, S_A\, S_B\, |\beta\alpha\rangle = \frac{1}{2}$$

$$\langle\beta\alpha|\, S_A\, S_B\, |\beta\alpha\rangle = -\frac{1}{4}$$

which gives

$$S_A\, S_B = \frac{1}{4}\begin{array}{c} \\ \langle\alpha\alpha| \\ \langle\alpha\beta| \\ \langle\beta\alpha| \\ \langle\beta\beta| \end{array}\begin{array}{cccc} |\alpha\alpha\rangle & |\alpha\beta\rangle & |\beta\alpha\rangle & |\beta\beta\rangle \\ \end{array}\begin{pmatrix} 1 & 0 & 0 & 0 \\ 0 & -1 & 2 & 0 \\ 0 & 2 & -1 & 0 \\ 0 & 0 & 0 & 1 \end{pmatrix} \tag{3.A.38}$$

## 3.A.6 Matrix Representation of the Hamiltonian Operator

The matrix representation of the Hamiltonian operator for spin systems can be obtained from the matrix representation of the containing operators.

For a single spin $S = \frac{1}{2}$ in a magnetic field $B_z$ the Hamiltonian operator and its matrix representation is

$$\mathcal{H} = -\gamma\, \hbar\, B_z\ S_z = -\ \frac{1}{2}\,\gamma\, \hbar\, B_z \begin{pmatrix} 1 & 0 \\ 0 & -1 \end{pmatrix} \tag{3.A.39}$$

or when we define $\nu = \frac{\gamma\, B_z}{2\pi}$ and express the Hamiltonian in frequency units (Hz)

$$\frac{\mathcal{H}}{h} = -\ \nu\, S_z = \begin{pmatrix} -\frac{1}{2}\nu & 0 \\ 0 & \frac{1}{2}\nu \end{pmatrix} \tag{3.A.40}$$

For two coupled spins A and B with resonance frequencies $\nu_A$ and $\nu_B$, the spin Hamiltonian in frequency units will be

$$\frac{\mathcal{H}}{h} = -\nu_A\, S_{Az} - \nu_B\, S_{Bz} + J\, S_A S_B \tag{3.A.41}$$

$$\frac{\mathcal{H}}{h} = -\nu_A\, S_{Az} - \nu_B\, S_{Bz} + J\, S_{Az} S_{Bz} + \frac{1}{2} J(S_{A+}S_{B-} + S_{A-}S_{B+}) \tag{3.A.42}$$

With the chosen basis set $\{ |\alpha\alpha\rangle, |\alpha\beta\rangle, |\beta\alpha\rangle, |\beta\beta\rangle \}$ the matrix representation of the spin Hamiltonian becomes

$$\frac{\mathcal{H}}{h} = \begin{pmatrix} -\frac{1}{2}\nu_A - \frac{1}{2}\nu_B + \frac{1}{4}J & 0 & 0 & 0 \\ 0 & -\frac{1}{2}\nu_A + \frac{1}{2}\nu_B - \frac{1}{4}J & \frac{1}{2}J & 0 \\ 0 & \frac{1}{2}J & \frac{1}{2}\nu_A - \frac{1}{2}\nu_B - \frac{1}{4}J & 0 \\ 0 & 0 & 0 & \frac{1}{2}\nu_A + \frac{1}{2}\nu_B + \frac{1}{4}J \end{pmatrix}$$

In the chosen basis set the Hamiltonian is non-diagonal.

The spin functions $\psi_1 = |\alpha\alpha\rangle$ and $\psi_4 = |\beta\beta\rangle$ are eigenfunctions of the Hamiltonian operator and the energies of these states are $E_1 = -\frac{1}{2}\nu_A - \frac{1}{2}\nu_B + \frac{1}{4}J$ and $E_4 = \frac{1}{2}\nu_A + \frac{1}{2}\nu_B + \frac{1}{4}J$, resp.

For weak coupling ($J \ll |\nu_A - \nu_B|$) the off-diagonal terms in the Hamiltonian matrix can be ignored and spin functions $\psi_2 = |\alpha\beta\rangle$ and $\psi_3 = |\beta\alpha\rangle$ are also eigenfunctions of the Hamiltonian operator and the energies of these states are $E_2 = -\frac{1}{2}\nu_A + \frac{1}{2}\nu_B - \frac{1}{4}J$ and $E_3 = \frac{1}{2}\nu_A - \frac{1}{2}\nu_B - \frac{1}{4}J$ , resp.

For strong coupling ($J \approx |\nu_A - \nu_B|$) the energies $E_2$ and $E_3$ can be only calculated by diagonalizing the central part of the Hamiltonian matrix and the eigenfunctions of the states spin functions $\psi_2$ and $\psi_3$ become linear combinations of $|\alpha\beta\rangle$ and $|\beta\alpha\rangle$. The solution for that can be found in several NMR textbooks (for example, [1]).

## 3.A.7 Density Operator, Density Matrix, and Observables

The wave function of a system composed of multiple states can be described by

$$\psi(t) = \sum_n c_n(t)\,\phi_n \tag{3.A.43}$$

The expectation values of measurable quantities represented by an operator A can be calculated as

$$\langle A \rangle = \int \psi^* A \psi \, d\tau = \sum_{i,j} c_i^*(t)\, c_j(t) \int \phi_i^* A \phi_j \, d\tau \tag{3.A.44}$$

Using the definition of the population matrix **P**

$$P_{nm} = c_n(t)\, c_m^*(t) \tag{3.A.45}$$

and the matrix representation, **A**, of the operator A

$$A_{nm} = \int \phi_n^* A \phi_m \, d\tau) \tag{3.A.46}$$

the expectation value of the operator A can be shown to be the trace of the product of the two matrixes

$$\langle A \rangle = \sum_{i,j} P_{ji} A_{ij} = \mathrm{Tr}\,(\mathbf{P}\,\mathbf{A}) \tag{3.A.47}$$

For an ensemble of spins, ensemble-average quantities will be observed. For this we use the density operator, defined as

$$\rho = \overline{|\psi\rangle\langle\psi|} \tag{3.A.48}$$

The density matrix, $\boldsymbol{\rho}$, is the matrix representation of the density operator. It has the matrix elements

$$\rho_{nm} = \langle\phi_n|\,\rho\,|\phi_m\rangle = \langle\phi_n\overline{|\psi\rangle\langle\psi|}\phi_m\rangle = \overline{c_n\, c_m^*} \tag{3.A.49}$$

Thus, the density matrix is the ensemble average of the population matrix

$$\rho_{nm} = \overline{P_{nm}} = \overline{c_n\, c_m^*} \tag{3.A.50}$$

The diagonal terms of the matrix $c_n^2(t)$ give the probability to find the system in state n. The off-diagonal terms $c_n(t)\, c_m^*(t)$ represent coherences.

The ensemble-averaged expectation value $\langle \overline{A} \rangle$ will be the trace of the product of the density matrix and matrix representation of the operator **A**:

$$\langle \overline{A} \rangle = \sum_{i,\,j} \overline{P_{ji}}\, A_{ij} = \sum_{i,\,j} \rho_{ji}\, A_{ij} = \mathrm{Tr}\,(\boldsymbol{\rho}\, \mathbf{A}) \tag{3.A.51}$$

where **A** is the same for all spins.

For further reading see [2–4].

## 3.A.8 Some Math

Here we summarize some mathematical expressions that are used in NMR and its quantum mechanical descriptions.

- Some trigonometric equations

$$e^{\,i\,\alpha} = \cos\alpha + i\ \sin\alpha \tag{3.A.52}$$

From this, one can easily derive

$$\begin{aligned}
\cos(\alpha+\beta) &= \cos\alpha\ \cos\beta - \sin\alpha\ \sin\beta \\
\cos(\alpha-\beta) &= \cos\alpha\ \cos\beta + \sin\alpha\ \sin\beta \\
\sin(\alpha+\beta) &= \sin\alpha\ \cos\beta + \cos\alpha\ \sin\beta \\
\sin(\alpha-\beta) &= \sin\alpha\ \cos\beta - \cos\alpha\ \sin\beta
\end{aligned} \tag{3.A.53}$$

and

$$\begin{aligned}
2\ \cos\alpha\ \cos\beta &= \cos(\alpha-\beta) + \cos(\alpha+\beta) \\
2\ \sin\alpha\ \sin\beta &= \cos(\alpha-\beta) - \cos(\alpha+\beta) \\
2\ \sin\alpha\ \cos\beta &= \sin(\alpha+\beta) + \sin(\alpha-\beta) \\
2\ \cos\alpha\ \sin\beta &= \sin(\alpha+\beta) - \sin(\alpha-\beta)
\end{aligned} \tag{3.A.54}$$

and

$$\begin{aligned}
\cos(2\,\alpha) &= \cos^2\alpha\ - \sin^2\alpha \\
\sin(2\,\alpha) &= 2\ \sin\alpha\ \cos\alpha
\end{aligned} \tag{3.A.55}$$

- Some properties of operators

$$\begin{aligned}
[A,\ B+C] &= [A,\ B] + [A,\ C] \\
[A,\ B\,C] &= [A,\ B]\,C + B\,[A,\ C] \\
[A\,B,\ C] &= A\,[B,\ C] + [A,\ C]\,B
\end{aligned} \tag{3.A.56}$$

- Exponential matrixes and operators

An exponential matrix is defined via a Taylor's expansion

$$e^{A} = \sum_{n=0}^{\infty} \frac{1}{n!} A^{n} \tag{3.A.57}$$

From this one can derive (using the Taylor's expansion for the cos and sin functions)

$$e^{\pm i\theta A} = E \cos\theta \pm iA \sin\theta \qquad \text{if } A^{2} = E \tag{3.A.58}$$

For a cartesian product operator, as used in NMR, this becomes

$$e^{\pm i\theta B_s} = E \cos\frac{\theta}{2} \pm 2iB_s \sin\frac{\theta}{2} \qquad \text{if } B_s^{2} = \frac{1}{4}E \tag{3.A.59}$$

- Other properties of exponential matrixes

$$\begin{aligned} &e^{A}\, e^{B} = e^{A+B} \qquad \text{if } [A,\, B] = 0 \\ &(e^{A})_{ij} = e^{\,A_{ij}} \end{aligned} \tag{3.A.60}$$

## References


1 Harris, R.K. (1986). *Nuclear Magnetic Resonance Spectroscopy: a Physicochemical View*. Essex, England: Longman Scientific & Technical. ISBN 0582446538.

2 Fano, U. (1957). Description of states in quantum mechanics by density matrix and operator techniques. *Rev. Mod. Phys.* 29: 74–93. doi: 10.1103/RevModPhys.29.74.

3 Slichter, C.P. (1990). *Principles of Magnetic Resonance*, 3e. Berlin,Heidelberg: Springer. doi: 10.1007/978-3-662-09441-9. ISBN 978-3-540-50157-2.

4 Levitt, M. (2008). *Spin Dynamics: Basics of Nuclear Magnetic Resonance*, 2e. Chichester, England: JohnWiley & Sons Ltd. ISBN 978-0-470-51117-6.

# 4

# Relaxation in NMR Spectroscopy

*Matthias Ernst*

*Physical Chemistry, ETH Zürich, Vladimir-Prelog-Weg 2, Zürich 8093, Switzerland*

## 4.1 Introduction

Relaxation in magnetic resonance describes the process of a spin system returning to the thermal-equilibrium state. In order to observe such a phenomenon, we have to generate a non-equilibrium state in the first place. In NMR, we can do this using radio-frequency (RF) pulses that manipulate the state of the spin system by rotations [1–3]. Combined with free-evolution periods, such experiments allow us to generate a large number of different non-equilibrium states and observe their return toward the thermal equilibrium. The complex path to thermal equilibrium involves coherent evolution of the density operator and incoherent auto- and cross-relaxation processes that lead to decay and transfer of coherences and populations. The versatility of Fourier-transform NMR allows us to measure the time evolution of different coherence and populations, thus allowing the detailed characterization of different relaxation processes as the density operator returns to the thermal-equilibrium state.

Phenomenologically, the relaxation process can be described by local fields at the location of the spin that fluctuate as a function of time on different time scales. Such fluctuating local fields can be generated by anisotropic interaction (dipolar coupling, chemical-shielding anisotropy [CSA], quadrupolar interaction for nuclei and g-tensor anisotropy, hyperfine interaction, and zero-field splitting for electrons) under reorientational (Brownian) motion of the molecule. Isotropic ($J$ coupling, isotropic chemical shift for nuclei and isotropic g value, Fermi-contact interaction for electrons) or anisotropic interactions can also be modulated due to exchange, chemical processes, or by fast relaxing spins (scalar relaxation). Since the stochastic modulation of the Hamiltonian is different for each spin in the sample, each spin will evolve on a different trajectory under the stochastic part of the Hamiltonian while they all evolve in the same way under the deterministic part of the Hamiltonian. As an example, such a time evolution is illustrated in Figure 4.1 for a homonuclear dipolar-coupled two-spin system undergoing rotational tumbling. As one can see, the expectation values, $\langle \hat{S}_{1z} \rangle$ and $\langle \hat{S}_{2z} \rangle$ evolve differently for each selected trajectory that represents a different series of stochastic reorientation of the molecule. The ensemble average over 300 such trajectories (Figure 4.1d) gives the well-known exponential transfer of polarization from spin 1 to spin 2 (dipolar cross relaxation) and at the same time the mono-exponential decay toward zero (dipolar auto-relaxation) with a time constant that depends on the strength of the square of the coupling and the correlation time of the rotational tumbling.

*Two-Dimensional (2D) NMR Methods*, First Edition. Edited by K. Ivanov, P.K. Madhu and G. Rajalakshmi.

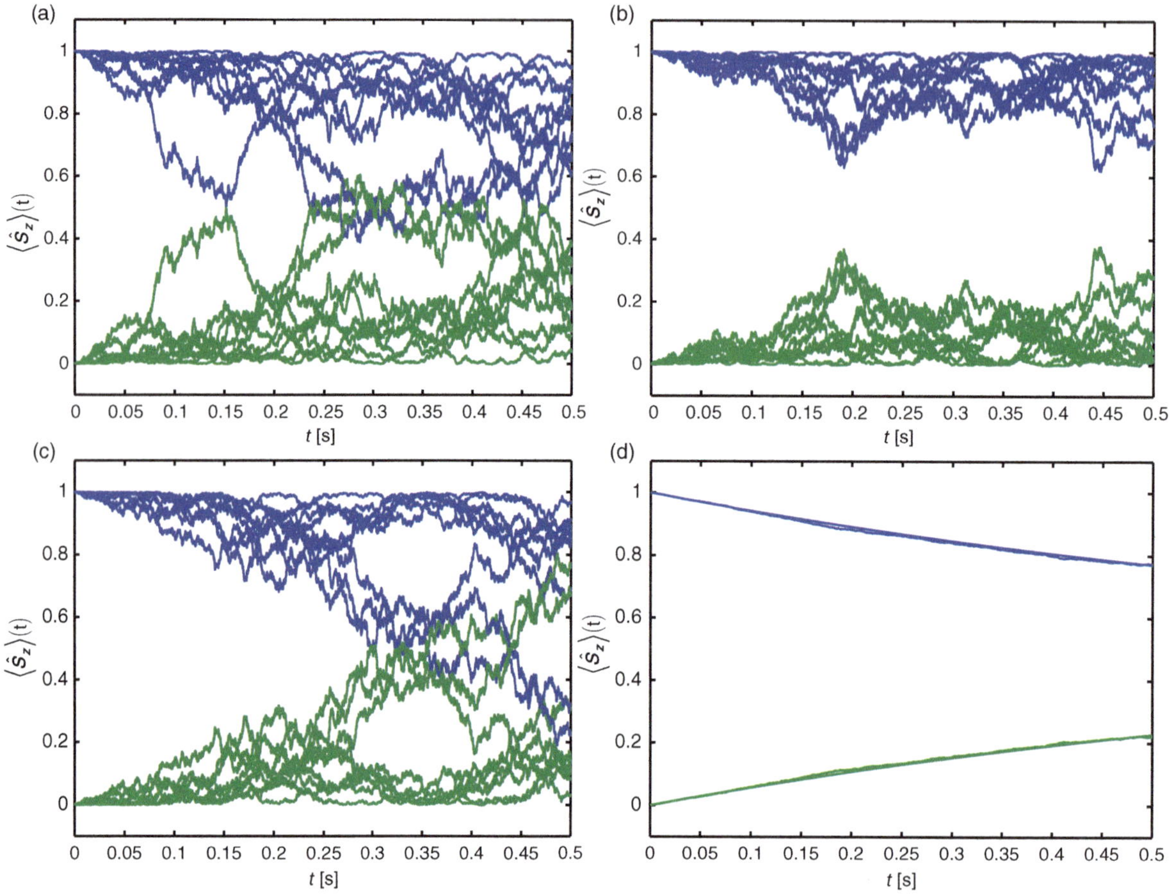


**Figure 4.1** Coherent numerical simulation of a homonuclear dipolar-coupled proton two-spin system in the laboratory frame ($\nu_0$ = 600 MHz, $\nu_1$ = 0 kHz, $\nu_2$ = -1 kHz, $\delta_D/(2\pi)$ = -8.9 kHz) undergoing stochastic molecular tumbling. One of the spins had an initial state of $\langle \hat{S}_{1z}\rangle(0) = 1$ while the other was assumed to be $\langle \hat{S}_{2z}\rangle(0) = 0$. The blue and green lines in panels (a)–(c) show the time evolution of the expectation values $\langle \hat{S}_{iz}\rangle(t)$ for different stochastic trajectories of the reorientation of the molecule containing the two-spin system. Each trajectory is different but the average of 300 such trajectories in panel (d) shows an almost perfect mono-exponential decay toward the thermal-equilibrium value with the expected theoretical time constant. This clearly illustrates that the experimentally observed exponential decay of magnetization is an ensemble property due to the many stochastic reorientations that the molecules undergo.

The description of relaxation processes was an important topic of magnetic resonance from the very beginning, starting with the initial work by Felix Bloch [4]. Over the years, a lot of effort has been put in continuously improving the description of relaxation processes as illustrated by many reviews [5–17] and books on this topic [18–24]. Even today, there are new developments as can be seen by the introduction of the Lindbladian formalism [25] into magnetic resonance to describe relaxation far from equilibrium and in highly polarized samples [26–28]. This adaptation was motivated by newly accessible experimental conditions that can be realized by dynamic-nuclear polarization (DNP) at temperatures around 1 K [29] and has started a debate whether the Lindbladian description is equivalent to the early quantum-mechanical description of relaxation in NMR [30, 31].

All relaxation treatments that are based on second-order perturbation theory require that the stochastic process happens on a much faster time scale than the evolution of the density operator. This requirement is often called the Redfield limit or the "weak collision" limit. If it is not fulfilled either because the stochastic process is slow or the density operator changes fast, we need to resort to other treatments of relaxation. In this case, the stochastic Liouville equation can be used [32–36] where we describe the spin system in a composite Liouville space with the stochastic process interconnecting the different states, in essence a generalized treatment similar to chemical exchange in the classical Bloch picture [37].

This chapter cannot cover all topics important in relaxation theory but focuses on the basics to give an introduction such that the reader can understand the original literature better. Of course, as always in such space-limited treatments, the selection is biased by the interests and knowledge of the author. For more comprehensive treatments, the interested reader is referred to the monographs about relaxation in magnetic resonance [18–24].

## 4.2 Theory

The theories derived to describe relaxation phenomena in magnetic resonance can be classified according to the way they describe the interaction with the environment. In the simplest approach (Bloch equations [4]), the magnetization and the environment are described classically and the relaxation has to be introduced phenomenologically. In the semi-classical approach [38–42], the spin system of interest is described by a quantum-mechanical density operator while the environment is described by classical spatial functions. The best-known example of such a semi-classical description is the Redfield theory [41, 42] that is based on second-order perturbation theory. In the semi-classical approach, the environment is at infinite temperature and the correct thermal-equilibrium value of the density operator has to be introduced ad hoc [42] or by a thermalization operator [9, 43–45]. In the quantum-mechanical description of relaxation [1, 25, 26, 46–48], both the spin system of interest and the environment are described by quantum-mechanical operators leading to the proper thermal-equilibrium density operator without any ad-hoc assumptions. In NMR, the semi-classical approach is still predominantly used to describe relaxation phenomena but there are some experimental conditions (high polarization levels and low temperatures) where a quantum-mechanical description is required [26].

### 4.2.1 Bloch Equations

The simplest model to describe magnetic-resonance experiments is the Bloch equations [4] where the time evolution of the magnetization is characterized by a classical magnetization vector $\vec{M}$ that interacts with the magnetic induction $\vec{B}$ (often colloquially called magnetic field):

$$\frac{d}{dt}\vec{M} = \gamma\vec{M} \times \vec{B} - \mathbf{R}\left(\vec{M} - \vec{M}_0\right) \tag{4.1}$$

with the magnetization vector $\vec{M} = \left(M_x, M_y, M_z\right)$ and the thermal-equilibrium magnetization $\vec{M}_0 = (0, 0, M_0)$ assuming that the external field is along the $z$ axis. The vector $\vec{B}$ describes the orientation of the magnetic induction. In the laboratory frame, $\vec{B} = (0, 0, B_0)$ is given by the static magnetic field, which, by convention, we put along the z-axis, while in the rotating frame, $\vec{B} = (B_1, 0, \Delta B_0)$ may contain the RF field amplitude $B_1$ and the resonance offset $\Delta B_0$. The relaxation matrix is diagonal,

$$\mathbf{R} = \begin{pmatrix} \frac{1}{T_2} & 0 & 0 \\ 0 & \frac{1}{T_2} & 0 \\ 0 & 0 & \frac{1}{T_1} \end{pmatrix} \tag{4.2}$$

with two phenomenological relaxation times $T_1$ and $T_2$ that describe the mono-exponential decay of the transverse and longitudinal components toward the thermal-equilibrium magnetization. The transverse relaxation time $T_2$ does not change the energy of the spin system and is sometimes referred to as the spin-spin relaxation-rate constant. The longitudinal relaxation time $T_1$ changes the length of the $M_z$ component, and therefore, the energy of the spin system and is sometimes referred to as the spin-lattice relaxation-rate constant. However, we should remember that these terms have historic connotations and should be used carefully or maybe better avoided [49]. We mention those terms here to provide context for understanding the older literature. Even in such a simple phenomenological model, certain limits are imposed on the relative ratio of the two relaxation times [50]. Based on the requirement that the length of the magnetization vector can never be longer than the length of the equilibrium magnetization, one can show that there is a limit imposed on the transverse relaxation times that is given by $T_2 \leq 2T_1$. To illustrate this, the magnetization trajectories from simulations based on the Bloch equations are shown in Figure 4.2. Using the parameters $-\gamma B_0/(2\pi) = \Omega/(2\pi) = 3$ Hz, $T_1 = T_2 = 0.75$ s (Figure 4.2a) one can clearly see that the length of the magnetization vector ($|M|$) is always smaller than one that corresponds to the equilibrium magnetization ($M_0 = 1$) used in the simulation. In Figure 4.2b, the transverse relaxation time $T_2$ was increased to a value of 2 s resulting in the length of the magnetization vector, $|M|$, to get larger than one which is physically not possible. The transition point for $|M|$ from a curve with a minimum that approaches asymptotically the equilibrium value for infinite times to a curve that has a minimum and a maximum above the equilibrium value happens for $T_2 > 2T_1$. Therefore, only values of $T_2 \leq 2T_1$ make physical sense for the relaxation times in the context of the Bloch equations. The Bloch picture does not provide a physical explanation for the magnitude of the relaxation-rate constants or the limitations discussed above.

Since the Bloch equations allow only the description of a classical magnetization interacting with the external field but not spin-spin interactions, we usually describe magnetic-resonance experiments based on quantum mechanics using the density-operator formulation [2, 3, 51]. We, therefore, also require a more complex theory for describing relaxation phenomena in such a system since the density operator has a larger basis and can assume more possible states for coupled spin systems.

### 4.2.2 Transition-rate Theory

A system of two coupled spins can be described by a Hilbert space of dimension four and has four energy levels and six transition probabilities. In such a coupled spin system, the transition-rate theory of Bloembergen, Pound, and Purcell [38] can be used, which was later formalized and generalized to larger spin systems in the Solomon equations [39]. Figure 4.3 shows the energy levels in a two-spin system and the corresponding transition-rate constants between the energy levels. Here, $W_\mathrm{I}$ and $W_\mathrm{S}$ are the one-quantum transition-rate constants where the I spin or S spin changes its state, respectively. The rate constant $W_0$ describes the zero-quantum transition where we have a flip-flop transition while $W_2$ describes the double-quantum rate constant where both spins flip in the same direction. The four populations are given by $P_{\kappa\mu}$ ($\kappa, \mu \in \{\alpha, \beta\}$) and the thermal-equilibrium populations by $P^0_{\kappa\mu}$. The deviation from the equilibrium population is characterized by $\Delta P_{\kappa\mu} = P_{\kappa\mu} - P^0_{\kappa\mu}$. We can set up a system of differential equations that describes the time evolution of the difference magnetization:

Figure 4.2 Simulated time evolution of the magnetization by solving the Bloch equations. The plots on the left show the three-dimensional visualization of the trajectory of the magnetization vector in space while the three components $M_x$, $M_y$, $M_z$ and the length of the magnetization $|M|$ are plotted on the right hand side. (a) The parameters are chosen as $-\gamma B_0/(2\pi) = \Omega/(2\pi) = 3$ Hz, $T_1 = T_2 = 0.75$ s. The length of the magnetization vector is always smaller or equal to one, which is the magnitude of the equilibrium magnetization. (b) The parameters are chosen as $-\gamma B_0/(2\pi) = \Omega/(2\pi) = 3$ Hz, $T_1 = 0.75$ s, $T_2 = 2$ s. The length of the magnetization vector becomes larger than one (red part of the plot of $|M|$), which is physically impossible and illustrates that there is a maximum allowed ratio of the two relaxation times given by $T_2 \leq 2T_1$.

(a) $T_1 = T_2 = 0.75$ s

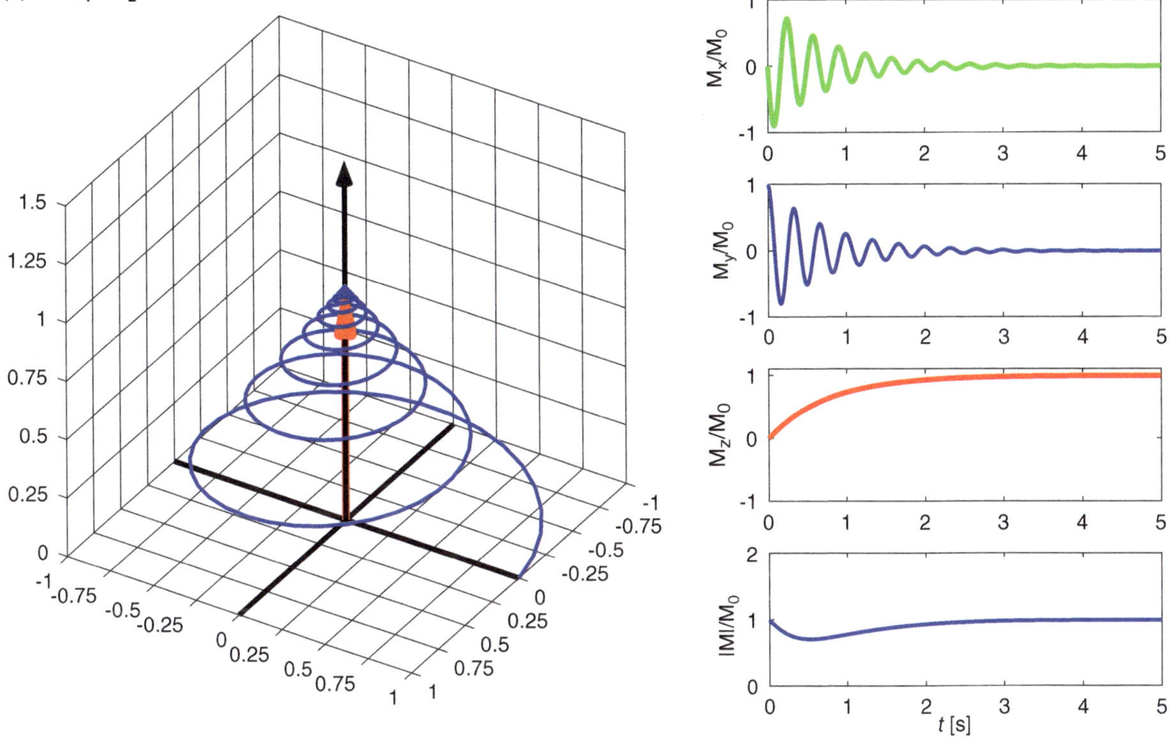


(b) $T_1 = 0.75$ s, $T_2 = 2$ s

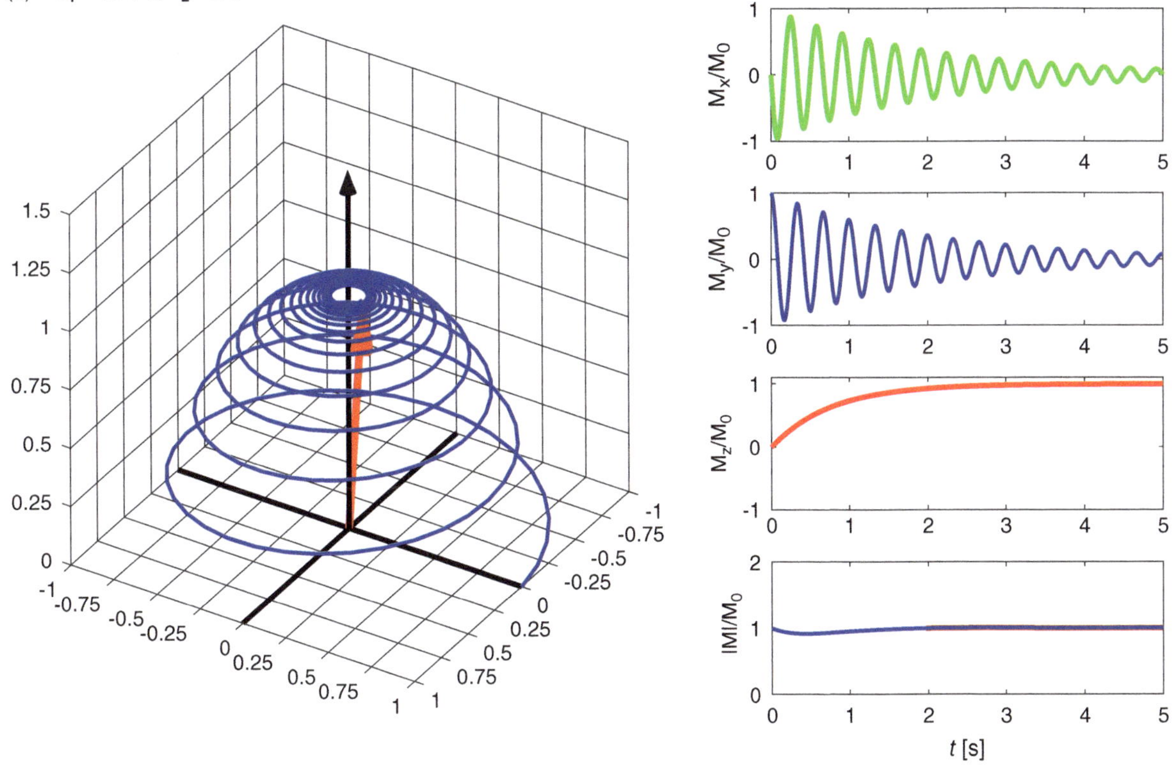

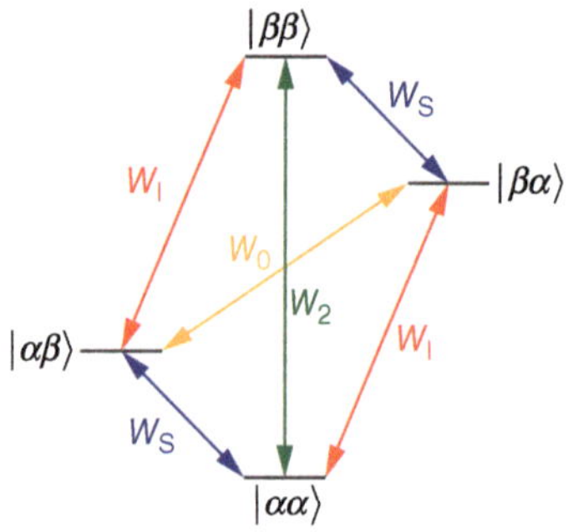


**Figure 4.3** Energy levels of an I-S two-spin system and transition-rate constants between the different energy levels.

$$\begin{aligned}
\frac{\mathrm{d}\Delta P_{\alpha\alpha}}{\mathrm{d}t} &= -(W_I + W_S + W_2)\Delta P_{\alpha\alpha} + W_S\Delta P_{\alpha\beta} + W_I\Delta P_{\beta\alpha} + W_2\Delta P_{\beta\beta} \\
\frac{\mathrm{d}\Delta P_{\alpha\beta}}{\mathrm{d}t} &= -(W_I + W_S + W_0)\Delta P_{\alpha\beta} + W_S\Delta P_{\alpha\alpha} + W_I\Delta P_{\beta\beta} + W_0\Delta P_{\beta\alpha} \\
\frac{\mathrm{d}\Delta P_{\beta\alpha}}{\mathrm{d}t} &= -(W_I + W_S + W_0)\Delta P_{\beta\alpha} + W_S\Delta P_{\beta\beta} + W_I\Delta P_{\alpha\alpha} + W_0\Delta P_{\alpha\beta} \\
\frac{\mathrm{d}\Delta P_{\beta\beta}}{\mathrm{d}t} &= -(W_I + W_S + W_2)\Delta P_{\beta\beta} + W_S\Delta P_{\beta\alpha} + W_I\Delta P_{\alpha\beta} + W_2\Delta P_{\alpha\alpha}.
\end{aligned} \tag{4.3}$$

The four populations can be connected to the expectation values of the three longitudinal spin operators by:

$$\begin{aligned}
\langle I_z\rangle(t) &= P_{\alpha\alpha} + P_{\alpha\beta} - P_{\beta\alpha} - P_{\beta\beta} \\
\langle S_z\rangle(t) &= P_{\alpha\alpha} - P_{\alpha\beta} + P_{\beta\alpha} - P_{\beta\beta} \\
\langle 2I_zS_z\rangle(t) &= P_{\alpha\alpha} - P_{\alpha\beta} - P_{\beta\alpha} + P_{\beta\beta},
\end{aligned} \tag{4.4}$$

which leads then to differential equations describing the relaxation of the three longitudinal spin operators:

$$\begin{aligned}
\frac{\mathrm{d}\langle\Delta I_z\rangle(t)}{\mathrm{d}t} &= -(W_0 + 2W_I + W_2)\langle\Delta I_z\rangle(t) - (W_2 - W_0)\langle\Delta S_z\rangle(t) \\
\frac{\mathrm{d}\langle\Delta S_z\rangle(t)}{\mathrm{d}t} &= -(W_0 + 2W_S + W_2)\langle\Delta S_z\rangle(t) - (W_2 - W_0)\langle\Delta I_z\rangle(t) \\
\frac{\mathrm{d}\langle\Delta 2I_zS_z\rangle(t)}{\mathrm{d}t} &= -2(W_I + W_S)\langle\Delta 2I_zS_z\rangle(t).
\end{aligned} \tag{4.5}$$

In principle, we also have a fourth longitudinal mode, the identity operator, but since it is invariant and does not couple to the other modes, we have neglected it here.

If we replace $\rho_I = W_0 + 2W_I + W_2$, $\rho_S = W_0 + 2W_S + W_2$ and $\sigma_{IS} = W_2 - W_0$, we obtain the original Solomon equations [39]:

$$\begin{aligned}
\frac{\mathrm{d}\langle\Delta I_z\rangle(t)}{\mathrm{d}t} &= -\rho_I\langle\Delta I_z\rangle(t) - \sigma_{IS}\langle\Delta S_z\rangle(t) \\
\frac{\mathrm{d}\langle\Delta S_z\rangle(t)}{\mathrm{d}t} &= -\rho_S\langle\Delta S_z\rangle(t) - \sigma_{IS}\langle\Delta I_z\rangle(t).
\end{aligned} \tag{4.6}$$

Here again, we have used the difference to the thermal equilibrium $\langle\Delta I_z\rangle(t) = \langle I_z\rangle(t) - \langle I_z^0\rangle$ and $\langle\Delta S_z\rangle(t) = \langle S_z\rangle(t) - \langle S_z^0\rangle$. The two-spin term $2I_zS_z$ is not coupled to the two other longitudinal spin operators as long as the two transition-rate constants for the $I$ spin ($W_I$) and the two rate constants for the $S$ spin ($W_S$) are identical and $\langle\Delta 2I_zS_z\rangle$ relaxes independently. The Solomon equations describe auto-relaxation (decay of magnetization, $\rho_I$ and $\rho_S$) and cross relaxation (transport of magnetization from one-spin species to another spin species, $\sigma_{IS}$) in a coupled system of two spins. They can be used to explain two-dimensional NOESY [52–54] spectroscopy, which is based on the cross relaxation between homonuclear spins. The Solomon equations are rate equations and can easily be extended to an $N$-spin system and one obtains a system of $N$ coupled differential equations:

$$\frac{\mathrm{d}\langle \Delta I_{kz}\rangle(t)}{\mathrm{d}t} = -\rho_k \langle \Delta I_{kz}\rangle(t) - \sum_{j\neq k} \sigma_{kj} \langle \Delta I_{jz}\rangle(t). \tag{4.7}$$

There are also other effects one can understand based on the Solomon equation, like the decoupling of coupled relaxation modes by saturating one of the spins [23] or the steady-state NOE [54] effect in heteronuclear spin systems. We will discuss them in the context of the semi-classical relaxation theory, which results in the same type of coupled differential equations for longitudinal relaxation.

### 4.2.3 Semi-classical Relaxation Theory

Semi-classical relaxation theory follows the ideas as formulated by Wangsness, Bloch, and Redfield [40–42] and is often referred to as WBR or Redfield theory. In the derivation given here, we follow closely the formulations used by Maurice Goldman [55]. The full time-dependent laboratory-frame Hamiltonian is split into two parts:

$$\mathcal{H}(t) = \mathcal{H}_0(t) + \mathcal{H}_1(t) \tag{4.8}$$

where $\mathcal{H}_0(t)$ describes the deterministic part of the Hamiltonian that can be time dependent due to RF irradiation or in solid-state NMR due to magic-angle spinning (MAS) [56–58] and is responsible for the coherent time evolution of the density operator. In the following we will assume that $\mathcal{H}_0$ is time independent to simplify the discussion. The Hamiltonian, $\mathcal{H}_1(t)$ describes the stochastic time-dependent part of the Hamiltonian, which causes the relaxation processes. We start from the Liouville-von Neumann equation:

$$\frac{\mathrm{d}\sigma(t)}{\mathrm{d}t} = -i\left[\mathcal{H}(t), \sigma(t)\right] \tag{4.9}$$

that describes the time evolution of the density operator in the laboratory frame. An analytical solution of the Liouville-von Neumann equation is only simple if the Hamiltonian is time independent. To simplify the Hamiltonian and focus on the effects of the stochastic part, we transform the system into an interaction frame with the deterministic Hamiltonian $\mathcal{H}_0$. The transformation of an operator $Q$ into the interaction frame is defined by:

$$\tilde{Q}(t) = e^{i\mathcal{H}_0 t} Q e^{-i\mathcal{H}_0 t} \tag{4.10}$$

where we use the tilde ($\tilde{\ }$) to symbolize an operator in the interaction frame. At the same time, the Liouville-von Neumann equation (Equation 4.9) needs to be corrected for the transformation into an accelerated frame of reference and we obtain:

$$\frac{\mathrm{d}\tilde{\sigma}(t)}{\mathrm{d}t} = -i\left[\tilde{\mathcal{H}}_1(t), \tilde{\sigma}(t)\right], \tag{4.11}$$

which does not contain the deterministic part of the Hamiltonian, $\mathcal{H}_0$. We can now formally integrate the Liouville-von Neumann equation:

$$\tilde{\sigma}(t) = \tilde{\sigma}(0) - i\int_0^t \left[\tilde{\mathcal{H}}_1(t'), \tilde{\sigma}(t')\right]dt' \tag{4.12}$$

and insert the formal solution into Equation 4.11 to obtain an iterative approximate solution, which corresponds to second-order perturbation theory:

$$\frac{\mathrm{d}\tilde{\sigma}(t)}{\mathrm{d}t} = -i\left[\tilde{\mathcal{H}}_1(t), \tilde{\sigma}(0)\right] - \int_0^t \left[\tilde{\mathcal{H}}_1(t), \left[\tilde{\mathcal{H}}_1(t'), \tilde{\sigma}(t')\right]\right]dt'. \tag{4.13}$$

In principle, we could go on and do the iterative solution multiple times to obtain a higher-order solution, but it turns out that the second-order solution is sufficient for most situations where Redfield theory is applicable. We do now a variable substitution by defining $t' = t - \tau$, $\mathrm{d}t' = -\mathrm{d}\tau$ and obtain by adjusting the integration boundaries:

$$\frac{\mathrm{d}\tilde{\sigma}(t)}{\mathrm{d}t} = -i\left[\tilde{\mathcal{H}}_1(t), \tilde{\sigma}(0)\right] - \int_0^t \left[\tilde{\mathcal{H}}_1(t), \left[\tilde{\mathcal{H}}_1(t-\tau), \tilde{\sigma}(t-\tau)\right]\right] d\tau. \tag{4.14}$$

So far we have made no assumptions except that we have a second-order perturbation solution of the Liouville-von Neumann equation. At this point, we introduce four modifications into Equation 4.14:

i) We take an ensemble average over all terms in Equation 4.14 despite the fact that the density operator already describes an ensemble of spin. This is required because different spins will have different stochastic trajectories (see Figure 4.1) and will evolve under different stochastic Hamiltonians. In each environment, the density operator will evolve in a different way and we are interested in the ensemble average of all these environments. We require that $\mathcal{H}_1(t)$ is a bias-free perturbation, i.e. the ensemble average $\overline{\mathcal{H}_1(t)} = 0$. Therefore, the first term of Equation 4.14 becomes zero and can be dropped.
ii) The thermal equilibrium that a system characterized by Equation 4.14 would reach corresponds to zero. This is a consequence of the fact that the forward and backward rate constants are equal. This assumption corresponds to a situation where the spin system is at infinite temperature but in reality we are interested in the system at a finite temperature. To correct this, we replace the density operator $\tilde{\sigma}(t-\tau)$ in Equation 4.14 by $\tilde{\sigma}(t-\tau) - \sigma_{\mathrm{eq}}$. This ad-hoc introduction of the correct thermal equilibrium makes the differential equation inhomogeneous.
iii) At this point, we will also make an *assumption* about the time scale of the stochastic process that generates the fluctuations of the Hamiltonian $\mathcal{H}_1(t)$. If we *assume* that the time scale is much shorter than the time scale on which the density operator typically evolves, i.e. $\tau \ll t$, then we can replace $\tilde{\sigma}(t-\tau)$ by $\tilde{\sigma}(t)$ in Equation 4.14. This assumption is known as the "weak collision" or Redfield limit since we assume that each stochastic event will change the state of the system only weakly and only the accumulation of many events will lead to a significant change of the state. Another equivalent formulation is the requirement that the magnitude of the anisotropic interaction is small compared to the inverse of the time scale of the stochastic process [23]. This is an important limitation of the Redfield relaxation theory that has to be considered when using it. In solution-state NMR, the overall tumbling is usually fast enough that we are safely within the Redfield limit. For slower tumbling, the lines become very broad and unobservable. In electron-paramagnetic resonance (EPR), where the interactions can be much larger, the question of the validity of the Redfield limit is less clear and alternate methods like the stochastic Liouville equation approach are frequently used [15, 59, 60].
iv) If we assume that the time $t$ is much longer than the time scale of $\tau$, we can replace the upper integration boundary $t$ of the integral by $\infty$.

Taking these assumptions and simplifications into account, we can simplify Equation 4.14 to:

$$\begin{aligned}\frac{\mathrm{d}\tilde{\sigma}(t)}{\mathrm{d}t} &= -\int_0^\infty \overline{\left[\tilde{\mathcal{H}}_1(t), \left[\tilde{\mathcal{H}}_1(t-\tau), \tilde{\sigma}(t) - \sigma_{\mathrm{eq}}\right]\right]} d\tau \\ &= -\hat{\hat{\Gamma}}\left(\tilde{\sigma}(t) - \sigma_{\mathrm{eq}}\right)\end{aligned} \tag{4.15}$$

where $\hat{\hat{\Gamma}}$ is the relaxation super operator in the interaction frame. We can now transform back into the laboratory frame with the inverse transformation of Equation 4.12 and obtain:

$$\begin{aligned}\frac{\mathrm{d}\sigma(t)}{\mathrm{d}t} &= -i\left[\mathcal{H}_0, \sigma(t)\right] - \int_0^\infty \overline{\left[\mathcal{H}_1(t), \left[e^{-i\mathcal{H}_0\tau}\mathcal{H}_1(t-\tau)e^{i\mathcal{H}_0\tau}, \sigma(t) - \sigma_{\mathrm{eq}}\right]\right]} d\tau \\ &= -i\left[\mathcal{H}_0, \sigma(t)\right] - \hat{\hat{\Gamma}}\left(\sigma(t) - \sigma_{\mathrm{eq}}\right)\end{aligned} \tag{4.16}$$

where the laboratory-frame relaxation super operator is defined by:

$$\hat{\Gamma}\{Q\} = -\int_0^\infty \overline{[\mathcal{H}_1(t), [e^{-i\mathcal{H}_0\tau}\mathcal{H}_1(t-\tau)e^{i\mathcal{H}_0\tau}, Q]]}d\tau. \tag{4.17}$$

Using the Hamiltonian commutator super operator $\hat{\mathcal{H}} = [\mathcal{H}, \quad]$, we can reformulate the master equation as:

$$\begin{aligned}\frac{d\sigma(t)}{dt} &= -i\hat{\mathcal{H}}_0\sigma(t) - \int_0^\infty \overline{\hat{\mathcal{H}}_1(t)e^{-i\mathcal{H}_0\tau}\hat{\mathcal{H}}_1(t-\tau)e^{i\mathcal{H}_0\tau}\sigma(t) - \sigma_{eq}}d\tau \\ &= -i\hat{\mathcal{H}}_0\sigma(t) - \hat{\Gamma}\left(\sigma(t) - \sigma_{eq}\right) = \hat{\mathcal{L}}\sigma(t) + \hat{\Gamma}\sigma_{eq}.\end{aligned} \tag{4.18}$$

So far, we have made no assumption about the form of $\mathcal{H}_1(t)$.

In semi-classical relaxation theory, we describe the spin system of interest quantum mechanically and the environment classically. This is typically implemented using a spherical-tensor notation [61] where:

$$\mathcal{H}_1(t) = \sum_{\mu}\sum_{\ell}\sum_{m=-\ell}^{\ell}(-1)^m A_{\ell,-m}^{(\mu)}(t)T_{\ell,m}^{(\mu)}. \tag{4.19}$$

Here, $\mu$ characterizes the interaction, $T_{\ell,m}^{(\mu)}$ are the spherical spin-tensor operators of rank $\ell$ and $A_{\ell,m}^{(\mu)}(t)$ are the spatial spherical tensors of rank $\ell$ characterizing the orientation dependence of the interaction $\mu$. Inserting Equation 4.19 into Equation 4.17 allows us to separate the ensemble average and the commutators:

$$\begin{aligned}\hat{\Gamma}\{Q\} = &\sum_{\mu,\mu'}\sum_{\ell,\ell'}\sum_{\substack{m=-\ell \\ m'=-\ell'}}^{\ell,\ell'}(-1)^{m+m'}\int_0^\infty \overline{A_{\ell,-m}^{(\mu)}(t)A_{\ell',-m'}^{(\mu')}(t-\tau)}d\tau \\ &\times\left[T_{\ell,m}^{(\mu)}, \left[e^{-i\mathcal{H}_0\tau}T_{\ell',m'}^{(\mu')}e^{i\mathcal{H}_0\tau}, Q\right]\right].\end{aligned} \tag{4.20}$$

In order to evaluate the correlation function, it is of advantage to substitute:

$$A_{\ell',-m'}^{(\mu')}(t-\tau)T_{\ell',m'}^{(\mu')} = A_{\ell',-m'}^{(\mu')}{}^{*}(t-\tau)T_{\ell',m'}^{(\mu')}{}^{\dagger} \tag{4.21}$$

where $^*$ indicates the conjugate complex and $^\dagger$ the adjoint of the operator. In addition, we can substitute $m'$ by $-m'$ since the summation runs over a symmetric range from $-\ell$ to $\ell$:

$$\begin{aligned}\hat{\Gamma}\{Q\} = &\sum_{\mu,\mu'}\sum_{\ell,\ell'}\sum_{\substack{m=-\ell \\ m'=-\ell'}}^{\ell,\ell'}(-1)^{m+m'}\int_0^\infty \overline{A_{\ell,-m}^{(\mu)}(t)A_{\ell',-m'}^{(\mu')}{}^{*}(t-\tau)}d\tau \\ &\times\left[T_{\ell,m}^{(\mu)}, \left[e^{-i\mathcal{H}_0\tau}T_{\ell',m'}^{(\mu')}{}^{\dagger}e^{i\mathcal{H}_0\tau}, Q\right]\right].\end{aligned} \tag{4.22}$$

We can now define a correlation function of the form:

$$C_{\ell,\ell',-m,-m'}^{(\mu,\mu')}(\tau) = \overline{A_{\ell,-m}^{(\mu)}(t)A_{\ell',-m'}^{(\mu')}{}^{*}(t-\tau)} \tag{4.23}$$

where we have assumed that we have stationary stochastic functions, which describe a time-independent process. In the case of $\mu = \mu'$, $\ell = \ell'$, and $m = m'$ we speak of an auto-correlation function, otherwise of a cross-correlation function.

We can now decompose the spherical-tensor operators in a sum of eigenoperators of $\mathcal{H}_0$ in order to evaluate the term $e^{-i\mathcal{H}_0\tau}T^{(\mu')}_{\ell',m'}e^{i\mathcal{H}_0\tau}$. We decompose the spherical-tensor operators as:

$$T^{(\mu)}_{\ell,m} = \sum_p V^{(\mu)}_p \tag{4.24}$$

where the eigenoperators fulfill the condition:

$$\hat{\mathcal{H}}_0 V^{(\mu)}_p = \left[\mathcal{H}_0, V^{(\mu)}_p\right] = \omega_p V^{(\mu)}_p \Longleftrightarrow e^{-i\mathcal{H}_0\tau}V^{(\mu)}_p e^{i\mathcal{H}_0\tau} = V^{(\mu)}_p e^{-i\omega_p\tau}. \tag{4.25}$$

Combining this all with Equation 4.22 we obtain:

$$\begin{aligned}\hat{\Gamma}\{Q\} = &\sum_{\mu,\mu'}\sum_{\ell,\ell'}\sum_{\substack{m=-\ell\\ m'=-\ell'}}^{\ell,\ell'}\sum_{p'}(-1)^{m+m'}\int_0^\infty C^{(\mu,\mu')}_{\ell,\ell',-m,-m'}(\tau)e^{i\omega_{p'}\tau}\mathrm{d}\tau\\ &\times\left[T^{(\mu)}_{\ell,m},\left[V^{(\mu')\dagger}_{p'},Q\right]\right].\end{aligned} \tag{4.26}$$

We can now defined the power spectral-density function as the Fourier transform of the correlation function:

$$J^{(\mu,\mu')}_{\ell,\ell',-m,-m'}(\omega) = \int_{-\infty}^\infty C^{(\mu,\mu')}_{\ell,\ell',-m,-m'}(\tau)e^{-i\omega\tau}\mathrm{d}\tau. \tag{4.27}$$

Since the integration in Equation 4.26 is only from 0 to $\infty$, we obtain also a complex part from the Fourier transformation and an additional factor of $\frac{1}{2}$:

$$\begin{aligned}\int_0^\infty C^{(\mu,\mu')}_{\ell,\ell',-m,-m'}(\tau)e^{-i\omega\tau}\mathrm{d}\tau =&\frac{1}{2}\int_{-\infty}^\infty C^{(\mu,\mu')}_{\ell,\ell',-m,-m'}(\tau)e^{-i\omega\tau}\mathrm{d}\tau\\ &+\frac{1}{2}\int_{-\infty}^\infty C^{(\mu,\mu')}_{\ell,\ell',-m,-m'}(\tau)\mathrm{sign}(\tau)e^{-i\omega\tau}\mathrm{d}\tau\\ =&\frac{1}{2}J^{(\mu,\mu')}_{\ell,\ell',-m,-m'}(\omega) + \frac{1}{2}iK^{(\mu,\mu')}_{\ell,\ell',-m,-m'}(\omega).\end{aligned} \tag{4.28}$$

The imaginary term $iK^{(\mu,\mu')}_{\ell,\ell',-m,-m'}(\omega)$ is responsible for dynamic-frequency shifts [62] due to stochastic processes and is often neglected. Our final result is then:

$$\begin{aligned}\hat{\Gamma}\{Q\} =&\frac{1}{2}\sum_{\mu,\mu'}\sum_{\ell,\ell'}\sum_{\substack{m=-\ell\\ m'=-\ell'}}^{\ell,\ell'}\sum_{p'}(-1)^{m+m'}\left(J^{(\mu,\mu')}_{\ell,\ell',-m,-m'}(\omega_{p'}) + iK^{(\mu,\mu')}_{\ell,\ell',-m,-m'}(\omega_{p'})\right)\\ &\times\left[T^{(\mu)}_{\ell,m},\left[V^{(\mu')\dagger}_{p'},Q\right]\right]\\ \approx&\frac{1}{2}\sum_{\mu,\mu'}\sum_{\ell,\ell'}\sum_{\substack{m=-\ell\\ m'=-\ell'}}^{\ell,\ell'}\sum_{p'}(-1)^{m+m'}J^{(\mu,\mu')}_{\ell,\ell',-m,-m'}(\omega_{p'})\left[T^{(\mu)}_{\ell,m},\left[V^{(\mu')\dagger}_{p'},Q\right]\right].\end{aligned} \tag{4.29}$$

We can rewrite the relaxation super operator completely in the terms of the eigenoperators $V^{(\mu)}_p$, which gives a more compact representation but hides some of the complexities of the summations:

$$\hat{\Gamma}\{Q\} =\frac{1}{2}\sum_{\mu,\mu'}\sum_{p,p'}\left[V^{(\mu)}_p,\left[V^{(\mu')\dagger}_{p'},Q\right]\right]J^{(\mu,\mu')}_{p,p'}(\omega_{p'})\delta\left(\omega_p,\omega_{p'}\right). \tag{4.30}$$

In order to calculate the relaxation-rate constants, we have to evaluate the double commutators that will give us the allowed relaxation pathways for each relaxation mechanism and we have to calculate the spectral-density functions that depend on the stochastic model of the motion and gives us the strength of the different relaxation pathway. We will discuss this in detail in Section 4.3.

### 4.2.4 Quantum-mechanical Relaxation Theory – Lindblad Formulation

Already in the early years of magnetic-resonance relaxation development, quantum-mechanical descriptions of NMR were developed [1, 46, 47]. However, compared to the semi-classical description of relaxation as formulated by Redfield [40–42], the quantum-mechanical description never gained wide-spread application in NMR. This can be mostly attributed to the fact that Redfield theory can explain most effects observed in NMR spectroscopy under typical conditions. Some shortcomings of Redfield theory can be avoided by using the homogeneous master equation [9, 43–45], which allows the proper calculation of relaxation under RF field irradiation and the implementation of effective Liouvillians to predict the steady-state density operator under a pulse sequence [43, 63–66]. The homogeneous master equation can be implemented in the high-temperature limit or for arbitrary temperatures but it has rarely been used. In operator terms, the homogeneous master equation implements a cross-relaxation term from the identity operator to the spin operators that have non-zero equilibrium values to achieve the proper thermal equilibrium.

Recent advances in DNP made high polarization levels and systems far from thermal equilibrium at low temperatures experimentally accessible [29]. Under such conditions, the problems associated with the Redfield approach become apparent and different approaches have to be utilized. In open-quantum systems, such an approach has been used for many years, the Lindblad thermalization [25, 26, 48]. It is based on a quantum-mechanical description of the environment leading to a different form of the thermalization operator. We can start with the master equation in the laboratory frame, Equation 4.17, which we can rewrite in a super-operator notation as:

$$\hat{\Gamma}\{Q\} = -\int_0^t \overline{\hat{\mathcal{H}}_1(t)e^{-i\hat{\mathcal{H}}_0\tau}\hat{\mathcal{H}}_1(t-\tau)e^{i\hat{\mathcal{H}}_0\tau}Q d\tau} \tag{4.31}$$

where $\hat{\mathcal{H}}_1(t)$ is the commutator super operator. In this way, the double commutator can be written in a more compact notation. Instead of expanding the time-dependent Hamiltonian as a product of the spherical spin-tensor operators and the spherical spatial tensors (see Equation 4.19), the time-dependent Hamiltonian is replaced by a fully quantum-mechanical operator of the form:

$$\mathcal{H}_1(t) = \sum_{\mu,p} V_p^{(\mu)} \otimes B_p^{(\mu)}(t) \tag{4.32}$$

where the $V_p^{(\mu)}$ are again the eigenoperators of $\hat{\mathcal{H}}_0$ as defined by Equation 4.25. In contrast to the semi-classical treatment, the environment is also described by quantum-mechanical operators $B_p^{(\mu)}(t)$ (bath operators). The total Hilbert space is defined as the direct product between the Hilbert space describing the spin system of interest and the Hilbert space describing the environment. We assume that the environment is big enough that it remains always close enough to the thermal equilibrium, even under energy exchange with the spin system. The description of the environment by quantum-mechanical operators instead of classical functions will change the double-commutator properties of Equation 4.31. If we evaluate the double commutators and apply a number of simplifications (see [26], page 12), we arrive at the so-called secular Lindbladian master equation:

$$\hat{\Gamma}\{Q\} = \sum_{\mu,\mu'}\sum_{p,p'}\left(V_p^{\mu}QV_{p'}^{\mu'\dagger} - \frac{1}{2}\left(V_p^{\mu}V_{p'}^{\mu'\dagger}Q + QV_{p'}^{\mu'\dagger}V_p^{\mu}\right)\right)J_{p,p'}^{\mathrm{QM}(\mu,\mu')}(\omega_{p'})\delta(\omega_p,\omega_{p'}). \tag{4.33}$$

This type of super operator is called a Lindbladian dissipator [25, 48] and is sometimes denoted as:

$$\hat{D}\left[A,B\right]Q = AQB - \frac{1}{2}\left(ABQ + QBA\right). \tag{4.34}$$

The quantum-mechanical correlation functions $J_{p,p'}^{\mathrm{QM}(\mu,\mu')}(\omega_{p'})$ are not symmetric with respect to an inversion of the frequency, i.e. $J_{p,p'}^{\mathrm{QM}(\mu,\mu')}(\omega) \neq J_{p,p'}^{\mathrm{QM}(\mu,\mu')}(-\omega)$. Since the quantum-mechanical spectral-density functions are often not known, they are approximated by a classical spectral-density function with a correction term to account for the asymmetry with respect to a sign change of the frequency:

$$J_{p,p'}^{\mathrm{QM}(\mu,\mu')}(\omega) \rightarrow J_{p,p'}^{(\mu,\mu')}(\omega)\exp\left(-\frac{1}{2}\beta_\Theta\omega\right). \tag{4.35}$$

where $\beta_\Theta = \hbar/(k_\mathrm{B}\Theta)$ and $\Theta$ is the temperature of the spin system. Putting this all together, we obtain the secular Lindbladian relaxation super operator in the laboratory frame as:

$$\hat{\Gamma}\{Q\} = \sum_{\mu,\mu'}\sum_{p,p'}\hat{D}[V_p^\mu, V_{p'}^{\mu'^\dagger}]QJ_{p,p'}^{(\mu,\mu')}(\omega_{p'})\exp\left(-\frac{1}{2}\beta_\Theta\omega_{p'}\right)\delta(\omega_p,\omega_{p'}). \tag{4.36}$$

This is the final result for the Lindbladian relaxation super operator and the results looks very similar to the Redfield result of Equation 4.26 except that the double commutator is replaced by the Lindbladian dissipator and the the additional $\exp\left(-\frac{1}{2}\beta_\Theta\omega_{p'}\right)$ term that makes the spectral-density function asymmetric in order to make forward and backward transition rates obey the detailed balance. At first glance, one might be tempted to introduce the thermal-correction term into the Redfield relaxation super operator in order to achieve the proper thermalization. This is, however, not possible because the contributions to the forward and backward rate constants are not properly separated in the double-commutator expression of Equation 4.26 [26].

## 4.3 Relaxation in Spin-1/2 Systems: Dipolar and CSA Relaxation

In this section we will discuss relaxation pathways enabled by dipolar couplings and CSA in a spin-1/2 two-spin system. We will use the Redfield formalism (see Section 4.2.3) [41, 42] since this is still the most commonly used formalism to describe relaxation processes in NMR. For simplicity, we will assume that the motional process is isotropic rotational tumbling leading to a correlation function of the form:

$$\begin{aligned}C_{\ell,\ell',-m,-m'}^{(\mu,\mu')}(\tau) =& \overline{A_{\ell,-m}^{(\mu)}(t)A_{\ell',-m'}^{(\mu')^*}(t-\tau)} \\ =& \delta_{\ell,\ell'}\delta_{m,m'}\frac{3}{2(2\ell+1)}\delta^{(\mu)}\delta^{(\mu')}e^{-\ell(\ell+1)D\tau}P_\ell\left(\cos\left(\Theta^{(\mu,\mu')}\right)\right).\end{aligned} \tag{4.37}$$

We assume a rigid molecule (no internal motion) and that both tensors are axially symmetric and can be characterized by a single parameter, the anisotropy of the tensor $\delta^{(\mu)}$. For the dipolar coupling we have $\delta^{(IS)} = -2\frac{\mu_0}{4\pi}\frac{\gamma_I\gamma_S\hbar}{r_{IS}^3}$ while for the CSA tensor we have $\delta^{(I)} = -\gamma_I\sigma_{zz}^{(I)}$. In the literature, we often find the span of the tensor $\Delta\sigma^{(I)} = \sigma_{zz}^{(I)} - \sigma_{xx}^{(I)} = \frac{3}{2}\sigma_{zz}^{(I)}$ used as a relaxation parameter instead of $\sigma_{zz}^{(I)}$, which leads to a difference in the prefactors of the relaxation-rate constants by a factor $\frac{3}{2}$ or $\frac{9}{4}$. The angle $\Theta^{(\mu,\mu')}$ describes the angle between the z-axes of the two tensors.

The spectral-density function can then be calculated as:

$$\begin{aligned} J^{(\mu,\mu')}_{\ell,\ell,-m,-m}(\omega) &= \int_{-\infty}^{\infty} C^{(\mu,\mu')}_{\ell,\ell,-m,-m}(\tau)e^{-i\omega\tau}\mathrm{d}\tau \\ &= \frac{3}{2}\delta^{(\mu)}\delta^{(\mu')}P_\ell\left(\cos\left(\Theta^{(\mu,\mu')}\right)\right)\frac{2}{2\ell+1}\frac{\tau}{1+(\omega\tau)^2} \\ &= \frac{3}{2}\delta^{(\mu)}\delta^{(\mu')}P_\ell\left(\cos\left(\Theta^{(\mu,\mu')}\right)\right)J_\ell(\omega), \end{aligned} \tag{4.38}$$

which is a Lorentz function described by $J_\ell(\omega)$ with some additional constants depending on the relaxation mechanism considered. For $\mu = \mu'$, $P_\ell(\cos(0)) = 1$ and we obtain an auto-correlated spectral density, for $\mu \neq \mu'$, we obtain a cross-correlated spectral density. For second-rank spatial tensors ($\ell = 2$) we obtain $J_2(\omega) = \frac{2}{5}\frac{\tau}{1+(\omega\tau)^2}$. If the molecule is not rigid but flexible also the internal motion can contribute to relaxation and modify the spectral-density function. Note that in the literature many different definitions for $J(\omega)$ are used (including or excluding the 2/5 pre factor or parts of it) leading to differences in the numerical factors in front of the relaxation-rate constants.

Besides the spatial part, we also have to calculate the double-commutator part $\left[T^{(\mu)}_{\ell,m}, \left[V^{(\mu')^\dagger}_{p'}, Q\right]\right]$ of the relaxation-rate constant that selects the allowed relaxation pathways. For the dipolar coupling, the spherical second-rank spin-tensor operators are given by:

$$\begin{aligned} T^{(IS)}_{2,0} &= \frac{1}{\sqrt{6}}\left(2I_zS_z - \frac{1}{2}(I^+S^- + I^-S^+)\right) \\ T^{(IS)}_{2,\pm1} &= \mp\frac{1}{2}(I^\pm S_z - I_zS^\pm) \\ T^{(IS)}_{2,\pm2} &= \frac{1}{2}I^\pm S^\pm, \end{aligned} \tag{4.39}$$

and in addition, we have to decompose these spherical spin-tensor operators into eigenoperators of $\hat{\mathcal{H}}_0$, which we assume to be the Zeeman Hamiltonian for simplicity. The eigenoperators can be found in Table 4.1 together with the relevant double-commutator expressions required for calculating the allowed relaxation pathways starting from $Q = I_z$. In addition, the eigenfrequencies are given for the different eigenoperators. From Table 4.1 we can see that starting from $I_z$ dipolar relaxation provides only two pathways: (i) auto-relaxation of the $I_z$ operator and (ii) cross relaxation from $I_z$ to $S_z$. No other relaxation pathways are possible.

**Table 4.1** Definition of eigenoperators of the dipolar-coupling Hamiltonian and double commutators for $Q = I_z$.

| $T^{(\mu)}_{\ell,m}$ | $V^{(\mu)}_p$ | $\omega_p$ | $\left[T^{(\mu)}_{\ell,m}, \left[V^{(\mu')^\dagger}_{p'}, Q\right]\right]$ |
|---|---|---|---|
| $T^{(IS)}_{2,0}$ | $\frac{2}{\sqrt{6}}I_zS_z$ | 0 | $\left[T^{(IS)}_{2,0}, \left[\frac{2}{\sqrt{6}}I_zS_z, I_z\right]\right] = 0$ |
| | $-\frac{1}{\sqrt{6}}I^+S^-$ | $\omega_I - \omega_S$ | $\left[T^{(IS)}_{2,0}, \left[-\frac{1}{\sqrt{6}}I^-S^+, I_z\right]\right] = \frac{1}{24}(I_z - S_z)$ |
| | $-\frac{1}{\sqrt{6}}I^-S^+$ | $\omega_S - \omega_I$ | $\left[T^{(IS)}_{2,0}, \left[\frac{1}{\sqrt{6}}I^+S^-, I_z\right]\right] = \frac{1}{24}(I_z - S_z)$ |
| $T^{(IS)}_{2,\pm1}$ | $\mp\frac{1}{2}I^\pm S_z$ | $\omega_I$ | $\left[T^{(IS)}_{2,\pm1}, \left[\mp\frac{1}{2}I^\mp S_z, I_z\right]\right] = \frac{1}{8}I_z$ |
| | $\mp\frac{1}{2}I_zS^\pm$ | $\omega_S$ | $\left[T^{(IS)}_{2,\pm1}, \left[\mp\frac{1}{2}I_zS^\mp, I_z\right]\right] = 0$ |
| $T^{(IS)}_{2,\pm2}$ | $\frac{1}{2}I^\pm S^\pm$ | $\omega_I + \omega_S$ | $\left[T^{(IS)}_{2,\pm2}, \left[\frac{1}{2}I^\mp S^\mp, I_z\right]\right] = \frac{1}{24}(I_z + S_z)$ |

In the same way, we can analyze the relaxation pathways provided by the $I$-spin CSA tensor where the spherical second-rank spin-tensor operators are given by:

$$
\begin{aligned}
T_{2,0}^{(IB)} &= \frac{1}{\sqrt{6}}(2I_zB_0) \\
T_{2,\pm1}^{(IB)} &= \mp\frac{1}{2}(I^{\pm}B_0) \\
T_{2,\pm2}^{(IB)} &= 0.
\end{aligned} \tag{4.40}
$$

In this case the eigenoperators, eigen frequencies, and the double-commutator expressions can be found in Table 4.2 and only a single auto-relaxation pathway of $I_z$ is possible. Cross-correlated relaxation pathways can be calculated in the same way but we have to use in the double commutator the spherical-tensor and eigen operators of two different interactions as shown in Table 4.3. For cross-correlated relaxation between the $I$-spin CSA and the dipolar coupling, we find only cross relaxation from the one-spin $I_z$ operator to the two-spin $2I_zS_z$ and the zero-quantum operators. In a fully analogous way, we can determine all the other allowed relaxation pathways in a two-spin system by using a different operator for $Q$.

Besides the commutator-allowed relaxation pathways, we also have to take the secular approximation for the relaxation super operator into account that generates a block-diagonal structure. Cross-relaxation rate constants between states that have different eigen frequencies under the coherent part of the Hamiltonian will average out to zero if the difference frequency is significantly larger than the cross-relaxation rate constant. This is known as the secular approximation in Redfield theory and leads to a block structure of the relaxation super operator. The structure of the relaxation super operator is illustrated in Figure 4.4 for different situations, i.e. a two-spin system

**Table 4.2** Definition of eigenoperators of the $I$-spin CSA Hamiltonian and double commutators for $Q = I_z$.

| $T_{\ell,m}^{(\mu)}$ | $V_p^{(\mu)}$ | $\omega_p$ | $\left[T_{\ell,m}^{(\mu)}, \left[V_{p'}^{(\mu')\dagger}, Q\right]\right]$ |
|---|---|---|---|
| $T_{2,0}^{(IB)}$ | $\frac{2}{\sqrt{6}}I_zB_0$ | 0 | $\left[T_{2,0}^{(IB)}, \left[\frac{2}{\sqrt{6}}I_zB_0, I_z\right]\right] = 0$ |
| $T_{2,\pm1}^{(IB)}$ | $\mp\frac{1}{2}I^{\pm}B_0$ | $\omega_I$ | $\left[T_{2,\pm1}^{(IB)}, \left[\mp\frac{1}{2}I^{\mp}B_0, I_z\right]\right] = \frac{1}{2}I_zB_0^2$ |

**Table 4.3** Definition of eigenoperators for CSAxDD cross relaxation and double commutators for $Q = I_z$.

| $T_{\ell,m}^{(\mu)}$ | $V_p^{(\mu')}$ | $\omega_p$ | $\left[T_{\ell,m}^{(\mu)}, \left[V_{p'}^{(\mu')\dagger}, Q\right]\right]$ |
|---|---|---|---|
| | $\frac{2}{\sqrt{6}}I_zS_z$ | 0 | $\left[T_{2,0}^{(IB)}, \left[\frac{2}{\sqrt{6}}I_zS_z, I_z\right]\right] = 0$ |
| $T_{2,0}^{(IB)}$ | $-\frac{1}{\sqrt{6}}I^+S^-$ | $\omega_I - \omega_S$ | $\left[T_{2,0}^{(IB)}, \left[-\frac{1}{\sqrt{6}}I^-S^+, I_z\right]\right] = \frac{1}{6}B_0I^-S^+$ |
| | $-\frac{1}{\sqrt{6}}I^-S^+$ | $\omega_S - \omega_I$ | $\left[T_{2,0}^{(IB)}, \left[\frac{1}{\sqrt{6}}I^+S^-, I_z\right]\right] = \frac{1}{6}B_0I^+S^-$ |
| $T_{2,\pm1}^{(IB)}$ | $\mp\frac{1}{2}I^{\pm}S_z$ | $\omega_I$ | $\left[T_{2,\pm1}^{(IB)}, \left[\mp\frac{1}{2}I^{\mp}S_z, I_z\right]\right] = \frac{1}{4}B_02I_zS_z$ |
| | $\mp\frac{1}{2}I_zS^{\pm}$ | $\omega_S$ | $\left[T_{2,\pm1}^{(IB)}, \left[\mp\frac{1}{2}I_zS^{\mp}, I_z\right]\right] = 0$ |
| $T_{2,0}^{(IS)}$ | $\frac{2}{\sqrt{6}}I_zB_0$ | 0 | $\left[T_{2,0}^{(IS)}, \left[\frac{2}{\sqrt{6}}I_zB_0, I_z\right]\right] = 0$ |
| $T_{2,\pm1}^{(IS)}$ | $\mp\frac{1}{2}I^{\pm}B_0$ | $\omega_I$ | $\left[T_{2,\pm1}^{(IS)}, \left[\mp\frac{1}{2}I^{\mp}B_0, I_z\right]\right] = \frac{1}{4}B_0\left(2I_zS_z - I^{\pm}S^{\mp}\right)$ |

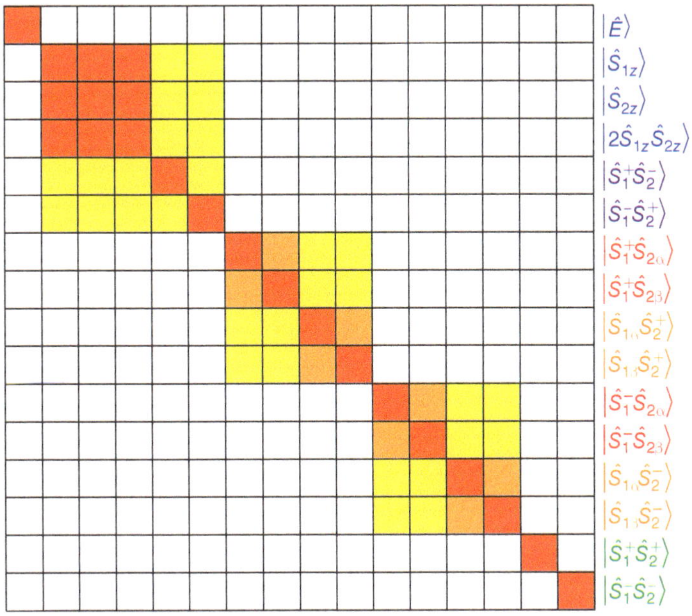


**Figure 4.4** Structure of the Redfield relaxation matrix for a spin-$\frac{1}{2}$ two-spin system under different conditions. (i) Red elements: heteronuclear spin system with $J$ coupling or homonuclear spin system with (large) chemical-shift difference and $J$ coupling. The only non-diagonal part are the populations that form a 3x3 sub block. This structure of the relaxation matrix is often called a Redfield kite. (ii) Orange elements: Heteronuclear spin system without $J$ coupling or homonuclear spin system with (large) chemical-shift difference but without $J$ coupling. In this case all single-quantum coherences are part of a 2x2 sub block. (iii) Yellow elements: homonuclear spin system with degenerate chemical shifts. All single-quantum coherences are now part of a 4x4 sub block and the populations and the zero-quantum coherences form a 5x5 sub block. The Liouville-state basis functions on the right are color coded for populations (blue), zero-quantum coherences (purple), plus-one-quantum coherences (red), minus-one-quantum coherences (orange), and double-quantum coherences (green).

with chemical-shift differences with $J$ couplings (red) and without $J$ couplings (orange) and a two-spin system with degenerate chemical shifts (yellow). The block structure of the Redfield relaxation matrix changes and the size of the sub blocks increase. Because of the block-diagonal structure of the Redfield matrix (red) with a 3×3 subblock for the populations and 1×1 blocks for all the coherences it is often called a Redfield kite.

### 4.3.1 Longitudinal Relaxation in a Two-spin System

From the structure of the Redfield relaxation matrix as shown in Figure 4.4, we see that the populations form a 3x3 sub block of coupled operators and in the case of a homonuclear spin system with degenerate chemical shifts, even a 5x5 sub block spanning not only the populations but also the zero-quantum coherences. Table 4.4 shows the commutator-allowed relaxation pathways in the longitudinal sub block of the Redfield matrix and the relaxation mechanisms that allow the various relaxation pathways. Two auto-correlated (dipolar and CSA relaxation) and three cross-correlated (dipolar/CSA and CSA/CSA) relaxation mechanisms are possible in such a spin system. Of course, as discussed above, the zero-quantum terms are only connected to the populations if the difference frequency of the two nuclei is small compared to the cross-relaxation rate constants, i.e. for spin systems that have degenerate or near-degenerate chemical shifts.

As we have seen in the previous section, we can use the double commutators to determine relaxation pathways. Based on Equation 4.22 and using the values from Tables 4.1 and 4.2, we can calculate the auto-relaxation-rate constant of the $I_z$ operator and the cross-relaxation rate constant between $I_z$ and $S_z$.

**Table 4.4** Allowed relaxation pathways in the longitudinal sub block of the Redfield relaxation matrix.

| | $I_z$ | $S_z$ | $2I_zS_z$ | $I^{\pm}S^{\mp}$ |
|---|---|---|---|---|
| $I_z$ | dipolar, CSA | dipolar | dipolar ⊗ CSA | dipolar ⊗ CSA |
| $S_z$ | | dipolar, CSA | dipolar ⊗ CSA | dipolar ⊗ CSA |
| $2I_zS_z$ | | | dipolar, CSA | dipolar, $CSA_I \otimes CSA_S$ |
| $I^{\pm}S^{\mp}$ | | | | dipolar, CSA, $CSA_I \otimes CSA_S$ |

$$
\begin{aligned}
\hat{\Gamma}\{I_z\} \approx& \frac{1}{2}\sum_{\substack{m=-2\\m'=-2}}^{2,2}\sum_{p'} J^{(IS,IS)}_{2,2,-m,-m'}(\omega_{p'})\left[T^{(IS)}_{2,m},\left[V^{(IS)\dagger}_{p'},I_z\right]\right]\\
&+\frac{1}{2}\sum_{\substack{m=-2\\m'=-2}}^{2,2}\sum_{p'} J^{(IB,IB)}_{2,2,-m,-m'}(\omega_{p'})\left[T^{(IB)}_{2,m},\left[V^{(IB)\dagger}_{p'},I_z\right]\right]\\
=&\frac{1}{2}\frac{3}{2}(\delta^{(IS)})^2\left(2\frac{1}{24}(I_z-S_z)J(\omega_I-\omega_S)+2\frac{1}{8}I_zJ(\omega_I)+2\frac{1}{24}(I_z+S_z)J(\omega_I+\omega_S)\right)\\
&+\frac{1}{2}\frac{3}{2}(\delta^{(I)})^2\left(2\frac{1}{2}I_zB_0^2J(\omega_I)\right)\\
=&\left(\frac{\delta^{(IS)}}{4}\right)^2((I_z-S_z)J(\omega_I-\omega_S)+3I_zJ(\omega_I)+6(I_z+S_z)J(\omega_I+\omega_S))\\
&+\frac{3}{4}\left(\delta^{(I)}B_0\right)^2 I_zJ(\omega_I)
\end{aligned}
\tag{4.41}
$$

To calculate the relaxation-rate constants, we take the trace with $I_z$ and $S_z$ over Equation 4.41 and obtain for the auto-relaxation-rate constant:

$$
\begin{aligned}
\Gamma_{I_z,I_z} = \frac{Tr\left(I_z\hat{\Gamma}\{I_z\}\right)}{Tr\left(I_zI_z\right)} =& \left(\frac{\delta^{(IS)}}{4}\right)^2(J(\omega_I-\omega_S)+3J(\omega_I)+6J(\omega_I+\omega_S))\\
&+\frac{3}{4}\left(\delta^{(I)}B_0\right)^2 J(\omega_I),
\end{aligned}
\tag{4.42}
$$

and for the cross-relaxation rate constant,

$$
\Gamma_{I_z,S_z} = \frac{Tr\left(S_z\hat{\Gamma}\{I_z\}\right)}{Tr\left(S_zS_z\right)} = \left(\frac{\delta^{(IS)}}{4}\right)^2(-J(\omega_I-\omega_S)+6J(\omega_I+\omega_S)).
\tag{4.43}
$$

Note that we will be using consistently the anisotropy ($\delta^{(\mu)}$) of the anisotropic interactions and not "dipolar couplings" or similar terms that are less well defined.

In a similar way, we can calculate the other auto- and cross-relaxation rate constants in the sub block of the populations, which leads to a set of eleven relaxation-rate constants that define the matrix elements in the 5×5 sub block:

$$\Gamma_{\text{pop}} = \begin{pmatrix} \Gamma_{I_z,I_z} & \Gamma_{I_z,S_z} & \Gamma_{I_z,2I_zS_z} & \Gamma_{I_z,I^+S^-} & \Gamma_{I_z,I^-S^+} \\ \Gamma_{S_z,I_z} & \Gamma_{S_z,S_z} & \Gamma_{S_z,2I_zS_z} & \Gamma_{S_z,I^+S^-} & \Gamma_{S_z,I^-S^+} \\ \Gamma_{2I_zS_z,I_z} & \Gamma_{2I_zS_z,S_z} & \Gamma_{2I_zS_z,2I_zS_z} & \Gamma_{2I_zS_z,I^+S^-} & \Gamma_{2I_zS_z,I^-S^+} \\ \Gamma_{I^+S^-,I_z} & \Gamma_{I^+S^-,S_z} & \Gamma_{I^+S^-,2I_zS_z} & \Gamma_{I^+S^-,I^+S^-} & \Gamma_{I^+S^-,I^-S^+} \\ \Gamma_{I^-S^+,I_z} & \Gamma_{I^-S^+,S_z} & \Gamma_{I^-S^+,2I_zS_z} & \Gamma_{I^-S^+,I^+S^-} & \Gamma_{I^-S^+,I^-S^+} \end{pmatrix}. \tag{4.44}$$

In general, the time evolution in this sub block is determined by solving the differential equation:

$$\frac{d}{dt}\begin{pmatrix} \langle I_z\rangle(t) \\ \langle S_z\rangle(t) \\ \langle 2I_zS_z\rangle(t) \\ \langle I^+S^-\rangle(t) \\ \langle I^-S^+\rangle(t) \end{pmatrix} = -\Gamma_{\text{pop}}\begin{pmatrix} \langle I_z\rangle(t) - \langle I_z\rangle_{\text{eq}} \\ \langle S_z\rangle(t) - \langle S_z\rangle_{\text{eq}} \\ \langle 2I_zS_z\rangle(t) - \langle 2I_zS_z\rangle_{\text{eq}} \\ \langle I^+S^-\rangle(t) \\ \langle I^-S^+\rangle(t) \end{pmatrix}, \tag{4.45}$$

which results in five coupled relaxation modes and in principle, in a time dependence of the five operators that is the sum of five exponentials. The matrix elements of the relaxation super operator of Equation 4.43 are given by:

$$\begin{aligned} \Gamma_{I_z,I_z} &= \left(\frac{\delta^{(IS)}}{4}\right)^2 (J(\omega_I - \omega_S) + 3J(\omega_I) + 6J(\omega_I + \omega_S)) + \frac{3}{4}\left(\delta^{(I)}B_0\right)^2 J(\omega_I) \\ \Gamma_{I_z,S_z} &= \left(\frac{\delta^{(IS)}}{4}\right)^2 (-J(\omega_I - \omega_S) + 6J(\omega_I + \omega_S)). \\ \Gamma_{S_z,S_z} &= \left(\frac{\delta^{(IS)}}{4}\right)^2 (J(\omega_I - \omega_S) + 3J(\omega_S) + 6J(\omega_I + \omega_S)) + \frac{3}{4}\left(\delta^{(S)}B_0\right)^2 J(\omega_S). \end{aligned} \tag{4.46}$$

The two-spin population term has auto- and cross-relaxation rate constants given by:

$$\begin{aligned} \Gamma_{I_z,2I_zS_z} &= \frac{3}{4}\delta^{(IS)}\delta^{(I)}B_0 J^{(IS,I)}(\omega_I) \\ \Gamma_{S_z,2I_zS_z} &= \frac{3}{4}\delta^{(IS)}\delta^{(S)}B_0 J^{(IS,S)}(\omega_S) \\ \Gamma_{2I_zS_z,2I_zS_z} &= \left(\frac{\delta^{(IS)}}{4}\right)^2 (3J(\omega_I) + 3J(\omega_S)) \\ &+ \frac{3}{4}\left(\delta^{(I)}B_0\right)^2 J(\omega_I) + \frac{3}{4}\left(\delta^{(S)}B_0\right)^2 J(\omega_S), \end{aligned} \tag{4.47}$$

and the auto- and cross-relaxation rate constants of the zero-quantum coherences are defined by the matrix elements:

$$
\begin{aligned}
\Gamma_{I_z,I^\pm S^\mp} &= -\frac{1}{4}\delta^{(IS)}\delta^{(I)}B_0J^{(IS,I)}(0) + \frac{1}{4}\delta^{(IS)}\delta^{(S)}B_0J^{(IS,S)}(0) \\
&\quad -\frac{3}{8}\left(\delta^{(IS)}\delta^{(I)}B_0J^{(IS,I)}(\omega_I) + \delta^{(IS)}\delta^{(S)}B_0J^{(IS,I)}(\omega_I)\right) \\
\Gamma_{S_z,I^\pm S^\mp} &= -\frac{1}{4}\delta^{(IS)}\delta^{(S)}B_0J^{(IS,S)}(0) + \frac{1}{4}\delta^{(IS)}\delta^{(I)}B_0J^{(IS,I)}(0) \\
&\quad -\frac{3}{8}\left(\delta^{(IS)}\delta^{(I)}B_0J^{(IS,I)}(\omega_S) + \delta^{(IS)}\delta^{(S)}B_0J^{(IS,I)}(\omega_S)\right) \\
\Gamma_{2I_zS_z,I^\pm S^\mp} &= -\frac{3}{2}\left(\frac{\delta^{(IS)}}{4}\right)^2 \left(J(\omega_I) + J(\omega_S)\right) \\
&\quad -\frac{3}{8}\delta^{(I)}\delta^{(S)}B_0^2\left(J^{(I,S)}(\omega_I) + J^{(I,S)}(\omega_S)\right) \\
\Gamma_{I^\pm S^\mp,I^\pm S^\mp} &= \left(\frac{\delta^{(IS)}}{4}\right)^2 \left(J(\omega_I-\omega_S) + \frac{3}{2}J(\omega_S) + \frac{3}{2}J(\omega_I)\right) \\
&\quad -\delta^{(I)}\delta^{(S)}B_0^2J^{(I,S)}(0) + \frac{1}{8}\left(\delta^{(S)}B_0\right)^2\left(4J(0) + 3J(\omega_S)\right) \\
&\quad +\frac{1}{8}\left(\delta^{(I)}B_0\right)^2\left(4J(0) + 3J(\omega_I)\right) \\
\Gamma_{I^\pm S^\mp,I^\mp S^\pm} &= -\left(\frac{\delta^{(IS)}}{4}\right)^2 J(\omega_I-\omega_S).
\end{aligned}
\tag{4.48}
$$

To discuss longitudinal relaxation in a two-spin system, we have to distinguish three different cases: (i) a heteronuclear spin system (AX spin system) where the zero-quantum terms are non-secular and the matrix has a 3x3 block structure (see Figure 4.4). (ii) A homonuclear spin system with a large chemical-difference (AX spin system) where the zero-quantum terms are also non-secular and the longitudinal relaxation is described by a 3×3 sub block. Compared to the heteronuclear spin system, the sampling of the spectral-density functions is different. (iii) A homonuclear spin system with degenerate chemical shifts ($A_2$ spin system) where the longitudinal relaxation is described by the full 5×5 matrix as described above. There is, of course, a continuous transition from the AX to the $A_2$ spin system through the intermediate cases (AB spin system) but these will not be discussed here [67–69].

#### 4.3.1.1 Heteronuclear Two-spin System

In a heteronuclear two-spin system, longitudinal relaxation is described by a 3×3 relaxation matrix that couples the operators $I_z$, $S_z$, and $2I_zS_z$. The time evolution in such a system is described by the differential equation:

$$
\begin{aligned}
\frac{d}{dt}\begin{pmatrix} \langle I_z\rangle(t) \\ \langle S_z\rangle(t) \\ \langle 2I_zS_z\rangle(t) \end{pmatrix} &= -\Gamma_{\text{pop}}\begin{pmatrix} \langle I_z\rangle(t) - \langle I_z\rangle_{\text{eq}} \\ \langle S_z\rangle(t) - \langle S_z\rangle_{\text{eq}} \\ \langle 2I_zS_z\rangle(t) - \langle 2I_zS_z\rangle_{\text{eq}} \end{pmatrix} \\
&= -\begin{pmatrix} \Gamma_{I_z,I_z} & \Gamma_{I_z,S_z} & \Gamma_{I_z,2I_zS_z} \\ \Gamma_{S_z,I_z} & \Gamma_{S_z,S_z} & \Gamma_{S_z,2I_zS_z} \\ \Gamma_{2I_zS_z,I_z} & \Gamma_{2I_zS_z,S_z} & \Gamma_{2I_zS_z,2I_zS_z} \end{pmatrix}\begin{pmatrix} \langle I_z\rangle(t) - \langle I_z\rangle_{\text{eq}} \\ \langle S_z\rangle(t) - \langle S_z\rangle_{\text{eq}} \\ \langle 2I_zS_z\rangle(t) - \langle 2I_zS_z\rangle_{\text{eq}} \end{pmatrix}.
\end{aligned}
\tag{4.49}
$$

In the general case, we will have three operators that are coupled leading to a time evolution of the magnetization that is characterized by three exponential functions. Formally, we can calculate the solution as:

$$\begin{pmatrix} \langle I_z\rangle(t) \\ \langle S_z\rangle(t) \\ \langle 2I_zS_z\rangle(t) \end{pmatrix} = e^{-\Gamma_{\text{pop}}t} \begin{pmatrix} \langle I_z\rangle(0) - \langle I_z\rangle_{\text{eq}} \\ \langle S_z\rangle(0) - \langle S_z\rangle_{\text{eq}} \\ \langle 2I_zS_z\rangle(0) - \langle 2I_zS_z\rangle_{\text{eq}} \end{pmatrix} + \begin{pmatrix} \langle I_z\rangle_{\text{eq}} \\ \langle S_z\rangle_{\text{eq}} \\ \langle 2I_zS_z\rangle_{\text{eq}} \end{pmatrix}. \tag{4.50}$$

To calculate the time evolution, we have to calculate the matrix exponential, which requires diagonalization of the relaxation matrix. This is analytically possible for a 3x3 matrix but gives quite complex expressions containing all six independent parameters of the relaxation matrix. Fitting such a multi-exponential time evolution is, in the presence of noise in experimental data, often not very stable. Therefore, one tries to do experiments that generate mono-exponential decays to simplify data evaluation. In this case, the typical modification is the saturation of one of the two-spin species during the relaxation delay, i.e. one irradiates the $I$ spins (protons) while one measure the return of the $S$ spins (carbon/nitrogen) to thermal equilibrium or vice versa. Complete saturation leads to $\langle I_z\rangle(t) = 0$ and $\langle 2I_zS_z\rangle(t) = 0$ resulting in a much simpler version of Equation 4.49 since we no longer have coupled differential equations and we obtain:

$$\frac{d}{dt}\langle S_z\rangle(t) = -\Gamma_{S_z,S_z}\left(\langle S_z\rangle(t) - \langle S_z\rangle_{\text{eq}}\right) + \Gamma_{S_z,I_z}\langle I_z\rangle_{\text{eq}} + \Gamma_{S_z,2I_zS_z}\langle 2I_zS_z\rangle_{\text{eq}}. \tag{4.51}$$

If we neglect the thermal-equilibrium value of the two-spin term because it is very small, we obtain a further simplified equation:

$$\frac{d}{dt}\langle S_z\rangle(t) = -\Gamma_{S_z,S_z}\left(\langle S_z\rangle(t) - \langle S_z\rangle_{\text{eq}}\right) + \Gamma_{S_z,I_z}\langle I_z\rangle_{\text{eq}} \tag{4.52}$$

with the mono-exponential solution:

$$\langle S_z\rangle(t) = \left(\langle S_z\rangle(0) - \langle S_z\rangle_{\text{SS}}\right) e^{-\Gamma_{S_z,S_z}t} + \langle S_z\rangle_{\text{SS}}. \tag{4.53}$$

In this case, the steady-state NOE value $\langle S_z\rangle_{\text{SS}}$ (assuming complete saturation and equilibration already before the start of the measurement) can be obtained by setting $\frac{d}{dt}\langle S_z\rangle(t) = 0$ leading to:

$$\begin{aligned} \langle S_z\rangle_{\text{ss}} &= \langle S_z\rangle_{\text{eq}} + \frac{\Gamma_{S_z,I_z}}{\Gamma_{S_z,S_z}}\langle I_z\rangle_{\text{eq}} = \langle S_z\rangle_{\text{eq}}\left(1 + \frac{\Gamma_{S_z,I_z}\langle I_z\rangle_{\text{eq}}}{\Gamma_{S_z,S_z}\langle S_z\rangle_{\text{eq}}}\right) \\ &= \langle S_z\rangle_{\text{eq}}(1+\eta). \end{aligned} \tag{4.54}$$

Measuring the steady-state NOE, $T_{1S} = 1/\Gamma_{S_z,S_z}$ and $T_{1I} = 1/\Gamma_{I_z,I_z}$ allows us to characterize the three most important relaxation-rate constants that determine the behavior of longitudinal relaxation in a two-spin system. The cross-correlated cross-relaxation rate constants to the two-spin term ($2I_zS_z$) as well as the auto-relaxation-rate constant of the two-spin term are more difficult to determine. In principle, we can use an INEPT experiment [70] to generate an initial state $2I_zS_z$ and then measure the cross-relaxation induced generation of $I_z$ or $S_z$ magnetization or use an INEPT back conversion step to measure the decay of the two-spin term. However, these measurements will lead to multi-exponential signals and depend on all six relaxation-rate constants in the 3x3 matrix. Since three of them can be measured independently (vide supra), data evaluation requires a simultaneous fit of the remaining three rate constants to the experimental measured data. Therefore, determining these rate constants is more complex than measuring the auto- and cross-relaxation rate constants of the two one-spin terms. Very often, cross-correlated cross relaxation in a two-spin system is neglected, and the complete relaxation matrix approximated by the two coupled modes $I_z$ and $S_z$.

We will now discuss how these relaxation-rate constants depend on the correlation time of the molecule and the static magnetic field $B_0$ that determines the values of the frequencies at which the spectral-density functions are sampled. Figure 4.5 shows the $T_1$ relaxation time of an isolated C–H two-spin system ($r_{\mathrm{CH}} = 1.09$ Å) as a function of the rotational correlation time $\tau_c$ and the static magnetic field $B_0$ for dipolar relaxation. We see a typical $T_1$ minimum for a correlation time in the nanosecond range with longer $T_1$ times for shorter or longer correlation times. The location of the $T_1$ minimum coincides roughly with a value of $\omega_S\tau_c = 1$. The dependence on the static magnetic field is less pronounced with an almost flat line for small molecules (short correlation times) but some field dependence for larger molecules with longer correlation times.

The same set of plots for CSA-induced longitudinal relaxation ($\delta^{(S)} = 50$ ppm) is shown in Figure 4.6. We see a very similar behavior of $T_1$ as a function of the rotational correlation time. The field dependence is different since now also the magnitude of the interaction itself depends on $B_0$. The relaxation-rate constant will scale with the square of the magnitude of the CSA tensor, i.e. doubling the CSA tensor will quadruple the rate constant. For large molecules (long correlation times), $T_1$ is almost independent of the field while for smaller molecules (shorter correlation times) there is a strong dependence of $T_1$ on $B_0$. Different relaxation mechanisms are always additive in the rate constant. For a C–H two-spin pair, the contribution of dipolar relaxation typically dominates

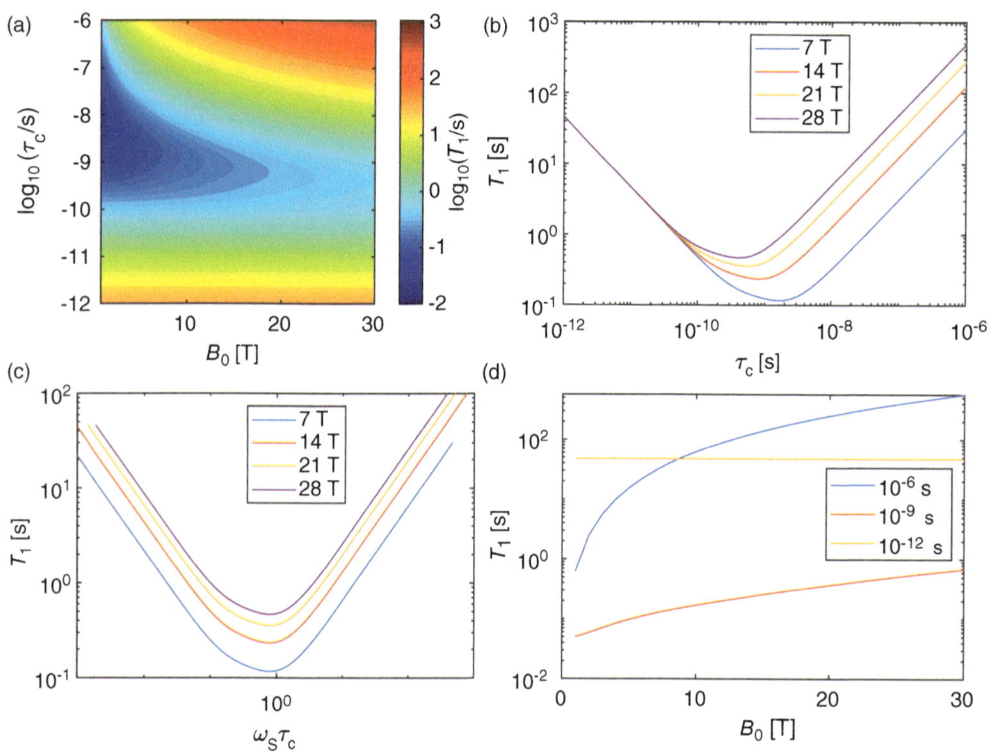


**Figure 4.5** (a) The dipolar-coupling induced $^{13}$C longitudinal relaxation time $T_1 = 1/\Gamma_{S_z,S_z}$ for a C–H two-spin system ($r_{\mathrm{CH}} = 1.09$ Å) as a function of the static magnetic field ($B_0$) and the rotational correlation time ($\tau_c$) of the molecular tumbling. One can clearly see that the $T_1$ time has a minimum for $\tau_c$ in the range of nanoseconds and is longer for shorter or longer correlation times. The $T_1$ minimum can also be seen in (b) and (c) where four slices corresponding to $B_0$ = 7, 14, 21, and 28 T are plotted as a function of the correlation time $\tau_c$ (b) and of $\omega_S\tau_c$ (c). The minimum is always roughly at $\omega_S\tau_c = 1$. The dependence on the static magnetic field $B_0$ is shown in (d) for correlation times of 1 μs, 1 ns, and 1 ps. For small molecules with short correlation times, we see almost no field dependence in the range of typical NMR frequencies while for larger molecules with longer correlation times a clear field dependence is observed.

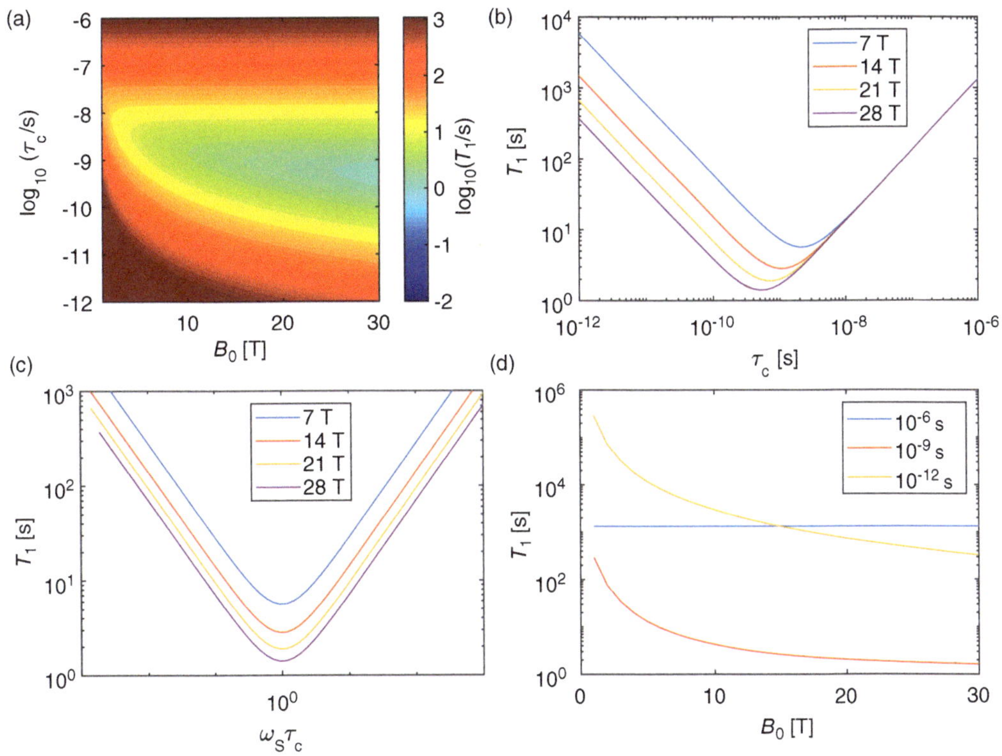


**Figure 4.6** (a) The CSA induced $^{13}$C ($\delta^{(S)}$ = 50 ppm) longitudinal relaxation time $T_1 = 1/\Gamma_{S_z,S_z}$ as a function of the static magnetic field ($B_0$) and the rotational correlation time ($\tau_c$) of the molecular tumbling. One can clearly see that the $T_1$ time has a minimum for $\tau_c$ in the range of nanoseconds and is longer for shorter or longer correlation times. The $T_1$ minimum can also be seen in (b) and (c) where four slices corresponding to $B_0$ = 7, 14, 21, and 28 T are plotted as a function of the correlation time $\tau_c$ (b) and of $\omega_S\tau_c$ (c). The minimum is always roughly at $\omega_S\tau_c = 1$. The dependence on the static magnetic field $B_0$ is shown in (d) for correlation times of 1 μs, 1 ns, and 1 ps. For large molecules with long correlation times, we see almost no field dependence in the range of typical NMR frequencies while for smaller molecules with shorter correlation times a clear field dependence is observed. Note that the absolute values of the rate constant is about one order of magnitude smaller than for dipolar relaxation.

the rate constant unless the CSA tensor is either very large or measurements are carried out at a very high static magnetic field. The situation is of course different for spins without directly bound protons like carbonyls where CSA relaxation can be the dominating mechanism.

The cross-relaxation rate constant $\Gamma_{I_z,S_z}$ is only influenced by dipolar relaxation that mediates the polarization transfer from $I_z$ to $S_z$ and is the source of the nuclear Overhauser effect [54]. The dependence of the cross-relaxation rate constant on the correlation time and the static magnetic field are shown in Figure 4.7. The cross-relaxation rate constant shows a very similar behavior as the dipolar $T_1$ relaxation time (see Figure 4.5). This is not surprising since both sample the spectral-density function at similar frequencies (Equation 4.46). There is a maximum of the rate constant around a correlation time of 1 ns and the field dependence is flat for fast tumbling molecules. The cross-relaxation rate constant determines the rate of polarization transfer and the magnitude of the steady-state NOE (see Equation 4.54). The steady-state NOE is often characterized by the value $\eta$ (see Equation 4.54), which is shown as a function of correlation time and static magnetic field in Figures 4.8 and 4.9 for a C–H and a N–H spin system, respectively. For small molecules (short correlation times), the NOE is field independent and has the maximum value of $\gamma_I/(2\gamma_S)$. Note that the sign of the NOE reflects the relative sign of the two gyromagnetic ratios, i.e. the NOE of $^{13}$C is positive while the NOE of $^{15}$N is negative under proton saturation. For large molecules

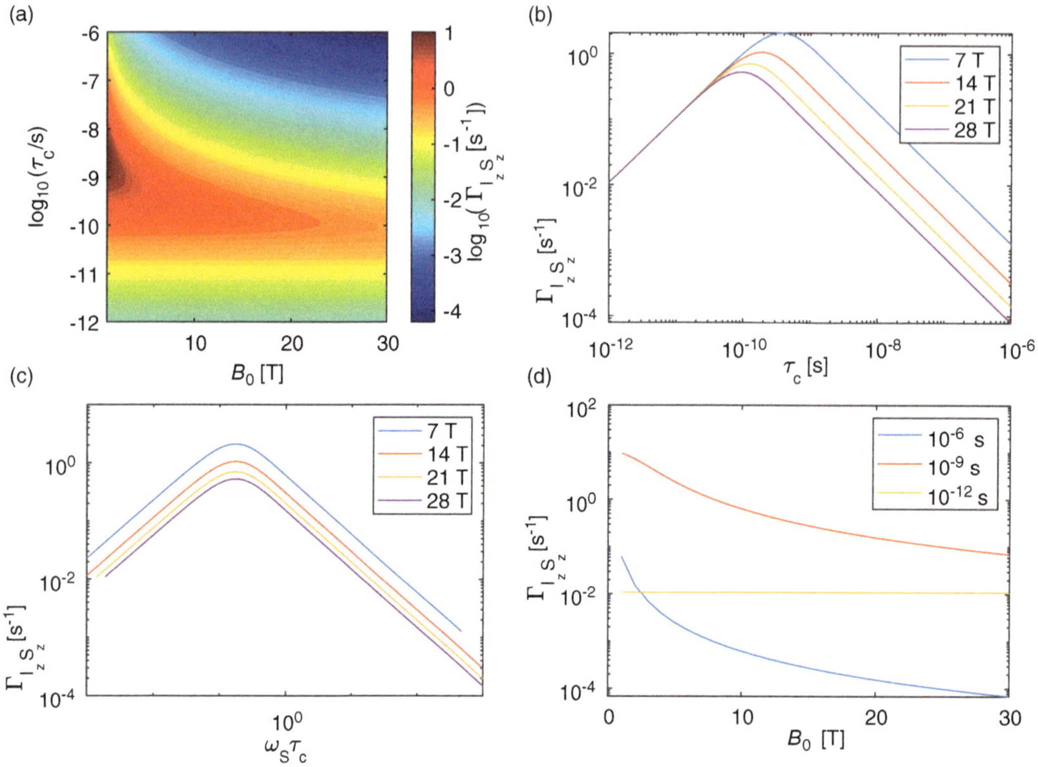


**Figure 4.7** (a) The dipolar-coupling induced $^{13}$C longitudinal cross-relaxation rate constant $\Gamma_{I_z,S_z}$ for a C–H two-spin system ($r_{\mathrm{CH}} = 1.09$ Å) as a function of the static magnetic field ($B_0$) and the rotational correlation time ($\tau_c$) of the molecular tumbling. One can clearly see that the cross-relaxation rate constant has a maximum for $\tau_c$ in the range of nanoseconds and is longer for shorter or longer correlation times. The maximum of $\Gamma_{I_z,S_z}$ can also be seen in (b) and (c) where four slices corresponding to $B_0$ = 7, 14, 21, and 28 T are plotted as a function of the correlation time $\tau_c$ (b) and of $\omega_S\tau_c$ (c). The maximum is always roughly at $\omega_S\tau_c = 1$. The dependence on the static magnetic field $B_0$ is shown in (d) for correlation times of 1 μs, 1 ns, and 1 ps. For small molecules with short correlation times, we see almost no field dependence in the range of typical NMR frequencies while for larger molecules with longer correlation times a clear field dependence is observed.

(long correlation times), the NOE becomes very small but in the transition area around 1 ns it is very sensitive to the correlation time. For small molecules, presaturation has been used to enhance the initial polarization on low-gamma nuclei through the NOE. However, the INEPT experiment [70] is usually the better choice for such a signal enhancement of low-$\gamma$ nuclei.

As discussed above, analysis of relaxation data in two-spin systems can be simplified by the saturation of the passive $I$ spin leading to a mono-exponential decay with the true $S$-spin $T_1$ relaxation time. This is very convenient and makes the determination of relaxation-rate constants in such spin systems simple and straightforward. As a note of caution, we would like to mention that saturation of the passive spins in order to obtain a mono-exponential decay with the true $T_1$ relaxation-rate constant does not work in arbitrary spin systems. Especially spin systems, which contain spins with degenerate chemical shifts, do not allow a complete saturation of all coupled magnetization modes by cw irradiation. In such spin systems, even under saturation, a system of coupled differential equations still exists leading to multi-exponential relaxation [7].

In the heteronuclear spin system, the zero-quantum terms are not coupled to the populations due to the secular approximation and relax mono exponentially. The zero-quantum relaxation time is plotted in Figure 4.10 as a

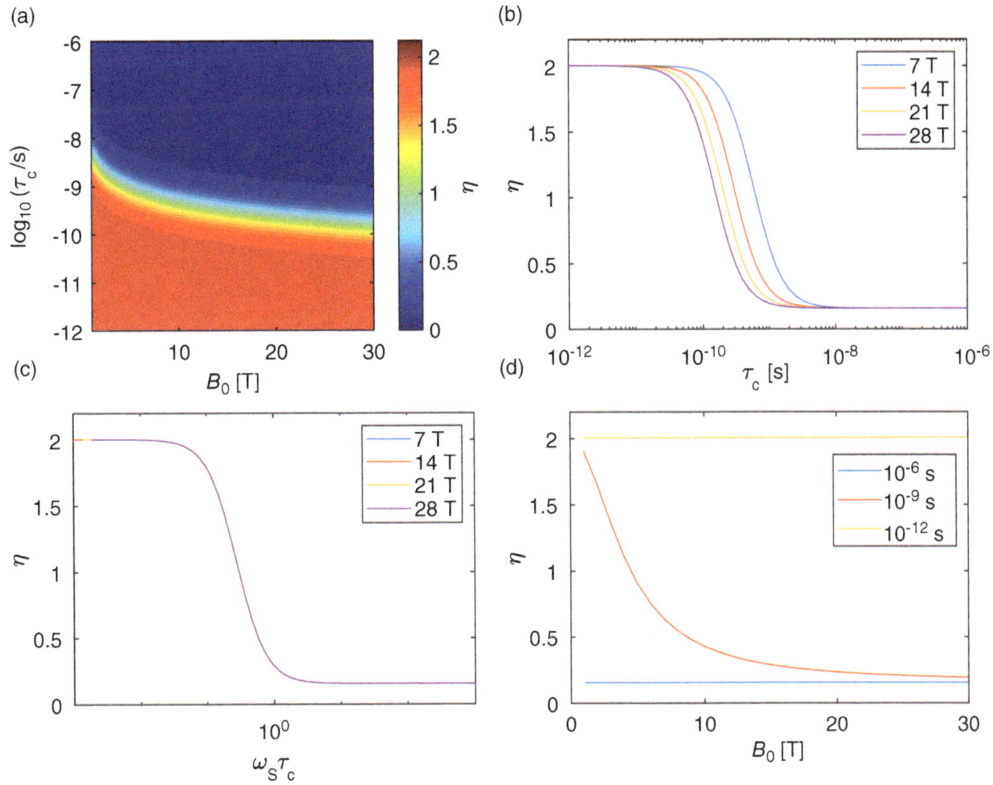


**Figure 4.8** (a) The steady-state NOE for a C–H two-spin system ($r_{CH} = 1.09$Å) as a function of the static magnetic field ($B_0$) and the rotational correlation time ($\tau_c$) of the molecular tumbling. For small molecules (short correlation times), the NOE is maximum and decreases with increasing correlation time. In (b) and (c) four slices of the steady-state NOE corresponding to $B_0$ = 7, 14, 21, and 28 T are plotted as a function of the correlation time $\tau_c$ (b) and of $\omega_S\tau_c$ (c). One can clearly see the transitions are around 1 ns where the NOE is most sensitive to changes in the correlation time. (d) Dependence of the steady-state NOE on the static magnetic field $B_0$ for correlation times of 1 μs, 1 ns, and 1 ps. For small molecules with short correlation times and large molecules with long correlation times, we see almost no field dependence but for intermediate-size molecules, the NOE can change significantly with $B_0$.

function of the static magnetic field and the rotational correlation time for a C–H two-spin system. There is a minimum of the relaxation time and the overall behavior is very similar to the $T_1$ relaxation time.

#### 4.3.1.2 Homonuclear Two-spin System with Chemical-Shift Difference

The homonuclear two-spin system with non-degenerate chemical shifts behaves essentially like a heteronuclear two-spin system and can be described in the same way. The sampling of the spectral-density function changes from $\omega_I - \omega_S$ to 0 and from $\omega_I + \omega_S$ to $2\omega_I$ resulting in a sampling of the spectral-density function at frequencies 0, $\omega_I$, and $2\omega_I$. These changes in the frequencies alter some of the behavior found in a homonuclear two-spin system compared to a heteronuclear spin system. Figure 4.11 shows the behavior of the $T_1$ time (Figure 4.11a and b) and the homonuclear NOE (Figure 4.11c and d) as a function of the static magnetic field $B_0$ and the correlation time $\tau_c$ of the overall tumbling of the molecule. In contrast to the heteronuclear case (Figure 4.5), the $T_1$ time decreases monotonously and has no minimum. This is due to the changes in the frequencies at which the spectral-density function is sampled. The appearance of the $J(0)$ term in the spectral density is responsible for this changed behavior which we will encounter again in transverse relaxation (vide infra).

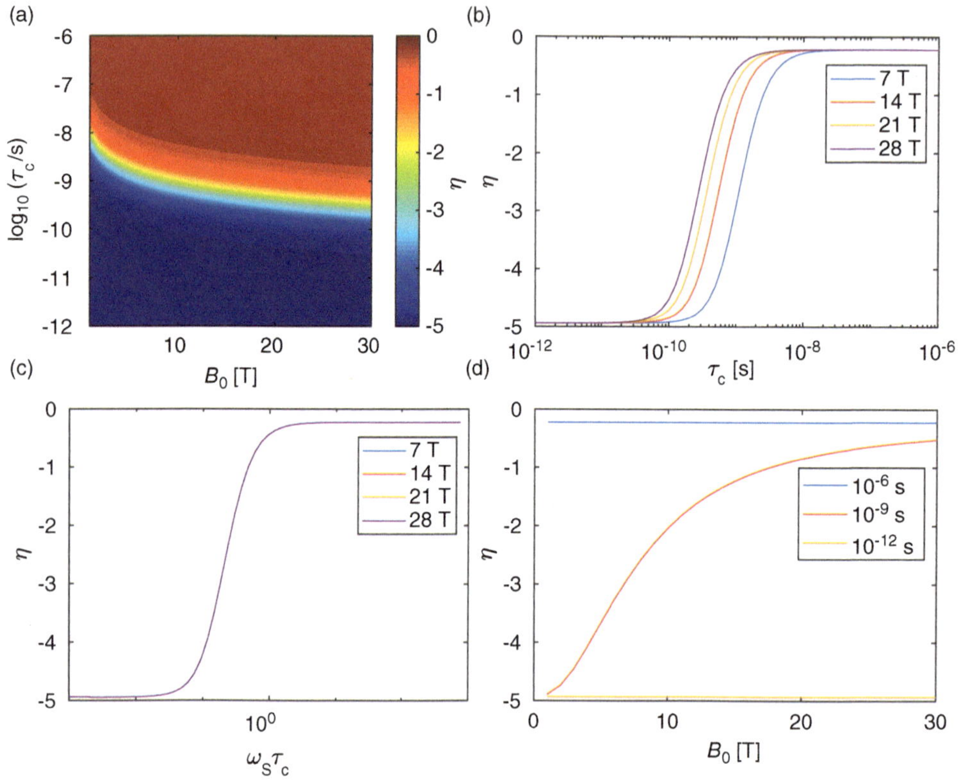


**Figure 4.9** (a) The steady-state NOE for a N–H two-spin system ($r_{NH}$ = 1.05Å) as a function of the static magnetic field ($B_0$) and the rotational correlation time ($\tau_c$) of the molecular tumbling. For small molecules (short correlation times), the NOE is maximum and decreases with increasing correlation time. In (b) and (c) four slices of the steady-state NOE corresponding to $B_0$ = 7, 14, 21, and 28 T are plotted as a function of the correlation time $\tau_c$ (b) and of $\omega_S\tau_c$ (c). One can clearly see the transitions are around 1 ns where the NOE is most sensitive to changes in the correlation time. (d) Dependence of the steady-state NOE on the static magnetic field $B_0$ for correlation times of 1 μs, 1 ns, and 1 ps. For small molecules with short correlation times and large molecules with long correlation times, we see almost no field dependence but for intermediate-size molecules, the NOE can change significantly with $B_0$. Note the negative sign of the steady-state NOE in the case of $^{15}$N due to the difference in sign of the $\gamma$ value compared to $^1$H.

The NOE also changes and shows a sign change and a zero crossing around a correlation time of 1 ns (Figure 4.11c and d). Such a zero crossing can only appear if the two frequencies are almost identical such that the difference-frequency term can dominate the sum-frequency term in the cross-relaxation rate constant (Equation 4.46). As a consequence of the zero crossing of the cross-relaxation rate constant $\Gamma_{I_z,S_z}$, intermediate-size molecules will not show homonuclear cross relaxation and have, therefore, no or very low-intensity cross peaks in two-dimensional NOESY spectra [53]. As an alternative, one can measure cross-relaxation in the rotating frame (ROESY) [54, 71, 72] where the magnetization is spin locked and also intermediate-size molecules will give rise to cross peaks in two-dimensional cross-relaxation spectra in the rotating frame.

#### 4.3.1.3 Homonuclear Two-spin System with Degenerate Chemical Shifts

For a homonuclear two-spin system with degenerate chemical shifts ($A_2$ spin system), the situation is different. First of all, the zero-quantum frequency is zero and the zero-quantum coherences couple to the populations and we have to deal with the full 5×5 matrix of coupled operators as given by Equations 4.44 and 4.45. In addition, we can only manipulate the two spins together, and experimentally, we can only observe the sum magnetization

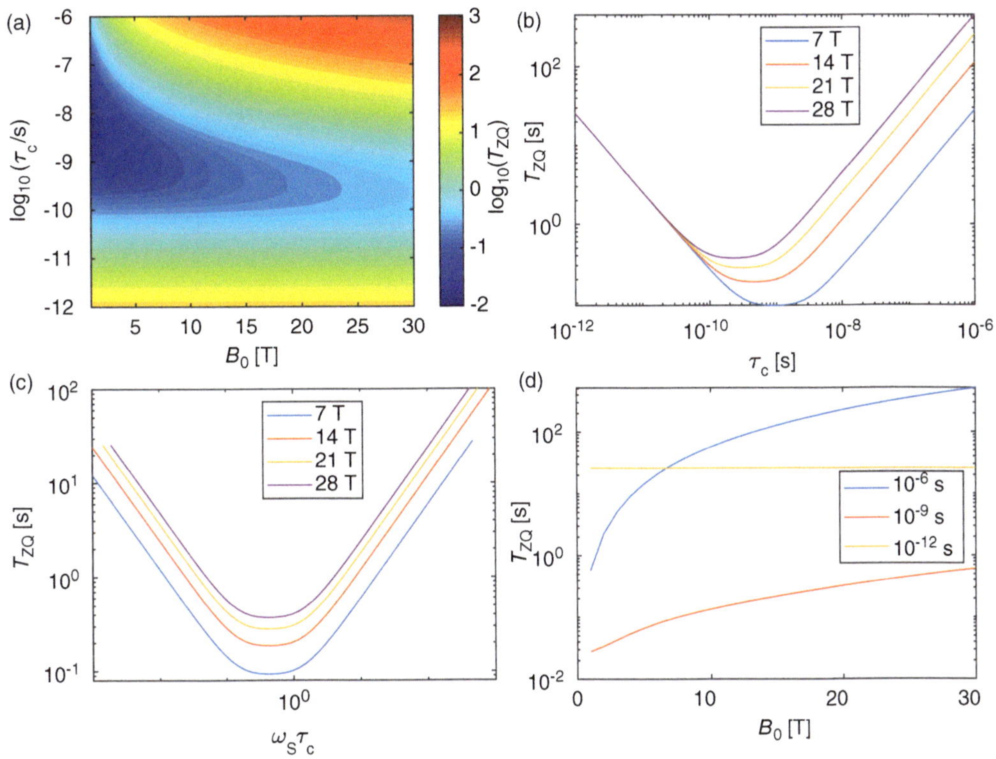


**Figure 4.10** (a) The dipolar-coupling induced zero-quantum relaxation time $T_{ZQ} = 1/\Gamma_{I\pm S\mp, I\pm S\mp}$ for a C–H two-spin system ($r_{CH} = 1.09\text{Å}$) as a function of the static magnetic field ($B_0$) and the rotational correlation time ($\tau_c$) of the molecular tumbling. One can clearly see that the $T_{ZQ}$ time has a minimum for $\tau_c$ in the range of nanoseconds and is longer for shorter or longer correlation times. The $T_{ZQ}$ minimum can also be seen in (b) and (c) where four slices corresponding to $B_0$ = 7, 14, 21, and 28 T are plotted as a function of the correlation time $\tau_c$ (b) and of $\omega_S\tau_c$ (c). The minimum is always roughly at $\omega_S\tau_c = 1$. The dependence on the static magnetic field $B_0$ is shown in (d) for correlation times of 1 μs, 1 ns, and 1 ps. For small molecules with short correlation times, we see almost no field dependence in the range of typical NMR frequencies while for larger molecules with longer correlation times a clear field dependence is observed.

$F_z = I_z + S_z$. This is the only accessible observable since there is no way to generate difference magnetization, two-spin order, or zero-quantum coherences in such a spin system. In an $A_2$ spin system, it is of advantage to describe relaxation using symmetry-adapted spherical-tensor operators. The relevant set of spherical-tensor operators is in this case $T_{1,0}^{(I)}$, $T_{1,0}^{(S)}$, $T_{0,0}^{(IS)}$, $T_{1,0}^{(IS)}$, and $T_{2,0}^{(IS)}$. Since only the sum magnetization is observable, we will use the symmetrized linear combinations $\frac{1}{2\sqrt{2}}(T_{1,0}^{(I)} + T_{1,0}^{(S)})$ and $\frac{1}{2\sqrt{2}}(T_{1,0}^{(I)} - T_{1,0}^{(S)})$. The transformation to the new relaxation modes is described by:

$$\begin{pmatrix} \frac{1}{2\sqrt{2}}(T_{1,0}^{(I)} + T_{1,0}^{(S)}) \\ \frac{1}{2\sqrt{2}}(T_{1,0}^{(I)} - T_{1,0}^{(S)}) \\ T_{0,0}^{(IS)} \\ T_{1,0}^{(IS)} \\ T_{2,0}^{(IS)} \end{pmatrix} = \begin{pmatrix} \frac{1}{2\sqrt{2}} & \frac{1}{2\sqrt{2}} & 0 & 0 & 0 \\ \frac{1}{2\sqrt{2}} & -\frac{1}{2\sqrt{2}} & 0 & 0 & 0 \\ 0 & 0 & -\frac{1}{2\sqrt{3}} & -\frac{1}{2\sqrt{3}} & -\frac{1}{2\sqrt{3}} \\ 0 & 0 & 0 & -\frac{1}{2\sqrt{2}} & \frac{1}{2\sqrt{2}} \\ 0 & 0 & \frac{1}{\sqrt{6}} & -\frac{1}{2\sqrt{6}} & -\frac{1}{2\sqrt{6}} \end{pmatrix} \begin{pmatrix} \langle I_z \rangle \\ \langle S_z \rangle \\ \langle 2I_zS_z \rangle \\ \langle I^+S^- \rangle \\ \langle I^-S^+ \rangle \end{pmatrix}. \tag{4.55}$$

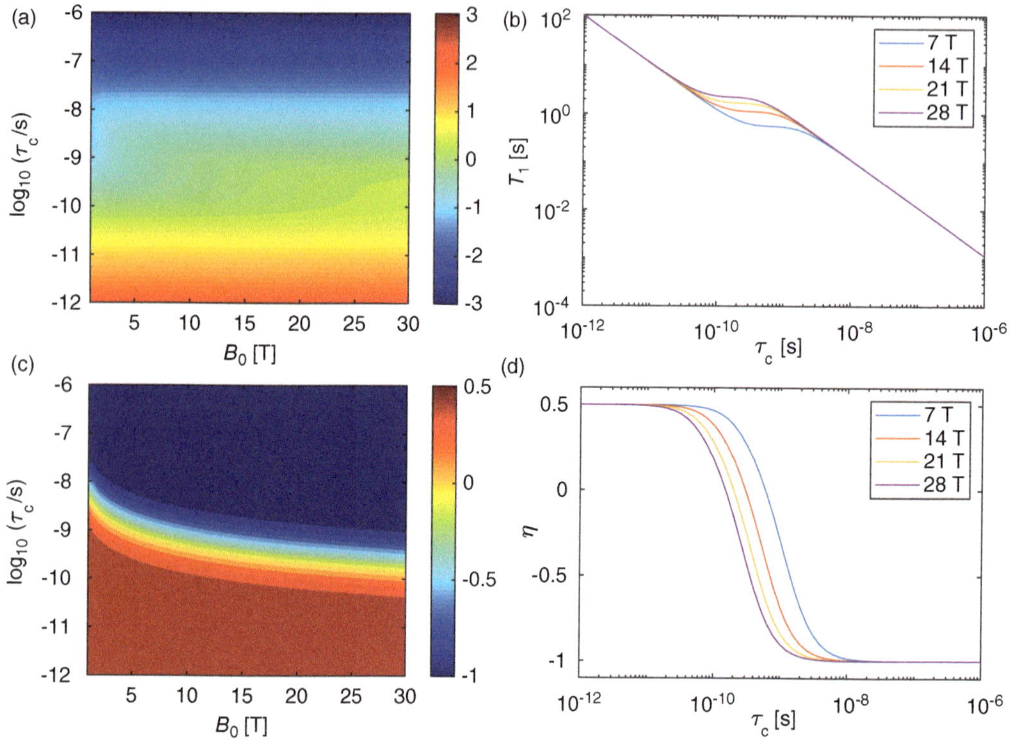


**Figure 4.11** (a) The dipolar-coupling induced $^1H$ longitudinal relaxation time $T_1 = 1/\Gamma_{I_z,I_z}$ for a H–H two-spin system ($r_{HH} = 1.26$Å) as a function of the static magnetic field ($B_0$) and the rotational correlation time ($\tau_c$) of the molecular tumbling. One can clearly see that the $T_1$ time has in this case no minimum and decreases monotonously. In (b) four slices corresponding to $B_0$ = 7, 14, 21, and 28 T are plotted as a function of the correlation time $\tau_c$ where the monotonous decrease of $T_1$ is clearly visible. (c) The steady-state NOE for a H–H two-spin system ($r_{HH} = 1.26$Å) as a function of the static magnetic field ($B_0$) and the rotational correlation time ($\tau_c$) of the molecular tumbling. For small molecules (short correlation times), the NOE is positive while for large molecules (long correlation time) the NOE is negative. In between, the NOE crosses zero around a correlation time of 1 ns, which explains why we do not see $^1H$-$^1H$ cross relaxation in intermediate-size molecules. In (d) four slices of the steady-state NOE corresponding to $B_0$ = 7, 14, 21, and 28 T are plotted as a function of the correlation time $\tau_c$. One can clearly see the zero crossing of the NOE around 1 ns.

We can use the transformation matrix of Equation 4.55 to transform the relaxation matrix of Equation 4.45 for longitudinal relaxation into the new basis. In the general case, we obtain a complete matrix that couples all five modes. If we neglect cross-correlated relaxation, however, the sum and difference magnetization are separated from the other modes and we obtain:

$$\frac{d}{dt}\begin{pmatrix}\langle I_z + S_z\rangle(t)\\ \langle I_z - S_z\rangle(t)\end{pmatrix} = -\frac{1}{2}\begin{pmatrix}2\Gamma_{I_z,I_z} + 2\Gamma I_z, S_z & 0\\ 0 & 2\Gamma_{I_z,I_z} - 2\Gamma I_z, S_z\end{pmatrix}\begin{pmatrix}\langle I_z + S_z\rangle(t)\\ \langle I_z - S_z\rangle(t)\end{pmatrix} \tag{4.56}$$

where we have assumed that the $T_1$ relaxation-rate constants of the $I$ and $S$ spin are the same since they are equivalent. Therefore, we will obtain a mono-exponential decay of the sum polarization assuming that the cross-correlated relaxation can be neglected.

Recently, the relaxation behavior in $A_2$ spin systems, has seen renewed interest in the context of long-lived spin states [73–76] that have lifetimes far exceeding the $T_1$ time of populations. Such long-lived states can be realized in the $T^{(IS)}_{0,0}$ operators of an $A_2$ spin system, which corresponds to the singlet state $|S_0\rangle$ (except for a part proportional to the identity operator) in a symmetry-adapted basis. Such states can be used to store magnetization for a longer time span than $T_1$, which is especially interesting in hyperpolarization applications where infrequently large amounts of polarization are generated. Using such a singlet state as a polarization storage system is hindered by the fact that they are inaccessible in a perfect $A_2$ spin system. Different strategies have been developed to interconvert almost degenerate AB spin systems into $A_2$ spin systems by changes in the static magnetic field, the use of RF irradiation, or changes in the chemical structure of the molecule [75, 76].

### 4.3.2 Transverse Relaxation in a Two-spin System

From the block structure of the Redfield matrix (Figure 4.4), we see that we have a maximum block size of the transverse single-quantum coherences of 4×4 and most of the time 2×2. Typically, transverse relaxation-rate constants are not given in the single-transition basis used in Figure 4.4 but for in-phase ($S^{\pm}$) and anti-phase ($2S^{\pm}I_z$) coherences. We can again obtain the relaxation pathways for the different relaxation mechanisms from the double commutators and use them to calculate the relaxation-rate constants. Table 4.5 shows the allowed relaxation pathways as well as the relaxation mechanisms that lead to auto- and cross relaxation in the transverse sub block of the Redfield relaxation matrix. Again, we have three auto-correlated relaxation mechanisms and three cross-correlated relaxation mechanisms that can appear. If the difference between the two chemical shifts is large, the secular approximation will reduce the block size to 2×2 sub blocks for the two spins, which will always be the case for heteronuclear spin systems.

In total, we have ten different relaxation-rate constants that need to be calculated. For the $I$ spin we find:

$$\begin{aligned}
\Gamma_{I^\pm,I^\pm} &= \frac{\left(\delta^{(IS)}\right)^2}{32}\left(4J(0)+J(\omega_I-\omega_S)+3J(\omega_I)+6J(\omega_S)+6J(\omega_I+\omega_S)\right)\\
&\quad+\frac{1}{8}\left(\delta^{(I)}B_0\right)^2\left(4J(0)+3J(\omega_I)\right)\\
\Gamma_{2I^\pm S_z,2I^\pm S_z} &= \frac{\left(\delta^{(IS)}\right)^2}{32}\left(4J(0)+J(\omega_I-\omega_S)+3J(\omega_I)+6J(\omega_I+\omega_S)\right)\\
&\quad+\frac{1}{8}\left(\delta^{(I)}B_0\right)^2\left(4J(0)+3J(\omega_I)\right)+\frac{3}{4}\left(\delta^{(S)}B_0\right)^2J(\omega_S)\\
\Gamma_{I^\pm,2I^\pm S_z} &= \frac{1}{8}\delta^{(IS)}\delta^{(I)}B_0\left(4J^{(IS,I)}(0)+3J^{(IS,I)}(\omega_I)\right),
\end{aligned} \tag{4.57}$$

**Table 4.5** Allowed relaxation pathways in the transverse sub block of the Redfield relaxation matrix.

| | $I^\pm$ | $2I^\pm S_z$ | $S^\pm$ | $2S^\pm I_z$ |
|---|---|---|---|---|
| $I^\pm$ | dipolar, CSA | dipolar ⊗ CSA | dipolar | dipolar ⊗ CSA |
| $2I^\pm S_z$ | | dipolar, CSA | dipolar ⊗ CSA | dipolar, $CSA_I \otimes CSA_S$ |
| $S^\pm$ | | | dipolar, CSA | dipolar ⊗ CSA |
| $2S^\pm I_z$ | | | | dipolar, CSA |

and equivalent expressions for the $S$ spin:

$$\begin{aligned}
\Gamma_{S^\pm,S^\pm} =& \frac{\left(\delta^{(IS)}\right)^2}{32}\left(4J(0)+J(\omega_I-\omega_S)+3J(\omega_S)+6J(\omega_I)+6J(\omega_I+\omega_S)\right)\\
&+\frac{1}{8}\left(\delta^{(S)}B_0\right)^2\left(4J(0)+3J(\omega_S)\right)\\
\Gamma_{2S^\pm I_z,2S^\pm I_z} =& \frac{\left(\delta^{(IS)}\right)^2}{32}\left(4J(0)+J(\omega_I-\omega_S)+3J(\omega_S)+6J(\omega_I+\omega_S)\right)\\
&+\frac{1}{8}\left(\delta^{(S)}B_0\right)^2\left(4J(0)+3J(\omega_S)\right)+\frac{3}{4}\left(\delta^{(I)}B_0\right)^2 J(\omega_I)\\
\Gamma_{S^\pm,2S^\pm I_z} =& \frac{1}{8}\delta^{(IS)}\delta^{(S)}B_0\left(4J^{(IS,S)}(0)+3J^{(IS,S)}(\omega_S)\right).
\end{aligned} \tag{4.58}$$

The cross-relaxation between the in-phase and anti-phase coherences of the two different spins is described by:

$$\begin{aligned}
\Gamma_{I^\pm,S^\pm} =& \frac{\left(\delta^{(IS)}\right)^2}{32}\left(2J(0)+2J(\omega_I-\omega_S)+3J(\omega_S)+3J(\omega_I)\right))\\
\Gamma_{I^\pm,2S^\pm I_z} =& \frac{1}{8}\delta^{(IS)}\delta^{(I)}B_0\left(2J^{(IS,I)}(0)-3J^{(IS,I)}(\omega_S)\right)\\
&+\frac{1}{8}\delta^{(IS)}\delta^{(S)}B_0\left(3J^{(IS,S}(\omega_I)+2J^{(IS,S)}(\omega_I-\omega_S)\right)\\
\Gamma_{S^\pm,2I^\pm S_z} =& \frac{1}{8}\delta^{(IS)}\delta^{(S)}B_0\left(2J^{(IS,S)}(0)-3J^{(IS,S)}(\omega_I)\right)\\
&+\frac{1}{8}\delta^{(IS)}\delta^{(I)}B_0\left(3J^{(IS,I)}(\omega_S)+2J^{(IS,I)}(\omega_I-\omega_S)\right)\\
\Gamma_{I^\pm S_z,2S^\pm I_z} =& \frac{\left(\delta^{(IS)}\right)^2}{32}\left(2J(0)+2J(\omega_I-\omega_S)\right))\\
&+\frac{3}{8}\delta^{(I)}\delta^{(S)}B_0^2\left(J^{(I,S)}(\omega_I)+J^{(I,S)}(\omega_S)\right).
\end{aligned} \tag{4.59}$$

To discuss the properties of transverse relaxation, we will again differentiate the three cases: (i) heteronuclear two-spin system (AX spin system) where we have to discuss in addition, the presence and magnitude of the $J$ coupling. (ii) Homonuclear spin system with distinct chemical shifts where we also have to discuss the role of the $J$ couplings. (iii) Homonuclear spin system with degenerate chemical shifts ($A_2$ spin system). Of course, it is again important to remember that the AX and $A_2$ spin systems are the limiting cases and that there is a continuous transition through an AB spin system in between the two cases.

#### 4.3.2.1 Heteronuclear Two-spin System

In a heteronuclear two-spin system, the secular approximation leads to two 2×2 sub blocks for the transverse relaxation consisting of the operators $I^\pm$ and $2I^\pm S_z$ and $S^\pm$ and $2S^\pm I_z$, respectively. If we measure the transverse relaxation in such a spin system, we expect a multi-exponential decay with two time constants. However, as in the case of longitudinal relaxation, we can saturate the $I$ spin, e.g. by on-resonance cw irradiation, and observe a mono-exponential decay of the $S$ spin with the $T_2$ time given by the auto-relaxation-rate constant $\Gamma_{S^\pm,S^\pm}$. Figure 4.12 shows the dependence of $T_2 = 1/\Gamma_{S^\pm,S^\pm}$ as a function of the static magnetic field $B_0$ and the rotational correlation time $\tau_c$. In contrast to the dependence of $T_1$ on the correlation time, $T_2$ shows no minimum but decreases monotonously with increasing molecular size (longer correlation times). This implies that the line width increases with increasing

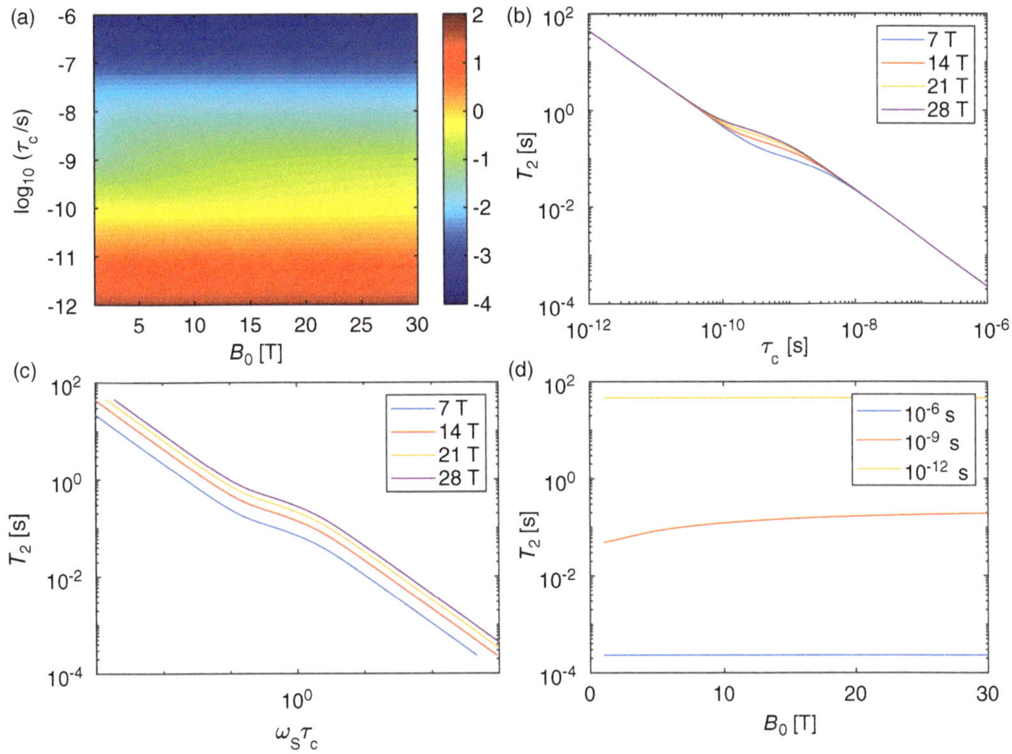


**Figure 4.12** (a) The dipolar-coupling induced $^{13}$C transverse relaxation time $T_2 = 1/\Gamma_{S^\pm,S^\pm}$ for a C–H two-spin system ($r_{\mathrm{CH}} = 1.09\,\text{Å}$) as a function of the static magnetic field ($B_0$) and the rotational correlation time ($\tau_c$) of the molecular tumbling. One can clearly see that the $T_2$ time decreases monotonously as a function of $\tau_c$ leading to an increase in the line width with increasing molecular size. The monotonous decrease of $T_2$ can also be seen in (b) and (c) where four slices corresponding to $B_0$ = 7, 14, 21, and 28 T are plotted as a function of the correlation time $\tau_c$ (b) and of $\omega_S\tau_c$ (c). The dependence on the static magnetic field $B_0$ is shown in (d) for correlation times of 1 μs, 1 ns, and 1 ps. The dependence on the static magnetic field is in all cases rather weak.

molecular size (slower tumbling, longer correlation times), which represents one of the limitations of solution-state NMR to larger molecules.

If $J$ couplings are present and the two multiplet lines are resolved, it is often more convenient to discuss the relaxation behavior in terms of single-transition operators $S^\pm I_\alpha$ and $S^\pm I_\beta$. The transformation is defined by:

$$\begin{pmatrix} S^\pm I_\alpha \\ S^\pm I_\beta \end{pmatrix} = \begin{pmatrix} \frac{1}{2} & \frac{1}{2} \\ \frac{1}{2} & -\frac{1}{2} \end{pmatrix} \begin{pmatrix} S^\pm \\ 2S^\pm I_z \end{pmatrix}. \tag{4.60}$$

and we can use this transformation matrix to calculate the Redfield relaxation matrix in the new basis:

$$\begin{pmatrix} \Gamma_{S^\pm I_\alpha, S^\pm I_\alpha} & \Gamma_{S^\pm I_\alpha, S^\pm I_\beta} \\ \Gamma_{S^\pm I_\alpha, S^\pm I_\beta} & \Gamma_{S^\pm I_\beta, S^\pm I_\beta} \end{pmatrix}$$

$$= \begin{pmatrix} \frac{1}{2} & \frac{1}{2} \\ \frac{1}{2} & -\frac{1}{2} \end{pmatrix} \begin{pmatrix} \Gamma_{S^\pm, S^\pm} & \Gamma_{S^\pm, 2S^\pm I_z} \\ \Gamma_{S^\pm, 2S^\pm I_z} & \Gamma_{2S^\pm I_z, 2S^\pm I_z} \end{pmatrix} \begin{pmatrix} 1 & 1 \\ 1 & -1 \end{pmatrix}. \tag{4.61}$$

The auto- and cross-relaxation rate constants of the two multiplet lines are, therefore, given by:

$$\Gamma_{S^\pm I_\alpha, S^\pm I_\alpha} = \frac{\Gamma_{S^\pm,S^\pm} + \Gamma_{2S^\pm I_z,2S^\pm I_z}}{2} + \Gamma_{S^\pm,2S^\pm I_z} = \Sigma + \Gamma_{S^\pm,2S^\pm I_z}$$
$$\Gamma_{S^\pm I_\beta, S^\pm I_\beta} = \frac{\Gamma_{S^\pm,S^\pm} + \Gamma_{2S^\pm I_z,2S^\pm I_z}}{2} - \Gamma_{S^\pm,2S^\pm I_z} = \Sigma - \Gamma_{S^\pm,2S^\pm I_z}$$
$$\Gamma_{S^\pm I_\alpha, S^\pm I_\beta} = \frac{\Gamma_{S^\pm,S^\pm} - \Gamma_{2S^\pm I_z,2S^\pm I_z}}{2} = \Delta. \tag{4.62}$$

The two multiplet lines relax with different $T_2$ relaxation-rate constants that differ by $2\Gamma_{S^\pm,2S^\pm I_z}$ and are coupled by the cross-relaxation rate constant $\Delta$. In addition, we have also the $J$ coupling that provides a frequency separation of the two operators and the time evolution of the two multiplet components can be described by:

$$\frac{d}{dt}\begin{pmatrix} \langle S^\pm I\alpha\rangle(t) \\ \langle 2S^\pm I_\beta\rangle(t) \end{pmatrix} =$$
$$-\begin{pmatrix} i\pi J_{IS} + \Sigma + \Gamma_{S^\pm,2S^\pm I_z} & \Delta \\ \Delta & -i\pi J_{IS} + \Sigma - \Gamma_{S^\pm,2S^\pm I_z} \end{pmatrix}\begin{pmatrix} \langle S^\pm I\alpha\rangle(t) \\ \langle 2S^\pm I_\beta\rangle(t) \end{pmatrix}. \tag{4.63}$$

If the separation by the $J$ coupling is much larger than the off-diagonal cross-relaxation rate constant $\Delta$, the cross relaxation is truncated according to the secular approximation and we find a mono-exponential decay of the two multiplet lines that differs by twice the cross-correlated cross-relaxation rate constant of the in-phase and anti-phase coherences.

The difference between the relaxation of the decoupled line, characterized by $T_2$, and the two multiplet lines is utilized in the TROSY [77] experiment. Figure 4.13 shows the transverse relaxation times in a C–H two-spin system ($r_{CH} = 1.09$ Å) with a CSA of $\delta^{(S)} = 50$ ppm as a function of the static magnetic field. The plots show the $T_2$ time of the decoupled line and the $T_2$ times of the two undecoupled multiplet components assuming an angle of 35° between the main axes of the dipolar and the CSA tensor. It is immediately obvious that one of the multiplet lines has a much longer $T_2$ (narrower line) than the decoupled line while the other multiplet line has a shorter $T_2$ (broader line). In TROSY spectroscopy, the experiment is set up such that only the narrow component is selected in the spectra, allowing the recording of spectra for molecules with larger molecular weight and slower rotational tumbling. In such experiments, it is important to preserve the state of the $I$ spins during $t_1$ to avoid a mixing of the narrow and broad component. The optimum frequency for TROSY spectroscopy depends on the magnitude of the CSA tensor in relation to the dipolar coupling and on the relative orientation of the two tensors. The TROSY technique was originally developed for N–H spin systems and has been extended also to aromatic C-H spin systems and $CH_3$ groups [78–80].

#### 4.3.2.2 Homonuclear Two-spin System with Distinct Chemical Shifts

For homonuclear two-spin systems with two distinct chemical shifts, we find the same behavior as in the heteronuclear two-spin system as long as the chemical-shift difference is large enough to ensure that the secular approximation is fulfilled and no cross relaxation between the two spins can happen. Of course, like in the case of longitudinal relaxation, the sampling of the spectral-density function is restricted to frequencies 0, $\omega_I$, and $2\omega_I$. Depending on the size of the $J$ coupling, there will be cross-correlated cross relaxation between the lines of the multiplet or not.

#### 4.3.2.3 Homonuclear Two-spin System with Degenerate Chemical Shifts

In the homonuclear case with degenerate chemical shifts, we face a similar situation as for the longitudinal relaxation. We have again an $A_2$ spin system with only a single observable, $F^\pm = I^\pm + S^\pm$ and we cannot manipulate

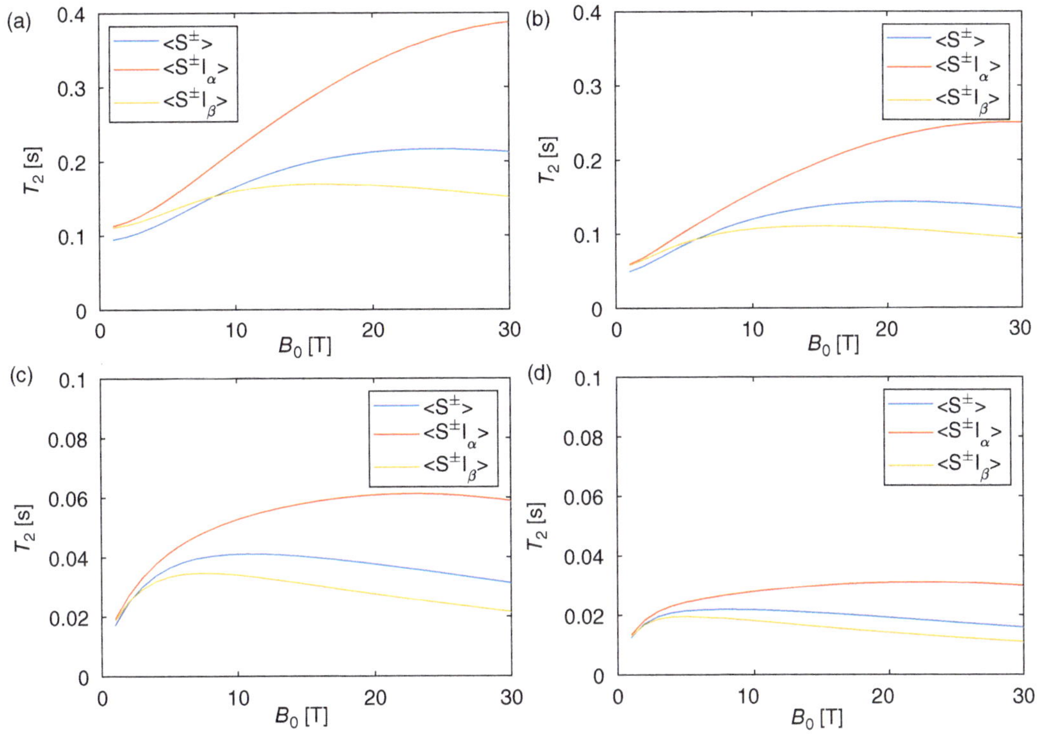


**Figure 4.13** The $T_2$ relaxation times of the decoupled line ($S^{\pm}$) and of the multiplet components ($S^{\pm}I_{\alpha}$ and $S^{\pm}I_{\beta}$) of a C–H two-spin system with an internuclear distance $r_{CH}$ = 1.09 Å and a CSA of $\delta^{(S)}$ = 50 ppm. The angle between the main axes of the dipolar coupling and the CSA tensor was set to 35°. The correlation times for the rotational tumbling were chosen to be (a) 0.5 ns, (b) 1 ns, (c) 5 ns, and (d) 10 ns. One can clearly see that one of the two multiplet lines has always a significantly longer transverse relaxation time (narrower line width) than the decoupled line while the second multiplet line has a shorter transverse relaxation time (broader line width) than the decoupled line.

the two spins independently. It is, therefore, again of advantage to use a different basis to describe relaxation in a homonuclear two-spin system with degenerate shift based on spherical-tensor operators. A typical basis set would consist of $T^{(I)}_{1,\pm1}$, $T^{(S)}_{1,\pm1}$, $T^{(IS)}_{1,\pm1}$, and $T^{(IS)}_{2,\pm1}$. In order to include the observable sum magnetization, it is of advantage to use again linear combinations of the two single-spin spherical-tensor operators, $T^{(I)}_{1,\pm1} \pm T^{(S)}_{1,\pm1}$. As in the longitudinal relaxation, neglecting cross-correlated relaxation leads to a mono-exponential decay of the sum magnetization with the relaxation-rate constant:

$$\begin{aligned}\Gamma_{F^{\pm},F^{\pm}} &= \Gamma_{I^{\pm},I^{\pm}} + \Gamma_{I^{\pm},S^{\pm}} \\ &= \frac{\left(\delta^{(IS)}\right)^2}{32}\left(9J(0) + 15J(\omega_I) + 6J(2\omega_I)\right) + \frac{1}{8}\left(\delta^{(I)}B_0\right)^2\left(4J(0) + 3J(\omega_I)\right).\end{aligned} \tag{4.64}$$

In the more general case, the complete 4×4 matrix is coupled and leads to a multi-exponential decay of the transverse relaxation with four distinct exponentials. Since we can only observe the sum magnetization, determining all relaxation-rate constants in such a spin system is probably not an easy task.

### 4.3.3 Double-quantum Relaxation

The final relaxation-rate constant that we have to consider in a two-spin system is the double-quantum relaxation-rate constant. The rate constant is given by:

$$\begin{aligned}\Gamma_{I^\pm S^\pm, I^\pm S^\pm} = & \frac{\left(\delta^{(IS)}\right)^2}{32}\left(3J(\omega_S) + 3J(\omega_I) + 12J(\omega_I + \omega_S)\right) \\ & -\delta^{(I)}\delta^{(S)}B_0^2 J^{(I,S)}(0) + \frac{1}{8}\left(\delta^{(S)}B_0\right)^2\left(4J(0) + 3J(\omega_S)\right) \\ & +\frac{1}{8}\left(\delta^{(I)}B_0\right)^2\left(4J(0) + 3J(\omega_I)\right).\end{aligned} \tag{4.65}$$

Figure 4.14 shows the behavior of the double-quantum relaxation time as a function of the static magnetic field and the overall rotational correlation time for a C–H two-spin system. The double-quantum coherence decays mono exponentially since it is not coupled to any other relaxation modes and behaves quite similar to the zero-quantum relaxation time.

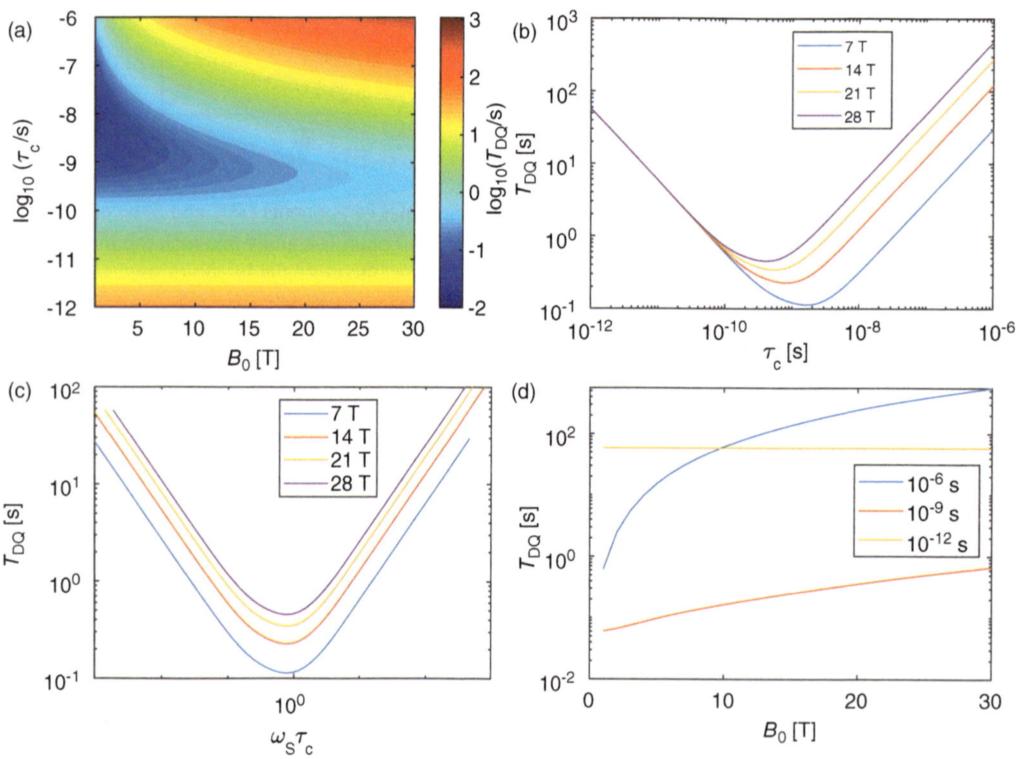


**Figure 4.14** (a) The dipolar-coupling induced double-quantum relaxation time $T_{DQ} = 1/\Gamma_{I^\pm S^\pm, I^\pm S^\pm}$ for a C–H two-spin system ($r_{CH} = 1.09$Å) as a function of the static magnetic field ($B_0$) and the rotational correlation time ($\tau_c$) of the molecular tumbling. One can clearly see that the $T_{DQ}$ time has a minimum for $\tau_c$ in the range of nanoseconds and is longer for shorter or longer correlation times. The $T_{DQ}$ minimum can also be seen in (b) and (c) where four slices corresponding to $B_0$ = 7, 14, 21, and 28 T are plotted as a function of the correlation time $\tau_c$ (b) and of $\omega_S\tau_c$ (c). The minimum is always roughly at $\omega_S\tau_c = 1$. The dependence on the static magnetic field $B_0$ is shown in (d) for correlation times of 1 μs, 1 ns, and 1 ps. For small molecules with short correlation times, we see almost no field dependence in the range of typical NMR frequencies while for larger molecules with longer correlation times a clear field dependence is observed.

### 4.3.4 Relaxation in Larger Spin Systems

Relaxation in larger spin systems like $CH_2$ or $CH_3$ groups is obviously more complex and we have to select suitable sets of basis operators. For longitudinal relaxation, this question has been covered in much detail in the literature in the so-called normal-mode approach to relaxation [7]. For transverse relaxation, a similar approach was used to analyze the relaxation behavior in different types of spin systems [8]. In these two seminal articles, cross-correlated relaxation is not included into the relaxation pathways. But for many of the more common spin systems, relaxation pathways including cross-correlated relaxation can be found in the literature. As mentioned above, saturation to isolate the relaxation decay in order to make the decay mono exponential does not always work in such larger spin systems especially in the case of spin systems with nuclei with degenerate chemical shifts where zero-quantum and homonuclear multi-spin terms are not saturated by cw irradiation. In these cases, a careful analysis of the multi-exponential decay is required in order to determine relaxation-rate constants precisely. Another commonly used approach is isotope substitution, i.e. the use of $CHD_2$ groups instead of methyl groups that behave essentially as an AX spin system if we neglect the influence of the deuterium atoms.

## 4.4 Other Relaxation Mechanisms

Besides relaxation through dipolar couplings and CSA quadrupolar couplings are another major source of relaxation in nuclei with a spin-quantum number $S > \frac{1}{2}$. Since the quadrupolar coupling is often much larger than all the other anisotropic interactions, quadrupolar relaxation will often dominate the relaxation of quadrupolar nuclei [81].

Scalar relaxation (a historic name which now also includes relaxation by non-scalar, anisotropic quantities) is another relaxation mechanism where the (scalar) interaction (often the $J$ coupling or the isotropic part of the hyperfine coupling) is modulated either by a chemical-exchange process (scalar relaxation of the first kind) or by a fast relaxing spin, i.e. a either a quadrupolar nucleus or an electron (scalar relaxation of the second kind). In solids also dipolar couplings can give rise to relaxation generated by the "scalar-relaxation mechanism" [1, 23, 82, 83].

Spin-rotation relaxation is the final mechanism, which is important in this context. The charges (nuclei and electrons) in a rotating molecules induce time-dependent magnetic fields that can interact with the nuclear and electron spins. Formally, this can be written as the interaction of the spins with the angular momentum of the molecule. An alternative formulation is a stochastic time-dependent magnetic field that interacts with the spins [84, 85].

### 4.4.1 Quadrupolar Relaxation

Relaxation in quadrupolar spins is more complex than in spin-$\frac{1}{2}$ systems due to the larger number of energy levels and the higher number of coupled relaxation modes. A complete basis set to describe the magnetization modes for relaxation is formed by the spherical-tensor operators. For a spin with a spin-quantum number $S$, we need the spin-tensor operators up to order $2S$, which is illustrated for a spin-$\frac{5}{2}$ by:

$$
\begin{array}{ll}
T^{(5/2)}_{0,0}, T^{(5/2)}_{1,0}, T^{(5/2)}_{2,0}, T^{(5/2)}_{3,0}, T^{(5/2)}_{4,0}, T^{(5/2)}_{5,0} & \text{populations} \\
T^{(5/2)}_{1,\pm1}, T^{(5/2)}_{2,\pm1}, T^{(5/2)}_{3,\pm1}, T^{(5/2)}_{4,\pm1}, T^{(5/2)}_{5,\pm1} & \text{single-quantum coherences} \\
T^{(5/2)}_{2,\pm2}, T^{(5/2)}_{3,\pm2}, T^{(5/2)}_{4,\pm2}, T^{(5/2)}_{5,\pm2} & \text{double-quantum coherences} \\
T^{(5/2)}_{3,\pm3}, T^{(5/2)}_{4,\pm3}, T^{(5/2)}_{5,\pm3} & \text{triple-quantum coherences} \\
T^{(5/2)}_{4,\pm4}, T^{(5/2)}_{5,\pm4} & \text{quadruple-quantum coherences} \\
T^{(5/2)}_{5,\pm5} & \text{quintuple-quantum coherences.}
\end{array}
\tag{4.66}
$$

Based on the secular approximation, we know that there will be no cross relaxation between different coherence orders leading to a block-diagonal representation of the Redfield relaxation matrix as shown in Figure 4.15 for a spin-$\frac{3}{2}$ nucleus. This block structure immediately tells us that we expect potentially multi-exponential relaxation behavior for populations as well as coherences except in the block with the highest coherence order. In liquid-state NMR, the number of observables is limited due to the efficient averaging of the quadrupolar coupling by the rotational tumbling of the molecule. This means that only the Zeeman polarization $T_{1,0} = S_z$ and the single-quantum coherence $T_{1,\pm1} = \mp\frac{1}{\sqrt{2}}S^{\pm}$ are accessible by experimental measurements. All other operators might get populated by cross relaxation but cannot be reconverted into detectable modes and are experimentally inaccessible. In solid-state NMR, where the quadrupolar coupling is not averaged, we have access to more transitions since we can excite and invert different transitions selectively, which gives us, in principle, access to all $T_{\ell,0}$ and $T_{\ell,\pm1}$ operators.

Quadrupolar nuclei are, of course, also relaxed by dipolar coupling and CSA interactions. Since the quadrupolar coupling is often several orders of magnitude larger than dipolar couplings or CSA tensors, they are often neglected and only the dominating quadrupolar relaxation will be considered. If we observe or are interested in cross relaxation to other spins, we have to include dipolar coupling-mediated relaxation since quadrupolar coupling will only lead to auto and cross relaxation between different modes of the quadrupolar spin.

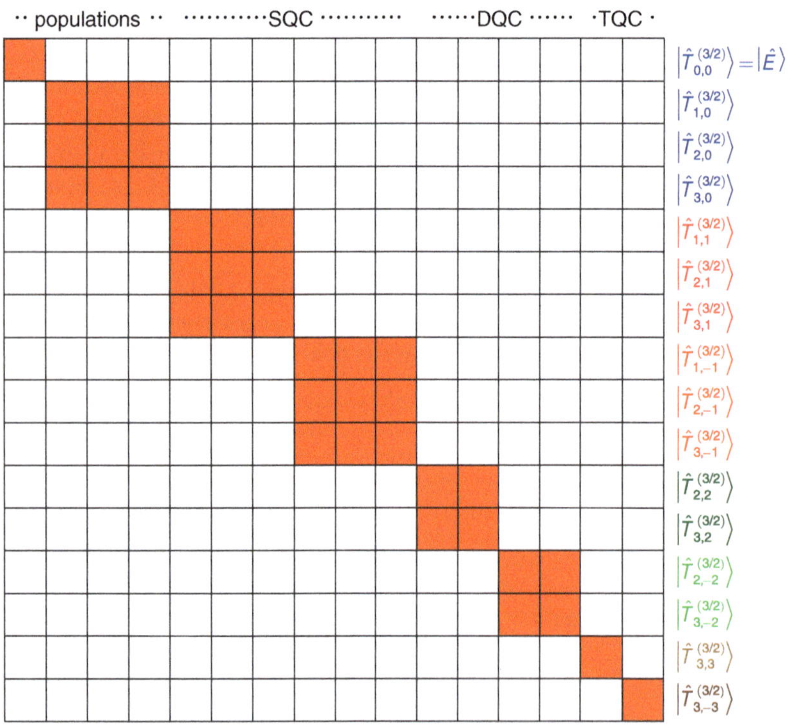


**Figure 4.15** Structure of the Redfield relaxation matrix for quadrupolar relaxation in a spin-$\frac{3}{2}$ nucleus. The different coherence orders are separated in sub blocks but within a sub block all operators are in principle coupled by cross relaxation. Therefore, we would expect multi-exponential relaxation behavior for the populations and the coherences except for the highest order coherence, which is described by a 1x1 block.

In order to calculate the relaxation pathways and the relaxation-rate constants for quadrupolar relaxation, we again have to evaluate the double commutators that require the eigenoperators of the Zeeman Hamiltonian. Tables 4.6 and 4.7 give the eigenoperators of the spherical-tensor operators used to describe the quadrupolar Hamiltonian and the double commutators for spin-quantum numbers $S = 1$ and $S = \frac{3}{2}$. Unfortunately, the double commutators depend on the spin-quantum number and cannot easily be given in a more general way. For the longitudinal relaxation, we see immediately, that quadrupolar relaxation couples the spherical-tensor operators $T_{\ell,0}$ and $T_{\ell\pm2,0}$ leading to two sub blocks, a block with even $\ell$ and a block with odd $\ell$. As a consequence of this, longitudinal relaxation of the Zeeman polarization in a spin-1 nucleus is mono exponential, which might be one of the reasons that $^{2}$H and $^{14}$N are nuclei that are quite often used for dynamics analysis. The longitudinal magnetization of quadrupolar nuclei with $S > 1$ will decay always multi-exponentially.

**Table 4.6** Definition of eigenoperators of the quadrupolar Hamiltonian and double commutators for $Q = T_{\ell,0}$ for $S = 1$.

| $T^{(\mu)}_{\ell,m}$ | $V^{(\mu)}_{p}$ | $\omega_p$ | $\left[T^{(\mu)}_{\ell,m},\left[V^{(\mu)\dagger}_{p},T_{\ell,0}\right]\right]$ |
|---|---|---|---|
| $T^{(Q)}_{2,0}$ | $T^{(Q)}_{2,0}$ | 0 | $\left[T^{(Q)}_{2,0},\left[T^{(Q)}_{2,0},T_{1,0}\right]\right]=0$ |
| | | | $\left[T^{(Q)}_{2,0},\left[T^{(Q)}_{2,0},T_{2,0}\right]\right]=0$ |
| $T^{(Q)}_{2,\pm1}$ | $T^{(Q)}_{2,\pm1}$ | $\omega_S$ | $\left[T^{(Q)}_{2,\pm1},\left[-T^{(Q)}_{2,\mp1},T_{1,0}\right]\right]=2T_{1,0}$ |
| | | | $\left[T^{(Q)}_{2,\pm1},\left[-T^{(Q)}_{2,\mp1},T_{2,0}\right]\right]=\frac{3}{2}T_{2,0}$ |
| $T^{(Q)}_{2,\pm2}$ | $T^{(Q)}_{2,\pm2}$ | $2\omega_S$ | $\left[T^{(Q)}_{2,\pm2},\left[-T^{(Q)}_{2,\mp2},T_{1,0}\right]\right]=\frac{1}{2}T_{1,0}$ |
| | | | $\left[T^{(Q)}_{2,\pm2},\left[-T^{(Q)}_{2,\mp2},T_{2,0}\right]\right]=0$ |

**Table 4.7** Definition of eigenoperators of the quadrupolar Hamiltonian and double commutators for $Q = T_{\ell,0}$ for $S = \frac{3}{2}$.

| $T^{(\mu)}_{\ell,m}$ | $V^{(\mu)}_{p}$ | $\omega_p$ | $\left[T^{(\mu)}_{\ell,m},\left[V^{(\mu)\dagger}_{p},T_{\ell,0}\right]\right]$ |
|---|---|---|---|
| | | | $\left[T^{(Q)}_{2,0},\left[T^{(Q)}_{2,0},T_{1,0}\right]\right]=0$ |
| $T^{(Q)}_{2,0}$ | $T^{(Q)}_{2,0}$ | 0 | $\left[T^{(Q)}_{2,0},\left[T^{(Q)}_{2,0},T_{2,0}\right]\right]=0$ |
| | | | $\left[T^{(Q)}_{2,0},\left[T^{(Q)}_{2,0},T_{3,0}\right]\right]=0$ |
| | | | $\left[T^{(Q)}_{2,\pm1},\left[-T^{(Q)}_{2,\mp1},T_{1,0}\right]\right]=\frac{6}{5}T_{1,0}+\frac{12}{5}T_{3,0}$ |
| $T^{(Q)}_{2,\pm1}$ | $T^{(Q)}_{2,\pm1}$ | $\omega_S$ | $\left[T^{(Q)}_{2,\pm1},\left[-T^{(Q)}_{2,\mp1},T_{2,0}\right]\right]=6T_{2,0}$ |
| | | | $\left[T^{(Q)}_{2,\pm1},\left[-T^{(Q)}_{2,\mp1},T_{3,0}\right]\right]=\frac{12}{5}T_{1,0}+\frac{24}{5}T_{3,0}$ |
| | | | $\left[T^{(Q)}_{2,\pm2},\left[-T^{(Q)}_{2,\mp2},T_{1,0}\right]\right]=\frac{24}{5}T_{1,0}-\frac{12}{5}T_{3,0}$ |
| $T^{(Q)}_{2,\pm2}$ | $T^{(Q)}_{2,\pm2}$ | $2\omega_S$ | $\left[T^{(Q)}_{2,\pm2},\left[-T^{(Q)}_{2,\mp2},T_{2,0}\right]\right]=6T_{2,0}$ |
| | | | $\left[T^{(Q)}_{2,\pm2},\left[-T^{(Q)}_{2,\mp2},T_{3,0}\right]\right]=\frac{12}{5}T_{1,0}+\frac{6}{5}T_{3,0}$ |

The quadrupolar relaxation-rate constants for a spin-1 nucleus are given by:

$$\begin{aligned}
\Gamma_{T^{(1)}_{1,0},T^{(1)}_{1,0}} &= \left(\frac{\delta^{(Q)}}{2}\right)^2 \left(1+\frac{1}{3}\left(\eta^{(Q)}\right)^2\right)(3J(\omega_S)+12J(2\omega_S))\\
\Gamma_{T^{(1)}_{2,0},T^{(1)}_{2,0}} &= \left(\frac{\delta^{(Q)}}{2}\right)^2 \left(1+\frac{1}{3}\left(\eta^{(Q)}\right)^2\right)9J(\omega_S)\\
\Gamma_{T^{(1)}_{1,1},T^{(1)}_{1,1}} &= \left(\frac{\delta^{(Q)}}{2}\right)^2 \left(1+\frac{1}{3}\left(\eta^{(Q)}\right)^2\right)\left(\frac{9}{2}J(0)+\frac{15}{2}J(\omega_S)+3J(2\omega_S)\right)\\
\Gamma_{T^{(1)}_{2,1},T^{(1)}_{2,1}} &= \left(\frac{\delta^{(Q)}}{2}\right)^2 \left(1+\frac{1}{3}\left(\eta^{(Q)}\right)^2\right)\left(\frac{9}{2}J(0)+\frac{3}{2}J(\omega_S)+3J(2\omega_S)\right).
\end{aligned} \tag{4.67}$$

As mentioned above, there are no cross-relaxation rate constants in a quadrupolar $S = 1$ spin system since cross relaxation can only connect spherical-tensor operators that differ in their rank by two. Such cross relaxation will only appear in a quadrupolar $S = 3/2$ system.

The same structure of connected even and odd values of $\ell$ can also be found for the transverse single-quantum relaxation of quadrupolar nuclei. An interesting feature of the cross-relaxation rate constants in $S = \frac{3}{2}$ quadrupolar nuclei is the fact that they all contain the spectral-density function as a difference of the form $k(J(\omega_S)–J(2\omega_S))$. This implies that in the extreme narrowing limit ($\omega_S\tau_c \ll 1$, small molecules) all cross-relaxation rate constants vanish and the Redfield relaxation matrix is diagonal. This turns out to be a general feature of quadrupolar relaxation independent of the spin-quantum number.

The properties of quadrupolar relaxation are shown in Figures 4.16 and 4.17 as a function of the static magnetic field and the correlation time of the overall rotational tumbling for a $^2$H spin with a quadrupolar-coupling constant of $C_{qcc} = 170$ kHz and $\eta = 0$. Note, that the anisotropy of the quadrupolar coupling that we use in the definition of relaxation-rate constants is related to the more commonly used quadrupolar-coupling constant by $\delta^{(Q)}/(2\pi) = C_{qcc}/(2S(2S-1))$. One can clearly see the typical features of $T_1$ and $T_2$ relaxation, which we have discussed previously. A $T_1$ minimum around $\omega_S\tau_c = 1$ and a monotonous decrease in $T_2$ leading to a broadening of the line with increasing molecular size (increasing correlation time). It is important to note that even for a relatively small quadrupolar-coupling constant like in $^2$H, the relaxation times are already quite short and typically in the sub-second range.

To illustrate the typical range of quadrupolar relaxation times, Figure 4.18 shows simulations for different quadrupolar nuclei that are common in biological macromolecules: $^2$H motionally averaged in fast rotating methyl groups, $^2$H in static environment, $^{14}$N and $^{17}$O in a static environment. The used quadrupolar parameters for the four simulations are given in the figure caption. One can clearly see that the relaxation times are short if the quadrupolar coupling is large. Relaxation times in the microsecond range and even below are possible. Such short $T_2$ relaxation times imply, on one hand, very broad lines in solution that are often difficult to observe and do not allow any spectral resolution since the chemical-shift range is typically smaller than these line widths. On the other hand, short $T_1$ relaxation times allow very fast repetition of experiments which allows efficient signal averaging.

### 4.4.2 Scalar Relaxation

Scalar relaxation originally described relaxation generated by a scalar interaction, e.g. the isotropic $J$ coupling or the isotropic part of the hyperfine coupling that is not modulated by the molecular tumbling. There are two

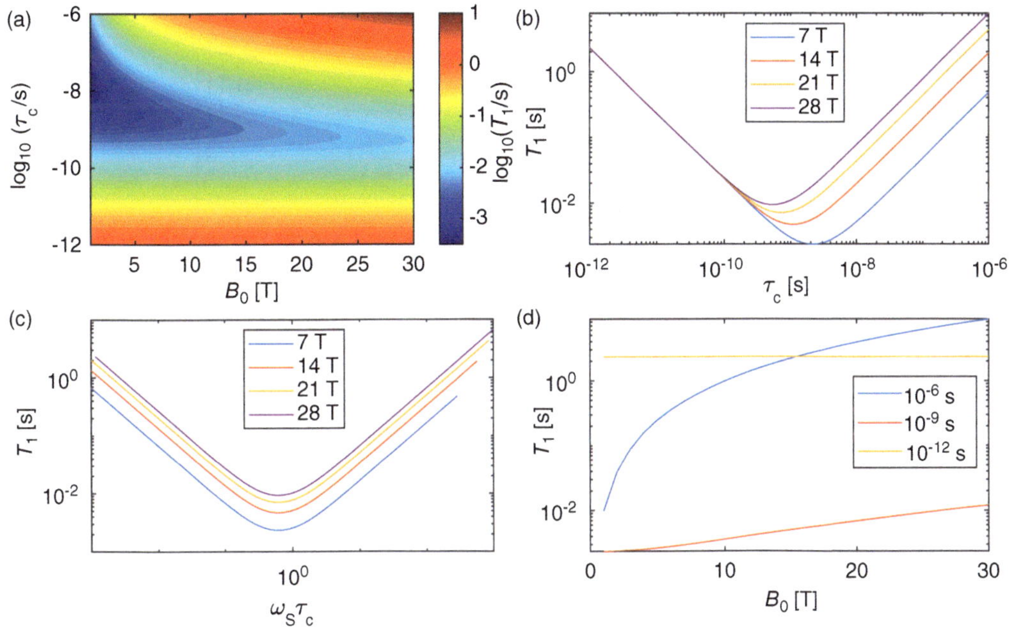


**Figure 4.16** (a) The quadrupolar longitudinal relaxation time $T_1 = 1/\Gamma_{S_z,S_z}$ for a $^2$H spin ($C_{qcc} = 170kHz$, $\eta = 0$) as a function of the static magnetic field ($B_0$) and the rotational correlation time ($\tau_c$) of the molecular tumbling. One can clearly see that the $T_1$ time has a minimum for $\tau_c$ in the range of nanoseconds and is longer for shorter or longer correlation times. The $T_1$ minimum can also be seen in (b) and (c) where four slices corresponding to $B_0$ = 7, 14, 21, and 28 T are plotted as a function of the correlation time $\tau_c$ (b) and of $\omega_S\tau_c$ (c). The minimum is always roughly at $\omega_S\tau_c = 1$. The dependence on the static magnetic field $B_0$ is shown in (d) for correlation times of 1 μs, 1 ns, and 1 ps. For small molecules with short correlation times, we see almost no field dependence in the range of typical NMR frequencies while for larger molecules with longer correlation times a clear field dependence is observed.

mechanisms that can lead to a stochastic modulation of a scalar interaction, a stochastic change in the magnitude of the coupling due to structural changes in the molecule for example by chemical exchange and a stochastic modulation of the spin state of the coupling partner by fast relaxing spins like quadrupolar nuclei (see previous section) or fast relaxing electrons. The first mechanism is called "scalar relaxation of the first kind" while the latter is called "scalar relaxation of the second kind."

If we consider scalar relaxation of the first kind, we can use the same methodology as discussed before and calculate the correlation function of the magnitude of the scalar interaction under the stochastic process. This will lead to a correlation function of an isotropic (rank zero) quantity and can lead to auto and cross-relaxation processes.

For scalar relaxation of the second kind, the magnitude of the scalar interaction is not modulated. The stochastic modulation happens through the relaxation of the coupled spin. Typically, one assumes that the fast relaxing spin relaxes mono exponentially with relaxation times $T_1$ and $T_2$ leading to different correlation functions for transverse and longitudinal spin components. In solids, this mechanism can also modulate anisotropic interactions like the dipolar coupling but is still referred to as scalar relaxation, which can lead to confusion.

As mentioned above, scalar relaxation is mostly important for nuclei coupled to a fast relaxing quadrupolar nucleus or in paramagnetic systems where couplings to an unpaired electron play an important role. For a more

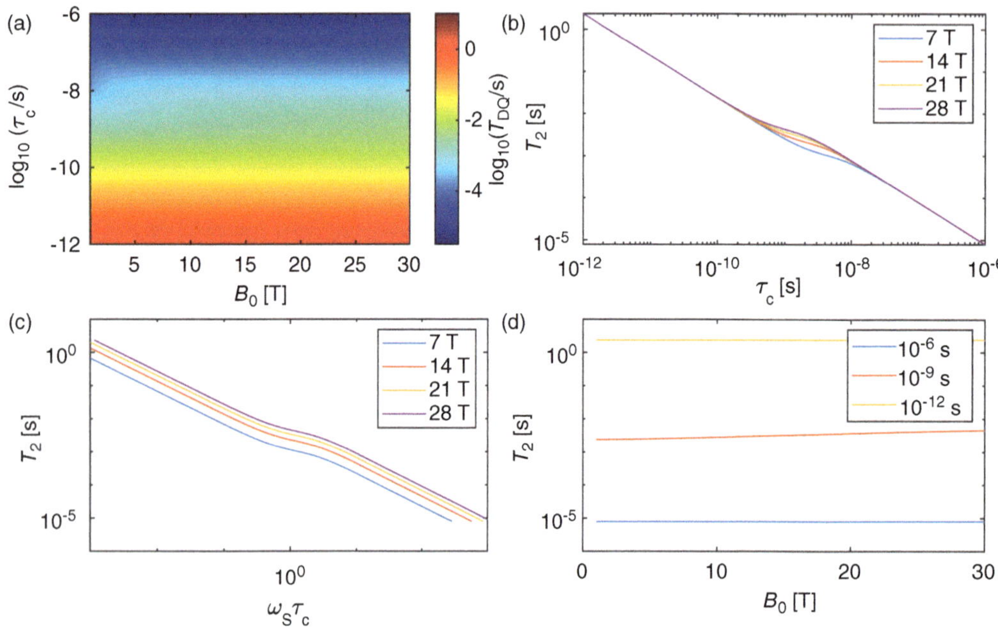


**Figure 4.17** (a) The quadrupolar longitudinal relaxation time $T_2 = 2/\Gamma_{S^\pm,S^\pm}$ for a $^2$H spin ($C_{qcc}$ = 170 kHz, $\eta$ = 0) as a function of the static magnetic field ($B_0$) and the rotational correlation time ($\tau_c$) of the molecular tumbling. One can clearly see that the $T_2$ time decreases monotonously as a function of $\tau_c$ leading to an increase in the line width with increasing molecular size. The monotonous decrease of $T_2$ can also be seen in (b) and (c) where four slices corresponding to $B_0$ = 7, 14, 21, and 28 T are plotted as a function of the correlation time $\tau_c$ (b) and of $\omega_S\tau_c$ (c). The dependence on the static magnetic field $B_0$ is shown in (d) for correlation times of 1 μs, 1 ns, and 1 ps. The dependence on the static magnetic field is in all cases rather weak.

detailed discussion of this relaxation mechanism, the reader is referred to the extensive literature about scalar relaxation [1, 23, 82, 83].

## 4.5 Concluding Remarks

Relaxation theory has evolved over the past 70 years from a phenomenological description to well established semi-classical and quantum-mechanical theories that give a robust description of relaxation processes in spin systems under almost all experimental conditions. Understanding the theoretical foundations of spin relaxation is important in order to select the correct theory applicable to a given experimental situation and to understand the limitations of different descriptions. The main application of relaxation measurements is the characterization of dynamic processes in molecules. Spin relaxation is the only measurable quantity that does not only give access to the amplitude of motion but in favorable cases also to the time scale of the motion. Especially in the interpretation of dynamics data, there are many pitfalls since the information content of relaxation data is often limited and different types of motional processes cannot be distinguished. Therefore, understanding the foundations of relaxation is important to recognize the limitations in order to avoid over interpretation of experimental data.

This short introduction can only present the basics and some simple examples of relaxation theory, but there is a vast literature where relaxation is discussed for different spin systems and under different conditions. It is not

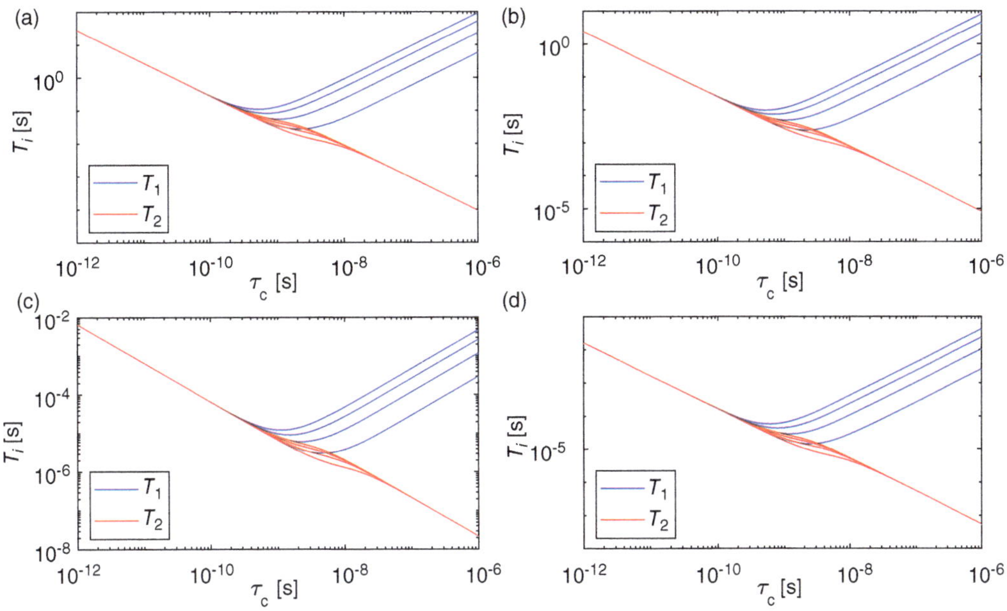


**Figure 4.18** The quadrupolar longitudinal ($T_1 = 2/\Gamma_{S^\pm,S^\pm}$) and transverse ($T_2 = 2/\Gamma_{S^\pm,S^\pm}$) relaxation times as a function of the rotational correlation time ($\tau_c$) for four different static magnetic fields, $B_0$ = 7, 14, 21, and 28 T. (a) $^2H$ with typical quadrupolar-coupling values for a $CH_3$ group ($C_{qcc}$ = 50 kHz, $\eta$ = 0), (b) $^2H$ with typical quadrupolar-coupling values for a static CH group ($C_{qcc}$ = 170 kHz, $\eta$ = 0), (c) $^{14}N$ with typical quadrupolar-coupling values of $C_{qcc}$ = 3.2 MHz and $\eta$ = 0.32, and (d) $^{17}O$ with typical quadrupolar-coupling values of $C_{qcc}$ = 8.2 MHz and $\eta$ = 0.

always easy to navigate this literature and one should be very careful to make sure that the description used is appropriate for the application to be discussed.

## References


1 Abragam, A. *The Principles of Nuclear Magnetism*. (1961). Oxford University Press.

2 Ernst, R.R., Bodenhausen, G., Wokaun, A. (1990). *Principles of Nuclear Magnetic Resonance in One and Two Dimensions*. Oxford: Oxford University Press.

3 Levitt, M.H. (2013). *Spin Dynamics*. Basics of Nuclear Magnetic Resonance. John Wiley & Sons.

4 Bloch, F. (1946). Nuclear induction. *Phys Rev.* 70 (7-8): 460–474.

5 Redfield, A.G. (1966). Intramolecular dipolar relaxation in multi-spin systems. *Adv. Magn. Reson.* 1: 1–32.

6 Kubo, R. (1977). A stochastic theory of line shape. *Adv. Chem. Phys.* 15: 101–127.

7 Werbelow, L.G. and Grant, D.M. (1977). Intramolecular dipolar relaxation in multi-spin systems. *Adv. Magn. Reson.* 9: 189–275.

8 Vold, R.L. and Vold, R.R. (1978). Nuclear magnetic relaxation in coupled spin systems. *Prog. NMR Spectr.* 12 (2): 79–133.

9 Jeener, J. (1985). Superoperators in magnetic-resonance. *Adv. Magn. Reson.* 10: 1–51.

10 Kowalewski, J., Nordenskiold, L., Benetis, N. and Westlund, P.O. (1985). Theory of nuclear-spin relaxation in paramagnetic systems in solution. *Prog. NMR Spectr.* 17: 141–185.

11 Bull, T.E. (1992). Relaxation in the rotating frame in liquids. *Prog. NMR Spectr.* 24: 377–410.

**12** Fischer, M.W.F., Majumdar, A. and Zuiderweg, E.R.P. (1998). Protein NMR relaxation: theory, applications and outlook. *Prog. NMR Spectr.* 33 (4): 207–272.

**13** Sharp, R., Lohr, L. and Miller, J. (2001). Paramagnetic NMR relaxation enhancement: recent advances in theory. *Prog. NMR Spectr.* 38 (2): 115–158.

**14** Luginbühl, P. and Wüthrich, K. (2002). Semi-classical nuclear spin relaxation theory revisited for use with biological macromolecules. *Prog. NMR Spectr.* 40 (3): 199–247.

**15** Schneider, D.J. and Freed, J.H. (2009). Spin relaxation and motional dynamics. *Adv. Chem Phys.* 73: 387.

**16** Nicholas, M.P., Eryilmaz, E., Ferrage, F., Cowburn, D. and Ghose, R. (2010). Nuclear spin relaxation in isotropic and anisotropic media. *Prog. NMR Spectr.* 57 (2): 111–158.

**17** Schanda, P. and Ernst, M. (2016). Studying dynamics by magic-angle spinning solid-state NMR spectroscopy: principles and applications to biomolecules. *Prog. NMR Spectr.* 96 (C): 1–46.

**18** Caspers, W.J. (1964). *Theory of Spin Relaxation*. Interscience Publishers.

**19** McConnell, J. (2009). The Theory of Nuclear Magnetic Relaxation in Liquids. Cambridge University Press.

**20** Gamliel, D. and Levanon, H. (1995). *Stochastic Processes in Magnetic Resonance*. World Scientific.

**21** Cowan, B. (2005). *Nuclear Magnetic Resonance and Relaxation*. Cambridge University Press.

**22** Bakhmutov, V.I. (2005). *Practical Nuclear Magnetic Resonance Relaxation for Chemists*. John Wiley & Sons.

**23** Kowalewski, J. and Mäler, L. (2006). *Nuclear Spin Relaxation in Liquids*. Theory, experiments, and applications. CRC Press.

**24** Kruk, D. (2007). *Theory of Evolution and Relaxation of Multi-Spin Systems*. Application to nuclear magnetic resonance and electron spin resonance. Suffolk: Arima Publishing.

**25** Lindblad, G. (1976). On the generators of quantum dynamical semigroups. *Commun. Math. Phys.* 48 (2): 119–130.

**26** Bengs, C. and Levitt, M.H. (2020). A master equation for spin systems far from equilibrium. *J. Magn. Reson.* 310: 106645.

**27** Bengs, C. (2021). Markovian exchange phenomena in magnetic resonance and the Lindblad equation. *J. Magn. Reson.* 322: 106868.

**28** Pell, A.J. (2021). A method to calculate the NMR spectra of paramagnetic species using thermalized electronic relaxation. *J. Magn. Reson.* 326: 106939.

**29** Kurzbach, D. and Jannin, S. (2019). Dissolution dynamic nuclear polarization methodology and instrumentation. *eMagRes.* 7.

**30** Barbara, T.M. (2021). The Lindbladian form and the reincarnation of Felix Bloch's generalized theory of relaxation. *Magn. Reson.* 2 (2): 689–698.

**31** Rodin, B.A. and Abergel, D. (2021). Spin relaxation: under the sun, anything new? *Magn. Reson. Discus.* 3, 1–22.

**32** Kubo, R. (1963). Stochastic liouville equations. *J. Math. Phys.* 4 (2): 174–183.

**33** Freed, J.H. and Fraenkel, G.K. (1963). Theory of linewidths in electron spin resonance spectra. *J. Chem. Phys.* 39 (2): 326–348.

**34** Vega, A.J. and Fiat, D. (1974). Stochastic liouville equation and approach to thermal equilibrium. *Pure Appl. Chem.* 40 (1-2): 181–192.

**35** Moro, G.J. and Freed, J.H. (1980). Efficient computation of magnetic resonance spectra and related correlation functions from stochastic Liouville equations. *J. Phys. Chem.* 84 (22): 2837–2840.

**36** Meirovitch, E., Shapiro, Y.E., Pohmeno, A. and Freed, J.H. (2010). Structural dynamics of bio-macromolecules by NMR: the slowly relaxing local structure approach. *Prog. NMR Spectr.* 56 (4): 360–405.

**37** Bain, A.D. (2003). Chemical exchange in NMR. *Prog. NMR Spectr.* 43 (3-4): 63–103.

**38** Bloembergen, N., Purcell, E.M. and Pound, R.V. (1948). Relaxation effects in nuclear magnetic resonance absorption. *Phys. Rev.* 73 (7): 679–712.

**39** Solomon, I. (1955). Relaxation processes in a system of 2 spins. *Phys. Rev.* 99 (2): 559–565.

**40** Wangsness, R.K. and Bloch, F. (1953). The dynamical theory of nuclear induction. *Phys. Rev.* 89 (4): 728–739.

41 Redfield, A.G. (1955). Nuclear magnetic resonance saturation and rotary saturation in solids. *Phys. Rev.* 98 (6): 1787–1809.

42 Redfield, A.G. (1957). On the theory of relaxation processes. *IBM J. Res. Dev.* 1: 19–31.

43 Levitt, M.H. and Di Bari, L. (1992). Steady-state in magnetic-resonance pulse experiments. *Phys. Rev. Lett.* 69 (21): 3124–3127.

44 Levitt, M.H. and Di Bari, L. (1994). The homogeneous master equation and the manipulation of relaxation networks. *Bull. Magn. Reson.* 16: 94–114.

45 Levante, T.O. and Ernst, R.R. (1995). Homogeneous versus inhomogeneous quantum-mechanical master-equations. *Chem. Phys. Lett.* 241: 73–78.

46 Bloch, F. (1957). Generalized theory of relaxation. *Phys. Rev.* 105 (4): 1206.

47 Hubbard, P.S. (1961). Quantum-mechanical and semiclassical forms of the density operator theory of relaxation. *Rev. Mod. Phys.* 33 (2): 249–264.

48 Breuer, H.P. and Petruccione, F. (2002). *The Theory of Open Quantum Systems.* Oxford University Press on Demand.

49 Gupta, A., Strait-Gardner, T. and Price, W.S. (2021). Is it time to forgo the use of the terms "Spin-Lattice" and "Spin-Spin" relaxation in NMR and MRI?. *J. Phys. Chem. Lett.* 12: 6305–6312.

50 Traficante, D.D. (1991). Relaxation. Can $T_2$, be longer than $T_1$? *Concepts. Magn. Reson.* 3: 171–177.

51 Keeler, J. (2011). *Understanding NMR Spectroscopy.* John Wiley & Sons.

52 Jeener, J., Meier, B.H., Bachmann, P. and Ernst, R.R. (1979). Investigation of exchange processes by two-dimensional NMR spectroscopy. *J. Chem. Phys.* 71 (11): 4546–4553.

53 Kumar, A., Ernst, R.R. and Wüthrich, K. (1980). A two-dimensional nuclear Overhauser enhancement (2D NOE) experiment for the elucidation of complete proton-proton cross-relaxation networks in biological macromolecules. *Biochem. Bioph. Res. Comm.* 95 (1): 1–6.

54 Neuhaus, D. and Williamson, M.P. (2000). *The nuclear Overhauser effect in structural and conformational analysis.* VCH Verlagsgesellschaft Mbh.

55 Goldman, M. (2001). Formal theory of spin-lattice relaxation. *J. Magn. Reson.* 149 (2): 160–187.

56 Andrew, E.R., Bradbury, A. and Eades, R.G. (1958). Nuclear magnetic resonance spectra from a crystal rotated at high speed. *Nature.* 182 (4650): 1659–1659.

57 Andrew, E.R., Bradbury, A. and Eades, R.G. (1959). Removal of dipolar broadening of nuclear magnetic resonance spectra of solids by specimen rotation. *Nature.* 183 (4678): 1802–1803.

58 Lowe, I.J. (1959). Free induction decays of rotating solids. *Phys. Rev. Lett.* 2 (7): 285–287.

59 Polimeno, A. and Freed, J.H. (1995). Slow motional ESR in complex fluids: the slowly relaxing local structure model of solvent cage effects. *J. Phys. Chem.* 99 (27): 10995–11006.

60 Polimeno, A., Barone, V. and Freed, J.H. (2011). Stochastic methods for magnetic resonance spectroscopies. In: *Computational Strategies for Spectroscopy: From Small Molecules to Nano Systems.* (ed. Barone, V.), 549-582. John Wiley & Sons, Ltd.

61 Mehring, M. (1983). *Principles of High Resolution NMR in Solids.* Springer-Verlag.

62 Werbelow, L.G. (2007). *Dynamic Frequency Shift.* Chichester, UK: John Wiley & Sons, Ltd.

63 Smith, S.A., Palke, W.E. and Gerig, J.T. (1994). Superoperator propagators in simulations of NMR spectra. *J. Magn. Reson. Ser. A.* 106 (1): 57–64.

64 Allard, P., Helgstrand, M. and Hard, T. (1997). A method for simulation of NOESY, ROESY, and off-resonance ROESY spectra. *J. Magn. Reson.* 129 (1): 19–29.

65 Allard, P., Helgstrand, M. and Hard, T. (1998). The complete homogeneous master equation for a heteronuclear two-spin system in the basis of Cartesian product operators. *J. Magn. Reson.* 134 (1): 7–16.

66 Ghose, R. (2000). Average Liouvillian theory in nuclear magnetic resonance—principles, properties, and applications. *Concepts Magn. Reson.* 12: 152–172.

67 Tropp, J.A. (1993). Dipolar cross correlation and the nuclear Overhauser effect in systems with strong scalar coupling. *J. Magn. Reson. Ser. A.* 103 (1): 90–91.

**68** Dorai, K. and Kumar, A. (2000). Cross correlations in the longitudinal relaxation of strongly coupled spins. *J. Magn. Reson.* 145 (1): 8–17.

**69** D'Silva, L., Pola, A., Dutta, P., Martinez, G.V., Sprenger, P., Gillies, R.J. et al. (2012). Slow relaxation of longitudinal multispin orders in weakly and strongly coupled two-spin systems. *Magn. Reson. Chem.* 50 (6): 443–448.

**70** Morris, G.A. and Freeman, R. (1979). Enhancement of nuclear magnetic-resonance signals by polarization transfer. *J. Am. Chem. Soc.* 101 (3): 760–762.

**71** Bothner-By, A.A., Stephens, R.L., Lee, J.M, Warren, C.D. and Jeanloz, R.W. (1984). Structure determination of a tetrasaccharide: Transient nuclear overhauser effects in the rotating frame. *J. Am. Chem. Soc.* 106 (3): 811–813.

**72** Bax, A. and Davis, D.G. (1985). Practical aspects of two-dimensional transverse NOE spectroscopy. *J. Magn. Reson.* 63 (1): 207–213.

**73** Carravetta, M., Johannessen, O.G. and Levitt, M.H. (2004). Beyond the $T_1$ limit: singlet nuclear spin states in low magnetic fields. *Phys. Rev. Lett.* 92 (15): 153003.

**74** Levitt, M.H. (2010). Singlet and other states with extended lifetimes. *eMagRes.*

**75** Pileio, G. (2017). Singlet NMR methodology in two-spin-1/2 systems. *Prog. NMR Spectr.* 98-99: 1–19.

**76** Teleanu, F., Sadet, A. and Vasos, P.R. (2021). Symmetry versus entropy: long-lived states and coherences. *Prog. NMR Spectr.* 122: 63–75.

**77** Pervushin, K., Riek, R., Wider, G. and Wüthrich, K. (1997). Attenuated $T_2$ relaxation by mutual cancellation of dipole–dipole coupling and chemical shift anisotropy indicates an avenue to NMR structures of very large biological macromolecules in solution. *Proc. Natl. Acad. Sci. USA.* 94 (23): 12366–12371.

**78** Fernández, C. and Wider, G. (2003). TROSY in NMR studies of the structure and function of large biological macromolecules. *Curr. Opin. Struct. Biol.* 13 (5): 570–580.

**79** Xu, Y. and Matthews, S. (2013). TROSY NMR spectroscopy of large soluble proteins. In: *Topics in Current Chemistry* 97–119.

**80** Schütz, S. and Sprangers, R. (2020). Methyl TROSY spectroscopy: a versatile NMR approach to study challenging biological systems. *Prog, NMR Spectr.* 116: 56–84.

**81** Werbelow, L.G. (2007). *Relaxation Theory for Quadrupolar Nuclei.* Chichester, UK: John Wiley & Sons, Ltd.

**82** Werbelow, L.G. and Kowalewski, J. (1997). Nuclear spin relaxation of spin one-half nuclei in the presence of neighboring higher-spin nuclei. *J. Chem. Phys.* 107 (8): 2775–2781.

**83** Werbelow, L.G. (2006). Effect of rapidly relaxed electron spin anisotropically coupled to nearby nuclear partners. *Magn. Reson. Chem.* 44 (3): 249–254.

**84** Hubbard, P.S. (1963). Theory of nuclear magnetic relaxation by spin-rotational interactions in liquids. *Phys. Rev.* 131 (3): 1155–1165.

**85** Matson, G.B. (1977). Methyl NMR relaxation: the effects of spin rotation and chemical shift anisotropy mechanisms. *J. Chem. Phys.* 67 (11): 5152–5161.

# 5

# Coherence Transfer Pathways

*David E. Korenchan*[1] *and Alexej Jerschow*[2]

[1] *Athinoula A. Martinos Center for Biomedical Imaging, Massachusetts General Hospital, Boston, Massachusetts, United States of America*
[2] *Department of Chemistry, New York University, New York, NY 10003, USA*

The product operator formalism discussed in Chapter 3 represents a powerful tool for a detailed description of the evolution of magnetization modes during an NMR pulse sequence. In this chapter, we discuss coherence transfer pathways (CTPs), which represent a more abstract concept for describing pulse sequences. CTP selection can be applied in many cases in which the use of the product operator formulation is too difficult or impossible. We discuss here the main methods used in coherence pathway selection, phase cycling and pulsed gradients, and we provide examples of how these are implemented in important classes of experiments.

## 5.1 Coherence Transfer Pathways: What and Why?

Directly detectable NMR signals arise only from very specific coherences between spin states. It was the discovery of controlled coherence conversions between detectable and undetectable coherences [1] that led to the rapid development of multi-dimensional and advanced NMR techniques [2]. Often, an important ingredient of experiments is the evolution of different magnetization modes under specific interactions, such that signal modulations can be read out indirectly even from modes that are not directly detectable. Such evolution is often conveniently represented via a CTP, which specifies the sequence of coherence orders during a pulse sequence. Radiofrequency (RF) pulses perform the conversions between coherence and population modes, but these conversions are in most cases not selective. As a result, very few pulse sequences would be complete without additional mechanisms for increasing the specificity of transitions. CTP selection by phase cycling and pulsed-field gradients has been the most successful and most universal tool for providing such functions.

While careful filtering of desired magnetization components is often required to make the experiment targeted for its intended use, a frequent additional implementation of CTP selection is directed toward eliminating artifacts. For example, non-ideal pulses often produce undesired (or unexpected) signals, and these can be reduced or eliminated with a proper CTP selection protocol. One of the first implementations of CTP selection arose in the elimination of quadrature and baseline artifacts for the single-pulse-acquire sequence [3]. Imperfect quadrature detection can be thought of as arising from the fact that the x- and y-axes are not exactly orthogonal to one another, which leads to the appearance of a mirror artifact of the signal with respect to the RF irradiation frequency. Such an artifact peak, like the one seen in Figure 5.1 below, could be easily mistaken for a true NMR peak. We will discuss this example below.

*Two-Dimensional (2D) NMR Methods*, First Edition. Edited by K. Ivanov, P.K. Madhu and G. Rajalakshmi.

CTP selection can also improve the appearance and facilitate the interpretation of NMR spectra. One prominent example is a COrrelated SpectroscopY (COSY) spectrum with and without double-quantum filtration (DQF), as seen in Figure 5.2. The cross-peaks of a DQF-COSY spectrum, unlike those of a COSY spectrum, appear in

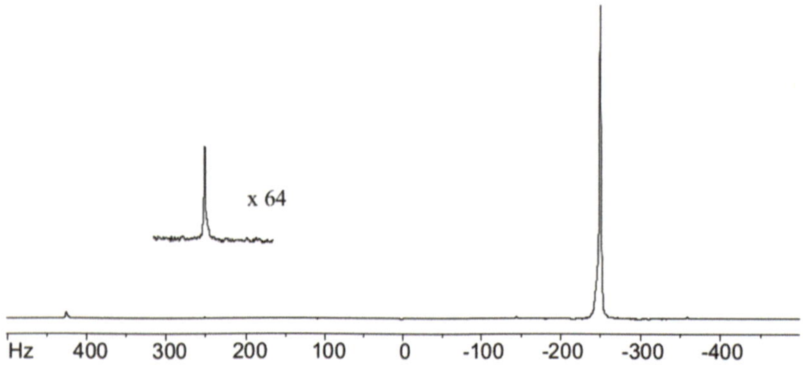


**Figure 5.1** $^{1}$H NMR spectrum of chloroform obtained on an AMX-500 spectrometer with slightly imbalanced quadrature detectors, leading to a small artifact (a "quadrature image" peak) opposite to the chloroform peak. A zoomed-in view of the quadrature image peak is shown directly above it. *Source:* [4], p. 81. Reproduced with permission of Wiley-VCH Verlag GmbH.

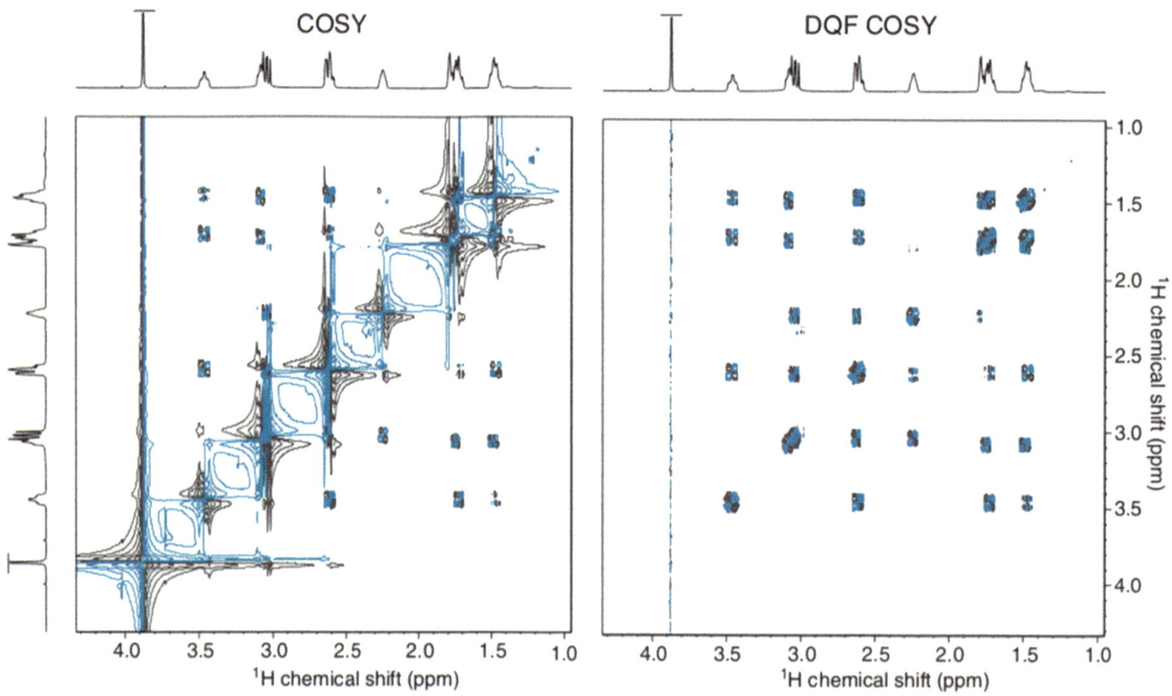


**Figure 5.2** Subsets of experimental COSY and DQF-COSY spectra of quinine, obtained on a 500 MHz NMR spectrometer. Unlike the COSY spectrum, which is dominated by dispersive lineshapes along the diagonal, the double-quantum filter in the DQF-COSY sequence removes some diagonal peaks, and causes the remaining diagonal peaks to be in absorptive mode. As a result, the cross-peaks are much easier to identify and to quantify. *Source:* [5], figure 8.17 (p. 202). Reproduced with permission of John Wiley & Sons, Ltd.

absorption mode, and strong singlet signals are also reduced or eliminated along the diagonal, making spectral analysis easier. Along similar lines, many multiple-quantum coherence (MQC) selection filters select for a given number of coupled spins in a cluster.

## 5.2 Principles of Coherence Selection

The mechanism of coherence selection is rooted in the properties of spherical tensor operators with respect to their rotation around the z-axis, and both phase cycling and pulsed-field gradients directly make use of these. As described below, MQCs of a given order can be differentiated based on these rotational properties. CTPs are typically classified by a sequence of coherence orders with RF pulses in between. The specification of a CTP further provides identification of the function of a given pulse sequence. For example, the differentiation between NOESY and DQF-COSY signals has to be made on the basis of proper coherence selection.

Although it is in principle possible to define CTPs with Cartesian operators, it is much more straightforward to do so with spherical tensor operators or raising and lowering operators. The simplest forms of these are the following:

$$\hat{I}_+ \equiv \hat{I}_x + i\hat{I}_y \tag{5.1}$$

$$\hat{I}_- \equiv \hat{I}_x - i\hat{I}_y. \tag{5.2}$$

These operators perform the interconversion between spin-up ($|\uparrow\rangle$) and spin-down ($|\downarrow\rangle$) states for a single spin. For this reason, they are often referred to as the "raising" ($\hat{I}_+$) and "lowering" ($\hat{I}_-$) operators. This property is expressed mathematically as follows:

$$\hat{I}_+|\downarrow\rangle = |\uparrow\rangle$$

$$\hat{I}_-|\uparrow\rangle = |\downarrow\rangle.$$

More importantly, rotation around the z-axis by an angle $\theta$, which we can represent by the action of a propagator $R_z(\theta) = e^{-i\theta\hat{I}_z}$, results in the spherical operator accumulating a phase factor, rather than converting to a different operator. The sign of the phase depends on whether we consider $\hat{I}_+$ or $\hat{I}_-$:

$$R_z(\theta)\hat{I}_+R_z^\dagger(\theta) = \hat{I}_+e^{-i\theta} \tag{5.3}$$

$$R_z(\theta)\hat{I}_-R_z^\dagger(\theta) = \hat{I}_-e^{i\theta}, \tag{5.4}$$

where the dagger symbol indicates the Hermitian conjugate operation. The rotational property can be shown in a straightforward manner by considering the matrix representations of these operators, especially because the rotation operators are diagonal in this case (owing to $\hat{I}_z$ being diagonal). We can also assign to each operator a coherence order, $p$, which designates the sign of the phase factor accrued by rotation around z. We assign $p = +1$ to $\hat{I}_+$ and $p = -1$ to $\hat{I}_-$. Equations (5.3) and (5.4) can then be combined compactly in terms of the coherence order, $p$, as follows:

$$R_z(\theta)\hat{I}_pR_z^\dagger(\theta) = \hat{I}_pe^{-i\theta p}. \tag{5.5}$$

Such a rotation about z can occur, for example, by precession in the rotating frame due to chemical shifts. For spins $>^1/_2$, the same rotational classification may be made according to the rotational properties of the spherical tensor operators, with the understanding that the coherence order can be larger than 1 for a single spin.

For a collection of spins in the spin-system, at a certain point in the pulse sequence, we may have products of Cartesian operators, which we may expand in terms of $I_+$, $I_-$, $I_z$, and $E$ (the identity operator) acting on individual spins. Each of these product operators can then be assigned a total coherence order $p$ as given by:

$$p \equiv (\textit{number of } I_+ \textit{ terms}) - (\textit{number of } I_- \textit{ terms}). \tag{5.6}$$

Note that $I_z$ and $E$ terms within the product operator do not contribute to the coherence order, as they have coherence-order zero (and do not change when rotated around z). Table 5.1 contains some spherical product operators and their corresponding (combinations of) Cartesian product operators and coherence orders.

During a pulse sequence, the application of an RF pulse will generate a set of new product operators. A perfect $\pi$-pulse is special in that it will simply flip the sign of the coherence order of any operator. Using techniques such as phase cycling and pulsed-field gradients, certain CTPs can be selected, and others suppressed. The *desired* CTPs selected by these techniques are often summarized in a *coherence transfer pathway diagram*. Figure 5.3 below shows an example for the DQF-COSY experiment.

The property of the operators regarding their rotation about z, as described in Equation 5.5 above, can now be used to describe their behavior in two contexts: (i) as the operator evolves under the action of a magnetic field along z and (ii) when the operator is converted into another one by an RF pulse of a given phase.

**Table 5.1** Some spherical product operators and their corresponding Cartesian product operator representations and coherence orders.

| Spherical product operator | Cartesian product operator | Coherence order ($p$) |
|---|---|---|
| $I_+$ | $I_x + iI_y$ | 1 |
| $I_-$ | $I_x - iI_y$ | −1 |
| $I_z$ | $I_z$ | 0 |
| $E$ | $E$ | 0 |
| $2I_+S_-$ | $2(I_xS_x + I_yS_y) - 2i(I_xS_y - I_yS_x)$ | 0 |
| $2I_-S_+$ | $2(I_xS_x + I_yS_y) + 2i(I_xS_y - I_yS_x)$ | 0 |
| $2I_+S_z$ | $2I_xS_z + 2iI_yS_z$ | 1 |
| $2I_zS_-$ | $2I_zS_x - 2iI_zS_y$ | −1 |
| $2I_+S_+$ | $2(I_xS_x - I_yS_y) + 2i(I_xS_y + I_yS_x)$ | 2 |
| $2I_-S_-$ | $2(I_xS_x - I_yS_y) - 2i(I_xS_y + I_yS_x)$ | −2 |

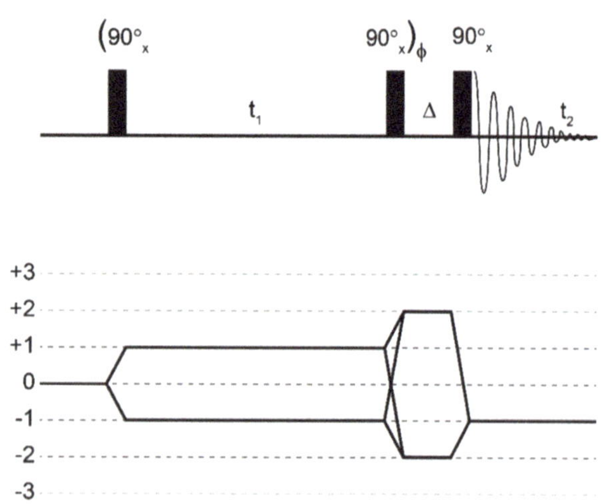


**Figure 5.3** Pulse sequence diagram for DQF-COSY, with its corresponding CTP diagram below. *Source:* [6], figure 5.1 (p. 41). Reproduced with permission of Oxford University Press.

### 5.2.1 Precession of a coherence about the z-component of a magnetic field

Consider the example product operator $P$, with overall coherence order $p$, below:

$$P = I_{1+}I_{2z}I_{3+}E_4I_{5-}\ \dots .$$

The identity operator, $\hat{E}$, is included for all spins without an x-, y-, or z-component contributing to the product operator. We can re-number the spins and group the different $\hat{I}_+$, $\hat{I}_-$, $\hat{I}_z$, and $\hat{E}$ operators together so that we now have the following representation of $P$:

$$\begin{aligned} P &= I_{1+}I_{2+}\dots I_{a+}I_{(a+1),-}\dots I_{(a+b),-}I_{(a+b+1),z}\dots I_{(a+b+c),z}E_{(a+b+c+1)}\dots E_{(a+b+c+d)} \\ &= \prod_{j=1}^{a} I_{j+}\prod_{k=1}^{b} I_{(a+k),-}\prod_{l=1}^{c} I_{(a+b+l),z}\prod_{m=1}^{d} E_{(a+b+c+m)}, \end{aligned}$$

where one finds $p = a - b$. The rotation of the product operator about a magnetic field pointing along z through an angle $\theta$ can be described by the action of the propagator $R_z(\theta) = e^{-i\theta\hat{I}_z}$, where $\hat{I}_z = \sum_i \hat{I}_{iz}$ is the sum over all individual $\hat{I}_z$ operators. The expression can be expanded into the individual spin operations by using the identity $R_z^\dagger(\theta)R_z(\theta) = \hat{1}$ to give:

$$\begin{aligned} &R_z(\theta)PR_z^\dagger(\theta) \\ &\quad = R_z(\theta)I_{1+}\left[R_z^\dagger(\theta)R_z(\theta)\right]\hat{I}_{2+}\left[R_z^\dagger(\theta)R_z(\theta)\right]\dots\left[R_z^\dagger(\theta)R_z(\theta)\right]E_{(a+b+c+d)}R_z^\dagger(\theta) \\ &\quad = \prod_{j=1}^{a} R_z(\theta)I_{j+}R_z^\dagger(\theta)\prod_{k=1}^{b} R_z(\theta)I_{(a+k),\,-}R_z^\dagger(\theta)\prod_{l=1}^{c} R_z(\theta)I_{(a+b+l),z}R_z^\dagger(\theta)\prod_{m=1}^{d} R_z(\theta)E_{(a+b+c+m)}R_z^\dagger(\theta). \end{aligned}$$

We find that rotation of $P$ is equivalent to the product of the rotations of its individual terms. We can use the property of spherical operators defined above (Equations 5.3 and 5.4) in order to write these rotations as accumulations of phase, then recombine the terms again (note that the terms $R_z(\theta)$, $I_{(a+b+l),z}$, $E_{(a+b+c+m)}$, and $R_z^\dagger(\theta)$ commute):

$$\begin{aligned} &\prod_{j=1}^{a} I_{j+}e^{-i\theta}\prod_{k=1}^{b} I_{(a+k),-}e^{i\theta}\prod_{l=1}^{c} I_{(a+b+l),z}\prod_{m=1}^{d} E_{(a+b+c+m)} \\ &\quad = e^{-i\theta(a-b)}\prod_{j=1}^{a} I_{j+}\prod_{k=1}^{b} I_{(a+k),-}\prod_{l=1}^{c} I_{(a+b+l),z}\prod_{m=1}^{d} E_{(a+b+c+m)} = Pe^{-i\theta p}. \end{aligned}$$

We thus arrive at a central result: At any time, the acquired phase angle of the coherence $P$ due to rotation around z is proportional to the overall coherence order, $p$:

$$R_z(\theta)PR_z^\dagger(\theta) = Pe^{-i\theta p}. \tag{5.7}$$

### 5.2.2 Effect of changing the phase of a radiofrequency pulse that converts one coherence order term into another

Consider a pulse along the +x-axis and a flip angle $\alpha$, which converts a spherical product operator $P_1$ into $P_2$, changing the coherence order from $p_1$ to $p_2$. We indicate this change in coherence order by the notation $(p_1 \to p_2)$. The pulse is described by the propagator $R_x(\alpha) = e^{-i\alpha\hat{I}_x}$. Increasing the phase of this pulse by an angle $\varphi$ corresponds to a rotation about the z-axis. This procedure produces a new propagator:

$$R_\varphi(\alpha) = R_z(\varphi)R_x(\alpha)R_z^\dagger(\varphi).$$

Note that by the properties of a unitary operator (such as those describing rotations),

$$R_\varphi^\dagger(\alpha) = R_z(\varphi)R_x^\dagger(\alpha)R_z^\dagger(\varphi).$$

Also note that in the two equations above, $R_z^\dagger(\varphi) = R_z(-\varphi)$ and $R_x^\dagger(\alpha) = R_x(-\alpha)$. Then the mathematical expression for the product operator $P_1$ under the influence of the RF pulse will be:

$$R_\varphi(\alpha)P_1R_\varphi^\dagger(\alpha) = R_z(\varphi)R_x(\alpha)R_z^\dagger(\varphi)P_1R_z(\varphi)R_x^\dagger(\alpha)R_z^\dagger(\varphi)$$

We can utilize Equation 5.7 to perform the substitution $R_z^\dagger(\varphi)P_1R_z(\varphi) = P_1e^{i\varphi p_1}$. Our expression thus becomes:

$$R_z(\varphi)R_x(\alpha)\left[P_1e^{i\varphi p_1}\right]R_x^\dagger(\alpha)R_z^\dagger(\varphi).$$

The term $R_x(\alpha)P_1R_x^\dagger(\alpha)$ (moving the phase term $e^{ip_1\varphi}$ to the end of the expression) describes the action of the RF pulse along the +x-axis on $P_1$. The pulse may create a number of terms with different coherence orders. Let's consider only one of such terms $P_2$, with coherence order $p_2$. For this component, the resulting expression will look as follows:

$$R_z(\varphi)P_2R_z^\dagger(\varphi)e^{i\varphi p_1}.$$

We can use again Equation 5.7 to perform the substitution $R_z(\varphi)P_2R_z^\dagger(\varphi) = P_2e^{-i\varphi p_2}$. In doing so, we arrive at the relationship between the change in pulse phase and the coherence orders of the desired CTP:

$$R_\varphi(\alpha)P_1R_\varphi^\dagger(\alpha) \rightarrow P_2e^{-i\varphi(p_2-p_1)} = P_2e^{-i\varphi\Delta p}. \tag{5.8}$$

This expression shows that a change in pulse phase, $\varphi$, introduces a phase term for $P_2$ that is proportional to the *difference* in coherence orders, $\Delta p = p_2 - p_1$ between the states before and after the pulse.

Based on these transformations, one can also conclude that if instead of a single pulse, we considered a sequence of pulses, which were all shifted in phase by the same amount, the same overall phase factor would be observed, which only depends on the initial and the final coherence orders. This observation can greatly simplify the design of phase cycling schemes, as we discuss below. Although the treatment here mainly focused on a collection of spin-$^1/_2$ nuclei, the extension to spins $>^1/_2$ is straightforward and follows the same pattern. Notably, the central expressions represented in Equations 5.7 and 5.8 apply for these as well.

## 5.3 Coherence Transfer Pathway Selection by Phase Cycling

Phase cycling has the longest history of utilization for CTP selection [7, 8]. The principle of CTP selection by phase cycling is based on the following strategy: Phases of individual or grouped RF pulses within the NMR pulse sequence can be changed in consecutive experiments such that when the signals are co-added, the signals arising from the desired pathways accumulate *constructively*, while the signals from undesired pathways *cancel* each other.

More formally, the requirements are: The phases of pulses or a sequence or block of pulses and the receiver phase must be iterated over a given number of scans, $N$, such that the following conditions are met:

1. For *desired* CTPs, the cumulative phase should change by an integer multiple of $2\pi$ between scans; and
2. For *undesired* CTPs, the cumulative phases from different scans should be evenly spaced around the unit circle when performing all phase cycle steps.

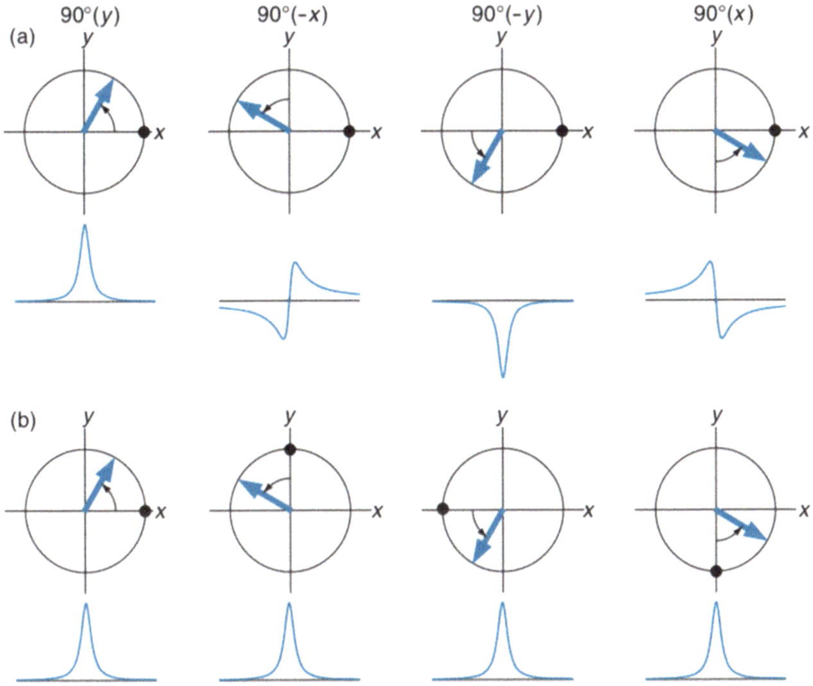


**Figure 5.4** Graphic depicting how the phase of the detected (−1)-coherence relates to the receiver phase. In each of the four phase cycles shown, the magnetization arising from the coherence is represented by a precessing vector, and the black dot represents the receiver phase. In (a), the changing excitation pulse phase leads to changes in the detected NMR peak between absorptive and dispersive modes. When summed together, these four spectra would lead to a net zero NMR signal. In (b), the receiver phase tracks the phase of the excitation pulse, thereby resulting in the same absorptive NMR peak mode in all four transients, which would sum to produce a non-zero NMR signal. *Source:* [5], figure 11.6 (p. 394). Reproduced with permission of John Wiley & Sons, Ltd.

This situation is illustrated in Figure 5.4. The principle of CTP selection via phase cycling is encapsulated in Equation 5.8, which relates a change in pulse phase to a coherence-order dependent change in phase of the product operator, and all phase cycling strategies derive from this property.

For a pulse sequence with $n$ pulses or $n$ cycled blocks, a given CTP with $n+1$ elements is denoted:

$$p = (p_0, p_1, p_2, \dots, p_n).$$

$p_i$ indicates coherence order after the $i$-th pulse, and $p_0$ refers to the coherence order before the first pulse. In the vast majority of cases, we assume $p_0 = 0$, if we start from thermal equilibrium, and the convention for detection operators is to detect coherence order $-1$, i.e. $p_n = -1$. One need not use these constraints in order to develop the general formalism. In fact, often one may wish to develop a phase cycle for only part of the pulse sequence, and in such a case, these conditions would not be used.

A phase cycle consists of $M$ steps with the phases of the pulses of the $m$-th step indicated as follows:

$$\boldsymbol{\varphi}^{(m)} = \left(\varphi_1^{(m)}, \varphi_2^{(m)}, \varphi_3^{(m)} \dots, \varphi_n^{(m)}\right).$$

We denote the coherence order jumps during the pulses by:

$$\boldsymbol{\Delta p} = (p_1 - p_0, p_2 - p_1, \dots, p_n - p_{n-1}). \tag{5.9}$$

Following the discussion of the previous section, one can write the accumulated phase for a given phase cycle step with index $m$ and a given phase setting $\boldsymbol{\varphi}^{(m)}$ as:

$$\varphi_{tot}^{(m)} = -\boldsymbol{\varphi}^{(m)} \cdot \boldsymbol{\Delta p}.$$

One may then further subtract from the accumulated phase a given receiver phase $\varphi_{rec}^{(m)}$. This procedure would amount to rotating the signal within the complex plane by the specified angle, $\varphi_{rec}^{(m)}$. As a result, the final signal phase recorded on the computer from the $m$-th phase cycling step would be:

$$\varphi_{acq}^{(m)} = -\left(\boldsymbol{\varphi}^{(m)} \cdot \boldsymbol{\Delta p}\right) - \varphi_{rec}^{(m)}. \tag{5.10}$$

The signal contribution from a given CTP will then be:

$$s_p^{(m)} = s_0 e^{-i\,\varphi_{acq}^{(m)}}.$$

In the equation above, $s_0$ is the portion of the signal independent of pulse phases, which includes factors related to coil efficiency, pulse flip angle effects, etc. After co-adding all signals from $M$ steps in a phase cycle, one has:

$$s_{tot} = s_0 \sum_{m=1}^{M} e^{-i\,\varphi_{acq}^{(m)}}.$$

The task of setting up a suitable phase cycle can be defined as finding the smallest number of $M$ steps and the associated $\varphi_i^{(m)}$ and $\varphi_{rec}^{(m)}$ values to achieve $\varphi_{acq}^{(m)} = const$ for all desired CTPs, so that these pathways are accumulated in the summation, and $\varphi_{acq}^{(m)}$ to cover the unit circle uniformly for undesired CTPs, so that these pathways are canceled.

A key consideration of phase cycling is that cycling a pulse (or block of pulses) through $M_i$ phase steps, with the phase being incremented by $\frac{2\pi}{M_i}$ in consecutive scans, can select more than one coherence order jump. If the receiver phase $\varphi^{(m)}$ is set to select one value of $\Delta p$, the sequence will also select $\Delta p \pm kM_i$ , where $k$ is an integer, since the $kM_i$ term will only contribute a phase of a multiple of $2\pi$. Say, for example, that one pulse undergoes a three-step phase cycle $\left[0, \frac{2\pi}{3}, \frac{4\pi}{3}\right]$, designed to select a coherence order change $\Delta p = +1$. The sequence will then also select all values of $\Delta p$ that are spaced apart by three, i.e. it will select $\Delta p = 1 + 3k = \ldots -5, -2, +1, \ +4, \ldots$ .

We can sum up these ideas in a general set of steps for designing a phase cycling scheme to perform CTP selection:

1. Identify the spin system: how many $J$-coupled or dipolar-coupled spins does it contain? What is the largest coherence order it could sustain?
2. Identify the desired CTPs, and calculate the coherence order jump at each pulse or pulse block ($\boldsymbol{\Delta p}$) for each CTP using Equation 5.9.
3. For each $i$-th (block of) pulse(s), set the number of phase cycle steps ($M_i$) to equal the difference in allowable coherence order jumps between the desired CTPs. For instance, if one CTP has $\Delta p_i = +1$ and another has $\Delta p_i = -3$, such as is the situation, for example after pulse sequence element A in Figure 5.5 below, set $M_i = 1-(-3) = 4$, that is, cycle the $i$-th pulse evenly across the unit circle in four steps as Table 5.2 shows. Choosing a large enough value of $M_i$ will hence minimize the selection of other potentially undesired pathways. If there is only one desired pathway, then the latter consideration becomes the deciding factor for the number of steps and consequently for the duration of the experiment.
4. Once you have determined each value of $M_i$, multiply these together to determine the total number of steps, $M$, of the phase cycle.
5. Create a table of values where the phase of each pulse (or block) cycles by incrementing the phase by the values calculated in Step 3. An example is shown below in Table 5.2, which has corresponding allowed and forbidden CTPs as shown in Figure 5.5. If multiple pulse sequence elements are phase cycled, one needs to nest one phase cycle within the other, i.e. change one pulse (or block) phase while holding all other phases constant. An

example of nested phase cycling is shown in Table 5.3, which provides a possible phase cycling scheme for a DQF-COSY sequence, where all three pulses are cycled independently.

6. For each phase cycle step $m$, calculate the value of $\varphi_{rec}^{(m)}$ such that the desired CTP is selected, according to Equation 5.10, such that $\varphi_{acq}^{(m)}$ is zero or constant for all phase cycling steps.

In order to simplify the phase cycling scheme as determined above, one could choose not to phase cycle the final pulse (or block). This is possible because CTPs generally end with order −1 during detection. However, the final pulse (or block) may need to be cycled with a Cyclically Ordered Phase Sequence (CYCLOPS) scheme, described below, in order to remove quadrature artifacts and/or the zero-frequency artifact.

Occasionally, it is important to consider the accumulated relative phase from symmetric pathways (i.e. pathways in which $p_i = -p'_i$), if one wishes to preserve these pathways. Consider the two symmetric CTP segments $(p_1 \rightarrow p_2)$ and $(-p_1 \rightarrow -p_2)$. Such segments occur often in multiple-quantum filtered sequences. A given pulse of zero phase and flip angle $\alpha$ performing these transformations would introduce flip angle factors that could differ by a minus sign, according to this Wigner rotation matrix relationship of rank $l$:

$$d^l_{p_1,p_2}(\alpha) = (-1)^{p_1-p_2} d^l_{-p_1,-p_2}(\alpha). \tag{5.11}$$

For example, in a triple-quantum filter segment, one may have a pathway of the type $(0 \rightarrow +3)$, but also $(0 \rightarrow -3)$. It is advantageous to keep both CTPs in order to maximize the signal. A six-step phase cycle would select both (+3)- and (-3)-coherences, but a relative sign would be incurred regardless of the flip angle, according to Equation 5.11. In such a case, one needs to take care to compensate for this sign. This can be achieved by shifting the pulse phase by an angle of $\pi/N$ (with $N = 6$), which, according to the phase cycling rules discussed above, would produce a relative negative sign to compensate for the aforementioned effect (see Equation 5.8, in which one would substitute

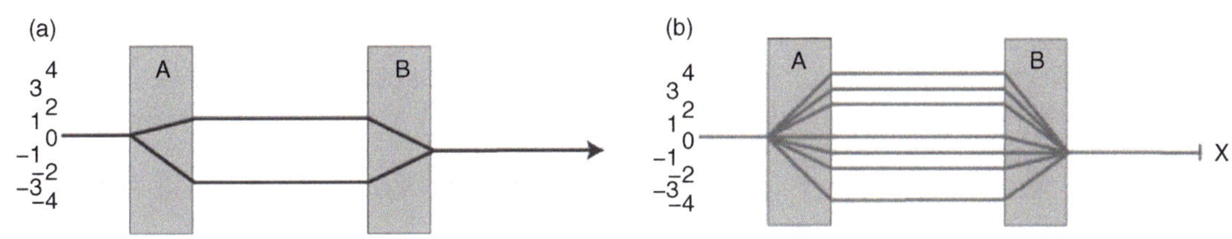


Figure 5.5 CTPs that are (a) allowed and (b) suppressed by the phase cycling scheme in Table 5.2. *Source:* [9], figures A.20 and A.21 (p. 636–37). Reproduced with permission of John Wiley & Sons, Ltd.

Table 5.2 Example four-step phase cycle for two pulse sequence elements A and B, each comprised of one or more RF pulses. The selected and suppressed CTPs are depicted in Figure 5.5.

| Cycle counter | $\varphi_A$ | $\varphi_B$ | $\varphi_{rec}$ |
|---|---|---|---|
| 0 | 0 | 0 | 0 |
| 1 | $\pi/2$ | 0 | $3\pi/2$ |
| 2 | $\pi$ | 0 | $\pi$ |
| 3 | $3\pi/2$ | 0 | $\pi/2$ |

*Source:* [9], equation A.68 (p. 636). Reproduced with permission of John Wiley & Sons, Ltd.

**Table 5.3** Nested phase cycling for a DQF-COSY sequence. The corresponding CTP diagram is shown in Figure 5.3.

| Cycle counter | $\varphi_1$ | $\varphi_2$ | $\varphi_3$ | $\varphi_{rec}$ |
|---|---|---|---|---|
| 0 | 0 | 0 | 0 | 0 |
| 1 | $\pi$ | 0 | 0 | $\pi$ |
| 2 | 0 | 0 | $\pi/2$ | $3\pi/2$ |
| 3 | $\pi$ | 0 | $\pi/2$ | $\pi/2$ |
| 4 | 0 | 0 | $\pi$ | $\pi$ |
| 5 | $\pi$ | 0 | $\pi$ | 0 |
| 6 | 0 | 0 | $3\pi/2$ | $\pi/2$ |
| 7 | $\pi$ | 0 | $3\pi/2$ | $3\pi/2$ |

*Source:* [9], p. 645. Reproduced with permission of John Wiley & Sons, Ltd.

$\Delta p\varphi = \left(3 * \frac{\pi}{6}\right) = \frac{\pi}{2}$, and $\Delta p\varphi = \left(-3 * \frac{\pi}{6}\right) = -\frac{\pi}{2}$ for the two different pathways). Thus, rather than using a six-step phase cycle of $\left[0, \frac{\pi}{3}, \frac{2\pi}{3}, 3\pi/3, \frac{4\pi}{3}, \frac{5\pi}{3}\right]$, one would use the phase-shifted cycle $\left[\frac{\pi}{6}, \frac{3\pi}{6}, \frac{5\pi}{6}, \frac{7\pi}{6}, \frac{9\pi}{6}, \frac{11\pi}{6}\right]$.

Selecting CTPs via phase cycling has some important advantages. To begin with, it is very easy to implement within a pulse sequence program on an NMR spectrometer. In addition, there are no special hardware requirements, such as pulsed-field gradients, which are less wide-spread (especially on many low-field instruments). However, there are disadvantages to phase cycling as well. First, because it requires multiple acquisitions in order to isolate the desired CTPs, the acquisition times scale with the complexity of the pathway selection tasks (i.e. generally, the more pulses, the larger the cycle). Furthermore, this property places strong requirements on good instrument stability and stable sample conditions. Lastly, more stringent CTP selection (i.e. fewer allowed pathways) requires more phase cycle steps, adding to the time and complexity of the sequence. Many of these disadvantages can be addressed by using pulsed gradient selection of CTPs.

Finally, we mention some specific phase cycling schemes that are pervasively used as components of more complex phase cycles.

### 5.3.1 CYCLOPS

Imperfections in the NMR detection setup can cause artifacts to appear in the acquired NMR spectrum, as seen earlier in Figure 5.1. For example, if the quadrature detection axes are misaligned (i.e. are not perfectly perpendicular to one another), then the NMR spectrum will show not only a peak corresponding with a particular nucleus in the sample, but also a smaller artifact peak at the negative frequency, that is, opposite the real peak from the center of the spectrum. A similar artifact is seen if the detectors are properly aligned but imbalanced in their detection sensitivities. This artifact is known as a "quadrature artifact" or "quadrature image" because it can arise in quadrature detection systems. In both cases, it can be shown that what is detected corresponds not only to $(-1)$-coherence ($I_-$), but also a small component of $(+1)$-coherence ($I_+$). Because typically every NMR experiment begins with $I_z$ operators (with coherence order $p = 0$), this undesired CTP is $(0 \rightarrow +1)$. [3, 10]. Quadrature artifacts are rare on modern NMR spectrometers, because these often employ digital filtering [11], which effectively removes the mirror artifact.

Another NMR artifact is observed when either of the detectors has a constant (or DC) voltage bias, resulting in a peak corresponding to zero frequency seen in the center of the NMR spectrum. Conceptually, because this signal does not precess nor change coherence order with the RF pulse, it can be thought of as an undesired $(0 \rightarrow 0)$ CTP.

The desired CTP can be represented as $(0 \rightarrow -1)$, corresponding to the coherence order difference through the pulse sequence $\Delta p = -1$. Using the rules of phase cycling described previously, one can identify the phase cycling scheme summarized in Table 5.4 to select this CTP:

This scheme is known as the CYCLOPS phase cycle scheme. It can be implemented either for a single-pulse sequence, or can be used as a 'supercycle' in a multi-pulse sequence. In the latter case all pulse and receiver phases are incremented by the values above. This approach is very effective for removing artifacts, but it will make any sequence four times as long. The shortest cycle that would achieve this selection would be a three-step cycle with $2\pi/3$ phase increments, and if one only wished to remove the spike at zero frequency, then a two-step cycle would be sufficient.

### 5.3.2 EXORCYCLE

The EXORCYCLE phase cycling scheme addresses artifacts due to imperfections in $\pi$-pulses, for example, in the implementation of refocusing elements. A perfect $\pi$-pulse reverses the sign of a coherence order; thus, in the case of $(\pm 1)$-coherences, it will select the CTPs $(-1 \rightarrow +1)$ and $(+1 \rightarrow -1)$. However, if the $\pi$-pulse is imperfect (i.e. the actual tip angle is greater than or less than $\pi$ radians), then other coherence orders may be generated and can eventually end up as detectable $(-1)$-coherences, giving rise to artifacts or undesired signal contributions in the acquired NMR spectrum. In the limit of small flip angle errors, undesired coherence order jumps generated by imperfect $\pi$-pulses will have a signal intensity proportional to the flip angle error raised to the power of the coherence order deviation [12].

A simple cycle is often used in such situations. The two desired CTPs mentioned above have coherence order changes $\Delta p = \pm 2$. Thus, one can choose the following phase cycling parameters, listed in Table 5.5 below.

**Table 5.4** Four-step CYCLOPS phase cycle for selecting only the $(0 \rightarrow -1)$ CTP.

| Cycle counter | $\varphi_{RF}$ | $\varphi_{rec}$ |
|---|---|---|
| 0 | 0 | 0 |
| 1 | $\pi/2$ | $\pi/2$ |
| 2 | $\pi$ | $\pi$ |
| 3 | $3\pi/2$ | $3\pi/2$ |

**Table 5.5** Four-step EXORCYCLE phase cycle for selecting only the $(-1 \rightarrow +1)$ and $(+1 \rightarrow -1)$ CTPs.

| Cycle counter | $\varphi_{RF}$ | $\varphi_{rec}$ |
|---|---|---|
| 0 | 0 | 0 |
| 1 | $\pi/2$ | $\pi$ |
| 2 | $\pi$ | 0 |
| 3 | $3\pi/2$ | $\pi$ |

This is the EXORCYCLE phase cycling scheme. Of course, this four-step phase cycle will also select higher coherence order jumps, including $\Delta p = \pm 6$ if the spin system can support them.

## 5.4 Cogwheel Phase Cycling

Cascading (or 'nesting') phase cycles is a straightforward way to set up a comprehensive phase cycle, but it comes with the notable disadvantage that this approach is exponential in the number of pulses. As a result, while the sequence will work well, the experiment can become prohibitively lengthy. Several strategies can be used to alleviate such effects. For example, one may choose to cycle pulses as blocks, whereby the phase of a group of pulses is incremented by the same amount between steps. This approach can work well if only the total coherence order change matters.

Cogwheel phase cycling is a type of concerted phase incrementation, but it had not been exploited until it was explicitly formulated by Levitt et al. [13]. The phase cycling scheme involves incrementing the phase of individual pulse sequence elements by different integer multiples of the same fundamental phase, such that each element performs a different number of full rotations of the phase about the unit circle. The analogy to meshing cogwheels of different sizes, where all rotate together but at different angular frequencies, is what gave this approach the name. The main power of this approach lies in the fact that it can avoid exponential scaling of the CTP selection problem.

A short example should illustrate this advantage: consider three $\pi$-pulses that would be used to invert coherence orders. If the desired CTP is $(+3 \rightarrow -3 \rightarrow +3 \rightarrow -3)$, and if all other pathways between the coherence orders of $\pm 3$ need to be suppressed, a nested phase cycle of $7^3$=343 steps would be required. Alternatively, by the cogwheel cycling approach one could consider shifting the first and the third pulse by $2\pi/19$, and the second pulse by $-2\pi/19$, and shifting the receiver phase by $2\pi * 18/19$ in consecutive steps of a 19-step cycle. This procedure would eliminate any other coherence transfers within the $\pm 3$ coherence order range.

Following the formalism for cogwheel phase cycling [13], we define the winding numbers $w_i$, which are integers and specify by how much the phase in a consecutive step will be incremented. Assuming a total number $M$ of phase cycling steps and considering the phase cycle counter $m$ to range from 0 to $M - 1$, the phases for a cogwheel cycle are set as follows:

$$\varphi^{(m)} = \left( w_1 m \frac{2\pi}{M}, w_2 m \frac{2\pi}{M}, w_3 m \frac{2\pi}{M} \ldots, w_N m \frac{2\pi}{M} \right).$$

Using Equation 5.10 one can then show that cycling the receiver phase with a winding number:

$$w_{rec} = -\sum_{i=1}^{N} w_i \Delta p_i$$

will produce an overall signal phase that cancels all accumulated phase for the pathway considered for all steps of the phase cycle. The key task here is to find the minimal $M$ and the winding numbers $w_i$, such that all undesired pathways are eliminated. This problem does not, at present, have a general solution, but numerical searches can be performed in a straightforward fashion if the number of pulses is not too large. Figure 5.6 shows a pictorial representation of an example of a cogwheel phase cycling scheme. In this example (corresponding to a FAM-RIACT-FAM sequence), pulse sequence block B is not cycled, whereas block A increments by $6\pi/23$ each scan, and block C increments by $2\pi/23$ each scan [13].

Select problems, such as a train of $\pi$-pulses with the goal of fully inverting the maximum coherence order at every step, can be solved exactly with cogwheel phase cycling. Such cases appear to show the largest benefit with respect to reducing the number of phase cycling steps compared to conventional cascaded or nested phase cycling.

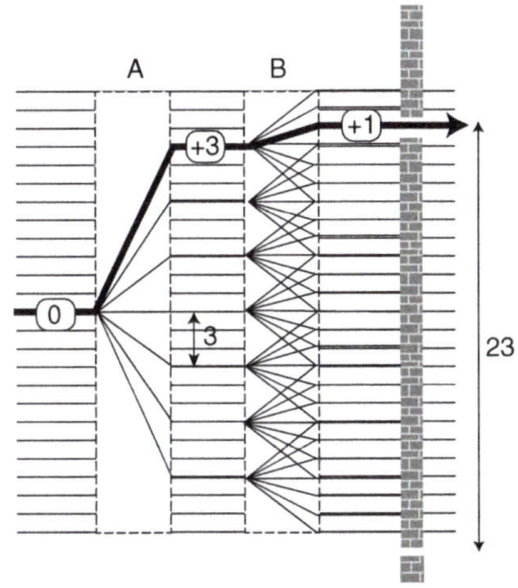


**Figure 5.6** Graphic illustrating the concept of cogwheel phase cycling with three pulse sequence elements A, B, and C (not shown). The difference in winding numbers between elements A and B is three times larger than that between B and C, leading to A phase cycling three times faster. The entire pulse sequence has 23 phase cycle steps, which is the minimum number to select the desired CTP ($0 \rightarrow +3 \rightarrow +1$). *Source:* [13], figure 2. Reproduced with permission of Academic Press.

Cogwheel phase cycles can be easily calculated using computer programs [14]. In addition to having several applications in liquid-state 2D and 3D NMR [15], it can also be used to efficiently reduce sidebands in solid-state NMR [16].

## 5.5 Coherence Transfer Pathway Selection by Pulsed-field Gradients

The use of pulsed-field gradients for CTP selection was first reported by Bax et al. [17], and today it represents an important approach. The technique can be implemented by itself, or in conjunction with phase cycling in order to use the advantages of both approaches. For more details than what will be covered here, we direct the interested reader to the following review [18]. The level of complexity found in phase cycle design can often be greatly reduced by utilizing pulsed-field gradients. The principle leveraged by gradients is the dependence of the precession frequencies of different coherences under the influence of the external magnetic field, and thus the same rotational principles illustrated in Equation 5.7 are employed here. By applying a magnetic field gradient, one can induce a spatially-dependent precession frequency, which will lead to an overall dephasing of the magnetization components. Essentially, the coherence selection principle is the following: *dephase* coherences by varying their precession frequencies across space using a linear pulsed-field gradient, and then after a pulse (or block of pulses) selectively *rephase* only the desired CTP(s).

If we apply a linear magnetic field gradient along the z-axis with a time-dependent amplitude $G_z(t)$ over a period of time, $T$, then the time-dependent z-component of the magnetic field across the z-axis ($B_z(z,t)$) can be written as:

$$B_z(z,t) = \gamma\,(B_0 + G_z(t)z)\,.$$

The Hamiltonian describing the interaction of all spherical product operators with the main magnetic field and gradient field is:

$$\hat{H}(z,t) = \hat{H}_0 + \hat{H}_G(z,t) = \omega_0 \hat{I}_z + \gamma G_z(t) z \hat{I}_z.$$

After completion of the time period $T$, the propagator can be written as:

$$e^{-i\omega_0 T \hat{I}_z} e^{-i\left(\gamma z \int_0^T G_z(\tau)d\tau\right)\hat{I}_z}.$$

Since all the terms are aligned with the static magnetic field, this propagator will simply lead to a rotation of all components around the z-axis, and, in particular, for a term $P$ of coherence order $p$, one obtains, in analogy to Equation 5.7:

$$e^{-i\left(\gamma z \int_0^T G_z(\tau)d\tau\right)\hat{I}_z} P e^{i\left(\gamma z \int_0^T G_z(\tau)d\tau\right)\hat{I}_z} = P e^{-i\left(\gamma z \int_0^T G_z(\tau)d\tau\right)p}.$$

The total phase, $\varphi(z)$, of the operator $P$ accumulated as a function of its location along the z-axis in space is:

$$\varphi(z) = p\left(\gamma z \int_0^T G_z(\tau)d\tau\right).$$

This situation leads to a helical distribution of phases across space, as can be seen in Figure 5.7. The larger the integral of $G_z(t)$ over time, the higher is the "spatial winding frequency" of this helix, and the lower the total NMR signal would be, were we to measure it as it precesses about the z-axis. Figure 5.8 demonstrates this fact by plotting the total signal attenuation as a function of the spatial frequency of the phase fluctuation across z.

Following some conversion between coherences by a RF pulse or a block of pulses, one can use the same rotational principles to refocus only the desired combination of coherence orders.

Consider that the first pulsed-field gradient, with profile $G_{z,A}(t)$ applied for total time $T_A$, dephased (among all other terms) a desired coherence term $P_A$ with coherence order $p_A$. The phase accumulation across space, as we have seen, is given by $\varphi_A(z) = \left(\gamma z \int_0^{T_A} G_{z,A}(\tau)d\tau\right) p_A$. A subsequent RF pulse (or series of RF pulses) converts the coherence term $P_A$ into a different coherence $P_B$, which now has coherence order $p_B$. If we are to detect the

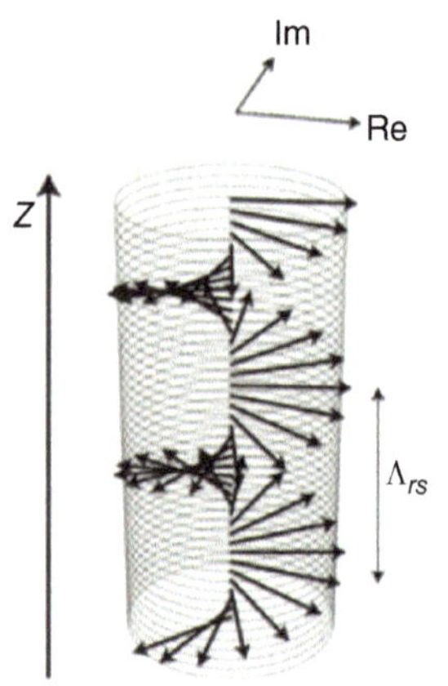


**Figure 5.7** Graphic depicting the xy-phase dispersion across space of a coherence when subjected to a linear field gradient. The parameter $\Lambda_{rs}$ denotes the spatial wavelength of the sinusoidal phase oscillation. *Source:* [9], figure A.32 (p. 651). Reproduced with permission of John Wiley & Sons, Ltd.

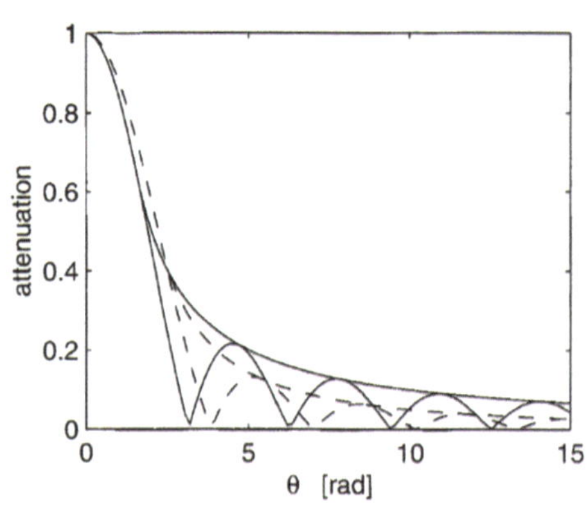


**Figure 5.8** Attenuation of the total detectable NMR signal summed across space as a function of $\theta = \gamma r \int_0^T G_r(\tau)d\tau$, where $\gamma$ is the gyromagnetic ratio, $G_r(t)$ is the gradient profile in time from $t = 0$ to $t = T$, and $r$ is the distance along the axis of the gradient. The solid and dashed lines indicate the attenuation function plus its envelope for gradients along the z-axis and along the x- or y-axis, respectively. *Source:* [19], figure 2. Reproduced with permission of Academic Press.

signal that may arise from $P_B$ (perhaps after an RF pulse that converts it to a detectable $(-1)$-coherence), we must apply another gradient pulse, with profile $G_{z,B}(t)$ applied for total time $T_B$, producing a spatially-varying phase $\varphi_B(z) = \left(\gamma z \int_0^{T_B} G_{z,B}(\tau)d\tau\right)$ that will undo the phase accrual across space due to the first gradient pulse. This requirement can be expressed as:

$$\varphi_A(z) + \varphi_B(z) = 0.$$

From this expression, one obtains the condition:

$$\frac{\int_0^{T_A} G_{z,A}(\tau)d\tau}{\int_0^{T_B} G_{z,B}(\tau)d\tau} = -\frac{p_B}{p_A}. \tag{5.12}$$

The most common gradient pulse shapes are either rectangular or a half-sine function, with the latter being often used simply to make the gradient pulse gentler on the hardware (although nowadays this consideration is becoming less important). The rectangular shape is the most time-efficient version, allowing the maximum dephasing per unit time. For rectangularly-shaped gradients, Equation 5.12 becomes:

$$\frac{G_{z,A}T_A}{G_{z,B}T_B} = -\frac{p_B}{p_A}. \tag{5.13}$$

The negative sign affects the relative polarity of the gradient pulses; if $p_A$ and $p_B$ have the same sign, then the gradients must be opposite in polarity from one another, and the gradients should have the same polarity if $p_A$ and $p_B$ have opposite signs. In order to match the gradient pulse integral ratio to the coherence order ratio, either the gradient pulse length, or amplitude, or both may be adjusted. Figure 5.9 below illustrates how CTPs may be selected using pulsed-field gradients.

Two very important points should be emphasized here. Because only the ratio of the coherences matter for the selection, the process is not unique. For example, if the ratio of the gradient pulses before and after the RF pulse are 1:2, then they will not only select a $(+2 \rightarrow -1)$ coherence transfer, but also a $(-2 \rightarrow +1)$ transfer, and also a $(+4 \rightarrow -2)$ coherence transfer if the spin system is large enough to sustain $(\pm 4)$-coherences. Therefore, it is essential to examine whether the selection process will be unique for the spin system in question. Another important aspect is that pulsed-field gradients can never be used to eliminate zero-order coherences (in a homonuclear system), but they could be used to eliminate all coherences other than zero-order ones. This last application is often a component of signal saturation (e.g. for a saturation recovery pulse sequence). It is also important to note that, *unlike phase cycling, pulsed-field gradient selection depends on the desired coherence orders before and after each*

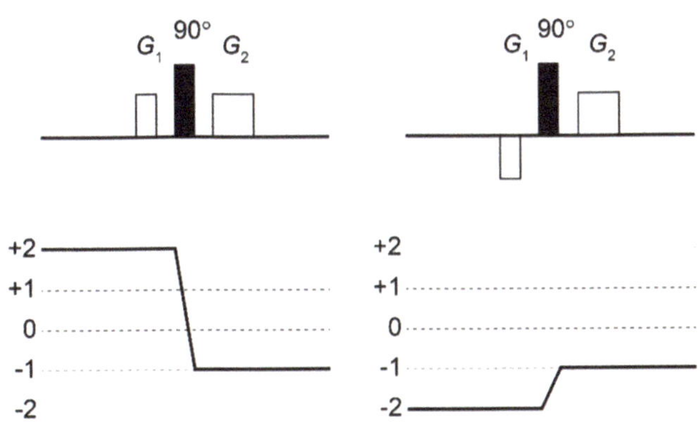


**Figure 5.9** Two examples of CTPs selected using pulsed-field gradients. Here, the gradient pulses are shown with similar amplitude but vary in duration. *Source:* [6], figure 5.2 (p. 43). Reproduced with permission of Oxford University Press.

*pulse, rather than the change in coherence order.* This means that the pulse sequence design process with pulsed-field gradients will be very different from the one using phase cycling.

We can now summarize the CTP selection considerations when performed by pulsed-field gradients:

1. Identify the spin system: how many *J*-coupled or dipolar-coupled spins does it contain? What is the largest coherence order it could sustain?
2. Identify the desired CTPs, and determine the coherence order(s) before and after each RF pulse (or a block of RF pulses)
3. For each RF pulse (or block of RF pulses) for which the desired coherence order changes, select pulsed-field gradient parameters (amplitude, polarity, and duration) such that the ratio determined in Equation 5.12 matches the desired selection. The gradient integral should be large enough to dephase coherence orders (see Figure 5.8) to a sufficient extent in order to limit signal leakage from unwanted pathways. On modern spectrometers, gradient pulse durations for this purpose typically range from 1-2 ms, and gradient amplitudes of 0.1-0.2 T/m are considered good choices.

We have only discussed here the case of two pulsed-field gradients working together, but a set of three or more gradients could be used to perform a given selection. In this case, the gradient ratios determine the ratios between three or more coherences in a similar way as discussed above [20]. It is much more common to use several pairs of pulsed field gradients to bracket different coherence conversion steps.

## 5.6 Comparison Between Phase Cycling and Pulsed-field Gradients

Both phase cycling and pulsed gradients can be combined together in a single pulse sequence for gradient selection. Therefore, if a desired CTP cannot be selected using pulsed-field gradients, especially in the case that higher multiples of the desired coherence orders would be selected, then phase cycling should be employed on that RF pulse in order to isolate the desired pathway.

Using pulsed-field gradients has distinct advantages over phase cycling. First, the design process becomes much simpler than with phase cycling, and it becomes much easier to select a single CTP. Additionally, pulsed-field gradients eliminate the need for multiple acquisitions and greater numbers of phase cycles in order to select a single CTP. Nevertheless, the pulsed-field gradient technique comes with its own disadvantages as well. The signal-to-noise ratio tends to be lower because it is harder to select more than one pathway if the two are asymmetric. Furthermore, pulsed-field gradients require significant additional evolution delays during pulse sequences, which are often not compatible with the goal of an experiment. A notable advantage of pulsed-field gradients is that coherence selection is often found to be cleaner, since the selection avoids relying on signal cancelation between scans. In many cases, the selection can be performed very well within a single scan. Water suppression such as WATERGATE [21] and excitation sculpting are good examples for robust use of pulsed-field gradients [22].

Most modern high-field spectrometers and probes have at least a single-axis pulsed-field gradient setup. Some have a triple-axis setup, which provides even further opportunities for clean coherence selection, because the gradients performed along different axes can be employed to perform independent selection steps, without the chance of undesired coherence pathways accidentally refocusing.

## 5.7 CTP Selection in Heteronuclear Spin Systems

We should note that the discussion of CTP selection primarily focused on the case of homonuclear spin systems. For heteronuclear spin systems, one needs to keep track of the individual coherence orders for each nucleus. This approach is more straightforward for phase cycling than it is for pulsed-field gradients. For phase cycling the accumulated phase is based only on coherence order *changes*, rather than *precession frequency* (compare Equations 5.7 and 5.8, noting that $\theta = \gamma B_0 t$ in Equation 5.7 depends on the nucleus). Hence, one can consider each pulse (performed on each nuclear species) separately, and apply the same rules as discussed. For pulsed-field gradients, one needs to consider the accumulated phase, which will involve the gyromagnetic ratios of all species involved in a coherence. Heteronuclear multiple-quantum coherence (HMQC) and heteronuclear single-quantum coherence (HSQC) experiments [23, 24] are often employed in such a way. Although the gradient ratios cannot strictly be represented as ratios of integers in such cases, integer approximations are often made, especially in the combination of $^1H$ and $^{13}C$ spins, because $\gamma_{1H} \approx 4\gamma_{13C}$.

## 5.8 Additional Approaches to Coherence Selection

It is worth mentioning an additional potential consideration of coherence selection, which mostly affects the pulsed-field gradient method. Motion in the form of diffusion and flow can produce additional dephasing under the influence of gradients, which generally increases with the separation between the defocusing and the refocusing gradients. Diffusion processes are used of course for diffusion measurements, but they could also be deliberately used for destroying signals arising from mobile components (as a diffusion filter). Special gradient arrangements and ratios are also important for the compensation of flow, and the method of convection compensation has found popularity in diffusion measurements that can be made robust against convection [25–28].

Both pulsed-field gradients and phase cycling fundamentally rely on the rotational properties of operators around z, from which all the rules for designing an effective selection protocol derive. It is, of course possible to consider more advanced protocols, which would select not only the coherence order but also the rank of operator terms. Such schemes require the combination of experiments with different flip angles, and these are much less frequently employed but can have important applications. One such example would be the use of zero-rank filters to select nuclear spin singlet states, which require three pulses and two field gradients [29]. Alternatively, a special phase cycling scheme without gradients can be employed, such as a polyhedral phase cycling approach [30]. Furthermore, a scheme called spherical tensor analysis was developed for specifically selecting both the rank and the order [31, 32].

## References

1 Hatanaka, H., Terao, T., and Hashi, T. (1975). Excitation and detection of coherence between forbidden levels in 3-Level spin system by multistep processes. *J. Phys. Soc. Japan* 39 (3): 835–836.
2 Munowitz, M. and Pines, A. (1986). Multiple-quantum nuclear-magnetic-resonance spectroscopy. *Science.* 233 (4763): 525–531
3 Hoult, D.I. (1973). The application of high field nuclear magnetic resonance. [Thesis (Ph.D.).].
4 Braun, S., Kalinowski, H-O., and Berger, S. (1998). 150 and more basic NMR experiments: a practical course, 2nd expanded ed. Weinheim: Wiley-VCH.
5 Keeler, J. (2010). *Understanding NMR spectroscopy*, 2e. Chichester, U.K.: John Wiley and Sons. 511.
6 Hore, P.J., Jones, J.A., and Wimperis S. (2000). NMR: the toolkit. Oxford; New York: Oxford University Press. 85.
7 Bain, A.D. (1984). Coherence levels and coherence pathways in NMR. A simple way to design phase cycling procedures. *J. Magn. Reson.* 56 (3): 418–427.
8 Bodenhausen, G., Kogler, H., and Ernst, R.R. (1984). Selection of Coherence-transfer pathways in NMR pulse experiments. *J. Magn. Reson.* 58 (3): 370–388.
9 Levitt, M.H. (2008). *Spin Dynamics: Basics of Nuclear Magnetic Resonance*, 2e. Chichester, England; Hoboken, NJ: John Wiley & Sons. 714.

**10** Reichert, D. and Hempel, Gn. (2002). Receiver imperfections and CYCLOPS: an alternative description. *Concepts Magn. Reson.* 14 (2): 130–139.

**11** Torres, A.M. and Price, W.S. (2016). Common problems and artifacts encountered in solution-state NMR experiments. *Concepts Magn. Reson. Part A* 45A (2).

**12** Jerschow, A. (2000). Nonideal rotations in nuclear magnetic resonance: estimation of coherence transfer leakage. *J. Chem. Phys.* 113 (3): 979–986.

**13** Levitt, M.H., Madhu, P.K., and Hughes, C.E. (2002). Cogwheel phase cycling. *J. Magn. Reson.* 155 (2): 300–306.

**14** Jerschow, A. and Kumar, R. (2003). Calculation of coherence pathway selection and cogwheel cycles. *J. Magn. Reson.* 160 (1): 59–64.

**15** Zuckerstätter, G., Müller, N. (2007). Coherence pathway selection by cogwheel phase cycling in liquid-state NMR. *Concepts Magn. Reson. A Part A* 30A (2): 81–99.

**16** Ivchenko, N., Hughes, C.E., and Levitt, M.H. (2003). Application of cogwheel phase cycling to sideband manipulation experiments in solid-state NMR. *J. Magn. Reson.* 164 (2): 286–293.

**17** Bax, A., Dejong, P.G., Mehlkopf, A.F., and Smidt, J. (1980). Separation of the different orders of NMR Multiple-Quantum transitions by the use of pulsed field gradients. *Chem. Phys. Lett.* 69 (3): 567–570.

**18** Kay, L.E. (1995). Field gradient techniques in NMR spectroscopy. *Curr. Opin. Struct. Biol.* 5 (5): 674–681.

**19** Jerschow, A. and Muller, N. (1998). Efficient simulation of coherence transfer pathway selection by phase cycling and pulsed field gradients in NMR. *J. Magn. Reson.* 134 (1): 17–29.

**20** Zhu, J.M. and Smith, I.C.P. (1995). Selection of coherence transfer pathways by Pulsed-field gradients in NMR spectroscopy. *Concepts Magn. Reson.* 7 (4): 281–291.

**21** Piotto, M., Saudek, V., and Sklenar, V. (1992). Gradient-tailored excitation for single-quantum NMR spectroscopy of aqueous solutions. *J Biomol. Struct. NMR.* 2 (6): 661–665.

**22** Hwang, T.L. and Shaka, A.J. (1995). Water suppression that works-excitation sculpting using arbitrary wave-forms and pulsed-field gradients. *J. Magn. Reson. Series A* 112 (2): 275–279.

**23** Parella, T., Sanchezferrando, F., and Virgili, A. (1995). Improved HMQC-Type and HSQC-Type 1D spectra using pulsed-field gradients. *J. Magn. Reson. Series A* 114 (1): 32–38.

**24** Mandal, P.K. and Majumdar, A. (2004). A comprehensive discussion of HSQC and HMQC pulse sequences. *J. Concepts Magn. Reson.* 20A (1): 1–23.

**25** Jerschow, A. and Muller, N. (1997). Suppression of convection artifacts in stimulated-echo diffusion experiments. Double-stimulated-echo experiments. *J. Concepts Magn. Reson.* 125 (2): 372–375.

**26** Jerschow, A. and Muller, N. (1998). Convection compensation in gradient enhanced nuclear magnetic resonance spectroscopy. *J. Concepts Magn. Reson.* 132 (1): 13–18.

**27** Jerschow, A. (2000). Thermal convection currents in NMR: flow profiles and implications for coherence pathway selection. *J. Concepts Magn. Reson.* 145 (1): 125–131.

**28** Nilsson, M. and Morris, G.A. (2005). Improving pulse sequences for 3D DOSY: convection compensation. *J. Concepts Magn. Reson.* 177 (2): 203–211.

**29** Thrippleton, M.J. and Keeler, J. (2003). Elimination of zero-quantum interference in two-dimensional NMR spectra. *Angew. Chem., Int. Ed. Engl.* 42 (33): 3938–3941.

**30** Pileio, G. and Levitt, M.H. (2008). Isotropic filtering using polyhedral phase cycles: application to singlet state NMR. *J. Concepts Magn. Reson.* 191 (1): 148–155.

**31** van Beek, J.D., Carravetta, M., Antonioli, G.C., and Levitt, M.H. (2005). Spherical tensor analysis of nuclear magnetic resonance signals. *J. Chem. Phys.* 122 (24): 244510.

**32** Chandra Shekar, S., Rong, P., and Jerschow, A. (2008). Irreducible spherical tensor analysis of quadrupolar nuclei. *Chem. Phys. Lett.* 464 (4–6): 235–239.

# 6

# Nuclear Overhauser Effect Spectroscopy

*P.K. Madhu*

*Tata Institute of Fundamental Research Hyderabad, Hyderabad 500 046, India*

## 6.1 Introduction

Nuclear magnetic resonance is an excellent spectroscopic tool to obtain the structure of molecules and elucidation of dynamics within them. Determination of internuclear distances is vital in both of these. Dipole-dipole (DD) couplings between nuclear spins are essential in this regard as they carry an $r^{-6}$ dependence in the expression for the DD Hamiltonian [1], where $r$ is the internuclear distance. Whilst normally in solution-state NMR all anisotropic (orientation-dependent) interactions that carry structural information are averaged to zero on account of rapid molecular tumbling, they lead to relaxation through higher-order effects. Nuclear Overhauser effect (NOE) is an elegant consequence of this that can yield distance information in molecules [2, 3]. Two-dimensional (2D) analogue of NOE, termed as nuclear Overhauser effect spectroscopy (NOESY), is often used for large molecules [3] in place of the one-dimensional (1D) NOE experiments. NOESY can also detect chemical and conformational exchange, the technique in this case named as exchange spectroscopy (EXSY) [4, 5]. The NOESY experiment can lead to distance information between nuclear spins even if they are not bonded. NOESY is also useful in determining protein-ligand binding contacts. This chapter will deal with principles of NOE, experimental realisation of NOE, details of NOESY, rotating-frame NOE and NOESY termed as ROE and ROESY, and certain illustrative applications.

The term Overhauser effect was first coined for polarisation of nuclei in a metal upon saturating the electrons by Overhauser in 1953 [6]. Experimental realisation of the Overhauser effect came soon by Carver and Slichter in 1953, itself [7] followed by a detailed article in 1956 [8]. The nuclear part of it was implemented by Solomon in 1955 [9]. The first 2D NOESY experiment was carried out on the protein BPTI by Anil Kumar in the research groups of Ernst and Wüthrich [10].

## 6.2 Nuclear Overhauser Effect

### 6.2.1 Qualitative Picture

NOE, by definition, is the change in the intensity of resonances in an NMR spectrum when any other single resonance is perturbed. We consider here, for simplicity, a two-spin system. For two spins I and S we assume

*Two-Dimensional (2D) NMR Methods*, First Edition. Edited by K. Ivanov, P.K. Madhu and G. Rajalakshmi.

observation of I-spin resonance intensity upon perturbation of S-spin resonance. The NOE, $\eta_I\{S\}$, is defined as a normalised fractional change in the intensity of the I-spin resonance upon perturbing the S spin as:

$$\eta_I\{S\} = \frac{(I - I^0)}{I^0}. \tag{6.1}$$

Here, $I^0$ corresponds to the equilibrium intensity of the I-spin resonance, before the perturbation of the S-spin resonance.

The origin of NOE is from cross-relaxation between the two spins mediated by DD couplings. Figure 6.1a shows the energy level diagram for a homonuclear two-spin system. The intensity of any I-spin transition is proportional to the difference between the populations of the concerned levels, namely, $(N_{\alpha\alpha} - N_{\beta\alpha})$ and $(N_{\alpha\beta} - N_{\beta\beta})$. Similarly, the intensity of any S-spin transition is proportional to $(N_{\alpha\alpha} - N_{\alpha\beta})$ and $(N_{\beta\alpha} - N_{\beta\beta})$. Here, $N_{ij}$ represents the population of the level $|ij\rangle$. Under thermal equilibrium, the energy level diagram for such a spin system may be depicted as shown in Figure 6.1b with the populations marked against each level. The I-spin transitions involve $|\alpha\alpha\rangle \leftrightarrow |\beta\alpha\rangle$ and $|\alpha\beta\rangle \leftrightarrow |\beta\beta\rangle$. The S-spin transitions involve $|\alpha\alpha\rangle \leftrightarrow |\alpha\beta\rangle$ and $|\beta\alpha\rangle \leftrightarrow |\beta\beta\rangle$. $\Delta$ refers to the population difference across which the transitions occur. The $W_0$, $W_1$'s, and $W_2$ correspond to the zero-, single-, and double-quantum (ZQ, SQ, and DQ) transition probabilities, respectively.

In the thermal equilibrium case, the population difference leading to both the I-spin and S-spin transitions is $\Delta$. Let us now consider the scenario leading to NOE when the S-spin is saturated, Figure 6.1c. There is no population difference, on account of saturation, across the S-spin transitions. The population difference across the I-spin transitions, however, remains at $\Delta$. This clearly says that, in case of saturation, considering only the SQ transitions, i.e. the $W_1$ transition probabilities, as the active relaxation pathways, saturating the S-spin transitions will not affect the intensity of the I-spin resonance. Hence, there is no NOE!

The significance of the $W_2$ and $W_0$ transition probabilities comes into play here as important playmakers in relaxation. They are brought about by the DQ and the ZQ terms in the DD Hamiltonian, which lead to NOE. Let us analyse the situation in the new light.

After saturation of the S-spin transitions, the following scenarios exist. We first consider the $|\alpha\beta\rangle \leftrightarrow |\beta\alpha\rangle$ transition. This is the case where $W_0$ is active.

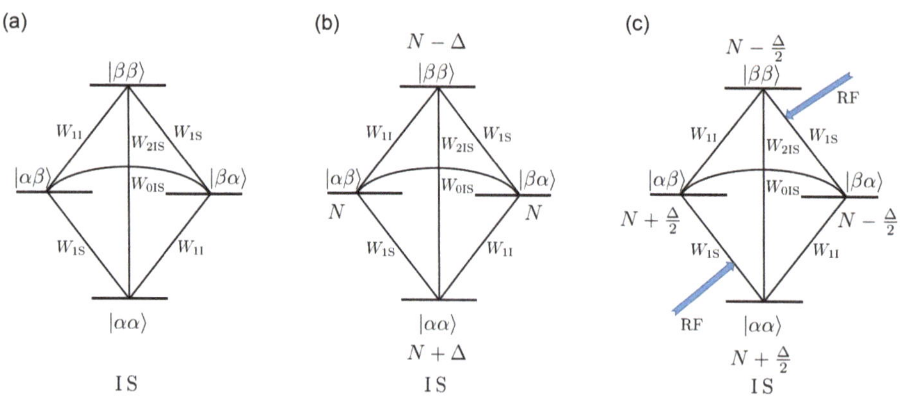


**Figure 6.1** (a) A homonuclear two-spin energy level schematic with the states marked against each level. Here, the first spin is I and the second spin is S. (b) A thermal equilibrium two-spin energy level schematic with populations marked against each level. Here, $\Delta$ refers to the population difference leading to a transition. (c) Same as (b), but upon saturation of the S-spin transitions. In the text, for both $W_2$ and $W_0$, the subscript *IS* is dropped as we only consider an isolated two-spin (*IS*) system.

- After saturation, the population difference across the considered transition becomes Δ, which was zero at thermal equilibrium.
- $W_0$ will ensure that this post-saturation population difference becomes zero, which can be done by transferring population from $|\alpha\beta\rangle \leftrightarrow |\beta\alpha\rangle$.
- This leads to a decrease in the population difference across the $|\alpha\alpha\rangle \rightarrow |\beta\alpha\rangle$ transition.
- This also leads to a decrease in the population difference across the $|\alpha\beta\rangle \rightarrow |\beta\beta\rangle$ transition.
- All the above imply that the $W_0$ effect is to effectively reduce the I-spin resonance intensity. This will be, however, compensated by $W_{1I}$ as before $W_0$ became effective, the I-spin populations were those at the thermal equilibrium.
- The net effect is that the I-spin resonance intensity would depend on the competition between $W_{1I}$ and $W_0$. Hence, in the case of $W_0$ being the dominant pathway, a negative NOE will result.

We now consider the second scenario with the $|\alpha\alpha\rangle \leftrightarrow |\beta\beta\rangle$ transition. This is the case where $W_2$ is active.

- After saturation, the population difference across the considered transition becomes Δ, which was 2Δ at thermal equilibrium.
- $W_2$ will ensure that this post-saturation population difference becomes 2Δ, which can be done by transferring population from $|\beta\beta\rangle$ to $|\alpha\alpha\rangle$ .
- This leads to an increase in the population difference across the $|\alpha\alpha\rangle \leftrightarrow |\beta\alpha\rangle$ transition.
- This also leads to an increase in the population difference across the $|\alpha\beta\rangle \rightarrow |\beta\beta\rangle$ transition.
- All the above imply that the $W_2$ effect is to effectively increase the I-spin resonance intensity. This will be, however, compensated by $W_{1I}$ as before $W_2$ became effective, the I-spin populations were those corresponding to the thermal equilibrium.
- The net effect is that the I-spin resonance intensity would depend on the competition between $W_{1I}$ and $W_2$. Hence, in the case of $W_2$ being the dominant pathway, a positive NOE will result.

To summarise the above discussion, if the $W_0$'s are the dominant pathways, a negative NOE results, and if instead the $W_2$'s are the leading pathways, a positive NOE results. The individual transitions of either the I or the S spins take place at well-defined frequencies. $W_1$'s correspond to either $\omega_I$ or $\omega_S$ with $\omega$ being the respective Larmor frequency. $W_0$'s will correspond to either zero for homonuclear spins or $\omega_I - \omega_s$ for heteronuclear spins. $W_2$'s will correspond to $2\omega$ for homonuclear spins or $\omega_I + \omega_s$ for heteronuclear spins.

These frequencies are important as relaxation is caused by the fluctuating local magnetic fields ($B_{\text{loc}}(t)$) around the nuclear spins. The effectiveness of a particular $W$ leading to a certain relaxation pathway depends on the strength of $B_{\text{loc}}(t)$ fluctuating at that particular frequency. For instance, if $B_{\text{loc}}(t)$ is predominantly at a frequency of $2\omega$ (for homonuclear spins), $W_2$'s will be dominant and so on for the other frequencies.

### 6.2.2 NOE: Quantitative Picture

We refer the reader to the derivation of the Solomon's equations in Chapter 4. These, which are the central part of NOE and relaxation, may be written as:

$$\frac{d}{dt}\begin{pmatrix} I_z(t) \\ S_z(t) \end{pmatrix} = -\begin{pmatrix} W_0 + 2W_{1I} + W_2 & W_2 - W_0 \\ W_2 - W_0 & W_0 + 2W_{1S} + W_2 \end{pmatrix}\begin{pmatrix} I_z - I_z^0 \\ S_z - S_z^0 \end{pmatrix}. \tag{6.2}$$

Here $I_z^0$ and $s_z^0$ are the equilibrium intensities of the I- and S-spin resonances. In the case of NOE discussed earlier, known as steady-state NOE (SSNOE), the I-spin resonance intensity is probed whilst saturating the S-spin transitions. At the steady state, the boundary conditions for the Solomon's equations are:

$$\frac{dI_z(t)}{dt} = 0 \quad \text{and} \quad S_z = 0. \tag{6.3}$$

Equation 6.2 then becomes:

$$0 = -\left(I_z - I_z^0\right)\left(W_0 + 2W_{1I} + W_2\right) + S_0^z\left(W_2 - W_0\right), \tag{6.4}$$

leading to:

$$\frac{I_z - I_z^0}{S_z^0} = \frac{W_2 - W_0}{W_0 + 2W_{1I} + W_2}. \tag{6.5}$$

Realising that $s_z^0 = \frac{\gamma_S}{\gamma_I} I_z^0$, the expression for NOE becomes:

$$\eta_I\{S\} = \frac{I_z - I_z^0}{I_z^0} = \frac{\gamma_S}{\gamma_I} \frac{W_2 - W_0}{W_0 + 2W_{1I} + W_2}. \tag{6.6}$$

From Equation 6.6, which has in the numerator $W_2 - W_0$, it is clear that positive NOE results if $W_2$'s are dominant and negative NOE results if $W_0$'s are dominant. $W_1$'s do not contribute to the NOE. $W_2 - W_0$ is the cross-relaxation rate constant denoted as $\sigma_{IS}$ and $W_0 + 2W_{1I} + W_2$ is the auto-relaxation rate constant denoted as $\rho_I$, with similar expression holding true for $\rho_S$.

As mentioned earlier, relaxation is caused by the fluctuating local dipolar fields and these fluctuations typically arise from molecular motions. This brings about a relationship between the overall molecular tumbling frequencies and rotational correlation timescales, $\tau_c$. Typically, $W_0$'s will be dominant if the molecules tumble at the difference of the Larmor frequencies $(\omega_I - \omega_s)$. Similarly, $W_1$'s will be dominant if the molecular tumbling is on the order of a few MHz (Larmor frequencies), and $W_2$'s will be active if the frequencies are at the sum of the Larmor frequencies $((\omega_I + \omega_s)$. Hence, qualitatively, small molecules lead to positive NOE, big molecules lead to negative NOE, and somewhere in between, the NOE can be even zero, all of these depending on the molecular correlation times.

In order to understand the dependence of NOE and $W$'s on $\tau_c$, it is important to find out how the energy available for causing relaxation depends on $\tau_c$. This is because the transition probability, $W_{ij}$, between two states $|i\rangle$ and $|j\rangle$ is given by:

$$W_{ij} = J(\omega)|\langle i|H_{\text{rel}}|j\rangle|^2. \tag{6.7}$$

Here, $H_{\text{rel}}$ is the appropriate relaxation Hamiltonian and $J(\omega)$ is the spectral density function that is closely associated with the molecular tumbling.

Correlation time is defined as the time taken by the molecule to rotate by one radian about any angle. A correlation function, $g(\tau)$, with a characteristic time constant of $\tau_c$ (for its decay) is required to describe the local fluctuating fields. $g(\tau)$ essentially describes how a parameter, $f(t)$, measured at an instant of time $t$ is correlated with the same parameter at another instant of time $t + \tau$ such that:

$$g(\tau) = \overline{f(t)f(t+\tau)} \tag{6.8}$$

where an ensemble time average is taken on the right-hand side to calculate the correlation function (e.g. free induction decay, [FID]). Note that $g(\tau)$ does not depend on where the observation starts, only the interval between the two time points matters. The desired properties for $g(\tau)$ are a fixed positive value at $t = 0$ (normally normalised to 1) and decaying to zero for long $\tau$. An easy functional form in case of NMR is an exponentially decaying function such that:

$$g(\tau) = exp\left(-\frac{\tau}{\tau_c}\right). \tag{6.9}$$

Figure 6.2a shows the behaviour of such a correlation function for three different $\tau_c$. In general, $\tau_c \ll 1$ ns corresponds to small molecules, $\tau_c \gg 1$ ns corresponds to big molecules, and otherwise is in the intermediate regime.

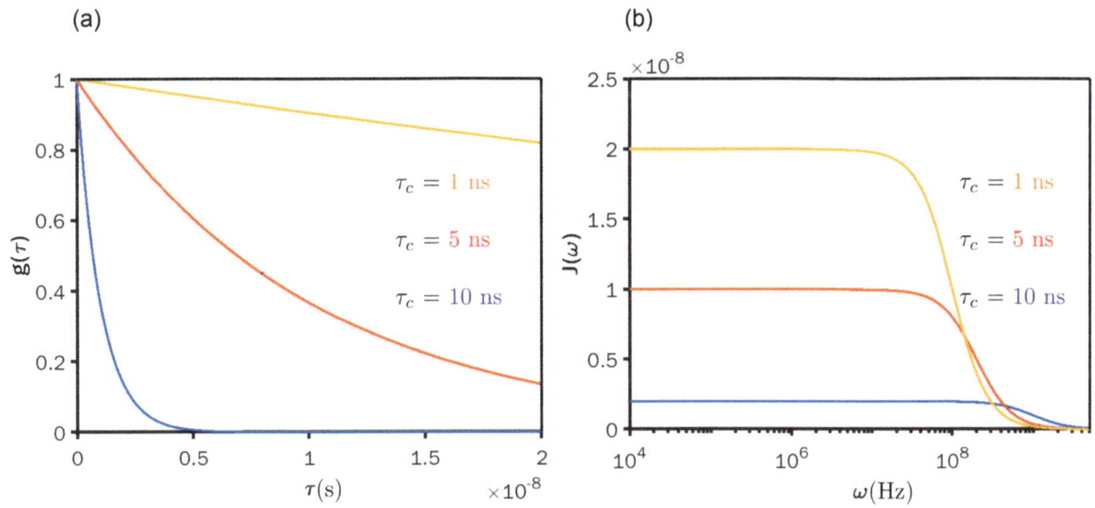


**Figure 6.2** Plot of the correlation function $g(\tau)$ for three different correlation times. (b) The spectral density $J(\omega)$ for each of the $g(\tau)$ in (a).

Correlation functions are statistically useful, however, for NMR purposes, a function that depends on a frequency is desired as this dependence is essential to characterise the $W$'s in terms of frequencies. Such a function, denoted as spectral density, can be generated from the correlation function as its Fourier transform and is given by:

$$J(\omega) = 2\int_0^{\infty} g(\tau)exp(-i\omega\tau)d\omega \tag{6.10}$$

As indicated earlier, $J(\omega)$'s are called spectral densities as they represent the available frequencies in a molecular system and the power that can be obtained from the molecular motion to effect relaxation at a particular $W$. Representative plots of $J(\omega)$ are given in Figure 6.2b. From an NMR parlance, $g(\tau)$ may be thought of as a single-resonance FID the Fourier transform of which is an absorption Lorentzian centred at $\omega$. The decay constants are typically on the order of $10^{-7}$ to $10^{-11}$ s and the corresponding line width of $J(w)$ is on the order of MHz. The functional form of $J(w)$ is given by:

$$J(\omega) = \frac{2\tau_c}{1+(\omega\tau_c)^2}, \tag{6.11}$$

and it has the following properties:

1. Total integrated area of an NMR spectrum is given by the first point of the fid, and since all $g(\tau)$ start at unity, all $J(\omega)$ would have the same area.
2. $J(\omega)$'s are initially flat, then they drop off at $\omega \approx \frac{1}{\tau_c}$ (which is equivalent to the full-width at half-height of a Lorentzian line) and then tend to zero.

In order to relate $W$'s, $J$'s, and $\tau_c$ it is necessary to consider the specific forms of the relaxation mechanisms. In this discussion, only DD interaction need to be considered. The expressions for the $W$'s are given by [11]:

$$
\begin{aligned}
W_{1I} &= \frac{3}{40}K^2 J(\omega_I) = \frac{3}{20}K^2 \frac{\tau_c}{1+(\omega_I\tau_c)^2} \\
W_{1S} &= \frac{3}{40}K^2 J(\omega_S) = \frac{3}{20}K^2 \frac{\tau_c}{1+(\omega_S\tau_c)^2} \\
W_0 &= W_{0IS} = \frac{1}{20}K^2 J(\omega_I-\omega_S) = \frac{1}{10}K^2 \frac{\tau_c}{1+(\omega_I-\omega_S)^2\tau_c^2} \\
W &= W_{2IS} = \frac{3}{10}K^2 J(\omega_I+\omega_S) = \frac{3}{5}K^2 \frac{\tau_c}{1+(\omega_I+\omega_S)^2\tau_c^2}.
\end{aligned}
\tag{6.12}
$$

Here, $K = \frac{\mu_0}{4\pi}\frac{\hbar\gamma_I\gamma_S}{r_{IS}^3}$ with $\gamma$ being the gyromagnetic ratio. For the fast tumbling case ($\omega\tau_c << 1$), $J(\omega) = 2\tau_c$, and this is independent of the spectrometer frequency. In this case, also denoted as the extreme narrowing limit, $W_0 : W_1 : W_2 = 1 : \frac{3}{2} : 6$. The auto-relaxation rate, $\rho$, is faster than the cross-relaxation (NOE) rate, $\sigma$, except for $\omega\tau_c \ll 1$. This is the spin-diffusion limit, or in other words, the spin-diffusion mechanism plays a larger role for $\omega\tau_c \gg 1$.

With all the above, Equation 6.6, for the homonuclear case, can be written as:

$$
\eta_I\{S\} = \frac{5+\omega^2\tau_c^2-4\omega^4\tau_c^4}{10+23\omega^2\tau_c^2+4\omega^4\tau_c^4}.
\tag{6.13}
$$

Figure 6.3 shows the behaviour of NOE with respect to the correlation time for a homonuclear case considering an isolated coupled two-spin system. This gives the range of $\tau_c$ for which positive and negative NOE's can be expected with the null NOE happening for $\omega\tau_c = \sqrt{\frac{5}{4}} \approx 1.12$. The maximum value of the NOE is 0.5, which is observed in small molecules whilst the minimum value is −1, which is observed in big molecules. This figure clearly says that irrespective of the distance between the two spins, the NOE factor only depends on the correlation time. NOE has no dependence on $r_{IS}$, which is also clear from Equation 6.13.

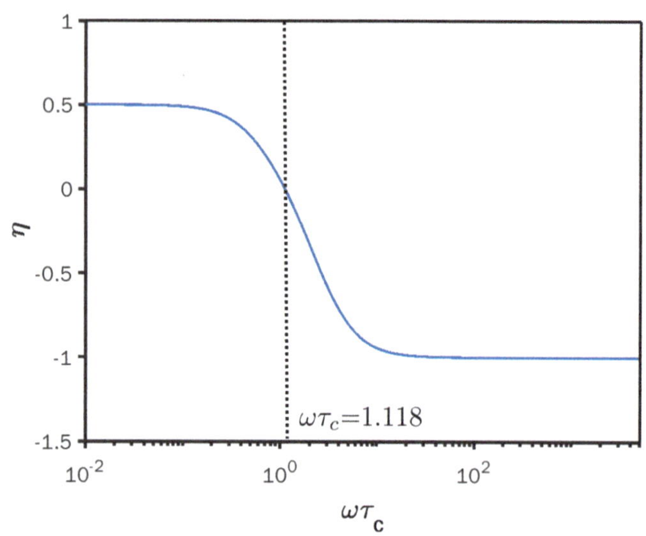


Figure 6.3 Plot of the NOE value as a function of $\omega\tau_c$ for a homonuclear two-spin system. Here, the Larmor frequency $\omega$ was assumed to be 500 MHz. The dashed line points to the value of $\omega\tau_c$ where the NOE goes to zero.

### 6.2.3 NOE and Distance Dependence: Many-spin System

The above considerations dealt with an isolated two-spin system for deriving the details of NOE. However, the observed I-spin can also relax through other mechanisms besides the ones considered earlier. In general, the auto-relaxation rate, $R_I$, of the I-spin is given by:

$$\begin{aligned} R_I &= \rho_I + \sum_X \rho_{IX} + \rho_I^\star \\ &= R_I^{DD} + \rho_I^\star . \end{aligned} \quad (6.14)$$

Here, $\rho_{IX}$ refers to the self-relaxation of the I-spin coupled to the other spins, X. $\rho_I^\star$ is the relaxation rate of the I-spin caused by mechanisms besides DD interactions, such as, chemical-shift anisotropy, quadrupole coupling, and scalar coupling. $R_I^{DD}$ corresponds to the overall I-spin relaxation caused by DD interactions between I and S spins and between I spin and other spins around it denoted as $X$. Hence, NOE in the case of a many spin system takes the form:

$$\eta_I\{S\} = \frac{\gamma_S}{\gamma_I} \frac{\sigma_{IS}}{\rho_I + \rho_I^\star + \rho_{IX}} \quad (6.15)$$

$\rho_I^\star$ and $\rho_{IX}$, hence, reduce the NOE effect ensuring that in a homonuclear case the NOE never reaches the theoretical maximum of 50%. These are referred to as leakage terms. Whilst both $\sigma_{IS}$ and $\rho_I$ are proportional to $\frac{1}{r^6{}_{IS}}$, the leakage terms have no dependence on the IS distance. Hence, if $r_{IS}$ decreases, $\sigma_{IS}$ and $\rho_I$ increase, and the contribution of the leakage terms to n becomes less significant and $n$ tends to the theoretical maximum of 0.5. This is shown in Figure 6.4a considering a two-spin system where $\rho_I^\star$ is the only leakage term. Two-spin NOE, hence, depends on the internuclear distance only when external relaxation mechanisms (besides the two- spin DD interaction) are present. This is clear from the plots in Figure 6.4a. Leakage terms are more significant for small molecules, and not so much for larger molecules due to spin-diffusion effects. This is because $\rho^\star$ does not normally scale with $\tau_c$ Figure 6.4b shows the dependence of NOE on $\omega\tau_c$ for various values of $\rho^\star$ for a given $r_{IS}$.

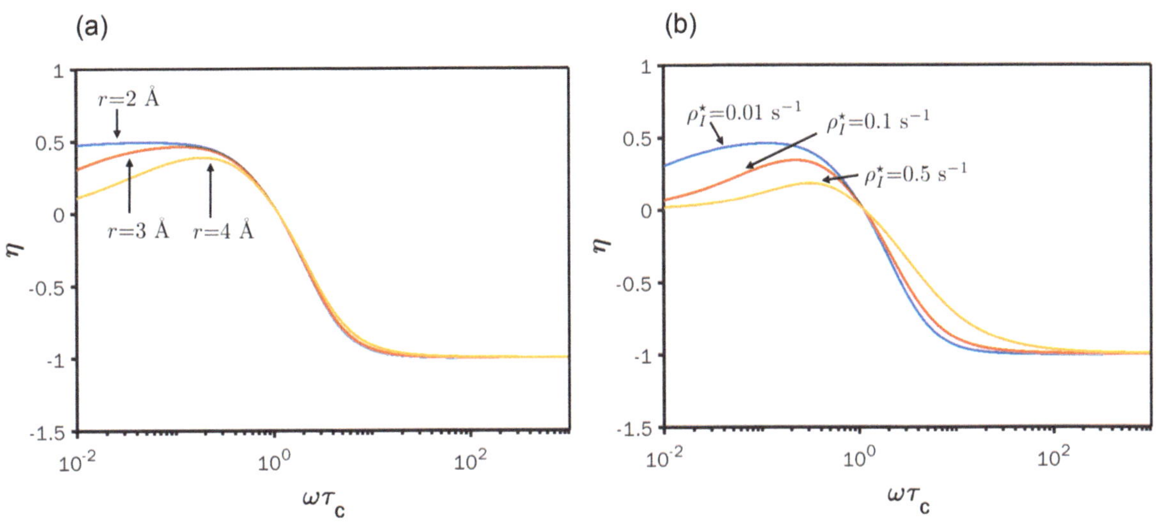


**Figure 6.4** (a) Plot of the NOE value as a function of $\omega\tau_c$ for a homonuclear spin system ($^1H$ in this case) for different values of $r_{IS}$ with $\rho_I{}^* = 0.1\ s^{-1}$. (b) Same as in (a) but for different values of $\rho_I{}^*$ with $r_{IS} = 3$ Å. The $^1H$ Larmor frequency was 500 MHz.

### 6.2.4 NOE Comparison and Distance Elucidation

Bell and Saunders did SSNOE for a series of small molecules and established the linear correlation between the SSNOE enhancement and $\frac{1}{r^6}$ [2]. However, this is far from general. The following conditions have to be met for this observation of NOE and distance dependence:

- The observed I-spin must not have an appreciable DD cross-relaxation with other spins, except the S spin, which is being saturated; that is I and S form a true two-spin system.
- $\frac{\rho_I^\star}{\tau_c}$ should be constant over the entire molecule.
- Extreme narrowing limit is assumed. Under this condition the following relations hold, such as:

$$\begin{aligned} \eta_I\{S\} &= \frac{\sigma_{IS}}{\rho_I + \rho_I^\star}, \quad \text{but } \frac{\sigma_{IS}}{\rho_I} = \frac{1}{2} \\ &= \frac{\rho_I}{2\left(\rho_I + \rho_I^\star\right)} \\ \frac{1}{\eta_{I\{S\}}} &= 2 + \frac{2\rho_I^\star}{\rho_I}. \end{aligned} \tag{6.16}$$

When $\omega\tau_c \ll 1$, $\rho_I \propto \frac{\tau_c}{r^6}$, leading to:

$$\frac{1}{\eta_I\{S\}} \propto \left(\frac{\rho_I^\star}{\tau_c}\right) r^6. \tag{6.17}$$

This is the $r^{-6}$ dependence of the NOE effect, which is true in case of small molecules characterised by the extreme narrowing limit.

### 6.2.5 Indirect NOE Effects

In case of an isolated two-spin system, saturating either of the spins and observing the other one will lead to $\eta = 0.5$ in both cases, irrespective of the distance between the two spins. Hence, observation of NOE between two spins does not necessarily mean that they are close.

Consider a linear arrangement of three spins, A, B, and C as in Figure 6.5a. Since $\sigma_{AB} \gg \sigma_{AC}$, irradiation of the A-spin leads to a large NOE enhancement at B, as B is closer to A. Irradiating the C-spin leads to a smaller enhancement at B for the same reason. These two experiments can be compared to get $r_{AB}$ and $r_{BC}$. Irradiating the B spin, on the other hand, will give similar enhancements at both A and C. This is because for both the A- and C-spins there is only one nearby spin, which is B. Hence, this measurement is not very useful. Here, we see that $\eta_B\{A\} = \eta_A\{B\}$. NOE is not symmetrical as it depends on the distribution of other spins around either A or B.

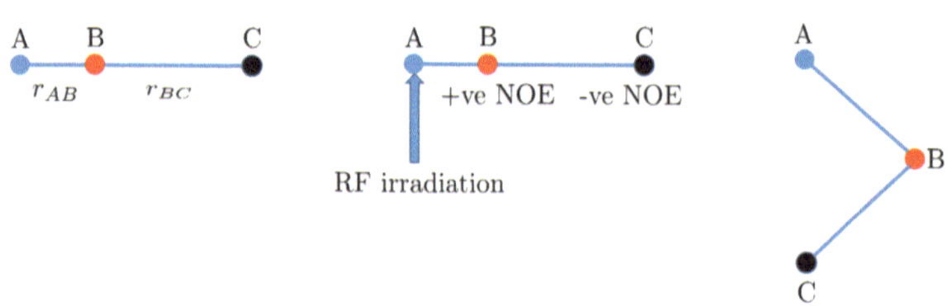


**Figure 6.5** (a) Linear arrangement of three spins, A, B, and C. (b) Schematic to demonstrate NOE effects on three spins in a linear arrangement. (c) A non-linear arrangement of three spins, A, B, and C.

What happens to the C-spin resonance intensity when the A-spin is irradiated? The A-spin has no significant direct effect on the C-spin, hence, the A-spin mainly experiences the population difference across the B-spin transitions. Irradiating the B- spin increases the population difference at B and decreases that at A and, hence, an increase at B will lead to a decrease at C. All these mean that saturation at A leads to a positive NOE at B and a negative NOE at C. This is called the three-spin effect, a configuration for which is shown in Figure 6.5b. It may be noted that observation of a negative homonuclear NOE is a clear indication of a linear arrangement if the experiment is performed under extreme narrowing conditions.

Now consider another geometry as shown in Figure 6.5c. Both the B- and C-spins are close to the A-spin. $\eta_C\{A\}$ depends on the direct effect from the A-spin and indirect effect from the B-spin. Since, they are opposite in sign, for some internuclear distances $\eta_C\{A\} \rightarrow 0$. However, the absence of NOE between the spins does not mean that they are "far apart." And also presence of NOE does not mean that the spins are "closer."

## 6.3 Measurement of NOE

The pulse sequence that may be used for the measurement of steady-state NOE is given in Figure 6.6a and 6.6b. The saturation pulse is on-resonance in Figure 6.6a whilst it is far off-resonance in Figure 6.6b, which serves as a control experiment. NOE effect will be the difference in the spectral intensities of the resonances between Figure 6.6a and Figure 6.6b.

## 6.4 Heteronuclear NOE

Denoting $^1$H as the I spin and $^{13}$C as the S spin, the heteronuclear NOE (for $^{13}$C spin) in the extreme narrowing limit becomes:

$$\eta_S\{I\} = \frac{\gamma_I}{2\gamma_S}. \tag{6.18}$$

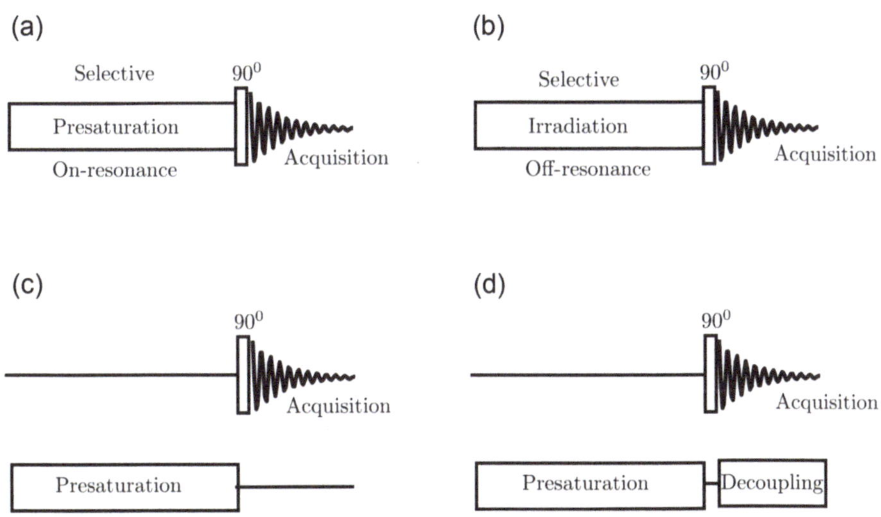


Figure 6.6 (a) Pulse scheme for measurement of NOE in a homonuclear spin system. (b) Control pulse scheme with irradiation performed at off-resonance. The NOE effect will be the difference in the intensities of the resonances with (a) and (b). Heteronuclear NOE measurement pulse methods for (c) only NOE and (d) for NOE with decoupling to remove the heteronuclear scalar couplings.

Hence, the maximum NOE in case of C is around 198.8% and that for $^5$N will be around −494%. The heteronuclear NOE effect on the S spin (e.g. $^{13}$C) upon irradiation of the I spin (e.g. $^1$H) may be expressed as:

$$\eta_S\{I\} = \frac{\gamma_I}{\gamma_S} \frac{\dfrac{6}{1+(\omega_S+\omega_I)^2\tau_c^2} - \dfrac{1}{1+(\omega_S-\omega_I)^2\tau_c^2}}{\dfrac{1}{1+(\omega_S-\omega_I)^2\tau_c^2} + \dfrac{3}{1+(\omega_S)^2\tau_c^2} + \dfrac{6}{1+(\omega_S+\omega_I)^2\tau_c^2}}. \tag{6.19}$$

Here, $\eta$ mostly stays positive unless $\frac{\gamma_I}{\gamma_S} < 2.38$. NOE becomes zero for:

$$\tau_c = \left[\frac{5}{(\omega_S+\omega_I)^2 - 6(\omega_S-\omega_I)^2}\right]^{\frac{1}{2}}. \tag{6.20}$$

Figure 6.7 shows the plot of NOE on $^{13}$C-spin upon irradiation of $^1$H-spin as a function of $\omega\tau_c$.

## 6.5 NOE Kinetics

As seen earlier, SSNOE gives a qualitative idea regarding distances between two spins. SSNOE is of not much use for large molecules as spin diffusion eventually ensures that the steady-state enhancement is 100%. The rate at which the steady state is reached is more important as this brings the kinetics into picture and enables estimation of distances even quantitatively. The main ideas exploited here are:

- In the initial stage of the NOE build-up, the NOE factor can be treated as coming from a two-spin system. This is the so-called initial-rate approximation.
- Kinetic effects can distinguish between direct and indirect effects.

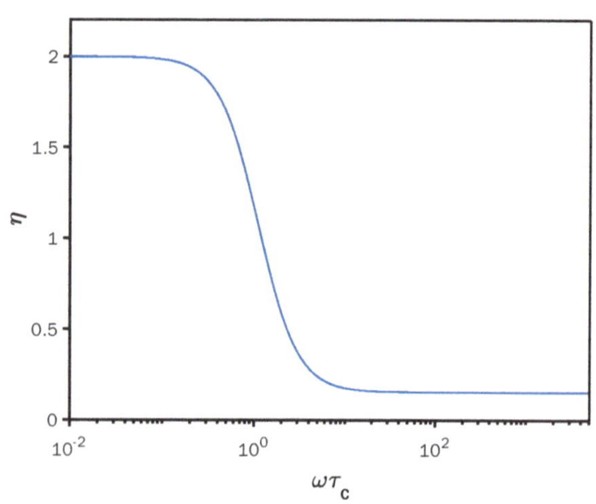


**Figure 6.7** Plot of the NOE value as a function of $ut_c$ for a heteronuclear two-spin system. $^1$H and $^{13}$C are considered here with $^1$H Larmor frequency as 500 MHz.

### 6.5.1 Initial-Rate Approximation

From the Solomon's equations, the rate of change of $I_z$ upon saturation of the S spin (for a two-spin system) is given by:

$$\frac{dI_z(t)}{dt} = -\rho_I\left(I_z - I_z^0\right) - \sigma_{IS}\left(S_z - S_z^0\right). \tag{6.21}$$

In case of SSNOE, the boundary conditions may be written as $\frac{dI_z(t)}{dt} = 0 = S_z$. $\eta_{max}$ is observed after full saturation of the S spin. In kinetic NOE experiments, $\eta$ evolves in the presence of the saturating field and is continuously driven toward its steady state. This means that NOE may be continuously observed during the saturation as a function of saturation pulse duration. This is called truncated driven NOE (TOE). Another way of performing kinetic NOE experiments is by a rapid perturbation of the S spin by a pulse. Here, the NOE evolves in the absence of any RF field. This is referred to as transient NOE (trNOE). Figure 6.8 shows the pulse schemes used for both TOE and trNOE.

In case of TOE, the initial conditions are $I_z = I_z^0$ and $S_z = 0$. The Solomon's equations then give:

$$\left.\frac{dI_z(t)}{dt}\right|_{t=0} = \sigma_{IS}S_z^0. \tag{6.22}$$

In case of trNOE, assuming inversion of $S_z$ rather than saturation, the initial conditions become $I_z = I_z^0$ and $S_z = -S_z^0$, leading to:

$$\left.\frac{dI_z(t)}{dt}\right|_{t=0} = 2\sigma_{IS}S_z^0. \tag{6.23}$$

Both TOE and trNOE equations imply that the moment the S spin is saturated or inverted, $\sigma_{IS}$ begins to grow and leads to NOE value at that rate, and $\rho_I$ or $R_I$ that opposes $\sigma_{IS}$ starts only later. Here, we analyse the trNOE experiment a bit more in detail.

Figure 6.9 shows the NOE build-up curves for a homonuclear isolated two-spin system as a function of the recovery time, $t$ (see Figure 6.8), for three different correlation times. They correspond to the low, intermediate, and high molecular weights.

To generate the trNOE curves, the Solomon's equations need to be solved. One way is to decouple both the equations by differentiating once more which will yield a second-order differential equation in both $I_z(t)$ and $S_z(t)$. For the NOE build-up, the relevant equation for $I_z(t)$ is given by (here, $\sigma = \sigma_{IS}$ and $\rho = \rho_I$):

$$\frac{d^2I_z(t)}{dt^2} + 2\rho\frac{dI_z(t)}{dt} - \left(\sigma^2 - \rho^2\right)I_z(t) = -\left(\sigma^2 - \rho^2\right)I_z^0. \tag{6.24}$$

The characteristic eigenvalues of this are:

$$\lambda_1 = -(\rho - \sigma), \quad \lambda_2 = -(\rho + \sigma). \tag{6.25}$$

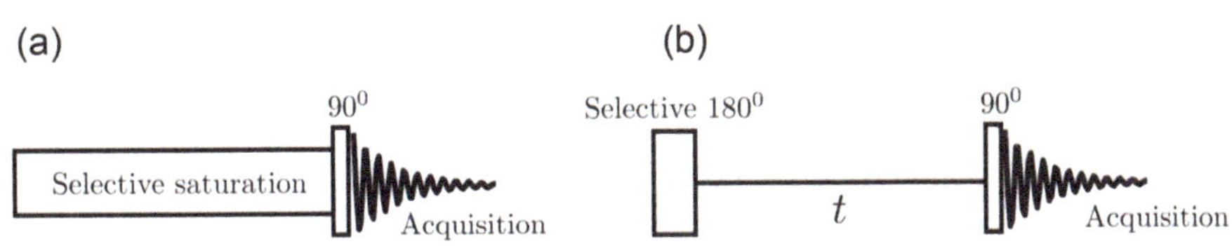


**Figure 6.8** Pulse schemes for measurement of (a) truncated driven NOE and (b) transient NOE, in a homonuclear spin system.

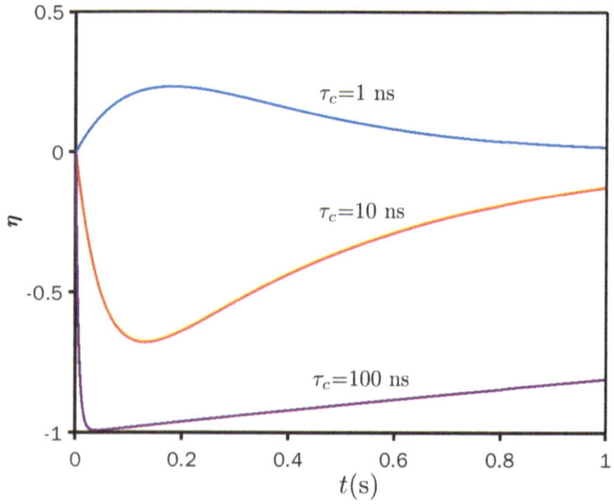


**Figure 6.9** The trNOE build-up curves for an isolated homonuclear two-spin system as a function of time, t, (see Figure 6.8) for the indicated correlation times. The distance between the two spins was 2 Å. The $^1$H Larmor frequency was 500 MHz.

The general solution for $I_z(t)$ becomes:

$$I_z(t) = c_{11}e^{-(\rho-\sigma)t} - c_{12}e^{-(\rho+\sigma)t} + I_z^0. \tag{6.26}$$

Using the initial conditions given by $I_z(t)|_{t=0} = I_z^0$ and $\frac{dI_z(t)}{dt}|_{t=0} = 2\sigma S_z^0$, the constants become:

$$c_{11} = S_z^0, \quad c_{12} = -S_z^0. \tag{6.27}$$

With all these, the general solution for $I_z(t)$ becomes:

$$I_z(t) = S_z^0 e^{-(\rho-\sigma)t} - S_z^0 e^{-(\rho+\sigma)t} + I_z^0. \tag{6.28}$$

For a homonuclear two-spin system, the NOE then becomes:

$$\begin{aligned} \frac{I_z(t)-I_z^0}{I_z^0} &= e^{-(\rho-\sigma)t} - e^{-(\rho+\sigma)t} \\ &= \left[1 - e^{-2\sigma t}\right] e^{-(\rho-\sigma)t}. \end{aligned} \tag{6.29}$$

This is the equation that is plotted in Figure 6.9.

## 6.6 Nuclear Overhauser Effect Spectroscopy, NOESY

### 6.6.1 NOESY Pulse Scheme

The pulse schemes for NOE have been already discussed. In the transient NOE scheme, one needs to invert a particular $^1$H (selectively) resonance to get all the necessary connectivities. This needs to be repeated for all the $^1$H resonances for a full distance elucidation. This becomes time consuming. The best approach, hence, is to non-selectively invert all the $^1$H resonances at the same time. Herein comes the concept of nuclear Overhauser effect spectroscopy (NOESY), the 2D analogue of the 1D NOE scheme.

The design of NOESY pulse scheme involves the following steps:

- During inversion the spins are chemical-shift and scalar-coupling labelled, that means no refocussing pulses.
- Then all the $^1$H spins are oriented along $-z$ (inversion) of whatever spin magnetisation is left.
- Let those along $z$ evolve during a $\tau_m$ mixing period, during which they experience NOE cross-relaxation.
- Read out the final state with another pulse, preferably a 90° pulse.

All the above may be accomplished by a three-pulse scheme as shown in Figure 6.10.

The resulting 2D spectrum has isotropic chemical shifts in both dimensions and cross peaks emanating from DD coupling. It may be noted that the NOESY pulse scheme looks exactly like multiple-quantum filtered COSY scheme except for the phase cycling. NOESY is a transient NOE experiment and the peak values are less than those from SSNOE experiments. For $\omega\tau_c > 1.12$, diagonal and cross peaks are of the same sign, otherwise they are of opposite sign. The same scheme can track chemical-exchange peaks, which are cross peaks with a negative sign (spin-diffusion limit).

### 6.6.2 NOESY Theory

Consider a homonuclear two-spin system having resonances at frequencies of $\Omega_1$ and $\Omega_2$. Here, we assume only spatial proximity between the spins (DD coupling) and no scalar coupling.

The analysis of the pulse scheme, shown along with the coherence pathways, in Figure 6.11, will be done here using product operator formalism. This will follow the fate of the density matrix, $p$, at various points marked in the figure and during the $t_1$, $\tau_m$, and $t_2$ periods.

The density matrix at ① and ② is:

$$\rho_{①} = I_{1z} + I_{2z} \xrightarrow{90^\circ_x} \rho_{②} - I_{1y} - I_{2y}. \tag{6.30}$$

The density matrix at ③ following $t_1$ time evolution becomes:

$$\rho_{③} = \left[-I_{1y}cos\left(\Omega_1 t_1\right) + I_{1x}sin\left(\Omega_1 t_1\right)\right]e^{-\lambda t_1} + \left[-I_{2y}cos\left(\Omega_2 t_1\right) + I_{2x}sin\left(\Omega_2 t_1\right)\right]e^{-\lambda t_1} \tag{6.31}$$

where $\lambda = \frac{1}{T_2}$ and assuming that any term created by scalar coupling is removed by phase cycling, if such couplings are present. Here, $T_2$ is the transverse relaxation time.

The density matrix at ④ after a $90^\circ_x$ pulse becomes:

$$\rho_{④} = [-I_{1z}cos(\Omega_1 t_1) + I_{1x}sin(\Omega_1 t_1)]e^{-\lambda t_1} + [-I_{2z}cos(\Omega_2 t_1) + I_{2x}sin(\Omega_2 t_1)]e^{-\lambda t_1}. \tag{6.32}$$

Performing phase cycling such that only signals corresponding to zero-quantum coherence pathway are retained, hence, removing all the transverse magnetisation components, $\rho_{④}$ can be written as:

$$\rho_{④} = [-I_{1z}cos(\Omega_1 t_1) - I_{2z}cos(\Omega_2 t_1)]e^{-\lambda t_1}. \tag{6.33}$$

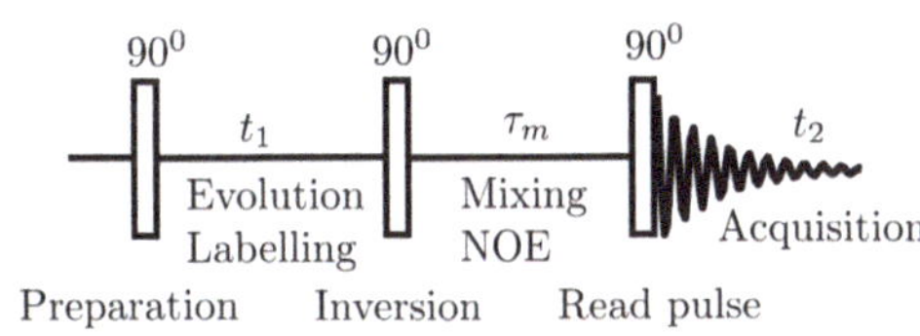


**Figure 6.10** The NOESY pulse sequence with the role of each pulse and delay mentioned.

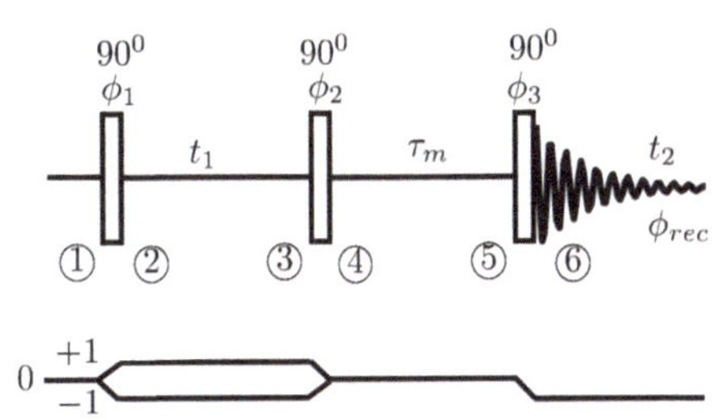


**Figure 6.11** The NOESY pulse sequence with the selected coherence transfer pathways depicted along with numbers indicating the positions where the density matrix is calculated as shown in the text. The phase cycling adopted here is as follows: $\phi_1 = x, -x, x, -x$, $\phi_2 = x$, $\phi_3 = x, x, -x, -x$, and receiver phase $\phi_{rec} = x, -x, -x, x$.

Note the inversion of the longitudinal components. The magnetisation now will evolve during $\tau_m$ according to the Solomon's equations such that during $\tau_m$ the following conversions happen given by:

$$\begin{aligned} I_{1z} &\xrightarrow{\tau_m} a_{1\to1}(\tau_m) I_{1z} + a_{1\to2}(\tau_m) I_{2z} \\ I_{2z} &\xrightarrow{\tau_m} a_{2\to1}(\tau_m) I_{1z} + a_{2\to2}(\tau_m) I_{2z} \end{aligned} \tag{6.34}$$

where the numerical values correspond to the entries in the 2 × 2 Solomon's equation matrix. In the above:

$$\begin{aligned} a_{1\to1} &= a_{2\to2} = \cosh(\sigma\tau_m) e^{-\rho\tau_m} \\ a_{1\to2} &= a_{2\to1} = -\sinh(\sigma\tau_m) e^{-\rho\tau_m}. \end{aligned} \tag{6.35}$$

With the above expression:

$$\begin{aligned} \rho_{⑤} = &-I_{1z} a_{1\to1}(\tau_m) cos(\Omega_1 t_1) e^{-\lambda t_1} - I_{1z} a_{2\to1}(\tau_m) cos(\Omega_2 t_1) e^{-\lambda t_1} \\ &-I_{2z} a_{2\to2}(\tau_m) cos(\Omega_2 t_1) e^{-\lambda t_1} - I_{2z} a_{1\to2}(\tau_m) cos(\Omega_1 t_1) e^{-\lambda t_1}. \end{aligned} \tag{6.36}$$

Whilst the first and third terms contribute to the diagonal peaks, the second and fourth contribute to the cross peaks in the 2D NOESY spectrum.

$\rho_{⑤}$ after the last $90^\circ_x$ pulse becomes.

$$\begin{aligned} \rho_{⑥} = &I_{1y} a_{1\to1}(\tau_m) cos(\Omega_1 t_1) e^{-\lambda t_1} + I_{1y} a_{2\to1}(\tau_m) cos(\Omega_2 t_1) e^{-\lambda t_1} \\ &+I_{2y} a_{2\to2}(\tau_m) cos(\Omega_2 t_1) e^{-\lambda t_1} + I_{2y} a_{1\to2}(\tau_m) cos(\Omega_1 t_1) e^{-\lambda t_1} \end{aligned} \tag{6.37}$$

The 2D signal is cosine modulated with the form given by:

$$\begin{aligned} S(t_1, t_2) = &a_{1\to1}(\tau_m) cos(\Omega_1 t_1) e^{-\lambda(t_1+t_2)} e^{i\Omega_1 t_2} + a_{2\to1}(\tau_m) cos(\Omega_2 t_1) e^{-\lambda(t_1+t_2)} e^{i\Omega_2 t_2} \\ &a_{2\to2}(\tau_m) cos(\Omega_2 t_1) e^{-\lambda(t_1+t_2)} e^{i\Omega_2 t_2} + a_{1\to2}(\tau_m) cos(\Omega_1 t_1) e^{-\lambda(t_1+t_2)} e^{i\Omega_1 t_2}. \end{aligned} \tag{6.38}$$

Amplitude of the diagonal peaks, hence, will be:

$$a_{1\to1} = a_{2\to2} = cosh(\sigma\tau_m) e^{-\rho\tau_m} \tag{6.39}$$

and that of the cross peaks will be:

$$a_{1\to2} = a_{2\to1} = sinh(\sigma\tau_m) e^{-\rho\tau_m}. \tag{6.40}$$

NOESY, hence, leads to pure-absorption line shapes in both the dimensions. At short mixing times of $\tau_m$, $a_{cross\text{-}peak}(\tau_m) \equiv \tau_m \propto r^{-6}$. Hence, estimates of distances are possible by performing a series of NOESY experiments as a function of $\tau_m$ and analysing the resulting build-up magnetisation curves.

For more details regarding the concepts outlined above, the reader is referred to [11–13].

## 6.7 Rotating-frame NOE, ROE

NOE involves relaxation between elements of longitudinal magnetisation. The null NOE aspect renders the method ineffective in measuring distances in intermediatesized molecules as indicated earlier. A way to avoid this is to mimic $\omega\tau_c \ll 1$ limit for all the molecules. This means creation of frequencies corresponding to the spectral density that causes relaxation on the order of tens to hundreds of kHz rather than the normal tens to hundreds of MHz (as in the NOE case). This may be easily accomplished in NMR by performing a rotating-frame experiment where the effective frequency is no more the Zeeman frequency but the nutation frequency of the spin-lock pulse. Hence, a positive NOE will result for every molecule. The resultant experiment is called a rotating-frame NOE or ROE where the relaxation involves transverse magnetisation components. The 2D version of ROE is the ROESY experiment. Figure 6.12 shows the pulse schemes for both ROE and ROESY along with the ROESY coherence pathways schematic.

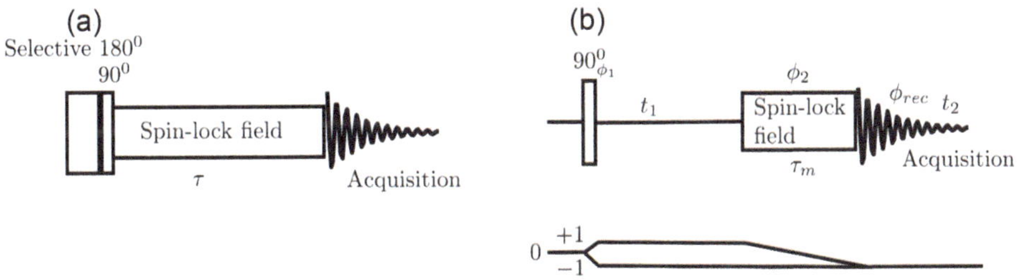


**Figure 6.12** The (a) ROE and (b) ROESY pulse sequence with coherence pathways shown.

We briefly analyse the ROE principle. Referring to Figure 6.12a, the spin-locked magnetisation, which here is assumed to be along the transverse x direction, of the two spins follows Solomon's equations as:

$$\begin{aligned}\frac{dI_x(t)}{dt} &= -\rho_I I_x(t) - \sigma_{IS} S_x(t)\\ \frac{dS_x(t)}{dt} &= -\rho_S S_x(t) - \sigma_{IS} I_x(t).\end{aligned} \tag{6.41}$$

For ROE, two experiments need to be carried out, one for reference (without the 180° pulse in Figure 6.12a and one for control, Figure 6.12a. For the reference experiment, the initial condition is $I_x(0) = S_x(0) = 1$. For the control experiment, this becomes $I_x(0) = 1$ and $S_x(0) = -1$. The solution for Equation 6.41 for the reference and control experiments may be written as (following a similar approach as in the case of trNOE solution, and denoting $\rho_I$ as $\rho$ and $\sigma_{IS}$ as $\sigma$ for simplicity):

$$\begin{aligned}I_x^{ref}(t) &= e^{-(\rho+\sigma)t}\\ I_x^{con}(t) &= e^{-(\rho-\sigma)t}.\end{aligned} \tag{6.42}$$

respectively. NOE, then is the difference of these two expressions and is given by:

$$\eta = e^{-(\rho-\sigma)t} - e^{-(\rho+\sigma)t}. \tag{6.43}$$

The maximum ROE, $\eta_{max}$, is given as:

$$\eta_{max} = \frac{(\rho+\sigma)}{(\rho-\sigma)}^{-\frac{(\rho-\sigma)}{2\sigma}} - \frac{(\rho+\sigma)}{(\rho-\sigma)}^{-\frac{(\rho+\sigma)}{2\sigma}}. \tag{6.44}$$

For homonuclear spins, $\rho$ and $\sigma$ are given by:

$$\begin{aligned}\rho &= \frac{3}{4}\gamma^4\hbar^2\left[\frac{3}{8}J(0) + \frac{15}{4}J(\omega) + \frac{3}{8}J(2\omega)\right]\\ \sigma &= \frac{3}{4}\gamma^4\hbar^2\left[\frac{1}{6}J(0) + \frac{3}{2}J(\omega)\right].\end{aligned} \tag{6.45}$$

The spectral densities, assuming isotropic Brownian motion, are given by:

$$\begin{aligned}J(0) &= \frac{1}{r^6}\frac{24}{15}\tau_c\\ J(\omega) &= \frac{1}{r^6}\frac{4}{15}\left(\frac{\tau_c}{1+\omega^2\tau_c^2}\right)\\ J(2\omega) &= \frac{1}{r^6}\frac{16}{15}\left(\frac{\tau_c}{1+4\omega^2\tau_c^2}\right).\end{aligned} \tag{6.46}$$

With all the above, $\rho$ and $\sigma$ become:

$$\rho = \frac{1}{20}\left(\frac{\gamma^4\hbar^2}{r^6}\right)\left(5\tau_c + \frac{9\tau_c}{1+\omega^2\tau_c^2} + \frac{6\tau_c}{1+4\omega^2\tau_c^2}\right)$$
$$\sigma = \frac{1}{20}\left(\frac{\gamma^4\hbar^2}{r^6}\right)\left(4\tau_c + \frac{6\tau_c}{1+\omega^2\tau_c^2}\right). \tag{6.47}$$

Figure 6.13 shows the plot of the maximum ROE for a homonuclear spin system as a function of $\omega\tau_c$.

As is clear from Figure 6.13, the ROE is always positive irrespective of the correlation time of the molecules. The minimum value in the case shown above is 38.5%, and the maximum value is 67.5%.

There are certain advantages and disadvantages for ROE or ROESY. ROE is desirable for large molecules as spin-diffusion effects are minimal, besides yielding positive NOE. ROE is, however, less sensitive than NOE. The sensitivity depends on the magnitude of the spin-lock field and the resonance offsets. ROE scheme can also lead to sample heating, particularly for salty samples, due to the spin-lock pulse that could be long enough. ROESY can also show coherence transfer due to J-coupling as in the case of TOCSY (total correlation spectroscopy). We also draw the attention of the readers to an observation by Claridge [14], "In view of the various complicating factors associated with the ROESY experiment, it is perhaps prudent to avoid using the technique whenever possible, and instead select a steady-state or conventional transient experiment as first choice."

Historically, ROE was termed as CAMELSPIN, which stands for cross-relaxation appropriate for minimolecules emulated by spin-locking [15]. (An anecdote says that CAMELSPIN, which is one of the three basic figure-skating spin positions, gave Bothner-By the inspiration to think of the rotating-frame NOE experiment, hence, the original name for the method.) More details of all the aspects outlined in the above sections may be found in Refs. [3, 11, 12, 16–18].

## 6.8 Relative Signs of Cross Peaks

Here, we briefly analyse the appearance of cross peaks in NOESY and ROESY spectra when chemical exchange can also happen between nuclear spins.

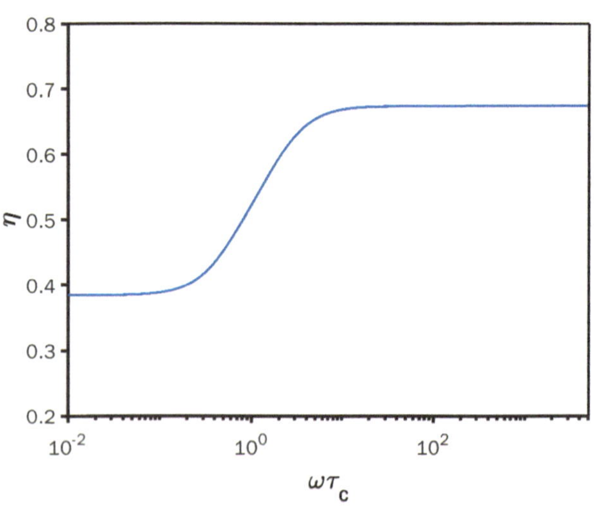


**Figure 6.13** The plot of the maximum ROE ($\eta$) as a function of $\omega\tau_c$ for a homonuclear two-spin system. This is always positive irrespective of the correlation time of the molecule unlike the NOE. The calculation is done for a pair of $^1H$-$^1H$ homonuclear spins separated by a distance of 2 Å. $^1H$ Larmor frequency was 500 MHz.

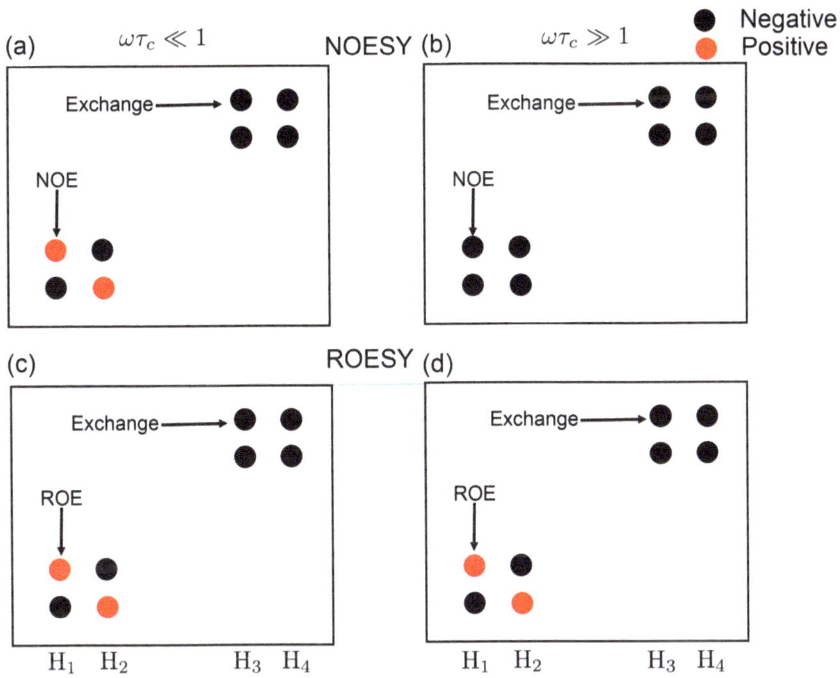


**Figure 6.14** Schematic representation of the signs of the diagonal and cross peaks in both NOESY and ROESY spectra for small and big molecules. (See https://nmr.chem.columbia.edu/sites/default/files/content/NOESY%20and%20ROESY%20experiments.pdf.)

We consider four $^1H$ spins, $H_1$, $H_2$, $H_3$, $H_4$; assume that NOE or ROE happens between $H_1$, $H_2$; and there is chemical exchange between $H_3$, $H_4$. Figure 6.14 depicts the phase pattern in both NOESY and ROESY spectra for small and big molecules of such a four-spin system.

In case of NOESY, for small molecules, the NOE cross peaks are opposite in sign to that of the diagonal peaks, if one is negative the other is positive. For large molecules, both of them will have the same sign. The sign, hence, is a clear indicator of the molecular size regime. Cross peaks due to chemical exchange have the same sign as the diagonal for all molecules in NOESY spectra.

In case of ROESY, the diagonal and cross peaks have opposite signs for all molecules as ROE is always positive. Cross peaks due to chemical exchange will have the same sign as the diagonal peaks for all molecules in ROESY spectra. Most importantly, cross peaks due to chemical exchange in ROESY spectra will be negative (as they are usually) whilst the ROE cross peaks will be positive. This is very significant in distinguishing cross-relaxation from chemical-exchange effects.

## 6.9 Generalised Solomon's Equation

In the scenario of having more than two spins that are coupled amongst themselves, the Solomon's equation Equation 6.2 may be written in a general form as:

$$\frac{dI_z(t)}{dt} = -\rho_i\left(I_{zi}(t) - I^0_{zi}\right) - \sum_{j\neq i}\left(I_{zj}(t) - I^0_{zj}\right). \tag{6.48}$$

For $i = 1 \dots n$ coupled spins, this results in $n$ coupled differential equations with the self relaxation of each spin given by $\rho_i$ and cross-relaxation with the other spins given by $\sigma_{ij}$. The solution yields in general a multiexponential time evolution of the magnetisation coupled to each other. Once the geometry of the spin configuration is known, it is possible to calculate the rate constants from Equation 6.48 and compute the expected diagonal and cross peak intensities in the NOESY spectra. These can be then fitted into the experimentally observed spectral intensities, and such a procedure carried out iteratively may converge on possible structure(s) consistent with the observed intensities. It is highly possible that there will be deviations from the computed and observed intensities, and this in turn is a measure of the internal motions. Another parameter that may be built into such calculations is the anisotropy of the reorientation of the molecules.

## 6.10 NOESY and ROESY: Practical Considerations and Experimental Spectra

In this section, we outline certain practical considerations that may be adopted whilst carrying out the NOESY and/or ROESY experiments together with experimental examples of the schemes.

NOESY spectra may contain artefacts in the form of zero-quantum coherences, which have the same coherence order as that of the desired and observed longitudinal magnetisation. The zero-quantum coherence will evolve during the mixing time as the difference between the offsets of the scalar coupled nuclear spins. These peaks will have the typical up-down nature of the double-quantum filtered COSY peaks, leading to zero total integral value. However, they can obscure the real NOE correlation peaks, making identification and quantification of such peaks an arduous task. Howe [19] and, later, Thrippleton and Keeler [20] have proposed ways to cancel out such peaks, and a summary of associated principles may be found in Ref. [21].

In ROESY, one of the issues is the appearance of the relayed TOCSY peaks. These have the same phase as the diagonal peaks and hence, similar to the exchange cross peaks. EASY-ROESY sequence could be an option here to circumvent the artefacts coming from relayed TOCSY effects [22].

From a practical point of view, the following points need attention. NOE effects may be reduced or totally quenched in the presence of dissolved oxygen or another paramagnetic species. Oxygen should be removed, especially in the case of small molecules, by methods like freeze-pump-thaw protocol. One way is to use special sample holders, such as Shigemi tubes, in case of low volume of the sample to avoid line shape issues. The temperature should be critically controlled. It is preferable not to spin the sample. Spinning may lead to modulation of the signals as a function of the evolution time leading to $t_1$ noise. Spinning also may lead to an increase in the diffusion rate in the sample which can cause a loss in the NOE signal. The current superconducting magnets are highly homogeneous, hence, spinning for improved homogeneity (as was the case earlier) really does not help.

Figure 6.15a and 6.15b depict NOESY ($\tau_m = 0.75$ s) and ROESY ($\tau_m = 0.5$ s) spectra, respectively, recorded on a sample of sucrose dissolved in $D_2O$ at 20 °C. Figure 6.15c and 6.15d show the aromatic region of NOESY ($\tau_m = 0.5$ s) and ROESY ($\tau_m = 0.1$ s) spectra, respectively, recorded on a sample of a small protein domain (A17G FF) dissolved in $D_2O$ at 20 °C. All the spectra were recorded at 700 MHz of $^1H$ Larmor frequency. Positive peaks are shown in red and negative peaks are shown in green. In the NOESY and ROESY spectra of the small molecule sucrose, the diagonal and cross peaks of the NOESY and ROESY spectra have opposite signs whilst in the larger protein molecule the diagonal and cross peaks have same signs in the NOESY and opposite signs in the ROESY spectra, showing that the sign of NOE enhancement varies with the rotational correlation time, as explained earlier.

## 6.11 Conclusions

This chapter has attempted a concise description of the principles of NOE, NOESY, and ROE, along with representative experimental spectra. The field is certainly exhaustive with NOE-related pulse methods used in the study of motions, conformational flexibility, elucidation of molecular reorientations and order parameters (through

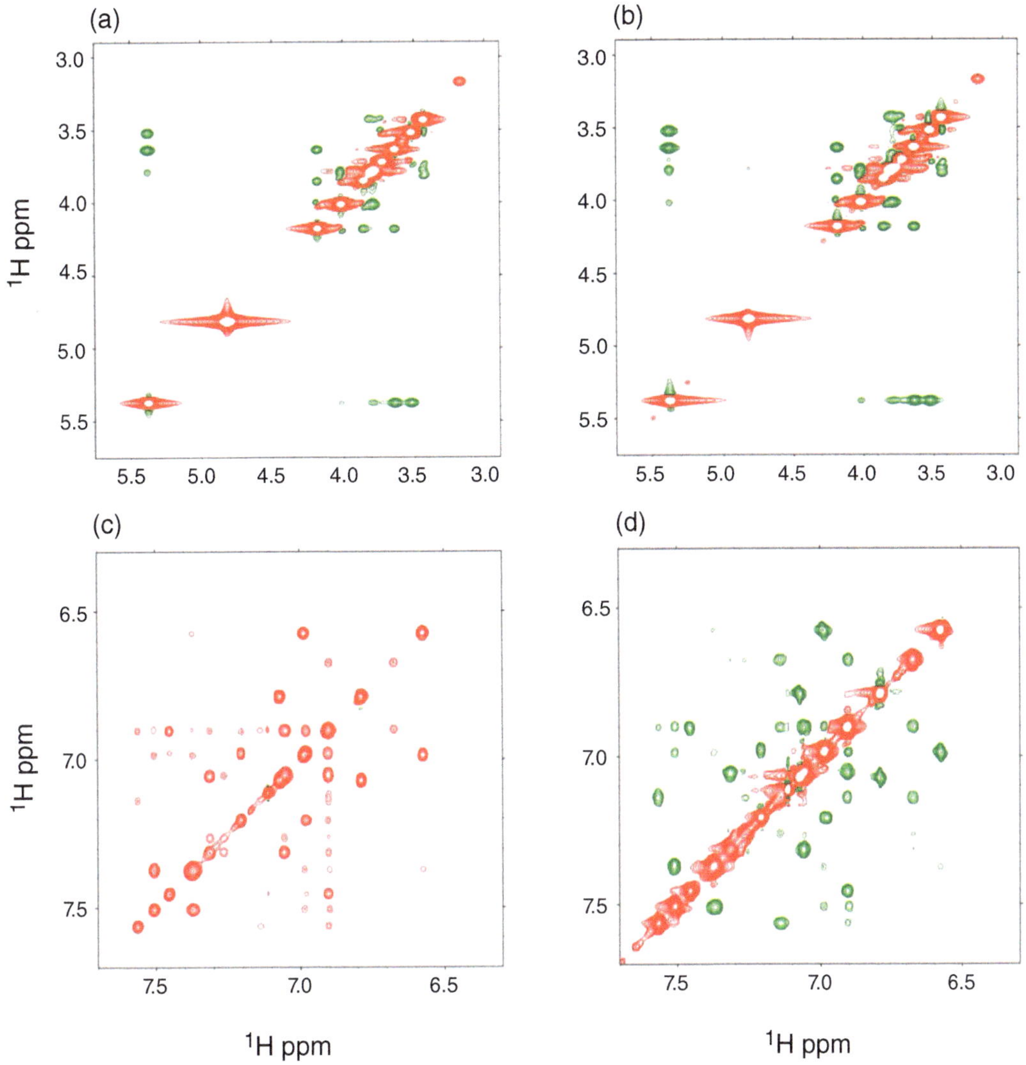


**Figure 6.15** (a) NOESY ($\tau_m = 0.75$ s) and (b) ROESY ($\tau_m = 0.5$ s) spectra recorded on a sample of sucrose dissolved in $D_2O$ at 20°C. (c) The aromatic region of NOESY ($\tau_m = 0.5$ s) and (d) ROESY ($\tau_m = 0.1$ s) spectra recorded on a sample of a small protein domain (A17G FF) dissolved in $D_2O$ at 20 °C. All the spectra were recorded at 700 MHz of $^1H$ Larmor frequency. Positive peaks are shown in red and negative peaks are shown in green. In the NOESY and ROESY spectra of the small molecule sucrose, the diagonal and cross peaks of the NOESY (a) and ROESY (b) spectra have opposite signs whilst in the larger protein molecule the diagonal and cross peaks have same sign in the NOESY (c) and opposite sign in the ROESY spectra (d), showing that the sign of NOE enhancement depends on the rotational correlation time.

the dipole-dipole coupling), exchange between chemically inequivalent sites, and indeed structural information. Different facets of NOE-related ideas have also emerged, like in the investigations of ligand binding with the feasibility of determining binding pose without protein-NMR resonance assignments [23], generation of protein-ligand costructures based on sparse NOE data [24], and elucidation of high-resolution small RNA structures [25], to name a few. The discussion here has completely ignored the effects of cross-correlated relaxation and its bearing on NOE for which the readers are referred to Ref. [26].

## Acknowledgements

I thank Prof. Anil Kumar, IISc, Bangalore, India, for the many discussions we had regarding NOE and relaxation. His insight was invaluable for me to gain a good understanding of these subjects. I acknowledge intramural funds at TIFR Hyderabad from the Department of Atomic Energy (DAE), India, under Project Identification Number RTI 4007; Pramodh Vallurupalli and Matthias Ernst for a careful reading of the manuscript; and thank Pramodh Vallurupalli for collecting the data pertaining to Figure 6.15.

## References

**1** Abragam, A. (1961). *Principles of nuclear magnetism*. Clarendon Press: Oxford.

**2** Bell, R.A. and Saunders, J.K. (1970). Correlation of the intramolecular nuclear Overhauser effect with internuclear distance. *Can. J. Chem.* 48: 414.

**3** Noggle, J.H. and Schirmer, R.E. (1971). *The Nuclear Overhauser Effect; Chemical Applications*. New York: Academic Press.

**4** Meier, B.H. and Ernst, R.R. (1979). Elucidation of chemical exchange networks by two-dimensional NMR spectroscopy: the heptamethylbenzenoniumion. *J. Am. Chem. Soc.* 101: 6441.

**5** Perrin, C.L., and Dwyer, T.J. (1990). Application of two-dimensional NMR to kinetics of chemical exchange. *Chem. Rev.* 90: 935.

**6** Overhauser, A.W. (1953). Polarization of nuclei in metals. *Phys. Rev.* 92: 411.

**7** Carver, T.R. and Slichter, C.P. (1953). Polarization of nuclear spins in metals. *Phys. Rev.* 92: 212.

**8** Carver, T.R. and Slichter, C.P. (1956). Experimental verification of the Overhauser nuclear polarization. *Phys. Rev.* 102: 975.

**9** Solomon, I. (1956). Relaxation processes in a system of two spins. *Phys. Rev.* 99: 559.

**10** Kumar, A., Ernst, R.R. and Wüthrich, K. (1980). A two-dimensional nuclear Overhauser enhancement (2D NOE) experiment for the elucidation of complete proton-proton cross-relaxation networks in biological macromolecules. *Biochem. Biophys. Res. Commun.* 95: 1.

**11** Neuhaus, D. and Williamson, M.P. (2000). *The Nuclear Overhauser Effect in Structural and Conformational Analysis*. New York: Wiely-VCH.

**12** Ernst, R.R., Bodenhausen, G. and Wokaun, A. (1990). *Principles of Nuclear Magnetic Resonance in One and Two Dimensions*. Oxford: Clarendon Press.

**13** Vincent, S.J.F., Zwahlen, C. and Bodenhausen, G. (1994). Selective measurement of the time-dependence of transient Overhauser effects in magnetic resonance applications to oligonucleotides. *Angew. Chem. Int. Ed. Engl.* 33: 343.

**14** Claridge, T.D.W. (2009). Correlations though space: the nuclear Overhauser effect. In High-Resolution NMR techniques in Organic Chemistry. Elsevier, Ch. 8.

**15** Bothner-By, A.A., Stephens, R.L., Lee, J.M., Warren, C.D. and Jeanloz, R.W. (1984). Structure determination of a tetrasaccharide: transient nuclear Overhauser effects in the rotating frame. *J. Am. Chem. Soc.* 106: 811.

**16** Williamson, M.P. (2005). Nuclear magnetic resonance spectroscopy techniques: nuclear Overhauser effect. Encyclopedia Anal. Sci. 342.

**17** Neuhaus, D. (2011). Nuclear Overhauser effect. In eMagRes (eds. R.K. Harris and R.L. Wasylishen). https://doi.org/10.1002/9780470034590.emrstm0350.pub2

**18** Kumar, A. and Grace, R.C.R. (2017). Nuclear Overhauser effect. Encyclopedia Spectrosc. Spectrom. 423.

**19** Howe, P.W.A. (2006). Removal of zero-quantum peaks from 1D selective TOCSY and NOESY spectra. *J. Magn. Reson.* 179: 217.

**20** Thrippleton, M.J. and Keeler, J. (2003). Elimination of zero-quantum interference in two-dimensional NMR spectra. *Angew. Chem. Int. Ed.* 42: 3938.

**21** Gil, R.R. and Navarro-Vazquez, A. (2017). Application of the nuclear Overhauser effect to the structural elucidation of natural products. Modern NMR approaches to the structural elucidation of natural products: Volume 2: Data acquisition and applications to compound clases. UK: Royal Society of Chemistry.

**22** Thiele, C.M., Petzold, K. and Jiirgen, S. (2008). EASY ROESY: Reliable cross-peak integration in adiabatic symmetrized ROESY. *Angew. Chem. Int. Ed.* 15: 585.

**23** Constantine, K.L., David, M.E., Metzler, W.J., Muller, L. and Claus, B.L. (2006). Protein-ligand NOE matching: a high-throughput method for binding pose evaluation that does not require protein NMR resonance assignments. *J. Am. Chem. Soc.* 128: 7252.

**24** Shah, D.M., Diercks, T., Hass, M.A.S., van Nuland, N.A.J. and Siegal, G. (2012). Rapid protein-ligand costructures from sparse NOE data. *J. Am. Chem. Soc.* 55: 10786.

**25** Nichols, P. J., Henen, M. A., Born, A., Strotz, D., Güntert, P., & Vögeli, B. (2018). High-resolution small RNA structures from exact nuclear Overhauser enhancement measurements without additional restraints. *Commun. Biol.* 1, 61. https://doi.org/10.1038/s42003-018-0067-x

**26** Kumar, A., Grace, R.C.R. and Madhu, P.K. (2000). Cross-correlations in NMR. *Prog. Nucl. Magn. Reson. Specy.* 37: 191.

# 7

# DOSY Methods for Studying Non-equilibrium Molecular and Ionic Systems

*Muslim Dvoyashkin*[1,*], *Monika Schönhoff*[2], *and Ville-Veikko Telkki*[3]

[1] *Institute of Analytical Chemistry, Leipzig University, Leipzig 04103, Germany*
[2] *Institute of Physical Chemistry, University of Münster, Münster 48149, Germany*
[3] *NMR Research Unit, University of Oulu, Oulu 90014, Finland*
* *Corresponding Author*

## 7.1 Introduction

The ability to "encode" and "decode" spin coherences via the superimposition of homogeneous and intentionally inhomogeneous magnetic fields enabled a plethora of powerful NMR techniques, including the **p**ulsed **f**ield-**g**radient (PFG) NMR (also frequently referred to as DOSY, PGSE/PGSTE NMR, or Affinity NMR), NMR imaging, rheo-NMR, electrophoretic-, and ultrafast NMR methods. This powerful toolbox of methods is daily utilized to investigate various biological (e.g. medical diagnostic by magnetic resonance tomography) and chemical systems. Among the relevant processes characteristic for both are molecular and ionic transport within various functional nanoenvironments, which can be studied directly by gradient NMR methods. Being omnipresent in nature, it remains at the core of many technological processes securing our daily life.

An extensive description of gradient NMR techniques and their application in diffusion studies exists in the literature [1–4]. In this context, the present chapter highlights the recent advances only with an emphasis on studying the non-equilibrium systems. This chapter is structured as follows: basic common principles of a gradient NMR are explained and followed by methodological aspects of electrophoretic and ultrafast NMR techniques for studying molecular and/or ionic transport in non-equilibrated systems.

## 7.2 Spatial Spin "Encoding" Using Magnetic Field Gradient

The nuclear spin, when placed in a magnetic field, undergoes precession with an angular frequency $\omega$ depending on the gyromagnetic ratio $\gamma$ of the nucleus

$$\omega(\vec{r}) = \gamma B(\vec{r}). \tag{7.1}$$

In the case of the external field $\overrightarrow{B_0}$ from the magnet, all spins precess with the same frequency $\omega_0$. The process of "encoding" implies a spectrum of $\omega$ depending on their position in space, i.e. according to (7.1) different strength of the magnetic field seen by spins sitting in different positions, i.e. artificially introduced in a controlled way by field inhomogeneity – the magnetic field gradient $\overrightarrow{B_G}$. Technically, it is convenient to produce it using a set of additional coils, for example, anti-Helmholtz coils in which the current is flowing in opposite directions (conventional design

*Two-Dimensional (2D) NMR Methods*, First Edition. Edited by K. Ivanov, P.K. Madhu and G. Rajalakshmi.

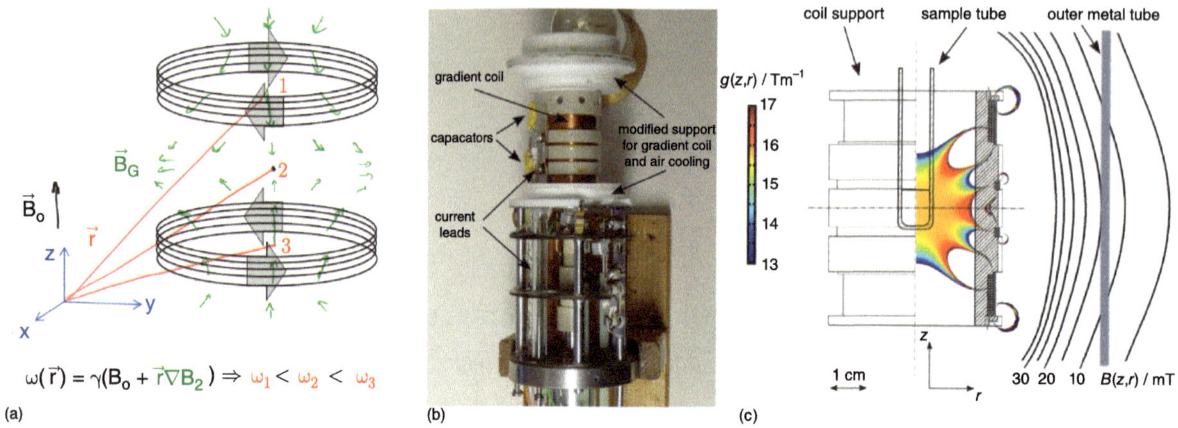


**Figure 7.1** (a) Illustration of lines of magnetic induction $\overrightarrow{B_G}$ resulting in anti-parallel- (point 1), zero- (point 2), and parallel (point 3) addition to the external magnetic field $\overrightarrow{B_0}$ caused by a pair of anti-Helmholtz gradient coils, (b) NMR probe with this type of gradient coils, and (c) calculated distribution of the resulting field gradient. Partially reproduced with permission from Ref. [5].

of high-gradient NMR systems), or by the saddle coils (used, e.g. in MRI systems) for the ability to generate field gradients in the orthogonal directions with respect to the orientation of $\overrightarrow{B_0}$ . In the former case, this results in the spatially variable lines of magnetic induction, as illustrated in Figure 7.1. In the points labeled as **1**, **2**, and **3**, one obtains different resultant field $\vec{B}$, which is a vector sum of the external field and the gradient field:

$$\vec{B} = \overrightarrow{B_0} + \overrightarrow{B_G}, \tag{7.2}$$

where $\overrightarrow{B_G} = \left(\vec{r}\frac{\partial B_x}{\partial x}\vec{i}\right)\vec{i} + \left(\vec{r}\frac{\partial B_y}{\partial y}\vec{j}\right)\vec{j} + \left(\vec{r}\frac{\partial B_z}{\partial z}\vec{k}\right)\vec{k}$.

Since only the magnitude of the total field is relevant for spins, while a local orientation of the total field is taken as a quantization axis and for respective precession around it, the resulting Larmor frequency follows Equation 7.1. Typically, the field generated by the magnet is notably larger than the one produced by gradient coils, even in high-gradient systems. Additionally, when the anti-Helmholtz coils are used in the gradient system, such as in all techniques described in the present chapter, the orthogonal x- and y-components of the total field and the higher-order magnetic field perturbations, the so-called concomitant terms, can be neglected. This simplifies Equation 7.1 to:

$$\omega(\vec{r}) = \gamma(B_0 + \vec{r}\frac{\partial B_z}{\partial z}\vec{k}). \tag{7.3}$$

According to Equation 7.3, in the mentioned spatial positions **1**, **2**, and **3**, having different z-coordinates, one obtains respective Larmor frequencies fulfilling the condition $\omega\left(\vec{r_1}\right) < \omega\left(\vec{r_2}\right) < \omega(\vec{r_3})$, i.e. they become position-dependent in the presence of a field gradient along the z-axis. Under the assumptions made, it is worth mentioning that Larmor frequencies would depend only on their z-coordinates and not on the x- and y-ones.

## 7.3 Formation of NMR Signal and Spin Echo in the Presence of Field Gradient

It is convenient to recall a simple phenomenological picture of the time-dependent evolution of nuclear magnetization in a single-pulse experiment to understand how the field gradient influences measured NMR signal. After application of a 90°-pulse, the signal is proportional to the transverse magnetization $M_{xy}$ represented by its real and imaginary parts and having a meaning of x- and y-components as follows (Figure 7.2):

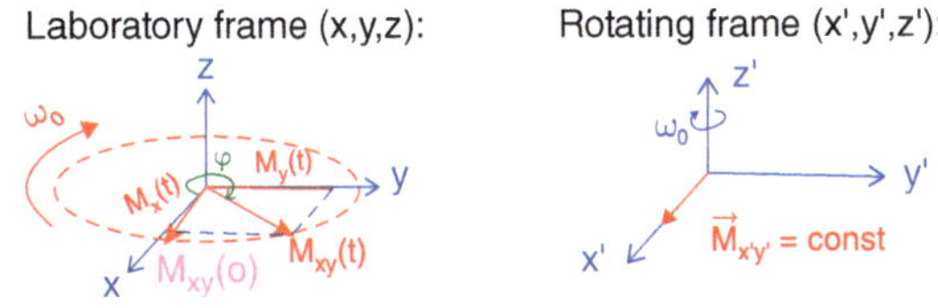


**Figure 7.2** Time evolution of magnetization in the laboratory and rotating frames in the *xy*-plane.

$$Signal \sim M_{xy} = M_x + iM_y. \tag{7.4}$$

In the laboratory frame, neglecting relaxation processes, the magnitude of magnetization vector remains constant, i.e. $\left|\overrightarrow{M_{xy}}(t)\right| = M_0 = const$, while its x- and y-components due to precession around z-axis can be expressed as follows:

$$M_x = \mathrm{Re}\left(M_{xy}\right) = M_0 \cos\varphi, \quad M_y = \mathrm{Im}\left(M_{xy}\right) = -M_0 \sin\varphi. \tag{7.5}$$

Using Euler's complex exponentials, Equation 7.4 can be represented as:

$$M_{xy} = M_0 \exp\left(-i\varphi\right) = M_0 \exp\left(-\gamma B_0 t\right), \tag{7.6}$$

where $\varphi$ is an angle counted from the initial magnetization position immediately after an r.f. pulse. Finally, in the rotating frame spinning at $\omega_0$, the magnetization simplifies to a constant and stays unchanged over time, i.e. $M'_{xy} = M_0$.

Turning the gradient on leads to the appearance of this additional term $\vec{r}\frac{\partial B_z}{\partial z}\vec{k}$ in the Larmor frequency, while the rest of the math is identical. This modifies expressions for magnetization accordingly to:

$$M_{xy} = M_0 \exp\left(-i\gamma\left(B_0 + \vec{r}\frac{\partial B_z}{\partial z}\vec{k}\right)t\right) \tag{7.7}$$

and

$$M_{x'y'} = M_0 \exp\left(-i\gamma\vec{r}\frac{\partial B_z}{\partial z}\vec{k}t\right) \tag{7.8}$$

in the laboratory and rotating frames, respectively.

The phase $\varphi' = \gamma\vec{r}\frac{\partial B_z}{\partial z}\vec{k}t$ in Equation 7.8 is an important parameter in the gradient NMR, directly influencing the amplitude of NMR signals. To illustrate, let us pick up five different z-positions $z_1..z_5$ and having the gradient field added to $B_0$ as it is represented in Figure 7.3. The total field thus monotonically increases from $z_1$ through $z_5$, so as the respective Larmor frequencies $\omega_1..\omega_5$. If now we follow evolutions of spin isochromats having these coordinates in the rotating frame, then the number 3 would stay fixed along x'-axis since $\omega_3 = \omega_0$, while numbers 1 and 2 would move anticlockwise because their frequencies are lower than $\omega_0$, and, finally, 4 and 5 would crawl clockwise. Finally, the gradient is turned off; all five spin isochromats will accumulate different phases. It leads to smaller transverse magnetization due to a loss of spin coherence (the so-called "dephasing").

One of the simple NMR pulse sequences capable of retrieving information about diffusion of spin-bearing species (atoms, molecules, ions) is the Stejskal-Tanner sequence, named after its pioneers who have introduced a pair of gradient pulses into the Hahn echo pulse sequence, as illustrated in Figure 7.4.

To understand an echo-formation containing loss of spin coherences caused by diffusion it is convenient to consider in contrast a hypothetical case of immobilized spins. In this thought experiment, two ensembles of spins having spatial coordinates $z_1$ and $z_2$ (shown in the inset in the upper left corner of Figure 7.4) in the presence of a field-gradient experience locally different total field $B_{\mathrm{tot}}$ and thus undergo a precession at different frequencies $\omega_1 < \omega_0 < \omega_2$ in the rotating frame. For clarity, all magnitudes are further considered in the rotating frame, and their primes denoting transition to it will be omitted. During the gradient duration $\delta$, according to Equation 7.8, the immobilized spins would accumulate position-dependent phases $\gamma z_1\frac{\partial B_z}{\partial z}|_{z_1}\delta$ and $\gamma z_2\frac{\partial B_z}{\partial z}|_{z_2}\delta$ (see blue and purple isochromats in Figure 7.4 moving anticlockwise and clockwise, respectively). The $\pi$-pulse applied along

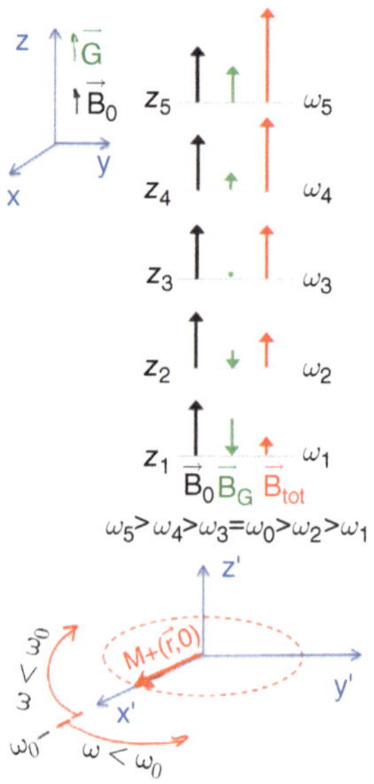


Figure 7.3 Representation of the total field in five arbitrary selected *z*-coordinates between the gradient coils 1 to 5 (top) and its result for magnetization rotation.

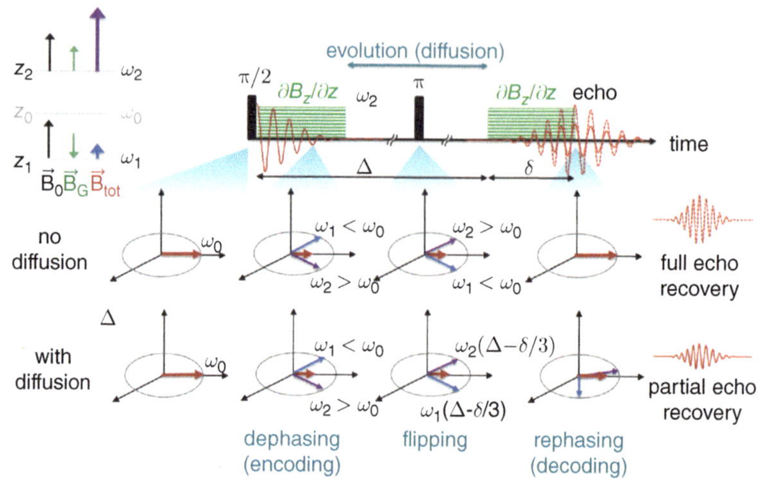


Figure 7.4 Illustration of basic steps in the Stejskal-Tanner pulse sequence leading to the spin-echo formation, accompanied by full (in case of negligibly slow diffusion) or partial (significant diffusion) signal recovery due to present field gradients.

the *y*-axis inverts the phases of isochromats, however preserving their magnitude, i.e. $\gamma z_i \frac{\partial B_z}{\partial z}|_{z_i}\delta \rightarrow -\gamma z_i \frac{\partial B_z}{\partial z}|_{z_i}\delta$. The second gradient pulse of the same length, magnitude, and direction will cause the rotation of isochromats with the same frequencies and in the same direction (blue – anticlockwise, purple – clockwise). At the end of the latter gradient, the spin isochromats will acquire precisely the same phases as during the former one, i.e. $\gamma z_1 \frac{\partial B_z}{\partial z}|_{z_1}\delta$ and $\gamma z_2 \frac{\partial B_z}{\partial z}|_{z_2}\delta$, leading to the full echo recovery. If some spins are allowed to diffuse away from their initial positions, and their z-coordinates may change, the phases after the first and the second gradient pulses may not coincide, thus leading to a non-zero total phase. This loss of spin coherence results in only partial echo recovery and measured NMR signal. Obviously, the larger the displacement of the spins, the stronger the observed NMR signal suppression due to applied gradients. Finally, the signal $\Psi$ is measured as a function of applied gradient, and, in the case of PGSE considered here, can be described as

$$\Psi = \frac{M(g,\delta)}{M(0,\delta)} = \exp\left(-\gamma^2 g^2 \delta^2 D\left(\Delta - \delta/3\right)\right), \qquad (7.9)$$

where $D$ is the constant known as the diffusion coefficient, or the self-diffusion coefficient when the experiment is conducted without a gradient of chemical potential in the detectable area.

## 7.4 NMR of Liquids in An Electric Field: Electrophoretic NMR

### 7.4.1 Measurement of Drift Velocities

Electrophoretic NMR (eNMR) is based on the same echo experiments with magnetic field gradients described above. In particular, it uses the spatial sensitivity introduced by gradient pulses to quantify the drift motion of charged species in an electric field. It is thus a specific case of the measurement of flow by gradient methods. An in-depth description of the phase evolution of a spin ensemble, which undergoes flow, is given in fundamental textbooks [3, 4, 6]. We will here explain the influence of flow on NMR signals in a simple consideration of the phase evolution.

If a molecular motion is not only random, but contains a coherent flow with a molecular drift velocity $v$, then the description of the phase evolution of each spin shown graphically in Figure 7.4 has to be modified. Figure 7.5 illustrates the phase evolution for the case of an electric field pulse applied in a simple Hahn echo for a system with and without flow.

Again, we consider two spin ensembles at different positions $z_1$ and $z_2$ along the gradient axis. The gradient strength is considered constant in space, $\frac{\partial B_z}{\partial z} = g$. The dephasing and rephasing during the gradient pulses are identical to the Hahn echo without an electric field. Thus, after the first gradient pulse, the phase angles are $\varphi_1 = \gamma z_1 g\delta$ and $\varphi_2 = \gamma z_2 g\delta$. During the electric field pulse, the spins are precessing with their Larmor frequency $\omega_0$, since no gradient is applied. Thus, no change of the phases occurs in the rotating frame (Figure 7.5); they are only exchanged by the flipping of the spins by the $\pi$ pulse. However, if the spins are drifting in the electric field, their positions at the time of the second gradient pulse are altered. We assume a plug flow with identical velocities $\boldsymbol{v}$ of all molecules bearing the observed spins. The mobility is defined as $\mu_i = v_z/E$, with $E$ denoting the electric field. With the joint $z$ component $v_z$, the respective new position is $z_i' = z_i + v_z\Delta$ for $i = 1, 2$. Note that here $\Delta$ denotes the length of the electric field pulse, which may differ slightly from the gradient spacing. The second gradient pulse now induces a phase shift according to $z_i'$, such that a phase $\varphi_i' = \gamma z_i' g\delta$ is added. The final phase shift can be described by $\Delta\varphi_i = -\gamma z_i g\delta + \gamma z_i' g\delta$, with the negative sign being a consequence of the flipping of the spins by the $\pi$ pulse. Expressing the difference in positions by the drift velocity, the result is $\Delta\varphi_i = \gamma v_z \Delta g\delta$ , which shows that the identity of drift velocity for all spins causes an additional phase shift $\Delta\varphi$, identical for all spins independent of their starting position. Flow thus has the effect of shifting the echo in phase, while it does not diminish the echo intensity, see Figure 7.5. However, the coherent molecular drift in a flow is often superimposed by self-diffusion; see the discussion further below.

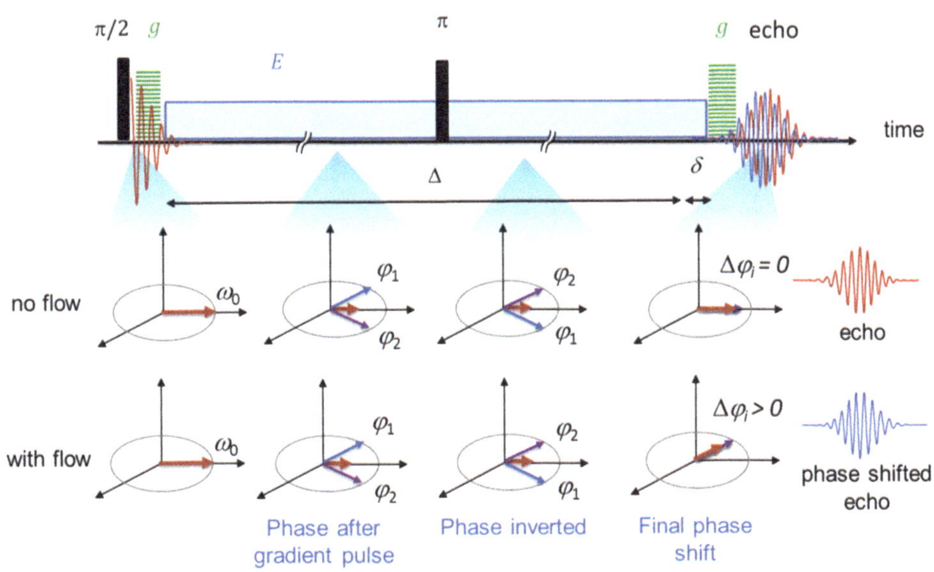


**Figure 7.5** Hahn echo with gradients and an applied electric field, showing the influence of the magnetic field-gradient pulses and molecular drift in the electric field on the phase evolution of spin ensembles starting at different positions $z_1$ and $z_2$.

In the Fourier transformed complex echo signal $S(\omega)$, the additional phase shift causes a modulation, which can be expressed using the absorptive and dispersive parts $A(\omega)$ and $D(\omega)$ of the complex signal:

$$S(\omega) = S_0[A(\omega) + iD(\omega)] \exp(i\Delta\varphi). \tag{7.10}$$

In a typical experiment, a series of spectra is taken and the phase shift $\Delta\varphi$ analyzed. Most commonly, the varied parameter is the electric field $E$ by applying different voltage values $U$ and thus varying the drift velocity as:

$$\nu_z = \mu E = \mu \frac{U}{d}, \tag{7.11}$$

where $\mu$ is the electrophoretic mobility and $d$ the electrode distance. The phase shift then results in:

$$\Delta\varphi = \gamma\mu\Delta g\delta\frac{U}{d}, \tag{7.12}$$

from which the electrophoretic mobility can be determined by a variation of $U$. Note that the gradient strength $g$ is a similarly suitable independent parameter, varied in some cases. However, as $g$ variation also reduces the echo signal due to diffusion, in most applications it is more appropriate to choose the voltage as an independent parameter, while $g$ is set to a fixed value which forms a compromise of a sufficient echo signal and large enough sensitivity of the phase shift.

Figure 7.6 shows an example of the eNMR spectra of a dilute salt solution, taken at different voltages as indicated. It is evident that the solvent signal, in this case residual protons in $D_2O$, is not affected by the electric field, while the trimethylammonium signal shows a phase shift, which changes with applied voltage according to Equation 7.12. In analogy to a PFG-NMR diffusion experiment, eNMR allows the evaluation of mobilities of all spectrally resolvable species.

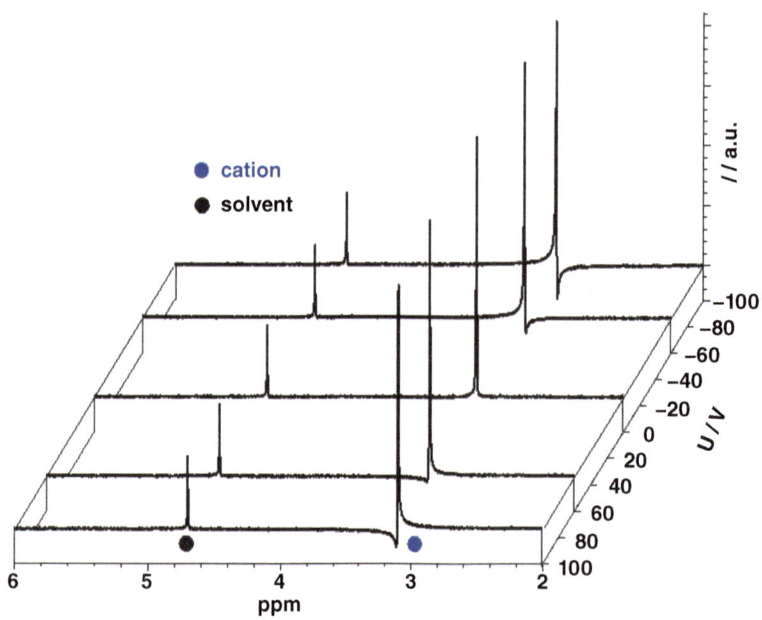


Figure 7.6 Selected $^1H$ spectra of 100mM trimethylammonium bromide in water ($D_2O$), taken by a double stimulated echo (see Figure 7.7) with pulses of different voltage. Spectra taken by P. Steinforth (unpublished data).

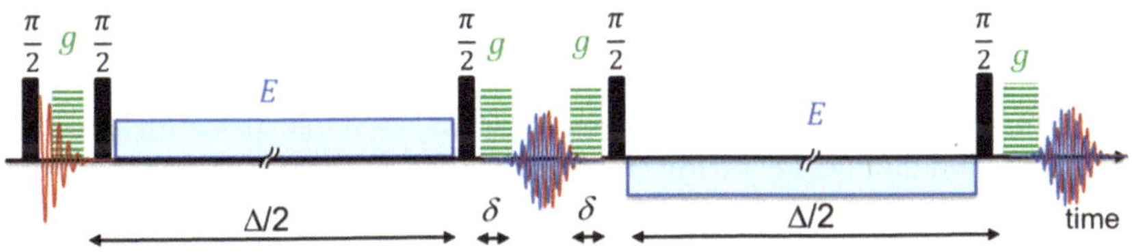


**Figure 7.7** Double stimulated echo with gradient pulses and electric field pulses: This sequence compensates the influence of a constant flow arising from convection and electroosmotic flow, while the migration of charged species in an electric field is inverted, leading to constructive addition of the phase contributions.

### 7.4.2 Technical Development

The history of eNMR dates back to the first observation of molecular flow by Packer [7]. Holz performed the first flow experiment under an electric field in a dilute aqueous salt sample [8]; Johnson and He introduced electric field pulses into a stimulated echo sequence [9], where signal modulations depending on the electric field as given by Equation 7.12 were observed [10]. Subsequently, 2D processing of the eNMR experiment was introduced involving Fourier transformations with respect to the FID time domain and E, allowing analysis of distributions of the electrophoretic mobility [11]. This type of experiment was termed MOSY (**m**obility **o**rdered 2D NMR **s**pectroscop**y**), in analogy to the acronym DOSY and is discussed further below [10, 12].

Since then, progress has been relatively slow, with instrumental optimizations covering several decades. Continuing challenges originate from non-electrophoretic flow artifacts, which may accompany the ion drift in the electric field. Consequently, a significant fraction of papers has dealt with setup improvements, including instrumentation and sample design, and developing optimized pulse sequences. A major problem was convection, induced by Joule heating of the conductive samples, which caused convective flow in the presence of temperature gradients. In addition, electroosmotic flow induced by a drag exerted on charged layers of liquid at glass tube interfaces may cause disturbances. A thorough discussion of flow artifacts is given by Pettersson et al., who also suggested a solution to suppress such undesired additional flow components [13]. For this purpose, a double stimulated echo sequence was introduced, see Figure 7.7. This is an extension of the previously suggested "back-to-back" echo sequence consisting of a double Hahn echo, which was introduced for convection compensation in eNMR by He and Wei [14]. Both sequences contain a second echo, where an electric field pulse with an inverted sign is applied. Consequently, the electrophoretic drift direction is inverted, while the perturbing non-electrophoretic bulk flow components persist. Since the phase is inverted at the center of the sequence, the electrophoretic phase modulation of both parts of the sequence is added constructively, while the effect of non-electrophoretic bulk flow is canceled [13].

Several reviews cover the apparative development for eNMR in (mostly aqueous) dilute solutions [13, 15]. In recent years the group of Furo et al. was leading the technical development, for example discussing different sample geometries and introducing pulse sequences such as an eNMR CPMG sequence for non-electrophoretic bulk flow compensation [16], and introducing current-controlled pulses for improved stability [17].

### 7.4.3 Application Areas: From Dilute to Concentrated Electrolytes

While the above experimental improvements were ongoing, the application of eNMR has for a long time focused on dilute solutions, targeting questions of ion condensation and self-assembly. Major achievements were made in the field of counter-ion condensation in polyelectrolyte solutions by the group of Scheler [18, 19]. eNMR also proved useful in the study of colloidal systems, such as the self-assembly of ionic surfactants or their aggregation with polyelectrolytes, as summarized in review articles [20, 21].

A key concept in interpreting eNMR mobilities in associating systems is the effective charge. In the ideal case of an electrolyte with completely uncorrelated ion motion, i.e. no ion condensation, aggregation, or pair formation, the ionic mobility $\mu_i^{NE}$ is given by the Nernst-Einstein equation:

$$\mu_i^{NE} = \frac{z_i\, e\, D_i}{k_B\, T}. \tag{7.13}$$

$D_i$ and $z_i$ are the diffusion coefficient and the nominal number of charges of the species $i$, respectively. Further, $k_B$ is the Boltzman constant, $T$ the temperature and $e$ the elementary charge. The effective charge relates the experimental electrophoretic mobility $\mu_i$ to this ideal value,

$$\varepsilon_i = \frac{\mu_i}{\mu_i^{NE}} = \frac{\mu_i k_B T}{z_i e D}. \tag{7.14}$$

In the case of complexation of ionic species or other ion correlations, the calculation from Equation 7.13 generally overestimates the mobility. The mismatch described by the effective charge $\varepsilon_i$ in Equation 7.14 is thus a measure for the number of charges of an ion complex, which is relevant for its transport. Note that in spite of the term "effective charge" $\varepsilon_i$ is dimensionless since it is not defined in units of charge, but in units of the nominal charge $z_i e$. By this concept, for example, the effective charge of proteins at different pH conditions was quantified [22], and an effective charge was attributed to the neutral polymer poly(ethyleneoxide), describing the binding of cations to the chain under different conditions [23].

Studies of highly concentrated electrolytes appeared more challenging due to the high ion concentrations enhancing the problem of Joule heating. For example, ionic liquids (ILs), of interest as poorly flammable electrolytes in Li-ion batteries, are liquid salts consisting solely of ionic species. First attempts of eNMR on such concentrated electrolytes lacked the linearity expected by Equations 7.11 and 7.12 and yielded extremely high mobilities, accompanied by enhanced diffusion [24]. eNMR studies of ILs became feasible only when two measures of convection compensation were applied, namely the double stimulated echo, and furthermore the introduction of a bundle of capillaries in the active volume. The latter is a method to suppress x- and y- components of non-electrophoretic flow, suggested by He and Wei [14]. The first eNMR studies of pure ILs with successful convection compensation appeared [25, 26] and opened the route for investigations of concentrated Li-conducting liquid battery electrolytes. Here, the power of multinuclear eNMR is tremendous, as it is a unique method to highlight dynamic correlations of different ion species by multinuclear eNMR. For example, in ILs with an added Li salt all ionic species bear suitable nuclei, and combined $^1$H (IL cations), $^{19}$F (anions) and $^7$Li (lithium ion) eNMR sheds light on the complete charge transport behavior. In this way, a vehicular transport of Li in anionic clusters was discovered in ILs, as the Li effective charge resulted in negative values, indicating the stoichiometry of the transport-relevant species being $\text{Li[Anion]}_n^{-(n-1)}$ with a negative net charge and $n$ depending on the type of anion [27–29]. Furthermore, the role of Li-coordinating solvents could be highlighted, and it could be distinguished whether the Li-ion either migrates jointly with a solvent shell, or the Li transport occurs via an ion hopping between non-migrating coordination sites [30–32].

### 7.4.4 Methods of Transformation and Processing: MOSY

Similar to standard PFG-NMR for diffusion measurements, eNMR is a 2D experiment, though in both methods the evaluation is often performed by a fit of data following a 1D transform. In eNMR, the phase shifts can be simply evaluated from single spectra like in Figure 7.6 and fitted according to Equation 7.12 to yield the mobility. A more elegant way is to perform a 2D transform, though this is more time consuming since it requires data acquisition for

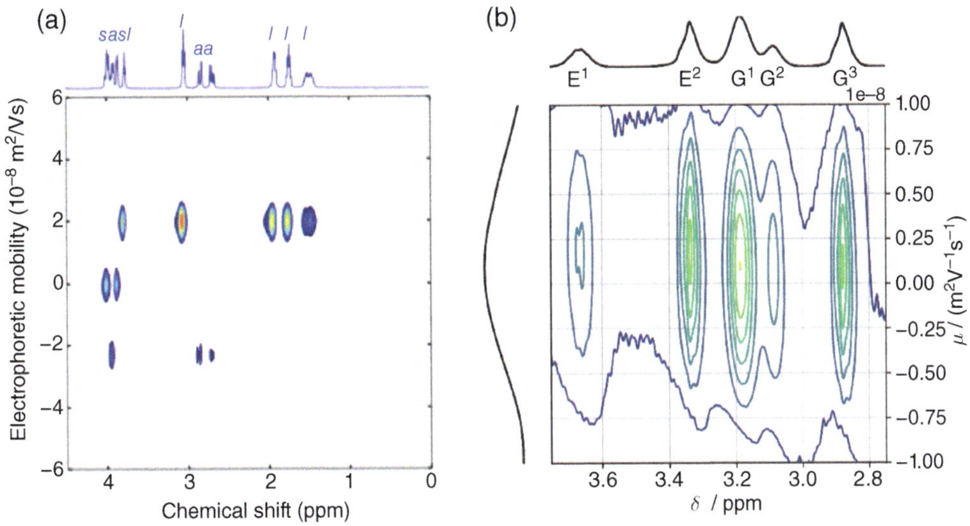


**Figure 7.8** (a) $^1$H MOSY spectrum of an equimolar aqueous mixture of L-aspartic acid (*a*), L-serine (*s*) L-lysine (*l*) in $D_2O$ at pD = 6.2. (b) $^1$H MOSY spectrum of LiTFSA:tetraglyme:$C_1C_2$ImTFSA in a molar ratio of 1:1:8, showing the spectral region with superimposed tetraglyme and ethylmethylimidazolium ($C_1C_2Im^+$) signals: $E^1$ and $E^2$ of $C_1C_2Im^+$, $G^1$, $G^2$ and $G^3$ of tetraglyme. Figures reproduced with permission from Refs. [33] and [30].

a larger number of voltage values. The second transform dimension is based on the phase term of Equation 7.10, where the modulation with the variable $\gamma\delta\Delta g\mu E$ is detailed as:

$$S(\omega) = S_0[A(\omega) + iD(\omega)][\cos(\gamma\delta\Delta g\mu E) + i\sin(\gamma\delta\Delta g\mu E)]. \tag{7.15}$$

Thus, a 2D Fourier transform will yield a 2D spectrum with the chemical shift and the mobility dimension. Figure 7.8a gives an example where the components of an amino acid mixture can be separated by MOSY, while their spectral separation is poor, highlighting the analytical potential of eNMR [33].

While MOSY 2D processing is a powerful analytical tool, its application is tied to the observable phase range. While that phase range can be large for small molecules in dilute, low viscosity solutions, it is typically far below a full period in concentrated electrolytes such as ILs, due to the far lower mobilities. This yields MOSY peaks, which are very broad in the mobility dimension with a width that is controlled by the limited phase range rather than the polydispersity of the sample, see Figure 7.8b. Especially for overlapping resonances, this prohibits the independent determination of the mobilities of different species. It was shown that this limitation in concentrated electrolytes can be overcome by spectral deconvolution of eNMR spectra via a set of Lorentz profiles with absorptive and dispersive components. The deconvolution procedure is superior, yielding a higher precision of the mobilities in cases of overlapping resonances of different species. It was employed to analyze the migration of uncharged, coordinating solvent molecules in concentrated lithium-conducting electrolyte [30].

### 7.4.5 Is eNMR a non-equilibrium experiment or a steady-state experiment?

It has often been questioned whether the high voltage of up to 100 V in eNMR is leading to electrolyte decomposition in the electric field, especially in comparison to typical battery cell voltages of just a few V. Another question is whether electrode polarization is causing relevant decays of the voltage applied to the bulk materials, which would require correction.

In the following, we discuss these questions in particular in the context of the relevant time and length scales, highlighting non-equilibrium aspects of the eNMR experiment. With ion-blocking metal electrodes, the eNMR cell can be considered a capacitor, where the applied voltage induces polarization and an electric double layer is formed at the electrode interfaces, see Figures 7.9a and b. While at $t = 0$ the full voltage drop occurs over the electrolyte, inducing ion drift velocities $v_+$ and $v_-$, the situation has changed after a time $t$ with $0 < t < \Delta$, since ions have accumulated at either electrode interface, forming an electric double layer. The consequence is a voltage drop $U'$ over the double layer, which reduces the voltage drop over the bulk electrolyte and thus the electric field in which the ions are migrating. In typical experiments on concentrated electrolytes, however, no dependence of the drift velocity on the pulse length is observed, and the phase shift obeys the linearity predicted by Equation 7.12.

In the following estimate, we consider a typical IL sample with mobilities on the order of $\mu = 10^{-9}$ m²/Vs (see EmImTFSI [26]). Experimental parameters are an observation time of $\Delta = 100$ ms and a voltage of 100 V applied over 2 cm, yielding $E = 50$ V/cm. This results in an ion drift velocity of $v_z = \mu \cdot E = 5 \cdot 10^{-6}$m/s, yielding a displacement of the ions by a distance of $\Delta z = v_z \cdot \Delta/2 = \mu \cdot E \cdot \Delta/2 = 0.25$ µm by one electric field pulse. This displacement is in fact very small, and it is instructive to compare it to the root-mean-square displacement resulting from diffusion. With $D = 3.4 \cdot 10^{-11}$ m/s² in the same IL [26], the diffusive displacement $\sqrt{\langle \Delta z^2 \rangle}$ amounts to about 2 µm, exceeding the field-driven displacement. Thus, in an eNMR experiment, a shuttling of ions between two close positions in space is superimposed by a substantial broadening of their distribution in space. This is illustrated in Figure 7.9c, where the development of the propagator for a slice of ions initially positioned at $z_0$ is considered. The propagator, often considered in the theory of NMR transport measurements, describes the time evolution of the spatial distribution. During the first electric field pulse the ions are displaced by $\Delta z_0$ to a new position. This shift of the propagator is accompanied by broadening due to diffusion (dashed line). In the second electric field pulse, which is applied with inverted polarity, the maximum of the propagator is shifted back to $z_0$, while diffusion-induced broadening continues, resulting in the distribution represented by the solid blue line. Successively applying an eNMR pulse sequence to a sample thus consists of shuttling ions back and forth by a sub-µm distance, while diffusion takes place.

While this shuttling is the bulk behavior of the ions, which is detected, thinghs are different at the electrodes. The electric double layer formed during an electric field pulse is characterized by the double layer capacitance at a conducting electrode. We assume a Helmholtz model consisting of a monolayer of ions (see Figure 7.9b), with a typical ionic molar volume of $v_m = 1.5 \cdot 10^{-4}$m³/mol, as calculated from density measurements (1,3 g/cm³) yielding a molecular dimension of $a = 6$ Å. With a relative permittivity $\varepsilon_r = 12$ (given for EMIm TFSI [34]), and an electrode area of A = 20 mm² the resulting capacitance is $C = \varepsilon_r \cdot \varepsilon_0 \cdot A/a \approx 4$ µF. This is in good agreement

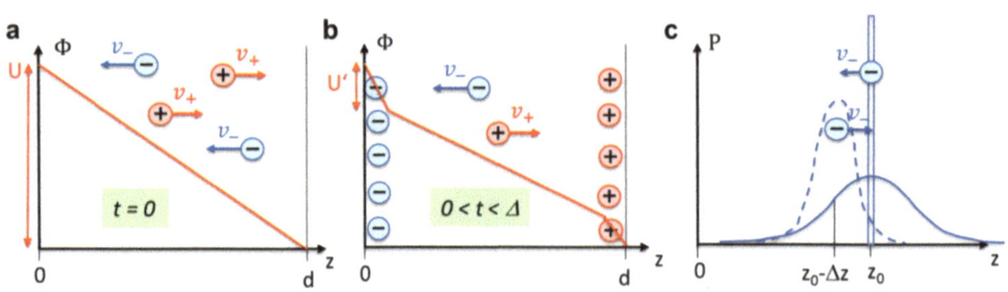


**Figure 7.9** Ion drift and electric potential (a) at the beginning and (b) during an electric field pulse of duration Δ in an eNMR experiment. Electrodes are positioned at z = 0 and z = *d*. (c) – Development of the propagator for a slice of ions initially positioned at $z_0$: Negative drift velocity in first E pulse shifting peak maximum to $z_0 - \Delta z$, superimposed by broadening due to diffusion (dashed line); second E pulse with inverted sign shifting peak maximum back to $z_0$, while diffusion-induced broadening continues, resulting in distribution shown by the solid blue line.

with experimental values determined for IL interfaces of $c \approx 2\ \mu F/cm^2$ [35], justifying the simplified assumption of a monolayer.

This layer bears a surface charge density at each electrode of $\sigma = (a \cdot e \cdot N_A)/v_m \approx 0.4\ C/m^2$, and with the above drift velocity, it can be built up in the time required for an ion displacement of $a$, which is $\tau = 0.1$ ms. It is thus clear that the polarization buildup is instantaneous and the equilibration time following the voltage increment is small compared to the observation time, i.e. $\tau << \Delta$. The next question is whether the potential drop over the polarization layer is causing a significant reduction of the electric field in the bulk volume, where the drift velocity is measured. Again making use of the capacitance of $\approx 4\ \mu F$, we obtain $U' = Q/C \approx 2$ V. Thus, in comparison to the externally applied voltage of $U = 100$ V, this potential drop is negligible. We can finally state that the effect of polarization is minimal, causing only a negligible reduction of the externally applied voltage.

However, a far larger amount of charge is transported during a complete voltage pulse. If an accumulation of all ions reaching the electrode within the time $\Delta$ is assumed, the surface charge density accumulated during an electric field pulse would reach values as large as $\sigma = (\Delta z \cdot e \cdot N_A)/v_m \approx 3 \cdot 10^2\ C/m^2$, far above that of a Helmholtz layer. This charge would thus cause a breakdown of the electric field in the bulk region. Apparently, this is not observed in the experiment. On the contrary, the linearity of the phase shift with $\Delta$, $g$ and in particular $E$, as predicted by Equation 7.12, can be tested and is valid under the considered conditions. It can be concluded that this amount of charge is thus transferred to the electrodes in the form of decomposition reactions of the electrolyte. Electrolyte decomposition is thus not an artifact in eNMR on concentrated electrolytes, but rather a prerequisite for steady-state conditions during the electric field pulse. To achieve this steady state, the charge transfer should not be rate-limiting, as this would cause a voltage breakdown in the bulk region due to an enhancement of $U'$.

It can be concluded that eNMR on concentrated electrolytes with voltages as high as up to 100 V is a non-equilibrium experiment with respect to electrolyte decomposition occurring at the electrodes. In the bulk of the active volume, however, ions are shuttled back and forth by sub-μm distances. It is essential to ensure that the measurement of the drift velocities in the bulk active volume is not affected by electrode reactions. This is best achieved by careful control over the linearity given by Equation (7.11), especially with respect to the applied voltage. In addition, following an eNMR experiment, the sample has to be discarded, and the sample cell carefully cleaned from any decomposition products.

Additional quality control of eNMR mobilities is easily possible, which consists of the comparison of the sum of the current carried by all ion species $i$, termed $\sigma_{tot}$, to an independent measurement by impedance spectroscopy, $\sigma_{IS}$:

$$\sigma_{tot} = \sum_i z_i\, c_i\, F\, \mu_i = \sigma_{IS}, \tag{7.16}$$

where $F$ is the Faraday constant. Excellent agreement of both quantities is confirmed in many different systems, proving the drift motion to be driven by an unperturbed electric field under steady-state conditions [28, 29, 36, 37].

With many precautions to be taken and experimental conditions to be verified against artifacts that might occur due to non-electrophoretic flow contributions or the mentioned potential chemical processes at the electrodes, eNMR remains a challenging technique. In addition, it should be mentioned that the experiment times and repetitions required make it a much slower technique as compared to PFG diffusion NMR. Experiment times are comparatively long, since a long relaxation delay is required to allow dissipation of the heat generated by Joule heating. Nevertheless, in many electrolyte systems, eNMR can give unique information, as the above examples have shown, and it may further prosper, in particular in the broad field of electrolytes for energy storage, where the species selectivity in the charge transport analysis is a property that makes eNMR unique in comparison to electrochemical techniques.

## 7.5 Ultrafast Diffusion Measurements

Conventional DOSY measurements are rather time-consuming because the experiment must be repeated many times with incremented gradient strength, as illustrated in Figure 7.10a. The number of repetitions, $N$, required for probing the diffusion decay curve well enough varies typically between 10 and 100. As the single experiment with one gradient value, including the relaxation delay $T_R$ for the recovery of the longitudinal magnetization, typically takes from 1 to 10 s, the total experiment time with a single scan varies between 10 and 1 000 s. This time must be multiplied by the number of scans needed for phase cycling and improving signal-to-noise ratio (SNR), which varies typically between 2 and 128. Therefore, the whole DOSY measurement usually lasts from minutes to hours.

The long DOSY experiment time may limit the throughput of NMR spectrometers. It also may hinder investigations of systems, which are changing in time. The changes may be caused by, e.g. sample degradation, chemical reactions, charge/discharge cycles of batteries, protein folding, metabolic processes, varying temperature, phase changes. Furthermore, the need for repetitions may prevent utilization of modern nuclear spin hyperpolarization techniques, which otherwise could improve the sensitivity of experiments by several orders of magnitude, and make, e.g. metabolites in physiological concentrations observable. For example, a dissolution dynamic nuclear polarization (dDNP) process requires a time scale from tens of minutes to hours [38]. Repetition of the polarization process from 10 to 100 times for the DOSY increments is not realistic time-wise, and varying the degree of hyperpolarization might induce additional challenges. There are several approaches to accelerate DOSY measurements, which are based on shortened relaxation delay, small flip angle pulses, trains of diffusion gradients, and compressed sensing [39].

One elegant and efficient method for accelerating DOSY measurements is based on the spatial encoding of diffusion data [40]. Spatial encoding is broadly exploited to accelerate 2D NMR experiments under the concept of ultrafast NMR spectroscopy [41, 42]. It allows single-scan 2D NMR measurements. The price of the speed

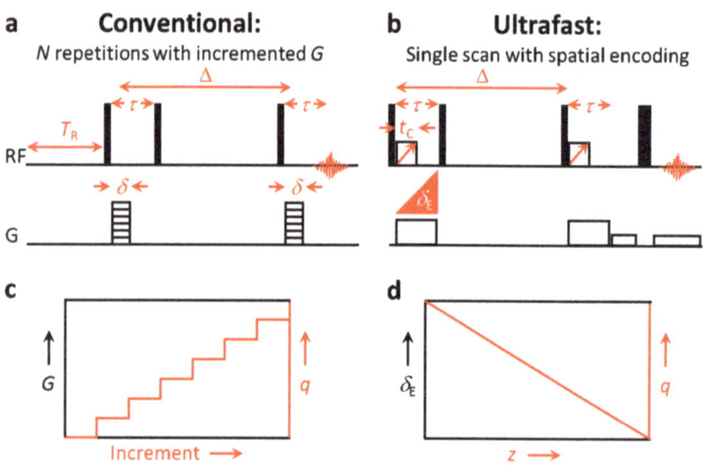


**Figure 7.10** (a) The pulse sequence for the conventional DOSY measurement. It requires $N$ repetitions with incremented gradient strength, which is represented by the ladder-like gradient pulse shape. (b) Ultrafast DOSY sequence includes additional frequency-swept 180° RF pulses, illustrated by black boxes with red arrows, which make the effective length of the diffusion gradient, $\delta_E$, linearly dependent on position. The sequence also contains an additional hard 180° RF pulse sandwiched by read gradients for reading the magnetization profile. (c) In the conventional DOSY experiment, various $q$ values are probed by changing the gradient strength, while (d) in the ultrafast DOSY experiment, a similar range of $q$ values are encoded in the magnetization profile in a single scan.

is a decrease in the sensitivity of the experiment, as, virtually, the sample is split into data layers. However, the single-scan approach makes it possible to overcome the sensitivity decrease by hyperpolarization; typically, the sensitivity improvement provided by hyperpolarization is much greater than the impairment due to spatial coding.

The pulse sequence for an ultrafast DOSY experiment is shown in Figure 7.10b. Compared to the conventional DOSY sequence in Figure 7.10a, it includes additional adiabatic 180° frequency-swept chirp pulses, which are switched on simultaneously with the diffusion gradients. The length of the chirp pulses is half of the length of the gradient pulses. The offset frequency (frequency in the rotating frame) of the chirp pulse, $\nu_C$, increases linearly with time from $-\Delta\nu/2$ to $+\Delta\nu/2$:

$$\nu_C(t) = \frac{\Delta\nu}{t_C}\left(t - \frac{t_C}{2}\right), \quad \text{when } 0 \leq t \leq t_C. \tag{7.17}$$

Here, $\Delta\nu$ and $t_C$ are the sweep width and length of the chirp pulse. On the other hand, the diffusion gradient makes the offset Larmor frequency of spins, $\nu_L$, linearly dependent on position (z) – it is the smallest at the bottom and the largest at the top:

$$\nu_L = \frac{\gamma g z}{2\pi}, \quad \text{when } -\frac{\Delta z}{2} \leq z \leq \frac{\Delta z}{2}. \tag{7.18}$$

Here, $\Delta z$ is the height of the spatial encoding region (center at $z = 0$). The chirp pulse rotates transverse magnetization by 180° when the frequency of the pulse matches with the Larmor frequency, and it acts as a refocusing pulse. Therefore, the time when the refocusing pulse is acting, $t_R$, becomes linearly dependent on position:

$$t_R(z) = \left(\frac{\gamma g z}{2\pi\Delta\nu} + \frac{1}{2}\right)t_C, \quad \text{when } -\frac{\Delta z}{2} \leq z \leq \frac{\Delta z}{2}. \tag{7.19}$$

The height of the spatial encoding region is defined by the combination of the chirp sweep width and gradient strength:

$$\Delta z = \frac{2\pi\Delta\nu}{\gamma G}. \tag{7.20}$$

According to Equations 7.19 and 7.20, the refocusing 180° chirp pulse acts at $t_R = 0$ and $t_R = t_C$ for the spins at the bottom and top encoding layers, respectively. The latter time instant is at the center of the gradient pulse. Therefore, the phases of spins are focused at the end of the gradient pulse, and, effectively, the spins at the top layer do not experience the gradient pulse at all. On the other hand, the spins at the bottom layer experience the full gradient. In the intermediate positions, the phases of the spins are refocused at $2t_R$, and therefore, effectively, the gradient pulse dephases spins only for time $2t_C - 2t_R$. Note that although the phase of the chirp pulse experienced by spins is dependent on the position, inducing an additional position-dependent phase of spins, this contribution is canceled by the second similar chirp pulse [39]. Hence, the effective length of the gradient, $\delta_E$, experienced by spins becomes linearly dependent on position (see Figure 7.10d):

$$\delta_E(z) = 2t_C - 2t_R(z) = \left(1 - \frac{\gamma g z}{\pi\Delta\nu}\right)t_C, \quad \text{when } -\frac{\Delta z}{2} \leq z \leq \frac{\Delta z}{2}. \tag{7.21}$$

As the magnetization wave vector $q$ is directly proportional to $\delta_E$ (see Figure 7.11d),

$$q(z) = \gamma g \delta_E(z), \tag{7.22}$$

also the signal amplitude in the ultrafast DOSY experiment becomes dependent on position $z$:

$$\Psi(z) = \exp\left[-D[q(z)]^2\Delta\right] \equiv \exp[-Db(z)], \tag{7.23}$$

where $b(z) = [q(z)]^2\Delta$. Consequently, the magnetization profile in the $z$-direction in the ultrafast DOSY experiment is equivalent to the magnetization decay curve observed in the conventional DOSY experiment (*cf.* Equation 7.9 under narrow gradient pulse approximation, i.e. $\delta \approx 0$).

In a conventional DOSY measurement, variable $q$ values are probed by incrementing the gradient strength in separate experiments (Figure 7.10c). In contrast, in an ultrafast DOSY measurement, the diffusion gradient remains constant, but the effective length of the gradient becomes linearly dependent on position, allowing probing a range of $q$ values similar to the conventional experiment (Figure 7.10d). The magnetization profile is read using the principles of 1D magnetic resonance imaging (MRI), by adding read gradients along with a hard 180° refocusing pulse to the end of the pulse sequence shown in Figure 7.10b (corresponds to a 1D spin-echo imaging block). Therefore, the ultrafast approach makes it possible to measure data equivalent to a conventional DOSY experiment in a single scan instead of 10 to 100 repetitions, significantly reducing experiment time.

Figure 7.11 shows an ultrafast DOSY magnetization profile of a doped water sample (red solid line) [43]. The single-scan experiment time was only 2 s. The sweep width of the chirp pulse was $\Delta\nu = 108$ kHz, and the diffusion gradient strength was $g = 0.220$ T/m. Therefore, according to Equation 7.20, the height of the spatial encoding region was $\Delta z = 1.2$ cm. The upper horizontal axis shows the z coordinates, while the lower horizontal axis indicates Larmor frequencies when the diffusion gradient was on. Horizontal gray dashed lines visualize the chirp sweep and spatial encoding region.

The spatially encoded diffusion decay curve, equivalent to the conventional diffusion decay curve measured with multiple repetitions, is visible within the vertical dashed lines. The curve has its maximum value close to the right edge of the spatial encoding region, as it corresponds to minimum $\delta_E$ and $q$. The curve decays when going to the left due to increased $\delta_E$ and $q$ (see also Figure 7.10d). In the vicinity of the edges of the spatial encoding region, the experimental curve deviates from the theoretically predicted curve due to the imperfect behavior of the chirp pulses at the beginning and end of the sweeps [39]. Therefore, those artificial regions and the data outside the spatial encoding region need to be excluded from the diffusion analysis. A 1D spin-echo image of the water

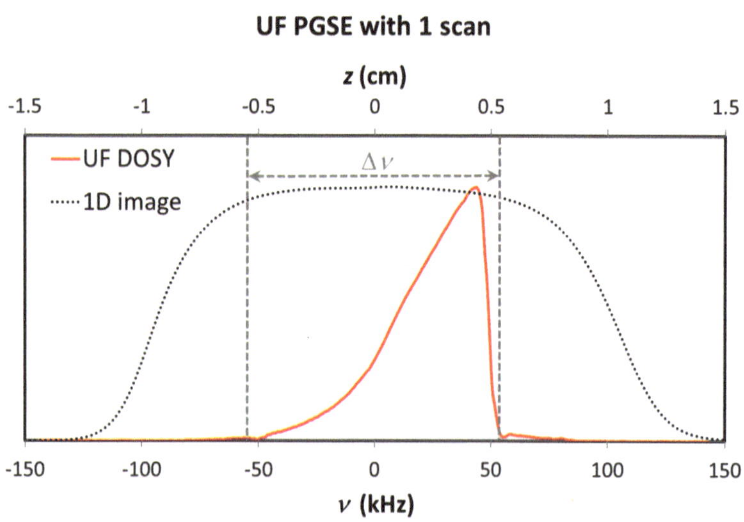


Figure 7.11 Ultrafast DOSY magnetization profile of a doped water sample measured by a single scan (solid red line). The spatial encoding and chirp sweep region are indicated by the gray vertical dashed lines. The black dotted line shows the coil excitation-detection sensitivity profile (1D MR image of the sample).

sample, representing the coil excitation-detection sensitivity profile [39], is also shown in Figure 7.11 (black dotted line). If the coil sensitivity profile is non-uniform within the spatial encoding region, it distorts the ultrafast DOSY magnetization profile. This effect can be removed by dividing the DOSY magnetization profile by the coil sensitivity profile [39]. The diffusion coefficient can be determined by fitting Equation 7.23 with the final DOSY magnetization profile.

It has been demonstrated that the ultrafast DOSY is compatible with all major nuclear spin hyperpolarization techniques, including parahydrogen induced polarization (PHIP) and related SABRE methods, dDNP, and spin-exchange optical pumping (SEOP) [44–49]. The high sensitivity of hyperpolarized ultrafast DOSY has been exploited in various applications, including dynamics of fluids in porous materials, identification of intra- and extracellular metabolites, analysis of complex mixtures, and monitoring chemical reactions [44–49]. The method is also feasible using low-field portable single-sided NMR spectrometers, which are much cheaper than the high-field instruments and can be used to explore surfaces of almost any kind of objects without such geometry and size restrictions imposed by the borehole of the high-field instruments [45].

A disadvantage of the ultrafast DOSY method is that chemical shift resolution is lost due to the application of the read gradients. This is not a problem if the spectrum includes only one peak, as it is often the case when, for example, the dynamics of a fluid in a porous material is studied. If the spectrum includes several peaks, a peak of interest can be excited by replacing one or several hard 90° pulses in the pulse sequence shown in Figure 7.10b by frequency-selective soft 90° pulses. Full chemical shift information can be achieved by the echo planar spectroscopic imaging (EPSI) type detection [48].

## 7.6 Ultrafast Diffusion Exchange Spectroscopy

Diffusion-exchange spectroscopy (DEXSY) is a useful method to investigate molecular exchange through diffusion contrast [50]. In addition to DOSY, it is also possible to significantly accelerate DEXSY measurements by the principles of ultrafast NMR spectroscopy. The pulse sequence of a conventional DEXSY experiment is shown in Figure 7.12a [50]. It includes two DOSY blocks (cf. Figure 7.10a) separated by a mixing time $\tau_M$. Therefore, the experiment enables one to correlate the diffusion coefficients ($D$) before and after the mixing time. If $D$ changes, it indicates that molecules have moved to a different physical or chemical environment characterized by a different $D$. The conventional DEXSY experiment is very time-consuming, as the gradients in the two DOSY blocks must be incremented independently in separate experiments. Therefore, if the number of increments in the first and second block is $N$ and $M$, altogether $N \times M$ repetitions are required (multiplied by the number of scans). If $N = M = 10$, the sequence must be repeated at least 100 times; if $N = M = 100$, the sequence must be repeated at least 10 000 times!

A pulse sequence for ultrafast DEXSY experiment, which allows a single-scan (instead of 100 to 10 000 repetitions) DEXSY measurement, is illustrated in Figure 7.12b [51]. The first DOSY block includes partially overlapping frequency-swept chirp 90° and 180° pulses applied together with diffusion gradients. The pulses form a double-spin echo, and due to the application of the diffusion gradient, the double spin-echo time, $t_1$, is linearly dependent on position:

$$t_1(z) = 2\left(1 - \frac{\gamma G_{\mathrm{SD}}}{\pi \Delta \nu} z\right) t_{\mathrm{C}}^{90^\circ}, \text{ when } -\frac{\Delta z}{2} \leq z \leq +\frac{\Delta z}{2}. \tag{7.24}$$

Here, $t_{\mathrm{C}}^{90^\circ}$ is the length of the 90° chirp pulse and $G_{\mathrm{SD}}$ is the gradient strength in the first DOSY block. The double echo time is zero at the top and maximum ($2t_{\mathrm{C}}^{90^\circ}$) at the bottom. If decay due to $T_2$ relaxation is negligible, after the first DOSY block, the longitudinal magnetization profile obeys the following equation [51]:

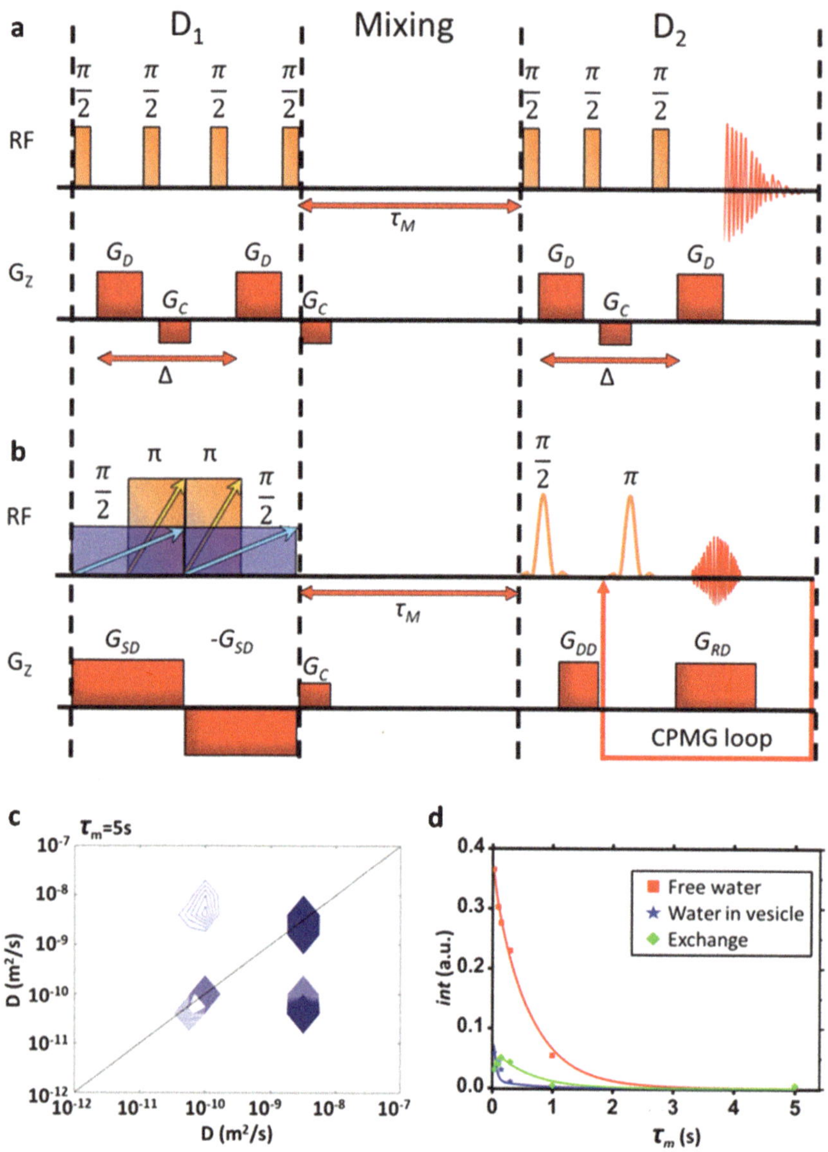


**Figure 7.12** Pulse sequence for (a) conventional and (b) ultrafast DEXSY experiments. (c) Ultrafast DEXSY map of water in aqueous sodium decanoate sample measured with a mixing time of $\tau_M = 5$ s. (d) Integrals of the ultrafast DEXSY maps as a function of mixing time. Partially reproduced with permission from Ref. [51].

$$E_1(z) = E_1^0 \exp\left[-\frac{\gamma^2 G_{SD}^2 [t_1(z)]^2}{48} D_1 t_1(z)\right], \tag{7.25}$$

where $D_1$ is the diffusion coefficient of molecules during the first DOSY block. After the mixing time, the signal is read by a single scan using a CPMG loop accompanied by gradient pulses. The gradient pulses have a double function: on the one hand, they serve as read gradients to read the spatially encoded diffusion data of the first dimension; on the other hand, they serve as diffusion gradients of the second dimension. During the second DOSY block, the signal decay obeys the following equation (if $T_2$ decay is negligible):

$$E_2(t_2) = E_0^2 \exp\left[-\frac{\gamma^2 G_{RD}^2 \tau^2}{3} D_2 t_2\right], \tag{7.26}$$

where $D_2$ is the diffusion coefficient, $2\tau$ is the echo time (the time between the 180° pulses), and $G_{RD}$ is the strength of the gradients in the CPMG loop. If there is a distribution of diffusion coefficients, instead of a single $D$, the signal observed in the ultrafast DOSY experiment is:

$$E\left[t_1(z), t_2\right] = E_0 \int P\left(D_1, D_2\right) \exp\left[-\frac{\gamma^2 G_{SD}^2 [t_1(z)]^2}{48} D_1 t_1(z)\right] \exp\left[-\frac{\gamma^2 G_{RD}^2 \tau^2}{3} D_2 t\right] dD_1 dD_2, \tag{7.27}$$

where $P(D_1, D_2)$ is the 2D diffusion coefficient distribution, which can be extracted by the inversion process which is often called a 2D inverse Laplace transform [39].

A DEXSY map of a water peak in an aqueous sodium decanoate sample is shown in Figure 7.12c. The diagonal peaks at around $D_1 = D_2 = 2 \cdot 10^{-9}$ and $6 \cdot 10^{-11}$ $m^2/s$ indicate that water molecules have two different physical environments characterized by different diffusion coefficients. The smaller $D$ originates from free water, while the higher $D$ arises from the water inside sodium decanoate vesicles. The off-diagonal cross-peaks reveal that there is an exchange of water molecules between these sites. Analysis of the ultrafast DEXSY peak intensities as a function of mixing time (Figure 7.12d) allowed the quantification of the exchange rate [51].

Theoretically, the ultrafast approach accelerates the DEXSY measurement by two to four orders of magnitude as compared to the conventional DEXSY measurements. Like in all ultrafast experiments, the spatial encoding lowers the sensitivity of ultrafast DEXSY, but, on the other hand, the single-scan nature significantly facilitates the use of hyperpolarization to achieve much higher sensitivity than in the conventional DEXSY experiment.

## References


1 Johnson, C.S. (1999). Diffusion ordered nuclear magnetic resonance spectroscopy: principles and applications. *Prog. Nucl. Magn. Reson. Spect.* 34 (3-4): 203–256. Available from: https://doi.org/10.1016/S0079-6565(99)00003-5.

2 Levitt, M.H. (2008). *Spin Dynamics: Basics of Nuclear Magnetic Resonance*. Chichester, Hoboken, NJ, San Francisco, CA: John Wiley and Sons Ltd.

3 Price, W.S. (2009). *NMR Studies of Translational Motion*. Cambridge: Cambridge University Press.

4 Callaghan, P.T. (2011). *Translational Dynamics and Magnetic Resonance: Principles of Pulsed Gradient Spin Echo NMR*. Oxford: Oxford University Press.

5 Schlayer, S., Pusch, A.-K., Pielenz, F., Beckert, S., Peksa, M. and Horch, C. et al. (2012). X-Nuclei NMR self-diffusion studies in mesoporous silica foam and microporous MOF CuBTC. *Mater. (Basel, Switzerland)* 5 (4): 617–633. Available from: https://doi.org/10.3390/ma5040617.

6 Stilbs, P. (2019). *Diffusion and Electrophoretic NMR*. Berlin, Boston: De Gruyter.

7 Packer, K.J. (1969). The study of slow coherent molecular motion by pulsed nuclear magnetic resonance. *Mol. Phys.* 17 (4): 355–368. Available from: https://doi.org/10.1080/00268976900101131.

8 Holz, M. and Müller, C. (1982). Direct measurement of single ionic drift velocities in electrolyte solutions. An NMR method. *Berichte der Bunsengesellschaft für physikalische Chemie* 86 (2): 141–147. Available from: https://doi.org/10.1002/bbpc.19820860210.

9 He, Q. and Johnson, C.S. (1989a). Stimulated echo electrophoretic NMR. *J. Magn. Reson. (1969)* 85 (1): 181–185. Available from: https://doi.org/10.1016/0022-2364(89)90334-X.

10 He, Q. and Johnson, C.S. (1989b). Two-dimensional electrophoretic NMR for the measurement of mobilities and diffusion in mixtures. *J. Magn. Reson. (1969)* 81 (2): 435–439. Available from: https://doi.org/10.1016/0022-2364(89)90079-6.

**11** He, Q., Hinton, D.P. and Johnson, C.S. (1991). Measurement of mobility distributions for vesicles by electrophoretic NMR. *J. Magn. Reson. (1969)* 91 (3): 654–658. Available from: https://doi.org/10.1016/0022-2364(91)90396-B.

**12** Morris, K.F. and Johnson, C.S. (1992). Mobility ordered 2D-NMR spectroscopy. *J. Am. Chem. Soc.* 114 (2): 776–777. Available from: https://doi.org/10.1021/ja00028a063.

**13** Pettersson, E., Furó, I. and Stilbs, P. (2004). On experimental aspects of electrophoretic NMR. *Concepts Magn. Reson. Part A* 22A (2): 61–68. Available from: https://doi.org/10.1002/cmr.a.20012.

**14** He, Q. and Wei, Z. (2001). Convection compensated electrophoretic NMR. *J. Magn. Reson. (San Diego, Calif. : 1997)*, 150(2): 126–131. Available from: https://doi.org/10.1006/jmre.2001.2321.

**15** Holz, M. (2005). Field-assisted diffusion studied by electrophoretic NMR. In: *Diffusion in Condensed Matter: Methods, Materials, Models.* (ed. P. Heitjans and J. Kärger), 717–742. Berlin/Heidelberg: Springer-Verlag.

**16** Fredrik Hallberg, István Furó, Pavel V. Yushmanov, Peter Stilbs, Sensitive and robust electrophoretic NMR: Instrumentation and experiments, Journal of Magnetic Resonance, Volume 192, Issue 1, 2008, Pages 69–77, ISSN 1090-7807, https://doi.org/10.1016/j.jmr.2008.02.001.

**17** Fang, Y., Yushmanov, P.V. and Furó, I. (2020). Improved accuracy and precision in electrophoretic NMR experiments. Current control and sample cell design. *J. Magn. Reson. (San Diego, Calif. : 1997)* 318: 106796. Available from: https://doi.org/10.1016/j.jmr.2020.106796.

**18** Scheler, U. (2007). Electrophoretic NMR. In: *Encyclopedia of Magnetic Resonance.* (ed. R.K. Harris), Chichester, UK: John Wiley & Sons, Ltd.

**19** Böhme, U. and Scheler, U. (2010). Counterion condensation and effective charge of poly(styrenesulfonate). *Adv. Colloid Interface Sci.* 158 (1-2): 63–67. Available from: https://doi.org/10.1016/j.cis.2010.02.010.

**20** Stilbs, P. and Furo, I. (2006). Electrophoretic NMR. *Curr. Opin. Colloid Interface Sci.* 11 (1): 3–6. Available from: https://doi.org/10.1016/j.cocis.2005.09.016.

**21** Griffiths, P.C. (2009). Chapter 5 Electrophoretic NMRŮIons, Molecules, Mixtures, Pores and Complexes, Annual Reports on NMR Spectroscopy, Academic Press, Volume 65, 2009, Pages 139–159, ISSN 0066-4103, ISBN 9780123747341, https://doi.org/10.1016/S0066-4103(08)00205-6.

**22** Böhme, U. and Scheler, U. (2007). Effective charge of bovine serum albumin determined by electrophoresis NMR. *Chem. Phys. Lett.* 435 (4-6): 342–345. Available from: https://doi.org/10.1016/j.cplett.2006.12.068.

**23** Fang, Y., Giesecke, M. and Furó, I. (2017). Complexing cations by poly(ethylene oxide): binding site and binding mode. *J. Phys. Chem. B.* 121 (9): 2179–2188. Available from: https://doi.org/10.1021/acs.jpcb.6b12381.

**24** Umecky, T., Saito, Y. and Matsumoto, H. (2009). Direct measurements of ionic mobility of ionic liquids using the electric field applying pulsed gradient spin-echo NMR. *J. Phys. Chem. A B* 113 (25): 8466–8468. Available from: https://doi.org/10.1021/jp9043946.

**25** Zhang, Z. and Madsen, L.A. (2014). Observation of separate cation and anion electrophoretic mobilities in pure ionic liquids. *J. Chem. Phys.* 140 (8): 84204. Available from: https://doi.org/10.1063/1.4865834.

**26** Gouverneur, M., Kopp, J., van Wüllen, L. and Schönhoff, M. (2015). Direct determination of ionic transference numbers in ionic liquids by electrophoretic NMR. *Phys. Chem. Chem. Phys.: PCCP* 17 (45): 30680–30686. Available from: https://doi.org/10.1039/c5cp05753a.

**27** Brinkkötter, M., Giffin, G.A., Moretti, A., Jeong, S., Passerini, S. and Schönhoff, M. (2018). Relevance of ion clusters for Li transport at elevated salt concentrations in Pyr12O1FTFSI ionic liquid-based electrolytes. *Chem. Comm. (Cambridge, England)* 54 (34): 4278–4281. Available from: https://doi.org/10.1039/c8cc01416g.

**28** Gouverneur, M., Schmidt, F. and Schönhoff, M. (2018). Negative effective Li transference numbers in Li salt/ionic liquid mixtures: does Li drift in the "Wrong" direction?. *Phys. Chem. Chem. Phys.: PCCP*, 20 (11): 7470–7478. Available from: https://doi.org/10.1039/c7cp08580j.

**29** Brinkkötter, M., Mariani, A., Jeong, S., Passerini, S. and Schönhoff, M. (2021). Ionic liquid in Li salt electrolyte: modifying the Li + transport mechanism by coordination to an asymmetric anion. *Adv. Energy Sustain. Syst.* 2 (2): 2000078. Available from: https://doi.org/10.1002/aesr.202000078.

**30** Schmidt, F., Pugliese, A., Santini, C.C., Castiglione, F. and Schönhoff, M. (2020). Spectral deconvolution in electrophoretic NMR to investigate the migration of neutral molecules in electrolytes. *Magn. Reson. Chem. MRC* 58 (3): 271–279. Available from: https://doi.org/10.1002/mrc.4978.

**31** Schmidt, F. and Schönhoff, M. (2020). Solvate cation migration and ion correlations in solvate ionic liquids. *J. Phys. Chem.* B, 124 (7): 1245–1252. Available from: https://doi.org/10.1021/acs.jpcb.9b11330.

**32** Ackermann, F. and Schönhoff, M. (2021). Chelating additives reversing the lithium migration direction in ionic liquid electrolytes. *J. Phys. Chem. C.* 125 (1): 266–274. Available from: https://doi.org/10.1021/acs.jpcc.0c09828.

**33** Fang, Y., Yushmanov, P.V. and Furó, I. (2017). Assessing 2D electrophoretic mobility spectroscopy (2D MOSY) for analytical applications. *Magn. Reson. Chem.: MRC*, 55 (6): 584–588. Available from: https://doi.org/10.1002/mrc.4558.

**34** Weingärtner, H. (2014). The static dielectric permittivity of ionic liquids. *J. Mol. Liq.* 192: 185–190. Available from: https://doi.org/10.1016/j.molliq.2013.07.020.

**35** Singh, M., Manoli, K., Tiwari, A., Ligonzo, T., Di Franco, C. and Cioffi, N. et al. (2017). The double layer capacitance of ionic liquids for electrolyte gating of ZnO thin film transistors and effect of gate electrodes. *J. Phys. Mater. A C*, 5 (14): 3509–3518. Available from: https://doi.org/10.1039/C7TC00800G.

**36** Rosenwinkel, M.P., Andersson, R., Mindemark, J. and Schönhoff, M. (2020). Coordination effects in polymer electrolytes: fast Li + transport by weak ion binding. *J. Chem. Phys.* C, 124 (43): 23588–23596. Available from: https://doi.org/10.1021/acs.jpcc.0c08369.

**37** Rosenwinkel, M.P. and Schönhoff, M. (2021). Polymer-induced inversion of the Li + drift direction in ionic liquid-based ternary polymer electrolytes. *Macromol. Chem. Phys.*, 2100320. Available from: https://doi.org/10.1002/macp.202100320.

**38** Ardenkjaer-Larsen, J.H., Fridlund, B., Gram, A., Hansson, G., Hansson, L. and Lerche, M.H. et al. (2003). Increase in signal-to-noise ratio of 10,000 times in liquid-state NMR. *Proc. Natl. Acad. Sci. U.S.A.* 100 (18): 10158–10163. Available from: https://doi.org/10.1073/pnas.1733835100.

**39** Telkki, V.-V., Urbańczyk, M. and Zhivonitko, V. (2021). Ultrafast methods for relaxation and diffusion. *Prog. Nucl. Magn. Reson. Spectr.* 126-127: 101–120. Available from: https://doi.org/10.1016/j.pnmrs.2021.07.001.

**40** Thrippleton, M.J., Loening, N.M. and Keeler, J. (2003). A fast method for the measurement of diffusion coefficients: one-dimensional DOSY. *Magn. Reson. Chem.* 41 (6): 441–447. Available from: https://doi.org/10.1002/mrc.1195.

**41** Frydman, L., Scherf, T. and Lupulescu, A. (2002). The acquisition of multidimensional NMR spectra within a single scan. *Proc. Natl. Acad. Sci. U.S.A.* 99 (25): 15858–15862. Available from: https://doi.org/10.1073/pnas.252644399.

**42** Pelupessy, P. (2003). Adiabatic single scan two-dimensional NMR spectrocopy. *J. Am. Chem. Soc.* 125 (40): 12345–12350. Available from: https://doi.org/10.1021/ja034958g.

**43** Ahola, S., Mankinen, O. and Telkki, V.-V. (2017). Ultrafast NMR diffusion measurements exploiting chirp spin echoes. *Magn. Reson. Chem. MRC* 55 (4): 341–347. Available from: https://doi.org/10.1002/mrc.4540.

**44** Ahola, S., Zhivonitko, V.V., Mankinen, O., Zhang, G., Kantola, A.M. and Chen, H.-Y. et al. (2015). Ultrafast multidimensional Laplace NMR for a rapid and sensitive chemical analysis. *Nat. Commun.* 6: 8363. Available from: https://doi.org/10.1038/ncomms9363.

**45** King, J.N., Fallorina, A., Yu, J., Zhang, G., Telkki, V.-V. and Hilty, C. et al. (2018). Probing molecular dynamics with hyperpolarized ultrafast Laplace NMR using a low-field, single-sided magnet. *Chem. Sci.* 9 (28): 6143–6149. Available from: https://doi.org/10.1039/c8sc01329b.

**46** Mankinen, O., Hollenbach, J., Ahola, S., Matysik, J. and Telkki, V.-V. (2018). Ultrafast Laplace NMR with hyperpolarized xenon gas. *Microporous Mesoporous Mater.* 269: 75–78. Available from: https://doi.org/10.1016/j.micromeso.2017.10.024.

**47** Zhang, G., Ahola, S., Lerche, M.H., Telkki, V.-V. and Hilty, C. (2018). Identification of intracellular and extracellular metabolites in cancer cells using 13C hyperpolarized ultrafast laplace NMR. *Anal. Chem.* 90 (18): 11131–11137. Available from: https://doi.org/10.1021/acs.analchem.8b03096.

**48** Guduff, L., Berthault, P., van Heijenoort, C., Dumez, J.-N. and Huber, G. (2019). Single-scan diffusion-ordered NMR spectroscopy of SABRE-hyperpolarized mixtures. *Chemphyschem : J. Phys. Chem. Chem. Phys.* 20 (3): 392–398. Available from: https://doi.org/10.1002/cphc.201800983.

**49** Tickner, B.J., Zhivonitko, V.V. and Telkki, V.-V. (2021). Ultrafast Laplace NMR to study metal-ligand interactions in reversible polarisation transfer from parahydrogen. *Phys. Chem. Chem. Phys.* PCCP, 23 (31): 16542–16550. Available from: https://doi.org/10.1039/d1cp02383g.

**50** Qiao, Y., Galvosas, P., Adalsteinsson, T., Schönhoff, M. and Callaghan, P.T. (2005). Diffusion exchange NMR spectroscopic study of dextran exchange through polyelectrolyte multilayer capsules. *J. Chem. Phys.* 122 (21): 214912. Available from: https://doi.org/10.1063/1.1924707.

**51** Mankinen, O., Zhivonitko, V.V., Selent, A., Mailhiot, S., Komulainen, S. and Prisle, N.L. et al. (2020). Ultrafast diffusion exchange nuclear magnetic resonance. *Nat. Commun.* 11 (1): 3251. Available from: https://doi.org/10.1038/s41467-020-17079-7.

# 8

# Multiple Acquisition Strategies

*Nathaniel J. Traaseth*

*Department of Chemistry, New York University, New York, NY, USA*

## 8.1 Introduction

Methodology advancements in NMR spectroscopy are typically classified as those involving improvements in sensitivity or resolution. The focus of this chapter is the detection of two or more free induction decays (FIDs) during a single experiment, referred to as *multiple acquisitions*. These detection schemes improve sensitivity of data collection since multiple datasets can be collected within the same time relative to the conventional experiment. Implementation of data collection requires suitable pulse sequences for detecting additional datasets and may involve the usage of two or more receivers for detecting FIDs on different nuclei. Several strategies and experiments are currently available, and this chapter classifies the approaches and underscores advantages relative to conventional experiments that detect a single FID.

## 8.2 Types of Multiple Acquisition Experiments

Acquisition of multiple datasets in one pulse sequence can be accomplished using sequential or simultaneous detection schemes (Figure 8.1). In the simplest form, two or more datasets can be collected with a single receiver, which requires detecting signals on the same nucleus concurrently or at different periods in the experiment (Figure 8.1a). The former method is called the time-shared experiment [1–3] and utilizes a single acquisition statement in the pulse sequence, whereas the latter requires two or more acquisition statements. In contrast to single receiver methods, spectrometers with two or more receivers can be configured to detect two or more nuclei at the same time [4]. Detection schemes with multiple receivers can be designed to record FIDs in a sequential or simultaneous manner (Figure 8.1b). The multiple receiver approach is also referred to as parallel acquisition NMR spectroscopy (PANSY) [4].

The simplest example of multiple receiver detection involves 90° excitation pulses on different nuclei followed by FID detection on each nucleus of interest. This can be implemented on as many channels and receivers available on the spectrometer and probe (i.e. most commonly $^{1}$H, $^{13}$C, $^{15}$N, $^{31}$P, and $^{19}$F). Since the spin relaxation times vary widely depending on the nucleus and sample, it is practical for designing such pulse sequences with optimal recycle delays for each nucleus of interest. Ultimately the sensitivity gained by detecting each experiment will depend on the relative abundance of a particular nucleus and the sensitivity of detection on the probe. A probe optimized for direct detection of the X-channel nucleus (e.g. $^{13}$C and $^{15}$N) would likely be the ideal configuration

*Two-Dimensional (2D) NMR Methods*, First Edition. Edited by K. Ivanov, P.K. Madhu and G. Rajalakshmi.

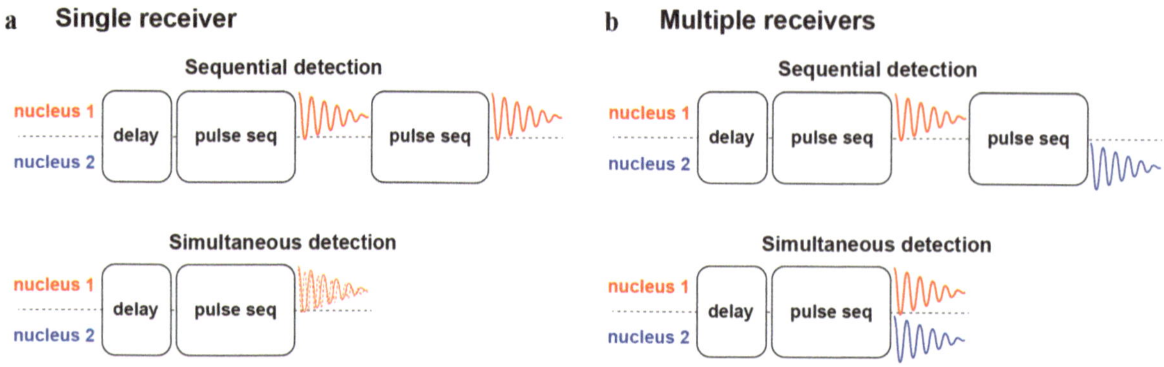


Figure 8.1 Schematic of multiple acquisition detection types. (a) Single receiver experiments that utilize sequential (top) or simultaneous (bottom) detection schemes to acquire two or more datasets. The simultaneous detection scheme with a single receiver displays two overlapped FIDs to represent the two datasets acquired. (b) Multiple receiver experiments that utilize sequential (top) or simultaneous (bottom) detection to acquire two or more datasets. Experiments using multiple receivers are also referred to as parallel acquisition and are the basis for the PANSY experiment [4]. In each panel, detection on nucleus 1 is in red, while detection on nucleus 2 is in blue.

with the goal of increasing signal-to-noise for the nucleus with the lowest inherent sensitivity. For example, the usage of cryogenic probes optimized for direct detection on $^{13}$C or $^{15}$N has been a key advancement to ensure sufficient sensitivity for multiple receiver experiments on small and large molecules.

A more significant benefit of multiple receiver experiments are multidimensional experiments establishing correlations among the nuclei. Several experiments commonly used for resonance assignments and structure determination have been designed, including various combinations of COSY, HMBC, HSQC, HMQC, and TOCSY [4–7]. The first implementation involved dual detection of $^{1}$H-$^{1}$H homonuclear COSY and $^{1}$H-$^{13}$C heteronuclear correlation spectra with subsequent detection on $^{1}$H and $^{13}$C channels [4]. For multiple acquisition types where complementary experiments are acquired, the first detected experiment within the group typically gives an identical signal-to-noise relative to acquisition types acquired in the conventional manner (i.e. one experiment at a time). The additional datasets may or may not have reduced signal-to-noise relative to experiments collected in the conventional manner. However, regardless of whether sensitivity is compromised in the secondary datasets, these spectra typically do not require additional experimental time when acquired in parallel since both detection periods share a common recycle delay between scans. Hence, if the signal-to-noise achieved is sufficient for the secondary datasets, there is an advantage of grouping experiments together using multiple receivers.

Currently, the state-of-the-art configuration for new commercial NMR spectrometers involves receivers for each channel, which makes the use of sequential or simultaneous detection possible for experiments that would benefit from multiple detection schemes. Pulse sequence libraries are also available on commercial spectrometers and setting up these sequences usually requires no advanced knowledge beyond setting up single detection experiments. Processing scripts are used to parse the FIDs which are usually stored in an interleaved manner and data analysis is performed in the conventional manner using commercial software or external software packages.

## 8.3 Utilization of *Forgotten* Spin Operators

Collection of multidimensional experiments requires establishing correlations to link spin systems for resonance assignments. During the process of selecting the desired coherence for spectral acquisition, unwanted coherences are often created and typically discarded using phase cycling and pulsed field gradients. This unutilized magnetization can be recovered and converted into signal for the detection of multiple datasets at the same time. Perhaps the earliest example of multiple acquisition experiments involved the collection of $^1$H-$^1$H COSY and $^1$H-$^1$H NOESY spectra in a single pulse sequence utilizing sequential detection with a single receiver [8, 9] (Figure 8.2). Subsequently, time-shared experiments enabled the collection of pairs of simultaneously acquired multidimensional experiments [1, 2, 10, 11], including $^1$H-$^{13}$C and $^1$H-$^{15}$N HSQC experiments using solution NMR [12] and $^{13}$C-$^{13}$C COSY and $^{13}$C-$^{13}$C DARR spectra using solid-state NMR experiment magic-angle-spinning (MAS) [13].

The concept of *afterglow* or residual magnetization was introduced by Kupče and Freeman for small molecules by employing the use of multiple receivers to acquire $^{13}$C-$^{13}$C INADEQUATE and $^1$H-$^{13}$C HMBC spectra in a single experiment using solution NMR [14]. This approach paved the way for *all-in-one* experiments, referred to as PANACEA (parallel acquisition NMR, an all-in-one combination of experimental applications), with the goal of obtaining spectral assignments from a single experiment employing multiple acquisitions (Figure 8.3) [15]. Related to this method is *NOAH*, which nests up to five multidimensional spectra for data collection during a single experiment and with a single receiver [16]. The reader is directed to the reviews by Kupče et al. for a comprehensive discussion of these techniques and their application to small molecules [6, 7].

It is important to note that sensitivity enhancement schemes that increase the signal-to-noise ratio of a single experiment have been differentiated from those that lead to additional spectral datasets. In some cases,

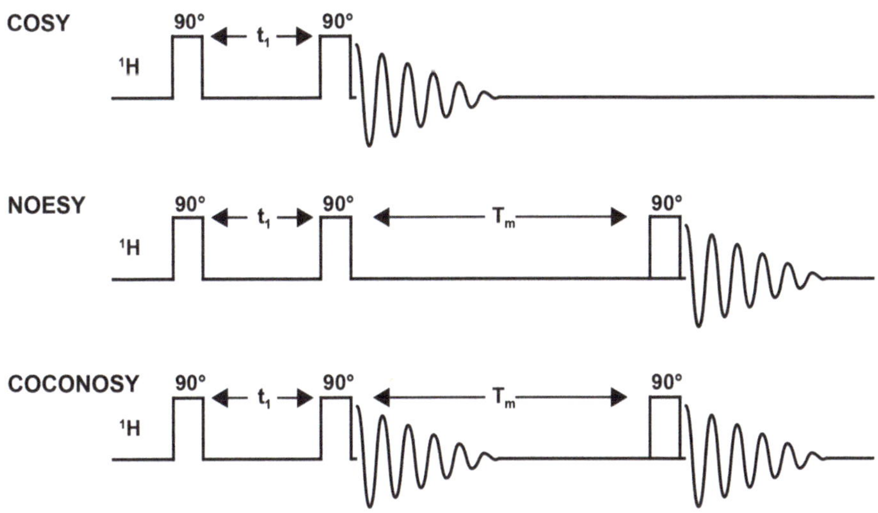


Figure 8.2 Combined two-dimensional pulse sequences of COSY and NOESY to give a single detected experiment named COCONOSY. The top and middle pulse sequences display schemes for conventional detection of COSY and NOESY spectra, where $T_m$ is the mixing period. The bottom sequence illustrates how these spectra can be detected in a single experiment with two sequential acquisition elements on the $^1$H channel. Adapted from Ref. [9], with permission from Elsevier.

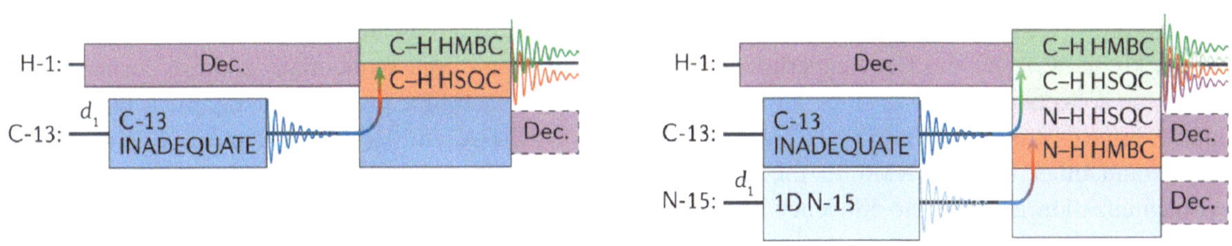


Figure 8.3 PANACEA (left) and extended PANACEA (right) methods utilizing multiple receiver detection to enable spectral assignments of small molecules in an *all-in-one* approach [15]. Modified from Ref. [6], with permission from Springer Nature.

these experiments precede those multiple acquisition techniques. For example, in HSQC and other multidimensional experiments, the sine component of the evolved chemical shift in the indirect dimension can be reconverted into observable magnetization and added with the cosine component to yield improvements in signal-to-noise [17–19]. This detection approach, referred to as *preservation of equivalent pathways (PEP)* [20], is common in biomolecular solution NMR spectroscopy and uses similar principles as afterglow experiments described above to recover lost coherences. Related approaches have also been introduced into solid-state NMR experiments [21, 22].

## 8.4 Application of Multiple Acquisition Techniques

A wide range of experiments have been developed that implement multiple acquisition periods in both solution and solid-state NMR spectroscopy. Below are examples of such methods grouped by sample type in the form of solution or solid-state NMR applications with a focus on biomolecular NMR applications.

### 8.4.1 Solution NMR Spectroscopy

Similar to solid-state NMR methods that will be discussed below, multiple acquisition methods in solution NMR rely on the efficient use of magnetization. Such repurposing of previously discarded magnetization can be used to increase the sensitivity of a single experiment (e.g. HSQC as described above) or through the creation of additional multidimensional datasets. The first occurrence of the latter for the detection of biomolecules was the collection of two-dimensional (HA)CACO and three-dimensional (HA)CA(CO)NNH experiments on the protein nuclease A inhibitor [23]. The acquisition of (HA)CACO utilized direct detection on $^{13}$C, which was made possible by sensitive cryogenic probes. Unlike the conventional (HA)CACO experiment, which ignored the residual $^{13}$C coherence at the end of the detection period, the afterglow approach refocused this magnetization, ultimately transferring it to $^{1}$H for collecting the (HA)CA(CO)NNH three-dimensional dataset. Although only ~10% of the magnetization remained on $^{13}$CO following the $^{13}$C directly detected dimension, $^{1}$H detection enabled sufficient sensitivity of the three-dimensional dataset with essentially no penalty in the overall experimental time. It is noteworthy that the use of $^{13}$C detection in the afterglow experiment converges with the popularity of direct detection methods for a range of biomolecules, including metalloproteins, intrinsically disordered proteins (IDPs), and membrane proteins [24, 25]. Hence, it is likely these methods will have continued usage from the biomolecular NMR community.

Several subsequent advancements have been made to perform sequential acquisition with a single receiver and parallel acquisitions using multiple receivers. For example, Kupče and Kay introduced an approach to measure two-dimensional $^{1}$H-$^{15}$N HSQC and $^{15}$N-$^{13}$CO HSQC spectra at the same time as well as three-dimensional versions of HNCA and NCACO spectra [26]. Chakraborty et al. adapted the three-dimensional HNN experiment used for sequentially walking from one amino acid to the next to collect two-dimensional NCA spectra using $^{13}$CA magnetization that was previously discarded [27]. This approach utilized multiple receiver detection on $^{1}$H and $^{13}$C and provided advantages of the NCA spectrum, such as identification of the directionality of the backbone walk for the HNN experiment. Bellstedt et al. introduced sequentially acquired experiments enabling the collection of pairs of three-dimensional experiments useful for resonance assignment [28]. These include HA(CA)NH and HA(CACO)NH, HA(CA)NH and H(N)CAHA, and H(N)CAHA and H(CC)NH. Overall, these and other solution NMR methods [29, 30] utilizing multiple receivers in conjunction with afterglow magnetization increase the efficiency of NMR spectrometer time relative to the conventionally detected experiments.

### 8.4.2 Solid-State NMR Spectroscopy

There are two key approaches utilized in multiple acquisition methods in biomolecular MAS solid-state NMR. The first is the use of afterglow magnetization and the second involves the creation of multiple polarization sources on $^{15}$N and $^{13}$C. The former method is based on the observation that the cross-polarization (CP) element [31], the most common polarization transfer mechanism in solid-state NMR, equilibrates the magnetization between the two nuclei involved in the transfer rather than mediating the complete transfer from one nucleus to the other. Indeed, CP can be thought of as a heat transfer process involving two reservoirs in thermal contact [32, 33]. Detection of the residual magnetization from CP can be observed in a relatively simple experiment that forms a building block for assignment of proteins using MAS. Figure 8.4a displays a pulse sequence corresponding to a one-dimensional NCA experiment [34] utilizing a SPECIFIC-CP element [35] for selective transfer between $^{15}$N and $^{13}$CA. In this experiment, $^{1}$H magnetization is initially transferred to $^{15}$N through CP and subsequently transferred to $^{13}$CA using SPECIFIC-CP. After these transfers, both $^{15}$N and $^{13}$C are detected using two receivers. Upon collecting a series of one-dimensional experiments where the field strength on $^{13}$CA is arrayed, a non-zero signal is observed in the $^{15}$N spectrum at the condition of the optimal polarization transfer (Figure 8.4b). This signal is direct evidence of afterglow magnetization [36] (also called orphan magnetization [37]) and can be used for the collection of two or more multidimensional experiments.

The first application of afterglow magnetization in solid-state NMR stemmed from simultaneously acquired NCA and NCO two-dimensional or NCACX and NCOCX three-dimensional experiments on microcrystalline ubiquitin (Figure 8.5) [36]. In this pulse sequence, magnetization is initially transferred from $^{1}$H to $^{15}$N, evolved on $^{15}$N in the indirect dimension, then followed by a SPECIFIC-CP transfer from $^{15}$N to $^{13}$CA. The magnetization on $^{13}$CA constitutes the first detected FID and gives the NCA or NCACX spectra. Following SPECIFIC-CP from $^{15}$N to $^{13}$CA, the afterglow $^{15}$N polarization is stored along the z-axis while the $^{13}$CA magnetization is

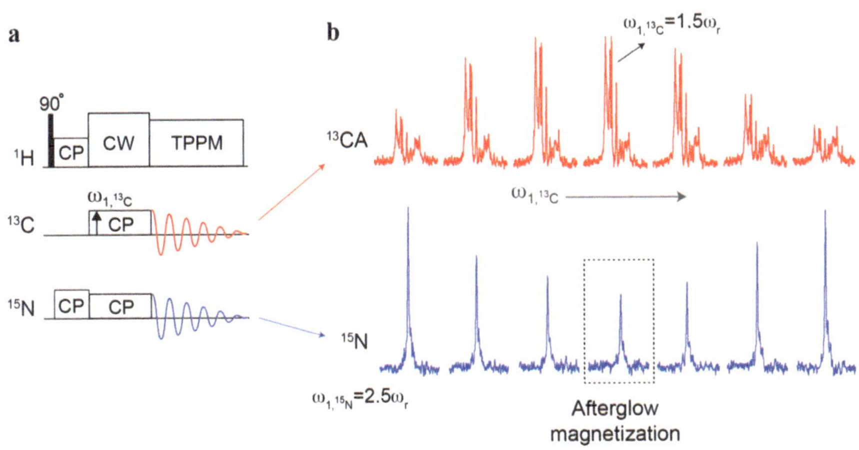


**Figure 8.4** Method for detecting afterglow $^{15}$N magnetization in solid-state NMR using SPECIFIC-CP from $^{15}$N to $^{13}$CA. (a) One-dimensional NCA pulse sequence of CP from $^{1}$H to $^{15}$N, followed by SPECIFIC-CP [35] from $^{15}$N to $^{13}$CA. Magnetization is detected on both $^{15}$N and $^{13}$C using two receivers. (b) $^{13}$CA (top) and $^{15}$N spectra (bottom) resulting from the pulse sequence in panel a. The top and bottom spectra correspond to a distinct experiment where the $^{15}$N field strength is fixed and the $^{13}$C field strength on $^{13}$CA is varied. At the optimal SPECIFIC-CP transfer from $^{15}$N to $^{13}$C, the $^{13}$CA signal is maximal while the $^{15}$N signal is minimal. The latter corresponds to afterglow magnetization and can be used for acquisition of additional datasets. Adapted from Ref. [34], with permission from Springer Nature.

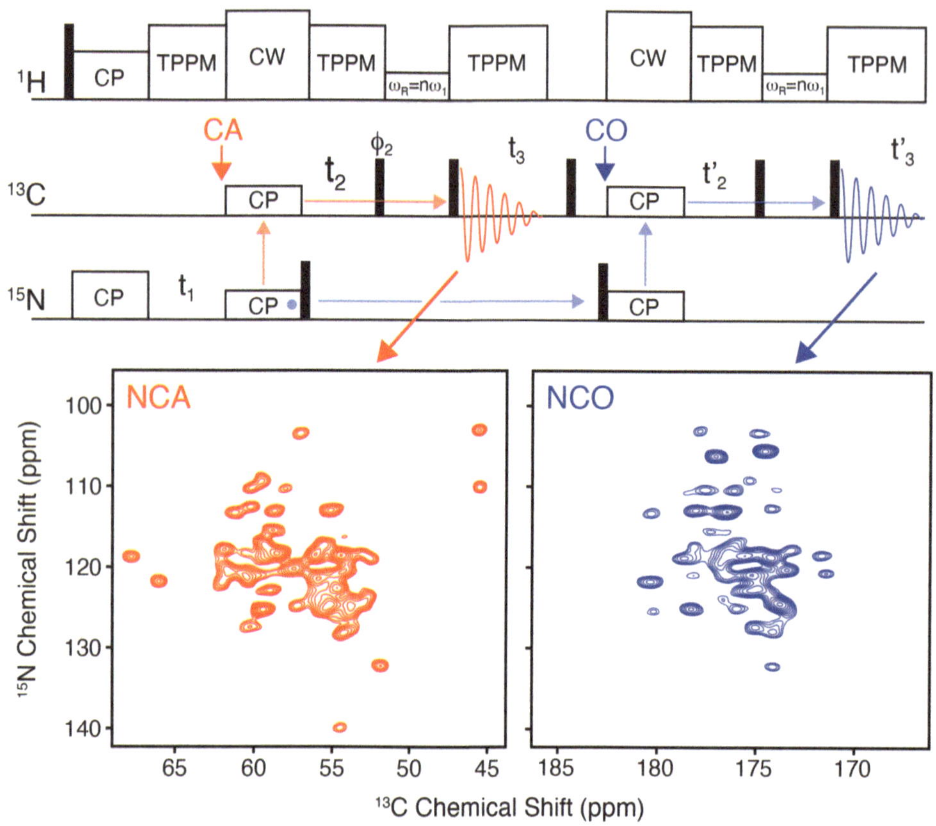


**Figure 8.5** Afterglow solid-state NMR experiment with simultaneous detection of two-dimensional NCA and NCO spectra or three-dimensional NCACX and NCOCX spectra. The red arrows trace the path of magnetization from SPECIFIC-CP leading to NCA or NCACX spectra, while the blue arrows trace the path of afterglow magnetization leading to NCO or NCOCX spectra. Adapted from Ref. [36], with permission from the American Chemical Society.

detected. The long $T_1$ of $^{15}N$ in solid-state NMR samples leads to a negligible loss of magnetization during the FID acquisition time [38]. After the FID is detected, afterglow polarization is rotated into the transverse plane and subjected to a second SPECIFIC-CP step to $^{13}CO$. Finally, the second FID is detected on $^{13}C$ to give the NCO or NCOCX spectra. In uniformly $^{13}C/^{15}N$ labeled proteins, the NCA (or NCACX) spectrum is essentially identical to the standard NCA (or NCACX), while the afterglow NCO (or NCOCX) spectrum gives ~32% of the signal relative to the standard NCO (or NCOCX). A larger sensitivity gain occurs for the afterglow NCO or NCOCX dataset when the protein is selectively or sparsely labeled [39, 40], such as those with 1,3-$^{13}C$ glycerol or 2-$^{13}C$ glycerol.

The second key advancement for multiple acquisition detection in solid-state NMR is the use of simultaneous CP from $^{1}H$ to both $^{13}C$ and $^{15}N$. This was initially implemented by Herbst et al. on RNA samples to detect two-dimensional CHHN and NHHC experiments at the same time using two receivers [41]. The simultaneous CP makes use of the abundant source of the $^{1}H$ polarization that can accommodate magnetization transfers to both $^{15}N$ and $^{13}C$ without significant losses in sensitivity compared to a single CP transfer element (i.e. $^{1}H$ to $^{13}C$ or $^{1}H$ to $^{15}N$). This method was advanced by Gopinath and Veglia to achieve sequentially acquired datasets using a single receiver [38]. This method, referred to as dual acquisition MAS (DUMAS), was introduced

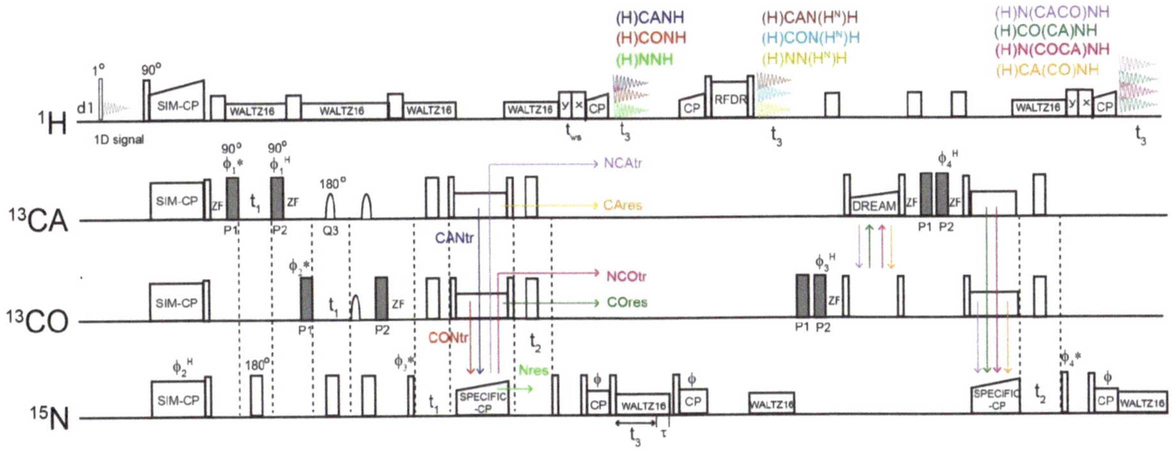


**Figure 8.6** PHRONESIS pulse sequence used to assign proteins through the detection of 10 multidimensional solid-state NMR experiments using fast MAS and $^1$H detection. Each multidimensional experiment is listed above the three $^1$H detection periods. Reprinted from Ref. [49], with permission from a Creative Commons license.

with two sets of sequentially detected two-dimensional experiments, DARR/NCA and DQSQ/NCO. Inclusion of three-dimensional experiments was made possible with bidirectional CP transfers that enabled the collection of NCACX/CANCO and NCOCX/CON(CA)CX, which provided several of the key experiments for protein assignments in only two experiments [42]. Such experiments have been broadly termed polarization optimized experiments (POE) [43, 44], which can also be collected for obtaining distance constraints [45]. The reader is directed to the review by Gopinath and Veglia for a comprehensive description of these multiple acquisition experiments [44].

Recent examples of solid-state NMR multiple acquisition technology have been developed for fast MAS techniques and $^1$H detection, both of which offer promising paths forward for studying proteins due to the sensitivity of $^1$H and resolution improvements from fast spinning rates [46]. Namely, Sharma et al. introduced a single experiment leading to the simultaneous collection of eight spectra datasets with demonstration on the model tripeptide f-MLF [47]. Stanek et al. introduced RAVASSA (redundant assignment via a single simultaneous acquisition) to assign the 42.5-kDa maltose-binding protein using a series of eight multidimensional experiments, at a spinning rate of 100 kHz [48]. And Gopinath et al. proposed PHRONESIS, a fast MAS experiment consisting of 10 three-dimensional experiments, which was applied to assign the resonances of GB1 using a single experiment at a spinning rate of 65 kHz (Figure 8.6) [49]. Overall, these experiments and others [50–54] provide suitable opportunities for acquiring multiple datasets that can be used to efficiently assign proteins in the solid-state, including microcrystalline samples, amyloid fibrils, and membrane proteins embedded in lipid bilayers.

## 8.5 Modularity of Multiple Detection Schemes and Other Novel Approaches

The multiple acquisition methods discussed above increase the effective sensitivity of NMR data collection since the additional spectra do not increase the overall experimental acquisition time. However, none of these methods directly address the second major hurdle in NMR spectroscopy: the rather long experimental times needed to resolve indirect dimensions. Fortunately, there has also been outstanding progress in methods capable of reducing time needed to sample these dimensions, including non-uniform sampling [55], covariance NMR [56], reduced dimensionality [57], projection reconstruction [58], and GFT methods [59]. For a thorough description of these methods, the reader is directed to literature cited therein. The salient feature of such methods is that most are compatible with multiple detection strategies and therefore can further decrease the total experimental time needed for data acquisition.

Another novel direction by Frydman and co-workers is the detection of multidimensional experiments in a single scan [60]. This fast acquisition approach utilizes spatial encoding of the indirectly evolved dimension along a particular axis that can be read out by a series of gradient pulses, similar to detection approaches in imaging experiments. This method enables multidimensional experiments to be collected in a matter of seconds. Furthermore, the approach is compatible with multiple acquisition techniques, as has been demonstrated in PANSY detection schemes to acquire $^{1}H$-$^{1}H$ and $^{1}H$-X correlation spectra in a single scan [61]. Such approaches have been named PUFSY (parallel ultrafast two-dimensional spectroscopy) and enable kinetics to be monitored on the second timescale and with atomic resolution.

Finally, reduction of time needed to sample pseudo three- or four-dimensional datasets encounter the same challenges as those three-dimensional experiments necessary for spectral assignment. One class of such experiments are chemical exchange methods that require a series of two- or three-dimensional experiments to derive insight into the intrinsic conformational dynamics of a system [62–64]. Among the toolkit of promising methods for probing slow conformational dynamics is the chemical exchange saturation transfer (CEST) method [65, 66]. Toward the goal of reducing the time necessary for sampling such dimensions, the use of multifrequency pulses have been employed to reduce the number of multidimensional experiments required by recording two-dimensional spectra in parallel [67, 68]. Depending on the nuclei to be sampled and the specifics of the experiment, this can reduce the overall experimental time by a factor of two- to four-fold. In principle, these advancements to chemical exchange methods are compatible with the growing arsenal of afterglow and PANSY methods and offer additional opportunities for reducing the time needed to collect pseudo three- or four-dimensional datasets that are often among the most time-consuming class of experiments applied to biomolecules.

## 8.6 Future of Multiple Acquisition Detection

The future of multiple detection schemes will continue to be commonplace as methods are further refined and implemented on commercial spectrometers and when spectrometers with multiple receivers become accessible to all NMR users. Beyond this, it is likely that the future of NMR will involve probes utilizing multiple coils that can accommodate several samples at the same time [69–72]. Data acquisition in a parallel manner would make more efficient use of expensive magnets and allow shared use of instrument time. Such developments have been applied for MAS of pharmaceutical solids that have long spin lattice relaxation times and incorporate multiple MAS stators within a single probe body [72]. More recent directions involve microcoils [73], which may be more easily accommodated into a detection apparatus for recording data using multiple receivers. Toward this goal, Wang et al. demonstrated the applicability of detecting up to eight separate radiofrequency coils with the use of switches and four receivers (Figure 8.7) [74]. Finally, the implementation of dynamic nuclear polarization methods [75], which boost sensitivity by significant factors, will further add to the reduction of time necessary for collecting atomic resolution data to probe the structure and dynamics for a range of chemical and biochemical applications.

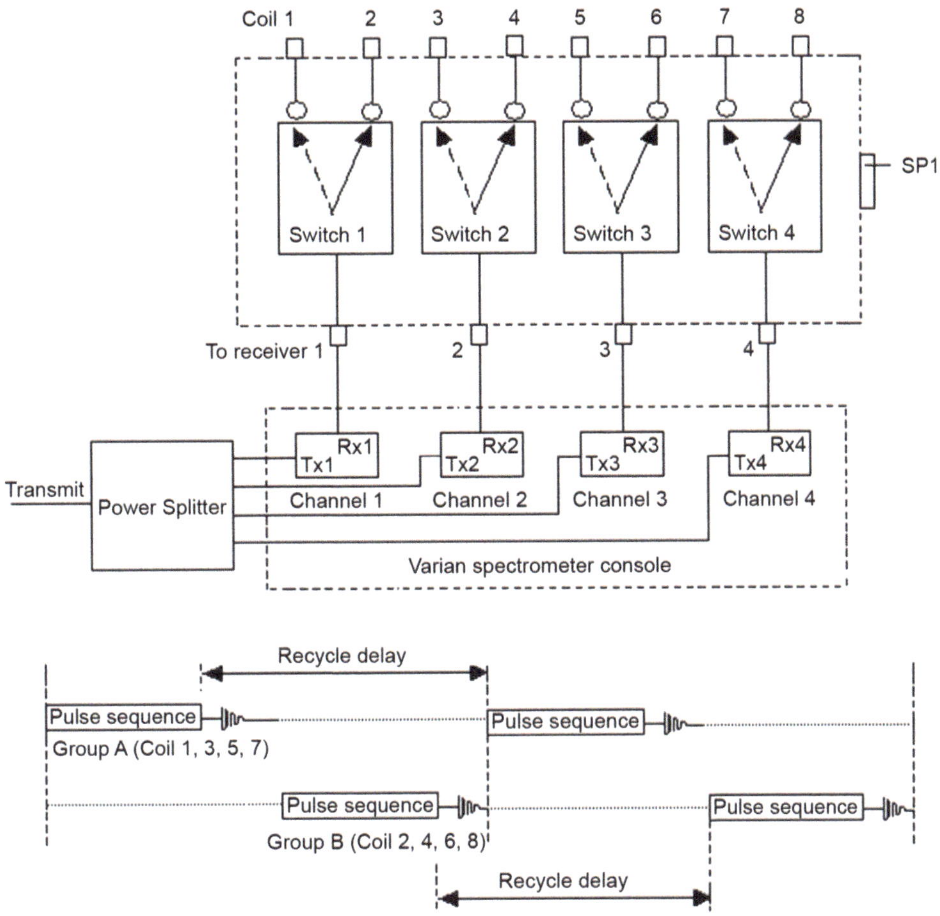


Figure 8.7 A schematic of the transmit and receive scheme for an eight-coil probehead design used to collect multiple FIDs from eight samples. Reprinted from Ref. [74], with permission from Elsevier.

## Acknowledgments

N.J.T. was supported by NIH (R01AI108889, R01AI165782) and NSF (MCB1902449) awards.

## References

1 Sorensen, O.W. (1990). Aspects and prospects of multidimensional time-domain spectroscopy. *J. Magn. Reson.* 89 (1): 210–216.

2 Farmer, B.T. (1991). Simultaneous [C-13, N-15]-Hmqc, a pseudo-triple-resonance experiment. *J. Magn. Reson.* 93 (3): 635–641.

3 Parella, T., Nolis, P. (2010). Time-shared NMR experiments. *Concept Magn. Reson.* A 36a (1): 1–23.

4 Kupce, E., Freeman, R. and John, B. K., (2006). Parallel acquisition of two-dimensional NMR spectra of several nuclear species. *J. Am. Chem. Soc.* 128 (30): 9606–9607.

5 Kupce, E. (2013). NMR with multiple receivers. *Top. Curr. Chem.* 335: 71–96.
6 Kupče, E., Frydman, L., Webb, A.G., Yong, J.R.J. and Claridge, T.D.W. (2021). Parallel nuclear magnetic resonance spectroscopy. *Nat. Rev. Methods Primers.* 1 (1): 27.
7 Kupce, E., Mote, K.R., Webb, A., Madhu, P.K. and Claridge, T.D.W. (2021). Multiplexing experiments in NMR and multi-nuclear MRI. *Prog. Nucl. Magn. Reson. Spectrosc.* 124: 1–56.
8 Gurevich, A.Z., Barsukov, I.L., Arseniev, A.S. and Bystrov, V.F. (1984). Combined cosy-noesy experiment. *J. Magn. Reson.* 56 (3): 471–478.
9 Haasnoot, C.A.G., Vandeven, F.M. and Hilbers, C.W. (1984). Coconosy – combination of 2d correlated and 2d nuclear Overhauser enhancement spectroscopy in a single experiment. *J. Magn. Reson.* 56 (2): 343–349.
10 Kay, L.E., Wittekind, M., Mccoy, M.A., Friedrichs, M.S. and Mueller, L. (1992). 4d Nmr triple-resonance experiments for assignment of protein backbone nuclei using shared constant-time evolution periods. *J. Magn. Reson.* 98 (2): 443–450.
11 Burgering, M., Boelens, R. and Kaptein, R. (1993). Observation of intersubunit noes in a dimeric P22 Mnt repressor mutant by a time-shared [N-15,C-13] double half-filter technique. *J. Biomol. NMR.* 3 (6): 709–714.
12 Sattler, M., Maurer, M., Schleucher, J. and Griesinger, C. (1995). A simultaneous N-15,H-1-Hsqc and C-13,H-1-Hsqc with sensitivity enhancement and a heteronuclear gradient-echo. *J. Biomol. NMR* 5 (1): 97–102.
13 Fukuchi, M. and Takegoshi, K. (2008). Combination of C-13-C-13 COSY and DARR (COCODARR) in solid-state NMR. *Solid State Nucl. Magn. Reson.* 34 (3): 151–153.
14 Kupce, E. and Freernan, R. (2008). Molecular structure from a single NMR experiment. *J. Am. Chem. Soc.* 130 (32): 10788–10792.
15 Kupce, E. and Freeman, R. (2008). Molecular structure from a single NMR experiment. *J. Am. Chem. Soc.* 130 (32): 10788–10792.
16 Kupce, E. and Claridge, T.D.W. (2017). NOAH: NMR supersequences for small molecule analysis and structure elucidation. *Angew. Chem. Int. Edit.* 56 (39): 11779–11783.
17 Palmer, A.G., Cavanagh, J., Wright, P.E. and Rance, M. (1991). Sensitivity improvement in proton-detected 2-dimensional heteronuclear correlation Nmr-spectroscopy. *J. Magn. Reson.* 93 (1): 151–170.
18 Cavanagh, J. and Rance, M. (1990). Sensitivity improvement in isotropic mixing (Tocsy) experiments. *J. Magn. Reson.* 88 (1): 72–85.
19 Kay, L.E., Keifer, P. and Saarinen, T. (1992). Pure absorption gradient enhanced heteronuclear single quantum correlation spectroscopy with improved sensitivity. *J. Am. Chem. Soc.* 114 (26): 10663–10665.
20 Cavanagh, J. and Rance, M. (1993). Sensitivity-enhancedNMR techniques for the study of biomolecules. *Annu. Rep. NMR Spectrosc.* 27, 1–52.
21 Gopinath, T. and Veglia, G. (2009). Sensitivity enhancement in static solid-state NMR experiments via single- and multiple-quantum dipolar coherences. *J. Am. Chem. Soc.* 131 (16): 5754–5756.
22 Fukuchi, M., Inukai, M., Takeda, K. and Takegoshi, K. (2008). Double-acquisition: utilization of discarded coherences in a 2D separation experiment using the states method. *J. Magn. Reson.* 194 (2): 300–302.
23 Kupce, E., Kay, L.E. and Freeman, R. (2010). Detecting the "afterglow" of 13C NMR in proteins using multiple receivers. *J. Am. Chem. Soc.* 132 (51): 18008–18011.
24 Bermel, W., Bertini, I., Felli, I.C., Piccioli, M. and Pierattelli, R. (2006). C-13-detected protonless NMR spectroscopy of proteins in solution. *Prog. Nucl. Magn. Reson.* 48 (1): 25–45.
25 Chhabra, S., Fischer, P., Takeuchi, K., Dubey, A., Ziarek, J.J., Boeszoermenyi, A., Mathieu, D., Bermel, W., Davey, N.E., Wagner, G. and Arthanari, H. (2018). (15)N detection harnesses the slow relaxation property of nitrogen: Delivering enhanced resolution for intrinsically disordered proteins. *Proc. Natl. Acad. Sci. USA* 115 (8): E1710–E1719.
26 Kupce, E. and Kay, L.E. (2012). Parallel acquisition of multi-dimensional spectra in protein NMR. *J. Biomolec. NMR* 54 (1): 1–7.

**27** Chakraborty, S., Paul, S. and Hosur, R.V. (2012). Simultaneous acquisition of C-13(alpha)-N-15 and H-1-N-15-N-15 sequential correlations in proteins: application of dual receivers in 3D HNN. *J. Biomolec. NMR* 52 (1): 5–10.

**28** Bellstedt, P., Ihle, Y., Wiedemann, C., Kirschstein, A., Herbst, C., Gorlach, M. and Ramachandran, R. (2014). Sequential acquisition of multi-dimensional heteronuclear chemical shift correlation spectra with H-1 detection. *Sci. Rep.* 2014, 4.

**29** Goradia, N., Wiedemann, C., Herbst, C., Gorlach, M., Heinemann, S.H., Ohlenschlager, O. and Ramachandran, R. (2015). An approach to NMR assignment of intrinsically disordered proteins. Chemphyschem 16 (4): 739–746.

**30** Furuita, K., Sugiki, T., Takamuku, M., Hattori, Y., So, M., Kawata, Y., Ikegami, T., Fujiwara, T. and Kojima, C. (2021). Sensitivity enhancement by sequential data acquisition for C-13-direct detection NMR. *J. Magn. Reson.* 322.

**31** Hartmann, S.R. and Hahn, E. L. (1962). Nuclear double resonance in rotating frame. *Phys. Rev.* 128 (5): 2042–2053.

**32** Demco, D.E., Tegenfeldt, J. and Waugh, J.S. (1975). Dynamics of cross relaxation in nuclear magnetic double-resonance. *Phys. Rev. B* 11 (11): 4133–4151.

**33** Xu, J., Struppe, J. and Ramamoorthy, A. (2008). Two-dimensional homonuclear chemical shift correlation established by the cross-relaxation driven spin diffusion in solids. *J. Chem. Phys.* 128 (5): 052308.

**34** Abramov, G. and Traaseth, N.J. (2018). Afterglow solid-state NMR spectroscopy. *Methods Mol. Biol* 1688, 55-66.

**35** Baldus, M., Petkova, A.T., Herzfeld, J.H. and Griffin, R.G. (1998). Cross polarization in the tilted frame: assignment and spectral simplification in heteronuclear spin systems. *Mol. Phys.* 95 (Journal Article): 1197–1207.

**36** Banigan, J.R. and Traaseth, N.J. (2012). Utilizing afterglow magnetization from cross-polarization magic-angle-spinning solid-state NMR spectroscopy to obtain simultaneous heteronuclear multidimensional spectra. *J. Phys. Chem. B* 116 (24): 7138–7144.

**37** Gopinath, T. and Veglia, G. (2013). Orphan spin operators enable the acquisition of multiple 2D and 3D magic angle spinning solid-state NMR spectra. *J. Phys. Chem.* 138 (18): 184201.

**38** Gopinath, T. and Veglia, G. (2012), Dual Acquisition Magic-Angle Spinning Solid-State NMR-Spectroscopy: Simultaneous Acquisition of Multidimensional Spectra of Biomacromolecules. *Angew. Chem. Int.* Ed., 51: 2731–2735. https://doi.org/10.1002/anie.201108132.

**39** Banigan, J.R., Gayen, A. and Traaseth, N.J. (2013). Combination of (1)(5)N reverse labeling and afterglow spectroscopy for assigning membrane protein spectra by magic-angle-spinning solid-state NMR: application to the multidrug resistance protein EmrE. *J. Biomol. NMR* 55 (4): 391–399.

**40** Good, D.B., Wang, S., Ward, M.E., Struppe, J., Brown, L.S., Lewandowski, J.R. and Ladizhansky, V. (2014). Conformational dynamics of a seven transmembrane helical protein Anabaena Sensory Rhodopsin probed by solid-state NMR. *J. Am. Chem. Soc.* 136 (7): 2833–2842.

**41** Herbst, C., Riedel, K., Ihle, Y., Leppert, J., Ohlenschlager, O., Gorlach, M. and Ramachandran, R. (2008). MAS solid state NMR of RNAs with multiple receivers. *J. Biomol. NMR* 41 (3): 121–125.

**42** Gopinath T. and Veglia G. (2012 July). 3D DUMAS: simultaneous acquisition of three-dimensional magic angle spinning solid-state NMR experiments of proteins. *J. Magn. Reson.* 220: 79–84. doi: 10.1016/j.jmr.2012.04.006. Epub 2012 Apr 26. PMID: 22698806; PMCID: PMC3487463.

**43** Gopinath, T., Weber, D.K. and Veglia, G. (2020). Multi-receiver solid-state NMR using polarization optimized experiments (POE) at ultrafast magic angle spinning. *J. Biomol. NMR* 74 (4-5): 267–285.

**44** Gopinath, T. and Veglia, G. (2018). Experimental aspects of polarization optimized experiments (POE) for magic angle spinning solid-state NMR of microcrystalline and membrane-bound proteins. *Methods Mol. Biol.* 1688: 37–53.

**45** Taware, P.P., Raran-Kurussi, S. and Mote, K.R. (2022). CURD: a single-shot Strategy to obtain assignments and distance restraints for proteins using solid-state MAS NMR spectroscopy. *J. Phys. Chem. B* 126 (17): 3269–3275.

**46** Andreas, L.B., Le Marchand, T., Jaudzems, K. and Pintacuda, G. (2015). High-resolution proton-detected NMR of proteins at very fast MAS. *J. Magn. Reson.* 253, 36–49.

**47** Sharma, K., Madhu, P.K., Agarwal, V. and Mote, K.R. (2020). Simultaneous recording of intra- and inter-residue linking experiments for backbone assignments in proteins at MAS frequencies higher than 60 kHz. *J. Biomol. NMR* 74 (4-5): 229–237.

**48** Stanek, J., Schubeis, T., Paluch, P., Guntert, P., Andreas, L.B. and Pintacuda, G. (2020). Automated backbone NMR resonance assignment of large proteins using redundant linking from a single simultaneous acquisition. *J. Am. Chem. Soc.* 142 (12): 5793–5799.

**49** Gopinath, T., Manu, V.S., Weber, D.K. and Veglia, G. (2022). PHRONESIS: a one-shot approach for sequential assignment of protein resonances by ultrafast MAS solid-state NMR spectroscopy. *Chemphyschem* 23 (13): e202200127.

**50** Klein, A., Vasa, S.K., Soldner, B., Grohe, K. and Linser, R. (2022). Unambiguous side-chain assignments for solid-state NMR structure elucidation of nondeuterated proteins via a combined 5D/4D side-chain-to-backbone experiment. *J. Phys. Chem. Lett.* 13 (7): 1644–1651.

**51** Takeda, K., Kusakabe, Y., Noda, Y., Fukuchi, M. and Takegoshi, K. (2012). Homo- and heteronuclear two-dimensional covariance solid-state NMR spectroscopy with a dual-receiver system. *Phys. Chem. Chem. Phys.* 14 (27): 9715–9721.

**52** Gallo, A., Franks, W.T. and Lewandowski, J.R. (2019). A suite of solid-state NMR experiments to utilize orphaned magnetization for assignment of proteins using parallel high and low gamma detection. *J. Magn. Reson.* 305: 219–231.

**53** Gopinath, T., Wang, S.L., Lee, J., Aihara, H. and Veglia, G. (2019). Hybridization of TEDOR and NCX MAS solid-state NMR experiments for simultaneous acquisition of heteronuclear correlation spectra and distance measurements. *J. Biomol. NMR* 73 (3-4): 141–153.

**54** Bellstedt, P., Herbst, C., Hafner, S., Leppert, J., Gorlach, M. and Ramachandran, R. (2012). Solid state NMR of proteins at high MAS frequencies: symmetry-based mixing and simultaneous acquisition of chemical shift correlation spectra. *J. Biomol. NMR* 54 (4): 325–35.

**55** Barna, J.C.J., Laue, E.D., Mayger, M.R., Skilling, J. and Worrall, S.J.P. (1987). Exponential sampling, an alternative method for sampling in two-dimensional Nmr experiments. *J. Magn. Reson.* 73 (1): 69–77.

**56** Bruschweiler, R. and Zhang, F.L. (2004). Covariance nuclear magnetic resonance spectroscopy. *J. Chem. Phys.* 120 (11): 5253–5260.

**57** Szyperski, T., Yeh, D.C., Sukumaran, D.K., Moseley, H.N.B. and Montelione, G.T. (2002). Reduced-dimensionality NMR spectroscopy for high-throughput protein resonance assignment. *Proc. Natl. Acad. Sci. U.S.A* 99 (12): 8009–8014.

**58** Kupce, E. and Freeman, R. (2003). Projection-reconstruction of three-dimensional NMR spectra. *J. Am. Chem. Soc.* 125 (46): 13958–13959.

**59** Kim, S. and Szyperski, T. (2003). GFT NMR, a new approach to rapidly obtain precise high-dimensional NMR spectral information. *J. Am. Chem. Soc.* 125 (5): 1385–1393.

**60** Frydman, L., Scherf, T. and Lupulescu, A. (2002). The acquisition of multidimensional NMR spectra within a single scan. Proc. *Natl. Acad. Sci. U.S.A* 99 (25): 15858–15862.

**61** Donovan, K. J., Kupce, E. and Frydman, L. (2013). Multiple parallel 2D NMR acquisitions in a single scan. *Angew. Chem. Int. Ed. Engl.* 52 (15): 4152–4155.

**62** Palmer 3rd, A.G. and Massi, F. (2006). Characterization of the dynamics of biomacromolecules using rotating-frame spin relaxation NMR spectroscopy. *Chem. Rev.* 106 (5): 1700–1719.

**63** Kay, L.E. (1998). Protein dynamics from NMR. *Nat. Struct. Biol.* 5 Suppl (Journal Article): 513–517.

**64** Liang, B.Y. and Tamm, L.K. (2016). NMR as a tool to investigate the structure, dynamics and function of membrane proteins. *Nat. Struct. Mol. Biol.* 23 (6): 468–474.

**65** Ward, K.M., Aletras, A.H. and Balaban, R.S. (2000). A new class of contrast agents for MRI based on proton chemical exchange dependent saturation transfer (CEST). *J. Magn. Reson.* 143 (1): 79–87.

**66** Vallurupalli, P., Bouvignies, G. and Kay, L.E. (2012). Studying "invisible" excited protein states in slow exchange with a major state conformation. *J. Am. Chem. Soc.* 134 (19): 8148–8161.

**67** Leninger, M., Marsiglia, W.M., Jerschow, A. and Traaseth, N.J. (2018). Multiple frequency saturation pulses reduce CEST acquisition time for quantifying conformational exchange in biomolecules. *J. Biomol. NMR* 71 (1): 19–30.

**68** Yuwen, T., Kay, L.E. and Bouvignies, G. (2018). Dramatic decrease in CEST measurement times using multi-site excitation. *Chemphyschem.* 19 (14): 1707–1710.

**69** MacNamara, E., Hou, T., Fisher, G., Williams, S. and Raftery, D. (1999). Multiplex sample NMR: an approach to high-throughput NMR using a parallel coil probe. *Anal. Chim. Acta.* 397 (1-3): 9–16.

**70** Li, Y., Wolters, A.M., Malawey, P.V., Sweedler, J.V. and Webb, A.G. (1999). Multiple solenoidal microcoil probes for high-sensitivity, high-throughput nuclear magnetic resonance spectroscopy. *Anal. Chem.* 71 (21): 4815–4820.

**71** Zhang, X., Sweedler, J.V. and Webb, A.G. (2001). A probe design for the acquisition of homonuclear, heteronuclear, and inverse detected NMR spectra from multiple samples. *J. Magn. Reson.* 153 (2): 254–258.

**72** Nelson, B.N., Schieber, L.J., Barich, D.H., Lubach, J.W., Offerdahl, T.J., Lewis, D.H., Heinrich, J.P. and Munson, E.J. (2006). Multiple-sample probe for solid-state NMR studies of pharmaceuticals. *Solid State Nucl. Magn. Reson.* 29 (1-3): 204–213.

**73** Olson, D.L., Peck, T.L., Webb, A.G., Magin, R.L. and Sweedler, J.V. (1995). High-resolution microcoil H-1-Nmr for mass-limited, nanoliter-volume samples. *Science* 270 (5244): 1967–1970.

**74** Wang, H., Ciobanu, L., Edison, A.S. and Webb, A.G. (2004). An eight-coil high-frequency probehead design for high-throughput nuclear magnetic resonance spectroscopy. *J. Magn. Reson.* 170 (2): 206–212.

**75** Maly, T., Debelouchina, G.T., Bajaj, V.S., Hu, K.N., Joo, C.G., Mak-Jurkauskas, M.L., Sirigiri, J.R., van der Wel, P.C., Herzfeld, J., Temkin, R.J. and Griffin, R.G. (2008). Dynamic nuclear polarization at high magnetic fields. *J. Chem. Phys.* 128 (5): 052211.

# 9

# Anisotropic One-dimensional/Two-dimensional NMR in Molecular Analysis

## Contributions and Opportunities of Anisotropic One-dimensional/Two-dimensional NMR to Recent Analysis of Small Organic Molecules

*Philippe Lesot*[1,*] *and Roberto R. Gil*[2]

[1] *Université Paris-Saclay, RMN en Milieu Orienté, ICMMO, UMR CNRS 8182, Site BPC, Bât. Henri Moissan (HM1), Orsay 91400, France*
[2] *Department of Chemistry, Carnegie Mellon University, Pittsburgh, PA, USA*
[*] *Corresponding Author*

## 9.1 Introduction

Due to its variety of analytical possibilities and methodologies developed over the last three decades in solid or liquid phase, NMR spectroscopy has become a powerful and indispensable tool in modern chemistry [1]. In this chapter, we will address an original aspect of modern NMR dedicated to the analysis of small (achiral, prochiral, and chiral) molecules using (weakly orienting) ordered media as alternative NMR solvents. Interestingly, this approach combines the use of residual anisotropic NMR interactions (tensorial properties) observed in oriented environments (liquid crystals, liquid-crystalline solution, ordered phases) and the spectral advantages of high-resolution liquid-state NMR. In the literature, there are numerous publications describing the analysis of mesogenic molecules forming generally nematics mesophases (thermotropics) or solutes aligned in oriented phases. Without wishing to be exhaustive, the readers may take a look at some "reference" books dealing with this subject [2–7]. In this chaper, we will use various terms to define ordered media, such as, anisotropic solvents, ordered liquids, aligning media, or a combination of them.

In isotropic solutions, small molecules tumble rapidily and randomly, adopting all possible orientations with respect to the external magnetic field $\mathbf{B}_0$ (of NMR spectrometer) with the same probability on the NMR timescale, i.e. it does not possess any orientational order. This molecular tumbling regime no longer exists in oriented media, leading to very informative molecular anisotropic NMR interactions. These new interactions, which no longer average out by molecular tumbling as in isotropic liquid phases, are: (i) the residual dipolar coupling (noted RDC or $D$), (ii) the chemical-shift ($\delta^{\text{aniso}}$) through the residual chemical-shift anisotropy (noted RCSA or $\Delta\sigma$), and (iii) finally the residual quadrupolar coupling (noted RQC or $\Delta\nu_Q$).

We will, first, present the specificities of anisotropic/oriented NMR using weakly aligning media as well as the tools (the spectral toolkit) and NMR approaches to extract and exploit the anisotropic information involved in the analysis of small organic molecules of synthetic or natural origin. Among weakly- aligning media as useful anisotropic NMR solvents (opposed to thermotropic liquid crystals), we will mainly present the case of polypeptide-based lyotropic liquid crystals (such as PBLG or PCBLL) and the polymeric aligning gels (such as PMMA or Poly-HEMA). Key experimental examples illustrating the principles of anisotropic NMR have been

*Two-Dimensional (2D) NMR Methods*, First Edition. Edited by K. Ivanov, P.K. Madhu and G. Rajalakshmi.

selected to show the diversity of analyzable compounds in these systems, associated with a description of spectral information visible on their anisotropic spectra.

Then we will describe the challenges and the analytical opportunities of NMR in oriented solvents (chiral and non-chiral), such lyotropic liquid crystals or stretched/compressed gels, through key illustrative applications. The examples proposed will show how anisotropic 1D and 2D NMR can solve many analytical problems encountered by a large community of chemists. Throughout this chapter we will show how 2D NMR methodologies provide solutions for simplifying complex anisotropic 1D spectra.

Finally, and beyond the description of NMR results in terms of spin system analysis (AX, AB, $A_3$, …) as well as magnetic or chemical equivalence, we will also refer to various stereochemical aspects encountered routinely in organic chemistry (chirality, prochirality, diastereotopic, and enantiotopic groups, etc.). While some aspects will be proposed in this chapter, a more precise description of these important concepts can be found in the "Bible" of stereochemistry written by Eliel and Wilen [8].

## 9.2 Advantages of Oriented Solvents

Discovered in 1888, liquid crystals (LCs) are fascinating molecules [9]. They show an interesting state of matter between solids and liquids (see Figure 9.1a), providing NMR spectroscopists with a wide variety of anisotropic environments (aligning media), with the advantages of combining solid-state NMR (SS-NMR) properties (through the detection of anisotropic observables) with those of liquid-state NMR (highly fluid solvents leading to high-resolution spectra) [4, 6, 7]. In other words, NMR parameters only observed in solid state can be measured with the resolution obtained in liquid state. This unique "anisotropy-fluidity" combination is at the origin of a strong analytical potential that can be successfully exploited in modern molecular analysis.

Regardless of how the molecular order is created, the use of NMR in oriented media provides a possible access to: (i) the determination the constitution, the configuration and the preferred conformation/s of the aligned guest molecules, (ii) the study the orientational behavior of solutes in relation to the solute-solvent interactions, and (iii) the evaluation of the enantiomeric purity of a mixture of chiral molecules (enantiomers or enantio-isotopomers) or the discrimination of enantiotopic element in prochiral molecules if the oriented system is chiral, for instance.

The analytical success of anisotropic NMR lies in the ability of the LC mesophases (chiral or non-chiral) or the strained (stretched or compressed) aligning gels to orient a solute homogeneously and uniformly inside the magnetic field ($\mathbf{B_0}$) of the NMR spectrometer. These unique properties lead to exploitable high-resolution NMR spectra that can be combined with the ability to spectrally discriminate enantiomers (chiral molecules) or enantiotopic elements (prochiral molecules) if the aligning system is homochiral (e.g. enantiopure). Herein, achiral and chiral LC will be denoted ALCs and CLCs, respectively.

A large collection of weakly aligning systems has been designed over the last two decades, both for water- and organo-soluble compounds [10, 12]. These molecular systems are very interesting because they produce a weak degree of solute alignment, leading to residual anisotopic NMR interactions of small magnitude, in the order of $10^{-3}$–$10^{-4}$ with respect to the values observed in solid state. Contrarily, thermotropic liquid-crystalline phases, another family (historical) of oriented media, show a degree of order ~0.15 with respect to solid state. Weak alignment reduces the complexity of NMR spectra and so the difficulty to record them. Among weakly aligning mesophases, a special attention has been paid to the design of many chiral polymer-based LC (see Figure 9.1b) as well as the development of mechanically controlled compressed or stretched achiral gels (see Figure 9.1c) [13–15]. Compared to thermotropics, the lyotropic systems allow an easy control of global orientation of solutes by adjusting: (i) the temperature of the sample, (ii) the concentration of components of the mesophase, (iii) the nature of the helical polymer, and (iv) the polarity of the co-solvent (organo-soluble polymer). These four parameters afford a full control and easy optimization of the properties of the mesophase in terms of solute orientation but also of difference in orientation (chiral molecules) if the system is enantiodiscriminating. For aligning gels, which are not

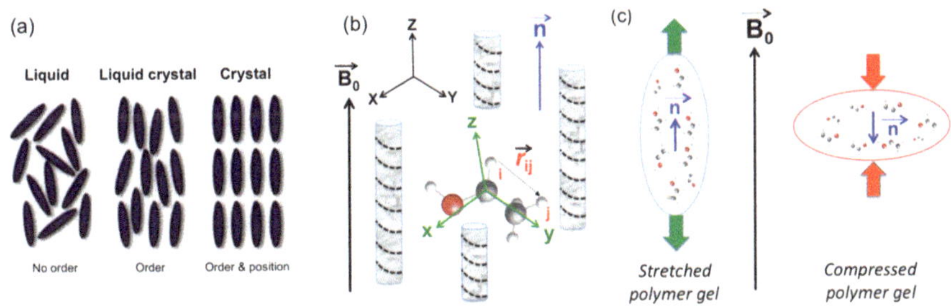


Figure 9.1 (a) Schematic description of the molecular orientational order in an isotropic liquid (no order, $S = 0$), a nematic LC (only order, $0 < S < 1$) and a crystalline solid (position and order with $S = 1$). (b) Example of a reference axis system (x, y, and z) attached to a molecule dissolved in a chiral polymer-based LLC, and used for the description of molecular order. (c) Model description of stretched or compressed gels inducing a molecular order. Figure partially adapted from Refs. [10, 11] with permission.

chiral, the degree of alignment generally depends on: (i) the cross-link density of the polymer gel [16, 17], and (ii) the degree of stretching or compression, which with the exception of polyacryamide gels, is reversible and can be tuned up and down at the user's whim [13–15].

### 9.2.1 Description of Orientational Order Parameters

Phenomenologically, molecular orientation can be described using the Saupe's order matrix, a second-rank tensor ($3 \times 3$) noted $S_{\alpha\beta}$ or $\{S_{\alpha\beta}\}$ or $\hat{S}$, associated with a molecular (orthogonal) axis system x, y, z, whose internal elements ($\alpha\beta \equiv$ x, y, z) can be individually expressed as [6]:

$$S_{\alpha\beta} = \frac{1}{2}\left\langle 3\cos\theta_Z^{\alpha}\cos\theta_Z^{\beta} - \delta_{\alpha\beta}\right\rangle. \tag{9.1}$$

In Equation 9.1, $\delta_{\alpha\beta}$ is the Kronecker function ($\delta = 1$ if $\alpha = \beta$, and $\delta = 0$ if $\alpha \neq \beta$) and $\theta_Z^{\alpha}$ are the angles between the axes, x, y, and z and the Z-axis of laboratory axis system (X, Y, Z). As a consequence, all order-dependent NMR interactions are tensorial properties related to these order parameters.

The Saupe's matrix has two important properties: (i) it is symmetric ($S_{\alpha\beta} = S_{\beta\alpha}$) since the orientation can have two opposite directions, (ii) it is traceless, i.e. the sum of its diagonal elements is null ($\sum S_{\alpha\alpha} = 0$), because no order exists in isotropic liquids. Expressed in an arbitrarily defined reference axis system (RAS) (see Figure 9.1b), the number of independent elements in $S_{\alpha\beta}$ is dependent on the symmetry of the analyte as well as whether the oriented solvent is chiral or not [18]. For instance, five (non-zero) order parameters are needed to describe the orientational behavior of chiral molecules of $C_1$ symmetry, both in ALCs and CLCs. In contrast, for $C_s$-symmetry prochiral (rigid) molecules, three (non-zero) order parameters are needed in ALCs to describe their orientational ordering, but five are needed in CLCs. As we will discuss in Section 9.3.4, three other molecular symmetries ($C_{2v}$, $D_{2d}$, and $S_4$) are also affected by the chirality of the solvent, thus leading to important spectral consequences.

From the matrix elements of $S_{\alpha\beta}$, we can derive any local order parameters, $S_{ij}$, associated with any internuclear direction, i–j, (the vector $\vec{r}_{ij}$) expressed in the molecular reference axis system (RAS) initially defined, (x, y, z), (see Figure 9.1b) on the basis of the so-called director cosines following Equation 9.2 [6, 10]:

$$S_{ij} = \sum_{\alpha,\beta=x,y,z} \cos\theta_{ij\alpha}\cos\theta_{ij\beta} S_{\alpha\beta}. \tag{9.2}$$

Interestingly, the local order term, $S_{ij}$, can be also described as a simple trigonometric function defined as:

$$S_{ij} = \frac{1}{2}\langle 3\cos^2\theta_{ij}^{B_0} - 1\rangle \tag{9.3}$$

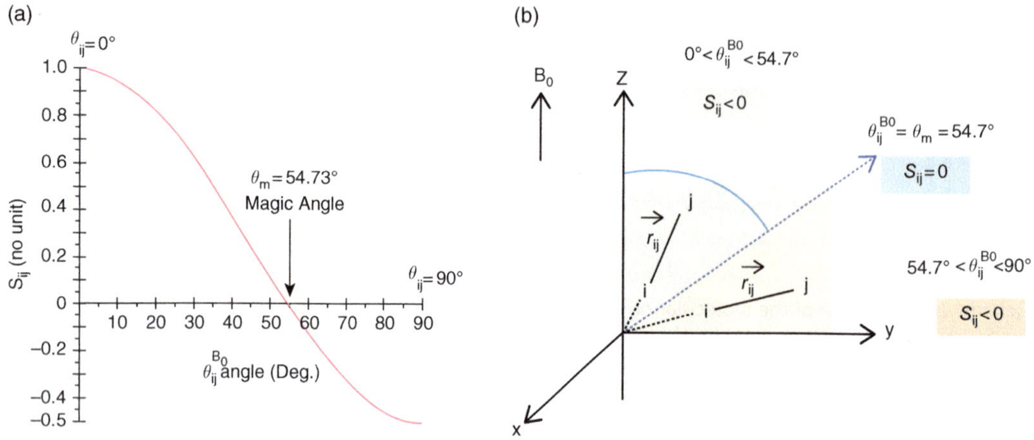


**Figure 9.2** (a) Variation of the trigonometric term, $(3\cos^2\theta_{ij}^{B_0} - 1)/2$, versus angle $\theta_{ij}^{B_0}$. Note the famous magic angle, $\theta_m$, for which the local order parameter, $S_{ij}$, is null. At $\theta_{ij}^{B_0} = \theta_m$, any ($^1$H-$^1$H)-RDCs or $^2$H-RQCs are equal to zero. (b) Vectorial description of the position of an internuclear vector i–j (noted also $\vec{r}_{ij}$) relative to the axis $\vec{B}_0$.

where $\theta_{ij}^{B_0}$ is the angle between a given i–j internuclear vector ($\vec{r}_{ij}$) and the $B_0$ axis. Figure 9.2a shows the variation of term, $3(\cos^2\theta_{ij}^{B_0} - 1)/2$, versus the $\theta_{ij}^{B_0}$ angle. As schematically described in Figure 9.2b, the $S_{ij}$ values are positive between $[0° \leq \theta_{ij}^{B_0} < 54.73°]$, negative between $[54.73° < \theta_{ij}^{B_0} \leq 90°]$, and null when $\theta_{ij}^{B_0} = 54.73°$ (the so-called magic angle). Remarkably, at $\theta_{ij}^{B_0} = \theta_m$, any RDC or RQC values are equal to zero (as in liquids), even if molecules are oriented in the field $\mathbf{B_0}$.

Note, finally, that the Saupe matrix, $S_{\alpha\beta}$, can be also diagonalized. This matrice, noted $S_{\alpha'\beta'}$, expressed in the unique principal axis system (PAS) of the molecule (noted x', y', z') contains only three parameters, $S_{x'x'}$, $S_{y'y'}$, and $S_{z'z'}$ and as previously described is traceless, $S_{x'x'} + S_{y'y'} + S_{z'z'} = 0$.

### 9.2.2 The GDO Concept

In 2001, the concept of generalized degree of order (GDO) in the PAS has been introduced to describe the orientational order of an analyte using a single value (scalar quantity). GDO is defined as follows [19, 20]:

$$\text{GDO} = \frac{|\mathbf{A}|}{|\mathbf{A}_{\max}|} = \sqrt{\frac{3}{2}}|\mathbf{A}| \tag{9.4}$$

where $|\boldsymbol{A}|$ is the norm of the alignment tensor, $A$, and $|\boldsymbol{A}_{max}|$ represents the maximum order for a static molecule (solid state). Note here that the relation between the order matrix, noted, $\hat{S}$, and the alignment tensor, noted $\hat{A}$, is simply defined as $\hat{A} = \frac{2}{3}\hat{S}$ [20]. Since in the PAS (x', y', z'), only the diagonal elements of $A$ are non-zero, the above equation is reduced to:

$$\text{GDO} = \sqrt{\frac{3}{2}}\sqrt{\left|A_{x'x'}^2 + A_{y'y'}^2 + A_{z'z'}^2\right|}. \tag{9.5}$$

For a static molecule, the GDO is equal to 1. The GDO is an indication of the degree of alignment induced by the orienting media. It cannot be calculated but it can be determined experimentally by first obtaining the alignment tensor A from fitting anisotropic data (RDCs, RCSAs, RQCs) to the correct 3D structure. It is important to highlight that the degree of alignment also depends on the shape and size of the molecule. Molecules that are very small and close to a spherical shape align poorly. However, for a molecule that aligns well, the GDO is a good indication of the strength of alignment of a particular orienting media. The bigger the GDO, the stronger the alignment is. Aligning gels have a GDO in the order of $10^{-4}$, while LLCs phase such PBLG it is in the order of $10^{-3}$, compared to solid state (static molecule). If we know the order of the GDO we can predict the maximum value of RDCs, RCSAs, and/or RQCs that we should expect to observe experimentally. For instance, the maximum possible $^{1}D_{CH}$ solid state for a CH bond is when the bond is parallel to the magnetic field $\mathbf{B_0}$. This value is −45.38 KHz. In a gel, with a GDO of 7.0 $10^{-4}$, a simple calculation of [−45.38 KHz × 7.0 $10^{-4}$] tells us that the maximum expected $^{1}D_{CH}$ would be of ~31.8 Hz with $r_{CH}$ = ~1.5 Å (see Table 9.1). As mentioned above, this value will also depend on the size and shape of the molecule. In conclusion, if we know the solid-state value of RDCs, RCSAs, and RQC, we can estimate their maximum value in solution if we know the GDO.

## 9.3 Description of Useful Anisotropic NMR Parameters

For any magnetically active nuclei ($I \neq 0$), chemical shifts (due to electronic shielding), $\delta^{iso}$ and (homo- or heteronuclear) scalar couplings between inequivalent nuclei (due to magnetic mutual interaction through bonds), $J_{ij}$, observed in isotopic liquids are two important informative sources of molecular data.

As in NMR in liquids, all non-zero spin nuclei of molecules dissolved in oriented media are also potential NMR probes for the three key anisotropic NMR interactions: (i) the RDC's associated with the mutual dipole-dipole interaction between magnetically active nuclei through space, (ii) the chemical shift ($\delta^{aniso}$) through the RCSA's originating from the specific electronic shielding of nuclei, and (iii) finally the RQC's resulting from the interaction between the electric quadrupolar moment of quadrupolar nuclei (spin $I > {}^1\!/_2$) and an electric field gradient. The value of these anisotropic parameters depends on the average orientation (at NMR time scale) of the molecules (and associated local internuclear directions) with respect to external magnetic field $\mathbf{B_0}$ (see Figures 9.1b and 9.1c).

Generally, the anisotropy of scalar coupling, $\Delta J$, the last anisotropic interaction is not considered because the anisotropic component of $J$ is not as strong as for the three other interactions ($\Delta\delta$, $D$, and $\Delta\nu_Q$), and its potential analytical use is only limited to very particular nuclei such as fluorine-19.

### 9.3.1 Residual Dipolar Coupling (RDC)

The homo- or heteronuclear dipolar coupling between two chemically non-equivalent coupled nuclei, i and j, (an AX or AB spin system) vanishes in isotropic solution because the bonded or non-bonded internuclear vectors ($r_{ij}$), e.g. "i–j" or "i···j," adopt all the possible orientations (defined by the angle $\theta_{ij}$) with respect to $\mathbf{B_0}$ with the same probability (molecular isotropic tumbling) (see Figure 9.3) [6, 21]. In contrast, in the presence of an anisotropic medium, only a fraction of the dipolar coupling is observed (about $10^{-3}$ – $10^{-4}$), hence the term "residual interaction." The resulting coupling is known as residual dipolar coupling (noted RDC) but contain the same information than the dipolar coupling.

Mathematical description of RDCs (whatever the pair of mutually interacting nuclei) is presented with key Equations 9.4 and 9.5. If an internuclear scalar coupling exists between two bonded nuclei, $i$ and $j$, ($J_{ij}$), the signal observed (for $i$ and $j$) shows a total splitting (or total spin-spin coupling) equal to $T_{ij} = J_{ij} + 2D_{ij}$ (see Figure 9.3). As $S_{ij}$ can be positive, negative or null (magic angle), the RDC values can be negative, positive, or null, respectively. Depending on the magnitude and the sign of $J_{ij}$ and $D_{ij}$, the final value of $T_{ij}$ can be positive, negative or null. Note that in the literature, the total coupling, $T_{ij}$, can also be defined as $J_{ij} + D_{ij}$; this difference comes from the definition of dipolar coupling used (see Equation 9.6). In this chapter, we adopt the notation "$2D_{ij}$" for the dipolar

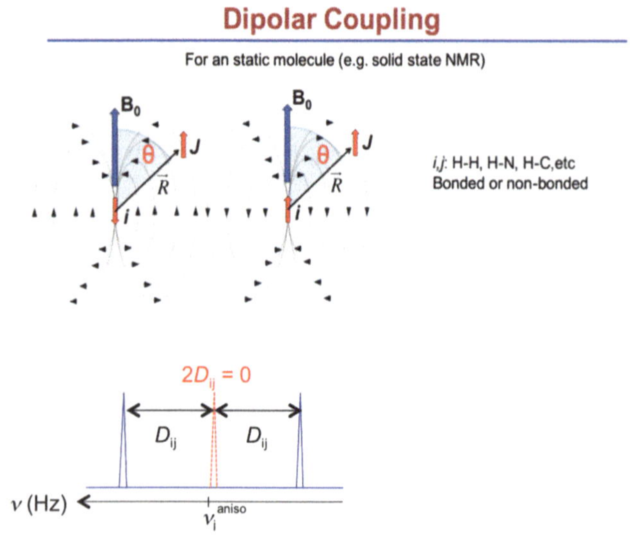


$$D_{\mathrm{ij}} = -k_{\mathrm{ij}} \left\langle \frac{S_{\mathrm{ij}}}{r_{\mathrm{ij}}^3} \right\rangle \tag{9.6}$$

with :

$$k_{\mathrm{ij}} = \left(\frac{\mu_0}{4\pi}\right) \times \left(\frac{\mathrm{h}\gamma_{\mathrm{i}}\gamma_{\mathrm{j}}}{4\pi^2}\right). \tag{9.7}$$

**Figure 9.3** The origin of the dipolar coupling, $D_{ij}$, between two magnetically active nuclei, $i$ and $j$, the spectral consequences for two inequivalent coupled nuclei (view on nucleus i), and associated equations to RDCs. In Equation 9.6, $S_{ij}$ and $r_{ij}$ are the local order parameter of the internuclear direction, i–j, and the distance separating the two nuclei $r_{ij}$. In Equation 9.7, $\mu_0$ is the permeability constant in vaccum, and $\gamma_i$ and $\gamma_j$ are the gyromagnetic ratios of $i$ and $j$. In the case of a mixture of enantiomers ($R/S$), a doubing of the spectral patterns (on *per* enantiomers) is expected to be observed if a spectral enantiodiscrimination occurs, while generally centered on the very close frequencies ($\delta^{\mathrm{aniso},R} \approx \delta^{\mathrm{aniso},S}$).

contribution to $J_{ij}$. RDCs provide information of the relative orientation of internuclear vectors between an atomic pair ($i$ and $j$) in a molecule. They can be bonded, such as C-H ($^1D_{CH}$), N-H ($^1D_{NH}$), C-F ($^1D_{CF}$), C-P ($^1D_{CP}$) bonds, but also non-bonded nuclei as geminal proton–proton pairs ($^{2,3}D_{HH}$), or two to three bond carbon-proton pairs ($^{2,3}D_{CH}$). From a practical standpoint, the most widely used RDC for structural analysis of small molecules is the one-bond proton-carbon ($^1D_{CH}$), because it is easy to measure from $F_2$- or $F_1$-coupled HSQC 2D experiments. Although less often, geminal proton–proton RDCs ($^{2,3}D_{HH}$) and long-range two to three bond proton-carbon RDCs ($^{2,3}D_{CH}$) were occasionally used in addition to $^1D_{CH}$ data.

Other RDC's can be exploited if they are detected: e.g., $^1D_{CF}$ can be straightforwardly measured from the fluorinated carbon signal splitting in the $^{13}C\text{-}\{^1H\}$ 1D NMR spectra as the difference between the isotropic and anisotropic spectra. From Equation 9.6, it can be noted that the magnitude of the RDC value, for a given $\theta$ angle and $r_{ij}$ distance, will strongly depend on the term $k_{ij}$, that corresponds to the product of gyromagnetic ratio for two interacting nuclei and the inverse of the cube of the distance between them, $(r_{ij})^{-3}$. Table 9.1 lists the most useful values of $k_{ij}$.

An interesting property of anisotropic NMR is that, different from the scalar coupling $J$, RDCs do not vanish under chemical equivalence. As a consequence, this unique property provides access to RDCs between equivalent nuclei, such as hydrogens inside a methyl group ($A_3$ spin system) and a methylene group ($A_2$ spin system), while their scalar coupling is not measured in NMR in liquids. For these both cases, purely dipolar splittings measured on the corresponding spectral patterns (a 1:2:1 triplet and a 1:1 doublet, respectively) is equal to $T_{ii} = 3D_{ii}$. These proton-proton RDCs within methyl or intramethylene groups can be easily measured in polypeptide/polyacetylenic-based systems but the situation is more difficult with aligning gels whose the degree of alignment is smaller (by at least a factor 10 or more) compared to previous systems. Indeed, if the spectral rules applied to analyze the anisotropic spectra are theoretically the same in the two families of orienting media, some practical differences can appear because all anisotropic interactions measured in alignment gels are generally one

**Table 9.1** Examples of $k_{ij}$ values (with sign) involving nuclei $^{1}H$, $^{2}H$, $^{13}C$, $^{19}F$, and $^{31}P$, and ranked in decreasing order

| Pair of interacting nuclei | $k_{ij}$ values (kHz.Å$^3$) | Sign[a] |
|---|---|---|
| $^{1}H$-$^{1}H$ | 120.07 | > 0 |
| $^{1}H$-$^{19}F$ | 112.99 | > 0 |
| $^{19}F$-$^{19}F$ | 106.30 | > 0 |
| $^{1}H$-$^{31}P$ | 48.62 | > 0 |
| $^{1}H$-$^{13}C$ | 30.20 | > 0 |
| $^{13}C$-$^{19}F$ | 28.41 | > 0 |
| $^{31}P$-$^{31}P$ | 19.68 | > 0 |
| $^{1}H$-$^{2}H$ | 18.44 | > 0 |
| $^{13}C$-$^{31}P$ | 12.23 | > 0 |
| $^{13}C$-$^{13}C$ | 7.59 | > 0 |
| $^{2}H$-$^{31}P$ | 7.46 | > 0 |
| $^{2}H$-$^{13}C$ | 4.64 | > 0 |
| $^{2}H$-$^{2}H$ | 2.83 | > 0 |

[a] $k_{ij}$ is negative when one of interacting atoms has a negative gyromagnetic ratio (e.g. $^{15}N$).

order of magnitude smaller than in PBLG and leading to RDC values smaller than the linewidths for these particular cases. In general, any pair of protons close in space that have the same chemical shift, either by coincidence or by symmetry ($A_2$, $A_3$, ....$A_n$), will show a measurable RDCs in anisotropic conditions.

Technically, one-bond ($^{13}C$-$^{1}H$)-RDCs, ($^{1}D_{CH}$), are by far the easiest to measure on any NMR spectrometers with the help or not of adapted homo- or heteronuclear 2D experiments to extract information, even with small molar amounts of analyte. Followed by geminal ($^{1}H$-$^{1}H$)-RDCs, ($^{2}D_{HH}$). Measuring other proton-proton RDCs is more challenging and barely used in structure elucidation of small molecules. One limitation exists, however, in the case of proton-deficient molecules. A peculiar situation in which $^{13}C$-RCSAs play a relevant role because they can be measured for every carbon, even those not protoned, as we will show below.

### 9.3.2 Residual Chemical-shift Anisotropy (RCSA)

In isotropic solutions, the chemical shift measured corresponds to the average value of the three main components of the chemical-shift anisotropic tensor (see Figure 9.4). In the presence of an anisotropic medium, this average value is shifted, and the shift is known as RCSA. Same as RDCs, the RCSA values can be positive, negative or null, leading to a low- or high-field shift of resonances compared to isotropic ones. Mathematical descriptions of RCSAs (whatever the nuclei) are presented with the key Equations 9.8 to 9.10 [6, 10].

According to IUPAC recommendations [22, 23], the absolute magnetic shielding, σ (expressed in ppm), is the difference in shielding between the frequency of the bare nucleus, $\nu_{\text{bare nucl.}}$, and the frequency of the same nucleus in the species under investigation, $\nu_{\text{spe.}}$:

$$\sigma(\text{ppm}) = 10^6 \times \left(\frac{\nu_{\text{bare nucl.}} - \nu_{\text{spe.}}}{\nu_{\text{bare nucl.}}}\right). \tag{9.11}$$

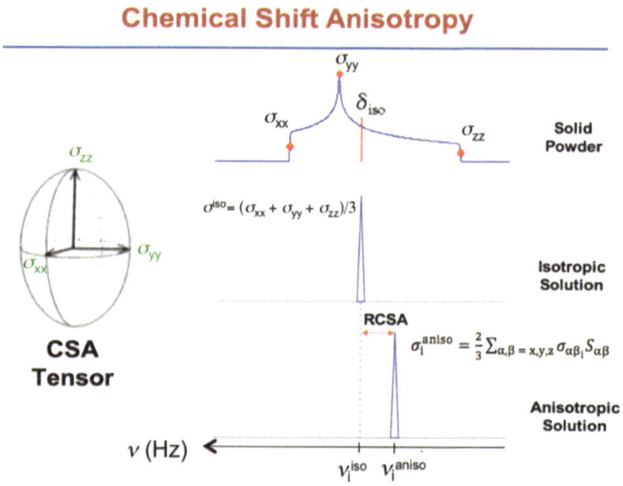


$$\nu_i^{aniso} = -\frac{\gamma}{2\pi}\left(1 - \sigma_i^{iso} - \sigma_i^{aniso}\right) B_0 \tag{9.8}$$

with :

$$\sigma_i^{iso} = \frac{1}{3}\left(\sigma_{xx_i}^{iso} + \sigma_{yy_i}^{iso} + \sigma_{zz_i}^{iso}\right) \tag{9.9}$$

and :

$$\sigma_i^{aniso} = \frac{2}{3} \sum_{\alpha,\beta=x,y,z} \sigma_{\alpha\beta_i} S_{\alpha\beta} \tag{9.10}$$

**Figure 9.4** The origin of chemical shift anisotropy, $\sigma^{aniso}$, related to anisotropic distribution of electrons (CSA tensor), spectral consequences and associated equations to RCSAs. In Equation (9.8), $\gamma_i$ is the gyromagnetic ratio of i, and $\sigma^{iso}$ and $\sigma^{aniso}$ are the isotropic and anisotropic terms contributing to the electronic shielding. In the case of a mixture of enantiomers ($R/S$), two shifted resonances are expected to be detected with $\nu^{aniso,R} \neq \nu^{aniso,S}$, if spectral enantiodiscrimination occurs. In the case of mixture of enantiomers, two resonances are expected to be detected, centered on $\nu^{aniso,R}$ and $\nu^{aniso,S}$ if a spectral enantiodiscrimination occurs.

This is the value that can be obtained when using chemical-shift calculations by DFT, and the chemical shift ($\delta$) is then determined using the $\sigma$ calculated for TMS, for instance. The well-known chemical shift, $\delta$(expressed in ppm), used by chemists in the, is the difference in shielding between the nucleus in the species under investigation, $\sigma_{spe.}$, and the shielding of the same nucleus in a reference compound, $\sigma_{ref.}$:

$$\delta(\text{ppm}) = 10^6 \times \left(\frac{\sigma_{ref.} - \sigma_{spe.}}{1 - \sigma_{ref.}}\right) = 10^6 \times \left(\frac{\nu_{spe.} - \nu_{ref.}}{\nu_{ref.}}\right). \tag{9.12}$$

Approximatively, we can write: $\delta(\text{ppm}) = 10^6 \times (\sigma_{ref.} - \sigma_{spe.})$.

When talking about chemical shift, as reported in the NMR spectral data given in database of in scientific literature, the quantity $\delta$(in ppm) is used, thus allows to compare the same information whatever the strength of the spectrometer (the Larmor frequency). The symbol $\sigma$ is reserved to the absolute magnetic shielding constant.

Mathematically, the chemical shift is a tensorial property. It is not a number, and it is represented by a second-rank tensor (3×3 matrix). The value of the chemical shift depends on the orientation of the molecule respect to the magnetic field $B_0$. This orientational dependence is known as chemical-shift anisotropy (CSA). In solid-state NMR, a powder pattern is observed, as shown in Figure 9.4b. As mentioned above, in solution, the isotropic chemical shift ($\delta^{iso}$) is observed, corresponding to the average value of the three principal components of the chemical shift tensor $(\delta_{xx} + \delta_{yy} + \delta_{zz})/3$.

For a given pair of spins (e.g. C-H bond) with same or similar internuclear distance, the maximum RDCS value at the same degree of order will be nearly the same. In order for this value to significantly chance, the internuclear distance has to change such as for the case of long-range RDCs ($^{2,3}D_{CH}$), which are one order of magnitude smaller that the corresponding one-bond RDC ($^1D_{CH}$). However, for RCSAs this is not the case. Its maximum value will depend on the anisotropy of the chemical shift tensor for each carbon in the molecule. If the value of the three components of the chemical shift tensor is the same ($\delta_{xx} = \delta_{yy} = \delta_{zz}$), the tensor will be spherical (zero anisotropy) and the RCSA will be null, no matter how strong is the anisotropy created by the alignment medium.

The anisotropy of a chemical shift tensor, in terms of the electronic shielding constant $\sigma$, is defined as:

$$\text{anisotropy} = \sigma_{zz} - \left(\frac{\sigma_{11} - \sigma_{22}}{2}\right). \quad (9.13)$$

In practice, carbons with $sp^3$ hybridization show very poor anisotropy while sp and $sp^2$ carbons normally show strong anisotropy, as shown in Figure 9.5.

In anisotropic media, the maximum RCSA values observed correspond to the product $(\sigma_{zz} - \sigma^{iso}) \times$ GDO (see 9.2.2 for the definition of GDO), because, $\sigma_{33}$ is the largest value of the CS tensor by convention. For example, a PMMA gel of 0.2 M% of cross-link density has a GDO value of $0.7\times10^{-4}$. For a $sp^2$ hybridized $^{13}$C nuclei where the difference $(\sigma_{zz} - \sigma^{iso})$ is equal to ~200 ppm (e.g. carbonyl group or aromatic carbon atom), the maximal $^{13}$C-RCSA value is equal to $200\times0.7\times10^{-4} = 0.14$ ppm namely ~17.5 Hz at 125 MHz (11.75 T). For a $CH_3$ group is $(\sigma_{zz} - \sigma^{iso})$ is about ~20 ppm, and then max value of RCSA is ~1.75 Hz. In conclusion, RCSAs are very small and are in the range of ppb's. Hence, in order to accurately measure $^{13}$C-RCSAs it is necessary to have a very high-resolution spectrum and the only way to achieve it is to run $^{13}$C-{$^1$H} 1D NMR spectra.

As we will see in the section of application of RCSAs to the structural analysis of small molecules (see Section 9.7.2), the larger the GDO the better. Thus PBLG-based LLCs are ideal orienting media to measure RCSAs, since they have GDO values with one order of magnitude larger than aligning gels. In fact, it is like having a "magnifying glass" to particularly enhance the $^{13}$C-RCSAs values of $sp^3$ carbons.

The same way that RDCs encode information about the relative orientation of internuclear vectors (e.g. C-H bonds), $^{13}$C-RCSAs encode information about the relative orientation of alignment tensors. It is easy to visualize the relative orientation of highly anisotropic CS tensors, because of their ellipsoidal shape (see aromatic carbon atoms in Figure 9.5). However, the poorer the anisotropy, the more spherical the CS tensor is, and if there is not anisotropy at all, there is no way to determine the relative orientation of two spherical objects (two balls). In addition, as for RDCs, parallel CS tensors do not provide additional orientational information. As you can see in Figure 9.5, the six CS tensors for the aromatic ring counts as one. In case of a double bond, they will count as two. The only advantage from the assignment standpoint is that if you have two double bonds ($sp^2 - sp^2$) with different

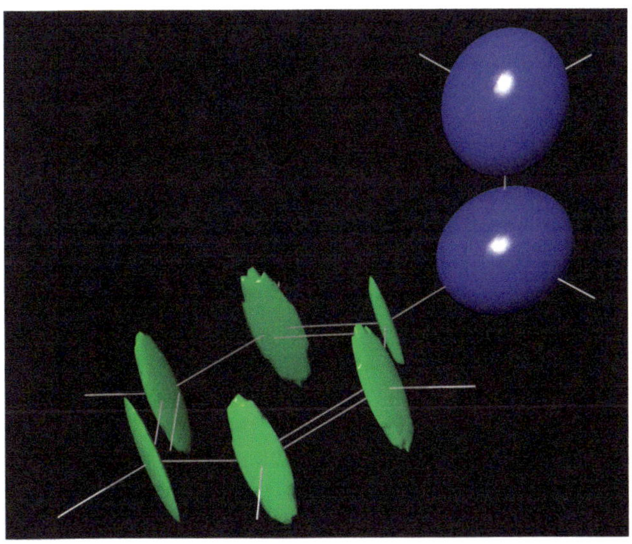

**Figure 9.5** Isosurface plot of the $^{13}$C chemical shift tensors of ethyl benzene ($sp^2$ in green and $sp^3$ in blue). The difference in size of the $sp^2$ and $sp^3$ carbons is because the CS tensor drawn in absolute chemical shielding units (ppm). The $sp^3$ carbon atoms are significantly more shielded than the $sp^2$ carbons, hence their larger size. Note the difference in shape between the aromatic carbons (flat ellipsoid) compared to the $CH_2$ and $CH_3$ carbons (more spherical ellipsoids).

relative orientation in space, the carbons that belong to the same double bond will have similar RCSAs values. Finally, judging from the orientation the CS tensors in ethyl benzene (Figure 9.5), the six aromatic carbons count as one (all parallel) since they provide the same orientational information, while the CS tensor of $CH_2$ and $CH_3$ carbon atoms show independent orientations of their $\sigma_{zz}$ components.

From an experimental point of view, the main problem when measuring RCSAs are the interferences from isotropic chemical shift changes. Indeed, many factors can affect the value of the chemical shift such as temperature, concentration, pH, changes in magnetic susceptibility of the medium (solvent changes), etc. The bottom line here is that if the isotropic and anisotropic spectra are not acquired in the same exact experimental conditions, the change in chemical shift value will not only be due to RCSA. Errors introduced by isotropic chemical shift changes can be high because RCSAs are very small compared to RDCs. The following are different experimental scenarios for combinations of isotropic and anisotropic chemical shift changes: (i) isotropic solvent A and then isotropic solvent B (isotropic shift), (ii) isotropic solvent and then solvent/gel (isotropic shift), (iii) isotropic solvent and then solvent/compressed gel (isotropic shift plus RCSAs), (iv) isotropic shift not observed in stretched gels, (v) same gel from relaxed to compressed (RCSA data and predictable isotropic shifts that can be corrected), and (vi) same gel from relaxed to stretched (only RCSAs). In the session of applications of RCSAs we will show how this problem was trackled in order to accurately measure them.

Among the common magnetically active nuclei present in the majority of organic molecules (C, H, O, N), the $^{13}C$ nucleus is the most efficient nuclear spy as far as RCSA is concerned. Indeed, $^{13}C$-RCSA can be measured for any type of carbons, even quaternary ($sp^3$), provided that the carbons show enough anisotropy. In practice, $sp^2$- and sp-hybridized carbon atoms (aromatic rings, double bonds, carbonyls) provide the largest values of $^{13}C$-RCSA. As we will see, $^{13}C$-RCSA is very useful for discriminating enantiomers on the basis of difference of $^{13}C$-RCSA as well as revealed to be very useful for proton-deficient molecules where the number of CH vectors is not enough to determine alignment tensors using only RDCs.

### 9.3.3 Residual Quadrupolar Coupling (RQC)

For the same reasons explained for the dipolar coupling above, the quadrupolar coupling vanishes in isotropic solutions. In the presence of an (weakly orienting) anisotropic medium, a fraction of it is observed ($10^{-3}$ – $10^{-4}$), and it is known as RQCs. As for RDCs, higher values are obtained with classical thermotropics.

Numerous quadrupolar nuclei exist and can be detected, but most of them present a strong quadrupolar moment, accelerating the relaxation process, leading in turn to low-resolution NMR signals. Interestingly, deuterium atoms ($I = 1$), naturally present in any organic molecules, as the second isotope of hydrogen possess a rather small quadrupolar moment (see Equation 9.15). They can be detected using modern NMR spectrometers with our without cryogenic probes. Evidently, using isotopically enriched molecules (deuterated analytes) is a spectral advantage in terms of sensitivity (100% instead of 1.5 $10^{-2}$%), but requires a molecular modification by chemical synthesis. Numerous synthetic approaches for introducing deuterium atoms (selectively or not) in achiral/prochiral/chiral molecules have been developed and are well documented. However, it is clear that they are not always possible or easy to make. Their description is over the scope of this chapter, but readers can see Refs [24–27].

Compared to the routine nuclei of spin one-half, the detection of deuterium nuclei gives access to the residual deuterium quadrupolar ($^2H$-RQC). These couplings originate from the interaction between the nuclear quadrupolar moment of a deuterium atom and an electric field gradient along the C-D bond.

In case of an isolated deuterium site, we observe one quadrupolar doublet ($^2H$-QD) in a non-chiral phase. For two monodeuterated enantiomers dissolved in a discriminating chiral LC, two quadrupolar doublets are expected. Interestingly, the quadrupolar interaction is very sensitive to small difference of molecular orientation. Mathematical description of $^2H$-RQCs is presented from key Equations. 9.14 to 9.16 (see Figure 9.6).

When preparing samples for ($^{1}$H-$^{1}$H)- and ($^{13}$C-$^{1}$H)-RDCs and $^{13}$C-RCSAs measurements in LLCs, the $^{2}$H 1D-NMR spectrum of the deuterated solvent can be used to verify: (i) if the sample is anisotropic, (ii) the degree of anisotropy by looking at the value of the $^{2}$H solvent signal splitting, and (iii) the quality of the anisotropy by looking at the shape of the solvent signal.

### 9.3.4 Spectral Consequences of Enantiodiscrimination

#### 9.3.4.1 Chirality and Prochirality

Among exciting domains of modern organic chemistry, the asymmetric synthesis of active substances used in human health has a special place (mainly from the terrible case of Thalidomide at the end of 1950s, a chiral drug with an enantiomer that was teratogenic). The development of new (NMR) tools to analyze and spectrally discriminate enantiomers of chiral molecules is a continuous challenge [28–30]. In this section, we describe the significant contribution of anisotropic NMR to this societal problem.

Enantiomers of chiral molecules are mirror-image objects by a plane of symmetry, but are not superimposable. Majority of chiral molecules possess a stereogenic center (as an asymmetric tetrahedral carbon atom, for instance) (see Figure 9.7), but it is not a prerequisite as observed in cases of planar chirality, axial chirality or atropoisomerism.

In the domain of chiral analysis by NMR spectroscopy, the determination of enantiomeric purity of chiral mixtures is one of the main analytical challenges [28, 31]. The enantiopurity of a sample of chiral molecules can be simply evaluated by the enantiomeric excess ($ee(\%)$) defined as [8]:

$$ee(\%) = 100 \times \frac{|A^R - A^S|}{|A^R + A^S|} \tag{9.17}$$

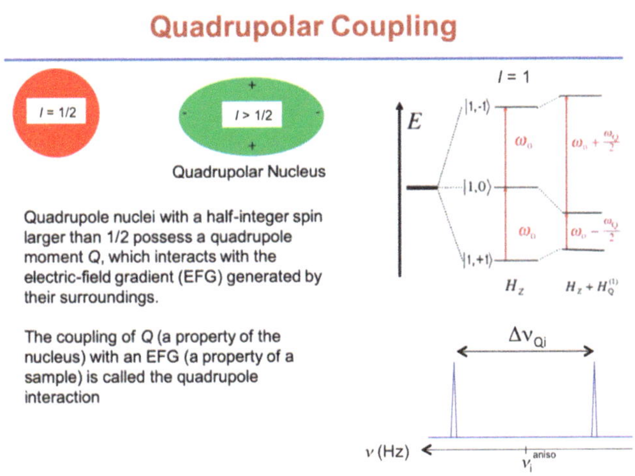


$$\Delta\nu_{Q_i} = \frac{3}{2} K_{\text{C-D}_i} S_{\text{C-D}_i} \tag{9.14}$$

with :

$$K_{\text{C-D}_i} = \frac{e^2 Q_{D_i} q_{\text{C-D}_i}}{h} \tag{9.15}$$

and :

$$S_{\text{C-D}_i} = \frac{1}{2}\left\langle 3\cos^2\theta_{\text{C-D}_i}^{B_0} - 1\right\rangle \tag{9.16}$$

**Figure 9.6** The origin quadrupolar coupling, $\Delta\nu_Q$, associated with a non-spherical distribution of electric charge inside of nuclei, energetic diagramm in case of spin $I = 1$, as deuterium, example of an $^{2}$H-QD, and associated equations to $^{2}$H-RQCs. In Equation 9.12, $K_{\text{C-D}_i}$ is the quadruplar coupling constant (QCC) for nucleus i (the C-D$_i$ bond). In Equation 9.13, $Q_D$ is the nuclear electric quadruplar moment and $q_{\text{C-Di}}$ is the electric field gradient of the C-D$_i$ bond. The anisotropic chemical shift, $\nu_i^{\text{aniso}}(^2\text{H})$, correspond to the contribution of $\nu_i^{\text{iso}}(^2\text{H})$ and $^{2}$H-RCSA terms. In the case of a mixture of monodeuterated enantiomers ($R/S$), two $^{2}$H-QDs are expected to be detected while centered on $\mu^R \approx \nu^{\text{aniso},S}$, if spectral enantiodiscrimination occurs.

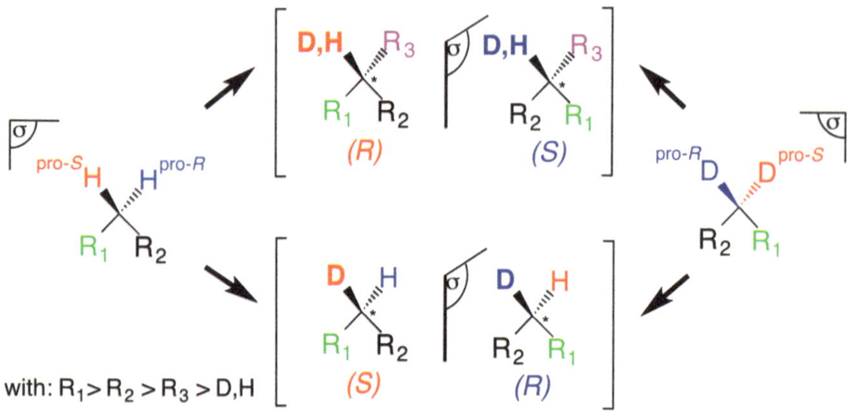


**Figure 9.7** Examples of *R*/*S*-enantiomeric pairs of a model $C_1$-symmetry chiral molecule ("classical" chirality and "[D/H] isotopic" chirality) with a single stereogenic center (tetrahedral carbon atom) deriving from their deuterated or hydrogenated prochiral precursors and characterized by methylene enantiotopic directions, pro-*R*/pro-*S*. Figure adapted from Ref. [32] with permission.

where $A^R$ and $A^S$ are the areas of resonances (or group of resonances) for the *R* and *S* enantiomers. Peak areas can be determined by peak integration or signal deconvolution as featured in any NMR processing software. When *ee* is equal to 0% and 100%, the mixture is called "racemic" and "enantiopure", respectively. Otherwise, it is called "scalemic."

The concept of enantiotopicity and prochirality in chemistry is another important aspect of the molecular enantiomorphism [8]. Unlike chirality, which involves two non-superimposable constitutionally identical molecular 3D objects, prochirality involves intramolecular enantiotopic elements (generally bonded to prostereogenic centers), such as nuclei, groups of nuclei or internuclear directions, which are exchangeable by an improper symmetry operation (see Figure 9.6) [8].

#### 9.3.4.2 NMR in CLCs

Contrarily to ALCs, distinct interactions of enantiomers (diastereomporphous interactions) with the enantiopure molecules of the chiral mesophase results in different average orientations of enantiomers with respect to the $\mathbf{B_0}$ magnetic field of the NMR spectrometer. Each orientation is described with the Saupe's order matrix, noted $S^S_{\alpha\beta}$ and $S^R_{\alpha\beta}$ (see Equation 9.1) [6, 33]. From an order-dependent NMR interaction point of view, each enantiomer presents a specific set of anisotropic observables (RDC$^{R \text{ or } S}$, RDC$^{R \text{ or } S}$, and RQC$^{R \text{ or } S}$) that (auto)consistently characterize each isomer (see Section 9.7). In practice, independent and distinct NMR spectra associated to each enantiomer in the mixture are expected to be observed in CLCs. Illustrative examples will be presented below. Interestingly, the intensity of resonances for each *R*- or *S*-spectrum is proportional to the concentration of each enantiomer in the mixture, and hence using different enantiomeric amounts ($[R] \neq [S]$) facilitate the assignment of spectra.

The peculiar case of spectral discrimination of enantiotopic elements of prochiral molecules is another interesting analytical challenge for anisotropic NMR. Using group theory, it has been demonstrated that for rigid molecules, the enantiotopic discrimination of prochiral solutes in CLCs for four molecular improper point groups, namely $C_s$, $C_{2v}$, $D_{2d}$, and $S_4$, originates from the reduction of their effective molecular symmetry when they interact with a chiral environment [18]. Interestingly, these four symmetries correspond to rigid molecules having enantiotopic faces, groups of atoms, or internuclear directions. This reduction of the effective symmetry increases the number of non-zero independent order parameters of $S_{\alpha\beta}$ as well as the changes in the location of the PAS of the orientational order matrixes, which allows enantiopic element to be discriminated using NMR in

CLCs. Experimentally, and contrarily to ALCs, the difference of orientation of enantiotopic elements (internuclear vectors) of $C_s$, $C_{2v}$, $D_{2d}$, and $S_4$ symmetry molecules in CLCs leads to distinct anisotropic NMR observables ($RDC^{pro-R\ or\ pro-S}$, $RDC^{pro\text{-}R\ or\ pro\text{-}S}$, and $RQC^{pro\text{-}R\ or\ pro\text{-}S}$). Hence, originating different spectral patterns for each enantiopic element but not for the different homotopic elements (exchangeable by a $C_n$-symmetry axis as in case of methyl groups) of the structure. Here again, some key examples will be presented below.

## 9.4 Adapted 2D NMR Tools

As in NMR of liquids (isotropic solvents), the analysis of complex anisotropic spectra can be simplified with the help of classical homo- or heteronuclear 2D NMR experiments (with or without specific adaptation) or specially designed anisotropic experiments. Numerous 2D experiments have been developed over the last two decades. In brief, we can mention the methodological developments around $^1$H and $^{13}$C NMR, such as the family of CLIP-CLAP HSQC or G-SERF 2D sequences) as well as for $^2$H NMR (the family of QUOSY 2D sequences), both recorded on poly- or perdeuterated molecules, or at natural abundance level (see examples in the following sections).

### 9.4.1 Spin-1/2 Based 2D Experiments

($^1$H-$^1$H)-RDCs and ($^1$H-$^{13}$C)-RDCs are two important sources of anisotropic information used in various applications such as structure determination but their measurement from complex 1D spectra are not always simple, and request 2D NMR sequences able to separate chemical shifts and homo- or heteronuclear couplings on both spectral dimensions. "$J$-resolved"-type 2D schemes or more sophisticated schemes as the $^1$H $J$-HSQC-BIRD correlation sequence [34] or CLIP-CLAP HSQC 2D sequence [35–37] for which only the (one bond) coupling interaction evolves during the $t_1$ dimension are interesting tools. Note that further spectral simplifications using a selective excitation of nuclear site of interest (family of homo- or heteronuclear G-SERF experiments) are obtained see Figure 9.6 [10, 38, 39]. A schematic description of these nD experiments and some other useful ones are depicted in Figure 9.8 [40].

An exhaustive description of all these sequences is out of the scope of this book chapter, but clearly one of the most popular and established heteronuclear 2D experiments for the measurement of one-bond scalar coupling ($^1J_{CH}$) or the total spin-spin coupling ($^1T_{CH} = {}^1J_{CH} + 2 \times {}^1D_{CH}$), are the $^{13}$C-$^1$H CLIP-HSQC 2D experiment [35], and some more sophisticated variations [37, 41].

#### 9.4.1.1 Extraction of Isotropic and Anisotropic Spin-1/2 Spectral Data

To determine ($^1$H-$^1$H)-RDCs, ($^1$H-$^{13}$C)-RDCs, or RDCs of any pair of homo- or heteronuclear magnetically inequivalent nuclei (X-X or X-Y), it is necessary to determine first the associated scalar couplings ($^nJ$($^1$H-$^1$H), $^nJ$($^1$H-$^{13}$C), $^nJ$(X-X), and $^nJ$(X-Y)). Under this condition, the spectral data of the analyte must be extracted using isotropic and anisotropic environments. From a practical point of view, this is generally achieved by preparing two disctint NMR samples, the first one using a liquid solvent, the second one using an anisotropic medium, both at the same sample temperature.

An interesting alternative to this two-step approach consists in measuring both types of data ($J$ and $D$) in a single sample. The first one relies on the so-called variable-angle spining sample (VASS) NMR technique that orients the axis of alignment of a LC (the director, $\vec{n}$) at varying angles to the magnetic field by sample rotation, including the magic angle ($\theta_m = 54.7°$) (see also Figure 9.45). For this specific angle, all anisotropic observables are averaged to zero and disappeared and scalar coupling can be measured (see Figure 9.2a) [42]. Another approach using a compressed aligning gel (see above) is also possible, as reported in 2016, with DMSO-compatible using cross-linked poly(2-hydroxylethyl methacrylate) (poly-HEMA) [43]. In this case, the significant difference in bulk magnetic susceptibility between the DMSO inside and outside the gel allows the simultaneous extraction of isotropic and

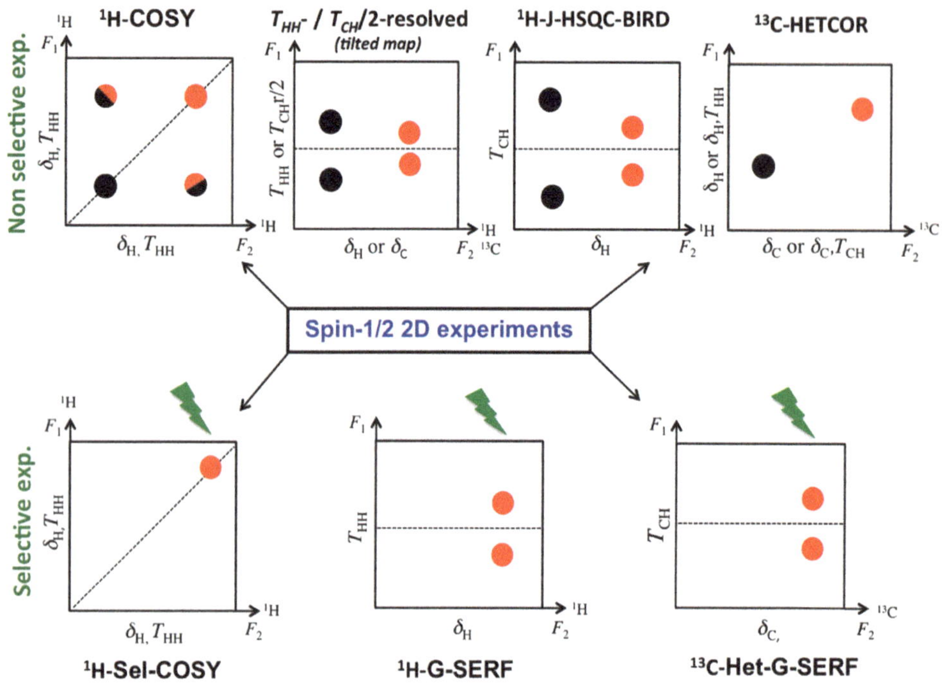


Figure 9.8 Schematic description of the main homo- and heteronuclear ($^1H$-$^1H$ or $^{13}C$-$^1H$ pairs) experiments used for simplifying the analysis of anisotropic spin-1/2 spectra ($^1H$/$^{13}C$) without (top) and with selective excitation (bottom). Figure adapted from Ref. [40] with permission.

anisotropic NMR data from the same spectra. Extraction of data is obtained with $F_1$ coupled HSQC 2D experiment. The concept and example of 2D map are shown in Figure 9.9.

$^1H$ *J*-HSQC-BIRD 2D experiment is a highly functional multipurpose HSQC 2D sequence with evolution of $^1J_{CH's}$ and $\delta(^1H)$ in $F_1$ and $F_2$ dimensions, respectively [44]. Using simple switches, the user can activate different blocks of the pulse program shown in Figure 9.10, including the option of having homonuclear decoupling (pure-shift) or not during the acquisition period. Block A controls the $^1J_{CH}$ evolution in $F_1$ using a scaling factor. A factor of zero corresponds to no $^1J_{CH}$ evolution ($^1H$ decoupling in $F_1$) and a regular HSQC experiment is collected. A value of 1 provides splitting in $F_1$ corresponding to the $^1J_{CH}$ value, and values other than one allows the user to collect experiments with *J*-scaling as it is normally done in the JBS-HSQC 2D experiment. Block C controls the chemical shift evolution in $F_1$ dimension via a chemical-shift scaling factor. A factor of zero produces a *J*-resolved experiment with no chemical shift information in $F_1$, just the $^1J_{CH}$ splitting in $F_1$ at the corresponding $^1H$ signal in $F_2$.

Figure 9.10b shows the *J*-resolved HSQC 2D spectrum of artemisinin [45]. As it can be seen, as long as there is good $^1H$ signals separation in $F_2$ dimension, $^1J_{CH}$ in isotropic conditions and $^1T_{CH}$ in anisotropic conditions permit the accurate extraction of $^1D_{CH}$ values. Since there is no chemical shift evolution in $F_2$, the experiment is much faster since there is no need for many increments in $F_1$ to obtain an excellent degree of digital resolution in $F_1$. Block B permits the measurement of long-range proton-carbon RDCs ($^{2,3}D_{CH}$) using the selective *J*-scaled (SJS) HSQC 2D experiments [46].

Finally, a third approach using stable biphasic liquid-crystalline phases (combination of anisotropic and nearly isotropic domains) in a single sample has been also proposed in 2017 with the using a lyotropic system based on a helically chiral polyisocyanide polymer [47]. In this last approach, spatially selectively excited $^{13}C$-$^1H$ CLIP-HSQC 2D experiments (along the z axis) are applied to the isotropic part and the anisotropic part of the sample, leading

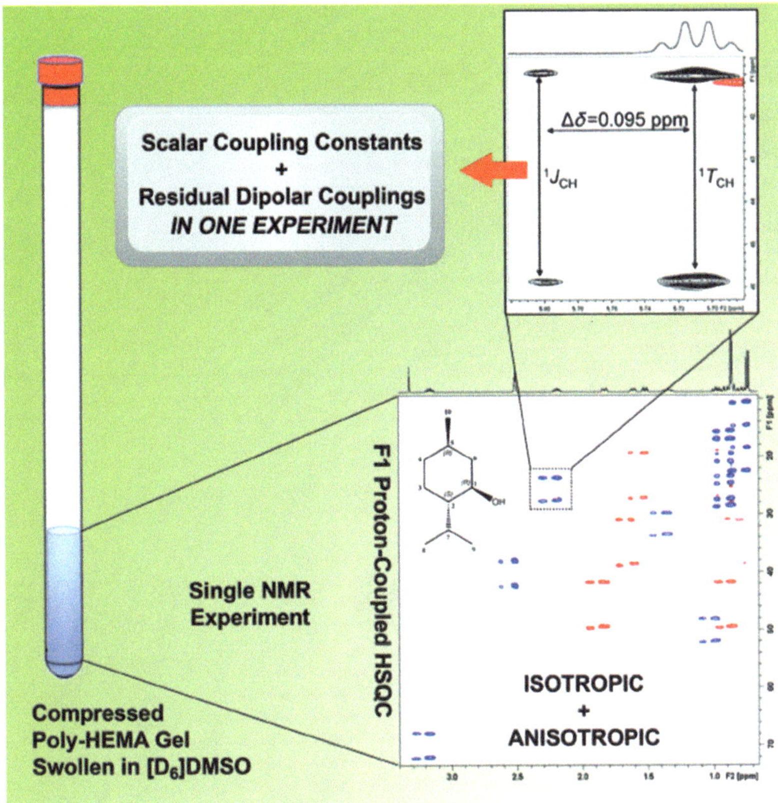


**Figure 9.9** Principle of the simultaneous detection of isotropic and anisotropic data in a single 2D experiment. Figure adapted from Ref. [43] with permission.

to 2D maps where ${}^{1}J_{CH}$ and ${}^{1}T_{CH}$ can be measured respectively. These specific selective experiments relies on the combined application of a pulsed field gradient along the z axis, and a shaped pulse for a spatial selection of a certain volume of the sample at a predefined position of the sample (the isotropic and anisotropic part).

### 9.4.2 Spin-1 Based 2D Experiments

${}^{2}$H-RQC is another powerful source of anisotropic information very useful in various applications involving enantiotopic or enantiomeric discriminations (enantiopurity determination or isotopic profile) as well as the structure determination as recently reported [10, 32, 40, 48]. The interest of ${}^{2}$H-RQCs lies in the magnitude of the observed ${}^{2}$H-quadrupolar splittings compared to (${}^{1}$H-${}^{1}$H)-RDCs or (${}^{13}$C-${}^{1}$H)-RDCs, for instance, and in the great sensitivity to small differences of molecular orientations (see Equation 9.14) [10, 40, 49]. This anisotropic interaction reveals to be extremely efficient for spectral enantiodisrimination when using monodeuterated analytes because we obtain simple deuterium spectra with an optimal sensitivity. However, this approach needs the chemical transformation of the molecule, which can be time consuming or difficult, and only one site can be spectrally observed. These two drawbacks vanished when detecting all deuterium sites by recording the ${}^{2}$H spectra at natural abundance level (or NAD NMR). Although of low sensitivity, NAD 1D/2D NMR is generally possible, even using routine NMR spectrometers operating at low field (5.87 T) [50]. The use of high-field NMR in combination with ${}^{2}$H cryogenic probes significantly its sensititivy.

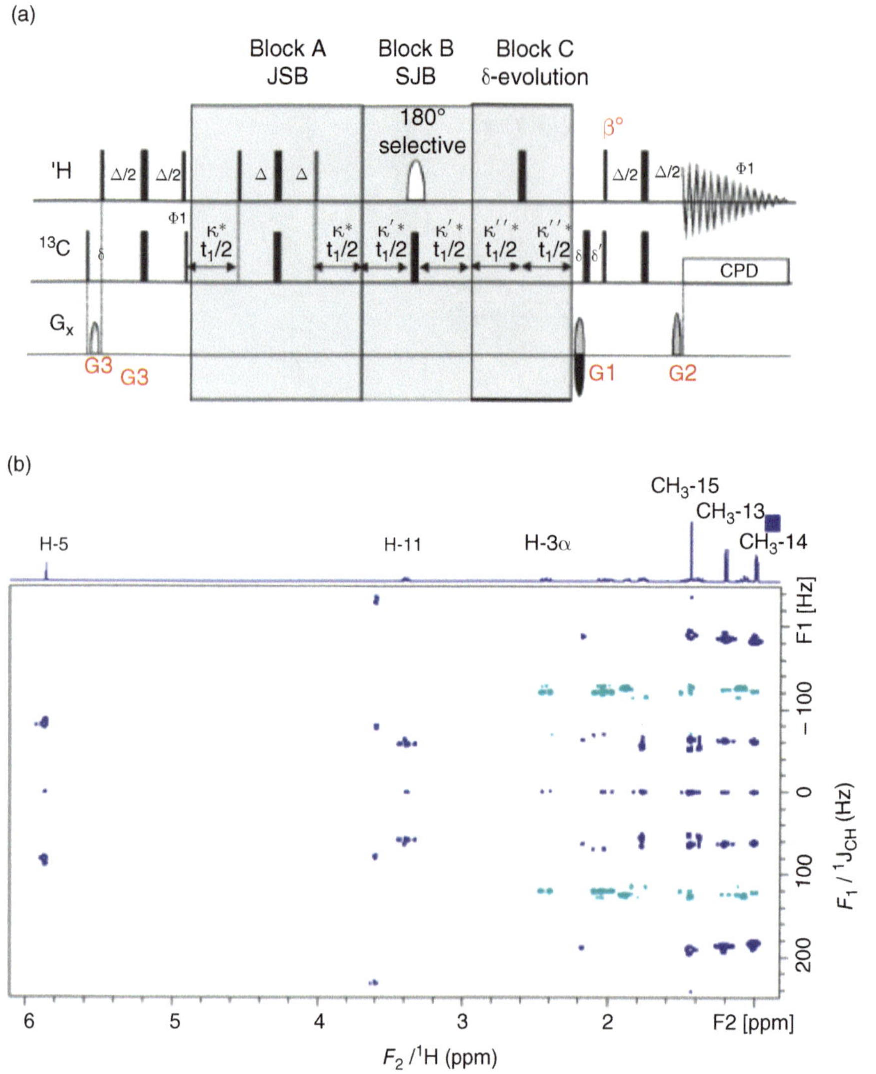


**Figure 9.10** (a) Pulse diagram of the multipurpose HSQC 2D. G1, G2, G3 are z-axis pulse gradients. The delays Δ are adjusted according to the magnitude average values of expected couplings (*J* or *T*). (b) Example of anisotropic *J*-resolved HSQC 2D spectrum of Artemisinin aligned in a PMMA/$CDCl_3$ gel using the compression device. Figure reproduced from Ref. [45] with permission.

As we simultaneously detect all monodeuterated species of the mixture molecular isotopomers, various $^2$H autocorrelation experiments, called "QUOSY" for QUadrupolar Order SpectroscopY, have been developed to simplify the analysis of complex NAD 1D spectra [50, 51]. Obviously, all these 2D experiments designed for recording anisotropic $^2$H spectra at natural abundance can be applied to study poly- or perdeuterated solutes as well. Among this class of specifically designed $^2$H 2D experiments, the most interesting one is the tilted *Q*-resolved 2D experiment (phased or not) that allows the full separation of $^2$H anisotropic chemical shifts, $\delta^{aniso}(^2H)$, and the RQCs, $\Delta\nu_Q(^2H)$, in two spectral dimensions of the map ($F_2$ and $F_1$). Homonuclear 3D experiments (as 3D *Q*-DQ) [52, 53] were also designed as well as heteronuclear versions of QUOSYs experiments. The latter correlates (one-bond scalarly or dipolarly coupled) $^{13}$C and $^2$H nuclei either in deuterated molecules ("CDCOM" 2D experiments: carbon-deuterium correlation in oriented media) [54, 55] or even at natural abundance level ("NASDAC"

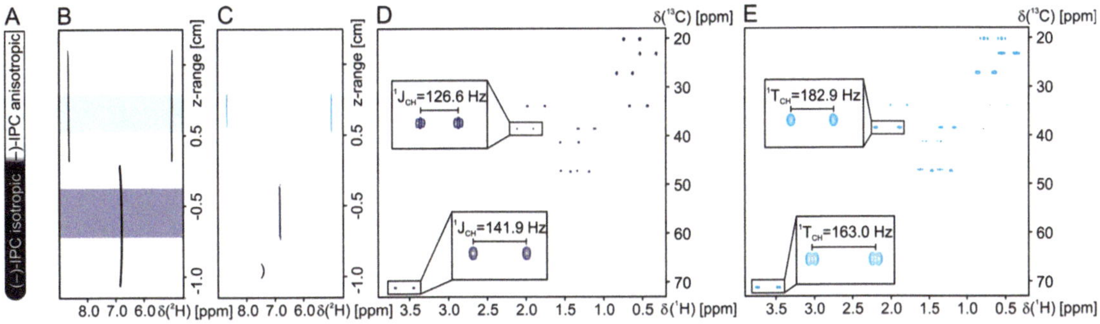

Figure 9.11 (a) Principle of biphasic LLC NMR samples where the anisotropic part (upper part) and quasi-isotropic (lower part) are clearly identified. (b and c) Phase encoded deuterium 2D spectrum selectively excited regions in a biphasic LLC. (d and e) Same as (b and c) but with the spatially selectively excited $^{13}C$-$^{1}H$ CLIP-HSQC 2D spectra applied to the quasi-isotropic part and the anisotropic part, respectively. Figure extracted from Ref. [47] with permission.

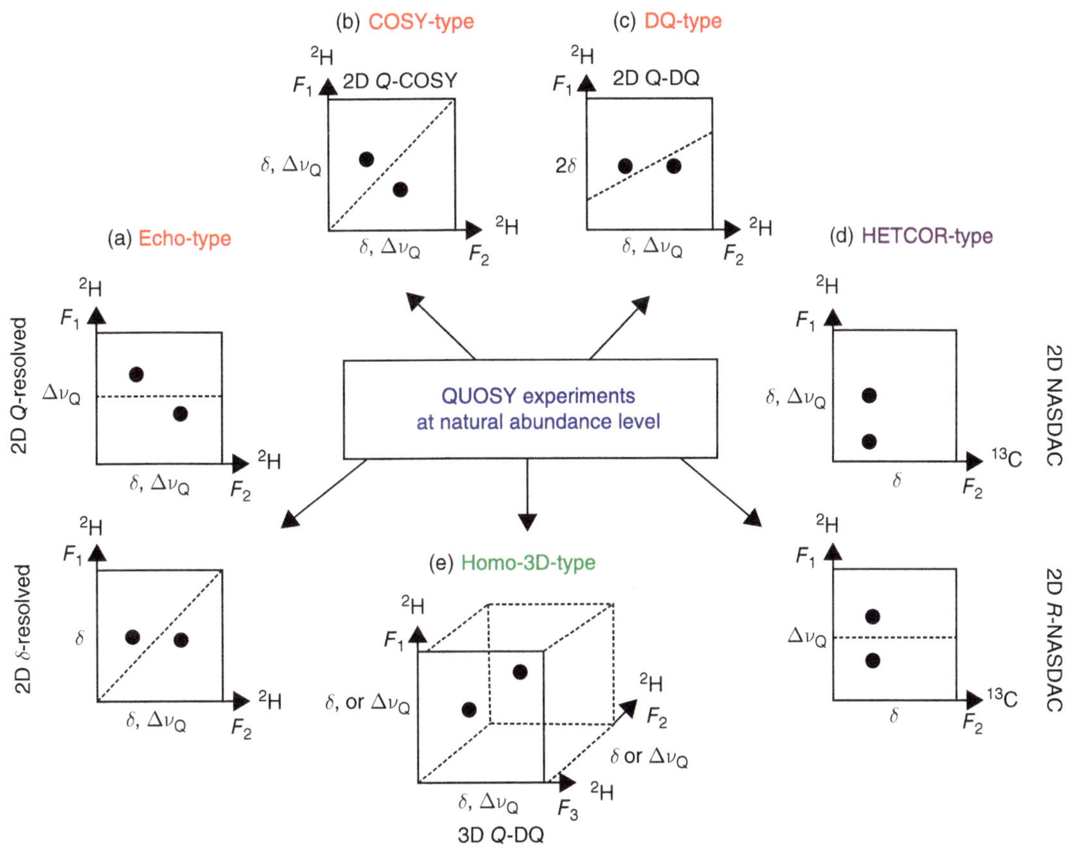


Figure 9.12 Schematic description of homo- and heteronuclear $^{2}H$ QUOSY 2D/3D experiments: (a) Echo-type scheme (*Q*-resolved and $\delta$-resolved sequences), (b) COSY-type scheme (*Q*-COSY), (c) Double quantum-type scheme (*Q*-DQ), (d) HETCOR-type 2D schemes (CDCOM, NASDAC and 2D Refocused-NASDAC sequences) and (e) homonuclear 3D-type scheme (3D *Q*-DQ sequence). Figure adapted from Ref. [52] with permission.

experiment: natural abundance spectroscopy deuterium and carbon), allowing the detection of $^{2}H$ and $^{13}C$ isotopologues without any isotropic enrichement [56]. A schematic description of these nD experiments and other also developed are depicted in Figure 9.12.

## 9.5 Examples of Polymeric Liquid Crystals

Anisotropic NMR using LCs as solvents has a long history since the first anisotropic NMR spectra recorded in achiral and chiral oriented systems have been described by Saupe and Englert in 1963 (nematic thermotropic) and Sackmann, Meiboom, and Snyder in 1968 (cholesteric thermotropic), respectively; namely in the early stage of NMR developments [57, 58].

From mid-1990, a paradigm shift occurred with the first uses of lyotropic (organo-soluble or water-compatible) weakly orienting (chiral or not) solvents [12, 59, 60]. Contrarily to thermotropics, these lyotropic phases are made up of molecular components that do not possess intrinsic mesogenic properties, but when mixed with suitable (organic) solvents under appropriate conditions (concentration, temperature, and pressure) lead to a uniform/homogeneous oriented medium in the magnetic field of the spectrometer. Interestingly, such aligning systems lead to weakly oriented solutes ($10^{-3}$ to $10^{-5}$) where the (high-resolution) 1D/2D NMR spectra ressembles, generally, ones recorded in isotropic solvents.

Among organo-compatible, weakly aligning enantiodiscriminating mesophases, a special attention has been paid to rod-like systems made $\alpha$-helicoidally chiral polymers (see Figure 9.1b) and stretched or compressed oriented gels (see Figure 9.1c).

### 9.5.1 Polypeptide or Polyacetylene-based Systems

Many LLCs are generally made by a helically chiral homopolymer, where $\text{Length}^{\text{poly}} \gg \text{Diameter}^{\text{poly}}$. The stereogenic center can be present in the central backbone (as polypeptidic polymers with achiral side chains), on the flexible side chain (as polyacetylenic or poly(arylisocyanide) polymers), or simultaneously on both these structural elements (polypeptides bearing a chiral side chain) (see Figure 9.9a).

Among the most investigated homopolypeptide polymers (used as efficient enantiodiscriminating CLCs) and applied in many differents applications (in particular at Orsay), we can mentionned the case of poly-$\gamma$-benzyl-*L*-glutamate (PBLG) or poly-$\gamma$-carbobenzyloxy-*L*-lysine (PCBLL) for each the absolute configuration (AC) is of *L*-type (see Figure 9.13a) [61]. Evidently, the enantiomers of these polymers (PBDG, PCBDL, ...) are also enantiodiscriminating systems but with inverted results in terms or enantio-orientation. Interestingly, mixtures prepared by mixing equal amounts by weight of polypeptides (close DP) of same nature but with opposite AC (e.g. PBLG and its enantiomer PBDG) lead to achiral liquid crystals (ALCs) where enantiodiscrimination phenomenon vanishes. In these achiral mesophases, noted « PBG », both enantiomers exchange rapidly, between the (*L*)- and (*D*)-helical vicinities of each polypeptide on the NMR time scale. These results in identical average magnetic interactions for enantiomers, and no differences in their NMR spectra are therefore expected [62]. Another source of new chiral LLCs consists simply in mixing two chiral polymers of the same family (with the same AC but with side chains chemically different, as polypeptides PBLG et PCBLL) [63]. Depending on the peptide unit ratio of each polymer in the mixture, a control of the orientation and enantiodiscrimination is possible. As previously, a mixture of two polymers with their respective enantiomers leads to an achiral oriented media [64]. Over the last decade, innovative and original enantiodiscriminating polypeptide-based polymers have been reported such as helical chiral systems made of (see Figure 9.13b): (i) poly-$\beta$-phenethyl-*L*-aspartate (PPLA) [65], (ii) poly-$\beta$-benzyl-*L*-aspartate (PBLA) [66], (iii) poly-$\gamma$-*p*-biphenylmethyl-*L*-glutamate (PBPMLG) [67], and (iv) poly-$\gamma$-(*S*)-2-methylbutyl-*L*-glutamate (PSMBLG) [68]. The idea related to the use of PSMBLG is to reinforce the enantiorecognition efficiency

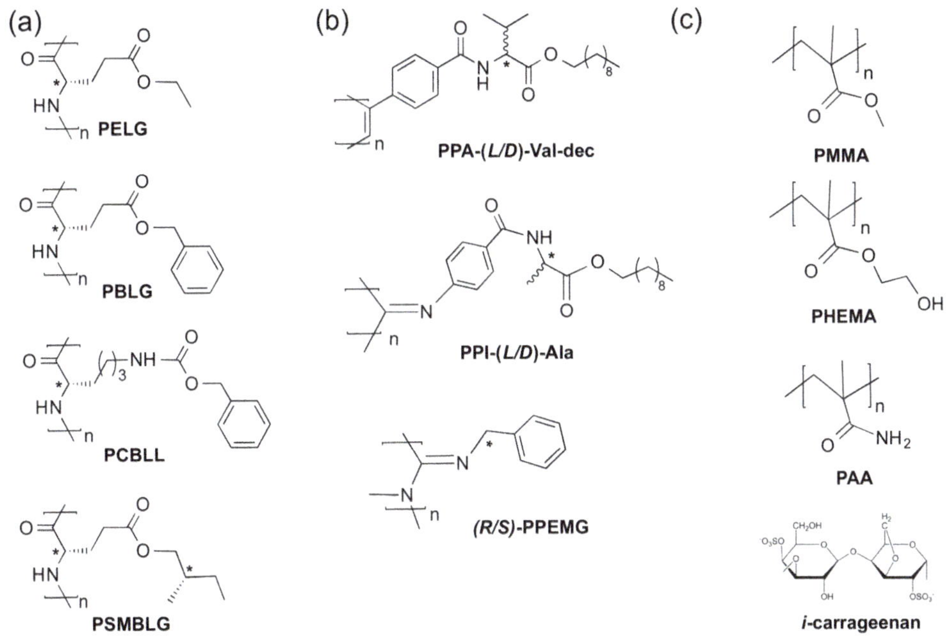


Figure 9.13 (a) Examples of structures of polypeptide-based helical polymers with a L-configuration for the repeating unit (PELG, PBLG, PCBLL, PSMBLG). (b) Other examples of non-polypeptide helical polymers (PPA-(*L*/*D*)-Val, PPI-(*L*/*D*)-Ala, (*R*/*S*)-PPEMG. In these polymers, the stereogenic center is located on the side chain. (b) Examples of polymers involved compressed/stretched gels (PMMA, PHEMA, PPA, i-carrageenan). These latter are compatible with chlorinated solvents but also DMSO, which is not the case with polypeptide-based phases.

and increase spectral enantio-separations by adding a second asymmetric center in the side chain. Finally, we can also mention the design of a polypeptide co-polymer (a polymer bearing two types of sidechain) made phenethyl and benzylpolyaspartate [65, 69]. Note that this list is not exhaustive but gives a brief overview of the diversity of chiral helical polymers.

### 9.5.2 Compressed and Stretched Gels

During the development of anisotropic media for proteins, methods to generate alignment by compressing a polyacrylamide (PA) gel were simultaneously proposed in 2000 [70, 71]. The method was referred to as strain-induced alignment (SAG). These approaches were inspired by an early work by Deloche and Samulski [72], which correlates the $^2$H RQC versus elasticity by stretching elastomers swollen in deuterated solvents in a solid-state NMR instrument. The only limitation of PA gels is that they are not elastic (no reversible compression) and they are only compatible with water. However, in 2013, the first report on the application of RDCs to the structure analysis of a small organic molecule in a PA gel was described using sodium cholate (see Figure 9.14a) as analyte [73].

In 2004, a very ingenious approach in which a cylindrical polystyrene gel stick (~3 mm of diameter and ~10 mm in length) was inserted in a regular 5-mm o.d. NMR tube was reported [74]. Then, $CDCl_3$ was added and the gel was allowed to swell. Once the gel touched the wall of the NMR tube, it continued swelling axially and over time it self-stretches inside the NMR tube generating anisotropy. Over a period of several days, the $^2$H RQC reached a maximum value of 25 Hz that did not change further. Once the maximum degree of alignment is

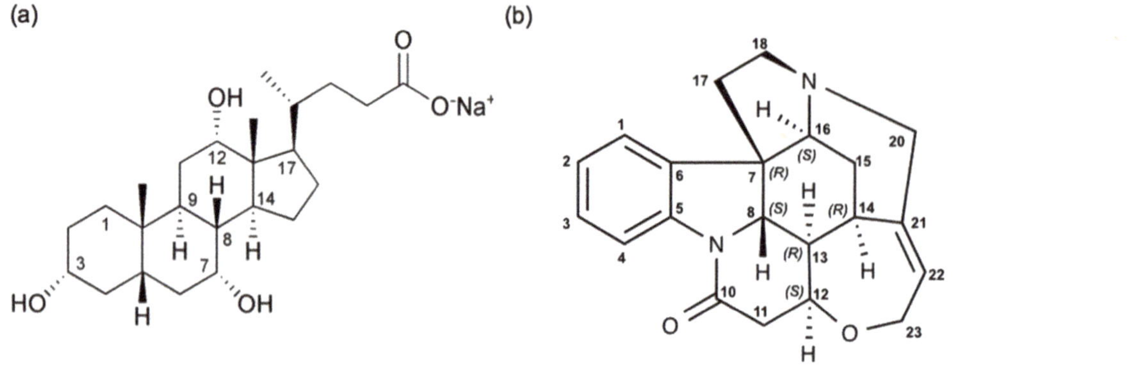


**Figure 9.14** (a) Structure of sodium cholate. (b) Structrure of strychnine, a (toxic) natural compounds abundantly used as model molecules in the framework of the structure/configuration analysis.

reached, it cannot be varied, but it was reported that the degree of alignment can be tuned by changing the cross-link density of the polymer gel. As a proof of concept, they aligned strychnine (see Figure 9.14b) and collected RDCs. This foundational piece of work opened many further developments in the field of NMR in anisotropic media. In a follow-up article, more detailed study on the aligning properties of the polystyrene gels was presented [16], continued with the development of other polymer gels compatible with different organic solvents, such as polydimethylsiloxane (PDMS) [75], poly(vinyl acetate) (PVAc) [76], polyacrylonitrile (PAN) [77], gelatin from commercial "gummy bears" with enantiodiscrimination properties [78], noncovalently and covalently cross-linked polyurethane gels [79], and poly(ethylene oxide) (PEO) [80]. Other interesting systems such as co-polymeric polyacrylamide gel, the first alignment medium compatible with DMSO have been proposed in 2005 [81], or chemically cross-liked poly(methyl methacrylate) (PMMA) compatible with $CDCl_3$ in 2008 [17] as a self-stretched gel as well as poly(2-hydroxyethyl methacrylate) (HEMA) gel compatible with DMS [43], poly[di(ethylene glycol) methyl ether methacrylate] (poly(DEGMEMA)) gel compatible with methanol [82], are all designed to work with the compression device, which was first developed for PMMA gels (see Figure 9.15a) [13].

As far as we know, these are the aligning gels developed so far since 2004. The PEO gels are compatible with solvents of any polarity, including water. In 2021, was developed cross-linked poly-(4-acrylomorpholine) (p-AM) blended with PMMA to add mechanical robustness [83]. This new gel is fully compatible with water and was designed to be used with the compression devices. Its analytical value was illustrated in the case of strychnine hydrochloride and a cyclic peptide.

The self-stretching method was not practical at all, and its limitation led to the design of stretching and compression devices. The first device reported in 2006 to generate anisotropy in stretched gelatin uses a silicon-rubber tube inside an open-ended NMR tube [14, 84]. The silicon rubber is compatible with polar solvents like water or DMSO but not with apolar organic solvents like chloroform; this device was then modified and replaced the silicon rubber by a Kalrez 8002 perfluoroelastomer [85]. Figure 9.15b shows a picture of the stretching apparatus. The only shortcoming of this device is the reluctancy of many users to insert an open-ended tube into the NMR probe for fear of accidental leaking. The degree of alignment can be tuned at the user's whim. Figure 9.16 shows an example of the CLIP-CLAP HSQC 2D experiment (case of the quinine in PDMS) with ${}^{1}T_{CH}$ values measured in $F_2$ at different degrees of stretching (see insets) [15].

In 2010, a fast and tunable alignment by reversible compression/relaxation of PMMA gels was reported using a compression device [13], which was later extended to other polymer gels, such as poly-HEMA [43], poly-EDGMEMA [82], and poly-(4-acrylomorpholine) [83].

The original (home-made) compression device consisted in swelling a cylindrical gel rod of ~2 mm in diameter and 25 – 30 mm in length inside a regular 5-mm o.d. NMR tube. The gel swells up to a diameter of ~4 mm and a

(a)

(b)

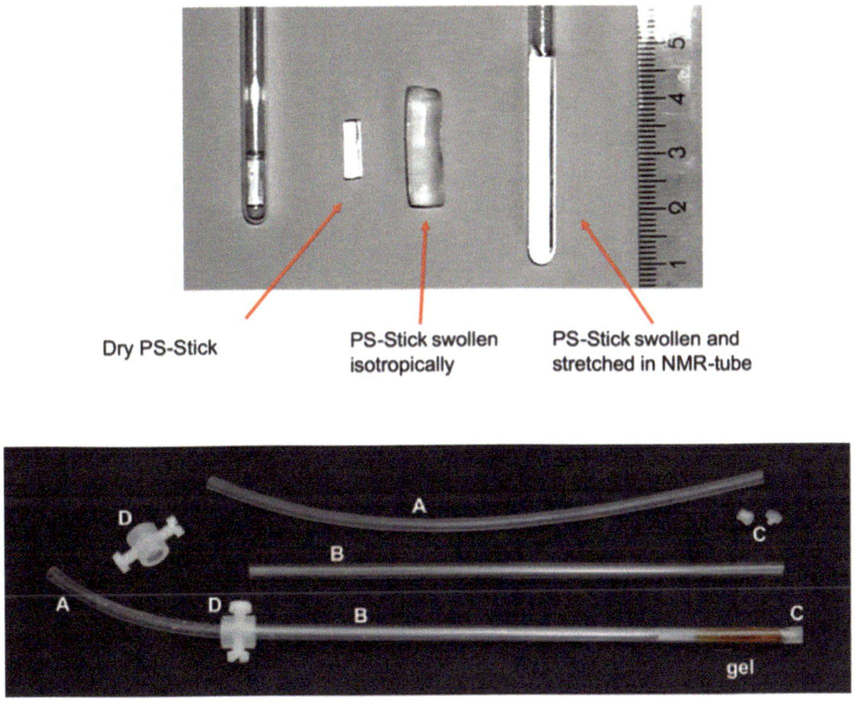


Figure 9.15 (a) Pictures of swollen gels (PS-stick) with and without stretching (b) Exemple of stretching apparatus for 5-mm o.d. standard high-resolution NMR probe heads. The flexible Kalrez 8002UP tube (A) is placed inside a cut-open 5-mm o.d. NMR tube (B) and fixed with a specially designed PCTFE screw (C) at the bottom. A PCTFE device with nylon screws (D) is used to fix the stretched tube at the top. Inside the assembled apparatus, a reddish-brown PAN/DMSO gel is ready to be stretched. Figure reproduced from reference [15, 74] with permission.

length of ~40 mm, but it does not touch the walls of the tube. Then, a Shigemi plunger is inserted, and the gel is compressed to the desired degree of anisotropy (see Figure 9.17).

The compression device evolved into a more sophisticated design to precisely lock the position of the plunger (see Figure 9.18a), and it is commercially available [86]. The Teflon piston, which is threaded, can be easily adjusted to any degree of compression with high accuracy. It is important to highlight that when the gel is fully relaxed, there is space between the gel and the wall of the NMR tube. That space is filled with solvent. The $^{2}$H 1D NMR spectrum of the relaxed gel shows signal of the solvent peak from outside and inside the gel. Both signals are from the solvent in isotropic conditions. The gel buckles upon compression, and the signal from outside gets reduced but never disappears. The compression generates a $^{2}$H-RDC, indicating that anisotropy was created inside the gel.

### 9.5.3 Polynucleotide-based Chiral Oriented Media

Helical polymers can be constituted by the repetition of a single monomeric unit (homopolymer) or of two different units (co-polymer) but also by the linking of various units (without repetition) as in the case of polynucleotides. The use of polynucleotide chiral polymers made of short chiral LLC made of short DNA-fragments was reported and showed an interesting enantiodicrimination potential when combined with $^{1}$H NMR [87] or $^{2}$H-{$^{1}$H} NMR of labeled molecules [88]. DNA strands are known to provide cholesteric oriented systems [89, 90]. From a practical point of view, if those systems might provide convenient oriented media for NMR studies, they are generally not

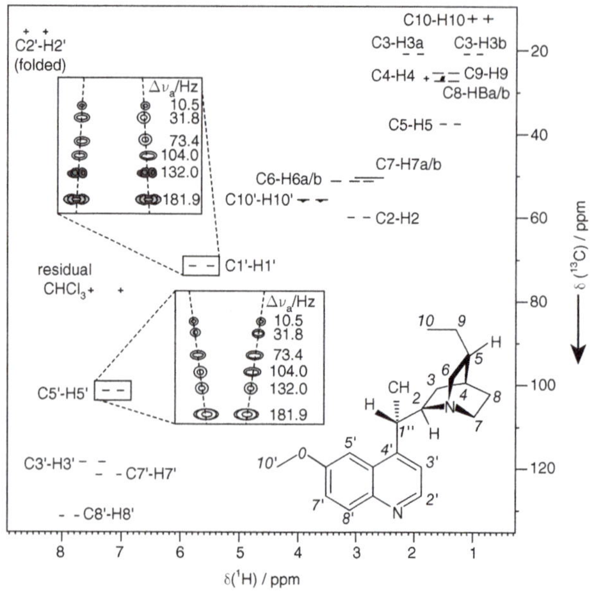


**Figure 9.16** Arbitrary scaling of RDCs in gels with halogenated solvents using a Kalrez 8002UP tubing and observed on CLIP-HSQC 2D spectrum of hydroquinidine in a PDMS/$CDCl_3$ gel. Insets show two signals (C1"–H1"and C5'–H5') at various stages of stretching with peaks separated in the vertical direction according to the corresponding quadrupolar splitting $\Delta\nu_Q$ of the solvent $CDCl_3$. The linear dependence between observed $^1T_{CH}$ couplings and quadrupolar splittings is evident. Figure extracted from Ref. [15] with permission.

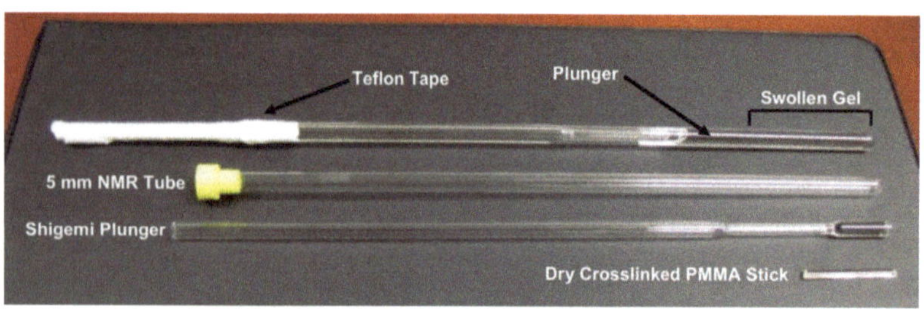


**Figure 9.17** First compression device. The plunger (here a Shigemi plunger) is hold tight inside the tube using Teflon tape. Figure reproduced from Ref. [13] with permission.

commercially available (fragmentation of DNA [200 – 300 base pairs] by sonication), and their preparation (such as control of pH, ionic balance, and sample homogeneity) is sometime tricky. An example of exciting applications of $^2$H-{$^1$H} NMR in DNA/water-based LLCs phase will be presented in Section 9.6.3 [91].

Without being exhaustive, it is is relevent to mention that the description of experimental spectral enantiodisciminations of small chiral molecules in water-based chiral LLCs were first discovered and reported by Tracey and Diehl in 1975 [92]. Despite this, the number of very efficient systems (easy to prepare) proposed so far remains rather limited. For instance, we can mention: (i) the potassium *N*-dodecanoyl-L-alaninate based system [93], (ii) the glucopon/hexanol/buffered water [94], and (iii) alanine-derivated system [95]. However, these systems did not meet a large success for routine applications, mainly due to their difficulties of preparation (their avalaibility/phase homogeneity) and their efficiency.

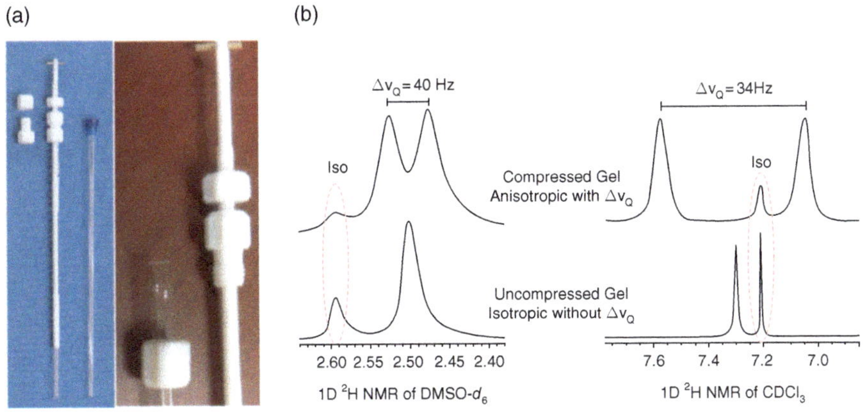


**Figure 9.18** (a) More recent version of the compression device. The plunger is made out of Teflon and it is locked into a special tube to maintain the anisotropy at a constant degree of alignment. (b) Example of $^2$H 1D NMR spectrum of DMSO-$d_6$ in a poly-HEMA gel (left) and of $CDCl_3$ in a PMMA gel (right). For each gel, the bottom trace corresponds to the fully relaxed gel and the top trace to the gels fully compressed. When the gel is relaxed, two signals are observed, one from inside and other from outside the gel. Both are in isotropic condition. The gel buckles upon compression and not matter how much is compressed, there are always pocket of solvent outside the gel, but the signal from inside the gel show the $^2$H-RQC, indicating that anisotropy was created upon compression.

### 9.5.4 Some Practical Aspects of Polymer-based LLCs Preparation

Classical oriented lyotropic solutions are prepared by mixing appropriate amounts of analyte and mesogen molecules for thermotropic systems and analyte and polymer for lyotropic phases. In all cases, the molar ratio analyte/mesogen or analyte/polymer/co-solvent must be determined to avoid biphasic systems (a mixture of oriented and non-oriented molecules), or destroy the liquid crystallinity of the sample. These ratios are very different with respect to the type of LCs and molecules involved.

Importantly, the oriented lyotropic solutions must be orientationally homogeneous (no concentration gradient) and the molecules oriented uniformly along the magnetic field $\mathbf{B_0}$ in order to record high-resolution NMR spectra. Typical linewidths in a range of 1–5 Hz can be obtained depending upon the choice of the type of LC (degree of order), the experimental conditions (concentration viscosity and temperature), and the quality of the sample preparation. For themotropics, homogenization can be reached by repeated cycles of heating the mixtures to the isotropic phase, physical shaking, and then cooling back to the nematic phase inside the magnetic field. For lyotropics, as polymeric aligned systems, a series of centrifugations of the sample at rather low speed (~500 rpm) during a dozen of seconds is useful (to accelerate the dissolution process or avoid matter gradients). Between each series, the 5-mm o.d. NMR tube in the centrifuge (up/down) is inverted. The thermic equilibration of the sample is then performed inside the magnet during a few minutes (see Figure 9.19a). Depending on the nature of the LCs, NMR samples can be spun at moderate rates (10 – 20 Hz).

At the last stage, the use of a simple polarizer (two polarized slabs) allows to controlling if the system is biphasic or not (dark zone) and uniformly oriented in the tube. Analysis of $^2$H signal(s) (lineshape and quadrupolar splitting of QD[s]) of deuterated co-solvents (organic or not) present in the oriented sample allows another simple control of the quality of aligning phase, and to verify if the system is not biphasic (presence of signal $\delta^{iso}(^2H)$). A fine approach based on the use of gradient NMR experiments to obtain a $^2$H "image" of the sample can also provide such information [96]. Depending on the nature of the LCs, NMR samples can be spun at moderate rates (10 – 20 Hz).

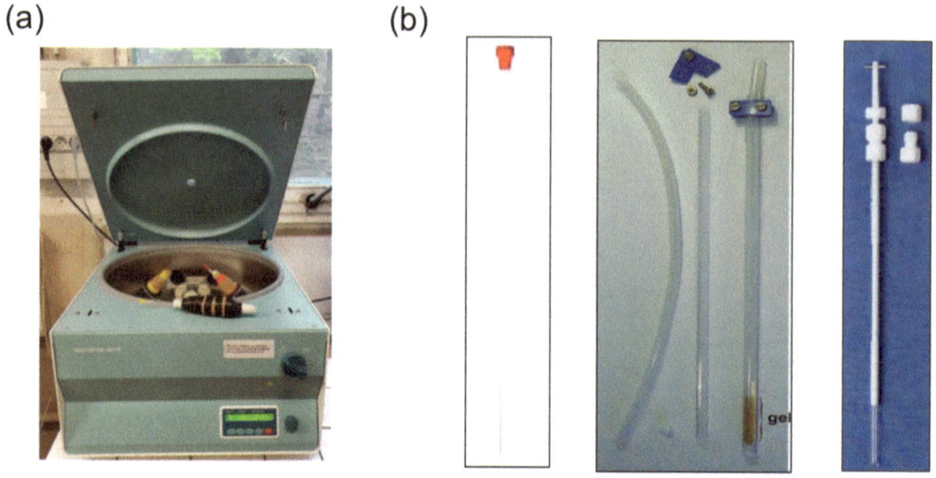


Figure 9.19 Example of modern (commercially available) centrifuge (a) used for preparing classical 5.mm o.d. NMR tubes (b) containing polymer-based LLCs (as PBLG, PELG, or PLA), for instance. The tube support (home-made) has been adapted to protect the NMR tube during its rotation (~500 rpm) in the centrifuge (Picture from P. Lesot).

Finally, whatever their nature, the samples can be then shimmed to reduce the homogeneities of the external magnetic field. Automatic shimming procedures (on a selective resonance of the $^{1}$H or $^{2}$H spectra) can even be applied.

## 9.6 Contribution to the Analysis of Chiral and Prochiral Molecules

Strictly imposed by the U.S. Food and Drug Administration (FDA) [97] or European control rules (registration, evaluation, and authorization of chemicals [REACH] [98]) this control is crucial for pharmaceutical laboratories, whose number one priority is to develop enantiopure bioactive drugs since two enantiomers can induce very different pharmacological effects/responses. Hence, enantiomers can be used to efficiently threat different diseases, as in the case of *Ritalin*. In other cases, the less active enantiomer can generate undesired side effects but remains harmless to human health. On the other hand, the simultaneous presence of the two forms in the drug composition can be highly dramatic as in the case of *Thalidomide* mentioned above. Indeed, while one of the enantiomers has a sedative effect and prevents nausea, the other has teratogenic properties leading to irreversible fetal malformations.

From the NMR spectroscopic point of view, enantiomers have identical magnetic properties. Therefore, this approach, alone, is unable to separate their signals when dissolved in a non-chiral media. In contrast, the magnetic properties of diastereoisomers are different in any environment, whether chiral or not. To discern enantiomers by NMR, it is therefore necessary to make them react (diastereoisomers) or interact (diastereomeric adducts) with an enantiomerically pure chiral entity or environment. In this field, NMR in chiral oriented environment provides new and original solutions, thus giving a suitable alternative to classical methods either chromatographic (GC, HPLC) or spectral (isotropic NMR) [28].

Enantiomorphism covers both the concept of chirality and prochirality, associated respectively with enantiomers of chiral compounds or enantiotopic elements in prochiral molecules. Prochiral molecules are simply defined as any molecule that can be transformed into a chiral one by a single isotopic substitution. Chirality can therefore originate from isotopic substitution around a stereogenic center or not and called isotopic chirality (see Figure 9.7) [8]. In this section, again, we describe the significant contribution of anisotropic NMR to the discrimination of enantiotopic elements in prochiral molecules, and its possible applications.

### 9.6.1 Analysis and Enantiopurity Determination of Chiral Mixtures

When the enantiodiscrimination mechanisms involved in CLCs are effective, a doubling of resonances (for each enantiomer or enantiotopic elements) is expected compared to NMR spectra recorded in ALCs. As a consequence of this additional spectral complexity in CLCs, all routine magnetically active nuclei (C, H, O, N) are more or less suitable NMR spies to use to reveal spectral differentiations in weakly aligning CLCs.

For almost all organic molecules, we can pay attention to $^{1}$H and its first isotope, $^{2}$H, as well as $^{13}$C nuclei, mainly; each of them having its own advantages and drawbacks depending on the spin $I$, the gyromagnetic ratio, $\gamma$, and the relative natural abundance. Quadrupolar nuclei, $^{14}$N and $^{17}$O, can be theoretically used. However, their quadrupolar properties and their spin (large $Q$-value (see Equation 9.13) reduce drastically the $T_1$, $T_2$ relaxation times and the spin 5/2 for $^{17}$O) make them rather poor analytical probe. When present in compounds, other nuclei of spin $I = \frac{1}{2}$, like $^{19}$F or $^{31}$P can be also successfully exploited. Below we discuss the potential of $^{1}$H, $^{2}$H, and $^{13}$C NMR.

#### 9.6.1.1 Proton NMR

A priori, hydrogen atoms appear to be the simplest nuclear spies for revealing spectral enantiodiscriminations based on differences in ($^{1}$H-$^{1}$H)-RDC values (see Equation 9.4), because $^{1}$H-RCSAs are generally rather small and rarely exploitable. The analytical advantages of this nucleus are its high isotopic abundance (99.985%), its large gyromagnetic constant ($\gamma$ and its spin $I = \frac{1}{2}$. However, even for small-size molecules, a dense homonuclear network made of (long- and short- range) ($^{1}$H-$^{1}$H)-RDCs exists, and the contribution of long- and short-range couplings ($J$ and $D$) generally leads to weakly resolved $^{1}$H spectral patterns, rather difficult to decipher compared to isotropic $^{1}$H spectra. Some exceptions can be found with (ideal) compounds, such as in case of $S$-enriched mixture of 3-butyn-2-ol dissolved in PBLG where all ($R$)- and ($S$)-resonances of the methyl group are nicely resolved and RDCs can be easily determined (see Figure 9.20).

Note finally that in case of compressed or stretched polymeric gels for which the degree of solute orientation is one of order magnitude less than in polypeptide-based oriented phases, it is sometimes difficult to obtain clear and resolved spectral patterns and so extract reliable proton dipolar information. In practice, using aligning gels for the extraction of ($^{13}$C-$^{1}$H)-RDCs from the total spin-spin coupling, $T$($^{13}$C-$^{1}$H), is much simpler.

To overcome the complexity of low-resolution proton 1D spectra, 2D approaches have been explored to improve the resolution of $^{1}$H NMR spectra of enantiomeric mixtures in CLCs. Most of these methods rely on the ability to separate the evolution of chemical shifts and spin-spin coupling interactions, combined with the selection of part of

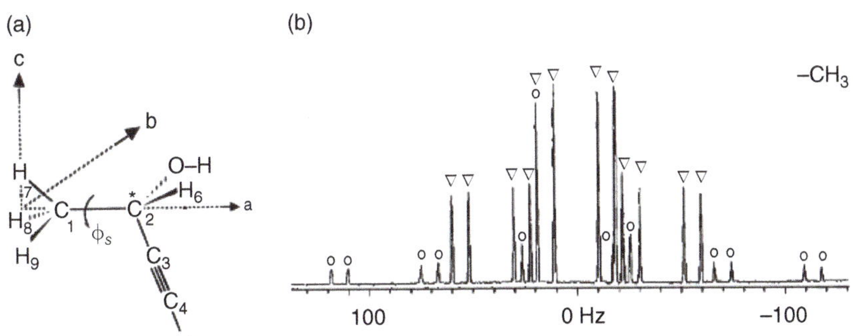


Figure 9.20 (a) Structure of 3-butyn-2-ol. (b) 400.1 MHz $^{1}$H 1D-NMR signal of methyl group ($ee(S)$ = 72%) in PBLG/CDCl$_3$ phase. For each enantiomer, the spectral pattern is made of two A$_3$MX spin system, namely a dipolar triplet (CH$_3$) coupled with methyne hydrogens 5 and 6. Note the difference of intensity of the two A$_3$MX systems due to the ee used. Figure adapted from Ref. [33] with permission.

spectral information using selective irradiation. All these anisotropic experiments derived from the first $^{1}H$ SERF 2D experiments introduced in 1995 [99]. Then, numerous Improvements to the original SERF pulse sequence were proposed to obtain cleaner and simpler spectra to analyze. Among them we can mention: (i) the experiments capable of providing pure absorption 2D spectra (SERFph) [100], (ii) the combination of these phased SERF with the VASS method [101] where the sample spins at a given angle [42], (iii) the implementation of spatially-encoded selective spin echoes in order to visualize the whole total coupling network involving a given proton site in a single $T$-edited 2D spectrum [102]. An example of spatially-encoded selective spin echoes 2D spectra is illustrated in Figures 9.21b, 9.21c, and 9.21d. The resulting maps show a series of multiplets that are often reduced to simple doublets and appear for each enantiomer at the resonance frequency of each coupling partner of the probed proton site. This allows for a straightforward assignment and measurement of the $^{1}H$-$^{1}H$ total couplings and a possible measurement of the ee of mixture [103].

#### 9.6.1.2 Carbon-13 NMR

Exploiting the potential of anisotropic interactions ($^{13}C$-RCSA or $^{13}C$-$^{1}H$ RDCs) associated with carbon-13 nuclei is at the origin of many applications: from the determination of enantiomeric purity [103–107], to the analysis of ($^{13}C/^{12}C$) isotopic ratios, as demonstrated in 2021 [107]. Although $^{13}C$ and $^{1}H$ are nuclei of spin 1/2, the main difference with $^{1}H$ is the low $^{13}C$ natural abundance of 1.1%. Interestingly, the detection of low abundance nuclei

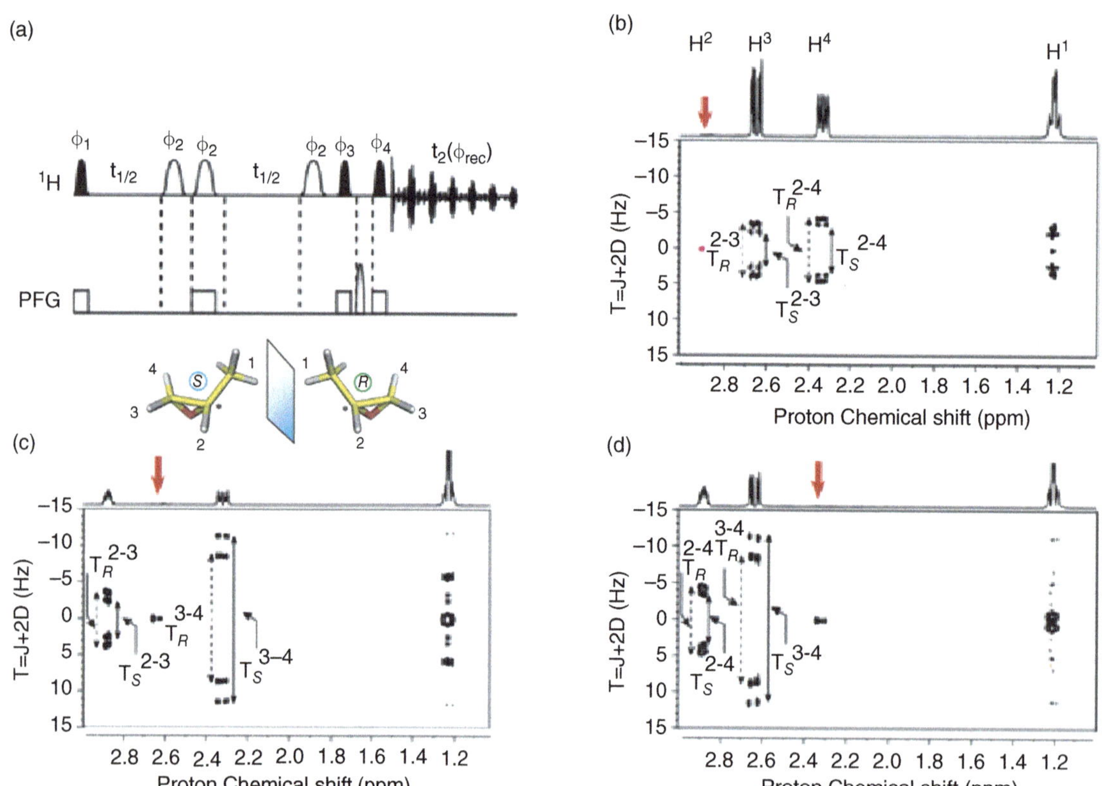


Figure 9.21 (a) Schematic pulse diagramm of the $^{1}H$ G-SERF 2D experiment. (b to c) Examples of $^{1}H$ G-SERF spectra applied on (±)-1,2-epoxypropane in PBLG/$CDCl_3$ for the edition of the total couplings involving proton sites, $H^2$ (b), $H^3$ (c), and $H^4$ (d). Figure partially adapted from Ref. [10, 103] with permission.

like $^{13}C$ leads to simpler NMR spectra (compared to $^{1}H$) to analyze, particularly when all protons are decoupled using classical composite pulse decoupling (CPD) sequences such as the WALTZ-16 sequence [104, 105]. In this case, the spectrum is a sum of independent resonances associated to each $^{13}C$-$^{12}C$ isotopologue of a molecule, while the probability to detect a pair of $^{13}C$-$^{13}C$ isotopologues (1.1% × 1.1% = 0.01%) is too low to be simply detected (INADEQUATE 2D NMR). Another specific advantage of $^{13}C$ nuclei is its large range of chemical shifts ($\delta^{\text{iso or aniso}}(^{13}C) = 0 - 250$ ppm, instead of a range of $\delta^{\text{iso or aniso}}(^{1}H) = 0$ to 15 ppm). This occurrence considerably enlarges the distribution of $^{13}C$ resonances or $^{13}C$ spectra patterns in the $^{13}C$-($^{1}H$) $^{13}C$ spectra, respectively, thus facilitating their assignment or analytical deciphering (coupling measurements).

Among drawbacks associated with this nucleus, it could be mentioned the lowest sensitivity compared to $^{1}H$ ($\gamma(^{13}C) = \gamma(^{1}H)/4$) as well as longer longitudinal relaxation times, $T_1^{\text{iso or aniso}}$ ($^{13}C$), notably with quaternary $^{13}C$ atoms. In fact, the relative sensitivity at natural abundance $^{1}H/^{13}C$ = 5,666 : 1. However, it should be noticed that: (i) with the anisotropic $^{13}C$ NMR, $T_1^{\text{aniso}}(^{13}C)$ are shorter than $T_1^{\text{iso}}(^{13}C)$ measured in isotropic NMR due to viscosity of the solvent and (ii) the use of new cryogenic $^{13}C$ NMR probes (that reduce the electronic noise) allows compensating for the problem of sensitivity due to low $\gamma$ and natural abundance, even with a small amount of analyte [109, 110].

When chiral-oriented solvents are used, spectral enantiodiscrimination can be detected by proton-coupled $^{13}C$ NMR on the basis of a difference of ($^{13}C$-$^{1}H$)-RDCs, and the *ee* easily measured on isolated sites in the molecule (few or no long-range RDCs). This situation is often observed with (more or less isolated) methyl groups as in case of 2-bromopropionic acid as illustrated in Figure 9.22a. Here for each enantiomer, the $^{13}C$ spectral pattern is associated with an $A_3MX$ spin system with X: $^{13}C$, and $A_3$ and M: $^{1}H$, but only one isomer shows a total non-zero $^{13}C$-$^{1}H$ coupling [104]. Figure 9.22b shows proton-coupled $^{13}C$ signal of diastereotopic (inequivalent sites in $\alpha$-position of the asymmetric carbon) hydrogens of the methylene group of $\beta$-trichloromethyl-$\beta$-propiolactone. The (*R*)- and (*S*)-spectral patterns are typically two ABX spin systems with X: $^{13}C$ and A and B: $^{1}H_a$ and $^{1}H_b$) with two distinct ($^{13}C$-$^{1}H$)-RDC values [111]. As we will see in the next section with other examples, the analysis of NAD signals of this $CH_2$ lead to the detection of four $^{2}H$-QDs, revealing de facto the inequivalence of diastereotopic $^{1}H$ sites through the detection of their four associated $^{2}H$ diastereo-isotopomers.

The $^{13}C$ NMR approach is often limited when a significant number of long-range ($^{13}C$-$^{1}H$)-RDCs contribute to signal of interest. As mentioned previously, the use of heteronuclear (selective or not) 2D experiments such as *J*-resolved, CLIP-CLAP HSQC or $^{1}H$ $F_1$-coupled *J*-scaled BIRD-filtered HSQC (JSB-HSQC) 2D experiments can overcome this problem [34, 35, 37].

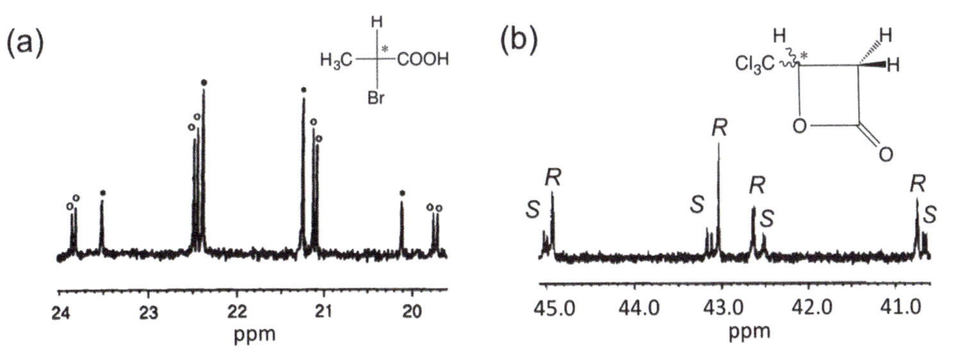


Figure 9.22 (a) 100.3 MHz $^{13}C$ spectral pattern of methyl group ($A_3X$ spin system with X =$^{13}C$) of (±)-2-bromopropionic acid showing. (b) $^{13}C$ spectral pattern (ABX spin system with X =$^{13}C$) of methylene group of $\beta$-trichloromethyl-$\beta$-propiolactone ((*ee*(*R*) = 40%) in PBLG. Both examples recorded in the PBLG phase show clear enantiodiscriminations. Figure adapted from Refs. [104, 111] with permission.

Simpler spectral analyses and successful results are generally obtained when all protons are decoupled ($^{13}$C-{$^1$H} NMR), because discrimination is detected only on the basis of differences in $^{13}$C-RCSA values. Thus, all inequivalent $^{13}$C-{$^1$H} signals of a chiral molecule give rise to a single resonance in ALCs and two resonances in a CLC, one for each enantiomer, if the enantiodiscrimination phenomenon occurs as seen in the simple case of (±)-phenethyl alcohol recorded at 16.5 T (176.1 MHz for $^{13}$C) (see Figure 9.23a) [107]. For this molecule, differences in $^{13}$C-RCSA values, Δ(RCSA), observed at this magnetic field strength (see Equation 9.8) vary within a range of 8 (sp$^3$) to 16 Hz (sp$^2$). Among other remarkable, more complex, examples of spectral $^{13}$C-{$^1$H} enantiodiscriminations, we can mention the case of enantiomers of planar chirality as chromium tricarbonyl complexes (($\eta^6$-arene)X(CO)$_3$ with X = Cr) obtained with new asymmetric synthesis reactions (see Figure 9.23b) [108].

Figure 9.23c presents a more atypical case of chirality (atropoisomerism) with the 1,1′-bi(2-naphthol) (BINOL), whose 80% of 10 $^{13}$C sites (sp$^2$ carbon atoms) in PBLG are enantiodiscriminated with a difference in $^{13}$C-RCSA values of about 10–17 Hz (9.4 T) [105]. Interesting results were also obtained with (±)-2,2'-dimethyl-1,1'-binaphthyl [112] as well as the possibility to differentiate enantiomers of chiral molecules having a heteronuclear stereogenic center, as it is the case of chiral sulfoxides (see Figure 9.23d).

In all cases, the large number of enantiodifferentiated $^{13}$C sites allows to select the optimal carbon site both in terms of spectral $R/S$ separations and signal-to-noise ratios (SNR) for a robust determination of the enantiopurity (*ee*) of chiral mixtures.

In collaboration with many chemists, this tool has been successfully applied in the evaluation of enantioselectivity in various asymmetric synthesis cases: (i) monofluorination reactions on propargylic compounds, (ii) double diastereoselection in [2 + 3] cycloaddition reactions of chiral oxazoline *N*-oxides and their application to the kinetic resolution of a racemic $\alpha,\beta$-unsaturated $\delta$-lactone [113], (iii) intramolecular hydroamination catalyzed by ate and neutral rare-earth complexes [114], (iv) aza-Michael additions of *O*-benzylhydroxylamine to *N*-alkenoyloxazolidinones catalyzed by samarium iodobinaphtholate [115].

Finally, it is interesting to note that by increasing the spectrometer's magnetic field strength by a factor of two (18.8 T), the enantiodifferences of all examples presented here will simply double (see Equation 9.8), facilitating the integration or the deconvolution of the $R/S$-enantiomer resonances.

#### 9.6.1.3 Deuterium NMR

Deuterium nuclei, the second isotope of hydrogen, is naturally present in all hydrogenated compounds at very low natural abundance level ($1.55\times10^{-2}$%) and resonate at a lower Larmor frequency ($\gamma(^2\text{H}) = \gamma(^1\text{H})/6.515$). However, same as for anisotropic $^{13}$C NMR, the use of high-field NMR spectrometers combined with $^2$H cryogenic probes compensate enough the very low sensitivity of this nucleus to be analytically exploited. This sensitivity limitation is no longer a problem for $^2$H isotopically enriched compounds.

Contrary to $^{13}$C and $^1$H, $^2$H nuclei ($I = 1$) possess a quadrupolar moment, $Q$, specific to "quadrupolar" nuclei with spin $I > \frac{1}{2}$ (see Equation 9.15), giving rise to a measurable $^2$H-RQC (see Equation 9.14) in oriented media. This new property represents an analytical advantage for various reasons. First, due to its spin number, any $^2$H signal is featured by a quadrupolar doublet ($^2$H-QD) that distribute the signal on only two components ($2I$ transitions), except if $\theta_{\text{CD}}^{\text{B}_0}$ is aligned on $\theta_{\text{m}}$ (see Equation 9.16). Second, the magnitude of the quadrupole moment, $Q$, that governs the efficiency of quadrupolar relaxation mechanisms, is small enough to lead to resolved and sharp spectral lines, unlike the majority of quadrupolar nuclei. Third, the range of magnitude of $^2$H-RQCs ($|\Delta\nu_{\text{Q}}| = 0$ to 1-2 kHz) in weakly aligning media compensate for the low distribution of $^2$H chemical shifts (0 to 15 ppm), thus limiting the peak overlaps. Last but not least, the $^2$H quadrupolar interaction is very sensitive to a difference of orientation of the C-D bond, since the $K_{\text{CD}}$ varies from 150 to 300 kHz depending on the hybridization state of the deuteron and the neighboring substituents, which one order of magnitude large than the corresponding ($^{13}$C-$^1$H)-RDCs for the same bond. All these reasons make proton-decoupled $^2$H NMR a very attractive tool. Finally, and contrary to $^{13}$C NMR, the possible exploitation of $^2$H-$^1$H scalar and dipolar couplings is extremely rare due to their low amplitude (0–1 Hz) measured in weakly aligned media, since they directly depend on the products of their corresponding gyromagnetic constants.

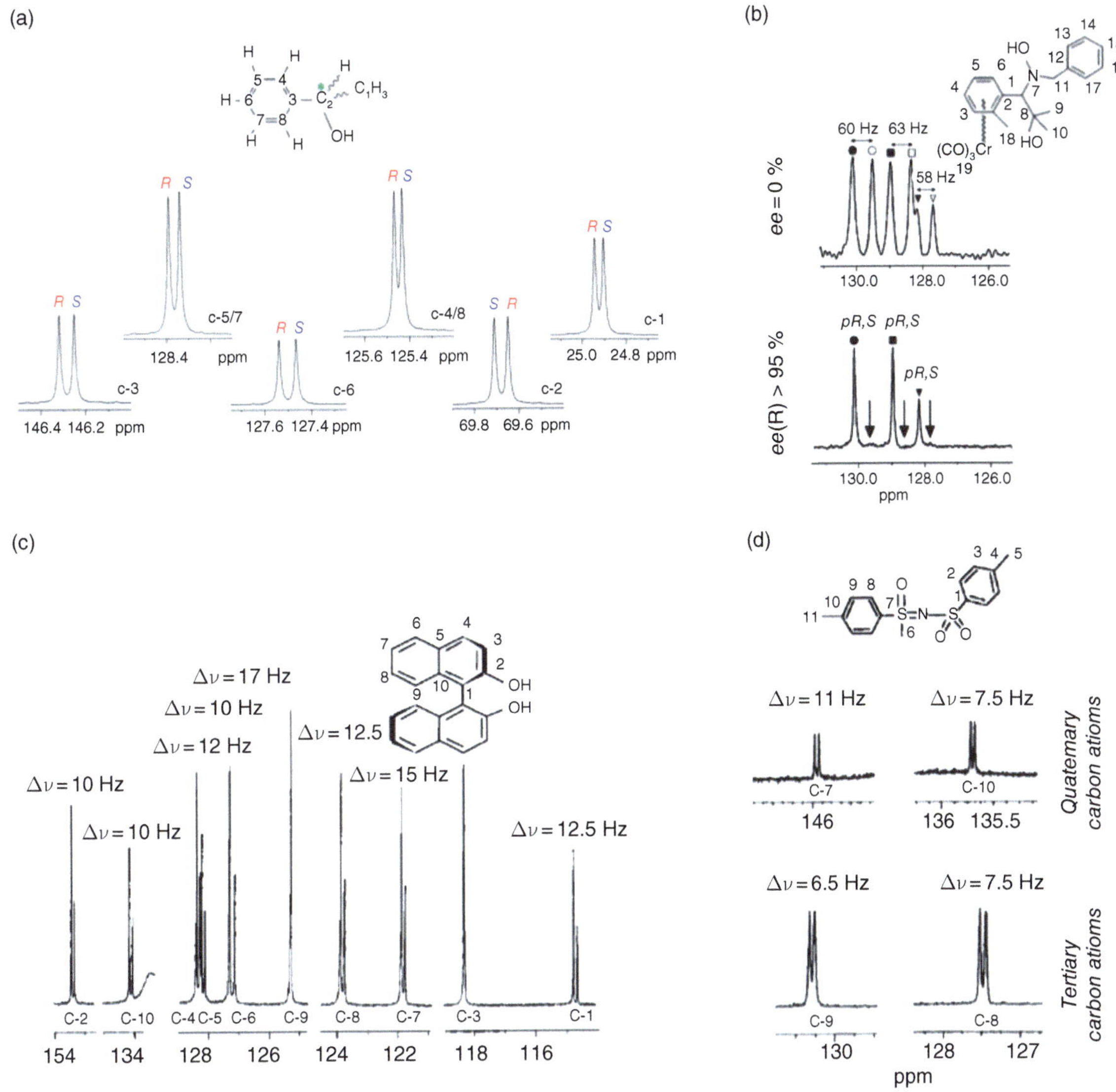


**Figure 9.23** Selected examples of $^{13}$C-{$^1$H} enantiodiscrimations (on sp$^2$ and sp$^3$ carbon atoms) associated with various types of chirality: (a) 176.1 MHz $^{13}$C-{$^1$H} 1D spectrum of (±)-phenethyl alcool in the PBLG/CDCl$_3$ phase showing an enantiodiscrimination on on all $^{13}$C sites. (b) Three differenciated aromatic $^{13}$C-{$^1$H} signals (100.3 MHz) of the *N*-(2-methyl-2-hydroxy-1-((2-methylphenyl) chromium tricarbonyl) propyl)-*N*-benzyl-hydroxylamine, a chiral ($\eta^6$-arene) chromium tricarbonyl complex recorded in PBLG, in racemic (top) and enantiopure series (bottom). (c) Eight (over ten) 100.3 MHz $^{13}$C-{$^1$H} 1D spectrum of 1,1′-bi(2-naphthol) (*ee*(*R*) = 31%) in PBLG/DMF-d$_7$. (d) Four (aromatic) enantiodisciminated $^{13}$C-{$^1$H} 1D signals of (±)-*S*-methyl-*S*-*P*-tolyl-*N*-tosylsulfoximine in PBLG. Figure adapted from Refs. [105, 107, 108] with permission.

From the first results reported in 1992 [59], a large collection of deuterated compounds have been tested to establish the analytical potential of $^2$H-{$^1$H} 1D NMR using polypeptide-based CLCs as enantiodiscriminating media. The experimental results were successful for almost all of the classes of organic chiral compounds tested, including the cases of chiral compounds by virtue of the isotopic substitution (D/H) as well as for enantiomers with a large variety of structures and functional groups (alcohol, amines, carboxylic acids, esters, ethers, epoxides, tosylates, chlorides, bromides). Even cyclic hydrocarbons were successfully discriminated [24, 112]. Clearly, the method revealed to be efficient regardless the molecular structure: rigid, semi-rigid, or flexible with however significant variations in RQC differences ($|\Delta\Delta\nu_Q|$), which generally depend on the distance of deuterated site to the sterogenic center in more or less flexible compounds [112].

Four typical examples of $^2$H-{$^1$H} 1D spectra of mono- or dideuterated chiral molecules aligned in PBLG are shown in Figure 9.23, clearly illustrating the $^2$H enantiodiscrimination power in a structuraly diverse group of chiral compounds (tetrahedral chirality, axial chirality, $C_3$ chirality, isotopic chirality).

As mentioned in Section 9.3.3, the introduction of a deuterium probe in a target molecule without a real guarantee of success can be seen by organic chemists as a severe technical limitation of $^2$H-NMR in CLCs applied. A first alternative consists on discriminating deuterated enantiomers prepared in situ by $^2$H NMR in CLCs, like in the case of compounds bearing a labile hydrogen (OH or NH) since their isotopic labeling can be achieved by a simple deuterium exchange in the presence of MeOD or $D_2O$ [116]. In the case of amides and amines the labile site is generally in a slow exchange regime at room temperature [117]. The case of chiral alcohols is more difficult. Indeed, at room temperature, the labile deuterons in -OD hydroxyl group are often in a fast exchange regime, from one enantiomer to the other (or with the media), preventing generally their visualization. This exchange can be significantly slowed down by decreasing the sample temperature [116].

To avoid any step of isotopic enrichment, ex- or in situ, the ideal alternative remains to be the detection of all natural monodeuterated isotopomers of a molecule by $^2$H-{$^1$H} NMR, as shown using achiral thermotropics or lyotropic chiral LC [121, 122]. The advent of high-field NMR spectrometers equipped with fully digital consoles and cryogenic $^2$H NMR probes (not a prerequisite) led to a successful application of anisotropic NAD NMR [109]. Combined with the use of $^2$H QUOSY-type 2D experiments (see Section 9.2), it possible to record NAD spectra for

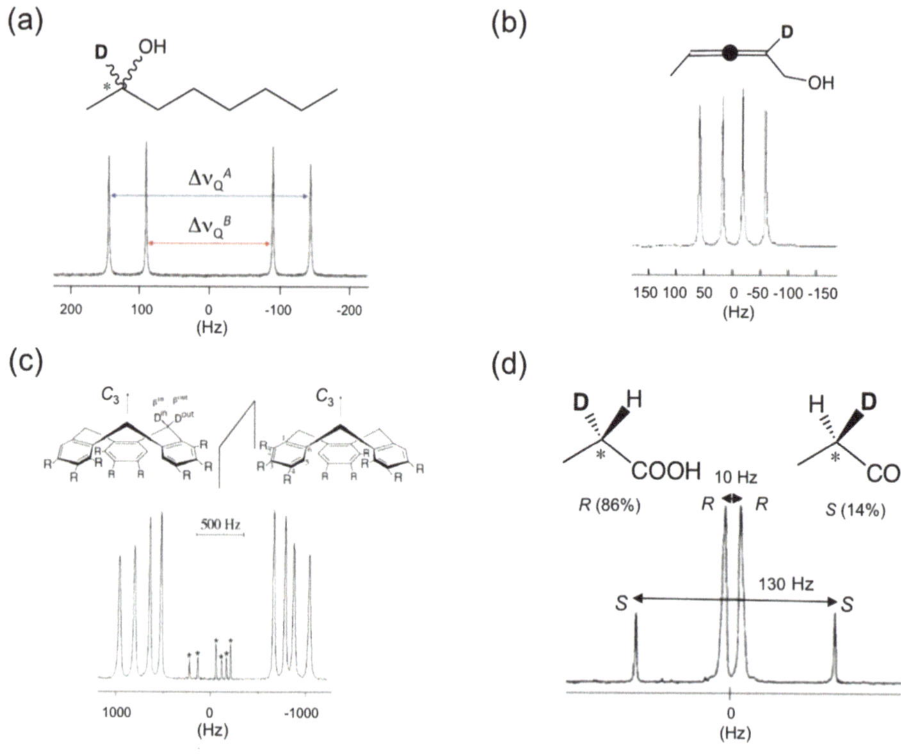


**Figure 9.24** Selected examples of enantiodiscriminations observed by $^2$H-{$^1$H} 1D NMR of deuterated chiral molecules dissolved in a PBLG phase: (a) (*R*/*S*)-2-deutero-octan-2-ol, (b) (*R*/*S*)-2-deutero-2,3-pentadiene-1-ol, (c) (*M*/*P*)-(±)-nonamethoxy-CTV hexadeuterated (methylene groups), and (d) (*R*)-enriched mixture of 2-deutero-propionic acid (*ee*(*R*) = 63%). Note that in panel (c) four $^2$H-QDs are observed due to the diasterotopicity of the deuterium sites in three equivalent methylene groups. Figure adapted from Refs. [27, 118–120] with permission.

the analysis of complex (chiral) molecules with significant MW, such as natural products (see Section 9.7.3) with reasonable amounts of solute ($4-6 \times 10^{-4}$ mol) [51, 123].

Figure 9.25a shows a typical example of NAD-{$^1$H} 1D spectrum recorded with the (±)-hept-3-yn-2-ol (seven non-chemically equivalent sites, except the hydroxyl group) in PBLG, where numerous components of 14 $^2$H-QDs of analyte are overlapped, leading to 1D spectra not trivially analyzable, even with this small molecule. The recording of a ANAD *Q*-resolved 2D spectrum followed by tilting the 2D map allows a simple analysis of various enantio-isotopomers by separating the $\delta^{aniso}(^2H)$ and $\Delta\nu_Q(^2H)$ in $F_2$ and $F_1$ dimension, respectively (see Figure 9.25a) [51, 124]. As it can be seen, the methyl group shows two $^2$H-QDs evidencing the discrimination of associated enantio-isotopomer. From practical aspects, the detection of enantiodiscriminations on methyl groups, where three equivalent enantio-isotopomer contribute to the signal ($A_3$), it is highly advantageous in terms of sensitivity and subsequent robustness of *ee* determination. As speculated for the methylene group of *β*-trichloromethyle-*β*-propiolactone, the diastereotopic sites in methylene 6,6' give rise to eight $^2$H-QDs, evidence of the discrimination of enantio- and diastereo-isotopomers associated with this $CH_2$ group.

Interestingly, these results permit the enantioselectivity of the alkyne zipper reaction, leading to hept-6-yn-2-ol from the hept-3-yn-2-ol, a possible chiral building block for preparing the dolatrienoic acid used to build the south fragment of dolastatine-14 [125]. The analyses of the NAD signals of the methyl of precursor and those associated with one of the diastereotopic deuterons of the methylene group 3 for product, both in racemic and enantioenriched series, have established unambiguously that the reaction was a racemization-free process (*ee* over 95%) (see Figure 9.25b). Other examples of NAD spectra will be presented in the applications presented in the next sections.

Note here that some attempts to measure $^2$H-RQCs in compressed gels compatible with chloroform (PMMA) and DMSO (poly-HEMA) [13, 43] failed due to strong polymer background signal of $^2$H NMR at natural abundance, and limited volume inside the gel (~300 mL) to dissolve large amounts of analyte. Finally, and contrarily to $^{13}$C-RCSAs, it is important to note that the magnitude of $^2$H-RQCs (as RDCs) is absolutely not affected by the strength of the magnetic field used.

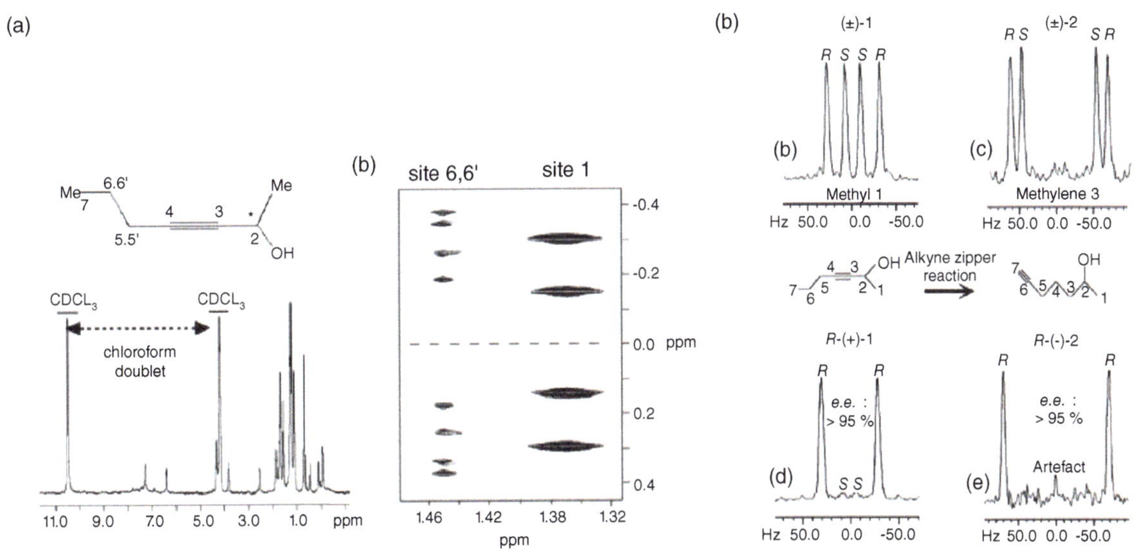


**Figure 9.25** (a) Example of NAD-{$^1$H} 1D spectrum of (±)-hept-3-yn-2-ol in PBLG/$CHCl_3$. Note the NAD signal of chloroform used as organic co-solvent. (b) Zoom on the *Q*-resolved 2D experiment of (±)-hept-3-yn-2-ol showing enantiodiscrimination on methyl and methylene group. (b) 61.4 MHz NAD signals of (b and c) methyl group 1 of **1** and (d and e) one of two deuterons in methylene group 3 of **2** in racemic (top) and enriched series (bottom) in PBLG/$CHCl_3$. Figure partially adapted from Refs. [124, 125] with permission.

#### 9.6.1.4 Combined Anisotropic $^2$H and $^{13}$C NMR

As demonstrated previously, the use of $^2$H homonuclear QUOSY 2D experiments greatly simplifies the analysis of overcrowded ANAD spectra. However, the assignment of $^2$H-QDs based on $^2$H chemical shifts is not always simple due to the rather low dispersion of $^2$H chemical shifts as in the complex case of triglycerides [126].

An interesting alternative approach to this limitation is to acquire anisotropic $^2$H-$^{13}$C heteronuclear correlation spectra using pulse sequences derived from the original $^2$H-$^{13}$C HETCOR experiment [54, 56, 123], but other schemes have been also proposed [54, 127], including in solid NMR [128]. Using polypeptide-based CLCs and deuterated analytes, the 2D experiment was renamed as carbon-deuterium correlation in oriented media (CDCOM) (see Figures 9.26a and 9.26c) [54]. In this experiment, the $^2$H magnetization is transferred to $^{13}$C so that one-bond $^2$H-$^{13}$C correlations appear in the direct domain at the chemical shift of each carbon nucleus. The resulting spectra provide significant resolution enhancements in the direct domain ($F_2$) as a result of the larger chemical shift dispersion of $^{13}$C nuclei (in ppm) compared to $^2$H.

It has been demonstrated that the anisotropic $^{13}$C-detected $^2$H-$^{13}$C HETCOR 2D sequence displays higher sensitivity than HMQC or HSQC-based 2D sequences, either detecting $^2$H or $^{13}$C nuclei [54, 56, 129].

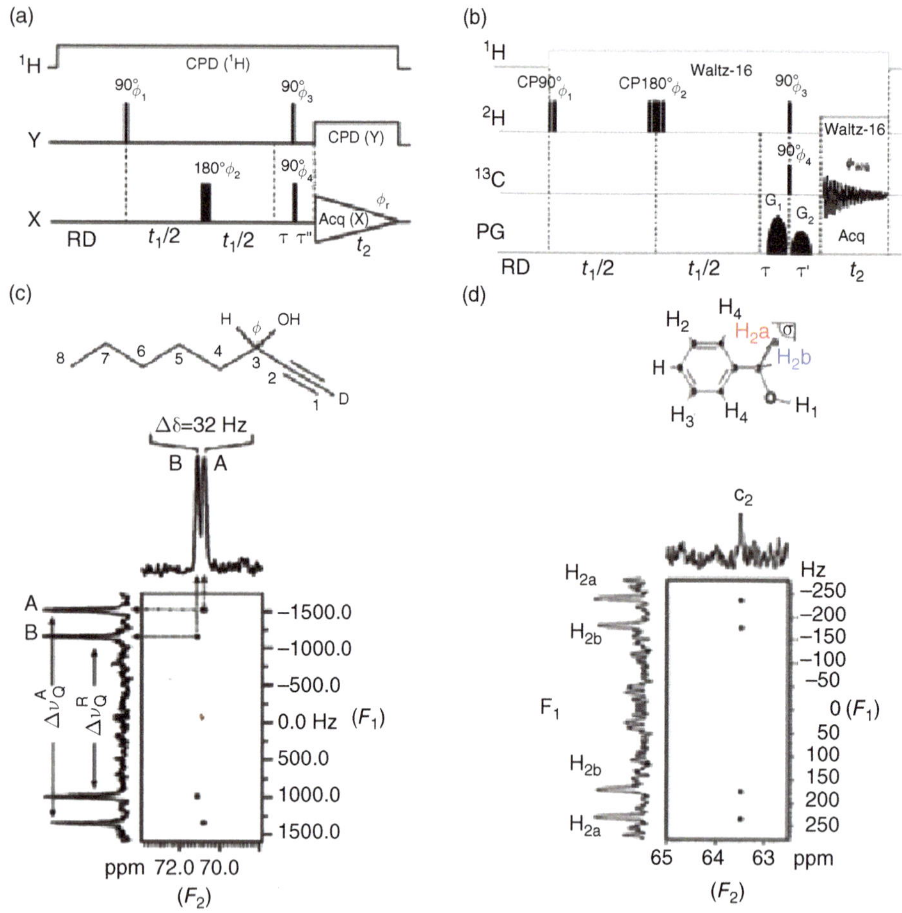


Figure 9.26 (a and b) Pulse diagram of the CDCOM (HETCOR-type scheme with X: $^{13}$C and Y = $^2$H) and *R*-NASDAC 2D experiments, respectively. (c) The CDCOM 2D spectrum of (±)-[1-$^2$H]-1-octyn-3-ol in PBLG phase recorded at 9.4 T (400.1 and 61.4 MHz) and showing the simultaneous $^{13}$C and $^2$H enantiodiscrimination ($F_2$ and $F_1$). (d) The *R*-NASDAC 2D spectrum (900.1 and 225.6 MHz) of benzyl alcohol in PBLG and showing the detection of $^2$H-X enantio-isotopomers with X: $^{13}$C. Figure adapted from Refs. [54, 56, 129] with permission.

In the same spirit, but experimentally more difficult, it is possible to detect isotopomers containing both deuterium and carbon nuclei at their natural abundance level ($1.55 \times 10^{-2}$ % $\times 1.1\%$) using NASDAC and *R*-NASDAC 2D experiments (see Figure 9.26b). This optimized version of the $^2$H-$^{13}$C correlation 2D experiments (HECTOR-type) are able to eliminate all $^2$H-$^{12}$C and $^1$H-$^{13}$C isotopomers, and so detecting 1 out of 600,000 molecules [56]. A first example was reported in the case of the $^2$H-$^{13}$C isotopomer of the small prochiral molecule benzyl alcohol in PBLG. The differentiation of two $^2$H-$^{13}$C enantio-isotopomers associated to the methylene group of benzyl alcohol is shown in Figure 9.26d. Finally, the comparison of the $F_2$ projection of anisotropic NASDAC 2D and $^{13}$C-{$^1$H} 1D spectra allows the determination of the $^2$H to $^{13}$C isotopic effect on $\delta(^{13}\text{C})$, without deuteration of the analyte.

### 9.6.2 Discrimination of Enantiotopic Elements in Prochiral Structures

So far, only a few isotropic NMR methodologies were proposed for resolving signals of enantiotopic elements in prochiral molecules. We can mention the use of cryptan ligands, leading to distinct isotropic chemical shifts in prochiral carboxylic acids [130]. When successful, differences in interactions between the enantiotopic elements and the enantiopure agent are usually too small to produce significant difference in terms of chemical shifts or scalar couplings. Here again, anisotropic NMR in CLCs has provided a new effective alternative for spectrally discriminating enantiotopic elements (see Section 9.3.4).

The spectral discrimination of enantiotopic elements can be achieved using various NMR approaches already described for enantiomers: $^{13}$C NMR, $^{13}$C-{$^1$H} NMR, and $^2$H-{$^1$H} NMR, the $^1$H NMR being the less adapted tool. Figures 9.27a and 9.27b show two examples of enantiotopic discrimination using $^{13}$C NMR and $^{13}$C-{$^1$H} NMR,

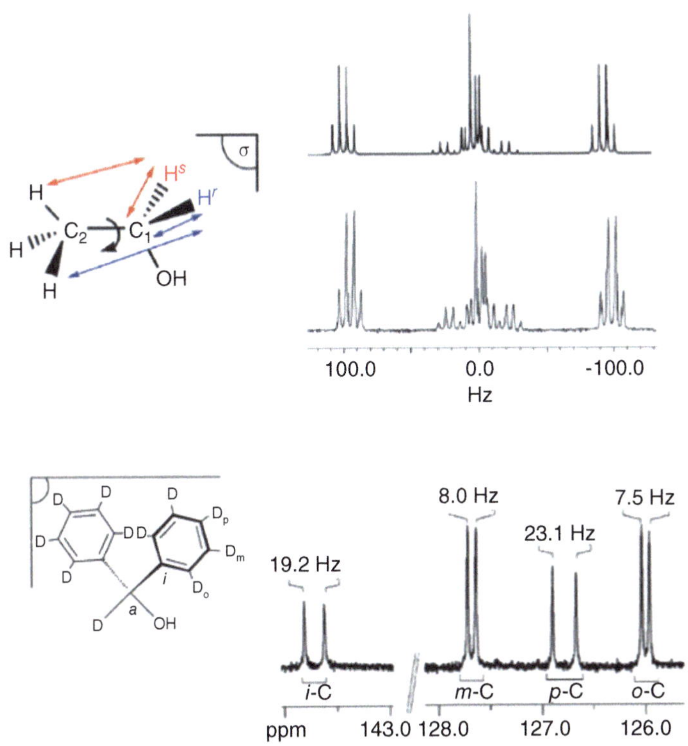


Figure 9.27 (a) Experimental (bottom) and simulated (top) proton-coupled $^{13}$C signal of the C-1 carbon of ethanol. The spectral pattern corresponds to a AXYM$_3$ spin system. Note the spectral desymmetry due to a second-order effect. (b) $^{13}$C-{$^2$H} 1D signals of aromatic carbons of perdeuterated diphenylmethanol ($C_s$ symmetry). Both spectra were recorded at 100.3 MHz in PBLG. Figure partially adapted from Refs. [54, 131] with permission.

respectively [131]. In the case of ethanol ($C_s$ symmetry), the proton-coupled observed $^{13}C$ spectral pattern corresponds to an $AXYM_3$ (with A = $^{13}C$ and $XYM_3$: $^{1}H$) spin system with two distinct ($^{13}C$-{$^{1}H$})-RDCs associated with the $^{13}C$-$^{1}H$ enantiotopic directions bonded on the prostereogenic center. Details can be found in Ref. [131]. The analysis of $^{13}C$-{$^{2}H$} spectrum of perdeuterated diphenylmethanol (another $C_s$-symmetry molecule) is simpler, since the doubling of $^{13}C$-{$^{2}H$} resonances of four aromatic sites revealed directly their spectral enantiodiscrimination. Outstandingly, up to 23 Hz of differences in $^{13}C$-RCSAs are observed. Using $^{13}C$-{$^{1}H$} NMR, an identical result would be obtained with a regular sample (not labeled) of diphenylmethanol or the isotopic chiral analogous when only one of aromatic groups is deuterated [132].

The use of $^{2}H$ 1D/2D NMR in CLCs is another excellent tool for revealing enantotopic discriminations in prochiral deuterated compounds. First tested on isotopically enriched $C_S$-symmetry molecules like benzyl alcohol [27], ethanol [131], or perdeuterated diphenylmethanol [54], but also compounds of $C_{2v}$ symmetry (acenaphten) [18]. This approach has been extended to the detection of deuterium atoms at natural abundance level. In this case, and contrary to enantiomers, only the monodeuterated enantio-isotopomers (isotopic chirality) of a mixture (associated with a pair of hydrogenated enantiotopic sites) will be discriminated, giving rise to two $^{2}H$-QDs observable when the enantiodiscrimination occurs (see Figure 9.7). It is important to understand here that the discrimination of enantiomers of isotopic chirality is a consequence of the discrimination of enantiotopic elements in the prochiral parent molecule. With this approach, it has been possible to experimentally validate the theoretical arguments predicting that all prochiral molecules of $C_s$, $C_{2v}$, $D_{2d}$, and $S_4$ possess enantiotopic directions discriminable in CLCs [18]. Figures 9.28a and 9.28b show two typical examples of enantiotopic discriminations in $D_{2d}$ and $S_4$ prochiral molecules using $^{2}H$-{$^{1}H$} NMR [133, 134]. Note that molecules of $S_4$ symmetry are rather rare in nature.

All examples shown so far were obtained with polypeptide-based LLCs co-dissolved in a broad range of organic solvents, but the enantiotopic discrimination phenomenon has been also revealed using other chiral, weakly orienting media. We can mention chiral systems made of: i) covalently cross-linked gelatin [21], ii) DNA-based water compatible LC in case of dideuterated glycine (see Figure 9.29a) [88], ii) stressed or compressed gelatin gels [136], iii) polysaccharide-based stretched gels (i-carrageenan) using DMSO-$d_6$ as prochiral probe (see Figure 9.29b). [137], or iv) polyacetylene-based LCs in case of perdeuterated $C_S$-symmetry analyte (ethanol, benzyl alcohol, DMSO) but also of $C_{2v}$ symmetry (acenaphthen) [138].

Before closing this section, it is noteworthy to examine the intriguing case of malononitrile ($H_2C(CN)_2$) (MLN), a rigid $C_{2v}$-symmetry molecule with a non-prostereogenic tetrahedral center [131]. For chemists, MLN could be defined as "pro-prochiral," because a priori two chemical steps would be needed to convert it into a chiral structure. From a symmetry point of view, MLN can be also defined as non-prochiral, because it may be superimposed upon itself by an overall rotation, thus producing a structure that is indistinguishable from the original. Finally, from a more stereochemical side, if the (bonded) $^{13}C$-$^{1}H$ internuclear directions are homotopic and cannot be differentiated in a CLC, the $^{13}C\cdots{}^{1}H$ (non-bonded) internuclear directions are enantiotopic (exchangeable by a plan) and so expected to be spectrally distinguished in a CLC. Experimentally, a typical second-order spectral pattern (AXX' spin system) was observed for $^{13}C$ signals (C-2/C-3) of nitrile group in the PBLG chiral phase showing the enantiodiscrimination of these directions. This pattern disappears in the PBG-based ALC. Amazingly, the spectral differentiation of enantiotopic directions in MLN, a $C_{2v}$ symmetry molecules without a prostereogenic tetrahedral center, has validated a stereochemical hypothesis made by Mislow and Raban in 1967 [139]. Indeed, it had been speculated that "for molecules of the type "CXXYY"…, the two X groups as well as the Y groups are equivalent and cannot be distinguished in chiral or achiral experimental conditions. However, the relationships between X and Y groups are not all equivalent. The four X-Y relationships may be ordered into two enantiotopic sets of two equivalent relationships." These results validate also a more recent concept of stereogenicity defined by Fujita in 1990, who considers that "the compounds ($CX_2Y_2$) can be regarded as prochirals,

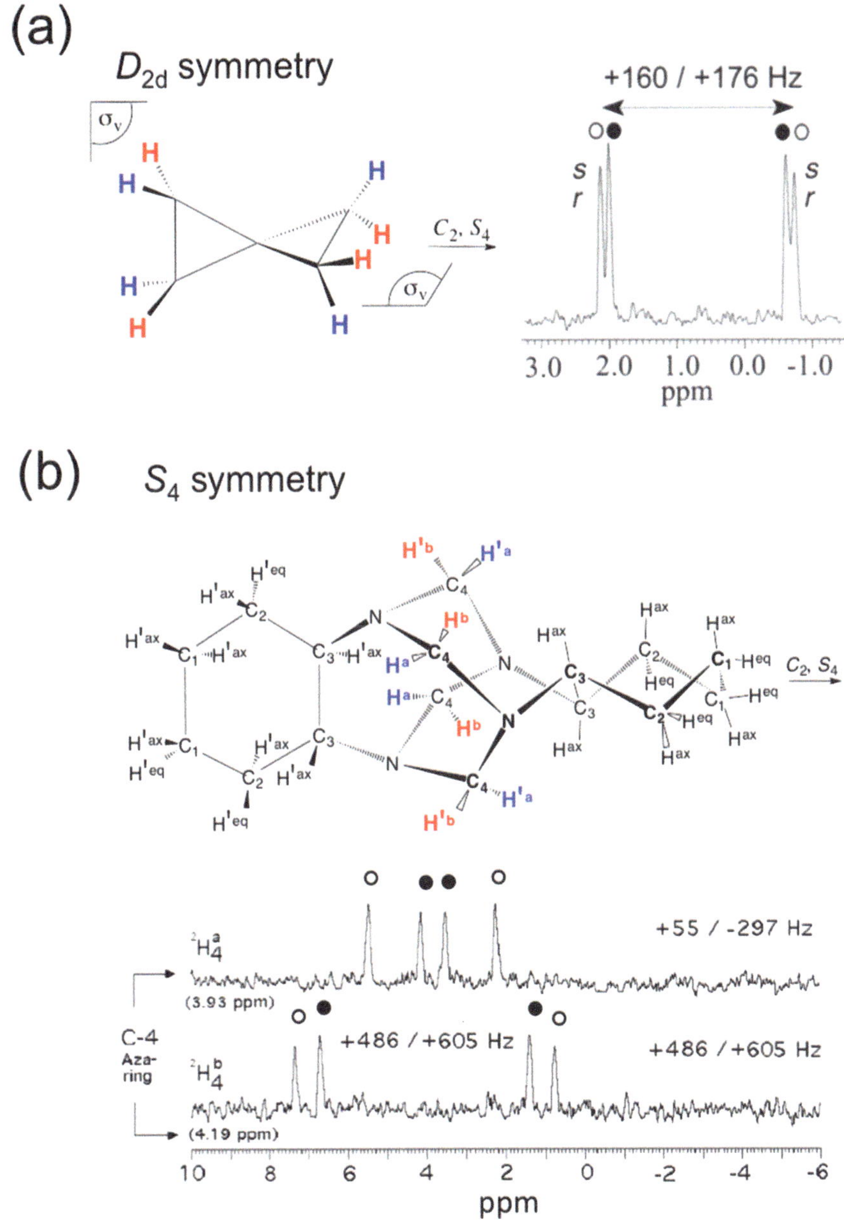


**Figure 9.28** Two examples of $D_{2d}$ and $S_4$ molecules possessing enantiotopic sites (red/blue) and their associated NAD-{$^1$H} enantiodiscriminated signal: (a) spiropentane ($D_{2d}$) and (d) icosane ($S_4$). The doubling of $^2$H-QDs indicates the discrimination of deuterated enantio-isotopomers. Figure from Refs. [133] and [134] with permission.

since the four edges (X-Y) construct an enantiospheric $C_{2v}(C_1)$ orbit" [140]. Clearly, a fundamental stereochemical issue about the definition of prochirality and the concept of enantiomorphism is raised here. Indeed, MLN can be regarded as a prochiral compound when interacting with the CLC, while for organic chemists, this molecule is not!

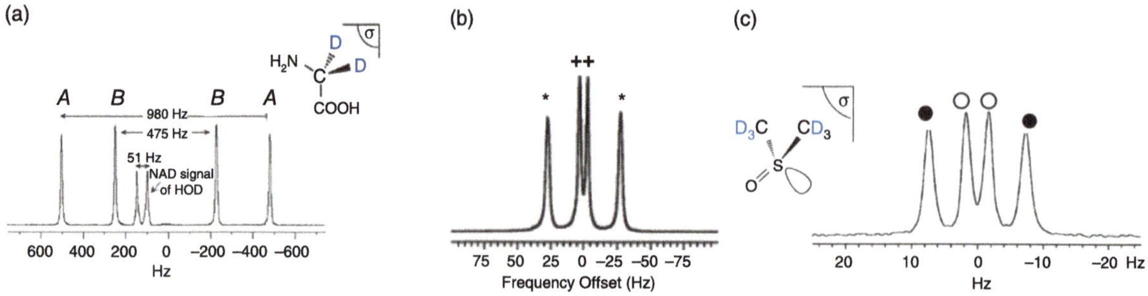


**Figure 9.29** Three examples of enantiotopic discriminations in $C_s$ symmetry prochiral molecule by $^2$H NMR: (a) glycine-d$_2$ in DNA/water-based CLC, (b) DMSO-$d_6$ in a stretched gelatin gel, and (c) DMSO-$d_6$ in a polysaccharide-based stretched gel (i-carrageenan). Figure from Refs. [135–137] with permission.

### 9.6.3 Dynamic Analysis by $^2$H NMR

The study of conformational dynamic processes (as intramolecular interconversion or exchange processes) in achiral or chiral, flexible molecules allows to understand their molecular internal motions [141, 142]. In particular, from the analysis of NMR spectra we can determine, for instance, exchange or interconversion rate constants, which themselves depend on the magnitude of the barrier to interconversion, $\Delta H^{\ddagger}$, and the sample temperature [143]. This can be done using isotropic or anisotropic NMR, chiral or not chiral-oriented media. However, the combination of NMR and chiral anisotropic environments makes it possible to study dynamic processes involving enantiomeric molecular forms or enantiotopic elements in prochiral molecules [144, 145], since in both cases their spectral discrimination is possible. These exchange processes are, obviously, invisible in achiral (oriented) media.

Among the NMR tools described above, $^2$H-{$^1$H} NMR possess three advantages for such investigations: (i) simple high-resolution spectra dominated by the $^2$H quadrupolar interaction, (ii) the phenomenon can be clearly identified, and (iii) spectral separations between exchanging $^2$H anisotropic signals can be much larger than those observed in isotropic $^2$H or $^{13}$C NMR, thus allowing a much wider dynamic process range to be studied. Two examples are proposed here: the case of (±)-*cis*-decalin and (±)-1-bromo-2-methyl-3-deuterio-5-(1'-naphthyl)benzene).

#### 9.6.3.1 Case of Cis-decalin

*Cis*-decalin (CDC) can be described as a chiral compound without any stereogenic center while possessing a two-fold rotational symmetry axis (see Figure 9.30a, top). Interestingly, this molecule is submitted to a possible interconversion between two limit conformers constituted by a pair of $C_2$-symmetry enantiomers. As a consequence, the orientational behavior of CDC and its associated NMR spectra recorded in a CLC is highly dependent on the sample temperature, while different structures are involved to qualitatively interpret the experimental data according to T (see Figure 9.30b, bottom).

Experimentally, CDC in PBLG phase can be studied at different temperatures, between 230 and 360 K, using deuterium (perdeuterated compound) [146]. In each case, the $^2$H NMR spectra obtained results from averaged NMR observables along the conformational pathway. In Figure 9.30b is presented the variation of $^2$H spectrum at three temperature (230, 300, and 360 K). As seen, the coalescence effect is obtained at 300 K. At this temperature, the single high-resolution doublet can be unambiguously assigned to the deuterated sites 9 and 10 located on the bridgehead of CDC because, whatever the temperature, they exhibit no kinetic averaging between the conformational forms 1a and 1b.

The analysis of $^2$H 1D NMR spectrum as well as $^2$H-$^{13}$C HETCOR 2D NMR map (Figure 9.29c) at low temperature (230 K) shows a chiral spectral differentiation between two $C_2$-symmetry invertomers while at high temperature (360 K) the spectrum with coherent with a high-energy, $C_{2v}$-symmetry conformation (in average).

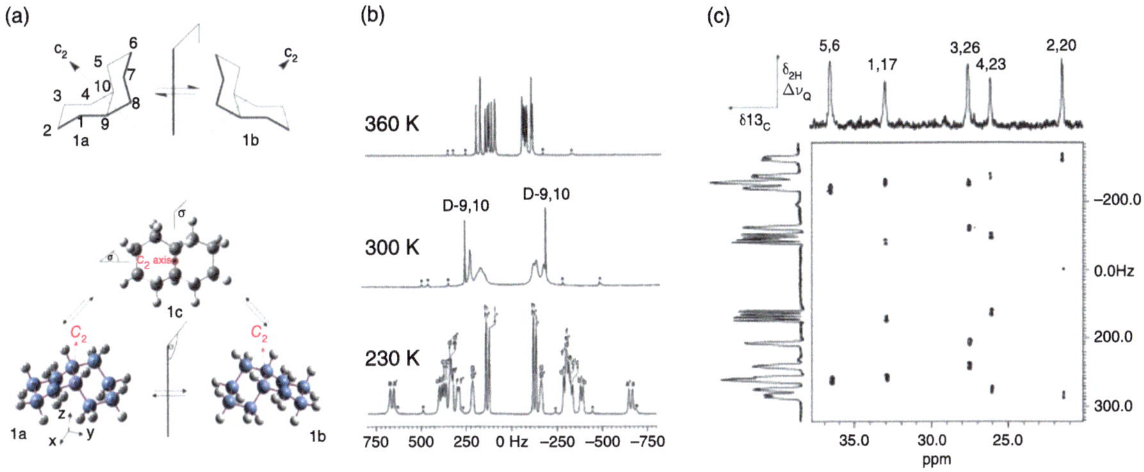


Figure 9.30 (a) (top) $C_2$-symmetry conformers of *cis*-decalin and associated position numbering, (bottom) $C_{2v}$-symmetry high-energy conformation of *cis*-decalin. (b) 61.4 MHz $^2$H 1D-NMR spectra of perdeuterated *cis*-decalin in PBLG recorded at three temperatures. Note the coalescence at 300 K. (c) $^2$H-$^{13}$C HETCOR 2D spectrum recoded at 243 K. Figure adapted from refs. [146, 147] with permission.

The analysis of CDC oriented in the PBLG-based ALC, where all enantiodiscriminations vanish, fully confirms the interpretation of results in the CLC. Thus a maximum of five $^2$H-QDs is expected in $^2$H NMR in the CLC, D-9/D-10, D-5, D-4, D-1, D-8, D-5', D-4', D-1', D-8', D-6, D-2, D-7, D-3, D-6', D-2', D-7', D-3, and nine of $^2$H-QDs in CLCs, two for each enantiotopic pairs and one for the homotopic pair D-9/D-10.

#### 9.6.3.2 Determination of the Activation Barrier Energy

As a second illustrative example, we consider the case of (±)-1-bromo-2-methyl-3-deuterio-5-(1'-naphthyl) benzene (BMNB), a monodeuterated atropoisomer orthosubstituted biaryl, investigated by $^2$H-{$^1$H} 1D NMR in PBLG, again (see Figure 9.31a). The spectral variation of BMNB-d$_1$ versus T is displayed in Figure 9.31b. Here, the coalescence phenomenon is observed at 250 K. This temperature indicates a weakly sterically crowded compound.

Determination of thermodynamic parameters from NMR data is possible from the Eyring's equation that described the dependence of the exchange/interconversion rate constant, $k$, on temperature [145]:

$$k = \frac{RT}{N_A h} exp\left(-\frac{\Delta G^{\ddagger}}{RT}\right) \text{ with } \Delta G^{\ddagger} = \Delta H^{\ddagger} - T \times \Delta S^{\ddagger}. \tag{9.18}$$

where $R = 8.32$ J.K$^{-1}$.mol$^{-1}$, $N_A = 6.02 \times 10^{23}$ mol$^{-1}$, and $h = 6.62 \times 10^{-34}$ J.s, and $T$ is expressed in K.

Using spectral data measured in simulated $^2$H NMR spectra and the analysis of the Eyring plot, namely the natural logarithm of $kN_Ah/RT$ against $1/T$ above and below the coalescence temperature [145]: the activation parameters $\Delta H^{\ddagger}$, $\Delta S^{\ddagger}$, and subsequently $\Delta G^{\ddagger}(T)$, can be extracted; the slope and the y-intercept of plot are equal to the ratio $(-\Delta H^{\ddagger}/R)$ and $(\Delta S^{\ddagger}/R)$, respectively.

Interestingly, the rate constant, $k$, and the free energy of activation, $\Delta G^{\ddagger}$, at the coalescence temperature, $T_C$, (denoted hereafter $\Delta G^{\ddagger}(T_C)$) can easily be deduced from the measurement of the half-difference of $^2$H quadrupolar splittings, $|\Delta\Delta\nu_Q| = |\Delta\nu_Q^A/2 - \Delta\nu_Q^B|/2$, in the $^2$H spectrum below $T_C$. At this particular temperature, assuming identical time constant ($T_2^*$) for the FID decay of both exchanging deuterons, we can write:

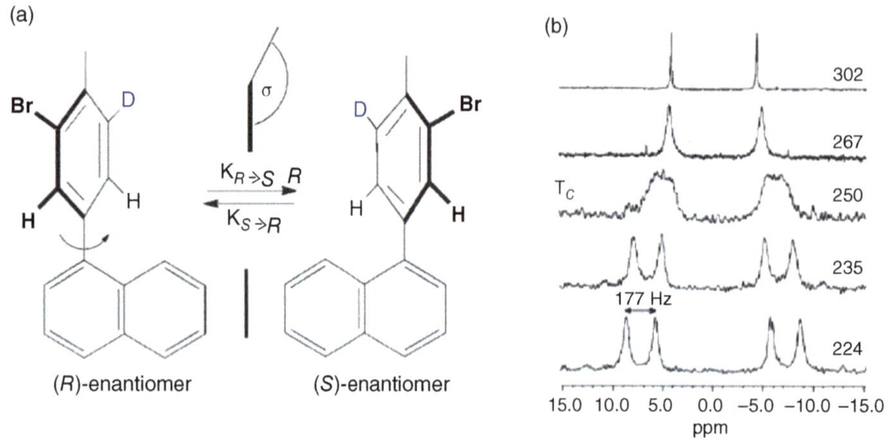


**Figure 9.31** (a) Enantiomeric conformers associated with the (±)-1-bromo-2-methyl-3-deuterio-5-(1'-naphthyl)benzene. (b) Associated experimental (left) and simulated (right) 61.4 MHz $^2$H-{$^1$H} 1D-NMR spectra in PBLG versus five temperatures. Figure adapted from Ref. [145] with permission.

$$k = \pi \times \left| \frac{\Delta\nu_Q^A}{2} - \frac{\Delta\nu_Q^B}{2} \right| \tag{9.19}$$

and

$$\Delta G^{\ddagger}(T_c) = RT_c \times ln\left( \frac{RT}{N_A h} \times \frac{2\sqrt{2}}{\left|\Delta\nu_Q^A - \Delta\nu_Q^B\right|} \right). \tag{9.20}$$

In practice, the anisotropic $^2$H-{$^1$H} or NAD-{$^1$H} spectra measured at $T_c$ can be very different depending on the signs of $\Delta\nu_Q$ for deuterons $A$ and $B$ as well as the magnitude of $\left|\Delta\nu_Q^A - \Delta\nu_Q^B\right|$ compared to $\Delta\nu_Q^A$ and $\Delta\nu_Q^B$ [144].

From data measured in simulated NMR spectra (not shown) and the analysis of the Eyring plot, namely the natural logarithm of $kN_Ah/RT$ against $1/T$ above and below the coalescence temperature, the activation parameters $\Delta H^{\ddagger}$, $\Delta S^{\ddagger}$, and subsequently $\Delta G^{\ddagger}(T)$, can be extracted from the slope $(-\Delta H^{\ddagger}/R)$ and the y-intercept of plot $(\Delta S^{\ddagger}/R)$ [143, 144]. Thus the coalescence reached at $T = 250$ K corresponds to a value of k equal to $3.2 \times 10^2$ s$^{-1}$. The activation parameters of Equation 9.20 derived from the Eyring plot analysis. [143, 148] were found to be equal to $\Delta H^{\ddagger} = 44.7 \pm 0.5$ kJ.mol$^{-1}$, $\Delta S^{\ddagger} = -18 \pm 2$ J.mol$^{-1}$.K$^{-1}$, and $\Delta G^{\ddagger}(T_c) = 49.0 \pm 0.5$ kJ.mol$^{-1}$.

### 9.6.3.3 Reaction Monitoring

Another important aspect of "dynamic" NMR spectroscopy is the chemical reaction monitoring by NMR, in situ and in real time. The idea is to follow the amount (concentration) of reactant and product(s) *versus* time when the reaction is performed in the NMR tube inside the magnet of the spectrometer. This can be done by integrating/deconvoluting the peak surfaces of reactant(s) and product(s) on spectra of any magnetically active nuclei showing a clear spectral signature (isolated peaks, for instance). Many investigations (inside the tube or in flow) using isotropic solvents and high-field NMR [149], as well as low-field NMR [165, 166] can be found in literature. Investigations using thermotropic LCs as reaction solvents have been also described [152, 153].

In 2013, an approach involving anisotropic $^2$H 1D NMR and DNA-based chiral LLCs was reported to follow an enzymatic racemization reaction (alanine racemase, an active enzyme) on a deuterated substrate (alanine-d$_3$) in a scalemic mixture in order to determine its kinetic parameters (time-dependent parameters) and understanding its mechanism reaction [91]. As this DNA-oriented system is enantiodiscriminating, it is possible to distinguish

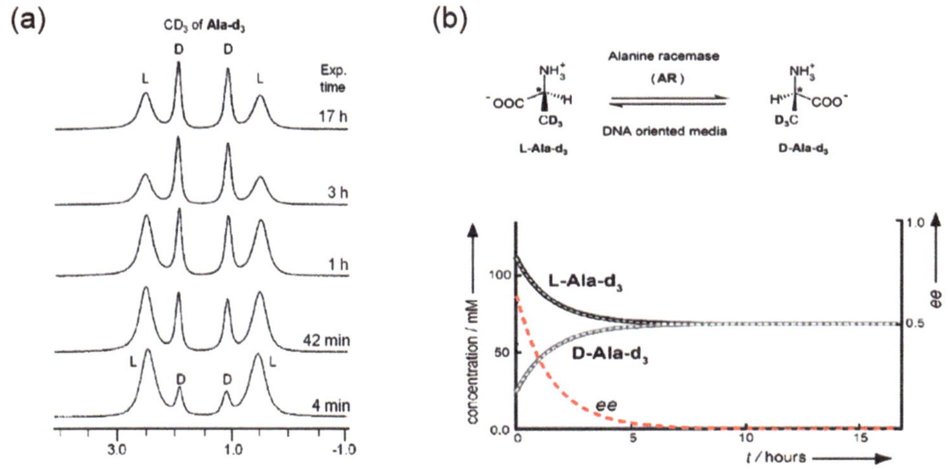


Figure 9.32 (a) Examples of $^{2}$H-{$^{1}$H} 1D signals (same vertical scale) of (*L*)- and (*D*)-d$_3$ recorded at different time intervals after the introduction of AR in the DNA LLC. (b) (top) Principle of racemization of alanine-d$_3$ by AR in DNA-based oriented media. (bottom) Variation of *ee* (red dashed line) and experimental (black and gray continuous lines) and fitted (white dashed lines) concentrations (in mM) of (*L*)- and (*D*)-Ala-d$_3$ as a function of time (in h). Figure adapted from Ref. [91] with permission.

between enantiomer signals of alanine-$d_3$, as shown in Figure 9.32a, and hence follow their respective concentrations *versus* time. Interestingly, $^{2}$H-{$^{1}$H} NMR of deuterated substrate provides simple analysis of signals. In this example, the larger linewidth for the (*L*)-isomer originates from internal $^{2}$H-$^{2}$H total couplings that are not equal to zero as for the (*D*)-isomer.

The time-dependent concentrations permit the evaluation of $k_{\text{catL}}$ and $k_{\text{catD}}$ (kinetic parameters) for a reversible Michaelis-Menten model, the Michaelis constants $K_{\text{M}}$ and $K_{\text{M'}}$ being determined independently. The values of $k_{\text{catL}}$ and $k_{\text{catD}}$ obtained with this tool are consistent with previous values using circular dichroism, thus showing the potential and robustness of the method proposed.

#### 9.6.3.4 Ultrafast 2D NMR

The main difficulty of chemical reaction monitoring in real time is the experimental time needed to record spectral data of all compounds in a molecular mixture. This situation can be drastic for extremely fast and/or multiple (cascade) chemical transformations for which the fast identification and reliable quantification of $^{2}$H signals of chiral (or not) products (reactants, [un]stable intermediates, and products) requires 2D-NMR experiments with sub-minute time resolutions or even sub-second time resolutions. Although non-uniform sampling (NUS) methods, which consist of sparse acquisitions of experimental data (Nyquist's condition not fulfilled), can reduce the experiment times of isotropic or anisotropic 2D-NMR experiments [154, 155]. The gain in time is not enough for reaching sub-second durations. To overcome these difficulties, ultrafast 2D-NMR experiments, first developed for solution-state NMR, provide an efficient alternative [156, 157]. First proposed in 2012 for recording ($^{1}$H-$^{13}$C)-coupled HSQC 2D spectra in a single PBLG phase [158], the approach has been extended to anisotropic $^{2}$H 2D NMR, and denoted as "ADUF" spectroscopy (see Figure 9.33) [159]. In this particular homonuclear anisotropic $^{2}$H ultrafast 2D-NMR experiments, the usual time ($F_1$) is replaced by a spatial encoding in QUOSY-like 2D sequences (*Q*-COSY, *Q*-resolved, and *Q*-DQ) [50], thus allowing recording anisotropic $^{2}$H spectra in sub-second experimental times. The choice of various ADUF 2D sequences enlarges the usefulness and applications of the proposed technique [160].

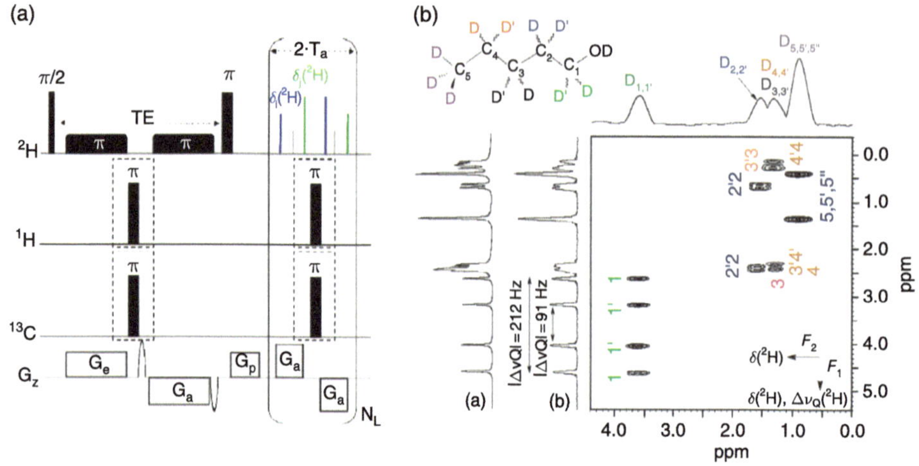

**Figure 9.33** (a) Pulse scheme of the $^2H$-{$^1H$} ultrafast-resolved constant-time pulse NMR sequence. The blocks for $^1H$ decoupling during the spatial encoding and acquisition periods are shown in dashed boxes. $G_e$ and $G_a$ correspond to the gradients applied during the same periods, respectively. (b) 107.5 MHz single-scan ADUF-{$^1H$} 2D map (magnitude mode) with $F_1$ and $F_2$ projections of [$^2H_{12}$]-1-pentanol-$d_{12}$ in the PBLG/$CHCl_3$ phase, recorded in ca. 400 ms. The 107.5 MHz $^2H$-{$^1H$} 1D spectrum is also shown as a top projection. Figures adapted from Ref. [159] with permission.

## 9.7 Structural Value of Anisotropic NMR Parameters

Among the analytical techniques used routinely in organic chemistry, 1D or 2D NMR spectroscopy is becoming an unavoidable tool for the structural analysis of small molecules, in particular, the determination of the constitution of chiral or non-chiral molecules [161]. The next challenge is the possibility to establish the 3D spatial structure of a molecule, or in other words determine its configuration and its preferred conformation/s. In isotropic solvents, this task has been historically achieved using scalar couplings (e.g. $^3J_{HH}$ and $^3J_{CH}$) [162–164], which follow Karplus-based equations to extract dihedral angle information, or the NOE effect to obtain proton-proton distances [165, 166]. These conventional NMR parameters provide structural information of local characters. If for some reason, the chain of $J$-couplings or interproton distances become disconnected there is no way to correlate the relative configuration of remotely located stereocenters.

For example, to determine the relative configuration of the stereocenters at carbons C-3 and C-17 in the compound in Figure 9.34a, a relay of $J$-couplings and NOE correlation from C-3 to C-17 is enough to achieve this goal. However, if those correlations are interrupted by a linker lacking any 3D structural information (flat in this example) such as that in the structure on the right, there is no way to perform it using $J$-couplings and NOE correlations only. A distance of 7.5 Å (structure on the right) is too far to observe an NOE correlation when the limit in NMR of small molecules in no more that 4 – 5 Å (see Figure 9.34b).

In this section, we show the power of anisotropic NMR for the determination of the relative configuration, namely the 3D structure of the molecule. An important number of articles have demonstrated the interest of extracting RDC and/or $^{13}C$-RCSAs in the determination of the configuration of molecule of natural compounds [11, 167–170], but the use of $^2H$-RQC has been only demonstrated in 2020 [10]. RDCs, RCSAs, and RQCs do not encode information of the distance between stereocenters, that's why it is necessary to first know how the atoms are connected in the molecule (molecular constitution). However, these anisotropic NMR parameters allow us to determine the relative orientation of stereocenter regardless of the distance between them. Here is where the power of these parameters reside by lifting the limitations imposed by conventional isotropic NMR parameters (scalar couplings and NOE).

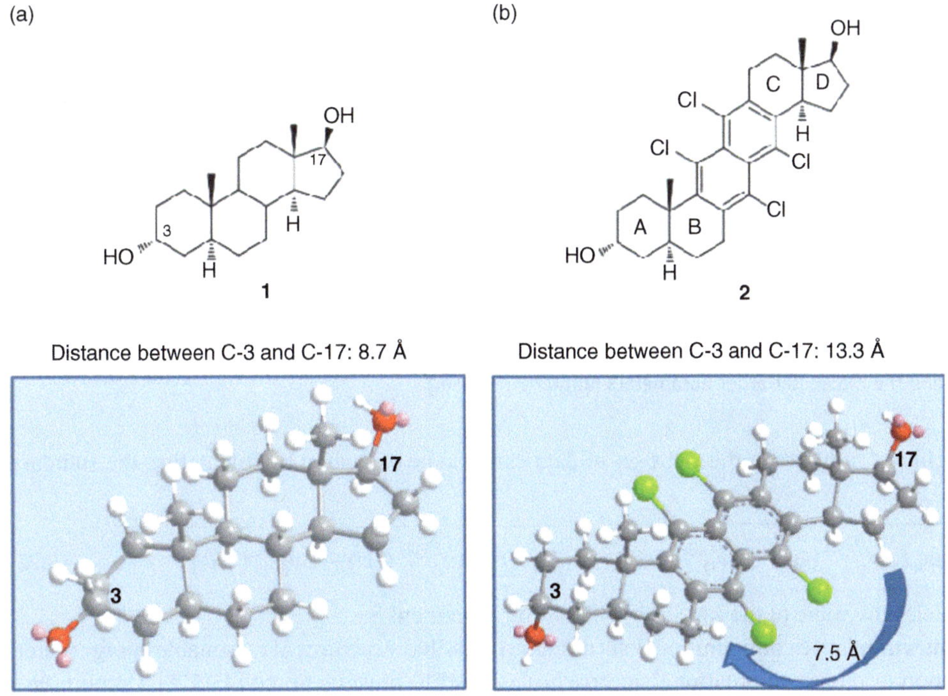


**Figure 9.34** 3D structures of (a) etiocholanediol (**1**) and (b) a hypothetical derivative (**2**). Figures adapted from Ref. [11] with permission.

## 9.7.1 From the Molecular Constitution to Configuration of Complex Molecules

### 9.7.1.1 Principle and Process

The main reason to use anisotropic NMR data is that it exists a univocal relationship between the order-dependent NMR observables (RCSA, RDC or RQC), the Saupe order matrix that describe the orientation of the molecule in the mesophase and the structure of the molecule (see Figure 9.34).

The fit between experimental and back-calculated anisotropic data (from the order matrix for a given [known] geometry) is obtained by varying the elements of Saupe order matrix, $S_{\alpha\beta}$, (using an algorithm based on the principle of singular value decomposition [SVD]) in order to minimize the difference between the two sets of data [171]. Figure 9.35 depicted the principle of the calculation. The agreement between the experimental ($\mathrm{Obs}_n^{Exptl.}$) and back-calculated ($\mathrm{Obs}_n^{Calc.}$) data during the SVD fitting procedure is evaluated by Cornilescu's quality factor, $Q$, calculated as follows [20, 172]:

$$Q = \sqrt{\frac{\sum w_n \left(\mathrm{Obs}_n^{Exptl.} - \mathrm{Obs}_n^{Calc.}\right)^2}{\sum w_n \left(\mathrm{Obs}_n^{Exptl.}\right)^2}} \tag{9.21}$$

where $w_n$ are normalized relative weighting factors. For uniform weighting, $w_n$ is equal to one when a single type of anisotropic data (RCSA, RDC, RQC) used. $W_n$ can be different if various anisotropic data are used simultaneously.

In practice, the smaller the value of $Q$, the better the agreement. The agreement between experimental and predicted value can be graphically presented simply by a correlation plot ($\mathrm{Obs}_n^{Exptl.}$ vs. $\mathrm{Obs}_n^{Calc.}$). One example of plot is shown in Section 9.7.4. In practice, $Q$-values below 0.05 means an excellent agreement between the two sets of data for a given geometry. $Q$-values over 0.1 indicate a smaller confidence between experimental data

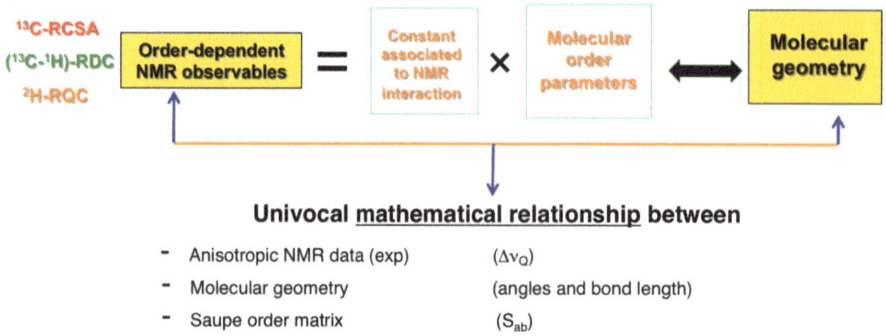


**Figure 9.35** (a) Schematic description of the univocal relationship between a set of experimental anisotropic observables, the molecular geometry and the molecular order parameters (alignment tensor).

and structure. The quality of fit between the twox set of data can also be evaluated by calculating the standard deviation as:

$$RMSD = \sqrt{\sum w_n \left(\Delta\nu_Q^{Exptl.}(^2\text{H}) - \Delta\nu_Q^{Calc.}(^2\text{H})\right)^2} = \sqrt{\frac{1}{N}\sum \left(\Delta\nu_Q^{Exptl.}(^2\text{H}) - \Delta\nu_Q^{Calc.}(^2\text{H})\right)^2}. \quad (9.22)$$

As for $Q$-factor, the smaller the value of the RMSD, the better the agreement is.

In the frame of (unknown) relative configuration determination, possible structures of reasonable energy (determined by DFT calculation or other computational approaches as MMFF) must be tested [45, 173]. Combining a given set of anisotropic experimental data for a set of diverse, possible relative configurations results in a variation of the $Q$-factor, as can be seen in Figure 9.36 [174]. In this case, two factors must be considered for a final selection of the configuration: (i) the smallest value of $Q$-factor and (ii) the difference to the second smaller $Q$-factor value.

Note that the case of flexible molecules is more complex to manage because it becomes necessary to consider various conformers with an individual weight in the calculation. In addition, different from isotropic NMR parameters, the value of the anisotropic NMR parameters (RDCs, RCSAs, and RQCs) for each conformer depends on their shape and orientation, as will be discussed more in detail in Section 9.7.2.

#### 9.7.1.2 Hyphenated Programs

From the experimental anisotropic data, various computational programs have been developed to exploit the analytical potential of anisotropic data. We can mention the program "SHAPE" initially developed by P. Diehl et al. that was suitably modified to handle RDCs and $^2$H-RQCs as input data [175, 176]. Two more recent important achievements were proposed with the program MSpin [177, 178] as well as the program ConArch+ [179–181]. Both programs use an algorithm based on an SVD process, using RDCs, RCSAs, and RQCs as input data. More interesting, they propose also simple graphical and interfaces that can combined various option such the PAS ($S_{a'a'}$, $S_{b'b'}$, $S_{c'c'}$) of the diagonalized Saupe matrix, the inertia tensor axes, ($I_{a'}$, $I_{b'}$, $I_{c'}$), and the Saupe tensor surface representation (see below).

### 9.7.2 Contribution of Spin-1/2 NMR

#### 9.7.2.1 Examples Using ($^{13}$C-$^1$H)-RDCs

We have selected a few examples where isotropic NMR parameters failed to provide a unique solution to the structural problem while RDCs lifted the limitation and their correct relative configuration was unambiguously determined. It is very important to highlight that different from isotropic parameter such as $J$-couplings and chemical shifts, which can be predicted using semi-empirical or DFT methods, anisotropic NMR parameter cannot be predicted by calculations. The alignment tensor is not known a priori and RDCs, RCSAs, and RQCs have to be fitted to a pool of candidate structures with the condition that the correct structure must be part of the pool. The

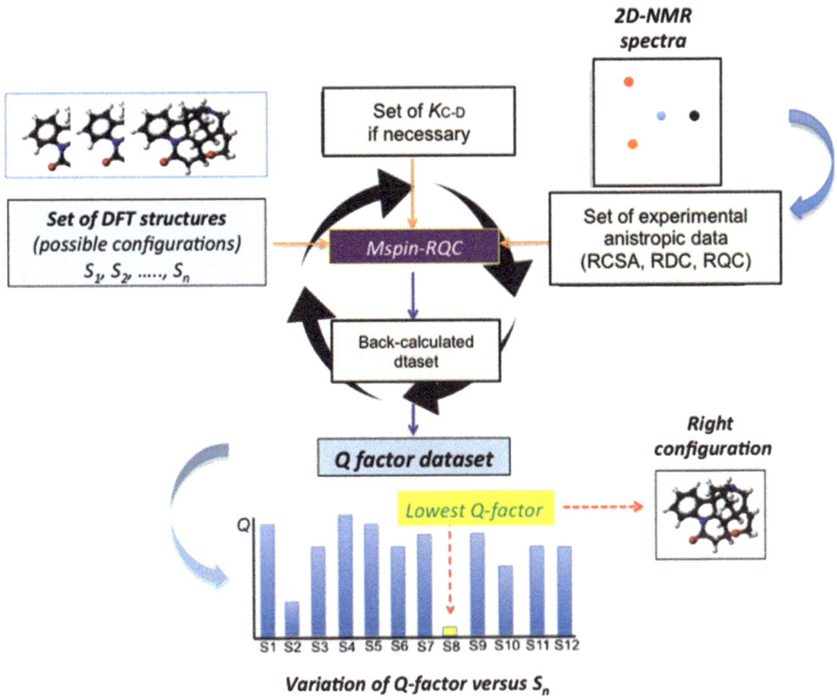


**Figure 9.36** Simplified principle of fitting of experimental anisotropic observable dataset to a set of possible structures. Figure adapted from Ref. [174] with permission.

fitting procedure will select the best fitting structure and if the correct structure is not present, the best fitting structure will be selected leading to a wrong solution.

Ludartin (see Figure 9.37) is a sesquiterpene lactone first isolated in 1972 from *Artemisia carruthii by* Geissman and Griffin as a mixture with it 11,13-dihydroderivative [182], and later isolated in pure form from *Stevia yaconensis var subeglandulosa*, a plant growing in the mountains of northwestern Argentina [183]. It has been shown to inhibit the aromatase enzyme activity in vitro, which is involved in hormone-dependent breast cancer [184]. In the original publication of 1989, it was pointed out that the chemical shift of H-6 alone was not enough to unambiguously determine the configuration of the 3,4-epoxyguainolide unless both $\alpha$- and $\beta$-epoxide are available for comparison [183].

Hence, both epoxide isomers were chemically prepared from ludartin to determine the correct configuration as 3$\alpha$, 4$\alpha$, based on the chemical shifts of H-5 and H-6. In 2008, ludartin was aligned in a self-stretched PMMA gel swollen in $CDCl_3$, and 10 proton-carbon ($^1D_{CH}$) RDCs were measured using an $F_2$ $^1$H-coupled HSQC 2D experiment. Singular value decomposition (SVD) fitting analysis of the RDC data to the 3D structures of both isomers of the epoxide group yielded *Q*-factors of 0.048 and 0.221 for the 3$\alpha$, 3$\beta$ and 3$\beta$, 4$\beta$, respectively. Hence, based on the lower value of the *Q*-factor, the correct configuration of the epoxide group of ludartin was determined as 3$\alpha$, 3$\beta$, clearly showing the structure elucidation power of anisotropic NMR [17]. Figure 9.37 also shows the $F_2$ traces of the associated $^1$H $F_2$-coupled HSQC 2D experiment. Those experiments were collected using a self-stretched PMMA gels. This experiment was performed at the beginning of the development of the stretched gels technology. It took nearly 20 days to fully swell and stabilize a PMMA rod of 1 cm in length and 0.4 cm of diameter. The homogeneity of the gel was not ideal and far from the quality of current gels. However, the data could be extracted and the RDC analysis done successfully.

Jaborosalactone 32 is a whitanolide (steroidal lactone) isolated from *Jaborosa rotaceae*. Its AC at C-23 was determined as *R* from the Cotton effect at 218 nm in the electronic circular dichroism (ECD) spectrum by the chromofore

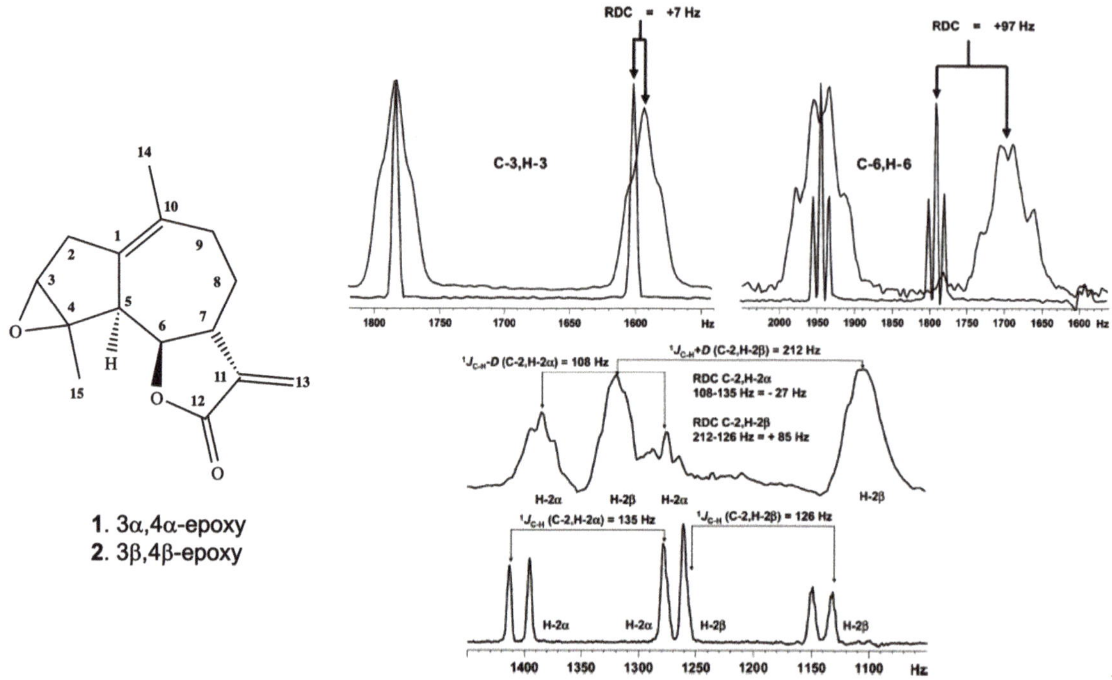


**Figure 9.37** (a) Two possible configurations of the epoxide ring of ludartin (**1** and **2**). (b) Selected overlapped $F_2$ traces of the $F_2$-$^1$H-coupled HSQC 2D spectrum (500 MHz for $^1$H and 125 MHz for $^{13}$C) in isotropic and anisotropic (self-stretched gel) conditions showing how RDCs were measured (here $^1D_{CH} = (^1T_{CH} - {}^1J_{CH})$). Partially reproduced from Ref. [17] with permission.

of the $\alpha,\beta$-insaturated $\gamma$-lactone group. Its structure was further confirmed by single-crystal X-ray diffraction analysis [185]. However, in jaborosalactol 24 (see Figure 9.38), isolated from *Jaborosa parviflora*, this chromophore is absent to perform an ECD analysis due to the conversion of the conjugated $\gamma$-lactone into an $\alpha,\beta$-epoxy-lactol by the natural chemical machine in the plant. A combined analysis using proton–proton $^3J$-coupling constants, NOE, and molecular modeling resulted in three possible solutions. Jaborosalactol 24 was aligned in a self-stretched PMMA gel swollen in $CDCl_3$ and 14 proton-carbon ($^1D_{CH}$) RDCs were measured using an $F_2$ $^1$H-coupled HSQC experiment. SVD fitting analysis of the RDC data was performed on the three possible candidates obtained by $^3J$-coupling and NOE analysis leading to a unique solution of 23*S*, 24*S*, 25*S*, and 26*S* relative to the AC of the main steroidal skeleton. The relative configuration was independently and almost simultaneously confirmed by powder X-ray diffraction analysis leading to a joint publication [186].

Based on the RDC analysis, the configuration of five other molecular analogs with different substitution pattern in the steroidal skeleton was determined, (see Figure 9.38b). This was the first report in which the structure a new natural product was determined using RDCs.

Among other interesting compounds studied, the phytochemical analysis of *Xylocarpus rumphii* yielded a series of tetranortriterpenoids that was isolated as hemiacetals, which could not be purified due to the side-chain ring opening and closing in solution, hence giving complex spectra. Purification was achieved after acetylation and subsequent separation of the epimeric mixtures of acetates. However, identifying which acetate derivative was the (*R*)- or the (*S*)-epimer at C-23 was not possible using isotropic NMR 1D/2D techniques. Proton H-23, with respect to all the nearby protons in the main skeleton, were beyond the maximum distance to observe viable NOE correlations. Acetylation and further separation of the C-23 epimers of mixture **5** (in the publication), yielded epimers **5a** and **5b** pure form. These two epimers were aligned in PMMA/$CDCl_3$ using the compression device leading to a $^2$H quadrupolar coupling of 17 Hz for chloroform. One-bond ($^{13}$C-$^1$H)-RDCs measured for carbons

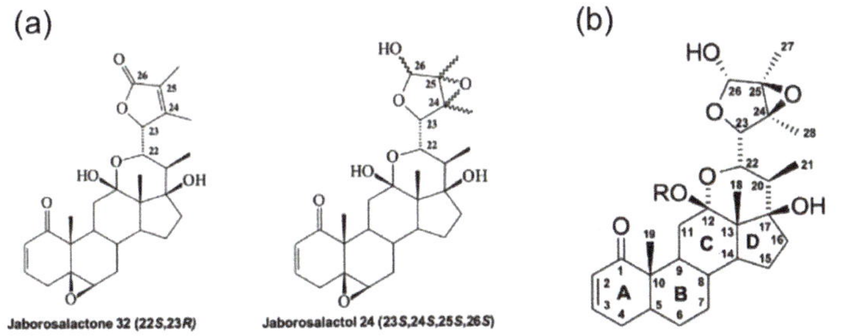


1 5β,6β-epoxy; R=H
2 5β,6β-epoxy; R=Ethyl
3 5α-Cl,6β-OH; R=H
4 5α-Cl,6β-OH; R=Ethyl
5 5α-Cl,6β-OH-2,3-dihydro; R=H
6 5α,6β-diOH; R=H

**Figure 9.38** (a) Structure of jaborosalactol 32 and jaborosalactol 24. (b) All the structures isolated from *Jaborosa parviflora*, including jaborosalactol 24 (**1**). Figures adapted from Ref. [186] with permission.

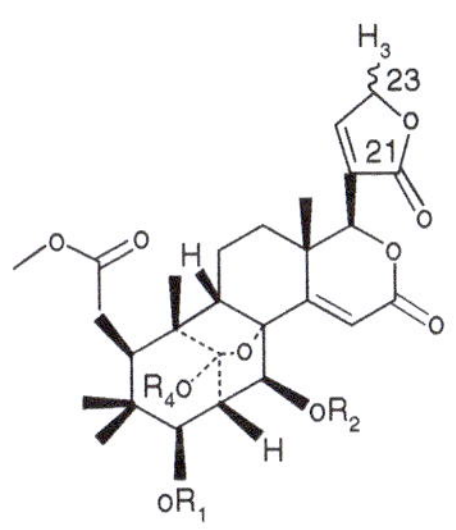

4 $R_1$ = 2*S*-methylbutyryl, $R_2$ = isobutyryl, $R_3$ = OH, $R_4$ = H
4a $R_1$ = 2*S*-methylbutyryl, $R_2$ = isobutyryl, $R_3$ = OAc (*S*), $R_4$ = H
4b $R_1$ = 2*S*-methylbutyryl, $R_2$ = isobutyryl, $R_3$ = OAc (*R*), $R_4$ = H
4c $R_1$ = 2*S*-methylbutyryl, $R_2$ = isobutyryl, $R_3$ = OAc (*S*), $R_4$ = Ac
4d $R_1$ = 2*S*-methylbutyryl, $R_2$ = isobutyryl, $R_3$ = OAc (*R*), $R_4$ = Ac
5 $R_1$ = isobutyryl, $R_2$ = 2*S*-methylbutyryl, $R_3$ = OH, $R_4$ = H
5a $R_1$ = isobutyryl, $R_2$ = 2*S*-methylbutyryl, $R_3$ = OAC (*S*), $R_4$ = H
5b $R_1$ = isobutyryl, $R_2$ = 2*S*-methylbutyryl, $R_3$ = OAC (*R*), $R_4$ = H
6 $R_1$ = $R_2$ = 2*S*-methylbutyryl, $R_3$ = OH, $R_4$ = H
6a $R_1$ = $R_2$ = 2*S*-methylbutyryl, $R_3$ = OAc (*S*), $R_4$ = H

**Figure 9.39** Structure of the compounds isolated from *Xylocarpus rumphii*. Figures adapted from Ref. [187] with permission.

C-2, C-3, C-5, C-15, C-22, C-23, C-29, and C-30. The structures of **5a** and **5b**, generated by DFT showed only one rotamer of the side chain for each configuration, which facilitated the SVD analysis of RDCs using just a single conformation per configuration. SVD analysis of the RDC data of 5a with the structure of both epimers yielded *Q*-factors of 0.081 and 0.177 for the C-23 *S* and C-23 *R* configurations, respectively. While the RDC data of **5b** yielded *Q*-factors of 0.110 and 0.055 for the C-23 *S* and C-23 *R* configurations, respectively [187]. It is very important to highlight that the 3D structure of both epimers is almost identical, except for the orientation of the acetyl group attached to C-23, and RDCs are able to detect that geometrical difference because they encode for the relative orientation of the C-H bonds involved in the SVD analysis.

Apart from these examples, the analysis of several others structures of natural products were reported in the literature [11] as well as the determination of molecular constitution by RDCs [188]. Again, for this purpose, the correct structure must be present in the pool of candidates to perform an accurate selection procedure. Thus the reaction of azide-containing 1,5-enyne (**1**) in the presence of electrophilic iodine sources yielded an unknown tricycliccompound (**4**) [188].

The potential proposed structures of different molecular constitution for **4** are shown in Figure 9.40. Compound **4** was aligned in stretched polystyrene gel swollen in $CDCl_3$. Twelve ($^{13}C$-$^{1}H$)-RDCs were measured using the $^{1}H$ $F_2$-coupled CLIP-HSQC experiment and additional five $^{2}D_{HH}$ proton–proton geminal RDCs measured with the P.E. HSQC 2D experiment [189]. SVD fitting of the RDCs data to the pool of proposed structures selected the constitution **Ba** (see Figure 9.41) for compound [188]. Thus far, we presented examples showing how powerful are RDCs to determine molecular configuration and constitution, particularly in cases where isotropic NMR parameters fails to provide a unique solution. These examples deal with rigid molecules, where if flexibility is present, only one conformation dominates the identity of the 3D structure. RDCs as well as the other two anisotropic parameters (RCSAs and $^{2}$H-RQCs) encode information of the relative position in space of all the atoms in the molecules. This relative position represents the constitution, the configuration and the conformation/s of the molecules as a whole. Determination of conformation by anisotropic NMR parameters is very tricky. For isotropic NMR parameters (*J*-couplings, $\delta$, and NOEs) in the presence of a fast exchange of conformers, the NMR response is the result of their

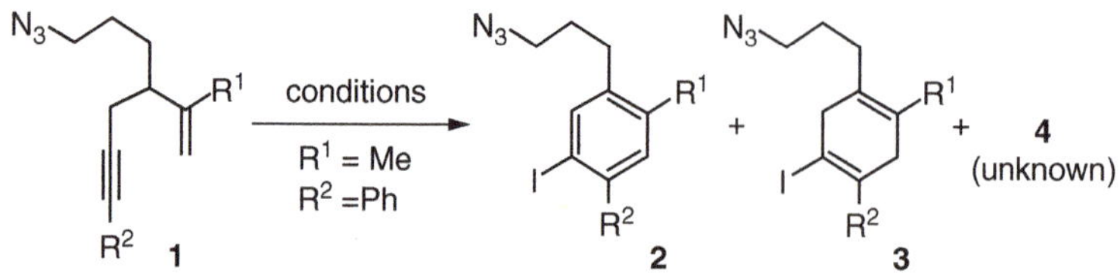


| | 2 : 3 : 4 yield [%] | | | |
|---|---|---|---|---|
| a) NIS, 50 °C, $CH_2Cl_2$ | 76 | 0 | 0 | Ref. [4] |
| b) $I_2$, $K_3PO_4$, 0 °C, $CH_2Cl_2$ | 0 | 99 | 0 | Ref. [4] |
| c) $I_2$, $K_3PO_4$, 23 °C, $CH_2Cl_2$ | 16 | 0 | 38 | |
| d) $I_2$, $K_3PO_4$, 50 °C, $CH_2Cl_2$ | 23 | 0 | 47 | |

**Figure 9.40** Experimental conditions leading to formation of products noted (**2**), (**3**), and (**4**) from acyclic compound (**1**). Figures adapted from Ref. [188] with permission.

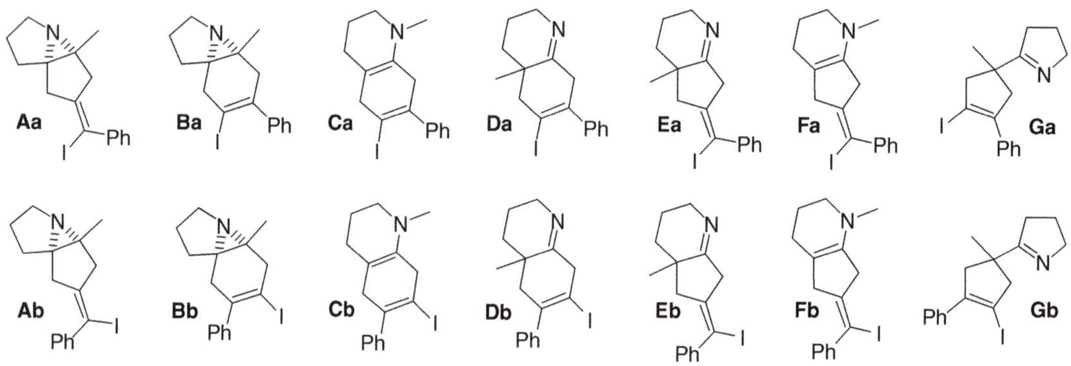


**Figure 9.41** Potential structures of product **4**. Figures adapted from Ref. [15] with permission.

population-weighted average, regardless of the orientation of the conformers; e.g., for a two-sites fast-exchange conformation system of conformers $A$ and $B$, the scalar $J$-coupling and the chemical shift values average as follows:

$$\delta^{aver} = P_A \times \delta_A + P_B \times \delta_B \tag{9.23}$$

and

$$J^{aver} = P_A \times J_A + P_B \times J_B \tag{9.24}$$

where $P_A$ and $P_B$ are the populations of conformers $A$ and $B$, respectively, with $P_A + P_B = 1$. But their corresponding RDC values average in multi-tensor approach, according to Equation 9.25.

$$\overline{D_{ij}^{A+B}} = P_A \left( \frac{k_{ij}}{r_{ij_A}^3} \right) \overrightarrow{r_{ij}}^T . \hat{A}_A . \vec{r_{ij}} + P_B \left( \frac{k_{ij}}{r_{ij_B}^3} \right) \overrightarrow{r_{ij}}^T . \hat{A}_B . \vec{r_{ij}} \tag{9.25}$$

where $\hat{A}$ is the alignment tensor, $\overrightarrow{r_{ij}}^T$ is the transposed of vector of $\vec{r_{ij}}$, and $k_{ij} = \left(-\frac{\mu_0}{4\pi}\right) \times \left(\frac{h\gamma_i\gamma_j}{4\pi^2}\right)$.

Conformers of significantly different shapes and rotational diffusion properties will tend to align differently. If a large portion of the molecule dominates the molecular tumbling properties, it is possible to assume that the alignment tensors for conformer $A$ and $B$ are the same. During the SVD fitting process, the structurally similar parts of the molecule are overlapped (as shown on the right part of the Figure 9.42) and the fittings are performed

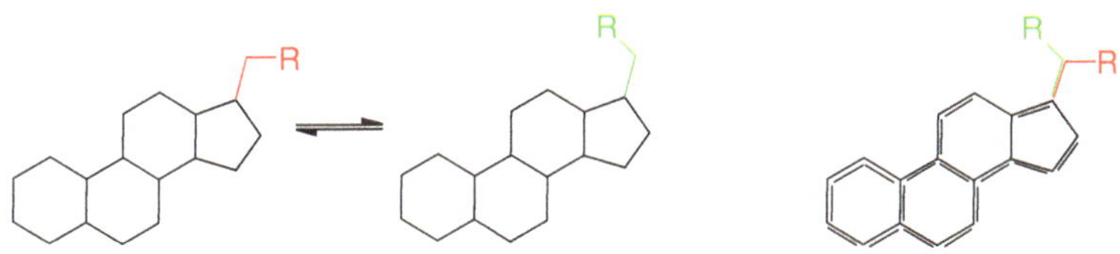


**Figure 9.42** Conformational exchange of two hypothetical compounds. Figures adapted from Ref. [190] with permission.

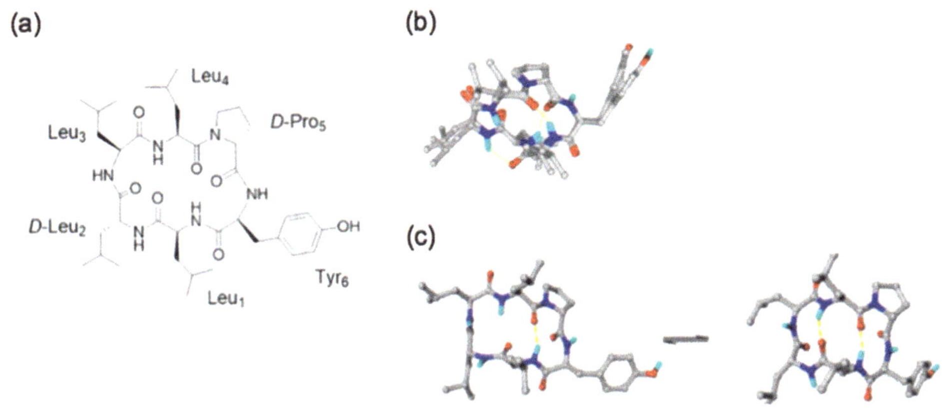


**Figure 9.43** (a) Chemical structure of cyclo-[Leu1-D-Leu2-Leu3-Leu4-D-Pro5-Tyr6]. Preferred conformations of cyclo-[Leu1-D-Leu2-Leu3-Leu4-D-Pro5-Tyr6] in (b) chloroform and (c) in DMSO. Figures adapted from Ref. [191] with permission.

using a population-weighted average [190]. If the populations are not known, it is possible to find the composition that produces the lowest quality factor value. This is known as the single-tensor approximation.

Among, many applications explored, this approximation has been successfully applied to the conformational analysis of cyclic peptides, since the cyclic structure of the backbone of the different conformations share very similar shapes.

The conformational analysis of peptide cyclo-[Leu1-D-Leu2-Leu3-Leu4-D-Pro5-Tyr6] (see Figure 9.43a) was carried out in $CDCl_3$ and DMSO-$d_6$ [191]. The analysis was performed using a combination of RDCs, intramolecular hydrogen bond (IMHB) analysis and $J$-couplings. The peptide was aligned in PMMA/$CDCl_3$ and Poly-HEMA/DMSO-$d_6$ using the gels reversible compression/relaxation method with the compression device. A total of six $^1D_{CH}$ and five $^1D_{NH}$ from the peptide backbone were measured using the $^1$H $F_1$-coupled $J$-scaled BIRD(JSB)HSQC 2D experiment. In addition, six $^3J_{NH\text{-}H\alpha}$ coupling constants and temperature coefficients for the five backbone NH groups were measure. The single-tensor approximation in combination with the Akaike information criterium (AIC) for model selection was used for SVD fitting of the RDC data to the conformational space of thousand DFT-optimized structures. The combined analysis, including the $J$-coupling constants led to the selection of only on conformation in chloroform with three IMHBs (see Figure 9.43b) and two conformations in DMSO with one and two IMHBs, respectively (see Figure 9.43c).

The following example is an interesting and very challenging case for many reasons. It is about an unexpected synthetic by-product. It is very small with oval overall shape and aligns very weakly in gels. Hence, the sample had to be aligned in a stronger aligning medium, such as the lyotropic liquid-crystalline phase (LLC) made of poly-$\gamma$-ethyl-$L$-glutamate (PELG). Attempts to prepare compound **3** from **1** (see Figure 9.44) failed and instead compounds **4** and **5** were produced [192].

Computer-assisted structure elucidation (CASE) analysis of the $^1$H and $^{13}$C 1D NMR spectra as well as HSQC, HMBC, 1,1-ADEQUATE, and 1,n-ADEQUATE 2D NMR spectra of unexpected compound **4** using the structure elucidator program ADC/Labs [193] resulted in only two molecular constitution structures satisfying the atoms

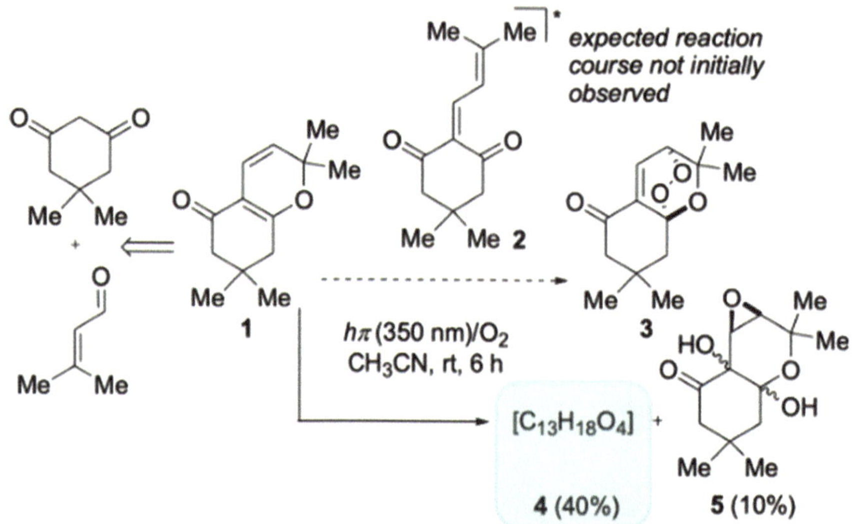


**Figure 9.44** Chemical transformation of (**1**) into unexpected by-product **4**. Figures adapted from Ref. [192] with permission.

connectivity maps, oxirane A, and oxetane B (see Figure 9.45a). These two compounds are constitutional isomers. The $^{1}J_{CH}$ values of the two oxygenated CH groups in A and B would be different enough to distinguish the oxirane from the oxetane structural feature. The DFT predicted values for these $^{1}J_{CH}$ coupling constants are 181.1 and 188.3 Hz for the oxirane CH bonds and 159.8 and 164.6 Hz for the oxetane CH bonds. The experimental values for compound **4** are 181.2 and 190.5 Hz, clearly indicating that compound 4 is an oxirane and not an oxetane. The sample was aligned in PELG dissolved in $CDCl_3$ and nine ($^{13}$C-$^{1}$H)-RDCs were measured using $^{1}$H $F_1$-coupled $J$-SB-HSQC 2D experiment.

There are two possible diastereoisomers for both the oxirane A and the oxetane B with two conformations each, as shown in Figure 9.45a above. SVD fitting of the RDCs date using the single-tensor approximation led to not only discriminate the oxirane from the oxetane, but also select the correct configuration (Figure 9.45b) with a $Q$-factor value of 0.030. This is a clear example of how RDCs can select in one-shot the constitution, configuration, and conformation of a small molecule. In fact, the fitting was performed to select the composition of conformers that produces the lowest $Q$-factor. In this particular case, a ratio of 0.83: 0.17 has been determined.

#### 9.7.2.2 Examples Using $^{13}$C-RCSA

As mentioned previously, the development of the application of RCSAs to the structural analysis of small organic molecules was delayed until scientists figured out how to deconvolute RCSAs from isotropic chemical shift contributions. In 2011, the successful measurement of $^{13}$C-RCSAs for strychnine (see Figure 9.14b) using a combination of a mechanically stretched polymer gel inside a variable-angle NMR probe was described [194]. To prevent interferences from isotropic sample, they used the same sample already anisotropic by stretching the gel inside the rotor. Then, they varied the angle of the director of the alignment in the gel respect to the magnetic field $\mathbf{B_0}$. They monitored the $^{2}$H RQC as function of this angle, scaling the anisotropy down to zero at 54.73° (the magic angle), as shown in Figure 9.46. At the magic angle, the equivalent of isotropic conditions, they measured one-bond proton-carbon $^{1}J_{CH}$ coupling constants for strychnine using the $F_2$-CLIP-HSQC. At angles $\Theta = 0°$ and 65°, they measure the total couplings $^{1}T_{CH}$ in order to obtain the $^{1}D_{CH}$ values. At the same angles, they also measured 1D $^{13}$C NMR spectra (see Figure 9.47).

Just by rotation, the anisotropy is changed while the sample maintains exactly the same experimental conditions, not giving room to contributions for isotropic chemical shifts. Since the correct 3D structure of strychnine is well known, the experimental RDCs were used to calculate an accurate alignment tensor. In turn, with this alignment

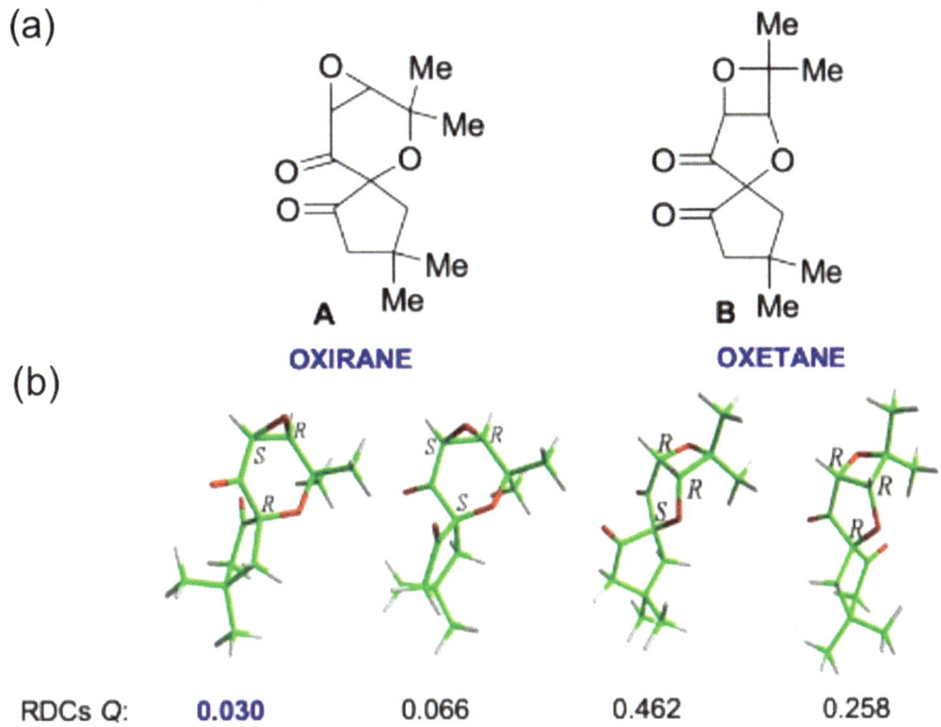


Figure 9.45 (a) Structure of oxirane (A) and oxetane (B). (b) 3D structure associated to the smallest $Q$-factors obtained. Figures adapted from Ref. [192] with permission.

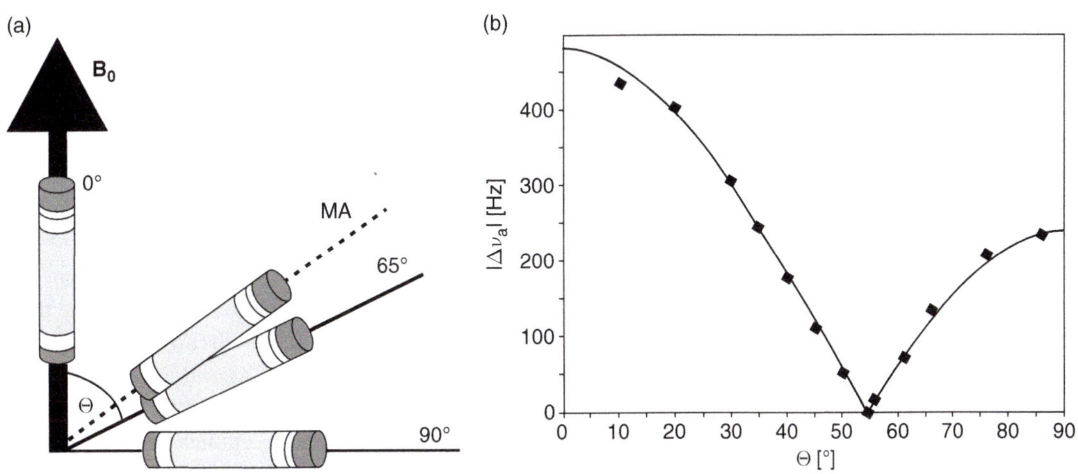


Figure 9.46 (a) Scaling the anisotropy by varing the angle Θ of a mechanically stretched gel with respect to the magnetic field $\mathbf{B_0}$, and without the need of sample spinning at the magic angle (MA) at 54.73°. Figures adapted from Ref. [194] with permission.

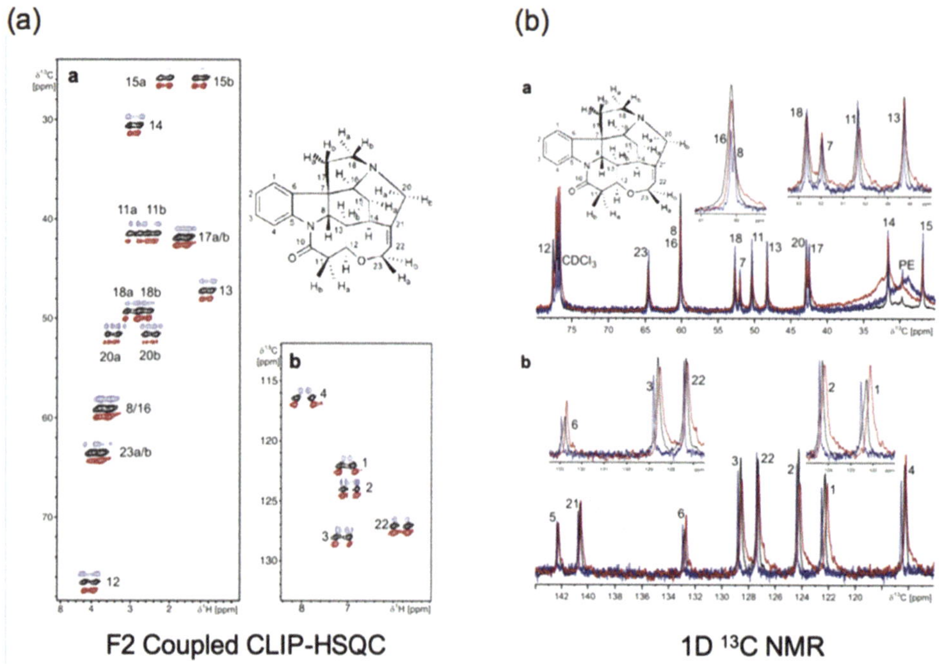


**Figure 9.47** (a) CLIP-HSQC 2D spectrum measured on strychnine in a mechanically stretched PDMS/$CDCl_3$ gel at $\theta$ angle values of 54.73° (black), 0° (blue, dark gray), and 65° (red, light gray). (b) Aliphatic (top) and aromatic (bottom) regions of $^{13}C$-{$^1H$} 1D spectra with same conditions (same color code). Figures adapted from Ref. [194] with permission.

tensor and the DFT calculated CS tensor, $^{13}C$-RCSAs were back-calculated for each carbon in the molecules. Then, experimental versus back-calculated $^{13}C$-RCSAs leading to an excellent correlation with an RMSD of 15.9 ppb, as shown in Figure 9.48.

Although this first attempt to accurately measure RCSAs for a small molecule was very successful, the experimental setup, which includes the use of a VASS NMR probe, made its application very impractical and not affordable to everyone.

The same year (2011), a collaborative effort led by several research groups attempted the differentiation of estrone and 13-*epi*-estrone using only RCSAs [195]. Estrone was aligned in a DMSO-compatible (*S*)-2-acrylamido-1-propanesulfonic acid gel (APS) and the gel was stretched [14]. $^{13}C$-{$^1H$} 1D NMR spectra were measured at two different degrees of stretching of the gel inside the stretching device. $^{13}C$-RCSAs were calculated as the difference in chemical shift for each carbon between the two degrees of alignment. Thus the authors of this article claimed that: "*This approach ensures that the sample composition changes only minimally, if at all, between the different alignment conditions, and the observed chemical shift changes are therefore caused mainly by the RCSAs and by changes in the overall magnetic susceptibility of the sample, which affects all resonances in the same way.*" However, the SVD analysis of fitting only $^{13}C$-RCSAs couldn't discriminate the structure of estrone from its C-13 epimer diastereoisomer. Only when combined with RDCs, the discrimination happened but with quite high *Q*-values (see Figure 9.48).

These results discourage NMR spectroscopists from using RCSAs for the determination of molecular configuration in small molecules for a while. It wasn't until five years later (2016) that a collaborative effort involving Merck & Co. (Rahway, NJ, USA), Carnegie Mellon University (Pittsburgh, PA, USA), Universidade Federal de Pernambuco (Recife, Brazil), and Max Planck Institute for Biophysical Chemistry (Göttingen, Germany) demonstrated using aligning gels and two different devices (compression and stretching) that it was possible to determine the relative configuration of small molecules using only $^{13}C$-RCSAs [168]. This is a foundational paper that sets the basis to a reliable application of $^{13}C$-RCSAs to the structural analysis of small molecules. It is a very complete

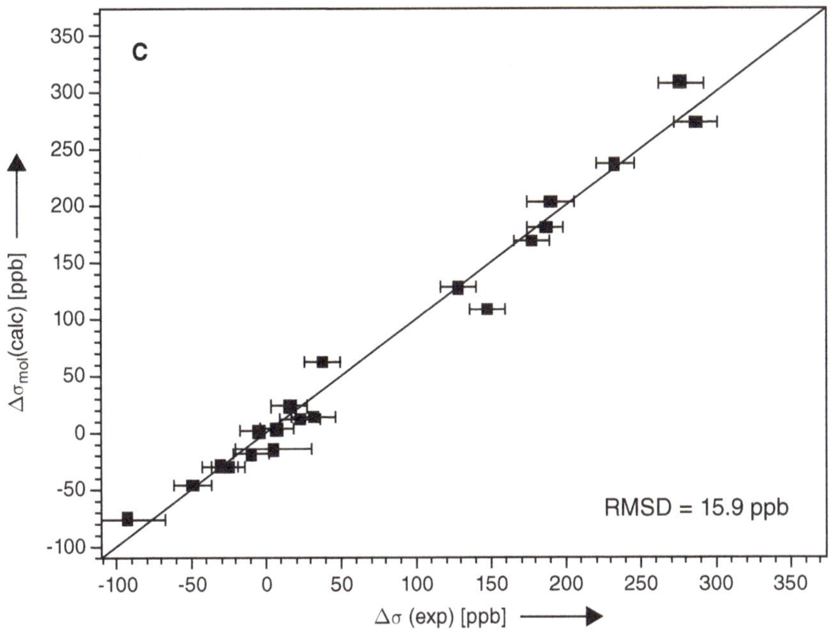


**Figure 9.48** Correlation plot of back-calculated $\Delta\sigma_{mol}$(calc) using the ab-initio $^{13}$C-CSA-tensors and the RDC-derived alignment tensor versus experimental $\Delta\sigma$(exp) residual chemical-shift anisotropies for strychnine in PDMS/$CDCl_3$ corrected by the average uncorrected RCSA value for C-13, C-14, and C-15 as the three nuclei with the smallest theoretically determined CSA tensor. Figure reproduced from Ref. [194] with permission.

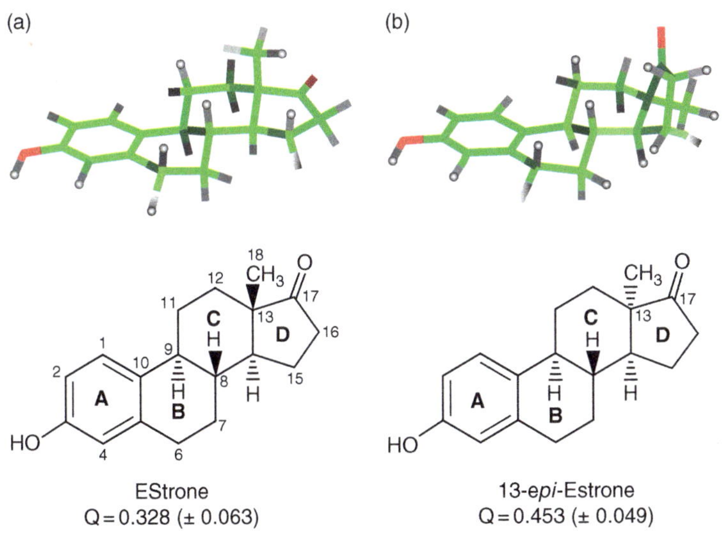


**Figure 9.49** The DFT 3D structure of (a) estone and (b) 13-*epi*-estrone. Figures adapted from Ref. [195] with permission.

and robust study in which ($^{13}$C-$^{1}$H)-RDCs and $^{13}$C-RCSAs were accurately measured for estrone, menthol, mefloquine, retrorsine, and strychnine in stretched and compressed (PMMA gels swollen in $CDCl_3$) (see Figure 9.50). We would like to highlight a few important aspects from this piece or work, but we strongly recommend further reading of this foundational article on $^{13}$C-RCSAs.

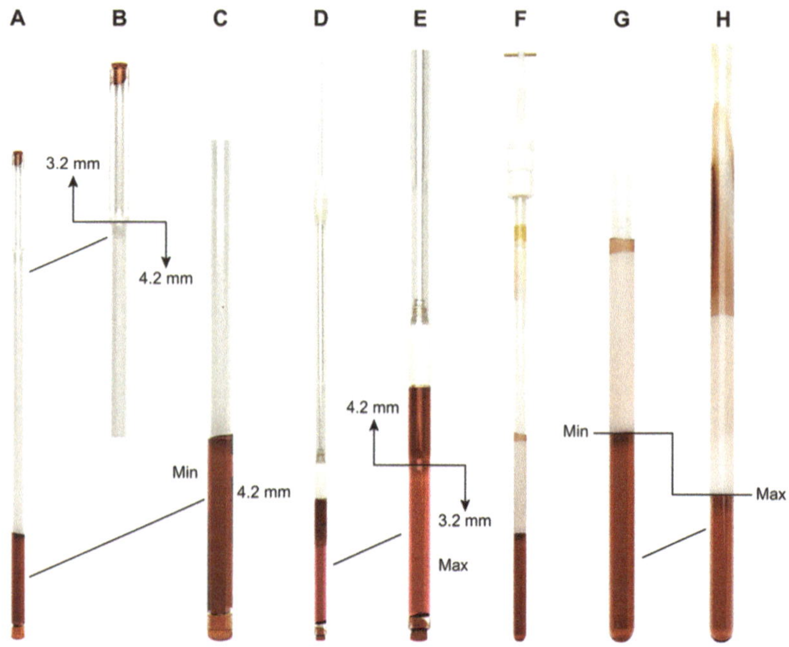


**Figure 9.50** Stretching (left) and compressing (right) device used in this study. Figures adapted from Ref. [168] with permission.

One important finding is that when the stretching device is used, the difference in $^{13}$C chemical shifts between the relaxed and the stretched states is solely due to RCSAs. While, the same difference between the relaxed and compressed states using the compression device is not only due to RCSAs but also to the addition of a predictable amount of isotropic shift that can be corrected [196]. During the compression, the same active volume is used for strong and weak alignment, yielding similar signal-to-noise for both conditions. However, due to solvent and gel composition changes experienced under the different alignment strengths, the isotropic shift must be compensated. This adjustment can be achieved by a robust post-acquisition correction done during the SVD fitting process [196]. The article clearly shows how the $Q$-factor significantly improves when the isotropic gel shifts is corrected when the compression device is used. This problem only exists when measuring RCSAs. RDCs are straightforwardly collected without problems at all.

Coming back to the analysis of estrone, Figure 9.51 shows $^{13}$C-RCSAs collected for estrone with both devices. Between the $^{13}$C-RCSAs collected with stretched gels (panels [a]–[d]) and collected with compressed gels (panels [e]–[h]), clear switch in sign is observed. This is because the alignment tensor rotates 90° between stretched and compressed gels and the all RCSAs values are scaled by the trigonometric function, $\cos^2\theta$, as seen above in Figure 9.46 for the quadrupolar coupling of the $^2$H NMR signal of $CDCl_3$ [194]. This change in sign by rotation of the alignment tensor was also observed experimentally when the compression method was reported in 2010 [13]. It is important to clarify that the stretching device used in this publication is not for variable degree of alignment. It has only two stages for the measurement of RDCs and RCSAs in proteins [197]. It was adapted for organic solvents. Its setup is not as user-friendly as the compression device. It uses an open-ended tube, which poses a risk of losing the stopper and leak the sample into the NMR probe. As shown in Figure 9.52, the structure of estrone and 13-epi-estrone were clearly differentiated by only RCSAs using stretched and compressed (with gel shift correction) PMMA gels.

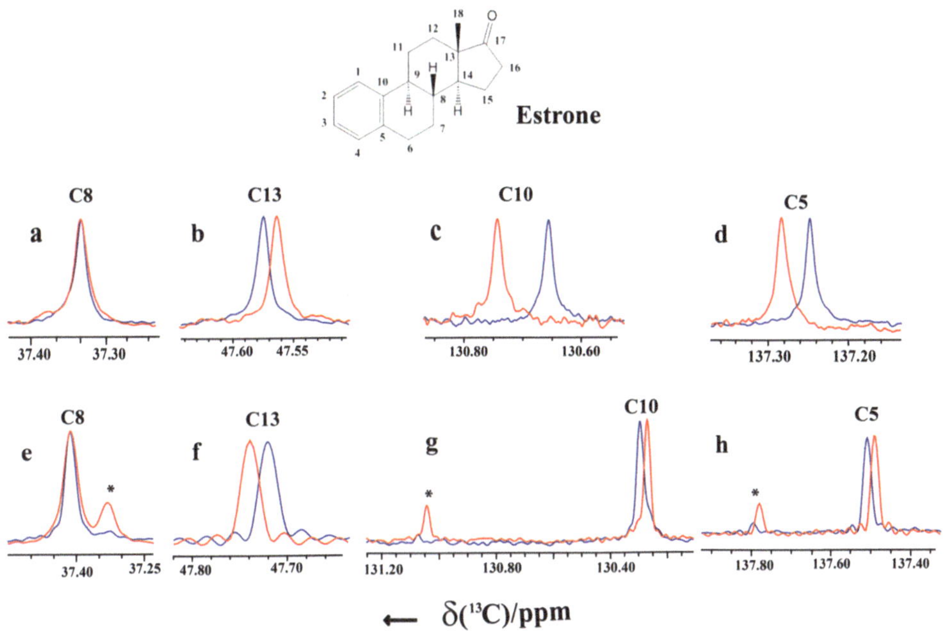


Figure 9.51 Variation of $^{13}$C-RCSAs extracted from 1D NMR spectra of estrone obtained with stretching device (top) at 150.0 MHz and compression device (bottom) at 225.0 MHz of estrone in the narrow-bore (blue) and wide-bore (red) sections of the tube. The C8 resonance shown in panel (e) was used as the reference resonance. Note the presence of both isotropic (marked with an asterisk) and anisotropic signals for some carbons. Figure reproduced from Ref. [168] with permission.

As mentioned previously, different from RDCs, the maximum value of RCSAs depends on the anisotropy of the CS Tensor, as well as on the GDO of the alignment medium (see above). This $^{13}$C-RCSA is significantly larger (~4 – 5 times) for sp$^2$ compared to sp$^3$ carbons. The best solution would be to measure the $^{13}$C-RCSAs in a stronger alignment medium, such as PBLG. However, at the time of this work in 2016, nobody had yet figured out how to compensate for isotropic shifts in PBLG. The GDO of PMMA-based gels is on the order of $10^{-3}$. Hence, an elegant solution was proposed with the introduction of a modified quality factor $Q_{CSA}$, which was based on the conventional $Q$-factor but takes into account the significantly varying CSA values of different carbon atoms, enhancing the capabilities of distinguishing different relative configurations based on $^{13}$C-RCSA data only. The $Q_{CSA}$ is calculates as follows: (i) the alignment tensor is derived by fitting all ΔRCSAs to the DFT-computed CSA tensor through the singular value decomposition (SVD) method and (ii) a new quality factor, $Q_{CSA}$, is calculated by scaling both experimental and calculated ΔRCSA values with the corresponding atom's chemical-shift anisotropies, using the formula given below, where $CSA_{i,ax}$ equals $\sigma_{33} - (\sigma_{22} + \sigma_{11})/2$ and the chemical-shielding eigenvalues $\sigma_{11} - \sigma_{33}$ are obtained from DFT. The $Q_{CSA}$ factor is defined as:

$$Q_{CSA} = \sqrt{\frac{\sum\left(\left(\Delta RCSA_{i,ax}^{Exptl.} - \Delta RCSA_{i,ax}^{Calc.}\right)/CSA_{i,ax}\right)^2}{\sum\left(\Delta RCSA_{i}^{Exptl.}/CSA_{i,ax}\right)^2}}. \tag{9.26}$$

This $Q_{CSA}$ was successfully used in structures containing both sp$^2$ and sp$^3$ carbons e.g. Figure 9.53 shows the $^{13}$C-RCSAs-assisted configurational analysis of strychnine in stretched and compressed PMMA/$CDCl_3$ gels comparing the conventional $Q$-factor with the new $Q_{CSA}$ quality factors. The diastereomers of strychnine were labeled via the $R$- or $S$-descriptors for the chiral carbons C-7, C-8, C-12, C-13, C-14, and C-16, respectively. Hence, *RSSRRS* represents the correct configuration 7$R$, 8$S$, 12$S$, 13$R$, 14$R$, 16$S$ selected by the SVD fitting procedure.

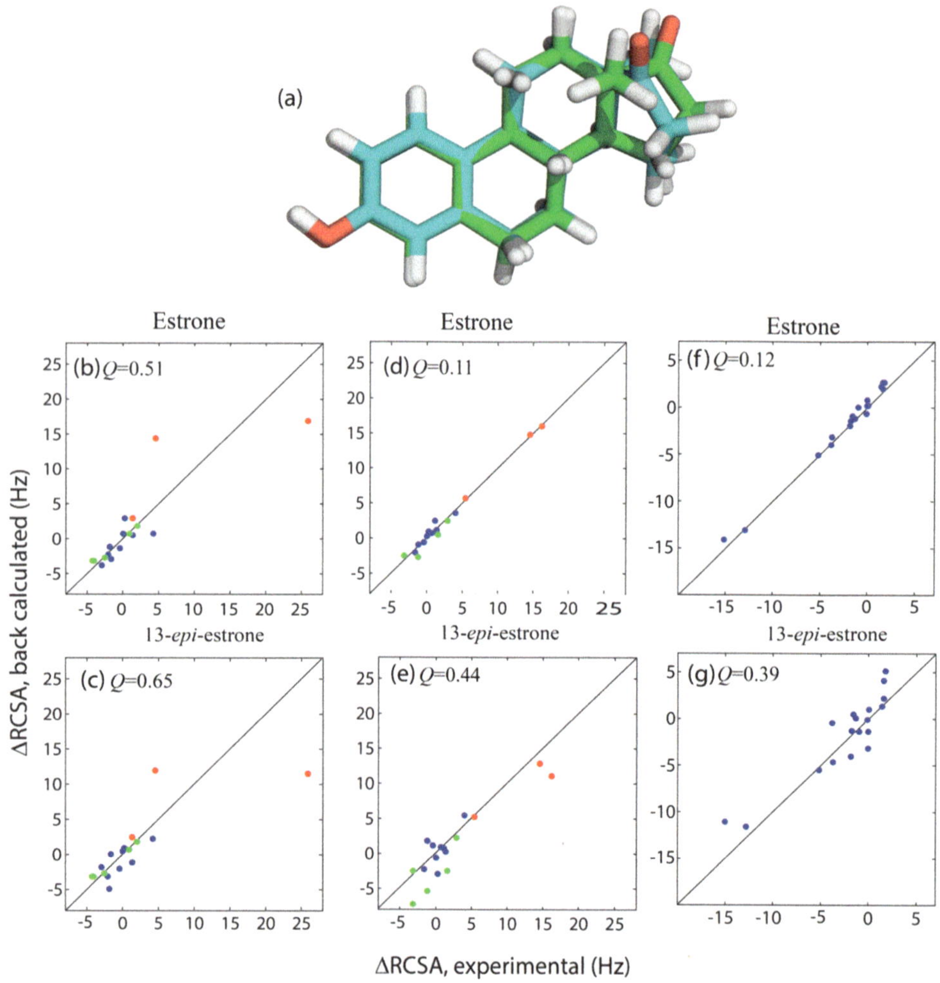


**Figure 9.52** (a) The DFT-computed 3D structural overlay of estrone and 13-*epi*-estrone. (b–g) Comparison of correlation plots $\Delta RCSA^{Exptl.}$ and $\Delta RCSA^{Calc.}$ values ($^{13}C$ data) for estrone and 13-*epi*-estrone using compression (b-e) and stretching (f–g) devices. Panels (b and c) and (d and e) show ΔRCSAs without and isotropic correction, respectively. Between panels b and c, the *Q*-values do favor the correct configuration ($Q = 0.51$ for estrone vs. $Q = 0.653$ for 13-*epi*-estrone). (d) Effect of the correction of the isotropic shift leading to a better differentiation with *Q*-values of 0.11 for estrone and (e) 0.44 for 13-*epi*-estrone. Figure reproduced with permission from reference [168].

Although not described here, to avoid redundancy, the correct configuration of menthol, mefloquine, and retrorsine was also determined successfully in this work. This work has opened new avenues to reliably applied RCSAs to the structural analysis of synthetic small organic molecules as well as natural compounds. Soon after this publication, another structure elucidation of the fungal metabolite Homodimericin A using CASE and a combination of isotropic and anisotropic NMR data, which included the first application of $^{13}C$-RCSAs to the structural analysis of a natural product [198].

Homodimericin A has a 20-carbon atom hexacyclic core with a carbon backbone containing 8 contiguous stereogenic carbons (see Figure 9.54a). Half of its carbon atoms don't have protons attached, presenting a significant challenge for NMR-based structural analysis. The presence of so many non-protonated carbons sets the perfect scenario to use $^{13}C$-RCSAs. RDCs and RCSAs were collected using a poly-HEMA gel swollen in DMSO-$d_6$ and stretched in the two-stages device [168, 197] described above. The combined SVD fitting of RDCs and RCSAs selected the correct configuration for Homodimericin A with a *Q*-factor of 0.162 (see Figure 9.54b).

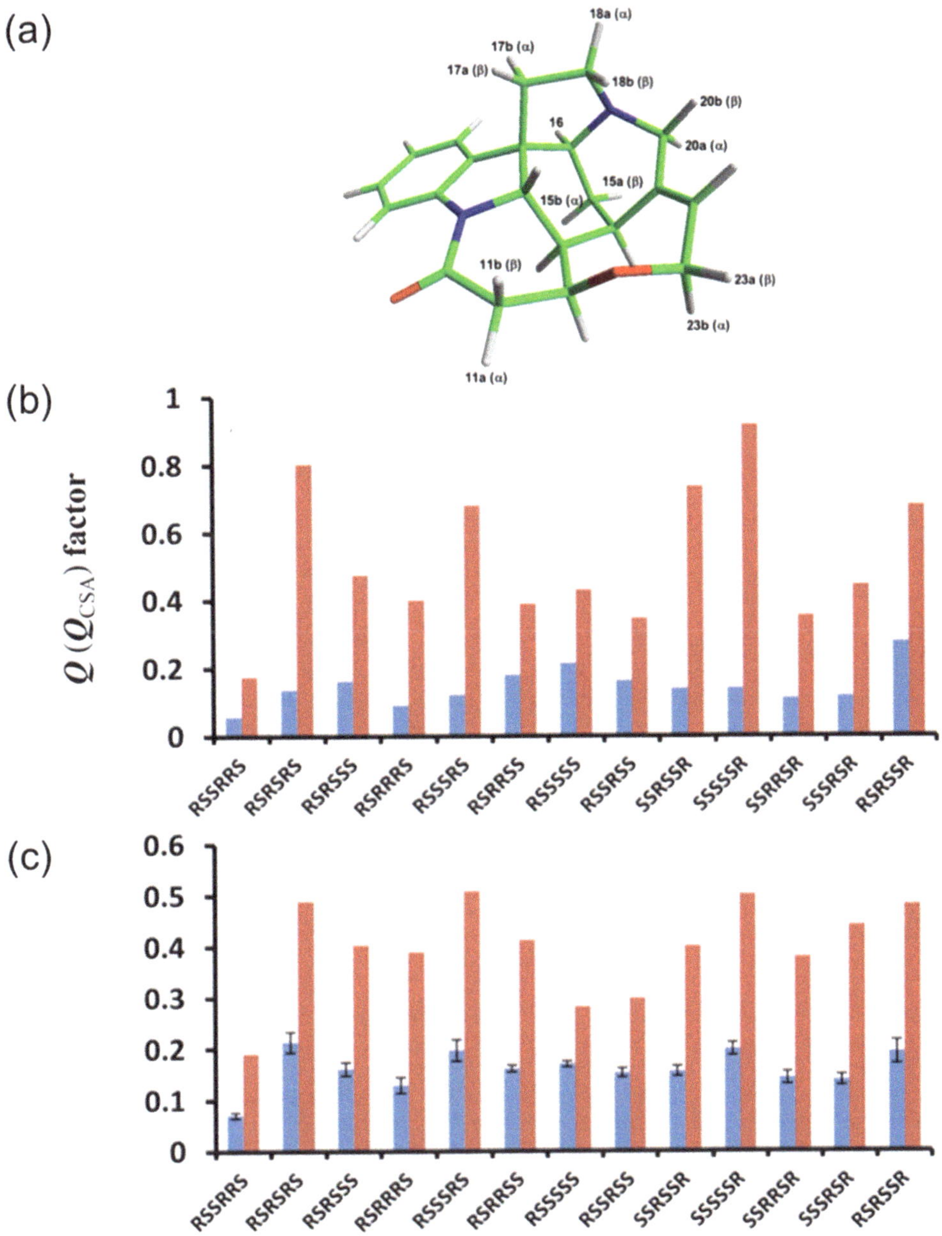


Figure 9.53 Variation of $Q$-factor obtained for strychnine with a ΔRCSA-based stereochemical analysis. (a) DFT 3D structure of strychnine. (b) Results from the stretching device: $Q$ (blue) and $Q_{CSA}$ (red) factors calculated for lowest-energy structures of different diastereoisomers from DFT calculation using only ΔRCSAs. (c) Results from the compression device: $Q$-factors obtained from the analysis of experimental ΔRCSAs and the 13 possible configurations. Figure reproduced from Ref. [168] with permission.

In 2018, a new method to accurately measure RCSAs in PBLG without interferences from isotropic chemical shifts was described [199]. It consists of adding small amounts of PBLG to the compound's solution (e.g. strychnine) and acquired a $^{13}C\{^1H\}$ spectrum after each addition (see Figure 9.55). The method assumes that adding small amounts of PBLG does not introduce isotropic chemical shifts. Tetramethyl silane (TMS) at 4% (v/v) was added for $^{13}C$ chemical shift referencing. The effect of the bulk susceptibility change at different PBLG concentrations is eliminated by TMS referencing. The method, by using PBLG, significantly enhances the $^{13}C$-RCSAs values of

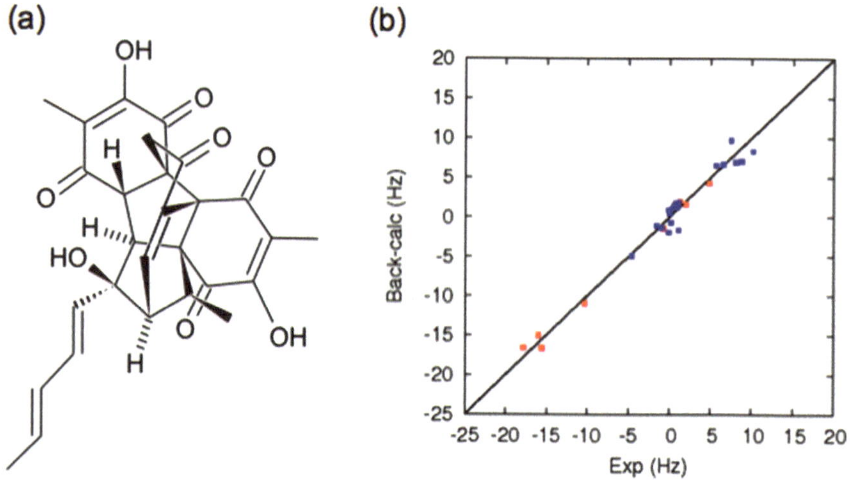


**Figure 9.54** (a) Structure of Homodimericin A. (b) Experimental versus back-calculated RDCs (red) and RCSAs (blue) [198] reproduced with permission.

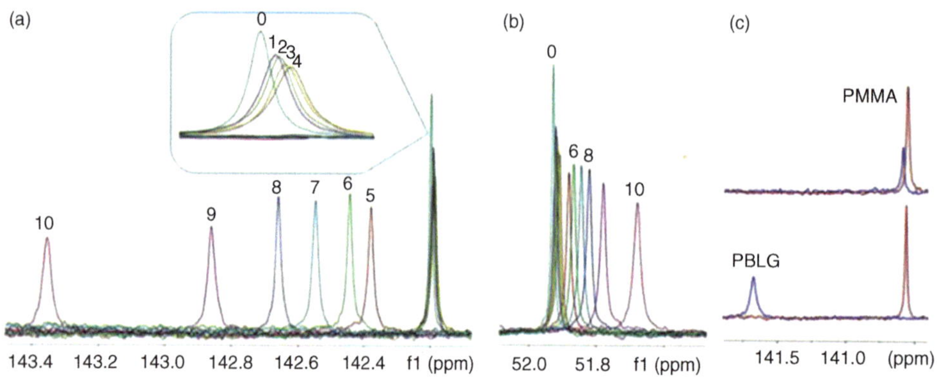


**Figure 9.55** Example of variations of $^{13}$C resonances used for the $^{13}$C-RCSA measurements. (a) Variation obtained when collected with solution made of 0, 2.1, 4.1, 6.6, 8.8, 11.4, 12.9, 15.5, 18.4, 22.9, and 34.5% (w/v) of PBLG in $CDCl_3$ (spectra labelled [0 – 10], respectively. Samples 0 – 4 were acquired with PBLG below its LC-forming critical concentration (isotropic solution spectra), whereas the data for samples 5 – 10 were collected with PBLG mesophase above its critical concentration ($C_{crit}$) and are increasingly anisotropic in nature. A quaternary sp$^2$ carbon of strychnine (1a) (C-21) is shown in (a), and the only quaternary sp$^3$ carbon in strychnine (C-7) is shown in (b). $^{13}$C-RCSA shifts in traces 5 – 10 are significantly larger than the change due to $\Delta\Delta_{iso}$ seen in traces 0 to 4, even for C7 (b), which has a very small DFT-computed CSA of 30 ppm. (c) Top traces: signal of C-21 carbon of strychnine in weakly (red) and strongly (blue) stretched PMMA gel. The RCSA values correspond to the separation between red and blue spectra. Bottom traces: same as top traces with 0% [red, "0" in (a)] and 34.5% PBLG [blue, "10" in (a)]; the actual $^{13}$C-RCSA value is actually slightly larger than the separation between red and blue spectra after correcting for $\Delta\Delta^{iso}$. Although 11 PBLG concentrations were used to study the trend of chemical shift changes, in practice only three concentrations are needed for $^{13}$C-RCSA data extraction. Figure reproduced from Ref. [199] with permission.

carbon atoms with poor anisotropy (low CSA), such as sp$^3$ carbons. Comparison of $^{13}$C-RCSAs measured in PBLG and PMMA gels are provided in Figure 9.55.

The configuration of strychnine was straightforwardly determined with this method (data not shown), though most strychnine carbons have enough CSA to measure $^{13}$C-RCSAs with PMMA gels, as shown above [168]. However, it is noteworthy to mention the power of this method to differentiate caulamidine A from its C-26 epimer

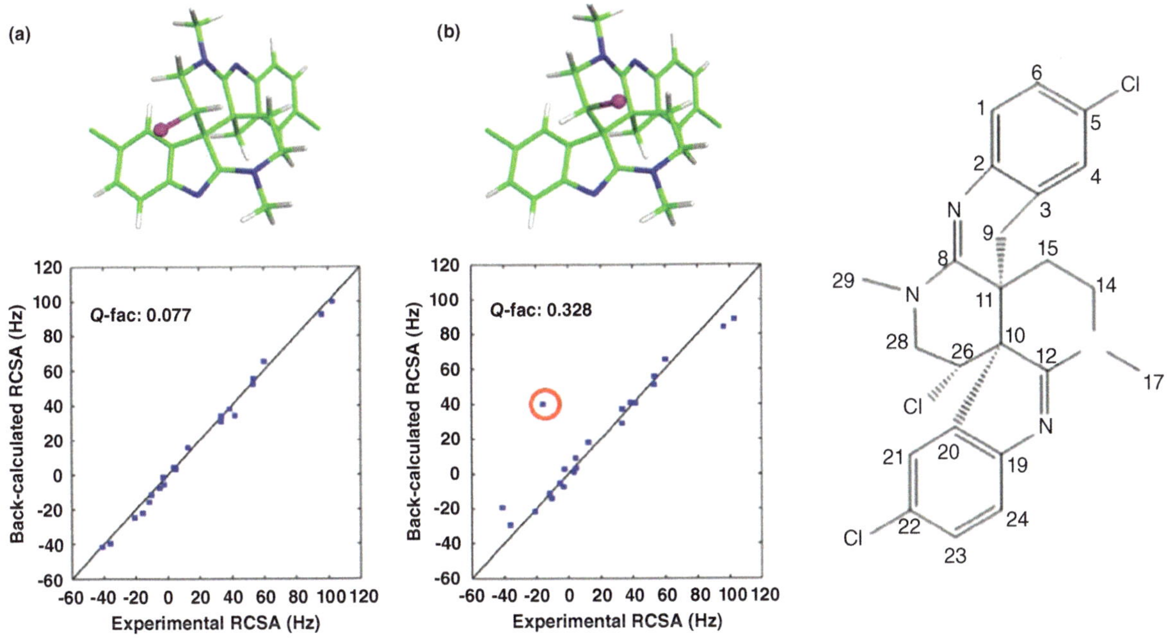


Figure 9.56 Stereochemical differentiation of caulamidine A using $^{13}$C-RCSA data collected in PBLG. (a) The DFT 3D structure of caulamidine A. (b) The energetically feasible C-26-inverted structure. The chlorine atom is represented as a magenta sphere. The outlying point circled in red is the RCSA for the inverted C-26. Figure reproduced from Ref. [199] with permission.

(see Figure 9.56). The structure of caulamidine A was previously reported using NMR data that included RDCs and RCSAs collected in a Poly-HEMA gel in DMSO [200].

This subsequent addition of the PBLG method was intensively used by various research groups. A few examples are described as follows. For complete NMR work, please check the original publications. The structure of dictyospiromide (see Figure 9.57), an antioxidant spirosuccinimide alkaloid from the marine alga, *Dictyota coriacea*, was determined using a combination of isotropic and anisotropic 1D/2D NMR experiments, complemented with chiroptical and computational methodologies. Anisotropic NMR parameters provided critical orthogonal verification of the configuration of the difficult to assign spiro carbon and the other stereogenic centers [201]. Particularly, the application of $^{13}$C-RCSAs led to select the 1*E*, 2*R* with a *Q*-factor of 0.097, as shown in Figure 9.57. As mentioned earlier, the determining of the configuration of proton-deficient molecules is a challenging problem when using conventional NMR methods (*J*-coupling constants and NOE analysis). Here again, determination of experimental $^{13}$C-RCSAs is a powerful alternative option as demonstrated to straightforwardly determine the double-bond configuration ($Z/E$) of a proton-deficient compounds as thiazolidinedione (see Figure 9.58) [202]. In this example, $^{13}$C-RCSAs selected the *E* configuration with a *Q*-factor of 0.068 versus a *Q*-factor of 0.404 for the *Z* isomer.

Another interesting example is provided for phormidolide A, a natural product isolated from the marine cyanobacterium *Leptolyngbya* sp. (strain ISB3NOV94-8A) that shows a mid-range toxicity in the brine shrimp model. It contains a 16-membered macrocycle linked to a pendant polyol side chain terminating in bromomethoxydiene (see Figure 9.59). In light of discordant results arising from recent synthetic and biosynthetic reports, a rigorous study of the configuration of phormidolide A was reported, which outlined a synergistic effort employing computational and anisotropic NMR investigation that provided orthogonal confirmation of the reassigned side chain as well as supporting a further correction of the C-7 stereocenter [203].

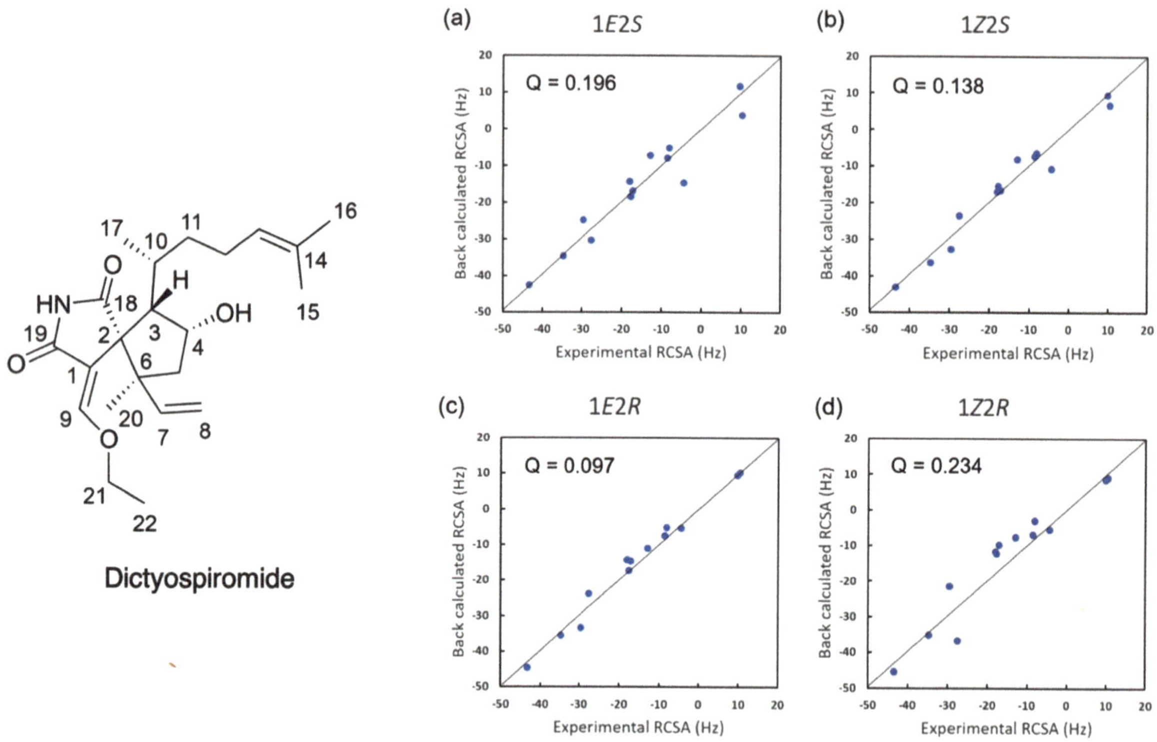


**Figure 9.57** Structure of dictyospiromide and correlation plots of back-calculated versus experimentally measured $^{13}$C-RCSA values for the (a) 1*E*, 2*S*, (b) 1*Z*, 2*S*, (c) 1*E*, 2*R*, and (d) 1*Z*, 2*R* isomers. **Partially reproduced from reference [201] with permission.**

At this point, it is very clear that the key to measure accurate $^{13}$C-RCSAs is to be able to experimentally remove the interferences from isotropic shifts. The stretching gels device described above [197], and the consecutive additions of PBLG method do a very good job to be free of isotropic chemical shift interferences. There was another device developed to remove isotropic chemical shift interferences reported in 2018 [204]. The method was intended as an extension of the compression method, stretching the same gel after it was used compressed to collect RDCs. In practice, the device pushes an already swollen gel through a funnel into the inner part of a 4-mm o.d. NMR tube and consists of a gel chamber, a funnel, a piston, and a piston driver. Details on the device are shown in Figure 9.59. The piston has a VitonS O-ring, which makes this apparatus compatible with $CDCl_3$.

The gel chamber, funnel, and piston are made out of Teflon. This is a modification of the original device developed to stretch polyacrylamide gels into open-ended 5-mm o.d. tube to align proteins in water [205]. Both devices are commercially available [86].

As illustrated in Figure 9.61a, the same PMMA gel stick used for the compression device [13] is first swollen in $CDCl_3$ in a regular 5-mm o.d. NMR tube. RDCs can be measured in the compressed stage, when anisotropy is created by compression. But, if $^{13}$C-RCSAs are collected, a post-acquisition correction done during the SVD fitting process is necessary, as mentioned earlier [196]. The already swollen gel can be further extruded from the gel chamber to a 4-mm o.d NMR tube. In this open-ended 4-mm o.d NMR tube, the gel is stretched and anisotropy is created. This method has the advantage that the 4-mm o.d NMR tube perfectly fits inside a conventional 5-mm o.d NMR tube and is held by a Shigemi tube's cap (see Figure 9.59c). In order to avoid solvent from diffusing into the space between the inner wall of the 5-mm o.d. tube and the outer wall of the 4-mm o.d tube, a tiny plug made of Teflon tape is inserted into the bottom open end of the 4-mm o.d. tube. No $CDCl_3$ isotropic peak is observed

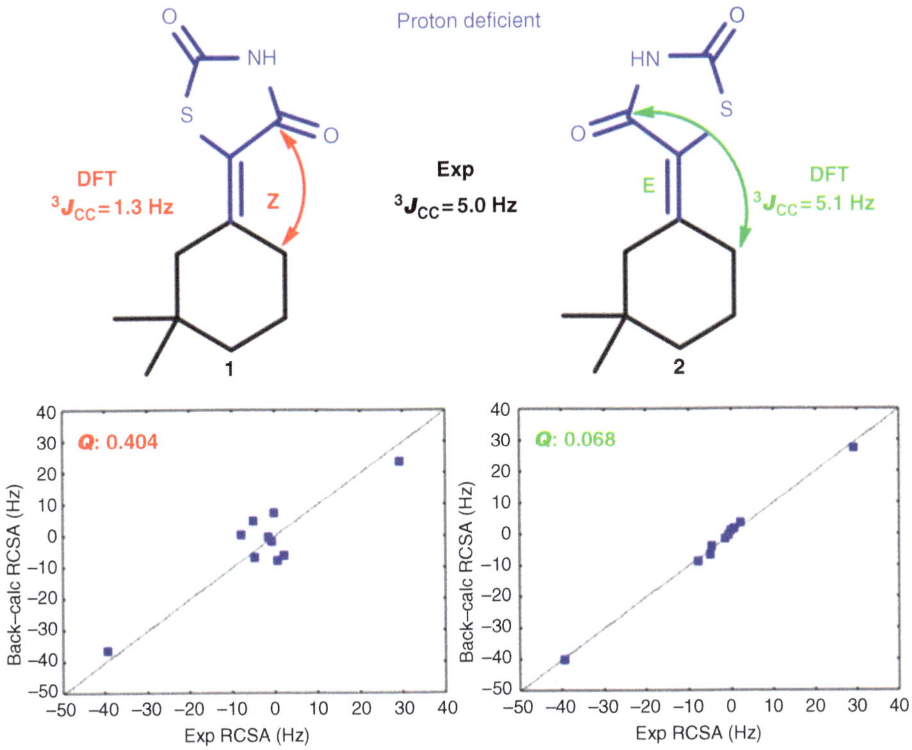


**Figure 9.58** Clear discrimination of the *E* isomer of a proton-deficient analyte, the thiazolidinedione, complemented with experimental versus calculated $^3J_{CC}$ data at $^{13}$C natural abundance. Reproduced from Ref. [202] with permission.

and pure $^{13}$C-RCSAs can be experimentally measured without isotropic chemical shift interferences. ($^{13}$C-$^{1}$H)-RDCs and $^{13}$C-RCSAs for 10-epi-8-deoxycumambrin B, strychnine, and yohimbine were measured and the correct configuration of each compound was successfully selected using only $^{13}$C-RCSAs without post-processing gel shift correct during the SVD fitting procedure [204].

The successful application of $^{13}$C-RCSAs, particularly to the structural analysis of proton-deficient molecules, created excitement in the community and led to the development of further creative methods to accurately measure $^{13}$C-RCSAs without the undesired contribution from isotropic chemical shifts. In this domain, the "one-shot" method, which consists in preparing a PBLG solution in $CDCl_3$ where both, isotropic and anisotropic conditions co-exist was recently explored and reported [206]. The concentration at which this so-called biphasic condition exists depends on the average molecular weight of the PBLG, as shown in Figure 9.62a.

As shown in Figure 9.61b, this is a very ingenious approach that permits the simultaneous collection of 1D $^{13}$C NMR spectra is isotropic and anisotropic condition in exactly the same experimental conditions (concentration, temperature, viscosity, etc.) and hence calculate the $^{13}$C-RDCs in "one-shot." The method was successfully applied to determine the structure of strychnine, neotricone, and excelsione. It is interesting to note that despite the recent report of this approach, it has already seen one application. The "one-shot" method permitted also the revision of the structure of a previously reported synthetic product proposed to be the 1*R*,2*S*-cannabidiol epoxide and was reassigned as cannabielsoin using $^{13}$C-RCSAs [207].

Discovery of new oriented phases is a continuous challenge. Recently, self-assembled oligopeptide nanotubes (noted AAKLVFF) as a new alignment medium for accurate RDC measurement in methanol was reported [208, 209]. A further step was accomplished with the application of the AAKLVFF-based LLCs phase to measure $^{13}$C-RCSAs without interferences from isotropic chemical shifts [210]. Since anisotropy does not develop immediately

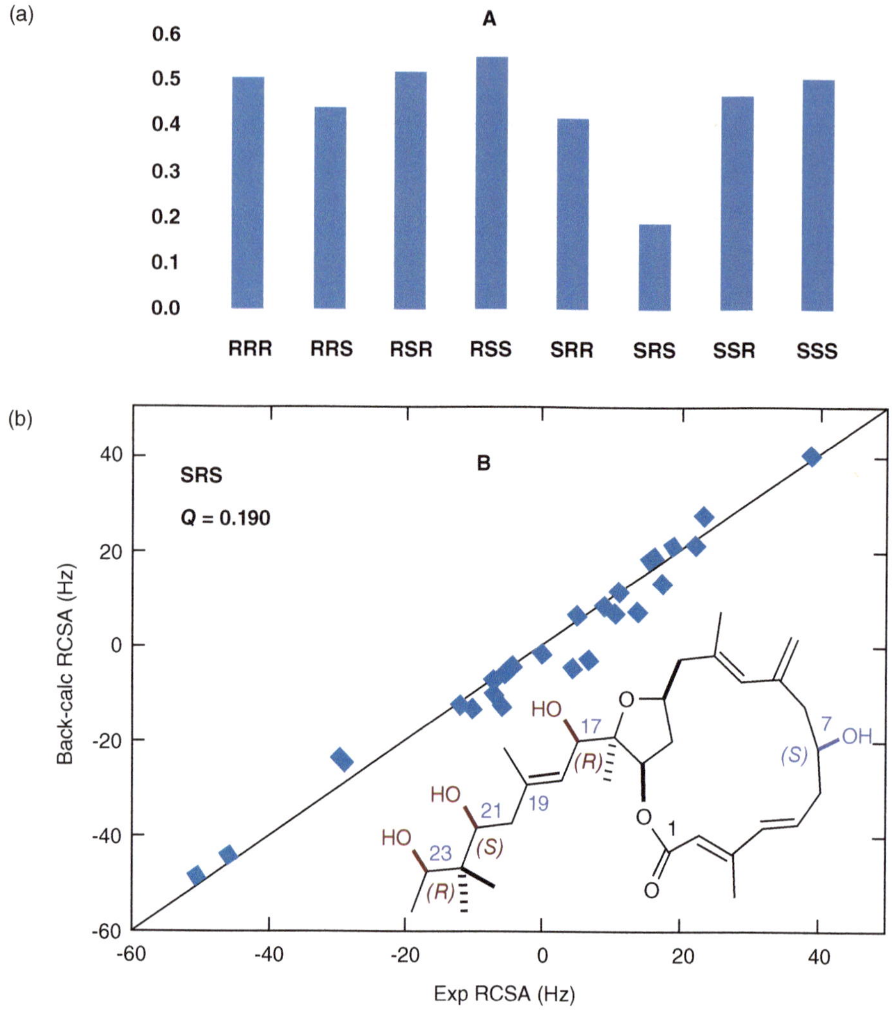


**Figure 9.59** (a) Bar charts of *Q*-factors of isomeric phormidolide A structures, involving permutational stereo-inversion at position C-7, C-17, and C-21/C-23 (highlighted inset, B) leading to structural revision of phormidolide A to *SRS*. (b) *Structure of the S,R,S* C24-truncated phormidolide A fragment, and correlation between $^{13}$C-RCSAs$^{\text{Exptl.}}$ versus $^{13}$C-RCSAs$^{\text{Back-calc.}}$ of the revised structure with *S,R,S* configuration for C-7, C-17 and C-21 stereocenters. Figure reproduced from Ref. [203] with permission.

after making the solution of AAKLVFF, it is possible to collect $^{13}$C-{$^{1}$H} NMR spectra in isotropic conditions right after the solution is prepared.

The anisotropy slowly develops in a time span of 30 days until a state of equilibrium of the LLC phase is achieved, strong RCSAs can measured after 10 days. During this period of time, since the experimental conditions are exactly the same, the change in chemical shift is purely due to RCSAs. Figure 9.63 shows the evolution of the anisotropy build-up as a function of time of the AAKLVFF phase. This approach was also successfully applied to the determination of relative configuration of the four known natural products ((-)-bilobalide, estrone, limonin, and $\beta$-artemether) belonging to different structural classes. In addition, the configuration of marine natural product spiroepicoccin A (see Figure 9.63), a rare thiodiketopiperazine whose configuration could not be assigned based on conventional NMR methods, was unambiguously determined.

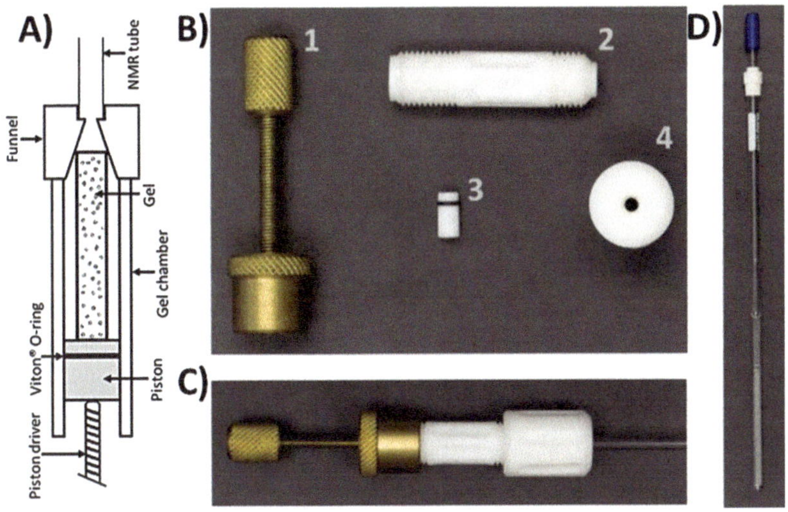


**Figure 9.60** Gels stretching device to measure RCSAs without interference from isotropic chemical shifts. Figure reproduced from Ref. [204] with permission.

Magnetic susceptibility-induced alignment in diamagnetic proteins is significantly small, as shown from 1996 [60]. However, in protein/DNA complexes, a significant anisotropic magnetic susceptibility builds-up by constructive addition of the individual magnetic susceptibility tensors of each nucleotide when they orient parallel in the DNA helical B-form. Thus RDCs of several Hertz for the backbone amides and $^{13}C_{\alpha}$-$^{1}H_{\alpha}$ sites have been measured [211]. Still, the magnetic susceptibility-induced alignment generally remains significantly smaller than what is desired to measure RDCs with high accuracy for proteins or nucleic acids [212].

Magnetic susceptibility-induced alignment scales with the square of the value of the magnetic field $\mathbf{B_0}$. Molecules self-align in solution without the need of any type of alignment media and the alignment can be predicted by DFT calculations. In 2020, this method was applied to measure RDCs and RCSAs (for the first time) on the new marine natural products gymnochrome G (see Figure 9.64a) isolated from the deep-sea crinoid *Hypalocrinus naresianus* featuring a large, proton-deficient aromatic system and two side chains with one stereocenter each [213]. Aromatic rings possess a strongly anisotropic magnetic susceptibility tensor, leading to a large degree of alignment using this approach, as experimentally observed for this compound but also for the strychnine.

Useful data (proton–proton couplings and extracted $^{13}C$ chemical shifts) were measured on $^{1}H$ and $^{13}C$-{$^{1}H$} 1D NMR spectra for each compound at two different magnetic-field strengths (400 and 950 MHz for $^{1}H$). For strychnine, they acquired $^{13}C$-$^{1}H$ and $^{1}H$-$^{1}H$ couplings from CLIP-HSQC 2D spectra [213].

As mentioned above, this type of alignment scales with the square of the magnetic field $\mathbf{B_0}$. While the isotropic component of the chemical shift and the *J*-couplings stay constant at all magnetic field values, any change in chemical shift and *J*-couplings observed on anisotropic spectra is due to the contribution of RCSAs and RDCs, respectively. It is a difference experiment, and the larger the difference in $\mathbf{B_0}$, the larger the values of RCSAs and RDCs are.

First RDCs measured for small molecules using this method were reported in 1986 for analyzing *o*-dichlorobenzene [214], for analyzing a porphyrin-quinone based molecular cage in 1988 [215], and hydrogenated fullerenes in 1997 [216]. However, this is the first time that RCSAs are reported for small molecules.

The beauty of this method, regardless of the fact that the alignment is very weak and it needs very high-field NMR instruments, resides in the fact that the magnetic susceptibility tensor can be predicted by DFT calculation

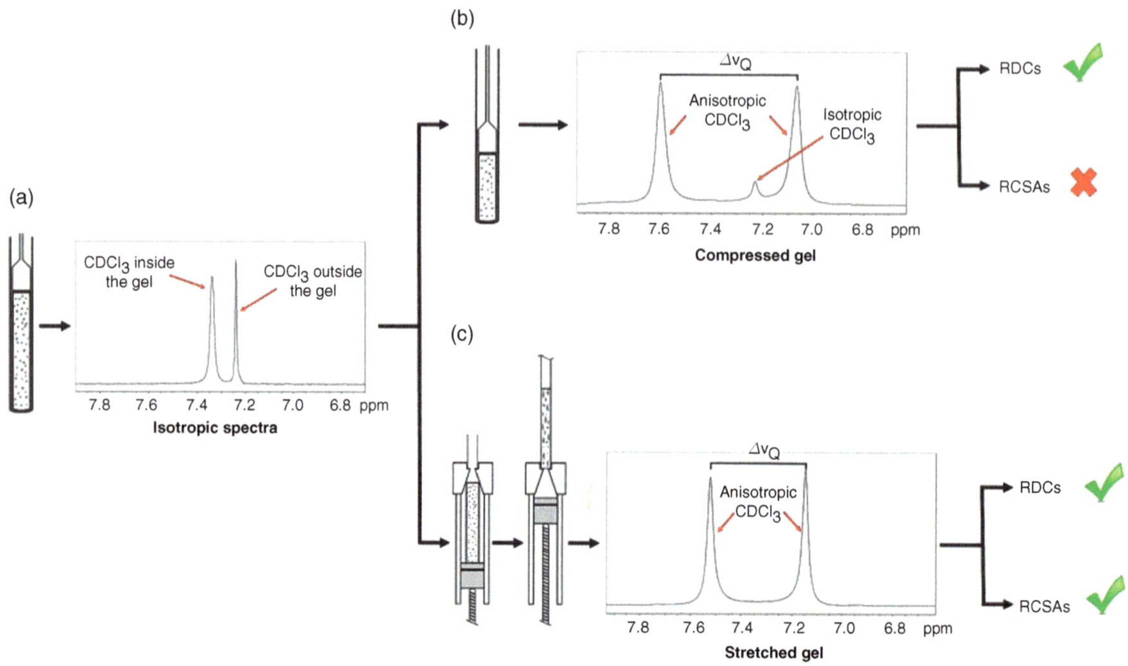


**Figure 9.61** Working flow for the method involving the device shown in Figure 9.59. (a) Swollen of gel in a 5-mm o.d. NMR tube. When relaxed, $CDCl_3$ shows $^2$H NMR isotropic signals from inside and outside the gel. (b) Compressed gel: the $^2$H NMR signal of $CDCl_3$ inside the gel splits into an $^2$H-QD while the signal from outside the gel stay unchanged (a single line) but reduced in intensity. In these conditions, it is possible to collect RDCs but not RCSAs without isotropic chemical shift interferences. (c) Extruded gel through the funnel and stretched inside a 4-mm o.d. NMR tube, no signals from outside the gel are observed. RDCs and RCSAs can be measured accurate without further corrections. Figure reproduced from Ref. [204] with permission.

and the align-ment can be predicted. Then, experimental versus predicted RDCs or RCSAs can be compared. If the molecule is flexible, the magnetic susceptibility tensor for each conformer can be calculated and Boltzmann averaging of RDCs or RCSAs can be predicted from the calculated energies, instead of having to perfume SVD fitting to the conformational ensemble using the single-tensor approximation. Figure 9.59 shows the *Q*-factors for all diastereomers of predicted and SVD-fitted RCSAs for gymnochrome G and strychnine. In both cases, the correct structure is selected.

At natural abundance, $^1$H NMR is 5 666 times more sensitive than $^{13}$C NMR. Hence, it would be more convenient to measure $^1$H- instead of $^{13}$C-RCSAs, but on the other side, carbons show stronger anisotropy than protons. It is evident that $^1$H-RCSA would be ideal for limited natural product samples at the microgram level. Let's see some facts about $^1$H-RCSAs. From DFT calculations, the most anisotropic proton in strychnine is H-18$\alpha$ with an anisotropy value of 11.900, while the less anisotropic one is H-8 with an anisotropy value of 3.227. LLC phase such as PBLG are a no go for $^1$H 1D NMR spectra due to the significant signal's line broadening observed due to their viscosity. The best option would be to use aligning gels. The GDO in aligning gels, such as compressed PMMA, is around $7 \times 10^{-4}$. Hence, the maximum $^1$H-RCSAs values for H-18$\alpha$ should be equal to $7 \times 10^{-4} \times (\sigma_{zz} - \sigma^{iso}) = 7 \times 10^{-4} \times (36.86 - 28.93) = 0.00555$ ppm. This value is a constant. It increases therefore in Hertz as the magnetic field $\mathbf{B_0}$ increases in value. To minimize the error, $^1$H-RCSAs should be collected at very high magnetic fields, e.g., at 800 MHz, this value corresponds to only 4.44 Hz. For H-8, the same calculations yield a value of only 1.20 Hz. Measuring such small values in Hz accurately and without isotropic shift interferences represents a very challenging task. However, in 2020, it has been reported the successful application of $^1$H-RCSAs for the structural

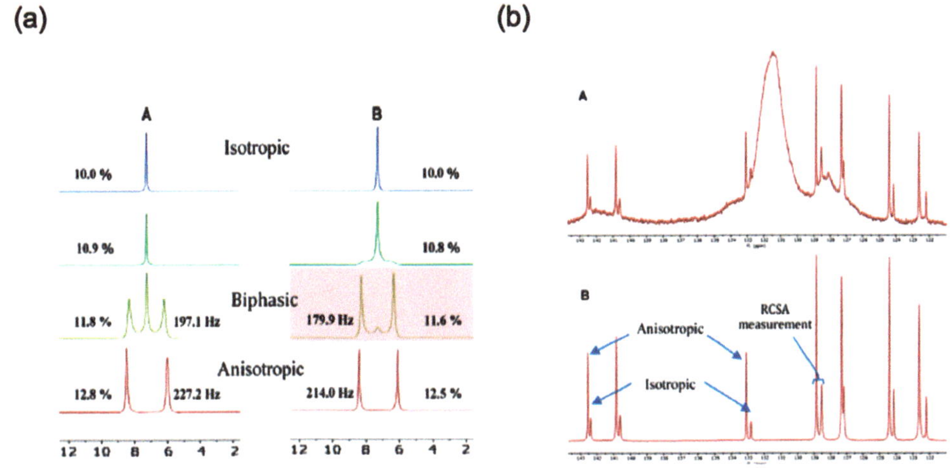


Figure 9.62 (a) 92.1 MHz $^2$H NMR spectra of varying concentrations of two different PBLG (w/v%) with differing molecular weight distributions in $CDCl_3$ at 300 K. ([left] PBLG MW of ~249 kDa, [right] PBLG MW of ~328.5 kDa). (b) Aromatic region expansion of the 150.9 MHz $^{13}$C-{$^1$H} 1D NMR spectrum of strychnine in PBLG (11.6%) showing NMR resonances for seven sp$^2$ carbons and their respective isotropic and anisotropic NMR resonances. The ratio of anisotropic to isotropic species is proportional to the highlighted $^2$H NMR data. Figures reproduced from Ref. [206] with permission.

analysis of natural products at microgram levels [217]. The study was conducted with NMR instruments operating at 700 and 800 MHz with several natural products and $^1$H-RCSAs measured with the samples oriented in aligning gels. Among a few known natural products, the relative configuration of a 35-μg sample of the new diterpenoid briarane B-3 isolated from the gorgonian *Briareum asbestinum* collected in the waters off the Yucatan Peninsula of Mexico (see Figure 9.65) was determined. The sample was aligned in a perdeuterated PMMA-$d_8$ gel stretched in a Hilgenberg's micro stretching device and the $^1$H 1D NMR spectra recorded at 800 MHz. Its AC was determined by ECD. A synergistic combination of anisotropic NMR with chiroptics.

### 9.7.3 Configuration Determination Using Spin-1 NMR Analysis

A powerful and recent alternative to RDCs and RCSAs is the use of RQCs, because this third anisotropic observable (specific to nuclei with spin $I > 1/2$) encodes also valuable 3D structural information and correlates the relative orientation of stereogenic centers, regardless of the distance between them in the molecule. As mentioned in Section 9.3.3, RQCs can be effectively exploited for analytical or structural purposes when the relaxation rates (governed by the quadrupole mechanisms) is relatively low, allowing for high-resolution NMR spectra. This characteristic is typically met with deuterium, with the advantage of being naturally present in all hydrogenated molecules, with $\delta^{\text{iso or aniso}}(^2\text{H}) \approx \delta^{\text{iso or aniso}}(^1\text{H})$, and probing the same molecular directions that the associated $^{13}$C-$^1$H bond vectors (see Equation 9.12).

If the analysis of oriented deuterated solutes was abundantly used, [4, 5, 218] the detection of first $^2$H spectra at natural abundance of a solute has been reported since 1964 [219], but isotropic NAD NMR finds interesting developments with the achievement of FT-NMR, as an aid to analysis and decipher complex $^1$H spectra, since $\delta^{\text{iso}}(^2\text{H}) \approx \delta^{\text{iso}}(^1\text{H})$ [220], as a simple nuclear probe to understand biosynthesis mechanisms [221] as well as the determination of site-specific ($^2$H/$^1$H) isotopic ratios in molecules by Martin and co-workers from 1981 (see Section 9.8) [222]. Disregarding rather examples of ANAD spectra of LCs (thermotropic) [223, 224], paradoxically, the first results on the utility of $^2$H-RQCs recorded at natural abundance level for analyzing solutes in oriented media have been only reported in 1998 [121] 35 years after the first anisotropic $^1$H 1D-NMR spectra

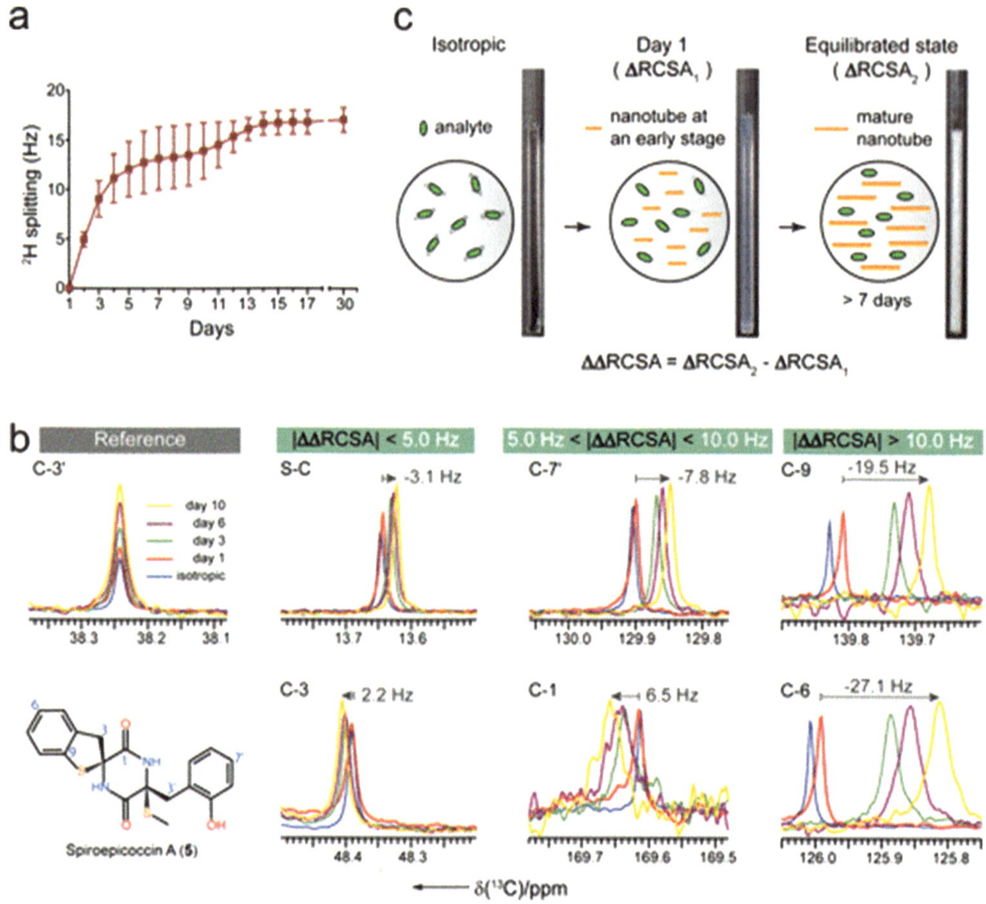


Figure 9.63 Experimental strategy for the measurement of ΔΔRCSAs in an AAKLVFF liquid-crystalline phase. (a) Time evolution of $^{2}$H quadrupolar splitting (Hz) measured using eight different samples. (b) Representative $^{13}$C-{$^{1}$H} signals of aliphatic, aromatic, and carbonyl carbons of spiroepicoccin A (5) measured under isotropic conditions (blue), day 1 (red), day 3 (green), day 6 (magenta), and day 10 (yellow). C-3′ was defined as the reference. (c) Schematic representation of the experimental procedure for the extraction of ΔΔRCSAs in an AAKLVFF phase. Figure reproduced from ref. [210] with permission.

of a solute [57], using also thermotropic systems (achiral nematics), the same year of the first enantiotopic and enantiomeric spectral discriminations in PBLG-based lyotropic CLCs [122]. Interestingly, in lyotropic systems, the significant difference of $T_2(^2H)$ values (due to their dynamical range) between the different deuterium sites in the polymer (such as polypeptide and polyacetylene) and in the solute leads to very low-resolution spectra (baseline) and high-resolution ANAD spectra, respectively, and no real interference between the two types of spectral signatures exits.

### 9.7.3.1 Analysis of Scalemic Mixtures by ANAD Two-dimensional NMR

Using ANAD QUOSY-type 2D experiments (see Section 9.4.2) to simply the identification and assignment of $^{2}$H-QDs, the analysis of enantio-isotopomeric mixture associated to prochiral compounds [131, 176, 225–227] is possible and can be extended to racemic or scalemic mixtures of two enantiomers [112, 177]. This is interesting

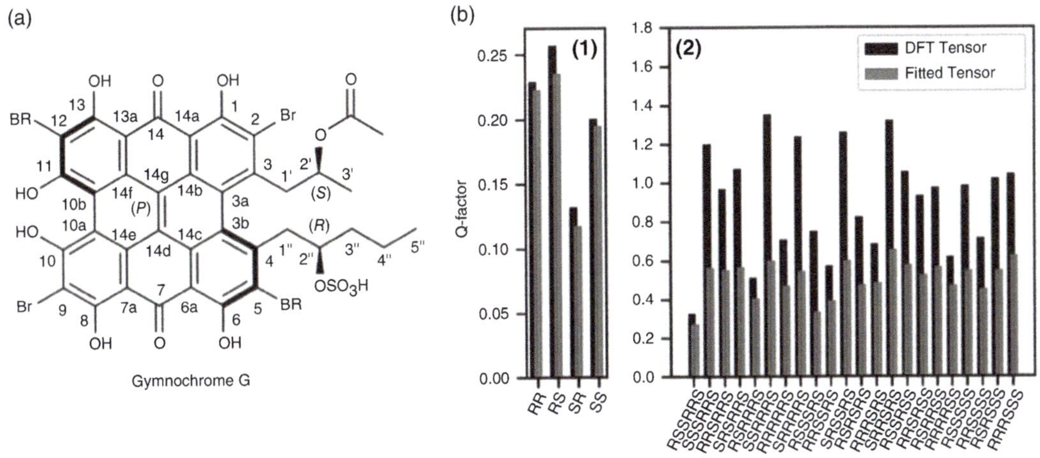


**Figure 9.64** (a) Structure of gymnochrome G. (b) $Q$-factors for the different configurations of gymnochrome G (left) and strychnine (right). The values obtained from predicting the alignment tensor with DFT are shown in black, the ones obtained from the fitted alignment tensor are shown in gray. In both cases, the true configuration is correctly identified by the lowest $Q$-factor. Figure reproduced from Ref. [213] with permission.

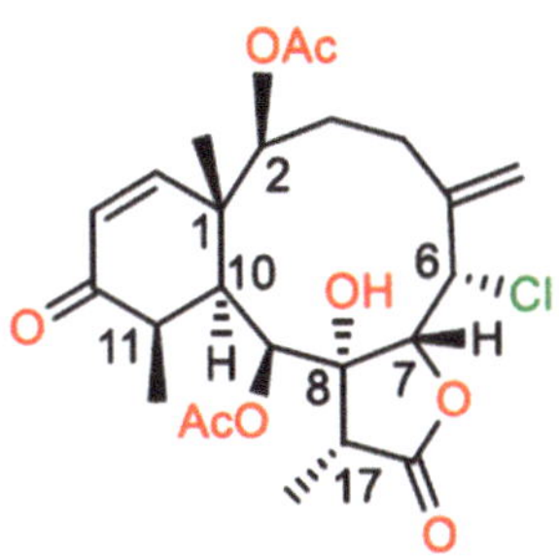


**Figure 9.65** Structure of briarane B-3 along with configuration.

because when RDCs are concerned, the data are generally extracted from two oriented samples, one for each enantiomer, assuming that all experimental conditions are the same.

Simplified schematic protocol showing the principle of molecular structural analysis using the $^2$H-RQCs extracted from ANAD 2D-NMR experiments as presented in Figure 9.66a [177]. The integrated computational program using SVD algorithm can be "MSpin-RQC" [177, 178] or "ConArch+" [129, 181]. Note here that the sign of $^2$H-RQCs for a given $^{13}$C-$^2$H bond, not directly accessible from ANAD 1D/2D spectra derived from a rapid analysis of the $^1J(^{13}\text{C-}^2\text{H})$ and $^1T(^{13}\text{C-}^2\text{H})$ values extracted from isotropic and anisotropic proton-coupled $^{13}$C 1D spectra or with the help of $J/D$-resolved 2D spectra.

As in case of RDCs and RCSAs, from the analysis of $^2$H-RQC data, we can evaluate the difference of average molecular orientation of each enantiomer and the comparison of their alignment tensors (see Figure 9.66b). Such information may provide a quantitive data on the enantiomeric discrimination ability (EDA) of various chiral polymers toward a given analyte or make a comparison of a series of chiral molecules toward a given chiral polymer [177, 181]. Such data are useful to have better insight on the enantiodiscrimination mechanisms in CLCs, with the possibility to correlate the enantiorecognition capabilities to the nature or the structure of chiral polymers (see Section 9.7.4).

In practice, this description is done by calculating the molecular Saupe matrix of each enantiomer, $\{S_{\alpha\beta}\}^{R,S}$ and then compared by evaluating the generalized 9D $\beta$ angle using Equation 9.27 [20, 177]:

$$\beta = \arccos\left(\frac{\sum_{\alpha\beta=x,y,z} S^{R}_{\alpha\beta} S^{S}_{\alpha\beta}}{\sqrt{\sum_{\alpha\beta=x,y,z}\left(S^{R}_{\alpha\beta}\right)^2}\sqrt{\sum_{\alpha\beta=x,y,z}\left(S^{S}_{\alpha\beta}\right)^2}}\right). \tag{9.27}$$

The $\beta$ angle value or the cosine($\beta$) value (GCB value) can be used to quantity the enantiodiscrimination efficiency for a enantiomeric pair. Thus, the closer this GCB value is to 1, the closer the positions of the alignment tensors are in space. Conversely, the closer this value is to 0, the more the positions of the axes of the alignment tensors are different, this second situation corresponds to the maximum enantiodifferentiation and so to an optimum EDA.

#### 9.7.3.2 First Analysis of Natural Drugs by ANAD 2D NMR

In the sections above, various examples of 3D structural elucidation of natural products or drugs as well as synthetic compounds of interest based on a unique set of RDCs or RCSAs (or a combination of both), with or without other data sources on the spatial arrangement of atoms in molecules (as nOe) have been proposed, clearly illustrating their analytical potentialities and their contribution to structural issues.

In 2020, a new door was opened with the demonstration that $^2$H-RQCs extracted from ANAD spectra could provide reliable and coherent set of anisotropic dataset and hence becoming a promising alternative to RDCs and RCSAs to solve non-trivial structural problems. The performance and scope of this tool was examined for two natural chiral compounds of pharmaceutical interest (artemisinin and strychnine). Figure 9.67a presents the molecular structure, atomic labels and $R/S$ descriptors of artemisinin as well as DFT-optimized 3D structures. This rigid molecule possesses seven stereogenic center, leading to 128 ($2^7$) isomers.

Using the fit procedure shown in Figure 9.66, it has been possible to identify the right molecular relative configuration (1$S$, 4$R$, 5$S$, 6$R$, 7$S$, 10$R$, 11$R$), using either the various $^2$H-RQCs and the average of $^2$H-RQCs for each methylene group and their respective prostereogenic methylene directions ($\alpha$, $\beta$), but also the $^2$H-RQC data of all

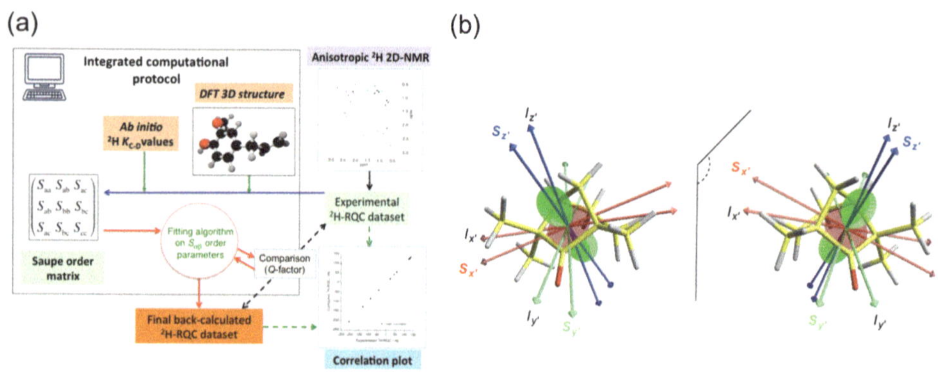


**Figure 9.66** (a) Simplified flowchart showing the principle of molecular structural analysis using the integrated computational program "MSpin-RQC" from the $^2$H-RQCs extracted from ANAD 2D-NMR experiments. The correlation plot ($^2$H-RQC$^{\text{Calc.}}$ vs. $^2$H-RQC$^{\text{Exptl.}}$) simply visualizes the quality of fit of experimental data. (b) The PAS ($S_{x'x'}$, $S_{y'y'}$, $S_{z'z'}$) of the diagonalized Saupe matrix, the inertia tensor axes, ($I_{x'}$, $I_{y'}$, $I_{z'}$), and the Saupe tensor surface representation (red and green surfaces indicate positive and negative $^2$H-RQCs, respectively) for (left) ($S$)-FCH and (right) ($R$)-FCH oriented in PBLG/$CHCl_3$. Figure adapted from ref. [177] with permission.

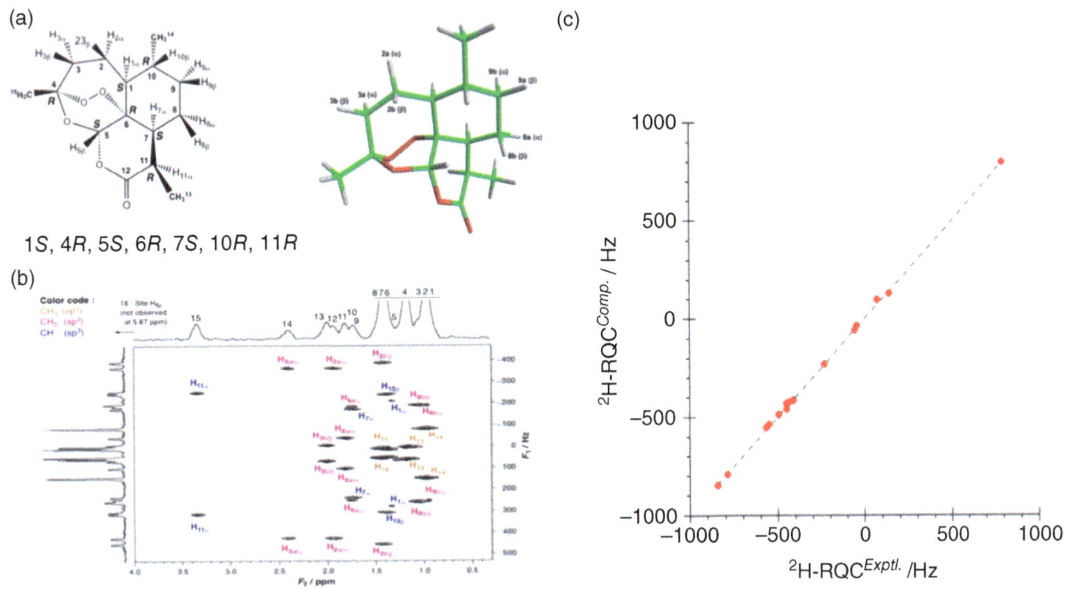


Figure 9.67 (a) Molecular structure, atomic labels and ($R/S$)-descriptors of artemisinin associated with the (known) AC, and DFT-optimized 3D structures labeling the $\alpha/\beta$ (a/b) positions of diastereotopic protons/deuterons. (b) Part of the NAD-{$^1$H} $Q$-resolved Fz 2D map (tilted and then symmetrized). (c) Correlation plot between experimental and back-calculated $^2$H-RQC values. Figure adapted from Ref. [174] with permission.

deuterium sites [174]. The smaller $Q$-value obtained dropped to 0.017 for all data, while a $Q$-value of 0.011 was calculated when the average $^2$H-RQCs for the diastereotopic site was used. The agreement between experimental and back-calculated RQCs for the smallest $Q$-factor as shown in Figure 9.67c demonstrated the quality of the coherence between the RQC dataset and the spatial arrangements associated to the 1$S$, 4$R$, 5$S$, 6$R$, 7$S$, 10$R$, 11$R$ configuration.

The final configuration obtained by the $^2$H-RQC protocol is fully consistent with previous reports using RDCs [45] as anisotropic data as well as with X-ray structure [228], consequently validating for the first time the robustness of the ANAD-NMR methodology.

Compared to RDC analysis, all monodeuterated isotopomers for a given molecule form independent dilute $^2$H-spin systems, and hence, in ANAD NMR the strong coupling effects that frequently hamper the accurate determination of $^1D_{CH}$ couplings in $CH_2$ groups no longer exist.

From an experimental point of view, the analysis of the ANAD 2D-NMR spectrum recorded in PBLG/$CHCl_3$ lead to the unambiguous extraction of fifteen signed $^2$H-RQCs out of sixteen possible sites (see Figure 9.67a). An important deuterium depletion effect (associated with a low $^2$H/$^1$H isotopic ratio) may lead to a drastic reduction of the $^2$H-QD intensity at certain sites not detected (see Section 9.8). However, the lack of some $^2$H-RQCs associated to a molecular structure does systematically not pre-empt the accomplishment of the structural analysis, if a sufficient number of anisotropic data are still experimentally available.

### 9.7.4 Determining the Absolute Configuration of Monostereogenic Chiral Molecules

Determining the AC of enantiomeric NMR signals ($R/S$, $M/P$, $D/L$, $\Delta/\Lambda$) with a single stereogenic center using anisotropic NMR in CLCs alone remains a real challenge and a difficult issue because no direct and simple experimental tools exist to do so [229]. Same situation arises for the signal assignment of enantiotopic elements (pro-*R*/pro-*S*) in prochiral molecules.

In 2007, an empirical approach combining NAD 2D-NMR and a PBLG-based mesophase was proposed for assigning the AC of small chiral molecules [230]. The key concept of this strategy lies on the fact that recognition phenomena of molecular shape play an important role in global ordering/differentiation mechanisms, and hence analogous molecular structures should be oriented similarly in the same aligned environment. By comparison with molecules in which the AC of enantiomeric signals is known and identified, it becomes possible to propose an AC for the enantiomeric signals of compounds of interest, with an unknown AC.

To date, the most promising alternative to is the development of ab initio methods (as molecular dynamic [MD] based computational simulations) able to predict molecular alignment. First descriptions were proposed in 1994 [231]. The main issue is to correctly evaluate all type of interactions (electronic distribution and molecular shape) between a given analyte and a helically chiral system from the evaluation of interactions between a given analyte and a helically chiral system [232–236]. Although some progresses were obtained for predicting orientation behavior of analytes interacting with achiral polypeptide-based achiral systems (PBG) [236], and in case of two enantiomers interacting with helical-polymer based CLCs [233, 235], the results obtained are rare and not always convincing in terms of reliability. Thus, MD simulation involving PBLG remains rather fragile because MD has to be performed with a sufficiently long time to correctly evaluate the effect of conformational dynamic of all side chains along the $\alpha$-helix on the enantiodiscriminating effects. This occurrence needs a formidable calculation power, even whether a small number of polymeric units is considered.

## 9.8 Conformational Analysis in Oriented Solvents

The presence of conformational dynamics in flexible molecules, i.e large, low frequency torsional molecular motions, may strongly complicate the interpretation and the correct use of the anisotropic NMR data (as RDCs) compared to the case of rigid compounds [237–239]. To solve this difficult problem, several (more or less complex) calculation approaches have been proposed so far. To help the reader to go further, several solutions were proposed to manage this important problem are briefly outlined below, but a more detailed overview can be found in the Ref. [40].

The simplest possible model, the rotational isomeric state (RIS) model assumes that only a restricted set of minimum-energy structures is populated [240]. For instance, the free rotation of methyl groups are described by the three statistically weighted staggered conformers. More realistic models allow for continuous bond rotations. Two main approaches, alone or combined, have been proposed in the past: (i) the additive potential (AP) method, an "approximate" approach whose aim is to reduce latest equations to manageable expressions by using some physically justifiable approximations [241] and (ii) the maximum entropy (ME) method, an "unbiased" method based on information theory [242–244]. Finally, in the last decade, hybrid strategies were developed, combining the AP method and a direct probability distribution, DPD, for describing directly the internal potential as sum of Gaussians [237, 238, 245]. These models were applied to evaluate the conformational dynamic of small flexible analytes as naproxen or flurbiprofen (enantiopure chiral drugs) in polypeptide (PBLG) mesophases [246–248] and a couple of enantiomers of (*R*/*S*)-1-(4-fluorophenyl)ethan-1-ol, for the first time in 2022 [249]. In this example, it was demonstrated that *R*/*S*-enantiomers show significant differences in their conformational behaviors, due to the difference of solute-solvent orientational molecular interactions experienced by the (*R*)- and (*S*)-isomers in a helical chiral environment as PBLG system. Understanding of chirality-dependent molecular interactions, often involved in a number of important chemical processes.

## 9.9 Anisotropic $^2$H 2D NMR Applied to Molecular Isotope Analysis

### 9.9.1 The Natural ($^2$H/$^1$H) Isotope Fractionation: Principle

The relative abundance of deuterium compared to proton is about 155 ppm, this specific value is known as Vienna-standard mean ocean water (V-SMOW) value [10, 22, 32]. However, in reality, the deuterium/proton isotopic ratio can vary from one site to another site of the molecule, thus leading to its molecular isotopic profile.

The chemical origin of the isotopic fractionation comes from a discrimination between light ($^1$H) and heavy atoms ($^2$H) that occurs during (bio)chemical processes [250]. The determination of isotopic fractionation can provide key data to: (i) understand the enzymatic mechanisms, (ii) validate the biosynthetic pathways of natural molecules, and (iii) authenticate the geographical/botanical origin of bioproducts [251].

The principle of deuterium isotope fractionation analysis consists in recording $^2$H-{$^1$H} 1D/2D NMR spectra under quantitative conditions (with recycling time $T_R(^2\text{H}) = 5 \times T_i^{(\text{the longest value})}(^2\text{H})$ in the presence of an internal reference (such as tetramethylurea, TMU), for which the ($^2$H/$^1$H) isotopic ratio is calibrated and well known [222]. The isotope ratio (expressed also in ppm) for each deuterium site is then calculated from the measurement of the peak surfaces in combination with masses of the product and reference, and the stoichiometric ratio according to Equation 9.28.

$$\left({}_1^2\text{H}/{}_1^1\text{H}\right)_i^{\text{Anal.}} = \left[\frac{A_i^{\text{Anal.}}}{A^{\text{Ref}}}\right] \times \left[\frac{P^{\text{Ref.}} \times m^{\text{Ref.}} \times M^{\text{Anal.}}}{P_i^{\text{Anal.}} \times m^{\text{Anal.}} \times M^{\text{Ref.}}}\right] \times \left({}_1^2\text{H}/{}_1^1\text{H}\right)_i^{\text{Ref.}}. \tag{9.28}$$

In this equation, $A_i^{\text{Anal.}}$ and $A^{\text{Ref.}}$ are the integrated area of the signals at site i for the analyte and that of the reference, both measured on the isotropic NAD 1D spectra, $P_i^{\text{Anal.}}$ and $P_i^{\text{Ref.}}$ are their stoichiometric numbers of hydrogen at site i (analyte or reference), $M^{\text{Anal.}}$, $M^{\text{Ref.}}$, $m^{\text{Anal.}}$, and $m^{\text{Ref.}}$ are the molecular weights and masses of the analyte and the reference, respectively. Finally, $(^2\text{H}/^1\text{H})^{\text{Ref.}}$ is the isotope ratio of chemical reference in the sample (at site i).

This method is well known as SNIF-NMR® and stands for site-specific natural isotopic fractionation first explored and then intensively developed by Martin and co-workers [222, 252]. It was an original development of the international company Eurofins. SNIF-NMR was successfully used in several analytical applications, particularly in food analysis [253–255]. However, the method suffers from two drawbacks: (i) Significant peaks overlapping due to the low $^2$H chemical shift dispersion (in Hz) compared to $^1$H ($\gamma(^2\text{H}) = \gamma(^1\text{H})/6.515$) and (ii) Impossibility of discriminating enantiotopic sites in prochiral molecules as we can see in the case of the methylene group of ethanol. Interestingly, these problems can be overcome by using NAD-{$^1$H} NMR in chiral LCs, because we detect a new NMR interaction, the deuterium quadrupolar (RQC). We can spectrally discriminate enantio-isotopomers in a CLC [33]. One illustrative example is discussed in Section 9.9.2.

### 9.9.2 Case of Prochiral Molecules: The Fatty Acid Family

The analytical potential of NAD 2D NMR in the framework of the isotope fractionation analysis was first explored in the case of a fragment of prochiral fatty acids, the 1,1′-bis(phenylthio)hexane (BPTH) [225], and then successfully demonstrated on a complete long-chain (C-18) unsaturated fatty acid, the methyl linolenate (ML) [226, 227]. Indeed, for such a prochiral analyte, we were able to spectrally separate all the monodeuterated enantio-isotopomers of the molecule in the PBLG/pyridine mesophase (see Figure 9.68b). As a consequence, it was possible to follow the variation of ($^2$H/$^1$H) isotopic ratios for enantiotopic positions in each methylene group.

As a new molecular isotopic information related to the discrimination of all ($R/S$)-enantio-isotopomers, it is possible to determine the bio enantio-isotopomeric excess, ($eie_i$(%)) for each $CH_2$ group i, according to the Equation 9.29 [227]:

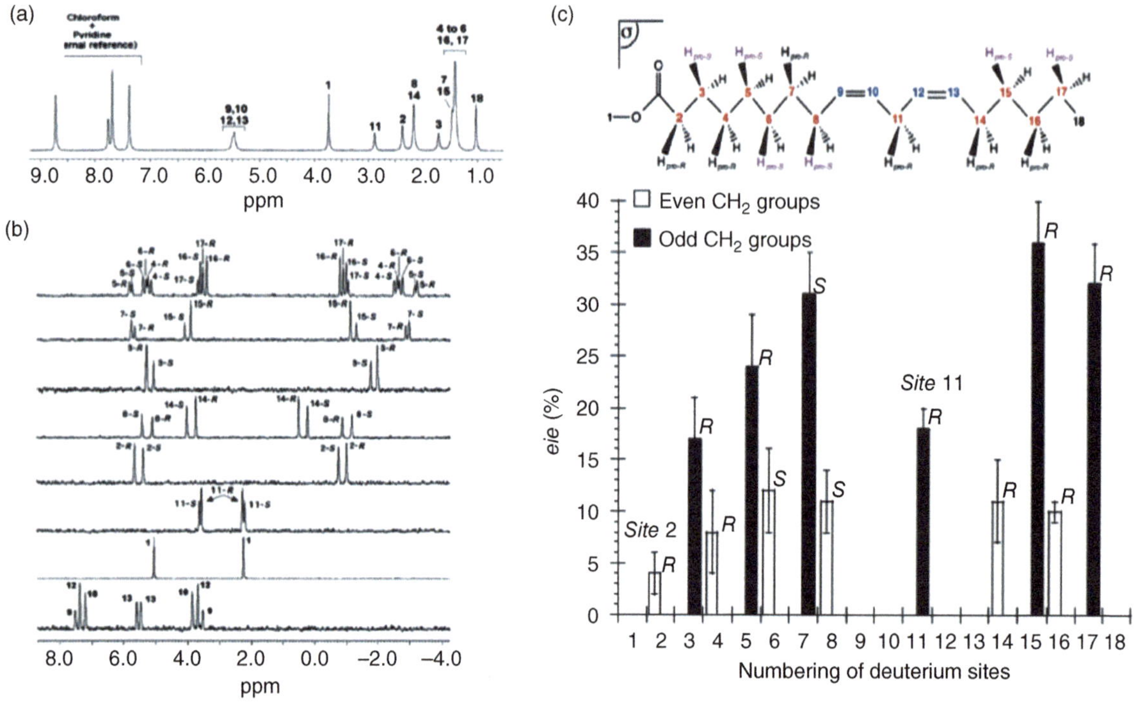


**Figure 9.68** (a) Isotropic 92.1 MHz NAD-{[1]H} 1D spectrum of methyl linoleate. Note the high rate of peak overlap. (b) Series of 92.1 MHz NAD-{[1]H} 1D sub-spectra of ML in PBLG/pyridine extracted from the $^2$H tilted *Q*-COSY Fz 2D spectrum. The $^2$H-QD of methyl 18 is not shown. (c) Variation of *eie*(%) versus the methylene groups. Figure adapted from Ref. [227] with permission.

$$eie_i\,(\%) = 100 \times \frac{\left|\left({}^2_1\mathrm{H}/{}^1_1\mathrm{H}\right)^R_i - \left({}^2_1\mathrm{H}/{}^1_1\mathrm{H}\right)^S_i\right|}{\left({}^2_1\mathrm{H}/{}^1_1\mathrm{H}\right)^R_i + \left({}^2_1H/{}^1_1\mathrm{H}\right)^S_i}. \tag{9.29}$$

It has been evident that the *eie*(%) values at the odd-numbered $CH_2$ positions are larger than those measured at even-numbered $CH_2$ groups all along the chain (see Figure 9.68c). This observation was explained by the different incorporation mechanisms of hydrogens into the chain during the elongation of fatty acids via the fatty acid synthetase (FAS) enzyme [256]. These new isotope data, only accessible by anisotropic $^2$H NMR allowed better understanding of the stereochemical aspects of the enzymatic reactions leading to methyl linoleate, and particularly those of FAS.

Taking advantage of NMR spectrometer operating at higher field (14.1 T, i.e. 92.1 MHz for $^2$H) equipped with an electronic modern and a selective $^2$H cryogenic probe (5-mm o.d.), the method was then successfully extended to the study of more complex long-chain prochiral FAMEs with longer chain, such as C-18 mono- or poly- unsaturated FAMEs as well as the conjugated unsaturated FAMEs (CUFAs) methyl oleate (MO), methyl linolenate (ML), methyl eleostearate (ME), methyl punicate (MP), including the chiral FAME methyl vernoleate (MV) [257]. With such a possibility in hand, it has been possible to experimentally evaluate the possible differences of bioconversion processes of ML to MV involved in the case *Euphorbia lagascae* and *Vernonia galamensis*, two plants that use different enzymatic systems [258].

The interest of this approach has been also successfully tested for the cases of C-14 to C-18 saturated FAMEs (SFA), namely the methyl myristate (MM), methyl palmitate (MP), and methyl stearate (MS) fatty acids. It has been demonstrated that chiral ANAD NMR could provide a robust means of accessing a larger number of $^2$H

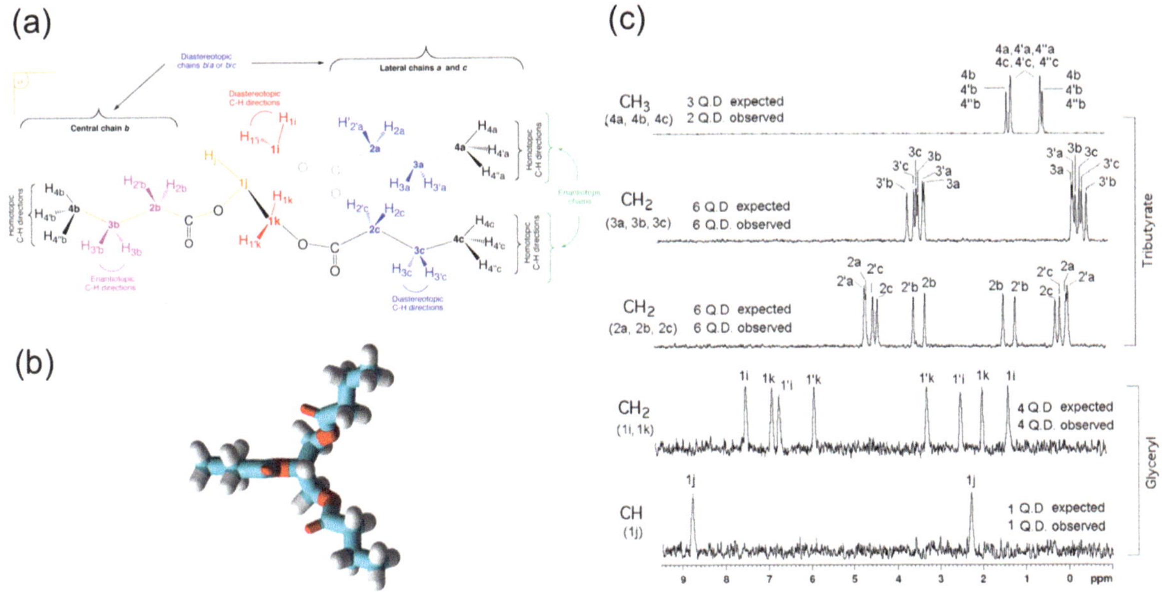


**Figure 9.69** (a) Atomic numbering and stereochemical relationships between the different hydrogenated sites of the glyceryl part (sites i, j, k) and methylene (2 and 3) and methyl (4) groups of the three hydrocarbon chains (a, b, and c) of TB. Note the different homotopic, enantiotopic and diastereotopic elements in this molecule. (b) 3D structure of TB, top view. (c) Series of 92.1 MHz NAD-{$^1$H} 1D sub-spectra of TB in PBLG/pyridine extracted from the tilted *Q*-COSY Fz 2D spectrum. Figure adapted from Ref. [126] with permission.

sites (compared to isotropic NMR) [236]. Finally as, a last frontier, to examine the ultimate potentialities of chiral ANAD 2D NMR, the complex case the homogeneous triglycerides (TGs), the last class of lipids was examined (see Figure 9.69) [126]. In accordance with Altmann's definition [259], homogeneous TGs are flexible molecules of $C_s$ symmetry on average containing a plane of symmetry $\sigma$. They have two enantiotopic side chains for which hydrogenated sites are diastereotopic and a central chain (that is diastereotopic relative to the two side chains) containing enantiotopic hydrogenated sites. For clarity, Figure 9.69 shows all stereochemical relationships (enantiotopicity and diastereotopicity) of trybutyrin (TB), a short-chain TG.

The analysis of the NAD-{$^1$H} *Q*-COSY Fz 2D map of TB in PBLG/Py indicates that 95% of the non-equivalent $^2$H sites (i.e. 19 $^2$H-QDs out of 20) are spectrally discriminated. In this example, the number of differentiated sites is in agreement with the theoretical number expected for a $C_s$-symmetry molecule (see Figures 9.69), which implies an efficient shape recognition phenomenon. This ability to discriminate all $^2$H sites was unfortunately not met in the case of trimyristate (TM), where the discrimination mechanisms that differentiate the central from the two side chains indicate that its no longer possible to distinguish the side chains (a and c) from the central chain (b) as is done for the case of TB. In other words, TM behaves more as a $C_{3v}$-symmetry molecule instead of a $C_s$-symmetry molecule on average. This situation occurs because the ratio ($V_{a,c}/V_b$) of the persistent molecular volume of the side chain ($V_{a,c}$) to the central chain ($V_b$) strongly tends toward unity. Here the molecular volumes, $V_{a,b,c}$, correspond to the 3D space explored by the moving atoms of each chain [126].

### 9.9.3 New Tools for Fighting Against Counterfeiting

The combined application of isotropic $^2$H and, more recently, $^{13}$C NMR techniques in a variety of domains of natural products has been intensively investigated [260, 261]. They were used in plant species differentiation, detection of adulteration, and bio-activity evaluation. In 2018, anisotropic $^2$H 2D-NMR has been involved in a new challenging analytical domain associated with molecular authenticity/traceability investigations in food products [262]. Among possible molecular targets, vanillin, the most widely used aroma molecule in food industry and an important component of perfumery has been examined for two reasons: (i) the price per kilo of natural and synthetic vanillin is extremely different, and therefore the temptation of selling synthetic vanillin at the price of vanillin from natural origin and (ii) the full $^2$H isotopic composition of the aromatic core cannot be evaluated by the isotropic irm-$^2$H 1D-NMR (two sites over three) because in the case of aromatic deuterium sites $^2H_a$ and $^2H_b$ we have $\delta(^2H_a) = \delta(^2H_b) \neq \delta(^2(H_c)$.

Using an optimized polypeptide-oriented phase (PBLG/$CHCl_3$/$CCl_4$), the $^2$H signal of both sites has been separated on the basis of a difference on the magnitude of the $^2$H-RQCs ($|\Delta\nu_Q(C\text{-}^2H_a)| \neq |(\Delta\nu_Q(C\text{-}^2H_b)|$) (see Figure 9.70a). Interestingly, the assignment for each $^2$H-QDs is possible, based on the collinearity of the C-D direction of sites in the aromatic ring of vanillin ($S_{C\text{-}Ha} = S_{C\text{-}Hc}$) leading to equal $^2$H-RQCs values for both sites ($|\Delta\nu_Q(C\text{-}^2H_a)| = |\Delta\nu_Q(C\text{-}^2H_c|$) (see Figure 9.70b).

To test the potential of ANAD NMR in the case of vanillin, we have analyzed different samples of vanillin, both from natural and industrial sources (synthetic or technological bioconversion), to see if we could discriminate them by only using isotopic data of the aromatic ring. Clearly, the access to the isotopic information for all aromatic sites could be an advantage to analytically control the synthetic or natural origin of vanillin.

For this, we have calculated the relative distribution coefficients, $k'_{^2H\,i(i=a,b,c)}$, of deuterium of three aromatic sites $H_a$, $H_b$ and $H_c$, corresponding to a fraction of peak area, $A$, and defined as:

$$k'_{^2H\,i(i=a,b,c)}(\%) = 100 \times \frac{A_{^2H_i}}{\sum_{i=a,b,c} A_{^2H_i}}. \tag{9.30}$$

In the series of vanillin samples analyzed, it appeared that the ratios vary as $H_a < H_b < H_c$ for the samples of natural origin (bean or from a natural precursor [ferulic acid or lignin]), whereas the two synthetic samples

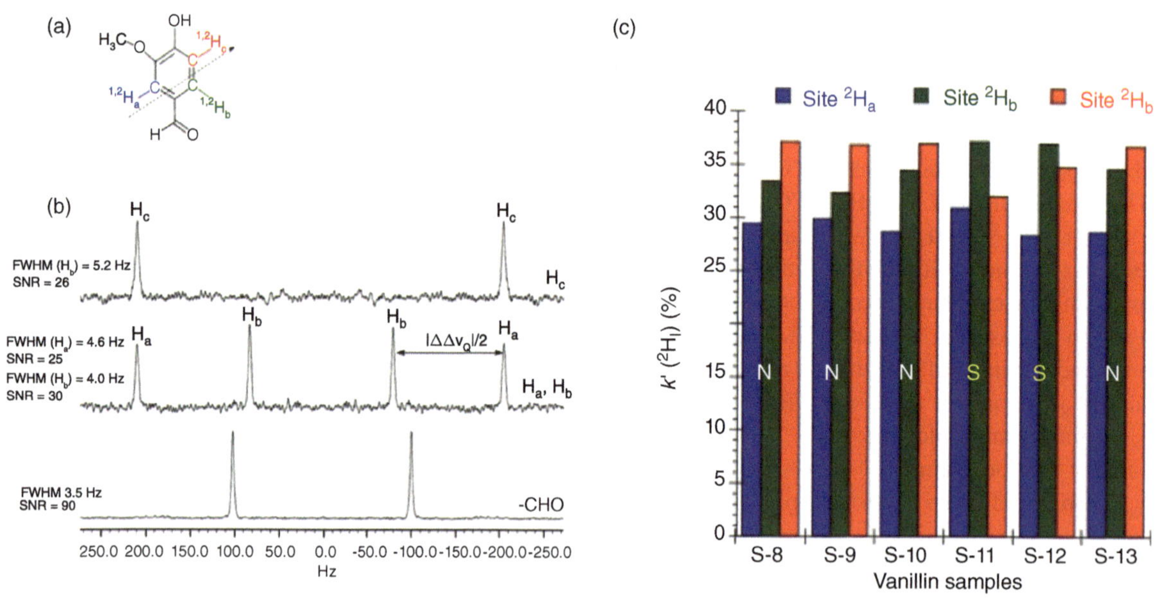


Figure 9.70 (a) Structure of vanillin recorded with TMU. (b) Series of 92.1 MHz NAD-{$^1$H} 1D sub-spectra of vanillin in PBLG extracted from the tilted $Q$-resolved Fz 2D spectrum. Compared to Figures 9.68a or 9.69c, all $^2$H-QDs are centered here on 0 Hz in $F_1$ dimension. Figure adapted from Ref. [261] with permission.

both show $H_a < H_b > H_c$. From these referent isotopic trends, it becomes possible to establish the origin of an unknown sample of vanillin by comparing its distribution coefficients on the three sites with the reference data obtained for known vanillin origin.

Access to new isotopic information, it is an important successful example. From a more theoretical standpoint, it has been shown from the analysis of all aromatic isotopic data how the $^{2}$H distribution might relate to the biosynthesis of vanillin [262]. From an applicative viewpoint, these results show that the evaluation of intramolecular $^{2}$H isotopic composition of the aromatic ring should enable discernment between natural and synthetic origin, information that can be used in combination with isotopic data of non-aromatic sites of the molecule.

#### 9.9.3.1 Exploring the ($^{13}$C/$^{12}$C) Isotope Fractionation

After a long analytical exploitation of $^{2}$H SNIF-NMR, the possibility of detecting the natural ($^{13}$C/$^{12}$C) isotope fractionation (with a precision of *per mil*) by quantitative $^{13}$C-{$^{1}$H} 1D NMR was explored and developed in 2000s [260, 263, 264]. In 2021, a new page of isotopic analysis by anisotropic NMR was opened with an exploratory experimental study to evaluate the possibility to establish enantiomeric $^{13}$C position-specific isotope fractionation (EPSIF), using quantitative $^{13}$C in PBLG-based CLC [107]. The idea is to provide a robust NMR tool to answer the intriguing and fundamental question related to chiral induction/amplification at the origin of homochirality in nature: "*Is there a relationship between enantiomeric and isotopic fractionation of carbon-13 in chiral molecules*?" The main challenge is the ability to record $^{13}$C-{$^{1}$H} spectra with sufficiently high signal-to-noise ratios to meet the required per-mil precision () in isotope analysis and hence to accurately determine the $^{13}$C/$^{12}$C isotope ratio, despite the doubling of $^{13}$C signals due to the spectral enantiodiscrimination.

Using phenethyl alcohol as molecule model (see Figure 9.16), it was possible to (i) separate the enantiomeric $^{13}$C signals in various carbon sites of the analyte, (ii) to maximize the spectral resolutions by optimization of the sample composition at 175.1 MHz, (iii) to work with a very good repeatability for a series up to seven consecutive replicates, and (iv) to obtain rather good long-term reproducible results.

## 9.10 Anisotropic NMR in Molecular Analysis: What You Should Keep in Mind

Anisotropic NMR using LLCs or alignment gels is a fascinating approach as they offer analytical possibilities and application opportunities that neither the liquid NMR nor the solid-state NMR has. As described here, we have highlighted how the three major anisotopic NMR observables, (RDCs, RSCAs, RQCs) were powerful tools to solve two important, modern problems for chemists: (i) the evaluation of enantiomeric purity of chiral mixtures and (ii) the determination of molecular structure of complex synthetic or natural compounds. To be brief, and as a toolkit for any reader, we list below ten important facts and specifics of anisotropic NMR (in weakly aligning media) that every NMR spectroscopists should keep in mind to assess its potential for routine uses:

(i) analytical richness induced by the molecular orientation inside the magnetic field $\mathbf{B_0}$;
(ii) access to order-dependent NMR interactions (RCSA, RDC, RQC) no longer averaged to zero as in liquids;
(iii) spectral advantages to work with weakly orienting phases such as LLC phases and gels, within a range of one order of magnitude in terms of molecular orientation between these two types of media;
(iv) full control of orientation properties by adjusting the molar composition of the sample, the choice/nature of polymer and/or the polarity of organic co-solvent, or simply optimize the sample temperature in $\mathbf{B_0}$;
(v) detection of very weakly abundant nuclei (as $^{2}$H) with reasonable experimental NMR times;
(vi) spectral discrimination enantiomers of chiral molecules to evaluate enantiomeric purity of mixtures as well as differentiate enantiotopic elements in prochiral compounds, when chiral oriented media are used;
(vii) possibility of NMR monitoring of dynamic processes (versus temperature or time) involving chiral objects (conformational equilibria, chemical transformations) with CLC;

(viii) analysis of site-specific isotopic profile using $^2$H or $^{13}$C NMR;
(ix) determination of molecular relative configuration because all anisotropic NMR observables encode valuable 3D structural information;
(x) and finally, the interest of diversifying the sources of molecular data using nD NMR for solving many challenging analytical problems met by chemists.

From the first pioneering work by Saupe and Englert, and later Sackmann and co-workers in the early 1960s, which first opened the doors of the anisotropic NMR spectroscopy in ALCs [57] and then in CLCs [58], respectively, numerous methodological and/or technological developments (NMR sequences, variable-angle spinning sample (VASS) NMR probes, new oriented systems, etc.) were proposed during three fertile decades of continuous progress. Combing with recent instrumental NMR achievements, they provide various keys to fully exploit the potential of NMR in oriented media and solve important analytical issues for a large community of chemists.

Clearly, after three decades dominated by thermotropic systems, an important rebirth of NMR in oriented solvents has occurred with the advent of (chiral or not) weakly alignment environments such as organic or water-compatible LLCs in the 1990s. Among future challenges associated with the next developments of anisotropic NMR in these specific systems, two important challenges can be mentioned. The first one concerns the increasing of sensitivity of anisotropic NMR, particularly the detection of low abundant nuclei as deuterium [265]. The second one concerns the possibility of proposing robust computational model to predict the molecular orientation of a solute and the complex mechanisms of enantiodiscrimination phenomena using molecular dynamics [229, 233, 235, 265–266]. The final objective of this second challenge would be the possibility of predicting the AC of chiral mono- or poly-stereogenic molecules without the help of any analytical tool other than NMR [229].

## References


1 Emsley, J.W. and Feeney, J. (1995). Milestones in the first fifty years of NMR. *Prog. Nucl. Magn. Reson. Spectrosc.* 2: 1–9.

2 Burnell, E.E., and Delange, C.A. (eds.) (2003). *NMR of Oriented Liquids.* Dordrecht: Springer Science.

3 Dong, R.Y. (2004). Advances in NMR studies of liquid crystals. *Ann. Rep. on NMR Spect.* 53: 67–155.

4 Dong, R.Y. (ed.) (2010). *NMR Spectroscopy of Liquid Crystals.* New Jersey: World scientific.

5 Dong, R.Y. (ed.) (2012). NMR spectroscopy in liquid crystalline and ordered phases. In: *Encyclopedia of Analytical Chemistry* (ed. R.A. Meyers), John Wiley & sons. doi: 10.1002/9780470027318.a9297.

6 Emsley, J.W. and Lindon, J.C. (eds.) (1975). Chap. 2, In *Book: NMR Spectroscopy Using Liquid Crystal Solvents.* Oxford: Pergamon Press.

7 Veracini, C. (1985). *In book: Nuclear Magnetic Resonance of Liquid Crystals.* NATO ASI Series.

8 Eliel, E.L. and Wilen, S.H. (1994). *Stereochemistry of Organic Compounds.* New York: John Wiley & Sons.

9 Reinitzer, P. (1888). Beiträge zur kenntnis des cholesterins, Monasth. *Chem.* 9: 421–441.

10 Lesot, P., Aroulanda, C., Berdagué, P., Meddour, A., Merlet, D., Farjon, J, Giraud, N., and Lafon, O. (2020-a). Three fertile decades of methodological developments and analytical challenges. *Prog. Nucl. Magn. Reson. Spectrosc.* 116: 85–154.

11 Gil, R.R. and Navarro-Vazquez, A. (2017-a). Application of residual dipolar couplings to the structural analysis of natural products. In: *book : From Modern NMR Approaches to the Structure Elucidation of Natural Products*, 2e, (ed. A.J. Williams, G.E. Martin, and D. Rovnyak), 117–176. Royal Society of Chemistry Cambridge (UK). ISBN: 978-1-84973-383-0.

**12** Tjandra, N. and Bax, A. (1997-a). Direct measurement of distances and angles in biomolecules by NMR in a dilute liquid crystalline medium. *Science* 278: 1111–1114.

**13** Gayathri, C., Tsarevsky N.V., and Gil, R.R. (2010). Residual dipolar couplings (RDCs) analysis of small molecules made easy: fast and tuneable alignment by reversible compression/relaxation of reusable PMMA gels. *Chem. Eur. J.* 16: 362–3626.

**14** Kuchel, P.W., Chapman, B.E., Boghda, Mueller, N., Bubb, W.A., Philp, D.J., and Torres, A.M. (2006). Apparatus for rapid adjustment of the degree of alignment of NMR samples in aqueous media: verification with residual quadrupolar splittings in $^{23}$Na and $^{133}$Cs spectra. *J. Mag. Res.* 180: 256–265.

**15** Kummerlöwe, G., McCord, E.F., Cheatham, S.F., Niss, S., Schnell, R.W., and Luy, B. (2010). Tunable alignment for all polymer gel/solvent combinations for the measurement of anisotropic NMR parameters. *Chem. Eur. J.* 16: 7087–7089.

**16** Luy, B., Kobzar, K., Knoer, S., Furrer, J., Heckmann, D., and Kessler, H. (2005). Orientational properties of stretched polystyrene gels in organic solvents and the suppression of their residual $^{1}$H NMR signals. *J. Am. Chem. Soc.* 127: 6459–6465.

**17** Gil, R.R., Gayathri, C., Tsarevsky, N.V., and Matyjaszewski, K. (2008). Stretched poly(methyl methacrylate) gel aligns small organic molecules in chloroform. Stereochemical analysis and diastereotopic proton NMR assignment in ludartin using residual dipolar couplings and $^{3}$ coupling constant. *J. Org. Chem.* 73: 840–848.

**18** Merlet, D., Emsley, J.W. Lesot, P., and Courtieu, J. (1999-a). The relationship between molecular symmetry and second-rank orientational order parameters for molecules in chiral liquid crystalline solvents. *J. Chem. Phys.* 111: 6890–6896.

**19** Tolman, J.R., Al-Hashimi, H.M., Kay, E. and Prestegard, J.H. (2001). Structural and dynamic analysis of residual dipolar coupling data for proteins. *J. Am. Chem. Soc.* 123: 1416–1424.

**20** Kramer, F., Deshmukh, M.V., Kessler, H., and Glaser, S.J. (2004). Residual dipolar coupling constants: An elementary derivation of key equations. *Concepts Magn. Reson.* 21A: 10–21.

**21** Kummerlöwe, G., Udaya Kiran, M., and Luy, B. (2009-a). Covalently cross-linked gelatin allows chiral distinction at elevated temperatures and in DMSO. *Chem. Eur. J.* 15: 12192–12195.

**22** IUPAC (1972). Recommendations for the presentation of NMR data for publication in chemical journals. *Pure Appl. Chem.* 29: 627–628.

**23** IUPAC (1976). Presentation of NMR data for publication in chemical journals - B. Conventions relating to spectra from nuclei other than protons. *Pure Appl. Chem.* 45: 217–220.

**24** Canet, I., Courtieu, J., Loewenstein, A., Meddour, A., and Péchiné, J.M. (1995). Enantiomeric analysis in a polypeptide lyotropic liquid crystal by deuterium NMR. *J. Am. Chem. Soc.* 117: 6520–6526.

**25** Gao, L., Perato, S., Garcia-Argote, S., Taglang, C., Miguel Martinez-Prieto, L., Chollet, C., Buisson, D. Dauvois, D., Lesot, P. Chaudret, B. Rousseau, B., Feuillastre, S., and Pieters. G. (2018). Ruthenium-catalyzed hydrogen isotope exchange of ($C_{sp3}$)-H bonds directed by a sulfur atom. *Chem. Commun.* 54: 2986–2989.

**26** Jackman, H., Marsden, S.P., Shapland, P., and Barrett, S. (2007). Isotopic labeling for determination of enantiomeric purity by $^{2}$H NMR spectroscopy. *Org Lett.* 9: 5179–5182.

**27** Meddour, A., Canet, I., Loewenstein, A., Péchiné, J.-M., and Courtieu, J. (1994). Observation of enantiomers, chiral by virtue of isotopic substitution, through deuterium NMR in a polypeptide liquid crystal. *J. Am. Chem. Soc.* 116: 9652–9656.

**28** Wenzel, T.J. (2007). *In Book: Discrimination of Chiral Compounds using NMR Spectroscopy.* (ed. H.D.C. Smyth and M.J. Donovan). New York: Wiley Interscience. ISBN: 978-0-471-76352-9, doi: 10.1080/03639040701662677.

**29** Wenzel, T.J. (2018). *In book: Differentiation of Chiral Compounds using NMR Spectroscopy*, 2e. Hoboken, USA.: John Wiley & sons. ISBN-10: 1119323916, ISBN-13: 978-1119323914

**30** Wenzel, T.J. and Chisholm, C.D. (2011). Assignment of the absolute configuration of polyfunctional compounds by NMR using chiral derivatizing agents. *Prog. Nucl. Magn. Reson. Spectrosc.* 59: 1–63.

**31** Sylva, M.S. (2017). Recent advances in multinuclear NMR spectroscopy for chiral recognition of organic compounds. *Molecules* 22: 247–268.

**32** Lesot, P. (2020). Determination of the natural deuterium distribution of fatty acids by application of $^{2}$H 2D-NMR in liquid crystals: fundamentals, advances, around and beyond. *Liq. Cryst.* 47: 1886–1910.

**33** Lesot, P., Merlet, D., Courtieu, J., Emsley, J.W., Rantala, T.T., and Jokisaari, J. (1997). Calculation of the molecular ordering tensors of (±)-3-butyn-2-ol in an organic solution of poly($\gamma$-benzyl-L-glutamate). *J. Phys. Chem. A* 101: 5719–5724.

**34** Thiele, C.M. and Bermel, W. (2012). Speeding up the measurement of one-bond scalar ($^{1}$J) and residual dipolar couplings ($^{1}$D) by using non-uniform sampling (NUS). *J. Magn. Reson.* 216: 134–143.

**35** Enthart, A., Freundenberger, J.C., Furrer, J., and Luy, B. (2008). The CLIP/CLAP-HSQC: pure absorptive spectra for the measurement of one-bond couplings. *J. Magn. Reson.* 192: 314–22.

**36** Reinsperger, T. and Luy, B. (2008). CLIP-ASAP-HSQC for fast and accurate extraction of one bond couplings from isotropic and partially aligned molecules. *J. Magn. Reson.* 239: 110–120.

**37** Reinsperger, T. and Luy, B. (2014). Homonuclear BIRD-decoupled spectra for measuring one-bond couplings with highest resolution: CLIP/CLAP-RESET and constant-time-CLIP/CLAP-RESET. *J. Magn. Reson.* 239: 110–120.

**38** Farjon, J. and Merlet, D. (2010). SERF-filtered experiments: new enantio-selective tools for deciphering complex spectra of racemic mixtures dissolved in chiral oriented media. *J. Magn. Reson.* 210: 24–30.

**39** Giraud, N., Beguin, L., Courtieu, J., and Merlet, D. (2010). Nuclear magnetic resonance using a spatial frequency encoding: application to J-edited spectroscopy along the sample. *Angew. Chem., Int. Ed.* 49: 3481–3484.

**40** Aroulanda, C. and Lesot, P. (2022). Molecular enantiodiscrimination by NMR in chiral oriented systems: Concept, tools and applications. *Chirality.* 34: 182–244.

**41** Becker, J. and Luy, B. (2015). CLIP-ASAP-HSQC for fast and accurate extraction of one-bond couplings from isotropic and partially aligned molecules. *Magn. Reson. Chem.* 53: 878–885.

**42** Courtieu, J., Bayle, J.-P., and Fung, B.M. (1994). Variable-angle sample-spinning NMR in liquid-crystals. *Prog. Nucl. Magn. Reson. Spectrosc.* 26: 141–169.

**43** Gil-Silva, L.F., Santamaria-Fernandez, R., Navarro-Vazquez, A., and Gil, R.R. (2016). Collection of NMR scalar and residual dipolar couplings using a single experiment. *Chem. Eur. J.* 22: 472–476.

**44** Souza, A.A., Gil, R.R., and Parella, T. (2017). Highly resolved HSQC experiments for the fast and accurate measurement of homonuclear and heteronuclear coupling constants. *J. Mag. Res.* 282: 54-61.

**45** Navarro-Vázquez, A, Gil, R.R., and Blinov, K. (2018). Computer-assisted 3D structure elucidation (CASE-3D) of natural products combining isotropic and anisotropic NMR parameters. *J. Nat. Prod.* 81: 203–210.

**46** Trigo-Mouriño, P., Navarro-Vazquez, A., Ying, J., Gil, R.R., and Bax, A. (2011). Structural discrimination in small molecules by accurate measurement of long-range proton-carbon NMR residual dipolar couplings. *Angew. Chem., Int. Ed.* 50: 7576–7580.

**47** Reller, M., Wesp, S., Koos, M.M.M., Reggelin, M., and Luy B. (2017). Biphasic liquid crystal and the simultaneous measurement of isotropic and anisotropic parameters by spatially resolved NMR spectroscopy. *Chem Eur J.* 23: 13351–13359.

**48** Lesot, P. (2019-b). Nuclear Magnetic Resonance Spectroscopy | Hydrogen Isotopes: $^{2}$H NMR, Encyclopedia of Analytical Science, 3e, 152–167. doi: 10.1016/B978-0-12-409547-2.14084-3.

**49** Luy, B. (2010). Distinction of enantiomers by NMR spectroscopy using chiral orienting media. *J. Indian Inst. Sci.* 90: 119–132.

**50** Merlet, D., Ancian, B., Smadja, W., Courtieu, J., and. Lesot, P. (1998). Analysis of the natural abundance deuterium NMR spectra of enantiomers in chiral liquid crystals through 2D auto-correlation experiments. *Chem. Commun.* 2301–2302.

**51** Merlet, D., Ancian, B., Courtieu, J., and Lesot, P. (1999). Two-dimensional deuterium NMR spectroscopy of chiral molecules oriented in a polypeptide liquid crystal: application for the enantiomeric analysis through natural abundance deuterium NMR. *J. Am. Chem. Soc.* 121: 5249–5258.

**52** Lesot, P. and Lafon, O. (2017). Natural abundance $^2$H NMR spectroscopy. In: *Book: Encyclopedia of Spectroscopy and Spectrometry*, 3e (ed. J. Lindon, G.E. Tranter, and D.W. Kopenaal), Vol. 3, 1–14. Elsevier / Academic Press.

**53** Lafon, O., and Lesot, P. (2005). Deuterium three-dimensional NMR experiments for analysing weakly aligned, isotopically enriched molecules. *Chem. Phys. Lett.*, 404: 90–94.

**54** Lafon, O., Berdagué, P., and Lesot, P. (2004). Use of two-dimensional correlation between $^2$H quadrupolar splittings and $^{13}$C CSA's for assignment of NMR spectra in chiral nematics. *Phys. Chem. Chem. Phys.* 6: 1080 –1084.

**55** Auger, C., Lesage A., Caldarelli, S., Hodgkinson P., and Emsley, L. (1998). Assignment and measurement of deuterium quadrupolar couplings in liquid crystals by deuterium-carbon NMR correlation spectroscopy. *J. Phys. Chem. B* 120: 3718–3723.

**56** Lesot, P. and Lafon, O. (2012). Experimental detection of achiral and chiral naturally abundant $^{13}$C-$^2$H isotopomers by 2D-NMR in liquids and chiral oriented solvents. *Anal. Chem.* 84: 4569–4573.

**57** Saupe, A. and Englert, G. (1963). High-resolution nuclear magnetic resonance spectra of orientated molecules. *Phys. Rev. Lett.* 11: 462–464.

**58** Sackmann, E., Meiboom, S., and Snyder, L.C. (1968). Nuclear magnetic resonance spectra of enantiomers in optically active liquid crystals. *J. Am. Chem. Soc.* 90: 2183–2184.

**59** Bayle, J.-P., Courtieu, J., Gabetty E., Loewenstein, A. and Péchiné, J.-M. (1992). Enantiomeric analysis in a polypeptide lyotropic liquid-crystal through proton decoupled deuterium NMR. *New J. Chem.* 16: 837–838.

**60** Tjandra, N., Grzeisek, S., and Bax, A. (1996). Magnetic field dependence of nitrogen-proton J splittings in $^{15}$Nenriched human ubiquitin resulting from relaxation interference and residual dipolar coupling. *J. Am. Chem. Soc.* 18: 6264–6272.

**61** Aroulanda, C., Sarfati, M., Courtieu, J., and Lesot, P. (2001). Investigation of enantioselectivity of three polypeptide liquid-crystalline solvents using NMR spectroscopy. *Enantiomer*. 6: 281–287.

**62** Canlet, C., Merlet, D., Lesot, P., Meddour, A. Loewenstein, A., and Courtieu, J. (2000). Deuterium NMR stereochemical analysis of threo-erythro isomers bearing remote chiral centres in racemic and non-racemic liquid crystalline solvents. *Tetrahedron: Asymmetry* 11: 1911–1918.

**63** Lesot, P., Lafon, O., Aroulanda, C., and Dong, R. (2008). $^2$H NMR studies of two-homopolypeptide lyotropic mesophases: toward the quantification of solute-fiber interactions. Chem. Eur. J. 14: 4082–4092.

**64** Serhan, Z., Aroulanda, C., and Lesot, P. (2016). Investigation of solute-fiber affinity and orientational ordering of norbornadiene interacting with two-polypeptide chiral liquid-crystalline solvents by NAD NMR. *J. Phys. Chem. A* 120: 6076–6088.

**65** Schwab, M., Herold, D., and Thiele, C.M. (2017). Polyaspartates as thermoresponsive enantiodifferentiating helically chiral alignment media for anisotropic NMR spectroscopy. *Chem. Eur.* 23: 14576–14584.

**66** Hirschmann, M., Schwab, M., and Thiele C.M. (2019). Molecular weights: The key of lyotropic liquid crystalline of poly-$\beta$-benzyl-L-aspartate. *Macromolecules* 52: 6025–6034.

**67** Jeziorowski, S. and Thiele, C.M. (2018). Poly-$\gamma$-p-biphenylmethyl-glutamate as enantiodifferentiating alignment medium for NMR spectroscopy with temperature-tunable properties. *Chem. Eur. J.* 24: 15631–15637.

**68** Hansmann, S., Schmidts, V., and Thiele, C.M. (2017). Synthesis of poly-$\gamma$-S-2-methylbutyl-L-glutamate and poly-$\gamma$-S-2-methylbutyl-D-glutamate and their use as enantiodiscriminating alignment media in NMR spectroscopy. *Chem. Eur. J.* 23: 9114–9121.

**69** Schwab, M., Schmidts, V., and Thiele, C.M. (2018). Thermoresponsive alignment media in NMR spectroscopy: helix reversal of a co-polyaspartate at ambient temperatures. *Chem. Eur. J.* 24: 14373–14377.

**70** Sass, H.-J., Musco, G., Stahl, S.J. Wingfield, P.T., and Grzesiek, S. (2000). Solution NMR of proteins within polyacrylamide gels: diffusional properties and residual alignment by mechanical stress or embedding of oriented purple membranes. *J. Biomol. NMR.* 18: 303–309.

**71** Tycko, R., Blanco, F.J., and Ishii, Y. (2000). Alignment of biopolymers in strained gels: a new way to create detectable dipole-dipole couplings in high-resolution biomolecular NMR. *J. Am. Chem. Soc.* 122: 9340–9341.

**72** Deloche, B. and Samulski, E.T. (1981). Short-range nematic-like orientational order in strained elastomers: a deuterium magnetic resonance study. *Macromolecules*. 14: 575-581.

**73** Mangoni, A., Esposito, V., and Randazzo, A. (2003). Configuration assignment in small organic molecules via residual dipolar couplings. *Chem. Commun.* 154–155.

**74** Luy, B., Kobzar, K., and Kessler, H. (2004). An easy and scalable method for the partial alignment of organic molecules for measuring residual dipolar couplings. *Angew. Chem., Int. Ed.* 43: 1092–1094.

**75** Freudenberger, J.C., Spiteller, P., Bauer, R., Kessler, H., and Luy, B. (2004). Stretched poly(dimethylsiloxane) gels as NMR alignment media for apolar and weakly polar organic solvents: an ideal tool for measuring RDCs at low molecular concentrations. *J. Am. Chem. Soc.* 126: 14690–14691.

**76** Freudenberger, J. C., Knoer, S., Kobzar, K., Heckmann, D., Paululat, T., Kessler, H., and Luy, B. (2005). Stretched poly(vinyl acetate) gels as NMR alignment media for the measurement of residual dipolar couplings in polar organic solvents. *Angew. Chem., Int. Ed.* 44: 423–426.

**77** Kummerlöwe, G., Auernheimer, J., Lendlein, A., and Luy, B. (2007). Stretched poly(acrylonitrile) as a scalable alignment medium for DMSO. *J. Am. Chem. Soc.* 129: 6080–6081.

**78** Kobzar, K., Kessler, H., and Luy, B. (2005). Stretched gelatin gels as chiral alignment media for the discrimination of enantiomers by NMR spectroscopy. *Angew. Chem., Int. Ed.* 44: 3145–3147.

**79** Kaden, P., Freudenberger, J.C., and Luy, B. (2012). Noncovalently and covalently cross-linked polyurethane gels as alignment media and the suppression of residual polymer signals using diffusion-filtered spectroscopy. *Mag. Reson. Chem.* 50 (S1): S22–S28.

**80** Merle, C., Kummerlöwe, G., Freudenberger, J. C., Halbach, F., Stoewer, W., von Gostomski, C.L., Hoepfner, J., Beskers, T., Wilhelm, M., and Luy, B. (2013). Crosslinked poly(ethylene oxide) as a versatile alignment medium for the measurement of residual anisotropic NMR parameters. *Angew. Chem., Int. Ed.* 52: 10309–10312.

**81** Haberz, P., Farjon, J., and Griesinger, C. (2005). A DMSO-compatible orienting medium: towards the investigation of the stereochemistry of natural products. *Angew. Chem., Int. Ed.* 44: 427–429.

**82** Garcia, M.E., Woodruff, S.R., Hellemann, E., Tsarevsky, N.V., and Gil, R.R. (2017). Di(ethylene glycol) methyl ether methacrylate (DEGMEMA)-derived gels align small organic molecules in methanol. *Magn. Reson. Chem.* 55: 206–209.

**83** Farley, K., Koos, M.R.M, Che, Y., Horst, R., Limberakis, C., Bellenger, J., Lira, R., Gil-Silva, L., and Gil, R.R. (2021). Cross-linked poly-4-acrylomorpholine: a flexible and reversibly compressible aligning gel for anisotropic NMR analysis of peptides and small molecules in water. *Angew. Chem. Int. Ed.* 60: 26314–26319.

**84** Naumann, C., Bubb, W.A., Chapman, B.E., and Kuchel, P.W. (2007). Tunable-alignment chiral system based on gelatin for NMR spectroscopy. *J. Am. Chem. Soc.* 129: 5340–5341.

**85** Dupont. Dupont elastomercompagny. http://www.dupontelastomers.com/products/kalrez/indsemi.asp.

**86** New Era Enterprise. http://newera-spectro.com/gel-presses.

**87** Sau, S.P. and Ramanathan, K.V. (1997). Visualization of enantiomers in the liquid-crystalline phase of a fragmented DNA solution. *J. Phys. Chem. B.* 113: 1530–1535.

**88** Lesot, P., Venkateswara Reddy, U., and Suryaprakash, N.S. (2011). Exploring the enantiodiscrimination potentialities of DNA-based orienting media using deuterium NMR spectroscopy. *Chem. Commun.* 47: 11736–11738.

**89** Brandes, R. and Kearns, D.R. (1986). Magnetic ordering of DNA liquid crystals. *Biochemistry* 25: 5890–5885.

**90** Strzelecka, T.E. and Rill, R.L. (1987). Solid State $^{31}$P NMR Studies of DNA liquid crystalline phases. The isotropic to cholesteric transition. *J. Am. Chem. Soc.* 109: 4513–4518.

**91** Chan-Huot, M., Lesot, P., Pelupessy, Duma, L., Duchambon, P., Bodenhausen, G., Toney, M.D., Reddy, V., and Suryaprakash, N. (2013). 'On-the-fly' kinetics of enzymatic racemization using deuterium NMR in DNA-based, oriented chiral solvents. *Anal. Chem.* 85: 4694–4697.

**92** Tracey, A.S., and Diehl, P. (1975). The interaction of D- and L-alanine with an optically active model membrane system. *FEBS Lett.* 59: 131–132.

**93** Tracey, A.S., and Radley, K. (1984). Effects of composition on cholesteric behavior in the lyotropic mesophase system of potassium N-dodecanoyl-L-alaninate. *J. Phys. Chem.* 88: 6044–6048.

**94** Solgadi, A., Meddour, A., and Courtieu, J. (2004). Enantiomeric discrimination of water-soluble compounds using deuterium NMR in a glucopon/buffered water/n-hexanol chiral lyotropic liquid crystal. *Tetrahedron: Asymmetry* 15: 315–1318.

**95** Baczko, K., Larpent, C., and Lesot, P. (2004). New amino acid based on anionic surfactants and their use as enantiodiscriminating lyotropic liquid crystalline NMR solvent. *Tetrahedron: Asymmetry* 15: 971–982.

**96** Trigo-Mouriño, P., Merle, C., Koos, M.R.M., Luy, B., and Gil, R.R. (2013). Probing spatial distribution of alignment by deuterium NMR imaging. *Chem. Eur. J.* 19: 7013–7019.

**97** FDA. FDA directives. https://www.fda.gov/.

**98** Reach directive. https://ec.europa.eu/environment/chemicals/reach/reach_en.htm

**99** Facke, T. and Berger, S. (1995). SERF, A new method for H,H spin-coupling measurement in organic-chemistry. *J. Magn. Reson. A* 113: 114–116.

**100** Beguin, L., Giraud, N., Ouvrard, J.-M., Courtieu, J., and, Merlet, D. (2009). Improvements to selective refocusing phased (SERFph). *J. Magn. Reson.* 199: 41–47.

**101** Beguin, L., Courtieu, J., Ziani, L., and Merlet, D. (2006). Simplification of the H-1 NMR spectra of enantiomers dissolved in chiral liquid crystals, combining variable angle sample spinning and selective refocusing experiments. *Magn. Reson. Chem.* 44: 10960–1101.

**102** Farjon J. and Giraud, N. (2018). Analyses of enantiomeric mixtures using chiral liquid crystals. *Curr. Opin. Colloid & Interface Sci.* 33: 1–8.

**103** Merlet, D. Beguin, L. Courtieu, J., and Giraud, N. (2011). Spin-spin coupling edition in chiral liquid crystal NMR solvent. *J. Magn. Reson.* 209: 315–322.

**104** Lesot, P., Merlet, D., Meddour, A., Courtieu, J, and Loewenstein, A. (1995). Visualization of enantiomers in a polypeptide liquid-crystal solvent through carbon-13 NMR spectroscopy. *J. Chem. Soc, Faraday Trans.* 91: 1371–1375.

**105** Meddour, A., Berdagué, P., Hedli, A., Courtieu, J., and Lesot, P. (1997). Proton-decoupled carbon-13 NMR spectroscopy in a lyotropic chiral nematic solvent as an analytical tool for the measurement of the enantiomeric excess, J. Am. Chem. Soc. 119: 4502-4508.

**106** Tzvetkova, P., Luy, B., and Simova, S. (2011). Chemical shift anisotropy-based measurement of enantiomeric excess for selected tetrasubstituted pyrrolidines. In: *Book: Topics in Chemistry and Material Science*, Vol. 5 (ed. R.D. Nikolova, S. Simova, P. Denkova, G.N. Vayssilov), 70–77. of Current Issues in Organic Chemistry. Birmingham: Heron Press Ltd.

**107** Lesot, P., Berdagué P., Silvestre V., and Remaud, G. (2021). Exploring the enantiomeric $^{13}C$ isotope fractionation: challenges and anisotropic NMR-based analytical strategy. *Anal. Bioanal. Chem.* 413: 6379–6392.

**108** Lafon, O., Lesot, P., Rivard, M., Chavarot, M., Rose-Munch, F., and Rose, E. (2005) Enantiomeric analysis of planar chiral ($\eta^6$-arene) chromium tricarbonyl complexes using NMR in oriented solvents organometallics. 24: 4021–4028.

**109** Kovacs, H., Moskau, D., and Spraul, M. (2005). Cryogenically cooled probes - a leap in NMR technology. *Prog. Nucl. Magn. Reson. Spectrosc.* 4: 131–155.

**110** Marathias, V.M., Tate, P.A., Papaioannou, N., and Massefski, W. (2010). An improved method for determining enantiomeric excess by $^{13}C$-NMR in chiral liquid crystal *media. Chirality* 22: 838 –843.

**111** Lesot, P., Gounelle, Y., Merlet, D. Loewenstein, A., and Courtieu, J. (1995). Measurement and analysis of the molecular ordering tensors of two enantiomers oriented in a polypeptide liquid crystalline system *J. Phys. Chem. A* 99: 14871–14875.

**112** Lesot, P., Sarfati, M., and Courtieu, J. (2003). Natural abundance deuterium NMR spectroscopy in polypeptide liquid crystals as a new and incisive means for enantiodifferentiation of chiral hydrocarbons. *Chem. Eur. J.* 9: 1724–1745.

**113** Dirat, O., Kouklovsky, C.Y., Langlois, P., Lesot, P., and Courtieu, J. (2010). Double diastereoselection in asymmetric [2+3] cycloadditions reactions of chiral oxazoline *N*-oxides and application to the kinetic resolution of a racemic $\alpha,\beta$-unsaturated $\delta$-lactone. *Tetrahedron: Asymmetry* 10: 3197–3207.

**114** Queffelec, C., Boeda, F., Pouilhès, A., Meddour, A., Kouklovsky, C., Hannedouche, J., Collin, J., and Schulz, E. (2011). Enantioselective intramolecular hydroamination of secondary amines catalyzed by easily accessible ate and neutral rare-earth complexes. *ChemCatChem*. 3: 122–126.

**115** Didier, D., Meddour, A., Bezzenine-Lafollée, S., and Collin, J. (2010). Samarium iodobinaphtholate: an efficient catalyst for enantioselective aza-michael additions of O-benzylhydroxylamine to N-alkenoyloxazolidinones. *Eur. J. Org. Chem.* 14: 2678–2684.

**116** Palomino, M., Khudr, H., Courtieu, J., Merlet, D., and Meddour, A. (2012). The use of exchangeable nuclei to observe enantiomers through deuterium NMR in chiral liquid crystalline solvents. *Magn. Reson. Chem.* 50: S12–S16.

**117** Phillips, A.R. and Sharman, G.J. (2004). The measurement of high enantiomeric excesses in chiral liquid crystals using $_{19}$F NMR and exchangeable protons in $^{2}$H NMR. *Chem. Commun.* 1330–1331.

**118** Lesot, P., Merlet, D., Sarfati, M., Courtieu, J., Zimmermann, H., and Luz, Z. (2002). Enantiomeric and enantiotopic analysis of cone-shaped compounds with $C_3$ and $C_{3v}$ symmetry using NMR spectroscopy in chiral anisotropic solvents. *J. Am. Chem. Soc.* 124: 10071–10082.

**119** Meddour, A., Atkinson, D., Loewenstein, A., and Courtieu, J. (1998). Enantiomeric analysis of homologous series of secondary alcohols by deuterium NMR spectroscopy in a chiral nematic liquid crystal: influence of molecular geometry on chiral discrimination. *Chem. Eur. J.* 4: 1142–1147.

**120** Smadja, W., Auffret, S., Berdagué, P., Merlet, D., Canlet, C., Courtieu, J., Legros, J.-Y., Boutros, A., and Fiaud, J.-C. (1997). Visualisation of axial chirality using $^{2}$H-$^{1}$H NMR in poly($\gamma$-benzyl-L-glutamate), a chiral liquid crystal solvent. *Chem. Commun.* 2031–2032.

**121** Khetrapal, C.L., Ramanathan, K.V., Suryaprakash, N., and Vivekanandan, S. (1998). Natural abundance $^{2}$H NMR spectra of molecules oriented in liquid crystals. *J. Magn. Reson.* 135: 265–266.

**122** Lesot, P., Merlet, D., Loewenstein, A., and Courtieu, J. (1998). Enantiomeric visualisation using proton-decoupled natural abundance deuterium NMR in poly($\gamma$-benzyl-L-glutamate) liquid crystalline solutions. *Tetrahedron: Asymmetry* 9: 1871–1881.

**123** Lesot, P. and Courtieu J. (2009). Natural abundance deuterium NMR spectroscopy: developments and analytical applications in liquids, liquid crystals and solid phases. *Prog. Nucl. Magn. Reson. Spectrosc.* 55: 128–159.

**124** Lesot, P. and Lafon, O. (2008). Enantiomeric analysis using natural abundance deuterium 3D NMR spectroscopy in polypeptide chiral oriented media. Chem. Phys. Lett. 458: 219–222.

**125** Parenty, A., Campagne, J.-M., Aroulanda, C., and Lesot, P. (2002). Routine use of natural abundance deuterium NMR in a polypeptidic chiral oriented solvent for determination of the enantiomeric composition of chiral building blocks. *Org. Lett.* 4: 1663–1666.

**126** Lesot, P., Serhan, Z., Aroulanda, C., and Billault, I. (2012). Analytical contribution of NAD 2D-NMR spectroscopy in polypeptide mesophases to the investigation of triglycerides. *Magn. Reson. Chem.* 5: S2–S11.

**127** Sandström, D. and Zimmermann, H. (2000). Correlation of deuterium quadrupolar couplings and carbon-13 chemical shifts in ordered media by multiple-quantum NMR. *J. Phys. Chem. B.* 104: 1490–1493.

**128** Robyr, P. and Meier, B.H. (2000). Deuterium-carbon NMR correlation spectroscopy: a tool for structural characterization of solids. *Chem. Phys. Lett.* 327: 319–324.

**129** Lesot, P., Lafon O., and Berdagué, P. (2014). Correlation 2D-NMR experiments involving both $^{13}$C and $^{2}$H Isotopes in oriented media: methodological developments and analytical applications. *Magn. Reson. Chem.* 52: 595–613.

**130** Bilz, A., Stork, T.G., and Helmchen, G. (1997). New chiral solvating agents for carboxylic acids: discrimination of enantiotopic nuclei and binding properties. *Tetrahedron: Asymmetry* 8: 3999–4002.

**131** Aroulanda, C., Merlet, D., Courtieu, J., and Lesot, P. (2001). NMR experimental evidence of the differentiation of enantiotopic directions in $C_s$ and $C_{2v}$ molecules using partially oriented, chiral media. *J. Am. Chem. Soc.* 123: 12059–12066.

**132** Lesot, P., Lafon, O., Courtieu, J., and Berdagué, P. (2004). Analysis of the $^{13}$C NMR spectra of molecules, chiral by isotopic substitution, dissolved in a chiral oriented environment: toward the absolute assignment of the pro-*R*/pro-*S* character of enantiotopic ligands in prochiral molecules. Chem. Eur. J. 10: 3741–3746.

**133** Aroulanda, C., Zimmermann, H., Luz, Z., and Lesot, P. (2011). Enantiotopic discrimination in the deuterium NMR spectrum of solutes with $S_4$ symmetry in chiral liquid crystals. *J. Chem. Phys.* 134: 134502–1/8.

**134** Lesot, P., Aroulanda, C., Zimmermann, H., and Luz Z. (2015). Enantiotopic discrimination in the NMR spectrum of prochiral solutes in chiral liquid crystals. *Chem. Soc. Rev.* 44: 230–275.

**135** Lesot, P., Venkateswara Reddy, U., and Suryaprakash, N.S. (2011). Exploring the enantiodiscrimination potentialities of DNA-based orienting media using deuterium NMR spectroscopy. Chem. Commun. 47: 11736–11738.

**136** Naumann, C. and Kuchel, P.W. (2008). Prochiral and chiral resolution in $^2$H NMR spectra: solutes in stretched and compressed gelatin gels. *J. Phys. Chem. A.* 112: 8659–8664.

**137** Naumann, C. and Kuchel, P.W. (2009). NMR (pro)chiral discrimination using polysaccharide gels. *Chem. Eur. J.* 15: 12189–12191.

**138** Lesot, P., Berdagué, P., Meddour, A., Kreiter, A., Noll, M., and Reggelin, M. (2019). $^2$H and $^{13}$C NMR-based enantiodetection using polyacetylene versus polypeptide aligning media: versatile and complementary tools for chemists. *ChemPlusChem.* 84: 144–155.

**139** Mislow, M. and Raban, M. (1967). In *Book: Topics in Stereochemistry*, Vol. 1, (eds. N.L. Allinger and E.L. Eliel) New York: John Wiley & sons.

**140** Fujita, S. (2010). Systematic characterization of prochirality, prostereogenicity, and stereogenicity by means of the sphericity concept. *Tetrahedron.* 56: 735–740.

**141** Luz, Z. (2003). *Dynamic NMR in Liquid Crystals and Liquid Crystalline Solutions* (ed. E.E. Burnell, C.A. de Lange). Dordrecht: Kluwer Academic Publisher. Chap. 19.

**142** Luz, Z. (1983). Dynamic of molecular processes by NMR in liquid crystalline solvents. Israel *J. Chem.* 23: 305–313.

**143** Bain, A.D. (2003). Chemical exchange in NMR. *Prog. Nucl. Magn. Reson. Spectrosc.* 43: 63–103.

**144** Lesot, P., Lafon, O., Kagan, H.B., and Fan, P. (2006). Study of molecular rotational isomerism using deuterium NMR in chiral oriented solvents. Chem. Commun. 389–391.

**145** Lafon, O., Lesot, P., Fan C.A., and Kagan, H.B. (2007). Analysis of intramolecular dynamic processes in enantiomeric diaryl atropoisomers and related derivatives through $^2$H NMR in polypeptide liquid crystals. *Chem. Eur. J.* 13: 3772–3786.

**146** Sarfati, M., Aroulanda, C., Courtieu, J., and Lesot, P. (2001). Enantiomeric recognition of chiral invertomers through NMR in chiral oriented solvents: a study of the *cis*-decalin. *Tetrahedron: Asymmetry.* 12: 73–744.

**147** Aroulanda, C., Lafon, O., and Lesot, P. (2009). Enantiodiscrimination of flexible cyclic solutes using deuterium NMR spectroscopy in polypeptide chiral mesophases: investigation of *cis*-decalin and THF. *J. Phys. Chem. B.* 113: 10628–10640.

**148** Eyring, H. (1935). Activated complex in chemical reactions. *Chem. Rev.* 17: 65–77.

**149** Maiwald, M., Holger H. Fischer, H.H., Kim Y.-K., Albert, K., and Hasse, H. (2004). Quantitative high-resolution online NMR spectroscopy in reaction and process monitoring. *J. Mag. Res.* 166: 135–146.

**150** Dalitz, F., Cudaj, M., Maiwald, M., and Guthau, G. (2012). Process and reaction monitoring by low-field NMR spectroscopy. *Prog. Nucl. Magn. Reson.* 60: 52–70.

**151** Gouilleux, N., Charrier, B., Akoka, S., Felpin, F.-X., Rodriguez-Zu-Biri, M., and Giraudeau P. (2016). Ultrafast 2D NMR on a benchtop spectrometer: applications and perspectives. *TRAC-Trends in Anal. Chem.* 83: 65–75.

**152** Leigh, W.J. (1991). Thermotropic liquid crystals as chemical reaction media. In: *book: Liquid Crystals - Applications and Uses*, 357–408. World Scientific Publishing Co. doi: 10.1142/9789814360449_0007.

**153** Leigh, W.J., Workentin, M.S., Demus, D., and Vill, V. (2008). Liquid crystals as solvents for spectroscopic, chemical reaction, and gas chromatographic applications. In: *book: Handbook of Liquid Crystals Set.* (ed. D. Demus, J. Goodby, G.W. Gray, H.-W. Spiess, and V. Vill). Wiley. doi: 10.1002/9783527619276.ch9da.

**154** Lesot, P., Kazimierczuk, K., Trebosc, J., Amoureux, J-P., and Lafon, O. (2015). Fast acquisition of multidimensional NMR spectra of solids and mesophases using alternative sampling methods. *Magn. Reson. Chem.* 53: 927–939.

**155** Mobli, M. and Hoch, J.C. (2014). Nonuniform sampling and non- Fourier signal processing methods in multidimensional NMR. *Prog. Nucl. Magn. Reson. Spectrosc.* 83: 21–41.

**156** Frydman L., Lupulescu, A., and Scherf, T. (2003). Principles and features of single scan two-dimensional NMR spectroscopy. *J. Am. Chem. Soc.* 125: 9204–9217.

**157** Giraudeau, P. and Frydman, L. (2014). Ultrafast 2D NMR: an emerging tool in analytical spectroscopy. *Ann. Rev. Anal. Chem.* 7: 129–161.

**158** Giraudeau, P., Montag, T., Charrier, B., and Thiele, C.M. (2012). Fast access to residual dipolar couplings by single-scan 2D NMR in oriented media. *Magn. Reson. Chem.* 50: S53–S57.

**159** Lesot, P., Berdagué, P., and Giraudeau, P. (2016). Detection of quadrupolar nuclei by ultrafast 2D-NMR: exploring the case of deuterated analytes aligned in chiral oriented solvents. *Chem. Commun.* 52: 2122–2125.

**160** Gouilleux, B., Meddour, A., and Lesot, P. (2020). $^{2}$H QUOSY 2D-NMR Experiments in weakly aligning systems: from the conventional to the ultrafast approach. *ChemPhysChem.* 21: 1548–1563.

**161** Claridge, T.D.W. (2016). *High-Resolution NMR Techniques in Organic Chemistry*, 3e. Amsterdam, Neth: Elsevier Science B.V. ISBN: 978-0-08-099986-9, DOI: 10.1016/b978-0-08-099986-9.00002-6.

**162** Haasnoot, C.A.G., De Leeuw, F.A.A.M., De Leeuw, H.P.M., and Altona, C. (1980). The relation between protonproton NMR coupling constants and substituent electronegativities. An empirical generalization of the Karplus equation. *Tetrahedron* 36: 2783–2792.

**163** Palermo, G., Riccio, R., and Bifulco, G. (2010). Effect of electronegative substituents and angular dependence on the heteronuclear spin-spin coupling constant $^{3}J_{C\text{-}H}$: an empirical prediction equation derived by density functional theory calculations. *J. Org. Chem.* 75: 1982–1991.

**164** Marshall, J.L. (1983). Carbon-carbon and carbon-proton NMR couplings: applications to organic stereochemistry and conformational analysis. In: *Book: Methods in Stereochemical Analysis*, Vol. 2. Weinheim, Germany: Verlag Chemie.

**165** Anet, F.A.L., and Bourn, A.J.R. (1965). Nuclear magnetic resonance spectral assignments from nuclear overhauser effects. *J. Am. Chem. Soc.* 87: 5250–5251.

**166** Neuhaus, D. (2012). *Nuclear Overhauser Effect. Encyclopedia of NMR.* (ed. R.K. Harris and R.E. Wasylishen) 6: 3020–3035. Wiley. doi: 10.1002/9780470034590.emrstm0350.pub2.

**167** Liu, Y., Navarro-Vázquez, A., Gil, R.R., Griesinger, C., Martin, G.E., and Williamsonn, R.T. (2018). Application of anisotropic NMR parameters to the confirmation of molecular structure. *Nat. Protocol.* 14: 217–247.

**168** Nath, N., Schmidt, M., Gil, R.R., Williamson, R.T., Martin, G.E., Navarro-Vazquez, A., Griesinger, C., and Liu, Y. (2016). Determination of relative configuration from residual chemical shift anisotropy. *J. Am. Chem. Soc.* 138: 9548–9556.

**169** Gil, R.R. Griesinger, C., Navarro-Vazquez, A., and Sun, H. (2014). Structural elucidation of small organic molecules assisted by NMR in aligned media. In: *From Structure Elucidation in Organic Chemistry*, (ed. M.M. Cid and J. Bravo), 279–323. Weinheim, Germany: Wiley-VCH Verlag GmbH & Co. KGaA. ISBN: 978-3-527-66461-0, doi: 10.1002/9783527664610.ch8.

**170** Boettcher, B. and Thiele, C.M. (2012). Stereochemistry of molecules determined from residual dipolar couplings. (Ed. Harris, R.K., and Wasylishen, R.E.) *Encyclopedia of NMR.* 8: 4736–4747.

**171** Losonczi, J.A., Andrec, M., Fischer, M.W.F., and Prestegard, J.H. (1999). Order matrix analysis of residual dipolar couplings using singular value decomposition. *J. Mag. Res.* 138: 334–342.

**172** Cornilescu, G. and Bax, A. (2000). Measurement of proton, nitrogen, and carbonyl chemical shielding anisotropies in a protein dissolved in a dilute liquid-crystalline phase. *J. Am. Chem. Soc.* 122: 10143–10154.

**173** Troche-Pesqueira, E., Anklin, C., Gil, R.R., and Navarro-Vazquez, A. (2017). Computer-assisted 3D structure elucidation of natural products using residual dipolar couplings. *Angew. Chem., Int. Ed.* 56: 3660–3664.

**174** Lesot, P., Gil, R.R., Berdagué, P., and Navarro-Vazquez, A. (2020). Residual quadruplar couplings: crossing the current frontiers in the relative configuration analysis of natural products. *J. Nat. Prod.* 83: 3141–3148.

**175** Diehl, P., Henrichs, P. M., and Niederberger, W. (1971). A study of the molecular structure and of the barrier to methyl rotation in o-chlorotoluene partially oriented in the nematic phase. *Mol. Phys.* 1: 139–145.

**176** Aroulanda, C., Lesot, P., Merlet, D., and Courtieu J. (2003). Structural ambiguities in bridged ring systems resolved using natural abundance deuterium NMR in chiral liquid crystals. *J. Phys. Chem. A.* 107: 10911–10918.

**177** Navarro-Vazquez, A., Berdagué, P., and Lesot, P. (2017). Integrated computational protocol for analyzing quadrupolar splittings from natural abundance deuterium NMR spectra in (chiral) oriented media. *ChemPhysChem.* 18: 1252–1266.

**178** Navarro-Vázquez, A. (2012). MSpin-RDC. A program for the use of residual dipolar couplings for structure elucidation of small molecules. *Magn. Reson. Chem.* 50: S73–S79.

**179** Immel, S., Köck, M., and Reggelin, M. (2018). Configurational analysis by residual dipolar coupling driven floating chirality distance geometry calculations. *Chem. Eur. J.* 24: 13918–13930.

**180** Köck, M., Reggelin, M., and Immel, S. (2020). The advanced floating chirality distance geometry approach - how anisotropic NMR parameters can support the determination of the relative configuration of natural products. *Mar. Drugs* 18: 330–352.

**181** Berdagué, P., Gouilleux, B. Nolls, M., Immel, S., Reggelin, M., and Lesot, P. (2021). Study and quantification of enantiodiscrimination power of four polymeric chiral LLCs using NAD 2D-NMR. *Phys. Chem. Chem. Phys.* 2022, 24: 7338–7348.

**182** Geissman, T.A. and Griffin, T.S. (1972). Sesquiterpene lactones of Artemisia Carruthii. *Phytochemistry* 11: 833–835.

**183** Sosa, V.E., Oberti, J.C., Gil, R.R., Ruveda, E.A., Goedken, V.L., Gutierrez, A.B., and Herz, W. (1989). 10-Epideoxycumambrin B and other constituents of Stevia yaconensis var. subeglandulosa. *Phytochemistry* 28: 1925–1929.

**184** Blanco, J.G., Gil, R.R., Alvarez, I., Patrito, L.C., Genti-Raimondi, and S., Flury, A. (1997). A novel activity for a group of sesquiterpene lactones: inhibition of aromatase. *FEBS Lett.* 409: 396–400.

**185** Nicotra, V.E., Ramacciotti, N.S., Gil, R.R., Oberti, J.C., Feresin, G.E., Guerrero, C.A., Baggio, R.F., Garland, M. T., and Burton, G. (2006). Phytotoxic Withanolides from Jaborosa rotacea. *J. Nat. Prod.* 69: 783–789.

**186** García, M.E., Pagola, S., Navarro-Vazquez, A., Phillips, D.D., Gayathri, C., Krakauer, H., Stephens, P.W., Nicotra V.E., and Gil, R.R. (2009). Stereochemistry determination by powder X-ray diffraction analysis and NMR spectroscopy residual dipolar couplings. *Angew. Chem., Int. Ed.* 48: 5670–5674.

**187** Waratchareeyakul, W., Hellemann, E., Gil, R.R., Chantrapromma, K., Langat, M.K., and Mulholland, D.A. (2017). Application of residual dipolar couplings and selective quantitative NOE to establish the structures of tetranortriterpenoids from Xylocarpus rumphii. *J. Nat. Prod.* 80: 391–402.

**188** Kummerlöwe, G., Crone, B., Kretschmer, M., Kirsch, Stefan F., and Luy, B. (2011). Residual dipolar couplings as a powerful tool for constitutional analysis: the unexpected formation of tricyclic compounds. *Angew. Chem., Int. Ed.* 50: 2643–2645. S2643/1-S2643/25.

**189** Tzvetkova, P., Simova, S., and Luy, B. (2007). P.E. HSQC: a simple experiment for simultaneous and signsensitive measurement of ($^1J_{CH} + D_{CH}$) and ($^2J_{HH} + D_{HH}$) couplings. *J. Mag. Res.* 186: 193–200.

**190** Gil, R.R. (2017). Residual dipolar couplings in small-molecules NMR. In: *Encyclopedia of spectroscopy and spectrometry*, 3e (ed. J.C. Lindon, G.E. Tranter, and D. Koppenaal), Vol. 3, 946–955.

**191** Farley, K.A., Che, Y., Navarro-Vazquez, A., Limberakis, C., Anderson, D., Yan, J., Shapiro, M., Shanmugasundaram, V., and Gil, R.R. (2019). Cyclic peptide design guided by residual dipolar couplings, *J*-couplings, and intramolecular hydrogen bond analysis. *J. Org. Chem.* 84: 4803–4813.

**192** Riveira, M.J., Trigo-Mourino, P., Troche-Pesqueira, E., Martin, G.E., Navarro-Vazquez, A., Mischne, M.P., and Gil, R.R. (2015). Self-sensitized photooxygenation of $^{2}$H-Pyrans: characterization of unexpected products assisted by computed structural elucidation and residual dipolar couplings. *J. Org. Chem.* 80: 7396–7402.

**193** ACD labs, https://www.acdlabs.com.

**194** Kummerlöwe, G., Grage, S.L.; Thiele, C.M., Kuprov, I., Ulrich, A.S., and Luy, B. (2011). Variable angle NMR spectroscopy and its application to the measurement of residual chemical shift anisotropy. *J. Magn. Reson.* 209: 19–30.

**195** Hallwass, F., Schmidt, M., Sun, H., Mazur, A., Kummerlöwe, G., Luy, B., Navarro-Vázquez, A., Griesinger, C., and Reinscheid, U.M. (2011). Residual chemical shift anisotropy (RCSA): A tool for the analysis of the configuration of small molecules. *Angew. Chem., Int. Ed.* 50: 9487–9490.

**196** Hallwass, F., Teles, R. R., Hellemann, E., Griesinger, C., Gil, R.R., and Navarro-Vazquez, A. (2018). Measurement of residual chemical shift anisotropies in compressed polymethylmethacrylate gels. Automatic compensation of gel isotropic shift contribution. *Mag. Reson. Chem.* 56: 321–328.

**197** Liu, Y. and Prestegard, J.H. (2010). A device for the measurement of residual chemical shift anisotropy and residual dipolar coupling in soluble and membrane-associated proteins. *J. Biomol.* NMR 47: 249–258.

**198** Mevers, E., Sauri, J., Liu, Y., Moser, A., Ramadhar, T.R., Varlan, M., Williamson, R.T., Martin, G.E., and Clardy, J. (2016). Homodimericin A: a complex hexacyclic fungal metabolite. *J. Am. Chem. Soc.* 138: 12324–12327.

**199** Liu, Y., Cohen, R.D., Gustafson, K.R., Martin, G.E., and Williamson, R.T. (2018). Enhanced measurement of residual chemical shift anisotropy for small molecule structure elucidation. *Chem. Commun.* 54: 4254–4257.

**200** Milanowski, D.J., Oku, N., Cartner, L.K., Bokesch, H.R., Williamson, R.T., Sauri, J., Liu, Y., Blinov, K.A., Ding, Y., Li, X-C., Ferreira, D., Walker, L.A., Khan, S., Davies-Coleman, M.T., Kelley, J.A., McMahon, J.B., Martin, G.E., and Gustafson, K.R. (2018). Unequivocal determination of caulamidines A and B: Application and validation of new tools in the structure elucidation toolbox. *Chem. Sci.* 9: 307–314.

**201** Yan, P., Li, G., Wang, C., Wu, J., Sun, Z., Martin, G.E., Wang, X., Reibarkh, M., Sauri, J., and Gustafson, K.R. (2019). Characterization by empirical and computational methods of dictyospiromide, an intriguing antioxidant alkaloid from the marine Alga Dictyota Coriacea. *Org. Lett.* 21: 7577–7581.

**202** Ndukwe, I.E., Brunskill, A., Gauthier, D.R., Zhong, Y-L., Martin, G.E., Williamson, R.T., Reibarkh M., and Liu, Y. (2019). $^{13}$C NMR-based approaches for solving challenging stereochemical problems. *Org. Lett.* 21: 4072–4076.

**203** Ndukwe, I.E., Wang, X., Lam, N.Y. S., Ermanis, K., Alexander, K.L.; Bertin, M.J., Martin, G.E., Muir, G., Paterson, I., Britton, R., Goodman, J.M., Helfrich, E.J.N., Piel, J., Gerwick, W.H., and Williamson, R.T. (2020). Synergism of anisotropic and computational NMR methods reveals the likely configuration of phormidolide A. *Chem. Commun.* 56: 7565–7568.

**204** Hellemann, E. and Gil, R.R. (2018). New stretching method for aligning gels: its application to the measurement residual chemical shift anisotropies (RCSAs) without the need for isotropic shift correction. *Chem. Eur. J.* 24: 3689–3693.

**205** Chou, J.J., Gaemers, S., Howder, B., Louis, J.M., and Bax, A. (2001). A simple apparatus for generating stretched polyacrylamide gels, yielding uniform alignment of proteins and detergent micelles. *J. Biomol. NMR.* 21: 377–382.

**206** Recchia, M.J.J., Cohen, R.D., Liu, Y., Sherer, E.C., Harper, J.K.; Martin, G.E., and Williamson, R.T. (2020). "Oneshot" measurement of residual chemical shift anisotropy using poly-$\gamma$-benzyl-L-glutamate as an alignment medium. *Org. Lett.* 22: 8850–8854.

**207** Monroe, A.Z., Gordon, W.H., Wood, J.S., Martin, G.E., Morgan, J.B., and Williamson, R.T. (2021). Structural revision of a Wnt/$\beta$-catenin modulator and confirmation of cannabielsoin constitution and configuration. *Chem. Commun.* 57: 5658–5661.

**208** Lei, X., Qiu, F., Sun, H., Bai, L., Wang, W.X., Xiang, W. and Xiao, H. (2017). A self-assembled oligopeptide as a versatile NMR alignment medium for the measurement of residual dipolar couplings in methanol. *Angew. Chem., Int. Ed.* 56: 12857–12861.

**209** Liu, H., Chen, P., Li, X.L., Sun, H., and Lei, X. (2020). Practical aspects of oligopeptide AAKLVFF as an alignment medium for the measurements of residual dipolar coupling of organic molecules. *Magn. Reson. Chem.* 58: 404–410.

**210** Li, X.-L., Chi, L.-P., Navarro-Vazquez, A., Hwang, S., Schmieder, P., Li, X.-M., Li, X., Yang, S.-Q., Lei, X. Wang, B.-G., and Sun, H. (2020). Stereochemical elucidation of natural products from residual chemical shift anisotropies in a liquid crystalline phase. *J. Am. Chem. Soc.* 142: 2301–2309.

**211** Tjandra, N., Omichinski, J.G., Gronenborn, A.M., Clore, G. M. and Bax, A. (1997). Use of dipolar $^{1}H$-$^{15}N$ and $^{1}H$-$^{13}C$ couplings in the structure determination of magnetically oriented macromolecules in solution. *Nat. Struc. & Mol. Biol.* 4: 732–738.

**212** Bax, A. (2003). Weak alignment offers new NMR opportunities to study protein structure and dynamics. *Protein Sci.* 12: 1–16.

**213** Karschin, N., Wolkenstein, K., and Griesinger, C. (2020). Magnetically induced alignment of natural products for stereochemical structure determination via NMR. *Angew. Chem., Int. Ed.* 59: 15862–15864.

**214** Anet, F.A.L. (1986). Magnetic alignment effects in the 500-MHz proton NMR spectrum of *o*-dichlorobenzene in acetone-$d_6$. *J. Am. Chem. Soc.* 108: 1354–55.

**215** Lisicki, M.A., Mishra, P.K., Bothner-By, A.A., and Lindsey, J.S. (1988). Solution conformation of a porphyrinquinone cage molecule determined by dipolar magnetic field effects in ultra-high-field NMR. *J. Phys. Chem.* 92: 3400–3403.

**216** Alemany, L.B., Gonzalez, A., Billups, W.E., Willcott, M.R., Ezell, E., and Gozansky, E. (1997). Alignment effects in high field proton NMR spectra of the hydrogenated fullerenes c60h2 and c60h4: Evidence for residual anisotropic dipole–dipole couplings. *J. Org. Chem.* 62: 5771–5779.

**217** Nath, N., Fuentes-Monteverde, J.C., Pech-Puch, D., Rodriguez, J., Jimenez, C., Noll, M., Kreiter, A., Reggelin, M., Navarro-Vazquez, A., and Griesinger, C. (2020). Relative configuration of micrograms of natural compounds using proton residual chemical shift anisotropy. *Nat. Commun.* 11: 4372–4381.

**218** Dong, R.Y. (1996). In book: *Encyclopedia of NMR.* (ed. Harris R.K., and Grant D.M.J.), 2752–2759. Chichester: Wiley & Sons.

**219** Diehl, P., and Leipert, T. (1964). Deuteronen- kernresonanzspektroskopie, *Helv. Chim. Acta.* 47: 545–557.

**220** Briggs, J.M., Farnell, L.F., and Randall, E.W. (1973). Proton-noise-decoupled deuterium resonance at natural abundance by Fourier transform. *Chem. Commun.* 70–71.

**221** Leopold, M.F., Epstein, W.W. and Grant, D.M. (1998). Natural abundance deuterium NMR as a novel probe of monoterpene biosynthesis: limonene. *J. Am. Chem. Soc.* 111: 616–617.

**222** Martin, G.J. and Martin, M.L. (1981). Deuterium labelling at the natural abundance level as studied by high-field quantitative H-2 NMR. Tetrahedron Lett. 22: 3525–3528.

**223** Prosser, R.S., Heaton, N.J., and Kothe, J. (1996). Application of the quadruplar Carr-Purcell-Meiboom-Gill pulse train for sensitivity enahancement in deuteron NMR of liquid Crystals. *J. Magn. Reson. B.* 112: 51–56.

**224** Tabayashi, K. and Akasaka, K. (1997). Natural abundance $^{2}H$ NMR for liquid crystal studies. application to 4‘-(hexyloxy)-4-cyanobiphenyl. *J. Phys. Chem. B.* 101: 5108–5111.

**225** Lesot, P . Aroulanda, C., and. Billault, I. (2004). Exploring the analytical potential of NMR spectroscopy in chiral anisotropic media for the study of the natural abundance deuterium distribution in organic molecules. *Anal. Chem.* 76: 2827–2835.

**226** Lesot, P., Baillif, V., and Billault, I. (2008). Combined analysis of four C-18 unsaturated fatty acids using natural abundance deuterium 2D-NMR spectroscopy in chiral oriented solvents. *Anal. Chem.* 80: 2963–2972.

**227** Serhan, Z, Martel, L., Billault, I., and Lesot, P. (2010). Complete determination of site-specific bio-enantiomeric excesses in linoleic acid using natural abundance deuterium 2D NMR in polypeptide mesophase. *Chem. Commun.* 46: 6599–6601.

**228** Guo, Z. (2016). Artemisinin anti-malarial drugs in China Zongru Guo. *Acta Pharm. Sin.* B 6: 115–124.

**229** Berger, R., Courtieu, J., Gil, R.R., Griesinger C, Köck M., Lesot, P., Luy, B., Merlet, D., Navarro-Vazquez, A., Reggelin, M, Reinscheid, U.M., Thiele, C.M., and Zweckstetter, M. (2012). Is the determination of absolute configuration possible by using residual dipolar couplings from chiral-non-racemic alignment media? - A critical assessment. *Angew. Chem., Int. Ed.* 51: 2–5.

**230** Ziani, L., Lesot, P., Meddour, A., and Courtieu, J. (2007). Empirical determination of the absolute configuration of small chiral molecules using natural abundance $^{2}$H NMR in chiral liquid crystals. *Chem Commun.* 45: 4735–4739.

**231** Helfrich, J., Hentschke, J., and Apel, U.M. (1994). Molecular-dynamics simulation study of poly(gamma-benzyl Lglutamate) in dimethylformamide. *Macromolecules* 2: 472–482.

**232** Alvira, E., Breton, J., and Plata, J. (1991). J. Chiral discrimination - a model for the interaction between a helicoidal system and an amino-acid molecule. *Chem. Phys.* 155: 7–18.

**233** Frank, A.O., Freudenberger, J.C., Shaytan A.K., Kessler H. and Luy, B. (2010). Direct prediction of residual dipolar couplings of small molecules in a stretched gel by stochastic molecular dynamics simulations. *Magn. Reson. Chem.* 53: 213–217.

**234** Nandi, N. (2004). Role of secondary level chiral structure in the process of molecular recognition of ligand: study of model helical peptide. *J. Phys. Chem. B.* 108: 789–797.

**235** Sager, E., Tzvetkova, P., Gossert, A.D., Piechon, P., and Luy, B. (2020). Determination of configuration and conformation of a reserpine derivative with seven stereogenic centers using molecular dynamics with RDCderived tensorial constraints. *Chem. Eur. J.* 26: 14435–14444.

**236** Serhan, Z., Billault, I., Borgogno, A., Ferrarini, A., and Lesot, P. (2012). Analysis of NAD 2D-NMR spectra of saturated fatty acids in polypeptide aligning media by experimental and modeling approaches. *Chem. Eur. J.* 18: 117–126.

**237** Celebre, G., De Luca, G., Emsley, J.W., Foord, E.K., Longeri, M., Lucchesini, F., and Pileio G. (2003). The conformational distribution in diphenylmethane determined by nuclear magnetic resonance spectroscopy of a sample dissolved in a nematic liquid crystalline solvent. *J. Chem. Phys.* 118: 6417–6426.

**238** Celebre, G. and Longeri, M. (2003b). *NMR Studies of Solutes in Liquid Crystals: Small Flexible Molecules* (ed. E.E. Burnell and C.A. de Lange). Dordrecht: Kluwer Academic Publisher, Chap. 14.

**239** Samulski, E.T. (2003). *Very Flexible Solutes: Alkyl Chains and Derivatives* (ed. E.E. Burnell, C.A. de Lange). Dordrecht: Kluwer Academic Publisher. Chap. 13.

**240** Flory, P.J. (1974). Foundations of rotational isomeric state theory and general methods for generating configurational averages. *Macromolecules* 7: 381–392.

**241** Emsley, J.W., Luckhurst, G.R., and Stockley, C.P. (1982). A theory of orientational ordering in uniaxial liquid crystals composed of molecules with alkyl chains. *Proc. R. Soc. London, Ser.* A 381: 117–138.

**242** Catalano, D., Di Bari, L., and Veracini C.A. (1991). A maximum- entropy analysis of the problem of the rotameric distribution for substituted biphenyls studied by $^{1}$H nuclear magnetic resonance spectroscopy in nematic liquid crystals. *J. Chem. Phys.* 94: 3928–3935.

**243** Di Bari, L., Forte, C., Veracini, C.A., and Zannoni C. (1988). An internal order approach to the investigation of intramolecular rotations in liquid crystals by NMR: 3-phenyl-thiophene in PCH and phase IV. *Chem. Phys. Lett.* 143: 263–269.

**244** Stevensson, B., Sandström, D., and Maliniak, A. (2003). Conformational distribution functions extracted from residual dipolar couplings: A hybrid model based on maximum entropy and molecular field theory. *J. Chem. Phys.* 119: 2738–2746.

**245** Celebre, G., De Luca, G., and Di Pietro, M.E. (2012). Conformational distribution of trans-stilbene in solution investigated by liquid crystal NMR spectroscopy and compared with in vacuo theoretical predictions. *J. Phys. Chem. B.* 116: 2876–2885.

**246** Di Pietro, M.E., Aroulanda C., Celebre, G., Merlet, D., and De Luca, G. (2015). The conformational behaviour of naproxen and flurbiprofen in solution by NMR spectroscopy. *New J. Chem.* 39: 9086–9097.

**247** Di Pietro, M.E., Celebre, G., Aroulanda C., Merlet, D., and De Luca, G. (2017). Assessing the stable conformations of ibuprofen in solution by means of residual dipolar couplings. *Eur. J. of Pharma. Sci.* 106: 113–121.

**248** Di Pietro, M.E., Sternberg, U., and Luy, B. (2019). Molecular dynamics with orientational tensorial constraints: a new approach to probe the torsional angle distributions of small rotationally flexible molecules. *J. Phys. Chem. B* 123: 8480–8491.

**249** Salvino, R.A. De Luca, G., and Celebre, G. (2022). Assessing the chirality-dependent conformational distribution of small flexible opposite enantiomers dissolved in weakly ordering enantiopure media by means of liquid crystal NMR techniques. *J. Mol. Liquids*, 347, 11794.

**250** Basov, A., Fedulova, L., Vasilevskaya, E., and Dzhimak, S. (2019). Possible mechanisms of biological effects observed in living systems during $^{2}H/^{1}H$ isotope fractionation and deuterium interactions with other biogenic isotopes. *Molecules*. 24: 4101–4123.

**251** Akoka, S. and Remaud, G.S. (2020). NMR-based isotopic and isotopomic analysis. *Prog. Nucl. Magn. Reson.* 120–121: 1–24.

**252** Martin, G.J., Martin, M.L., Mabon F., and Bricout, J. (1982). A new method for the identification of the origin of natural products. Quantitative $^{2}H$ NMR at the natural abundance level applied to the characterization of anetholes. *J. Am. Chem. Soc.* 104: 2658–2659.

**253** Duan, J.-R., Billault, I., Mabon, F., and Robins, R.J. (2002). Natural deuterium distribution in fatty acids isolated from peanut seed oil: A site-specific study by quantitative $^{2}H$ NMR spectroscopy. *ChemBioChem.* 3: 752–759.

**254** Remaud, G.S., Martin, Y.-L., Martin, G.G., and Martin G.J. (1997). Detection of sophisticated adulterations of natural vanilla flavors and extracts: application of the SNIF-NMR method to vanillin and p-hydroxybenzaldehyde. *J. Agric. Food Chem.* 45: 859–866.

**255** Remaud, G., Debon, A.A., Martin, Y.-L., Martin G.G., and Martin, G.J. (1997-b). Authentication of bitter almond oil and cinnamon oil: application of the SNIF-NMR method to benzaldehyde. *J. Agric. Food Chem.* 45: 4042–4048.

**256** Baillif, V., Robins, R.J., Le Feunten, Lesot, P., and Billault, I. (2009). Investigation of fatty acid elongation and desaturation steps in Fusarium Lateritium by quantitative two-dimensional deuterium NMR spectroscopy in Chiral Oriented *Media. J. Biol. Chem.* 284: 10783–10792.

**257** Lesot, P., Serhan, Z., and Billault, I. (2011-b). recent advances in the analysis of the site-specific isotopic fractionation of metabolites such as fatty acids using anisotropic natural abundance 2H NMR spectroscopy: application on conjugated linolenic methyl esters. *Anal. Bioanal. Chem.* 399: 1187–1200.

**258** Billault, I., Ledru, A., Ouetrani, M., Serhan, Z., Lesot, P., and Robin, R.J. (2012). Probing substrate-product relationships by natural abundance deuterium 2D NMR spectroscopy in liquid-crystalline solvents: the case of the epoxidation of linoleate to vernoleate by two different plant enzymes. *Anal. Bioanal. Chem.* 402: 2985–2998.

**259** Altmann, S.I. (1967). The summary of nonrigid molecules: the Schrödinger supergroup. *Proc Roy Soc. (London)* A298: 184–203.

**260** Jézéquel, T., Joubert, V., Giraudeau, P., Remaud, G.S., and Akoka, S. (2017). The new face of isotopic NMR at natural abundance. *Magn. Reson. Chem.* 55: 77–90.

**261** Zhao, J., Wang, M., Saroja, S.G., and Khan, I.A. (2021). NMR technique and methodology in botanical health product analysis and quality control. *J. of Pharm. and Biomed. Anal.* 207: 114376, 1–27.

**262** Texier-Bonniot, T., Berdagué, P., Robins, R., Remaud, G., and Lesot, P. (2018). Analytical contribution of deuterium 2D-NMR in oriented solvents to $^{2}H/^{1}$ isotopic characterization: the case of Vanillin. *Flavour Frag J.* 34: 217–229.

**263** Portaluri, V., Thomas F., Jamin, E., Akoka, S., and Remaud G. (2021). Authentication of agave products through isotopic intramolecular $^{13}C$ content of ethanol: optimization and validation of $^{13}C$ quantitative NMR methodology. *ACS Food Sci. Technol.* 1: 1316–1322.

**264** Tenailleau, E., Lancelin, P., Robins, R.J., and Akoka, S. (2004). Authentication of the origin of vanillin using quantitative natural abundance C-13 NMR. *J. Agric. Food Chem.* 52: 7782–7787.

**265** Griesinger, C., Bennatia, M., Viethn, H.M, Luchinat, C., Parigi, G., Höferd, P., Engelked, F., Glasere, S.J., Denysenkov, V., and Prisne, T.F. (2012). Dynamic nuclear polarization at high magnetic fields in liquids. *Prog. Nucl. Magn. Reson. Spectrosc.* 64: 4–28.

**266** Celebre, G. (2001). On the anisotropic intermolecular potential of biaxial apolar solutes in nematic solvents: Monte Carlo predictions and experimental data. *J. Chem. Phys.* 115: 9552–9556.

**267** Celebre, G., De Luca, G., D'Urso, C., and Di Pietro, M.E. (2019). Helical solutes orientationally ordered in anisotropic media composed of helical particles: Formulation of a mean torque potential sensitive to P and M chirality as a tool for the assignment of the absolute configuration of enantiomers. *J. Mol. Liq.* 288 (1–5): 111044.

**268** Ibanez de Opakua, A., Klama, F., Ndukwe, I.E., Martin, G.E., Williamson, R.T., and Zweckstetter, M. (2020). Determination of complex small-molecule structures using molecular alignment simulation. *Angew. Chem., Int. Ed.* 59: 617–6176.

**269** Tzvetkova, P., Sternberg, U., Gloge, T., Navarro-Vázquez, A., and Luy, B. (2019). Configuration determination by residual dipolar couplings: accessing the full conformational space by molecular dynamics with tensorial constraints. *Chem. Sci.* 10: 8774–8791.

# 10

# Ultrafast 2D methods

*Boris Gouilleux*

*Université Paris-Saclay, laboratoire ICMMO, ERMN 17 Av. des Sciences, Orsay 91400, France*

## 10.1 Introduction

Much of the power of NMR spectroscopy relies on the ability to correlate information in a multi-dimensional (nD) pattern delivering valuable structural and dynamic insights on the probed compounds. Moreover, the gain in resolution yielded by spreading out the signals over several dimensions makes nD NMR spectroscopy a central method to address analytical challenges such as structural and dynamic studies of large biomolecules or the analysis of smaller compounds within complex matrixes.

The important benefits of the multi-dimensional approach come with a long experimental duration when the 2D spectra are recorded in respect to the conventional scheme introduced by Jeener [1]. In its original form, a 2D interferogram is obtained by performing a series of similar experiments with a linearly incremented $t_1$-evolution period (Figure 10.1a). This multi-experiment scheme is in essence a time-consuming process, all the more so several scans may be averaged per $t_1$-increment to complete a phase-cycling or for sensitivity reasons. The overall experiment duration $T_{exp}$ pertained to 2D experiments is expressed by the equation below:

$$T_{exp} = T_R N_S N_1. \tag{10.1}$$

In this equation, $N_1$ is the number of necessary $t_1$-increments to sample the indirect dimension, $N_S$ is the number of scans averaged per $t_1$-increment and $T_R$ is the repetition time from scan to scan. Over the past 30 years, a set of methods have been developed to speed up the record of nD spectra [2]. The most intuitive idea is to retain the original scheme, but with a reduced number of increments $N_1$ or shorter repetition times $T_R$ (or both at the same time), accompanied by ingenious changes in the pulse sequences and data processing to counteract the decrease in resolution and sensitivity that this would entail.

A first set of fast nD methods consists of dramatically shortening the recycling delay, together with an excitation at the Ernst flip angle. Usually, the sensitivity of such fast-pulsing experiments are further improved by an enhancement of the $T_1$-relaxation of spins of interest [3]. Meanwhile, efforts have been paid to reduce the number of $t_1$-increments $N_1$ while maintaining (or even improving) the resolution along the indirect dimension F1. Non-uniform sampling (NUS) techniques are widely used in that respect [4].

Despite impressive acceleration, these strategies face a conceptual limitation, as $N_1$ and $T_R$ can only be reduced to a certain level to preserve a usable 2D interferogram. To push forward this acceleration, an alternative concept

*Two-Dimensional (2D) NMR Methods*, First Edition. Edited by K. Ivanov, P.K. Madhu and G. Rajalakshmi.

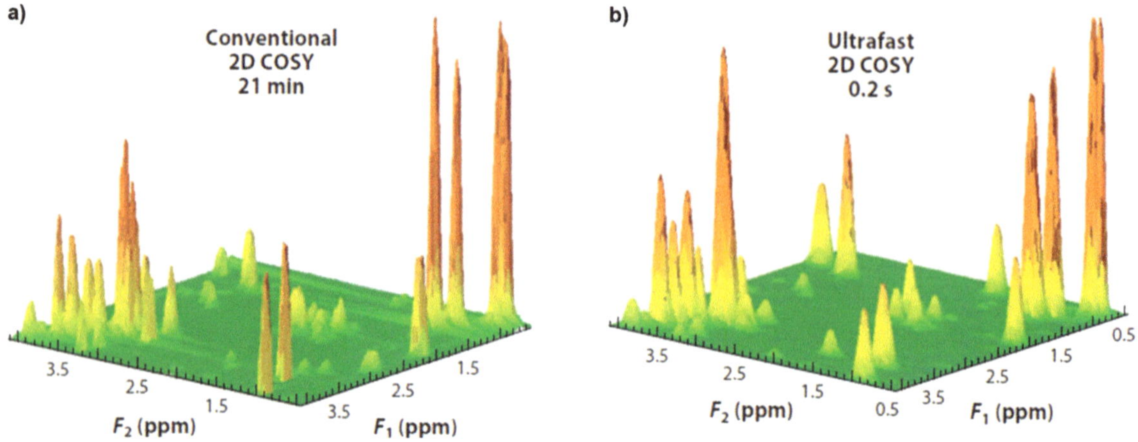


Figure 10.1 (a) Conventional and (b) $^1$H UF 2D COSY spectra recorded on an aqueous mixture of metabolites with a 500 MHz spectrometer equipped with a cryogenic probe. Note the dramatic acceleration with the UF method (0.2 s versus 21 min).

for the record of 2D spectra has to be considered. Frydman and coworkers addressed this bottleneck with the development of ultrafast (UF) 2D NMR, which enables the recording of any multi-dimensional NMR spectra within a single scan (Figure 10.1) [5]. This concept proposed about 20 years ago fits well the current state-of-the-art of NMR where hardware improvements (very high magnetic field, cryogenic probe) and the rise of hyperpolarization techniques, warrant in many instances sufficient signal-to-noise ratio (SNR) with a single-scan acquisition.

The key feature of UF 2D NMR is the replacement of the time-consuming $t_1$-incrementation by a spatiotemporal manipulation of the spins making their $t_1$-evolution dependent on their spatial position. This spatial parallelization is a much faster process than the collection of $N_1$ successive experiments. This concept is presented in more detail in Figure 10.2. In conventional 2D NMR, all the spins experience the same incremented $t_1$-evolution period at each transient, whereas in the UF approach spins undergo different $t_1$-evolutions at the same time according to their spatial position: $t_1(z) = Cz$.

From this guiding concept, UF 2D NMR requires two specific elements, which are not present in its conventional counterpart: (i) a spatial encoding step (SPEN) to entangle the evolution time and the spatial position; (ii) a detection period, namely echo planar spectroscopic imaging (EPSI), capable of reading out the spatially encoded signals. This is followed by a specific data processing to retrieve 2D correlation maps similar to those afforded by conventional techniques. These two key elements, SPEN and EPSI, rely on gradient-based methods inspired from the NMR-imaging methodology.

Several reviews about UF 2D NMR have already been published, providing the reader with a comprehensive overview of this methodology both from a theoretical point of view and the practical applications that follow [6–9].

The herein chapter aims at surveying the fundamental aspects of UF 2D NMR as well as the current methods used in practical applications. In a first section, the key elements necessary to understand UF 2D NMR are described in detail, such as the spatial encoding step, and the decoding of this information during the detection period. The specificities and limits of this ultrafast methodology are then discussed in a second section, while the most significant improvements that remedy the initial limitations are presented in a third section. The fourth section aims at demonstrating the versatility of this ultrafast concept to accelerate many types of 2D spectroscopic

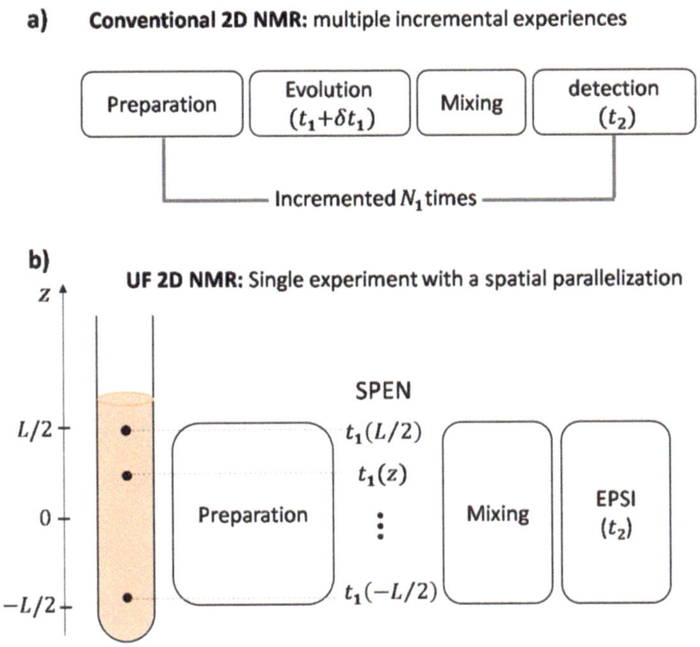


**Figure 10.2** Comparison between conventional 2D NMR based on the Jeener's scheme and the ultrafast (UF) 2D NMR. (a) in the conventional version, experiments with similar preparation, mixing and detection steps are repeated $N_1$ times along with an incrementing of the $t_1$-evolution period. (b) in UF 2D NMR, this series of experiments is replaced by a spatial parallelization of the $t_1$-evolution time: $t_1(z) = Cz$ carried out in a single scan.

methods (homo- and hetero-nuclear), as well as pseudo-2D dynamic experiments such as inversion recovery (IR) or diffusion ordered spectroscopy (DOSY). Finally, an overview of UF 2D NMR applications is presented in the last section, with representative examples from the fields of reaction monitoring, hyperpolarization, oriented media and high-resolution NMR in inhomogeneous magnetic fields.

## 10.2 UF 2D NMR Principles: Entangling the Space and the Time

### 10.2.1 Spatial Encoding

The spatial encoding (SPEN) is the necessary element to link the $t_1$-evolution period of the spins with their spatial location. This can be achieved in practice with different schemes, as further detailed, which share the combined use of magnetic field gradient with selective pulses or frequency-swept pulses.

When a magnetic field gradient is applied along the z-axis, the resonance frequency pertained to a chemical environment $i$ becomes proportional to the z-position:

$$\omega_i(z) = \Omega_i + \gamma G z. \tag{10.2}$$

This leads to magnetization vectors that rotate at different angular rates in the rotating frame along the sample length. This can be visualized as a z-axis helix whose pitch depends on the amplitude of the pulse-field-gradient $G$ and the gyromagnetic ratio $\gamma$.

Thereafter, the combined used of a magnetic field gradient with a selective pulse allows one to select a slice of the sensitive volume of the probe, where the center $z_c$ and the thickness $\Delta z$ are tunable according to the pulse length of the selective pulse (corresponding to a frequency bandwidth $BW$), the RF-frequency offset $O_{RF}$ as well as with the gradient amplitude $G$ (Figure 10.3a):

$$\Delta z = \left|\frac{BW}{\gamma G}\right|. \tag{10.3}$$

$$z_c = \frac{\Omega_i - O_{RF}}{\gamma G} \tag{10.4}$$

Hence, applying successive selective pulses with different $O^j_{RF}$, rotates the magnetization (excitation or refocusing) from slice to slice at distinct instants. From this capability, a first SPEN was proposed that consists of a train of $N_1$ selective pulses with a linear increase of $O^j_{RF}$, applied together with bipolar pulse-field gradients (Figure 10.3b). This scheme is referred to as "discrete spatial encoding" and was used as a proof of concept for UF 2D NMR published in 2002 [5]. This leads to the desired array of $N_1$ evolution periods (one per slice) in a single scan. Yet, a large number of selective pulses is necessary to reach a descent resolution, which is demanding for the hardware, especially when it comes with bipolar gradient pulses. Furthermore, the presence of "ghost-peaks" arising from the discrete nature of this encoding limits further its use. Soon after this discrete encoding has been superseded by the emergence of continuous SPEN, which improve the performance while being less demanding for the hardware.

The main change in continuous SPEN is the use of a single broad-band frequency-swept pulse, rather than a train of selective pulses. Most frequently, "chirp-pulse" are used for that purpose, providing a linear variation of the RF-frequency offset $O_{RF}$ in the course of the pulse application:

$$O_{RF}(t) = O^0_{RF} + Rt \text{ with } 0 < t < T. \tag{10.5}$$

In this equation, $T$ is the frequency-swept pulses duration, $O^0_{RF}$ is the initial frequency offset and $R$ corresponds to the sweep rate defined as $BW/T$ where $BW$ is the bandwidth fixed by the operator. Under the combined application of a gradient $G_e$ and a chirp-pulse, the nuclear spins from a chemical site $i$ experience a rotation whenever the

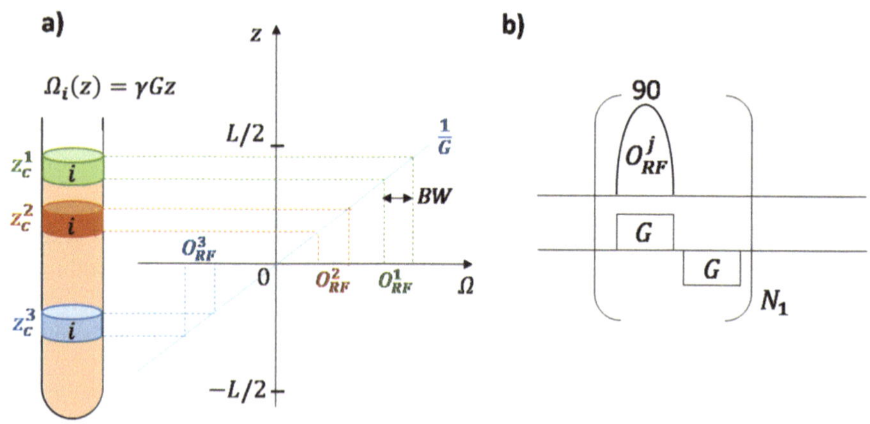


Figure 10.3 (a) Slice selection by the combined use of a selective pulse and a magnetic-field gradient. Location and thickness of the slice depend on the bandwidth $BW$ of the pulse, the RF offset $O_{RF}$ and the strength of the gradient $G$. (b) a pulsesequence for a discrete spatial encoding. The train of selective pulses with a linear increase of the RF offset allows the rotation of spins present in each slice in a sequential way. The second gradient of opposite amplitude is necessary for rephrasing the magnetization after each slice selection.

$O_{RF}(t)$ of the pulse matches with their z-dependent frequency offset $\omega_i(z)$. By combining Equations 10.2 and 10.5, this occurs at a time $t_{rot}$ expressed as follows:

$$t_{rot}(\mathrm{z}) = \frac{\Omega_i - \gamma G_e z - O_{RF}^0}{R}. \tag{10.6}$$

The rotation is assumed to be instantaneous, which is ensured in practice by using frequency-swept pulses so that the product $T \times BW$ exceeds 40. Equation 10.6 shows clearly the opportunity to rotate nuclear spins of a same frequency offset $\Omega_i$ at different instants $t_{rot}(\mathrm{z})$, enabling a continuous spatial parallelization of the evolution time (Figure 10.4). This might be seen as manipulating infinitesimal slices of the sensitive volume, in contrast to the discrete SPEN involving finite slices of thickness $\Delta z$ (see Equation 10.3)

However, the combined use of a chirp-pulse and a magnetic field gradient involves a complex dephasing of the transverse magnetization. Equations 10.7 and 10.8 express the resulting phase shift for a 90° excitation and a refocusing 180° chirp-pulse, respectively [6, 9]. In both cases, the dephasing exhibits the desired linear phase term in z to enable the spatial parallelization, nonetheless, there is a quadratic contribution (i.e. right-hand term) that highly complicate the subsequent signal detection. Repeating the same combination with an opposite gradient amplitude leads to an overall phase shift $\Phi(z)$, in which the quadratic phase term is canceled. From this base, several continuous SPEN schemes have been considered and proposed.

$$\Phi^{90}(z) = \frac{T}{2}\left(\Omega_i + \frac{\gamma G_e L}{4}\right) + \frac{T}{L}\left(\frac{\gamma G_e L}{2} - \Omega_i\right) z - \frac{\gamma G_e T}{2L} z^2 \tag{10.7}$$

$$\Phi^{180}(z) = \frac{\gamma G_e L T}{4} - \left(2\gamma G_e T - \frac{2T}{L}\Omega_i\right) z + \frac{\gamma G_e T}{L} z^2 \tag{10.8}$$

A first one consists of a pair of 90° chirp-pulses applied together with opposite encoding gradients $G_e$ (Figure 10.5a) [10]. The first 90° chirp-pulse flips the spins $i$ from equilibrium onto the transverse plane at different times $t_{rot}^{+}(z)$ according to their z-position in the tube while the second chirp-pulse flips back the magnetization at a time $t_{rot}^{-}(z)$ along the z-axis (Figure 10.5a). This pulse sequence, noted (90–90) leads to an amplitude modulation of the stored magnetization (regardless relaxation and diffusion):

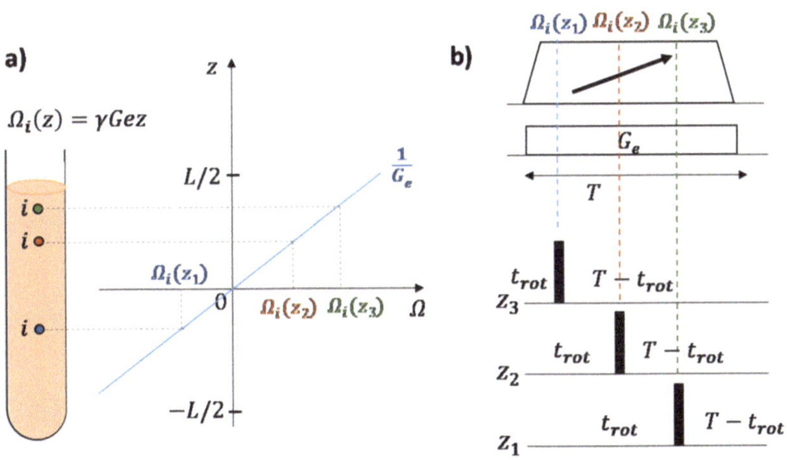


Figure 10.4 (a) The spatial dependence of resonance frequency under the application of magnetic field gradient. (b) The combined use of a gradient $G_e$ and a frequency-swept pulse, e.g. chirp-pulse, leads to spin rotations at different instants according to the z-position as it is during the discrete encoding, but in a continuous fashion.

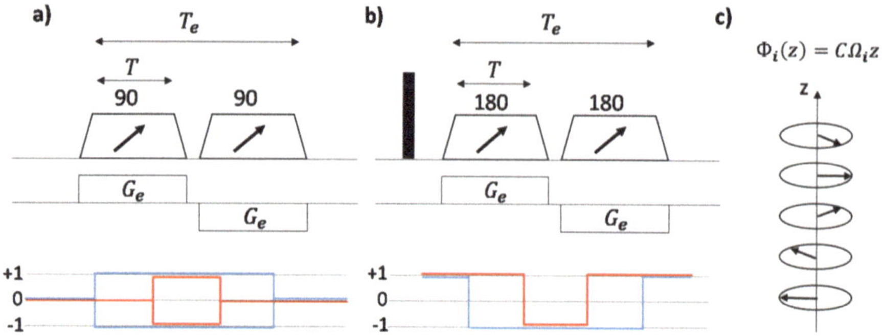


**Figure 10.5** Pulse sequences for two spatial encoding. Black thin rectangle is a 90° hard pulse. (a) The real-time amplitude-modulated spatial encoding (90–90) and (b) the constant-time phase-modulated spatial encoding (180–180). At the bottom is displayed, for both encodings, the magnetization trajectory for two spins of identical frequency offset $\Omega_i$, but with different initial spatial positions along the NMR tube. This difference of trajectories arising from spins of similar $\Omega_i$ corresponds to the spatial parallelization of the evolution period. Note that the two spins spend equivalent duration into the transverse plane (i.e. coherence order ±1) in the (180–180) SPEN in contrast to the (90–90) one. (c) The resulting magnetization helix at the end of SPEN. Whatever the pulse sequence, the overall dephasing is linear in $z$ and $\Omega_i$.

$$M_z(z) \propto M_0 \sin\left(\frac{2T}{L}\Omega_i z + K\right). \tag{10.9}$$

With $M_0$ the initial magnetization and $K$ a constant term. This encoded longitudinal magnetization will undergo mixing sequence before tipped back to the transverse plane for detection. This (90–90) scheme is qualified as a real-time amplitude-modulated SPEN. The term "real-time" reflects the different durations spent by the spins $i$ in the transverse plane relatively to their initial z-position.

Another continuous SPEN, noted (180–180) starts with a classical 90° hard pulse followed by a pair of 180° refocusing chirp-pulses applied in combination with opposite encoding gradients $G_e$ (Figure 10.5b) [11]. The transverse magnetization is here phase-modulated while coherences are inverted twice at dates $t^+_{rot}(z)$ and $t^-_{rot}(z)$. The spatial dependence of the phase shift at the end of this encoding is (regardless of relaxation, diffusion, and $J$-coupling):

$$M_{xy}(z) \propto M_0 \exp\left(i\frac{4T}{L}\Omega_i z\right). \tag{10.10}$$

In contrast to the (90 – 90) SPEN, all the spins spend the same amount of time in the transverse plane whatever their initial spatial position (Figure 10.5b). This (180–180) scheme is thus referred as a constant-time (CT) phase-modulated encoding.

Regardless of the SPEN used, a linear spatial dephasing $\Phi_i(z) = C\Omega_i z$ is induced where the spatial encoding constant $C$ characterizes the resulting magnetization helix (Figure 10.5c). Note that a particular winding of the magnetization helix is obtained for each frequency offset $\Omega_i$ leading to a specific spatial encoding for each chemical shift.

The (180–180) SPEN has been found to be preferable for different reasons: (i) for an identical encoding time $T_e$, the constant $C$ is twice as large in this SPEN ($C = 4T/L$) as in the (90–90) one ($C = 2T/L$). This is an important feature since the line-width of the subsequent signal is inversely proportional to $C$ (see in Equation 10.15); (ii) this SPEN does not implies a dispersive component into the resulting line-shape (see Section 10.4.1); and (iii) 180° chirp-pulses are in practice easier to calibrate and more robust than the 90° ones [6, 12, 13]. A last important point is that this CT spatial encoding shares pros and cons of conventional 2D CT experiment (CT-COSY, CT-HSQC, etc...), such as a homo-decoupled indirect dimension – which is a useful boost in sensitivity and resolution –

and the tedious peak amplitude modulation according to the nature of $J$-couplings ($J$ constant values and spin-systems). Yet, this drawback may be alleviated by the average of a few single-scan 2D spectra recorded with an incremented delay placed prior to the SPEN [14].

### 10.2.2 Reading Out the Spatially Encoded Signal

As described in the previous section, the spatially encoded magnetization exhibits a dephasing that depends on the frequency offset $\Omega_i$ and on the spatial position. Considering the helix shown in Figure 10.5c, the vector sum of the transverse magnetization vectors $M_{xy}(z)$ along the z-axis is null, so that no signals can be detected at this level. It is, in turn, necessary to unwind the helix until the vectors $M_{xy}(z)$ can be summed in a coherent way. This is achieved by applying a magnetic field gradient – so called acquisition gradient ($G_a$) – while the receiver is open. The effect of $G_a$ on the spatially encoded magnetization is illustrated in Figure 10.6. This progressively unwinds the helix until the $M_{xy}(z)$ be in coherence leading to an echo that arises at a time proportional to the frequency offset $\Omega_i$. A more formal vision consists in introducing the wave number $k$ defined as below:

$$k(t_2) = \gamma \int_0^{t_2} G_a(t)dt. \tag{10.11}$$

Where $t_2$ is the detection time when the receiver is open. Under the application of $G_a$, the transverse magnetization undergoes an additional phase $k(t_2)z$, resulting in overall dephasing $\Phi(k, t_2) = C\Omega_i z + k(t_2)z$. Vectorial sum along the sample length $L$ is maximal whenever the overall dephasing $\Phi(k, t_2)$ is null. Therefore, the echo occurs at a time $t_{echo} = C\Omega_i/\gamma G_a$ and at a position $k = -C\Omega_i$ along the $k$-domain (Figure 10.6). Consequently, distinct chemical shifts give echoes with distinct locations with respect to $k$. This series of echoes reflecting the chemical sift information corresponds to the indirect dimension – also called "UF dimension" – of the 2D spectrum. The wave number, $k$ is expressed in $m^{-1}$, which is a quite unusual unit in NMR spectroscopy. This axis may be rescaled in frequency units by dividing the $k$-values by the constant $C$. The shift between two successive echoes as well as the maximum dephasing that the acquisition gradient is able to refocus for a given duration $T_a$ are linked to the spatial encoding constant $C$ and $G_a$. These parameters thus determine the spectral width and the resolution in this dimension (see Section 10.3.2).

The second dimension of the UF 2D experiment is obtained through the repetition of this gradient-based detection period while reversing the gradient amplitude every other time. This train of $N_L$ bipolar pulse-field gradient $\pm G_a$ pairs is analogous to the EPSI (Figure 10.7a), a detection technique that originated in the world of magnetic resonance imaging [15, 16]. Here, the EPSI periodically refocus and defocus the spatially encoded magnetization, giving rise to a series of mirror-image 1D spectra along the $k$-axis (Figure 10.6). During this "zig-zag" trajectory

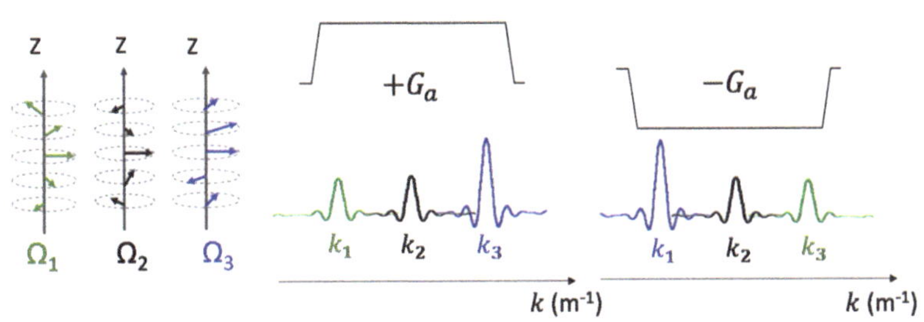


Figure 10.6 Cartoon showing the reading out of spatially encoded signals. The acquisition gradient refocus the magnetization phase shift until the rise of an echo along the $k$-domain. The location of the echo is specific to the frequency offset, reflecting the chemical shift information. A mirror series of echoes is obtained with a similar acquisition gradient but with an opposite amplitude.

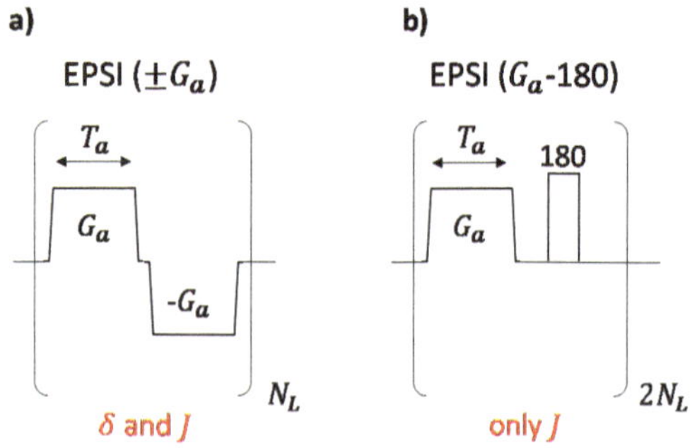


**Figure 10.7** Two types of EPSI scheme. Through the EPSI (±$G_a$) the signal is monitored by the chemical shift and the *J*-coupling (a) whereas in the ($G_a$-180) variant the chemical shift is refocused (b) This second EPSI is mainly used for UF *J*-resolved experiments.

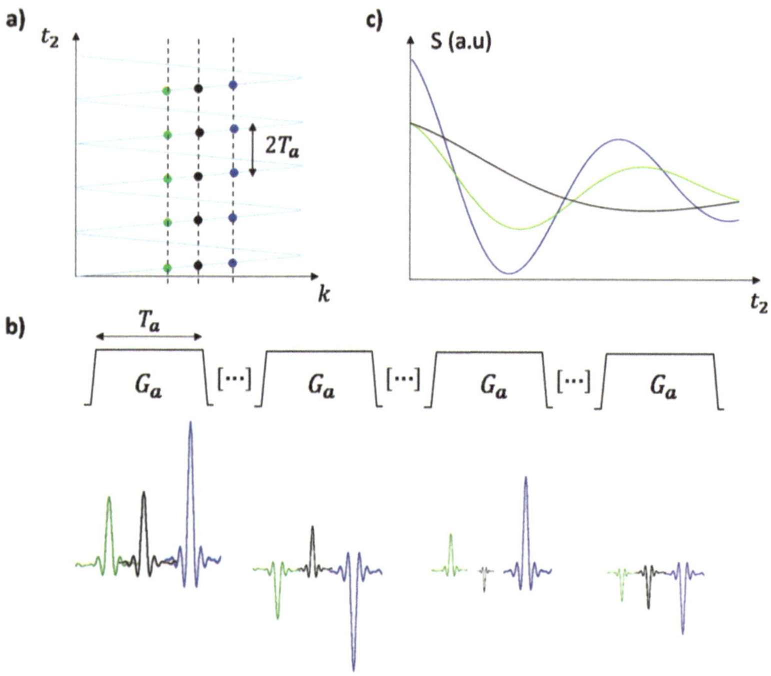


**Figure 10.8** (a) Scheme representing the "zig-zag" trajectory onto the ($k$, $t_2$) space during EPSI. As dataset from positive and negative acquisition gradients are split and processed separately, the effective dwell time is equal to $2T_a$. (b) Example of echoes arising from positive $G_a$ and experiencing free precession. The evolution of echo amplitudes in the course of the EPSI leads to free induction decay curves (c), which are then Fourier transformed to yield the second dimension of the UF 2D spectrum.

into the $(k, t_2)$ plane (Figure 10.8a), the system evolves under conventional NMR parameters, according to the EPSI type (chemical shift, *J*-couplings…) (Figure 10.8a and c). After Fourier transform along the detection time $t_2$, the second dimension is yielded, called "conventional dimension," corresponding to the direct dimension in conventional 2D experiments. This dimension is sampled with an effective dwell time of $2T_a$ (Figure 10.8a), with $T_a$ the acquisition gradient duration, providing a spectral width equal to $1/2T_a$ and a resolution inversely proportional of the overall duration $2T_aN_L$.

In contrast to the conventional detection whereby the signal evolves under both chemical shift and *J*-coupling, two types of EPSI detection scheme are available that allow one to monitor signal by free precession (Figure 10.7a) or by the J-coupling only (Figure 10.7b).

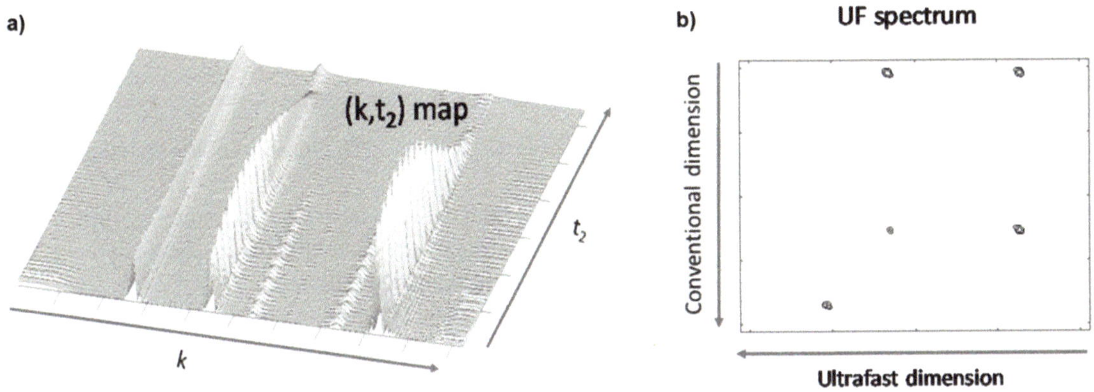


Figure 10.9 (a) Example of a $(k, t_2)$ map obtained after rearrangement of data extracted from positive gradients. The k-axis gives the chemical shift information without the need of a Fourier Transform while the peak amplitude evolves under chemical shift and J-coupling. The Fourier transform with respect to $t_2$ delivers the single-scan UF 2D spectrum (b). This maps were gathered from a sample of ethanol dissolved in $D_2O$.

### 10.2.3 Processing Workflow in UF Experiments

Owing to use of a EPSI block, UF 2D NMR requires a specific data processing workflow to build a 2D correlation map similar to what is provided in conventional 2D methods. As a first step, data arising from positive and negative acquisition gradients are split and rearranged, so that gathering two mirror-image $S(k, t_2)$ maps (Figure 10.9a). Fourier transform is subsequently applied with respect to $t_2$, resulting in two symmetric $S(k, F_2)$ maps (from $\pm G_a$, respectively). The map related to negative gradients is then inverted, and added to the other one giving the final ultrafast 2D spectrum (Figure 10.9b). The $k$-axis, i.e. the UF dimension, may be further converted to frequency units by dividing it by the spatial encoding constant $C$ to get a $S(F_1, F_2)$ like 2D spectrum. In this convention, F1 corresponds to the UF dimension while F2 to the conventional one. It could be noted there is not a consensual way of displaying the UF 2D spectra, as the UF dimension is arbitrarily plotted horizontally or vertically in the literature.

## 10.3 Specific Features of UF 2D NMR

### 10.3.1 Line-shape of the Signal

The 2D peak line-shape observed in UF 2D spectra is quite different of those from their conventional counterparts. In what follows, we focus on the line-shape from the CT (180–180) SPEN, which is today the most frequently used scheme. Similar conclusions are reported for real-time encodings, nonetheless with few differences [6, 12].

The signal $S_i$ for a chemical site $i$ at a time $t_2$ under the application of the reading gradient $G_a$ is proportional to the vector sum – from $-L/2$ to $L/2$ – of the projections of the magnetization vectors onto the transverse plane (x'y') in the rotating frame. Expressed in a complex form, without considering either the relaxation or the diffusion, this leads to:

$$S_i(k(t_2)) \propto M_0 \int_{-L/2}^{L/2} e^{iC\Omega_i z}.e^{ik(t_2)z} dz. \tag{10.12}$$

In this equation, $M_0$ is the initial magnetization while $k(t_2)$ is the wave number as previously defined. The expression of $S_i$ (Equation 10.12) might be visualized as a Fourier transform along the z-axis of a rectangular function of width $L$. This leads in turn to a *sinc*-function:

$$S_i(k(t_2)) \propto M_0 L \,\mathrm{sinc}\left(\frac{L(C\Omega_i + k(t_2))}{2}\right). \tag{10.13}$$

Equation 10.13 involves important features of the signal $S_i$ along the $k$-domain:

(i) The signal $S_i$ is maximal for $k(t_2) = -C\Omega_i$ and the height is equal to $M_0L$. The location of the signal depends on the frequency offset of the spins $i$ as expected.
(ii) The line-width of the signal is equal to $1/L$, or $1/CL$ whether the axis is viewed in frequency units. The line-width depends only on the overall encoding time $T_e$: the longer is $T_e$, better is the resolution. Hence, an analogy can be made between $T_e$ and the $t_{1max}$ value in conventional 2D NMR.
(iii) The *sinc*-shape derives from the rectangular spatial profile of the probe, together with the assumption of a perfect homogenous sample in space. Wiggles associated with the *sinc*-function may cause some signal overlaps, especially in case of high dynamic range. However, spatial apodization applied at the processing level tackle this issue (further details in Section 10.4.1.1). In practice, diffusion effects during the spatial encoding contribute at various extents to the final line-shape. Then, in the case of the amplitude-modulated (90-90) SPEN, the $T_2$-relaxation decay is z-dependent giving an asymmetric envelope about $z = 0$. The resulting signal pattern exhibits an undesired dispersive component [6]. The (180–180) SPEN is not concerned by this issue since all the spins experience the same amount of time into the transverse plane whatever their z-position, leading to a uniform $T_2$-decay.

The signal $S_i$ also evolves under free precession during $t_2$ through the EPSI block, which leads to a standard Lorentzian function after Fourier transform in this conventional dimension. Thus, 2D-peak from the $S(k, F_2)$ map is a combination of a *sinc* and Lorentzian function. In practice the UF 2D spectrum is plotted in magnitude.

### 10.3.2 Resolution and Spectral Width

We survey here the key equations that feature the UF spectra in both dimensions to highlight the specificities of this approach in comparison with conventional 2D NMR. First, the spectral width in the UF dimension is linked to the ability of the reading gradient $G_a$ to refocus the magnetization helix. The maximal wave number reached for a given gradient $G_a$, applied during $T_a$, is $k_{max} = \gamma G_a T_a$. In frequency units, the spectral width is obtained by dividing $k_{max}$ by the constant $C$ whereas the line-width is given in Equation 10.15:

$$SW_{UF} = \frac{\gamma G_a T_a}{C}, \tag{10.14}$$

$$\Delta\nu_{UF} \propto \frac{1}{CL}. \tag{10.15}$$

The signal in the conventional dimension has the classical characteristics of Fourier transform NMR. The spectral width is determined by the effective dwell-time, $2T_a$, while the resolution is driven by the total EPSI duration: $2T_aN_L$:

$$SW_{conv} = \frac{1}{2T_a}, \tag{10.16}$$

$$\Delta\nu_{conv} \propto \frac{1}{2T_aN_L}. \tag{10.17}$$

Combining Equations 10.14–10.16 leads to the equation Equation 10.18 below, which highlights how spectral width and resolution in both dimensions are entangled. Increasing the encoding period improve the resolution while limiting the spectral width in the UF dimension. Meanwhile, a high value of $T_a$ increases the $SW_{UF}$, but at the expense of $SW_{conv}$. There is therefore a trade-off between resolution and spectral widths when performing UF 2D NMR, which is sorted out with the analytical needs. Yet, it is worth mentioning that working with a strong $G_a$

alleviates significantly this compromise as it ensures a high $SW_{UF}$ without concessions. In practice, the maximum value of $G_a$ depends on the probe and one should keep in mind that trains of intense bipolar pulse-field-gradient applied at high rates is very demanding for the hardware, in particular when signal averaging is necessary:

$$\gamma G_a L = \frac{SW_{UF} SW_{conv}}{\Delta\nu_{UF}}. \tag{10.18}$$

### 10.3.3 Sensitivity Considerations

Sensitivity is a central feature when performing UF 2D experiments as this method involves a SNR decrease in comparison with its conventional counterpart. Thus, recording a whole 2D spectrum in a single scan is suitable only if the sensitivity is sufficient, otherwise signal averaging is required. The main source of sensitivity loss is due to the application of an intense magnetic field gradient $G_a$ while the receiver is open [6–9, 17]. The signal is thus spread out over a wide range of frequencies that involves the use of a broad digital filtering (FW). A broad FW leads to a significant increase of the noise root-mean-square. Another source of sensitivity decrease comes with the translational diffusion effect during the spatial encoding [17, 18]. Indeed, the diffusion causes a signal decrease during and between the encoding gradients $G_e$. While this is a rather small effect in most of UF pulse sequences with encoding times of 30–50 ms, this becomes an increasing penalty for experiments as UF $J$-resolved, which relies on a longer SPEN duration. In overall, one should consider a SNR penalty of a factor 5–10 compared to conventional 2D experiments for a given number of scans [8].

## 10.4 Advanced UF Methods

### 10.4.1 Improving the Sensitivity

#### 10.4.1.1 Spatial Apodization

Apodization along the conventional dimension can be applied as it is usual done in Fourier transform NMR. It is particularly relevant since the signal is highly truncated in $t_2$ because of the limited number $N_L$ of bipolar gradient pairs in EPSI.

Moreover, apodization functions may also be applied in the $k$-domain, i.e. along the UF dimension, in order to modify and optimize the line-shape [19], as shown in Figure 10.10. To carry out this processing in practice, a Fourier transform is applied along the $k$-domain, converting the wave number domain into a spatial domain. The resulting spatial profile is then multiplied by an apodization window, usually Gaussian, to eliminate the signals arising from the edge of the sensitive volume (Figure 10.10). Besides the line-shape improvement, this process also yields a significant sensitivity enhancement by suppressing the noise outside the spatial profile. An inverse Fourier transform is then applied to get back the signals in the $k$-domain. This spatial apodization processing is today routinely used.

#### 10.4.1.2 Reduction of the Diffusion Losses

As previously discussed, translational diffusion leads to a sensitivity decrease, especially when a long SPEN is used, as it is in UF $J$-resolved experiments. This effect can be alleviated in first step by reducing physically the diffusion, by lowering the sample temperature or by using viscous solvents such as DMSO. Furthermore, for biological fluids, specific sample preparations based on medium-size phospholipid vesicles are available to encapsulate the analytes, de facto limiting the diffusion [20].

A more general approach is to redesign the spatial encoding to be more immune against the diffusion losses. A multi-echo variant of the (180–180) SPEN has been proposed in that respect (Figure 10.11) [13, 21]. This consists of replacing the pair of chirp-pulses by multiple ones with shorter durations. For a same spatial encoding time $T_e$,

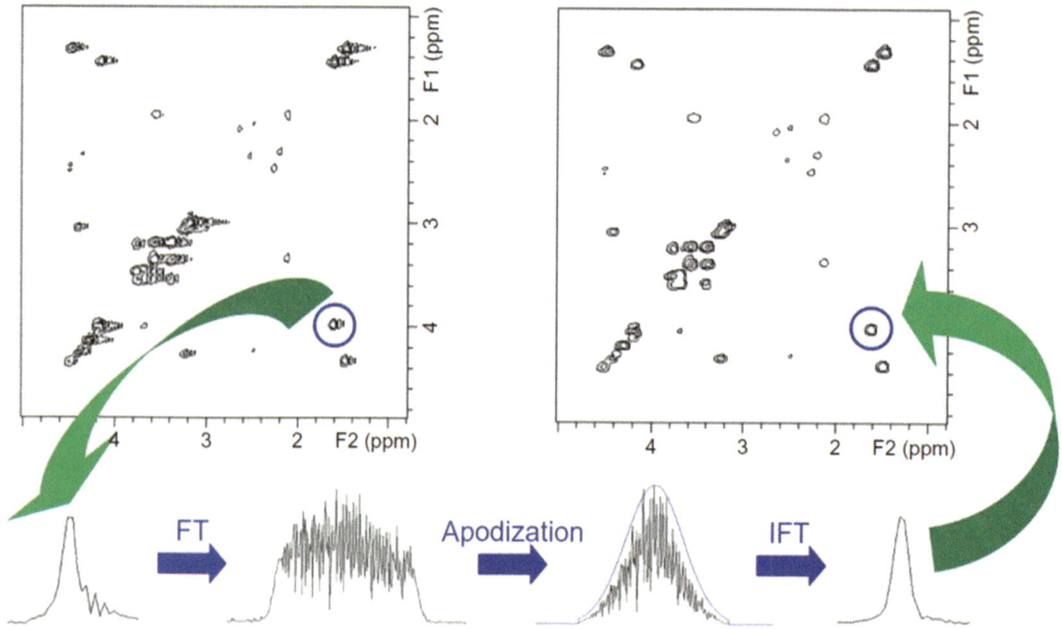


Figure 10.10 Spatial apodization of a UF 2D spectrum. A Fourier transform is applied in the *k*-domain and the resulting spatial profile is multiplied by a Gaussian window. After inverse Fourier transform, the resulting signal profits from more symmetric line-shapes along a significant of the SNR as seen on the 2D spectrum on the right. Reproduced from Ref. 19 with permission from John Wiley and Sons.

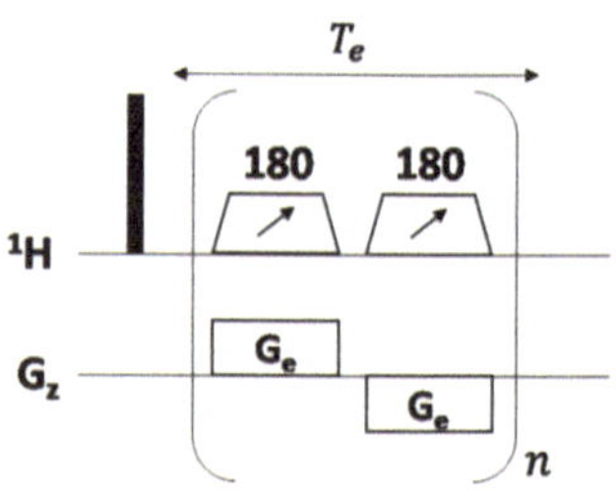


Figure 10.11 Pulse sequence of the multi-echo (180-180) SPEN. This variant is more immune against the diffusion losses when long encoding time $T_e$ are used. Black thin rectangle is a 90° hard pulse.

the diffusion impact is reduced owing to the use of shorter encoding gradients $G_e$. Yet, one should bear in mind that chirp-pulses can be shorten only up to certain point, for a given bandwidth $BW$, otherwise the adiabaticity behavior is no longer warranted (i.e. $T \times BW$ must exceed 40).

## 10.4.2 Improving Spectral Width and Resolution

### 10.4.2.1 Folding Gradients

In UF 2D NMR, aliasing occurs in the conventional dimension, i.e. the resonances are folded with respect to the extremity of the spectral window. Contrary, there is no aliasing in the ultrafast dimension since no Fourier transform is applied along the *k*-domain. However, applying a 90° mixing pulse splits the spatially encoded signals in two mirrors components with opposite dephasing. Adding pulse-field-gradients at the both sides of the mixing pulse allow the two mirror signals to be partially superimposed so that all the relevant signals can be observed in a reduced spectral width. This feature can be finely adjusted thanks to the parameters (amplitude and/or duration) of these "folding" gradients [22].

This gradient-controlled folding is suitable in all UF 2D experiments but is particularly useful for hetero-nuclear ones, generally characterized by a sparse indirect dimension with a large spectral width. This is even more true for experiments performed at very high magnetic field.

#### 10.4.2.2 Interleaved Acquisitions

Interleaved acquisitions are basically designed to reach higher spectral widths in both UF and conventional dimensions. Moreover, this can also be used to reduce the demand on the acquisition gradient amplitude $G_a$, or to improve the resolution in the conventional dimension [23]. In this method, multiple UF experiments are carried out while a pre-acquisition delay $\tau$ is incremented as shown in Figure 10.12a. For $N_i$ interleaved acquisitions, $\tau$ is incremented by a factor $2T_a/N_i$ between successive scans. The $N_i$ acquisitions undergo interleaved trajectories in the

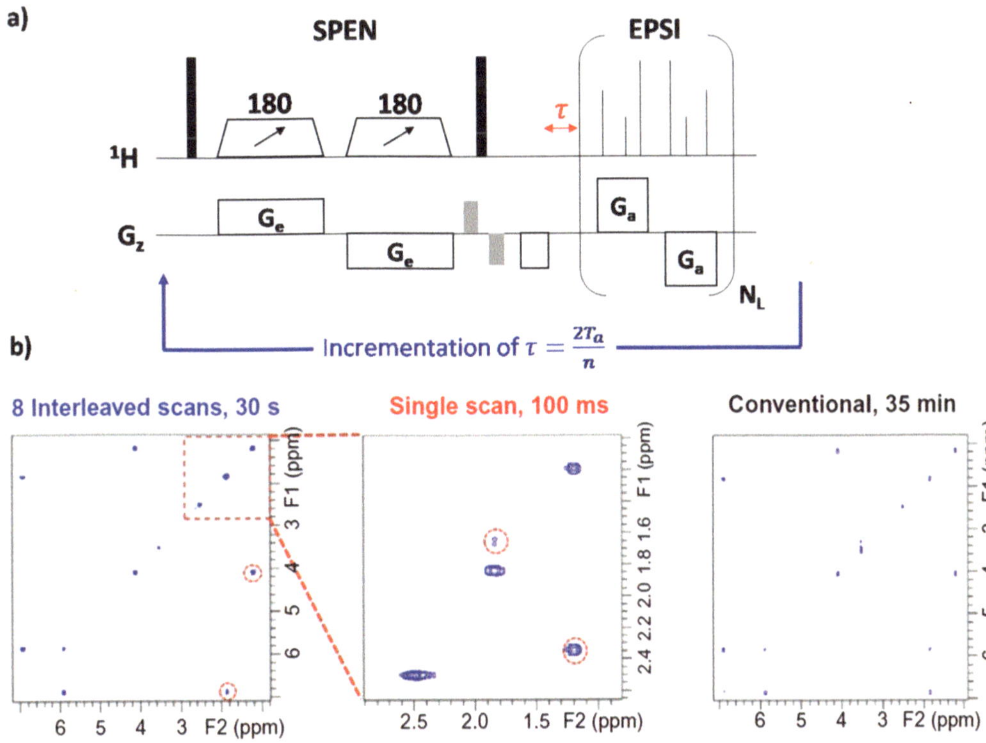


Figure 10.12 (a) Example of a UF COSY pulse-sequence with interleaved acquisitions. To do so, an incremented delay $\tau$ is added just prior to the EPSI. Black and thin rectangles are 90° hard pulses; gray filled boxes are pulse-field-gradients for coherence selection ; the open before EPSI is a "prephasing" gradient used to adjust the location of the echo train along the *k*-axis. (b) Comparison of 700 MHz UF 2D COSY spectra collected from a sample of ethyl crotonate with eight interleaved acquisitions (left), with a single-scan acquisition (middle), and the conventional COSY spectra (right). Two-dimensional peaks in the red circles come from aliasing in the conventional dimension. Note the significant increase of spectral widths with the interleaving approach. The experimental duration of this multi-scan method remains much shorter that than that of the conventional experiments (30 s *versus* 35 min). Reproduced from S. Akoka, P. Giraudeau, Fast hybrid multi-dimensional NMR methods based on ultrafast 2D NMR, *Magn. Reson. Chem.* 53 (2015) 986–994 with permission from John Wiley and Sons.

$(k, t_2)$ plane. After Fourier transform in $t_2$, followed by data processing, the spectral width in the conventional dimension increases as:

$$SW_{conv} = \frac{N_i}{2T_a}. \tag{10.19}$$

Interleaved acquisitions enables the use of a high $T_a$ value to access broader $SW_{UF}$ while the subsequent reduction of $SW_{conv}$ is counterbalanced by the increase of $N_i$ (Equation 10.19). This alleviates the comprise between the both spectral widths described in Section 10.3.2. Yet, this method may lead to artifacts along the conventional dimension. Several processing strategies are available to remove theses artifacts efficiently [24].

At the cost of few scans, this interleaving approach provides users UF 2D spectra with an important increase of the spectral widths (Figure 10.12b). This an even more relevant approach when working at very high magnetic fields where the accessible frequency range with a single-scan UF experiments is rather poor (see further in Table 10.1).

#### 10.4.2.3 Non-uniform Sampling of Data in EPSI

A non-uniform sampling approach can be used in the conventional dimension through the use of pseudo-random oscillating gradients in the ESPI detection [25]. Consequently, the "zig-zag" trajectory into the $(k, t_2)$ plane becomes non-periodic making possible a non-uniform sampling of the free induction decay in $t_2$ (Figure 10.13). With this scheme, the amplitude of the acquisition gradients $G_a$ may be significantly reduced thereby alleviating the technical demand while increasing the sensitivity since the digital filter bandwidth $FW$ can be reduced. The 2D spectrum is then reconstructed with a compressed sensing algorithm. This coupled NUS-UF experiment is particularly suitable in the case of a high SNR and a sparse direct domain.

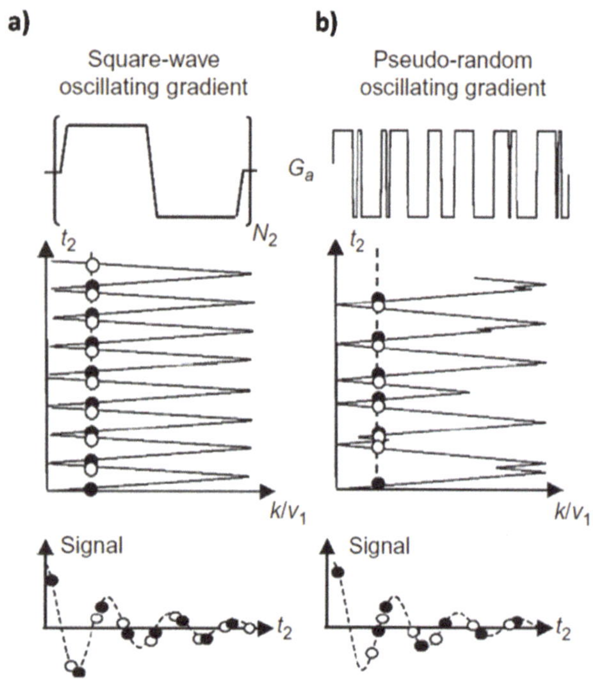


Figure 10.13 (a) Standard ($\pm G_a$) ESPI scheme leading to a periodic trajectory into the $(k, t_2)$ plane and so a uniform sampling of the signal in $t_2$. (b) A non-uniform sampling of the same signal can be performed by using bipolar acquisition gradients with pseudo-random durations. Reproduced from Ref. 25 with permission from Elsevier.

## 10.5 UF 2D NMR: A Versatile Approach

### 10.5.1 Accelerating 2D NMR Spectroscopy Experiments

While many fast methods target a specific set of 2D experiments, UF 2D NMR is in principle suitable for any kind of 2D homo- or hetero-nuclear correlations as long as ample SNR is available with a single or few of scans. This is carried out by replacing the $t_1$-evolution and the detection periods by a spatial encoding and a EPSI block, respectively. Preparation and mixing steps do not generally require modifications compared to the conventional version. UF variants of two classical 2D correlation experiments, namely COSY and HSQC, are illustrated in Figure 10.14. Due to the use of the CT (180 – 180) SPEN, the exact conventional counterparts are CT-COSY and CT-HSQC. This implies an indirect dimension (i.e. UF dimension) without J-coupling multiplicity, along

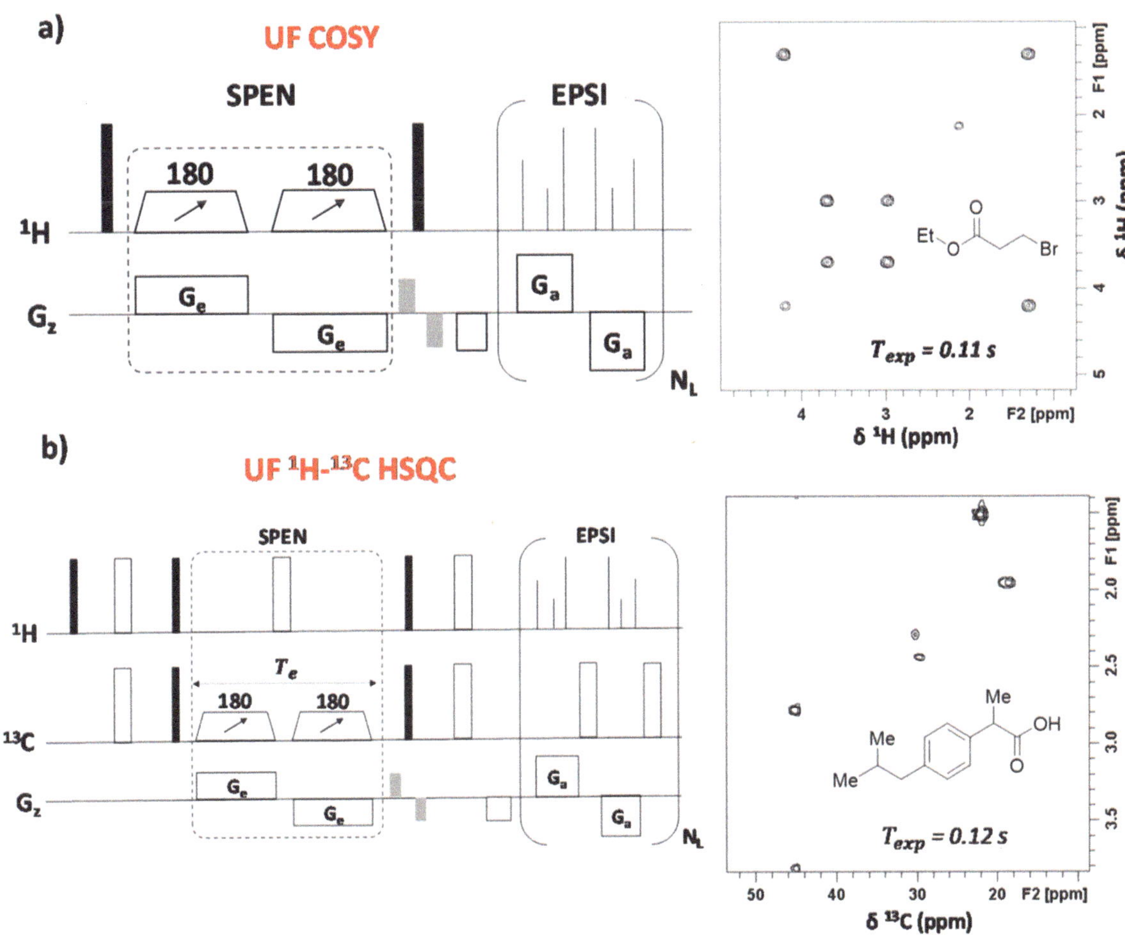


Figure 10.14 Examples of UF 2D pulse sequences accompanied with the resulting 2D spectra. (a) UF COSY pulse sequence performed on a sample of ethyl-3-bromopropionate in acetone-$d_6$. The 2D spectrum is collected in 0.11 s at 400 MHz. (b) UF $^1H$-$^{13}C$ HSQC pulse sequence applied to a sample of ibuprofen in acetone-$d_6$. This hetero-nuclear 2D spectrum (aliphatic range only) is recorded in 0.12 s with a 500 MHz spectrometer equipped with a cryogenic probe. The thin black rectangles are 90° hard pulses, while the wide open rectangles are 180° hard pulses; gray boxes are gradients for coherence selection and the gradient applied just prior to the EPSI is a "prephasing" gradient used to adjust the position of the echo train along the *k*-domain.

with a $J$-modulation of 2D-peak volumes [14]. While this last feature may cause a tedious cancelation of some 2D peaks, one may use this property to determine tiny $J$-coupling constants by monitoring the diagonal/cross-peak ratio as a function of the overall echo time $T_e$ [26]. The concept of UF 2D NMR has been extended to other widely used 2D experiments such as COSY, HSQC, TOCSY, HMBC, or homo- and hetero-nuclear $J$-resolved [5, 23, 27–29].

Moreover, it is also possible to turn 2D multi-quantum-single-quantum (MQ-SQ) experiments in a ultrafast fashion. Dumez, Caldarelli, and collaborators have reported the first UF DQ-SQ experiments applied to a metabolite samples [30]. This method has been then extended to higher coherence orders. When spatially encoded, MQ coherences exhibit a dephasing linear in $z$ and proportional to the sum of the frequencies offsets $\Omega_i$ of the coupled spins [31]:

$$\Phi_{MQ}(z) = \varphi_0 - Cz\sum_{i=1}^{p} \Omega_i.$$

In Equation 10.20, $p$ is the coherence order, $\varphi_0$ the phase shift experienced at the end of the MQ-buildup and $C$ is the spatial encoding constant mentioned previously in Section 10.2.1.

This dephasing $\Phi_{MQ}(z)$ is then transferred into observable single-quantum coherences (SQC) by the action of a mixing pulse. Through the EPSI block, these SQC lead to echoes in the k-domain at a location reflecting the MQ-chemical shift pertained to the coupled spins while evolving under free precession. After Fourier transform with respect to $t_2$, the subsequent UF 2D spectrum correlates the MQ and SQ chemical shifts in a similar way than proposed in conventional MQ-SQ NMR. The coherence order monitored along the indirect dimension (i.e. UF dimension) is selected by a pair of pulse-field gradients flanking the mixing pulses. An example of UF MQ-SQ pulse sequence as well as MQ-SQ 2D spectra recorded on a mixture of aromatic compounds are presented in Figure 10.15. This approach is capable of recording MQ-SQ spectra in less than one second up to $p = 5$, which corresponds, in this example, to the maximum quantum spectrum. The ultrafast approach enables the collection a MQ-SQ 2D spectra of various coherence orders in a reasonable experimental duration. This UF MQ-SQ approach has found promising applications in the analysis of complex mixture [31], as well as for absolute quantifications [32].

The ultrafast concept can also be combined with other fast 2D methods. The Section 10.4.2.3 above shows how a non-uniform sampling strategy may be applied at the detection stage. Then, UF 2D NMR is compatible with fast-pulsing methods as SOFAST (selective optimized flip-angle short-transient) [33], resulting in the UltraSOFAST 2D experiment that allows signal averaging at high rate [34]. Further details and example of applications are given at the end of the Section 10.6.1. Finally, spatial encoding may be also used to speed up experiments based on a Hadamard encoding [35]. In this UF version, the use of spatial/spectral pulses allow to encode one raw of the Hadamard matrix into one spatial region of the sensitive volume. The use of UF 2D Hadamard is particularly interesting when the UF dimension is sparse since the sensitivity enhancement (compared to standard UF 2D NMR) is inversely proportional to the size of the Hadamard matrix (correlated to the number of spectral regions to be encoded). This technique is however irrelevant in case of crowded spectra and requires to know beforehand the 1D spectrum of the studied sample.

While UF 2D NMR dramatically speeds up the acquisition of any 2D spectra, this method also comes with drawbacks regarding the analytical performance. As exposed in Section 10.3, a trade-off between the spectral widths and the resolution has to be considered, along with a SNR decreases because of the wide filter bandwidth $BW$ used during the detection. Moreover, the typical spectral widths, resolution, and sensitivity accessible in UF 2D NMR depend on the hardware (probe, gradient coil, and static magnetic field $B_0$). To give the reader an order of magnitude, Table 10.1 surveys the typical performance of two types of UF experiments (homo- and hetero-nuclear) with two different hardware configurations. For a constant resolution, this table underlines how the increase of the magnetic field $B_0$ leads to a better sensitivity, albeit at the cost of a reduced spectral width. This last feature

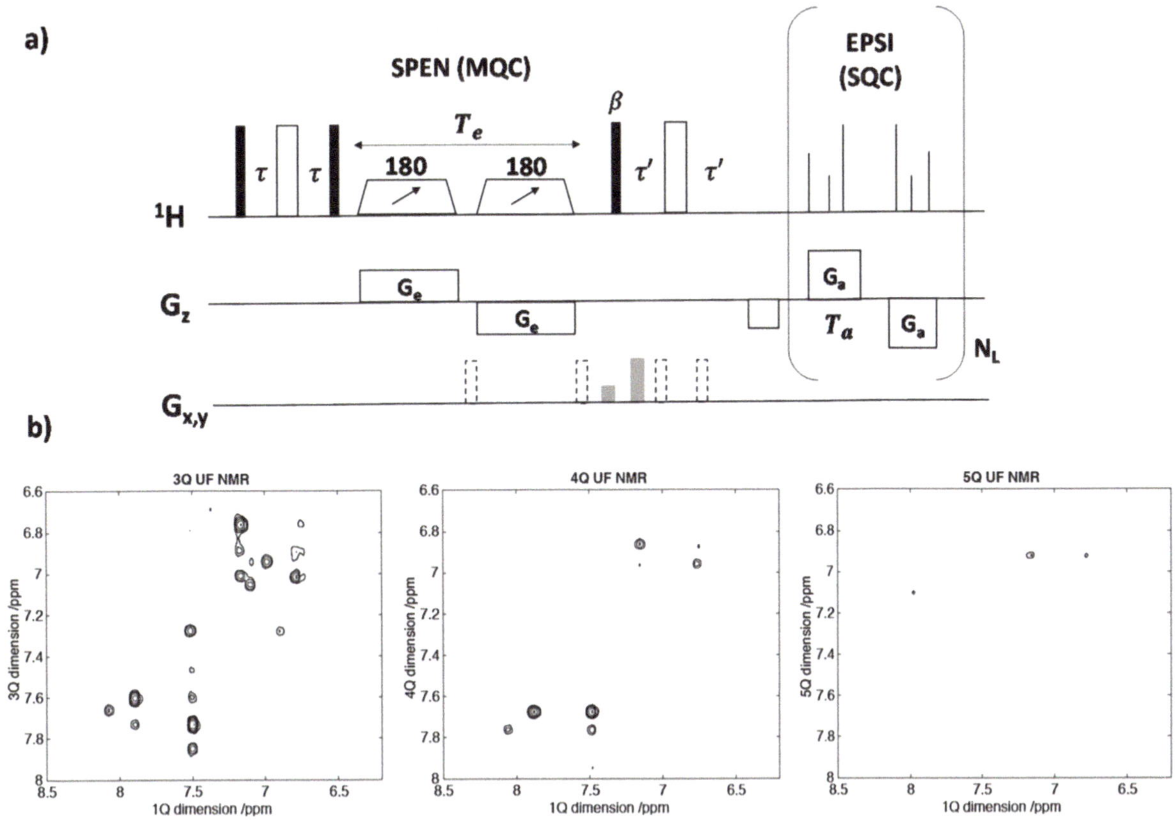


**Figure 10.15** (a) UF MQ 2D pulse-sequence. The delay $\tau$ is fixed at $1/4J$ for an efficient MQ-building up. The optional $2\tau'$ spin echo allows the conversion from anti-phase to in-phase magnetizations before the detection period. The coherence order monitored in the UF dimension is selected with the amplitude/duration ratio of gradients (gray filled boxes) flanking the mixing pulse. $\beta$ is the mixing pulse flip angle (often 120° or 90°); dashed box are spoiler gradients; thin black rectangles are 90° hard pulses, while the wide open rectangles are 180° hard pulses; the gradient applied just prior to the EPSI is a "prephasing" gradient used to adjust the position of the echo train along the $k$-domain. (b) UF 2D MQ spectra of various coherence orders: 3Q (left), 4Q (middle), and 5Q (right), recorded on a mixture of aromatic compounds. Here, the UF 5Q 2D spectrum reaches the maximum quantum 2D spectrum. Reproduced from Ref. 31, with permission from John Wiley and Sons.

can be counterbalanced by a multi-scan strategy with interleaved acquisitions (Section 10.4.2.2). Averaging signal in this way until one minute ensure an appealing performance while remaining fast compared to conventional 2D NMR.

### 10.5.2 Accelerating Dynamic Experiments (UF pseudo-2D)

So far, we have discussed how UF 2D NMR allows the record of a large variety of 2D NMR spectra in a single-scan manner. Besides the spatial encoding of spectroscopic observables (chemical shifts, $J$-couplings, etc..), this versatile approach may be also used to encode dynamic parameters such as $T_1$-relaxation times or diffusion coefficients ($D$). Popular pseudo-2D experiments such as diffusion ordered spectroscopy (DOSY) or inversion recovery (IR) can be acquired in a UF fashion. Note that the term "pseudo-2D" refers to the fact that the resulting 2D maps are not derived from a double Fourier transform.

A first UF pseudo-2D experiment is the single-scan measurement of the longitudinal relaxation times $T_1$ of a molecule at the atomic level. Generally, the measurement of $T_1$ values – with chemical shift resolution – is carried

**Table 10.1** Typical analytical performances of homo and hetero-nuclear UF 2D experiments carried out in a pure single-scan fashion or with 1 min of averaging. Two hardware configurations are also compared.

| | Pure single scan (0.2 s) | | Multi-scans (60 s) | |
|---|---|---|---|---|
| | 400 MHz RT probe | 700 MHz Cryoprobe | 400 MHz RT probe | 700 MHz Cryoprobe |
| **LOD** | 100 mM | 1 mM | 50 mM | 0.5 mM |
| **SW in COSY ($^1$H x $^1$H)** | 6 × 6 ppm | 2.5 × 2.5 ppm | 12 × 12 ppm | 5 × 5 ppm |
| **SW in HSQC ($^{13}$C x $^1$H)** | 40 × 4 ppm | 15 × 3 ppm | 80 × 8 ppm | 30 × 6 ppm |
| **FWHM (UF dim.)** | 50 Hz | 50 Hz | 50 Hz | 50 Hz |
| **FWHM (Conv. dim.)** | 35 Hz | 25 Hz | 35 Hz | 25 Hz |

LOD: Limit Of Detection; SW: spectral width; FWHM: Full Width at Mid-Height in the indirect dimension (UF dim.) or in the conventional dimension (Conv. dim.).

out by IR [36], which is a multistep experiment whereby an inversion delay is incremented. Moreover, the inter-scan delay must be long enough ($>5 \times T_1$) to vouch full longitudinal relaxation leading to a long experimental duration. This process would be highly accelerated with the spatial parallelization of the inversion delay within a single scan. Such an approach has been developed and used for a real-time monitoring of the $T_1$ relaxation times during a xylose-borate reaction [37]. The pulse sequence of this UF experiment is given in Figure 10.16a. The spatial parallelization of the inversion period is achieved with the combined use of a 180° chirp-pulse and a magnetic field gradient. Thus, spins *i* from a same chemical shift experience various inversion delays according to their position along the NMR tube before excitation. To read out the spatially encoded signal, several acquisition schemes have been proposed. One may use a long and weak acquisition gradient, followed by a Fourier transform applied in the *k*-domain, which leads to spatial profiles for each chemical shift. The effect of the longitudinal relaxation is "printed" onto the spatial profiles showing variations of magnetization along the length of the sample. This detection based on a weak acquisition gradient suffers from the compromise between the accuracy of $T_1$-measurment (requiring large profiles) and the chemical shifts resolution (narrow profiles to avoid overlaps). This trade-off can be alleviated by the use of a EPSI with strong bipolar gradients. After 2D Fourier transform onto the $(k, t_2)$ plane, a 2D map is obtained that correlates chemical shift and spatial profiles in two orthogonal dimensions (Figure 10.16b). Once again, the $T_1$-relaxation is reflected through the modification of the spatial profiles that it induces (Figure 10.16c). This UF method is quite efficient for the determination of short $T_1$ values, but becomes unsuitable when large values are involved. Such a single-scan $T_1$ measurement is a relevant tool for dynamic studies over a wide range of timescales that are inaccessible with standard IR experiments.

Another development with growing interest is the transposition of the UF concept in the field of diffusion NMR. DOSY encodes the effect of random Brownian motion in the indirect dimension while the chemical shift information is retained in the direct dimension [38]. This discriminates the compounds via their diffusion coefficients within a same sample without any physical separation, making DOSY a well-recognized tool for the analysis of complex mixtures. In its original form, this experiment performs a series of transients recorded with incremented gradient amplitudes. Once again, such a multistep process involves a significant experimental duration. Moreover, the original DOSY pulse sequence involves phase-cycling increasing even more the experiment duration. Hence, speeding up the record of 2D DOSY map has been a long standing concern. Among various strategies, the spatial parallelization of the diffusion encoding is one of the fastest approach. This concept, introduced by Keeler and coworkers [39] and then reintroduced by Frydman et al. [40], makes it possible to acquired 2D DOSY data in a single-scan manner. In the UF version, the incremented pulse-field-gradient-stimulated-echo, commonly

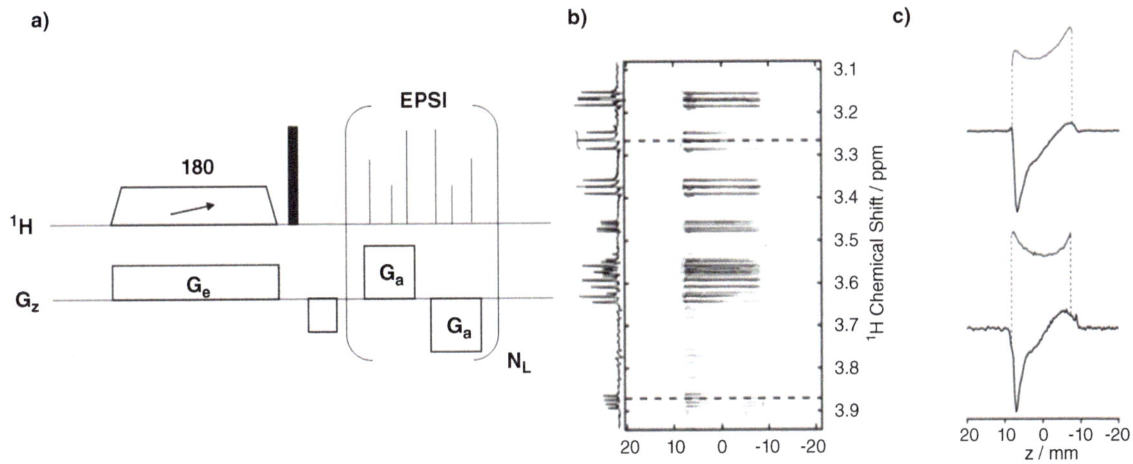


**Figure 10.16** Single-scan measurement of $T_1$ values. (a) UF IR pulse sequence. The thin black rectangle is a 90° hard pulse while the gradient applied just prior to the EPSI is a "prephasing" gradient. (b) UF IR pseudo-2D spectrum, acquired on 100 mM of D-xylose dissolved in $D_2O$, obtained after a 2D FT transform applied onto the (k, t2) plane. (c) Slices taken at 3.3 ppm and 3.9 ppm (dashed lines) portray the variation of magnetization along the length of the sample. Shown in gray are profiles arising from the same sites after full recovery. These are used as reference for an accurate reconstruction of the IR curves. Reproduced from Ref. 37, with permission from John Wiley and Sons.

encountered in 2D DOSY, is replaced by the concurrent application of chirp-pulses and magnetic field gradients. This leads to a spatial dependence of the diffusion weighting. As in the UF $T_1$-measurment described above, the desired dynamic information for each chemical shift is recovered from the shape of spatial profiles gathered by either applying a long weak gradient [39], or by using an EPSI [40]. The latter option is generally preferred in UF 2D DOSY because of its better resolution of the chemical shift. The accuracy of the diffusion coefficients from this single-scan method has been significantly improved by Dumez and coworkers leading to 2D DOSY map with accurate $D$ values (Figure 10.17). Furthermore, the same authors highlight how this spatial encoded DOSY variant is suitable for the extension to 3D DOSY experiments [41].

In several applications, there is a need of mapping dynamic parameters together, instead of retaining the chemical shift in the direct dimension [42]. These $T_1-T_2$ and D-$T_2$ correlation experiments are generally referred as 2D Laplace NMR due to the 2D inverse Laplace transformation involved in the data processing of such experiments. For the past few years, Telkki and coworkers have adapted the UF concept to Laplace 2D NMR to accelerate the experimental duration by one to two orders of magnitude for $T_1-T_2$ [43] and $D$-$T_2$ correlation experiments [44]. This UF 2D Laplace NMR methods have been implemented on high-field spectrometers [43, 44] as well as on low-field single-sided NMR system [45].

Another example of such UF pseudo-2D method is the acceleration of the chemical exchange saturation transfer (CEST) experiment [46]. CEST is a widely used technique for generating MRI contrast in vivo and in vitro. To screen relevant candidates as contrast agents, the polarization of the reporter signal (typically water) is measured as a function of the frequency offset of the saturating radiofrequency irradiation. The resulting curve, refereed as "Z-spectrum," is yielded point by point, involving the repetition of a large number of experiments. Jerschow and coworkers has designed a UF variant to collect a Z-spectrum over a large range of frequency offsets from only two scans [47]. This relies on a saturation with a radiofrequency pulse applied together with a magnetic field gradient during the preparation step. After excitation, the signal is read out with a weak acquisition gradient while the receiver is open. The saturation is turned "off" and "on" for respectively the first and the second scan, leading to a whole Z-spectrum by a simple comparison of the two scans. This ultrafast method enables the high-throughput screening of CEST contrast agents with various experimental conditions [48].

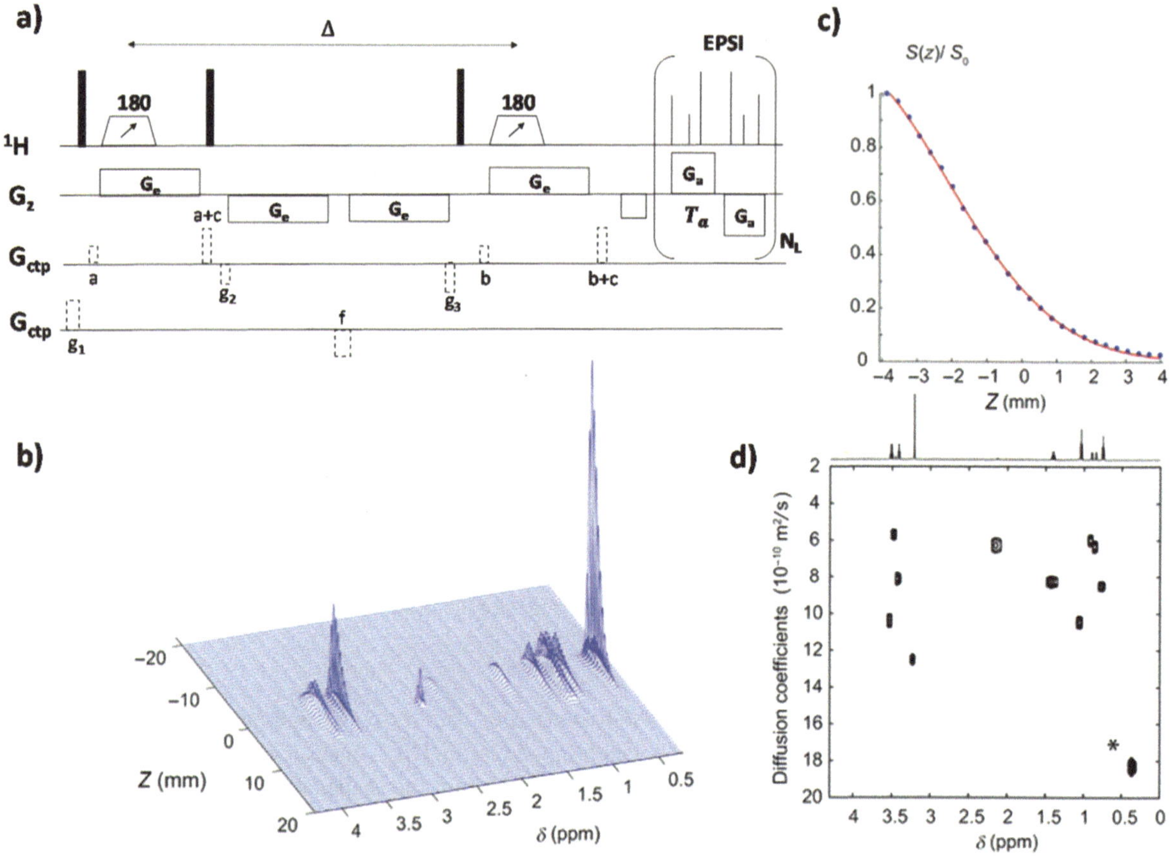


Figure 10.17 Single-scan 2D DOSY experiment. (a) Pulse sequence for spatial encoded DOSY. Gradients "a" and "b" are crushers surrounding the refocusing chirp-pulses; gradient "c" selects the anti-echo pathway for the stimulated echo; gradient "f" is a spoiler during longitudinal storage; "g1," "g2," and "g3" are compensating gradients. (b) 2D spectroscopic imaging data set obtained with the spatial encoding 2D DOSY experiment on a mixture of three alcohols (methanol, ethanol, propanol) and an amino-acid (L-valine), at a concentration of ~100 mM in $D_2O$. (c) Diffusion decay curve obtained from the data set shown in (b), for the methanol $CH_3$ resonances at 3.2 ppm. (d) 2D DOSY display obtained from the data set shown in (b). The water peak is folded and indicated by an asterisk. The $^1H$ pulse-acquire spectrum is shown above the DOSY display. The experiment was carried out with a 600 MHz spectrometer equipped with a triple-axis gradient high-resolution probe. Reproduced from Ref. 41, with permission from Royal Society of Chemistry.

## 10.6 Overview of UF 2D NMR Applications

By delivering exploitable 2D spectra on timescales that cannot be reached with conventional methods, UF 2D NMR has played a central role in applications where the time constraint is predominant, such as monitoring of organic reactions, real-time dynamic studies of biomolecules, 2D NMR on hyperpolarized substrates, coupling with other analytical techniques. Some of representative applications drawn from the literature are presented here.

### 10.6.1 Reaction Monitoring

The first use of UF 2D NMR for real-time reaction monitoring was reported by Frydman and coworkers where time series of UF $^{15}N$-$^1H$ 2D HSQC spectra were collected at a high rate to monitor site-specifically the proton/

deuterium exchange of amide groups after the dissolution of a lyophilized protonated protein ubiquitin in $D_2O$ [49]. From this pioneering work, UF 2D NMR has been used to probe multi-component chemical reactions whose the resulting 1D spectra are overcrowded and unmanageable. This is well illustrated by the works of Herrera and coworkers [50]. In a first study, the mechanisms involved in the formation of alkyl pyrimidines is investigated with UF homo-nuclear 2D NMR [51]. Several hundreds of 2D TOCSY spectra are recorded in a high-throughput manner (every 10 s) to monitor 1 h 30 min of reaction. These 2D spectra collected on-the-fly delivers structural insights matching with the different species involved in the mechanism (Figure 10.18). The evolution of the cross-peak amplitude in the course of the reaction gives access to the kinetic rates of both the starting material and the final products while confirming the transient nature of some species. Such studies can be complemented by real-time hetero-nuclear 2D experiments to further characterize the transient species [52]. Other reaction monitoring based on UF 2D TOCSY and UF 2D HMBC highlight the complementarity of fast homo and hetero-nuclear 2D experiments for a deep understanding of reaction mechanisms [53, 54].

UF 2D NMR is also compatible with flow NMR systems. This enables the real-time monitoring of batch reactions taking place outside the spectrometer in conditions much more realist (temperature, mechanical agitation, etc...) than in-situ monitoring within the NMR tube. Giraudeau and coworkers reported the first real-time monitoring by UF 2D NMR on a flow system, consisting of a closed loop linking the reactor with a benchtop NMR

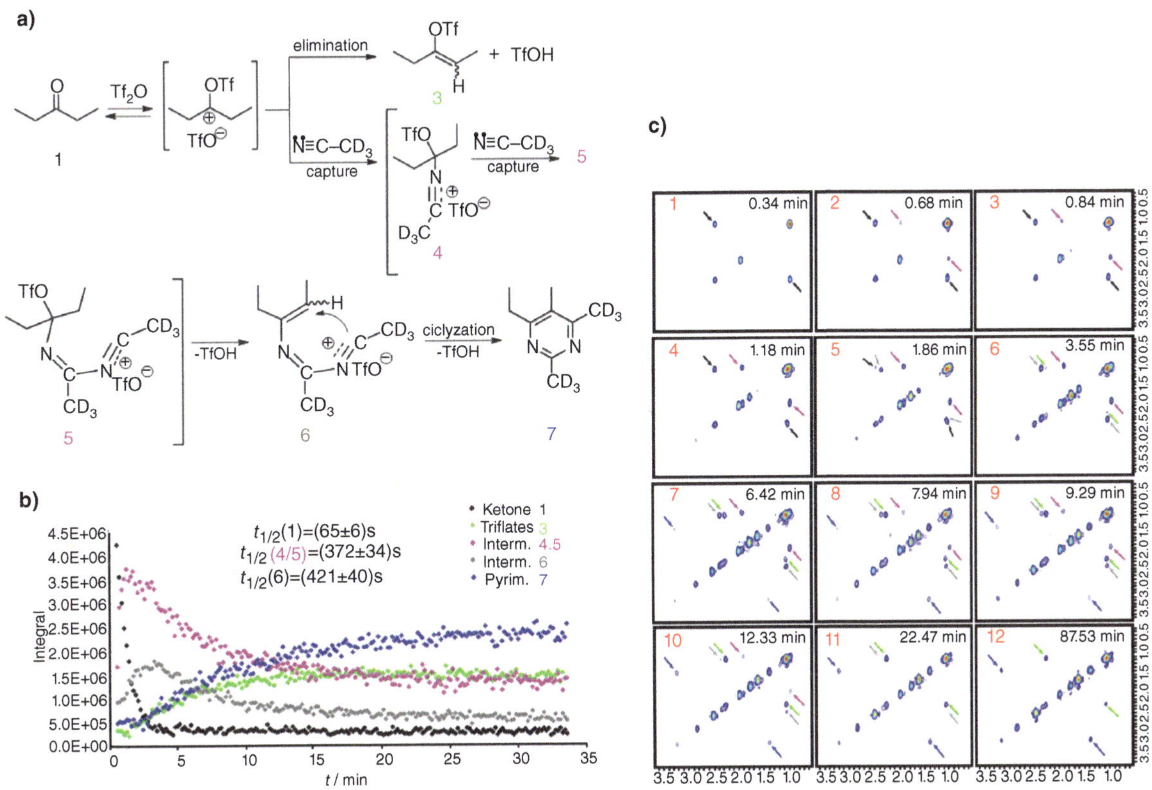


Figure 10.18 UF 2D TOCSY monitoring of a complex organic reaction. (a) Proposed mechanism for the reaction between a symmetric ketone, triflic anhydride, and deuterated acetonitrile. (b) Plots of the averaged integrated peak intensity as a function of the reaction time. (c) Examples of UF 2D TOCSY spectra taken from different instants in the course of the reaction. Reproduced from Ref. 52, with permission from American Chemical Society.

spectrometer [55]. The authors have taken advantage that the spatial encoding direction is orthogonal to the flow in such a benchtop system, so that the spatial encoding is little disturbed by the sample motion. Furthermore, Dumez and coworkers have characterized the interference between flow and spatial encoding in other configurations, including the one where the flow is collinear to the spatial encoding direction, which is the situation most often encountered in flow systems coupled with high-field spectrometers. From this study, pulse sequence- and hardware-based solutions have been proposed [56].

UF 2D NMR has also been exploited for the studies of biomolecule samples. To boost the sensitivity while maintaining a high time-resolution, the UF experiments are carried out in a SOFAST manner. This well-known approach in the protein NMR community dramatically reduces the inter-scan delay upon gathering a large amount of longitudinal magnetization [33]. Such a gain in sensitivity per unit time is achieved by selectively exciting a subset of protons that are of interest (typically amid groups for protein samples) while leaving all other protons unperturbed. This allows for efficient relaxation of the protons of interest, significantly reducing their effective $T_1$ relaxation time. Short recycling delay (e.g. 100–300 ms) can therefore be applied leading to high repetition rates. When possible, the flip angle is also adjusted (Ernst angle) to further improve the sensitivity. This combined use of UF 2D NMR with SOFAST leads to an UltraSOFAST HMQC experiment. This method delivers 2D hetero-nuclear spectra in about 1 s on protein samples at 2 mM (Figure 10.19) enabling the real-time monitoring of protein kinetics occurring on timescales down to a few seconds [34].

Besides applications in (bio)chemical process monitoring, the UF approach has also benefited for hyphenating 2D NMR with other analytical techniques, such as high-performance liquid chromatography (HPLC) combined with single-scan 2D NMR detection [57, 58], or even in-situ electrochemistry process monitored in real time by UF 2D experiments [59].

### 10.6.2 Single-scan 2D Experiments on Hyperpolarized Substrates

Hyphenating UF 2D NMR with hyperpolarization techniques is probably one of the most important development in the UF methodology. Hyperpolarization techniques tackle the sensitivity limit of NMR – caused by the weak bulk polarization observed at room-temperature (about $10^{-5}$) – with metastable spin states that exhibit a much higher polarization (nearby unity values for some techniques). Hyperpolarized samples therefore generate super-signals with an increase of several orders of magnitude of the SNR. A set of methods are available to provide these hyperpolarized states in liquids, such as the optical pumping of noble gases [60, 61], the use of parahydrogen [62] and microwave-driven dynamic nuclear polarization (DNP) from unpaired electrons [63, 64]. In any case, hyperpolarization is a short-lived state that remains on the $T_1$-relaxation time scale of the nuclei. Thus, hyperpolarization methods are hardly compatible with multi-scan experiments, which hampers the acquisition of 2D spectra on such samples with the conventional scheme. UF 2D NMR is in turn an appealing solution to deliver hyperpolarized 2D NMR spectra.

The first acquisition of a 2D spectrum from a hyperpolarized substrate by UF 2D NMR was reported by Frydman and Blazina in 2007 using a dissolution DNP system [65]. This work demonstrates that 2D spectra of hyperpolarized liquid samples at submicromolar concentrations can be acquired in a single-scan fashion, i.e. in 0.1 s. These results shed light on the high complementarity of the two methods: UF 2D NMR is capable of delivering 2D spectra from short lived samples while hyperpolarization remedies the low sensitivity of this spatial encoded method. Another complication with dissolution DNP is the time spent (second time scale) to transfer the sample from the polarizer to the spectrometer during which the hyperpolarized state vanishes through $T_1$-relaxation. This implies the storage of the hyperpolarized state onto slow relaxing nuclei to maintain a sufficient amount of super-signal. In this vein, UF 2D HMBC exploiting long distance correlations between protons and hyperpolarized quaternary $^{13}C$ or $^{15}N$ has been successfully performed [66]. Thereafter, a band-selective variant of this UF experiment may be carried out to probe several correlations in a single shot on mixtures of hyperpolarized natural products at millimolar

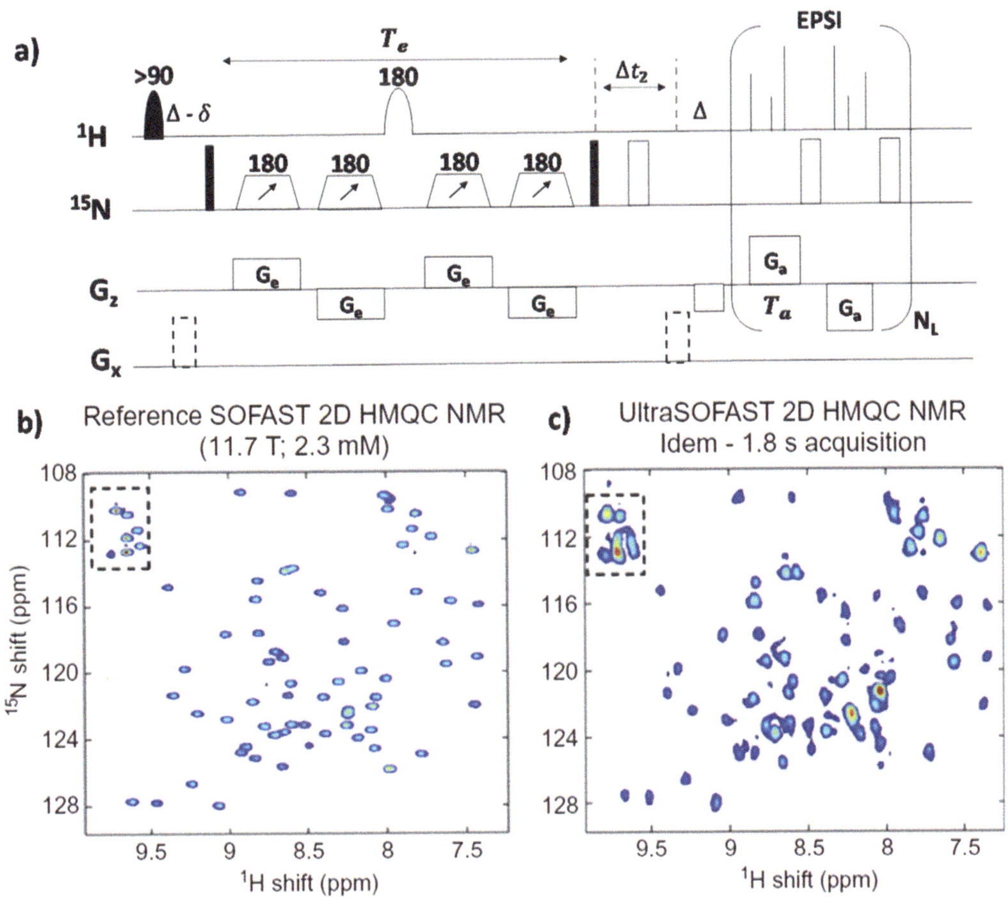


**Figure 10.19** Real-time monitoring of protein kinetics occurring on timescales down to a few seconds. (a) scheme of the UltraSOFAST HMQC sequence. A CT phase-modulated spatial encoding is performed to monitor the $^{15}$N indirect dimension. This sequence incorporates a SOFAST excitation where both excitation and refocusing $^1$H pulses are band-selective on the protons of interest, i.e. amide protons. Note that the selective excitation is applied with a tuned flip angle (Ernst angle) to further enhance the sensitivity per time units. (c) Resulting UltraSOFAST HMQC spectrum versus (b) its reference SOFAST counterpart of $^{15}$N-labeled ubiquitin at mM concentration. Dashed boxes contain folded peaks arising from amine groups. Reproduced from Ref. 34, with permission from American Chemical Society.

concentrations [67]. UF 2D HMBC is also suitable in biochemical applications such as the study of hyperpolarized plant and cancer cell extracts at natural abundance with a 500 MHz spectrometer equipped with a cryogenic probe (Figure 10.20a) [68]. In this example, the 2D spectra are collected in a few seconds, after 30 min of polarization, with a sufficient performance to probe and identify metabolites, while it takes hours with conventional 2D experiments. This coupled approach is undergoing progresses, and homo-nuclear UF experiments, namely 2D TOCSY and 2D MQ-SQ have been recently proposed in the sake of hyperpolarized mixture analysis (Figure 10.20b) [69]. Overall, these results demonstrate the ability of dissolution DNP coupled with UF 2D NMR of delivering 2D spectra in the second time-scale on complex substrates at millimolar concentrations. These developments pave the way for future applications in authentication studies as well as in "omics sciences."

Besides the coupling with dissolution DNP, UF 2D NMR has been used to record single-scan 2D spectra from samples hyperpolarized with parahydrogen techniques, such as the widely used signal amplification by reversible

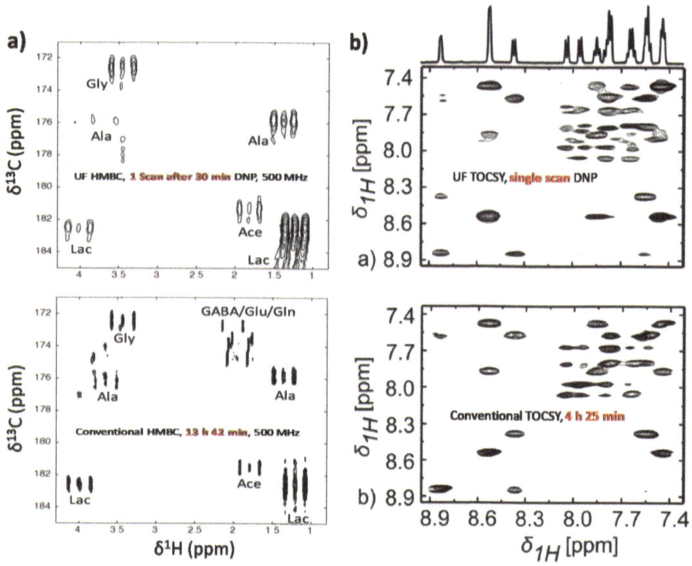


**Figure 10.20** Hyperpolarized UF 2D experiments *versus* conventional 2D techniques. (a) $^{1}H$-$^{13}C$ HMBC spectra of $^{13}C$-enriched extracts of human breast cancer cell lines. Hyperpolarized single-scan HMBC spectrum at 500 MHz after 30 min of polarization (top) and conventional HMBC spectrum, recorded in 13 h 42 min at 500 MHz (bottom). (b) Hyperpolarized UF $^{1}H$–$^{1}H$ 2D TOCSY spectrum collected in a single scan on a model mixture of quinolone, benzophenone, and pyridine (top) and its conventional counterpart recorded on the same sample after rethermalization and shimming, in 4 h and 25 min using 128 $t_1$ increments with 8 scans per increment. Reproduced from Ref. 68 and 69, with permission from Royal Society of Chemistry.

exchange (SABRE) [70, 71]. Lloyd et al., have reported the use of UF 2D COSY experiments to detect SABRE-based hyperpolarization of quinoline at a 10 mM concentration [72]. Moreover, another study used a similar coupled approach to collect COSY spectra in less than 1 s on a complex mixture of analytes at sub-millimolar concentrations (Figure 10.21) [73].

Furthermore, the UF experiments developed to encode relaxation and diffusion parameters could also benefit from hyperpolarization techniques. Dumez and coworkers have reported a UF $^{13}C$ DOSY experiments on a model mixture polarized by dissolution DNP [74], while Telkki and coworkers have demonstrated the possibility of collecting a 2D $D$-$T_2$ map in a ultrafast manner on hyperpolarized substrates [75]. From a proof of concept on a model sample of hyperpolarized DMSO in $D_2O$, the concept of UF hyperpolarized 2D Laplace has been used to identify intracellular and extracellular metabolites in cancer cells (Figure 10.22) [76] as well as to study metal–ligand interactions in reversible polarization transfer from parahydrogen [77].

### 10.6.3 Quantitative UF 2D NMR

Applied under certain conditions, NMR spectroscopy enables quantitative measurements. From an analytical chemistry point of view, NMR is defined as a primary method since the measured signal of a compound is directly proportional to its concentration. Consequently, the measurement of a NMR signal against a reference compound provides the concentration of the probed molecule [78]. Quantitative NMR (qNMR) has been widely used in a broad range of applications such as pharmaceuticals, food science, authentication studies [79–81]. However, $^{1}H$ qNMR suffers from severe peak-overlaps when complex samples are targeted that is detrimental for the analytical performance in terms of accuracy and precision. Among different strategies to address this issue, the use of multidimensional NMR has been particularly effective as it offers a much better signal discrimination than 1D NMR.

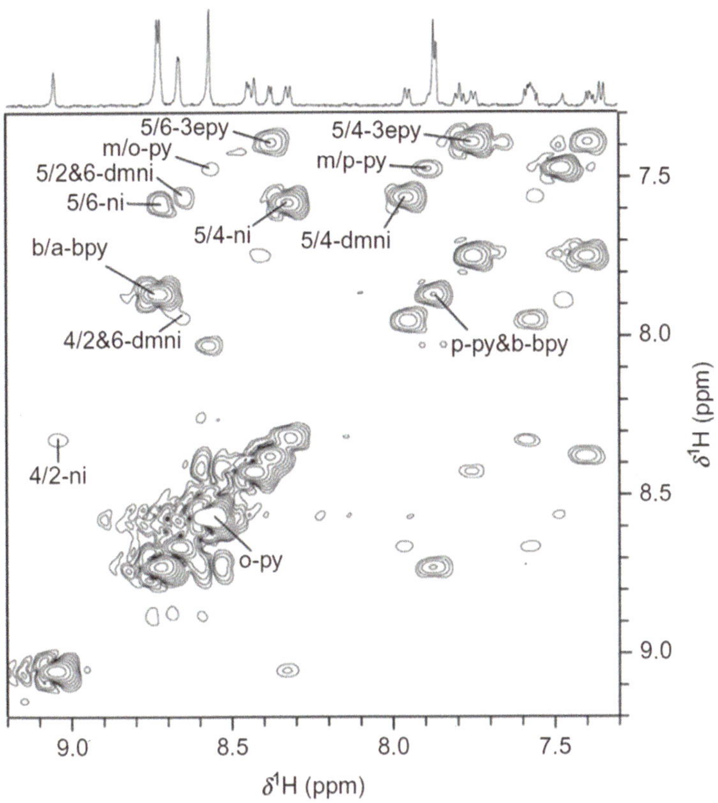


**Figure 10.21** Example of UF 2D NMR coupled with SABRE-enhancement. Single-scan UF 2D COSY spectrum of a mixture consisting of small molecules at concentrations on the order of 0.5 mM, with SABRE hyperpolarization. In addition to the substrates, the sample contained 0.13 mM of complex precursor [Ir(COD)(IMes)Cl], 10 mM of deuterated pyridine as a co-substrate and 4 bar of p- $H_2$ in methanol-$d_4$. Reproduced from Ref. 73, with permission from Royal Society of Chemistry.

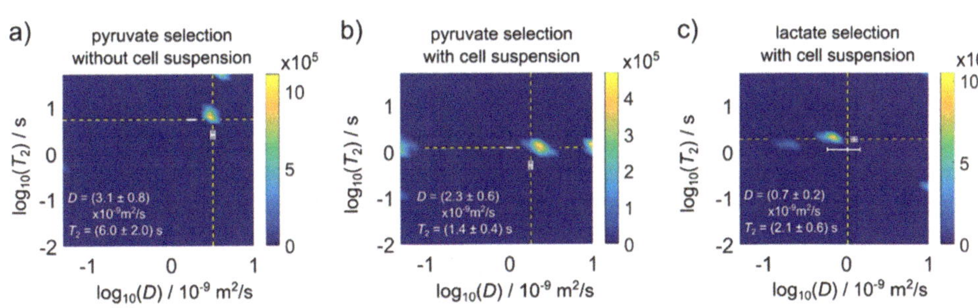


**Figure 10.22** Chemically selective ultrafast $D$–$T_2$ map of (a) pyruvate without cell suspension, (b) pyruvate in cell suspension, and (c) lactate in cell suspension. The $D$ and $T_2$ values shown in the figures correspond to the maxima of the peaks. Reproduced from Ref. 76, with permission from American Chemical Society.

Yet, the response factor of the 2D-peak volume is no longer homogenous for all the probed nuclei in contrast to 1D experiments [82]. A calibration procedure is thus required involving a series of 2D experiments. The use of UF 2D NMR allows to expedite such calibration procedure making quantitative 2D NMR more accessible and suitable in a high-throughput context [83].

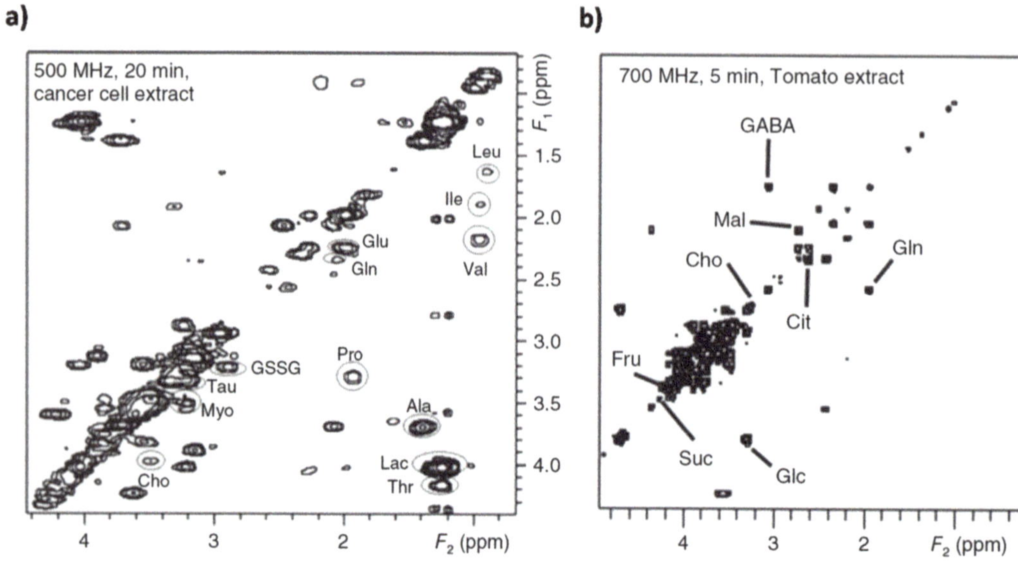


**Figure 10.23** Example of UF 2D COSY spectra recorded on metabolic extracts for absolute quantification purposes. Major metabolites (circled on the figure) are quantified with an accuracy of a few percent. (a) Signal averaged UF 2D COSY spectrum collected in 20 min at 500 MHz with a cryogenic probe on a breast cancer cell extract. (b) Signal averaged UF 2D COSY spectrum obtained in 5 min at 700 MHz with a cryogenic probe using interleaved acquisitions on a tomato fruit pericarp extract. Reproduced from Ref. 88 with permission from American Chemical Society.

The potential of this quantitative approach was firstly assessed with UF 2D *J*-resolved and TOCSY experiments. A precision and a long-term stability lower than 1% and 4%, respectively, (values corresponding to relative standard deviations) were reported on a model mixture of alcohols (>100 mM) in DMSO-$d_6$ [84]. Both methods ensured an excellent linearity, which is an important property when using a calibration procedure. In quantitative applications, a single-scan experiment is rarely sufficient as the precision is inversely proportional to SNR. Signal averaging is thus required, nonetheless, UF experiments do not suffer from "$t_1$-noise" observed in conventional 2D spectra arising from hardware instabilities during the construction of the interferogram. This feature provides UF 2D NMR a better precision than 2D experiments for a similar experiment duration [85]. With this advantages – higher precision and fast calibration procedure – Giraudeau and coworkers have used this protocol to measure site-specific isotopic enrichment in complex biological mixtures [86]. The method has even been extended to 3D experiments, namely UF 3D COSY-*J*-resolved in case of remaining signal overlaps in the 2D spectra [87]. Quantitative UF experiments have gradually become established in the field of metabolomics. UF 2D COSY experiments have been performed to determine the absolute concentration of 14 metabolites in three breast cancer cell line extracts [88] and 8 metabolites in tomato extracts [89].

### 10.6.4 UF 2D NMR in Oriented Media

The ultrafast NMR concept has also been transposed for 2D experiments in oriented media, such as lyotropic liquid crystals (weakly aligning systems) or even solid-sate samples. Giraudeau, Thiele and coworkers have reported the use of UF 2D HSQC in a liquid crystal to measure $^1$H-$^{13}$C residual dipolar couplings (RDC) on a small chiral molecule, e.g. (+)-isopinocampheol, in high concentration at natural abundance [90]. The RDC values extracted were in good agreement with the ones measured with a conventional 2D HSQC. UF 2D NMR is also suitable to probe quadrupolar nulcei in weakly aligning systems. Lesot, Giraudeau, and coworkers designed the first deuterium UF 2D experiment in order to extract $^2$H residual quadrupolar couplings (RQC) on deuterated analytes dissolved in a chiral polypeptide liquid crystal [91]. This pioneering work has recently been followed

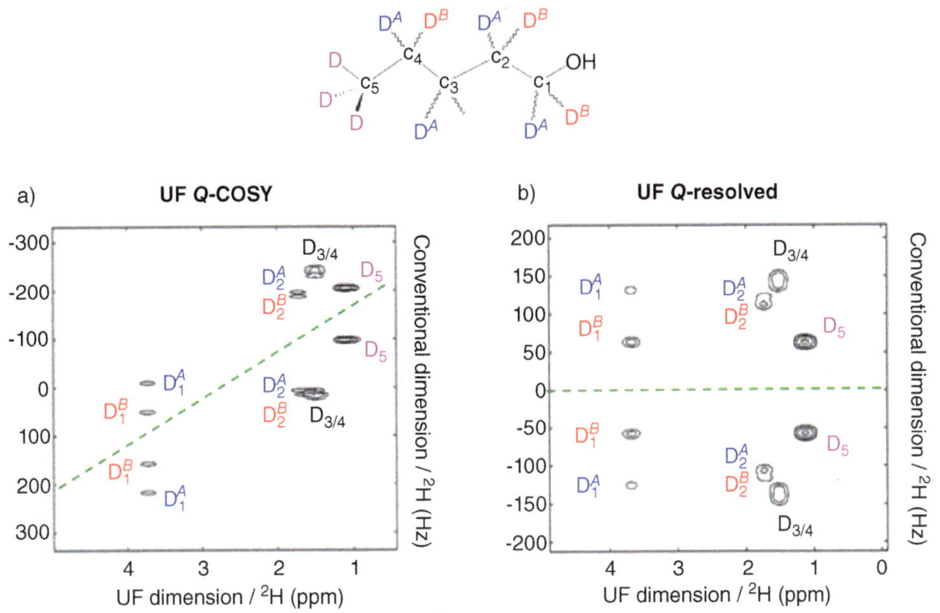


**Figure 10.24** Two single-scan 92.1 MHz UF $^2H$ QUOSY 2D experiments performed on pentanol-$d_{12}$ dissolved in a chiral lyotropic liquid crystal (PBLG/$CHCl_3$). (a) UF 2D Q-COSY spectrum recorded in 0.15 s, correlating the RQC values with the chemical shift position. In this experiment, only the anisotropic chemical shift is encoded in the UF dimension while both anisotropic chemical shift and quadrupolar coupling are monitored along the conventional one. (b) UF 2D Q-resolved spectrum recorded in 0.20 s. In this variant, anisotropic chemical shift and quadrupolar coupling information are encoded in two orthogonal directions. The pro-R/pro-S assignment shown here is arbitrary. Reproduced from Ref. 92 with permission from John Wiley and Sons.

by the development of various UF $^2H$ 2D quadrupolar ordered spectroscopy (QUOSY) methods (Figure 10.24), including single and double-quantum experiments. [92].

The ability of UF methods to probe solid-state samples has also been demonstrated. The proof of concept has been achieved on a high-resolution magic angle spinning (HR-MAS) set up [93]. In this work, UF spectra are collected on banana slopes while spinning the sample at the magical angle (Figure 10.25a and 10.25b). Besides this development in HR-MAS, the UF methodology can be used in the solid-state at very high field (17.6 T), using a double resonance MAS probe together with external micro-imaging gradients [94]. Two widely used 2D solid-state NMR experiments: double-quantum correlation and RF-driven proton spin diffusion (PSD), can be revisited in a UF manner. This has been highlighted on elastomers samples under magic-angle spinning (Figures 10.25c and 10.25d). This UF approach is particularly relevant to expedite series of 2D experiments needed, for instance, to collect spin-diffusion build-up curves.

### 10.6.5 UF 2D NMR in Spatial Inhomogeneous Fields

The spatial encoding of the NMR interactions has been exploited in experiments that aim at gathering high-resolution spectra from samples submitted to an inhomogeneous $B_0$ magnetic field. In many circumstances, the spatial and temporal homogeneities of $B_0$ are degraded as in the studies of heterogeneous biological tissues in vivo, complex food matrices, or in the case of in-situ NMR spectro-electrochemistry [95]. Numerous strategies have been proposed to address this challenge. One family of methods relies on 2D experiments, in which intra- or intermolecular multiple-quantum coherences (iMQCs) that are immune to the magnetic field inhomogeneity, are monitored in the indirect dimension [96]. The UF concept can be exploited here to yield such 2D datasets in a single-scan fashion, which is a determining advantage for practical applications involving spatiotemporal variations of $B_0$, e.g.

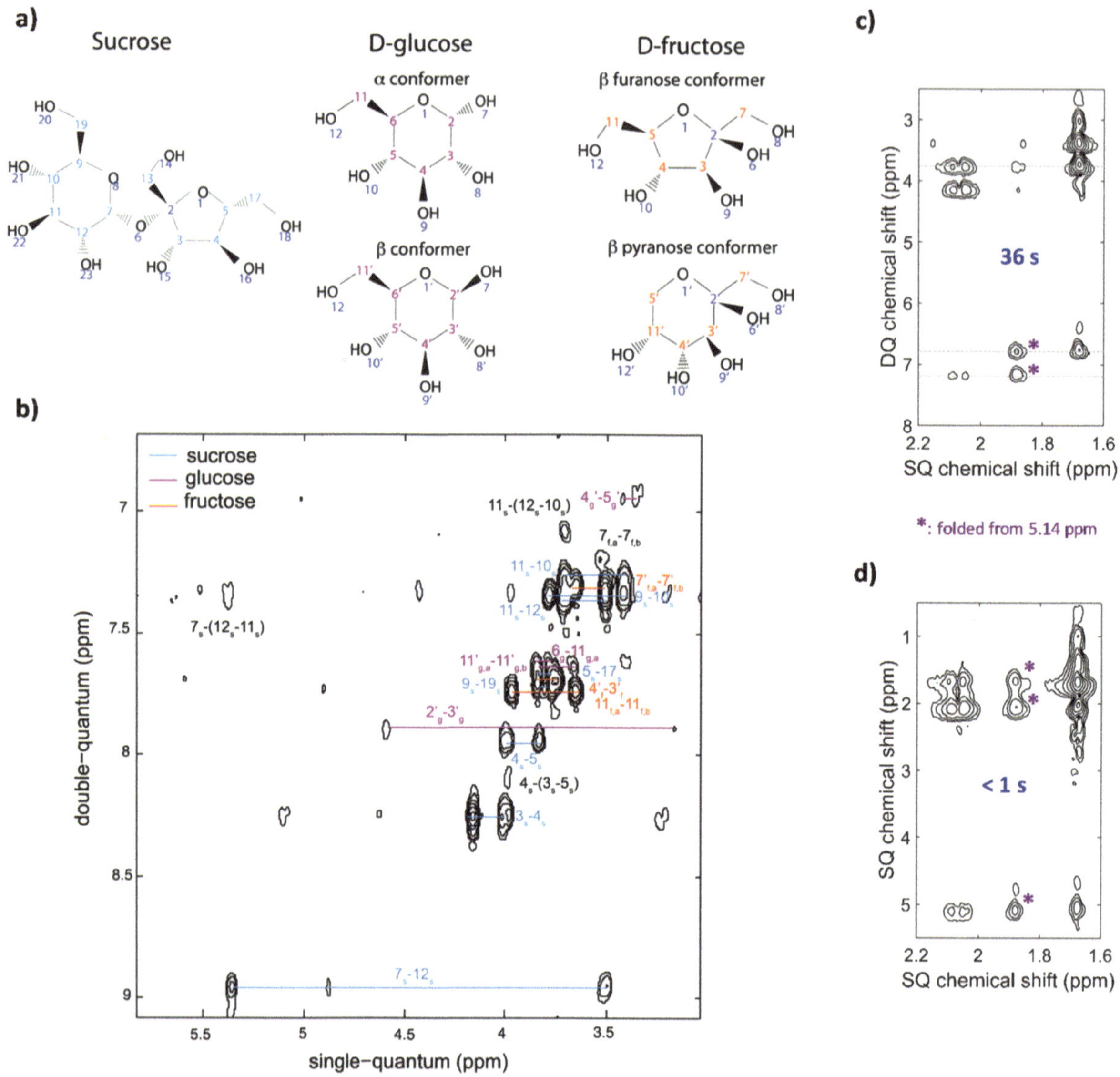


**Figure 10.25** UF 2D NMR applied in HR-MAS and solid-state. (a) Chemical structure of sucrose, D-glucose, and D-fructose. (b) UF HR-MAS DQS spectrum of a 10 mg sample of fresh banana, spinning at a frequency of 4.53 kHz. Pairs of correlation peaks for the three most abundant sugars in the sample (glucose, fructose and sucrose) are indicated by colored lines. (c) UF DQ spectra of a natural rubber sample recorded in 36 s. (d) UF PSD spectra of the same sample, recorded in less than 1 s. Reproduced from Ref. 93 with permission from Royal Society of Chemistry.

in vivo experiments. About 10 years ago, Pelupessy et al. proposed in 2009 a spatially encoded method to yield high-resolution 2D spectra in a single scan in inhomogeneous fields [97]. The key feature of this method is to spatially encode the chemical shift differences for pairs coupled spins with two similar spatial encoding schemes flanking a mixing pulse. This leads to the differential evolution of two single-quantum coherences while the unknown spatially dependent frequency induced by the inhomogeneous B0 field is eliminated by simple subtraction. As a result, the overall phase before detection is only dependent on the chemical shift differences between coupled spins and proportional to the spatial encoding constant. $J$-coupling can be then monitored trough a ($G_a$-180°) EPSI scheme similar to the one used by Giraudeau et al. for recording UF $J$-resolved spectra. The same authors also proposed an improved version of the pulse sequence consisting of an odd number of spatial encoding step at both sides of the mixing pulse to refocus the $B_0$ inhomogeneity effects at the rise of the echo, rather than at the beginning of the detection. By the way, these UF experiments lead to a 2D spectrum where the $J$-coupling in the direct dimension is correlated with the chemical shift differences from coupled spins along the indirect dimension. This F1 dimension that is reminiscent of a zero-quantum (ZQ) evolution makes direct spectral assignments challenging and the uncoupled spins (singlets) are lacking. Further efforts have been made so far to tackle these drawbacks. Chen and coworkers proposed a new experiment, which combines a similar spatial encoding approach but with another coherence transfer pathway based on intermolecular zero-quantum coherences (iZQC) [98]. This leads to a 2D spectrum akin the common 2D $J$-resolved with chemical shifts and $J$-coupling along two orthogonal dimensions (Figure 10.26b). UF 2D NMR based on other iMQC have also been reported such as UF iSQC [99, 100], or UF iDQS [101]. Finally, the same research group extended this approach to 2D correlation spectroscopy through the development of UF SECSY (spin-echo spectroscopy) and SETOCSY (spin-echo TOCSY) experiments. After a specific mathematical manipulation of the datasets, these deliver 2D COSY- and TOCSY-like spectra in a single scan

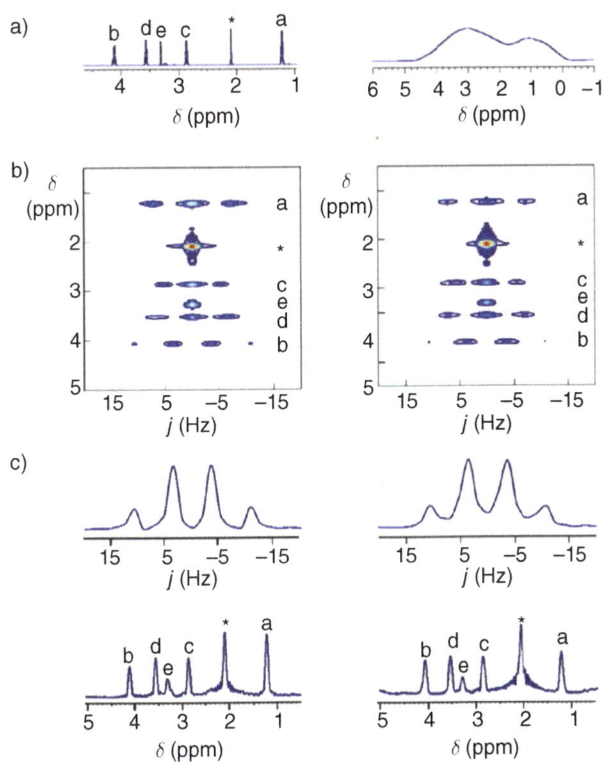


Figure 10.26 2D $J$-resolved experiment in inhomogeneous fields combining spatial encoding and iZQC coherence transfer scheme. (a) 1D $^1H$ NMR spectrum of a sample of ethyl 3-bromopropionate and methanol in acetone recorded in a homogeneous field (left) and in the presence of 1.8 kHz field inhomogeneity artificially introduced by detuning the $\{z_1, x_1, y_1\}$ shim coils. The peaks with asterisk correspond to the solvent. (b) 2D iZQC $J$-resolved spectra, obtained in the homogeneous (left) and inhomogeneous fields (right). (c) Projections along the F2 dimension of the quadruplet at 4.16 ppm from spectra in (b). (d) Sum of the projections along the F1 dimension from spectra in (b). Reproduced from Ref. 9 with permission from American Chemical Society.

[102]. This method is suitable through linear inhomogeneities along the orientation of encoding and acquisition gradients. Besides these developments on homo-nuclear experiments, Zhang et al. extended this UF methodology to 2D hetero-nuclear correlation experiments [103].

## 10.7 Conclusion

Ultrafast NMR is the fastest way to collect any multi-dimensional spectra whenever ample SNR is available from a single or few scan experiment. Since the original version was proposed about 20 years ago, many improvements have allowed this ultrafast method to deliver usable 2D spectra in a few seconds, opening the field of 2D NMR to previously inaccessible applications. This versatile concept can, in principle, benefit any kind of 2D spectroscopic methods but also dynamic-based experiments. This has led to a multitude of fundamental and practical applications, from the real-time reaction monitoring, to the study of dynamic processes occurring on timescales down to a few seconds, or even for the record of high-resolution spectra in inhomogeneous magnetic fields. Moreover, this ability to deliver 2D spectra from short lived-state species is perfectly in line with the emerging hyperpolarization techniques. Further developments and exciting applications are expected along this direction.

## References


1 Jeener, J. (1971). *Lecture Presented at Ampere International Summer School II*. Yugoslavia: Basko Polje.

2 Rouger, L., Gouilleux, B. and Giraudeau, P. (2017). Fast n-Dimensional Data Acquisition Methods, in: *Encyclopedia of Spectroscopy and Spectrometry*, 3e (ed. J.C. Lindon, G.E. Tranter, D.W. Koppenaal), 588–596. Oxford: Academic Press.

3 Schanda, P. (2009). Fast-pulsing longitudinal relaxation optimized techniques: enriching the toolbox of fast biomolecular NMR spectroscopy. *Prog. Nucl. Magn. Reson. Spectrosc.* 55: 238–265.

4 Orekhov, V.Y. and Jaravine, V.A. (2011). Analysis of non-uniformly sampled spectra with multi-dimensional decomposition. *Prog. Nucl. Magn. Reson. Spectrosc.* 59: 271–292.

5 Frydman, L., Scherf, T. and Lupulescu, A. (2002). The acquisition of multidimensional NMR spectra within a single scan. *Proc. Natl. Acad. Sci. U.S.A.* 99 (2002) 15858–15862.

6 Tal, A. and Frydman, L. (2010). Single-scan multidimensional magnetic resonance. *Prog. Nucl. Magn. Reson. Spectrosc.* 57: 241–292.

7 Gal, M. and Frydman, L. (2015). Multidimensional NMR spectroscopy in a single scan. *Magn. Reson. Chem.* 53: 971–985. doi:10.1002/mrc.4271.

8 Giraudeau, P. and Frydman, L. (2014). Ultrafast 2D NMR: an emerging tool in analytical spectroscopy. *Annu. Rev. Anal. Chem.* 7: 129–161.

9 Gouilleux, B., Rouger, L. and Giraudeau, P. (2018). Ultrafast 2D NMR: methods and applications. *Annu. Rep. NMR Spectrosc.* 93: 75–144.

10 Shrot, Y., Shapira, B. and Frydman, L. (2004). Ultrafast 2D NMR spectroscopy using a continuous spatial encoding of the spin interactions. *J. Magn. Reson.* 171: 163–170.

11 Pelupessy, P. (2003). Adiabatic single scan two-dimensional NMR spectroscopy. *J. Am. Chem. Soc.* 125: 12345–12350.

12 Shrot, Y. and Frydman, L. (2008). Spatial encoding strategies for ultrafast multidimensional nuclear magnetic resonance. *J. Chem. Phys.* 128: 052209.

13 Giraudeau, P. and Akoka, S. (2008). Resolution and sensitivity aspects of ultrafast J-resolved 2D NMR spectra. *J. Magn. Reson.* 190: 339–345.

**14** Gouilleux, B., Rouger, L., Charrier, B,. Kuprov, I., Akoka, S., Dumez, J.-N. and Giraudeau, P. (2015). Understanding J-modulation during spatial encoding for sensitivity-optimized ultrafast NMR spectroscopy. *ChemPhysChem.* 16: 3093–3100.

**15** Mansfield, P. (1977). Multi-planar image formation using NMR spin echoes. *J. Phys. C: Solid State Phys.* 10: 55–58.

**16** Mansfield, P. and Morris, P.G. (1982). *NMR Imaging in Biomedicine*. New-York: Academic Press.

**17** Dumez, J.N. (2018). Spatial encoding and spatial selection methods in high-resolution NMR spectroscopy. *Prog. Nucl. Magn. Reson. Spectrosc.* 109: 101–134.

**18** Giraudeau, P. and Akoka, S. (2008). Sources of sensitivity losses in ultrafast 2D NMR. *J. Magn. Reson.* 192: 151–158.

**19** Giraudeau, P. and Akoka, S. (2011). Sensitivity and lineshape improvement in ultrafast 2D NMR by optimized apodization in the spatially encoded dimension. *Magn. Reson. Chem.* 49: 307–313.

**20** Rouger, L., Loquet, D., Massou, S., Akoka, S. and Giraudeau, P. (2012). Limitation of diffusion effects in ultrafast 2D nuclear magnetic resonance by encapsulation of analytes in phospholipidic vesicles. *ChemPhysChem.* 13: 4124–4127.

**21** Giraudeau, P. and Akoka, S. (2008). Sensitivity losses and line shape modifications due to molecular diffusion in continuous encoding ultrafast 2D NMR experiments. *J. Magn. Reson.* 195: 9–16.

**22** Giraudeau, P. and Akoka, S. (2010). A new gradient-controlled method for improving the spectral width of ultrafast 2D NMR experiments. *J. Magn. Reson.* 205: 171–176.

**23** Frydman, L., Lupulescu, A. and Scherf, T. (2003). Principles and features of single-scan two-dimensional NMR spectroscopy. *J. Am. Chem. Soc.* 125: 9204–9217.

**24** Rouger, L., Charrier, B., Pathan, M., Akoka, S. and Giraudeau, P. Processing strategies to obtain clean interleaved ultrafast 2D NMR spectra. *J. Magn. Reson.* 238: 87–93.

**25** Shrot, Y. and Frydman, L. (2011). Compressed sensing and the reconstruction of ultrafast 2D NMR data: principles and biomolecular applications. *J. Magn. Reson.* 209: 352–358.

**26** Hamdoun, G., Gouilleux, B., Sebban, M., Barozzino-Consiglio, G., Harrison-Marchand, A., Giraudeau, P., Maddaluno, J. and Oulyadi, H. (2017). 2D 7 Li ultrafast CT-COSY: a new tool for the rapid measurement of tiny homonuclear lithium scalar couplings. *Chem. Commun.* 53: 220–223.

**27** Herrera, A., Fernández-Valle, E., Gutiérrez, E.M., Martínez-Álvarez, R., Molero, D., Pardo, Z.D. and Sáez, E. (2010). 2D ultrafast HMBC: a valuable tool for monitoring organic reactions. Org. *Lett.* 12: 144–147.

**28** Giraudeau, P. and Akoka, S. (2007). A new detection scheme for ultrafast 2D J-resolved spectroscopy. *J. Magn. Reson.* 186: 352–7.

**29** Pathan, M., Akoka, S. and Giraudeau, P. (2012). Ultrafast hetero-nuclear 2D J-resolved spectroscopy. *J. Magn. Reson.* 214: 335–339.

**30** Le Guennec, A., Giraudeau, P., Caldarelli, S. and Dumez, J.-N. (2015). Ultrafast double-quantum NMR spectroscopy. *Chem. Commun.* 51: 354–357.

**31** Jacquemmoz, C., Boyarskaya, D., Masson, G. and Dumez, J.N. (2018). Ultrafast maximum-quantum NMR spectroscopy for the analysis of aromatic mixtures. *ChemPhysChem.* 19: 3310–3317.

**32** Rouger, L., Gouilleux, B., Pourchet-Gellez, M., Dumez, J.-N. and Giraudeau, P. (2016). Ultrafast double-quantum NMR spectroscopy with optimized sensitivity for the analysis of mixtures. *Analyst.* 141: 1686–1692.

**33** Schanda, P. and Brutscher, B. (2005). Very fast two-dimensional NMR spectroscopy for real-time investigation of dynamic events in proteins on the time scale of seconds. *J. Am. Chem. Soc.* 127: 8014–8015.

**34** Gal, M., Schanda, P., Brutscher, B. and Frydman, L. (2007). UltraSOFAST HMQC NMR and the repetitive acquisition of 2D protein spectra at Hz rates. *J. Am. Chem. Soc.* 129: 1372–1377.

**35** Tal, A., Shapira, B. and Frydman, L. (2009). Single-scan 2D Hadamard NMR spectroscopy, *Angew. Chem. Int. Ed.* 48: 2732–2736.

**36** Vold, R.L., Waugh, J.S., Klein, M.P. and Phelps, D.E. (1968). Measurement of spin relaxation in complex systems. *J. Chem. Phys.* 48: 3831–3832.

**37** Smith, P.E.S., Donovan, K.J., Szekely, O. Baias, M. and Frydman, L. (2013). Ultrafast NMR T1 relaxation measurements: probing molecular properties in real time. *ChemPhysChem.* 14: 3138–3145.

**38** Barjat, H., Morris, G.A., Smart, S., Swanson, A. and Williams, S.C.R. (1995). High-resolution diffusion-ordered spectroscopy (HR-DOSY) - a new tool for the analysis of complex mixtures. *J. Magn. Reson. Ser. B.* 108: (1995) 170–172.

**39** Loening, N.M. Keeler, J. and Morris, G.A. One-dimensional DOSY. *J. Magn. Reson.* 153: 103–112. doi:10.1006/jmre.2001.2423.

**40** Shrot, Y. and Frydman, L. (2008). Single-scan 2D DOSY NMR spectroscopy. *J. Magn. Reson.* 195: 226–231.

**41** Guduff, L., Kuprov, I., van Heijenoort, C. and Dumez, J.-N. (2017) Spatially encoded 2D and 3D diffusion-ordered NMR spectroscopy. *Chem. Commun.* 53: 701–704.

**42** Godefroy, S. and Callaghan, P.T. (2003). 2D relaxation/diffusion correlations in porous media. *Magn. Reson. Imaging.* 21: 381–383.

**43** Ahola, S. and Telkki, V.-V. (2014) Ultrafast two-dimensional NMR relaxometry for investigating molecular Processes in real time, *ChemPhysChem.* 15: 1687–1692.

**44** Ahola, S., Zhivonitko, V.V., Mankinen, O., Zhang, G., Kantola, A.M., Chen, H.-Y., Hilty, C., Koptyug, I.V. and Telkki, V.-V. (2015). Ultrafast multidimensional Laplace NMR for a rapid and sensitive chemical analysis. *Nat. Commun.* 6: 8363.

**45** King, J.N., Lee, V.J., Ahola, S., Telkki, V.-V. and Meldrum, T. (2016). Ultrafast multidimensional laplace NMR using a single-sided magnet. *Angew. Chem. Int. Ed.* 128: 5124–5127.

**46** Wolff, S.D. and Balaba, R.S. (1990). NMR imaging of labile proton exchange. *J. Magn. Reson.* 86: 164–169.

**47** Xu, X., Lee, J.-S. and Jerschow, A. (2013). Ultrafast scanning of exchangeable sites by NMR spectroscopy. *Angew. Chem. Int. Ed.* 52: 8281–8284.

**48** Xu, X., Yadav, N.N., Song, X., McMahon, M.T., Jerschow, A., van Zijl, P.C.M. and Xu, J. (2016). Screening CEST contrast agents using ultrafast CEST imaging. *J. Magn. Reson.* 265: 224–229.

**49** Gal, M., Mishkovsky, M. and Frydman, L. (2006). Real-time monitoring of chemical transformations by ultrafast 2D NMR spectroscopy. *J. Am. Chem. Soc.* 128: 951–956.

**50** Herrera, A., Fernández-Valle, E., Martínez-Álvarez, R., Molero-Vílchez, D., Pardo-Botero, Z.D. and Sáez-Barajas, E. (2015). Monitoring organic reactions by UF-NMR spectroscopy. *Magn. Reson. Chem.* 53: 952–970.

**51** Herrera, A., Fernández-Valle, E., Martínez-Álvarez, R., Molero, D., Pardo, Z.D., Sáez, E. and Gal, M. (2009). Real-time monitoring of organic reactions with two-dimensional ultrafast TOCSY NMR spectroscopy13. *Angew. Chem. Int. Ed.* 48: 6274–6277.

**52** Pardo, Z.D., Olsen, G.L., Fernández-Valle, M.E., Frydman, L., Martínez-Álvarez, R. and Herrera, A. (2012). Monitoring mechanistic details in the synthesis of pyrimidines via real-time, ultrafast multidimensional NMR spectroscopy. *J. Am. Chem. Soc.* 134: 2706–2715.

**53** Fernández, I., Fernández-Valle, M.E., Martinez-Alvarez, R., Molero-Vilchez, D., Pardo, Z.D., Sáez-Barajas, E., Sanchez, A. and Herrera, A. (2014). Discovering mechanistic insights by application of tandem ultrafast multidimensional NMR techniques. *J. Org. Chem.* 79: 8086–8093.

**54** Queiroz Jr, L.H., Giraudeau, P., dos Santos, F.A., de Oliveira, K.T. and Ferreira, A.G. (2012). Real-time mechanistic monitoring of an acetal hydrolysis using ultrafast 2D NMR. *Magn. Reson. Chem.* 50: 496–501.

**55** Gouilleux, B., Charrier, B., Akoka, S., Felpin, F.X., Rodriguez-Zubiri, M. and Giraudeau, P. (2016). Ultrafast 2D NMR on a benchtop spectrometer: applications and perspectives, TrAC Trends Anal. *Chem. Part A* 83: 65–75.

**56** Jacquemmoz, C., Giraud, F. and Dumez, J.N. (2020). Online reaction monitoring by single-scan 2D NMR under flow conditions. *Analyst.* 145: 478–485.

**57** Shapira, B., Karton, A., Aronzon, D. and Frydman, L. (2004). Real-time 2D NMR identification of analytes undergoing continuous chromatographic separation. *J. Am. Chem. Soc.* 126: 1262–1265.

**58** Queiroz, L.H.K., Queiroz, D.P.K., Dhooghe, L., Ferreira, A.G. and Giraudeau, P. (2012). Real-time separation of natural products by ultrafast 2D NMR coupled to on-line HPLC. *Analyst.* 137: 2357–2361.

**59** Boisseau, R., Bussy, U., Giraudeau, P. and Boujtita, M. (2015). In situ ultrafast 2D NMR spectroelectrochemistry for real-time monitoring of redox reactions. *Anal. Chem.* 87: 372–375.

**60** Albert, M.S., Cates, G.D., Driehuys, B., Happer, W., Saam, B., Springer, C.S. and Wishnia, A. (1994). Wishnia, Biological magnetic resonance imaging using laser-polarized 129Xe, *Nature.* 370: 199–201.

**61** Navon, G., Song, Y.Q., Room, T., Appelt, S.T.R.E., Taylor, R.E. and Pines, A. (1996). Pines, enhancement of solution NMR and MRI with laser-polarized xenon. *Science.* 271.

**62** Green, R.A., Adams, R.W., Duckett, S.B., Mewis, R.E., Williamson, D.C. and Green, G.G. (2012). The theory and practice of hyperpolarization in magnetic resonance using parahydrogen. *Prog. Nucl. Magn. Reson. Spectrosc.* 67: 1–48.

**63** Wolber, J., Ellner, F., Fridlund, B., Gram, A., Johannesson, H., Hansson, G., Hansson, L.H., Lerche, M.H., Månsson, S., Servin, R. and Thaning, M. (2004). Generating highly polarized nuclear spins in solution using dynamic nuclear polarization. *Nucl. Instr. and Meth. in Phys. Research, Section A: Accelerators, Spectrometers, Detectors and Associated Equipment* 526: 173–181.

**64** Griesinger, C., Bennati, M., Vieth, H.M., Luchinat, C., Parigi, G., Höfer, P., Engelke, F., Glaser, S.J., Denysenkov, V. and Prisner, T.F.(2012). Dynamic nuclear polarization at high magnetic fields in liquids. *Prog. Nucl. Magn. Reson. Spectrosc.* 64: 4–28.

**65** Frydman, L. and Blazina, D. (2007). Ultrafast two-dimensional nuclear magnetic resonance spectroscopy of hyperpolarized solutions. *Nat. Phys.* 3: 415–419.

**66** Mishkovsky, M. and Frydman, L. (2008). Progress in hyperpolarized ultrafast 2D NMR spectroscopy. *ChemPhysChem.* 9: 2340–2348.

**67** Giraudeau, P., Shrot, Y. and Frydman, L. (2009). Multiple ultrafast, broadband 2D NMR spectra of hyperpolarized natural Products. *J. Am. Chem. Soc.* 131: 13902–13903.

**68** Dumez, J.N., Milani, J., Vuichoud, B., Bornet, A., Lalande-Martin, J., Tea, I., Yon, M., Maucourt, M., Deborde, C., Moing, A. and Frydman, L. (2015). Hyperpolarized NMR of plant and cancer cell extracts at natural abundance. *Analyst.* 140: 5860–5863.

**69** Singh, K., Jacquemmoz, C., Giraudeau, P., Frydman, L. and Dumez, J.N. (2021). Ultrafast 2D $^{1}H$–$^{1}H$ NMR spectroscopy of DNPhyperpolarised substrates for the analysis of mixtures. *Chem. Commun.* 57: 8035–8038.

**70** Adams, R.W., Aguilar, J.A., Atkinson, K.D., Cowley, M.J., Elliott, P.I., Duckett, S.B., Green, G.G., Khazal, I.G., Lóрez-Serrano, J. and Williamson, D.C. (2009). Reversible interactions with para-hydrogen enhance NMR sensitivity by polarization transfer. *Science.* 323: 1708–1711.

**71** Cowley, M.J., Adams, R.W., Atkinson, K.D., Cockett, M.C., Duckett, S.B., Green, G.G., Lohman, J.A., Kerssebaum, R., Kilgour, D. and Mewis, R.E. (2011). Iridium N-heterocyclic carbene complexes as efficient catalysts form magnetization transfer from para-hydrogen. *J. Am. Chem. Soc.* 133: 6134–6137.

**72** Lloyd, L.S., Adams, R.W., Bernstein, M., Coombes, S., Duckett, S.B., Green, G.G., Lewis, R.J., Mewis, R.E. and Sleigh, C.J. (2012). Utilization of SABRE derived hyperpolarization to detect low concentration analytes via 1D- and 2D NMR methods. *J. Am. Chem. Soc.* 134: 12904–12907, 120719120603004.

**73** Daniele, V., Legrand, F.X., Berthault, P., Dumez, J.N. and Huber, G. (2015). Single-scan multidimensional NMR analysis of mixtures at sub-millimolar concentrations by using SABRE hyperpolarization. *ChemPhysChem.* 16: 3413–3417.

**74** Guduff, L., Kurzbach, D., Van Heijenoort, C., Abergel, D. and Dumez, J.N. (2017). Single-scan 13C diffusion-ordered NMR spectroscopy of DNP-hyperpolarised substrates. *Chem. Eur. J.* 23: 16722–16727.

**75** Ahola, S., Zhivonitko, V.V., Mankinen, O., Zhang, G., Kantola, A.M., Chen, H.Y., Hilty, C., Koptyug, I.V. and Telkki, V.V. (2015). Ultrafast multidimensional Laplace NMR for a rapid and sensitive chemical analysis. *Nat. Commun.* 6: 8363.

**76** Zhang, G., Ahola, S., Lerche, M. H., Telkki, V. V., & Hilty, C. (2018). Identification of intracellular and extracellular metabolites in cancer cells using 13c hyperpolarized ultrafast Laplace NMR. *Anal. Chem.* 90 (18): 11131–11137.

**77** Tickner, B.J., Zhivonitko, V.V. and Telkki, V.V. (2021). Ultrafast Laplace NMR to study metal–ligand interactions in reversible polarisation transfer from parahydrogen. *Phys. Chem. Chem. Phys.* 23 (31): 16542–16550.

**78** Evilia R.F. (2001). Quantitative NMR spectroscopy. *Anal. Letter.* 34: 2227–2236.

**79** Holzgrabe, U. (2004). Quantitative NMR spectroscopy in pharmaceutical applications. *Prog. Nucl. Magn. Reson.* Spectrosc. 57: 229–240.

**80** Tenailleau, E., Lancelin, P., Robins, R.J. and Akoka, S. (2004). Authentication of the origin of vanillin using quantitative natural abundance 13C NMR. *J. Agric. Food Chem.* 52: 7782–7787.

**81** Mannina, L., Sobolev, A.P. Viel, S. (2012). Liquid state 1H high field NMR in food analysis. *Prog. Nucl. Magn. Reson.* Spectrosc. 66: 1–39.

**82** Giraudeau, P. (2014). Quantitative 2D liquid-state NMR. *Magn Reson Chem.* 52: 259–272.

**83** Giraudeau, P. (2017). Challenges and perspectives in quantitative NMR. *Magn. Reson. Chem.* 55: 61–69.

**84** Giraudeau, P., Remaud, G.S. and Akoka, S. (2009). Evaluation of ultrafast 2D NMR for quantitative analysis. *Anal. Chem.* 81: 479–484.

**85** Pathan, M., Akoka, S., Tea, I., Charrier, B. and Giraudeau, P. (2011). "Multi-scan single shot" quantitative 2D NMR: a valuable alternative to fast conventional quantitative 2D NMR. *Analyst* 136: 3157–3163

**86** Massou, S., Nicolas, C., Letisse, F. and Portais, J.-C. (2007). Application of 2D-TOCSY NMR to the measurement of specific 13C-enrichments in complex mixtures of 13C-labeled metabolites. *Metab. Eng.* 9: 252–257.

**87** Boisseau, R., Charrier, B., Massou, S., Portais, J.C., Akoka, S. and Giraudeau, P. (2013). Fast spatially-encoded 3D NMR strategies for $^{13}$C-based metabolic flux analysis. *Anal. Chem.* 85: 9751–9757.

**88** Le Guennec, A., Tea, I., Antheaume, I., Martineau, E., Charrier, B., Pathan, M., Akoka, S. and Giraudeau, P. (2012). Fast determination of absolute metabolite concentrations by spatially-encoded 2D NMR: application to breast cancer cell extracts. *Anal. Chem.* 84: 10831–10837.

**89** Jézéquel, T., Deborde, C., Maucourt, M., Zhendre, V., Moing, A. and Giraudeau, P. (2015). Absolute quantification of metabolites in tomato fruit extracts by fast 2D NMR, *Metabolomics.* 11: 1231–1242.

**90** Giraudeau, P., Montag, T., Charrier, B. and Thiele, C.M. (2012) Fast access to residual dipolar couplings by single-scan 2D NMR in oriented media. *Magn. Reson. Chem.* 50: 53–S57.

**91** Lesot, P., Berdagué, P. and Giraudeau, P. (2016). Detection of quadrupolar nuclei by ultrafast 2D NMR: exploring the case of deuterated analytes aligned in chiral oriented solvents. *Chem. Commun.* 52: 2122–2125.

**92** Gouilleux, B. Meddour, A. Lesot, P. (2020). $^2$H QUOSY 2D-NMR experiments in weakly aligning systems: from the conventional to the UF approach. *ChemPhysChem.* 21: 1548–1563.

**93** Andre, M., Piotto, M., Caldarelli, S. and Dumez, J.-N. (2015). Ultrafast high-resolution magic-angle-spinning NMR spectroscopy. *Analyst.* 140: 3942–3946.

**94** Rouger, L., Yon, M., Sarou-Kanian, V., Fayon, F., Dumez, J.-N. and Giraudeau, P. (2017). Ultrafast acquisition of $^1$H-$^1$H dipolar correlation experiments in spinning elastomers. *J. Magn. Reson.* 277: 30–35.

**95** Chen, Z., Cai, S., Huang, Y. and Lin, Y. (2015). High-resolution NMR spectroscopy in inhomogeneous fields. *Prog. Nucl. Magn. Reson. Spectrosc.* 90: 1–31.

**96** Vathyam, S. Lee, S. and Warren, W.S. (2009). Homogeneous NMR spectra in inhomogeneous fields. *Science.* 272: 92–96.

**97** Pelupessy, P., Renella, E. and Bodenhausen, G. (2009). High-resolution NMR in magnetic fields with unknown spatiotemporal variations. *Science.* 324: 1693–1697.

**98** Lin, Y., Zhang, Z., Cai, S. and Chen, Z. (2011). High-resolution 2D J-resolved spectroscopy in inhomogeneous fields with two scans. *J. Am. Chem. Soc.* 133: 632–7635.

**99** Cai, H.H., Chen, H., Lin, Y.L., Feng, J.H., Cui, X.H. and Chen, Z. (2015). Feasibility of ultrafast intermolecular single-quantum coherence spectroscopy in analysis of viscous-liquid foods. *Food Anal. Methods.* 8: 1682–1690.

**100** Lin, Y., Zhang, Z., Cai, S. and Chen, Z. (2015). Spatially-encoded intermolecular single-quantum coherence method for high-resolution NMR spectra in inhomogeneous fields. *Chem. Phys. Lett.* 634: 11–15.

**101** Wang, K.Y., Chen, H., Zhang, Z.Y., Huang, Y.Q. and Chen, Z. (2016). Two-dimensional J-resolved NMR analyses of fish and its products via spatially encoded intermolecular double-quantum coherences. *Food Anal. Methods.* 9: 1502–1511.

**102** Zhang, Z., Chen, H., Wu, C., Wu, R., Cai, S. and Chen, Z. (2013). Spatially encoded ultrafast high-resolution 2D homonuclear correlation spectroscopy in inhomogeneous fields. *J. Magn. Reson.* 227: 39–45.

**103** Zhang, Z., Huang, Y., Smith, P.E., Wang, K., Cai, S. and Chen, Z. (2014). High-resolution heteronuclear multi-dimensional NMR spectroscopy in magnetic fields with unknown spatial variations. *J. Magn. Reson.* 242: 49–56.

# 11

## Multi-dimensional Methods in Biological NMR

*Tobias Schneider[1,2] and Michael Kovermann[1,2]*

[1] *Department of Chemistry, Universitätsstrasse 10, Universität Konstanz, DE-78457 Konstanz*
[2] *Graduate School Chemical Biology KoRS-CB, Universitätsstrasse 10, Universität Konstanz, DE-78457 Konstanz*

## 11.1 Introduction

High-resolution NMR spectroscopy represents a key method to probe biological molecules. This is due to its ability to acquire experimental data on structural, dynamical, and functional characteristics of the molecule under study at atomic resolution. In this respect, NMR active nuclei of high interest are represented by proton ($^{1}$H), carbon ($^{13}$C), nitrogen ($^{15}$N), fluorine ($^{19}$F), and phosphorus ($^{31}$P), which all are spin $^1/_2$ nuclei. Whereas natural abundance of $^{1}$H is about 99.99% and of $^{31}$P is 100%, efficient and targeted labeling of desired molecules using $^{13}$C, $^{15}$N, and $^{19}$F nuclei is achieved by established protocols [1–11]. As a general rule, the first important characterization of the biomolecule of interest should be done by acquiring a highly resolved 1D NMR spectrum reporting on $^{1}$H and $^{19}$F (holding for fluorinated peptides, fluorinated proteins) or $^{1}$H and $^{31}$P (holding for nucleic acids) resonance signals. For this purpose, a magnetic field strength of about $B_0$ = 14.1 T is sufficient in most instances. Advantageously, the gyromagnetic ratios of $^{1}$H, $^{19}$F, and $^{31}$P nuclei permit to employ a sample concentration in the range of 10 to 100 $\mu$M (suitable molar mass provided). If quality of spectral data can be rated as good enough, more time-consuming multi-dimensional NMR spectroscopic experiments can be subsequently conducted increasing spectral resolution. Here, the most commonly used multi-dimensional NMR spectroscopic experiment, which is applied on peptides and proteins is the 2D heteronuclear $^{1}$H-$^{15}$N HSQC spectrum. This spectrum reports on backbone correlations between proton and nitrogen (one single cross-peak for every proteinogenic amino acid, except proline, can be expected) as well as on side chain proton-nitrogen correlations of asparagine, glutamine, and arginine. The power of this experiment is based on its sensitivity and the relatively narrow spectral width within the heteronuclear dimension, which makes an acquisition on, e.g. $^{15}$N-labeled ubiquitin (possessing a concentration of about 10 $\mu$M only) in about 10 minutes possible (applying a cryo probe at a high field magnet of $B_0$ = 18.8 T). Another popular two-dimensional heteronuclear correlation spectrum is represented by an $^{1}$H-$^{13}$C HSQC especially when it comes to large molecular dimensions, which goes along with significant slower overall molecular tumbling. Here, predominantly the analysis of cross-peaks comprising methyl groups enables the analysis even of large molecular machines.

The aim of this chapter "Multi-dimensional Methods in Biological NMR" lies in presenting the large potential of high-resolution NMR spectroscopic methods employing multiple dimensions, which have been recently shown on a variety of biological molecules. The final part of this chapter is devoted to applications, which can be classified

*Two-Dimensional (2D) NMR Methods*, First Edition. Edited by K. Ivanov, P.K. Madhu and G. Rajalakshmi.

as non-standard. We acknowledge that here has been a wealth of studies published in the recent years significantly advancing the field of NMR spectroscopy. Saying that, we apologize that we cannot mention all of these studies.

## 11.2 Experimental Approaches

### 11.2.1 NMR Spectroscopic Information on Structural Features

The numerical value of the chemical shift of a resonance signal arising in an NMR spectrum gives precise information regarding the chemical environment of the nucleus, which has been probed. This can then be used to characterize structural features of the biomolecule under study. Thus, the acquisition of an ordinary 1D proton NMR spectrum of a protein which can be done in seconds already reports on its overall structural characteristics (its "fold"). Chemical shifts arising from aliphatic protons which point into upfield-direction possessing $\omega^{1H} < 0$ ppm and a ratio in signal height of about 1:5 relating upfield-shifted and methyl-group protons, which sense non-secondary structural motifs indicate a properly folded protein in most cases (Figure. 11.1a). Contrary, absence of resonance signals in the upfield range of the proton dimension goes along with non-folded conformations of the protein under study (Figure. 11.1b). It is highly recommended to probe structural features of a protein in the beginning of a project applying such simple experimental setup on non-labeled samples before time-consuming tailor-made labeling schemes for subsequent multi-dimensional NMR experiments are applied. The same holds true for the structural characterization of biomolecules which carry $^{19}$F or $^{31}$P nuclei. Simple, fast recorded 1D spectral information should be obtained before sophisticated experimental approaches are conducted.

Multi-dimensional NMR spectroscopic approaches are of up-to-date relevance when the three-dimensional structural determination of a biomolecule is intended. Several NMR parameters like the NOE [12–14], PCS [15–18], RDC [19–21], J-coupling [13, 22, 23], PRE [24–26], and others can then be used to obtain the 3D structure utilizing suitable software packages like, e.g. X-Plor NIH [27, 28], CNS [29, 30], Dyana [31] or Aria [32]. Notably, a broad repertoire of adapted multi-dimensional NMR experiments targeting individual research objectives is available to obtain the desired NMR parameter: IPAP approaches to obtain information on RDC [33], NOESY-edited HSQC

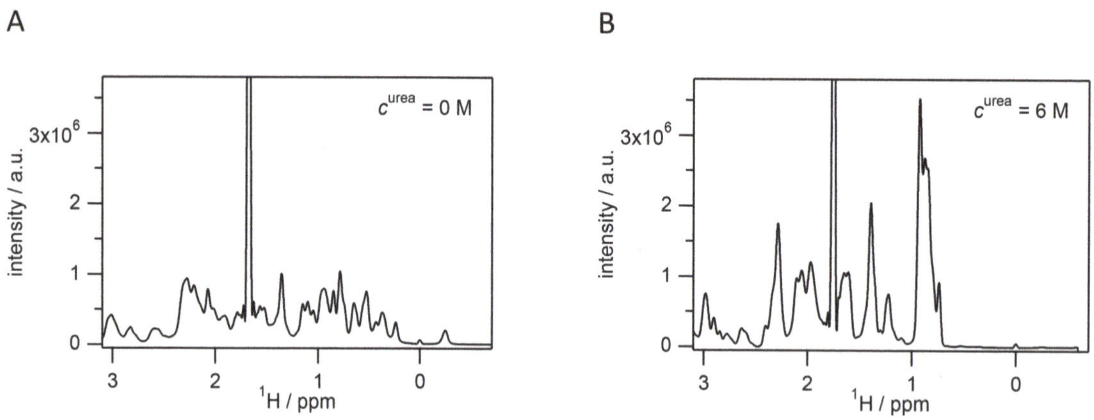


**Figure 11.1** Pattern of one-dimensional $^1$H NMR spectra that have been acquired for a protein at different experimental conditions. A native environment generates values of chemical shifts reporting on a properly folded protein that are significantly below 0.5 ppm and that possess a reasonable signal height (a). A change to an unfolded environment leads to a disappearance of these chemical shifts, clearly indicating an unfolded protein ensemble (b). The NMR spectra shown here have been acquired for cold shock protein B from *Bacillus subtilis* comprising 67 amino acids at $T = 298$ K and $B_0 = 14.1$ T in 20 mM sodium cacodylate, pH 7.0. The protein concentration was set to 30 $\mu$M.

methodology [34, 35] or a read-out of HSQC spectra acquired in absence and presence of paramagnetic agents to name only a few.

### 11.2.2 Spectroscopic Information on Dynamical Features

The large power high-resolution NMR spectroscopy inherently owns is prominently expressed when it comes to applications tackling dynamic features within the (bio)molecule under study. Thereby, the time scale of adapted NMR experiments is extremely broad: ranging from fast picosecond-to-nanosecond motions up to dynamic processes, which take place on the slow seconds-to-minutes-to-hours time scale. There exist numerous applications of multi-dimensional NMR methods on biomolecules to probe such dynamic events. The following part of this chapter gives an overview of this wide range of experimental possibilities.

(1) Spin-relaxation methods focusing predominantly on $^{13}C$ or $^{15}N$ nuclei report on motional events taking place on the fast picosecond-to-nanosecond time scale. It has been very popular to link relaxation parameters as longitudinal relaxation rate constant, $R_1$, transversal relaxation rate constant, $R_2$, or the heteronuclear NOE, $\{^1H\}$-$^{13}C$ or $\{^1H\}$-$^{15}N$, with motional parameters: (i) the amplitude of motion as expressed by an order parameter, $S^2$; (ii) the rate of motion as expressed by an internal as well as an overall correlation time, $\tau_e$ and $\tau_c$; and (iii) a term considering chemical exchange, $R_{ex}$. Once relaxation data of the biomolecule of interest have been acquired and analyzed at a residue-by-residue basis, software packages as, e.g. MODELFREE [36, 37] can be used to obtain motional parameters of interest. This experimental strategy enables characterizing fast time scale–motions of the biomolecule under study in a sequence-dependent manner [38–41].

(2) The determination of the dispersion of the transversal relaxation rate constant, $R_{2,eff}$, of either $^1H$ , $^{13}C$, or $^{15}N$ nuclei enables to obtain dynamic information on the micro-to-millisecond time scale of the biomolecule under study [42–45]. Predominantly, 2D $^1H$-$^{13}C$ or $^1H$-$^{15}N$ correlation spectra are acquired and the signal height of cross-peaks is then used to determine the particular relaxation rate constant dependent on the field strength, which has been applied (typically given per Hz). Finally, the regression of the general solution of a two-site exchange process gives quantitative information on its structural (difference in chemical shifts, $\Delta\omega$), dynamic (transversal relaxation rate constant in absence of exchange, $R^A_{2,0}$ and $R^B_{2,0}$), thermodynamic (populations, $p^A$ and $p^B$), and kinetic features (kinetic rate constant of exchange, $k_{ex}$) [46–49]. Another possibility for probing dynamic exchange processes taking place on the millisecond time scale is the application of chemical exchange saturation transfer (CEST) methodology. Here, a weak $B_1$ field is applied along x- or y-axis for a certain period of time, $T_{ex}$, allowing occurrence of chemical exchange between high and low populated states of the molecule under study within $T_{ex}$. Varying the offset frequency of $B_1$ typically in a range between 100 and 135 ppm (for $^{15}N$ nuclei) leads to an intensity profile reporting on the signal height of individual cross-peaks observed in heteronuclear 2D NMR spectra. This experimental strategy enables the determination of quantities reporting on structural, thermodynamic, and kinetic features of the underlying exchange process [50–53].

(3) High-resolution NMR spectroscopy also offers a variety of suitable methods for the targeted observation of dynamic processes comprising the slow millisecond-to-second-to-minute-to-hour time scale. For example, experiments monitoring the exchange of protons between the (bio)molecule under study and the solvent is one possibility. Thus, approaches like MEXICO (measurement of exchange of isotopically labeled compounds) [54] or CLEANEX (clean chemical exchange spectroscopy) [55, 56] enable to obtain valuable insights into, e.g. the thermodynamic stability of two-domain prolyl-peptidyl *cis/trans* isomerase SlyD [57] at a residue-by-residue level. Also, classical hydrogen-to-deuterium exchange depicts an experimental possibility to follow the replacement of a proton comprising the molecule under study and a deuteron provided by the solvent in a time-dependent manner. Here, ordinary 1D or 2D NMR spectra are acquired to follow this exchange process in real time. Analyzing the integral (in the case of 1D NMR spectroscopy) or the signal height of crosspeaks (in the case of 2D NMR spectroscopy) allows then to obtain the time constant quantifying the exchange process. Finally,

this quantity reports on the thermodynamic stability of the biomolecule under study – if a suitable exchange regime is present [57–59]. It should be also noted that kinetic processes of biomolecules that take place on the second-to-minute-to-hour time scale can be followed by the serial acquisition of NMR spectra. Here, the number of dimensions is adjusted to the kinetic process that is observed. Analyzing the quantity of integrals, signal heights, line width, chemical shifts, or also relaxation parameters gives then access to the kinetic rate constant(s) and contains, in parallel, valuable information regarding the amplitude of the underlying kinetic process [60–64].

### 11.2.3 NMR Spectroscopic Information Obtained from Interaction Studies

Advantageously, the toolbox of an NMR spectroscopist is not limited to the characterization of biomolecules that are present in the free form. Specific isotopic labeling schemes enable selectively obtaining spectroscopic information of a (bio)molecule that is present in a complex environment. Thus, quantifying the interaction between, e.g. protein and oligonucleotides, becomes feasible not only applying in vitro conditions but also in presence of a high density of macromolecular agents [65] as even when experimental work is done in cell lysate [66]. The large capacity of multi-dimensional NMR spectroscopy unfolds greatly when such interaction studies are carefully analyzed. Identifying the interface of interaction depicts thereby one major objective. The experiments are generally designed such that the target biomolecule is isotopically labeled (e.g. using sources comprising $^{13}$C, $^{15}$N, or $^{19}$F nuclei) whereas the ligand molecule is non-labeled. Adding non-labeled ligand molecules to the isotopically labeled molecule under study enables then to exclusively monitor desired resonance signals by applying suitable NMR pulse sequences. If an intermolecular interaction takes place, changes in chemical shift, in line width, in signal height, or a combination thereof will be monitored. It should be mentioned that interaction processes can take place on different NMR time scales referred to as slow, intermediate, and fast exchange [67]. As a rule of thumb, tight binding processes take place on the slow time scale whereas weak binding is mainly represented by fast exchange. Finally, the assignment of resonance signals enables the identification of the site of the intermolecular interaction. Moreover, high-resolution NMR spectroscopy can provide quantitative insights into intermolecular interactions. Thus, following the change in chemical shift during an NMR titration experiment enables quantifying the dissociation constant, $K_D$, as well as the stoichiometry, $n$, of the interaction. The endpoint of an NMR titration experiment should be designed such that an about three times molar excess of ligand molecules regarding the target molecule exist.

### 11.2.4 Quench Flow Methodology in Combination with NMR – Hydrogen-to-deuterium Exchange

Following the exchange from a hydrogen to a deuteron offers the possibility to obtain thermodynamic information of the biomolecule under study at atomic resolution as outlined under the third point in Section 11.2.2 ("Spectroscopic Information on Dynamical Features"). Thus, valuable information can be received when monitoring this time-dependent process by using multi-dimensional NMR spectroscopic methods. Beside this "classical" H/D exchange [68–70], a so-called quench flow technique can be applied to obtain structural information of a biomolecule. Pioneering work has been performed in this regard by unraveling intermediate structures residing on the folding path of cytochrome C [71] and ribonuclease A [72]. Both studies have monitored homonuclear $^1$H-$^1$H COSY spectra enabling to obtain site-specific spectroscopic information on folding intermediates on a time scale between 4 ms and 60 s. Later, folding processes of several other biomolecules have been determined including heteronuclei into the experimental setup as exemplarily shown by [73, 74]. Also, the complexity of molecules that are studied has been expanded to molecules comprising multi domains as, e.g. represented by apo myoglobin [75]. The strength of the combination between a hydrogen-to-deuterium exchange NMR setup with quench flow lies in the possibility to obtain structural information of kinetic intermediates that are potentially populated on the millisecond time scale when the refolding process of a protein takes place. The targeted application of multi-dimensional NMR spectroscopic methods allows then to characterize structural features of these transient species at atomic resolution.

### 11.2.5 Expanding Multi-dimensional NMR Spectroscopy from in vitro to in vivo Applications

As life scientists are in particular interested in the functional characterization of biomolecules, the focus of high-resolution NMR spectroscopy is gradually expanding from classical in vitro to in vivo applications meaning that the biomolecule of interest is studied directly within living cells. This research domain is often termed as in-cell NMR spectroscopy [76].

It is obvious that the ambient conditions of a biomolecule dramatically differ when it is placed within a cell. Most importantly, the concentration of the surrounding molecules increases up to 200 g/l in prokaryotic and even 400 g/l in eukaryotic organisms [77, 78]. This goes often along with consequences for the overall thermodynamic stability of the biomolecule and its interaction affinity toward ligands. The underlying molecular mechanisms have been extensively investigated by biophysical studies in the presence of (macro)molecular crowding agents or cell lysates mimicking distinct aspects of the cytosolic environment [65, 66, 79–81]. With in-cell NMR spectroscopy the next step forward is ventured fully accounting for the complexity of native cellular conditions. Thereby, isotopic labeling of NMR active heteronuclei is mandatory to obtain highly resolved structural information from multi-dimensional NMR experiments. For a protein, for instance, two different strategies are principally feasible to reach this goal.

The first one relies on overexpression of the desired protein from isotopically enriched medium directly in the host organism, which is usually *Escherichia coli* [76] but has also been demonstrated on yeast [82], insect [83], and mammalian cell lines [84]. This approach generally requires high expression levels of the protein to be discriminated from the background [85]. Embarking on the second strategy, the protein of interest is recombinantly expressed and purified in presence of isotopic sources first and is then inserted into the cell prior to NMR measurement [85, 86]. This can be done either by microinjection into *Xenopus laevis* oocytes [87, 88] or by temporarily increasing membrane permeability of mammalian cells through electroporation or pore-forming toxins [89]. More ingeniously, it was shown that a protein can be shuttled over the membrane into the cytosol of human cells by fusing it to the cell-penetrating peptide (CPP) from the Human-immunodeficiency virus (HIV) Tat protein. After successful translocation, together with CPP, the tag can then optionally be cleaved off [90]. Consequently, from a technical point of view a broad variety of biological systems is already accessible by in-cell NMR spectroscopy.

Key knowledge about the thermodynamic stability and the folding/unfolding kinetics of proteins obtained from numerous in vitro experiments can thus be evaluated under near native conditions. This means that the generality of fundamental concepts like the excluded volume effect need to be reappraised under the simultaneous influence of, for example, extreme viscosity, posttranslational modification, accumulation of proteins in distinct subcellular compartments, and permanent interactions with a variety of specific and unspecific binding partners spanning a broad range of affinities [91, 92]. Moreover, in-cell NMR spectroscopy was proven to be subject of protein structure determination. Despite the short lifetime of cells within the NMR tube, even 3D NMR spectra for resonance assignment and NOE collection could be acquired by applying a nonlinear sampling scheme and replacing the samples every 3–4 h [93]. In this way a high-resolution structure of the protein TTHA1718 form *Thermus thermophilus* HB8 could be solved within living *E. coli* cells [94].

### 11.2.6 Multi-Dimensional NMR Spectroscopy as an Integrated Approach in Structural Biology

Today's structural biology research provides a broad repertoire of techniques capable to obtain several kinds of conformational restraints for characterizing biomolecules. However, different methodologies rely on different physical phenomena and often require to examine a biomolecule under specific experimental conditions. The full potential of all these techniques including multi-dimensional NMR spectroscopy is thus best clear when combined in a synergistic manner. This is, in particular, true for molecular structure determination, when data obtained by a single technique do not lead to a converging model due to intrinsic properties of the biomolecule itself and experimental limitations in that respect. One example where this problem was defeated by an integrated approach is presented below in Section 11.3.6. Therein, the structure of the half-megadalton enzyme complex TET2 was solved by intertwining cryo-electron microscopy with solution and solid-state high-resolution NMR spectroscopy [95].

Another prominent example of the fruitful combination between NMR spectroscopy and adjacent experimental as theoretical approaches that are applied in structural biology is represented by the three-domain enzyme adenylate kinase. Thus, complementing high-resolution NMR spectroscopic data with X-ray crystallography, single-molecule fluorescence resonance energy transfer (smFRET) and molecular dynamics (MD) simulations methodologies enabled to obtain valuable insights into intrinsic protein dynamics which led to quantitatively understand the process of enzymatic catalysis [96]. Adenylate kinase has also been used to obtain precise insights into both conformational selection [97] as well as induced fit mechanism [98] that describe two fundamental scenarios of ligand binding to a target molecule. Thus, the well-designed application of isothermal titration calorimetry (ITC), stopped flow fluorescence spectroscopy and X-ray crystallography in combination with NMR spectroscopic approaches (e.g. probing RDCs, MEXICO, TALOS, line shape analysis, spin relaxation) permitted the investigation of the energy landscape of substrate binding for adenylate kinase at atomic resolution [98]. It should be also noted that NMR spectroscopic-derived RDC values have been successfully used to refine an ensemble of 46 ubiquitin structures obtained from X-ray crystallography that have been relaxed by MD simulations to get information regarding recognition dynamics of this protein to eventually form complexes [21]. Noteworthy, the variety of different methods provided by high-resolution NMR spectroscopy play also a key role to better understand the molecular process of aggregation and fibrillation, e.g. of alpha-synuclein [99, 100] that is related to Parkinson's disease or A$\beta$ [101] that is related to Alzheimer's disease to name just two prominent examples. The synergy between experimental methods provided predominantly by NMR spectroscopy as well as electron microscopy has significantly supported to promote the understanding of aggregation driven processes in an integrated fashion.

## 11.3 Case Studies

This chapter aims to provide an overview into the broad area of different applications high-resolution multi-dimensional NMR spectroscopy can do. This comprises, e.g. using samples in natural abundance, heteronuclear correlation spectra, and paramagnetic relaxation enhancements, PREs, that arise from the solvent.

### 11.3.1 Determining Thermodynamic Stability of Biomolecules at Atomic Resolution

As discussed at the beginning of this chapter, NMR spectroscopy is a convenient way to monitor the overall fold of a protein structure. This is not only beneficial for quality control of the sample integrity, but also allows sophisticated studies on protein folding and unfolding events reporting on the thermodynamic stability. Owing to multi-dimensional NMR spectroscopy this can be done with atomistic detail. The main observable for determining the protein folding state is the chemical shift, as it is a sensitive probe for the chemical environment of a certain spin. In a properly folded protein this environment is quite individual, as several diverse residues are in close proximity, resulting in a strong signal dispersion with minor degree of signal overlap. This is in particular true for spins that are part of secondary structural elements such as $\beta$ sheets where through-space distances to other residues are relatively short. In contrast, for a protein sensing unfolding conditions this is not the case and the chemical environment of, for example, an amide proton spin is essentially defined by the side chain of its own residue and the adjacent amino acids [102].

Which impact the global conformation can have on the chemical-shift value was illuminated in a biophysical study employing a peptide called Trpzip2 [103]. Trpzip2 consists of 12 amino acids including four tryptophan residues, which induce the formation of a hairpin motif through side chain interactions. In this study, it acts as a template to elucidate the effects of disease-causing polyglutamine (polyQ) elements, which were introduced systematically into the sequence. The low molecular weight as well as high peptide concentration of up to 3 mM enable the acquisition of 2D $^{1}H/^{15}N$ HSQC spectra at natural abundance of the $^{15}N$ isotope without artificial enrichment. Due to the hairpin fold, which is adapted by Trpzip2 the dispersion of backbone amide resonances in the spectrum is quite broad spanning proton chemical shifts in a range of 2.8 ppm (Figure. 11.2). This is not significantly altered upon the insertion of one glutamine residue into each cross strand of the hairpin ($\Delta\delta$ = 2.9 ppm). However, when using longer polyQ elements containing six or even ten glutamine residues (three or five in a row per strand), the range of proton chemical shifts decreases down to 1.4 ppm. This is a strong indicator that the hairpin conformation of Trpzip2 is disrupted when longer polyQ elements are inserted and a disordered structure is predominantly adopted instead [103].

Another way to enforce protein unfolding is the excessive addition of denaturing agents such as urea or guanidinium salts. This was applied to the cold shock protein B from *Bacillus subtilis* to determine its overall thermodynamic stability in absence and presence of (macro)molecular crowding agents [81]. As the cold shock protein B belongs to the class of β barrel proteins, an excellent signal dispersion is apparent in the $^{1}H/^{15}N$ TROSY-HSQC spectrum of the native polypeptide (Figure. 11.3a). However, in consequence of the stepwise addition of urea crosspeaks originating from the folded ensemble gradually disappear, whereas crosspeaks originating from the unfolded ensemble arise to the same extent within a narrow spectral region around 8.3 ppm regarding proton chemical shift (Figure. 11.3b–11.3g). Signal heights of resonances originating from both states were taken into account to determine the transition midpoint which is reached at a urea concentration of 3 M. Focusing on the spectrum at that concentration one can visually anticipate that both the folded state and the unfolded ensemble are

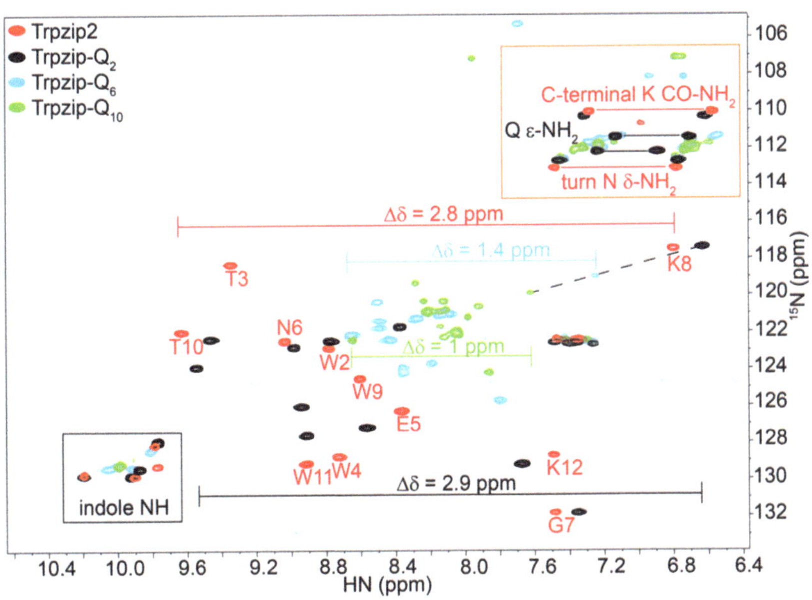


**Figure 11.2** Superimposition of 2D $^{1}H$-$^{15}N$ HSQC NMR spectra acquired for Trpzip2 at natural abundance without insertion of polyQ motif (colored in red) as well as in presence of two glutamine (colored in red), six glutamine (colored in blue), and ten glutamine residues that have been inserted (colored in green), respectively [103]. The corresponding signal dispersion in the proton dimension is indicated by $\Delta\delta$.

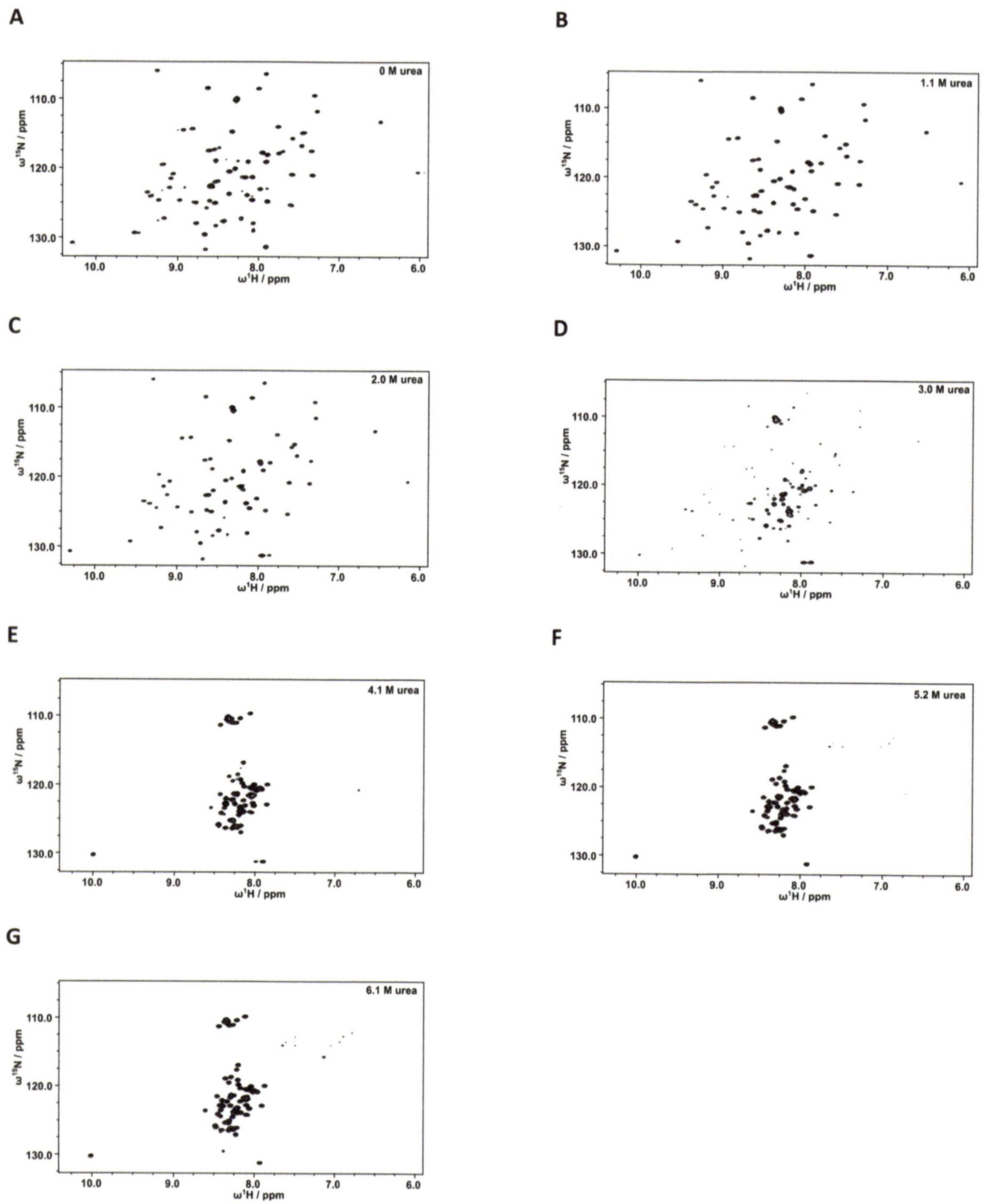


**Figure 11.3** Progression of chemical denaturation of the cold shock protein B from *Bacillus subtilis* by using increasing amounts of urea [81]. Two-dimensional $^{1}H$-$^{15}N$ TROSY-HSQC NMR spectra acquired under dilute conditions (a) and in presence of $c = 1.1$ M (b), $c = 2.0$ M (c), $c = 3.0$ M (d), $c = 4.1$ M (e), $c = 5.2$ M (f), and $c = 6.1$ M urea (g), respectively.

populated almost equally at that point (Figure. 11.3d). This highlights that multi-dimensional NMR spectroscopy is an exquisite tool for thermodynamic studies on protein folding and unfolding, as it allows to distinguish distinct transitional states, which are populated in thermal equilibrium. This study could further show that the transition between folded state and the unfolded ensemble is significantly shifted to higher levels, if (macro)molecular crowding agents like PEG1 (polyethylene glycol possessing molar mass of 1 kDa) or dextran 20 (possessing molar mass of 20 kDa) are supplemented. Remarkably, the effect seems to be irrespective of intrinsic properties of the crowding agents such as polarity and size. Owing to the systematic approach applying crowding agents of different types and monitoring resonances originating from main chain as well as from side chain groups warrant the conclusion that the thermodynamic stabilization is an entropically driven event caused by an excluded volume effect [81].

### 11.3.2 Exotic Heteronuclear NMR Spectroscopy Correlating $^{31}$P with $^{13}$C

The manipulation of the transfer of coherence in multi-dimensional NMR spectroscopy opens the way for correlating spins of virtually every type of nuclei with each other. It "only" requires that both spins are NMR active and can be excited and detected by the NMR spectrometer, as it is equipped with corresponding radiofrequency channels. Numerous pulse sequences have been developed to recover such correlations by applying different spin coupling mechanisms [104]. This is typically employed by experiments intended for resonance assignment such as the COSY, TOCSY, and NOESY experiments, for example, but is also the basis for the HMQC and HSQC experiments that are prominently used in the context of protein NMR spectroscopy [105, 106]. Therein, chemical shifts are correlated between protons and either one-bound coupled $^{13}$C or $^{15}$N nuclei to provide an informative fingerprint spectrum of the protein, which can be exploited for a variety of structural and dynamic studies.

When focusing on biomolecules other than proteins which are relevant for structural biologists such as ribonucleic acids (RNA), classical heteronuclear $^{1}$H-$^{13}$C and $^{1}$H-$^{15}$N correlations turned out to be less beneficial. RNA molecules generally suffer from relatively low proton density within the nucleotide building blocks, poor signal dispersion due to the structural similarity between the nucleotides, and unfavorable relaxation and exchange properties of the protons [107, 108]. For that reason, more exotic heteronuclear correlations become eligible. The 2D (H)CPC experiment for instance correlates $^{13}$C and $^{31}$P spins along the phosphodiester bond of the ribose-phosphate backbone within RNA molecules (Figure. 11.4a) [108]. The coherence is transferred first from position H4' to C4' on the ribose scaffold by an initial INEPT block and is further transferred to the $^{31}$P spins of the phosphate groups belonging to the own and the succeeding nucleotide via $^{3}J$(C4',$P_i$) and $^{3}J$(C4',$P_{i+1}$) coupling, respectively. Acquisition is then accomplished by $^{13}$C-direct detection after coherence back transfer to the original C4' [109]. In this way, sequential assignment of consecutive nucleotides can be achieved from the 5'- to the 3'-end of the RNA molecule by following the connections in the resulting (H)CPC spectrum (Figure. 11.4b) [108, 109]. A clear advantage of this experiment is that the $^{31}$P nucleus is a natural component of RNA molecules. However, this setup can also be transferred to other non-native types of nuclei such as $^{19}$F, if they are introduced artificially with coupling to $^{13}$C spins [108, 110].

### 11.3.3 Following Biomolecular Dynamics by Homonuclear and Heteronuclear ZZ Exchange

As discussed, in Section 11.2.2 ("Spectroscopic Information on Dynamical Features") NMR spectroscopy is ideally suited to investigate dynamic processes on a broad range of time scales. In this context, the power of multi-dimensional experiments is in particular exploited by the ZZ exchange or the so-called EXSY experiment, probing conformational changes that occur in the slow millisecond-to-second time window. Under these circumstances, each conformational state gives rise to a single peak in the NMR spectrum provided that the states are populated to somewhat similar extent. The exchange event may take place during a mixing time where the coherence

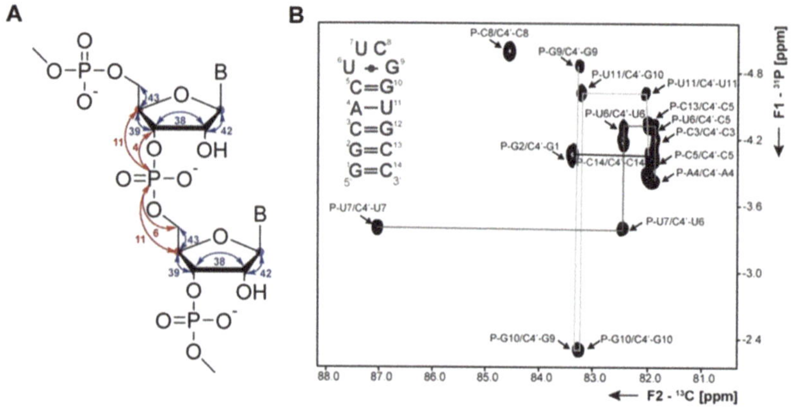

Figure 11.4 Two-dimensional (H)CPC NMR experiment targeting on the sequential assignment of RNA oligonucleotides [108]. (a) Schematic representation of the coherence transfer executed by the (H)CPC experiment along the ribose-phosphate backbone. The $^3J(C4',P_i)$ and $^3J(C4',P_{i+1})$ couplings used for the sequential walk are highlighted in red and indicated by a coupling constant of 11 Hz. Other $J$(C,C) and $J$(C,P) couplings, which are also used for coherence transfer in related experiments, are additionally shown. The nuclei that are addressed for $^{13}$C-direct detection are indicated by circles colored in red. "B" acts as a placeholder for arbitrary nucleobases. (b) Two-dimensional (H)CPC NMR spectrum of the oligonucleotide motif depicted in the upper left corner. The sequential assignment can be followed by the solid lines.

information is preserved in the form of z-magnetization. Within the detection period, cross-peaks appear in the NMR spectrum that connect the diagonal auto-peaks with each other that represent the original states due to conformational interconversion. Exchange kinetics can then be determined by fitting the signal heights of both auto- and cross peak to an appropriate model [111, 112].

In a homonuclear variant of the experiment, the dynamic process underlying the gating mechanism of the 20S core particle proteasome from *Archaea* was investigated [113]. The 20S core particle proteasome consists of four staggered rings with seven homologous subunits, each and facilitates efficient protein degradation to maintain cellular proteostasis [114]. Thereby, the N-terminal residues of the seven $\alpha$-subunits, which form the outer ring act as a gatekeeper to the proteasome by moving between an inside and outside position at the entry [113]. The study is focused on methyl side chain groups of methionine residues located at the N-terminus of the $\alpha$-subunits. These side chain groups can be monitored exclusively by preparing a perdeuterated sample of an outer ring construct with selective $^{13}CH_3$-methionine labeling [113, 115]. With this sample a 2D ZZ exchange experiment was performed selecting only the hydrogen spins of the methionine methyl groups. In consequence, a conformational equilibrium in slow chemical exchange has been revealed and quantified undergoing continuous interconversion between three distinct states ("A", "B" and "C") (Figure. 11.5). By means of PRE distance measurements "A" could be ascribed to the open gate state, whereas "B" and "C" are associated with closed gate states. Strikingly, the dynamics regulating gate opening and closing turned out to be directly coupled to the proteolysis activity of the proteasome [113].

In structural biology, the ZZ exchange experiment is often performed in a heteronuclear variant utilizing the $^1$H-$^{15}$N spin system of the protein backbone amide group [116]. This setup was used, for instance, to characterize an unknown conformation adapted by the polypeptide ubiquitin after phosphorylation of Ser65 in which the C-terminal tail is retracted into the ubiquitin core by two amino acids [117]. This retracted conformation is in slow chemical exchange with a so-called relaxed conformation resembling the structure of wild-type ubiquitin (Figure. 11.6). Both conformations are well populated at nearly neutral pH and interchange with a rate constant of about 2 s$^{-1}$ at 25 °C [117]. From a structural point of view an alternating leucine pattern (L67-X-L69-X-L71-X-L73) comprising the C-terminal tail and the preceding $\beta$5 strand is a prerequisite for the retraction of the

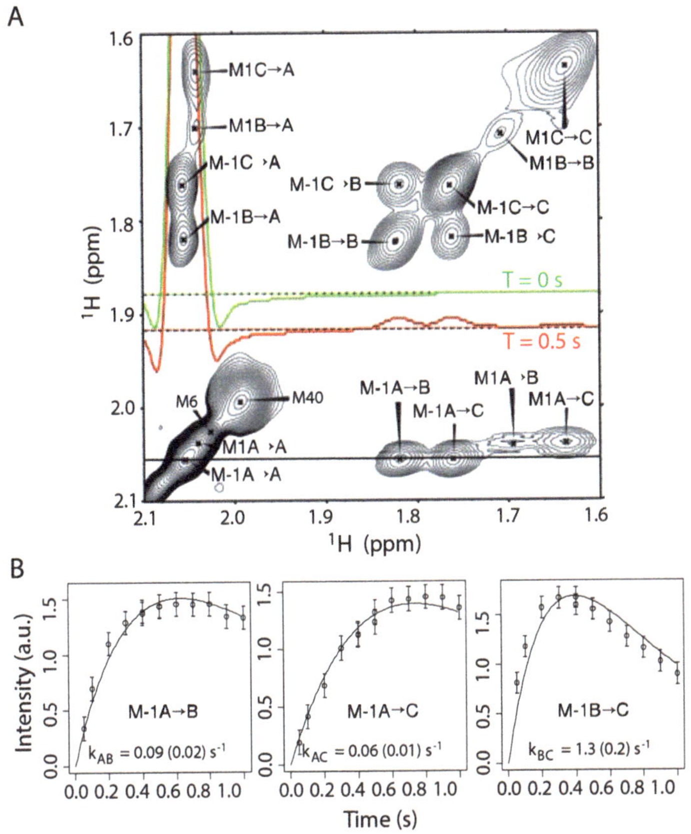


**Figure 11.5** Homonuclear 2D $^1$H-$^1$H ZZ exchange NMR spectroscopy applied on an outer ring construct of the archaeal 20S core particle proteasome [113]. (a) ZZ exchange NMR spectrum acquired using a mixing time of 0.5 ms. Only resonances originating from protons of methionine methyl groups are visible due to the specific labeling strategy used. M-1 denotes a methionine residue, which is artificially introduced at the N-terminus as an additional reporter. Auto-peaks on the diagonal are assigned to the different conformational states termed "A", "B" and "C" and are connected vertically and horizontally by corresponding exchange peaks. One-dimensional traces of the 2D experiment at definite $^1$H frequency indicated by the horizontal black line ($\delta = 2.05$ ppm) are also enclosed using a mixing time of 0 s (colored in green) and 0.5 s (colored in red), respectively. (b) Build-up curves used to quantify the rate constants of exchange between the different conformational states on basis of Met-1.

C-terminus, as it allows slippage of the leucine side chains along corresponding pockets in the core [52]. Interestingly, the same pattern together with Ser65 is also conserved in the primary sequence of the ubiquitin-like modifier NEDD8 (Figure. 11.7a) [118]. In analogy to ubiquitin, NEDD8 exhibits a conformational equilibrium between similar relaxed and retracted states after phosphorylation at Ser65 that interconverts with a rate constant of about 8 $s^{-1}$ [119]. From Figure. 11.7b it becomes apparent that for the ZZ exchange experiment, which has been applied in this study a multi-dimensional acquisition technique is mandatory to get both auto- and cross peaks properly resolved. This is in particular true as additional resonances originating from non-phosphorylated wild-type NEDD8 comprising the sample under study further complicate the NMR spectrum under observation. However, in such case the spectral connection provided by the ZZ exchange cross peaks contains additional information to successfully conduct the assignment of resonance signals. We note that for the quantification of the rate of interconversion only a subset of six residues was sufficient (Figure. 11.7c) [119].

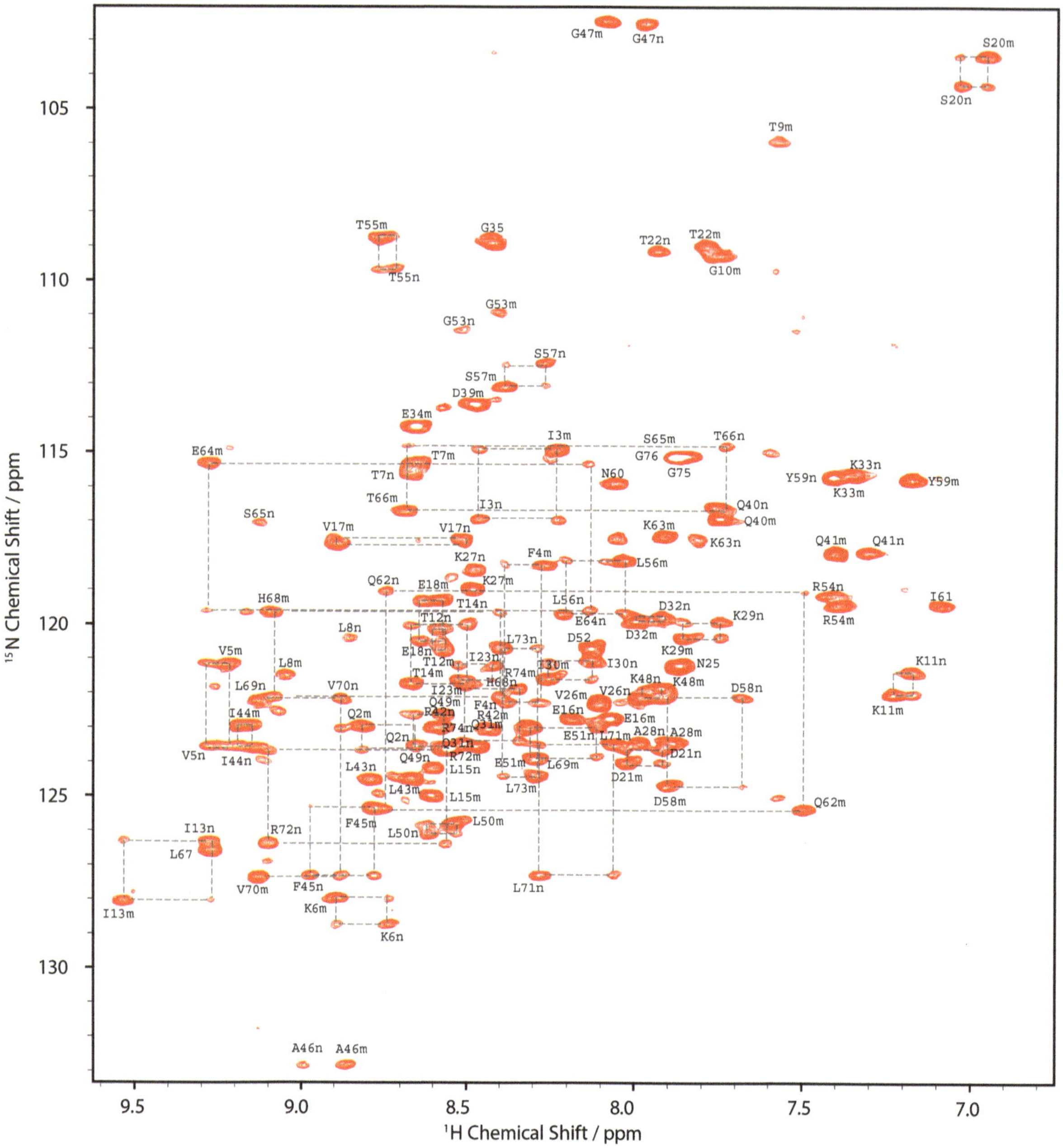


**Figure 11.6** Two-dimensional $^1$H-$^{15}$N HSQC NMR spectrum of the ZZ exchange experiment performed on ubiquitin, which is phosphorylated at position Ser65 using a mixing time of 92 ms [117]. The relaxed conformation is denoted with "m" and the retracted conformation with "n". Auto- and cross peaks of corresponding residues are connected by dashed lines.

### 11.3.4 Probing Structural Features by Solvent PREs

Long-range distance information are valuable restraints for NMR structure calculation. This kind of restraint is frequently provided by the paramagnetic relaxation enhancement (PRE) effect commonly exerted by spin labels, which are introduced into the biomolecule under investigation by using approaches provided by chemical biology [120, 121]. The principle is that a paramagnetic center, which is installed at a defined position on the molecule accelerates relaxation of spins, which are spatially near. This results in line-broadening of associated resonance cross peaks that correlates with the distance between a certain spin and the paramagnetic center [26]. Typical spin labels which, are employed for this purpose are nitroxide radicals and metal-chelating tags that bear an unpaired electron [122]. The PRE effect is known to decrease with the inverse sixth power to the distance between the unpaired electron and the observed spin [26]. Besides paramagnetic tags, which are covalently attached to the molecule, soluble paramagnetic probes have also been proven a useful tool for studying structure and dynamics in biomolecules and provide numerous applications in combination with multi-dimensional NMR spectroscopy [123–126]. Thereby, paramagnetic compounds such as TEMPOL or Gd(DTPA-BMA) are added as a cosolute to the biomolecule enabling to adjust the sensitivity of the probe by tuning its concentration. As spins presented on the surface are more densely surrounded by the paramagnetic cosolute than spins, which are buried in the core, they are expected to experience a stronger PRE [127].

The PRE approach may be used in an intelligent way for editing of multi-dimensional NMR spectra, which suffer from annoying peak overlap [124]. As a proof of principle, the paramagnetic cosolute Gd(DTPA-BMA) was added to a sample of ubiquitin from yeast, which originally consists of 76 amino acids but is modified with an additional N-terminal extension of 23 amino acids in this study. Since the cosolute diffuses around the protein a strong signal attenuation was primarily caused for residues, which are exposed to the solvent in the $^{1}$H-$^{15}$N HSQC spectrum. This is applicable especially for residues that are located at the unstructured N-terminal extension or the flexible C-terminal tail but not for core residues. The result is a simplified NMR spectrum with a reduced number of resonances, which is related to unexposed residues only (Figures 11.8a and 11.8b). Strikingly, subtraction of the spectrum with Gd(DTPA-BMA) from the spectrum without Gd(DTPA-BMA) gives the opposite result showing only resonance signals originating from solvent-exposed residues (Figure. 11.8c) [124]. In consequence, the initial spectral complexity could be diminished according to the criterion of solvent accessibility.

It is worth, to mention that this approach is not only suitable to resolve crowded spectral regions but at the same time provides valuable information about the location of a certain residue within the protein structure. This is emphasized in a study on the maltose-binding protein [123, 127]. Herein, the solvent PRE is determined in an explicitly quantitative manner allowing an implementation of structure calculation protocols. Thereby, $^{1}$H longitudinal relaxation rate constants ($R_1$) were obtained for backbone amide protons as well as for carbon-bound protons of the maltose-binding protein at different concentrations of Gd(DTPA-BMA). The exact solvent PRE value is given by the slope of a linear fitting curve (Figure. 11.9a) and is significantly higher for protons of residues that reside on the protein surface than for residues that are trapped in the interior (Figure. 11.9b) [123, 127]. The solvent PRE is thus a convenient and precise option to investigate solvent accessibility without the need of further modifications on the biomolecule itself.

In the case of multi-domain proteins, large-scale interdomain motions can directly affect solvent accessibility. They are characterized by a heterogeneous conformational ensemble comprising different arrangements where distinct sites on the protein surface of one domain can either be covered by another domain or left vacant. Since both situations are sampled in solution, the solvent PRE contains time-averaged information about the interdomain dynamics [128]. This could be demonstrated for example for the calcium sensor protein calmodulin, which is known to undergo a conformational switch from an open to a closed conformation upon $Ca^{2+}$-ligand binding [129, 130]. Residue-specific solvent PRE values were back-calculated from the crystal structure of calmodulin in its unbound state representing the open conformation and were compared to experimentally obtained values (Figure. 11.10) [131]. Although a strong correlation was observed within the N-terminal and C-terminal domain (Figure. 11.10a), the α-helical linker region in between revealed large deviations (Figure. 11.10b). To compensate the discrepancies a two-state equilibrium was assumed taking into account two structures, which were selected from an ensemble generated by MD simulations. Importantly, the structures comprise an open and a closed conformation and particularly differ in the helix-character and the solvent exposure of the linker region. Indeed, the overall correlation coefficient could be significantly improved from 0.71 to the best of 0.87, when the relative populations were optimized to 55% and 45%, respectively, regarding the open and closed conformation (Figure. 11.10b) [131]. The static picture given by the crystal structure of $Ca^{2+}$-free calmodulin is thus not sufficient to account for the dynamic character of the protein even in the absence of ligand.

### 11.3.5 Discerning Protein Dynamics by Probing Fast Amide Proton Exchange

Amide proton-exchange experiments, and in particular the MEXICO approach, also provide an attractive opportunity for probing the solvent exposure of proteins by means of multi-dimensional NMR spectroscopy on a residue-by-residue basis. In this sense, the MEXICO experiment is complementary to the solvent PRE approach, although different observables are addressed. Whereas this is $^1$H spin relaxation, in case of the solvent PRE [131], the magnetization recovery after replacement of irradiated amide protons by unaffected water hydrogens is recognized by the MEXICO approach [54]. The valid time regime of the MEXICO approach is thus defined by the rate of exchange events taking place during a defined mixing time and is in the range of microseconds to milliseconds. It is noteworthy to mention that the time scale of amide proton exchange within proteins is much broader in general (up to days/weeks) meaning that under physiological conditions the MEXICO experiment is preferentially suited to monitor fast exchanging protons, which are commonly located in loop and turn regions or at the termini [132].

This feature is exemplified by the polypeptide ubiquitin, which is characterized by its compact β-grasp fold [133]. Most backbone amide groups are involved in a global hydrogen-bonding network hampering exchange with protons from surrounding water molecules [134]. Consequently, pronounced rates of exchange could only be observed in three distinct regions, which are the β1-β2 loop at the N-terminus (Leu8-Thr12), a short linker connecting the β3 and β4 strand (Ala46), and the unanchored C-terminal tail (Leu73-Gly75) [135]. Exactly those regions have been reported to be most flexible within ubiquitin [21]. The well-balanced response of the MEXICO experiment for corresponding residues was utilized to detect linkage-specific interdomain dynamics sampled by two distinct ubiquitin dimers [136]. The dimers are designed in a way that the C-terminus of one ubiquitin molecule (termed distal) is conjugated to the side chain of either Lys11 or Lys27 of another ubiquitin molecule (termed proximal). Although only the proximal moiety is isotopically labeled, relative domain motions of the distal moiety that take place on an effective time scale are sensed and manifested by a decrease of the amide proton-exchange rate at the affected sites (Figure. 11.11). The elusive conformational ensembles of the Lys11- and Lys27-linked ubiquitin dimers could then be described in great detail by combining MD simulations with complementary NMR approaches [136].

The observation that the localization of an amide group within a protein structure is crucial for its individual property of undergoing hydrogen exchange was also exploited in a study focusing on the cold shock protein B from *B. subtilis* [79]. As a model organism it is used to investigate the effect of (macro)molecular crowding on protein stability. In this way it is intended to understand the thermodynamics a protein is subjected to in a cellular environment. A consistent decrease of amide proton-exchange rates in the presence of all crowding agents under

**Figure 11.7** Heteronuclear $^1$H-$^{15}$N ZZ exchange experiment on NEDD8, which is phosphorylated at position Ser65 [119]. (a) Two-dimensional $^1$H-$^{15}$N HSQC NMR spectrum of phosphorylated NEDD8 presenting the backbone amide cross peaks originating from the relaxed conformation (denoted with "m") colored in orange and from the retracted conformation (denoted with "n") colored in cyan. Unassigned resonance signals comprising both phosphorylated and non-phosphorylated wild-type NEDD8 are colored in gray. (b) Excerpts taken from the NMR spectrum of the ZZ exchange experiment using a mixing time of 40 ms. Auto- and cross peaks originating from slow conformational exchange between the relaxed and the retracted conformation are connected by dashed lines. Resonance signals associated with corresponding residues of non-phosphorylated Nedd8, which is contained as impurity in the sample, are denoted with "WT". (c) Build-up curves from the composite ratios of signal heights of associated auto- and cross peaks. The rate constant of exchange is calculated by a global fitting procedure including the six residues that are presented in (b).

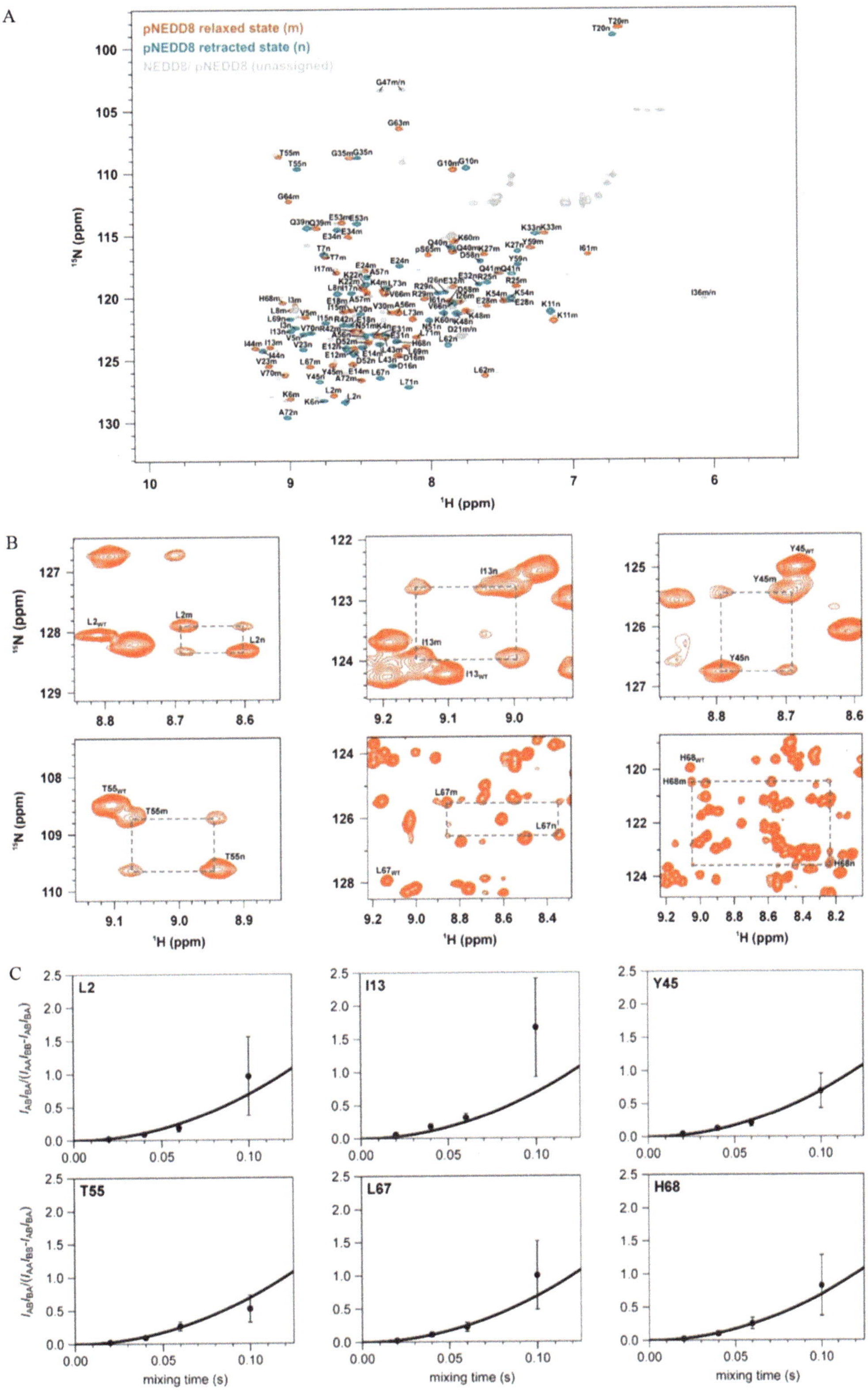

A
pNEDD8 relaxed state (m)
pNEDD8 retracted state (n)
NEDD8/ pNEDD8 (unassigned)
¹⁵N (ppm)
¹H (ppm)
B
¹⁵N (ppm)
¹H (ppm)
C
L2
I13
Y45
T55
L67
H68
$I_{AB}I_{BA}/(I_{AA}I_{BB}-I_{AB}I_{BA})$
mixing time (s)

investigation (PEG8 and dextran 20) were compliant with their stabilizing effect observed by unfolding experiments using complementary fluorescence and CD spectroscopic approaches. However, only the MEXICO NMR approach was capable of revealing that electrostatic interactions are a relevant factor in this context, as the effect turned out to be higher for polar dextran 20 than for PEG8 (Figure. 11.12a). Moreover, amide groups in flexible loop regions are experiencing better protection from exchange with increasing concentrations of the crowding agent (Figures 11.12b and 11.12c). It is thus conclusive that protein stabilization by (macro)molecular crowding is accompanied by an enhanced rigidity of especially non hydrogen-bonded residues [79].

### 11.3.6 Integrated Approaches Utilizing Structural Information from NMR Spectroscopy

Besides X-ray crystallography and cryo-EM, NMR spectroscopy is one of the leading approaches in structural biology capable of providing macromolecular structure information in atomistic detail. Although the main obstacle of NMR spectroscopy in this context is the size of the molecule under investigation, as the overall molecular tumbling directly affects spin relaxation and thus line width of resonance signals, continuous improvements in terms of instrumental power, pulse sequence developments, and isotopic labeling strategies are still raising the limits. However, for structure determination of molecules comprising molecular masses higher than 40 kDa, X-ray crystallography is the foremost method of choice [137]. Cryo-EM, on the other hand, is best suited when it comes to very large macromolecular assemblies but offers somewhat smaller resolution [138]. It is noteworthy to mention that the physical basis of the restraints obtained by x-ray crystallography, cryo-EM, and NMR spectroscopy is fundamentally different. Whereas the former ones rely on molecular replacement of an electron density arising from heavy atoms, it is the positioning of the hydrogen nuclei relative to each other that is sensed by the latter one [139]. Moreover, only NMR spectroscopy can be accurately performed in solution and is thus capable of taking dynamic events into account that are often essential to understand molecular function. The goal of integrative structural biology is to overcome the intrinsic limitations of single methodologies by combining restraints originating from different approaches. The most realistic representation is given when the information used to refine or to converge to a structural model is – to our mind – best diversified.

One elegant protocol to combine structural data from cryo-EM and multi-dimensional NMR spectroscopy in a synergistic manner is schematically shown in Figure. 11.13 and was initially developed for structure determination of the aminopeptidase TET2 from *Pyrococcus horikoshii* [95]. TET2 is an almost half-megadalton enzyme complex consisting of twelve homologous subunits with 353 residues each [140]. Three- and even four-dimensional experiments were conducted by solution and solid-state NMR spectroscopy to assign most of the backbone and side chain resonances of the TET2 subunit and to collect distance and dihedral restraints for structure determination. However, due to the lack of long-range distance restraints (313 restraints with $|i - j| \geq 4$) from NMR and the modest resolution of existing cryo-EM maps (4.5–8 Å), neither NMR nor cryo-EM data alone were sufficient to solve the structure. Nevertheless, it was possible to identify α-helical parts with high confidence within the electron density. Stretches of residues that are ascribed to α-helices on the basis of chemical shifts by using TALOS were then placed into the corresponding EM densities, thereby testing the possibilities in a combinatory manner (Figure. 11.13a). Each combination was used to perform a structure calculation run with CYANA in which the three-dimensional positions of the α-helices were fixed and the correct arrangement was unraveled in this way (Figure 11.13b and

**Figure 11.8** Two-dimensional $^1$H-$^{15}$N HSQC NMR spectra acquired for ubiquitin from yeast possessing an N-terminal extension illustrating the beneficial use of solvent PREs for spectral editing [124]. (a) $^1$H-$^{15}$N HSQC NMR spectrum acquired in absence of the paramagnetic cosolute Gd(DTPA-BMA). (b) $^1$H-$^{15}$N HSQC NMR spectrum acquired in presence of 5 mM Gd(DTPA-BMA) showing mainly resonances originating from core residues. (c) Difference spectrum resulting from the subtraction of (b) from (a). Resonance signals mainly originating from solvent-exposed residues are apparent.

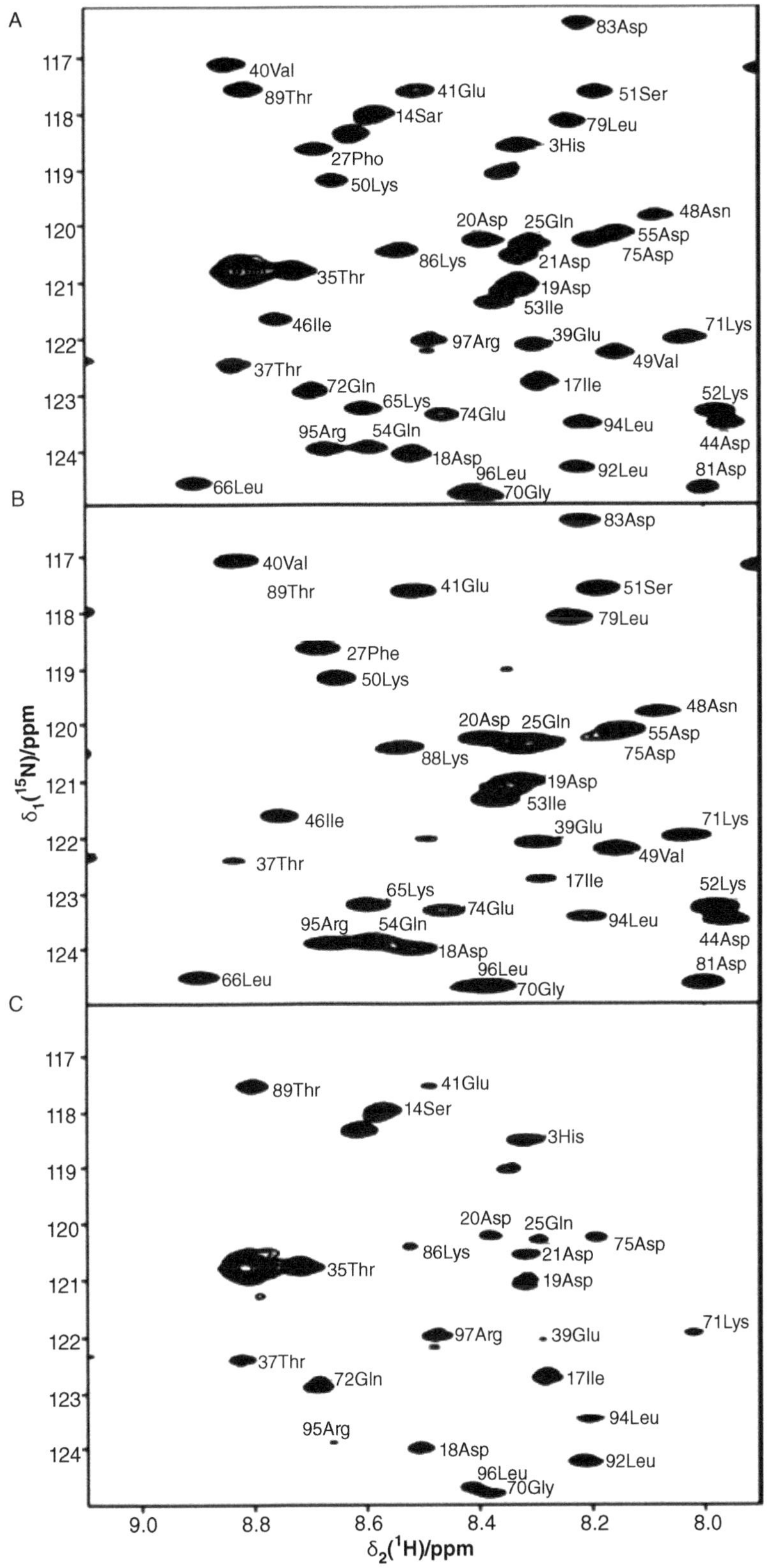

A
83Asp
40Val
89Thr
41Glu
51Ser
14Sar
79Leu
3His
27Pho
50Lys
20Asp
25Gln
48Asn
55Asp
86Lys
21Asp
75Asp
35Thr
19Asp
53Ile
46Ile
71Lys
39Glu
97Arg
49Val
37Thr
17Ile
72Gln
65Lys
52Lys
74Glu
94Leu
95Arg
54Gln
44Asp
18Asp
96Leu
92Leu
81Asp
66Leu
70Gly
B
83Asp
40Val
89Thr
41Glu
51Ser
79Leu
27Phe
50Lys
20Asp
25Gln
48Asn
55Asp
88Lys
75Asp
19Asp
53Ile
46Ile
71Lys
39Glu
49Val
37Thr
17Ile
65Lys
52Lys
74Glu
94Leu
95Arg
54Gln
44Asp
18Asp
81Asp
96Leu
66Leu
70Gly
C
89Thr
41Glu
14Ser
3His
20Asp
25Gln
75Asp
86Lys
21Asp
35Thr
19Asp
71Lys
97Arg
39Glu
37Thr
72Gln
17Ile
94Leu
95Arg
18Asp
92Leu
96Leu
70Gly
117
118
119
120
121
122
123
124
9.0
8.8
8.6
8.4
8.2
8.0
$\delta_1(^{15}N)$/ppm
$\delta_2(^1H)$/ppm

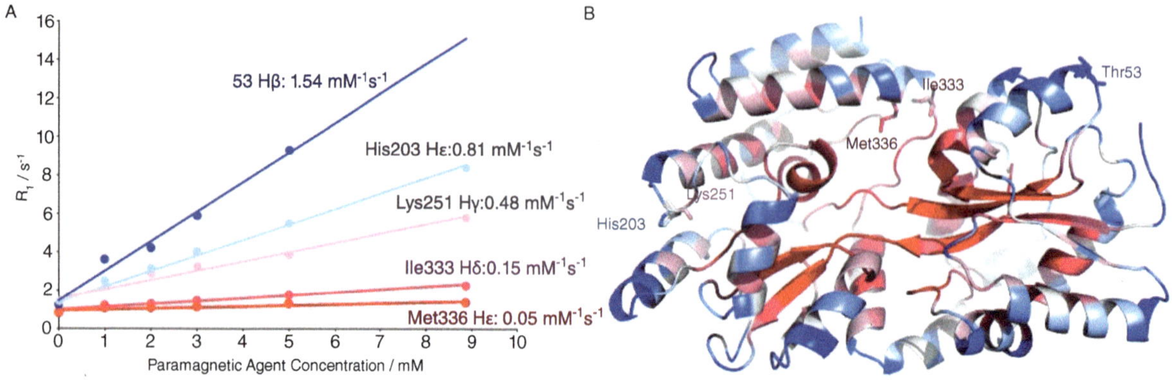


**Figure 11.9** Solvent PRE studies on the maltose-binding protein [127]. (a) Longitudinal relaxation rate constants ($R_1$) of carbon-bound side chain protons plotted against the concentration of Gd(DTPA-BMA). The corresponding solvent PRE values are given by the slope in $mM^{-1}s^{-1}$. (b), Experimental solvent PRE values as determined in (a) are mapped on the structure of the maltose-binding protein using colors ranging from red (low solvent PRE) to blue (high solvent PRE).

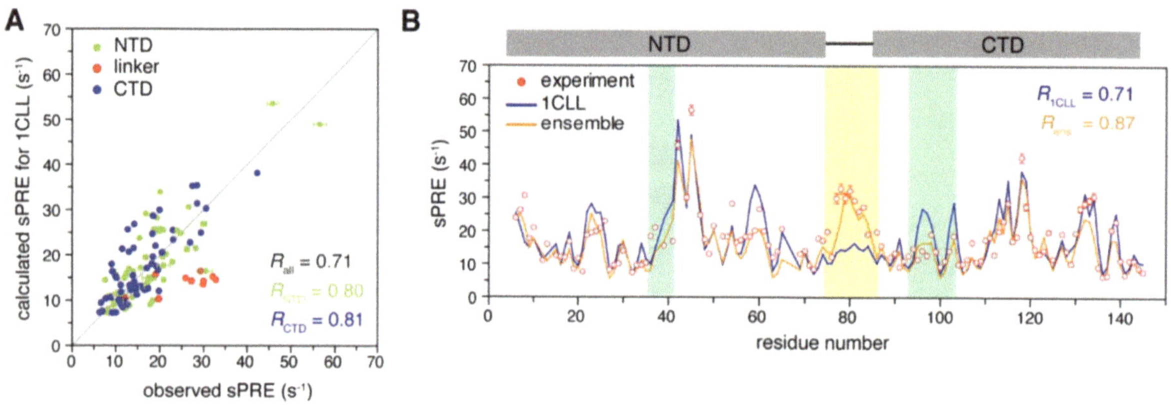


**Figure 11.10** Characterization of protein dynamics of $Ca^{2+}$-free calmodulin by means of solvent PRE measurements [131]. (a) Experimental solvent PRE values are correlated with values that have been back-calculated from the crystal structure of calmodulin in its unbound state representing the open conformation. Solvent PRE values of residues in the N-terminal domain (NTD) are colored in green, in the C-terminal domain (CTD) are colored in blue, and in the linker region are colored in red. Correlation coefficients, $R$ values, are calculated for both domains separately as well as for the whole protein. (b) Experimental solvent PRE values are plotted against the primary sequence of calmodulin as red open dots. The blue line represents solvent PRE values that have been back-calculated from the crystal structure and the orange line represents values originating from the optimized conformational ensemble. Background colors in green and yellow highlight significant differences between experimentally obtained solvent PRE values and PRE values that have been back-calculated from the crystal structure.

11.13c). On that basis, remaining flexible segments were next fitted into the EM map and several rounds of CYANA runs were iterated with an increasing number of NMR restraints included. After convergence, a final refinement step was implemented resulting in a structural ensemble with a backbone RMSD of 0.29 Å calculated using 20 structures possessing lowest energy (Figure. 11.13d). Starting from an EM map with 4.1 Å resolution, a significant improvement and subsequently, a high-resolution structure could thus be achieved by combining complementary NMR and cryo-EM data [95].

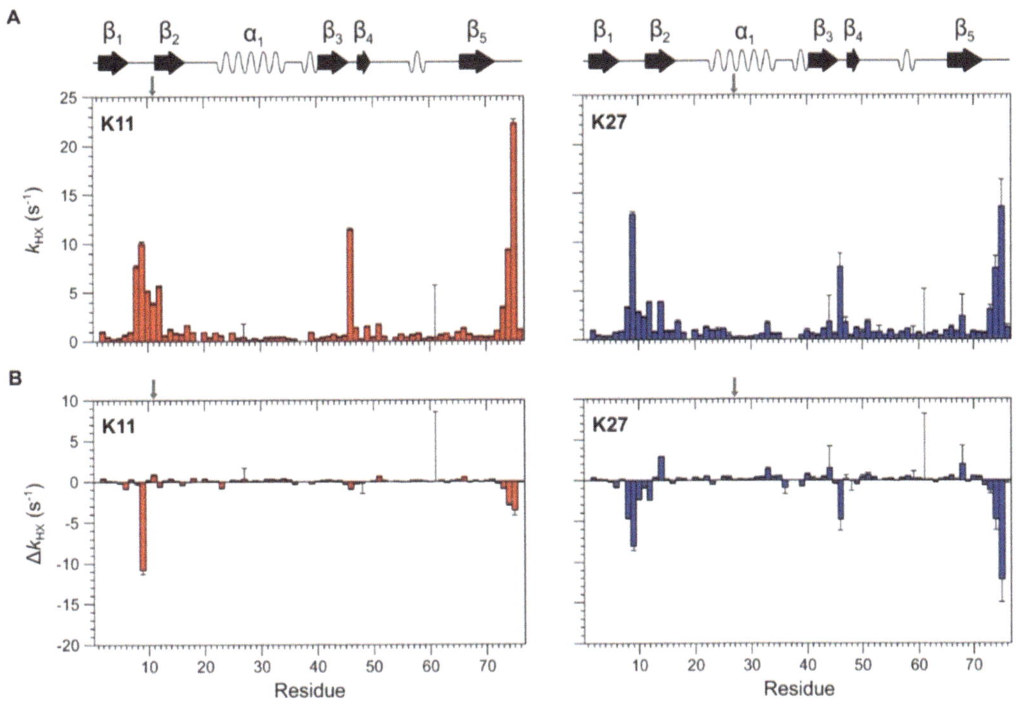


Figure 11.11 Modified MEXICO experiment used to investigate interdomain dynamics within ubiquitin dimers [136]. (a) Amide proton-exchange rates ($k_{HX}$) determined for the proximal moiety of the K11- (colored in red) and K27-linked ubiquitin dimer (colored in blue). (b) Differences of amide proton-exchange rates ($\Delta k_{HX}$) calculated by subtracting the values from (a) from the values determined for monomeric wild-type ubiquitin.

### 11.3.7 Multi-dimensional NMR Spectroscopy on ex vivo Samples

Multi-dimensional NMR spectroscopy is capable of expanding the spatial resolution of commonly crowded 1D spectra originating from biomolecules by the introduction of one or more dimensions. Yet, to solve biophysical and structural biology issues, the corresponding samples have to satisfy high quality criteria. Consequently, NMR measurements are usually performed on isolated molecules that are prepared in a buffer at constant pH and saline environment. To circumvent sensitivity limitations, high molecular concentrations and isotope enrichment are often required. In the case of proteins, for example, this can be achieved by recombinant expression and extensive purification. As all these steps are elaborate, expensive, and only partially suited for high-throughput applications, it is a reasonable objective to perform NMR spectroscopy also on samples in more complex media such as those obtained ex vivo.

Indeed, established protocols are now available to investigate samples directly from biofluids such as blood serum and urine without the need of time-consuming pretreatment [141]. In particular, this is utilized in metabolomics studies in the fields of biomedical, food, and environmental sciences but is also applied in clinical practice allowing, for example, disease diagnosis by screening for biomarkers [142]. Thereby, metabolites that are present in the sample as a mixture can be identified by comparison with reference spectra from databases containing the pure compounds [143].

Although most NMR metabolomics studies are based on 1D experiments detecting different types of nuclei ($^1$H, $^{13}$C, $^{15}$N, $^{31}$P), clear benefits in terms of resolution can also be gained in this context from 2D homonuclear $^1$H-$^1$H (COSY, TOCSY, NOESY) and heteronuclear $^1$H-$^{15}$N or $^1$H-$^{13}$C correlation experiments (HSQC, HMQC) [144]. This is exemplarily shown in Figure. 11.14 where – despite of numerous overlapping resonances in the 1D $^1$H spectrum – a plenty of compounds could be verified without ambiguity in a plant extract from *Arabidopsis* by means of 2D $^1H$-$^{13}C$ HSQC spectroscopy [145].

While metabolomics studies essentially focus on small organic compounds such as amino acids, sugars, phospholipids, and nucleotides, the potential of high-resolution NMR spectroscopy on ex vivo protein samples is impressively demonstrated in a case study with $\alpha$1-antitrypsin (AAT) [146]. This represents a glycoprotein possessing a molecular mass of 52 kDa and acts as an endogenous inhibitor of the serine-protease elastase. Distinct single point mutations of ATT (E264V and E342K) are known to be prone to misfolding and polymerization leading to accumulation in liver cells and cirrhosis on the one hand [147] and dysregulation of elastase and pulmonary disorder on the other hand [148]. However, an existing crystallographic structure of the E342K mutant does not

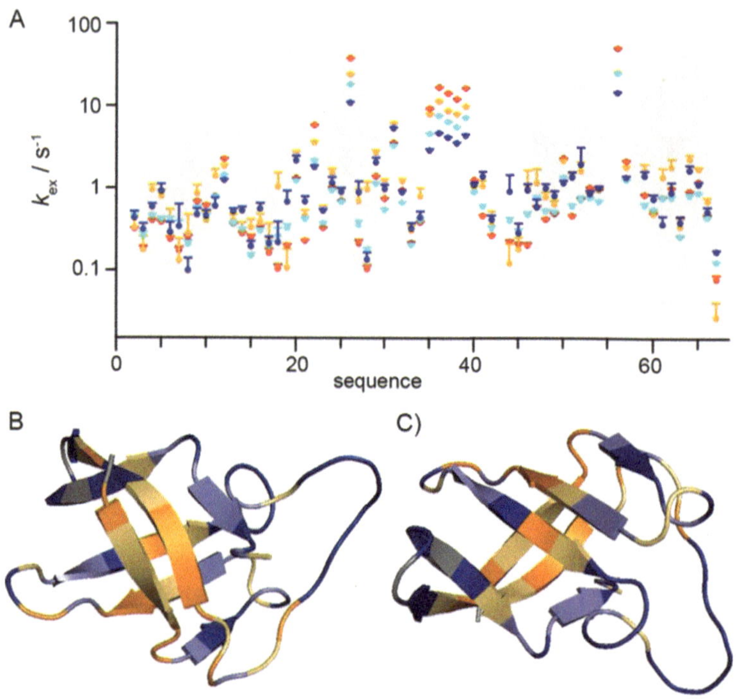


**Figure 11.12** Modified MEXICO NMR experiment applied on the cold shock protein B from *B. subtilis* to study the effects of macromolecular crowding [79]. (a) Amide proton-exchange rates ($k_{ex}$) determined under dilute conditions (colored in red), supplemented with 15% (*w*/*v*) PEG8 (colored in orange), 12% (*w*/*v*) dextran 20 (colored in cyan) and 24% (*w*/*v*) dextran 20 (colored in purple). (b, c) Differences of amide proton-exchange rates ($\Delta k_{ex}$) comparing values obtained under dilute conditions and in presence of 12% (*w*/*v*) dextran 20. A decrease of $k_{ex}$ upon addition of dextran 20 is indicated by blue color, whereas an increase is indicated by orange color.

**Figure 11.13** Schematic workflow developed for integrated structure elucidation of TET2 by combining structural information from NMR spectroscopy and cryo-EM [95]. (a) All possible arrangements are identified in which $\alpha$-helical stretches as indicated by TALOS analysis potentially fit to appropriate electron densities in the cryo-EM map of TET2. (b) The correctness of the various arrangements from (a) is evaluated by the CYANA target score of a structure calculation run. (c) Examples are presented for the correct arrangement of $\alpha$-helical stretches (Example 2) showing a high CYANA target score and a good agreement with the cryo-EM map and for an incorrect arrangement (Example 1) showing a low CYANA target score and a poor agreement with the cryo-EM map. (d) Several rounds of structure calculation with CYANA were performed thereby iteratively adding additional unambiguous NMR restraints. If an acceptable level of convergence is reached, the structure is subjected to XPLOR-NIH for further refinement.

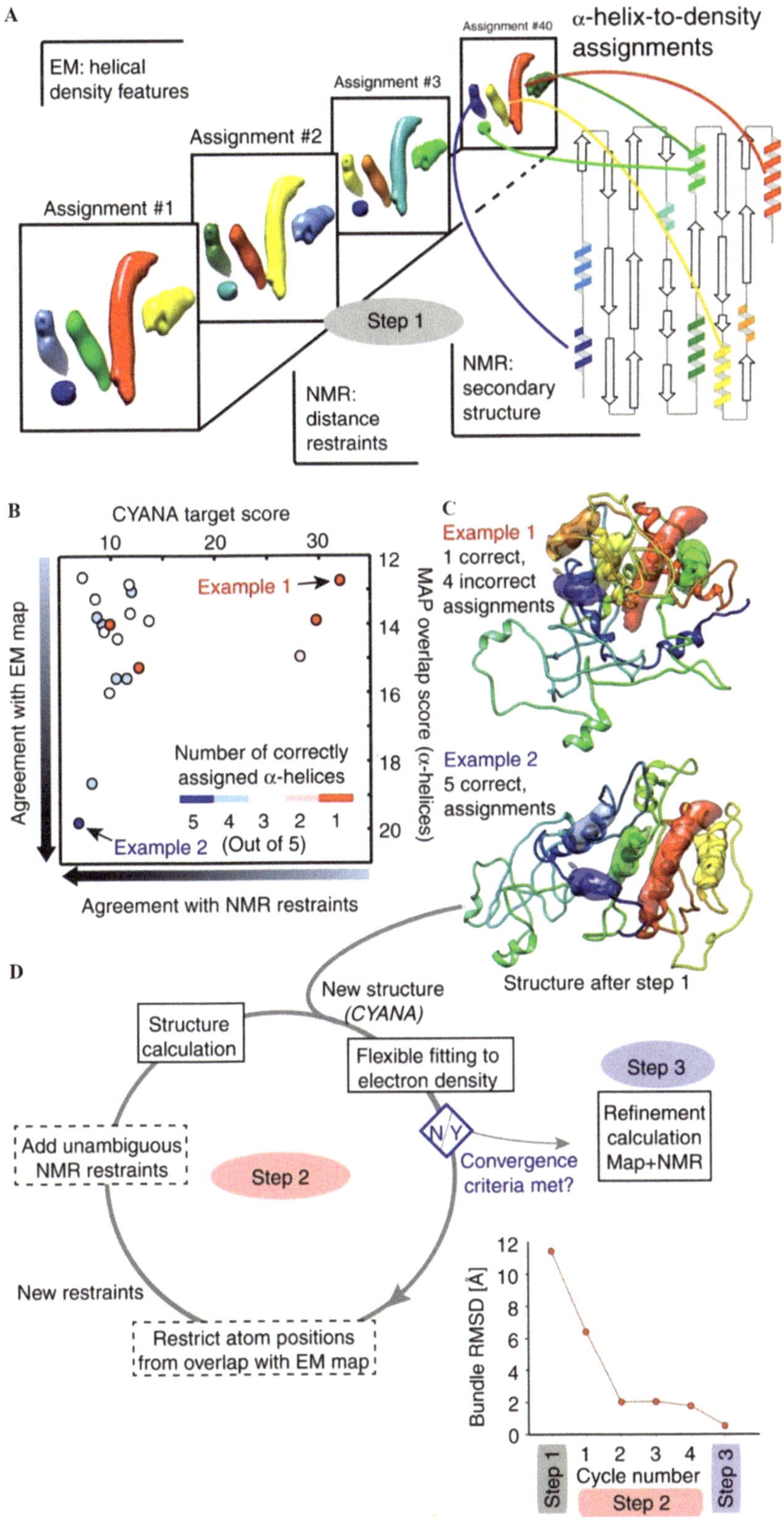

A
EM: helical density features
Assignment #1
Assignment #2
Assignment #3
Assignment #40
α-helix-to-density assignments
Step 1
NMR: distance restraints
NMR: secondary structure
B
CYANA target score
10
20
30
12
14
16
18
20
Example 1
Agreement with EM map
MAP overlap score (α-helices)
Number of correctly assigned α-helices
5 4 3 2 1
Example 2 (Out of 5)
Agreement with NMR restraints
C
Example 1
1 correct,
4 incorrect
assignments
Example 2
5 correct,
assignments
Structure after step 1
D
New structure (CYANA)
Structure calculation
Flexible fitting to electron density
Step 3
Refinement calculation Map+NMR
Add unambiguous NMR restraints
Step 2
N/Y
Convergence criteria met?
New restraints
Restrict atom positions from overlap with EM map
Bundle RMSD [Å]
12
10
8
6
4
2
0
Step 1
1 2 3 4
Cycle number
Step 2
Step 3

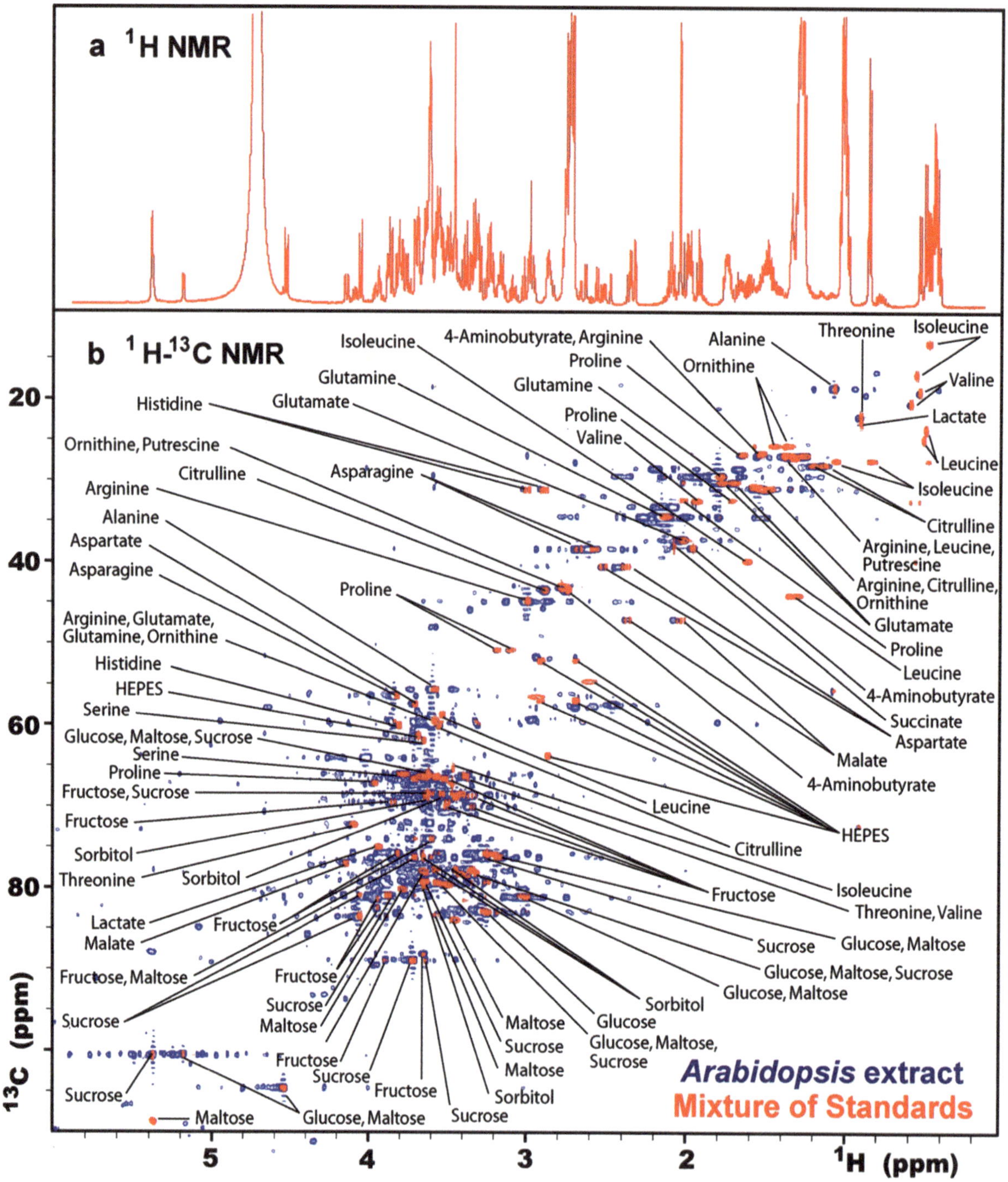


**Figure 11.14** Metabolomics NMR study performed on an aqueous whole-plant extract from *A. thaliana* [145]. (a) One-dimensional $^{1}H$ spectrum of a mixture containing equimolar amounts of 26 small-molecule standards. (b) Superimposition of two-dimensional $^{1}H$-$^{13}C$ HSQC NMR spectra acquired for the plant extract from *A. thaliana* colored in blue and the reference mixture from (a) colored in red. Cross peaks arising from the reference spectrum are assigned to the corresponding metabolite.

provide an explanation for this behavior [146, 149]. A destabilizing intermediate could be confirmed by using NMR spectroscopy performing an analysis of chemical-shift perturbations. This intermediate is well populated within the conformational ensemble of the E342K mutant in solution, thus triggering protein misfolding [146].

Strikingly, as conventional protocols for recombinant expression of ATT failed in the case of the E342K mutant due to reduced stability in the absence of glycosylation, purified ex vivo samples were used throughout the study that were obtained from human plasma. Despite the high molecular mass, the intrinsic instability of ATT mutants, and the lack of isotope enrichment, 2D $^{1}H$-$^{13}C$ HMQC spectra addressing side chain methyl groups could be acquired in this way with remarkable quality (Figure. 11.15b). Since this was achieved over a measurement time of 80 h using a protein concentration of 400 $\mu$M, a quality control strategy was pursued as depicted in Figure. 11.15a. Accordingly, the aspired two-dimensional NMR spectrum was recorded in an interleaved mode with a simple one-dimensional $^{1}H$ experiment and a translational diffusion experiment starting every 5 h. The integrity of the sample could thus be ensured at any time, even over a period of days. Finally, the assignment of methyl groups in the spectrum of wild-type ATT was accomplished on the basis of recombinant samples and could be transferred directly to the spectra of the ATT mutants (Figure. 11.16) [146]. In conclusion, this case study demonstrates that multi-dimensional NMR techniques are not per se limited to in vitro samples but are also applicable to more complex

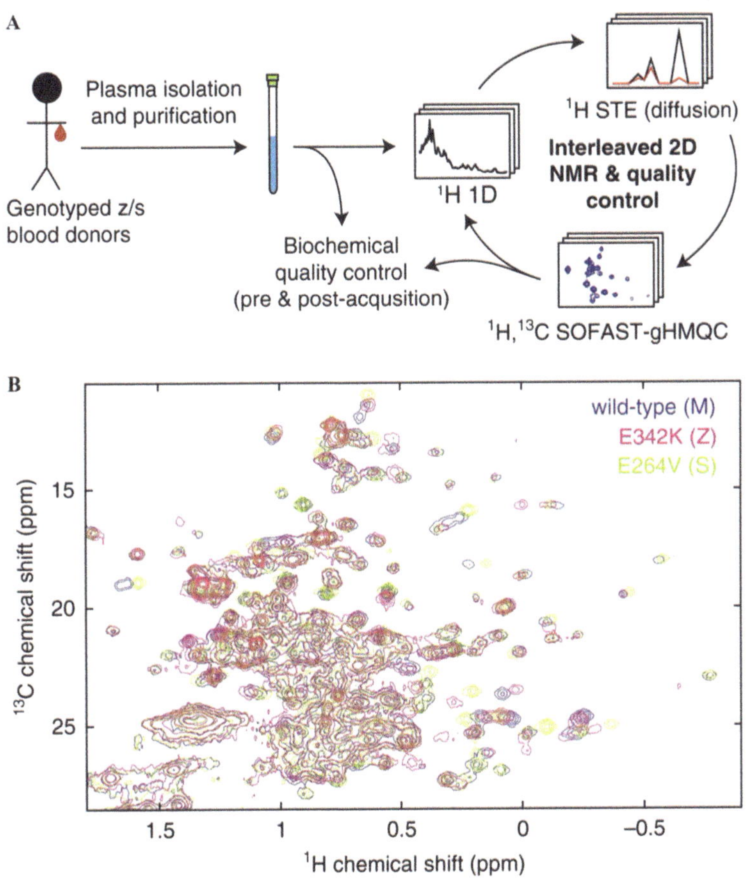


**Figure 11.15** NMR spectroscopy on ex vivo $\alpha$1-antitrypsin (ATT) samples [146]. (a) Experimental setup for sample recovery from human plasma and quality control for long-term NMR measurement. (b) Methyl group region of the $^{1}H$-$^{13}C$ SOFAST HMQC NMR spectra acquired with gradient selection for ex vivo samples of wild-type AAT (M) colored in purple and the disease-causing E342K (Z) and E264V (S) single mutants colored in pink and green, respectively.

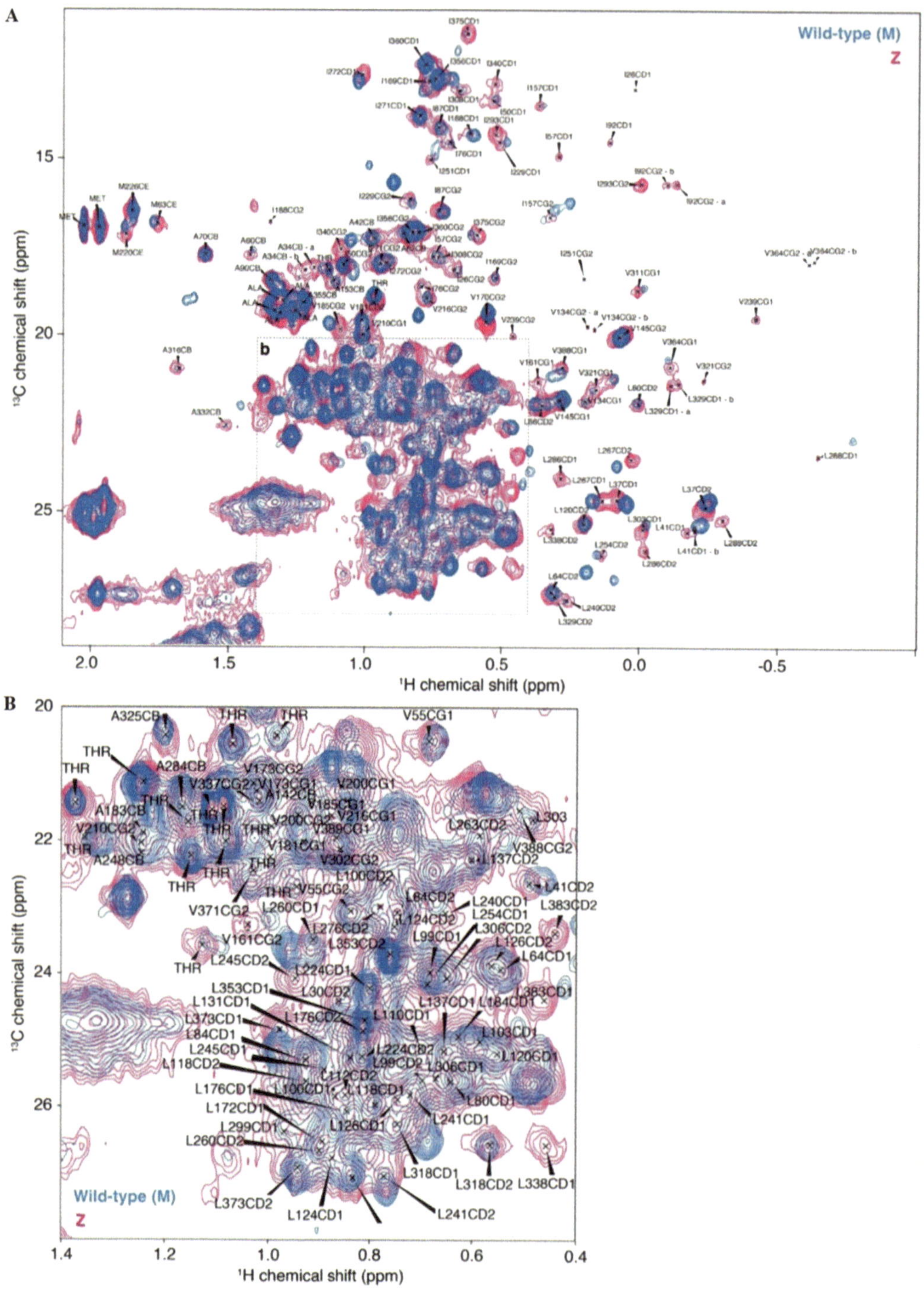


**Figure 11.16** Comparison of ex vivo α1-antitrypsin (AAT) samples [146]. (a), Superimposition of $^{1}H$-$^{13}C$ SOFAST HMQC spectra acquired with gradient selection on wild-type ATT (M) colored in blue and the disease-causing E342K (Z) single mutant colored in pink. Due to moderate peak shifts, methyl group assignments can be transferred with less ambiguity from wild-type ATT to the mutant with minor ambiguity. (b) Close-up view of the corresponding spectral region from (a).

media, if the experimental setup is well adapted. Thereby, we note that also the pulse sequence was optimized in this study regarding the ex vivo AAT samples to account for the low sensitivity and the large levels of $T_1$ noise the measurements are suffering from [146, 150].

## References


1 McIntosh, L.P., and Dahlquist, F.W. (1990). Biosynthetic incorporation of 15N and 13C for assignment and interpretation of nuclear magnetic resonance spectra of proteins. *Q. Rev. Biophys.* 23: 1–38.

2 Marley, J., Lu, M., and Bracken, C. (2001). A method for efficient isotopic labeling of recombinant proteins. *J. Biomol. NMR.* 20: 71–75.

3 Tugarinov, V., and Kay, L.E. (2003). Ile, Leu, and Val methyl assignments of the 723-residue malate synthase G using a new labeling strategy and novel NMR methods. *J. Am. Chem. Soc.* 125: 13868–13878.

4 Frieden, C., Hoeltzli, S.D., and Bann, J.G. (2004). The preparation of 19F-labeled proteins for NMR studies. *Methods Enzymol.* 380: 400–415.

5 Studier, F.W. (2005). Protein production by auto-induction in high-density shaking cultures. *Protein Expr. Purif.* 41: 207–234.

6 Kainosho, M., Torizawa, T., Iwashita, Y., Terauchi, T., Ono, A.M., and Güntert, P. (2006). Optimal isotope labelling for NMR protein structure determinations. *Nature.* 440: 52–57.

7 Lu, K., Miyazaki, Y., and Summers, M.F. (2010). Isotope labeling strategies for NMR studies of RNA. *J. Biomol. NMR.* 46: 113–125.

8 Otten, R., Chu, B., Krewulak, K.D., Vogel, H.J., and Mulder, F.A.A. (2010). Comprehensive and cost-effective NMR spectroscopy of methyl groups in large proteins. *J. Am. Chem. Soc.* 132: 2952–2960.

9 Nelissen, F.H.T., Tessari, M., Wijmenga, S.S., and Heus, H.A. (2016). Stable isotope labeling methods for DNA. *Prog. Nucl. Magn. Reson. Spectrosc.* 96: 89–108.

10 Cai, M.L., Huang, Y., Yang, R., Craigie, R., and Clore, G.M. (2016). A simple and robust protocol for high-yield expression of perdeuterated proteins in Escherichia coli grown in shaker flasks. *J. Biomol. NMR.* 66: 85–91.

11 Welte, H., and Kovermann, M. (2019). Targeted expression and purification of fluorine labelled cold shock protein B by using an auxotrophic strategy. *Protein Expr. Purif.* 157: 86–91.

12 Williamson, M.P., Havel, T.F., and Wüthrich, K. (1985). Solution conformation of proteinase inhibitor IIA from bull seminal plasma by 1H nuclear magnetic resonance and distance geometry. *J. Mol. Biol.* 182: 295–315.

13 Antuch, W., Güntert, P., and Wüthrich, K. (1996). Ancestral βγ-crystallin precursor structure in a yeast killer toxin. *Nat. Struct. Biol.* 3: 662–665.

14 Liu, G., Shen, Y., Atreya, H.S., Parish, D., Shao, Y., Sukumaran, D.K., Xiao, R., Yee, A., Lemak, A., Bhattacharya, A., Acton, T.A., Arrowsmith, C.H., Montelione, G.T., and Szyperski, T. (2005). NMR data collection and analysis protocol for high-throughput protein structure determination. *Proc. Natl. Acad. Sci. U. S. A.* 102: 10487–10492.

15 Gochin, M., and Roder, H. (1995). Use of the pseudocontact shift as a structural constraint for macromolecules in solution. *Bull. Magn. Reson.* 17: 298–299.

16 Banci, L., Bertini, I., Bren, K.L., Cremonini, M.A., Gray, H.B., Luchinat, C., and Turano, P. (1996). The use of pseudocontact shifts to refine solution structures of paramagnetic metalloproteins: Met80Ala cyano-cytochrome c as an example. *J. Biol. Inorg. Chem.* 1: 117–126.

17 Bertini, I., Luchinat, C., and Parigi, G. (2002). Paramagnetic constraints: an aid for quick solution structure determination of paramagnetic metalloproteins. *Concepts Magn. Reson.* 14: 259–286.

18 Bahramzadeh, A., Huber, T., and Otting, G. (2019). Three-dimensional protein structure determination using pseudocontact shifts of backbone amide protons generated by double-histidine $Co^{2+}$-binding motifs at multiple sites. *Biochemistry* 58: 3243–3250.

**19** Tjandra, N., Omichinski, J.G., Gronenborn, A.M., Clore, G.M., and Bax, A. (1997). Use of dipolar 1H-15N and 1H-13C couplings in the structure determination of magnetically oriented macromolecules in solution. *Nat. Struct. Biol.* 4: 732–738.

**20** Tjandra, N., and Bax, A. (1997). Direct measurement of distances and angles in biomolecules by NMR in a dilute liquid crystalline medium. *Science* 278: 1111–1114.

**21** Lange, O.F., Lakomek, N.A., Fars, C., Schröder, G.F., Walter, K.F.A., Becker, S., Meiler, J., Grubmüller, H., Griesinger, C., and de Groot, B.L. (2008). Recognition dynamics up to microseconds revealed from an RDC-derived ubiquitin ensemble in solution. *Science* 320: 1471–1475.

**22** Pardi, A., Billeter, M., and Wüthrich, K. (1984). Calibration of the angular-dependence of the amide proton-Ca proton coupling constants, $^3$JHN$\alpha$, in a globular protein: Use of $^3$JHN$\alpha$ for identification of helical secondary structure *J. Mol. Biol.* 180: 741–751.

**23** Salvador, P. (2014). Dependencies of J-couplings upon dihedral angles on proteins. *Annu. Rep. NMR Spectrosc.* 81: 185–227.

**24** Battiste, J.L., and Wagner, G. (2000). Utilization of site-directed spin labeling and high-resolution heteronuclear nuclear magnetic resonance for global fold determination of large proteins with limited nuclear overhauser effect data. *Biochemistry* 39: 5355–5365.

**25** Iwahara, J., Schwieters, C.D., and Clore, G.M. (2004). Ensemble approach for NMR structure refinement against 1H paramagnetic relaxation enhancement data arising from a flexible paramagnetic group attached to a macromolecule. *J. Am. Chem. Soc.* 126: 5879–5896.

**26** Clore, G.M., and Iwahara, J. (2009). Theory, practice, and applications of paramagnetic relaxation enhancement for the characterization of transient low-population states of biological macromolecules and their complexes. *Chem. Rev.* 109: 4108–4139.

**27** Schwieters, C.D., Kuszewski, J.J., Tjandra, N., and Clore, G.M. (2003). The Xplor-NIH NMR molecular structure determination package. *J. Magn. Reson.* 160: 65–73.

**28** Schwieters, C.D., Kuszewski, J.J., and Clore, G.M. (2006). Using Xplor-NIH for NMR molecular structure determination. *Prog. Nucl. Magn. Reson. Spectrosc.* 48: 47–62.

**29** Brünger, A.T., Adams, P.D., Clore, G.M., DeLano, W.L., Gros, P., Grosse-Kunstleve, R.W., Jiang, J.S., Kuszewski, J., Nilges, M., Pannu, N.S., Read, R.J., Rice, L.M., Simonson, T., and Warren, G.L. (1998). Crystallography & NMR system: a new software suite for macromolecular structure determination. *Acta Cryst.* D54: 905–921.

**30** Brunger, A.T. (2007). Version 1.2 of the crystallography and NMR system. *Nat. Protoc.* 2: 2728–2733.

**31** Güntert, P., Mumenthaler, C., and Wüthrich, K. (1997). Torsion angle dynamics for NMR structure calculation with the new program DYANA. *J. Mol. Biol.* 273: 283–298.

**32** Rieping, W., Habeck, M., Bardiaux, B., Bernard, A., Malliavin, T.E., and Nilges, M. (2007). ARIA2: automated NOE assignment and data integration in NMR structure calculation. *Bioinformatics* 23: 381–382.

**33** Ottiger, M., Delaglio, F., and Bax, A. (1998). Measurement of J and dipolar couplings from simplified two-dimensional NMR spectra. *J. Magn. Reson.* 131: 373–378.

**34** Marion, D., Driscoll, P.C., Kay, L.E., Wingfield, P.T., Bax, A., Gronenborn, A.M., and Clore, G.M. (1989). Overcoming the overlap problem in the assignment of 1H NMR spectra of larger proteins by use of three-dimensional heteronuclear 1H-15N Hartmann-Hahn-multiple quantum coherence and nuclear Overhauser multiple quantum coherence spectroscopy: application to interleukin 1ß. *Biochemistry* 28: 6150–6156.

**35** Talluri, S., and Wagner, G. (1996). An optimized 3D NOESY-HSQC. *J. Magn. Reson. B* 112: 200–205.

**36** Palmer III, A.G., Rance, M., and Wright, P.E. (1991). Intramolecular motions of a zinc finger DNA-binding domain from xfin characterized by proton-detected natural abundance 13C heteronuclear NMR spectroscopy. *J. Am. Chem. Soc.* 113: 4371–4380.

**37** Mandel, A.M., Akke, M., and Palmer, III, A.G. (1995). Backbone dynamics of Escherichia coli ribonuclease HI: correlations with structure and function in an active enzyme. *J. Mol. Biol.* 246: 144–163.

38 Volkman, B.F., Lipson, D., Wemmer, D.E., and Kern, D. (2001). Two-state allosteric behavior in a single-domain signaling protein. *Science* 291: 2429–2433.

39 Yan, J., Liu, Y., Lukasik, S.M., Speck, N.A., and Bushweller, J.H. (2004). CBF allosterically regulates the Runx1 Runt domain via a dynamic conformational equilibrium. *Nat. Struct. Mol. Biol.* 11: 901–906.

40 Ma, D., Tillman, T.S., Tang, P., Meirovitch, E., Eckenhoff, R., Carnini, A., and Xu, Y. (2008). NMR studies of a channel protein without membranes: structure and dynamics of water-solubilized KcsA. *Proc. Natl. Acad. Sci. U. S. A.* 105: 16537–16542.

41 Wang, W., Weng, J., Zhang, X., Liu, M., and Zhang, M. (2009). Creating conformational entropy by increasing interdomain mobility in ligand binding regulation: a revisit to N-terminal tandem PDZ domains of PSD-95. *J. Am. Chem. Soc.* 131: 787–796.

42 Carr, H.Y., and Purcell, E.M. (1954). Effcts of diffusion on free precession in nuclear magnetic resonance experiments. *Phys. Rev.* 94: 630–638.

43 Meiboom, S., and Gill, D. (1958). Modified spin-echo method for measuring nuclear relaxation times. *Rev. Sci. Instrum.* 29: 688–691.

44 Luz, Z., and Meiboom, S. (1963). Nuclear magnetic resonance study of protolysis of trimethylammonium ion in aqueous solution-order of the reaction with respect to solvent. *J. Chem. Phys.* 39: 366–370.

45 Loria, J.P., Rance, M., and Palmer III, A.G. (1999). A relaxation-compensated Carr-Purcell-Meiboom-Gill sequence for characterizing chemical exchange by NMR spectroscopy. *J. Am. Chem. Soc.* 121: 2331–2332.

46 Wolf-Watz, M., Thai, V., Henzler-Wildman, K., Hadjipavlou, G., Eisenmesser, E.Z., and Kern, D. (2004). Linkage between dynamics and catalysis in a thermophilic-mesophilic enzyme pair. *Nat. Struct. Mol. Biol.* 11: 945–949.

47 Sugase, K., Dyson, H.J., and Wright, P.E. (2007). Mechanism of coupled folding and binding of an intrinsically disordered protein. *Nature* 447: 1021–1025.

48 Whittier, S.K., Hengge, A.C., and Loria, J.P. (2013). Conformational motions regulate phosphoryl transfer in related protein tyrosine phosphatases. *Science* 341: 899–903.

49 Karamanos, T.K., Tugarinov, V., and Clore, G.M. (2020). An S/T motif controls reversible oligomerization of the Hsp40 chaperone DNAJB6b through subtle reorganization of a β sheet backbone. *Proc. Natl. Acad. Sci. U. S. A.* 117: 30441–30450.

50 Long, D., Bouvignies, G., and Kay, L.E. (2014). Measuring hydrogen exchange rates in invisible protein excited states. *Proc. Natl. Acad. Sci. U. S. A.* 111: 8820–8825.

51 Deshmukh, L., Tugarinov, V., Louis, J.M., and Clore, G.M. (2017). Binding kinetics and substrate selectivity in HIV-1 protease-Gag interactions probed at atomic resolution by chemical exchange NMR. *Proc. Natl. Acad. Sci. U. S. A.* 114: E9855–E9862.

52 Gladkova, C., Schubert, A.F., Wagstaff, J.L., Pruneda, J.N., Freund, S.M.V., and Komander, D. (2017). An invisible ubiquitin conformation is required for efficient phosphorylation by PINK1. *EMBO J.* 36: 3555–3572.

53 Xie, T., Saleh, T., Rossi, P., and Kalodimos, C.G. (2020). Conformational states dynamically populated by a kinase determine its function. *Science* 370: eabc2754.

54 Gemmecker, G., Jahnke, W., and Kessler, H. (1993). Measurement of fast proton-exchange rates in isotopically labeled compounds. *J. Am. Chem. Soc.* 115: 11620–11621.

55 Hwang, T.L., Mori, S., Shaka, A.J., and van Zijl, P.C.M. (1997). Application of Phase-Modulated CLEAN chemical EXchange spectroscopy (CLEANEX-PM) to detect water-protein proton exchange and intermolecular NOEs. *J. Am. Chem. Soc.* 119: 6203–6204.

56 Hwang, T.L., van Zijl, P.C.M., and Mori, S. (1998). Accurate quantitation of water-amide proton exchange rates using the Phase-Modulated CLEAN chemical EXchange (CLEANEX-PM) approach with a Fast-HSQC (FHSQC) detection scheme. *J. Biomol. NMR* 11: 221–226.

57 Haupt, C., Weininger, U., Kovermann, M., and Balbach, J. (2011). Local and coupled thermodynamic stability of the two-domain and bifunctional enzyme SlyD from Escherichia coli. *Biochemistry* 50: 7321–7329.

**58** Chandak, M.S., Nakamura, T., Makabe, K., Takenaka, T., Mukaiyama, A., Chaudhuri, T.K., Kato, K., and Kuwajima, K. (2013). The H/D-exchange kinetics of the Escherichia coli co-chaperonin GroES studied by 2D NMR and DMSO-quenched exchange methods. *J. Mol. Biol.* 425: 2541–2560.

**59** Bhunia, A., Ilyas, H., and Bhattacharjya, S. (2020). Salt dependence conformational stability of the dimeric SAM domain of MAPKKK Ste11 from budding yeast: a native-state H/D exchange NMR study. *Biochemistry* 59: 2849–2858.

**60** Balbach, J., Forge, V., van Nuland, N.A.J., Winder, S.L., Hore, P.J., and Dobson, C.M. (1995). Following protein folding in real time using NMR spectroscopy. *Nat. Struct. Biol.* 2: 865–870.

**61** Feige, M.J., Groscurth, S., Marcinowski, M., Yew, Z.T., Truffault, V., Paci, E., Kessler, H., and Buchner, J. (2008). The structure of a folding intermediate provides insight into differences in immunoglobulin amyloidogenicity. *Proc. Natl. Acad. Sci. U. S. A.* 105: 13373–13378.

**62** Haupt, C., Patzschke, R., Weininger, U., Gröger, S., Kovermann, M., and Balbach, J. (2011). Transient enzyme-substrate recognition monitored by real-time NMR. *J. Am. Chem. Soc.* 133: 11154–11162.

**63** Reining, A., Nozinovic, S., Schlepckow, K., Buhr, F., Fürtig, B., and Schwalbe, H. (2013). Three-state mechanism couples ligand and temperature sensing in riboswitches. *Nature* 499: 355–359.

**64** Macek, P., Kerfah, R., Erba, E.B., Crublet, E., Moriscot, C., Schoehn, G., Amero, C., and Boisbouvier, J. (2017). Unraveling self-assembly pathways of the 468-kDa proteolytic machine TET2. *Sci. Adv.* 3: e1601601.

**65** Köhn, B., Schwarz, P., Wittung-Stafshede, P., and Kovermann, M. (2021). Impact of crowded environments on binding between protein and single-stranded DNA. *Sci. Rep.* 11: 17682.

**66** Welte, H., Sinn, P., and Kovermann, M. (2021). Fluorine NMR spectroscopy enables to quantify the affinity between DNA and proteins in cell lysate. *Chembiochem* 22: 2973–2980.

**67** Williamson, M.P. (2013). Using chemical shift perturbation to characterise ligand binding. *Prog. Nucl. Magn. Reson. Spectrosc.* 73: 1–16.

**68** Bai, Y., Milne, J.S., Mayne, L., and Englander, S.W. (1993). Primary structure effects on peptide group hydrogen exchange. *Proteins* 17: 75–86.

**69** Connelly, G.P., Bai, Y., Jeng, M.F., and Englander, S.W. (1993). Isotope effects in peptide group hydrogen exchange. *Proteins* 17: 87–92.

**70** Bai, Y., Milne, J.S., Mayne, L., and Englander, S.W. (1994). Protein stability parameters measured by hydrogen exchange. *Proteins* 20: 4–14.

**71** Roder, H., Elöve, G.A., and Englander, S.W. (1988). Structural characterization of folding intermediates in cytochrome c by H-exchange labelling and proton NMR. *Nature* 335: 700–704.

**72** Udgaonkar, J.B., and Baldwin, R.L. (1988). NMR evidence for an early framework intermediate on the folding pathway of ribonuclease A. *Nature* 335: 694–699.

**73** Schulenburg, C., Löw, C., Weininger, U., Mrestani-Klaus, C., Hofmann, H., Balbach, J., Ulbrich-Hofmann, R., and Arnold, U. (2009). The folding pathway of onconase is directed by a conserved intermediate. *Biochemistry* 48: 8449–8457.

**74** Di Paolo, A., Balbeur, D., De Pauw, E., Redfield, C., and Matagne, A. (2010). Rapid collapse into a molten globule is followed by simple two-state kinetics in the folding of lysozyme from bacteriophage λ. *Biochemistry* 49: 8646–8657.

**75** Dyson, H.J., and Wright, P.E. (2017). How does your protein fold? Elucidating the apomyoglobin folding pathway. *Acc. Chem. Res.* 50: 105–111.

**76** Serber, Z., and Dötsch, V. (2001). In-cell NMR spectroscopy. *Biochemistry* 40: 14317–14323.

**77** Zimmerman, S.B., and Trach, S.O. (1991). Estimation of macromolecule concentrations and excluded volume effects for the cytoplasm of Escherichia coli. *J. Mol. Biol.* 222: 599–620.

**78** Cheung, M.C., LaCroix, R., McKenna, B.K., Liu, L., Winkelman, J., and Ehrlich, D.J. (2013). Intracellular protein and nucleic acid measured in eight cell types using deep-ultraviolet mass mapping. *Cytometry A* 83A: 540–551.

**79** Köhn, B., and Kovermann, M. (2019). Macromolecular crowding tunes protein stability by manipulating solvent accessibility. *Chembiochem* 20: 759–763.

**80** Welte, H., and Kovermann, M. (2020). Insights into protein stability in cell lysate by 19F NMR spectroscopy. *Chembiochem* 21: 3575–3579.

**81** Köhn, B., and Kovermann, M. (2020). All atom insights into the impact of crowded environments on protein stability by NMR spectroscopy. *Nat. Commun.* 11: 5760.

**82** Bertrand, K., Reverdatto, S., Burz, D.S., Zitomer, R., and Shekhtman, A. (2012). Structure of proteins in eukaryotic compartments. *J. Am. Chem. Soc.* 134: 12798–12806.

**83** Hamatsu, J., O'Donovan, D., Tanaka, T., Shirai, T., Hourai, Y., Mikawa, T., Ikeya, T., Mishima, M., Boucher, W., Smith, B.O., Laue, E.D., Shirakawa, M., and Ito, Y. (2013). High-resolution heteronuclear multidimensional NMR of proteins in living insect cells using a baculovirus protein expression system. *J. Am. Chem. Soc.* 135: 1688–1691.

**84** Banci, L., Barbieri, L., Bertini, I., Luchinat, E., Secci, E., Zhao, Y., and Aricescu, A.R. (2013). Atomic-resolution monitoring of protein maturation in live human cells by NMR. *Nat. Chem. Biol.* 9: 297–299.

**85** Nishida, N., Ito, Y., and Shimada, I. (2020). In situ structural biology using in-cell NMR. *Biochim. Biophys. Acta Gen. Subj.* 1864: 129364.

**86** Luchinat, E., and Banci, L. (2017). In-cell NMR: a topical review. *IUCrJ* 4: 108–118.

**87** Selenko, P., Serber, Z., Gadea, B., Ruderman, J., and Wagner, G. (2006). Quantitative NMR analysis of the protein G B1 domain in Xenopus laevis egg extracts and intact oocytes. *Proc. Natl. Acad. Sci. U. S. A.* 103: 11904–11909.

**88** Sakai, T., Tochio, H., Tenno, T., Ito, Y., Kokubo, T., Hiroaki, H., and Shirakawa, M. (2006). In-cell NMR spectroscopy of proteins inside Xenopus laevis oocytes. *J. Biomol. NMR* 36: 179–188.

**89** Ogino, S., Kubo, S., Umemoto, R., Huang, S., Nishida, N., and Shimada, I. (2009). Observation of NMR signals from proteins introduced into living mammalian cells by reversible membrane permeabilization using a pore-forming toxin, streptolysin O. *J. Am. Chem. Soc.* 131: 10834–10835.

**90** Inomata, K., Ohno, A., Tochio, H., Isogai, S., Tenno, T., Nakase, I., Takeuchi, T., Futaki, S., Ito, Y., Hiroaki, H., and Shirakawa, M. (2009). High-resolution multi-dimensional NMR spectroscopy of proteins in human cells. *Nature* 458: 106–109.

**91** Freedberg, D.I., and Selenko, P. (2014). Live cell NMR. *Annu. Rev. Biophys.* 43: 171–192.

**92** Luchinat, E., and Banci, L. (2016). A unique tool for cellular structural biology: in-cell NMR. *J. Biol. Chem.* 291: 3776–3784.

**93** Ikeya, T., Sasaki, A., Sakakibara, D., Shigemitsu, Y., Hamatsu, J., Hanashima, T., Mishima, M., Yoshimasu, M., Hayashi, N., Mikawa, T., Nietlispach, D., Wälchli, M., Smith, B.O., Shirakawa, M., Güntert, P., and Ito, Y. (2010). NMR protein structure determination in living E. coli cells using nonlinear sampling. *Nat. Protoc.* 5: 1051–1060.

**94** Sakakibara, D., Sasaki, A., Ikeya, T., Hamatsu, J., Hanashima, T., Mishima, M., Yoshimasu, M., Hayashi, N., Mikawa, T., Wälchli, M., Smith, B.O., Shirakawa, M., Güntert, P., and Ito, Y. (2009). Protein structure determination in living cells by in-cell NMR spectroscopy. *Nature* 458: 102–105.

**95** Gauto, D.F., Estrozi, L.F., Schwieters, C.D., Effantin, G., Macek, P., Sounier, R., Sivertsen, A.C., Schmidt, E., Kerfah, R., Mas, G., Colletier, J.P., Güntert, P., Favier, A., Schoehn, G., Schanda, P., and Boisbouvier, J. (2019). Integrated NMR and cryo-EM atomic-resolution structure determination of a half-megadalton enzyme complex. *Nat. Commun.* 10: 2697.

**96** Henzler-Wildman, K.A., Thai, V., Lei, M., Ott, M., Wolf-Watz, M., Fenn, T., Pozharski, E., Wilson, M.A., Petsko, G.A., Karplus, M., Hübner, C.G., and Kern, D. (2007). Intrinsic motions along an enzymatic reaction trajectory. *Nature* 450: 838–844.

**97** Kovermann, M., Grundström, C., Sauer-Eriksson, A.E., Sauer, U.H., and Wolf-Watz, M. (2017). Structural basis for ligand binding to an enzyme by a conformational selection pathway. *Proc. Natl. Acad. Sci. U. S. A.* 114: 6298–6303.

**98** Kovermann, M., Ådén, J., Grundström, C., Sauer-Eriksson, A.E., Sauer, U.H., and Wolf-Watz, M. (2015). Structural basis for catalytically restrictive dynamics of a high-energy enzyme state. *Nat. Commun.* 6: 11792.

**99** Pálmadóttir, T., Malmendal, A., Leiding, T., Lund, M., and Linse, S. (2021). Charge regulation during amyloid formation of α-synuclein. *J. Am. Chem. Soc.* 143: 7777–7791.

**100** Yang, X., Wang, B., Hoop, C.L., Williams, J.K., and Baum, J. (2021). NMR unveils an N-terminal interaction interface on acetylated-α-synuclein monomers for recruitment to fibrils. *Proc. Natl. Acad. Sci. U. S. A.* 118: e2017452118.

**101** Törnquist, M., Cukalevski, R., Weininger, U., Meisl, G., Knowles, T.P.J., Leiding, T., Malmendal, A., Akke, M., and Linse, S. (2020). Ultrastructural evidence for self-replication of Alzheimer-associated Aβ42 amyloid along the sides of fibrils. *Proc. Natl. Acad. Sci. U. S. A.* 117: 11265–11273.

**102** Wishart, D.S., Bigam, C.G., Holm, A., Hodges, R.S., and Sykes, B.D. (1995). 1H, 13C and 15N random coil NMR chemical shifts of the common amino acids. I. Investigations of nearest-neighbor effects. *J. Biomol. NMR* 5: 67–81.

**103** Siu, H.W., Heck, B., Kovermann, M., and Hauser, K. (2021). Template-assisted design of monomeric polyQ models to unravel the unique role of glutamine side chains in disease-related aggregation. *Chemical Science* 12: 412–426.

**104** Bax, A. (1989). Two-dimensional NMR and protein structure. *Annu. Rev. Biochem.* 58: 223–256.

**105** Fesik, S.W., and Zuiderweg, E.R.P. (1990). Heteronuclear three-dimensional NMR spectroscopy of isotopically labelled biological macromolecules. *Q. Rev. Biophys.* 23: 97–131.

**106** Mandal, P.K., and Majumdar, A. (2004). A comprehensive discussion of HSQC and HMQC pulse sequences. *Concepts Magn. Reson. Part A* 20A: 1–23.

**107** Ebrahimi, M., Rossi, P., Rogers, C., and Harbison, G.S. (2001). Dependence of 13C NMR chemical shifts on conformations of RNA nucleosides and nucleotides. *J. Magn. Reson.* 150: 1–9.

**108** Schnieders, R., Keyhani, S., Schwalbe, H., and Fürtig, B. (2020). More than proton detection-new avenues for NMR spectroscopy of RNA. *Chem. Eur. J.* 26: 102–113.

**109** Richter, C., Kovacs, H., Buck, J., Wacker, A., Fürtig, B., Bermel, W., and Schwalbe, H. (2010). 13C-direct detected NMR experiments for the sequential J-based resonance assignment of RNA oligonucleotides. *J. Biomol. NMR* 47: 259–269.

**110** Boeszoermenyi, A., Chhabra, S., Dubey, A., Radeva, D.L., Burdzhiev, N.T., Chanev, C.D., Petrov, O.I., Gelev, V.M., Zhang, M., Anklin, C., Kovacs, H., Wagner, G., Kuprov, I., Takeuchi, K., and Arthanari, H. (2019). Aromatic 19F-13C TROSY: a background-free approach to probe biomolecular structure, function, and dynamics. *Nat. Methods.* 16: 333–340.

**111** Palmer, III, A. G., Kroenke, C.D., and Loria, J.P. (2001). Nuclear magnetic resonance methods for quantifying microsecond-to-millisecond motions in biological macromolecules. *Methods Enzymol.* 339: 204–238.

**112** Kleckner, I.R., and Foster, M.P. (2011). An introduction to NMR-based approaches for measuring protein dynamics. *Biochim. Biophys. Acta* 1814: 942–968.

**113** Religa, T.L., Sprangers, R., and Kay, L.E. (2010). Dynamic regulation of archaeal proteasome gate opening as studied by TROSY NMR. *Science* 328: 98–102.

**114** Baumeister, W., Walz, J., Zühl, F., and Seemüller, E. (1998). The proteasome: paradigm of a self-compartmentalizing protease. *Cell* 92: 367–380.

**115** Gelis, I., Bonvin, A.M.J.J., Keramisanou, D., Koukaki, M., Gouridis, G., Karamanou, S., Economou, A., and Kalodimos, C.G. (2007). Structural basis for signal-sequence recognition by the translocase motor SecA as determined by NMR. *Cell* 131: 756–769.

**116** Farrow, N.A., Zhang, O., Forman-Kay, J.D., and Kay, L.E. (1994). A heteronuclear correlation experiment for simultaneous determination of 15N longitudinal decay and chemical exchange rates of systems in slow equilibrium. *J. Biomol. NMR* 4: 727–734.

**117** Wauer, T., Swatek, K.N., Wagstaff, J.L., Gladkova, C., Pruneda, J.N., Michel, M.A., Gersch, M., Johnson, C.M., Freund, S.M.V., and Komander, D. (2015). Ubiquitin Ser65 phosphorylation affects ubiquitin structure, chain assembly and hydrolysis. *EMBO J.* 34: 307–325.

**118** Whitby, F.G., Xia, G., Pickart, C.M., and Hill, C.P. (1998). Crystal structure of the human ubiquitin-like protein NEDD8 and interactions with ubiquitin pathway enzymes. *J. Biol. Chem.* 273: 34983–34991.

**119** Stuber, K., Schneider, T., Werner, J., Kovermann, M., Marx, A., and Scheffner, M. (2021). Structural and functional consequences of NEDD8 phosphorylation. *Nat. Commun.* 12: 5939.

**120** Gillespie, J.R., and Shortle, D. (1997). Characterization of long-range structure in the denatured state of staphylococcal nuclease. I. Paramagnetic relaxation enhancement by nitroxide spin labels. *J. Mol. Biol.* 268: 158–169.

**121** Gillespie, J.R., and Shortle, D. (1997). Characterization of long-range structure in the denatured state of staphylococcal nuclease. II. Distance restraints from paramagnetic relaxation and calculation of an ensemble of structures. *J. Mol. Biol.* 268: 170–184.

**122** Keizers, P.H.J., and Ubbink, M. (2011). Paramagnetic tagging for protein structure and dynamics analysis. *Prog. Nucl. Magn. Reson. Spectrosc.* 58: 88–96.

**123** Madl, T., Bermel, W., and Zangger, K. (2009). Use of relaxation enhancements in a paramagnetic environment for the structure determination of proteins using NMR spectroscopy. *Angew. Chem. Int. Ed.* 48: 8259–8262.

**124** Kellner, R., Mangels, C., Schweimer, K., Prasch, S.J., Weiglmeier, P.R., Rösch, P., and Schwarzinger, S. (2009). SEMPRE: spectral editing mediated by paramagnetic relaxation enhancement. *J. Am. Chem. Soc.* 131: 18016–18017.

**125** Hartlmüller, C., Günther, J.C., Wolter, A.C., Wöhnert, J., Sattler, M., and Madl, T. (2017). RNA structure refinement using NMR solvent accessibility data. *Sci. Rep.* 7: 5393.

**126** Kooshapur, H., Schwieters, C.D., and Tjandra, N. (2018). Conformational ensemble of disordered proteins probed by solvent paramagnetic relaxation enhancement (sPRE). *Angew. Chem. Int. Ed.* 57: 13519–13522.

**127** Hocking, H.G., Zangger, K., and Madl, T. (2013). Studying the structure and dynamics of biomolecules by using soluble paramagnetic probes. *Chemphyschem* 14: 3082–3094.

**128** Gong, Z., Gu, X.H., Guo, D.C., Wang, J., and Tang, C. (2017). Protein structural ensembles visualized by solvent paramagnetic relaxation enhancement. *Angew. Chem. Int. Ed.* 56: 1002–1006.

**129** Zhang, M., Tanaka, T., and Ikura, M. (1995). Calcium-induced conformational transition revealed by the solution structure of apo calmodulin. *Nat. Struct. Biol.* 2: 758–767.

**130** Kuboniwa, H., Tjandra, N., Grzesiek, S., Ren, H., Klee, C.B., and Bax, A. (1995). Solution structure of calcium-free calmodulin. *Nat. Struct. Biol.* 2: 768–776.

**131** Gong, Z., Schwieters, C.D., and Tang, C. (2018). Theory and practice of using solvent paramagnetic relaxation enhancement to characterize protein conformational dynamics. *Methods* 148: 48–56.

**132** Mori, S., Abeygunawardana, C., Berg, J.M., and van Zijl, P.C.M. (1997). NMR study of rapidly exchanging backbone amide protons in staphylococcal nuclease and the correlation with structural and dynamic properties. *J. Am. Chem. Soc.* 119: 6844–6852.

**133** Vijay-Kumar, S., Bugg, C.E., and Cook, W.J. (1987). Structure of ubiquitin refined at 1.8 Å resolution. *J. Mol. Biol.* 194: 531–544.

**134** Cordier, F., and Grzesiek, S. (2002). Temperature-dependence of protein hydrogen bond properties as studied by high-resolution NMR. *J. Mol. Biol.* 317: 739–752.

**135** Jurt, S., and Zerbe, O. (2012). A study on the influence of fast amide exchange on the accuracy of 15N relaxation rate constants. *J. Biomol. NMR* 54: 389–400.

**136** Schneider, T., Berg, A., Ulusoy, Z., Gamerdinger, M., Peter, C., and Kovermann, M. (2019). Conformational and functional characterization of artificially conjugated non-canonical ubiquitin dimers. *Sci. Rep.* 9: 19991.

**137** Sugiki, T., Kobayashi, N., and Fujiwara, T. (2017). Modern technologies of solution nuclear magnetic resonance spectroscopy for three-dimensional structure determination of proteins open avenues for life scientists. *Comput. Struct. Biotechnol. J.* 15: 328–339.

**138** Kühlbrandt, W. (2014). The resolution revolution. *Science* 343: 1443–1444.

**139** Schirò, A., Carlon, A., Parigi, G., Murshudov, G., Calderone, V., Ravera, E., and Luchinat, C. (2020). On the complementarity of X-ray and NMR data. *J. Struct. Biol. X* 4: 100019.

**140** Durá, M.A., Receveur-Brechot, V., Andrieu, J.P., Ebel, C., Schoehn, G., Roussel, A., and Franzetti, B. (2005). Characterization of a TET-like aminopeptidase complex from the hyperthermophilic archaeon Pyrococcus horikoshii. *Biochemistry* 44: 3477–3486.

**141** Beckonert, O., Keun, H.C., Ebbels, T.M.D., Bundy, J., Holmes, E., Lindon, J.C., and Nicholson, J.K. (2007). Metabolic profiling, metabolomic and metabonomic procedures for NMR spectroscopy of urine, plasma, serum and tissue extracts. *Nat. Protoc.* 2: 2692–2703.

**142** Wishart, D.S. (2016). Emerging applications of metabolomics in drug discovery and precision medicine. *Nat. Rev. Drug Discov.* 15: 473–484.

**143** O'Shea, K., and Misra, B.B. (2020). Software tools, databases and resources in metabolomics: updates from 2018 to 2019. *Metabolomics* 16: 36.

**144** Emwas, A.H., Roy, R., McKay, R.T., Tenori, L., Saccenti, E., Gowda, G.A.N., Raftery, D., Alahmari, F., Jaremko, L., Jaremko, M., and Wishart, D.S. (2019). NMR spectroscopy for metabolomics research. *Metabolites* 9: 123.

**145** Lewis, I.A., Schommer, S.C., Hodis, B., Robb, K.A., Tonelli, M., Westler, W.M., Sussman, M.R., and Markley, J.L. (2007). Method for determining molar concentrations of metabolites in complex solutions from two-dimensional 1H-13C NMR spectra. *Anal. Chem.* 79: 9385–9390.

**146** Jagger, A.M., Waudby, C.A., Irving, J.A., Christodoulou, J., and Lomas, D.A. (2020). High-resolution ex vivo NMR spectroscopy of human Z α1-antitrypsin. *Nat. Commun.* 11: 6371.

**147** Lomas, D.A., Evans, D.L., Finch, J.T., and Carrell, R.W. (1992). The mechanism of Z α1-antitrypsin accumulation in the liver. *Nature* 357: 605–607.

**148** Wewers, M.D., Casolaro, M.A., Sellers, S.E., Swayze, S.C., McPhaul, K.M., Wittes, J.T., and Crystal, R.G. (1987). Replacement therapy for alpha1-antitrypsin deficiency associated with emphysema. *New Engl. J. Med.* 316: 1055–1062.

**149** Huang, X., Zheng, Y., Zhang, F., Wei, Z., Wang, Y., Carrell, R.W., Read, R.J., Chen, G.Q., and Zhou, A. (2016). Molecular mechanism of Z α1-antitrypsin deficiency. *J. Biol. Chem.* 291: 15674–15686.

**150** Amero, C., Schanda, P., Dura, M.A., Ayala, I., Marion, D., Franzetti, B., Brutscher, B., and Boisbouvier, J. (2009). Fast two-dimensional NMR spectroscopy of high molecular weight protein assemblies. *J. Am. Chem. Soc.* 131: 3448–3449.

# 12

# TROSY

Principles and Applications

*Harindranath Kadavath[1,2] and Roland Riek[1,*]*

[1] *Laboratory of Physical Chemistry, ETH Zurich, Switzerland*
[2] *Department of Structural Biology, St Jude Children's Research Hospital, Memphis, TN, USA*
* *Corresponding Author*

## 12.1 Introduction

Nuclear magnetic resonance (NMR) spectroscopy is a powerful and versatile experimental tool in structural and molecular biology, that allows the structure determination of biomolecules, study of biomolecular interactions and dynamics at atomic-resolution and near-physiological conditions. Understanding the structure-dynamics-function paradigm of biomacromolecules requires high quality data. Structural biology is an emerging area of research with developments in the technical and experimental methods used. In the early phase of structural studies, all atomic-resolution structures of biomolecules were solved either by X-ray diffraction of single protein crystals or by NMR in solution [1, 2]. Recently the revolution in the field of cryo-electron microscopy single particle analysis provided an alternative to X-ray crystallography for large (>100 kDa) molecules with significant gain in resolution [3, 4]. Progress in solid-state NMR methods enabled the determination of high-resolution 3D structures of amyloid fibrils at atomic resolution [5, 6]. Although the developments in multidimensional NMR methods allowed the structure determination of small proteins with size up to ~30 kDa, it is harder to achieve high-quality NMR spectra with enough resolution and sensitivity [7]. In other words, conventional solution NMR-based investigation of biomolecules and macromolecular assemblies is limited by two major challenges. First, the signal overlap caused by a large number of resonances makes the spectral analysis very difficult. Second, in the case of larger molecules faster relaxation of resonances leads to line broadening, poor spectral sensitivity, and hence, much fewer visible NMR peaks (Figure 12.1, and 12.6).

These problems with larger molecules are directly reflected in the scarcity of NMR-based structural studies of biomolecules larger than 25 kDa. With increasing size of the molecule of interest the basic requirements to have an ideal NMR spectrum are difficult to be achieved. However, during the early twenty-first century, significant number of NMR methods were developed devoting to extend the applications of solution NMR to larger molecular systems [8, 13]. In addition, several isotope-labeling schemes were proposed to overcome the overlap of NMR resonances in the spectra [14–16]. Despite these developments, line broadening caused by transverse relaxation remained a major challenge.

To this end, the transverse relaxation-optimized spectroscopy (TROSY) [8] was introduced to reduce transverse relaxation to such an extent to achieve relatively narrow line widths and sensitivity in NMR experiments with larger proteins. TROSY reduces transverse relaxation by spectroscopic means and has significantly extended the

*Two-Dimensional (2D) NMR Methods*, First Edition. Edited by K. Ivanov, P.K. Madhu and G. Rajalakshmi.

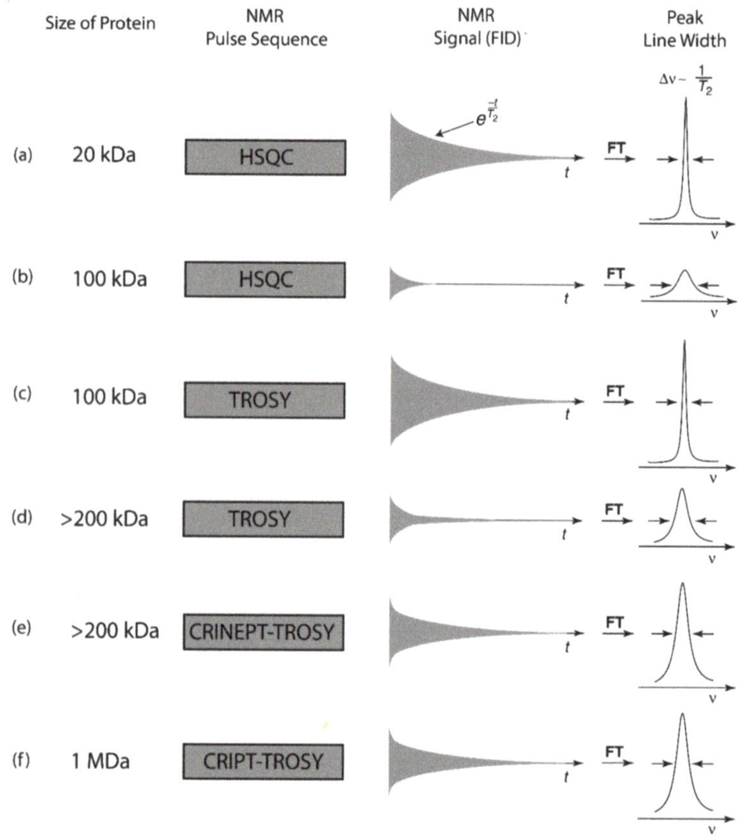


**Figure 12.1** Sketch on the various solution NMR experiments with small and large proteins. (a) The NMR signal obtained from an HSQC spectrum of small proteins (~20 kDa) relaxes slowly with a long transverse relaxation time ($T_2$). Fourier transformation (FT) of the FID with a large $T_2$ value leads to narrow line widths ($\Delta\nu$) in the NMR spectrum. (b) For larger molecules (~100 kDa), the decay of the NMR signal is faster, with shorter $T_2$ values. The HSQC experiment results in a weaker and broad signal in the spectrum. (c) For larger molecules (~100 kDa), using TROSY technique the transverse relaxation can be substantially reduced. This results in an improved spectral resolution and sensitivity with narrow line widths. (d) For very large molecules (~200 kDa), the TROSY experiment yields broad signals. (e and f) In the case of very large molecules (~200 kDa to 1 MDa), the TROSY experiment combined with more efficient polarization transfer methods such as CRINEPT (e) and CRIPT (f) yields the same broad signals as in the TROSY, but with more signal intensities.

size limit of biomacromolecules. Following the introduction of TROSY combined with deuteration, a wide range of new applications in solution NMR is obtained. Transverse relaxation-optimized polarization transfer methods such as cross-correlated relaxation-enhanced polarization transfer (CRINEPT) and cross-correlated relaxation-induced polarization transfer (CRIPT) further increase the sensitivity of NMR experiments and improved line width of NMR peaks of very large systems (Figure 12.1) [17, 18]. Taken together, solution NMR methods combined with TROSY enabled the studies of molecular systems with masses of up to 1 MDa [17, 19–21].

In this chapter, we describe the physical picture and concepts of TROSY, followed by a detailed description of the theoretical principles and important applications. Various developments in the TROSY-based pulse sequences and its applications are described in a separate section. The important applications of TROSY in structural and functional studies of large biological macromolecules are further discussed.

## 12.2 The Principles of TROSY

TROSY (transverse relaxation-optimized spectroscopy) is based on the concept of cross-correlated relaxation rates associated with the interferences between chemical shift anisotropy (CSA) and dipole-dipole interactions that can be significantly reduced. Transverse relaxation of nuclear spins is dominated by dipole-dipole (DD) coupling and CSA. In order to reduce the transverse relaxation rates during the frequency labeling period and acquisition, TROSY exploits constructive interference between the aforementioned relaxation mechanisms. This interference between CSA and DD coupling is termed cross-correlated relaxation [22]. TROSY works best with deuterated proteins and with high-field NMR spectrometers and is ideal for applications to apo and holo forms of proteins with protonated amide groups.

### 12.2.1 The Physical Picture of TROSY

Before providing the theoretical description of the principle of TROSY it is important to have a simple physical picture of a two-spin system, which can be described by a semi classical relaxation theory [8, 22] as described in Equation 12.1. For example, an isolated scalar coupled spins of magnitude $\frac{1}{2}$, such as $^{1}H$ ($I$) and $^{15}N$ ($S$), with a scalar coupling constant of $J_{HN}$ between them can be considered [23]. The transverse relaxation of this spin system is dominated by the DD coupling between the two spins and by the CSA of each individual spin. As shown in Figure 12.2 the relaxation rates of the individual multiplet components of $^{15}N$ spin can be discussed by assuming an axially symmetric $^{15}N$ CSA tensor with the axial principal component parallel to the $^{15}N$–$^{1}H$ vector. The CSA of $^{15}N$ induces a motion-influenced time-dependent magnetic field $B^{CSA}(t)$ on spin $^{15}N$ [24].

$$B^{CSA}(t) \propto \gamma_N B_0 \Delta\sigma_N \left[3\cos^2\theta(t) - 1\right] \tag{12.1}$$

where $\theta(t)$ is the angle between the magnetic field $B_0$ and the axial principal component of the CSA tensor, $\gamma_N$ is the gyromagnetic ratio of $^{15}N$, and $\Delta\sigma_N$ is the CSA part of the chemical shift tensor. The angle $\theta(t)$ and concomitantly $B^{CSA}(t)$ are modulated with time, since the molecule tumbles in the solution as a result of the Brownian motion. In addition, $B^{CSA}(t)$ is dependent on the magnetic field strength $B_0$. This motion-influenced magnetic field $B^{CSA}(t)$ couples to the precessional motion of the nuclear spin and leads to transverse relaxation and line broadening. Similarly, the DD coupling between $^{1}H$ and $^{15}N$ spins induces a motion-influenced time-dependent magnetic field [24],

$$B^{DD}(t) \propto \gamma_H \gamma_N / r_{HN}^3 \left[3\cos^2\theta(t) - 1\right] \tag{12.2}$$

where $r_{HN}$ is the internuclear distance. As discussed above, the tumbling of the molecule modulates $B^{DD}(t)$, which leads to transverse relaxation and line broadening, very similar to $B^{CSA}(t)$. However, unlike $B^{CSA}(t)$, $B^{DD}(t)$ is independent of $B_0$ and the sign of $B^{DD}(t)$ depends on whether the two spins $^{1}H$ and $^{15}N$ are parallel or antiparallel as indicated in Figure 12.2.

Both the time-dependent magnetic fields $B^{CSA}(t)$ and $B^{DD}(t)$ simultaneously influence the relaxation of spin $^{15}N$ and show the same angular and time dependence [23]. Thus, depending on whether the $^{1}H$ spin is parallel or antiparallel with respect to spin $^{15}N$, the two fields either add or subtract, as demonstrated in Figure 12.2. For the multiplet component $S^{12}$, $B^{DD}(t)$ opposes $B^{CSA}(t)$ and results in the favorable narrow linewidth component as shown in Figure 12.3. The other multiplet component $S^{34}$ relaxes faster as the two time-dependent magnetic fields $B^{DD}(t)$ and $B^{CSA}(t)$ adds up together. An optimal compensation of $B^{CSA}(t)$ with $B^{DD}(t)$ to get minimal transverse relaxation rates can be achieved adjusting the size of $B^{CSA}(t)$, which is possible by choosing an optimal magnetic field strength $B_0$ of ~1 GHz [8]. The main principle of TROSY is the selection of the most favorable multiplet component $S^{12}$ and $I^{13}$ by considering transverse relaxation-optimization along both dimensions, $^{15}N$ evolution period and $^{1}H$ acquisition [8]. The half difference between the two relaxations of $S^{34}$ and $S^{12}$ is termed cross-correlated relaxation between DD coupling and CSA. As both the mechanisms interfere, the described effect between DD coupling and CSA is termed cross-correlation between CSA and DD coupling.

In conventional NMR experiments, the multiplet pattern (Figure 12.3) is generally collapsed by decoupling the protons. The $^{1}H$ decoupling flips the spin $^{1}H$ and concomitantly the sign of the local magnetic field produced by $B^{DD}(t)$. Thus, during $^{15}N$-evolution each $^{15}N$ spin is perturbed by the term $B^{CSA}(t)$+$B^{DD}(t)$ during the first half

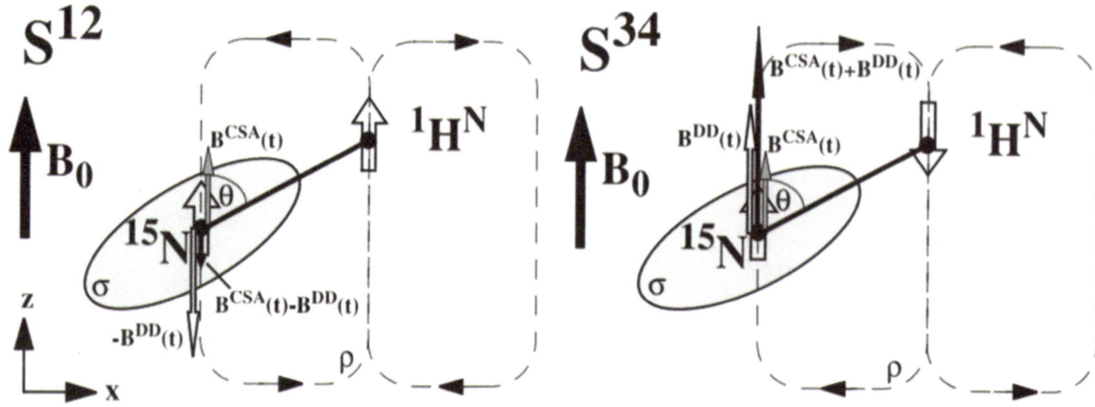


Figure 12.2 Interactions of the local magnetic fields $B^{DD}$(t) with $B^{CSA}$(t). $B_0$ is the static magnetic field. The CSA tensor $\sigma$ ($\Delta\sigma_N$ of Equation 12.1) is displayed by an ellipse. $\theta$ designates the angle between the $^{15}$N–$^{1}$H bond and $B_0$. $\rho$ designates the dipole-dipole interaction between the spins $^{1}$H and $^{15}$N. Figure adapted from Ref. [25] with permission.

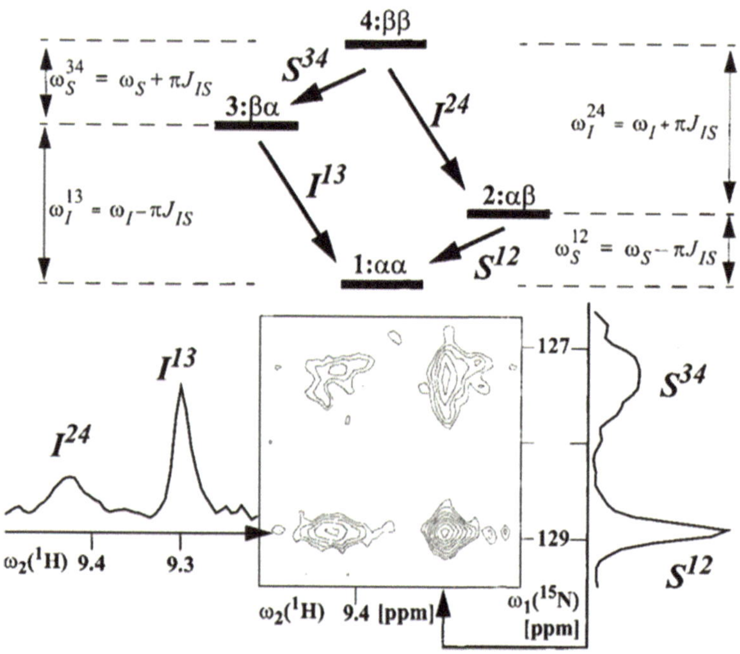


Figure 12.3 Energy level diagram of a two-spin $^{1}/_{2}$ system $I$ and $S$ showing the identification of components of the 2D multiplet expressed via single-transition basis operators, $I^{13}$, $I^{24}$, $S^{12}$, $S^{34}$. Contour plots of a $^{1}$H-$^{15}$N backbone moiety with cross sections of a conventional [$^{1}$H,$^{15}$N] correlation spectrum without decoupling during evolutions. The spectra were measured with a $^{2}$H, $^{15}$N labeled 110-kDa DHNA at 750 MHz at 20 °C. Figure adapted from Ref [25] with permission.

of the evolution time and by $B^{CSA}$(t)–$B^{DD}$(t) during the other half of the evolution time, which leads to a less favorable relaxation when compared with the relaxation of the component $S^{12}$ selected by TROSY.

Thus in a simple and technical perspective, the transverse relaxation-optimization is performed by eliminating the inversion (decoupling) of the scalar-coupled spin $^{1}$H during $^{15}$N evolution and by eliminating the inversion (decoupling) of $^{15}$N during $^{1}$H acquisition.

### 12.2.2 Theory of TROSY

The concept of TROSY is described here on the basis of semi-classical relaxation theory [24]. Consider a system of two scalar-coupled spins $^1/_2$, $I$ and $S$, with a scalar-coupling constant of $J_{\mathrm{IS}}$ and located in a protein molecule. Conventionally $I$ and $S$ stands for $^1$H and $^{15}$N in a $^1$H–$^{15}$N moiety. As explained in the previous section, transverse relaxation of this spin system is dominated by the DD coupling between $I$ and $S$ and by CSA of each individual spin. We should also take into account the additional relaxation mechanism of DD coupling with a few of remote protons, $I_{\mathrm{k}}$. Hence, the relaxation rates of the individual multiplet components in a single-quantum spectrum may be widely different (Figure 12.3) [8, 22]. They can be described using the single-transition basis operators $I^{\pm}_{13}, I^{\pm}_{24}, S^{\pm}_{12}$, and $S^{\pm}_{34}$, which refer to the transitions 1→2, 1→3, 2→4, and 3→4 in the standard energy-level diagram for a system of two spins $^1/_2$ (Figure 12.3), with the corresponding precession frequency [8, 22, 26, 27]:

$$
\begin{aligned}
I^{\pm}_{13} &= \frac{1}{2}I^{+}(1-2S_z) & \omega_I^{13} &= \omega_I - \pi J_{IS},\\
I^{\pm}_{24} &= \frac{1}{2}S^{\pm}(1+2S_z) & \omega_I^{24} &= \omega_I + \pi J_{IS},\\
S^{\pm}_{12} &= \frac{1}{2}S^{\pm}(1-2I_z) & \omega_S^{12} &= \omega_S - \pi J_{IS},\\
S^{\pm}_{34} &= \frac{1}{2}S^{\pm}(1+2I_z) & \omega_S^{34} &= \omega_S + \pi J_{IS}.
\end{aligned}
$$

Then, the first-order relaxation matrix results in an uncoupled system of differential equations with the diagonal form:

$$
\frac{d}{dt}\begin{bmatrix}\langle I^{\pm}_{13}\rangle\\ \langle I^{\pm}_{24}\rangle\\ \langle S^{\pm}_{12}\rangle\\ \langle S^{\pm}_{34}\rangle\end{bmatrix} = -\mathrm{diag}\left(\begin{bmatrix}\pm i\omega_I^{13}\\ \pm i\omega_I^{24}\\ \pm i\omega_S^{12}\\ \pm i\omega_S^{34}\end{bmatrix} + 4J(0)\left(\begin{bmatrix}p^2-2C_{p\delta_r}p\delta_I+\delta_I^2\\ p^2+2C_{p\delta_r}p\delta_I+\delta_I^2\\ p^2-2C_{p\delta_s}p\delta_S+\delta_S^2\\ p^2-2C_{p\delta_s}p\delta_S+\delta_S^2\end{bmatrix} + \begin{bmatrix}\frac{1}{T_{2I}}+\frac{1}{2T_{1S}}\\ \frac{1}{T_{2I}}+\frac{1}{2T_{1S}}\\ \frac{1}{T_{2S}}+\frac{1}{2T_{1I}}\\ \frac{1}{T_{2S}}+\frac{1}{2T_{1I}}\end{bmatrix}\right)\right)\cdot\begin{bmatrix}\langle I^{\pm}_{13}\rangle\\ \langle I^{\pm}_{24}\rangle\\ \langle S^{\pm}_{12}\rangle\\ \langle S^{\pm}_{34}\rangle\end{bmatrix}. \tag{12.3}
$$

In the slow tumbling limit in the absence of radiofrequency pulses, only terms with $\mathrm{J}(\omega = 0) = \frac{2\tau_c}{5}$ need to be retained. The contribution of the DD coupling is,

$$
p = \frac{1}{2\sqrt{2}}\gamma_I/\gamma_S\hbar/r_{IS}^3, \tag{12.4}
$$

and the contributions of the CSAs of spin $I$ and $S$ are

$$
\delta_S = \frac{1}{3\sqrt{2}}\gamma_S B_0 \Delta\sigma_{\mathrm{S}} \tag{12.5}
$$

and

$$
\delta_I = \frac{1}{3\sqrt{2}}\gamma_I B_0 \Delta\sigma_{\mathrm{I}}. \tag{12.6}
$$

where $\gamma_I$ and $\gamma_S$ are the gyromagnetic ratios of $I$ and $S$, $\hbar$ is the Planck constant divided by $2\pi$, $r_{IS}$ the distance between spins $S$ and $I$, $B_0$ the polarizing magnetic field, and $\Delta\sigma_{\mathrm{S}}$ and $\Delta\sigma_{\mathrm{I}}$ are the differences between the axial and the perpendicular principal components of the axially symmetric chemical shift tensors of spins $S$ and $I$, respectively. $C_{p\delta_s} = 0.5(\cos(\nu_{p\delta_s})^2 - 1)$ where $\nu_{p\delta_S}$ is the angle between the tensor axes of the CSA of spin $S$ and the $I$-$S$ vector. Correspondingly, $C_{p\delta_I} = 0.5(\cos(\nu_{p\delta_I})^2 - 1)$ where $\nu_{p\delta_I}$ is the angle between the tensor axes of the CSA of spin $I$ and the $I$–$S$ vector. $1/T_{2I}$, $1/T_{2S}$ and $1/T_{1I}$, $1/T_{1S}$ account for the transverse relaxation and the longitudinal relaxation of spin $I$ and spin $S$ due to DD coupling with remote protons $I_k$.

In the following, we will discuss the transverse relaxation of the different single transitions on the basis of Equation 12.3. The two brackets multiplied by $J(0)$ in Equation 12.3 are the terms of interest. The dominant relaxation mechanisms of spin $I$ and $S$ are the DD coupling and the CSA, which are listed in the first bracket. The relaxation due to DD coupling with remote protons $I_k$ is encountered in the second bracket and will be discussed below. Because of the terms containing $C_{p\delta_I}$ and $C_{p\delta_S}$ the relaxation rates of the individual multiplet components of spin $S$ and spin $I$ are different, as can be inferred also by the different linewidths of the multiplet components in the [$^1$H,$^{15}$N]-HSQC spectrum of Figure 12.3. These terms are due to cross-correlated relaxation between DD coupling and CSA. The name cross-correlated relaxation comes from calculations using a different set of eigenvectors ($I_x$, $S_x$, $2I_xS_z$, $2S_xI_z$). On the basis of these eigenvectors, the relaxation matrix of Equation 12.3 is containing off-diagonal elements with the terms $C_{p\delta_I}$ and $C_{p\delta_S}$. Thus, the term cross-correlation is anticipated.

Whenever CSA and DD coupling are comparable, i.e., $p \approx \delta_S$ or/and $p \approx \delta_I$ and the angles $\nu_{p\delta_N} \approx 0°$ and $\nu_{p\delta_H} \approx 0°$, the first and third row of the bracket of interest would be close to zero, and concomitantly, the transverse relaxation due to DD coupling and CSA would be small for resonances at frequencies $\omega_S^{12}$ and $\omega_I^{13}$. TROSY is just selecting out of the four multiplet components the multiplet with frequencies $\omega_S^{12}$ and $\omega_I^{13}$ (Figure 12.3).

Since DD coupling is field independent, whereas CSA increases proportionally to the field strength (Equations 12.3 to 12.6), there is actually a "magic field" at which for a specific multiplet component, the so-called TROSY component, the transverse relaxation due to DD and CSA will be near zero. For $^1$H–$^{15}$N groups, one approaches this situation at the highest $^1$H frequencies of 800 MHz to 1.2 GHz, and a minimum of transverse relaxation is expected in the $^1$H frequency range from about 950 to 1050 MHz [1, 8]. The ideal situation, where transverse relaxation would be completely quenched, will foreseeably not be attained in practice, for the following reasons: (i) It appears that the CSA is slightly variable depending on the residue type, sequence, and possibly three-dimensional structure, so that there is no common "magic field" for all residues in a protein. The $\Delta\sigma_{HN}$ of backbone amide protons are in the range from 3 to 15 ppm [28, 29] and the $\Delta\sigma_N$ is between −125 and −216 ppm, respectively. (ii) For complete cancellation of transverse relaxation, the CSA tensor would need to be collinear with the $^1$H–$^{15}$N bond, which is not strictly valid. For example, the $\nu_{p\delta_N}$ range from 6-26° [30, 31]. (iii) The remaining relaxation of the single transitions $I_{13}^{\pm}$ and $S_{12}^{\pm}$ is dominated by the DD coupling to remote protons, represented in Equation 12.3 by $1/T_{1I}$, $1/T_{2I}$, $1/T_{1S}$, and $1/T_{2S}$. To evaluate these contributions we identify $I$ and $S$ as the $^1$H and $^{15}$N spins in a $^1$H–$^{15}$N moiety. The relaxation of $^{15}$N is then mainly determined by the CSA and DD interactions with the directly attached proton [26] so that the contributions with remote protons $I_k$, $1/T_{1S}$, and $1/T_{2S}$, can to a good approximation be neglected. However, the DD interactions between spin $I$ and remote protons $I_k$ have to be accounted for [26]:

$$1/T_{1I} = \sum_k \left(\gamma_I^2 \hbar / 2r_k^3\right)^2 J(0), \tag{12.7}$$

$$1/T_{2I} = \sum_k \left(\gamma_I^2 \hbar / 2r_k^3\right)^2 \frac{5}{2} J(0). \tag{12.8}$$

This relaxation pathways cannot be influenced by TROSY. Only with the replacement of non-labile protons with deuterons the transverse relaxation is significantly reduced further, as can be inferred from Table 12. A.1 in Appendix 12.A, where the transverse relaxation rates of $^1$H and $^{15}$N are predicted for a 23-kDa protein. In a conventional [$^1$H,$^{15}$N]-HSQC experiment of a 23-kDa protein, deuteration reduces the $^1$H relaxation rates 2.5-fold and 1.6-fold for $\beta$-sheets and $\alpha$-helices, respectively, and deuteration yields only a small reduction in the $^{15}$N relaxation rate by less than a factor of 1.3 [26]. For [$^1$H,$^{15}$N]-TROSY, deuteration has approximately the same absolute effects on the transverse relaxation, but because of the much smaller relaxation rates the relative improvement is larger, up to 6.5 for $^1H$ and up to 2.9 for $^{15}N$. Conclusively, the combination of TROSY with deuteration dramatically reduces the transverse relaxation.

## 12.3 Practical Aspects of TROSY

The TROSY approach is based on the following technique: in heteronuclear two-spin systems, such as $^1H$–$^{15}N$ and aromatic $^1H$–$^{13}C$ moieties, the NMR signal of each nuclear spin is split into two distinct components by the scalar spin-spin coupling. Thus, a four-line fine structure can be observed in 2D correlation experiments (Figure 12.3). Several of the modern multidimensional NMR experiments use broad-band decoupling during the evolution and detection periods in order to obtain a simplified spectrum with improved sensitivity. As a result, the four-line pattern has been collapsed into a single, central peak. However, it has been known for long time that individual multiplet components have different transverse relaxation rates and hence different line widths [8]. These are mixed by the aforementioned decoupling in several of the classical NMR measurements (Figure 12.3). In contrast, the TROSY approach uses a different technique, which instead of decoupling the multiplet structure, retains only the narrowest and most slowly relaxing line of each multiplet.

In the experimental point of view, the most slowly relaxing multiplet component is specifically selected using pulse sequence elements. The ST2-PT element (single-transition-to-single-transition polarization transfer) [27] in the [$^1H$,$^{15}N$]-TROSY experiment is an example for this approach (Figure 12.4). Absence of any radiofrequency pulses on the attached spin prevents the mixing of the multiplet components during evolution periods. For example, during the $t_1$-evolution period of the [$^1H$,$^{15}N$]-TROSY (Figure 12.4), $^{15}N$ evolves and no other pulses on $^1H$ are applied (please note: radio frequency pulses on $^{15}N$ or a short 360° pulse on $^1H$ during the $^{15}N$-evolution would not destroy the TROSY effect).

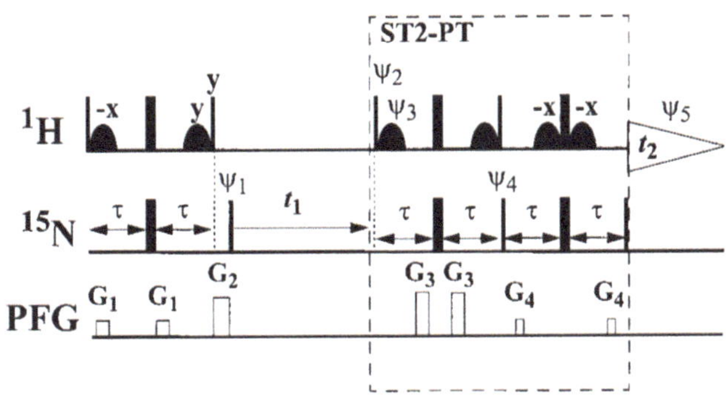


Figure 12.4 Experimental scheme for the 2D [$^1H$,$^{15}N$]-TROSY using single-transition-to-single-transition polarization transfer (box labeled ST2-PT). On the lines marked $^1H$ and $^{15}N$, narrow and wide bars stand for nonselective 90° and 180° radiofrequency pulses, respectively. The delay $\tau$ = 2.7 ms (see text). The line marked PFG indicates the pulsed magnetic field gradients applied along the z-axis: G1, amplitude 30 G/cm, duration 1 ms; G2, 40 G/cm, 1 ms; G3, 40 G/cm, 1 ms; G4; 48 G/cm, 1 ms. The following two-step phase cycling scheme was used: $\Psi_1 = \{y, -x\}$, $\Psi_2 = \{-y\}$), $\Psi_3 = \{y\}$, $\Psi_4 = \{-y\}$, $\Psi_5 = \{y, -x\}$; x on all other pulses. To obtain a complex interferogram a second FID is recorded for each t1 delay, with $\Psi_1 = \{y, x\}$, $\Psi_2 = \{y\}$, $\Psi_3 = \{-y\}$, $\Psi_4 = \{y\}$. The use of ST2-PT thus results in a 2D [$^1H$,$^{15}N$]-correlation spectrum that contains only the most slowly relaxing component of the 2D $^1H$-$^{15}N$ multiplet. The data are processed as described by Kay et al. [33] in an echo/antiecho manner. Water saturation is minimized by keeping the water magnetization along the z-axis during the entire experiment, which is achieved by the application of the water-selective 90° RF pulses indicated by curved shapes on the line $^1H$. It was reported that on some NMR instruments the phase cycle mentioned above does select the desired multiplet component. On these instruments, the replacements of $\Psi_1$ with $\Psi_1 = (y, x)$ for the first FID and $\Psi_1 = (y, -x)$ for the second FID select the desired multiplet component. Figure modified from Ref [25] with permission.

To further explain this approach, the pulse sequence of the [$^1$H,$^{15}$N]-TROSY experiment is discussed below in detail. The evolution of the density operator [8, 27, 34, 35] can be schematically represented as:

$$\frac{1}{2}(uI_z + vS_z) \rightarrow \frac{u+v}{2}S_{12}^{+}\exp(-iw_s^{12}t_1) \rightarrow \frac{u+v}{2}I_{13}^{-}\exp(-iw_s^{12}t_1) \quad (12.9)$$

and

$$\frac{1}{2}(uI_z + vS_z) \rightarrow \frac{u+v}{2}S_{34}^{+}\exp(-iw_s^{34}t_1) \rightarrow \frac{u+v}{2}I_{24}^{-}\exp(-iw_s^{34}t_1). \quad (12.10)$$

The first arrow of Equation 12.9 and 12.10 corresponds to the coherence transfer starting from the steady-state magnetizations $^1$H ($I$) and $^{15}$N ($S$) to $^{15}$N followed by the $^{15}$N chemical shift evolution during the delay $t_1$. This is followed by the ST2-PT element and the second arrow represents the coherence transfer from $^{15}$N to $^1$H using a train of pulses in ST2-PT element. The positive constant factors $u$ and $v$ represents the relative magnitude of the steady-state $^1$H and $^{15}$N magnetization. It can be noticed that only in TROSY-experiments the $^1$H and $^{15}$N steady-state magnetizations can be merged synergistically (Equation 12.9 and 12.10) [27, 36, 37]. This results in a signal increase of 10% for $^1$H–$^{15}$N [27] and 50–100% in the case of $^1$H–$^{13}$C aromatic moieties [36].

After retaining both pathways indicated in Equation 12.9 and 12.10, two diagonally shifted signals per $^1$H–$^{15}$N moiety are observed representing two out of four $^1$H–$^{15}$N multiplet components in the final [$^1$H,$^{15}$N]-correlation spectrum. The undesired polarization transfer pathway, $S_{34}^{\pm} \rightarrow I_{34}^{\pm}$, is suppressed by a two-step phase cycling of $\Psi_1$ and $\Psi_5$ (Figure 12.4). The remaining antiecho polarization transfer pathway of $S_{12}^{+} \rightarrow I_{13}^{-}$ connects single transitions of spin $S$ with spin $I$, and in alternate scans with inversion of the phases $\Psi_2$ and $\Psi_4$, the corresponding echo transfer, $S_{12}^{+} \rightarrow I_{13}^{-}$, is recorded. The sensitivity loss due to the use of only one of the multiplet components can be recovered when working with larger molecules. On the basis of experimental data and theoretical calculations, it was demonstrated that relaxation-induced imbalances between the coherence transfer pathways utilized in the ST2-PT element (Figure 12.4) leads to additional signals at the positions of the broader multiplet components [38]. The intensities of these unanticipated multiplet components are very little compared to the intensities of the TROSY component, however, it is desirable to suppress them. To this end, the "clean TROSY" was proposed, which significantly suppresses the aforementioned artifacts by modifying the ST2-PT element of the [$^1$H,$^{15}$N]-TROSY [39].

### 12.3.1 Field Strength Dependence of TROSY for $^1$H–$^{15}$N Groups

As described in the previous sections, the amide proton of a $^1$H–$^{15}$N moiety relaxes as a result of DD interaction with nitrogen and its own CSA. These interfering relaxation mechanisms results in different relaxation rates for the two multiplet components of the proton, where one rate is smaller and the other one is larger than the average relaxation rate. This effect is very much dependent on the external magnetic field as only CSA, and not DD, relaxation is field dependent. The optimal TROSY effect can thus be obtained by choosing the appropriate field strength, where its relaxation rate will be negligible. For amide protons present in proteins, this "magic field" is ~23.5 T, which correspond to a proton resonance frequency of ~1000 MHz. In a $^1$H–$^{15}$N moiety the $^{15}$N shows a similar interference between $^1$H–$^{15}$N DD interaction and its CSA, which minimizes the $^{15}$N relaxation at the "magic field."

However, since the CSA varies depending on the exact geometry of the amide moieties, in reality significant deviations can be expected from the "magic field" calculated for an isolated two-spin system. In addition, the residual DD couplings with remote protons induces relaxation that cannot be compensated by the TROSY scheme, however can be minimized by sample deuteration. In general, while using deuterated proteins in aqueous solutions, the optimal TROSY effect for $^1$H–$^{15}$N moieties with optimal resolution and sensitivity can be achieved at the high field magnets of presently available 900–1200 MHz spectrometers [8].

### 12.3.2 Peak Pattern of $^1H$-$^{15}N$ TROSY Spectrum

The TROSY scheme is based on the interference of different relaxation mechanisms of a particular nucleus [8], where the interference can be additive or subtractive as explained above. In addition to the ubiquitous relaxation due to DD coupling, CSA of $^1H$, $^{15}N$, and $^{13}C$ nuclei can significantly contribute to the transverse relaxation at the high magnetic fields. This interference of both relaxation mechanisms has been nicely illustrated in a $^1H$–$^{15}N$ correlation spectrum of the original literature (Figure 12.5). Figure 12.5 shows a selected region from $^1H$–$^{15}N$ correlation spectra of the *ftz* homeodomain-DNA complex with the resonance of the indole $^1H$–$^{15}N$ group of Trp-48 buried in the core of the protein [40]. $^1H$ couples to its directly attached $^{15}N$ via scalar coupling and the $^1H$ NMR spectrum of an amide moiety consists of two peaks corresponding to the $^{15}N$ attached protons with spin up and down, respectively, where the effect is observed for the $^{15}N$ nucleus as well. Therefore, in a non-decoupled 2D correlation experiment a four-line fine spectrum is observed (Figure 12.5). However, in conventional heteronuclear correlation experiments, decoupling averages the relaxation rates, and hence the four multiplet peaks are collapsed into a single resonance. For smaller molecules, with all four multiplet components having comparable linewidths, a simplified spectrum with improved sensitivity is observed.

However, in the case of large molecules the multiplet components have variable linewidths (Figure 12.5b) and as a result of decoupling a broad line is observed, which is broader and less intense than the narrowest multiplet component (Figure 12.5c). To this end, TROSY technique selects the slowest relaxing component (the narrowest component) of the four-line pattern (Figure 12.5c) giving rise to sharp signals. Although TROSY neglects part of the potential signal from other multiplet components, it is compensated in large molecules by the slower relaxation and gain in spectral resolution.

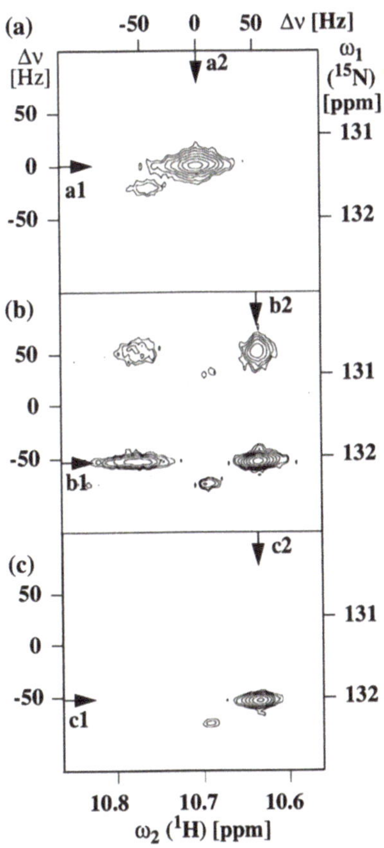


Figure 12.5 Contour plots of $^1H$-$^{15}N$ heteronuclear correlation experiment of a selected resonance. Correlation peak extracted from three distinct types of 2D [$^1H$,$^{15}N$] correlation experiments are shown. (a) Conventional broad-band decoupled [$^1H$,$^{15}N$]-HSQC, (b) conventional [$^1H$,$^{15}N$]-HSQC recorded without decoupling during the experiment and (c) [$^1H$,$^{15}N$]-TROSY spectrum. In both dimensions, chemical shifts in ppm and shifts in Hz relative to the center of the multiplet are shown. Figure adapted from Ref [8], copyright (1997) National Academy of Sciences.

## 12.4 Applications of TROSY

As discussed in the sections above, conventional multidimensional NMR spectroscopy-based structure-function studies are limited to biomolecules smaller than 50 kDa [41, 42]. To this end for a wide range of biomacromolecules the TROSY scheme is highly effective providing workable correlation spectra [27, 34, 43, 44], series of triple-resonance spectra for sequential assignment [26, 38, 45, 46], and nuclear Overhauser effect spectroscopy (NOESY) spectra for collecting conformational restraints [47–49].

The TROSY approach is, in particular, powerful for $^{1}H$–$^{15}N$ backbone moieties, $^{1}H$–$^{13}C^{methyl}$ and $^{1}H$–$^{13}C$ aromatic moieties of proteins [8, 36], and $^{1}H$–$^{15}N$ and $^{1}H$–$^{13}C$ base moieties of oligonucleotides [37, 50–52]. To this end, selected applications of multidimensional TROSY-based [$^{1}H$,$^{15}N$]- and [$^{1}H$,$^{13}C$]-experiments are briefly discussed and compared with the conventional biomolecular NMR experiments.

### 12.4.1 Two-Dimensional [$^{1}H$,$^{15}N$]-TROSY

In the case of larger proteins, 2D [$^{1}H$,$^{15}N$]-TROSY provides high-resolution spectra with enhanced sensitivity as highlighted by a comparison with the corresponding conventional NMR spectra. As an example, a 2D [$^{1}H$,$^{15}N$]-TROSY correlation spectrum of the $^{2}H$,$^{15}N$-labeled homo-octameric protein 7,8-dihydroneopterin aldolase of 110 kDa, from *Staphylococcus aureus* (DHNA) measured with the pulse sequence of Figure 12.4 [27] is shown in Figure 12.6. The optimal sensitivity is achieved by customizing the polarization transfer $\tau$ in Figure 12.4 (3 ms $< 2\tau <$ 5.4 ms) [10].

Considerable spectral resolution and sensitivity enhancements can be achieved for $^{2}H$, $^{15}N$-labeled, or $^{2}H$,$^{13}C$,$^{15}N$-labeled proteins and protein complexes with molecular weights of ~100 kDa [1, 17, 19, 53, 54] and at magnetic fields between 600 and 1200 MHz. Due to the transverse relaxation induced by the DD coupling with remote hydrogens, the enhancements in sensitivity and resolution of non-deuterated large proteins are less pronounced [25]. For proteins with a molecular weight below ~15 kDa, the TROSY is generally less sensitive than the conventional [$^{1}H$,$^{15}N$]-correlation experiment, however, it still has the advantage of higher resolution. It can also be noticed that while using the pulse sequence of Figure 12.4, the side-chain $NH_2$-moieties of Gln and Asn residues are suppressed (Figure 12.6), however, they can be retained using $\tau$ = 1.7 ms, compromising the imperfect suppression of the unwanted multiplet components [55]. An important example of the successful application of TROSY to larger protein, above the "magic field" is the well-resolved resonances obtained for the 419-residue long $^{15}N$-labeled protein PGK1 (phosphoglycerate kinase 1) with a molecular weight of 45 kDa, where the [$^{1}H$,$^{15}N$]-TROSY was measured at the highest presently commercially available field strength of 28.2 T with a $^{1}H$ frequency of 1200 MHz (Figure 12.7, unpublished data). The extremely narrow line width of the PGK1 resonances is attributed to the minimum of transverse relaxation beyond the "magic field" as described in Section 12.2.2

Overall, the gain in spectral resolution and sensitivity extends the applications of 2D [$^{1}H$,$^{15}N$]-heteronuclear correlation experiments to larger systems. Application of TROSY to large proteins and protein complexes yielded valuable quantitative information such as ligand binding based on chemical shift mapping, amide proton-exchange rates, residual dipolar couplings (RDCs), $^{15}N$ spin-relaxation measurements for dynamic elucidation, etc. and are briefly discussed in the upcoming sections.

### 12.4.2 [$^{1}H$,$^{15}N$]-TROSY for Backbone Resonance Assignments in Large Proteins

The assignment of the chemical shifts to individual nuclei is indispensable as a basis for detailed structural studies. Sequential assignment is achieved with "triple-resonance experiments" [56] and obtaining good quality spectra is an essential requirement. Variety of triple-resonance experiments, including the amide moiety, are usually applied for backbone resonance assignments of $^{13}C$,$^{15}N$-labeled proteins [57–63]. Usually, these experiments need to transfer magnetization between $^{1}H_N$, $^{15}N$, and $^{13}C$, and they are commonly applied with molecular sizes up to about

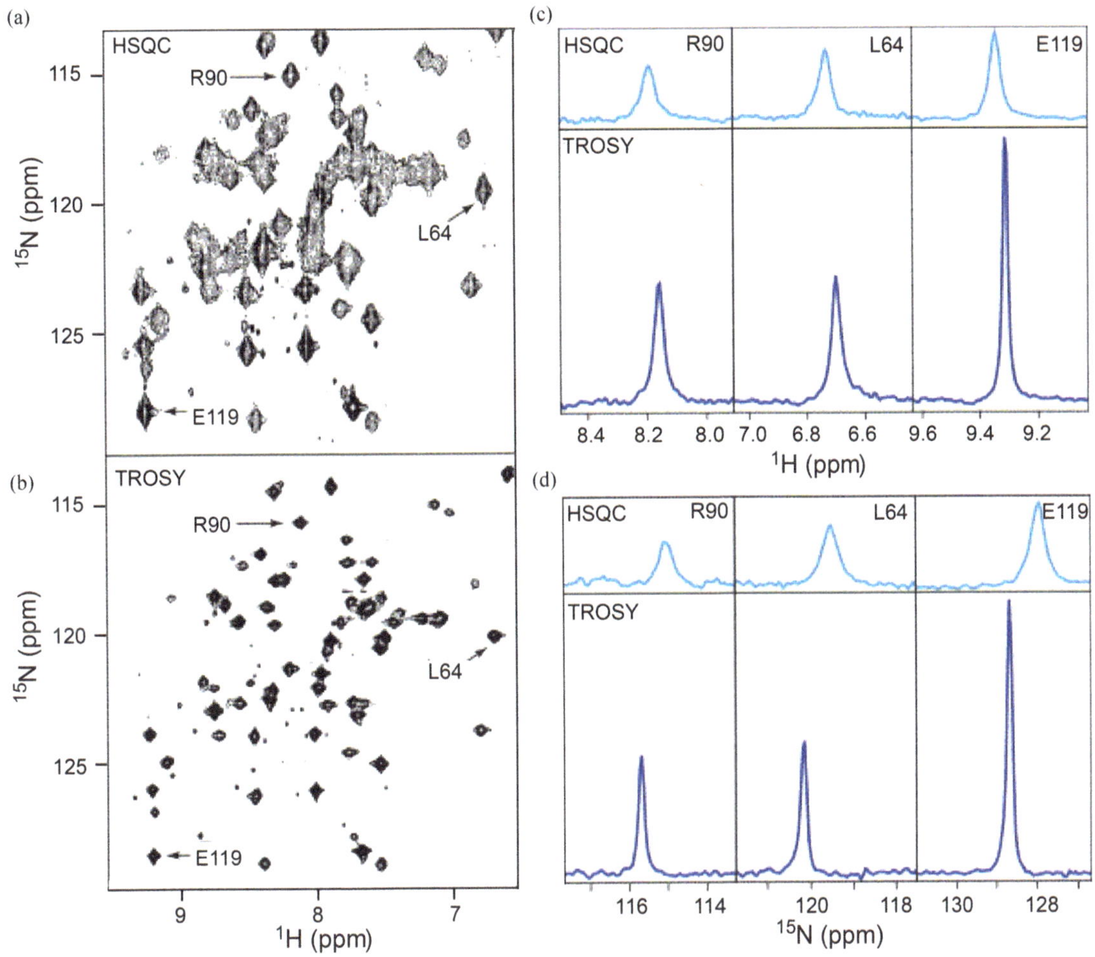


**Figure 12.6** TROSY-type and conventional [$^1$H,$^{15}$N]-correlation spectra of the $^2$H,$^{15}$N-labeled 110 kDa protein DHNA. [$^1$H,$^{15}$N]-correlation experiments afford a "fingerprint" of the protein, which is highly sensitive to changes in the protein environment and thus presents a many-parameter NMR probe for studies of intermolecular interactions and ensuring conformational changes. Comparison of conventional [$^1$H,$^{15}$N]-HSQC (a) and the corresponding [$^1$H,$^{15}$N]-TROSY (b) spectra are shown as contour plots. The spectra were measured at 750 MHz $^1$H frequency and at 20°C. Three selected resonances are marked with an arrow and are labeled with the one-letter amino acid code. One-dimensional cross sections of these residues along the $\omega_2$($^1$H)-dimension (c) and the $\omega_1$($^{15}$N)-dimension (d) are shown for both HSQC and TROSY peaks. Cross sections through individual peak maxima in the contour plots of (a) and (b) provide a 'side view' of the NMR signals that enables a straightforward assessment of the relative peak intensities. In the cross sections, with respect to TROSY, the peaks in HSQC can be seen to be displaced along both frequency axes by ~45 Hz, which corresponds to 0.6 ppm and 0.06 ppm along $\omega_1$($^{15}$N) and $\omega_2$($^1$H), respectively. Figure modified from Ref [9], with permission.

25 kDa. These triple-resonance experiments usually contain extended time periods with transverse amide nitrogen magnetization and when combined with TROSY, it shows considerable signal enhancement.

The use of TROSY in triple-resonance experiments [26, 45, 46, 54, 65] enabled backbone resonance assignments of very large proteins, such as $^2$H,$^{13}$C,$^{15}$N-labeled 67-kDa p53 dimer and the $^2$H,$^{13}$C,$^{15}$N-labeled homo-oligomeric 110-kDa DHNA (Figure 12.8). TROSY effect yielded in ~20-fold signal enhancements for individual residues in the structured segments of the DHNA polypeptide chain, whereas the highly flexible C-terminal residue Lys 121 gave comparable results with both TROSY and non-TROSY experiments [65] (Figure 12.8). Thus, TROSY-based triple-resonance experiments with $^2H,^{13}C,^{15}N$-labeled proteins of various size are superior to conventional triple-resonance experiments. A series of TROSY-based 3D and 4D triple-resonance experiments for $^{13}C,^{15}N$-labeled and $^2H,^{13}C,^{15}N$-labeled proteins can be found in literature and a list of these experiments can be found in Appendix 12. A (Table 12. A.2). The readers are recommended to the given corresponding references for the pulse sequences and technical details. The conversion of conventional triple-resonance experiments into TROSY-based pulse schemes is straightforward and can be done by replacing the [$^1H,^{15}N$]-HSQC module by the [$^1H,^{15}N$]-TROSY of Figure 12.4 (with $\Psi_1 = (y, x)$ for the first FID and $\Psi_1 = (y, -x)$ for the second FID), and the $^1H$- and $^{15}N$-decoupling is removed from the whole pulse sequence.

Most relevant TROSY-based triple-resonance experiments, in general, gain sensitivity of more than one order of magnitude for proteins in the molecular weight range from 25 to 150 kDa [17, 45, 65, 66]. Visibly, (Figure 12.8) shows dramatic improvement in spectral quality where the corresponding TROSY version shows clear cross peaks (Figure 12.8 a) unlike the conventional spectrum, which cannot yield the desired correlations (Figure 12.8 b).

### 12.4.3 [$^1H,^{15}N$]-TROSY for Assignment of Protein Side-chain Resonances

In addition to the backbone resonance assignments of proteins, [$^1H,^{15}N$]–TROSY experiments were developed for assigning $^1H$ and $^{13}C$ chemical shifts of methyl groups in deuterated, selectively methyl-protonated large proteins. As an example for this application, the membrane protein OmpX in 60-kDa DHPC micelles was selectively labeled with protons at the Val, Leu, and Ile (δ1) methyl groups [67]. The correlation of methyl groups with the backbone amide groups was performed using the [$^1H,^{15}N$]-TROSY-based $^{13}C$–$^{13}C$ TOCSY experiments, allowing complete sequence-specific assignments of the protonated methyl groups [68]. This unambiguous assignment procedure had a significant impact to calculate more precise NMR structure of the integral membrane protein OmpX [69]. Another application was to obtain nearly complete assignments of Val, Leu, and Ile (δ1)

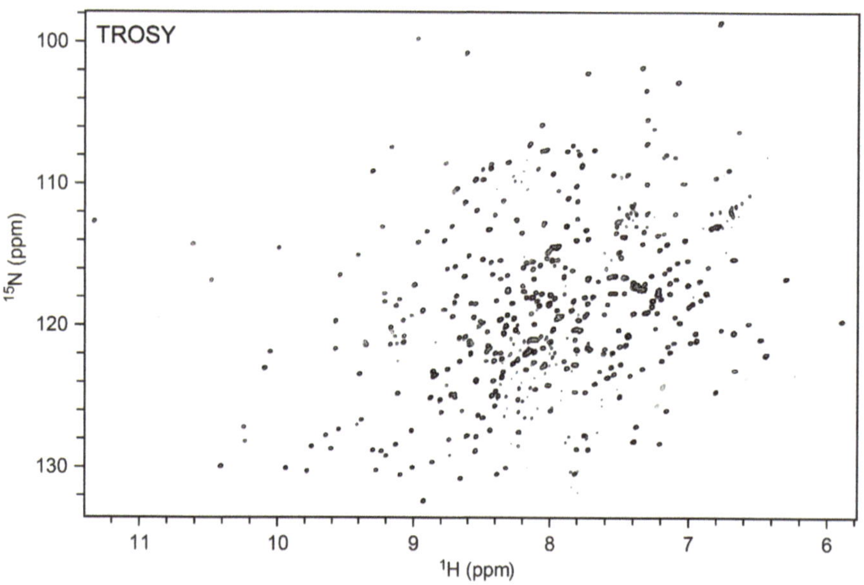


Figure 12.7 [$^1H,^{15}N$]-TROSY spectrum of the $^{13}C$, $^{15}N$-labeled 45-kDa protein PGK1 acquired above the "magic field." The 402 non-proline resonances expected out of the 419 residues are well resolved. The spectrum was measured at 1200 MHz $^1H$ frequency and at 25 °C (unpublished data).

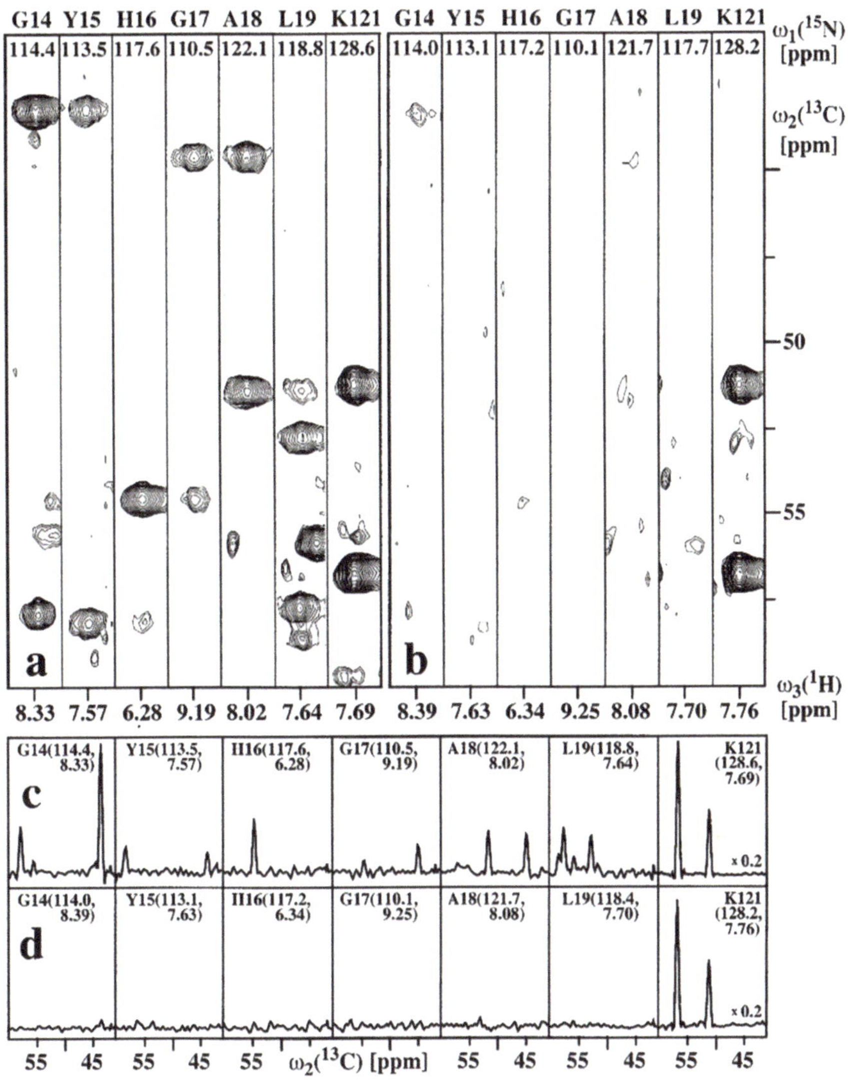


**Figure 12.8** Comparison of corresponding [$\omega_2(^{13}C)$, $\omega_3(^1H)$] strips from two 3D HNCA experiments recorded with a 0.5 mM solution of the uniformly $^2H,^{13}C,^{15}N$-labeled octameric 110-kDa *Staphylococcus aureus* 7,8-dihydroneopterin aldolase (DHNA). (a) [$^{15}N,^1H$]-TROSY-HNCA [11] and (b) conventional HNCA adapted to the scheme employed for panel a by using water flip-back pulses [64]. The strips were taken at the $^{15}N$ chemical shifts assigned to the residues 14-19 and the C-terminal Lys 121. They are centered about the corresponding amide proton chemical shifts and have a width of 131 Hz along $\omega_3(^1H)$. The panels c and d show cross sections along the $\omega_2(^{13}C)$ axis through the peaks in panels (a) and (b), respectively. In each panel, the intraresidual peak is identified by the one-letter amino acid symbol and the residue number, and the $\omega_1(^{15}N)$ and $\omega_3(^1H)$ chemical shifts are indicated in parentheses. Figure adapted from Ref [65].

methyl groups for the 81-kDa protein malate synthase G using new labeling strategies in combination with new TROSY-type NMR experiments designed for side-chain resonances where the Val and Leu isopropyl groups were labeled with $^1H$ and $^{13}C$ in only one of the two methyl groups [70–73] and were assigned using TROSY-based experiments.

### 12.4.4 Application of [1H,15N]-TROSY for RDC Measurements

RDCs are excellent NMR probes to study protein structure and dynamics [74–79]. This is, in particular, important for large, deuterated molecules as only a limited number of constraints can be obtained from NOEs [32, 80]. Dipolar couplings can usually be observed as a splitting of the NMR peak of a given spin that is dipolar coupled to another spin. Unlike low molecular-weight proteins, it is challenging to measure RDCs for biomacromolecules in an aligned medium. Applications of the various TROSY-based experiments developed for measuring RDCs [81–83] showed that precise RDCs can be obtained for proteins with a 30–40 kDa molecular weight [82, 83]. TROSY-based experiments were performed to measure RDCs in larger molecules such as the 723-residue malate synthase G [84], in the 41-kDa maltose-binding protein [85], and in a 53-kDa homomultimeric trimer from mannose-binding protein [86]. However, RDC measurements of large proteins and supramolecular complexes of ~100 kDa and above demands more efficient and sensitive TROSY scheme, and to this end a new solution NMR experiment was proposed named 2D SE2 J-TROSY that measures N-H RDCs for proteins and supramolecular complexes of ~200 kDa [87]. Accurate N-H RDCs were measured for 11-mer 93-kDa $^{2}$H,$^{15}$N-labeled Trp RNA-binding attenuator protein [87].

The HMQC-based methyl TROSY described in detail in Section 12.7.1 can also be used to collect RDC data. There are several limitations to measure conventional N–H RDCs for larger proteins. To gain more adequate sensitivity and resolution, unique properties of $^{1}$H–$^{13}$C cross peaks from methyl-TROSY spectra can be exploited [11, 88, 89]. It is difficult to use RDC data coming from methyl groups of Ile, Leu, or Val residues due to the number of degrees of freedom introduced by side-chain torsion angles. However, RDCs from $^{13}$C-labeled alanine methyl groups can be easily interpreted as they are directly attached to the protein backbone [90]. As an example, a strategy for acquiring RDC data from sparsely isotopically labeled large proteins was illustrated with a 145-kDa dimer of the *Esherichia* heat-shock protein, HtpG [91]. $^{13}$C-alanine methyl-labeled, perdeuterated protein provided backbone-centered structural information from $^{1}$H–$^{13}$C RDCs of alanine methyl groups.

### 12.4.5 [1H,15N]-TROSY-based NOESY Experiments

NMR-based protein structures use interproton distance restraints obtained from $^{15}$N- and $^{13}$C-edited 2D, 3D, and 4D NOESY data sets [92, 93] together with several other structural information such as RDC, dihedral angles, etc. NOE restraints are most essential to define the structure at high resolution which are usually obtained from routine, 3D or 4D $^{15}$N - and $^{13}$C - edited, NOESY experiments [94]. In order to collect structural constraints in large biomolecules, the TROSY principle can further be incorporated into the 3D and 4D NOESY experiments [47–49, 95]. Both, the $^{1}$H–$^{13}$C and $^{1}$H–$^{15}$N TROSY effects have been exploited in 3D $^{13}$C - and $^{15}$N-resolved [$^{1}$H,$^{1}$H]-NOESY experiments [35, 47, 52]. In these experiments, instead of the conventional HSQC or HMQC building blocks, TROSY-based chemical shift correlation schemes were used. The TROSY scheme is used in all three frequency labeling periods, and as an advantage, the otherwise very intense diagonal peaks are extraordinarily suppressed and the weak resonances close to the diagonal are better resolved for analysis, alleviating a limitation of conventional NOE spectroscopy. In addition, the resolution for NH-NH cross peaks is enhanced in all dimensions. The advantages of this approach were demonstrated for the 110-kDa protein aldolase as an example [48]. The CH-NH NOEs in large proteins can be collected by using the 3D $^{15}$N-resolved [$^{1}$H,$^{1}$H]-NOESY experiment [47], where TROSY can be applied during the $^{15}$N-evolution and the acquisition. However, there is minimal benefit of TROSY during the $^{15}$N-evolution period due to the small maximal chemical shift evolution.

The major drawback of the aforementioned experiments is the lesser sensitivity compared to conventional 2D NOESY and 3D NOESY-HSQC/HMQC experiments. For very large systems, the sensitivity loss is compensated by the narrower line widths [8]. The TROSY effect can be enhanced by partial deuteration of proteins, reducing the contributions of $^{1}$H–$^{1}$H dipolar interactions to the relaxation of the $H^{N}$ (and $^{15}$N).

### 12.4.6 Studies of Dynamic Processes Using the [$^1$H,$^{15}$N]-TROSY Concept

In addition to structural studies, NMR can also be used to investigate protein dynamic processes at near atomic resolution over a wide range of time scales [96–99] to exploit the structure-dynamics-function paradigm [100–103]. The most important probes for dynamic studies are T1 and T2 relaxation times and $^1$H–$^{15}$N heteronuclear NOEs of the $^{15}$N nuclei in amide groups. In order to perform these experiments with large biomolecules, pulse sequences using [$^1$H-$^{15}$N]-TROSY were developed [104–106]. Later on, relaxation experiments based on methyl TROSY were implemented in order to study the slow (millisecond) dynamic processes by exploiting the relaxation properties of methyl groups [102, 107].

## 12.5 Transverse Relaxation-optimization in the Polarization Transfers

In heteronuclear NMR experiments, magnetization is usually transferred between the different nuclei via scalar spin-spin couplings using INEPT polarization transfer elements (insensitive nuclei-enhanced polarization transfer [108] (Figure 12.4). Since the slow and fast relaxing transitions are usually mixed, the TROSY scheme is not active during INEPT transfers. Instead, the transverse relaxation-optimization is usually applied only during the evolution and detection periods [8]. In the case of very large biomolecules with molecular weights above approximately ~200 kDa, fast transverse relaxation during the INEPT transfers tends to be a limiting factor in H–N correlation experiments that leads to a complete loss of most signals. Hence, alternative polarization transfer methods are required and to this end various polarization transfer methods such as CRINEPT (Cross-correlated relaxation enhanced polarization transfer) [9, 10, 17] and CRIPT (cross-correlated relaxation induced polarization transfer) [17, 109] were proposed to alleviate the inefficient INEPT transfer. In the CRINEPT-based approach, INEPT and CRIPT schemes are combined [17]. In contrast to INEPT, the efficiency of CRIPT-based polarization transfer increases proportional to the size of the molecule, and it is shown to be an efficient magnetization transfer mechanism for molecules above 200 kDa [10, 17]. CRINEPT and CRIPT were shown to overcome the aforementioned limitations by using cross-correlated relaxation between DD coupling and CSA, which with the use of TROSY becomes a highly efficient transfer mechanism for molecular weights above 200 kDa [10, 110]. Thus, while studying very large biomolecules, significant sensitivity can be achieved by substituting INEPT by CRIPT or CRINEPT in H–N correlation experiments [17].

To allow resonance assignments of macromolecules above ~100 kDa the combination of TROSY and CRIPT or CRINEPT as [$^1$H,$^{15}$N]-CRIPT-TROSY or [$^1$H,$^{15}$N]-CRINEPT-TROSY experiments results in fully transverse relaxation-optimized experiments. It therefore allows the detection of $^1$H-$^{15}$N-fingerprints of macromolecules. As an example, [$^1$H,$^{15}$N]-CRIPT-TROSY spectrum of the heptameric $^2$H,$^{15}$N-labeled GroES in complex with a $^2$H-labeled single-ring variant of GroEL is shown in Figure 12.9. The molecular weight of this complex is 480 kDa and more than 85% of the expected resonances of GroES are observed in the CRIPT-TROSY spectrum [17, 19]. As in the case of TROSY-based experiments, the CRINEPT scheme can be used to study relatively small proteins, ~ 200 amino-acid residues, in complex with large macromolecules, or in large detergent/lipid micelles. The potential of the CRINEPT and CRIPT experiments was also demonstrated for a 900 kDa complex formed by GroES with GroEL (Figure 12.9) [19]. More interestingly, [$^1$H,$^{15}$N]-CRINEPT-HMQC-TROSY can be used to identify the completely bound state of the disordered proteins bound to macromolecular assemblies. It was applied to detect the amino-acid residues of the large (441-residue long) intrinsically disordered Tau protein bound to cytoskeletal microtubule, where the size of the protein complex is of the order of MDa [111, 112].

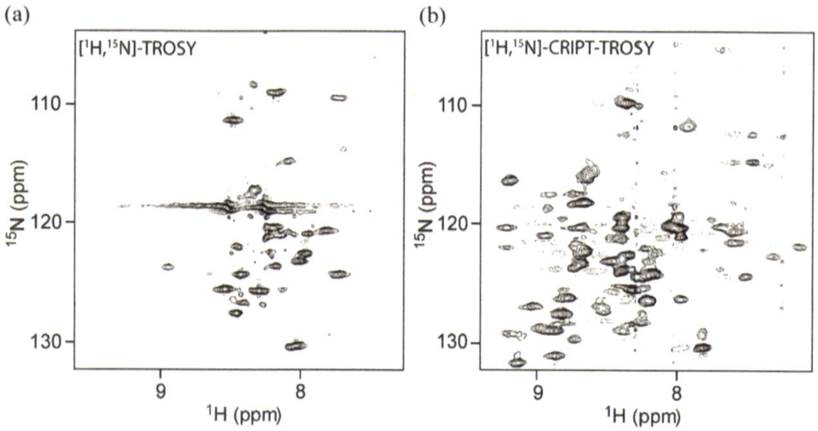


**Figure 12.9** Two-dimensional [$^1$H,$^{15}$N]-TROSY (a) and [$^1$H,$^{15}$N]- CRIPT-TROSY (b) spectra of uniformly $^2$H,$^{15}$N-labeled co-chaperonin GroES bound to SR1 with a molecular weight of 480 kDa. GroES is a homo-heptameric protein with a molecular weight of 78 kDa, and SR1 is a homoheptameric single ring variant of GroEL with a molecular weight of 400 kDa. The cross peaks observed in the [$^1$H,$^{15}$N]-TROSY spectrum are residuals of free GroES. Figure modified from Ref [25], with permission.

## 12.6 $^{15}$N Direct Detected TROSY

Recently, heteronuclear NMR experiments detecting nuclei with low gyromagnetic ratio ($\gamma$) have been proposed, utilizing the slower relaxation properties of $^{13}$C and $^{15}$N. Due to the low gyromagnetic ratio ($\gamma$) of $^{15}$N and low sensitivity of such experiments $^{15}$N detected multidimensional NMR of proteins is sparsely used. To expand the utility of NMR in structural studies of biomolecules, a variety of $^{13}$C- and $^{15}$N-direct detected experiments have been proposed [113–118]. Although $^{15}$N is a low $\gamma$ NMR active nucleus and thus its direct detection shows low sensitivity, $^{15}$N-direct detection can yield highly narrow NMR resonances, which resolves the degeneracy of resonances in high molecular weight and intrinsically disordered proteins (IDPs) and yields good signal to noise in particular at ultra-high magnetic fields [119–121].

Using this approach, a $^{15}$N-detected TROSY spectrum of a 1 mM protein of 67 kDa was recorded in 2 hours with the additional advantage in resolution [120]. The $^{15}$N TROSY is highly magnetic-field dependent and the narrowest linewidth for the TROSY $^{15}$N$_H$ component was obtained at 900 MHz, however, the maximum sensitivity is at 1.2 GHz. In addition, in contrast to conventional TROSY $^1$H detection, deuteration is not necessary for TROSY with $^{15}$N detection. This is, in particular, important because in certain cases of large proteins expressed in deuterated media, amide proton back exchange could be incomplete, which hampers the detection of amide groups at the core of the protein. Thus, the TROSY $^{15}$N detection supports a wider range of proteins such as those that can only be expressed in mammalian or insect cells or even proteins that cannot be refolded for amide back exchange after deuteration where proteins can be expressed in a $^1$H medium without detrimental effects on the $^{15}$N NMR.

## 12.7 [$^1$H,$^{13}$C]-TROSY Correlation Experiments

The most popular TROSY applications mainly include $^1$H–$^{15}$N moieties. However, the TROSY principle is also very effective for $^1$H–$^{13}$C groups in aromatic side chains of proteins and in DNA and RNA nucleotides (see also Section 12.3) [36, 51, 52]. In aromatic spin systems, the $^1$H and $^{13}$C transverse relaxation mechanisms are $^1$H-$^{13}$C DD coupling and $^{13}$C-CSA. The large CSA of $^{13}$C can efficiently compensate the transverse relaxation of $^{13}$C by dipolar coupling to the attached proton. However, the small CSA values of aromatic protons makes the application of TROSY to $^1H$ inefficient and protons are decoupled from $^{13}C$ during acquisition. The optimal TROSY effect of aromatic $^1H$–$^{13}C$ groups is observed at 600 MHz $^1H$ frequency. Thus, for $^{13}C$-labeled biological macromolecules, unlike the conventional HSQC-based experiments, significant sensitivity enhancement can be obtained in [$^1H,^{13}C$]-TROSY spectra of aromatic spin systems.

The basic experiment that exploits the [$^1H,^{13}C$]-TROSY effect in aromatic systems is the 2D constant-time-[$^1H,^{13}C$]-TROSY [36]. With the application of 2D ct-[$^1H,^{13}$Caromatic]-TROSY spectra, a 4- to 10-fold signal enhancement was achieved in $^{13}C,^{15}N$-labeled 18-kDa cyclophilin, when compared with conventional ct-[$^1H,^{13}$Caromatic]-HSQC [36]. Similar signal enhancements were achieved for RNA and DNA molecules [51, 52]. It is interesting to note that merging both the $^1H$ and $^{13}C$ steady-state magnetization resulted in ~2-fold signal gain [36]. Based on the [$^1H,^{13}C$]-TROSY building block, specific 3D pulse sequences can be developed to assist assignments of aromatic spin systems in proteins and nucleic acids [36].

### 12.7.1 Methyl-TROSY NMR

Very interestingly, it was demonstrated that the TROSY effect is not limited to simple AX spin systems but can also be applied to the $^{13}CH_3$ spin system as well, where different dipolar interactions compensate each other in very large molecules [13, 122–126]. Interestingly, the conventional [$^1H,^{13}C$]-HMQC experiment is an optimized TROSY experiment for methyl carbons, maintaining 50% of the carbon magnetization in slowly relaxing states. It was shown that 2D [$^1H,^{13}C$]-HMQC spectra of a perdeuterated, selectively methyl-$^1H,^{13}C$-labeled protein with a rotational correlation time of ~450 ns (equivalent to a ~800-kDa globular molecule at 37°C) gained large enhancements in sensitivity (up to a factor of 3) and resolution relative to [$^1H,^{13}C$]-HSQC-based data sets. These enhancements are derived from a TROSY effect where complete cancellation of intra-methyl $^1H$–$^1H$ and $^1H$–$^{13}C$ dipolar interactions occurs for 50% of the signal in the case of HMQC [11, 70]. It is important to note that the dipolar interaction is field independent, and hence the methyl TROSY can be applied at all field strengths for large molecules. The three identical methyl protons, by default, provide additional sensitivity when recorded with proton detection. This effect was first demonstrated for isoleucine $\delta_1$ methyl groups in a highly deuterated 82-kDa protein, malate synthase G [124]. As discussed in the case of [$^1H,^{15}N$]-TROSY, high levels of deuteration are critical for maximizing the [$^1H,^{13}C$]-TROSY effect. In this context it is important to note that the methyls of Ile ($\delta$1 only), Leu, and Val can be specifically protonated in deuterated proteins.

Overall, both the theoretical and experimental evidence show that TROSY can be very efficiently applied to methyl groups, allowing study of the structure and dynamics of high molecular-weight proteins [67, 126–129]. In particular, since methyls are often localized to hydrophobic cores and molecular interfaces, methyl-methyl distances provide valuable restraints in structural studies [127]. It is even possible to determine exact NOE (eNOE) derived distance restraints [130–133] at ~500 kDa with methyl TROSY-based build up curves. As an example, we demonstrated the applicability of eNOE investigations of the 360-kDa 2 × 7-mer half proteasome from *Thermoplasma acidophilium.* [126] Using the methyl-TROSY approach, eNOEs were measured for the perdeuterated and selectively methyl-labeled Ile, Leu, and Val residues of 360-kDa 2 × 7-mer half proteasome (Figure 12.10). Using the full relaxation matrix approach [134], we corrected spin-diffusion for a NOESY buildup series and extracted several eNOEs [126]. Thus, it was demonstrated that rather accurate distance restraints between methyl-labeled residues can be collected for very large systems, because of the nature of the NOE being a relaxation-based entity, extending the applicability of methyl TROSY to eNOEs.

To elucidate dynamics of a protein relaxation of methyl $^{13}C$, which originates due to its dipolar coupling with the three methyl protons, are measured. Interesting to note is a CPMG-based multiple-quantum relaxation dispersion experiment that measures millisecond dynamic processes at side-chain methyl positions [107].

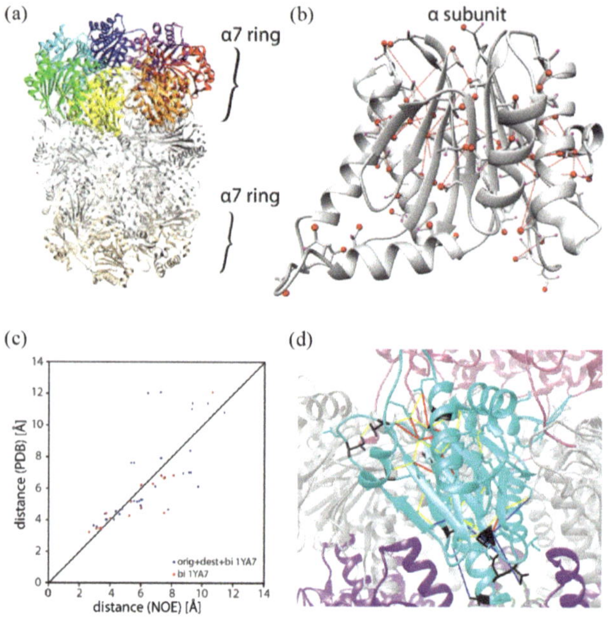


**Figure 12.10** Exact NOEs measured on a 360-kDa proteasome with the NOE-sensitive methyl groups on the α ring of 20S proteasome used for distance measurements. (a) Ribbon representation of the backbone of PDB 1PMA consisting of two heptameric α rings are colored and two heptameric β rings are shown in white. The seven α subunits in the uppermost α ring are distinguished by rainbow colors. (b) The side chains of residues Ile, Leu, and Val are plotted on an individual α subunit. The NOE-sensitive $^{1}H,^{13}C$-labeled methyl groups such as Ileδ1, Valγ2, and Leuδ2 are shown with red spheres at the carbon positions, and the NOE-insensitive $^{2}H,^{12}C$-labeled methyl groups are highlighted as sticks with pink shading. Methyl group pairs for which buildups can be measured are connected by red lines. For more details see detailed description in reference [126]. Figure reproduced from Ref [126] with permission.

## 12.8 Applications to Nucleic Acids

In comparison to proteins, nucleic acids have very little protons available as sources for structural information. NMR spectroscopy of large RNAs is often difficult due to poor signal-to-noise ratio (S/N), resonance overlap, and relaxation properties associated with the larger molecular weight of the RNA of interest. To this end, TROSY offers considerable signal enhancements and resolution for NMR structural studies of nucleic acids [135]. Several of the TROSY-based experiments discussed in the previous sections are of considerable importance for the structure determination of nucleic acids as well. Direct detection of hydrogen bonds [136] and the measurements of RDCs and NOEs are valuable sources for structural restraints. In addition to the increased sensitivity and resolution in correlation experiments for nucleic acids, TROSY further extends the application of state-of- the-art NMR methods to much larger oligonucleotides. Many applications of TROSY can be found in literature and we refer to the following references for further reading [51, 137].

Further TROSY-based NMR experiment use the $^{13}C$ nuclei with its large chemical shift dispersion and favorable relaxation properties with $^{13}C$-detection [135, 138]. In addition, the $^{15}N$-detected TROSY introduced for the investigation of proteins [119] was exploited in RNA to detect H–N correlations. In particular, the recently established, highly sensitive $^{15}N$-detected BEST-TROSY can be applied to RNA molecules ranging in size from 5 to 100 kDa [139, 140].

## 12.9 Intermolecular Interactions and Drug Design

Intermolecular interactions of proteins with other protein partners, nucleic acids, small molecule ligands, etc. provide important insights into the physiological roles of a protein of interest. Therefore, a detailed investigation of all these aspects is of primary requisite in structural biology and drug discovery. As discussed in the previous sections conventional correlation experiments have limitations when the size of the proteins or protein complexes are larger than 30–50 kDa. To this end, TROSY supports a wide range of NMR measurements related to the structural and functional properties of larger macromolecular complexes. Several of the commonly used NMR experiments to study intermolecular interaction can take advantage of TROSY, which include chemical shift mapping, spin-relaxation studies to determine the relaxations constants, intermolecular magnetization transfer studies such as cross-saturation, and hydrogen–deuterium exchange measurements, etc.

Using chemical shift mapping studies, putative contact regions at amino-acid residues resolution can be identified in the protein complex. To study the binding event of verge complexes, TROSY is an ideal method. Many examples of TROSY-based protein–protein and protein–ligand interaction studies can be found in the literature [141, 142]. In this context the first application of TROSY was to study the interactions of pilus chaperone FimC with adhesin FimH from *E. coli* [53]. Since then, a similar approach has been applied to various macromolecular complexes and interaction studies. CRINEPT-based TROSY NMR studies are highly valuable in the case of protein complexes larger than ~200 kDa as discussed in Section 12.5 for GroEL-GroES complex. Another example is the application of CRINEPT-TROSY to the predominantly unfolded p53 core domain bound to Hsp90, which revealed the molecular mechanism of binding and interaction [143]. Furthermore, CRINEPT-HMQC-TROSY was applied to study the interaction of the large, 441-residue long and intrinsically disordered Tau protein binding to microtubules [111] to elucidate the molecular mechanism of interaction of Tau with cytoskeletal microtubules, where CRINEPT transfer allowed the detection of amino-acid residues of Tau tightly bound to microtubules [111, 112].

In addition, several other experiments were proposed to identify the interfaces of large protein–protein complexes in which TROSY and isotope-labeling techniques are coupled. These include the chemical shift mapping, measurements of amide proton exchange rates, and, more importantly, cross-saturation NMR [144]. Cross-saturation phenomena in combination with TROSY detection can be used to identify the contact residues in a large protein–protein complex more accurately, in an optimally deuterium labeled system. This approach was applied to study a 64-kDa immunoglobulin complex with the B domain of protein A (FB), which binds specifically to the Fc fragment of immunoglobulin G. Two-dimensional [$^{1}$H,$^{15}$N]-TROSY experiments were used to identify the binding sites of the FB–Fc complex formed between $^{2}$H,$^{15}$N-labeled FB and unlabeled Fc [144].

## 12.10 Conclusion

An overview of the principles and applications of the TROSY-based methods for the characterization of very large biological macromolecules by solution NMR are provided in this chapter. The development of TROSY, CRINEPT, CRIPT-based high-resolution NMR spectroscopy of macromolecules, and de novo isotope-labeling schemes advanced the applications of solution NMR to biological macromolecules beyond 100 kDa. The recent advances in NMR instrumentation such as high performance cryo probes and high-field superconducting magnets up to 1.2 GHz field strength allowed the implementation of TROSY to macromolecules up to 1000 kDa and above. TROSY combined with conventional resonance assignment strategy and TROSY-based NOESY experiments allow to obtain resonance assignments and NOE restraints for very large biomolecules. Also, it is now possible to collect

NOE restraints to side-chain resonances such as methyl and aromatic protons. This information, combined with various alternative structural constraints such as RDCs, opened an avenue to the de novo 3D structure determination of very large systems by NMR. The TROSY scheme is sufficient to perform detailed studies on intermolecular interactions and investigations of dynamic processes. These developments have been demonstrated in various studies dealing with NMR resonance assignment, structural investigation, intermolecular interactions involving large structures, investigation of integral membrane proteins in solution, drug discovery, and protein dynamic processes both related to function and disease [145]. These data have and will contribute to our mechanistic understanding of many biologically relevant problems. With the prospect of further advances in NMR methods, instrumentation, improved protein isotopic labeling schemes and implementation of advanced machine learning applications solution NMR applied to biological systems both in vitro as well as in cells [146] will contribute significantly to the next frontiers in structural biology as a highly versatile method. We invite the reader to join on this journey.

## 12.A Appendix

**Table 12.A.1** Transverse $^{1}H^{N}$ and $^{15}N$ relaxation rates in [$^{1}H$,$^{15}N$]-TROSY and in [$^{1}H$,$^{15}N$]-HSQC spectra[a], predicted for a 23 kDa protein at 750 MHz. Table modified from Ref [25], with permission.

| $^{1}H$-$^{15}N$ moiety | Transverse relaxation of $^{15}N$ [$s^{-1}$] | | Transverse relaxation of $^{1}H^{N}$ [$s^{-1}$] | |
|---|---|---|---|---|
| | TROSY | HSQC | TROSY | HSQC |
| Isolated | 3.0 | 20.9 | 3.2 | 20.3 |
| $\beta$-sheet of a $^{13}C$,$^{15}N$ labeled protein[b] | 10.6 | 28.5 | 41.1 | 58.2 |
| $\alpha$-helix of a $^{13}C$,$^{15}N$ labeled protein[c] | 8.7 | 26.6 | 31.5 | 48.6 |
| $\beta$-sheet of a $^{2}H$,$^{13}C$,$^{15}N$ labeled protein[d] | 3.7 | 21.6 | 6.3 | 23.5 |
| $\alpha$-helix of a $^{2}H$,$^{13}C$,$^{15}N$ labeled protein[e] | 5.0 | 22.9 | 13.2 | 30.3 |

a) $^{1}H^{N}$ and $^{5}N$ relaxation rates were calculated using Equation 12.5. The values listed for the [$^{1}H$,$^{15}N$]-HSQC are the average relaxation rates of both components of the $^{1}H^{N}$ and the $^{15}N$ doublets. The following parameters were used: $r_{HN} = 1.04$ Å, $\Delta\sigma_{N} = -155$ ppm, $\Delta\sigma_{H} = -15$ ppm, $\nu_{p\delta N} = 15°$, $\nu_{p\delta N} = 10°$ and $\tau_{c} = 15$ ns.
b) Remote protons considered are $^{1}H^{N}(i-1)$, $^{1}H^{N}(i+1)$, $^{1}H^{N}(j)$, $^{1}H(i)$, $^{1}H(i-1)$, $^{1}H(j)$, $^{1}H(i)$, and $^{1}H(i-1)$ at distances of 4.3, 4.3, 3.3, 2.8, 2.2, 3.2, 2.5, and 3.2 Å, respectively, which are typical for an antiparallel $\beta$- sheet. (i is the observed residue, (i – 1) and (i + 1) are sequential neighboring residues and j indicates a long-range contact across the $\beta$-sheet).
c) Remote protons considered are $^{1}H^{N}(i-1)$, $^{1}H^{N}(i+1)$, $^{1}H^{N}(i-2)$, $^{1}H^{N}(i+2)$, $^{1}H(i)$, $^{1}H(i-1)$, $^{1}H(i-2)$, $^{1}H(i-3)$, $^{1}H(i-4)$, $^{1}H(i)$, at distances of 2.8, 2.8, 4.2, 4.2, 2.6, 3.5, 4.4, 3.4, 4.2, and 2.5 Å, respectively.
d) Remote protons considered are $^{1}H^{N}(i-1)$, $^{1}H^{N}(i+1)$, $^{1}H^{N}(j)$ at distances of 4.3, 4.3, and 3.3 Å.
e) Remote protons are $^{1}H^{N}(i-1)$, $^{1}H^{N}(i+1)$, $^{1}H^{N}(i-2)$, $^{1}H^{N}(i+2)$ at distances 2.8, 2.8, 4.2 and 4.2 Å.

**Table 12.A.2** Selected TROSY-triple resonance experiments for proteins. Table modified from Ref [25], with permission.

| Experiment | Reference |
|---|---|
| *For $^{2}H$,$^{13}C$,15N-labeled proteins*: | |
| 3D TROSY-HNCA | Salzmann et al., 1999 [147] |
| 3D TROSY-HNCO | Loria et al., 1999 [38] |
| 3D TROSY-HNCAC*B*[a] | Salzmann et al., 1999 [45] |
| 3D TROSY-CT-HNCA[a] | Salzmann et al., 1999 [148] |
| 3D TROSY-HN(CA)C*B*[b] | Yang and Kay, 1999 [46] |
| 3D TROSY-HN(CO)CA[a] | Salzmann et al., 1999 [45] |
| 3D TROSY-HN(CO)CAC*B*[a] | Salzmann et al., 1999 [45] |
| 3D TROSY-HN(COCA)C*B*[b] | Yang and Kay, 1999 [46] |
| 3D TROSY-HN(CA)CO | Loria et al., 1999 [38] |
| 3D MP-CT-HNCA | Permi and Annila, 2001 [149] |
| 3D sequential HNCAC*B*[c] | Meissner and Sorensen, 2001 [150] |
| 3D HNCAN[d] | Löhr et al., 2000 [151] |
| 4D TROSY-HNCACO[b] | Yang and Kay, 1999 [46] |
| 4D TROSY-HNCOCA[b] | Yang and Kay, 1999 [46] |
| 4D TROSY-HNCOi-1CA[b] | Konrat et al., 1999 [66] |
| 3D SEA-TROSY[e] | Pellecchia et. al., 2001 [152] |
| 3D TROSY-XY-HNCA[f] | Pervushin et. al., 2001 [153] |
| *for $^{13}C$,$^{15}N$-labeled proteins*: | |
| 3D TROSY-HNCA[g] | Eletsky et al., 2001 [154] |
| 3D TROSY-HNCACB | Meissner and Sorensen, 2001, [150] Eletsky et al., 2001 [154] |

a) Improved sensitivity can be gained by concatenating the ST2-PT element [27] with the $^{15}N$ constant-time period as demonstrated for the TROSY-HNCA experiment [38, 147].
b) A different TROSY-component selection is used [155] with similar properties to those of the ST2-PT element (Figure 12.4) [27].
c) A TROSY-HNCACB experiment detecting only the sequential cross peaks.
d) For sequential backbone resonance assignment across proline residues.
f) TROSY-type triple resonance experiments for the detection of solvent-accessible loops in large proteins.
g) TROSY-HNCA experiment with suppression of conformational exchange-induced relaxation.

## Acknowledgement

The authors would like to thank Matthias Bütikofer and Riccardo Cadelbert for providing the PGK1 sample to acquire the TROSY spectrum at the 1.2 GHz spectrometer at ETH Zurich.

## References

1 Wüthrich, K. (1998). The second decade - Into the third millenium. *Nat. Struct. Biol.* 5: 492–495.
2 Greenwald, J., Buhtz, C., Ritter, C., Kwiatkowski, W., Choe, S., Maddelein, M.L., Ness, F., Cescau, S., Soragni, A., Leitz, D., Saupe S.J. and Riek, R. (2010). The mechanism of prion inhibition by HET-S. *Mol. Cell.* 38: 889–899.

3 Amunts, A., Brown, A., Bai, X.C., Llácer, T.J., Hussain, L., Emsley, P., Long, F., Murshudov, G., Scheres, S., H. W. and Ramakrishnan, V. (2014). Structure of the yeast mitochondrial large ribosomal subunit. *Science.* 343: 1485–1489.

4 Scheres, S.H., Zhang, W., Falcon, B. and Goedert, M. (2020). Cryo-EM structures of tau filaments. *Curr Opin. Struct. Bio.* 64: 17–25.

5 Meier, B.H., Riek, R. and Böckmann, A. (2017). Emerging Structural Understanding of Amyloid Fibrils by Solid-State NMR. *Trends. Biochem. Sc.* 42: 777–787.

6 Wälti, M.A., Ravotti, F., Arai, H., Glabe, C.G., Wall, J.S., Böckmann, A., Güntert, P., Meier, B.H. and Riek, R. (2016). Atomic-resolution structure of a disease-relevant Aβ(1-42) amyloid fibril. *Proc. Natl. Acad. Sci. U.S.A.* 113: E4976–4984.

7 Tzakos, A.G., Grace, C.R.R., Lukavsky, P.J. and Riek, R. (2006). NMR techniques for very large proteins and RNAs in solution. *Annu. Rev. Biophys. Biomol. Struct.* 35: 319–342.

8 Pervushin, K., Riek, R., Wider, G. and Wüthrich, K. (1997). Attenuated T2 relaxation by mutual cancellation of dipole-dipole coupling and chemical shift anisotropy indicates an avenue to NMR structures of very large biological macromolecules in solution. *Proc. Natl. Acad. Sci. U.S.A.* 94: 12366–12371.

9 Riek, R. Pervushin, K. and Wüthrich, K. (2000). TROSY and CRINEPT: NMR with large molecular and supramolecular structures in solution. *Trends. Biochem. Sci.* 25: 462–468.

10 Riek, R., Wider, G., Pervushin, K. and Wüthrich, K. (1999). Polarization transfer by cross-correlated relaxation in solution NMR with very large molecules. *Proc. Natl. Acad. Sci. U.S.A.* 96: 4918–4923.

11 Tugarinov, V., Hwang, P.M., Ollerenshaw, J.E. and Kay, L. E. (2003). Cross-correlated relaxation enhanced 1H-13C NMR spectroscopy of methyl groups in very high molecular weight proteins and protein complexes. *J. Am. Chem. Soc.* 125: 10420–10428.

12 Wider, G. and Wüthrich, K. (1999). NMR spectroscopy of large molecules and multimolecular assemblies in solution. *Curr. Opin. Struct. Bio.* 9: 594–601.

13 Huang, C. and Kalodimos, C.G. (2017). Structures of Large Protein Complexes Determined by Nuclear Magnetic Resonance Spectroscopy. *Annu. Rev. Biophys.* 46: 317–336.

14 Gardner, K.H. and Kay, L.E. (1998). in *Annu. Rev. Biophys. Biomol. Struct.* 27: 357–406.

15 Otomo, T. Teruya, K. Uegaki, K. Yamazaki, T. and Kyogoku, Y. (1999). Improved segmental isotope labeling of proteins and application to a larger protein. *J. Biomol. NMR.* 14: 105–114.

16 Kim, I., Lukavsky, P.J. and Puglisi, J.D. (2002). NMR study of 100 kDa HCV IRES RNA, using segmental isotope labeling. *J. Am. Chem. Soc.* 124: 9338–9339.

17 Riek, R., Fiaux, J., Bertelsen, E.B., Horwich, A. L. and Wüthrich, K. (2002). Solution NMR Techniques for Large Molecular and Supramolecular Structures. *J. Am. Chem. Soc.* 124: 12144–12153.

18 Wider, G. (2005). NMR techniques used with very large biological macromolecules in solution. *Methods Enzymol.* 394: 382–398.

19 Fiaux, J., Bertelsen, E.B., Horwich, A.L. and Wüthrich, K. (2002). NMR analysis of a 900K GroEL-GroES complex. *Nature.* 418: 207–211.

20 Fernández, C. and Wider, G. (2003). TROSY in NMR studies of the structure and function of large biological macromolecules. *Curr. Opin. Struct. Biol.* 13: 570–580.

21 Jiang, Y. and Kalodimos, C.G. (2017). NMR Studies of Large Proteins. *J. Mol. Biol.* 429: 2667–2676.

22 Goldman, M. (1984). Interference effects in the relaxation of a pair of unlike spin-12 nuclei. *J. Magn. Reson.* 60: 437–452.

23 Farrar, T.C. and Stringfellow, T.C. (2007). in *eMagRes.*

24 Abragam, A. (1961). *The Principles of Nuclear Magnetism.* Oxford: Clarendon Press.

25 Riek, R. (2002). in *BioNMR in Drug Res.* 227–241.

26 Salzmann, M., Pervushin, K., Wider, G., Senn, H. and Wüthrich, K. (1998). TROSY in triple-resonance experiments: new perspectives for sequential NMR assignment of large proteins. *Proc. Natl. Acad. Sci. U.S.A.* 95: 13585–13590.

27 Pervushin, K.V., Wider, G. and Wüthrich, K. Single Transition-to-single Transition Polarization Transfer (ST2-PT) in [15N,1H]-TROSY. *J. Biomol. NMR.* 12: 345–348.

28 Tjandra, N. and Bax, A. (1997). Solution NMR Measurement of Amide Proton Chemical Shift Anisotropy in 15N-Enriched Proteins. *J. Am. Chem. Soc.* 119: 8076–8082.

29 Tessari, M. and Vuister, G. W. (2000). A novel experiment for the quantitative measurement of CSA(1H(N))/CSA(15N) cross-correlated relaxation in 15N-labeled proteins. *J. Biomol. NMR.* 16: 171–174.

30 Fushman, D. and Cowburn, D. (1998). Model-Independent Analysis of 15N Chemical Shift Anisotropy from NMR Relaxation Data. *J. Am. Chem. Soc.* 120: 7109–7110.

31 Kroenke, C.D., Rance, M. and Palmer, (1999). Variability of the 15N Chemical Shift Anisotropy in Escherichia coli Ribonuclease H in Solution. *J. Am. Chem. Soc.* 121, 10119–10125.

32 Choy, W.-Y., Tollinger, M. Mueller, G.A. and Kay, L.E. (2001). Direct structure refinement of high molecular weight proteins against residual dipolar couplings and carbonyl chemical shift changes upon alignment: an application to maltose binding protein. *J. Biomol. NMR.* 21: 31–40.

33 Kay, L., Keifer, P. and Saarinen, T. (1992). Pure absorption gradient enhanced heteronuclear single quantum correlation spectroscopy with improved sensitivity. *J. Am. Chem. Soc.* 114: 10663–10665.

34 Rance, M., Loria, J.P. and Palmer, A.G. (1999). Sensitivity Improvement of Transverse Relaxation- Optimized Spectroscopy. *J. Magn. Reson.* 136: 92–101.

35 Zhu, G., Xia, Y., Sze, K.H. and Yan, X. (1999). 2D and 3D TROSY-enhanced NOESY of 15N labeled proteins. *J. Biomol. NMR.* 14: 377–381.

36 Pervushin, K., Riek, R., Wider, G. and Wüthrich, K. (1998). Transverse Relaxation-Optimized Spectroscopy (TROSY) for NMR Studies of Aromatic Spin Systems in 13C-Labeled Proteins. *J. Am. Chem. Soc.* 120: 6394–6400.

37 Riek, R. (2001). Enhancement of the steady-state magnetization in TROSY experiments. *J. Biomol. NMR.* 21: 99–105.

38 Loria, J.P., Rance, M. and Palmer, A.G. (1999). 3rd, Transverse-relaxation-optimized (TROSY) gradient-enhanced triple-resonance NMR spectroscopy. *J. Magn. Reson.* 141: 180–184.

39 Schulte-Herbrüggen, T. and Sorensen, O. W. (2000). Clean TROSY: compensation for relaxationinduced artifacts. *J. Magn. Reson.* 144: 123–128.

40 Qian, Y.Q., Furukubo-Tokunaga, K., Resendez-Perez, D., Müller, M., Gehring W.J. and Wüthrich, K. (1994). Nuclear Magnetic Resonance Solution Structure of the fushi tarazu Homeodomain from Drosophila and Comparison with the Antennapedia Homeodomain. *J. Mol. Biol.* 238: 333–345.

41 Bax, A. (1994). Multidimensional nuclear magnetic resonance methods for protein studies. *Curr. Opin. Struct. Biol.* 4: 738–744.

42 Gardner, K.H., Zhang, X., Gehring, K. and Kay, L.E. (1998). Solution NMR Studies of a 42 KDa Escherichia Coli Maltose Binding Protein/β-Cyclodextrin Complex: Chemical Shift Assignments and Analysis. *J. Am. Chem. Soc.* 120: 11738–11748.

43 Zhu, G., Kong, X., Yan, X. and Sze, K. (1998). Sensitivity Enhancement in Transverse Relaxation Optimized NMR Spectroscopy. *Angew. Chem. Int. Ed. Engl.* 37: 2859–2861.

44 Weigelt, J. (1998). Sensitivity- and Gradient-Enhanced TROSY for Multidimensional NMR Experiments. *J. Am. Chem. Soc.* 120: 10778–10779.

45 Salzmann, M., Wider, G., Pervushin, K., Senn, H. and Wüthrich, K. (1999). TROSY-type triple-resonance experiments for sequential NMR assignments of large proteins. *J. Am. Chem. Soc.* 121: 844–848.

46 Yang, D. and Kay, L. E. (1999). TROSY triple-resonance four-dimensional NMR spectroscopy of a 46 ns tumbling protein. *J. Am. Chem. Soc.* 121: 2571–2575.

**47** Zhu, G., Xia, Y., Lin, D. and Gao, X. (2004). TROSY-based correlation and NOE spectroscopy for NMR structural studies of large proteins. *Methods Mol. Biol.* 278: 57–78.

**48** Pervushin, K.V., Wider, G., Riek, R. and Wüthrich, K. (1999). The 3D NOESY-[(1)H,(15)N,(1)H]-ZQTROSY NMR experiment with diagonal peak suppression. *Proc. Natl. Acad. Sci. U.S.A.* 96: 9607–9612.

**49** Xia, Y., Sze, K. and Zhu, G. (2000). Transverse relaxation optimized 3D and 4D 15N/15N separated NOESY experiments of 15N labeled proteins. *J. Biomol. NMR.* 18: 261–268.

**50** Pervushin, K., Ono, A., Fernández, C., Szyperski, T., Kainosho, M. and Wüthrich, K. (1998). NMR scalar couplings across Watson-Crick base pair hydrogen bonds in DNA observed by transverse relaxation-optimized spectroscopy. *Proc. Natl. Acad. Sci. U.S.A.* 95: 14147–14151.

**51** Riek, R., Pervushin, K., Fernández, C., Kainosho, M. and Wüthrich, K. (2001). [13C,13C]- and [13C,1H]-TROSY in a Triple Resonance Experiment for Ribose-Base and Intrabase Correlations in Nucleic Acids1. *J. Am. Chem. Soc.* 123: 658–664.

**52** Brutscher, B., Boisbouvier, J., Pardi, A., Marion, D. and Simorre, J.-P. (1998). Improved Sensitivity and Resolution in 1H-13C NMR Experiments of RNA. *J. Am. Chem. Soc.* 120: 11845–11851.

**53** Pellecchia, M., Sebbel, P., Hermanns, U., Wüthrich, K. and Glockshuber, R. (1999). Pilus chaperone FimC-adhesin FimH interactions mapped by TROSY-NMR. *Nat. Struct. Biol.* 6: 336–339.

**54** Mulder, F.A., Ayed, A., Yang, D., Arrowsmith, C.H. and Kay, L.E. (2000). Assignment of 1H(N), 15N, 13C(alpha), 13CO and 13C(beta) resonances in a 67 kDa p53 dimer using 4D-TROSY NMR spectroscopy. *J. Biomol. NMR.* 18: 173–176.

**55** Pervushin, K., Braun, D., Fernández, C. and Wüthrich, K. (2000). [15N,1H]/[13C,1H]-TROSY for simultaneous detection of backbone 15N-1H, aromatic 13C-1H and side-chain 15N-1H2 correlations in large proteins. *J. Biomol. NMR.* 17: 195–202.

**56** Bax, A. and Grzesiek, S. (1993). Methodological advances in protein NMR. *Accounts Chem. Res.* 26: 131–138.

**57** Ikura, M., Kay, L.E., Krinks, M. and Bax, A. (1991). Triple-resonance multidimensional NMR study of calmodulin complexed with the binding domain of skeletal muscle myosin light-chain kinase: indication of a conformational change in the central helix. *Biochemistry.* 30: 5498–5504.

**58** Ikura, M., Krinks, M., Torchia, D.A. and Bax, A. (1990). An efficient NMR approach for obtaining sequence-specific resonance assignments of larger proteins based on multiple isotopic labeling. *FEBS. Lett.* 266: 155–158.

**59** Grzesiek, S. and Bax, A. (1993). Amino acid type determination in the sequential assignment procedure of uniformly 13C/15N-enriched proteins. *J. Biomol. NMR.* 3: 185–204.

**60** Grzesiek, S. and Bax, A. (1997). A three-dimensional NMR experiment with improved sensitivity for carbonyl-carbonyl J correlation in proteins. *J. Biomol. NMR.* 9: 207–211.

**61** Sattler, M., Maurer, M., Schleucher, J. and Griesinger, C. (1995). A simultaneous (15)N, (1)H- and (13)C, (1)H-HSQC with sensitivity enhancement and a heteronuclear gradient echo. *J. Biomol. NMR.* 5: 97–102.

**62** Sattler, M., Schwendinger, M.G., Schleucher, J. and Griesinger, C. (1995). Novel strategies for sensitivity enhancement in heteronuclear multi-dimensional NMR experiments employing pulsed field gradients. *J. Biomol. NMR.* 6: 11–22.

**63** Schleucher, J., Schwendinger, M., Sattler, M., Schmidt, P., Schedletzky, O., Glaser, S.J., Sørensen, O.W. and Griesinger, C. (1994). A general enhancement scheme in heteronuclear multidimensional NMR employing pulsed field gradients. *J. Biomol. NMR.* 4: 301–306.

**64** Grzesiek, S. and Bax, A. (1992). Correlating backbone amide and side chain resonances in larger proteins by multiple relayed triple resonance NMR. *J. Am. Chem. Soc.* 114: 6291–6293.

**65** Salzmann, M., Pervushin, K., Wider, G., Senn, H. and Wüthrich, NMR assignment and secondary structure determination of an octameric 110 kDA protein using TROSY in triple resonance experiments. *K. J. Am. Chem. Soc.* 122: 7543–7548.

**66** Konrat, R., Yang, D. and Kay, L. E., (1999). A 4D TROSY-based pulse scheme for correlating 1HNi,15Ni,13Calphai,13C'i-1 chemical shifts in high molecular weight, 15N,13C, 2H labeled proteins. *J. Biomol. NMR.* 15: 309–313.

**67** Goto, N.K., Gardner, K.H., Mueller, G.A., Willis, R.C. and Kay, L.E. (1999). A robust and cost-effective method for the production of Val, Leu, Ile (ä1) methyl-protonated 15N-, 13C-, 2Hlabeled proteins. *J. Biomol. NMR.* 13: 369–374.

**68** Hilty, C., Fernández, C., Wider, G. and Wüthrich, K. (2022). Side chain NMR assignments in the membrane protein OmpX reconstituted in DHPC micelles. *J. Biomol. NMR.* 23: 289–301.

**69** Fernández, C., Hilty, C., Wider, G., Güntert, P. and Wüthrich, K. (2004). NMR structure of the integral membrane protein OmpX. *J. Mol. Biol.* 336: 1211–1221.

**70** Tugarinov, V. and Kay, L.E. (2004). An Isotope Labeling Strategy for Methyl TROSY Spectroscopy. *J. Biomol. NMR.* 28: 165–172

**71** Tugarinov, V. and Kay, L.E. (2004). Stereospecific NMR Assignments of Prochiral Methyls, Rotameric States and Dynamics of Valine Residues in Malate Synthase G. *J. Am. Chem. Soc.* 126: 9827–9836.

**72** Neri, D., Szyperski, T., Otting, G., Senn, H. and Wüthrich, K. (1989). Stereospecific nuclear magnetic resonance assignments of the methyl groups of valine and leucine in the DNA-binding domain of the 434 repressor by biosynthetically directed fractional 13C labeling. *Biochemistry.* 28: 7510–7516.

**73** Szyperski, T., Neri, D., Leiting, B., Otting, G. and Wüthrich, K. (1992). Support of 1H NMR assignments in proteins by biosynthetically directed fractional 13C-labeling. *J. Biomol. NMR.* 2: 323–334.

**74** Meiler, J., Prompers, J.J., Peti, W., Griesinger, C., and Brüschweiler, R. (2001). Model-Free Approach to the Dynamic Interpretation of Residual Dipolar Couplings in Globular Proteins. *J. Am. Chem. Soc.* 123: 6098–6107.

**75** Lakomek, N.-A., Walter, K. F. A., Farès, C., Lange, O.F., de Groot, B.L., Grubmüller, H., Brüschweiler, R., Munk, A., Becker, S., Meiler, J. and Griesinger, C. (2008). Self-consistent residual dipolar coupling based model-free analysis for the robust determination of nanosecond to microsecond protein dynamics. *J. Biomol. NMR.* 41: 139.

**76** Tolman, J.R. (2002). A Novel Approach to the Retrieval of Structural and Dynamic Information from Residual Dipolar Couplings Using Several Oriented Media in Biomolecular NMR Spectroscopy. *J. Am. Chem. Soc.* 124: 12020–12030.

**77** Lakomek, N.A., Carlomagno, T., Becker, S., Griesinger, C. and Meiler, J. (2006). A Thorough Dynamic Interpretation of Residual Dipolar Couplings in Ubiquitin. *J. Biomol. NMR.* 34: 101–115.

**78** Zweckstetter, M. and Bax, A. (2001). Single-step determination of protein substructures using dipolar couplings: aid to structural genomics. *J. Am. Chem. Soc.* 123: 9490–9491.

**79** Zweckstetter, M. and Bax, A. (2002). Evaluation of uncertainty in alignment tensors obtained from dipolar couplings. *J. Biomol. NMR.* 23: 127–137.

**80** Mueller, G.A., Choy, W.Y., Yang, D., Forman-Kay, J.D., Venters, R.A. and Kay, L.E. (2000). Global folds of proteins with low densities of NOEs using residual dipolar couplings: application to the 370-residue maltodextrin-binding protein. *J. Mol. Biol.* 300: 197–212.

**81** Prestegard, J. H., Bougault, C.M. and Kishore, A.I. (2004). Residual Dipolar Couplings in Structure Determination of Biomolecules. *Chem. Rev.* 104: 3519–3540.

**82** Fitzkee, N.C. and Bax, A. (2010). Facile measurement of $^{1}$H-$^{15}$N residual dipolar couplings in larger perdeuterated proteins. *J. Biomol. NMR.* 48: 65–70.

**83** Bhattacharya, A., Revington, M. and Zuiderweg, E.R. (2010). Measurement and interpretation of 15N-1H residual dipolar couplings in larger proteins. *J. Magn. Reson.* 203: 11–28.

**84** Tugarinov, V. and Kay, L.E. (2003). Quantitative NMR studies of high molecular weight proteins: application to domain orientation and ligand binding in the 723 residue enzyme malate synthase G. *J. Mol. Biol.* 327: 1121–1133.

**85** Evenäs, J., Mittermaier, A., Yang, D. and Kay, E. (2001). Measurement of 13Cá.13Câ Dipolar Couplings in 15N,13C,2H-Labeled Proteins: Application to Domain Orientation in Maltose Binding Protein. *J. Am. Chem. Soc.* 123: 2858–2864.

**86** Jain, N.U., Noble, S. and Prestegard, J.H. (2003). Structural characterization of a mannose-binding protein-trimannoside complex using residual dipolar couplings. *J. Mol. Biol.* 328: 451–462.

**87** Eletsky, A., Pulavarti, S.V.S.R.K., Beaumont, V., Gollnick, P. and Szyperski, T. (2015). Solution NMR Experiment for Measurement of 15N-1H Residual Dipolar Couplings in Large Proteins and Supramolecular Complexes. *J. Am. Chem. Soc.* 137: 11242–11245.

**88** Rosenzweig, R. and Kay, L.E. (2014). Bringing dynamic molecular machines into focus by methyl- TROSY NMR. *Annu. Rev. Biochem.* 83: 291–315.

**89** Liao, X., Godoy-Ruiz, R., Guo, C. and Tugarinov, V. (2011). Simultaneous measurement of $^{1}$H-$^{15}$N and methyl $^{1}$Hm-$^{13}$Cm residual dipolar couplings in large proteins. *J. Biomol. NMR.* 51: 191–198.

**90** Godoy-Ruiz, R., Guo, C. and Tugarinov, V. (2010). Alanine Methyl Groups as NMR Probes of Molecular Structure and Dynamics in High-Molecular-Weight Proteins. *J. Am. Chem. Soc.* 132: 18340–18350.

**91** Pederson, K., Chalmers, G. R., Gao, Q., Elnatan, D., Ramelot, T.A., Ma, L.-C., Montelione, G.T., Kennedy, M.A., Agard, D.A. and Prestegard, J.H. (2017). NMR characterization of HtpG, the E. coli Hsp90, using sparse labeling with 13C-methyl alanine. *J. Biomol. NMR.* 68: 225–236.

**92** Clore, G.M. and Gronenborn, A.M. (1991). Structures of larger proteins in solution: three- and fourdimensional heteronuclear NMR spectroscopy. *Science.* 252: 1390–1399.

**93** Wagner, G. (1989). Heteronuclear nuclear magnetic resonance experiments for studies of protein conformation. *Methods Enzymol.* 176: 93–113.

**94** Kay, L.E., Clore, G.M., Bax, A. and Gronenborn, A. M. (1990). Four-dimensional heteronuclear tripleresonance NMR spectroscopy of interleukin-1 beta in solution. *Science.* 249: 411–414.

**95** Meissner, A. and Sørensen, O.W. (1999). Suppression of diagonal peaks in TROSY-type 1H NMR NOESY spectra of 15N-labeled proteins. *J. Magn. Reson.* 140: 499–503.

**96** Kay, L.E., Torchia, D.A. and Bax, A. (1989). Backbone dynamics of proteins as studied by nitrogen-15 inverse detected heteronuclear NMR spectroscopy: application to staphylococcal nuclease. *Biochemistry.* 28: 8972–8979.

**97** Palmer, A.G. (2004). NMR Characterization of the Dynamics of Biomacromolecules. *Chem. Rev.* 104: 3623–3640.

**98** Kay, L.E. (2005). NMR studies of protein structure and dynamics. *J. Magn. Reson.* 173: 193–207.

**99** Griesinger, C., Peti, W., Meiler, J. and Brüschweiler. R. (2004). Projection angle restraints for studying structure and dynamics of biomolecules. *Methods Mol. Biol.* 278: 107–121.

**100** B. Vögeli, S. Olsson, R. Riek and P. Güntert, Complementarity and congruence between exact NOEs and traditional NMR probes for spatial decoding of protein dynamics. *J. Struct. Biol.* 2015, 191, 306-317.

**101** Mulder, F.A., Mittermaier, A., Hon, B., Dahlquist, F. W. and Kay, L. E. (2001). Studying excited states of proteins by NMR spectroscopy. *Nat. Struct. Biol.* 8: 932–935.

**102** Vahidi, S., Ripstein, Z.A., Juravsky, J.B., Rennella, E., Goldberg, A.L., Mittermaier, A.K., Rubinstein, J.L. and Kay, L.E. (2020). An allosteric switch regulates Mycobacterium tuberculosis ClpP1P2 protease function as established by cryo-EM and methyl-TROSY NMR. *Proc. Natl. Acad. Sci. U.S.A.* 117: 5895–5906.

**103** Eisenmesser, E.Z., Bosco, D.A., Akke, M. and Kern, D. (2002). Enzyme dynamics during catalysis. *Science.* 295: 1520–1523.

**104** Lakomek, N.A., Ying, J. and Bax, A. (2012). Measurement of $^{15}$N relaxation rates in perdeuterated proteins by TROSY-based methods. *J. Biomol. NMR.* 53: 209–221.

**105** Ishimá, R. (2012). Recent developments in (15)N NMR relaxation studies that probe protein backbone dynamics. *Top. Curr. Chem.* 326: 99–122.

**106** O'Brien, P.A. and Palmer, A.G. (2018) 3rd, TROSY pulse sequence for simultaneous measurement of the (15)N R(1) and (1)H-(15)N NOE in deuterated proteins. *J. Biomol. NMR.* 70: 205–209.

**107** Korzhnev, D.M., Kloiber, K., Kanelis, V., Tugarinov, V. and Kay, L. E. (2004). Probing Slow Dynamics in High Molecular Weight Proteins by Methyl-TROSY NMR Spectroscopy: Application to a 723-Residue Enzyme. *J. Am. Chem. Soc.* 126: 3964–3973.

**108** Morris, G.A. and Freeman, R. (1979). Enhancement of nuclear magnetic resonance signals by polarization transfer. *J. Am. Chem. Soc.* 101: 760–762.

**109** Dalvit, C. (1992). 1H to 15N polarization transfer via 1H chemical-shift anisotropy — 1H-15N dipole-dipole cross correlation. *J. Magn. Reson.* 97: 645–650.

**110** Brüschweiler, R. and Ernst, R.R. (1992). Molecular dynamics monitored by cross-correlated cross relaxation of spins quantized along orthogonal axes. *J. Chem. Phys.* 9: 1758–1766.

**111** Kadavath, H., Hofele, R.V., Biernat, J., Kumar, S., Tepper, K., Urlaub, H., Mandelkow, E. and Zweckstetter, M. (2015). Tau stabilizes microtubules by binding at the interface between tubulin heterodimers. *Proc. Natl. Acad. Sci. U.S.A.* 112: 7501–7506.

**112** Kadavath, H., Jaremko, M., Jaremko, Ł., Biernat, J., Mandelkow, E. and Zweckstetter, M. (2015). Folding of the Tau Protein on Microtubules. *Angew. Chem. Int. Ed. Engl.* 54: 10347–10351.

**113** Felli, I.C. and Brutscher, B. (2009). Recent Advances in Solution NMR: Fast Methods and Heteronuclear Direct Detection. *ChemPhysChem.* 10: 1356–1368.

**114** Hsu, S.-T.D., Bertoncini, C.W. and Dobson, C.M. (2009). Use of Protonless NMR Spectroscopy To Alleviate the Loss of Information Resulting from Exchange-Broadening. *J. Am. Chem. Soc.* 131: 7222–7223.

**115** Serber, Z., Richter, C. and Dötsch, V. (2001). Carbon-Detected NMR Experiments To Investigate Structure and Dynamics of Biological Macromolecules. *Chembiochem.* 2: 247–251.

**116** Gal, M., Edmonds, K.A., Milbradt, A.G., Takeuchi, K. and Wagner, G. (2011). Speeding up direct 15N detection: hCaN 2D NMR experiment. *J. Biomol. NMR.* 51: 497–504.

**117** Vasos, P.R., Hall, J.B., Kümmerle, R. and Fushman, D. (2006). Measurement of 15N relaxation in deuterated amide groups in proteins using direct nitrogen detection. *J. Biomol. NMR.* 36: 27–36.

**118** Levy, G.C. (1979). *Spectroscopy.*

**119** Takeuchi, K., Arthanari, H., Imai, M., Wagner, G. and Shimada, I. (2016). Nitrogen-detected TROSY yields comparable sensitivity to proton-detected TROSY for non-deuterated, large proteins under physiological salt conditions. *J. Biomol. NMR.* 64: 143–151.

**120** Takeuchi, K., Arthanari, H., Shimada, I. and Wagner, G. (2015). Nitrogen detected TROSY at high field yields high resolution and sensitivity for protein NMR. *J. Biomol. NMR.* 63: 323–331.

**121** Chhabra, S., Fischer, P., Takeuchi, K., Dubey, A., Ziarek, J.J., Boeszoermenyi, A., Mathieu, D., Bermel, W., Davey, N.E., Wagner, G. and Arthanari, H. (2018). (15)N detection harnesses the slow relaxation property of nitrogen: Delivering enhanced resolution for intrinsically disordered proteins. *Proc. Natl. Acad. Sci. U.S.A.* 115: E1710–e1719.

**122** Kreishman-Deitrick, M., Egile, C., Hoyt, D. W., Ford, J.J., Li, R., and Rosen, M.K. (2003). NMR Analysis of Methyl Groups at 100-500 kDa: Model Systems and Arp2/3 Complex. *Biochemis.* 42: 8579–8586.

**123** Guo, C., Zhang, D. and Tugarinov, V. (2008). An NMR Experiment for Simultaneous TROSY-Based Detection of Amide and Methyl Groups in Large Proteins. *J. Am. Chem. Soc.* 130: 10872–10873.

**124** Tugarinov, V. and Kay, L. E. (2003). Assignments of the 723-Residue Malate Synthase G Using a New Labeling Strategy and Novel NMR Methods. *J. Am. Chem. Soc.* 125: 13868–13878.

**125** Rossi, P., Monneau, Y.R., Xia, Y., Ishida, Y. and Kalodimos, C.G. (2019). Toolkit for NMR Studies of Methyl-Labeled Proteins. *Methods Enzymol.* 614: 107–142.

**126** Chi, C.N., Strotz, D., Riek, R. and Vogeli, B. (2018). NOE-Derived Methyl Distances from a 360 kDa Proteasome Complex. *Chem. Eur. J.* 24: 2270–2276.

**127** Gardner, K.H., Rosen, M.K. and Kay, L.E. (1997). Global folds of highly deuterated, methylprotonated proteins by multidimensional NMR. *Biochemistry.* 36: 1389–1401.

**128** Ollerenshaw, J.E., Tugarinov, V., Skrynnikov, N.R. and Kay, L.E. (2005). Probing Slow Time Scale Dynamics at Methyl-Containing Side Chains in Proteins by Relaxation Dispersion NMR Measurements: Application to Methionine Residues in a Cavity Mutant of T4 Lysozyme. *J. Biomol. NMR.* 33: 25–41.

**129** Skrynnikov, N.R., Mulder, F.A.A., Hon, B., Dahlquist, F.W. and Kay, L.E. (2001). Comparison of 13CH3, 13CH2D, and 13CHD2 methyl labeling strategies in proteins. *J. Am. Chem. Soc.* 123: 4556–4566.

**130** Vogeli, B., Kazemi, S., Guntert, P. and Riek, R. (2012). Spatial elucidation of motion in proteins by ensemble-based structure calculation using exact NOEs *Nat. Struct. Mol. Biol.* 19: 1053–U1110.

**131** Vogeli, B., Olsson, S., Guntert, P. and Riek, R. (2016). The Exact NOE as an Alternative in Ensemble Structure Determination. *Biophys. J.* 110: 113–126.

**132** Vögeli, B., Orts, J., Strotz, D., Chi, C., Minges, M., Wälti, M.A., Güntert, P. and Riek, R. (2014). Towards a true protein movie: a perspective on the potential impact of the ensemble-based structure determination using exact NOEs. *J. Magn. Reson.* 241: 53–59.

**133** Kadavath, H. and Riek, R. (2021). *in NMR Spectroscopy for Probing Functional Dynamics at Biological Interfaces. R. Soc. Chem.*

**134** Orts, J., Vogeli, B. and Riek, R. (2012). Relaxation Matrix Analysis of Spin Diffusion for the NMR Structure Calculation with eNOEs. *J. Chem. Theory. Comput.* 8: 3483–3492.

**135** Schnieders, R., Knezic, B., Zetzsche, H., Sudakov, A., Matzel, T., Richter, C., Hengesbach, M., Schwalbe, H. and Fürtig, B. (2020). NMR Spectroscopy of Large Functional RNAs: From Sample Preparation to Low-Gamma Detection. *Curr. Protoc. Nucleic. Acid. Chem.* 82: e116.

**136** Dallmann, A. and Sattler, M. (2014). Detection of hydrogen bonds in dynamic regions of RNA by NMR spectroscopy. *Curr. Protoc. Nucleic. Acid. Chem.* 59: 7.22.21–19.

**137** Brutscher, B., Boisbouvier, J., Kupče, E., Tisné, C., Dardel, F., Marion, D. and Simorre, J.-P. (2001). Base-type-selective high-resolution 13C edited NOESY for sequential assignment of large RNAs. *J. Biomol. NMR.* 19: 141–151.

**138** Richter, C., Kovacs, H., Buck, J., Wacker, A., Fürtig, B., Bermel, W. and Schwalbe, H. (2010). 13C-direct detected NMR experiments for the sequential J-based resonance assignment of RNA oligonucleotides. *J. Biomol. NMR.* 47: 259–269.

**139** Schnieders, R., Richter, C., Warhaut, S., de Jesus, V., Keyhani, S., Duchardt-Ferner, E., Keller, H., Wöhnert, J., Kuhn, L.T., Breeze, A.L., Bermel, W., Schwalbe, H. and Fürtig, B. (2017). Evaluation of 15N-detected H-N correlation experiments on increasingly large RNAs. *J. Biomol. NMR.* 69: 31–44.

**140** Schnieders, R., Keyhani, S., Schwalbe, H. and Fürtig, B. (2020). H. Schwalbe, B. Fürtig, More than Proton Detection-New Avenues for NMR Spectroscopy of RNA. *Chemistry.* 26: 102–113.

**141** Gao, G., Williams, J.G. and Campbell, S.L. (2004). Protein-protein interaction analysis by nuclear magnetic resonance spectroscopy. *Methods Mol. Biol.* 261: 79–92.

**142** Bonvin, A.M., Boelens, R. and Kaptein, R. (2005). NMR analysis of protein interactions. *Curr. Opin. Chem. Biol.* 9: 501–508.

**143** Rudiger, S., Freund, S.M., Veprintsev, D.B. and Fersht, A.R. (2002). CRINEPT-TROSY NMR reveals p53 core domain bound in an unfolded form to the chaperone Hsp90. *Proc. Natl. Acad. Sci. U.S.A.* 99: 11085–11090.

**144** Takahashi, H., Nakanishi, T., Kami, K., Arata, Y. and Shimada, I. (2000). A novel NMR method for determining the interfaces of large protein-protein complexes. *Nat. Struct. Biol.* 7: 220–223.

**145** Schuetz, A.K. and Kay, L.E. (2016). A Dynamic molecular basis for malfunction in disease mutants of p97/VCP. *eLife.* 5: e20143.

**146** Burmann, B.M., Gerez, J.A., Matečko-Burmann, I., Campioni, S., Kumari, P., Ghosh, D., Mazur, A., Aspholm, E.E., ulskis, D., Wawrzyniuk, M., Bock, T., Schmidt, A., Rüdiger, S.G.D., Riek, R. and Hiller S. (2000). Regulation of á-synuclein by chaperones in mammalian cells. *Nature.* 577: 127–132.

**147** Salzmann, M., Wider, G., Pervushin, K. and Wüthrich, K. (1999). Improved sensitivity and coherence selection for [15N,1H]-TROSY elements in triple resonance experiments. *J. Biomol. NMR.* 15: 181–184.

**148** Salzmann, M., Pervushin, K., Wider, G., Senn, H. and Wüthrich, K. (1999). [13C]-constant-time [15N,1H]-TROSY-HNCA for sequential assignments of large proteins. *J. Biomol. NMR.* 14: 85–88.

**149** Permi, P. and Annila, A. (2001). A new approach for obtaining sequential assignment of large proteins. *J. Biomol. NMR.* 20: 127–133.

**150** Meissner, A. and Sørensen, O.W. (2001). Sequential HNCACB and CBCANH protein NMR pulse sequences. *J. Magn. Reson.* 151: 328–331.

**151** Löhr, F., Pfeiffer, S., Lin, Y.J.,and Hartleib, J., Klimmek, O. and Rüterjans, H. (2000). HNCAN pulse sequences for sequential backbone resonance assignment across proline residues in perdeuterated proteins. *J. Biomol. NMR.* 18: 337–346.

**152** Pellecchia, M., Meininger, D., Shen, A.L., Jack, R., Kasper, C.B. and Sem, D.S. (2001). SEA-TROSY (solvent exposed amides with TROSY): a method to resolve the problem of spectral overlap in very large proteins. *J. Am. Chem. Soc.* 123: 4633–4634.

**153** Pervushin, K., Gallius, V. and Ritter, C. (2001). Improved TROSY-HNCA experiment with suppression of conformational exchange induced relaxation. *J. Biomol. NMR.* 21: 161–166.

**154** Eletsky, A., Kienhöfer, A. and Pervushin, K. (2001). TROSY NMR with partially deuterated proteins. *J. Biomol. NMR.* 20: 177–180.

**155** Yang, D. and Kay, L.E. (1999). Improved 1HN-detected triple resonance TROSY-based experiments. *J. Biomol. NMR.* 13: 3–10.

# 13

# Two-Dimensional Methods and Zero- to Ultralow-Field (ZULF) NMR

*K. Ivanov[1], John Blanchard[2], Dmitry Budker[3], Fabien Ferrage[4], Alexey Kiryutin[1], Tobias Sjolander[5], Alexandra Yurkovskaya[1], and Ivan Zhukov[1]*

[1] *International Tomography Center SB RAS, 630090 Novosibirsk, Russia*
[2] *Quantum Technology Center, University of Maryland, College Park, MD 20742, USA*
[3] *Helmholtz Institut Mainz, Johannes Gutenberg-Universität Mainz, 55128 Mainz, Germany; Department of Physics, University of California, Berkeley, California 94720-300, USA*
[4] *Laboratoire des Biomolécules, LBM, Département de chimie, École normale supérieure, PSL University, Sorbonne Université, CNRS, 75005 Paris, France*
[5] *Department of Physics, University of Basel, Klingelbergstrasse 82, Basel CH-4056, Switzerland*

## 13.1 Introduction and Motivation

Two-dimensional NMR spectroscopy, first proposed by Jean Jeener [1] and implemented by Aue, Bartholdi, and Ernst [2], transformed NMR into a powerful tool for chemistry, materials science, structural biology, and medicine. Two-dimensional NMR has increased both the resolution and the information content of NMR spectra, revealing correlations between nuclei through chemical bonds or through space. At the heart of 2D-NMR is the ability to control which interactions govern the evolution of the spin system, whether to transfer polarization or select which information is displayed in each spectral dimension. The control of nuclear spin systems is usually obtained by application of well-defined sequences of radiofrequency (RF) pulses and delays [3, 4].

A completely different way to select how the spin system evolves, introduced by Pines and coworkers for solids [5] and later extended to solutions [6, 7], consists in reducing the magnetic field down to strengths small enough that the interaction with the magnetic field is negligible compared to internal interactions, such as dipolar and scalar couplings. Nowadays, magnetic fields in the nanotesla range (or smaller) can be obtained so that proton homonuclear scalar couplings may become larger than the Zeeman interaction in the zero- to ultralow-field (ZULF) regime. In this chapter, we review recent literature that uses evolution in ZULF conditions to tailor the evolution of spin systems in 2D-NMR.

A 2D NMR experiment can be schematically represented with four consecutive elements (Figure 13.1):

- Preparation: converting the equilibrium polarization into a desired coherence;
- $t_1$ time evolution: the coherence obtained after the preparation evolves under one or more interactions;
- Mixing: selected coherences at the end of the delay $t_1$ are converted into an observable operator;
- $t_2$ time evolution: detection.

It is possible to select and let evolve during the delay $t_1$, in the indirect dimension, coherences that cannot be detected at high-field, such as multiple-quantum coherences. The effective Hamiltonian operator during $t_1$ can

*Two-Dimensional (2D) NMR Methods*, First Edition. Edited by K. Ivanov, P.K. Madhu and G. Rajalakshmi.

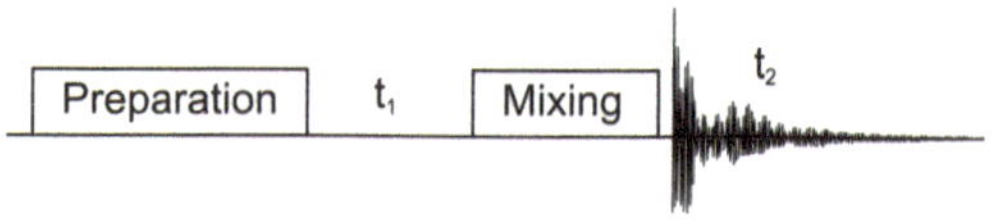


**Figure 13.1** Principle of a 2D-NMR experiment. The preparation is the part of the sequence before the delay $t_1$. The density operator evolves under possibly tailored interactions during the delay $t_1$. Between delays $t_1$ and $t_2$, the density operator changes during mixing before detection during the time $t_2$.

also be designed so as to obtain information on suited parts of the static Hamiltonian operator: scalar couplings, chemical shifts, or their combinations. In addition, the evolutions during $t_1$ and $t_2$ can involve coherences of like spins, in homonuclear correlations, or unlike spins, in heteronuclear correlations.

At high magnetic fields, a key difference between heteronuclear spin systems composed of unlike spins and homonuclear spin systems composed of like spins is that the former can be manipulated independently by hard RF pulses with no Bloch-Siegert effects. On the other hand, RF pulses applied on homonuclear spin systems can affect all spins. Zero- and ultralow-field conditions are defined by the fact that Zeeman interactions are negligible with respect to scalar-coupling interactions. Thus, even in heteronuclear systems of unlike spins, it is not possible to apply a pulsed DC field on only one kind of nuclear spin. However, the precession under a DC pulse will take place with different frequencies for nuclei with different gyromagnetic ratios.

In principle, 2D-NMR experiments can be designed with the use of field-cycling and ZULF conditions at any of the four steps shown in Figure 13.1:

- **Preparation** at high field in a ZULF experiment. Although an entire 2D-NMR pulse sequence can be performed under ZULF conditions, the **preparation** step usually involves the polarization of nuclear spins at high or moderate fields for sensitivity purposes.
- **Preparation** and **mixing** during field cycling. The transition between high-field and ZULF conditions often involves the non-adiabatic passage through level anti-crossings [8], which leads to another essential part of the **preparation**: the excitation of multiple-quantum coherences that will evolve at ZULF conditions.
- $t_1$ evolution in ZULF conditions and **detection** at high field. Performing the evolutions at ZULF in the indirect dimension and at high field in the direct dimension offers the possibility to correlate scalar-coupling quantities (at ZULF) and chemical shifts at high field, in a manner reminiscent of J-spectroscopy. Detection at high field offers chemical-shift resolution, which allows the investigations of complex molecules or mixtures, combined with high sensitivity.
- **Mixing** at ZULF with $\mathbf{t_1}$ and $\mathbf{t_2}$ evolutions at high field. Field cycling between high-field and ZULF may also be used in experiments with evolutions in the direct $\mathbf{t_1}$ and indirect $\mathbf{t_2}$ dimensions at high field but **mixing** under the particular Hamiltonian operator obtained in ZULF conditions.

## 13.2 Early Work

It is worth noting that the first experimental realizations of zero-field NMR [5, 9] were carried out as 2D experiments, with preparation and detection carried out at high magnetic field. The general concept is illustrated in Figure 13.2a. For these experiments, the initial state was prepared in a high-field (4 T) spectrometer, after which the sample was shuttled adiabatically out of the magnet into a fringe field of 10 mT. Spin evolution was then initiated by lowering the field suddenly to zero using a pair of resistive coils – after allowing the spins to evolve for a time $t_1$, the 10 mT field was reintroduced, selectively preserving only those components of the nuclear magnetization aligned with the magnetic field. The sample was then shuttled back into the high-field spectrometer for readout. Using such schemes, high-field–zero-field correlation spectroscopy was reported in [9].

The first "true" 2D zero-field NMR experiment – having two variable delays at zero magnetic field – was proposed in [9] and demonstrated by Thayer et al. in 1986 [10]. Suter et al. then expanded upon this concept, designing a four-pulse exchange experiment that allowed for the observation of spin diffusion [11] at zero magnetic field. The experimental protocol is shown in Figure 13.2b, and features adiabatic demagnetization to zero field rather than the sudden lowering of the field shown in Figure 13.2a. By varying the mixing time, the authors were able to extract

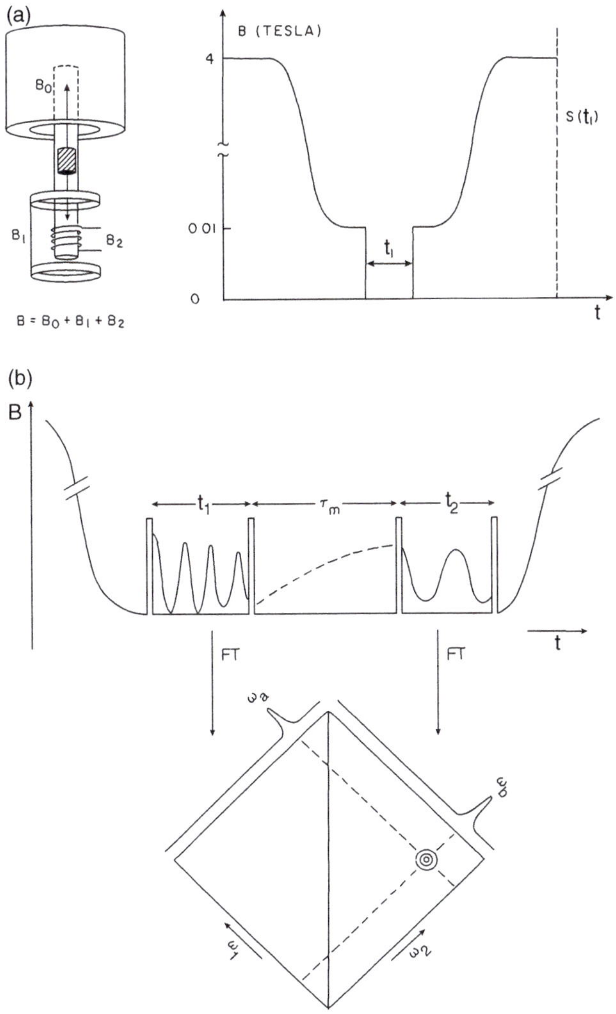


**Figure 13.2** (a) Schematic diagram of the field-cycling apparatus and a time-dependent field profile used for early zero-field NMR experiments [9]. (b) Illustration of the 2D zero-field exchange experiment reported in Ref. [11].

information about the time scale of spin diffusion. Because there is no preferred direction at zero magnetic field, all crystallites in the sample were equivalent, eliminating line broadening due to anisotropic power averaging. The authors of [11] were thus able to obtain "solution-like" resolution in disordered solid samples.

The indirect point-by-point acquisition used for these early zero-field NMR experiments was rather time consuming, especially for multidimensional experiments, which limited early propagation of the technique. In the past few decades, however, advances in non-inductive detectors – such as superconducting quantum interference devices (SQUIDs) and atomic magnetometers – have enabled direct acquisition of NMR signals at zero field, making 2D techniques involving such fields more attractive.

## 13.3 Two-dimensional NMR Measured at Zero Magnetic Field

The first 2D experiments performed completely in zero-field conditions were reported by Sjolander et al. [12], where coherent spin decoupling was used to obtain $^{13}$C-decoupled, $^{1}$H-coupled NMR spectra. The resulting spectra were fully determined by the proton-proton J-coupling network. While detection of NMR signals in zero magnetic field requires at least two different nuclear spin species, the proton J-spectrum is independent of isotopomer, thus potentially simplifying the spectra and improving the analytical capabilities of zero-field NMR. The protocol uses a distant third spin, which is subsequently decoupled, to selectively address the two-proton groups and prepare the proton-proton coherence. This allows for direct determination of J-coupling constants between chemically equivalent spins.

In an isotropic system (liquid or gas) at zero magnetic field, the scalar spin-spin interactions are the only (and thus, dominant) terms in the spin Hamiltonian. While temporal evolution of a heteronuclear system under the scalar-coupling Hamiltonian generally results in observable time-dependent magnetization, magnetization of a homonuclear spin system is independent of time (apart from the overall decay) and thus does not produce observable nonzero-frequency NMR signals.

To better understand why homonuclear systems do not generate nonzero-frequency NMR signals at zero field, consider the zero-field spin Hamiltonian $H = \sum_{i>j} J_{ij} I_i \cdot I_j$ where $I_{i,j}$ are spin angular-momentum operators. This Hamiltonian is isotropic and the eigenstates must, therefore, also be eigenstates of the total angular momentum, $F$, and can be labeled with the quantum numbers $F^2$ and $F_z$. Meanwhile, the observable in a ZULF NMR experiment is the total sample magnetization along some axis in the laboratory frame, often taken to be the z-axis. The corresponding operator is $M_z = \sum_i \gamma_i I_{z,i}$, where $\gamma_i$ is the gyromagnetic ratio for each spin. For a homonuclear spin system, all $\gamma_i$ are identical. For this case, the operator for observable magnetization becomes $M_z = \gamma \sum_i I_{z,i} = \gamma F_z$ where $F_z$ is the operator for the projection of total spin angular-momentum onto the z-axis. But the eigenstates of the zero-field Hamiltonian are also eigenstates of the operator $F_z$ and it follows that all transition matrix elements of the $M_z$ operator are zero by symmetry for the homonuclear case, since $F_z$ cannot couple different eigenstates. A similar consideration for a heteronuclear system shows that nonzero-frequency signals can, indeed, result in this case.

Even though homonuclear systems cannot generate time-dependent magnetization at zero field, it is still possible to detect static magnetization from such samples. The timescale for the decay of static magnetization can provide useful information in ZULF-relaxometry experiments as demonstrated by Ganssle et al. [13], Tayler et al. [14], and Put et al. [15]. Like spectroscopy, relaxometry can also be conducted across several time dimensions [16]. For example, Ganssle et al. [13] presented 2D correlations of $T_2$ versus $T_1$. Later on, Tayler et al. [14] noted that increased relaxation dispersion at ZULF makes it possible to quantify adsorption affinities for small molecules in porous media. Put et al. [15] showed that the rate of decay of static magnetization from bulk water at zero field depends on the concentration of dissolved D-glucose, so relaxometry can be used as an indirect probe of small biologically relevant molecules in solutions.

Sjolander et al. [12] introduced an indirect 2D approach for obtaining $^{13}$C-decoupled $^{1}$H-coupled J-spectra in ZULF NMR. The presence of a secondary spin species in the molecule, e.g. $^{13}$C, is still required, but it is made "invisible" by averaging out its couplings to the other nuclei while the pure proton J-spectrum is encoded. Such "homonuclear J-spectroscopy" is useful in ZULF NMR because the J-spectrum corresponding to isotopomers of the same molecule can be entirely different, while proton J-spectrum is independent of the location of the label such as $^{13}$C. Homonuclear J-spectroscopy therefore offers significant practical advantages for the study of molecules that are not isotopically enriched and thus are made up of mixtures of isotopomers. Decoupling of heteronuclear interactions also simplifies crowded ZULF-NMR spectra and allows accurate determination of homonuclear scalar-coupling constants in the presence of heteronuclear couplings even if they are of similar magnitude.

With 2D spectra, one retains information on the location of the $^{13}$C label and it is possible to directly determine the scalar-coupling constants between chemically equivalent protons, which is a problem that has attracted interest in the past [17, 18].

How can one perform decoupling of a particular spin species at zero field, where spins are typically manipulated with DC pulses affecting all magnetic moments non-selectively? The "trick" is in taking advantage of specific gyromagnetic ratios ($\gamma$) of the species involved. For instance, $\gamma_p/\gamma_{13C} \approx 3.98$, so performing, for example, a $\pi$-pulse on $^{13}$C, leaves the spin state of protons nearly unchanged as the pulse is nearly a $4\pi$-pulse for them. The residual effect can be corrected by designing decoupling pulse sequences compensating the "imperfections." One such sequence, the so-called XY-16 [19], was used by Sjolander et al. [12].

As a model system with which to demonstrate spin decoupling, Sjolander et al. used $^{13}$C-labeled propionic acid. The energy levels of this molecule in zero field are given by the J-coupling Hamiltonian and are shown in Figure 13.3, both with and without considering couplings to the $^{13}$C nucleus. In the figure, S, K, and L are angular momentum quantum numbers. S is the $^{13}$C spin, and K and L correspond to the total angular-momentum of the methylene and methyl groups, respectively. The colored arrows indicate observable transitions. Only transitions that conserve both K and L are allowed, since they correspond to magnetically equivalent groups of spins – and nothing in this experiment could break the equivalence. Figure 13.4 shows an experimental ZULF J-spectrum of $^{13}$C-labeled propionic acid, with 15 resolvable peaks. The stick spectrum shows the transitions predicted by numerical diagonalization of the J-coupling Hamiltonian.

In order to observe purely homonuclear transitions in ZULF, a 2D scheme was necessary since, per the argument given above, such coherences do not correspond to observable magnetization. The scheme used by Sjolander et al. [12] is outlined in Figure 13.5. The spin system was prepared in a state that corresponds to a coherence between the L and K proton groups. This was done by letting the system evolve under the full J-coupling Hamiltonian, since the coupling to the S spin is different for the two groups. In this scheme, the system is then allowed to evolve under spin decoupling for a variable time $t_1$ during which it accumulates phase solely due to proton–proton coherences. Finally, a fully coupled and therefore directly observable ZULF spectrum is acquired as a function of $t_2$. FFT with respect to both $t_1$ and $t_2$ should then reveal a homonuclear ZULF spectrum along F1.

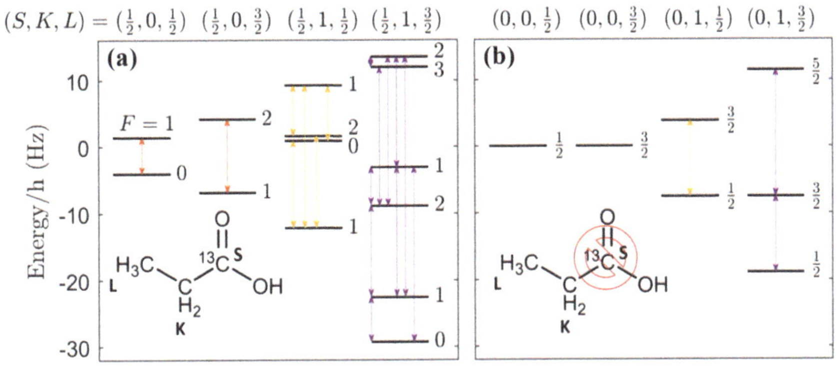


Figure 13.3 (a) Energy level diagram for $^{13}$C-propionic acid in ZULF conditions. (b) Effective energy level diagram under $^{13}$C decoupling. Reproduced from [12].

Figure 13.4 ZULF J-spectrum of $^{13}$C-propionic acid. The stick spectrum corresponds to the indicated transitions in Figure 13.3a. Reproduced from [12].

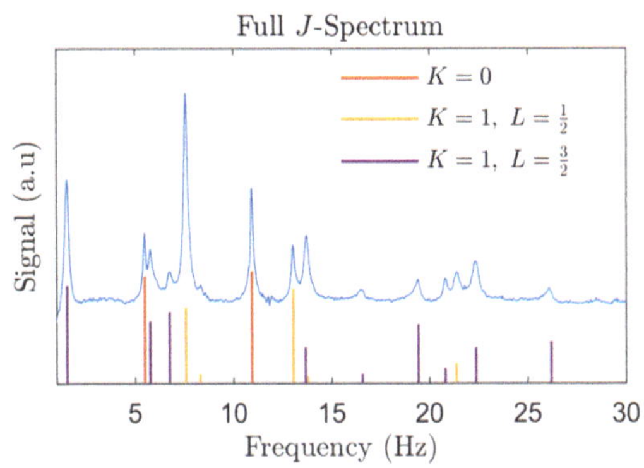

Sjolander et al. [12] used an XY-16 sequence of DC pulses calibrated to obtain a $\pi$ rotation of the carbon spin in order to decouple the carbon from the protons, while leaving the proton–proton coupling unchanged. Since each $\pi$ pulse on carbon is approximately $4\pi$ on proton, the proton spin is unchanged by this pulse sequence. The small residual action on the proton spin due to $\gamma_p/\gamma_{13C} \approx 3.98$ and not being exactly four was compensated for by the XY-16 sequence.

Figure 13.6 shows the resulting 2D ZULF spectrum of $^{13}$C-labeled propionic acid, with the $^{13}$C decoupled dimension along the y-axis. Since K and L, the quantum numbers for the total angular momentum of the proton groups are conserved throughout the experiment, cross peaks are only observed between states that share K and L values. Note that there are no cross peaks corresponding to K = 0, since the pure proton J-coupling Hamiltonian goes to zero for this case, see Figure 13.3b. The red shaded region in Figure 13.6 has been projected on the F1 axis in

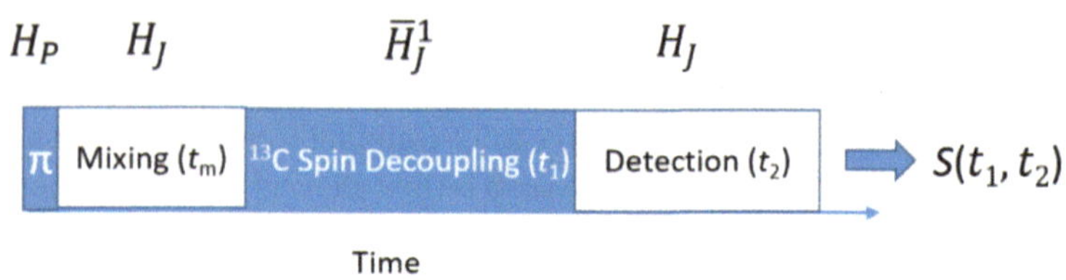


**Figure 13.5** Schematic for acquisition of $^{13}$C-decoupled ZULF spectra. Adapted from [12].

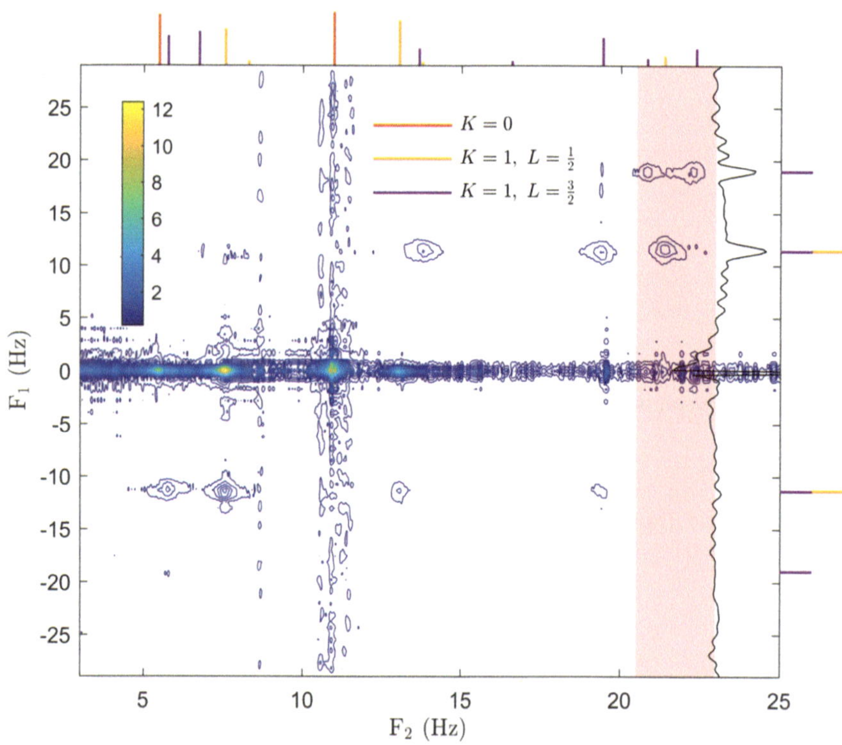


**Figure 13.6** 2D ZULF spectrum of $^{13}$C-propionic acid. F1 corresponds to the $^{13}$C decoupled dimension. The stick spectrum corresponds to the transitions indicated in Figure 13.3b. F2 corresponds to the fully coupled dimension, the stick spectrum matches Figure 13.4 and Figure 13.3a. The data in the shaded red region have been projected on the F1-axis in order to reproduce a $^{13}$C decoupled 1D J-spectrum of propionic acid. Reproduced from [12].

order to generate the 1D pure proton J-spectrum plotted in the figure. This spectrum shows the two distinct resonances indicated in Figure 13.3b, as expected for the ZULF J-spectrum of an effective spin-1 particle coupled to an effective spin-3/2 particle.

Multiple-quantum correlation J-spectroscopy detected in zero magnetic field was presented by Sjolander et al. [20]. They used correlation spectroscopy to separate the zero-field spectra of two $^{13}$C isotopomers of ethanol appearing in a mixture as shown in Figure 13.7. As expected, cross peaks were observed only between peaks belonging to the same isotopomer. The simple pulse sequence consisted of an initial excitation pulse, followed by $t_1$ evolution and finally, reconversion with a second pulse. The pulses were calibrated to perform a $\pi$ rotation on $^{13}$C and approximately $4\pi$ on proton, meaning the effect was to invert the relative orientation of the carbon and proton polarizations. This is a standard way of generating coherences in ZULF NMR when the spin order is the result of adiabatic transport from a pre-polarizing magnet. This work also showed that 2D spectroscopy allows separation of overlapping resonances into distinct cross peaks, improving spectral resolution. Like for the case of propionic acid above, the 2D ZULF J-spectrum of ethanol can be analyzed in terms of the quantum numbers for the total angular momentum of the methyl and methylene groups, which are conserved. The right panel of Figure 13.7 shows a single 1-$^{13}$C ethanol multiplet. The large peak at 211 Hz consists of two overlapping resonances belonging to the K = 1/2 and K = 3/2 spin manifolds, respectively, where K denotes the total angular momentum of the three hydrogens in the methyl group. In the 1D spectrum, the two peaks cannot be resolved, however, distinct cross peaks show up for both K = 1/2 and K = 3/2 peaks, thus confirming the presence of two overlapping resonances at 211 Hz.

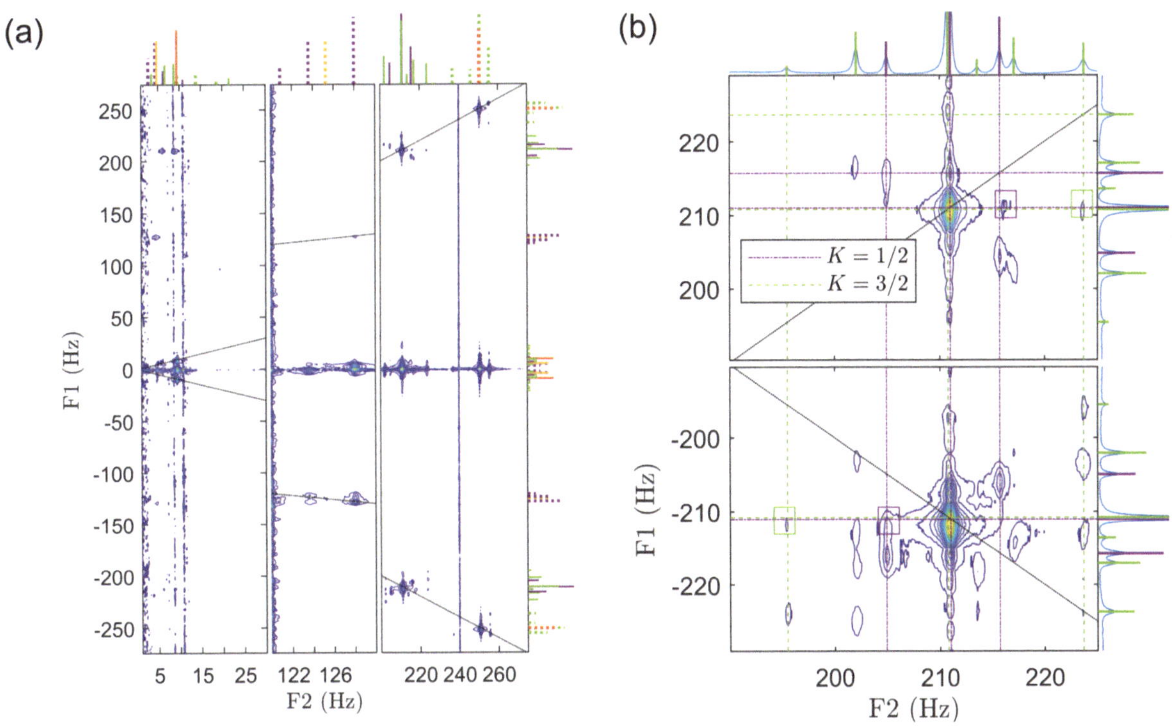


Figure 13.7 (a) 2D ZULF J-spectrum of a mixture of 1- and 2-$^{13}$C ethanol. The stick spectra are simulated ZULF J-spectra for 1-$^{13}$C ethanol (solid) and 2-$^{13}$C ethanol (dotted). Cross peaks occur only between peaks belonging to the same isotopomer. (b) Zoomed in on the high-frequency region of the 1-$^{13}$C ethanol spectrum, showing how 2D ZULF spectroscopy can be used to resolve overlapping peaks in the 1D J-spectrum, if the peaks belong to different angular-momentum manifolds. Adapted from [20].

In addition, Sjolander et al. [20] observed the zero-field equivalent of a double-quantum transition in $^{13}C_2$-acetic acid. In ZULF, dipole allowed transitions are those for which the total angular momentum of the spin system changes by 0 or $+/-1$. The ZULF equivalent of a multiple-quantum transition is then one for which F changes by more than 1. Such coherences do not correspond to observable magnetization and have to be detected indirectly. Sjolander et al. [20] used a pulse sequence consisting of two $\pi$ pulses on $^{13}C$ (a $\pi$ pulse on $^{13}C$ is almost a $4\pi$ pulse on $^{1}H$) interspersed with a free evolution period to excite the double-quantum transition, followed by $t_1$ evolution. The reconversion was done by another $^{13}C$ $\pi$ pulse. Since this pulse sequence did not feature any filtering or coherence selective excitation all single-quantum coherences were also excited in addition to the double-quantum coherence. The resulting 2D spectrum shown in Figure 13.8 therefore contained cross peaks between all coherences of matching K value, including the single-quantum ones.

Both homo- and heteronuclear multiple-quantum coherences were also investigated via a 2D technique at somewhat higher fields on the order of 0.1 mT by Buckenmaier et al. [21]. These authors used signal amplification by reversible exchange (SABRE) hyperpolarization of a proton-$^{19}F$ system and detection using a superconducting quantum interference device (SQUID). The technique relied on heteronuclear correlation spectroscopy together with phase-cycling.

ZULF methods can also be exploited beyond the spectroscopic characterization of systems relevant to chemists. For example, some modern theories of dark matter predict that so-called ultralight bosonic dark matter could interact with nuclear spins, which could be detected by ZULF NMR. Garcon *et al.* suggested a pseudo-2D ZULF-NMR method to probe such interaction [22], which is analogous to the two-dimensional one-pulse (TOP) spectroscopy described by Blümich et al. [23].

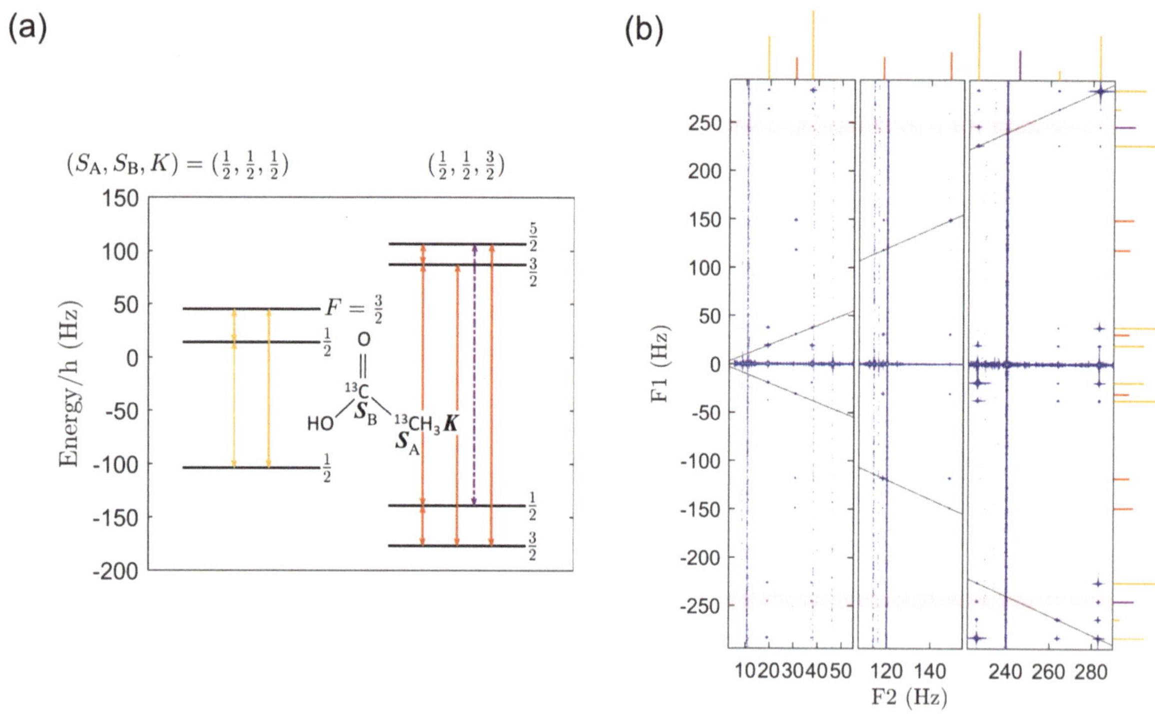


**Figure 13.8** (a) Energy level diagonal for $^{13}C_2$-acetic acid with dipole allowed transitions indicated by solid arrows. This spin system also supports the ZULF equivalent of a double-quantum transition, where the total spin angular momentum F changes by $+/-2$, indicated by a dotted arrow. (b) 2D detected multiple-quantum spectra of $^{13}C_2$-acetic acid. The double-quantum transition is observed as a cross peak within the K = 3/2 manifold. Adapted from [20].

## 13.4 Nuclear Magnetic Resonance at Millitesla Fields Using a Zero-Field Spectrometer

Tayler et al. [23] performed the first experiments where spins were manipulated under high-field conditions but detection was performed in zero field (opposite to the situation described in Section 13.2). They used coils to apply a millitesla bias field to the spins while inside a ZULF spectrometer in order to temporarily impose high-field conditions, defined as the Larmor frequencies being much larger than any couplings in the system. In principle, this technique opens up the entire library of high-field pulse sequences for use in ZULF. In particular it enables selectively addressing different spin species based on different resonance frequencies in the temporary bias field. Meanwhile, DC pulses in ZULF rotate all spin species proportionally to their gyromagnetic ratios. Such DC pulses were used to generate the 2D ZULF spectra in Section 13.3. Figure 13.9 shows a simulation of a pulse sequence for spin-species selective inversion in ZULF introduced by Tayler et al. [24]. Notably this pulse sequence does not rely on any particular relationship between the gyromagnetic ratios of the two spins species. After shuttling into the ZULF spectrometer the spins are adiabatically remagnetized along the z-axis using an applied DC field. Once the spins are in the high-field regime, a half-sine frequency-swept pulse inverts only those spins whose Larmor frequency falls within the sweep window. After inversion, the bias field is suddenly turned off and spin evolution is monitored under ZULF conditions. The ability to invert one spin species with respect to another is important in ZULF spectroscopy as relative spin inversion is one of the primary means by which one can excite ZULF coherences. In the case of sudden (as opposed to adiabatic) transport from the high-field regime to ZULF an inversion pulse is technically not required since the spin system in this case will automatically contain

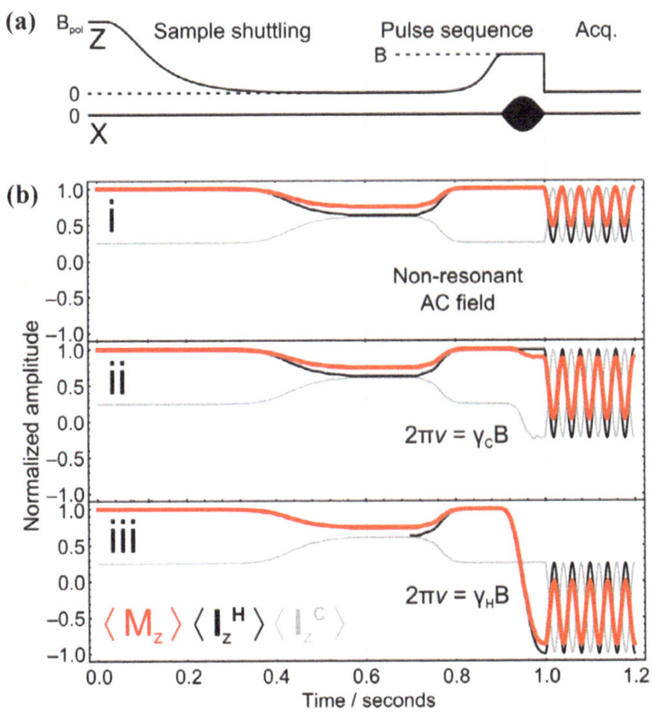


**Figure 13.9** (a) Experimental protocol for the frequency-swept pulse experiments. An initial polarizing field is applied along the z-axis. This field is adiabatically tuned to zero as the sample is shuttled into the ZULF spectrometer. A bias field along the z-axis is then adiabatically turned on using a coil. Once the bias field reaches a set value a frequency-swept adiabatic inversion pulse is applied along the x-axis of the laboratory frame. (b) Simulations of a proton/carbon spin system trajectory during the pulse sequence when the inversion pulse is (i) off-resonant, (ii) resonant with the carbon spin, and (iii) resonant with the proton spin. The carbon-13 z polarization is shown in gray, the proton z polarization in black and the total z magnetization in red. Reproduced from [24].

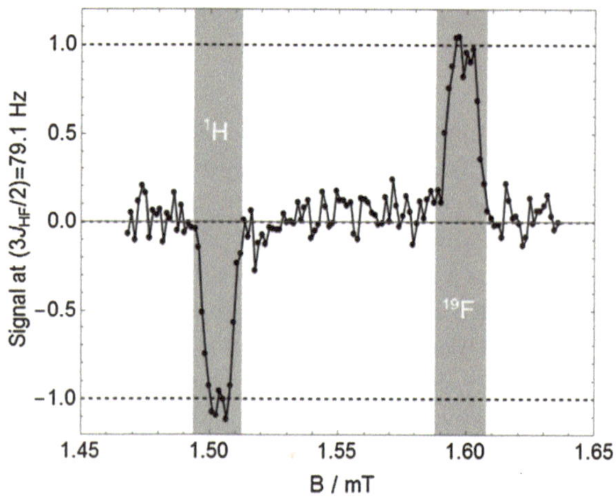


**Figure 13.10** Demonstration of selective inversion of the proton and fluorine spins in difluoroacetic acid in ZULF. Each data point corresponds to the amplitude of the ZULF spectrum for a particular value of the z-axis bias field during the inversion pulse. The center frequency of the AC inversion pulsed is the same for each data point. Reproduced from [24].

coherences. However, inverting the spins of a given spin species with respect to another is still useful since it increases the coherence amplitude. For a two-spin system, an inversion pulse improves the signal by a factor of $(\gamma_I + \gamma_S)/(\gamma_I - \gamma_S)$. For a carbon/hydrogen spin system, this ratio is ~5/3, meanwhile for a fluorine-19/hydrogen spin system the improvement is a factor of ~41.

An experimental demonstration of spin-species selective inversion in ZULF is shown in Figure 13.10. Here, Tayler et al. [24] used the pulse sequence presented in Figure 13.9 to selectively invert the proton and fluorine spins in difluoroacetic acid ($CHF_2COOH$), something that is very difficult to do using DC pulses in zero field as the gyromagnetic ratios of protons and fluorine are nearly the same, $\gamma_H/\gamma_F \sim 1.06$. By using frequency-swept inversion pulses, which are insensitive to resonance offset, they circumvented the problems of the pulsed bias field being relatively inhomogeneous, and the excitation coils not being tuned to a particular frequency. Band-selective composite DC pulses, developed later by Tayler [25], can also excite specific spins in systems of multiple species in ZULF. These are an order of magnitude faster than adiabatic sweeps, and do not require tuned coils or exceptionally homogeneous fields. The magnetic resonance spectrum of difluoroacetic acid in zero field consists of a single peak at 79.1 Hz. In Figure 13.10 the amplitude of this peak is plotted against the strength of the bias field applied to bring the spins to the high-field regime. The frequency-swept AC pulse was applied at a fixed center frequency (64 kHz) and the resonances occur as the proton and fluorine Larmor frequencies end up inside the sweep window (+/− 400 Hz from the center frequency). When either spin is inverted, the signal amplitude increases by x41 and a J-spectrum is observed, indicated by non-zero value for the peak amplitude in Figure 13.10. Outside this window neither spin is inverted, and the signal was too weak to be detected with the instrument used for this experiment.

While these results were not presented in a 2D fashion, the idea relied on spin evolution and manipulation in both ZULF and high field during the course of a single experiment, paving the way for true 2D experiments leveraging both high- and low-field conditions.

## 13.5 Field Cycling NMR and Correlation Spectroscopy

Zero and ultra-low magnetic fields in solutions lead to conditions where scalar-coupling interactions dominate the Hamiltonian that drive the evolution of the spin system. Such conditions are sought after in a precise type of high-field NMR: total correlation spectroscopy (TOCSY) [26]. In TOCSY experiments, one obtains correlations between nuclear spins that belong to a scalar-coupling network. These correlations are achieved by transitive polarization transfer: polarization originating from one nuclear spin flows toward the other nuclear spins throughout the coupled network. Unlike in correlation spectroscopy (COSY) [2] or insensitive nuclei enhanced by polarization transfer experiments [27], where polarization transfer between two nuclei requires the existence of a direct non-zero scalar coupling between the two nuclei, in TOCSY, polarization transfer happens coherently among all coupled nuclei, in a manner reminiscent of spin-diffusion experiments. Such polarization transfer is achieved at high magnetic fields under isotropic-mixing sequences [28–30]. Under these pulse sequences, the offset terms in the average Hamiltonian are canceled and the isotropic scalar coupling is preserved, with a scaling factor [29, 30]. During isotropic mixing, the zero-quantum part of the scalar-coupling Hamiltonian, which is averaged out in weak-coupling conditions, drives the polarization transfer of longitudinal polarization.

We have recently introduced a series of TOCSY experiments where isotropic mixing is performed at low or ultra-low magnetic field, while the evolution under chemical shifts and detection take place at high magnetic field to benefit from high resolution and sensitivity [31, 32]. In this manner, conventional NMR spectra recorded at high magnetic fields are correlated through isotropic mixing at magnetic fields lower by orders of magnitude.

Interestingly, magnetic fields just low enough to reach strong-coupling conditions for homonuclear scalar-coupling interactions may not be low enough if additional heteronuclear scalar-coupling interactions are present. We have performed $^{13}$C TOCSY experiments with isotropic mixing at a field of 0.33 T (resonance frequency of 3.5 MHz for $^{13}$C nuclei) on a mixture of two uniformly $^{13}$C-labeled amino acids: phenylalanine and leucine [31]. Cross-peaks were obtained in the aliphatic region of leucine and the aromatic ring of phenylalanine in the absence of RF irradiation (see Figure 13.11a). A close inspection of spin dynamics at 0.33 T in uniformly labeled leucine showed that polarization transfer occurred from cross-relaxation and not strong scalar coupling (Figure 13.11b). Indeed, when two spins in a homonuclear spin system are coupled to a heteronucleus with different scalar couplings, the heteronuclear scalar couplings perturb the strong scalar-coupling regime in the homonuclear spin system [33]. Efficient polarization transfer was retrieved with the use of radiofrequency pulses at low field on a two-field NMR spectrometer [34, 35]. Composite-pulse decoupling applied to protons leads to efficient polarization transfer between $^{13}$C nuclei in the isopropyl moiety of leucine (Figure 13.11c) [33] while an isotropic mixing sequence applied to $^{13}$C nuclei suppresses the effects of $^{13}$C offsets and heteronuclear scalar couplings alike, leading to efficient broadband isotropic mixing within a coupled $^{13}$C spin system [31].

The detrimental role of heteronuclear scalar couplings at low or moderate fields can be fully alleviated in ZULF conditions. At 100 nT, the difference of the Larmor frequencies of proton and $^{13}$C nuclei falls down to ~3 Hz, leading to energy levels and eigen states defined by strong scalar-coupling interactions for homo- and heteronuclear spin systems. Differences of polarization generated at high field, by selective inversion or by the evolution

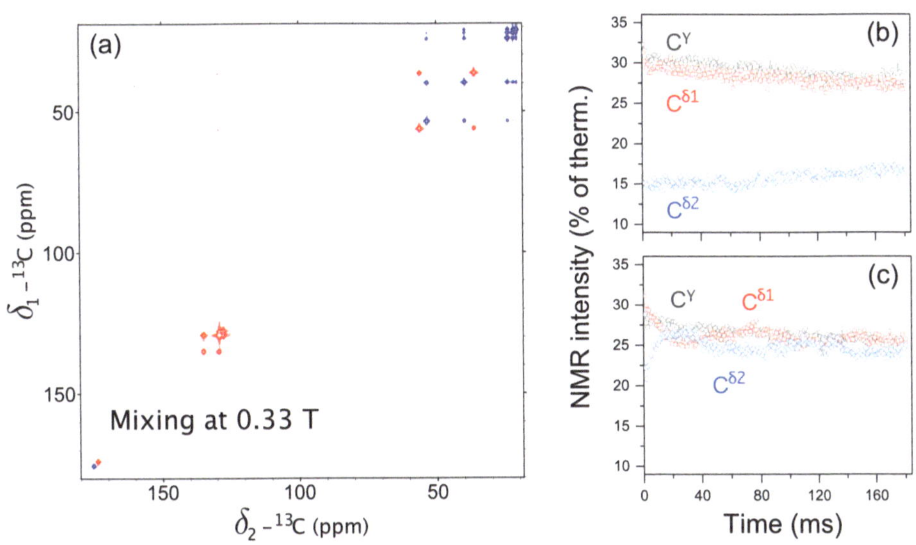


**Figure 13.11** 2D NMR with mixing at low magnetic field. (a) Adaptation of a two-field TOCSY experiment with mixing at 0.33 T but no pulse applied at low field. The rest of the pulse sequence takes place at 14.1 T. (b,c) Evolution $^{13}$C the polarization in the isopropyl moiety of uniformly $^{13}$C-labeled leucine at 0.33 T after the selective inversion of $^{13}$C $\delta 2$ at high field. (b) No pulse is applied at 0.33 T as in (a). (c) Composite-pulse decoupling is applied on the protons at 0.33 T.

under different chemical shifts, are transferred in a non-adiabatic manner to an ultralow field, where they are converted to coherences that evolve under scalar-coupling interactions, homo- and heteronuclear alike. Thus, correlations between all NMR-active nuclei can be obtained, provided relaxation is not too efficient at ultralow fields.

We recently introduced the ZULF-TOCSY experiment [32] (Figure 13.12a) where most of the pulse sequence is performed at high field (preparation, $t_1$ evolution, detection), but mixing is obtained by transferring the sample to ZULF conditions with the use of a sample-shuttle apparatus coupled to a field-compensation magnet [36]. In this

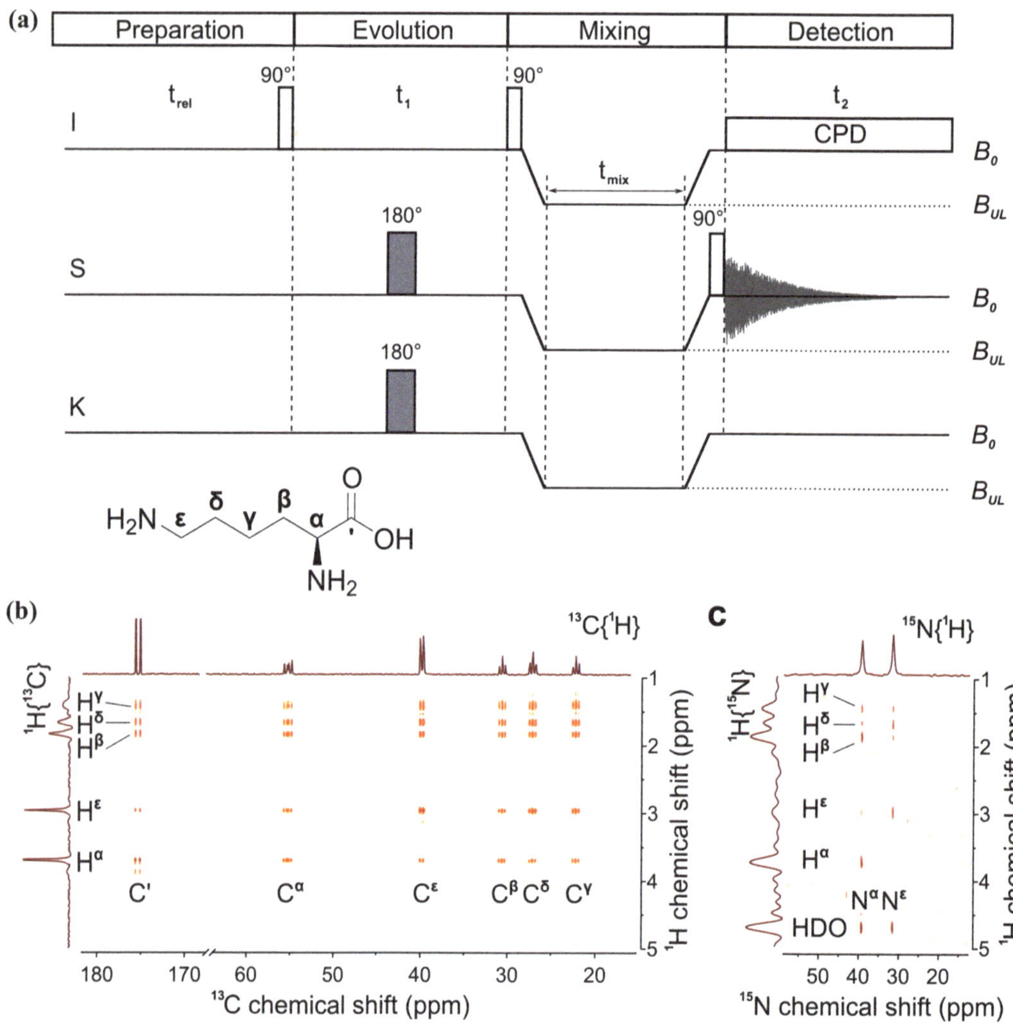


**Figure 13.12** (a) Protocol of the heteronuclear ZULF-TOCSY experiment. A 90° pulse after a suitable relaxation delay $t_{rel}$ generates the transverse coherence of the I spins (here, protons). This single-quantum coherence evolves under the chemical-shift Hamiltonian during the period $t_1$, mapping the indirect spectrum dimension. The evolutions under the IS and IK J-couplings are refocused by the simultaneous application of 180° pulses on the S and K channels, positioned in the middle of the $t_1$ evolution period. Here, as S and K spins we used $^{13}C$ and $^{15}N$ nuclei; one of them was used for detection. At $t = t_1$, the transverse polarization of the I spin is converted back into longitudinal polarization. The mixing block consists of a field switch $B_0 \leftrightarrow B_{UL}$, lasting for $t_{sw} \approx 0.4$ s and isotropic mixing under ZULF conditions for $t_{mix}$ of several tens of milliseconds. When the FID signal of S spins is acquired (i.e., the time domain signal in the direct dimension), composite-pulse decoupling on the I channel is applied. (b) $^{13}C$-$^1H$ ZULF-TOCSY spectrum of $^{13}C$- and $^{15}N$-labeled L-lysine. Experimental parameters: $B_{UL}$ = 100 nT, $t_{mix}$ = 2 ms, 128 transients in indirect direction, 4 scans per transient, relaxation delay 6 s, and total experiment duration ca. 70 min. (c) $^{15}N$-$^1H$ ZULF-TOCSY spectrum of $^{13}C$- and $^{15}N$ labeled L-lysine. Experimental parameters: $B_{UL}$ = 100 nT, $t_{mix}$ = 50 ms, 64 transients in indirect direction, 128 scans per transient, relaxation delay 23 s, and total experiment duration ca. 62 h. Reproduced from [32].

manner, longitudinal polarizations of nuclear spins selected at high field are transferred to all NMR-active nuclei during mixing at ZULF conditions. Detection can be performed on any chosen nucleus. We first demonstrated ZULF-TOCSY by recording proton-$^{13}$C and proton-$^{15}$N correlations on a sample of uniformly $^{13}$C-and $^{15}$N-labeled lysine (Figures 13.12b and 13.12c). All expected correlations between $^{13}$C nuclei and protons can be observed as well as a majority of the possible proton-$^{15}$N correlations.

The information content in a 2D spectrum displaying correlations between all NMR-active nuclei is notably different from conventional multidimensional heteronuclear correlations, obtained in uniformly labeled biomolecules, where well-controlled flows of polarization lead to the observation of selected correlations [37]. We showed that ZULF-TOCSY could be used efficiently for the analysis of a mixture of isotopically labeled molecules, such as the ISOGRO rich medium used for the bacterial expression of uniformly labeled proteins. ZULF-TOCSY led to the identification of the signals of several metabolites in this culture medium [32] (Figure 13.13).

The use of ZULF-TOCSY is not limited to the investigation of isotopically labeled molecules. In molecules with natural isotopic abundance, in particular for $^{13}$C, heteronuclear couplings through two or three bonds can be large enough to dominate Zeeman interactions so that ZULF-TOCSY can also be used to obtain heteronuclear correlations through multiple bonds, in a manner reminiscent of the HMBC experiment [38]. We have used the ZULF-TOCSY to obtain ($^1$H,$^{13}$C) 2D correlations through multiple bonds in a small peptide: Boc-Met-enkephalin, at natural isotopic abundance [39] (Figure 13.14). Although correlations between all protons and all $^{13}$C nuclei within three bonds could be expected in principle, only a selection of these correlations were observed. In particular, all intra-residue correlations between the α-proton and the carbonyl $^{13}$C nuclei ($H_i^\alpha, CO_i$) were observed. Importantly, all inter-residue correlations ($H_i^\alpha, CO_{i-1}$) were also observed, allowing the straightforward sequential assignment of all α-proton and carbonyl $^{13}$C resonances in this small peptide, making use of the known amino-acid sequence.

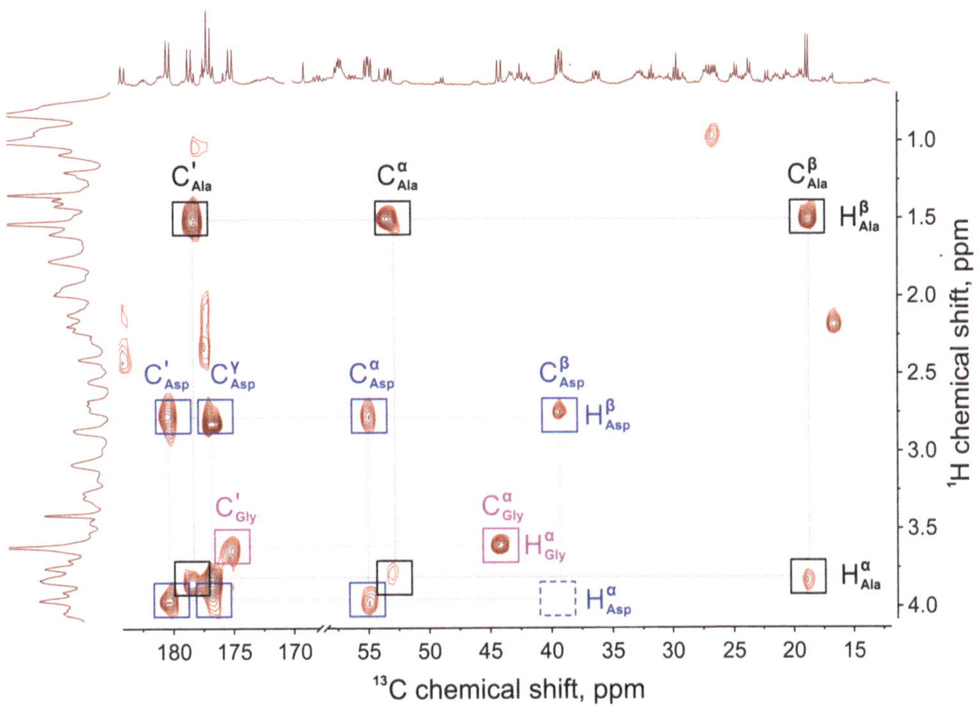


**Figure 13.13** ZULF-TOCSY spectrum of the ISOGRO supernatant in $D_2O$ (pH 6.5). Experimental parameters: $B_{UL}$ = 50 nT, $t_{mix}$ = 40 ms, 64 transients in the indirect direction, 960 scans per transient, relaxation delay 3 s, and total experiment duration ca. 69 hours. Reproduced from [31].

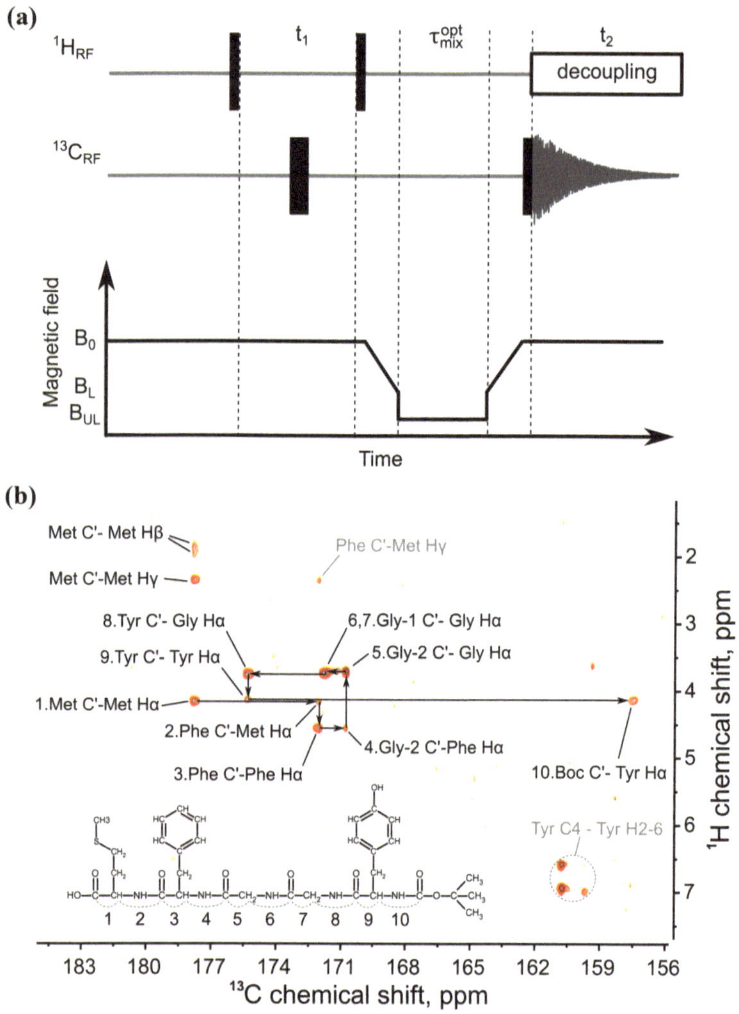


**Figure 13.14** (a) Protocol of ZULF-TOCSY experiment, see caption to Figure 13.12 for the detailed description. (b) $^{13}C$-$^{1}H$ ZULF-TOCSY spectrum of Boc-Met-enkephalin. Only the signals of the carbonyl carbons and $H^{\alpha}$ protons are shown here. Relevant parameters: $B_0 = 9.4$ T, $B_{UL} = 50$ nT; 80 increments of $t_1$ were used to sample the indirect dimension, 256 scans per transient, the relaxation delay was 6 s, the field switching time was 403 ms in each direction, the total experimental time was ca. 58 hours. All possible - correlations are visible and the connectivity of the peaks for sequential assignment is indicated by arrows. Reproduced from [39].

Other correlations, such as one-bond $(H_i^{\alpha}, C_i^{\alpha})$ correlations were not observed. We interpret this lack of signal as a result of fast relaxation of the $\alpha$-$^{13}C$ nucleus, mostly due to the dipole-dipole interaction with the $\alpha$-proton. Indeed, as shown by investigations of complex systems by $^{13}C$ relaxometry [40, 41], $^{13}C$ relaxation rates increase with decreasing field due to dipole-dipole interactions, usually with protons, that fluctuate with nanosecond correlation times. Given the duration of the sample transfer (~400 ms) between high-field and ZULF conditions of the current apparatus, only nuclei with long longitudinal relaxation rates, even at low magnetic fields, can be observed in ZULF-TOCSY experiments. Under these limitations, ZULF-TOCSY results obtained on Boc-Met-enkephalin are promising and demonstrate that some complex molecules such as small peptides are amenable to ZULF-TOCSY. Yet, macromolecules, such as proteins or nucleic acids typically investigated by NMR are out of reach of current methods. Relaxation in ZULF conditions can be dramatically reduced in long-lived states (LLS) [42] in both homonuclear [43] and heteronuclear spin systems [44]. Yet, the combination of LLS and isotropic mixing in ZULF conditions would be a very challenging endeavor.

## 13.6 ZERO-Field - High-Field Comparison

ZULF spectra allow J-couplings measurements with unprecedented precision, which can be as low as 20 mHz. Usually, the analysis of ZULF-NMR spectra in more complex cases than a lone $IS_n$ (I and S are sundry spin-1/2 nuclei) group requires numerical methods to simulate spectra as well as assumptions on the magnitudes of J-coupling constants involved. The complexity of ZULF-NMR spectra makes signal assignment and interpretation difficult, especially for mixtures of several compounds. Since in a ZULF-NMR experiment the external magnetic field is made as low as possible, these spectra do not contain any information about chemical shift of nuclei, which is the merit of shielding of the external magnetic field by the surrounding electrons. On the contrary, in high-field NMR, chemical shifts of NMR signals readily provide valuable information on the chemical environment of the nucleus. However, the limits on possible spectral resolution in high-field NMR spectra of small molecules are imposed by inhomogeneity of the external magnetic field, and by short transverse relaxation time values for larger molecules. Thus, high-field NMR and zero-field NMR are complementary approaches for extracting information on molecular structure.

Recently, we proposed an approach to combine high-field NMR information on chemical shifts and ZULF-NMR spectra of individual sample components in one 2D NMR spectrum [45]. In this spectrum, information about chemical shifts is encoded along the direct dimension, while spectra of oscillations at zero- to ultralow field are presented along the indirect dimension.

The protocol of the experiment is the following (see Figure 13.15): the spins are allowed to relax to near-equilibrium polarization at high field, then non-adiabatic field switch from high field to ZULF generates zero-quantum coherences in heteronuclear spin system, which are allowed to evolve for a time interval $t_1$. Next, the second non-adiabatic field switch converts the coherences into non-equilibrium high-field spin-state populations, which determine amplitudes and phases of NMR signals in the spectrum obtained by detecting FID of one of

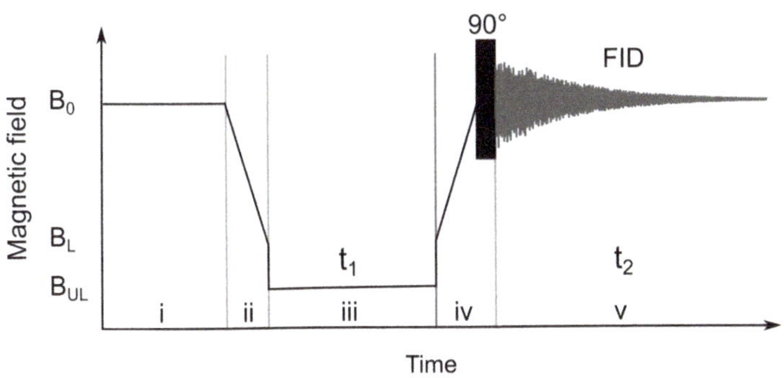


**Figure 13.15** Protocol of a field-cycling experiment used to measure the coherent spin evolution under ZULF conditions. The protocol comprises the following steps: (i) relaxation of the spin system to thermal equilibrium at the high magnetic field $B_0$, (ii) non-adiabatic field switching: mechanical sample transfer into the magnetic shield with a pre-set low magnetic field $B_L$ inside, which is non-adiabatically switched to an ultralow field $B_{UL}$ immediately after sample arrival. The field jump $B_L \rightarrow B_{UL}$ creates a heteronuclear zero-quantum coherence, which evolves in step (iii) during a free-evolution period $t_1$ of variable length; (iv) non-adiabatic field switching $B_{UL} \rightarrow B_L \rightarrow B_0$, followed by the application of an RF pulse at the resonance frequency of the nuclei to be observed; and (v) free induction decay (FID) acquisition during $t_2$. The NMR spectrum is given by the Fourier transform of the FID. Reproduced from [44].

the spins, either I or S, after application of a resonant 90° RF pulse at high field, in the presence of broadband composite-pulse decoupling on the other RF channel. The cycle is repeated with systematically incrementing evolution time $t_1$, and the application of a 2D Fourier transform of the signal $S(t_1, t_2)$ rendering a 2D NMR spectrum, with ZULF-NMR spectrum for high-field NMR signals in the indirect dimension and the chemical shifts of these signals in the direct dimension.

In Figure 13.16, an example of such a 2D correlation spectrum is shown. The sample investigated here is made of equal volumes of ethanol, methanol, acetonitrile, acetic acid, and DMSO-d6. All components of the mixtures have natural isotope abundance, except DMSO-$d_6$, which was added to maintain deuterium lock signal. Along the direct spectral dimension F2 (horizontal axis) one can observe the conventional high-field proton-decoupled $^{13}C$ NMR spectrum of the mixture, where each signal corresponds to a particular $^{13}C$ nucleus in one of the compounds in the mixture. Along the indirect spectral dimension, F1, a frequency spectrum of oscillations of proton-carbon-13 zero-quantum coherences at zero- to ultralow field is obtained. These coherences make the populations of eigenstates in ZULF depending on evolution time $t_1$. While usually in ZULF NMR zero-quantum coherences are detected directly by an atomic magnetometer, since the time dependence of eigenstate populations modulate sample magnetization, here the second non-adiabatic field switching projects the transient ZULF state populations onto the populations of high-field Zeeman states. If neglecting relaxation effects during field switching, these two approaches to monitor zero-quantum coherences at ZULF are fully equivalent, but technically the detection of magnetization oscillations by an atomic magnetometer seems to be superior to tedious point-by-point acquisition of a high-field NMR spectrum for each value of $t_1$. However, up to now, the sensitivity of inductive detection used in high-field NMR is drastically higher than that of the most sensitive atomic magnetometers. In ZULF NMR, one usually needs to use isotopically enriched neat compounds to obtain spectra with good signal-to-noise ratio,

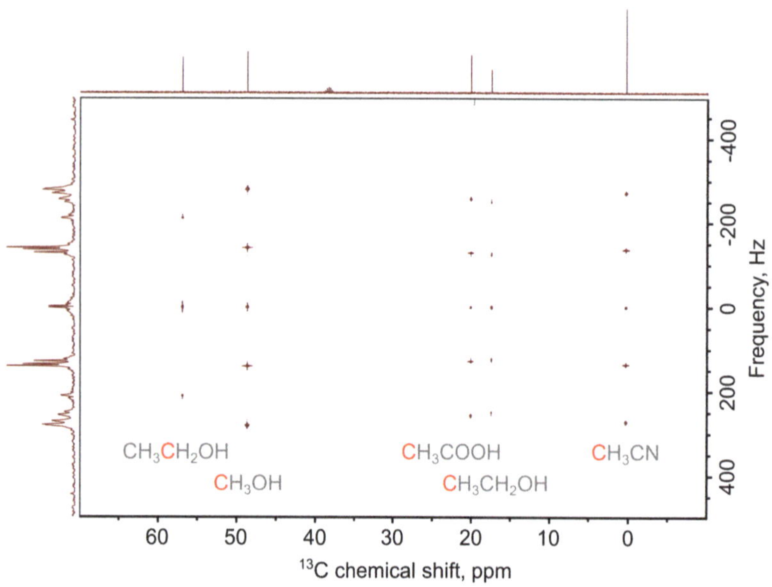


**Figure 13.16** 2D HF-ZULF-spectrum of the mixture. In this spectrum, the horizontal axis represents chemical-shift values of $^{13}C$ NMR signals, while the vertical axis shows frequencies derived from the coherent spin evolution under ZULF conditions. The spectrum presents a magnitude-mode Fourier transform of the $S(t_1,t_2)$ signal, shown by $1500 \times 4096$ points. Apodization of the acquired data with decaying exponential functions (10 Hz in the direct dimension and 1 Hz in the indirect dimension) was applied prior to the 2D Fourier transform. The 1D spectrum on the left shows the projection of the 2D spectrum on the indirect domain, i.e. the ZULF-NMR spectrum. The 1D spectrum on the top shows the standard $^{13}C$ NMR spectrum acquired with $^{1}H$ decoupling. Reproduced from [45].

or average a vast number of acquisitions. In turn, the sensitivity of modern high-field NMR spectrometers allows to obtain spectra with good signal-to-noise ratio with one scan at natural isotopic abundance for substances with concentrations ca. 1 M, corresponding to 10 mM concentration of individual $^{13}C$ isotopomers. Another advantage of the field-cycling approach is the resolution of chemical shifts, which potentially allows the analysis of dozens of analytes in one experiment.

The comparison of ZULF spectra measured directly at ZULF by an atomic magnetometer and indirectly, at high field, by field cycling and inductive detection approach is shown in Figure 13.17. One can note the close similarity between them. There are two main discrepancies: the first one is the artifact signals in ZULF-NMR spectra coming from an AC current modulation, which are marked in Figure 13.17 by stars. The second discrepancy is in much narrower high-frequency spectral lines in ZULF-NMR spectra if they are compared to field cycling ones. The high-field ZULF-NMR spectral lines are sensitive to the residual magnetic field ~100 nT present in field-cycling experiments. Apart from these, the ZULF-NMR spectra of the same compounds taken by different methodologies match well one another. Simplicity of signal assignments in 2D spectra correlating high-field NMR properties, namely, chemical shifts and ZULF-NMR spectra, is paving the way for the future applications of ZULF NMR as an analytical tool.

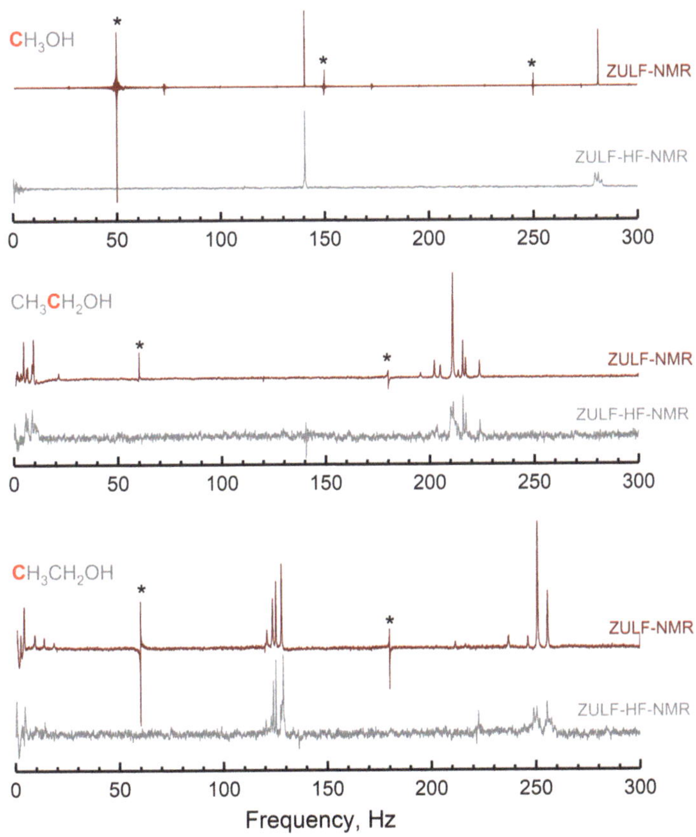


**Figure 13.17** Comparison of J-spectra measured by optical magnetometry at zero field and using field cycling and high-field NMR detection. The spectra shown in brown (ZULF-NMR spectra) are the spectra of $^{13}C$-enriched neat liquids detected using a ZULF-NMR spectrometer. The spectra shown in gray (ZULF-HF-NMR spectra) have been obtained in 4 hours for the mixture of substances with natural isotopic abundance. To obtain ZULF-HF-NMR spectra, field cycling was used, and the $B_{UL}$ field was about 100 nT. The asterisks in ZULF-NMR spectra indicate artifacts coming from the AC line. Reproduced from [45].

## 13.7 Conclusion and Outlook

While 2D methods involving ZULF NMR are far less ubiquitous than they are in high-field NMR, it is the author's hope that this chapter demonstrates that introducing them directly into ZULF NMR or in experiments involving field cycling between ZULF and high field clearly opens a new dimension (pun intended). For example, 2D methods enable ZULF measurements of homonuclear (e.g. proton-proton) J-couplings, observation of multiple-quantum coherences, and performing the ZULF version of correlation spectroscopy.

Field cycling over an ultra-wide range of magnetic fields allows to combine all the powerful capabilities of a modern high-field NMR spectrometer with the unique opportunity of switching between strong and weak-coupling regimes, even for heteronuclear spin systems. Although point-by-point acquisition of spin-system evolution in the ZULF regime may seem tedious, currently, conventional detection of FID with high-field NMR spectrometers provides higher sensitivity than detection with atomic magnetometers, allowing to work with samples with natural isotopic abundance. Field-cycling ZULF NMR approaches being realized in a 2D manner provide original correlations between standard FID detected NMR spectra and ZULF-NMR spectra. In addition, it allows straightforward implementation of coherence transfer simultaneously for all magnetic nuclei. We expect that the synergies between the two drastically different regimes of high-field and ZULF NMR will open the way for the design of a new generation of advanced NMR methods.

Some directions that it would be interesting to explore in future work include:

- The characterization of mixtures by ZULF-NMR.
- Applications of ZULF-TOCSY in the assignment strategy of small molecules.
- Implementation of ZULF-TOCSY in multidimensional methods, including with parallel detection.

## Acknowledgments

The authors are grateful to Danila Barskiy and Kirill Sheberstov for helpful suggestions. The work of AK, AY, and IZ was supported by the Russian Fund for Basic Research (grant #21-53-12023) and by Ministry of Science and Higher Education of the Russian Federation (contract #075-15-2021-580). The work of DB was supported in part by German Research Foundation (DFG), project # 465084791.

## References

1 Jeener, J. and Alewaeters, G. (2016). "Pulse pair technique in high resolution NMR" a reprint of the historical 1971 lecture notes on two-dimensional spectroscopy. *Prog. Nucl. Magn. Resn. Spec.* 94: 75.

2 Aue, W.P., Bartholdi, E., and Ernst, R.R. (1976). 2-Dimensional spectroscopy - application to nuclear magnetic-resonance. *J. Chem. Phys.* 64 (5): 2229.

3 Günther, H. (2013). *NMR Spectroscopy : Basic Principles, Concepts, and Applications in Chemistry*, 3e. New York: Wiley, Chichester.

4 Keeler, J. (2005). *Understanding NMR spectroscopy*. England; Hoboken, NJ: Wiley, Chichester.

5 Weitekamp, D.P., Bielecki, A., Zax, D., Zilm, K., and Pines, A. (1983). Zero-field nuclear magnetic-resonance. *Phys. Rev. Lett.* 50 (22): 1807.

6 Ledbetter, M.P., Theis, T., Blanchard, J.W., Ring, H., Ganssle, P., Appelt, S., Blümich, B., Pines, A., and Budker, D. (2011). Near-zero-field nuclear magnetic resonance. *Phys. Rev. Lett.* 107 (10): 107601.

7 McDermott, R., Trabesinger, A.H., Mück, M., Hahn, E.L., Pines, A., and Clarke, J. (2002). Liquid-state NMR and scalar couplings in microtesla magnetic fields. *Science* 295 (5563): 2247.

8 Ivanov, K.L., Pravdivtsev, A.N., Yurkovskaya, A.V., Vieth, H.-M., and Kaptein, R. (2014). The role of level anti-crossings in nuclear spin hyperpolarization. *Prog. Nucl. Magn. Reson. Spectrosc.* 81: 1.

9 Zax, D.B., Bielecki, A., Zilm, K.W., Pines, A., and Weitekamp, D.P. (1985). Zero-field NMR and NQR. *J. Chem. Phy.* 83 (10): 4877.

10 Suter, D., Jarvie, T.P., Sun, B., and Pines, A. (1987). Observation of spin diffusion in zero-field magnetic resonance. *Phys. Rev. Lett.* 59 (1): 106.

11 Thayer, A.M., Millar, J.M., and Pines, A. (1986). Two-dimensional zero-field NMR and NQR. *Chem. Phys. Lett.* 129 (1): 55.

12 Sjolander, T.F., Tayler, M.C.D., Kentner, A., Budker, D., and Pines, A. (2017). C-13-Decoupled J-coupling spectroscopy using two-dimensional nuclear magnetic resonance at zero-field. *J. Phy. Chem. Lett.* 8 (7): 1512.

13 Ganssle, P.J., Shin, H.D., Seltzer, S.J., Bajaj, V.S., Ledbetter, M.P., Budker, D., Knappe, S., Kitching, J., and Pines, A. (2014). Ultra-low-field NMR relaxation and diffusion measurements using an optical magnetometer. *Angew. Chem. Int. Ed.* 53 (37): 9766.

14 Tayler, M.C.D. and Gladden, L.F. (2018). Scalar relaxation of NMR transitions at ultralow magnetic field. *J. Mag. Res.*

15 Put, P., Pustelny, S., Budker, D., Druga, E., Sjolander, T.F., Pines, A., and Barskiy, D.A. (2021). Zero- to ultralow-field NMR spectroscopy of small biomolecules. *Anal. Chem.* 93 (6): 3226.

16 English, A.E., Whittall, K.P., Joy, M.L.G. and Henkelman, R.M. (1991). Quantitative Two-dimensional time Correlation Relaxometry. *Magn. Reson. Med.* 22: 425–434.

17 Tayler, M.C.D. and Levitt, M.H. (2011). Singlet nuclear magnetic resonance of nearly-equivalent spins. *Phys. Chem. Chem. Phys.* 13 (13): 5556.

18 Turanov, A. and Khitrin, A. (2016). Measurement of J-couplings between chemically equivalent nuclei using off-resonance decoupling. *J. Magn. Reson.* 270: 31.

19 Gullion, T., Baker, D.B., and Conradi, M.S. (1990). New, compensated Carr-Purcell sequences. *J. Magn. Reson.* (1969) 89 (3): 479.

20 Sjolander, T.F., Blanchard, J.W., Budker, D., and Pines, A. (2020). Two-dimensional single-and multiple-quantum correlation spectroscopy in zero-field nuclear magnetic resonance. *J. Magn. Reson.* 318: 106781.

21 Buckenmaier, K., Scheffler, K., Plaumann, M., Fehling, P., Bernarding, J., Rudolph, M., Back, C., Koelle, D., Kleiner, R., Hovener, J.B., and Pravdivtsev, A.N. (2019). Multiple quantum coherences hyperpolarized at ultra-low fields. ChemPhysChem 20 (21): 2823.

22 Garcon, A., Blanchard, J.W., Centers, G.P., Figueroa, N.L., Graham, P.W., Kimball, D.F.J., Rajendran, S., Sushkov, A.O., Stadnik, Y.V., Wickenbrock, A., Wu, T., and Budker, D. (2019). Constraints on bosonic dark matter from ultralow-field nuclear magnetic resonance. *Sci. Adv.* 5 (10): eaax4539.

23 Blümich, B., Blümler, P., and Jansen, J. (1992). Presentation of sideband envelopes by two-dimensional one-pulse (TOP) spectroscopy. *Nucl. Magn. Reson.* 1 (2): 111.

24 Bodenstedt, Sven and Mitchell, Morgan W. and Tayler, Michael C. D. (2022 September). Meridional composite pulses for low-field magnetic resonance. *Phys. Rev. A* 106 (3): 033102. American Physical Society. doi: 10.1103/PhysRevA.106.033102. https://link.aps.org/doi/10.1103/PhysRevA.106.033102.

25 Tayler, M.C.D., Sjolander, T.F., Pines, A., and Budker, D. (2016). Nuclear magnetic resonance at millitesla fields using a zero-field spectrometer. *J. Magn. Reson.* 270: 35.

26 Braunschweiler, L. and Ernst, R.R. (1983). Coherence transfer by isotropic mixing-application to proton correlation spectroscopy. *J. Magn. Reson.* 53 (3): 521.

27 Morris, G.A. (2007). INEPT. In: *eMagRes* (ed. R.K. Harris and R.L. Wasylishen).

28 Rucker, S.P. and Shaka, A.J. (1989). Broad-Band homonuclear cross polarization in 2D NMR using DIPSI-2. *Mol. Phys.* 68 (2): 509.

29 Kadkhodaie, M., Rivas, O., Tan, M., Mohebbi, A., and Shaka, A.J. (1991). Broad-band homonuclear cross polarization using flip-flop spectroscopy. *J. Magn. Reson.* 91 (2): 437.

30 Levitt, M.H. (1986). Composite pulses. *Prog. Nucl. Magn. Reson. Spectrosc.* 18: 61.

**31** Kaderavek, P., Strouk, L., Cousin, S.F., Charlier, C., Bodenhausen, G., Marquardsen, T., Tyburn, J.M., Bovier, P.A., Engelke, F., Maas, W., and Ferrage, F. (2017). Full correlations across broad NMR spectra by two-field total correlation spectroscopy. *ChemPhysChem* 18 (19): 2772.

**32** Zhukov, I.V., Kiryutin, A.S., Ferrage, F., Buntkowsky, G., Yurkovskaya, A.V., and Ivanov, K.L. (2020). Total Correlation Spectroscopy across All NMR-Active Nuclei by mixing at zero field. *J. Phys. Chem. Lett.* 11 (17): 7291.

**33** Zhukov, I. V., Kiryutin, A.S., Wang, Z., Zachrdla, M., Yurkovskaya, A.V., Ivanov, K.L., and Ferrage, F. (2020). Surprising absence of strong homonuclear coupling at low magnetic field explored by two-field nuclear magnetic resonance spectroscopy. *Magn. Reson.* 1 (2): 237.

**34** Cousin, S.F., Charlier, C., Kaderavek, P., Marquardsen, T., Tyburn, J.M., Bovier, P.A., Ulzega, S., Speck, T., Wilhelm, D., Engelke, F., Maas, W., Sakellariou, D., Bodenhausen, G., Pelupessy, P., and Ferrage, F. (2016). High-resolution two-field nuclear magnetic resonance spectroscopy. *Phys. Chem. Chem. Phys.* 18 (48): 33187.

**35** Cousin, S.F., Kaderavek, P., Haddou, B., Charlier, C., Marquardsen, T., Tyburn, J.M., Bovier, P.A., Engelke, F., Maas, W., Bodenhausen, G., Pelupessy, P., and Ferrage, F. (2016). Recovering invisible signals by two-field NMR spectroscopy. *Angew. Chem. Int. Ed.* 55 (34): 9886.

**36** Zhukov, I.V., Kiryutin, A.S., Yurkovskaya, A.V., Grishin, Y.A., Vieth, H.M., and Ivanov, K.L. (2018). Field-cycling NMR experiments in an ultra-wide magnetic field range: relaxation and coherent polarization transfer. *Phys. Chem. Chem. Phys.* 20 (18): 12396.

**37** Cavanagh, J. (2007). Protein NMR Spectroscopy : Principles and Practice, 2e. Boston: Academic Press, Amsterdam.

**38** Bax, A. and Summers, M.F. (1986). H-1 and C-13 assignments from sensitivity-enhanced detection of heteronuclear multiple-bond connectivity by 2D multiple quantum NMR. *J. Am. Chem. Soc.* 108 (8): 2093.

**39** Kiryutin, A.S., Zhukov, I.V., Ferrage, F., Bodenhausen, G., Yurkovskaya, A.V., and Ivanov, K.L. (2021). Sequential assignment of NMR spectra of peptides at natural isotopic abundance with zero- and ultra-low-field total correlation spectroscopy (ZULF-TOCSY). *Phys. Chem. Chem. Phys.* 23 (16): 9715.

**40** Cousin, S.F., Kaderavek, P., Bolik-Coulon, N., Gu, Y., Charlier, C., Garber, L., Bruschweiler-Li, L., Marquardsen, T., Tyburn, J.M., Bruschweiler, R., and Ferrage, F. (2018). Time-Resolved Protein Side-Chain Motions Unraveled by High-Resolution Relaxometry and Molecular Dynamics Simulations. *J. Am. Chem. Soc.* 140 (41): 13456.

**41** Sivanandam, V.N., Cai, J.F., Redfield, A.G., and Roberts, M.F. (2009). Phosphatidylcholine "Wobble" in vesicles assessed by high-resolution C-13 field cycling NMR spectroscopy. *J. Am. Chem. Soc.* 131 (10): 3420.

**42** Carravetta, M., Johannessen, O.G., and Levitt, M.H. (2004). Beyond the T 1 limit: singlet nuclear spin states in low magnetic fields. *Phys. Rev. Lett.* 92 (15): 153003.

**43** Pravdivtsev, A.N., Ivanov, K.L., Kaptein, R., and Yurkovskaya, A.V. (2013). Theoretical study of dipolar relaxation of coupled nuclear spins at variable magnetic field. *Appl. Magn. Reson.* 44 (1): 23.

**44** Kiryutin, A.S., Zhukov, I.V., Yurkovskaya, A.V., Budker, D., and Ivanov, K.L. (2020). *Chapter 23 Singlet Order in Heteronuclear Spin Systems. Long-lived Nuclear Spin Order: Theory and Applications*, 418 The Royal Society of Chemistry.

**45** Zhukov, I.V., Kiryutin, A.S., Yurkovskaya, A.V., Blanchard, J.W., Budker, D., and Ivanov, K. L. (2021). Correlation of high-field and zero- to ultralow-field NMR properties using 2D spectroscopy. *J. Chem. Phys.* 154 (14).

# 14

# Multidimensional Methods and Paramagnetic NMR

*Thomas Robinson, Kevin J. Sanders, Andrew J. Pell, and Guido Pintacuda**

*CRMN, Centre de RMN à Très Hauts Champs de Lyon (UMR 5082 CNRS / Ecole Normale Supérieure de Lyon / Université Claude Bernard Lyon 1), Université de Lyon, 69100, Villeurbanne, 5 rue de la Doua, France*
** Corresponding Author*

## 14.1 Introduction

Paramagnetic centers are atoms or ions with unpaired electrons, including transition $d$ or $f$ metals but also organic radicals. They have a central role in many fields within life sciences, medicine, and industry. Unpaired electrons interact with the surrounding nuclear spins and change the appearance of their NMR spectrum in several ways, by altering chemical shifts, shift anisotropies, and increasing relaxation rates. NMR of paramagnetic samples (paramagnetic NMR, pNMR) thus provides a direct probe of the electronic structures in many important compounds. It can disclose unique information in the study of the atomic-level properties of metal complexes, clusters, magnetic frameworks, and metalloproteins, constituting essential steps for the design of new catalysts and new materials and for understanding biochemical processes inside cells.

The impact of NMR on paramagnetic molecules and materials, however, is hampered by the very same enhanced paramagnetic shifts, shift anisotropies, and relaxation rates. These hinder the acquisition of the NMR experiments and the following spectral assignment and interpretation, and require the spectroscopist to modify the approaches that are optimal for diamagnetic substrates. The electronic properties of paramagnetic centers are widely different and depending on the nature of the paramagnetic center, on the coordination environment, and on the nucleus investigated, NMR shifts, and linewidths are affected to different extents, and different scenarios may be encountered. Figure 14.1 illustrates this point for some ions across the $d$ and $f$ transitions [1], presenting the expected contributions to nuclear relaxation and chemical shifts for $^{1}H$ spins in close proximity of a metal center. Rapid electronic relaxation (with electronic spin-lattice relaxation times typically ranging between $10^{-13}$ and $10^{-8}$ s) enhance nuclear relaxation rates, which on the one hand contributes to broaden the lines up to hundreds of thousands of Hz and on the other hand, reduces the spin-lattice relaxation times of the order of ms or lower. Large hyperfine coupling constants (up to several MHz for atoms in the bonding environment of a paramagnetic center) can result in contributions to NMR shifts of several hundreds of ppm. In materials, with a dense network of paramagnetic centers, these effects add up and result of shift and shift anisotropies of up to several thousands of ppm.

In the presence of large paramagnetic effects, 1D NMR experiments provide the most straightforward and sensitive fingerprint to probe the surrounding of a paramagnetic center. Assignment of these spectra can often be attempted exploiting the geometrical dependence of the paramagnetic effects, for example by correlating spin-lattice relaxation times to the metal-nucleus distances from a structural model or (when available) a single crystal X-ray structure. As an alternative, resonances can be tentatively interpreted on the basis of the paramagnetic shifts

*Two-Dimensional (2D) NMR Methods*, First Edition. Edited by K. Ivanov, P.K. Madhu and G. Rajalakshmi.

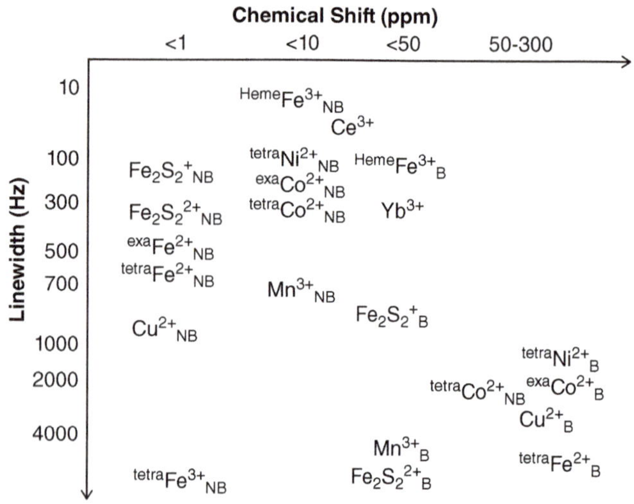


**Figure 14.1** Expected hyperfine shifts versus expected paramagnetic line broadening in $^1$H resonances for a number of metal centers for directly metal bound (3.5 Å distance from the metal, subscript B) and non-bound (6 Å distance from the metal, subscript NB) residues in tetrahedral ('tetra') and octahedral ('exa') coordination geometries. Reproduced with permission from [2]. Copyright (2014) Springer Nature.

predicted by rigorous quantum chemistry/density functional theory (DFT) calculations. Given the potential complexity of such spectra, the large linewidths, the overlap of the resonances, and the uncertainty of DFT methods in the reliable quantification of contact hyperfine interactions, however, 2D correlation spectra are clearly essential for a more extensive characterization.

In this chapter, we survey the experimental 2D NMR methods that have either been used on or are applicable to paramagnetic systems. These comprise many NMR pulse sequences that have been designed for diamagnetic systems, including small- to medium-sized molecules in solution [3], large biomolecules in solution [4], solid materials [5], and biosolids [6]. We will discuss the modifications required in order to apply these methods to paramagnetic systems, and additionally present the tailored schemes, designed specifically to evaluate quantitatively paramagnetic effects. A more extensive discussion of these topics can be found in references [1, 7].

## 14.2 NMR Methods for Paramagnetic Systems in Solution

### 14.2.1 Homonuclear Correlations

#### 14.2.1.1 Paramagnetic COSY

The correlation spectroscopy (COSY) sequence is is the method of choice to identify coupled protons within each spin system [3]. In its most simple implementation, the sequence is composed of only two pulses, which is an advantage in that this minimizes the potential for problems with non-ideal excitation to occur. Each diagonal-peak multiplet appears as an in-phase array, while crosspeaks are antiphase in each dimension. This form of the spectrum has a disadvantage when the linewidth is comparable to or larger than the *J*-coupling, as the components of the crosspeak multiplet cancel to an extent, resulting in a low-intensity multiplet, whereas the diagonal peaks reinforce. Therefore, it is often observed in rapidly-relaxing systems that COSY spectra exhibit very intense diagonal-peak multiplets that dominate the crosspeak multiplets. This problem is overcome by the double-quantum-filtered COSY (DQF-COSY) experiment (Figure 14.2a), which yields anti-phase square arrays for both the diagonal- and crosspeak multiplets and hence, a more equal balance of intensities. The double-quantum

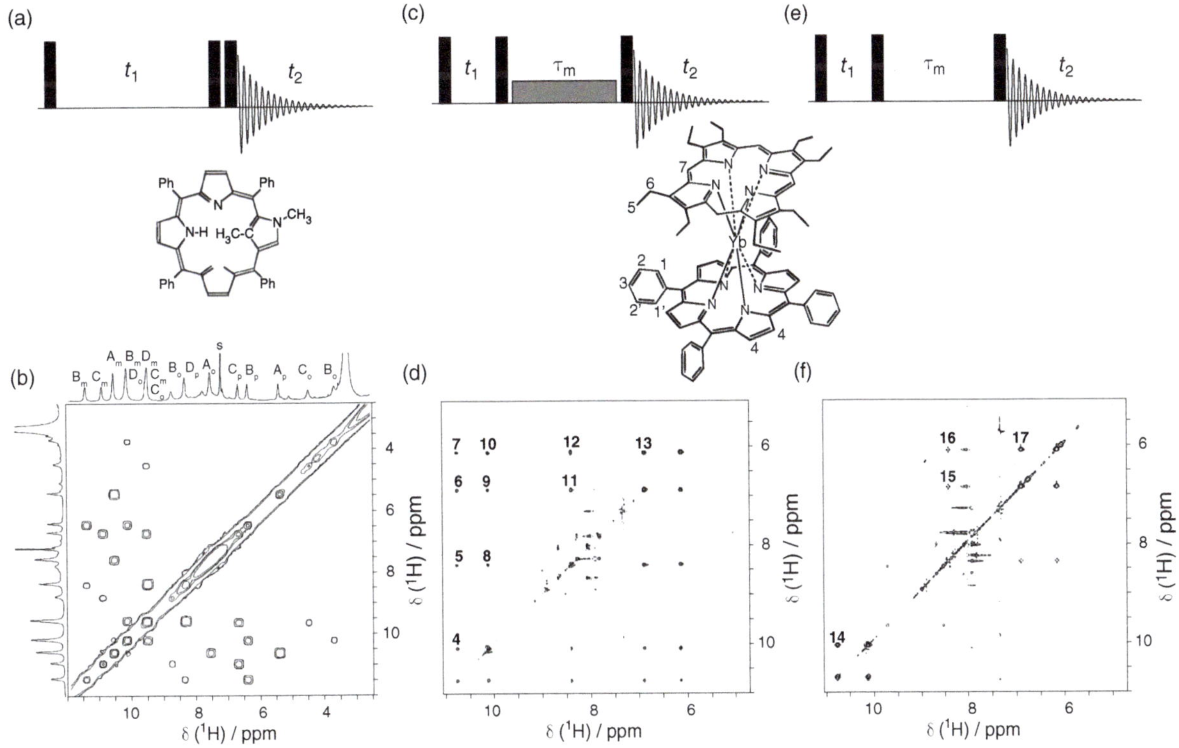


**Figure 14.2** Pulse sequences for the 2D DQF-COSY (a), TOCSY (c) and NOESY (e) experiments. Filled rectangles indicate pulses with a nominal flip angle of 90° and phase *x*, while the gray rectangle in the TOCSY experiment represents continuous irradiation during an isotropic mixing sequence. (b) Meso-phenyl region of the experimental 2D $^1H$–$^1H$ DQF-COSY spectrum of (2-$NCH_3$-21-$CH_3$CTPP)-NiCl in $CDCl_3$ (the structure of the porphyrin ring 2-$NCH_3$-21-$CH_3$CTPPH is shown above the spectrum). (d, f) Experimental 2D $^1H$–$^1H$ homonuclear correlation spectra of YbH(oep)(tpp) complex acquired in solution at 11.74 T. The spectra shown are (e) a TOCSY with an isotropic mixing time of 40 ms and (f) a NOESY spectrum with a mixing time of 40 ms. Adapted with permission from [8] and from [10]. Copyright (1996) and (2010) American Chemical Society.

filter also insures that only $J-$ coupled spins contribute to the spectrum forbidding intense singlet resonances, though the overall sensitivity is reduced by a factor of two with respect to the simple COSY.

In both variants of the experiment, the signal intensity varies as $\sin(\pi J t_1)$ and so the typical evolution time ($t_1$) for a maximum crosspeak signal is roughly about $1/(2J)$. For $^1H$–$^1H$ correlation experiments, $t_1$ should thus extend to about 25 ms. Clearly, this may present a problem for systems with large transverse PREs. In order to maximize the signal-to-noise ratio of the cross-peaks, it is therefore important to tailor the time-domain weighting functions to the expected $T_2$ time constants. However, for paramagnetic systems where $JT_2 \ll 1$, a weighting function is required with maxima at approximately $t_1 = t_2 = T_2$, and data points in the $t_1$ dimension should be acquired only out to a maximum time of $2T_2$.

When transverse PREs are relatively modest, as in the case of some 3*d* metal ions such as $Ni^{2+}$, $Co^{2+}$, or $Ln^{3+}$, the COSY experiment yields a substantial number of $^1H$–$^1H$ through-bond correlations despite larger dispersions of the signals. As an example, a $^1H$–$^1H$ COSY spectrum of (2-$NCH_3$-21-$CH_3$CTPP)-NiCl in $CDCl_3$ is shown in Figure 14.2d [8]. Interestingly, reduction of the recycle delays and acquisition times due to the fast-relaxation times enable the acquisition of more data in a shorter amount of time using these optimized paramagnetic COSY parameters (typically 5–10 mins for 4–8 scans) compared to the standard COSY parameters (10–30 min for 1–2 scans) [9].

Great care must be taken in interpreting the COSY spectra of paramagnetic molecules with slow rotational diffusion, where additional relaxation-allowed crosspeaks may be observable complicating assignment. These additional crosspeaks, first described by Wimperis and Bodenhausen [11] as well as Bertini [12], are not the result of a through-bond coupling but rather of cross-correlation from through-space (NOE) coupling between the nuclei as well as between the nuclei and the paramagnetic center. The cross-correlation results in the two components of the doublet having different linewidths in the two dimensions, and so for zero $J$-coupling we do not obtain the perfect cancellation that we would see without cross-correlation. The result is that with cross-correlation, even in the limit of zero $J$-coupling, we can both generate an antiphase coherence, which is transferred to the second spin by the mixing period, and observe an antiphase doublet in the crosspeak multiplet.

#### 14.2.1.2 Paramagnetic TOCSY

The total correlation spectroscopy (TOCSY) sequence provides an alternative, which compensates some of these weaknesses but is vulnerable to other ones. The form of the spectrum has a distinct advantage over COSY and DQF-COSY as *all* the multiplets are in phase. This means the crosspeak multiplets due to $J$-couplings that are comparable to the linewidth have a larger intensity and so are easier to observe. In this sequence, however, mixing is accomplished by a continuous low-power RF irradiation with duration $\tau_m$ (Figure 14.2b). This element may result in low sensitivity due to (i) the length of the mixing time giving large relaxation losses and (ii) inefficient transfer due to the low RF-field amplitude of the irradiation being insufficient to cover the spectral range of the paramagnetic shifts. The former problem is mitigated by compromizing on the length of the mixing time, and ensuring that it is no longer than $T_2$. A successful example of a $^1$H–$^1$H TOCSY spectrum recorded on the Yb complex YbH(oep)(tpp) is shown in Figure 14.2e [10].

#### 14.2.1.3 Paramagnetic NOESY

Either NOESY (Figure 14.2c) or steady-state NOE experiments are recommended to complete the $^1$H NMR spectrum assignment. These two experiments provide similar structural information but NOESY has the advantage that the exchange and NOE crosspeaks can be observed in a single 2D spectrum compared with multiple 1D spectra for steady-state NOE experiments.

The standard NOESY pulse program needs to be modified for application to paramagnetic complexes. Mixing times need to be chosen as a compromize between the short relaxation times and comparatively long NOE cross-relaxation rates. Given the range of $T_1$ relaxation times within the complex, optimization of the mixing time is an essential preliminary step. For protons with $T_1$ relaxation times significantly longer than the mixing time, the exchange integral approached a maximum as the mixing time increased. However, for protons with shorter $T_1$ relaxation times, the exchange integral reached a maximum before decreasing as relaxation began competing with exchange when the mixing time increased. Mixing times of 10–20 ms are usually found to be a good compromize for maximizing the exchange cross-peak of all protons despite their differing $T_1$ relaxation times. A successful example of a $^1$H–$^1$H NOESY spectrum recorded on the Yb complex YbH(oep)(tpp) is shown in Figure 14.2f [10].

#### 14.2.1.4 Exchange Spectroscopy

NOE intensities are, in general, small for small molecules and reduced even further by the fast relaxation from coupling to the paramagnetic center. In case some excess ligand is also present in the complex solution, however, the same sequence can be exploited to observe exchange crosspeaks between the metal ligands in the complex and the excess ligands. If the exchange is sufficiently rapid, EXSY crosspeak intensities can be close to 100% even for short mixing times, allowing the assignment of protons in the complex using the free ligand assignments (Figure 14.3a).

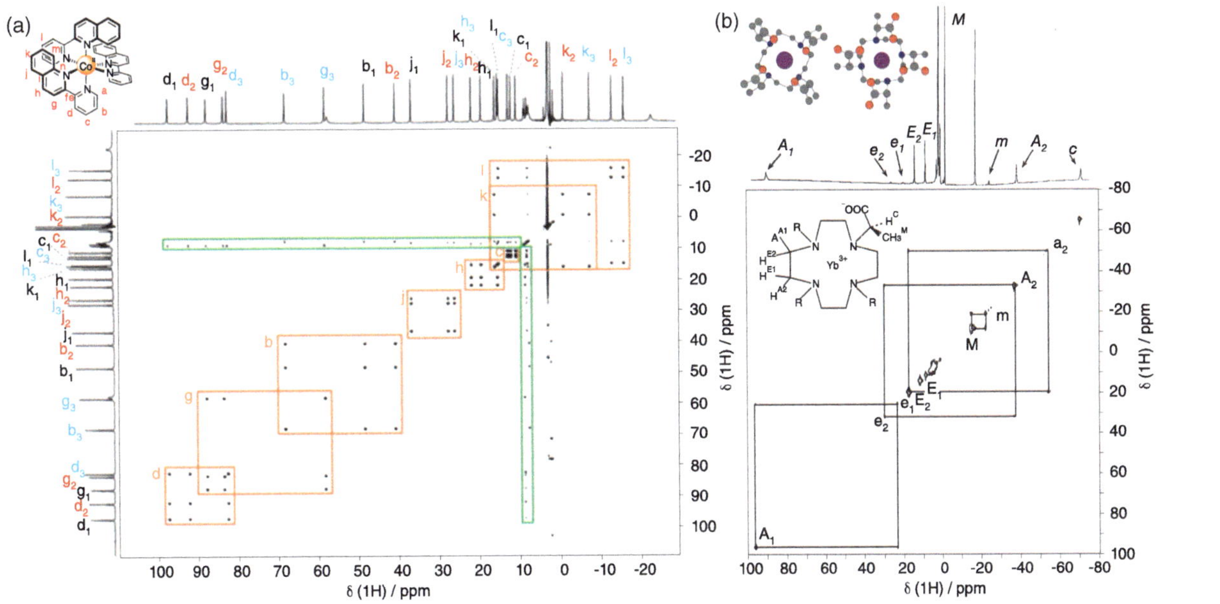


**Figure 14.3** (a) Experimental 2D $^{1}H$ EXSY spectrum of *mer*-$[Co(pq)_3](BF_4)_2$. Exchange crosspeaks are represented by the orange boxes and exchange crosspeaks between the complex and excess ligand are shown by the green boxes. (b) Experimental 2D $^{1}H$ EXSY spectrum of $YbDOTA^{-}$ acquired with a mixing time of 5 ms. The geometries of the two isomers of $YbDOTA^{-}$ are depicted above the spectrum. Adapted with permission from [9] and from [13]. Copyright (2019) and (2001) Wiley.

Two-dimensional $^{1}H$ EXSY has been used to study the slow-exchange dynamics of the two isomeric forms of molecules derived from the MRI contrast agent GdDOTA [14] and YbDOTMA [13] (Figure 14.3b).

### 14.2.2 Heteronuclear Correlations

Two experiments for one-bond correlations that have been successfully applied to paramagnetic molecules are the heteronuclear multiple-quantum correlation (HMQC) and the heteronuclear single-quantum correlation (HSQC) sequences. These experiments correlate the in-phase single-quantum coherence of the $I$-spin in $t_2$ against respectively multiple-quantum coherences of the $I$ and $S$-spin of total coherence orders −2, 0, and +2 (HMQC, Figure 14.4a) and single-quantum coherences of the $S$-spin in $t_1$ (HSQC, Figure 14.4b). Heteronuclear coherences are generated and reconverted to $I$-spin coherences by a pair of heteronuclear INEPT sequences, with half-echo delays $\tau$.

When applied to paramagnetic systems, the most significant weaknesses in these sequences are the narrow bandwidth associated with the 180° pulses and the coherence losses due to paramagnetic relaxation during the echo delays $\tau$.

The effects of the small bandwidth of conventional 180° pulses can be mitigated by the implementation of broadband inversion pulses (BIPs) of Smith et al [15]. In the HMQC, the small number of pulses (four) reduces the effects of imperfections.

As for the duration of the INEPT delays, the peak intensity is proportional respectively to $\sin^2(\pi J_{IS}\tau)$ and $\sin^2(2\pi J_{IS}\tau)$, where $J_{IS}$ is the one-bond heteronuclear $J$-coupling constant, and so the optimum value for $\tau$ is $1/(4J_{IS})$. For one-bond correlations between $^{1}H$ and $^{13}C$ the coupling constant is typically 140 Hz, resulting in an optimum $\tau$ of 1.8 ms. However for short $T_2$ it is usually found that the optimum delay is shorter, as shown in

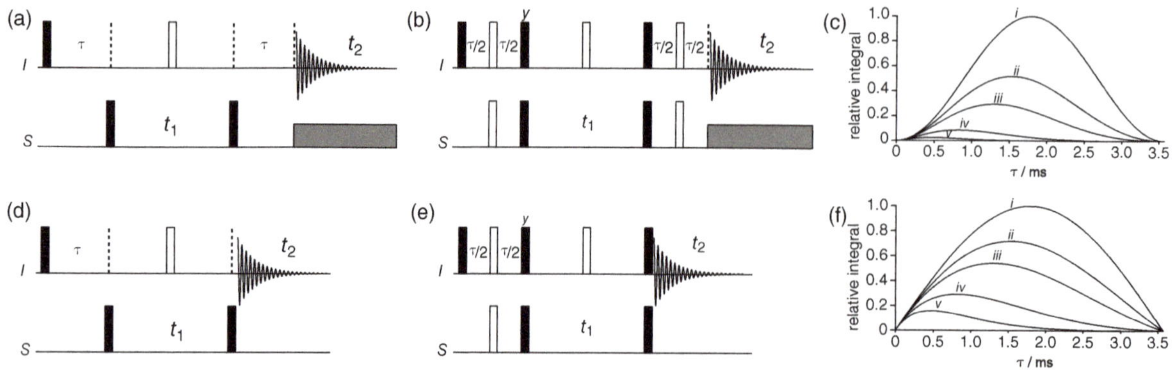


**Figure 14.4** (a, b) Pulse sequences for the 2D HMQC (a) and HSQC (b) experiments. The filled rectangles indicate pulses with a nominal flip angle of 90°, and unfilled rectangles indicate pulses with a nominal flip angle of 180°, and the gray rectangle represents broadband heteronuclear decoupling. The delays $\tau$ have an optimum value of $1/(2J_{IS})$ in the absence of relaxation. (c) Efficiency of the HMQC and HSQC transfers with $J = 140$ Hz at different $^1$H $T_2$ values: (i) no relaxation; (ii) 10 ms, (iii) 5 ms, (iv) 2 ms, ( v) 1 ms. (d, e) Pulse sequences for the 2D AP-HMQC (d) and AP-HSQC (e) experiments. (f) Efficiency of the AP-HMQC and AP-HSQC transfers with $J = 140$ Hz at the different $^1$H $T_2$ values of panel (c).

Figure 14.4c where, e.g. a $T_2$ of 5 ms reduces the optimum value of $\tau$ to below 1.5 ms. The maximum peak intensity is reduced to $\simeq$ 30% of the value expected with no relaxation.

In order to further mitigate sensitivity losses due to transverse relaxation in systems with short $T_2$ times, the inverse INEPT block, during which coherences are back-converted into observable, in-phase $H$ coherences immediately prior to acquisition can be omitted. This produces AP-HMQC and AP-HSQC sequences where an antiphase splitting is detected in the direct $^1$H dimension in the absence of $^{15}$N decoupling (Figures 14.4d and 14.4e). Sensitivity is thus significantly improved as compared to the standard sequence. Figure 14.4f shows, e.g. that with the same $T_2$ of 5 ms, the optimum $\tau$ is again reduced to below 1.5 ms, but that the intensity of the peak is reduced only to $\simeq$ 60%. When signal linewidth is larger than doublet separation, the components of in phase doublets partly overlap and cancel signal. However, when the doublet is phased in dispersion mode, it gives rise to a "pseudosinglet" originated by the sum of the two dispersive components of the doublet. This contributes to the identification of broad, fast-relaxing peaks and also to discriminate signals more severely affected by paramagnetic relaxation with respect to the others.

These detection schemes can be additionally combined via an inversion-recovery (IR) excitation block, which discriminates nuclear spins according to their $T_1$ relaxation times and enables a more convenient identification of residues at a short distance from a metal center. In the resulting IR-$^{15}$N-HSQC-AP sequence (Figure 14.5), a suitable choice of the recovery delay causes a sign difference between fast-relaxing and slow-relaxing signals. Combined with a short duty cycle, this allows a significant suppression of the bulk envelope of the diamagnetic signals, easing the observation of paramagnetic $^1$H resonances.

### 14.2.3 Long-Range Paramagnetic Effects

Nuclear species at large distances from a paramagnetic metal center can be observed by employing the standard NMR experiments for diamagnetic proteins. Nonetheless, the presence of a fast-relaxing paramagnetic ion in the protein changes the appearance of the spectrum of the surrounding nuclei in several ways, which are manifested as changes in chemical shifts, line widths, and one-bond coupling constants (Figures 14.6a–14.6d):

- The **pseudocontact shift (PCS)** changes the observed chemical shifts due to interaction with the anisotropic susceptibility tensor of the paramagnetic center, and depends on the position of a given nucleus with respect to

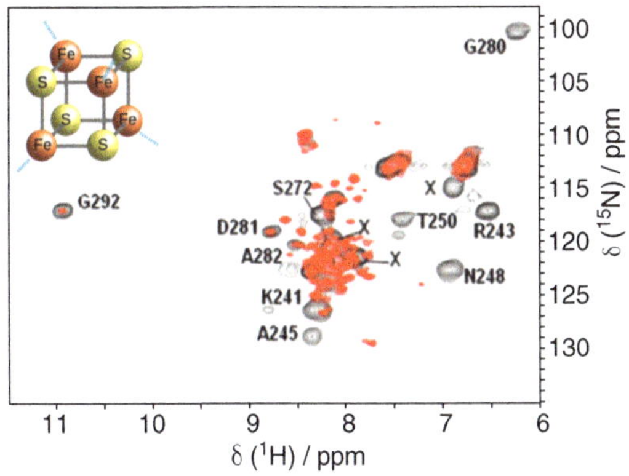


**Figure 14.5** Overlay of a standard $^{15}$N-HSQC (red contours) and IR-$^{15}$N-HSQC-AP (black contours) experiments acquired on the $[Fe_2S_2]^{2+}$-CIAPIN1 domain. Peaks are labeled according to their assignment. Reproduced with permission from [2]. Copyright (2014) Springer Nature.

the magnetic susceptibility tensor of the unpaired electrons, via the distance between the nucleus and paramagnetic center $r$, and the angles $(\theta, \phi)$ that define the orientation of the electronic–nuclear vector with respect to the principal axis system (PAF) of the susceptibility tensor.

- The **nuclear paramagnetic relaxation enhancement (PRE)**, due to a combination of the Solomon and the Curie contributions, depends on the distance $r$ of the nucleus from the paramagnetic center.
- The **cross-correlated relaxation (CCR)** between the Curie spin and dipole-dipole relaxation both the distance $r$ between the paramagnetic center and the observed nucleus $n$, and the angle $\vartheta$ between the electronic–$n$ vector and the bond vector defining an internuclear dipolar coupling.
- Finally, the **residual dipolar coupling (RDC)**, due to alignment of the molecule with respect to the magnetic field, depends on the angles $(\Theta, \Phi)$ that describe the orientation of an internuclear vector and the PAF of the magnetic susceptibility tensor.

Figure 14.6e illustrates the information available from the different paramagnetic effects for an N–H amide bond of a paramagnetic protein, the 30 kDa complex between the N-terminal domain of the $\epsilon$ subunit and the $\theta$ subunit of *Escherichia coli* DNA polymerase III [16]. All four paramagnetic parameters are readily measured by comparison of undecoupled $^{1}$H–$^{15}$N HSQC spectra recorded in the presence and absence of a paramagnetic metal ion. Line shape fitting of the anti-phase doublets observed in the $^{1}$H dimension yields the line width and frequency positions of each doublet component and hence PCS, PRE, CCR, and RDC values. Given the explicit dependence on the 3D structure of the molecule (once the electronic chi tensor anisotropy and the correlation time $\tau_r$ are known), these effects can be used in combination with conventional restraints, such as those obtained from NOESY, to obtain the 3D protein structure.

### 14.2.4 Heteronuclear Detection Strategies

Conventional sequences exploit the larger gyromagnetic ratio of $^{1}$H nuclei and use initial excitation and detection of $^{1}$H to maximize the sensitivity. In paramagnetic systems, however, a larger gyromagnetic ratio also translates into larger transverse PREs, which in turn increases the line broadening observed in the $^{1}$H dimension and accelerates coherence decay during transfer steps that involve $^{1}$H. This leads to a reduction in sensitivity, countering the advantage associated to the initial excitation and detection of $^{1}$H. In these cases, a solution is offered by protonless NMR experiments, where better sensitivity is observed for resonances from the close proximities of a metal center by performing both initial excitation and observation on $^{13}$C or on $^{15}$N. For example, the use of $^{13}$C COSY, $^{13}$C multiple-quantum experiments between CO and CA (COCAMQ), and $^{13}$C NOESY techniques was shown to be efficient for the detection of signals from nuclei as close as 4 Å to the $Cu^{2+}$ ion in $Cu^{2+}/Zn^{2+}$ superoxide dismutase (SOD), or as close as 8 Å to the lanthanide ion in $Tb^{3+}$-substituted human oncomodulin (OM) (in these molecules,

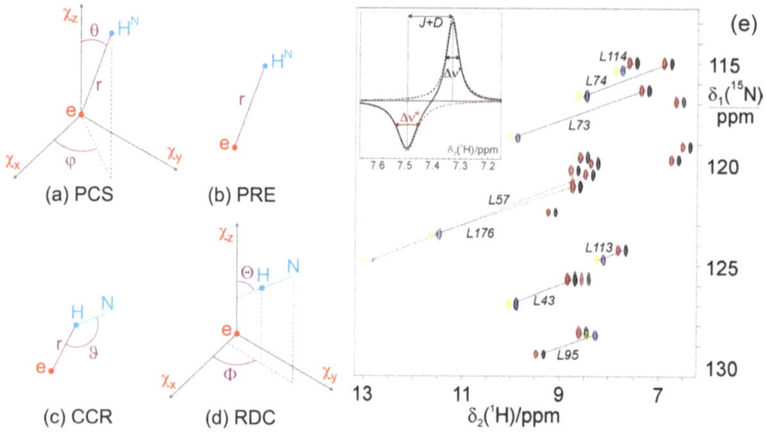


**Figure 14.6** (a–d) Dependence of paramagnetic effects on local geometry: (a) pseudocontact shifts (PCS), (b) paramagnetic relaxation enhancements (PRE), (c) cross-correlated relaxation (CCR) between the Curie spin of the paramagnetic protein and the dipole-dipole interaction between two nuclear spins, (d) residual dipolar couplings (RDC). (e) $^1H-^{15}N$ HSQC spectrum of the 30 kDa complex between the N-terminal domain of the $\epsilon$ subunit and the $\theta$ subunit of *Escherichia coli* DNA polymerase III in the absence and presence of a paramagnetic metal ion. The spectrum contains the anti-phase doublets of the two proteins, one binding paramagnetic $Dy^{3+}$ (yellow and blue peaks) and the other with no metal ion (red and blue peaks). Straight dotted lines connect each paramagnetic resonance with the diamagnetic equivalent. Also shown is a cross section through a crosspeak, with two Lorentzian lineshapes fitted to the doublet components (dashed lines). The PCS is given by the overall change in chemical shift of the resonance, the transverse PRE is given by the overall line broadening, the CCR effects are measured from the difference in linewidth of the two components of the antiphase doublet, and the RDC is calculated from the change in the splitting across the doublet. Reproduced with permission from [16]. Copyright (2004) American Chemical Society.

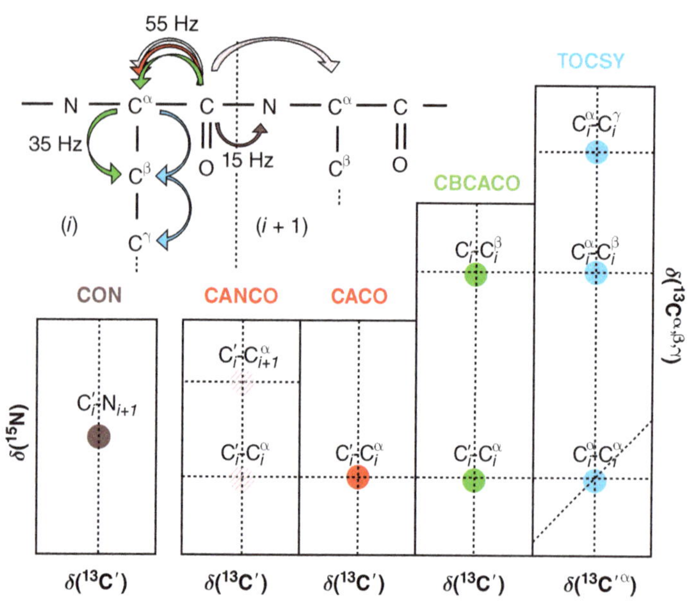


**Figure 14.7** Illustration of the sequential assignment procedure using $^{13}C/^{15}N$ experiments. The relevant one-bond *J*-coupling constants are given. Reproduced with permission from [18]. Copyright (2005) John Wiley and Sons.

$^{1}$H-detection experiments are inefficient respectively from 10 and 16 Å from the metal ion). In the case of paramagnetic proteins, a large palette of 2D and 3D protonless NMR experiments has been designed for the sequential assignment of the backbone and sidechains (Figure 14.7). These include a combination of $^{13}$C/$^{15}$N experiments such as CON, CANCO, CACO, CBCACO, and $^{13}$C TOCSY [17, 18].

## 14.3 NMR Methods for Paramagnetic Systems in Solids

### 14.3.1 Adiabatic Pulses

In solids, in addition to isotropic shifts, the high density of paramagnetic centers causes extremely large paramagnetic shift anisotropies, which can be on the order of 100–1 000 ppm and are thus larger than the practicable RF-amplitudes. Under these conditions, square pulses do not provide the necessary bandwidth, and alternative pulse schemes have to be used to ensure broadband inversion or refocusing. Adiabatic pulses offer a solution to this problem, featuring an impressive ratio between the RF-power used and the achieved bandwidth, and providing a performance that is virtually independent from the resonance offsets.

Classical adiabatic pulses in solution are based on long low-power irradiations, while in paramagnetic solids, under MAS efficient inversion is best achieved with the so-called short high-power adiabatic pulse (SHAPs) [19]. These elements are endowed with high power, which induces a larger effective field overcoming the modulation associated to the shift anisotropy, a wide and fast frequency sweep, which provides a uniform amplitude modulation for all the crystallites, and a short duration, which minimizes signal losses due to the PRE.

Once optimized, SHAPs can be incorporated as inversion elements in more complex pulse sequences, or used for refocusing through a double spin echo, where a second SHAP compensates the frequency-dependent phase errors on the coherences induced by the first SHAP and leading to a dephasing of the signal in the powder [20].

### 14.3.2 Homonuclear Correlations

#### 14.3.2.1 Spin Diffusion

The most important mechanism for homonuclear correlation spectroscopy in solids is spin diffusion, in which we transfer longitudinal magnetization from one spin to another via a coherent homonuclear dipolar coupling interaction[21]. This mechanism is particularly effective for $^{1}$H nuclei, where it is referred to as proton spin diffusion (PSD), as the large gyromagnetic ratio gives large dipolar coupling constants, and therefore relatively rapid transfer. Spin diffusion exhibits good efficiency for systems showing moderate PRE, and at moderate spinning speed. Nevertheless the rate of transfer decreases inversely to the spinning speed, resulting in reduced transfer yields at fast MAS and/or for systems withe pronounced PRE.

#### 14.3.2.2 Homonuclear Recoupling

The rate of transfer can be increased by employing a homonuclear dipolar-recoupling sequence during the mixing time. The most widely used technique for this in solid-state NMR is the radio frequency-driven dipolar-recoupling (RFDR) experiment, which mixes nuclear coherences by way of a train of $\pi$ pulses [22]. Figures 14.8a and 14.8b show the corresponding pulse sequence, together with a RFDR spectrum correlating highly paramagnetic $^{1}$Hs at the core of the iron-sulfur EhHiPIP protein [23]. Despite the large hyperfine shifts (up to 100 ppm), the sequence provides a broadband, offset-insensitive route to efficient 2D correlations.

Due to the limited excitation bandwidth of $\pi$ pulses in general, a long train of 50 or more pulses over 100s or 1 000s of $\mu$s may result, however, not only in spectral biasing due to accumulated pulse bandwidth issues, but potentially significant relaxation losses due to $T_1$ relaxation during this period. A method using $\pi/2$ pulses, which

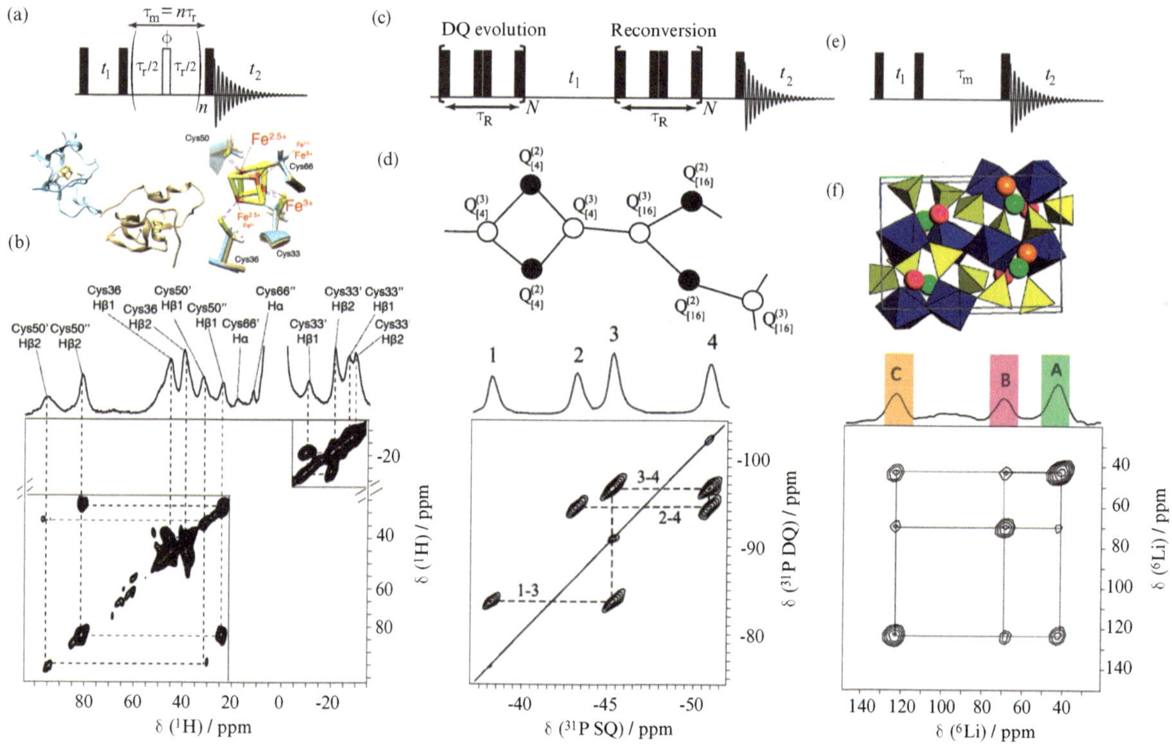


**Figure 14.8** (a) Pulse sequence for the RFDR homonuclear correlation experiment. The phases $\phi$ of the 180° RFDR pulses vary according to the XY-8 supercycle *xyxyyxyx* [24]. (b) Experimental two-dimensional $^{1}H$–$^{1}H$ dipolar correlation spectrum acquired with a microcrystalline, $^{13}C$- and $^{15}N$-labelled sample of the protein EhHiPIP I with RFDR mixing at 60 kHz MAS. Reproduced with permission from [23]. Copyright (2017) American Chemical Society. (c) Pulse sequence for the BABA homonuclear correlation experiment. (d) Experimental 2D BABA spectrum of $MgP_4O_{11}$ recorded using an MAS rotation rate of 14.2 kHz. It exhibits three correlated pairs of $^{31}P$ nuclei, consistent with the structure composed of corrugated sheets of linked rings containing 4 or 16 $PO_4$ tetrahedra. Reproduced with permission from [25]. Copyright (1996) Elsevier. (e) Pulse sequence for EXSY. (f) Experimental 2D $^{6}Li$ EXSY spectra acquired with monoclinic $Li_3Fe_2(PO_4)_3$ at 25 kHz MAS.
The 1D spectrum contains three distinct resonances for the three Li sites A, B, and C, and the EXSY spectrum, recorded with a long mixing time of 3 ms, exhibits crosspeaks due to exchange between all three sites. Reproduced with permission from [26]. Copyright (2010) American Chemical Society.

have much greater bandwidth than $\pi$ pulses, over a shorter recoupling period is the single-quantum (SQ) – double-quantum (DQ) back-to-back experiment, with the pulse sequence given in Figure 14.8c. This experiment correlates the SQ resonance of two nuclei by way of the evolution of a DQ resonance of the pair; as such, the two SQ resonances with frequencies $\omega_1$ and $\omega_2$ in the direct dimension will both have a frequency of $\omega_1 + \omega_2$ in the indirect dimension. This is illustrated on the 2D $^{31}P$ BABA spectrum of $MgP_4O_{11}$ in Figure 14.8d [25]. When applied to protons in paramagnetic organometallic complexes, the BABA experiment is a powerful method to resolve and assign $^{1}H$ peaks.

#### 14.3.2.3 Exchange Spectroscopy

The same 2D EXSY sequence as for solution NMR can be used to obtain correlations between nuclear sites experiencing slow chemical exchange 14.8 e. Also in this case, the mixing time is required not to exceed values of the order of $T_1$. EXSY in paramagnetic solids has been mainly used to investigate the dynamics of lithium exchange in lithium-ion-conducting materials. Figure 14.8f illustrates an example where 2D $^{6}Li$ EXSY spectra were acquired on monoclinic $Li_3Fe_2(PO_4)_3$ and used to measure exchange between the three distinct Li sites in the material [26].

### 14.3.3 Heteronuclear Correlations

#### 14.3.3.1 TEDOR

Because CP-based dipolar polarization transfers for paramagnetic materials are typically inefficient, it is necessary to achieve transfer of polarization from protons via short dipolar-recoupling sequences that use high-powered short RF pulses, such as the HETCOR variant of the TEDOR sequence, known also as dipolar insensitive nucleus enhanced by polarization transfer (DINEPT) sequence [27, 28]. This sequence uses rotor-synchronized $\pi$ pulses to reintroduce, during a relatively short interval $\tau$, the $^{1}$H-$^{13}$C dipolar couplings, otherwise averaged out by MAS, and is usually acquired as a $^{13}$C-detected experiment at intermediate MAS rates in the 30 kHz range. The 2D TEDOR sequence in Figure 14.9 has been proven to be particularly effective both on paramagnetic organometallic complexes and on metalloproteins. An example 2D TEDOR spectrum of the Fe(II) (S = 2) containing complex DIAD-Fe(II) dichloride (DIAD: 2,3-dimethyl-1,4-[(2′,6′)-diisopropylphenyl]-N,N′- diazediene) is given in Figure 14.9c. Eight major resonances corresponding to eight of the nine C–H pairs in the molecule are observed as well as a few additional weaker resonances corresponding to more distant coupled C–H pairs. Heteronuclear correlations achieved by 2D TEDOR or similar techniques represent a major step toward assignment of resonances in small- to medium-sized paramagnetic complexes, and indeed, with the aid of computational techniques may permit full assignment of the observed resonances.

#### 14.3.3.2 HSQC-TEDOR

With the use of faster MAS it becomes possible to observe the $^{1}$H spectrum directly, via a version of the TEDOR sequence with indirect detection, which benefits from the higher sensitivity of observing the nucleus with the larger $\gamma_I$ (HSQC-TEDOR) [31, 32]. The uniformly short $T_1$ relaxation times observed in a paramagnetic solid sample, together with the absence of the requirement for high-power RF irradiations during the experiment, allow the use of very short recycle delays between acquisitions. Combined with the high sensitivity of $^{1}$H detection, this produces a sensitivity boost, and for example in a 1.3 mm rotor at 60 kHz MAS, this scheme provides comparable sensitivity to the $^{13}$C-detected spectrum acquired in a 2.5 mm at 30 MAS, with a five-fold reduction in the sample volume.

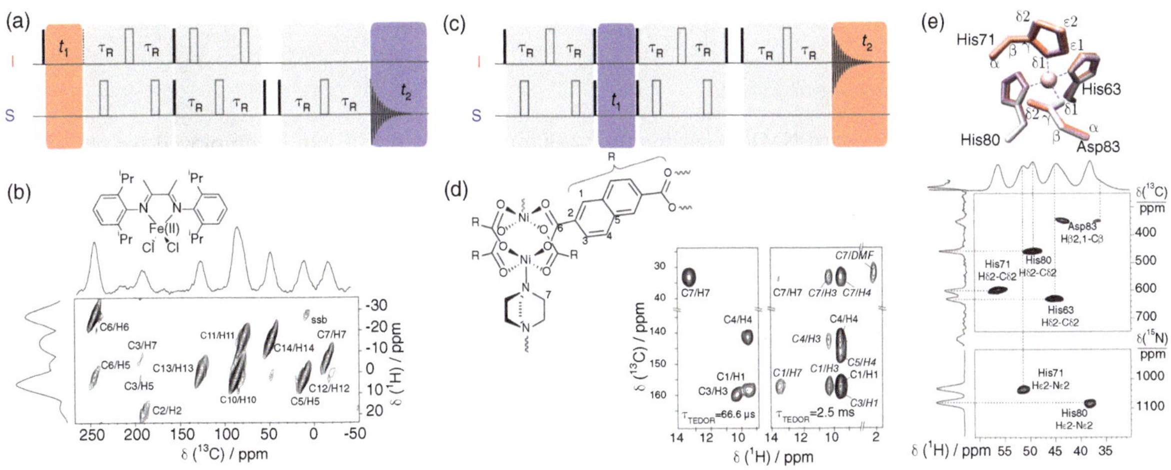


Figure 14.9 (a, c) Pulse sequences for the TEDOR experiments with direct- (a) and indirect-detection (c). (b) $^{1}$H−$^{13}$C TEDOR spectrum of the compound Fe(II)-DIAD acquired at 33 kHz MAS. (d) $^{1}$H−$^{13}$C HSQC-TEDOR spectra of open-pore Ni-DUT acquired with short (66.6 μs, left) and long (2.5 ms, right) $\tau_{\mathrm{TEDOR}}$ recoupling at 60 kHz MAS. Long-range correlations are labeled in italics. (e) $^{1}$H−$^{13}$C and $^{1}$H−$^{15}$N HSQC-TEDOR correlation spectra of Co-SOD acquired at 100 kHz MAS. Reproduced with permission from refs [28–30]. Copyright (2006) and (2020) American Chemical Society, (2021) Wiley.

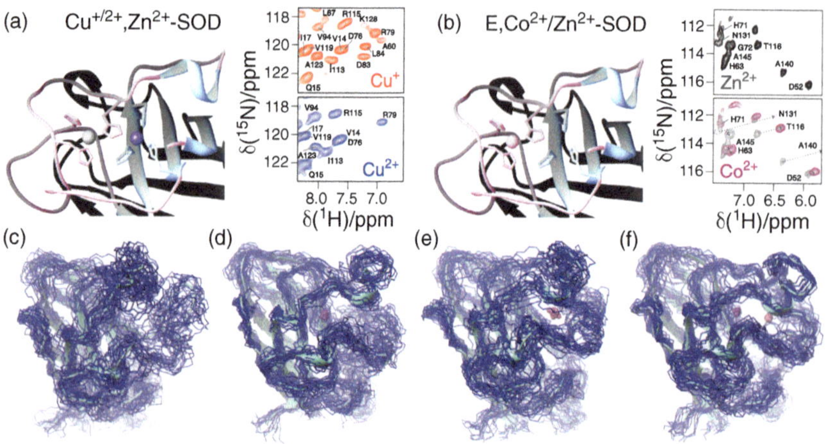


**Figure 14.10** $^{1}$H-detected measurement of $^{15}$N PREs from the comparison of $^{1}$H,$^{15}$N CP-HSQC spectra of $Cu^{2+}$,$Zn^{2+}$-SOD and $Cu^{+}$,$Zn^{2+}$-SOD. (b) $^{1}$H-detected measurement of $^{1}$H and $^{15}$N PCS from the comparison of $^{1}$H,$^{15}$N CP-HSQC spectra of $Co^{2+}$-SOD and Zn-SOD. (c–f) Bundles of SOD structures calculated (a) without paramagnetic restraints, (b) with $^{15}$N and $^{13}$C PREs, (c) with $^{1}$H, $^{15}$N, and $^{13}$C PCSs, and (d) with both PREs and PCSs. The Cu and Co ions are represented by violet and pink spheres, and the mean NMR structure is depicted as an aquamarine ribbon. Reproduced with permission from [35]. Copyright (2013) American Chemical Society.

The HSQC-TEDOR was recently employed for the characterization of a Fe(II) organometallic complex [32] and a metallo-organic framework containing Ni(II) paddle-wheel units [29], where 2D correlation can be acquired within 30 minutes on 1–2 mg of samples at natural $^{13}$C abundance. Depending on the recoupling period employed, $^{1}$H–$^{13}$C correlations can be observed either over short distances corresponding to a single bond, or over longer distances, including correlations between $^{1}$Hs and quaternary $^{13}$Cs (Figure 14.9d). These HSQC-TEDOR spectra resolve and assign the individual frequencies of all the $^{1}$H and $^{13}$ nuclei in the sample. The same approach was also used to elucidate the coordination sphere of a large paramagnetic Co(II) metalloenzyme, using correlations acquired on less than 0.5 mg of a large, uniformly $^{15}$,$^{13}$-labeled and fully protonated protein dimer (Figure 14.9e) [30].

### 14.3.4 Long-Range Paramagnetic Effects

As for solution NMR, the standard NMR experiments for diamagnetic proteins can be employed for observing nuclear species at large distances from a paramagnetic metal center, allowing the measurement of PRE and of pseudocontact shifts. In deuterated protein at 60 kHz MAS, 2D and 3D $^{1}$H-detected and CP-based correlations such as $^{1}$H-$^{15}$N CP-HSQC and (H)CANH correlations result in highly sensitive experiments capable to provide more than one hundred PREs [33] and PCS [34] between 10 and 24 Å from a paramagnetic metal ion in Cu(II)- and Co(II)-substituted superoxide dismutase. These data can be easily implemented as restraints in combination with internuclear distances and dihedral angles, improving the quality of an NMR-determined structure, in particular in the proximity of a metal ion.

### 14.3.5 Separation of Shift and Shift-anisotropy Interactions

Due to large anisotropic effects induced by paramagnetic centers on the NMR resonances of nuclei in paramagnetic complexes, a significant number of sidebands or some or all of the resonances may be observed. This often results in spectra consisting of many isotropic resonances overlapping with spinning sidebands. Of course, this problem can

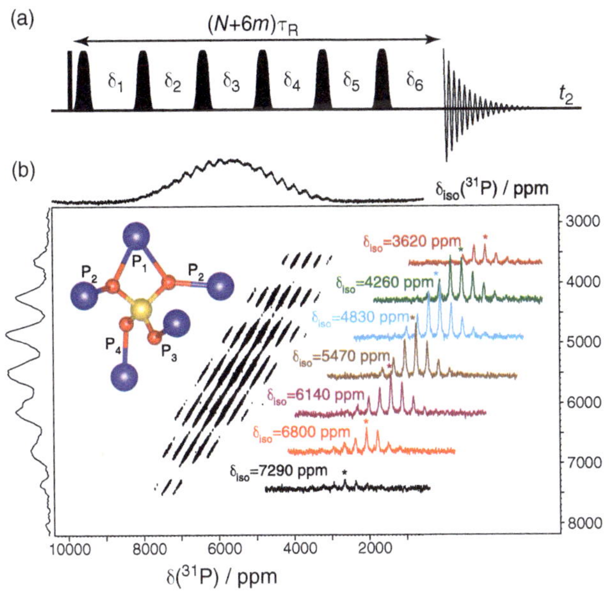


**Figure 14.11** (a) aMAT pulse sequence and (b) 2D aMAT spectrum of LiFe0.5Mn0.5PO4 acquired with a MAS rotation rate of 60 kHz. Reproduced with permission from [38]. Copyright (2012) American Chemical Society.

be addressed simply by rotating at a faster rate, but in extreme cases even MAS rates up to 60 kHz cannot resolve this issue. 2D sideband separation experiments like the magic-angle turning (MAT) sequence [36] or the phase-adjusted spinning-sidebands (PASS) sequence [37] address this issue by suppressing spinning sidebands in the direct (PASS) or indirect (MAT) dimensions, thus achieving a 2D spectrum correlating pure isotropic frequencies to mixed isotropic/anisotropic frequencies, the latter of which can be transformed to pure anisotropic frequencies with an appropriate shearing transformation. These experiments use a train of $\pi$ refocusing pulses with unique timings in order to result in pure isotropic evolution, but the aforementioned bandwidth problems with $\pi$ pulses can cause these experiments to not work properly for paramagnetic samples. Clément et al. addressed this problem in the MAT experiment by the inclusion of SHAP refocusing pulses, which was dubbed the adiabatic MAT (aMAT) sequence, and is given in Figure 14.11a [38]. The authors applied this methodology to untangle the $^{31}$P resonances in the olivine-type mixed phase lithium-ion battery cathode materials with the stoichiometries $LiFe_xMn_{1-x}PO_4$. The $^{31}$P MAS NMR spectra of these mixed-phases are composed of up to 32 unique isotropic resonances, each with potentially unique principle components of the anisotropy of their shift tensors. To make matters worse, lines were severely affected by significant anisotropic bulk susceptibility broadening, resulting an a 1D MAS spectrum completely devoid of any appreciable resolution. The use of aMAT (shown in Figure 14.11b on the $LiFe_{0.5}Mn_{0.5}PO_4$ composition) greatly enhanced the spectral resolution by suppression of spinning sidebands. Subsequent fitting of the isotropic projection of this spectrum permitted the identification of isotropic shifts for each of the 32 possible sites.

### 14.3.6 Separation of Shift-anisotropy and Quadrupolar Interactions

In this section, we present experimental schemes that are designed to separate the contributions to the spectral lineshape from the paramagnetic SA and quadrupolar interactions. Though quadrupolar nuclei in paramagnetic systems have complicated spectra, the tensor parameters are useful indicators of the structural and electronic properties. There are two important types of quadrupolar nucleus (with integer- and half-integer-spins respectively) characterized by their different observable signal components, with half-integer spins exhibiting both central and satellite transitions signals, and integer spins only having ST. These two types of nuclei, therefore, need to be treated separately with specifically-tailored pulse sequences.

#### 14.3.6.1 Integer-Spin Quadrupolar Nuclei

Under static conditions, we can in principle measure the effects on the spectrum of two different interactions: the isotropic shift and SA both evolve in the same way with frequencies that are proportional to the coherence order $p$, and the first-order quadrupolar interaction evolves with a frequency that is proportional to the satellite order $d$. The first pulse sequence that was developed to separate these two interactions is the shifting $p$-echo experiment of Antonijevic and Wimperis and is shown in Figure 14.12a [39].

The $p$- and $d$-symmetry pathways are also shown. Therefore, there are two echoes formed during the pulse sequence: the $p$-echo is formed at points where $t_1 = t_2$, and the $d$-echo is formed at $t_2 = 0$. The quadrupolar interaction echo ridge is therefore aligned along the $t_1$-axis, whereas the shift/SA ridge echo is along $t_1 = t_2$. After an appropriate 2D Fourier transform combined with a shear, we obtain a spectrum in which the quadrupolar broadening is present only in one dimension whereas the shift/SA appears in the other.

There is room for improvement in this experiment as the coherence-order selection in each solid echo only retains half the total observable signal components, and so the sequence gives a maximum sensitivity that is one quarter of that theoretically attainable in the 1D experiment. An improvement to the shifting $p$-echo experiment was proposed by Walder et al. who developed the shifting $d$-echo pulse sequence shown in Figure 14.12b [40]. Here, the solid echo during the evolution period is replaced with a spin echo, with the result that the $p$- and $d$-symmetry pathways are altered so that it is the $p$-echo that is now formed at $t_2 = 0$ and the $d$-echo that is formed at $t_1 = t_2$. As for the shifting $p$-echo experiment, complete separation of the two interactions is achieved using the same shearing and scaling transformations. The use of the spin echo is advantageous for two reasons: firstly, the zero-frequency spike in the quadrupolar dimension is eliminated, and secondly, the spin echo does not suffer from the reduction in signal intensity of the solid echo, with the result that the shifting $d$-echo experiment has double the sensitivity of the shifting $p$-echo experiment. The disadvantage of the shifting $d$-echo sequence is the lower bandwidth of the 180° pulse compared to the 90° pulse; this bandwidth issue has recently been tackled using adiabatic pulses. In order to improve the phase cancellation from the shift interaction and phase acquired due to the time-dependent adiabatic pulse phase, a pair of adiabatic pulses have been added by Aleksis et al. [41] before and after the $\frac{\pi}{2}$ pulse as illustrated in Figure 14.13a. The quadruple adiabatic shifting $d$-echo sequence appears

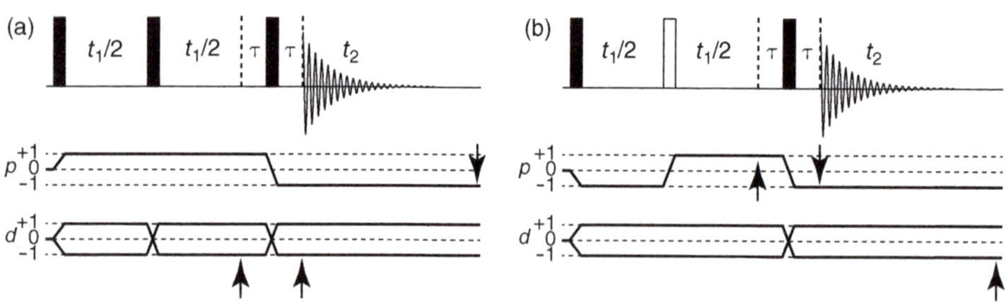


**Figure 14.12** Pulse sequences and symmetry-transfer pathways for correlating the paramagnetic shift anisotropy and quadrupolar lineshapes for spin $I = 1$ nuclei. Also shown are the transfer pathways for the coherence order $p$ and satellite order $d$. The sequence in (a) is for the shifting $p$-echo experiment of Antonijevic and Wimperis in which the first-order quadrupolar interaction is refocused at the end of $t_1$ [39]. The sequence for the shifting $d$-echo sequence of Walder et al. is shown in (b) [40], where the SA is refocused at the end of $t_1$. The time points at which the shift/shift anisotropy, and the quadrupolar interactions are refocused are indicated on the relevant symmetry pathways with arrows. Filled rectangles indicate pulses with a nominal flip angle of 90° and unfilled rectangles indicate pulses with a nominal flip angle of 180°. All pulses have phase $x$.

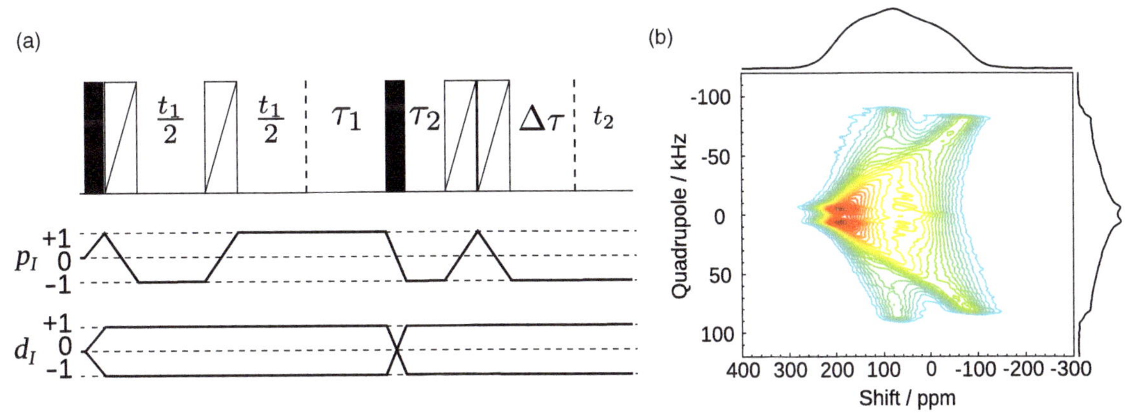


**Figure 14.13** Pulse sequences for and quadruple adiabatic shifting d- (a) echo sequences and the according experimental 2D spectra of $^2H$ in $CuCl_2 \cdot D_2O$ (b). Here the filled rectangle represent $\frac{\pi}{2}$ pulses and rectangle with a diagonal represent adiabatic pulses. $\tau_1$, $\tau_2$ and $\Delta\tau$ are free precession periods to facilitate correct echo formation. $p_1$ and $d_1$ are the symmetry pathways for the experiments. Adapted with permission from [41]. Copyright (2019) Elsevier.

to be superior since the spectrum of Figure 14.13b shows no trace of any artifacts and yields a clear lineshape characteristic of the two interactions. Therefore this sequence is beneficial for complex materials with large shift anisotropies and more than one local environment where the artifact would degrade the spectrum to the point of interfering with its assignment and interpretation.

While the shifting $p$- and $d$-echo experiments have proven useful for characterizing $^2H$ environments in paramagnetic systems, the experiments have the disadvantage of only being applicable to static samples, which result in spectra that suffer from low resolution and sensitivity. Magic-angle spinning alleviates these problems but does require different pulse sequences in order to obtain a separation of shift/SA and quadrupolar effects. Such separation and correlation has recently been achieved by Aleksis and Pell using the PASS experiment. [42] The first order quadrupolar nuclei appears along an axis parallel to the direct dimension $\omega_2$ while the shift anisotropy does so along the anti-diagonal axis $\omega_2 = -\omega_1$. Hence the shift anisotropy only appears along the indirect dimension $\omega_1$ while both quadrupolar interaction and shifts appear along the direct dimension $\omega_2$. A shearing transformation similar to that required for a shifting $d$-echo has to be applied to the spectrum in order to remove the shift anisotropy from the $\omega_2$ dimension. It has been demonstrated that PASS shows high sensitivity and allows total separation of both interactions.

#### 14.3.6.2 Half-Integer-Spin Quadrupolar Nuclei

If central transition (CT) is not broaden by the first-order quandrupolar interaction, the satellite transitions (ST) are, and both CT and ST are broadened by the (C)SA and second-order quadrupolar interaction therefore one must be especially cautious about their separation. Since the introduction of the multiple-quantum magic-angle spinning experiment (MQMAS) and the satellite transition magic-angle (STMAS) experiment, the spectroscopist can rely on an efficient method to record resolved NMR spectra of half-integer quadrupolar nuclei. However, in paramagnetic systems, half-integer quadrupolar nuclei experience the inhomogeneous paramagnetic broadening from various sources. This includes contributions from the system itself, the most important of which is the anisotropic bulk magnetic susceptibility (ABMS), which arises for systems exhibiting an anisotropic magnetic susceptibility and is due to the nature of the packing of the crystallites in the sample container and contributions from the instrumental setup, such as temperature gradient across the sample as a result of frictional heating due to MAS and, which is particularly problematic for paramagnetic systems due to the strong Curie–Weiss temperature dependence of the shifts. These paramagnetic broadening effects

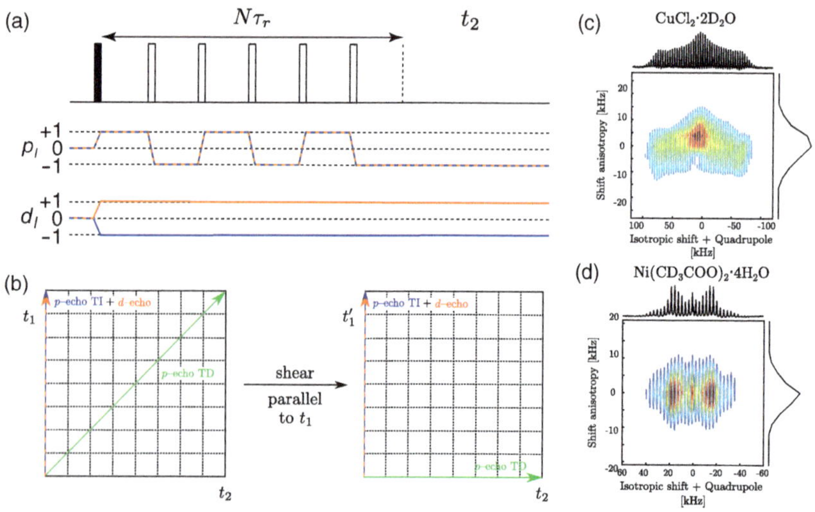


**Figure 14.14** (a) The PASS sequence, unless otherwise specified, the filled and empty rectangles represent RF pulses with flip angles $\frac{\pi}{2}$ and $\pi$, respectively. (b) Schematic representation of the time domain of the PASS experiment before and after shearing parallel to pseudo-$t_1$ dimension with a shear ratio of $\kappa = -1$ The blue, orange and green arrows represent time-independent (TI) p-echo ridge, the d-echo ridge and the time-dependent (TD) p-echo ridge. Experimental $^2$H PASS spectra of two paramagnetic (c) $CuCl_2 \cdot 2D_2O$ t and (d) $Ni(CD_3CO_2)_2$,$4H_2O$. Both spectra have been sheared. The projection of the isotropic shift/quadrupolar dimension shows the spinning-sideband manifold, while the projection of the shift-anisotropy dimension shows the profile of the spinning-sideband intensities. Adapted with permission from [42]. Copyright (2021) American Institute of Physics.

broaden the individual sidebands due to MAS and thus broaden the CT resonance. When combined with the broadening due to the second-order quadrupolar interaction, this results in a CT lineshape that is particularly difficult to interpret, thus requiring careful separation of the paramagnetic and second-order quadrupolar contributions.

Initially used separate the STs and enhance the CT signal intensity by a factor of $I+\frac{1}{2}$ the rotor assisted population transfer (RAPT) was used to estimate quadrupolar couplings on systems broaden by paramagnetic effects [43] but still cannot separate shift/SA from quadrupolar coupling. The correlation of anisotropies separated through echo refocusing (COASTER)[44] method has been successfully used but requires sample spinning at the unconventional angle of 70.18° therefore this experiment remains difficult to implement on commercial ssNMR probes.

Carvalho et al. recently proposed a new way of processing STMAS and MQMAS to separate in orthogonal dimensions the second-order fourth rank quadrupolar interaction tensor and the inhomogeneous paramagnetic broadening[45]. This method is based on the 2D one-pulse (TOP) transformation. A first shearing is applied after which a new 2D time-domain system is constructed in which shift and ABMS broadening are separated along two separated axes. A second shear transformation is applied yielding a pure quadrupolar and a pure shift dimension. Figure 14.15a shows the structure of an yttrium-alumium garnets (YAG) (b) and (c) show TOP-STMAS spectra acquired on samples doped with europium and neodymium, respectively.

The use of RAPT-enhanced MAT experiment (or RAPT-MAT) allows gaining the needed resolution to plot the RAPT enhancement for a given resonance versus the repetition frequency of RAPT pulses. [46] The maximum enhancement can be determined from such buildup and be used to extract the magnitude and range of the quadrupolar couplings. Figure 14.16a shows the resolved, RAPT-enhanced MAT spectra of $NaMnO_2$. The shift resolution allowed by this experiment is then used to obtain the buildup curves shown in Figure 14.16c.

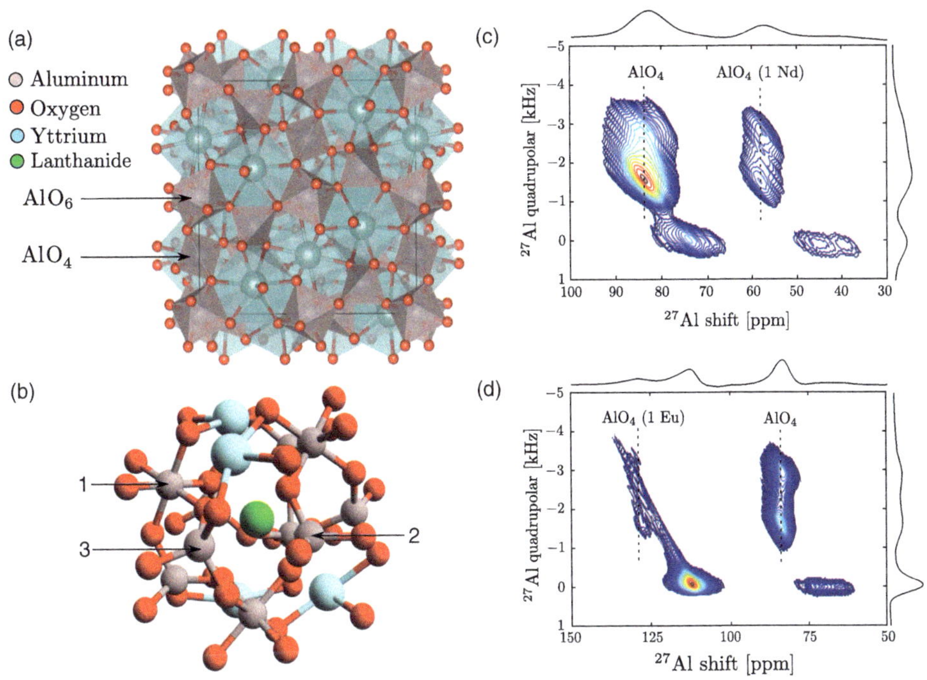


**Figure 14.15** (a) Structure of an Ln:Yttrium-alumium garnet (YAG) and (b) a molecular cluster defined by the atoms present in the Ln-Y coordination shell. Three types of alumium environments are represented : six coordinates (1) and four coordinates (2,3). (c, d) TOP-STMAS spectra acquired on samples doped with neodymium and europium, respectively. Adapted with permission from [45]. Copyright (2020) John Wiley and sons.

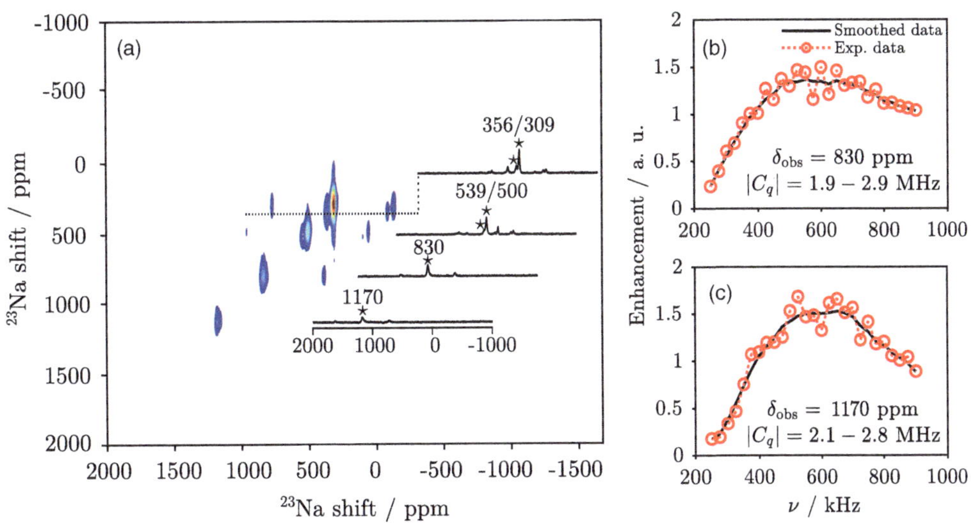


**Figure 14.16** (a) MAT spectrum of $NaMnO_2$ exhibiting six resonances with distinct observed shifts. Inset spectra correspond to extracted direct-dimension slices, and the observed shifts correspond to the centerbands marked by stars. RAPT pulses were applied before the MAT experiment to enhance the signal-to-noise ratio and suppress the signals arising from the satellite transitions. (b, c) Build up of the RAPT enhancement obtained for two of the $^{23}Na$ resonance of the MAT spectrum in (a) versus repetition frequency of RAPT pulses. Red circles represent data points while black lines are representation of the moving average. Adapted with permission from [46]. Copyright (2022) Elsevier.

## Acknowledgments

The development of paramagnetic NMR at CRMN Lyon has been funded over the past few years by the European Research Council (ERC) under the European Union's Horizon 2020 research and innovation programme (ERC-2015-CoG GA 648974), by the European Union's FP7 research and innovation programme (FP7-PEOPLE-2012-ITN GA 317127), by the Agence Nationale de la Recherche (ANR-15-CE29-0025-01 MrCAT and ANR-21-CE29-0010-01 MatPNMR), by joint research activities and transnational access within the consortium PANACEA (A Pan-European Solid-State NMR Infrastructure for Chemistry-Enabling Access, GA 101008500), by the Institut de Chimie de Lyon (FR3023) and by the CNRS (IR-RMN-THC FR3050 and Infranalytics FR2054).

## References


**1** Bertini, I., Luchinat, C., Parigi, G. and Ravera, E. (2016). *NMR of Paramagnetic Molecules: Applications to Metallobiomolecules and Models*. Elsevier Science Ltd: Amsterdam.

**2** Cio-Baoni, S., Gallo, A., Muzzioli, R. and Piccioli, M. (2014). The IR-$^{15}$N-HSQC-AP experiment: a new tool for NMR spectroscopy of paramagnetic molecules. *J. Biomol. NMR*, 58: 123–128.

**3** Keeler, J. (2010). *Understanding NMR Spectroscopy Second Edition*. Wiley and Sons, Ltd: Chichester.

**4** Cavanagh, J., Fairbrother, W. J., Palmer III, A.G., Rance, M. and Skelton, N. J. (2007). *Protein NMR spectroscopy: principles and practice*. London: Academic Press.

**5** Duer, M.J. (2005). *Introduction to Solid-State NMR Spectroscopy*. Oxford: Blackwell.

**6** Le Marchand, T., Schubeis, T., Bonaccorsi, M., Paluch, P., Lalli, D., Pell, A..J., Andreas, L.B., Jaudzems, K., Stanek, J. and Pintacuda, G. (2022). $^{1}$H-detected biomolecular NMR under fast magic-angle spinning. *Chem. Rev.* 122: 9943–10018.

**7** Pell, A..J., Pintacuda, J. and Grey, C.P. (2019). Paramagnetic NMR in solution and in the solid state. *Prog. Nucl. Magn. Reson. Spectrosc.*, 111: 1–271.

**8** Chmielewski, P.J., Latos-Grazynski, L. and Glowiak, T. (1996). Reactions of nickel(II) 2-Aza-5,10,15,20-tetraphenyl-21- carbaporphyrin with methyl iodide. The first structural characterization of a paramagnetic organometallic nickel(II) complex. *J. Am. Chem. Soc.*, 118: 5690–5701.

**9** Lehr, M., Paschelke, T., Trumpf, E., Vogt, A.-M., Naether, C., Soennichsen, F.D. and McConnell, A.J. (2020). A paramagnetic nmr spectroscopy toolbox for the characterisation of paramagnetic/spin-crossover coordination complexes and metal-organic cages. *Angew. Chem. Int. Ed.* 59: 19344–19351.

**10** Bertini, I., Coutsolelos, A., Dikiy, A., Luchinat, C., Spyroulias, G.A. and Troganis, A. (1996). Structural and dynamic information on double-decker $Yb^{3+}$ and $Dy^{3+}$ porphyrin complexes in solution through $^{1}$H NMR. *Inorg. Chem.* 35: 6308–6315.

**11** Wimperis, S. and Bodenhausen, G. (1989). Relaxation-allowed cross-peaks in two-dimensional NMR correlation spectroscopy. *Mol. Phys.* 66: 897–919.

**12** Bertini, I., Luchinat, C. and Tarchi, D. (1993). Are true scalar proton–proton connectivities ever measured in COSY spectra of paramagnetic macromolecules? *Chem. Phys. Lett.* 203: 445–449.

**13** Di Bari, L., Pintacuda, G. and Salvadori, P. (2000). Solution equilibria in YbDOTMA, a chiral analogue of one of the most successful contrast agents for MRI, GdDOTA. *Eur. J. Inorg. Chem.* 75–82.

**14** Aime, S., Botta, M. and Ermondi, G. (1992). NMR study of solution structures and dynamics of lantbanide(III) complexes of DOTA. *Inorg. Chem.*, 31: 4291–4299.

**15** Smith, M.A., Hu, H. and Shaka, A.J. (2001). Improved broadband inversion performance for NMR in liquids. *J. Magn. Reson.* 151: 269–283.

**16** Pintacuda, G., Keniry, M.A., Huber, T., Park, A.Y., Dixon, N.E. and Otting, G. (2004). Fast structure-based assignment of $^{15}$N HSQC spectra of selectively $^{15}$N-labeled paramagnetic proteins. *J. Am. Chem. Soc.* 126: 2963–2970.

**17** Bermel, W., Bertini, I., Duma, L., Felli, I.C., Emsley, L., Pierattelli, R. and Vasos, P. R. (2005). Complete assignment of heteronuclear protein resonances by protonless NMR spectroscopy. *Angew. Chem. Int. Ed.* 44: 3089–3092.

**18** Bertini, I., Luchinat, C., Parigi, G. and Pierattelli, R. (2005). NMR spectroscopy of paramagnetic metalloproteins. *ChemBioChem* 6: 1536–1549.

**19** Kervern, G., Pintacuda, G. and Emsley, L. (2007). Fast adiabatic pulses for solid-state NMR of paramagnetic systems. *Chem. Phys. Lett.* 435: 157–162.

**20** Pell, A.J., Kervern, G., Emsley, L., Deschamps, M., Massiot, D., Grandinetti, P.J. and Pintacuda, G. (2011). Broadband inversion for MAS NMR with single-sideband-selective adiabatic pulses. *J. Chem. Phys.* 134: 024117.

**21** Bloembergen, N. (1949). On the interaction of nuclear spins in a crystalline lattice. *Physica* 15: 386–426.

**22** Bennett, A.E., Ok, J.H., Griffn, R.G. and Vega, S. (1992). Chemical shift correlation spectroscopy in rotating solids: radio frequency-driven dipolar recoupling and longitudinal exchange. *J. Chem. Phys.* 96: 8624–8627.

**23** Bertarello, A., Schubeis, T., Fuccio, C., Ravera, E., Fragai, M., Parigi, G., Emsley, L., Pintacuda, G. and Luchinat, C. (2017). Paramagnetic properties of a crystalline iron–sulfur protein by magic-angle spinning NMR spectroscopy. *Inorg. Chem.* 56: 6624–6629.

**24** Gullion, T., Baker, D.B. and Conradi, M.S. (1990). New, compensated Carr–Purcell sequences. *J. Magn. Reson.* 89: 479–484.

**25** Feike, M., Demco, D.E., Graf, R., Gottwald, J., Hafner, S. and Spiess, H.W. (1996). Broadband multiple-quantum NMR spectroscopy. *J. Magn. Reson. A* 122: 214–221.

**26** Davis, L.J.M., Heinmaa, I. and Goward, G. R. (2010). Study of lithium dynamics in monoclinic $Li_3Fe_2(PO_4)_3$ using $^6$Li VT and 2D exchange MAS NMR spectroscopy. *Chem. Mater.* 22: 769–775.

**27** Wickramasinghe N.P., and Ishii, Y. (2006). Sensitivity enhancement, assignment, and distance measurement in C-13 solid-state NMR spectroscopy for paramagnetic systems under fast magic angle spinning. *J. Magn. Reson.* 181: 233–243.

**28** Kervern, G., Pintacuda, G., Zhang, Y., Oldeld, E., Roukoss, C., Kuntz, E., Herdtweck, E., Basset, J.M., Cadars, S. and Emsley, L. (2006). Solid-state NMR of a paramagnetic DIAD-Fe-II catalyst: Sensitivity, resolution enhancement, and structure-based assignments. *J. Am. Chem. Soc.* 128: 13545–13552.

**29** Blahut, J., Lejeune, A.L., Ehrling, S., Senkovska, I., Kaskel, S., Wisser, F.M. and Pintacuda, G. (2021). Monitoring dynamics, structure, and magnetism of switchable metal-organic frameworks via h-1-detected mas nmr. *Angew. Chem. Int. Ed.* 60: 21778–21783.

**30** Bertarello, A., Benda, L., Sanders, K.J., Pell, A.J., Knight, M.J., Pelmenschikov, V., Gonnelli, L., Felli, I.C., Kaupp, M., Emsley, L., Pierattelli, R. and Pintacuda. (2020). Picometer resolution structure of the coordination sphere in the metal-binding site in a metalloprotein by NMR. *J. Am. Chem. Soc.* 142: 16757–16765.

**31** Swamy, S.K.K, Karczmarska, A., Makowska-Janusik, M., Kassiba, A. and Dittmer, J. (2013). Solid-state NMR correlation experiments and distance measurements in paramagnetic metalorganics exemplified by Cu-Cyclam. *ChemPhysChem* 14: 1864–1870.

**32** Blahut, J., Benda, L., Lejeune, A.L., Sanders, K.J., Burcher, B., Jeanneau, E., Proriol, D., Catita, L., Breuil, P.A.R., Quoineaud, A.A., Pell, A.J. and Pintacuda, G. (2021). Proton-detected fast-magic-angle spinning NMR of paramagnetic inorganic solids. *RSC Adv.* 11: 29870–29876.

**33** Knight, M.J., Pell, A.J., Bertini, I., Felli, I.C., Gonnelli, L., Pierattelli, R., Herrmann, T., Emsley, L. and Pintacuda, G. (2012). Structure and backbone dynamics of a microcrystalline metalloprotein by solid-state NMR. *Proc. Natl. Acad. Sci. USA* 109: 11095–11100.

**34** Knight, M.J., Felli, I.C., Pierattelli, R., Bertini, I., Emsley, L., Herrmann, T. and Pintacuda, G. (2012). Rapid measurement of pseudocontact shifts in metalloproteins by proton-detected solid-state NMR spectroscopy. *J. Am. Chem. Soc.* 134: 14730–14733.

**35** Knight, M.J., Felli, I.C., Pierattelli, R., Emsley, L. and Pintacuda, G. (2013). Magic angle spinning NMR of paramagnetic proteins. *Acc. Chem. Res.* 46: 2108–2116.

**36** Hu, J.Z., Alderman, D.W., Ye, C.H., Pugmire, R.J. and Grant, D.M. (1993). An isotropic chemical shift–chemical shift anisotropy magic-angle slow-spinning 2D NMR experiment. *J. Magn. Reson. Ser. A* 105: 82–87.

**37** Antzutkin, O.N., Shekar, S.C., and Levitt, M. H. (1995). Two-dimensional sideband separation in magic-angle-spinning NMR. *J. Magn. Reson. Ser. A* 115: 7–19.

**38** Clément, R.J., Pell, A.J., Middlemiss, D.S., Strobridge, F.C., Miller, J.K., Whittingham, M.S., Emsley, L., Grey, C.P. and Pintacuda, G. (2012). Spin-transfer pathways in paramagnetic lithium transition metal phosphates from combined broadband isotropic solid-state MAS NMR spectroscopy and DFT calculations. *J. Am. Chem. Soc.* 134: 17178–17185.

**39** Antonijevic, S. and Wimperis, S. (2005). Separation of quadrupolar and chemical/paramagnetic shift interactions in two-dimensional $^{2}$H $I = 1$ nuclear magnetic resonance spectroscopy. *J. Chem. Phys.* 122: 044312.

**40** Walder, B.J., Dey, K.K., Davis, M.C., Baltisberger, J.H. and Grandinetti, P.J. (2015). Two-dimensional NMR measurement and point dipole model prediction of paramagnetic shift tensors in solids. *J. Chem. Phys.* 142: 014201.

**41** Aleksis, R., Carvalho, J.P., Jaworski, A. and Pell, A.J. (2019). Artefact-free broadband 2d NMR for separation of quadrupolar and paramagnetic shift interactions. *Solid State Nucl. Magn. Reson.* 101: 51–62.

**42** Aleksis, R. and Pell, A. J. (2021). Separation of quadrupolar and paramagnetic shift interactions in high-resolution nuclear magnetic resonance of spinning powders. *J. Chem. Phys.* 155: 094202.

**43** Yao, Z., Kwak, H.T., Sakellariou, D., Emsley, L. and Grandinetti, P.J. (2000). Sensitivity enhancement of the central transition nmr signal of quadrupolar nuclei under magic-angle spinning. *Chem. Phys. Lett.* 327: 85–90.

**44** Ash, J.T., Trease, N.M., and Grandinetti, P.J. (2008). Separating chemical shift and quadrupolar anisotropies via multiple-quantum NMR spectroscopy. *J. Am. Chem. Soc.* 130: 10858–10859.

**45** Carvalho, J.P., Jaworski, A., Brady, M.J. and Pell, A.J. (2020). Separation of quadrupolar and paramagnetic shift interactions with TOP-STMAS/MQMAS in solid-state lighting phosphors. *Mag. Reson. Chem.* 58: 1055–1070.

**46** Carvalho, J.P., Papawassiliou, W. and Pell, A. J. (2022). Half-integer-spin quadrupolar nuclei in magic angle spinning paramagnetic NMR: The case of namno$_2$. *J. Magn. Reson.* 340: 107235.

# 15

# Chemical Exchange

*Ashok Sekhar*[1] *and Pramodh Vallurupalli*[2]

[1] *Molecular Biophysics Unit, Indian Institute of Science, Bengaluru - 560012, Karnataka, India*
[2] *Tata Institute of Fundamental Research Hyderabad, 36/P, Gopanpally Village, Serilingampally Mandal, Ranga Reddy District, Hyderabad 500046*

## 15.1 Introduction

The chemical shift of a nucleus in a molecule is extremely sensitive to its electronic environment and consequently the chemical shift of the nucleus changes if the chemical environment around it changes. The change in chemical environment can be due to chemical reactions which involve the making and breaking of bonds or due to conformational changes that involve dihedral rotations. Hence, this phenomenon is called "Chemical Exchange" in the NMR literature as the nucleus of interest is exchanging between different chemical environments. Solution NMR spectra often report on the chemical shifts of various sites in the molecule and hence chemical exchange can have a large effect on the NMR spectrum of the molecule. This is illustrated in Figure 15.1 for an azapropazone derivative where the two N-methyl groups exchange with each other due to rotation about the C–N bond. At very low temperatures (223 K), when this interconversion is essentially nonexistent, the N-methyl $^1$H spectrum consists of two sharp peaks with one arising from each of the N-methyl groups. As the temperature is increased and the rate of interconversion increases, the peaks first broaden, then coalesce into a single broad peak, and finally give rise to a sharp narrow peak at the average position. This sensitivity of the NMR spectrum to chemical exchange makes it a very powerful method to study reactions and conformational exchange in solution. In this chapter we start by discussing the effect of various exchange parameters like the exchange rate, chemical shifts of the different states, etc. on the NMR spectrum of the molecule, following which we the describe various experiments that one can use to extract the underlying exchange parameters. Only chemical exchange in solution under equilibrium conditions for isolated spin $^1/_2$ sites is discussed here. Further, the discussion is largely restricted to two-state exchange. Readers should consult the literature [1–8] for a more detailed and in-depth discussion of chemical exchange effects on NMR spectra.

We start by considering a simple two-state reaction $A \underset{k_{BA}}{\overset{k_{AB}}{\rightleftharpoons}} B$ between states A and B. The kinetic rate equations for the system are given by:

$$\frac{d}{dt}\begin{bmatrix} [A] \\ [B] \end{bmatrix} = \begin{bmatrix} -k_{AB} & k_{BA} \\ k_{AB} & -k_{BA} \end{bmatrix}\begin{bmatrix} [A] \\ [B] \end{bmatrix}. \tag{15.1}$$

Here [A] and [B] are concentrations of the species in states A and B respectively, $k_{AB}$ is the forward rate constant and $k_{BA}$ is the reverse rate constant. The fractional population of state B, $p_B = k_{AB}/k_{ex}$ where $k_{ex} = k_{AB}+k_{BA}$. $\varpi_A$

*Two-Dimensional (2D) NMR Methods*, First Edition. Edited by K. Ivanov, P.K. Madhu and G. Rajalakshmi.

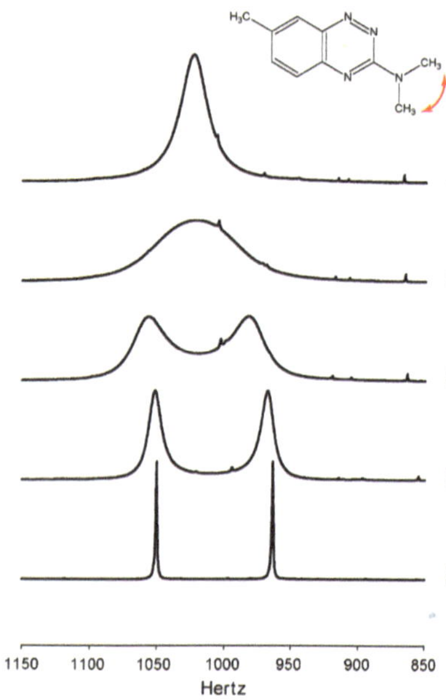


**Figure 15.1** The N-methyl region from a 300 MHz $^1$H NMR spectrum of an azapropazone derivative shown on top recorded at 223, 243, 253, 263, and 273 K. Due to rotation about the C–N bond, the two N-methyl groups interconvert with each other. At 223 K, where there is no exchange, there are two peaks arising from each of the methyl groups while at 273K, where the exchange is rapid, there is single narrow line at the average position. *The figure is adapted from [8].*

and $\varpi_{\mathrm{B}}$ are the chemical shifts (ppm) of the site of interest in states A and B, respectively, while $\omega_{\mathrm{A}}$ and $\omega_{\mathrm{B}}$ are the corresponding resonance frequencies in angular frequency units (rad/s). $\Delta\varpi_{\mathrm{AB}} = \varpi_{\mathrm{B}} - \varpi_{\mathrm{A}}$ and $\Delta\omega_{\mathrm{AB}} = \omega_{\mathrm{B}} - \omega_{\mathrm{A}}$ are differences in resonance frequencies of the two sites in ppm and rad/s respectively. Hence $\Delta\omega_{\mathrm{AB}}$ scales with static field strength while $\Delta\varpi_{AB}$ does not. In this chapter we will use $\Delta\omega_{AB}$ and $\Delta\omega$ interchangeably. Similarly, we will use $\Delta\varpi_{AB}$ and $\Delta\varpi$ interchangeably.

As the molecule jumps randomly between states A and B, the chemical-shift Hamiltonian is time dependent with $\omega(t) = \omega_A + \Delta\omega_{AB}h(t)$. Here, $h(t) = 0$ when the molecule is in state A and $h(t) = 1$ when the molecule is in state B. The observed NMR signal $M_+ = M_x + iM_y$ is given by $M_+(t) \propto \langle e^{i\int_0^t \omega(t)dt}\rangle$ where the integral is over a single molecule while the outer average ($\langle\rangle$) is over all the molecules in the sample. Solving this equation for most cases is quite involved and the effects of exchange on the NMR spectrum can be conveniently studied using the Bloch-McConnell formalism [9] discussed in the next section.

## 15.2 Bloch-McConnell Equations

For a spin $^1/_2$ particle the Zeeman Hamiltonian has two eigenstates $|\alpha\rangle$ and $|\beta\rangle$. We start by writing out the two-state reaction in terms of these eigenstates:

$$\begin{aligned} A_\alpha &\underset{k_{BA}}{\overset{k_{AB}}{\rightleftharpoons}} B_\alpha, \\ A_\beta &\underset{k_{BA}}{\overset{k_{AB}}{\rightleftharpoons}} B_\beta. \end{aligned} \tag{15.2}$$

$X_i$ specifies that the spin of interest is in the i-th eigenstate in state (species) X with $i\epsilon(\alpha,\beta)$ and $X\epsilon(A,B)$. In writing down the above equations, we have assumed that the kinetic-rate equations do not depend on the spin-state of the nucleus of interest as the chemical species are the same. We have also assumed that the spin-state of the nucleus does not change during the barrier crossing event that leads to the reaction. This is a reasonable assumption as barrier crossing occurs on the microsecond to picosecond time-scale while the nuclear $T_1$ values are on the order of a second. As in Equation 15.1 we can write the rate equations for the four species $A_\alpha$, $B_\alpha$, $A_\beta$ and $B_\beta$ as:

$$\frac{d}{dt}\begin{bmatrix}[A_\alpha]\\ [B_\alpha]\\ [A_\beta]\\ [B_\beta]\end{bmatrix}=\begin{bmatrix}-k_{AB} & k_{BA} & 0 & 0\\ k_{AB} & -k_{BA} & 0 & 0\\ 0 & 0 & -k_{AB} & k_{BA}\\ 0 & 0 & k_{AB} & -k_{BA}\end{bmatrix}\begin{bmatrix}[A_\alpha]\\ [B_\alpha]\\ [A_\beta]\\ [B_\beta]\end{bmatrix}. \tag{15.3}$$

The z-component of magnetization arising from molecules in state A, $M_z^A = c\left([A_\alpha]-[A_\beta]\right)$. Similarly, the z-component of magnetization arising from molecules in state B, $M_z^B = c\left([B_\alpha]-[B_\beta]\right)$. Here, c is a constant. Combining these expressions for $M_z^A$ and $M_z^B$ with Equation 15.3 results in the following equation for the evolution of $M_z^A$ and $M_z^B$ due to chemical exchange:

$$\frac{d}{dt}\begin{bmatrix}M_z^A\\ M_z^B\end{bmatrix}=\begin{bmatrix}-k_{AB} & k_{BA}\\ k_{AB} & -k_{BA}\end{bmatrix}\begin{bmatrix}M_z^A\\ M_z^B\end{bmatrix}. \tag{15.4}$$

Similar equations can be derived for the evolution of x and y magnetization arising from states A and B resulting in:

$$\frac{d}{dt}\begin{bmatrix}M_x^A\\ M_x^B\end{bmatrix}=\begin{bmatrix}-k_{AB} & k_{BA}\\ k_{AB} & -k_{BA}\end{bmatrix}\begin{bmatrix}M_x^A\\ M_x^B\end{bmatrix}, \tag{15.5}$$

$$\frac{d}{dt}\begin{bmatrix}M_y^A\\ M_y^B\end{bmatrix}=\begin{bmatrix}-k_{AB} & k_{BA}\\ k_{AB} & -k_{BA}\end{bmatrix}\begin{bmatrix}M_y^A\\ M_y^B\end{bmatrix}. \tag{15.6}$$

These can then be combined to obtain:

$$\frac{d\vec{M}}{dt}=\hat{K}\vec{M} \tag{15.7}$$

where,

$$\vec{M}=\begin{bmatrix}E/2\\ M_x^A\\ M_y^A\\ M_z^A\\ M_x^B\\ M_y^B\\ M_z^B\end{bmatrix} \text{ and } \hat{K}=\begin{bmatrix}0&0&0&0&0&0&0\\ 0&-k_{AB}&0&0&k_{BA}&0&0\\ 0&0&-k_{AB}&0&0&k_{BA}&0\\ 0&0&0&-k_{AB}&0&0&k_{BA}\\ 0&k_{AB}&0&0&-k_{BA}&0&0\\ 0&0&k_{AB}&0&0&-k_{BA}&0\\ 0&0&0&k_{AB}&0&0&-k_{BA}\end{bmatrix}. \tag{15.8}$$

It is clear from the Equations 15.4-15.8 that reaction kinetics does not interconvert the different components (x,y,z) of magnetization. Additionally for reasons that will come clear soon, we have included the identity operator $E$, which does not interconvert with the x-, y-, and z-components due to exchange. Ignoring exchange, the effects of chemical-shift evolution and any applied radiofrequency (RF) field on the spin $^1\!/_2$ nucleus of interest can be calculated in the usual manner resulting in the Bloch equations for evolution of magnetization in the rotating frame shown below for magnetization arising from species A.

$$\frac{d}{dt}\begin{bmatrix}E/2\\ M_x^A\\ M_y^A\\ M_z^A\end{bmatrix}=\begin{bmatrix}0&0&0&0\\ 0&-R_{2A}&-\Omega_A&\omega_{1y}\\ 0&\Omega_A&-R_{2A}&-\omega_{1x}\\ 2R_{1A}M_{zeq}^A&-\omega_{1y}&\omega_{1x}&-R_{1A}\end{bmatrix}\begin{bmatrix}E/2\\ M_x^A\\ M_y^A\\ M_z^A\end{bmatrix} \tag{15.9}$$

$R_{1i}$ and $R_{2i}$ are the longitudinal and transverse relaxation rates respectively for the site of interest in state i. $M_{zeq}^i$ is equilibrium z-magnetization in state i. $\omega_{1x}$ and $\omega_{1y}$ (rad/s) are the x- and y-components, respectively, of the applied radio frequency field while $\Omega_i = \omega_i - \omega_{RF}$, is the offset between the resonance frequency of the nucleus in state i, $\omega_i$ (rad/s) and the frequency, $\omega_{RF}$ (rad/s) at which RF the field is applied. The identity operator ($E$) lets us express

Equation 15.9 as a homogenous differential equation. Similar equations are obtained for magnetization arising from state B and the Bloch equations for the evolution of x, y, and z-magnetization from two independent states (A and B) that do not exchange with each other are shown below.

$$\frac{d\vec{M}}{dt} = \hat{B}\vec{M} \tag{15.10}$$

where,

$$\hat{B} = \begin{bmatrix} 0 & 0 & 0 & 0 & 0 & 0 & 0 \\ 0 & -R_{2A} & -\Omega_A & \omega_{1y} & 0 & 0 & 0 \\ 0 & \Omega_A & -R_{2A} & -\omega_{1x} & 0 & 0 & 0 \\ 2R_{1A}M^A_{zeq} & -\omega_{1y} & \omega_{1x} & -R_{1A} & 0 & 0 & 0 \\ 0 & 0 & 0 & 0 & -R_{2B} & -\Omega_B & \omega_{1y} \\ 0 & 0 & 0 & 0 & \Omega_B & -R_{2B} & -\omega_{1x} \\ 2R_{1B}M^B_{zeq} & 0 & 0 & 0 & -\omega_{1y} & \omega_{1x} & -R_{1B} \end{bmatrix}. \tag{15.11}$$

The time-dependent evolution of magnetization under the joint influence of the chemical-shift Hamiltonian, RF irradiation, and chemical exchange can be obtained by combining Equations 15.7 and 15.10 to obtain the Bloch-McConnell equations:

$$\frac{d\vec{M}}{dt} = (\hat{B} + \hat{K})\vec{M}. \tag{15.12}$$

In Equation 15.12, $\hat{B}$ (Equation 15.11) describes the evolution of the magnetization arising from molecules in states A and B in the absence exchange while the effect of exchange is introduced via $\hat{K}$ (Equation 15.8). To understand how $\hat{K}$ modifies the Bloch equations while introducing the effect of exchange, we look at the evolution of transverse $M_+$ magnetization from states A and B in the absence of RF irradiation:

$$\frac{d}{dt}\begin{bmatrix} M_+^A \\ M_+^B \end{bmatrix} = \left\{\begin{bmatrix} -R_{2A} + i\Omega_A & 0 \\ 0 & -R_{2B} + i\Omega_B \end{bmatrix} + \begin{bmatrix} -k_{AB} & k_{BA} \\ k_{AB} & -k_{BA} \end{bmatrix}\right\}\begin{bmatrix} M_+^A \\ M_+^B \end{bmatrix}. \tag{15.13}$$

Equation 15.13 is obtained directly from Equation 15.12 by retaining only the $M_x$ and $M_y$ terms and combining them to obtain $M_+$. The first $2 \times 2$ matrix on the right-hand side is the Bloch matrix for the evolution of $M_+$ magnetization. It is clear that in the absence of exchange ($k_{AB} = k_{BA} = 0$), the $M_+^A$ and $M_+^B$ components evolve independently at frequencies $\Omega_A$ and $\Omega_B$ respectively, while simultaneously relaxing with transverse relaxation rate constants $R_{2A}$ and $R_{2B}$. The second 2×2 matrix on right hand side introduces the effect of exchange. Focusing on the evolution of $M_+^A$ magnetization, we need to consider two exchange processes: (i) conversion of molecules from state A to state B, which results in decrease of A-state magnetization that is accounted by the $-k_{AB}$ term on the diagonal and (ii) conversion of molecules from state B to state A, which results in a transfer of B-state magnetization to A that is accounted by the off-diagonal $k_{BA}$ term. Similarly, for $M_+^B$ magnetization, we need to consider two exchange processes: (i) conversion of molecules from state B to state A, which results in a decrease of B-state magnetization that is accounted by the $-k_{BA}$ term on the diagonal and (ii) conversion of molecules from state A to state B, which results in a transfer of A-state magnetization to B that is accounted by the off-diagonal $k_{AB}$ term. Another thing to note is that even if $M_+^A$ and $M_+^B$ are aligned at $t = 0$ they will not remain aligned as the $M_+^A$ and $M_+^B$ components evolve at different frequencies. In a typical one pulse experiment, the NMR spectrum is obtained from the Fourier transform of $M_+(t)$. For the two-state system being considered here,

$M_+(t) = M_+^A(t) + M_+^B(t)$ and $M_+^A(t)$ and $M_+^B(t)$ can be obtained by integrating Equation 15.13 to obtain:

$$\begin{bmatrix} M_+^A(t) \\ M_+^B(t) \end{bmatrix} = e^{\left\{\begin{bmatrix} -R_{2A} + i\Omega_A & 0 \\ 0 & -R_{2B} + i\Omega_B \end{bmatrix} + \begin{bmatrix} -k_{AB} & k_{BA} \\ k_{AB} & -k_{BA} \end{bmatrix}\right\}t} \begin{bmatrix} M_+^A(0) \\ M_+^B(0) \end{bmatrix}. \tag{15.14}$$

Solutions of Equation 15.14 are given by:

$$\begin{bmatrix} M_+^A(t) \\ M_+^B(t) \end{bmatrix} = \begin{bmatrix} a_{11}(t) & a_{12}(t) \\ a_{21}(t) & a_{22}(t) \end{bmatrix} \begin{bmatrix} M_+^A(0) \\ M_+^B(0) \end{bmatrix} \tag{15.15}$$

with

$$\begin{aligned} a_{11} &= \frac{1}{2}\left[\left(1 - \frac{i\Delta\omega_{AB} - \Delta R_{2AB} + k_{AB} - k_{BA}}{(\lambda_+ - \lambda_-)}\right)e^{-\lambda_- t} + \left(1 + \frac{i\Delta\omega_{AB} - \Delta R_{2AB} + k_{AB} - k_{BA}}{(\lambda_+ - \lambda_-)}\right)e^{-\lambda_+ t}\right], \\ a_{22} &= \frac{1}{2}\left[\left(1 + \frac{i\Delta\omega_{AB} - \Delta R_{2AB} + k_{AB} - k_{BA}}{(\lambda_+ - \lambda_-)}\right)e^{-\lambda_- t} + \left(1 - \frac{i\Delta\omega_{AB} - \Delta R_{2AB} + k_{AB} - k_{BA}}{(\lambda_+ - \lambda_-)}\right)e^{-\lambda_+ t}\right], \\ a_{12} &= \frac{k_{BA}}{(\lambda_+ - \lambda_-)}\left(e^{-\lambda_- t} - e^{-\lambda_+ t}\right), \\ a_{21} &= \frac{k_{AB}}{(\lambda_+ - \lambda_-)}\left(e^{-\lambda_- t} - e^{-\lambda_+ t}\right), \\ \lambda_\pm &= \frac{1}{2}\left[(-i\Omega_A - i\Omega_B + R_{2A} + R_{2B} + k_{ex}) \pm \left\{(i\Delta\omega_{AB} - \Delta R_{2AB} + k_{AB} - k_{BA})^2 + 4k_{AB}k_{BA}\right\}^{1/2}\right]. \end{aligned} \tag{15.16}$$

The imaginary parts of the eigenvalues ($\lambda_+$ and $\lambda_-$) give the frequencies at which components evolve and the real part determines the rates at which the components decay. Here $\Delta R_{2AB} = R_{2B} - R_{2A}$. In this chapter we will use $\Delta R_{2AB}$ and $\Delta R_2$ interchangeably.

Assuming that the system starts at equilibrium, we have $M_{z,eq}^A \propto p_A$ and $M_{z,eq}^B \propto p_B$ and consequently an ideal pulse that acts on $M_z$ magnetization generates $M_+^A(0) \propto p_A$ and $M_+^B(0) \propto p_B$. We can then obtain $M_+(t)$ using either Equation 15.14 or Equation 15.15 and then calculate the spectrum of a system undergoing exchange as shown in Figure 15.2 for various $k_{ex}$ values and two different $p_B$ values. In Figure 15.2, $R_{2A} = R_{2B}$ for all the cases. In Figure 15.2a spectra are calculated for various $k_{ex}$ values when states A and B are equally populated. As in Figure 15.1, we find in Figure 15.2a that when $k_{ex}/\Delta\omega = 0$ we have two narrow distinct peaks that begin to broaden as $k_{ex}$ increases, coalescing and disappearing as $k_{ex}/\Delta\omega \sim 1$, following which a narrow peak emerges at the average frequency that further narrows when $k_{ex}/\Delta\omega$ increases. The more interesting case is shown in Figure 15.2b where one state has a significantly higher population than the other state ($p_A \gg p_B$). In Figure 15.2b, $p_A = 0.93$ and $p_B = 0.07$. When there is no exchange, we have two peaks, one at $\varpi_A$ arising due to molecules in state A and the other at $\varpi_B$ arising due to molecules in state B. The heights of peaks at $\varpi_A$ and $\varpi_B$ are proportional to the equilibrium populations of states A and B, respectively. When the rate of exchange begins to increase, the minor B-state peak broadens more than the major peak and cannot be detected ($k_{ex}/\Delta\omega = 0.1$). The major peak, on the other hand, can be seen even if the minor peak cannot and initially continues to broaden as $k_{ex}$ increases and then begins to become sharper as $k_{ex}$ further increases. Finally, when exchange is very fast, we have a single peak at the average chemical shift ($p_A\varpi_A + p_B\varpi_B$). The exact positions and linewidths of the peaks can be obtained from the imaginary and real parts of eigenvalues (Equation 15.16) of the matrix in Equation 15.13. Next, we try to understand the line shapes in Figure 15.2 by considering two limiting cases.

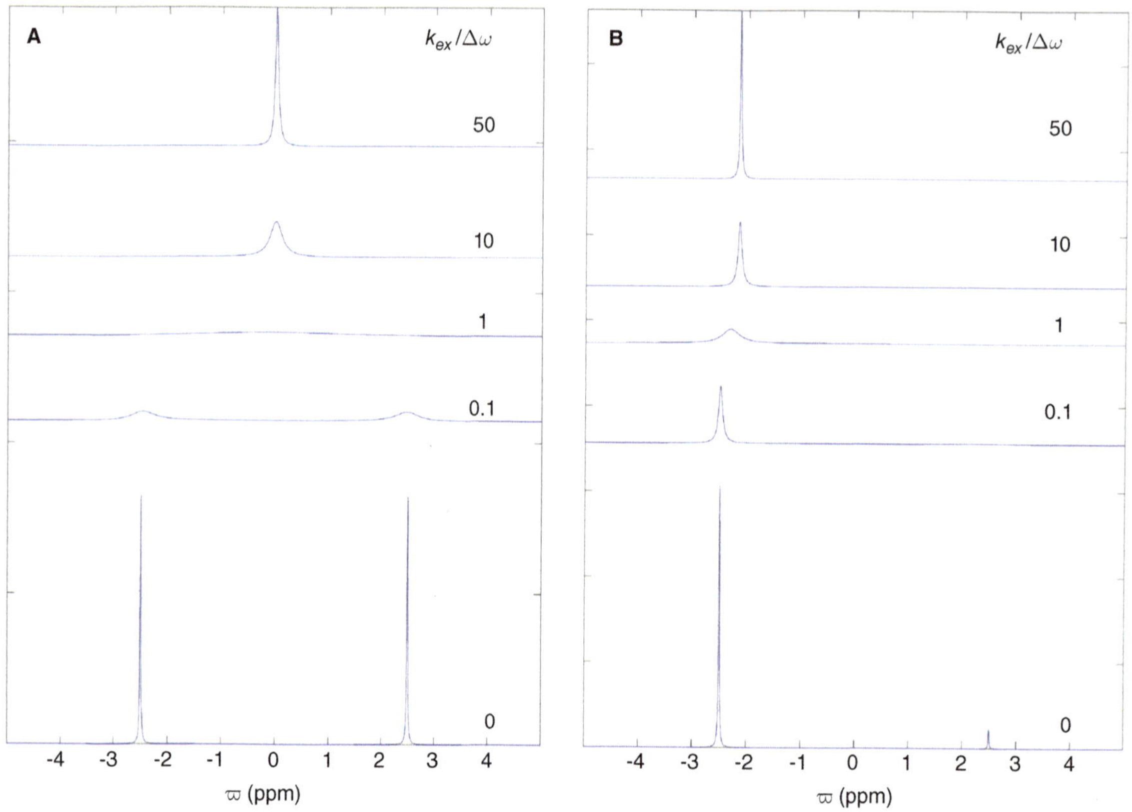


**Figure 15.2** Spectra of a $^{15}N$ nucleus undergoing chemical exchange in a 11.74 T magnet, calculated using the Bloch-McConnell equations (15.12) for different $k_{ex}/\Delta\omega$ values. $\Delta\varpi = 5$ ppm ($\Delta\omega \sim 1570$ rad/s) and $p_B = 0.5$ on the left (a) and $p_B = 0.07$ on the right (b).

### 15.2.1 Slow Exchange

In Equation 15.13 without any loss in generality we can set $\Omega_A = 0$, which means that $\Omega_B = \Delta\omega_{AB}$ leading to:

$$\frac{d}{dt}\begin{bmatrix} M_+^A \\ M_+^B \end{bmatrix} = \begin{bmatrix} -R_{2A} - k_{AB} & k_{BA} \\ k_{AB} & -R_{2B} + i\omega_{AB} - k_{BA} \end{bmatrix}\begin{bmatrix} M_+^A \\ M_+^B \end{bmatrix}. \tag{15.17}$$

The condition for slow exchange is $k_{ex}/|i\Delta\omega_{AB} + \Delta R_{2AB}| \ll 1$. As $|\Delta R_{2AB}|$ is often ~0, the slow exchange condition is usually written as $|k_{ex}/\Delta\omega_{AB}| \ll 1$. When exchange is slow, the off-diagonal terms in Equation 15.17 can be neglected, which means the two components evolve independently with $M_+^A(t) = M_+^A(0)e^{-(R_{2A}+k_{AB})t}$ and $M_+^B(t) = M_+^B(0)e^{-(R_{2B}+k_{BA}+i\Delta\omega_{AB})t}$ resulting in two lines in the spectra, one at $\omega_A$ and one at $\omega_B$. The lines are, however, broadened due to chemical exchange and the effective transverse relaxation rate ($R_{2,\mathrm{eff}}$) for the peak at $\omega_A$ is given by $R_{2A} + k_{AB}$, while that for the peak at $\omega_B$ is given by $R_{2B} + k_{BA}$. This occurs because once a molecule jumps from state A to B, it spends a considerable amount of time (on average $\sim 1/k_{BA}$) in the new state B, and when it jumps back to A it has lost phase with the molecules in state A and does not contribute to A-state magnetization. Hence, slow exchange gives rise to a leaking effect and the rate of loss of magnetization from A is determined by $k_{AB}$ and from B by $k_{BA}$. When $p_A = p_B$, $k_{AB} = k_{BA}$ and exchange broadens both peaks by the same amount. However, when $p_A \gg p_B$, $k_{BA} \gg k_{AB}$ and the peak at $\omega_B$ is broadened significantly more than peak at $\omega_A$ as can be seen in Figure 15.2b. The important consequence of this is that it is very difficult to observe the states with low populations (minor states) in spectra even under slow exchange. Further a major state peak that is not severely exchange broadened in the absence of a minor peak does not mean that it is in fast exchange with other peaks.

### 15.2.2 Fast Exchange

When exchange is fast, $k_{ex}/|i\Delta\omega_{AB} + \Delta R_{2AB}| \gg 1$ (or $|k_{ex}/\Delta\omega_{AB}| \gg 1$ if $|\Delta R_{2AB}| \sim 0$), Equation 15.14 reduces to:

$$M_{+}(t) \propto e^{\left(i\overline{\Omega} - \overline{R}_2 - \left(p_A p_B \Delta\omega_{AB}^2/k_{ex}\right)\right)t}. \tag{15.18}$$

Here, $\overline{\Omega} = p_A\Omega_A + p_B\Omega_B$. Hence, there is a peak at the population weighted chemical shift $\overline{\varpi} = p_A\varpi_A + p_B\varpi_B$ whose linewidth is determined by the population weighted transverse relaxation rate $\overline{R}_2 = p_A R_{2A} + p_B R_{2B}$ and an additional contribution $\left(p_A p_B \Delta\omega_{AB}^2/k_{ex}\right)$ due to exchange that becomes smaller as exchange becomes faster leading to the narrowing seen in Figure 15.1 and 15.2 as $|k_{ex}/\Delta\omega_{AB}|$ becomes very high. Equation 15.18 was obtained by simplifying $\lambda_-$ using the fact that $k_{ex}/|i\Delta\omega_{AB} + \Delta R_{2AB}| \gg 1$ and retaining only the terms that evolve according to the $\lambda_-$ eigenvalue as terms that evolve due to $\lambda_+$ decay rapidly. In both the slow exchange and fast exchange limits, the total magnetization for the system is given by $M_+(t) = M_+^A(t) + M_+^B(t)$, and the spectrum is obtained from the Fourier transform of $M_+$. However, in slow exchange, the molecule spends a significant amount of time in each state. When the molecule is in state A, there is sufficient time for the chemical shift of A to evolve; similarly, when the molecule jumps to state B, there is ample time for phase evolution to occur due to $\omega_B$. This time evolution provides the magnetization of states A and B with independent identities and results in two separate peaks after Fourier transformation. In contrast, in the fast exchange limit, the molecule jumps rapidly between states A and B and the lifetimes in each state are not large enough for a phase difference to build-up due to chemical-shift evolution. As a result, the distinction between the two states is blurred and a single peak is observed at the average chemical shift with a linewidth given by the population-weighted average of the transverse relaxation rate constants. The fact that $R_{ex} = \left(p_A p_B \Delta\omega_{AB}^2/k_{ex}\right)$ can be obtained from the random phase approximation [1].

### 15.2.3 Dependence of the Linewidth On Magnetic Field Strength

Since $\Delta\omega_{AB}$ scales with magnetic field strength ($B_0$), chemical exchange will affect NMR spectra recorded at different field strengths differently. As we have seen above, chemical exchange results in line broadening. Here, we look at how the exchange contribution to relaxation ($R_{ex}$) scales with $B_0$. We restrict ourselves to the case where $p_A > p_B$ and focus on the major peak that arises largely due to state A. The parameter $\alpha$ relates $R_{ex}$ to $B_0$ according to [10]:

$$\frac{\delta R_{ex}}{R_{ex}} = \alpha\frac{\delta B_0}{B_0} \Rightarrow \frac{dlnR_{ex}}{dln\Delta\omega_{AB}} = \alpha. \tag{15.19}$$

When $p_A \gg p_B$ and $R_{2A} = R_{2B}$ it can be shown from Equation 15.15 that:

$$R_{ex} \approx \frac{p_A p_B k_{ex}}{1 + \left(k_{ex}/\Delta\omega_{AB}\right)^2}. \tag{15.20}$$

Hence, we have:

$$\alpha = \frac{2\left(k_{ex}/\Delta\omega_{AB}\right)^2}{1 + \left(k_{ex}/\Delta\omega_{AB}\right)^2}. \tag{15.21}$$

Hence $\alpha$ is only a function of $|k_{ex}/\Delta\omega_{AB}|$ that can take up values between 0 and 2, and the chemical exchange time-scale can be defined as:

$$\begin{aligned} &0 \leq \alpha < 1 \text{ Slow exchange,} \\ &\alpha = 1 \text{ Intermediate exchange,} \\ &1 < \alpha \leq 2 \text{ Fast exchange.} \end{aligned} \tag{15.22}$$

Equation 15.22 shows that the linewidth does not depend on $B_0$ in the slow exchange limit and scales quadratically with field strength in the fast exchange limit. When the exchange parameters ($k_{ex}$, $p_B$ and $\Delta\varpi$) are fixed the exchange contribution to linewidth will increase with increasing field strength reaching a maximum value of $k_{AB}$ as discussed earlier when $|k_{ex}/\Delta\omega_{AB}| \ll 1$. Just as the linewidth, the peak positions also depend on the strength of the magnetic field strength ($B_0$) and expressions similar to the ones above (Equation 15.19 to Equation 15.22) have been obtained for the dependence of the peak positions on $B_0$ field strength [11].

### 15.2.4 Exchange in the Absence of Chemical-Shift Differences

So far, our discussion has focused on cases in which $\Delta R_{2AB} \sim 0$. Interestingly, it turns out that exchange between two species (A and B) with $\Delta\omega_{AB} = 0$ but with $|\Delta R_{2AB}| \gg 0$ can also have a large effect on the spectra as shown in Figure 15.3 for various $k_{ex}$ values with $R_{2A} = 3\ s^{-1}$ and $R_{2B} = 253\ s^{-1}$ and $|\Delta R_{2AB}| = 250\ s^{-1}$. As expected, there is a single line in all the spectra at $\omega_A = \omega_B = 0$, that broadens when $k_{ex}$ increases. When $p_A \gg p_B$ it can be shown from Equation 15.15 that:

$$R_{ex} \approx k_{AB} \frac{k_{BA}\Delta R_{2AB} + \Delta R_{2AB}^2 + \Delta\omega_{AB}^2}{\left(k_{BA} + \Delta R_{2AB}\right)^2 + \Delta\omega_{AB}^2}. \tag{15.23}$$

To focus solely on the effect of differential transverse relaxation rates, Equation 15.23 can be further simplified by setting $\Delta\omega_{AB} = 0$ to obtain:

$$R_{ex}\left(\Delta\omega = 0\right) \approx \frac{k_{AB}\Delta R_{2AB}}{\left(k_{BA} + \Delta R_{2AB}\right)}. \tag{15.24}$$

Equation 15.24 shows that even when $\Delta\omega = 0$, exchange can lead to line broadening. For simplicity, here we assume that $R_{2B} > R_{2A}$. When $k_{ex} = 0\ s^{-1}$, molecules in states A and B do not interconvert with each other, hence we have two independent components that relax with very different rates, and as shown in Figure 15.3, we see only the narrow component that arises from state A. As $k_{ex}$ increases molecules from state A jump to state B where the magnetization decays rapidly leading to a broadening of the observed line. When $k_{ex} \gg \Delta R_{2AB}$, the observed signal will decay with an $R_{2,eff}$ value of $p_A R_{2A} + p_B R_{2B}$. In the limit of very large $\Delta R_{2AB}$ values ($k_{ex} \ll \Delta R_{2AB}$), the observed signal will decay with an $R_{2,eff}$ value of $R_{2A} + k_{AB}$ for reasons analogous to those discussed in Section 15.2.1. In the NMR literature, exchange between sites with different $R_2$ values is referred to as R-type exchange while exchange between sites with different chemical shifts is referred to as C-type exchange [12].

### 15.2.5 Multi-State Exchange

Following the steps outlined above, Bloch-McConnell type equations can be written down for more complicated reaction schemes involving more than two species. Sometimes, apparently more complicated schemes can be reduced to an effective two-state process under equilibrium conditions as in the reaction scheme shown below:

$$A + R \underset{k_r}{\overset{k_f}{\rightleftharpoons}} B + P. \tag{15.25}$$

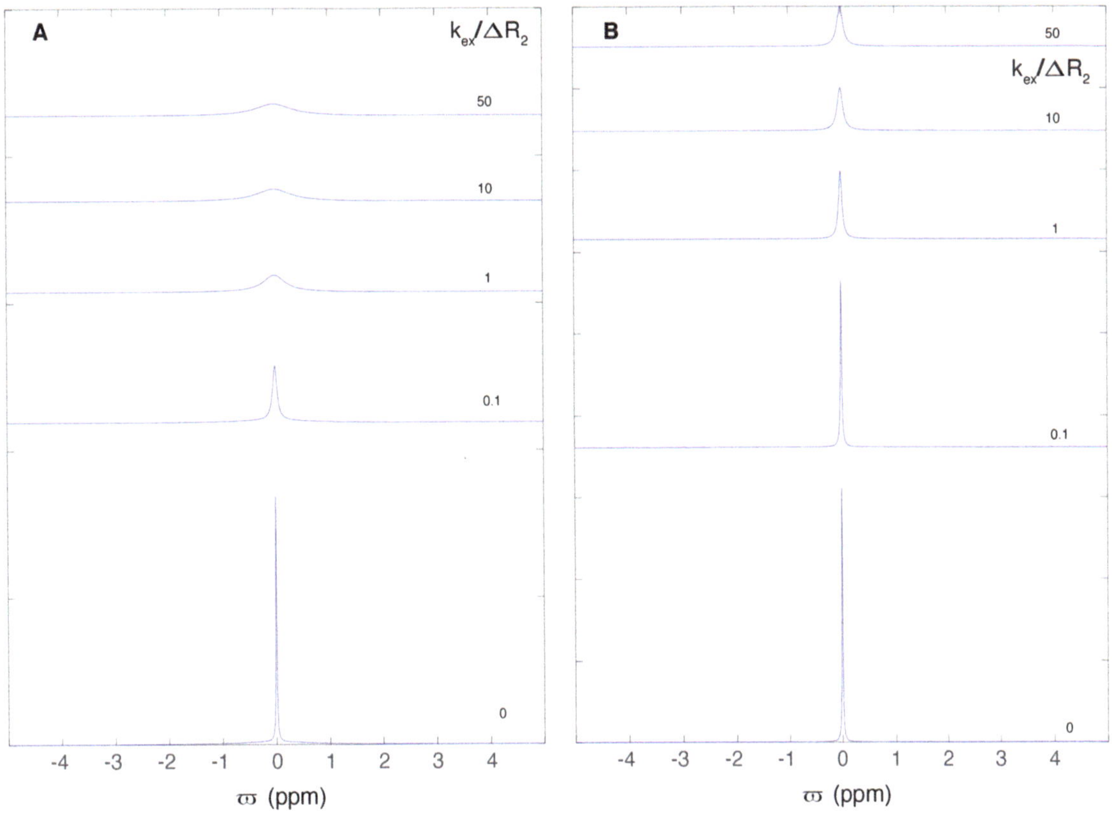


**Figure 15.3** Spectra of a $^{15}$N nucleus undergoing chemical exchange in a 11.74 T magnet, calculated using the Bloch-McConnell equations (15.12) for different $k_{ex}/\Delta R_2$ values. Here, $R_{2A} = 3\ s^{-1}$ and $R_{2B} = 253\ s^{-1}$, $p_B = 0.5$ on the left (a) and $p_B = 0.07$ on the right (b).

If the nucleus of interest is present only in species A and B, we can rewrite the rate equations as:

$$\frac{d}{dt}\begin{bmatrix}[A]\\ [B]\end{bmatrix} = \begin{bmatrix}-k'_{AB} & k'_{BA}\\ k'_{AB} & -k'_{BA}\end{bmatrix}\begin{bmatrix}[A]\\ [B]\end{bmatrix}. \tag{15.26}$$

Where $k'_{AB} = k_f[R]$ and $k'_{BA} = k_r[P]$. Hence, we can use the two-state Bloch-McConnell equations that were described earlier. Multi-state exchange can also lead to interesting effects; for example, unlike two-state exchange where $R_{ex}$ only depends on $|\Delta\omega|$ and not the sign of $|\Delta\omega|$, in the case of multi-state exchange, $R_{ex}$ can depend on the relative sign of the $\Delta\omega$ values between different sites.

## 15.3 Studying Exchange Between Visible States

As discussed earlier, exchange can occur among states that are sufficiently populated to be visible in the NMR spectrum and also between a highly populated visible state and a marginally populated 'invisible' state. In this section we describe methods to study exchange between visible states.

### 15.3.1 Lineshape Analysis

As described in Section 15.2, the trajectory of magnetization in the presence of chemical or conformational exchange can be described satisfactorily within the framework of the Bloch-McConnell equations, which depend on parameters defining the exchange process such as the populations of the different states, the rate constants coupling them, and the chemical shifts of nuclei of interest in the various states. As discussed in Sections 15.1 and 15.2, the lineshapes of NMR resonances in a system undergoing chemical exchange are very sensitive to the exchange parameters, and the chemical shifts and linewidths depend on $k_{ex}$, $p_B$, and $\Delta\omega$ (Figures 15.1 and 15.2). This sensitivity of the NMR lineshape to exchange parameters provides a natural method for characterizing conformational exchange, and the theory and practice of 1D NMR lineshape analysis has flourished for over six decades [13, 14]. One-dimensional lineshape analysis is typically carried out by systematically varying $k_{ex}/\Delta\omega$ values and sampling NMR spectra from the slow exchange regime through coalescence to fast exchange, as this protocol facilitates the robust estimation of $k_{ex}$, $p_B$, and $\Delta\omega$. For small molecules, temperature has been the variable of choice to increase $k_{ex}$. Spectra acquired over a significant range of temperatures are generally modeled using analytical equations or by numerically propagating the Bloch-McConnell equations. The dependence of $\omega_A$ and $\omega_B$ with temperature is evaluated in the slow exchange regime, where exchange is not fast enough to affect the chemical shifts of the two states. This dependence is assumed to be linear and extrapolated across the temperature range in which lineshapes are measured. In the case of biomolecules such as proteins, the dependence of folding/unfolding rate constants with urea has also been leveraged to generate 1D exchange spectra.

The 1D NMR lineshape analysis of unimolecular conformational exchange processes, particularly in biomolecules, is impaired by the fact that the thermodynamic and kinetic parameters are not true global variables but differ at each temperature/urea concentration. Data fitting is thus done separately for 1D spectra acquired at each condition, and this limits the number of parameters that can be reliably estimated. For this reason, 1D lineshape analysis has been used in conjunction with other methods like circular dichroism that can provide a measure of $p_B$ at every temperature or urea concentration, making $k_{ex}$ the only variable that is estimated by fitting each spectrum. It is also possible to impose models derived from Arrhenius or transition-state theories for rate constants and therefore employ activation energies as temperature-independent global parameters, but such models have limited relevance for biomolecular processes that have large $\Delta C_p$ values, as enthalpic and entropic activation free energy contributions vary with temperature. One-dimensional lineshape analysis has largely been superseded by techniques such as chemical exchange saturation transfer (CEST), dark-state exchange saturation transfer (DEST), Carr-Purcell-Meiboom-Gill (CPMG), and $R_{1\rho}$, which can probe conformational exchange at a single temperature or urea concentration. However, 1D and 2D lineshape analyzes affords a number of advantages for studies of protein-protein and protein-ligand interactions, where the exchange between free (state A) and bound (state B) conformations occur at or near the intermediate exchange timescale. First, unlike in unimolecular processes, 1D or 2D NMR spectra spanning a range of chemical shifts and $R_{ex}$ values can be acquired at a single temperature simply by titrating the ligand concentration (L), which determines both the $p_B$ and the $k_{ex}$ (Equation 15.25, Figure 15.4). Second, global modeling of lineshapes acquired across different ligand concentrations is possible because $k_{on}$, $k_{off}$, and $\Delta\omega$ remain constant for the exchange process. Finally, NMR-detected ligand binding titrations are powerful reporters of the stoichiometry and mechanism of binding (Figure 15.4), information that is not easily obtained from other biophysical methods.

### 15.3.2 ZZ-Exchange Experiment

In instances where all the states participating in conformational exchange are observable as distinct resonances in the NMR spectra, the magnetization exchange or ZZ-exchange (ZZ-ex) experiment provides a robust estimate of the rate constants coupling the various states. ZZ-ex is intrinsically a 2D NMR experiment in which the existence of conformational exchange can be discerned by the presence of cross peaks connecting diagonal peaks belonging to different exchanging species (Figure 15.5).

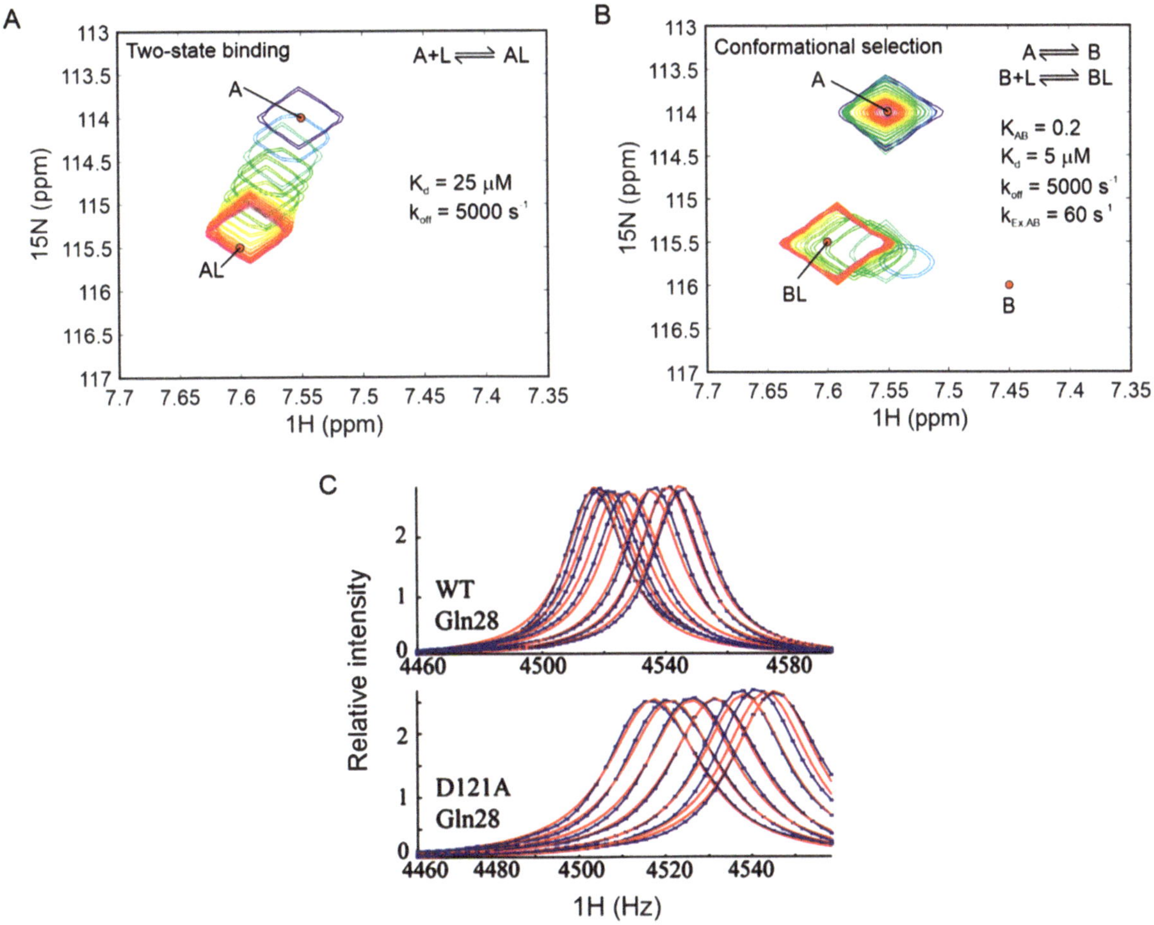


**Figure 15.4** Deciphering mechanism from 1D and 2D lineshape analysis. Simulated peak trajectories during a ligand binding titration that follows the two-state (a) or the conformational selection (b) mechanism. The effective thermodynamic dissociation constants for A↔AL in panel (a) or A↔BL in panel (b) are the same (25 μM). Other values used in the simulation are given in the figure. The chemical-shift positions of all species are indicated by red dots. The ligand in panel (b) binds to species B, which is a minor species. Clear indications of this binding are observed in 2D HSQC series in the form of resonances that trace a linear path from B to BL. In contrast, the titration in panel (a) follows a simple two-state mechanism, reflected by the linear trajectory of peaks from A to AL. (c) Use of 1D lineshape analysis to extract equilibrium and kinetic dissociation constants for the binding of 3′-CMP to wild-type and D121A RNase A. The Kds for the wild-type and mutant proteins are 210 and 78 μM, while the $k_{off}$ values are 1 700 and 2 700 $s^{-1}$, respectively. *Panel (c) adapted from [26].*

In a typical ZZ-ex pulse sequence (Figure 15.5a) a $t_1$ evolution period precedes the exchange period $T_{EX}$ and the chemical shifts of states A and B are recorded during this $t_1$ period, following which the magnetization is placed along the z-axis for the duration of $T_{EX}$. During $T_{EX}$, the two states exchange z-magnetization with each other, so that a fraction (f) of state-A magnetization labeled with the chemical shift $\varpi_A$ of state (a) during $t_1$ is converted to z-magnetization of state (b):

$$M_z^A cos(\Omega_A t_1) \xrightarrow{T_{EX}} (1-f)M_z^A cos(\Omega_A t_1) + f M_z^B cos(\Omega_A t_1) \tag{15.27}$$

and vice-versa. Subsequent to the $T_{EX}$ period, transverse magnetization is generated for detection. While the A-state magnetization that remains on state A (first term on the right-hand side of Equation 15.27) generates a diagonal peak at $(\varpi_A, \varpi_A)$ in the final 2D correlation map, magnetization from molecules that transition to state B during $T_{EX}$ (second term on the right-hand side of Equation 15.27) contributes to a cross peak between states A and

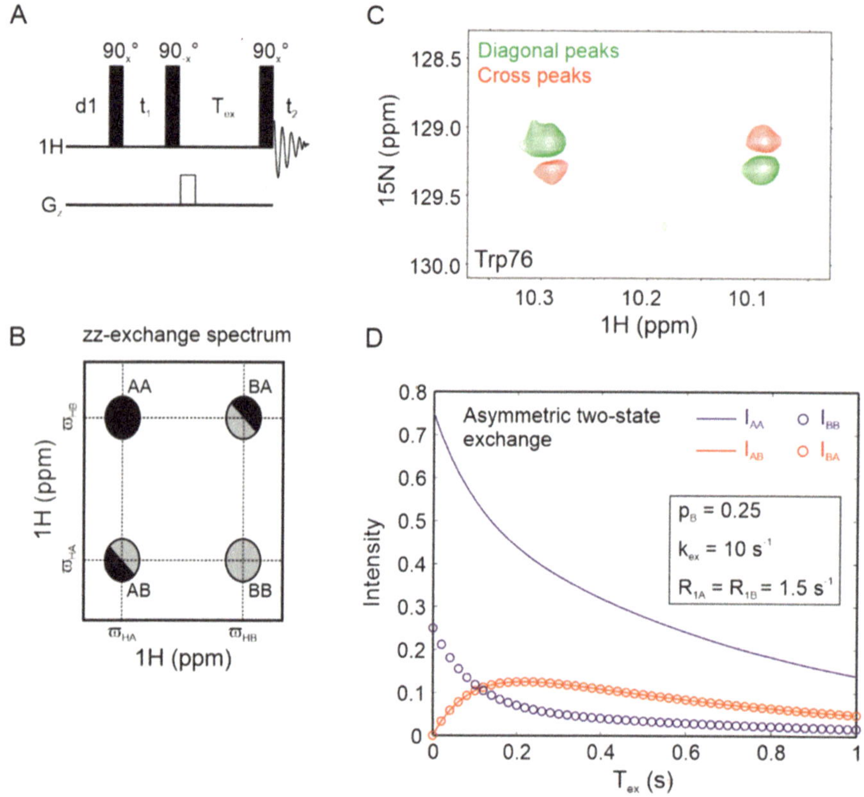


**Figure 15.5** (a) Pulse sequence for acquiring $^1$H-based 2D ZZ-ex spectra. $^1$H magnetization is first labeled with its chemical shift during $t_1$ and subsequently placed along the z-axis during the mixing time $T_{EX}$ for chemical exchange to occur. Then, $^1$H signals are detected during $t_2$. Phase cycling is not shown. Residual transverse magnetization after the second 90° pulse is dephased by a magnetic field gradient. (b) Schematic representation of a 2D $^1$H-$^1$H plane from a magnetization exchange experiment, showing diagonal (black and gray) and cross peaks (gray-black). (c) Experimental magnetization spectra of the iron-sulfur cluster assembly scaffold protein IscU, which exists in two conformations. Magnetization exchange is detected via a $^{15}$N-$^1$H correlation map and diagonal and cross peaks for Trp76 in the two conformations are shown in green and red, respectively. *Panel (c) adapted from [27]* (d) Decay of diagonal and cross-peak intensities as a function of $T_{ex}$ for asymmetric populations.

B at $(\varpi_A, \varpi_B)$ with the chemical shift of state A during $t_1$ and the chemical shift of state B during $t_2$ (Figures 15.5b and 15.5c). Similarly, the diagonal peak at $(\varpi_B, \varpi_B)$ and the cross peak at $(\varpi_B, \varpi_A)$ are generated from magnetization that originated on state B at the start of $T_{EX}$ labeled with the chemical shift of state B ($M_z^B cos\,(\Omega_B t_1)$) (Figures 15.5b and 15.5c).

The evolution of $\Delta M_z^A$ and $\Delta M_z^B$ during $T_{EX}$ is described by the following Bloch-McConnell equations:

$$\frac{d}{dt}\begin{bmatrix} \Delta M_z^A \\ \Delta M_z^B \end{bmatrix} = \left\{\begin{bmatrix} -R_{1A} & 0 \\ 0 & -R_{1B} \end{bmatrix} + \begin{bmatrix} -k_{AB} & k_{BA} \\ k_{AB} & -k_{BA} \end{bmatrix}\right\}\begin{bmatrix} \Delta M_z^A \\ \Delta M_z^B \end{bmatrix} \tag{15.28}$$

with the following solution:

$$\begin{bmatrix} \Delta M_z^A(t) \\ \Delta M_z^B(t) \end{bmatrix} = \begin{bmatrix} a_{11}(t) & a_{12}(t) \\ a_{21}(t) & a_{22}(t) \end{bmatrix}\begin{bmatrix} \Delta M_z^A(0) \\ \Delta M_z^B(0) \end{bmatrix} \tag{15.29}$$

where:

$$a_{11}(t) = \frac{1}{2}\left[\left(1 - \frac{-\Delta R_{1AB} + k_{AB} - k_{BA}}{(\lambda_+ - \lambda_-)}\right)e^{-\lambda_- t} + \left(1 + \frac{-\Delta R_{1AB} + k_{AB} - k_{BA}}{(\lambda_+ - \lambda_-)}\right)e^{-\lambda_+ t}\right],$$

$$a_{22}(t) = \frac{1}{2}\left[\left(1 + \frac{-\Delta R_{1AB} + k_{AB} - k_{BA}}{(\lambda_+ - \lambda_-)}\right)e^{-\lambda_- t} + \left(1 - \frac{-\Delta R_{1AB} + k_{AB} - k_{BA}}{(\lambda_+ - \lambda_-)}\right)e^{-\lambda_+ t}\right],$$

$$a_{12}(t) = \frac{k_{BA}}{(\lambda_+ - \lambda_-)}\left(e^{-\lambda_- t} - e^{-\lambda_+ t}\right),$$

$$a_{21}(t) = \frac{k_{AB}}{(\lambda_+ - \lambda_-)}\left(e^{-\lambda_- t} - e^{-\lambda_+ t}\right),$$

$$\lambda_\pm = \frac{1}{2}\left[(R_{1A} + R_{1B} + k_{ex}) \pm \left\{(-\Delta R_{1AB} + k_{AB} - k_{BA})^2 + 4k_{AB}k_{BA}\right\}^{1/2}\right]. \quad (15.30)$$

In the above equations, $\Delta M_z^A$ $(=M_z^A - M_{z,eq}^A)$ and $\Delta M_z^B$ $(=M_z^B - M_{z,eq}^B)$ are the deviations of the z components of magnetization arising from species A and B, respectively, from their equilibrium values while $\Delta R_{1AB} = R_{1B} - R_{1A}$. The experiments are often carried out in a difference manner so that $\Delta M_z^A$ and $\Delta M_z^B$ can be replaced by $M_z^A$ and $M_z^B$, respectively. From Equation 15.29, the intensities of the diagonal and cross peaks at the end of $T_{EX}$ period can be obtained according to:

$$\begin{bmatrix} I_{AA}(T_{EX}) & I_{AB}(T_{EX}) \\ I_{BA}(T_{EX}) & I_{BB}(T_{EX}) \end{bmatrix} = \begin{bmatrix} a_{11}(T_{EX}) & a_{12}(T_{EX}) \\ a_{21}(T_{EX}) & a_{22}(T_{EX}) \end{bmatrix}\begin{bmatrix} I_A(0) & 0 \\ 0 & I_B(0) \end{bmatrix}. \quad (15.31)$$

where $I_{ij}(T_{EX})$ is the intensity of a peak that started in state $j$ at the beginning of $T_{EX}$ and ended in state $i$ at the end of the $T_{EX}$ period and $I_A(0)$ and $I_B(0)$ are proportional to the amount of longitudinal magnetization associated with states A and B at the start of $T_{EX}$. As the system is at equilibrium, $I_A(0)$ and $I_B(0)$ are often proportional to $p_A$ and $p_B$, respectively. Since the intensities of the diagonal and cross peaks depend not only on the longitudinal relaxation rate constants of the two states, but also on the exchange rate constants $k_{AB}$ and $k_{BA}$, the experiment is performed for different $T_{EX}$ values and the $T_{EX}$ dependent intensity values are fit to the above equations to extract $k_{AB}$ and $k_{BA}$. To get a qualitative picture of how the peak intensities vary with the $T_{EX}$ values, we simplify Equation 15.31 by assuming that $R_{1A} = R_{1B} = R_1$, which results in:

$$a_{11}(t) = \left[p_A + p_B e^{-k_{ex}t}\right]e^{-R_1 t},$$

$$a_{22}(t) = \left[p_B + p_A e^{-k_{ex}t}\right]e^{-R_1 t},$$

$$a_{12}(t) = p_A\left[1 - e^{-k_{ex}t}\right]e^{-R_1 t},$$

$$a_{21}(t) = p_B\left[1 - e^{-k_{ex}t}\right]e^{-R_1 t}, \quad (15.32)$$

Hence, the intensities of the diagonal peaks decrease monotonically with $T_{EX}$ because of longitudinal relaxation and because of loss magnetization due to exchange (Figure 15.5d). However, the cross-peak intensities initially build-up at rates proportional to $k_{ex}$, reach a maximum and subsequently decay to zero because of longitudinal relaxation (Figure 15.5d). The value of $T_{EX}$ at which the cross-peak intensity is maximum is given by:

$$T_{EX}^{Max} = \frac{ln\left(\frac{\lambda_-}{\lambda_+}\right)}{(\lambda_- - \lambda_+)}. \quad (15.33)$$

In practice, $T_{EX}$ values smaller and larger than $T_{EX}^{Max}$ must be chosen in order to sample a substantial fraction of both the build-up and decay of the cross-peak profile. $T_{EX}^{Max}$ thus provides a useful guideline for choosing the $T_{EX}$ array while setting up a ZZ-ex experiment in cases where approximate values of $k_{AB}$, $k_{BA}$, $R_{1A}$, and $R_{1B}$ are known.

## 15.4 Studying Exchange Between a Visible State and Invisible State(s)

In several chemical processes of interest, exchange occurs between states with very unequal populations. In the case of two-state exchanges discussed here, this means $p_A \gg p_B$ and the peak near $\varpi_B$ will not be visible in spectra recorded on such a system, because the peak has very low intensity to begin with as $p_B$ is very small and exchange will further broaden this small peak making it 'invisible' (Figure 15.2). Hence, these 'invisible' minor states have to be detected and characterized by manipulating the visible major state magnetization. Carr-Purcell-Meiboom-Gill (CPMG), chemical exchange saturation transfer (CEST), dark-state exchange saturation transfer (DEST), and $R_{1\rho}$-type experiments are routinely used to detect minor states populated to as low as ~1%. The choice of experiment depends on the timescales of the exchange processes. Naively, one might expect that the effect of exchange on the major state signal will be on the order of $p_B$ making it very difficult to detect these minor states. However, central to all these experiments is an exchange period $T_{EX}$ during which the visible state magnetization is manipulated. $T_{EX}$ varies between ~10 and ~1 000 ms and amplifies the effect of exchange on the visible state signal making it possible to detect minor states with very low populations. Consider an exchange process where $k_{ex} = 50\ s^{-1}$ and $p_B = 1\%$. Such a process is usually studied using a CEST experiment with $T_{EX} = 0.5$ s. As $k_{AB} = 0.5\ s^{-1}$, ~22% ($p_A(1 - e^{-k_{AB}T_{EX}})$) of the molecules visit state B from state A during the $T_{EX}$ period. Hence if the magnetization of molecules in state B can be destroyed during $T_{EX}$, exchange will have a ~22% and not a ~1% effect on the visible state signal. The various experiments used to study exchange between a visible major state and invisible minor state(s) are described below.

### 15.4.1 CPMG Experiments

The Carr-Purcell-Meiboom-Gill (CPMG) experiment [15] is used to study exchange between a major visible state and 'invisible' minor states occurring on the microsecond to millisecond timescale. In the constant time (CT) version of the CPMG experiment [16] transverse magnetization is allowed to evolve for a time $T_{EX}$ during which a variable number of refocusing $\pi$ pulses are applied (Figure 15.6a). The effective relaxation $R_{2,eff}$ is characterized as function of the frequency ($\nu_{CPMG}$) at which the refocusing $\pi$ pulses are applied. Here, $\nu_{CPMG} = 1/(4\tau_{CP})$ where $2\tau_{CP}$ is the spacing between the refocusing $\pi$ pulses (Figure 15.6a) and $R_{2,eff}(\nu_{CPMG}) = \frac{-1}{T_{EX}} ln\left(\frac{I(\nu_{CPMG})}{I_0}\right)$ where $I_0$ is the intensity of the visible peak in the absence of the $T_{EX}$ delay and $I(\nu_{CPMG})$ is the intensity of the visible peak when the $\pi$ pulses are applied at $\nu_{CPMG}$. To reduce off-resonance artifacts an even number of $\pi$ pulses are applied, hence the experiment consists of 2N repeating [$\tau_{CP}$-$\pi$-$\tau_{CP}$] Hahn-echo elements so that $4N\tau_{CP} = T_{EX}$ with N being any positive integer.

In the absence of exchange the transverse magnetization is fully refocused at the end of each [$\tau_{CP}$-$\pi$-$\tau_{CP}$] Hahn-echo element. This is illustrated in Figure 15.6b when $\nu_{CPMG} = 50$ Hz where magnetization starts off along the x-axis ($M_x$, phase=0) and evolves during the $\tau_{CP}$ period so that phase builds up ($\omega_A > 0$) before the first $\pi$ pulse is applied. Here, phase refers to the angle that the magnetization vector makes with respect to the x-axis. The effect of the $\pi$ pulse is to invert the sense of precession of the magnetization and the phase starts to decrease returning to zero at time $2\tau_{CP}$. By comparing plots for $\nu_{CPMG} = 50$, 400, and 1000 Hz it is clear that the maximum value of the phase decreases as $\nu_{CPMG}$ increases because $\tau_{CP}$ reduces. However, as the magnetization is fully refocused irrespective of $\nu_{CPMG}$, $R_{2,eff}(\nu_{CPMG})$ adopts a constant value of $R_{2A}$ (Figure 15.6c). In Figure 15.6 $R_{2A} = R_{2B} = 0\ s^{-1}$.

In the presence of chemical exchange, the magnetization is not fully refocused as the evolution frequency of the different spins fluctuates randomly between $\omega_A$ and $\omega_B$. For example, if a spin evolves at $\omega_A$ in the first $\tau_{CP}$ period (before the $\pi$ pulse) and then the molecule jumps to state B at some point during the second $\tau_{CP}$ period after the first $\pi$ pulse the phase will not return to zero at $2\tau_{CP}$ as $\omega_B$ differs from $\omega_A$. This is illustrated in Figure 15.6d where 250 phase trajectories have been calculated for a site undergoing exchange. The different trajectories do not overlap as individual molecules undergo exchange at different times and many of them do not return to their

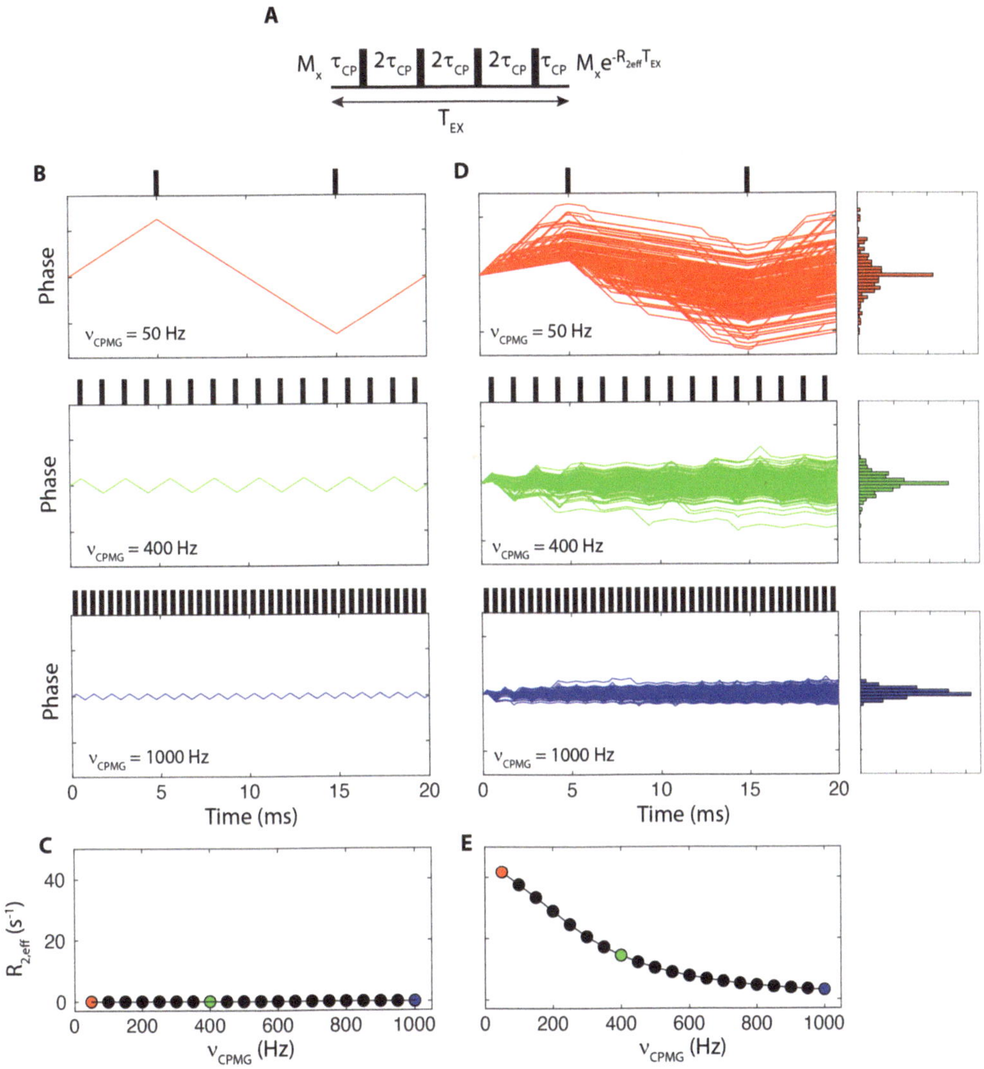


**Figure 15.6** Principle behind the CPMG experiment to quantify chemical exchange. (a) In the CT-CPMG experiment, transverse magnetization evolves for a time $T_{EX}$ during which it is subjected to varying number of refocusing $\pi$ pulses. $2\tau_{CP}$ is the spacing between the refocusing pulses and the frequency at which refocusing $\pi$ pulses are applied is defined as $\nu_{CPMG} = 1/4\tau_{CP}$. (b) In the absence of exchange, the build-up of phase due to chemical-shift evolution during the $T_{EX}$ delay is plotted for three different $\nu_{CPMG}$ values. The $\pi$ pulses are indicated by bars above the graphs. In the absence of exchange, there is complete refocusing of magnetization and $R_{2,eff}$ is independent of $\nu_{CPMG}$ (c). (d) In the presence of exchange, the build-up of phase due to chemical-shift evolution during the $T_{EX}$ delay is plotted for three different $\nu_{CPMG}$ values. Each plot, consists of 250 trajectories. On the right of each plot, the histogram shows the distribution of phase values. The distribution becomes sharper around zero as $\nu_{CPMG}$ increases leading to an increase in detected x magnetization and a concomitant decrease in $R_{2,eff}$ as $\upsilon_{CPMG}$ increases (e). Here, $T_{EX}$ was set to 20 ms, $\omega_A > 0$, $\Delta\omega_{AB} > 0$, and $R_{2A} = R_{2B} = 0\ s^{-1}$.

starting orientation (phase = 0) along the x-axis leading to a loss of detected x magnetization at the end of the $T_{EX}$ period. As $\nu_{CPMG}$ increases the maximum value of the phase that builds up decreases and more signal is detected along x at the end of the experiment (compare $\nu_{CPMG}$ = 50, 400, and 1000 Hz in Figure 15.6d). Hence, in the presence of exchange $R_{2,\text{eff}}$ can decrease as a function of $\nu_{CPMG}$ leading to the relaxation dispersion curve

(Figure 15.6e). Having recognized that the effect of the $\pi$ pulse is to invert the sense of precession, we can use the Bloch-McConnell equations to quantitatively obtain the relaxation dispersion curves.

$$\begin{bmatrix} M_+^A(\nu_{CPMG}) \\ M_+^B(\nu_{CPMG}) \end{bmatrix} = \left[ e^{\begin{bmatrix} -R_{2A}+i\Omega_A-k_{AB} & k_{BA} \\ k_{AB} & -R_{2B}+i\Omega_B-k_{BA} \end{bmatrix}\tau_{CP}} e^{\begin{bmatrix} -R_{2A}-i\Omega_A-k_{AB} & k_{BA} \\ k_{AB} & -R_{2B}-i\Omega_B-k_{BA} \end{bmatrix}2\tau_{CP}} \right. \\ \left. e^{\begin{bmatrix} -R_{2A}+i\Omega_A-k_{AB} & k_{BA} \\ k_{AB} & -R_{2B}+i\Omega_B-k_{BA} \end{bmatrix}\tau_{CP}} \right]^N \begin{bmatrix} M_+^A(0) \\ M_+^B(0) \end{bmatrix} \tag{15.34}$$

Here we have assumed ideal $\pi$ pulses whose effect is implicitly taken into account by inverting the sign of $\Omega_A$ and $\Omega_B$. $R_{2,eff}(\nu_{CPMG})$ is calculated from $M_+^A(\nu_{CPMG})$ and $M_+^A(0)$ as described earlier. Usually, $\begin{bmatrix} M_+^A(0) \\ M_+^B(0) \end{bmatrix} \propto \begin{bmatrix} p_A \\ p_B \end{bmatrix}$. Analytical solutions to Equation 15.34 are extremely cumbersome, however, when $p_A \gg p_B$ we have:

$$R_{2,eff}(\nu_{CPMG}) = R_2(\infty) + \frac{p_A p_B \Delta\omega_{AB}^2 k_{ex}}{k_{ex}^2 + \left(p_A^2 \Delta\omega_{AB}^4 + 36864\nu_{CPMG}^4\right)^{\frac{1}{2}}}. \tag{15.35}$$

It is clear that $R_{2,eff}(\nu_{CPMG})$ decreases as $\nu_{CPMG}$ is increased reaching its minimum value $R_2(\infty) = p_A R_{2A} + p_B R_{2B}$, when the effect of exchange has been quenched by pulsing infinitely fast. Further, by setting the second term in Equation 15.35 to half of its maximum value ($\nu_{CPMG} = 0$ Hz), we can estimate the rate of pulsing ($\nu_{CPMG\frac{1}{2}}$) at which the exchange contribution to relaxation has been reduced to half of its maximum value:

$$\nu_{CPMG\frac{1}{2}} = \frac{1}{4}\left(\frac{1}{12}\right)^{\frac{1}{2}} \left[\left(k_{ex}^2 + 2p_A\Delta\omega_{AB}^2\right)^2 - p_A^2\Delta\omega_{AB}^4\right]^{\frac{1}{4}}. \tag{15.36}$$

It is clear from Equation 15.36 that increasing $k_{ex}$ or $|\Delta\omega_{AB}|$ necessitates larger $\nu_{CPMG}$ values to quench effect of exchange. Equation 15.34 assumes that the starting magnetization is of the $M_+$ variety. A similar equation for $M_-$ is obtained by negating the signs of the $\Omega$ in Equation 15.34 resulting in a dispersion curve that is identical to the one obtained for the $M_+$ coherence. Hence, the shape of the relaxation dispersion curve depends on $|\Delta\omega|$ and not $\Delta\omega$, which is also confirmed by Equation 15.35 and Equation 15.36 that depend only on even powers of $\Delta\omega$. The discussion above has conveyed the impression that $R_{2,\mathrm{eff}}$ monotonously decreases with increasing $\nu_{CPMG}$. However, this is not always the case. When exchange is slow and $\nu_{CPMG}$ is low, interference effects between chemical-shift evolution and the CPMG pulse train can lead to an increase in $R_{2,\mathrm{eff}}$ leading to oscillations in the early part (low $\nu_{CPMG}$ values) of the relaxation dispersion curve [17]. As the shape of the relaxation dispersion curve depends on the exchange parameters and $|\Delta\omega|$, an analysis of the relaxation dispersion curve can give the exchange parameters, and $|\Delta\omega|$ and CPMG experiments are routinely used to study the kinetics and thermodynamics of various exchange processes (Figure 15.7). To obtain robust estimates of the exchange parameters, it is often useful to analyze CPMG relaxation dispersion profiles recorded at two well-separated $B_0$ field strengths as all exchange parameters remain the same except for $\Delta\omega$, which scales with field strength. It is clear from Equations 15.35 and 15.36 that when exchange is fast, high $\nu_{CPMG}$ values have to be used to quench out the effects of exchange. This may not be possible due to hardware limitations putting an upper limit on the $k_{ex}$ values that can be studied. Further, when $k_{ex}$ is large the size of the dispersion will be very small making it once again difficult to study fast processes. The length of the $T_{EX}$ period in CPMG experiments is chosen such that the signal of interest does not decay significantly and is determined largely by $R_{2,\mathrm{eff}}$ and varies between 20 and 40 ms. When $k_{ex}$ is very low there may not be enough number of exchange events during $T_{EX}$ leading to a very small loss in magnetization due to exchange making it difficult to study slow processes.

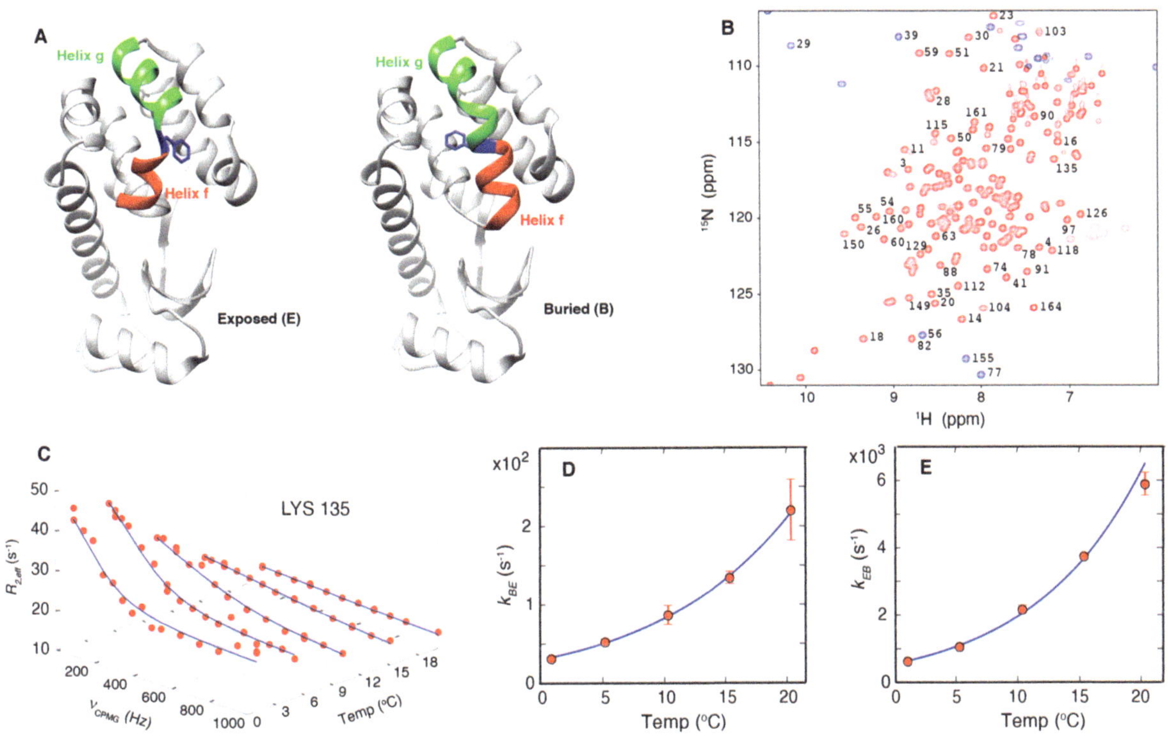


**Figure 15.7** Studying conformational exchange between a visible major state and an "invisible" minor state using CPMG experiments. (a) The L99A, G113A, R119P mutant of T4 lysozyme (T4Ltm) interconverts between two conformations the exposed form (E) in which the sidechain of Phe114 is exposed to solvent and the buried form (B) in which the Phe114 is buried in the core of the protein. At ambient temperature the buried form is the major state ($p_{Buried}$ ~96.5% at 15 °C) and the $^{15}$N - $^{1}$H (HSQC) correlation plot (b) consists of correlations only from the B form. The correlations are labeled according the residue from which they arise and aliased peaks are in blue. Amide $^{15}$N CPMG experiments were performed by quantifying the intensity of these peaks for varying $\nu_{\mathrm{CPMG}}$ values shown in panel (c) for Lys 135 at different temperatures. In (d) and (e) the CPMG-derived exchange rates are shown as red circles while the best-fit Arrhenius curves are shown as blue lines. *Figure adapted from [28].*

#### 15.4.1.1 Determining the Sign of $\Delta\omega$

As is clear from the discussion above, the sign of $\Delta\omega_{AB}$ cannot be obtained from the CPMG experiment. However, the sign of $\Delta\omega$ is required to obtain the minor state chemical shift from the major state chemical shift. Hence, one has to obtain the information regarding the sign of $\Delta\omega_{AB}$ independently from the visible major state resonance. As we can see in Figure 15.2b, the visible major state resonance that is at $\varpi_{\mathrm{A}}$ in the limit of slow exchange moves toward $\varpi_{\mathrm{B}}$ as $k_{\mathrm{ex}}$ increases, eventually reaching the population weighted average position ($p_{\mathrm{A}}\varpi_{\mathrm{A}} + p_{\mathrm{B}}\varpi_{\mathrm{B}}$) in the fast exchange limit. Thus, to obtain the sign of $\Delta\omega_{AB}$ one might consider tracking the visible peak position by varying the temperature and modulating $k_{\mathrm{ex}}$, however, attempts to modulate $k_{\mathrm{ex}}$ by varying the temperature will have other undesirable effects like affecting the minor state population. The sign of $\Delta\omega$ can, however, be obtained by recording spectra at different field strengths under identical conditions (Figure 15.8). As all the kinetic parameters remain constant, while $|k_{\mathrm{ex}}/\Delta\omega_{AB}|$ varies between the spectra recorded at different fields because $\Delta\omega$ scales with field strength, the exchange regime changes from slow to fast as the field strength is decreased. Hence the visible peak position will shift toward the minor peak position as the field strength is decreased and the sign of $\Delta\omega_{AB}$ can be obtained by recording spectra at two well-separated field strengths (Figure 15.8) [18]. Quantitatively, when $p_{\mathrm{B}} \ll 1$ and $\Delta R_2 = 0$, the exchange induced shift in the visible peak position from $\omega_A$ denoted by $\delta_{\mathrm{ex}}$ is given by:

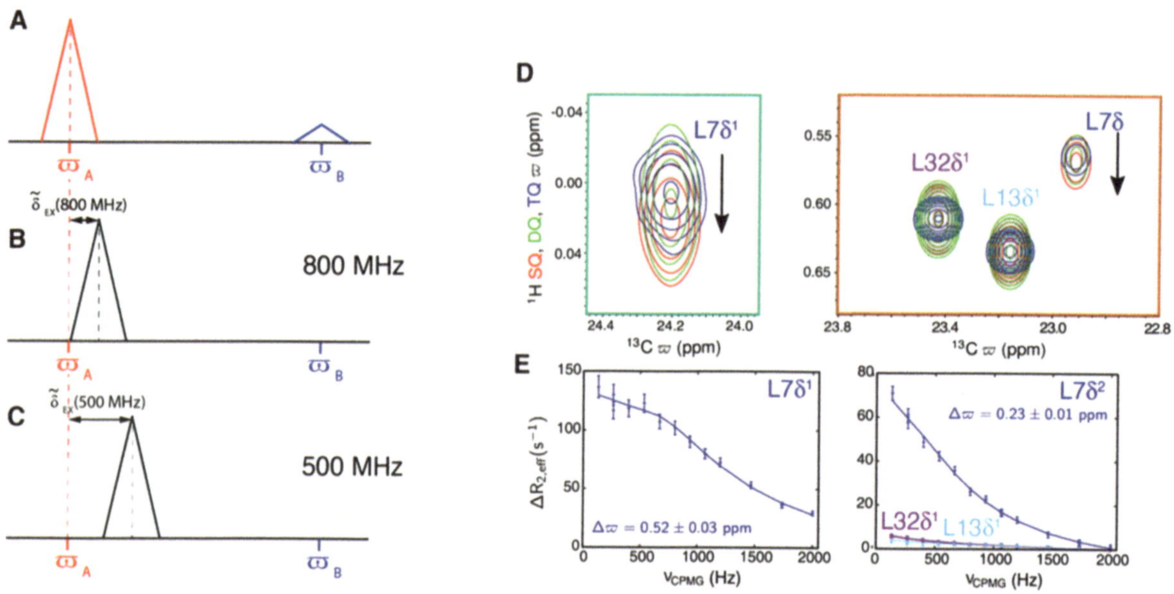


**Figure 15.8** The sign of $\Delta\varpi$ can be obtained by recording spectra at two well-separated fields (500 and 800 MHz). (a) Schematic illustration of the spectrum in the absence of exchange with the major site A resonating at $\varpi_A$ and the minor-site B resonating at $\varpi_B$. (b) Schematic illustration of the spectrum recorded in the presence of exchange at 800 MHz. The minor state is not visible and visible peak is shifted from $\varpi_A$ toward $\varpi_B$ by $\hat{\delta}_{ex}$ *(800 MHz)*. (c) Schematic illustration of the spectrum recorded in the presence of exchange at 500 MHz. The minor state is not visible and visible peak is shifted from $\varpi_A$ toward $\varpi_B$ by $\hat{\delta}_{ex}$*(500 MHz)*. As $|\hat{\delta}_{ex}$ *(500 MHz)*$|$ is greater than $|\hat{\delta}_{ex}$ *(800 MHz)*$|$ the visible state peak is shifted more toward $\varpi_B$ at 500 MHz than at 800 MHz and the sign of $\Delta\varpi$ can be obtained by comparing spectra recorded at 500 and 800 MHz, respectively, and the visible peak in the spectrum recorded at the lower field (500 MHz) will be shifted more toward the minor state than the visible peak in the spectrum recorded at the higher field (800 MHz). (d) Methyl $^1$H, single-quantum (SQ), double-quantum (DQ), and triple-quantum (TQ) spectra of T4Ltm (see Figure 15.7) recorded at 700 MHz. The DQ and TQ spectra can be thought of as spectra recorded and 1 400 and 2 100 MHz, respectively. The peaks from sites that show CPMG dispersions (e) are not on top of each other with the peak in the SQ spectrum (akin to the lowest $B_0$) shifted most toward the minor state and the one from TQ spectrum (akin to the highest $B_0$) shifted least toward the minor state, hence, providing information regarding the sign of $\Delta\varpi$. *Panels (d) and (e) adapted from [29].*

$$\delta_{ex} \approx \frac{p_B \Delta\omega_{AB}}{1 + \left(\Delta\omega_{AB}/k_{ex}\right)^2}. \tag{15.37}$$

Equation 15.37 can be obtained from Equation 15.15 and as expected we can see that in the slow exchange limit $\delta_{ex}$ is zero and in the fast exchange limit it is $p_B\Delta\omega_{AB}$. Hence, the exchange-induced shift in the visible peak position in ppm ($\hat{\delta}_{ex}$) between spectra recorded at two field strengths ($B_{0,1}$ and $B_{0,2}$) is given by:

$$\hat{\delta}_{ex}\left(B_{0,1}\right) - \hat{\delta}_{ex}\left(B_{0,2}\right) \approx \frac{p_B \Delta\varpi_{AB}}{1 + \left(\gamma\Delta\varpi_{AB}B_{0,1}/k_{ex}\right)^2} - \frac{p_B \Delta\varpi_{AB}}{1 + \left(\gamma\Delta\varpi_{AB}B_{0,2}/k_{ex}\right)^2} \tag{15.38}$$

Equation 15.38 shows us that when both $(\gamma\Delta\varpi_{AB}B_{0,1}/k_{ex})$ and $(\gamma\Delta\varpi_{AB}B_{0,2}/k_{ex})$ are very large (slow exchange) or very small (fast exchange) the visible peak will be at the same position (ppm) in spectra recorded at the two fields and will not give us the desired sign experiments and one will have to record spectra at different fields or modulate kinetic parameters to move exchange into the desired timescale. Here $\gamma$ is the gyromagnetic ratio of the nucleus of interest.

### 15.4.2 CEST and DEST Experiments

The CEST [19] and DEST [20] experiments were originally developed to study slow chemical exchange processes ($\sim$10 to $\sim$300 $s^{-1}$). In these experiments, longitudinal magnetization is subjected to weak to moderate $B_1$ irradiation ($\sim$5 to 300 Hz) during a $T_{EX}$ period and magnetization that is along the z-axis at the end of the $T_{EX}$ period is detected (Figure 15.9a). The effect of the magnitude and offset of the $B_1$ irradiation on the visible state magnetization is quantified and analyzed to study the underlying exchange processes. In the CEST class of experiments the chemical shift of the minor state of interest differs from the major state, while in the DEST class of experiments there is a large difference in the intrinsic transverse relaxation rates between major (a) and minor (b) states. Since longitudinal magnetization whose decay is governed by $R_1$ ($R_1 \ll R_2$) is present during the $T_{\rm EX}$ period in these experiments, $T_{\rm EX}$ can be quite large ($\sim$0.5s) compared to the $T_{\rm EX}$ delay in CPMG experiments, which is determined $R_2$ relaxation. Hence these experiments can be used to study slower exchange processes.

As mentioned earlier, in the CEST class of experiments the chemical shift of the minor state is different from that of the major state and this difference is used to detect the minor state. We start with a qualitative description of the CEST experiment assuming that $R_{2,\rm A} = R_{2,\rm B}$ and that $\Delta\varpi_{\rm AB}$ is very large ($\gg B_1$). When the $B_1$ irradiation during the $T_{\rm EX}$ period is applied at $\varpi_{\rm B}$, spins in molecules that are in state B will precess around the $B_1$ magnetization and will not be along the z-axis when the molecule returns to state A (Figure 15.9b) resulting in a net loss in A-state z magnetization. On average the molecules in state B will precess around the $B_1$ magnetization by angle $\langle\phi\rangle \sim \frac{2\pi B_1}{k_{BA}}$. Hence in some sense, in the CEST experiment $2\pi B_1$ is similar to $|\Delta\omega_{AB}|$ in a free precession experiment. As a

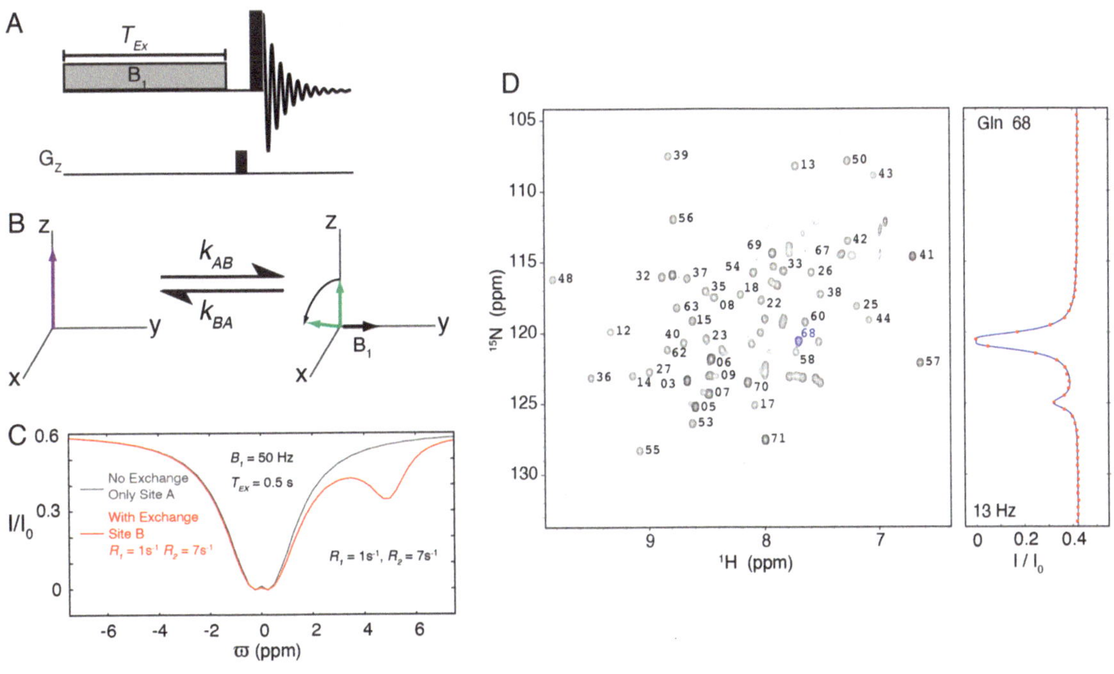


**Figure 15.9** a) Schematic illustration of the CEST experiment. (b) Weak irradiation at $\varpi_{\rm B}$ causes spins in the minor state to precesses around the y-axis. (c) CEST profile calculated for a $^{15}$N spin $1/2$ nuclei undergoing (red, $k_{ex} = 50\ s^{-1}$, $p_B = 0.02$, $\varpi_A = 0$ ppm, and $\varpi_B = 5$ ppm) and not undergoing (black) exchange at 16.45 T. The CEST profile for the site undergoing exchange clearly shows a dip at the resonance position of the minor state. (d) Studying the folding of the A39G FF domain at 1 °C. The spectrum corresponds to the folded form, however, by quantifying the peak intensities as a function of offset at which $B_1$ irradiation is applied, the unfolded state populated to just 1.65% (kex = 51.6 $s^{-1}$) can be detected. This is illustrated for Gln68 where two dips can be seen in the CEST profile, the large dip corresponds to the folded state while the small one corresponds to the unfolded state. *Panel (d) adapted from [30].*

large number of molecules visit state B from state A and then return to state A during $T_{EX}$, the net loss in A-state magnetization due to exchange can be considerable (if $B_1$ is large enough compared to $k_{BA}$). Consequently, a plot of $I(\varpi)/I_0$ versus $\varpi$, in which $I(\varpi)$ is the intensity of the visible state magnetization at the end of $T_{\mathrm{EX}}$ when the $B_1$ is applied at an offset $\varpi$, will have a dip at $\varpi_{\mathrm{B}}$ in addition to the expected one at $\varpi_{\mathrm{A}}$ (Figure 15.9c). $I_0$ is the intensity of the visible state magnetization in the absence of the $T_{\mathrm{EX}}$ delay.

In the absence of exchange, the effect of weak $B_1$ irradiation on the detected magnetization can be obtained by evolving the Bloch equations. Under saturating conditions ($\frac{\omega_1^2}{R_1R_2} \gg 1$), we obtain the following solution for $M_{z\mathrm{A}}$ at end of a long $T_{\mathrm{EX}}$ period:

$$M_{zA} = \frac{(\omega_A - \omega_{RF})^2}{(\omega_A - \omega_{RF})^2 + \frac{\omega_1^2 R_2}{R_1}}. \tag{15.39}$$

Equation 15.39, shows that there is a loss of state-A magnetization when the offset at which $\mathrm{B}_1$ irradiation is applied ($\omega_{RF}$) is coincident with $\omega_A$. In the presence of exchange, expressions have been derived for the evolution magnetization in the presence of $\mathrm{B}_1$ irradiation. Zaiss and Bachert [21] have shown that when $R_{1\mathrm{A}} = R_{1\mathrm{B}} = 0\ \mathrm{s}^{-1}$ and $\Delta\varpi_{AB} \to \infty$, the minor state dip can be described by:

$$\frac{I\,(\Delta\omega_{RF}, B)}{I_0} = 1 - e^{-R_{EX}(\Delta\omega_{RF,B})T_{EX}} \tag{15.40}$$

where,

$$R_{EX}\left(\Delta\omega_{RF,B}\right) = \frac{R_{ex}^{max}\mathrm{T}^2}{\mathrm{T}^2 + \Delta\omega_{RF,B}^2},$$

$$R_{ex}^{max} = \frac{k_{ex}p_B\omega_1^2}{\omega_1^2 + k_{ex}\left(k_{ex} + R_{2B}\right)},$$

$$\mathrm{T} = \sqrt{\frac{k_{ex} + R_{2B}}{k_{ex}}\omega_1^2 + \left(k_{ex} + R_{2B}\right)^2}. \tag{15.41}$$

Hence, when the $\mathrm{B}_1$ irradiation in coincident with $\varpi_{\mathrm{B}}$ ($\Delta\omega_{RF,B} = 0$), we only need to consider $R_{ex}^{max}$, which is very similar to Equation 15.20 if one replaces $\omega_1$ with $\Delta\omega$ consistent with the picture that $\omega_1$ is now analogous to $\Delta\omega$ in the case of free precession. More general expressions, which we shall not reproduce here, have been derived for the shape of the CEST profile. CEST experiments are carried out usually at least two $\mathrm{B}_1$ values, preferably with $2\pi B_1$ less than and greater than $k_{\mathrm{ex}}$. The CEST profiles are analyzed using the standard Bloch-McConnell equations to obtain the exchange rate(s), minor state population(s), and chemical shift(s). Unlike CPMG experiments that require additional experiments to obtain the minor-state chemical shifts (see Section 15.4.1.1), the minor-state chemical shift can be directly obtained from the CEST profile making them a very powerful technique to study exchange between a visible major state and invisible minor state(s) (Figure 15.9d). Although CEST experiments were initially developed to study relatively slow exchange (~10 to ~300 $\mathrm{s}^{-1}$) processes, they are now used to study chemical exchange occurring on the ~10 to ~10 000 $\mathrm{s}^{-1}$ timescale [22, 23].

As we have seen earlier in Section 15.2.4, chemical exchange can have an effect on the NMR spectrum even when $\Delta\varpi = 0$ ppm so long as $|\Delta R_2| \gg 0\ \mathrm{s}^{-1}$. The DEST experiments, which are experimentally identical to the CEST experiment, take advantage of this different physical effect to study exchange between a major visible conformer (A) and a minor invisible conformer (B) with $R_{2\mathrm{B}} \gg R_{2\mathrm{A}}$. This is illustrated in Figure 15.10 where DEST profiles have been calculated for various $R_{2\mathrm{B}}$ values with $\Delta\varpi = 0$ ppm and it is clear that when the $R_{2\mathrm{B}}$ values are very large the DEST profiles differ considerably from the case when $R_{2\mathrm{B}} = R_{2\mathrm{A}}$ allowing one to detect minor states with

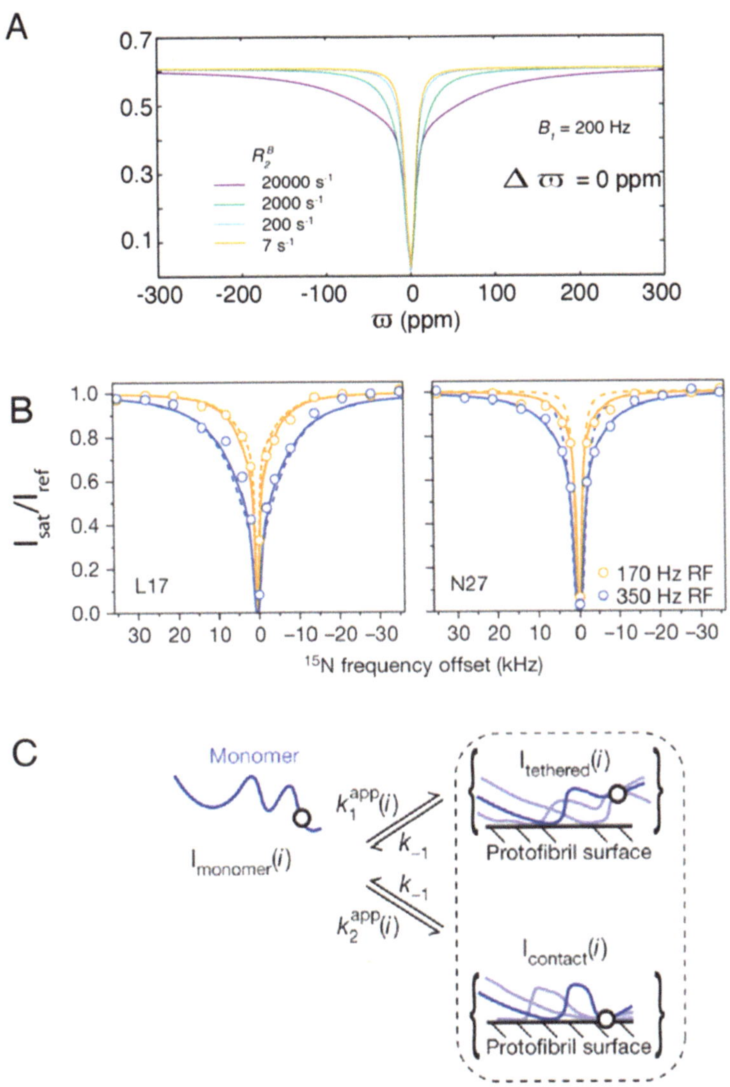


**Figure 15.10** (a) Large $R_{2B}$ values can have dramatic effects on the CEST profiles even when $\Delta\varpi = 0$. DEST profiles (16.45 T) were calculated for different $R_{2B}$ values with $R_{2A} = 7\ s^{-1}$, $T_{EX} = 0.5$ s, $p_B = 2\%$, and $k_{ex} = 50\ s^{-1}$. (b) Amide $^{15}N$ DEST profiles for A$\beta$ peptide clearly show that the peptide is in exchange with a larger species. (c) A detailed analysis of the DEST profiles showed that the peptide was associating with the (large) protofibrils in a wide variety of conformations. *Panels b and c adapted from [20].*

$R_{2B} \gg R_{2A}$. One can think of the DEST profile as a CEST profile with a very large $R_{2B}$ value so that the minor state dip is very broad and merged with the major state. The minor invisible state B is often a large aggregate and DEST experiments for example were used to detect and characterize A$\beta_{40/42}$ peptides that were bound to protofibrils (Figures 15.10b and 15.10c) by studying exchange between free A$\beta_{40/42}$ molecules (visible, small $R_2$) and those that were bound to the very large protofibrils that have very large $R_2$ values and are invisible in regular solution state NMR spectra.

### 15.4.3 $R_{1\rho}$ Relaxation Dispersion Experiment

The $R_{1\rho}$ relaxation-dispersion experiment ($R_{1\rho}$ RD) is another method [3, 4] for probing transiently and sparsely populated conformations in exchange with a major state with $k_{ex}$ values as high as ~35 000 $s^{-1}$. The $R_{1\rho}$ experiment is very similar to the CEST/DEST experiments described in the previous section in that a radiofrequency (RF) field is applied along the transverse plane at a specific frequency offset ($\omega_{RF}$) and amplitude ($\omega_1 = \gamma B_1$) for a duration $T_{EX}$ and its effect on the visible state magnetization is monitored. In the rotating frame (in the absence of exchange), it is useful to think of an effective magnetic field ($B_{1,eff}$) with a z-component ($\omega_A$-$\omega_{RF}$) and an x-component $\omega_1$ ($B_1$ is applied along x). However, unlike the CEST/DEST experiments, the magnetization of the nucleus of interest is aligned along the effective magnetic field prior to the start of $T_{EX}$ and its rotating-frame relaxation-rate constant ($R_{1\rho}$) is measured. The applied RF field is also called a spin-lock field as it 'locks' the magnetization along the direction of the effective $B_{1,eff}$ field. Since the offsets for the two states are different, the effective fields $B_1$ experienced by the ground and excited-state magnetization are also different with the direction ($\vec{\omega}_{eff,i}$) and magnitude ($\omega_{eff,i}$) of the effective field experienced by the spin given by $\vec{\omega}_{eff,i} = \vec{\omega}_1 + \vec{\Omega}_i$ and $\omega_{eff,i} = \sqrt{\omega_1^2 + \Omega_i^2}$, respectively, with $i\epsilon\{A, B\}$. For systems in slow exchange, the magnetization of states A and B are aligned along $\vec{\omega}_{eff,A}$, while for systems in intermediate to fast exchange, initial alignment occurs along the average effective field, whose magnitude and direction is given by $\vec{\omega}_{eff} = \vec{\omega}_1 + \vec{\overline{\Omega}}$ and $\omega_{eff} = \sqrt{\omega_1^2 + \overline{\Omega}^2}$, respectively, ($\overline{\Omega} = p_A\Omega_A + p_B\Omega_B$). $R_{1\rho}$ measurements are classified as on-resonance or off-resonance $R_{1\rho}$ based on the magnitude of the chemical-shift offsets and as weak-field ($B_1$ ~ 100 Hz) or strong-field $R_{1\rho}$ ($B_1$ > 1000 Hz) depending on the amplitude of the RF field. In the on-resonance $R_{1\rho}$ RD experiment (Figure 15.11a), $R_{1\rho}$ values for a particular nucleus are measured as a function of $B_1$ when the RF irradiation is on-resonance to this nucleus. In contrast, off-resonance $R_{1\rho}$ RD data (Figure 15.11b–15.11d) describe the variation of $R_{1\rho}$ or $R_{2,eff}$ as a function of the offset at which the RF field is applied. Conformational exchange is visible in the on-resonance $R_{1\rho}$ RD profile as a reduction in $R_{1\rho}$ values with increasing spin-lock field strength ($B_I$) (Figure 15.11a). In order to understand why this happens, let us consider a system in two-state slow exchange where the $B_1$ field is on-resonance to the ground state A. As long as the molecule stays in state A, there is no dephasing of magnetization. When the molecule jumps to state B, with a chemical shift $\varpi_B$ different from $\varpi_A$, the efficacy of the spin-lock field wanes and the spin acquires a phase difference in the transverse plane that leads to a net decrease in intensity when the molecule returns to state A, causing exchange broadening. At higher $B_1$ field strengths, the spin-lock field is able to suppress the dephasing of spins more effectively when the molecule resides in state B. Therefore, the net dephasing due to occasional jumps to state B decreases and exchange broadening is quenched at stronger spin-lock fields. For the same $B_1$ field, the efficiency of spin-locking is higher at lower $B_0$ fields because the chemical shift of the excited state (in rad/s) is smaller; this is why on-resonance $R_{1\rho}$ values for the same $B_1$ field are lower at a $B_0$ field of 14.1 T compared to 18.8 T (Figure 15.11a). On the other hand, off-resonance $R_{1\rho}$ experiments measure the variation in $R_{1\rho}$ values at constant $B_1$ as a function of the offset at which the RF field is applied. Several equations have been derived for $R_{1\rho}$ in various exchange regimes and population ratios. An equation for $R_{1\rho}$ that is valid for all timescales provided magnetization decay is single-exponential is [3]:

$$R_{1\rho} = R_1 \cos^2\theta + R_{2,eff} \sin^2\theta \tag{15.42}$$

with $\tan\theta = \omega_1/\overline{\Omega}$ and $R_{2,eff} = R_2 + R_{ex}$. Here, for simplicity, we have assumed that $R_{1A} = R_{1B}$ and $R_{2A} = R_{2B}$. Simplified expressions have been derived for $R_{2,eff}$ as shown below for the case when $p_A \gg p_B$:

$$R_{ex} = \frac{p_A p_B \Delta\omega_{AB}^2 k_{ex}}{\Omega_B^2 + \omega_1^2 + k_{ex}^2} = \frac{p_A p_B \Delta\omega_{AB}^2 k_{ex}}{\omega_{eff,B}^2 + k_{ex}^2}. \tag{15.43}$$

As expected from the discussion above, $R_{ex}$ decreases as $B_1$ increases (Figure 15.11) but interestingly $R_{ex}$ is also strongly dependent on the offset at which the $B_1$ field is applied (Figure 15.11b–15.11d) with a maximum when it is applied at $\omega_B$. Consequently, $R_{1\rho}$ data are sensitive not only to the magnitude of $\Delta\omega_{AB}$, but also to its sign, unlike CPMG datasets, which depend only on the magnitude. An exhaustive discussion of the spin physics underlying the off-resonance $R_{1\rho}$ experiment can be found in review articles [3, 4]. The offset-dependencies of $R_{1\rho}$ and $R_{2,eff}$, and the observed patterns, are best understood by considering the various effective fields operating during $T_{EX}$. Consider the case of slow two-state exchange, where the magnetization is initially aligned along $\omega_{eff,A}$. When the offset value is large compared to $B_1$, the direction of $\omega_{eff,A}$ approaches the z-axis and magnetization relaxes according to $R_{1A}$. As the offset becomes smaller, the effective field tilts more and more toward the transverse plane and $R_{1\rho}$ has significant contributions from $R_{2,eff}$, which includes the exchange broadening term. Therefore, $R_{1\rho}$ is highest around offset values close to $\omega_A$ (Figure 15.11b). As mentioned above, an important feature of off-resonance $R_{1\rho}$ profiles is that $R_{2,eff}$ is maximum at the chemical-shift offset of the minor state (Figure 15.11c,d). The offset-dependence of $R_{2,eff}$ can be understood as follows: magnetization is initially aligned along $\omega_{eff,A}$ and does not precess around this field. When a sub-ensemble of molecules transitions to state B, its effective field switches to $\omega_{eff,B}$ and its magnetization is no longer aligned with this effective field. Therefore, the magnetization precesses about $\omega_{eff,B}$ and acquires a phase difference with the magnetization from the rest of the ensemble. Once this sub-ensemble returns back to state A, the resulting phase difference depletes the magnetization of state A and causes exchange broadening. While such phase differences in CPMG experiments can be understood in terms of precession in two dimensions and a single phase angle ($\phi$, rotation around z), precession in $R_{1\rho}$ experiments occurs along $\omega_{eff,B}$, which is tilted with respect to the z-axis. Therefore, precession

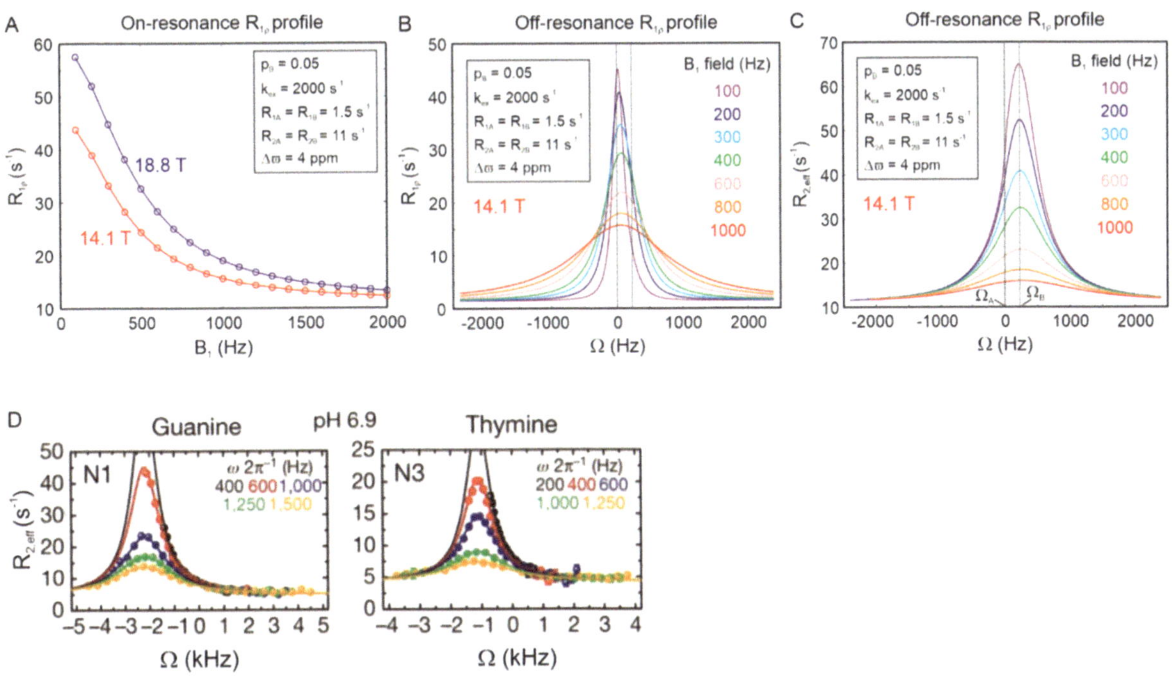


**Figure 15.11** Simulated on-resonance (a) and off-resonance (b,c) $R_{1\rho}$ relaxation-dispersion profiles depicting the variation of $R_{1\rho}$ (a,b) or $R_{2,eff}$ (c) with the amplitude of the spin-lock field (a) or the offset (b,c). (d) Experimental off-resonance $R_{1\rho}$ relaxation dispersion profiles of the imino protons (dG-N1 and dG-N3) of a hairpin DNA duplex, which contains a dG-dT wobble base pair, transiently exchanging with a tautomeric form that adopts Watson-Crick-like geometry. The pB for the tautomeric form is 0.17% and its lifetime is ~380 μs. *Figure adapted from [31].*

causes changes in both $\theta$ and $\phi$ polar angles. Phase differences, which lead to exchange broadening, do not accumulate as long as magnetization remains in state A aligned along $\omega_{\text{eff,A}}$. Accordingly, the value of $R_{2,\text{eff}}$ is maximum at the chemical-shift offset of B (Figure 15.11c, d), where magnetization arriving at state B relaxes primarily due to $R_{2,\text{eff}}$. On the other hand, when the offset $\Omega_B$ is large and irradiation occurs far from state B, $\omega_{\text{eff,B}}$ is close to the z-axis and precession of molecules that have jumped to state B is reduced leading to a small $R_{2,\text{eff}}$ value.

A key practical aspect of obtaining $R_{1\rho}$ relaxation dispersion profiles is the need to align magnetization of a particular nucleus to its specific effective field. This is a challenging task when measuring $R_{1\rho}$ for all residues of a protein or nucleic acid simultaneously because of the typically large dispersion in chemical-shift offsets. A number of pulse sequence modules that employ hard pulses as well as adiabatic ramps have been proposed to address this challenge [4]. Adiabatic ramps are very effective for $B_1$ field strengths greater than 500 Hz. However, at lower $B_1$ fields, the duration of the adiabatic sweep causes significant relaxation losses and limits the 2D $R_{1\rho}$ method to $B_1 > 500$ Hz; this, in turn, restricts the rate constants that can be measured to $\sim k_{ex} > 1000\ \text{s}^{-1}$. An elegant solution to this problem has been the development of a selective 1D-based $R_{1\rho}$ pulse sequence, which uses selective cross-polarization between a $^1$H–$^{13}$C or $^1$H–$^{15}$N scalar-coupled spin pair to measure $R_{1\rho}$ for spins one at a time [4, 24]. The advantage of this method is that multiple spins do not have to be simultaneously aligned, and alignment of the single nucleus of interest along its effective field is easily accomplished with hard pulses. Finally, the $R_{1\rho}$ experiment has many parallels with CEST/DEST, as all these experiments involve irradiating magnetization using variable $B_1$ fields at different offsets spanning the chemical-shift range of the target nucleus. The $R_{1\rho}$ values measured in an off-resonance $R_{1\rho}$ experiment are related to the CEST profile of the system through the relation:

$$\frac{I}{I_0} = \cos^2\theta e^{-R_{1\rho}T_{EX}}. \tag{15.44}$$

The factor of $\cos^2\theta$ accounts for the projection of the z-magnetization on the effective field in the CEST experiment, with one factor of $\cos\theta$ at the beginning (alignment) and one at the end (de-alignment) of $T_{\text{EX}}$ [25]. For example, the following approximation for $R_{\text{ex}}$:

$$R_{ex} = k_{ab}\frac{k_{ba}\left(\Delta\omega_{AB}^2 + \Delta R_{2AB}^2\right) + \Delta R_{2AB}\left(\omega_{1,eff}^2 + k_{BA}^2\right)}{k_{ba}\left(\omega_{2,eff}^2 + k_{BA}^2 + \Delta R_{2AB}^2\right) + \Delta R_{2AB}\omega_1^2}. \tag{15.45}$$

that accounts for $\Delta R_2$ and $\Delta\omega$ can be used to generate very good approximations to both CEST and DEST profiles showing that the $R_{1\rho}$, CEST, and DEST experiments are all related to one another.

## 15.5 Summary

In this chapter, we have discussed how chemical exchange can have conspicuous effects on NMR spectra and then showed how these effects can be predicted by the Bloch-McConnell equations. We also described various experiments that can be used to characterize chemical exchange and determine the mechanism of the underlying processes. Here, we restricted ourselves to a single spin ½ nucleus. When experiments are carried out in spin-systems containing more than one spin, care must be taken to supress artifacts that arise due to J couplings, cross-correlated relaxation, and cross relaxation. On the other hand, multi-spin systems also open up possibilities unavailable in an isolated spin ½ system. For example, one can generate multiple-quantum coherences and study their relaxation properties under exchange or J couplings between nuclei may be conformation dependent, opening up another avenue to characterize exchange. NMR methods are actively being developed to study exchange between multiple states and to extend existing experiments to more complicated spin systems. The dramatic impact of chemical exchange on NMR spectra has been leveraged for amplifying the spectral signatures of states that are present in low populations (<2%) and not readily detectable by other biophysical methods. The experiments described here have been applied to study conformational exchange processes occurring over a wide range of timescales and have not only provided insights into the reaction mechanisms of small molecules but have also helped in characterizing the conformational dynamics of proteins and nucleic acids that are often linked to function and disease.

## Acknowledgments

AS was funded by a DBT/Wellcome Trust India Alliance Fellowship (grant no.: IA/I/18/1/503614) and a DST/SERB Core Research Grant (no. CRG/2019/003457) as well as a start-up grant from IISc. PV was supported by intramural funds at TIFR Hyderabad from the Department of Atomic Energy (DAE), India, under Project Identification Number RTI 4007 and by funds from DBT via project BT/INF/22/SP42684/2021.

## References

1 Cavanagh, J. et al. (2006). *Protein NMR Spectroscopy, Principles and Practice*. 2e. Academic Press.

2 Palmer, A.G. (2004). *NMR characterization of the dynamics of biomacromolecules. Chem. Rev.* 104 (8): 3623–3640.

3 Palmer, A.G. and Massi, F. (2006). *Characterization of the dynamics of biomacromolecules using rotating-frame spin relaxation NMR spectroscopy*. *Chem. Rev.* 106 (5): 1700–1719.

4 Rangadurai, A. et al. (2019). Characterizing micro-to-millisecond chemical exchange in nucleic acids using off-resonance $R_{1\rho}$ relaxation dispersion. *Prog. Nucl. Magn. Reson. Spectrosc.* 112–113: 55–102.

5 Sekhar, A. and Kay, L.E. (2019). An NMR view of protein dynamics in health and disease. *Annu. Rev. Biophys.* 48: 297–319.

6 Tugarinov, V. and Clore, G.M. (2019). Exchange saturation transfer and associated NMR techniques for studies of protein interactions involving high-molecular-weight systems. *J. Biomol. NMR.* 73 (8–9): 461–469.

7 Anthis, N.J. and Clore, G.M. (2015). Visualizing transient dark states by NMR spectroscopy. *Q. Rev. Biophys.* 48 (1): 35–116.

8 Bain, A.D. (2003). Chemical exchange in NMR. *Prog. Nucl. Magn. Reson. Spectrosc.* 43 (3–4): 63–103.

9 McConnell, H.M. (1958). Reaction rates by nuclear magnetic resonance. *J. Chem. Phys.* 28 (3): 430–431.

10 Millet, O. et al. (2000). The static magnetic field dependence of chemical exchange linebroadening defines the NMR chemical shift time scale. *J. Am. Chem. Soc.* 122 (12): 867–2877.

11 Vallurupalli, P., Bouvignies, G. and Kay, L.E. (2011). Increasing the exchange time-scale that can be probed by CPMG relaxation dispersion NMR. *J. Phys. Chem. B.* 115 (49): 14891–900.

12 Allerhand, A. and Thiele, E. (1966). Analysis of carr purcell spin- echo NMR experiments on multiple- spin systems. II. the effect of chemical exchange. *J. Chem. Phys.* 45: 902–916.

13 Kovrigin, E.L. (2012). NMR line shapes and multi-state binding equilibria. *J. Biomol. NMR.* 53 (3): 257–270.

14 Rao, B.D. (1989). Nuclear magnetic resonance line-shape analysis and determination of exchange rates. *Methods Enzymol.* 176: 279–311.

15 Loria, J.P., Rance, M. and Palmer, A.G. (1999). A relaxation-compensated Carr-Purcell-Meiboom-Gill sequence for characterizing chemical exchange by NMR spectroscopy. *J. Am. Chem. Soc.* 121 (10): 2331–2332.

16 Mulder, F.A. et al. (2001). Measurement of slow (micros-ms) time scale dynamics in protein side chains by (15)N relaxation dispersion NMR spectroscopy: application to Asn and Gln residues in a cavity mutant of T4 lysozyme. *J. Am. Chem. Soc.* 123 (5): 967–975.

17 Tollinger, M. et al. (2001). Slow dynamics in folded and unfolded states of an SH3 domain. *J. Am. Chem. Soc.* 123 (46): 11341–11352.

**18** Skrynnikov, N.R., Dahlquist, F.W. and Kay, L.E. (2002). Reconstructing NMR spectra of "invisible" excited protein states using HSQC and HMQC experiments. *J. Am. Chem. Soc.* 124 (41): 12352–12360.

**19** Forsen, S. and Hoffman, R.A. (1963). Study of moderately rapid chemical exchange reactions by means of nuclear magnetic double resonance. *J. Chem. Phys.* 39 (11): 2892–2901.

**20** Fawzi, N.L. et al. (2011). Atomic-resolution dynamics on the surface of amyloid-beta protofibrils probed by solution NMR. *Nature* 480 (7376): 268–272.

**21** Zaiss, M. and Bachert, P. (2013). Chemical exchange saturation transfer (CEST) and MR Z-spectroscopy in vivo: a review of theoretical approaches and methods. *Phys. Med. Biol.* 58 (22): R221–R269.

**22** Rangadurai, A., Shi, H. and Al-Hashimi, H.M. (2020). Extending the sensitivity of CEST NMR spectroscopy to micro-to-millisecond dynamics in nucleic acids using high-power radio-frequency fields. *Angew. Chem. Int. Ed. Engl.* 59 (28): 11262–11266.

**23** Tiwari, V.P. et al. (2021). The A39G FF domain folds on a volcano-shaped free energy surface via separate pathways. *Proc. Natl. Acad. Sci. U.S.A.* 118 (46).

**24** Korzhnev, D.M., Orekhov, V.Y. and Kay, L.E. Off-resonance R(1rho) NMR studies of exchange dynamics in proteins with low spin-lock fields: an application to a Fyn SH3 domain. *J. Am. Chem. Soc.* 127 (2): 713–721.

**25** Palmer, 3rd A.G. and Koss, H. (2019). Chemical exchange. *Methods Enzymol.* 615: 177–236.

**26** Kovrigin, E.L. and Loria, J.P. (2006). Enzyme dynamics along the reaction coordinate: critical role of a conserved residue. *Biochemistry.* 45 (8): 2636–2647.

**27** Kim, J.H. et al. (2009). Structure and dynamics of the iron-sulfur cluster assembly scaffold protein IscU and its interaction with the cochaperone HscB. *Biochemis.* 48 (26): 6062–6071.

**28** Vallurupalli, P. et al. (2016). Atomistic picture of conformational exchange in a T4 lysozyme cavity mutant: an experiment-guided molecular dynamics study. *Chem. Sci.* 7 (6): 3602–3613.

**29** Gopalan, A.B. and Vallurupalli, P. (2018). Measuring the signs of the methyl $^1$H chemical shift diferences between major and 'invisible' minor protein conformational states using methyl $^1$H multiquantum spectroscopy. *J. Biomol. NMR.* 70 (3): 187–202.

**30** Vallurupalli, P., Bouvignies, G. and Kay, L.E. (2012). Studying "invisible" excited protein States in slow exchange with a major state conformation. *J. Am. Chem. Soc.* 134 (19): 8148–8161.

**31** Kimsey, I.J. et al. (2015). Visualizing transient Watson-Crick-like mispairs in DNA and RNA duplexes. *Nature* 519 (7543): 315–320.

# Appendix A

## Proton-Detected Heteronuclear and Multidimensional NMR

*Christian Griesinger, Harald Schwalbe, Jürgen Schleucher, and Michael Sattler*

This chapter describes proton-detected (also called inverse) heteronuclear two- and three-dimensional NMR spectroscopy. These techniques make it possible to answer biologically interesting questions and to derive solution structures of proteins, nucleic acids, glycoproteins, and other biomolecules with high precision.

The goal of the chapter is to give an overview of the bewildering variety of techniques used for heteronuclear correlation spectroscopy, compare the various basic building blocks, and show the advantages and disadvantages of those techniques. Three-dimensional techniques in particular can be used to overcome assignment problems and to measure homonuclear and heteronuclear $^{n}J$-coupling constants that could not be derived using conventional 2D NMR experiments.

This contribution is focused on methods rather than on applications, which are the topic of other chapters. It is split into five sections of varying length and complexity. Familiarity with the product operator formalism will be helpful, [1–3] although a short introduction will be given. Product operator notations of the state of the spin system at important points in the pulse sequence will be presented as well as a more general description of the transfer pathway in the course of the particular experiments.

## A.1 Introduction

### A.1.1 Sensitivity

The sensitivity of NMR is inherently low [3] because the energy differences between nuclear spin states are of the order of $10^{-5}$ of the thermal energy at room temperature. Therefore sensitivity enhancement is one of the major goals in NMR spectroscopy. The sensitivity (signal-to-noise ratio, S/N) of a one-dimensional experiment is proportional to

$$\mathrm{S/N} \sim N\gamma_{\mathrm{exc}}\, \gamma_{\mathrm{det}}^{3/2} B_0^{3/2}\, (\mathrm{NS})^{1/2} \frac{T_2}{T}$$

$N$ is the number of molecules in the active sample volume, $\gamma_{\mathrm{exc}}$ is the gyro- magnetic ratio of the excited spin, $\gamma_{\mathrm{det}}$ is the gyromagnetic ratio of the detected spin, $B_0$ is the static magnetic field, NS is the number of experiments, $T_2^{-1}$ is the homogeneous line width, and $T$ is the temperature. This formula immediately shows how to increase the S/N:

**Note:** Throughout this chapter we will use $I$ for proton and $S$ or $T$ for heteronuclear spins. Since in most of the examples $^{13}$C is the heteronuclear spin, we sometimes use $S$ and $^{13}$C synonymously.

*Two-Dimensional (2D) NMR Methods*, First Edition. Edited by K. Ivanov, P.K. Madhu and G. Rajalakshmi.

1. Increase the number of molecules $N$ by increasing the concentration of the sample at fixed volume, increasing the number of magnetically active molecules at fixed volume by labeling of low natural abundance spins or increasing the volume of the sample at fixed concentration.
2. Increase the static magnetic field $B_0$.
3. Increase NS. Simply increasing the number of experiments lengthens the measurement. Alternatively, decreasing the duration of one experiment, that is, decreasing the repetition time makes it possible to complete more experiments in the same total time (see Section A.3.2.1).
4. Increase $T_2$ by decreasing the viscosity of the sample.
5. Choose the gyromagnetic ratio of the excited and detected nucleus to be as high as possible.

For a given sample and magnetic field strength $B_0$, item (5) especially can be optimized by the selection of the most advantageous pulse scheme.

Table A.1a lists the relative sensitivity of some heteronuclear combinations. The relative measurement times to achieve identical S/N are given in Table A.1b. The gain in sensitivity is the more dramatic the larger the difference in gyromagnetic ratio between the proton and the heterospin. Experiments that start with proton magnetization and that detect proton magnetization are in principle the most sensitive.

### A.1.2 Resolution

Heteronuclear two-dimensional correlation experiments are asymmetric with respect to reflection at $\omega_1 = \omega_2$ because the nucleus that evolves during $t_1$ is not identical to the nucleus that is detected in $t_2$. It is necessary to decide which nucleus will take advantage of the high resolution that can be achieved in the detection period. During an evolution time, the resolution is limited by the number of experiments one can perform in a given measurement time. On the other hand, the digitization in $\omega_2$ is essentially limited only by disk space and has little influence on the measurement time, because the recycle time of the experiment does not need to be changed when $t_2^{max}$ is increased. Thus it would be desirable to detect, for example, $^{13}C$ in samples with natural abundance of $^{13}C$, because the resolution of $^{13}C$ in such samples under proton decoupling is very high due to the large $T_2$ and the lack of homonuclear C,C couplings. Evolution of proton magnetization during an evolution time would be adequate. The proton spectrum requires smaller digitization because the lines are broader due to smaller $T_2$ and the presence

**Table A.1a** Relative Sensitivity S/N.

| Heteronuclear Combination | X exc X det | H exc X det | X exc H det | H exc H det |
|---|---|---|---|---|
| H/P | 1/10 | 1/4 | 2/5 | 1 |
| H/C | 1/32 | 1/8 | 1/4 | 1 |
| H/N | 1/300 | 1/30 | 1/10 | 1 |

**Table A.1b** Relative Measurement Time to Achieve Identical S/N.

| Heteronuclear Combination | X exc X det | H exc X det | X exc H det | H exc H det |
|---|---|---|---|---|
| H/P | 100 | 16 | 6.25 | 1 |
| H/C | 1024 | 64 | 16 | 1 |
| H/N | 100000 | 1000 | 100 | 1 |

of homonuclear couplings and because the spectral width (in hertz) is smaller. This acquisition scheme, however, has dramatic sensitivity disadvantages.

Therefore in proton-detected methods one tries to increase the resolution during the evolution time for the well-resolved heteronuclear resonances. Folding (see Section A.3.1) in the indirectly sampled frequency domain ($\omega_1$, for 2D NMR), or the application of selective heteronuclear pulses can be used to reduce the spectral width for detection of the heteronucleus. Processing with non-FT methods like linear prediction (see Section A.3.2) can be used to increase the number of points in the $t_1$ dimension. The loss in resolution one has to face in the case of indirectly sampled heteronuclear spectra is most dramatic for natural abundance samples. In completely labeled samples, on the other hand, the homonuclear couplings between the heteronuclei introduce multiplet structure, so that the loss in resolution (see Section A.2.6) is not as dramatic as with the natural abundance samples. Constant time methods to remove homonuclear couplings in indirectly sampled heterospin spectra will be discussed in this chapter.

### A.1.3 Suppression of Unwanted Signals

Heteronuclear experiments with heteronuclear detection normally are not affected by solvent signals or by signals from protons not bound to the NMR active heteronucleus. However, in proton-detected pulse sequences, solvent suppression, especially of water, is a major issue. Conventional presaturation of the water signal can be employed. Making use of heteronuclear correlation, water protons can also be efficiently suppressed by selecting protons that are bound to carbon, nitrogen, and so on. Such a suppression scheme is based on connectivity rather than on the chemical shift of the solvent.

For natural abundance samples, suppression of signals of protons that arise from isotopomers that do not carry the heteronucleus at the specific site responsible for a cross peak is an additional problem. Suppression factors of the inverse of the natural abundance are necessary, namely 100 for $^{13}C$ and 300 for $^{15}N$. Three strategies exist, which will be discussed:

1. Cancellation of undesired signals by phase cycling
2. Topology-sensitive presaturation of undesired signals (BIRD sequences)
3. Suppression of undesired signals with inhomogeneous $B_1$ pulses (spinlock pulses) or $B_0$ pulses ($B_0$ gradient pulses)

Finally, it is possible to select or suppress protons that are bound to an NMR active heteronuclear spin. Such accessories to a pulse sequence are called high-pass or low-pass $J$ filters (described in Section A.3.5).

### A.1.4 Product Operator Formalism

We will use the so-called product operator formalism, which is the most appropriate tool for the description of complicated experiments for weakly coupled spin systems. Product operator formalism provides a handy description of the states of the spin system under conditions that can be described by a density matrix. Like any quantum mechanical state, the state of the density matrix has certain observables, such as magnetization, that have to be extracted by mathematical procedures.

We will not present here a mathematical introduction, starting from the Liouville von Neumann equation and introducing the observables as traces of the product of density matrix and the corresponding operator to the observable. This approach is nicely covered in many references. [1–3] Instead we present a rather phenomenological introduction based on analogy to the easy-to-grasp vector model (see ref. 4). In vector formalism, the magnetization of a spin is composed of the components $M_x$, $M_y$, and $M_z$. These magnetizations precess around an externally applied magnetic field tracing out the surface of a cone. This external field may either be the static magnetic field along the $z$ axis or a transverse field that is generated by an rf pulse and is static in the rotating frame. The axis of the cone on whose outer surface the magnetization precesses is given by the orientation of the external field. The precession frequency $\omega$ is given by the product of the field strength and the gyromagnetic ratio: $\omega = -\gamma B$.

The precession of magnetization in the rotating frame due to its chemical shift is understood as the precession about a static field along $z$ with strength $\Omega$. This is illustrated in Figure A.1a for the evolution of $x$ magnetization. The transformation properties for the other orientations of magnetizations are:

$$M_x \xrightarrow{\Omega t} M_x \cos\Omega t + M_y \sin\Omega t \quad M_y \xrightarrow{\Omega t} M_y \cos\Omega t - M_x \sin\Omega t \quad M_z \rightarrow M_z$$

The application of rf pulses can be understood as the application of a magnetic field that is static in the rotating frame and lies in the transverse plane. A pulse about the $x$ axis originates from a field $B_1$ along the $x$ axis in the rotating frame. Any magnetization will then precess about this axis with the frequency $\omega_1 = -\gamma B_1$. Figure A.1b gives a graphic representation of this precession for the evolution of $z$ magnetization. $z$ magnetization is present, for example, at the beginning of any pulse sequence.

$$M_y \xrightarrow{\omega_1 t} M_y \cos\omega_1 t + M_z \sin\omega_1 t = M_y \cos\beta + M_z \sin\beta$$

$$M_z \rightarrow M_z \cos\omega_1 t - M_y \sin\omega_1 t = M_z \cos\beta - M_y \sin\beta$$

$$M_x \rightarrow M_x$$

The flip angle $\beta$ is given by the product of rotation frequency $\omega_1$ and the duration $t$ of the pulse. A special duration is $t = \pi/(2\omega_1)$, which defines a 90° pulse. A $90^\circ_y$ pulse followed by evolution of chemical shift will lead to the following transformations:

$$M_z \xrightarrow{90_y} M_x \xrightarrow{\Omega t} M_x \cos\Omega t + M_y \sin\Omega t$$

A phase-sensitive detector records $M_x$ and $M_y$ as a function of time and stores them in different memory locations A and B, respectively. A complex signal is reconstructed from A and B, of the form:

$$\exp(i\Omega t) = \cos\Omega t + i \sin\Omega t$$

by adding the contents of A and $i$ times the contents of B. Fourier transformation of this complex FID yields a complex Lorentzian line provided the FID decays with a single time constant $T_2$.

The transition to product operator formalism is achieved by the following correspondence principle. The magnetization **M** generated by the ensemble of microscopic spins is replaced by operators **I** that are indexed by the Cartesian coordinates. $I_x$ then represents a state of an ensemble of spins $\boldsymbol{I}$ that carries $x$ magnetization. The transformations and properties indicated previously are exactly the same as for the magnetizations.

Thus precession of a state of the ensemble that carries $I$-spin magnetization about a static magnetic field along $z$ with frequency $\Omega$ in the rotating frame gives the following transformations:

$$I_x \rightarrow I_x \cos\Omega t + I_y \sin\Omega t \quad I_y \rightarrow I_y \cos\Omega t - I_x \sin\Omega t \quad I_z \rightarrow I_z$$

A transverse $B_1$ field, for example, from the $x$ direction, will lead to a precession about the $x$ axis with the frequency $\omega_1 = -\gamma B_1$:

$$I_y \rightarrow I_y \cos\omega_1 t + I_z \sin\omega_1 t \quad I_z \rightarrow I_z \cos\omega_1 t - I_y \sin\omega_1 t \quad I_x \rightarrow I_x$$

Thus a $90^\circ_y$ pulse applied on the equilibrium state of the spins followed by chemical shift precession will lead to the following transformations:

$$I_z \xrightarrow{90_y} I_x \xrightarrow{\Omega t} I_x \cos\Omega t + I_y \sin\Omega t$$

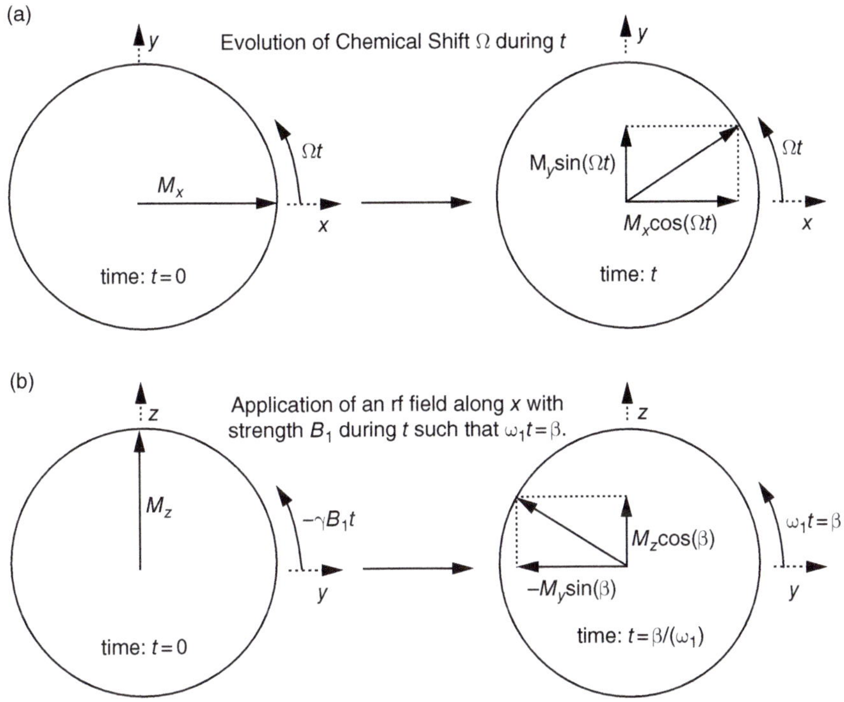


**Figure A.1** Representation of the evolution of chemical shift in a vector diagram, (a) $x$ magnetization evolves under the influence of chemical shift by rotation with frequency $\Omega$ in the rotating frame. After a time $t$, the phase of the magnetization is given by $\Omega_t$. The magnetization is composed of the two orthogonal components $M_x$ and $M_y$. (b) Rotation of a magnetization vector under the action of a pulse. The pulse induces a rotation of the magnetization about its direction with the frequency $\omega_1$. The flip angle $\beta$ is given by the duration of the pulse $t$ and the strength of the field: $\beta = -\gamma B_1 t$.

#### A.1.4.1 Coupling

The advantage of product operator formalism is that it allows the description of coupled spin systems, where states that do not carry observable magnetization play the central role. Although the transformation of these nonobservable states under pulses cannot be described by vector formalism, we will introduce them relying as long as possible on vector formalism. Suppose we have a spin that is coupled to another spin with a coupling constant $J = 10$ Hz and therefore appears in the spectrum as a doublet. When transverse magnetization of this spin is excited it disappears at odd multiples of 50 ms and then reappears again. In the vector model this is rationalized by two magnetization vectors that rotate in opposite directions, each with a frequency of 5 Hz (see Figure A.2). The two vectors are oriented antiparallel (antiphase) after one-fourth of a full revolution, that is, after 50 ms. The state of the spin system at 50 ms is devoid of any macroscopic magnetization, yet it is very different from the state that is reached after a long irradiation of the spin of interest, which destroys all magnetization. States without observable magnetization turn out to be the crucial states of a spin system that make possible most of the heteronuclear transfer experiments we are about to discuss in this chapter. We come back to our two-spin system.

A spin $I_1$ that is coupled to another spin $I_2$ (spin-1/2) appears as a doublet in the spectrum. The doublet structure of the signal of $I_1$ arises from two different types of molecules, namely one type of molecule in which $I_2$ is in the $\alpha$ state and a second type of molecule in which $I_2$ is in the $\beta$ state. Suppose the chemical shift of the spin $I_1$ is $\Omega_1$. Then the frequency of the $I_1$ line in the $I_{2\alpha}$ spin isomers is $\Omega_l + \pi J$, and the frequency of the $I_1$ line in the $I_{2\beta}$ spin isomers is $\Omega_1 - \pi J$. The spectrum of the $I_1$ spin is the superposition (sum) of the spectra originating from the two

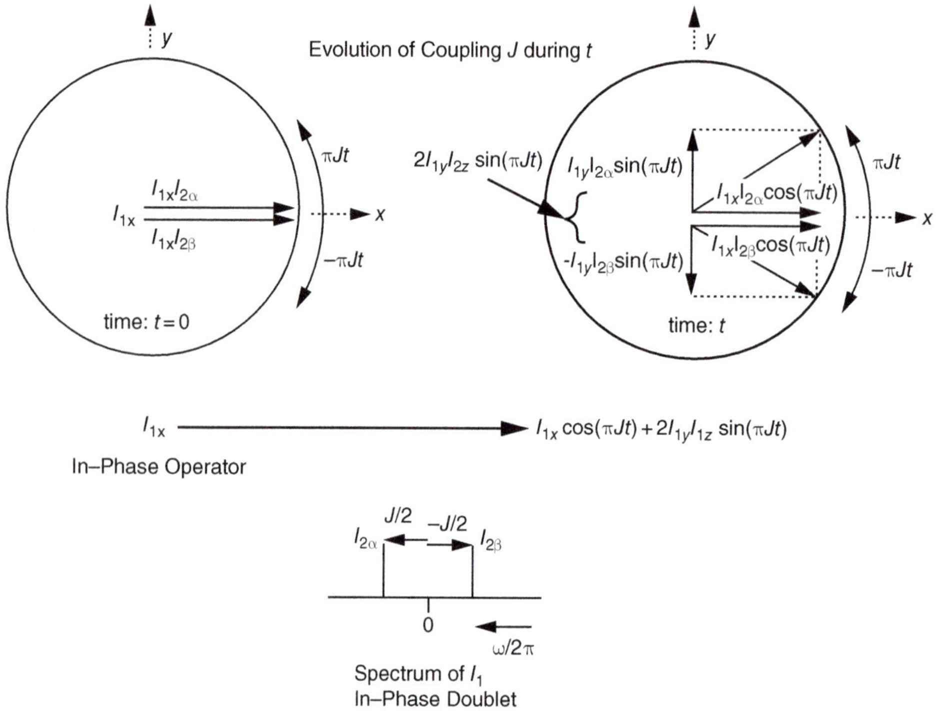


**Figure A.2** Vector diagram representation of the evolution of coupling of spin $I_1$ to spin $I_2$ in the rotating frame. The chemical shift is assumed to be zero. The two lines in the spectrum, one for the spin isomer with $I_2$ in the $\alpha$ state and one for $I_2$ in the $\beta$ state, correspond to two vectors that rotate in the rotating frame with identical frequency $\pi J$ but in opposite directions. At time zero the two vector components are parallel, resulting in full observable magnetization. At time $t$ they have acquired a phase shift of $\pi Jt$ and $-\pi Jt$, respectively. The two parallel vector components $I_{1x}I_{2\alpha}$ and $I_{1x}I_{2\beta}$ are combined to yield the magnetization $I_{1x}$. The two vectors that lie antiparallel $I_{1y}I_{2\alpha}$ and $I_{1y}I_{2\beta}$ are combined to $2I_{1y}I_{2z}$. This product operator is modulated with $\sin(\pi Jt)$.

spin isomers (Figure A.2). Therefore we can write down the transformations in the following way:

$$I_{1x}I_{2\alpha} \rightarrow I_{1x}I_{2\alpha}\cos(\Omega_1 + \pi J)t + I_{1y}I_{2\alpha}\sin(\Omega_1 + \pi J)t$$
$$I_{1x}I_{2\beta} \rightarrow I_{1x}I_{2\beta}\cos(\Omega_1 - \pi J)t + I_{1y}I_{2\beta}\sin(\Omega_1 - \pi J)t$$

We need to introduce two rules:

1. $I_\alpha + I_\beta = I_z$

This equation can be rationalized by considering $I_\alpha$ and $I_\beta$ as the probability of finding a spin in these states or the population of these states. Since there are just these two states for a spin-1/2, the sum over these probabilities is 1.

2. $I_\alpha - I_\beta = 2I_z$

This can be rationalized by the observation that a population difference between $\alpha$ state and $\beta$ state carries $z$ magnetization. The factor of 2 is just a normalization constant. This rule actually describes a transformation from a single element basis set for the description of longitudinal magnetization ($I_\alpha$, $I_\beta$) to a Cartesian basis set ($2I_z$, identity operator 1) [2].

With these two rules for the calculation and some trigonometric transformations we arrive at:

$$I_{1x} \rightarrow I_{1x} \cos \Omega_1 t \cos \pi J t + I_{1y} \sin \Omega_1 t \cos \pi J t + 2I_{1y} I_{2z} \cos \Omega_1 t \sin \pi J t$$
$$- 2I_{1x} I_{2z} \sin \Omega_1 t \sin \pi J t$$

The transformation properties of $I_y$ can be obtained by cyclic permutation: $x \rightarrow y \rightarrow -x \rightarrow -y$. The state $2I_{1y}I_{2z}$ is a product of two operators. It does not carry magnetization; in fact it is a general rule that only states of the spin system that can be represented by a single operator carry magnetization. The $2I_{1y}I_{2z}$ operator is called an antiphase operator (Figure A.3). This becomes plausible if one considers this operator in detail:

$$2I_{1y}I_{2z} = I_{1y}I_{2\alpha} - I_{1y}I_{2\beta}$$

Obviously, $2I_{1y}I_{2z}$ is the difference spectrum of the spin isomers with $I_2$ in the $\alpha$ and in the $\beta$ state. This spectrum has two lines at $\Omega_1 + \pi J$ and at $\Omega_1 - \pi J$, respectively; the two lines have different sign. In the absence of pulses,

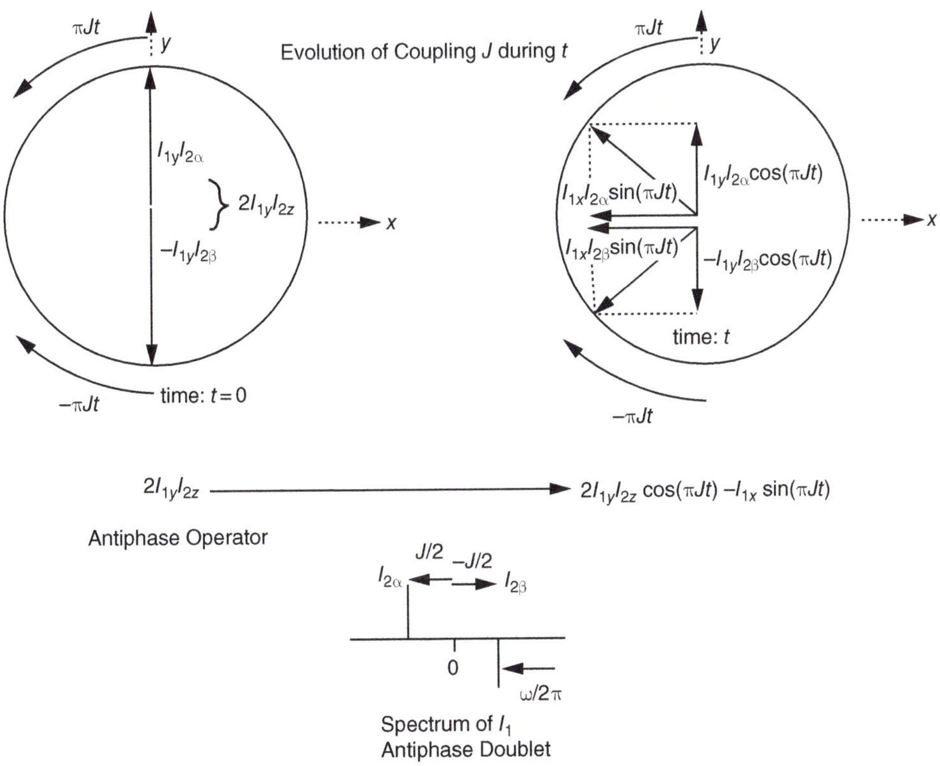


**Figure A.3** Evolution of the coupling during a time $t$ starting from antiphase coherence of spin $I_1$. This state of the spin system contains no observable magnetization at time $t = 0$. Evolution of coupling leads to a refocusing of the two antiphase components and detectable magnetization again appears. Since the state of the spin system in the vector diagram at time $t = 0$ is indistinguishable from two magnetization vectors originating from two spins with a difference in chemical shift of $2\pi J$ that lie in opposite directions, a spectrum results in which the two lines have opposite sign (one line in emission, the other in absorption).

such an antiphase operator evolves as follows:

$$2I_{1y}I_{2z} \rightarrow 2I_{1y}I_{2z}\cos\Omega_1 t\cos\pi Jt - 2I_{1x}I_{2z}\sin\Omega_1 t\cos\pi Jt$$
$$- I_{1x}\cos\Omega_1 t\sin\pi Jt - I_{1y}\sin\Omega_1 t\sin\pi Jt$$

The transformation properties of product operators under pulses are as follows:

- A pulse is applied to all operators individually.
- Chemical shift evolution and evolution of coupling need not be applied simultaneously in periods of free evolution that may be interrupted by 180° pulses. Often one achieves a simpler description by consecutive application of the interactions to the spin state (*vide infra*).

To conclude this rather cursory introduction to product operators we discuss the polarization transfer process that is essential for the majority of experiments in high-resolution NMR that use $J$-coupling for transfer of magnetization from one spin to another spin. Using product operator formalism (and assuming for the moment that spin $I_1$ is on-resonance), polarization transfer can be described very simply. An in-phase operator $I_{1x}$ evolves during a delay $\Delta$ completely into an antiphase operator $2I_{1y}I_{2z}$, provided $\sin\pi J\Delta = 1$, that is, $\Delta = 1/(2J)$. Application of a $90_x$ pulse to $I_1$ and $I_2$ transforms this operator into $-2I_{1z}I_{2y}$, which after another delay $\Delta = l/(2J)$ forms $I_{2x}$; this state again carries observable magnetization. Thus transverse magnetization of spin $I_1$ is transferred to transverse magnetization of spin $I_2$.

### A.1.5 Amplitude and Phase Modulation

We want to introduce some additional notions that are frequently used throughout this text:

Consider a spin $I_1$ with its characteristic chemical shift $\Omega_1$. Assume that this spin has an interaction with another spin $I_2$, with its characteristic chemical shift $\Omega_2$, such that we can transfer magnetization of spin $I_1$ to $I_2$. Now let us consider a pulse sequence that first excites spin $I_1$ and then allows for evolution of chemical shift during $t_1$:

$$I_{1x} \rightarrow I_{1x}\cos(\Omega_1 t_1) + I_{1y}\sin(\Omega_1 t_1)$$

Now we apply the specific mixing scheme that transfers magnetization from $I_1$ to $I_2$ in a 2D correlation experiment. There are two possibilities. Either magnetization is transferred from both magnetization components of $I_1$ to $I_2$ or only from one. We first consider transfer from both components:

$$I_{1x} \rightarrow I_{2x} \qquad \text{and} \qquad I_{1y} \rightarrow I_{2y}$$

In this case the following 2D FID is obtained after evolution of chemical shift during $t_2$:

$$I_{1x} \xrightarrow{t_1} I_{1x}\cos(\Omega_1 t_1) + I_{1y}\sin(\Omega_1 t_1) \xrightarrow{\text{magnetization transfer}}$$
$$I_{2x}\cos(\Omega_1 t_1) + I_{2y}\sin(\Omega_1 t_1) \xrightarrow{t_2}$$
$$I_{2x}\left[\cos(\Omega_1 t_1)\cos(\Omega_2 t_2) - \sin(\Omega_1 t_1)\sin(\Omega_2 t_2)\right] +$$
$$I_{2v}\left[\sin(\Omega_1 t_1)\cos(\Omega_2 t_2) + \cos(\Omega_1 t_1)\sin(\Omega_2 t_2)\right]$$

The 2D FID that the phase-sensitive receiver records by complex addition of $x$ magnetization plus $i$ times $y$ magnetization is therefore given by: $\exp(i\Omega_1 t_1)\exp(i\Omega_2 t_2)$. Such a 2D FID is said to be phase modulated in $t_1$, because $t_1$ enters the FID in $t_2$ solely by a phase factor. It yields after complex Fourier transformation so-called mixed phases (a mixture of absorptive and dispersive signals) in the 2D spectrum. The phase-modulated FID cannot be Fourier transformed such that pure phases result in $\omega_1$ and $\omega_2$ and at the same time the sign of the chemical shift is recognized. This is undesirable because purely absorptive peaks have minimum linewidth. It is obvious that the phase-modulated signal allows the differentiation of the sign of frequencies in $t_1$. This is always the case if the phase modulation is due to the chemical shift. The FID $\exp(i\Omega_1 t_1)\exp(i\Omega_2 t_2)$ gives rise to the antiecho part of a spectrum (see Chapter 2 in this volume). Antiecho pathways are characterized by the fact that the sign of the chemical shift is the same for $t_1$ $(+\Omega_1)$ and $t_2$ $(+\Omega_2)$. The echo part of the signal is characterized by the fact that the sign of the chemical shift is opposite in $t_1$ $(-\Omega_1)$ and in $t_2$ $(+\Omega_2)$: $\exp(-i\Omega_1 t_1)$ $\exp(i\Omega_2 t_2)$. The echo part of the spectrum could be obtained in the foregoing sequence by introducing a $180^\circ_x$ pulse before the transfer of magnetization from $I_1$ to $I_2$. The echo part of the signal derives its name from the fact that $B_0$ inhomogeneity behaves like chemical shift and is refocused after $\gamma(I_1)t_1 = \gamma(I_2)t_2$, where $\gamma(I_n)$ is the gyromagnetic ratio of $I_n$. Refocusing of $B_0$ inhomogeneities gives rise to the formation of an echo. There is a product operator basis set that allows one to describe echo and antiecho signals in a simple way. Using the equations for the evolution of chemical shift for $I_x$ and $I_y$ it is easily found that the two single-element operators: $I^+ = I_x + iI_y$ and $I^- = I_x - iI_y$ evolve chemical shift in the following way:

$$I^+ \rightarrow I^+ \exp(-i\Omega t) \qquad \text{and} \qquad I^- \rightarrow I^- \exp(i\Omega t)$$

The antiecho transfer therefore comes about from $I_1^- \rightarrow I_2^-$, whereas the echo transfer originates from $I_1^+ \rightarrow I_2^-$.

If magnetization is transferred only from either of the two components: $I_{lx}$ to $I_{2x}$ or $I_{ly}$ to $I_{2y}$, the signal that is obtained in the hypothetical experiment just described is given by

$$I_{1x} \xrightarrow{t_1} I_{1x}\cos(\Omega_1 t_1) + I_{1y}\sin(\Omega_1 t_1) \xrightarrow{\text{magnetization transfer}}$$

$$I_{2x}\cos(\Omega_1 t_1) \xrightarrow{t_2} I_{2x}\cos(\Omega_1 t_1)\cos(\Omega_2 t_2) + I_{2y}\cos(\Omega_1 t_1)\sin(\Omega_2 t_2)$$

The signal the phase-sensitive receiver records is now given by $\cos(\Omega_1 z_1)$ $\exp(i\Omega_2 z_2)$. The amplitude of the FID recorded in $t_2$ is modulated by $\cos(\Omega_1 z_1)$. The signal is amplitude modulated in $t_1$. Sign discrimination in one experiment is no longer possible since cosine is an even function. Therefore a second experiment is recorded that modulates the $t_2$ FID with $\sin(\Omega_1 t_1)$. This can easily be done by starting with $I_{1y}$ magnetization instead of $I_{1x}$ at the beginning of $t_1$, hence by a phase shift by 90° of the pulses that precede the evolution time. The Fourier transformation of these two amplitude-modulated signals yields pure phases in the spectrum (cf. Section A.3.1).

The reader should note that the two amplitude-modulated parts of the spectrum:

$$\cos(\Omega_1 t_1)\exp(i\Omega t_2) \qquad \text{and} \qquad \sin(\Omega_1 t_1)\exp(i\Omega_2 t_2)$$

can be obtained from the echo and the antiecho by linear combination and vice versa:

$$\cos(\Omega t_1)\exp(i\Omega t_2) = (1/2)\left[\exp(i\Omega_1 t_1)\exp(i\Omega_2 t_2) + \exp(-i\Omega_1 t_1)\exp(i\Omega_2 t_2)\right]$$

and

$$\sin(\Omega_1 t_1)\exp(i\Omega_2 t_2) = (1/2i)\left[\exp(i\Omega_1 t_1)\exp(i\Omega_2 t_2) - \exp(-i\Omega_1 t_1)\exp(i\Omega_2 t_2)\right]$$

We will in the following discuss sequences that yield pure absorption spectra after Fourier transformation by recording separately the echo and the antiecho part (see Sections A.2.4 and A.3.3).

### A.1.6 Phase Cycles

Phase cycles have been discussed in chapter 5 and we recall just a few rules that will be helpful for this chapter. In this chapter a pulse sequence is considered to consist of pulses (thin bars: 90°, thick bars: 180°; pulses deviating from these two values are designated $\theta$ in the context of editing and elsewhere with $\beta$) with fixed phases (these fixed phases are given with the pulse sequence and should not be changed in most of the cases), delays (we use only $t_n$, $\Delta$, and $\tau$), and pulses with phases that change in the course of a phase cycle. Pulses with variable phase are normally indicated by Greek symbols that come later in the Greek alphabet than or include $\phi$. Phases that have to be incremented in the course of the experiment to allow for sign discrimination of the frequencies in $t_1$ are designated $\phi$.

Phase cycles cancel out certain coherence transfers that are undesired and should therefore be suppressed. Throughout the text we use axial peak suppression by phase cycling a 90° pulse between the two phase values 0° and 180°. This can be understood in the following way. Excitation of transverse magnetization from longitudinal magnetization before an evolution time may be incomplete, so that in addition to the desired pathway there is also an undesired pathway:

$$\begin{array}{lcccll} & \text{excitation } (90^\circ_y) & & \text{evolution} & & \\ I_{1z} & \rightarrow & I_{1x} & \rightarrow & I_{1x}\cos(\Omega_1 t_1) + I_{1y}\sin(\Omega_1 t_1) & \text{desired pathway} \\ I_{1z} & & I_{1z} & & I_{12}\cos(0t_1) & \text{axial peak, undesired} \end{array}$$

The undesired pathway leads to axial peaks, since a peak with frequency 0 appears in the spectrum after Fourier transformation on the $\omega_1 = 0$ axis. The axial peak can be suppressed by recording a second experiment in which the phase of the 90° excitation pulse is changed by 180° to $(-y)$ and the phase of the receiver is changed by 180° simultaneously:

$$\begin{array}{lcccll} & \text{excitation } (90^\circ_{-y}) & & \text{evolution} & & \\ I_{1z} & \rightarrow & -I_{1x} & \rightarrow & -I_{1x}\cos(\Omega_1 t_1) - I_{1y}\sin(\Omega_1 t_1) & \text{desired pathway} \\ I_{1z} & & I_{1z} & & I_{12}\cos(0t_1) & \text{axial peak, undesired} \end{array}$$

The signal from the desired pathway is inverted whereas the undesired axial peaks do not change their phase in response to the phase of the excitation pulse. Adding the two signals with receiver phase 0° for the first and receiver phase 180° for the second experiment suppresses the axial peak. Two experiments have to be recorded to achieve the suppression. We will see in the following that the undesired signals of, for example, non-$^{13}$C-bound protons in natural abundance samples contribute exclusively to axial peaks.

### A.1.7 Multiple Quantum Filters

Multiple-quantum filters select multiple-quantum coherences. Suffice to say here that a product operator that contains $n$ transverse operators (e.g., $I_{lx}I_{2y}I_{3x}$), contains among others $n$-quantum (in the example given triple-quantum) coherence. In a single element basis set as introduced in the preceding, triple-quantum coherence is described by the operators $I_1^+I_2^+I_3^+$ and $I_1^-I_2^-I_3^-$. The observation that triple-quantum coherence, for example, can only be formed by at least three spins makes these filters a handy tool for selecting, for example, spin systems that contain at least three spins. The dependence of the phase of such coherences on a phase change $\phi$ of a pulse is obtained by summation over the "pluses" and the "minuses" in the product operator. Thus for a phase shift by 60° the coherences $I_1^+I_2^+I_3^+$ and $I_1^-I_2^-I_3^-$ change their phase by 180°. This means that they invert their sign. Then the implementation of a triple-quantum filter is an easy task. All pulses preceding the formation of the triple quanta are incremented in steps of 60°. The sign change of the desired coherences is compensated by simultaneous phase shift of the receiver in steps of 180°. Summing the phase-shifted experiments cancels undesired coherences but retains triple-quantum coherences.

## A.2 Basic Experiments

Correlation experiments between heteronuclei make use of the coupling constants between these nuclei; values for some heteronuclear coupling constants are summarized in Table A.2. Heteronuclear $^1J$ couplings are at least an order of magnitude larger than the homonuclear $^2J_{HH}$ and $^3J_{HH}$ couplings. The $^2J-$ and $^3J$-couplings tend to be smaller and of the order of proton–proton coupling constants. Since the time required to transfer magnetization from one nucleus to another is of the order of $J^{-1}$, the use of large couplings is advantageous for molecules with short $T_2$ times. This concept is important in the studies of completely $^{13}$C-labeled proteins.

Consider a correlation experiment between two protons separated by three bonds (Figure A.4). In a homonuclear experiment the time required to transfer in-phase magnetization between two protons in a H-C-C-H moiety, with the protons sharing a rather large $^3J$-coupling of 10 Hz, is $1/J = 100$ ms. The double heteronuclear relayed transfer, however, H→$^{13}$C→$^{13}$C→H requires only 36 ms for $^1J_{HC} = 140$ Hz and $^lJ_{CC} = 35$ Hz (Figure A.4). In addition, the $^3J_{HH}$ coupling is strongly conformation dependent whereas the $^1J_{CC}$- and $^1J_{CH}$-couplings are not. Furthermore, every relay step involving a carbon atom can be directly transformed to an evolution period, leading to multidimensional NMR experiments that spread out overlapping proton resonances of large molecules.

Heteronuclear sequences are constructed almost exclusively from 90° and 180° pulses and delays (Heteronuclear TOCSY experiments are not covered in this Section). The essential building blocks are:

1. *Simultaneous application of 90° pulses to both nuclei.* Assume that we have transverse magnetization of a spin at a given time during the course of a pulse sequence. After a suitable delay $\Delta$, there will be antiphase magnetization of spin $I$ present, of the form $2I_yS_z$ (Figure A.5a) as well as $2I_xS_z$ (not shown), allowing for evolution of chemical shift and heteronuclear coupling. Two $90^\circ_x$ pulses perform coherence transfer to $-2I_zS_y$, which represents antiphase magnetization of spin S. This transfer was already described for COSY [5, 6] in Chapter 2 and applies to the heteronuclear analog INEPT [7–9] (Figure A.5a,b). Thus transverse magnetization of spin $I(-I_y$ at the beginning of $\Delta$) can be transferred to transverse antiphase coherence of spin $S$:$-2I_zS_y$ (a) or vice versa (b). This method of transfer forms the basis for so-called transfer experiments via scalar coupling in the majority of pulse sequences. Chemical shift of spin $I$ evolves before the transfer and chemical shift of spin $S$ evolves after the transfer. Longitudinal two-spin order (represented by the product operator $2I_zS_z$) is obtained between the two 90° pulses when they are applied consecutively rather than simultaneously. The longitudinal two-spin order $2I_zS_z$ is not detectable, but it is oriented along $z$ and therefore invariant under rotations about the $z$ axis. Therefore $B_0$ gradients (Section A.3.3) can be inserted in coherence transfer segments between the 90° ($I$) pulse and the 90°($S$) pulse without affecting the coherence transfer from $I$ coherence to $S$ coherence.

**Table A.2** Ranges of Heteronuclear Coupling Constants[a].

| Nuclei | H,C | H,N | H,P | C,C | C,N | C,P |
|---|---|---|---|---|---|---|
| $^1J$ [Hz] | 125–280 | | 110–1200 | −20–200 | 0–(−77) | −53–476 |
| | (125–180) | (86–95) | (700) | (35–55) | (11–15) | (18) |
| $^2J$ [Hz] | −20–66 | 0–10 | 0–200 | 0–\|20\| | 0–\|14\| | −15–86 |
| | (0–\|7\|) | (0-5) | | (0–5) | (0–8) | |
| $^3J$ [Hz] | 0–16 | 0–10 | 0–50 | 0–16 | 0–(−7) | 0–47 |
| | (0–8) | (0–5) | (0–20) | (0–5) | (0–(−7)) | |

[a] Typical values for biomolecules appear in parentheses.

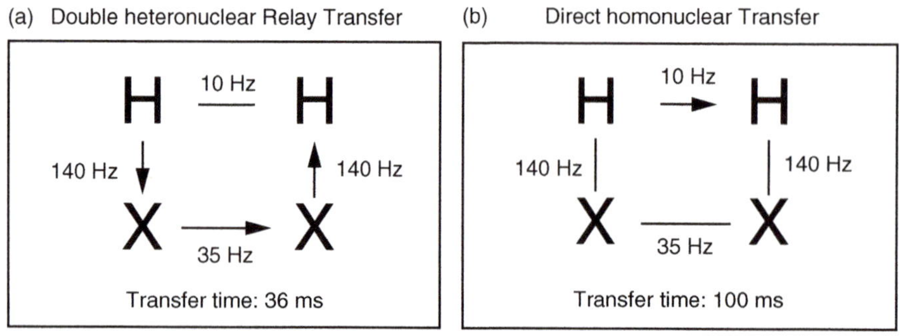


**Figure A.4** HXXH spin system: (a) Double relayed transfer from H to H via the HX, XX, and HX coupling. Since each transfer takes about $1/J$, for X $= ^{13}$C the relay transfer: H→C→C→H takes $[2(140\,\text{Hz})^{-1} + (35\,\text{Hz})^{-1}] = 42$ ms. The transfer time can be reduced to $[(140\,\text{Hz})^{-1} + (35\,\text{Hz})^{-1}] = 36$ ms because H↔X and X→X transfer can be performed simultaneously, in part. (b) Transfer via the homonuclear coupling of 10 Hz takes $(10\,\text{Hz})^{-1} = 100$ ms.

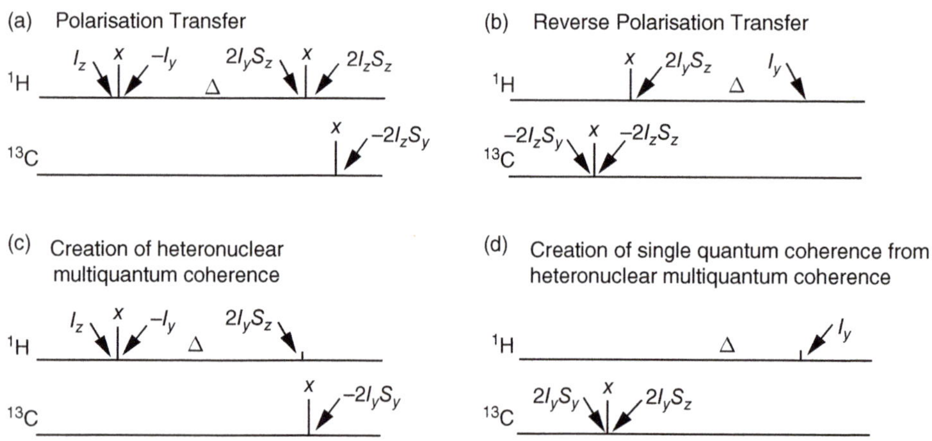


**Figure A.5** Building blocks in heteronuclear spectroscopy: Evolution of chemical shift and heteronuclear coupling in an $I, S$ spin system is considered. (a) Coherence transfer from antiphase $I$ ($2I_yS_z$) to antiphase $S$ coherence ($2I_zS_y$) is effected by two 90° pulses on $I$ and $S$. Sequential application of the pulses creates intermediate longitudinal two-spin order $2I_zS_z$. (b) Reverse coherence transfer from antiphase $S$ coherence ($2I_zS_y$) to antiphase $I$ coherence ($2I_yS_z$) which is refocused to in-phase $I$ coherence ($I_y$) at the end of $\Delta$. (c) Creation of heteronuclear multiquantum coherence ($2I_yS_y$) by application of an $S$ pulse to antiphase Z magnetization, (d) Reverse of (c).

2. *Selective application of a 90°(S) pulse.* This turns antiphase magnetization of spin $I$: $2I_yS_z$ into two-spin coherence that is transverse on both spins: $2I_yS_y$ (Figure A.5c, d). Two spin coherence evolves according to the chemical shift of both spins $I$ and $S$. The coupling $J_{IS}$ between the two spins is, however, switched off. Subsequent insertion of 180° pulses can refocus chemical shift evolution of either $I$ or $S$ (not shown).

3. *Application of 180° pulses in the middle of a delay to refocus chemical shift and/or heteronuclear couplings.* Four different situations must be distinguished (Figure A.6a–d): A 180° pulse applied to a spin $I$ in the middle of a delay $2\Delta$ (Figure A.6b) refocuses its chemical shift at the end of the delay $2\Delta$. The heteronuclear coupling between $I$ and $S$ is refocused after $2\Delta$ if either a 180°($I$) or a 180°($S$) pulse is applied in the middle of the delay (Figure A.6b, c).

Another way to think about the action of 180° pulses is the following [10]: Each 180° pulse inverts the evolution of chemical shift of the nucleus to which it is applied (jumping from the $\Omega$ line to the $-\Omega$ line or the reverse,

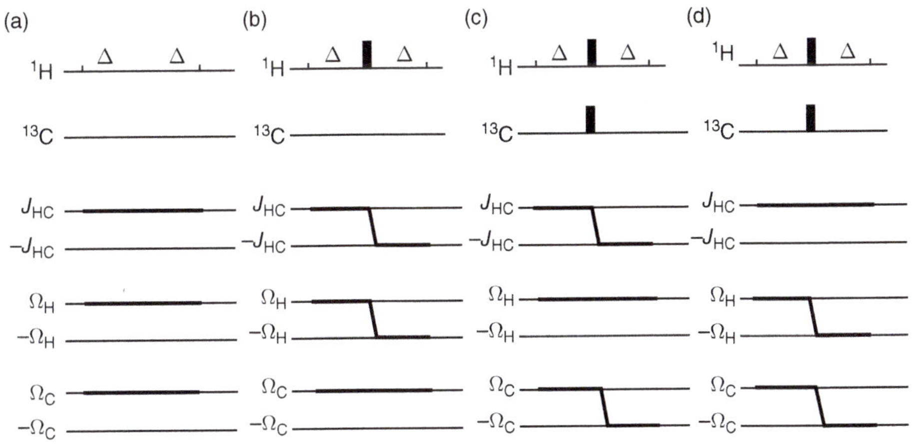


**Figure A.6** 180° pulses in the middle of a delay $2\Delta$. The evolution of coupling, chemical shift of H, and chemical shift of C is graphically represented in the manner often used for coherence orders, (a) With no 180° pulse all three interactions evolve, (b) Application of a 180°($I$) pulse refocuses chemical shift of $I$ and heteronuclear $J_{IS}$-coupling, (c) Application of a 180°($S$) pulse refocuses heteronuclear $J_{IS}$ and chemical shift of $S$. (d) Application of 180°($I$, $S$) pulses refocuses chemical shift of $I$ and $S$ but not $J_{IS}$-coupling.

Figure A.6). Each 180°($I$) or 180°($S$) pulse also inverts evolution of all heteronuclear couplings of the spin to which the pulse is applied (jumping from the $J_{HC}$ line to the $-J_{HC}$ line or the reverse). The evolution of heteronuclear coupling and chemical shifts in arbitrary sequences of 180° pulses can thus be visualized, as shown in Figure A.6.

As an example, the more complicated sequence in Figure A.7a behaves as follows:

- Evolution of heteronuclear $J_{HC}$-coupling (Figure A.7b) during $\Delta_1 - \Delta_2 + \Delta_3 - \Delta_4$
- Evolution of the chemical shift of $^1$H (Figure A.7c) during $\Delta_1 - \Delta_2 - \tau(180°^{13}C) - \Delta_3 + \Delta_4$
- Evolution of the chemical shift of $^{13}$C (Figure A.7d) during $\Delta_1 + \Delta_2 - \Delta_3 - \Delta_4$

With this graphical representation of the evolution of interactions, sequences that achieve a certain desired behavior with respect to evolution of all three possible interactions can easily be designed. If, for example, the coherence at the beginning of $\Delta_1$ contains only longitudinal proton operators, proton chemical shift evolution need not be taken into account. The two 180°(H) pulses can then be concatenated to one 180°(H) pulse at $\Delta_1 + 2\Delta_2 + \Delta_4$ provided $\Delta_3 > \Delta_2$ (Figure A.7e). Net evolution of $J_{HC}$ and $\Omega_C$ is the same as for the original sequence. If on the other hand the coherence at the beginning of $\Delta_1$ contains only longitudinal carbon operators, carbon chemical shift evolution need not be taken into account. The two 180°(H) pulses can then be concatenated at the position $\Delta_1 + \Delta_4$ and the 180°($^{13}$C) pulse is located at $\Delta_1 - \Delta_2$ (Figure A.7f).

As a rule of thumb, 180° pulses can be concatenated if the number of 180° pulses is larger than the number of interactions one has to consider.

We now discuss some basic sequences.

### A.2.1 HMQC (Heteronuclear Multiple-Quantum Correlation), HMBC (Heteronuclear Multiple-Bond Correlation)

HMQC [11–14] is one of the oldest sequences employed (Figure A.8a). Disregarding for the moment the decoupling with the GARP sequence during $t_2$ it is also a very simple sequence, consisting of only four pulses. Transverse proton magnetization is excited by the first pulse and is present during the whole sequence. Without the heteronuclear

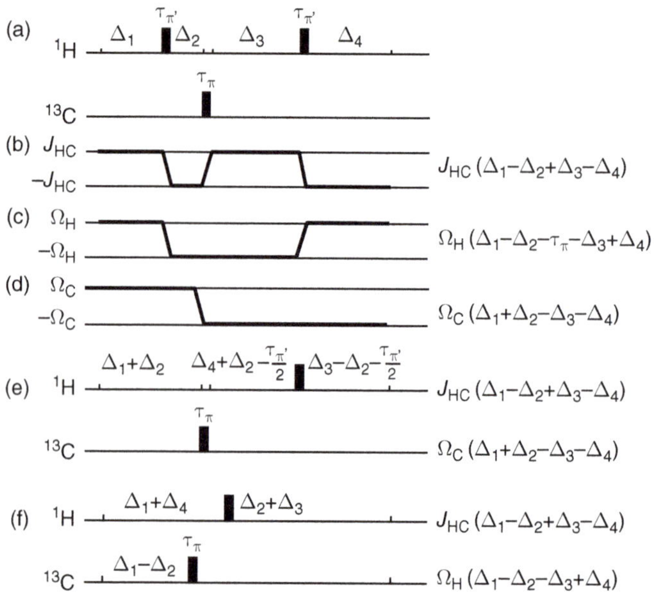


**Figure A.7** (a) Pulse sequence containing two 180°(*I*) pulses and a 180°(*S*) pulse. The evolution of interactions is given below: (b) for the heteronuclear H,C coupling, (c) for the *I* chemical shift, and (d) for the *S* chemical shift, (e) If all *I* spins are longitudinal, the *I* chemical shift need not be taken into account. Then the two *I* pulses can be concatenated, (f) If the chemical shift of *S* need not be taken into account, the two proton pulses are concatenated in a different way.

pulses, a 2D $J$ spectrum [15] would result. The 90°($S$) pulse turns the carbon "operator" also into the transverse plane. Heteronuclear double- and zero-quantum coherences ($2I_{x,y}S_y$) evolve during $t_1$. Proton chemical shift is refocused by the 180° proton pulse during $t_1$. Therefore it evolves during $t_2$ only. Homonuclear $J_{\mathrm{HH}}$-couplings evolve during the whole sequence $t_1 + t_2 + 2\Delta$, carbon chemical shift and homonuclear carbon-carbon couplings evolve during $t_1$, and the heteronuclear coupling evolves during $\Delta$ and again during $\Delta + t_2$. The transfer amplitude of this pulse sequence for a H,C cross peak in a H,C,H$_j$,C$_k$ spin system (where $j$ enumerates passive proton spins and $k$ enumerates passive carbon spins coupled to the carbon of interest) is therefore

$$\sin(\pi J_{\mathrm{HC}}\Delta)\sin(\pi J_{\mathrm{HC}}(\Delta + t_2))\prod_j \cos(\pi J_{\mathrm{HH}_j}(t_1 + t_2 + 2\Delta))$$

$$\prod_k \cos(\pi J_{\mathrm{CC}_k} t_1)\exp(i\Omega_{\mathrm{H}} t_2)\cos(\Omega_{\mathrm{C}} t_1)$$

The heteronuclear zero- and double-quantum coherence evolving during $t_1$ is selected by the phase cycle $\phi = 0$, 180; $\psi = 0, 0, 180, 180$; receiver phase $= 0, 180, 180, 0$.

Note that suppression of protons not bound to $^{13}$C is achieved only after two experiments by subtraction of unwanted coherences. Without decoupling in $t_2$, the signal that is canceled by the phase cycle is at least 200 times stronger than the desired signal given the 1% natural abundance of $^{13}$C and the doublet character of the carbon-bound proton. Therefore the dynamic range of the digitizer must be high for proton-detected heteronuclear spectroscopy of natural abundance samples. If the desired signal is a factor of 200 smaller than the undesired signal, almost half the number of bits of a 16-bit digitizer are used for signal that is canceled away by the phase-cycling procedure (8 bits correspond to a factor of 256). To obtain good subtraction the phases of the proton pulses are *not* changed during the sequence. Change of proton phases generally leads to poorer cancellation of undesired signals because a new equilibrium state is established between scans.

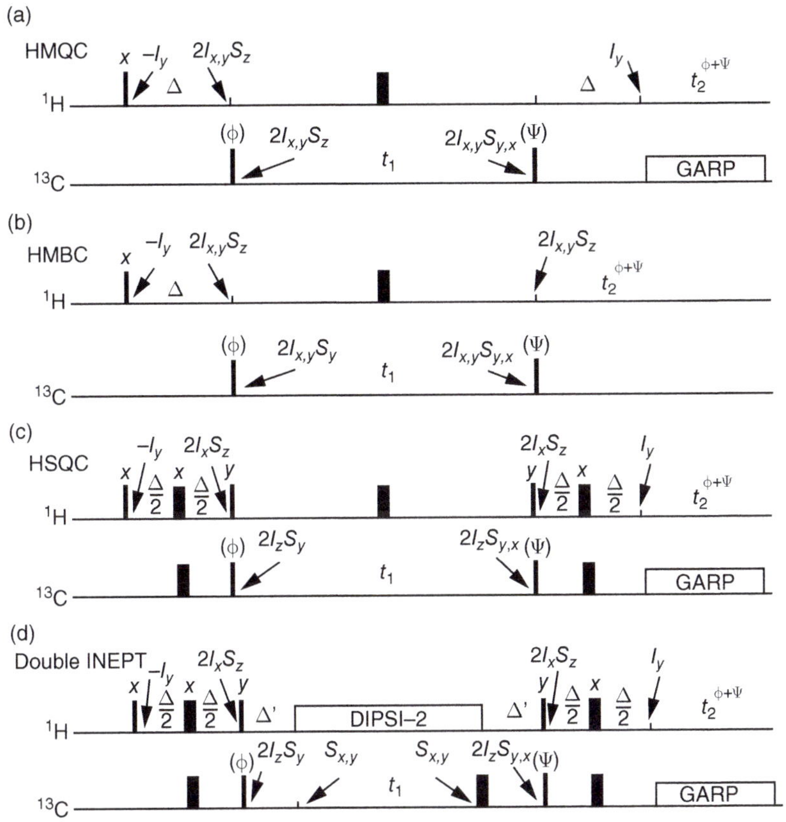


**Figure A.8** (a) Pulse sequence of the HMQC experiment with refocusing of the proton chemical shift in $t_1$ and decoupling in $t_2$. In absence of the carbon pulses a 2D $J$-resolved experiment results. The phases $\phi$ and $\psi$ are cycled according to 0,180 and 0,0,180,180, respectively. The receiver phase is cycled accordingly: 0,180,180,0. TPPI, RSH, or States-TPPI (cf Section A.3.1) is performed on $\phi$. (b) Pulse sequence of the HMBC experiment. Only the refocusing period is missing compared with the HMQC. Consequently, proton decoupling must be avoided and the active heteronuclear coupling is in antiphase. (c) Pulse sequence of the HSQC experiment. Defocusing of the heteronuclear $J_{HC}$ coupling occurs during $\Delta$. The $S$-spin antiphase coherence evolves during $t_1$. The 180° pulse refocuses heteronuclear coupling in $t_1$. The efficiency of the sequence is independent of the multiplicity. (d) Double INEPT sequence with refocusing during $\Delta'$. In-phase heteronuclear magnetization at the beginning of $t_1$ evolves under proton decoupling. Since spin polarization is lost during the $I_n \rightarrow S$ and $S \rightarrow I_n$ transfer the sequence leads to lower signal to noise ratio than HSQC/HMQC for multiplicities $n > 1$.

As we have done in this example we will throughout this chapter represent the state of the spin system by product operators at crucial points. Phase cycles should be applied in such a way as to select the specified transformations. We will denote the phases to be incremented according to the TPPI or States–Haberkorn–Ruben procedure by $\phi$ (for references see Section A.3.1).

In the HMQC experiment, heteronuclear decoupling (see Section A.3.5) can be applied during $t_2$ to gain sensitivity due to the reduction of the multiplet structure from a doublet to a singlet. The transfer amplitude is simplified:

$$\sin^2(\pi J_{\mathrm{HC}}\Delta)\prod_j \cos(\pi J_{\mathrm{HH}_j}(t_1 + t_2 + 2\Delta))$$

$$\exp(i\Omega_{\mathrm{H}}(t_2))\prod_k \cos(\pi J_{\mathrm{CC}_k}t_1)\cos(\Omega_{\mathrm{C}}t_1)$$

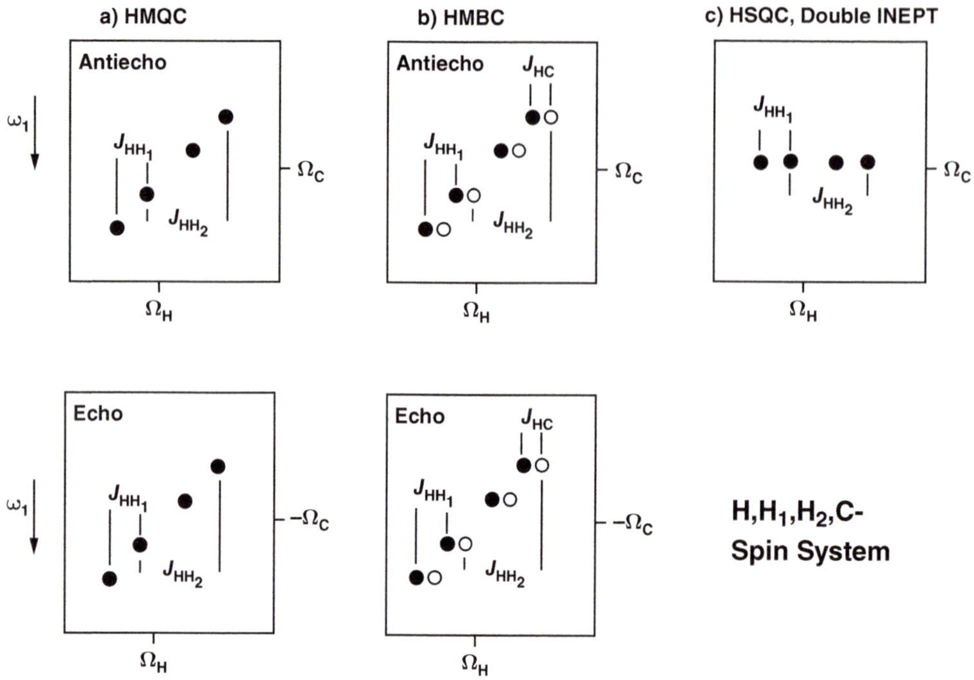


**Figure A.9** (a) Schematic multiplet structure in a HMQC experiment. Modulation by the homonuclear coupling in $t_1$ and $t_2$ leads to the tilted multiplet structure in the echo and antiecho spectra. The two spectra do not match on folding at $\omega_1 = 0$, which leads to mixed phases in HMQC. (b) HMBC multiplet pattern. In contrast to the HMQC pattern, the HMBC pattern is convoluted with an antiphase splitting in $\omega_2$ due to the active heteronuclear coupling not being refocussed before the beginning of $t_2$. (c) Multiplet pattern in HSQC and double INEPT: Modulation with the homonuclear coupling in $\omega_1$ is removed. Because the echo and antiecho part superimpose exactly on folding at $\omega_1 = 0$, spectra with pure phases can be obtained.

Fourier transformation of the decoupled experiment yields a multiplet as shown schematically in Figure A.9a (echo part). The antiecho multiplet structure at $\omega_1 = -\Omega_C$ is shown as well.

In natural abundance samples, the number of multiplet lines is minimal, namely just the number of lines in the proton multiplet, irrespective of the size of the heteronuclear coupling. Pure phases can only approximately be achieved, because the echo and the antiecho part do not exactly match after folding at $\omega_1 = 0$ (Figure A.9a). The deviation from amplitude modulation in $t_1$ is small if the resolution is of the order of, or less than, the width of the proton multiplets. Then one can Fourier transform an HMQC in the standard way. If the resolution in $\omega_1$ approaches the width of the proton multiplets, this recipe will yield mixed phases.

The HMQC sequence can be used for transfer via long-range heteronuclear couplings. In this so-called HMBC experiment [16] the second refocusing period is normally omitted and no decoupling is applied. This approach turns out to have a higher S/N than the refocused and $\omega_2$ decoupled HMBC. [17] The HMBC sequence is shown in Figure A.8b. The transfer amplitude is identical to the transfer amplitude of the fully coupled HMQC experiment with preacquisition delay $\Delta = 0$. (The evolution of homonuclear couplings of the heterospin is not considered further here because in fully labeled molecules the transfer via $^1J$-couplings is much faster than via heteronuclear long-range couplings, making other experimental approaches to obtain H,X-correlation more attractive.)

$$\sin(\pi J_{HC}\Delta)\sin(\pi J_{HC}t_2)\prod_j \cos(\pi J_{HH_j}(t_1+t_2+\Delta))\exp(i\Omega_H(t_2-\Delta))\cos(\Omega_C t_1)$$

This gives rise to a schematic multiplet in the echo part as shown in Figure A.9b. The difference between the carbon-decoupled HMQC and the HMBC multiplet consists of the additional convolution of the 2D multiplet by the heteronuclear antiphase splitting $J_{HC}$ in $\omega_2$, due to the modulation of the 2D signal with $\sin(\pi J_{HC} t_2)$. As in the HMQC experiment, the signal is phase modulated due to H,H-coupling in $t_1$. However, if $t_1^{max}$ is short such that the resolution achieved in $\omega_1$ is lower than the multiplet width, the signal is approximately amplitude modulated in $t_1$. Therefore the spectrum is usually recorded in the standard way (e.g., with TPPI), and Fourier transformed in such a way as to obtain pure phases. The $\omega_1$ dimension is then phased to pure absorption. Since the phasing in $\omega_2$ is impossible due to the evolution of homonuclear coupling (during $\Delta$ in the nondecoupled and during $2\Delta$ in the decoupled variant) and chemical shift in $\Delta$ (in the nondecoupled variant), the absolute value of the signal in $\omega_2$ is taken. This is obtained by combining the $R_1R_2$ and the $R_1I_2$ parts according to: $[(R_1R_2)^2 + (R_1I_2)^2]^{1/2}$. [18] The resulting spectrum has pure absorption phase in $\omega_1$ (X-nucleus) but shows the absolute value of the signal in $\omega_2$ ($^1$H). The abbreviations $R_1R_2$ and $R_1I_2$ refer to real (R) and imaginary (I) parts of the spectrum in $\omega_1$ (index 1) and $\omega_2$ (index 2), respectively.

An experimental spectrum of cyclo-(D-Pro$^6$-Phe$^{11}$-Thr$^{10}$-Ala$^9$-Trp$^8$-Phe$^7$-) is shown in Figure A.10. The HMBC experiment can be used to connect proton spin systems that are interrupted by nonprotonated atoms. In peptides, for example, the carbonyl groups constitute such an interruption of the proton spin system. The detection of cross peaks between the $H_{i+1}^N, C_i'$ and $H_{\alpha i}, C'$ permits the sequential assignment of a sequence of amino acids. This method for sequential assignment does not produce nonsequential cross peaks, in contrast to NOE-based sequential assignment. However, the size of the $^2J_{HC'}$-coupling constants depends on the conformation ($\phi$ and $\psi$ angles) [19].

The HMBC spectrum shown in Figure A.10 was acquired with a selective pulse (Gaussian 90° pulse [20]; see also Chapter 2 in this volume) in the carbonyl region to avoid folding problems in $\omega_1$ and to achieve high resolution in the indirectly sampled frequency domain. A 270° Gaussian pulse[21] or a G4 Gaussian pulse cascade [22] would be more up to date. In Figure A.10 only the echo part of the spectrum is shown. The sequential assignment of the cyclic hexapeptide is indicated in the figure [23].

Due to the $\exp(i\Omega_H(t_2 - \Delta))$ factor in the transfer amplitude of the HMBC experiment, a large first-order phase correction would be required in $\omega_2$. Magnitude calculation of the spectrum in $\omega_2$ is therefore usually performed as described previously. However, when a quantitative evaluation of heteronuclear coupling constants is required, the distorted phases are retained (Section A.5).

The abundance of connectivity information available in HMBC spectra is demonstrated (Figure A.11) with another variation of the HMBC experiment applied to a protected disaccharide, ethyl-6-*O*-(2,3,4-tri-*O*-benzyl-$\alpha$-L-fu-copyranosyl)-(l,6)-3-*O*-acetyl-4-*O*-(p-methoxybenzyl)-2-desoxy-2-phthalim-ido-1-thio-$\beta$-D-glucopyranoside. Nonselective carbon pulses were used.

1

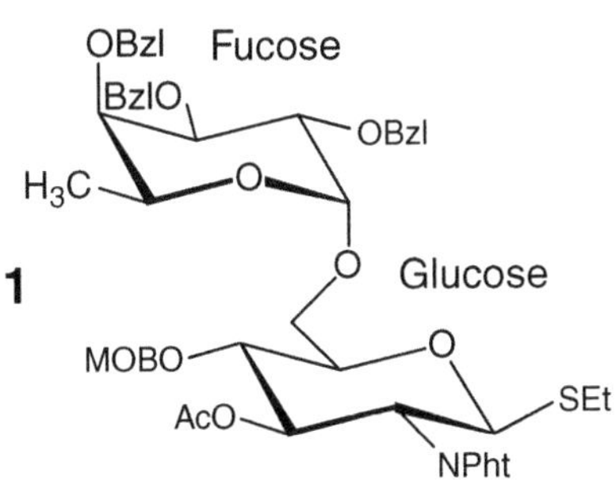


Folding (see Section A.3.1) was applied to obtain sufficient resolution. The spectrum shows nonprotonated carbon resonances (e.g., the carbonyl carbon of the acetyl group) as well as resonances of protonated carbons that are detected also in the HSQC spectrum (Figure A.23). Due to the intensity modulation of the cross peaks in HMBC

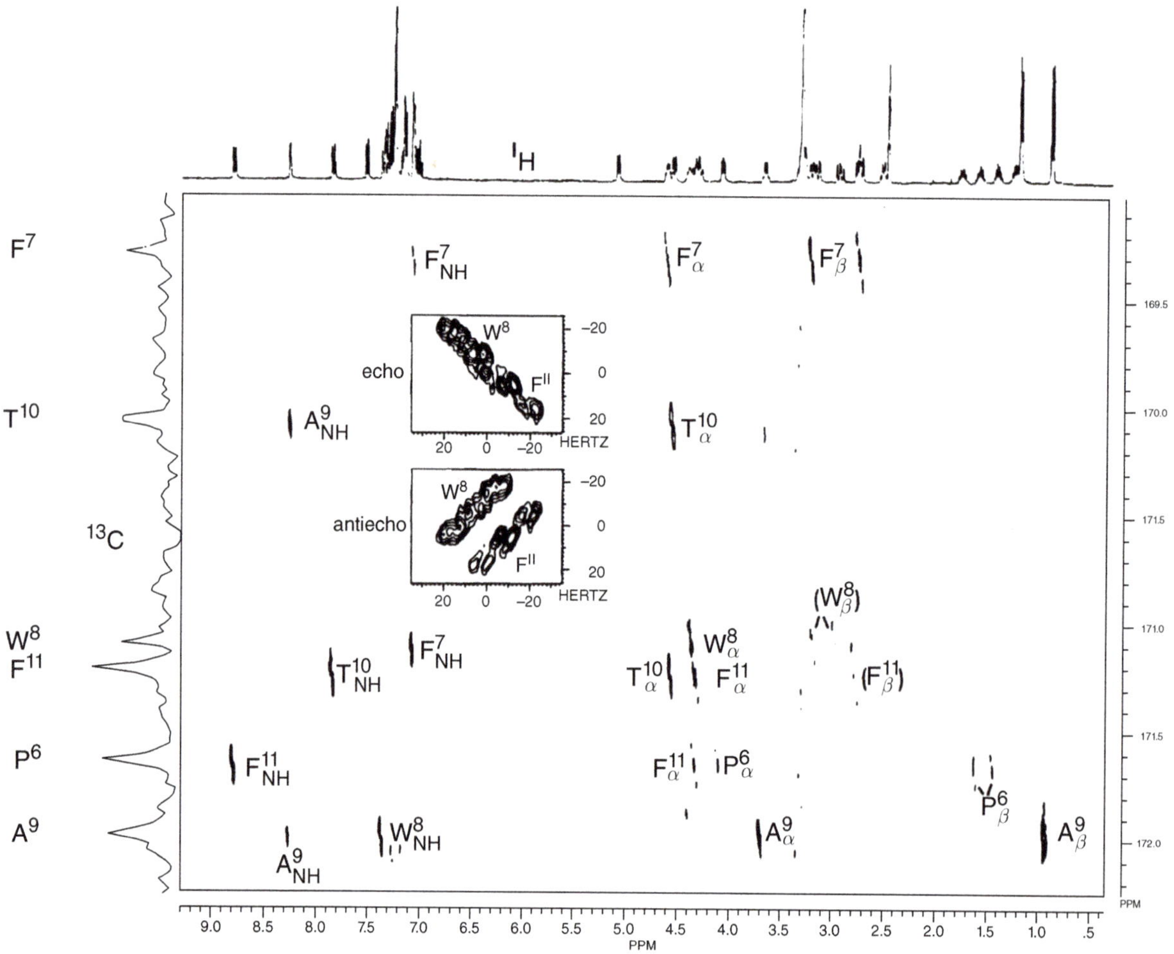


**Figure A.10** HMBC spectrum with a 90° Gaussian carbon pulse applied to carbonyl resonances of the cyclic hexapeptide cyclo-(D-Pro$^6$-Phe$^{11}$-Thr$^{10}$-Ala$^9$-Trp$^8$-Phe$^7$-). The connectivities between the neighboring amino acids are visible from the $H^N_{i+1}, C'_i$ cross peaks. $\Delta = 70$ ms, 128 $t_1$ experiments, the Gaussian pulse had a duration of 3.5 ms. The inserts show the tilted multiplet structure of the W$^8$(H$_\alpha$,C′) and F$^{11}$(H$_\alpha$,C′) cross peaks.

spectra with $\sin \pi J_{HC}\Delta$, some connectivities via $^lJ_{CH}$-couplings are missing if $^lJ_{HC}\Delta$ is close to a multiple of 1. For example, the direct connectivity of the $\underline{CH_2}$ of S-CH$_2$-CH$_3$ ($\omega_1 = 66.3$ ppm, $\omega_2 = 2.4$ ppm) is missing in the spectrum.

The HMBC spectrum shows the sugar linkage via the glucose-C(6) ($\omega_1 = 65.8$ ppm), fucose-H(l) ($\omega_2 = 4.9$ ppm) cross peak (G$_6$,F$_1$). The assignment of the benzyl groups can be derived from the Ph-$\underline{C}$H$_2$-O-C$\underline{H}$ as well as from the Ph-C$\underline{H_2}$-O-$\underline{C}$H cross peaks. The assignment of the benzyl protons and carbons can be derived from the $\underline{C_0}$,C$\underline{H_2}$ as well as from the $\underline{C_i}$,C$\underline{H_2}$ cross peak ($C_i$, $C_o$, $C_m$, and $C_p$ for ipso, ortho, meta, and para position in the aromatic ring). These cross peaks are indicated in the spectrum for the p-Meth- oxybenzyl (MOBO) group. The connectivities within the aromatic ring are derived from the $C_m$ and $C_p$ resonances in $\omega_1$, observing that cross peaks due to $^3J_{CH}$-couplings are much stronger than those due to $^2J_{CH}$-couplings in aromatic systems. Note the direct $^1J_{CH}$ as well as remote $^3J_{CH}$ connectivities between ortho protons and ortho carbons, which lead to an apparent triplet structure of these cross peaks [e.g., at $\omega_1 = 61$ ppm, $\omega_2 = 7.1$ ppm for the direct C$_o$,H$_o$ (doublet) and the remote C$_o$,H$_o$, (singlet) peaks].

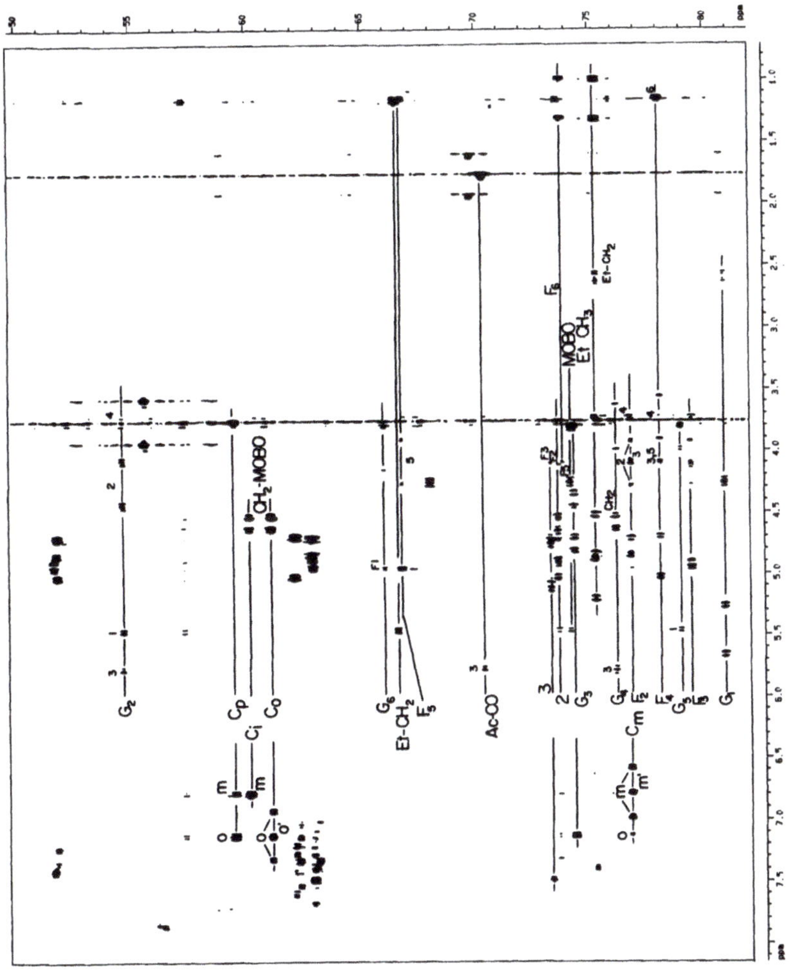

**Figure A.11** HMBC experiment on ethyl-6-*O*-(2,3,4-tri-*O*-benzyl-αL-fucopyranosyl)-(1,6)-3-O-acetyl-4-*O*-(p-methoxybenzyl)-2-desoxy-2-phthalimido-1-thio-β-D-glucopyranoside **1**. $\Delta = 50$ ms; nonselective pulses were used. The spectral width in $\omega_1$ was limited to 50 ppm. Peaks with F refer to the fucosyl resonances; MOBO refers to the (p-methoxybenzyl) protecting group. The carbons assigned $C_i$, $C_o$, $C_m$, $C_p$ refer to the aromatic carbons of this group. The protons are referred to as o, o′, m, m′. 2,3, and MOBO are the methylene carbons of the benzyl/methoxybenzyl protection groups at positions 2 and 3 of the fucose and 4 of glucose. Et-$CH_3$ and Et-$CH_2$ refer to the l-thioethyl group of the glucose. The direct $CH_2$-peak of the ethyl group is missing.

In HMQC and HMBC, the resolution in $\omega_1$ is limited and the S/N is reduced by the proton multiplet width because the cross peaks are modulated with the homonuclear proton couplings in $t_1$. In addition, for macromolecules, the heteronuclear zero- and double-quantum coherence containing transverse proton magnetization relaxes quickly due to the rather short $T_2$ of protons. [24, 25] Both of these problems are solved in the next basic experiment.

### A.2.2 HSQC (Heteronuclear Single-Quantum Correlation)

The HSQC [26, 27] sequence can be derived from the HMQC sequence by rotating the transverse proton magnetization to the longitudinal plane at the beginning of $t_1$ and then rotating it back to the transverse plane after $t_1$. To refocus chemical shift modulation during $\Delta$, a 180°$(I, S)$ pulse is introduced in the middle of each of the two delays A. Because the proton magnetization is longitudinal during $t_1$, no homonuclear couplings of the active spin H to the passive spins $H_j$ ($J_{HH}$) evolve during $t_1$. The evolution of heteronuclear couplings is refocused by the 180°$(I)$ pulse in the middle of $t_1$. The pulse sequence is shown in Figure A.8(c).

The transfer amplitude for the desired coherence transfer (proton magnetization longitudinal during $t_1$) is

$$\sin^2(\pi J_{HC}\Delta)\prod_j \cos(\pi J_{HH_j}\Delta)\exp(i\Omega_H t_2)\prod_j \cos[\pi J_{HH_j}(\Delta + t_2)]$$

$$\prod_k \cos(\pi J_{CC}t_1)\cos(\Omega_C t_1)$$

This formula is only true for a very large difference between $J_{HC}$ and $J_{HH}$, which is the situation for transfer via $^1J_{HC}$-couplings. If the heteronuclear coupling used for the polarization transfer is about the same size as the homonuclear coupling, the transfer becomes inefficient due to the excitation of proton multiple-quantum coherences after the second 90°$(I)$ pulse. Therefore, in contrast to HMQC, HSQC is used only for transfer via $^1J_{HC}$-couplings.

The schematic multiplet structure of an HSQC correlation peak is shown in Figure A.9b for a $HCH_1H_2$ spin system. The sensitivity of the experiment is comparable to HMQC for transfer via $^1J_{HC}$-couplings. The resolution in $\omega_1$ is no longer limited by the homonuclear proton couplings. For biomolecules HSQC has a higher S/N than HMQC because the fast decay due to proton transverse relaxation during $t_1$ is absent [24, 25].

So far we have seen that chemical shift information about a heteronuclear spin can be obtained by evolution of two types of operators:

| Experiment | Initial operator | Couplings | Relaxation rate |
|---|---|---|---|
| HMQC | $I_xS_x$ | Homonuclear couplings of $I$ and $S$ | $1/T_{2I} + 1/T_{2S}$ |
| HSQC | $I_zS_x, S_x$ | Homonuclear couplings of $S$ | $1/2T_{1I} + 1/T_{2S}$ |

In the HSQC experiment the heteronuclear chemical shift evolution is obtained from operators of the type: $2I_zS_x$ and $S_x$. The $2I_zS_x$ antiphase operator relaxes faster than the in-phase operator $S_x$ since the former also decays due to proton longitudinal relaxation. This effect is especially strong in larger molecules. To circumvent this problem, one may refocus the heteronuclear antiphase operator prior to $t_1$. This leads to the next experiment, the double INEPT sequence. Here in-phase heteronuclear single-quantum coherence $S_x$ evolves under decoupling of protons during $t_1$.

### A.2.3 Double INEPT

The double INEPT sequence [8, 24, 25, 28] introduces refocusing and defocusing periods for the heteronuclear coupling before and after $t_1$. Therefore pure in phase heteronuclear magnetization $S_x$, $S_y$ encodes heteronuclear chemical shifts during $t_1$ (Figure A.8d). The sequence is applied exclusively for transfer via $^1J$-couplings for the same reasons cited for the HSQC experiment.

The transfer amplitude of this sequence is

$$\sin^2(\pi J_{HC}\Delta)\sin^2(\pi J_{HC}\Delta')\cos^{2(n-1)}(\pi J_{HC}\Delta')$$

$$\cos(\Omega_C t_1)\exp(i\Omega_H t_2)\prod_j \cos(\pi J_{HH_j}\Delta)\prod_j \cos\left[\pi J_{HH_j}(\Delta + t_2)\right]$$

$n$ is the number of protons bound to the heteronucleus. The optimal delays to refocus heteronuclear antiphase coherence $2I_zS_y$ before $t_1$ and to defocus heteronuclear in-phase coherence after $t_1$ depend on the multiplicity of the heteronucleus. Maximum transfer is obtained with $1/(2J)$, $1/(4J)$, and $0.2/J$ for $IS$, $I_2S$, and $I_3S$ moieties, respectively. Consequently the sequence has a S/N identical to HMQC or HSQC only for $IS$ moieties. For low natural abundance nuclei, however, the transfer efficiency can be improved by incorporating composite bilinear rotations [2].

Since the heteronuclear coherence relaxes during $t_1$ only with the heteronuclear transverse relaxtion time $T_{2S}$, double INEPT is optimal for macromolecules, where longitudinal self-relaxation of protons is fast. Double INEPT is therefore used with advantage for H,N correlations, where NH groups are of main interest. Figure A.12 shows the comparison of HMQC (a), HSQC (b), and double INEPT (c) cross sections through the $^{15}$N,$^{1}$H cross peak of Gln$^{74}$-NH in $^{15}$N-labeled ribonuclease A.

### A.2.4 Sensitivity-Enhanced Correlations

Sensitivity can be enhanced in all three proton-detected correlation experiments via $^1J$-couplings. [29, 30] The idea will be demonstrated using the HSQC experiment. In the conventional experiment (Figure A.8c) either $2I_zS_x$ or $2I_zS_y$ coherence is selected after $t_1$ depending on the relative phases of the two $90°(S)$ pulses that surround $t_1$. Assume that the phase of the $90°(S)$ pulse at the end of $t_1$ is $x$. Then $2I_zS_y$ is transferred to $2I_yS_z$ and is refocused to $I_x$ at the beginning of $t_2$. $2I_zS_x$, however, turns into two-spin coherence that is not transferred to observable magnetization in the conventional experiment. This means that in each scan half of the magnetization is lost. More precisely, half of the power spectrum is lost leading to a sensitivity loss by $2^{1/2}$ compared with the theoretical optimum.

By introducing an additional refocusing period, the $2I_zS_x$ coherence can be recovered, which leads to an increase in S/N by $2^{1/2}$ (Figure A.13). $2I_zS_y$ at the end of $t_1$ is "stored" as $I_z$ at the beginning of the second delay whereas the two-spin coherence $2I_zS_x$ at the end of $t_1$ is converted then into $2I_yS_z$ coherence. This coherence is refocused in the second period $\Delta$. A final $90^\circ_x(I)$ pulse converts the stored $I_z$ as well as the refocused $I_x$ magnetization into detectable magnetization. Since $2I_zS_x$ is modulated with $\cos \Omega_S t_1$ and $2I_zS_y$ is modulated with $\sin \Omega_S t_1$, the following detectable operators are recorded at the beginning of $t_2$:

$$I_x \cos(\Omega_s)t_1 + I_y \sin(\Omega_s)t_1$$

The signal detected in $t_2$ is then $\exp(i\Omega_S t_1)\exp(i\Omega_1 t_2)$. This is a phase-modulated antiecho signal ($S^- \rightarrow I^-$) that leads to cross peaks that cannot be phased to pure absorption. However, recording a second experiment in which the third $90°(S)$ pulse is inverted leads to

$$I_x \cos(\Omega_s)t_1 - I_y \sin(\Omega_s)t_1$$

at the beginning of $t_2$. Adding or subtracting the two data sets yields amplitude-modulated signals:

$$\mathbf{R}_1 = I_x \cos(\Omega_s)t_1 \qquad \text{and} \qquad I_1 = I_y \sin(\Omega_s)t_1$$

that are stored at different memory locations in the computer. The enhancement of sensitivity in these experiments is due to creating a phase-modulated signal in each scan. It will be shown in Section A.3.3 that $B_0$ gradients applied in evolution times yield phase-modulated signals. Since echo as well as antiecho signals can be created, spectra with pure phases can be restored.

There are two possibilities to obtain sensitivity-enhanced correlations along these lines:

1. A Ruben–States–Haberkorn data set can be constructed from the two signals by observing that the $\mathbf{I}_1$ FID has to be phased in $\omega_2$ by an additional phase correction of 90° with respect to the $\mathbf{R}_1$ FID. The maximum enhancement for an $IS$-system is $2^{1/2}$ and may be less, since both components relax during the additional delay $\Delta$. Figure A.14 shows $\omega_1$ slices of an NH resonance of ribonuclease A obtained from enhanced HSQC compared with the usual HSQC-experiment. For a $I_1I_2S$ moiety, antiphase magnetization of the type $2I_zS_y$ at the end of

$t_1$ is not converted to observable magnetization. Therefore no enhancement is observed for such moieties for $\Delta = l/(2J)$. In addition, for $I_1I_2S$ the $\cos \Omega_S t_1$ and $\sin \Omega_S t_1$ parts do not have the same signal amplitude that leads to quadrature image formation in $\omega_1$. Therefore quadrature images are observed in $\omega_1$ for the $NH_2$ resonances. (Quadrature images are resonances that appear at minus the chemical shift at which they are supposed to appear.)

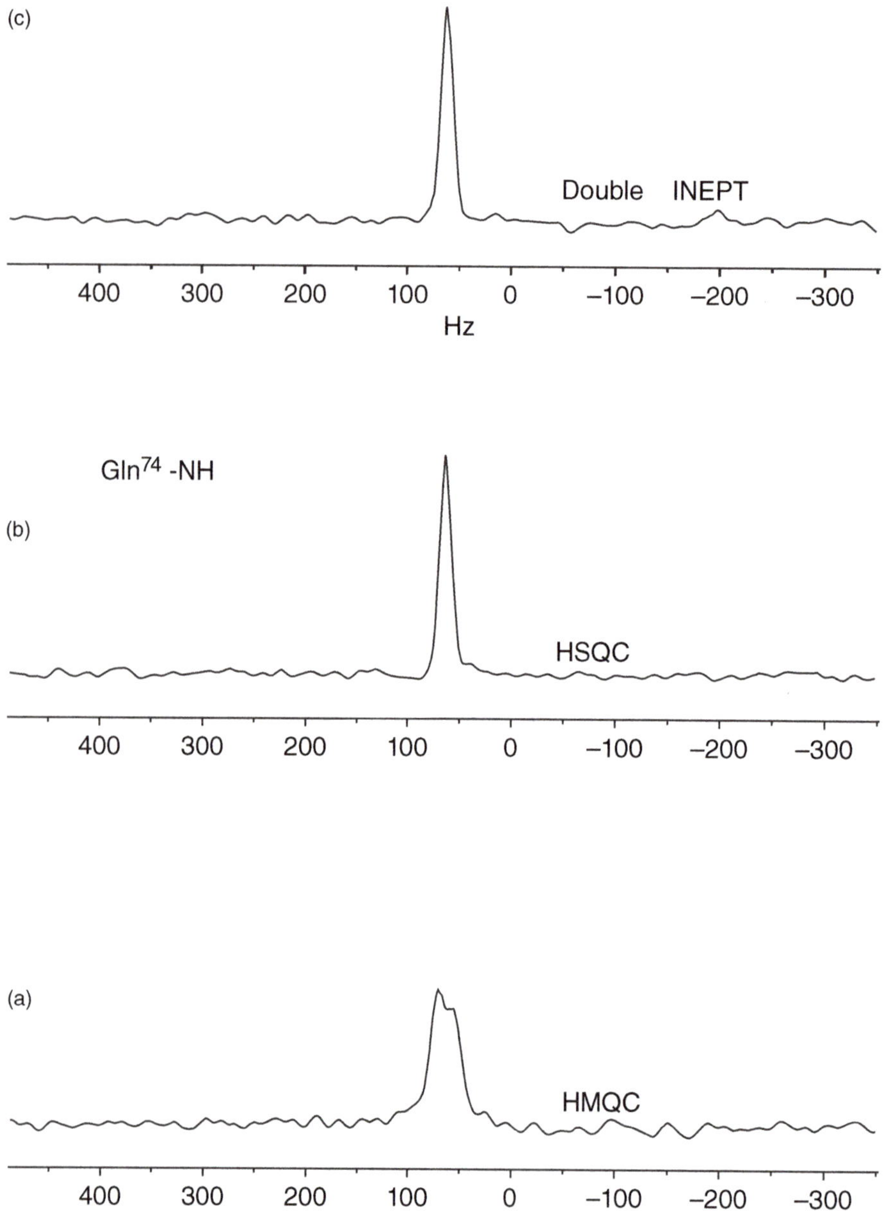


**Figure A.12** Comparison of (a) HMQC, (b) HSQC, and (c) double INEPT at Gln[74] of $^{15}$N-labeled ribonuclease A by taking $\omega_1$ traces through the respective spectra. The intensity loss of the HMQC due to the $H_\alpha$,$H^N$ coupling is clearly visible. No relaxation effect is visible at this molecular weight as an intensity difference between HSQC and double INEPT.

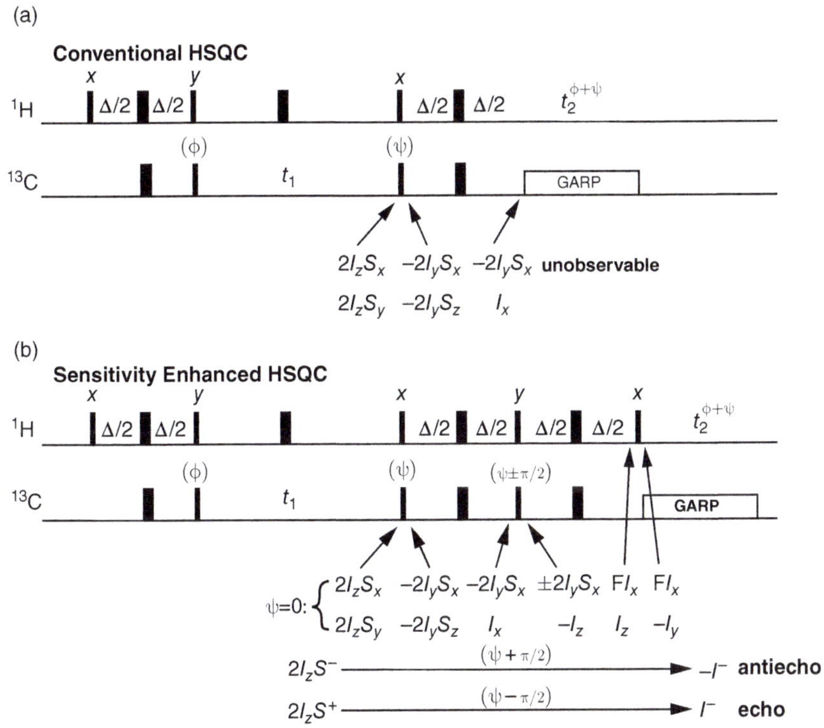


Figure A.13 Comparison of conventional HSQC (a) with sensitivity-enhanced HSQC (b). $\phi$ and $\psi$ are cycled in steps of 180° to suppress axial peaks. When the phase of the last 90°($S$) pulse in sequence b is $\psi + \pi/2$, the antiphase coherence $2I_zS^+$ is transferred exclusively to $I^-$ (heteronuclear antiecho), whereas the phase of the last 90°($S$) pulse being $\psi - \pi/2$ the antiphase coherence $2I_zS^+$ is transferred exclusively to $I^-$ (heteronuclear echo). Both components are needed to derive from the two phase-modulated 2D FIDs the amplitude-modulated 2D FIDs normally used to obtain pure phases in $\omega_1$ and $\omega_2$.

2. The normal TPPI, RSH, or States-TPPI procedure is applied to record both data sets. They are combined either before or after Fourier transformation, but with the application of a 90° zero-order phase correction in $\omega_1$ and $\omega_2$ for one of the signals. The speed of recording the spectrum is not increased but the S/N is larger for $IS$ moieties. No quadrature images are observed for $I_nS$ moieties with $n > 1$.

### A.2.5 Simultaneous Correlation to Several Heteronuclear Spins

The heteronuclear correlation experiments discussed so far can be expanded in such a way that correlations to several heteronuclei can be obtained simultaneously. Since each proton participates in at most one $^1J$ coupling to one heteronucleus, the evolution of heteronuclear coupling to the different heteronuclear spins is independent. Therefore the pulse sequence in Figure A.15 yields a correlation of protons with heteronuclei $S$ and $T$ each with optimal sensitivity. [31, 32] The product operators in the description of the pulse sequence refer to an $IS$ and an $IT$ spin system. To distinguish between the two $I$ spins, we call $I$ bound to $S$:$I_1$ and $I$ bound to $T$:$I_2$. The delays are optimized for complete defocusing and refocusing of the potentially different heteronuclear coupling constants $\Delta' = (2^1J(I_1S))^{-1}$ and $\Delta = (2^lJ(I_2T))^{-1}$. Four transients differing in the phases $\phi_l$ and $\phi_2$ have to be selected to get either the $I, S$ or the $I,T$ correlation spectrum. The result for each $(\phi_1, \phi_2)$ pair is stored separately. After the experiment the data are added and subtracted in appropriate combinations to yield the two correlation spectra.

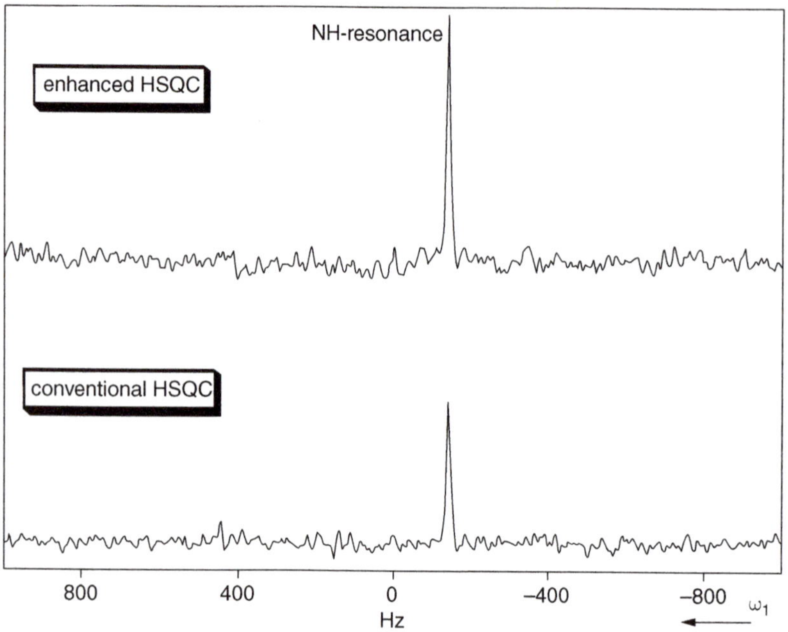


**Figure A.14** Comparison of the intensity of a $^{15}$N,H correlation peak in a conventional HSQC and an enhanced HSQC. The intensity gain in the enhanced spectrum is approximately a factor of $2^{1/2}$. The bar in the upper trace indicates the intensity of the peak in the lower trace.

The phase cycle is given in Figure A.15. To account for different chemical shift ranges of the spin types $S$ and $T$, scaling of the chemical shift range can be performed as indicated in Figure A.15. For the sequence in Figure A.15, the chemical shift range of the $S$ ($^{13}$C) spin is given by $[\Delta t_1(1\text{-}\kappa)]^{-1}$; the chemical shift range of the $T$ ($^{15}$N) spin is given by $(\Delta t_1)^{-1}$.

Although the experiment is elegant, since two correlations can be recorded in one, it has so far not gained widespread application.

### A.2.6 Constant-Time Experiments

The experiments described thus far have increasing overall duration with increasing evolution time. In constant-time experiments the chemical shift evolves during a fixed delay that is interrupted by a 180° refocusing pulse whose position steps through the fixed delay [33, 34] (Figure A.16).

Since the time for evolution of homonuclear $^{13}$C,$^{13}$C-coupling is independent of the evolution time, the couplings contribute an amplitude factor rather than a modulation in the evolution time. The heteronuclear coupling does not evolve between the two 90°($S$) pulses [$J_{HC}$ : $t_1/2 - \tau/2 + (\tau/2 - t_1/2) = 0$]. The carbon chemical shift evolves during $t_1/2 + \tau/2 - (\tau/2 - t_1/2) = t_1$. The homonuclear couplings to the nucleus active in the evolution time are removed from the indirectly sampled frequency domain due to $\tau = n/{}^1J_{CC}$. The resolution as well as the S/N is therefore normally increased. The 180°($S$) pulse should be placed in the middle of $\tau$ in the first experiment. It does not matter whether it moves toward the first or the second 90°($S$) pulse except for the direction of the $\omega_1$ frequency axis.

To be sure what the direction of the indirectly sampled frequency dimension is, the following rule is given: We call a COSY spectrum that is obtained by Fourier transformation along $t_1$, when TPPI is applied before $t_1$ in the usual way (phase increment 90°), a normal spectrum. If the TPPI phases are applied for all pulses before

Simultaneous $^{15}$N,H and $^{13}$C,H–HSQC: $I_1S$ and $I_2T$ Spin System

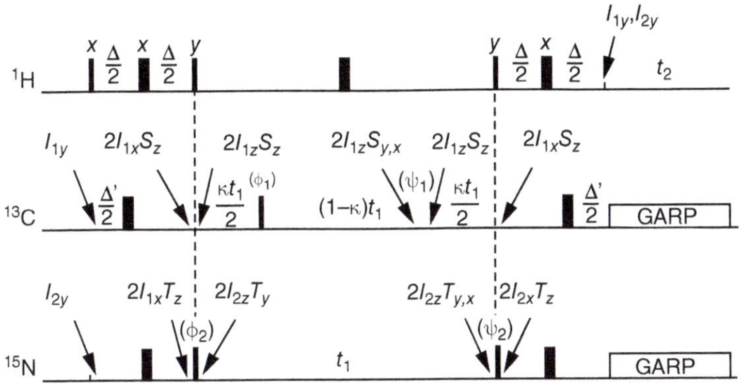


| ($\phi_1$) | ($\phi_2$) | $^{13}$C,H–correlation (rec.) | $^{15}$N,H–correlation (rec.): |
|---|---|---|---|
| 0 | 0 | 0 | 0 |
| 2 | 0 | 2 | 0 |
| 0 | 2 | 0 | 2 |
| 2 | 2 | 2 | 2 |

TPPI simultaneously on $\phi_1$ and $\phi_2$

**Figure A.15** HSQC with simultaneous correlation of protons to carbons and nitrogen. $\Delta' = 1/(2J_{HC})$ and $\Delta = 1/(2J_{HN})$. The $180°(I)$ pulse in the middle of $t_1$ ensures that heteronuclear couplings to the proton are refocused. The spectral range of the carbon spectrum is $SW = [(1-\kappa)DW]^{-1}$. The coupling between the heteronuclei $^{13}$C and $^{15}$N is not refocused in this sequence and cannot be in an experiment with simultaneous evolution of the chemical shifts of the two heteronuclei. It develops during $(l-\kappa)t_1$ for $^{13}$C and during $t_1$ for $^{15}$N. The experiments with the four phase settings for $\phi_1$ and $\phi_2$ given in multiples of 90° are stored separately. Additive combination of the four experiments with the second and fourth FID inverted yields the $^{13}$C,$^1$H correlation, whereas additive combination of the four experiments with the third and fourth FID inverted yields the $^{15}$N,$^1$H correlation. TPPI is applied simultaneously to $\phi_1$ and $\phi_2$. Axial peak suppression can be applied in addition to $\psi_1$ and $\psi_2$.

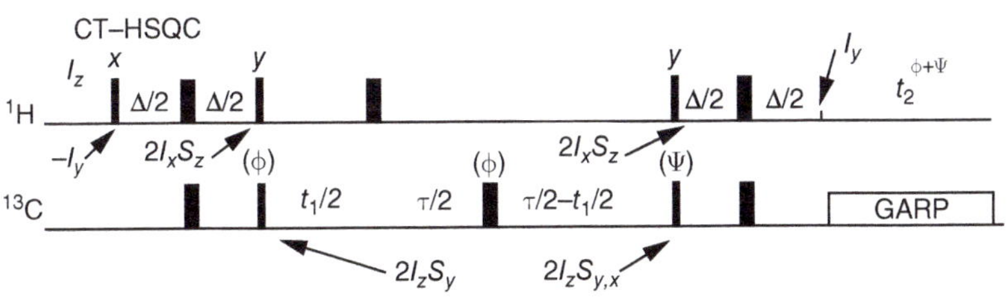


**Figure A.16** Constant-time HSQC. The delay $\tau = n/^1J_{CC}$ is chosen such that the homonuclear C,C couplings are refocused. The phases $\phi$ and $\psi$ are cycled in steps of 180°. The phase for the differentiation of signs of the chemical shifts in $\omega_1$ is either $\phi$ or $\psi$.

an incremented/decremented evolution time, the direction of the frequency axis $\omega_1$ will be normal/inverted. If the TPPI phases are applied to all pulses including the receiver phase after an incremented/ decremented $t_1$, the direction of $\omega_1$ will be inverted/normal. If in Figure A.16, TPPI is applied on all pulses [$90°(S)$ and $180°(S)$] before the decremented $t_1/2$ delay the application of the pulse sequence in Figure A.16 will lead to a spectrum that is inverted in $\omega_1$.

Constant-time experiments are useful in cases where there are homonuclear couplings for the S spin. This is the case, for example, in HSQC of completely $^{13}$C-labeled molecules. The transfer amplitude is given by

$$\sin^2(\pi J_{HC}\Delta)\prod_j \cos(\pi J_{HH_j}\Delta)\exp(i\Omega_H t_2)\prod_j \cos\left[\pi J_{HH_j}(\Delta + t_2)\right]$$

$$\cos(\Omega_C t_1)\prod_i \cos(\pi J_{CC_i}\tau)$$

As an example, the constant time HSQC [35–38] of completely labeled glutamic acid is shown in Figure A.17a. The $S$-pulses are selective to the aliphatic carbons and do not affect the carbonyl resonances. Therefore the couplings between the aliphatic carbons are removed but not the couplings between the carbonyl carbons and the aliphatic carbons, $\tau$ was set to $1/{}^1J_{CC}$ between aliphatic carbons, which limits the resolution obtained in the carbon dimension, because $t_1^{max}$ cannot be larger than $\tau$. More generally, $\tau$ can be set to $n/{}^1J_{CC}$, allowing for better resolution at the expense of sensitivity due to relaxation. Because there is one aliphatic carbon directly bound to $C_\alpha$ and $C_\gamma$ $(\cos(\pi J_{C_\alpha C_\beta}\tau) = -1, \cos(\pi J_{C_\gamma C_\beta}\tau) = -1)$, but two aliphatic carbons directly bound to $C_\beta(\cos(\pi J_{C_\beta C_\alpha}\tau)\cos(\pi J_{C_\beta C_\gamma}\tau = 1)$, the $C_\beta$ resonance is inverted in sign compared with the two other cross peaks. The two carbonyl carbons ($C'_\alpha$ and $C'_\gamma$) are not touched in the experiment. Therefore the ${}^1J_{C'_\gamma C_\gamma}$-coupling is clearly visible in $\omega_1$ (Figure A.17a). The ${}^1J_{C'_\alpha C_\alpha}$ coupling cannot be easily seen due to incomplete $C'_\alpha$ labeling. The increased resolution of the constant-time experiment versus the normal HSQC (Figure A.17b) due to the removed homonuclear C,C couplings is also visible in the spectra. The $\omega_2$ displacements of the two peak parts in the $C_\alpha$,$H_\alpha$ and $C_\gamma$, $H_\gamma$ cross peaks in the CT HSQC are the ${}^2J_{C'H_\alpha}$-coupling and the ${}^2J_{C'H_\gamma}$ coupling, respectively [39].

(a)

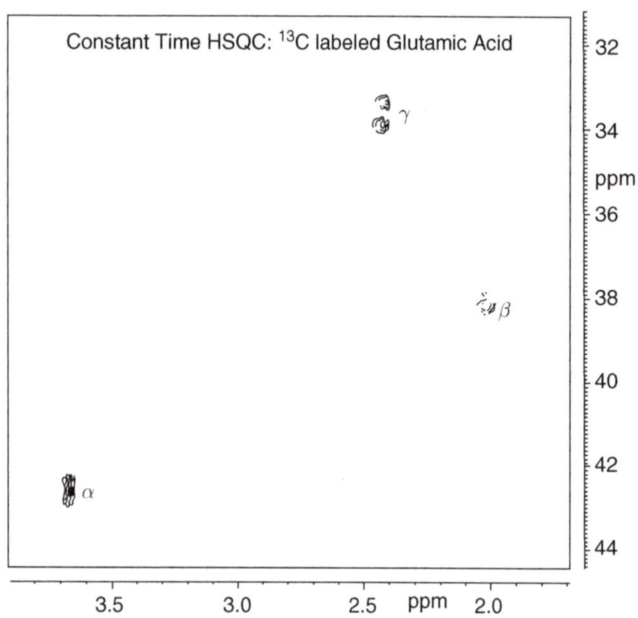


**Figure A.17** Comparison of constant-time HSQC (a) and conventional HSQC (b) of $^{13}$C-labeled glutamic acid. The ${}^1J_{CC}$-couplings between aliphatic carbons have not evolved in $t_1$ and therefore are removed from the spectrum in (a), whereas they have developed in $t_1$ and are therefore visible as line broadening in (b). The resolution enhancement can be clearly seen at the $C_\beta$ resonance. No carbonyl decoupling was applied (e.g., by a 180°(C′) pulse simultaneously with the 180°(H) pulse). Therefore in spectrum (a) the ${}^1J_{C_\alpha C'}$ and the ${}^1J_{C_\gamma C'}$-couplings lead to splittings in $\omega_1$.

*(Continued)*

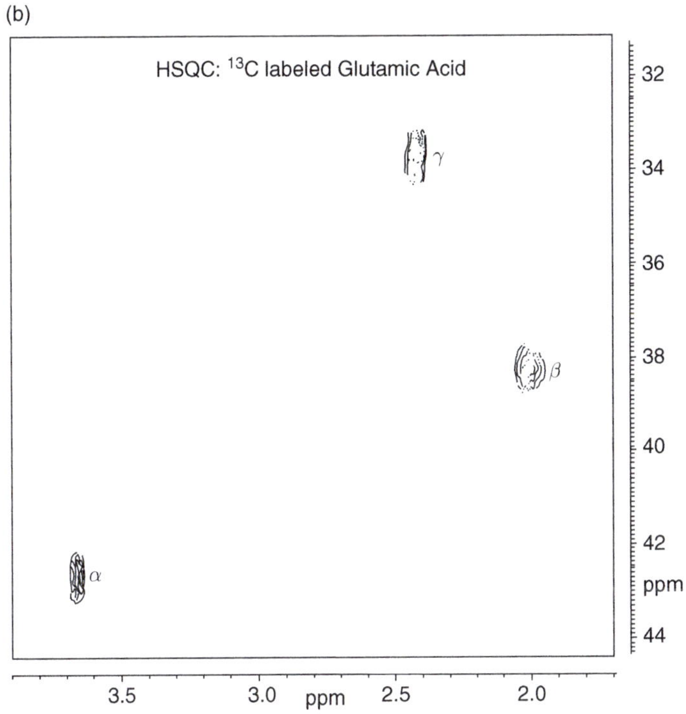


**Figure A.17(b) (Cont'd)**

Constant-time experiments are especially useful when a relatively small coupling should be refocused or defocused at the end of the constant-time delay. This will be discussed further in connection with 3D experiments on $^{13}$C-labeled biomolecules.

## A.2.7 Editing

We have just seen an editing effect in the constant time HSQC spectrum of $^{13}$C-labeled glutamic acid. The $C_\beta$ has a different multiplicity with respect to directly bound aliphatic carbons (namely 2) compared with the $C_\alpha$ and the $C_\gamma$ (namely 1) and appears inverted.

Editing usually refers to the selection of heteronuclear spins on the basis of their proton multiplicities, that is, on the number of attached protons. Discrimination of heteronuclei bound to a varying number of protons can be achieved in sequences based on HMQC or on HSQC. To achieve editing of proton multiplicities the magnetization of the heteronuclear spin must be in the transverse plane. Therefore introduction of an additional delay $1/J$ between the two heteronuclear 90° pulses opens up editing possibilities (Figure A.18). Given a $CH_n$ group, the state of the spin system before the $\theta$ pulse is described as $n2^nI_x(I_z)^{n-1}S_{x,y}$ for HMQC (Figure A.18a) and $S_{x,y}2^{n-1}\,nI_z^{n-1}$ for HSQC (Figure A.18b).

Insertion of the editing pulse must not change the type of the proton operator because only 180° pulses are applied later. Therefore we need only consider how the $\theta$ pulse transfers the initial operator onto itself. For a $\theta_x$ pulse in HMQC, [40, 41] this transfer is: $nS_{x,y}2^nI_x(I_z)^{n-1} \rightarrow (\cos^{n-1}\theta)nS_{x,y}2^nI_x(I_z)^{n-1}$. For a $\theta_x$ (or $\theta_y$) pulse in HSQC [42]: $nS_{x,y}(2I_z)^{n-1} \rightarrow (\cos^{n-1}\theta)nS_{x,y}(2I_z)^{n-1}$. Thus identical editing factors result for edited HSQC and

HMQC with a $\theta_x$ editing pulse. On the other hand, application of a $\theta_y$ pulse in HMQC [43–45] produces

$$nS_{x,y}2^nI_x(I_z)^{n-1} \rightarrow \left[\cos^n\theta - (n-1)\sin^2\theta\cos^{n-2}\theta\right] nS_{x,y}2^nI_x(I_z)^{n-1} \qquad \text{(Figure A.18a)}$$

As already stated, for HSQC (Figure A.18b) the phase of the editing pulse has no effect on the transfer amplitude.

Complete editing and selection of multiplicites in 1D experiments can be accomplished in HSQC by the application of at least three different $\theta$ flip angles. Often it is sufficient to distinguish $CH_2$ ($NH_2$) from $CH_3$ and CH (NH), which requires recording of $\theta = 0$ and $\theta = 180$ data [40].

Separation of all three multiplicities is made possible by recording a third experiment with $\theta = 90°$, which will detect only CH groups. Editing with a $\theta_x$ pulse scales the signal for the $CH_n$ groups by a factor of $\cos^{n-1}\theta$ and yields

$$\text{CH: } \theta = 90°$$

$$\text{CH}_2\text{: } (\theta = 0°) - (\theta = 180°)$$

$$\text{CH}_3\text{: } (\theta = 0°) + (\theta = 180°) - 2 \cdot (\theta = 90°)$$

No signal is lost for the CH and $CH_2$ groups; for $CH_3$ the double-quantum filtering eliminates half of the signal. Distinction of CH and $CH_3$ versus $CH_2$ is possible without any loss of signal.

For the more complicated editing function after a $\theta_y$ editing pulse in the edited HMQC, also called DEPT-HMQC [43–45]: $\cos^n\theta - (n - l)\sin^2\theta\cos^{n-2}$ we obtain

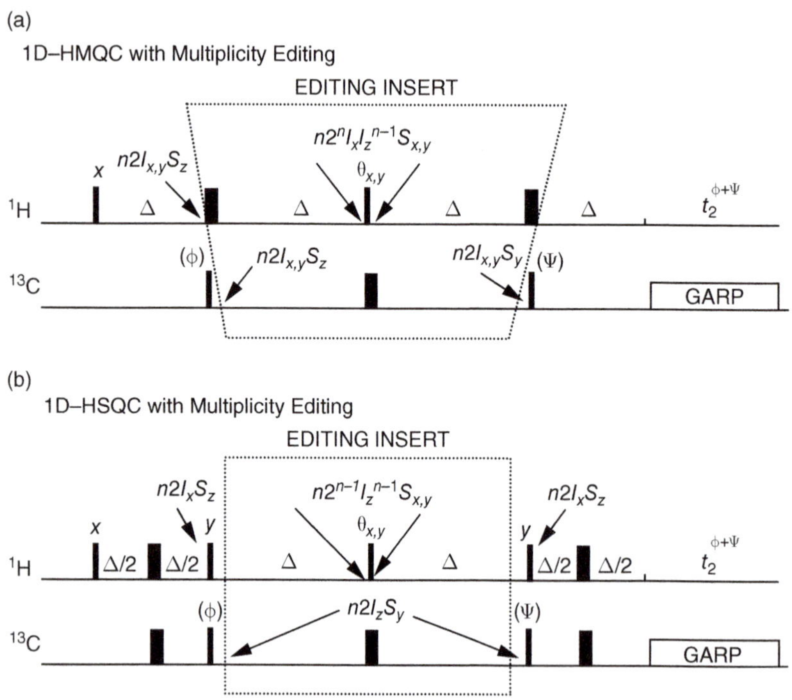


**Figure A.18** Editing delay insertion in the 1D versions of HMQC (a) and HSQC (b). The state of the spin system for the two pulse sequences before the editing pulse is different only in that the $I_x$ operator is present for the HMQC sequence. For magnetization to pass through the editing sequence and be detected, the state of the operator must not be changed during the editing insert. Using this rule the editing factors for the different multiplicities may be derived as described in the text.

$$\text{CH: } \cos\theta\text{: } 5\cdot(\theta = 0) + 4\cdot(\theta = 60) - 4\cdot(\theta = 120) - 5\cdot(\theta = 180)$$
$$- 4\cdot(\theta = 240) + 4\cdot(\theta = 300)$$
$$\text{CH}_2\text{: } \cos 2\theta\text{: } (\theta = 90°)$$
$$\text{CH}_3\text{: } \cos^3\theta - 2\sin^2\theta\cos\theta\text{: } (\theta = 0) - (\theta = 60) + (\theta = 120)$$
$$- (\theta = 180) + (\theta = 240) - (\theta = 300)$$

Again $\theta = 0°$ and $\theta = 180°$ is sufficient to distinguish CH and $CH_3$ from $CH_2$. $CH_3$ is selected by a triple-quantum filter process, in contrast to the double-quantum filter employed for the former editing function. The advantage of using this editing principle is that $CH_2$ can be recorded with a single $\theta = 90°$ experiment. $CH_3$ and $CH_2$ are selected with the same S/N as with the editing scheme of the HSQC or HMQC using $\theta_x$ editing pulse. For CH selection only 70% of the signal is retained.

2D correlation experiments are obtained by insertion of $t_1$ either before or after the "editing insert" (Figure A.19 for HSQC, Figure A.20a,b for HMQC). Taking advantage of the fact that the editing pulse $\theta_y$ can be split into two pulses à la POMMIE [46] $\theta_y = (90_\theta\ 90_{-x})$, a sequence has also been proposed [47] that inserts $t_1$ between those pulses (Figure A.20c).

Summarizing the results obtained so far: Coherences of the form $2S_xI_z$, $S_x$, or $2S_xI_y$ can be used for heteronuclear correlation experiments. Editing is achieved by the evolution of heteronuclear coupling during additional delays.

Selection of methyl groups has been applied in 3D HSQC-TOCSY experiments of samples in natural abundance based on the triple-quantum editing just described [48, 49] All sequences discussed so far require a total time of $1/J_{HC}$ to achieve editing. For specific applications, shorter editing sequences are possible. Multiplicity selection of $I_2S$ groups can be achieved in an edited HSQC with an editing delay of only $1/(2J)$ instead of $1/J$ used in the foregoing. The $2I_zS_y$ state after the first polarization transfer is then converted to $(2I_z)^{n-1}S_y$, which after a second $S$-$I$ polarization transfer yields proton single-quantum coherence only for $n = 2$. Application of the reverse INEPT sequence will yield only $I_2S$ groups in the spectrum [50].

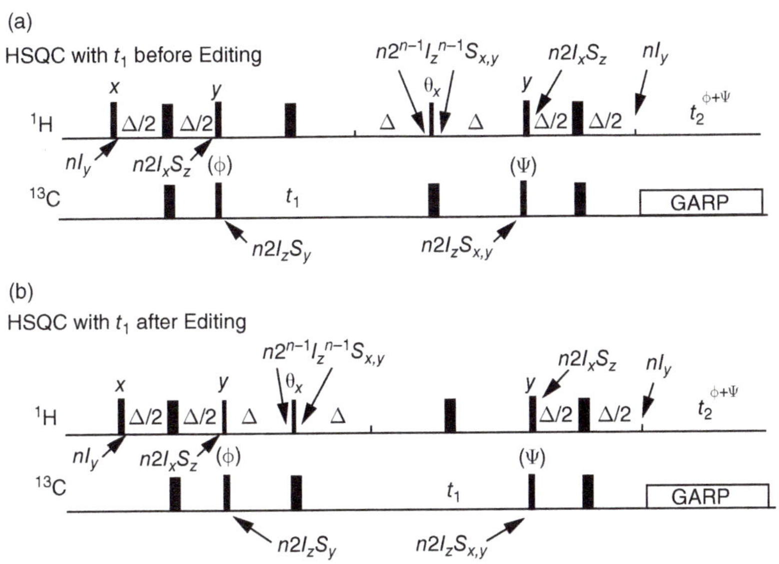


Figure A.19 Expansion of the ID HSQC sequences with editing to 2D sequences. The $t_1/2$ $180°(I)t_1/2$ segment that allows for evolution of heteronuclear chemical shift with refocusing of the heteronuclear coupling can be inserted before editing (a) or after editing (b).

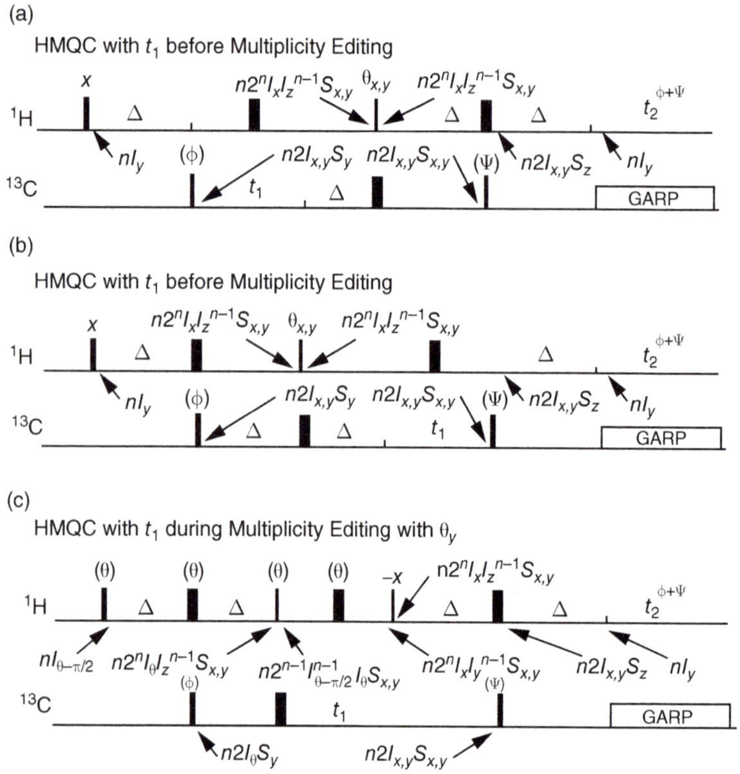


Figure A.20 Expansion of the ID HMQC editing sequences to 2D sequences. Inclusion of $S$-chemical shift evolution can be done in the HMQC in three different ways: (a) before the editing segment, (b) after the editing segment, and (c) within the editing segment. The identity used here is $\theta_y = 90_\theta\ 90_{-x}$. The magnetization of the protons is transverse between the two 90° pulses. Refocusing of proton chemical shift is achieved as usual with a 180°($I$) pulse.

For $I_nS$ systems a HMQC sequence with an additional $1/(2J)$ evolution period for heteronuclear coupling inserted between the two 90°($S$) pulses will yield the state $I_y(2I_z)^{n-1}$ for the proton part of the spin system at the beginning of the detection. Application of a $90^\circ_y(I)$ pulse yields detectable magnetization only for $n = 1$ and thus produces a spectrum containing $IS$ groups. Antiphase magnetization resulting exclusively from $I_2S$ groups can be produced by application of a $90^\circ_x(I)$ pulse [50].

## A.3 Accessories

In the following section some helpful accessories are discussed: Sequences that suppress undesired signals for low natural abundance heteronuclei and solvent(s), methods to increase the limited resolution achieved for the heteronucleus by the application of selective pulses, folding or non-FT data processing, filters, and special tricks.

### A.3.1 Folding

Due to the narrow linewidths common for heteronuclear spectra, methods to increase resolution in indirectly detected frequency domains are highly desirable. This can be achieved by undersampling in the evolution time, or by choosing a dwell time larger than the reciprocal chemical shift range of the nuclei of interest. This leads to a reduced spectral width and thus to an increased resolution with the same number of experiments. It also leads to folding of signals in the frequency domain corresponding to the evolution time. There are two basic folding patterns. [51] Data derived from a real FID fold differently from data derived from a complex FID.

### A.3.1.1 Folding of Data Derived from a Real FID

The folding characteristics of signals in a frequency domain derived from a real FID (usually recorded using TPPI) [52] can be understood by analyzing the real time domain signal: $\cos[\Omega(k/(2\text{SW})+\Delta)+\phi]$. $\Omega$ is the frequency of the nucleus under consideration. The running index for sampling in the time domain is $k$. The sampling rate or dwell time DW is $\text{DW} = \Delta t = l/2\text{SW}$ and the elapsed time consequently is $t = k\Delta t$. $\phi$ is the zero-order and $\Omega\Delta$ the first-order phase of the signal. If $\Omega$ does not fall within the spectral range covered by such an acquisition scheme, we use the periodicity of the trigonometric function to derive the following identity:

$$\cos\left[\Omega\left(\frac{k}{2\text{SW}}+\Delta\right)+\phi\right] = \cos\left[-\Omega\left(\frac{k}{2\text{SW}}+\Delta\right)-\phi\right]$$
$$= \cos\left[(4\pi\text{SW}-\Omega)\left(\frac{k}{2\text{SW}}+\Delta\right)-\phi-\frac{2\pi\Delta}{\Delta t}\right]$$

Thus a signal at $\Omega/2\pi > \text{SW}$ appears in the spectrum with the frequency $2\text{SW}-\Omega/2\pi$ (Figure A.21a). The relative phase of a genuine signal and a folded signal is: $-2\phi - 2\pi\Delta/\Delta t$. Provided there is no first-order phase correction to be applied in $\omega_1$ ($\Delta = 0$), the phase difference is $-2\phi$ between a signal within the spectral range and a signal that is folded once. Setting $\Delta = 0$ can be fulfilled by carefully tuning the delays during which the heteronuclear magnetization is transverse. [53, 54] This is achieved by introducing an additional 180°($S$) pulse in order to refocus all chemical shift evolution for the first increment. Experimental ways to verify this will be given in the following.

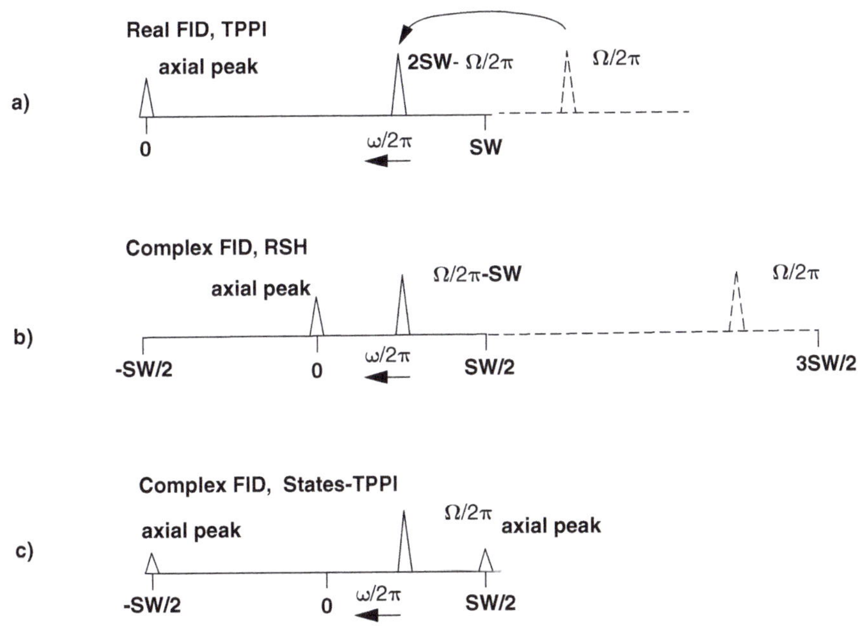


Figure A.21 Folding characteristics of a spectrum obtained (a) from a real FID (TPPI), (b) from a complex FID (Ruben, States, Haberkorn), and (c) from a complex FID according to the States-TPPI method, (a) Peaks outside the spectral range in either direction fold at the proximal (closest) edge. The axial peak is at the low field edge, (b) Peaks outside the spectral width fold by aliasing (they wrap around). The axial peaks are in the middle of the spectrum, (c) The folding characteristic is the same as for (b); the axial peaks, however, lie at both edges of the spectrum.

The choice of zero order phase $\phi = 90°$ (in an HSQC spectrum this corresponds to a phase difference of 90° between the first and last $90°(S)$ pulse) will yield positive sign for genuine signals and signals that are folded an even number of times. Negative sign is obtained for signals that are folded an odd number of times (phase difference $2\phi = 180°$). The two types of signals can therefore be distinguished. For the sake of completeness, the position of axial peaks is also indicated in the schematic Figure A.21a).

#### A.3.1.2 Folding of Data Derived from a Complex FID

Complex data acquired with the method of Ruben, States, Haberkorn, [55] on the other hand, fold by aliasing (Figure A.21b). This can be understood again in the time domain (sampling rate or dwell time is $\mathrm{DW} = \Delta t = 1/\mathrm{SW}$, $t = k\Delta t$):

$$\exp[i\Omega\left(\frac{k}{\mathrm{SW}} + \Delta\right) + i\phi] = \exp[i(\Omega - 2\pi\mathrm{SW})\left(\frac{k}{\mathrm{SW}} + \Delta\right) + i\phi + 2\pi i\mathrm{SW}\Delta]$$

Differentiation between folded and non-folded signals cannot be accomplished by means of a frequency-independent phase difference $\phi$ between the 90°(S) pulses. Only a frequency-dependent phase shift introduced by delaying the start of evolution by a time $\Delta$ allows for this differentiation. The phase difference between genuine and folded signals is given by: $2\pi\mathrm{SW}\Delta$. Setting the first increment to half the dwell time applied for the sampling in $\omega_1$: $\Delta = 0.5\ \mathrm{DW} = (2\mathrm{SW})^{-1}$, introduces phase alterations by 180° [53, 54, 56] between folded and nonfolded signals. To accomplish this, exact tuning of the delays with the rules just given is mandatory. As an example, an HSQC spectrum of the disaccharide 1 with editing and folding, recorded with TPPI and a BIRD inversion pulse to suppress undesired $^{12}$C-H signals (pulse sequence in Figure A.22) (see Section A.3.3), is shown in Figure A.23. The phase difference between the two 90° $^{13}$C pulses is 90°. This ensures that the phase of signals changes every time they are folded. Signals that are folded twice have the same sign as signals that are not folded. Two data sets are acquired, one with the sequence of Figure A.22a and one with the sequence in Figure A.22b. The sum and the difference of the two data sets yield two new data sets, each of which contains positive and negative signals. When positive and negative signals are plotted separately, four spectra result, which contain the following signals: Figure A.23a, CH, $CH_3$ folded an even number of times; Figure A.23b, $CH_2$ folded an even number of times; Figure A.23c, CH, $CH_3$ folded an odd number of times; Figure A.23d, $CH_2$ folded an odd number of times. The spectral width in the carbon dimension is chosen to be 50 ppm (41–91 ppm) although the chemical shift range is

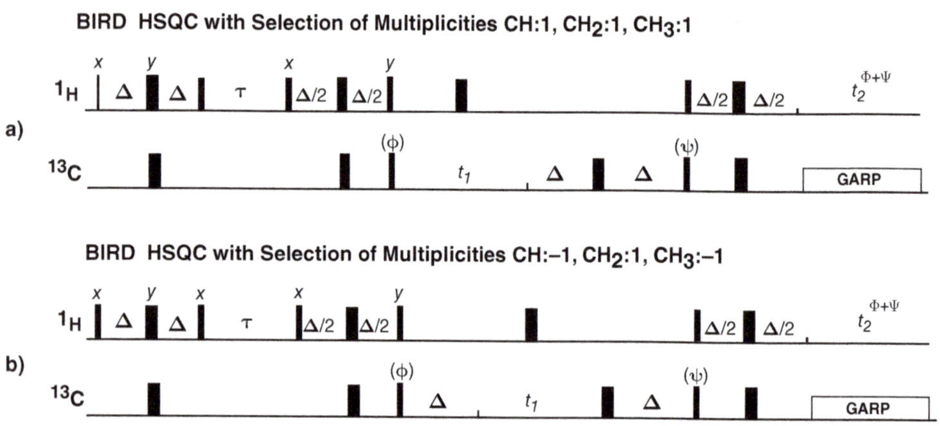


Figure A.22 BIRD-HSQC pulse sequence with discrimination of CH, $CH_2$, and $CH_3$. Heteronuclear coupling does not evolve during the $t_1 + 2\Delta$ duration in (a), whereas it evolves during $2\Delta$ in (b). This leads to a sign inversion $\cos(\pi J\Delta) = -1$ of the carbons carrying an odd number of protons. Sum and difference of (a) and (b) then yield the $CH_2$ and the $CH/CH_3$ spectrum, respectively. Setting the initial phase $\phi = 90°$ yields a sign change of folded peaks an odd number of times versus peaks that are folded an even number of times.

three times as large. Signals between −8 ppm and 42 ppm and signals between 92 and 142 ppm are folded once and therefore have opposite sign to signals that fall within the spectral width. The two folded ranges have the opposite order of resonances due to folding. The assignment of folded peaks is not difficult, because proton and carbon chemical shifts run parallel for most CH moieties. Aromatic protons are directly coupled to low field folded carbons, aliphatic to high field folded carbons. The separation of multiplicities resolves some potential overlap, for example between H(5′) and H(6′/6″).

A third folding method, "states-TPPI", combines the complex acquisition of RSH with the apparent shifting of the spectrum as in the TPPI method. [57] The folding characteristics are identical to the RSH method; the axial peaks, however, appear on both edges of the spectrum (Figure A.21c).

### A.3.2 Non-FT Methods for Processing

To increase the resolution in an indirectly sampled frequency domain, we have so far seen the application of folding and the use of selective pulses. The first method relies on the fact that the spectrum is sparse enough so that overlap between folded and genuine signals is avoided. This is the case for $^1J_{CH}$ correlations because the chemical shifts of protons and carbons correlate very well with each other. The situation is very different for $^1J_{NH}$ correlations. Peaks are distributed throughout the spectrum. Therefore folding cannot be applied in this case. Selective pulses are also not useful if all resonances are of interest. In these circumstances non-FT processing methods can be employed to

(a)

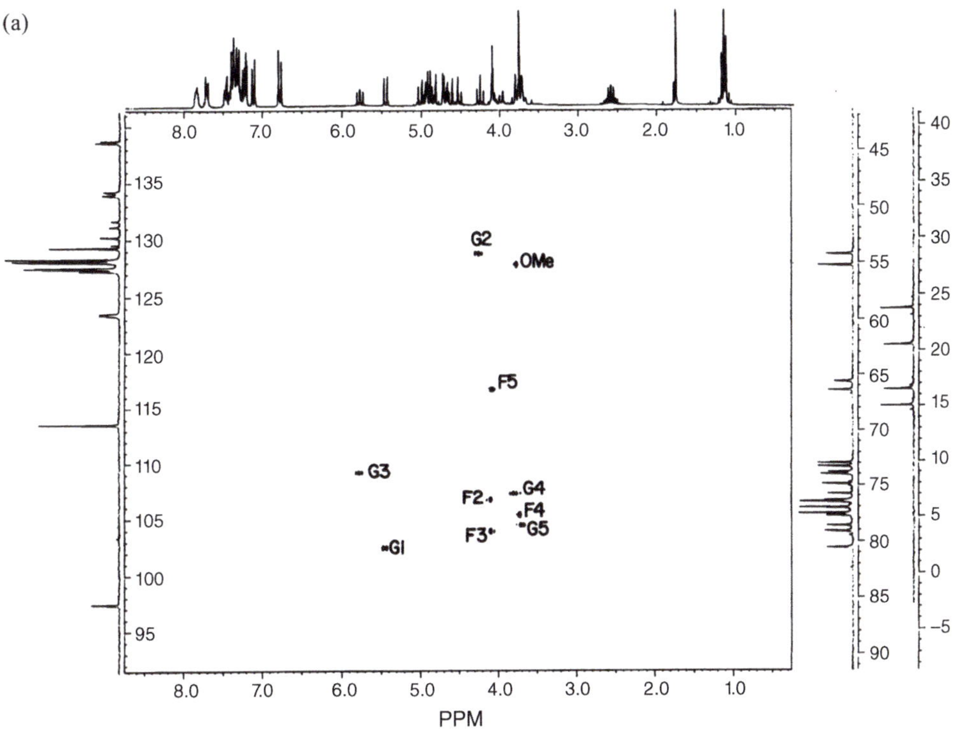


Figure A.23 Four spectra obtained from the pulse sequence of Figure A.22 for ethyl-6-O-(2,3,4-tri-O-benzyl-α-L-fucopyranosyl)-(l,6)-3-O-acetyl-4-O-(p-methoxy-benzyl)-2-desoxy-2-phthalimido-l-thio-β-D-glucopyranoside **1**. $\phi = x, -x$; $\psi = y, y, -y, -y$ ensures that the signals that are folded an odd number of times have inverted sign, (a) CH/$CH_3$ "even-folded"; positive peaks in the difference of Figure A.22a and Figure A.22b. (b) $CH_2$ "even folded": positive peaks in the sum of Figure A.22a and Figure A.22b. (c) CH/$CH_3$ "odd-folded": negative peaks in the difference of Figure A.22a and Figure A.22b. (d) $CH_2$ "odd folded": negative peaks in the sum of Figure A.22a and Figure A.22b.

*(Continued)*

(b)

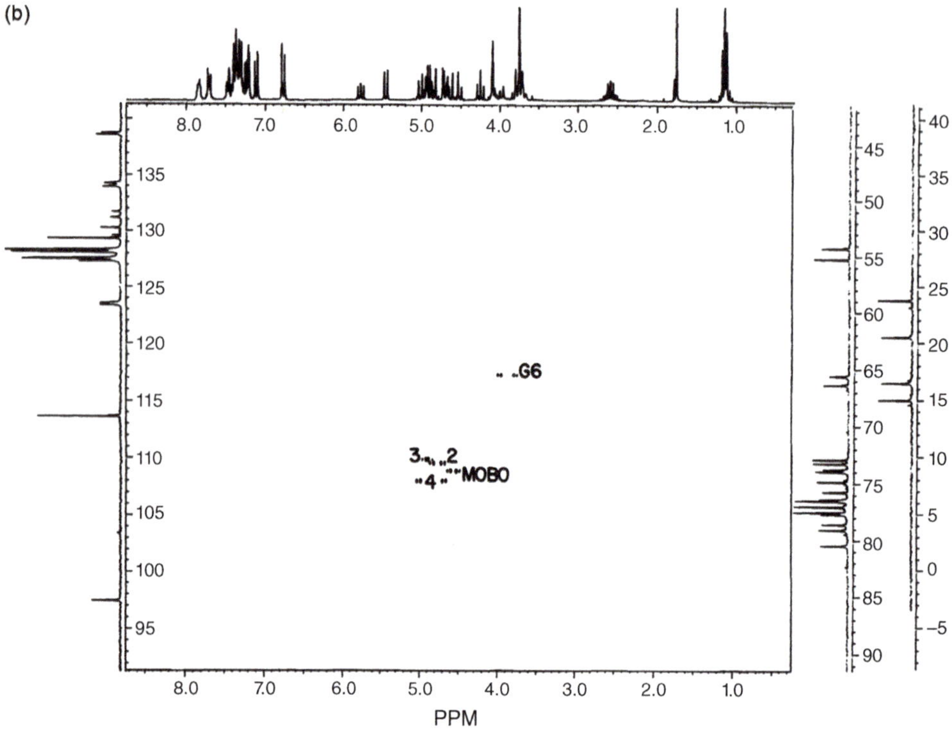


(c)

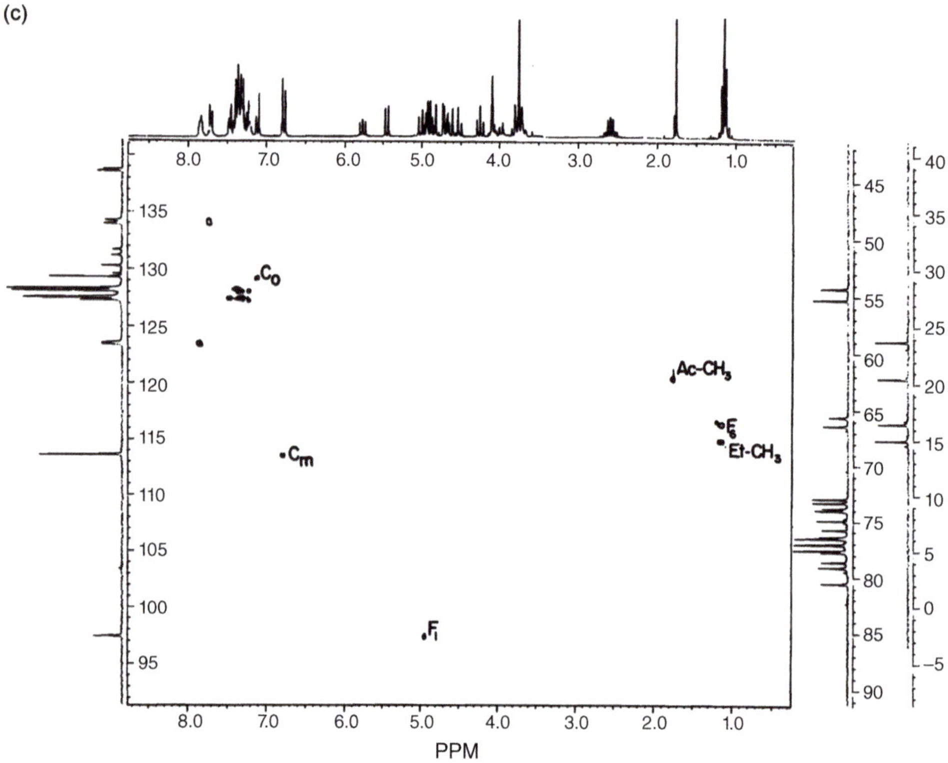


**Figure A.23(b,c) (Cont'd)**

*(Continued)*

(d)

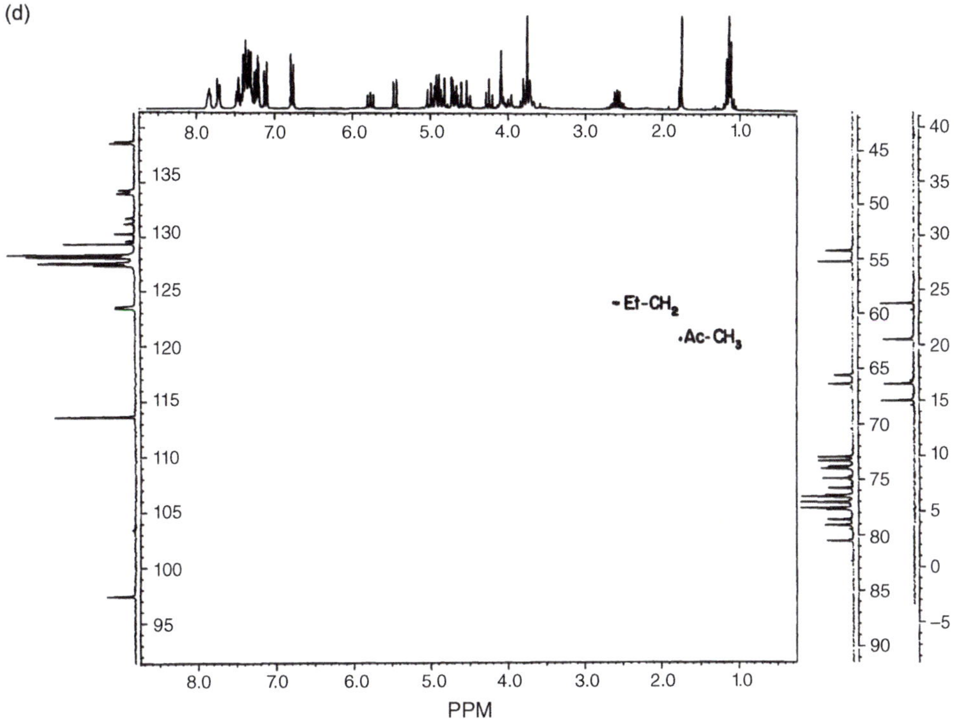


**Figure A.23(d) (Cont'd)**

increase the resolution. Although a whole array of such methods exists, including maximum entropy [58–63] and linear prediction, [64–69] the discussion is confined to the application of linear prediction to increase $t_1^{\max}$ of the FID in the evolution period. [70–73] Linear prediction of the FID is implemented in most commercial software packages and it is rather fast and easy to use. 2D linear prediction has been described recently [74].

A FID sampled at positions $k\Delta t$ containing $p$ Lorentzian lines with frequencies $\omega_j$, line widths $\lambda_j$, and amplitudes $A_j$ is described by the following equation:

$$f_k = f(k\Delta t) = \sum_{j=1}^{p} A_j \exp((i\omega_j - \lambda_j)k\Delta t)$$

This FID also fulfills the following equation for every point $k$, which allows one to predict signal at time $k\Delta t$ from points that were recorded earlier:

$$f_k = \sum_{m=1}^{p} V_m f_{k-m}$$

The $V_m$ are the linear prediction coefficients.

An analogous equation allowing one to express points earlier in time at $k\Delta t$ from signal recorded later $(k+m)\Delta t$ holds also. The relevant equation is again independent of $k$:

$$f_k = \sum_{m=1}^{p} R_m f_{k+m}$$

To predict a FID forward (backward), $p$ coefficients must be determined: $V_1, V_2, \dots, V_p$ (or $R_1, R_2, \dots, R_p$). Using the first $2p$ points of the FID, the following $p$ equations for the determination of the coefficients $V_m$ are obtained:

$$f_{p+1} = \sum_{m=1}^{p} V_m f_{p+1-m}$$

$$f_{p+2} \sum_{m=1}^{p} V_m f_{p+2-m}$$

$$\vdots$$

$$f_{p+k} = \sum_{m=1}^{p} V_m f_{p+k-m}$$

$$\vdots$$

$$f_{2p} = \sum_{m=1}^{p} V_m f_{2p-m}$$

$$f(k\Delta t) = \sum_{m=1}^{p} A_j \exp[(i\omega_j - \lambda_j)k\Delta t]$$

This system of linear equations can be solved for the coefficients. New points beyond the end of the acquired FID can be generated (predicted) by applying the prediction equation to the second half of the FID.

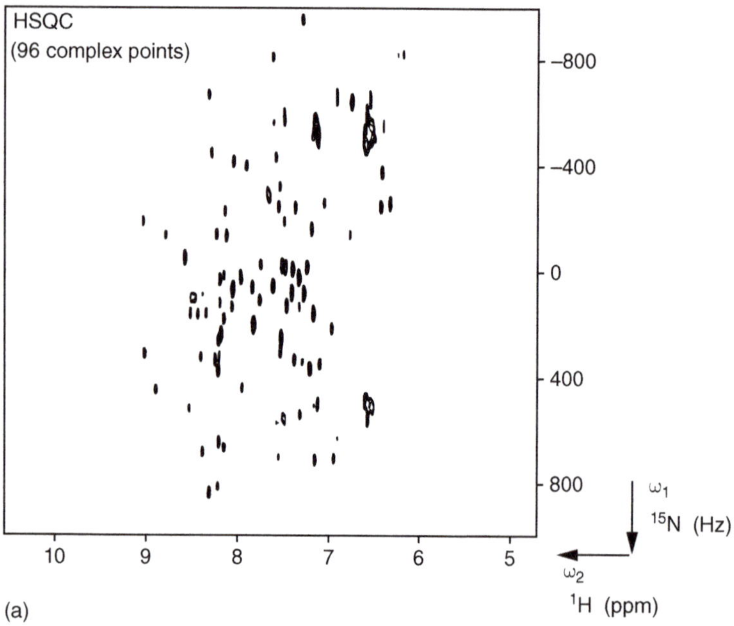


**Figure A.24** HSQC of $^{15}N$-labeled ribonuclease A. Ninety-six complex points were recorded in $t_1$. The number of real points in the $\omega_1$ direction is 256 in all spectra, (a) Ninety-six experimental complex points were Fourier transformed, (b) Thirty-two complex points were Fourier transformed. The resolution is rather low. There is considerable overlap of resonances in $\omega_1$. (c) Sixty-four new complex points were linearly predicted along $t_1$ from 32 complex points. The resolution is now sufficient to remove the overlap in $\omega_1$. This Spectrum compares well with the spectrum in (a).

*(Continued)*

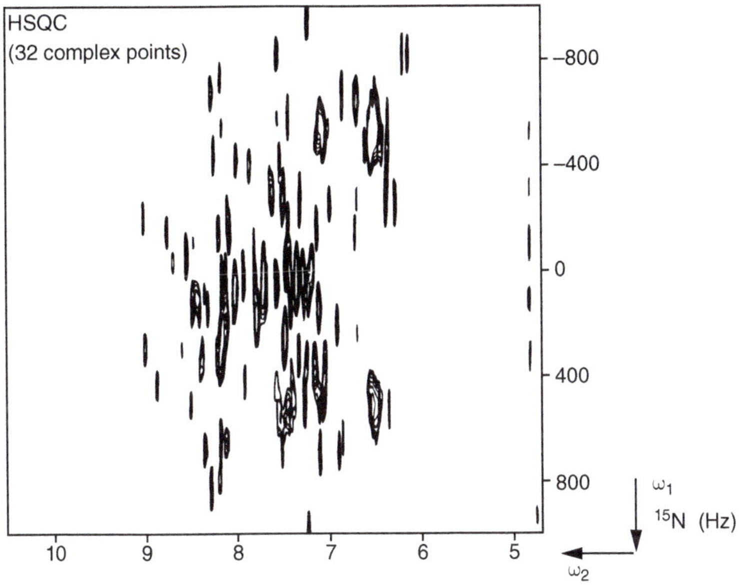


(b)

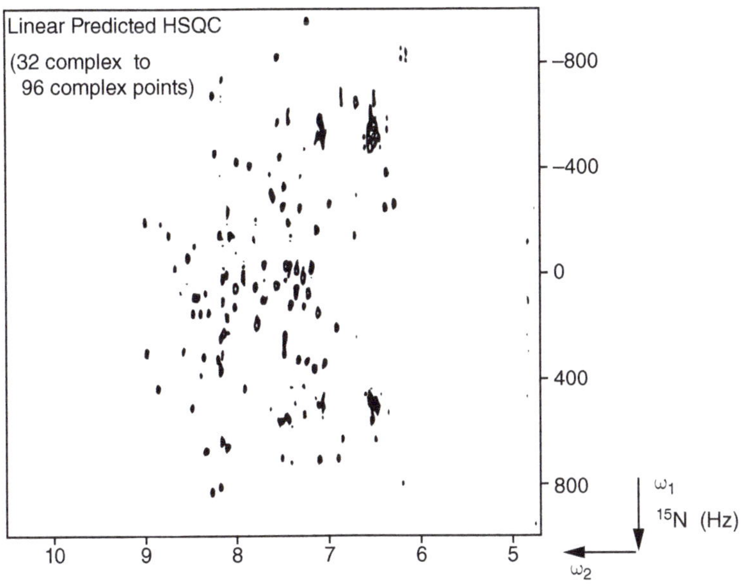


(c)

**Figure A.24(b,c) (Cont'd)**

As an example, $^{1}H$,$^{15}N$ HSQC spectra of $^{15}N$-labeled ribonuclease A all with identical digitization in $\omega_1$ are shown in Figure A.24. The spectra were recorded with 96 complex points in $t_1$. Conventional Fourier transformation including all $t_1$ points leads to the spectrum in Figure A.24a. Fourier transformation of only 32 complex points yields the spectrum in Figure A.24b. Application of linear prediction to the first 32 complex $t_1$ points generates 64 additional complex points. Fourier transformation of the predicted time-domain data yields a spectrum with dramatically increased resolution (Figure A.24c).

Linear prediction may be successfully applied if the S/N is high, the FID is truncated, and the spectrum contains fewer resonances than the number of recorded points in the frequency dimension whose resolution is enhanced by linear prediction. This requires, for ID linear prediction in multidimensional data, that all dimensions orthogonal to the predicted dimension be Fourier transformed before the application of linear prediction. The number of faithfully predictable points depends strongly on the signal to noise.

#### A.3.2.1 Mirror Image LP [75]

In constant-time experiments, the FID in $t_1$ does not decay. The decay constants $\lambda_j$ are zero. Provided the FID in $t_1$ is sampled at positions: $k/\mathrm{SW}$ or $(k+0.5)/\mathrm{SW}$, no linear phase correction is required and the FID for negative times can be obtained from the FID at positive times by complex conjugation:

$$f(-k\Delta t) = \sum_{j=1}^{p} A_j \exp((i\omega_j)(-k\Delta t)) = \sum_{j=1}^{p} A_j \exp((i\omega_j)k\Delta t)^*$$
$$= (\mathrm{f}(k\Delta t))^*$$

The number of basis points for the linear prediction is therefore increased by a factor of 2, which improves the results. [75] An example in connection with the determination of coupling constants will be given later (Figure A.68).

## A.3.3 Bilinear Rotations

Pulses are rotations that act on magnetization vectors or the individual components of each spin in the density matrix. Bilinear rotations consist of pulses and delays effecting rotations that can be described as rotations about axes given by bilinear product operators. The evolution of coupling is a very simple example of a bilinear rotation (Figure A.2).

#### A.3.3.1 BIRD Pulse

The simplest bilinear pulse is the BIRD pulse: $(90_x(I)-\Delta-180_x(I,S)-\Delta-90_x(I, \Delta = 1/(2J))$. [76] To understand what this pulse accomplishes, note that $180_x(I,S)$ refocuses chemical shift in the middle of $2\Delta$, so only the coupling needs to be taken into account. Since coupling evolves irrespective of the $180°(I,S)$ pulses, these two pulses can be thought of as if they occurred at the beginning of the sequence, yielding $180_x(I,S)\,90_x(I) - 2\Delta - 90_x(I) = 180_x(S)\,90_{-x}(I) - 2\Delta - 90_x(I)$. This sandwich acts on single spin magnetization, that is, on spins that do not have a $^1J_{IS}$ coupling, like a $180_x(S)$, since the effect of the two $90°(I)$ pulses is canceled. However, for magnetization of a proton bound to a heteronucleus the effect is different: Consider transverse $x$ - magnetization of spin $I$. It is not affected by the $90_{-x}(I)$ pulse, changes its sign due to the evolution of the heteronuclear coupling during $1/J$ and is left unaffected by the pulse $90_x(I)$, too. So in contrast to the magnetization of an $I$-spin that is not coupled to an $S$ spin, $I_x$ magnetization of an $I$-spin bound to an $S$ spin is inverted by the BIRD pulse. $I_y$ magnetization on the other hand is converted to $I_z$ by the first pulse. This magnetization does not evolve any heteronuclear coupling and therefore is reconstituted at the end of the BIRD pulse. Finally $I_z$ magnetization is converted to $-I_y$ magnetization, that is, is inverted due to evolution of heteronuclear coupling during $2\Delta$ and is therefore converted to $-I_z$ after the BIRD pulse. In conclusion, $\mathrm{BIRD}_x$ acts on $I$ magnetization in a coupled $IS$ spin system like a $180_y$ inversion pulse, whereas it acts as a $0°$ pulse on $I$ magnetization in an uncoupled $I$ spin system. Since even more input magnetizations can be conceived, in the following we try to describe the bilinear rotation executed by the BIRD pulse. This description is more abstract and needs to be read only by the reader interested in mathematical excursions.

Starting again with the pulse $180_x(S)\,90_{-x}(I) - 2\Delta - 90_x(I)$ we try to express the evolution of coupling during $2\Delta$. The Hamiltonian for the heteronuclear coupling is $2\pi J I_z S_z$. This coupling applied during $2\Delta = 1/J_{IS}$ executes a bilinear rotation by $\pi$ around the $(2I_zS_z)$ axis. Now, due to the $90_{-x}(I)\ 90_x(I)$ sandwich around the free evolution, the rotation axis of the bilinear rotation for spin $I$ is rotated by 90° about the $x$-axis from $z$ to $-y$, and the $180_x(S)$ pulse is applied in the beginning. So the BIRD pulse rotates by $\pi$ around the bilinear axis $-2I_yS_z$. (This concept has been used by Sørensen [2] for the derivation of optimum coherence transfer $I \rightarrow S$ in $I_nS$ systems). The transformations of some representative operators are given in Figure A.25.

As is obvious from the transformations and the explanations just given, the BIRD$_x$ [76] pulse $(90_x(I) - \Delta - 180_x(I,S) - \Delta - 90_x(I))$ with $\Delta = l/(2J)$ inverts $z$ magnetization of protons bound to carbon, whereas magnetization of protons not bound to carbon is untouched. The reverse holds for BIRD$_y$ $(90_x(I) - \Delta - 180_y(I,S) - \Delta - 90_x(I))$.

The BIRD$_y$ element can be used to selectively annihilate magnetization of protons that are not bound to $^{13}C$ or $^{15}N$ by selective inversion of $S$-bound $I$ spin magnetization. Due to $T_1$ relaxation these protons have zero magnetization after a suitably chosen delay [77]. The magnetization of protons bound to $^{13}C$ or $^{15}N$ is untouched. Application to a spin system with $z$-magnetization of $I$ spins inverts the non-$S$-bound $I$ spins (protons), whereas the $S$-bound protons are untouched. After an appropriate delay $\tau$ the $z$-magnetization of the non-$S$-bound protons is zero and the pulse sequence is started with only the $S$-bound protons contributing Figure A.26.

Although the proton relaxation times $T_1$ are not uniform for all protons in a molecule, good results are achieved by taking the smallest $T_1$ and fulfilling the equation:

$$1 - \exp\left[-\frac{(t_2^{\max} + T)}{T_1}\right] = -\left[1 - \exp\left(\frac{\tau}{T_1}\right)\right]$$

where $T$ and $\tau$ are defined in Figure A.26. These equations make the realistic assumption that there is no longitudinal proton magnetization at the beginning of $t_2$ and impose the requirement that there is no longitudinal proton magnetization at the beginning of the pulse sequence for the non $S$-bound $I$ spins. In principle, all values for $t_2^{\max} + T$ can be chosen. However, the range of $T_1$ times for which suppression is effective and the S/N per unit time depend on the choice of $t_2^{\max} + T$ Good results are obtained for $t_2^{\max} + T = 0.82\,T_1$, $\tau = 0.48\,T_1$. This corresponds to rather rapid repetition of the pulse sequence, which is fine from the point of sensitivity. Of course, $t_2^{\max} + T$ should be longer than $T_2$. Otherwise, $t_1$ noise will result. The spectra in Figure A.23 were obtained with an acquisition time $t_2^{\max} = 400$ ms ($T = 0$) and a relaxation delay $\tau = 228$ ms.

Further applications of bilinear $\pi$ pulses are "homonuclear" decoupling in an evolution time, using the fact that a BIRD$_y$ pulse inverts transverse magnetization of spin $I$ in an $IS$ spin system and also inverts $z$-magnetization of an $I$ spin not bound to $S$. [78] An analogous application allows the distinction of $^2J_{CH}-$ and $^3J_{CH}$-couplings in long-range correlation experiments with protonated carbons [79].

**Figure A.25** Operator transformations under the action of a bilinear inversion pulse for an $IS$ and a pure $I$ spin system, (a) BIRD$_x$ inverts, for example $z$-magnetization of an $I$ spin that is bound to an S spin ($\Delta = 1/2J_{IS}$) whereas the $z$-magnetization of a spin that is not bound to an S spin is not inverted. The operator transformations are derived from the identity BIRD$_x = \pi_x(S)\pi(2I_yS_z)$. $\pi_x(S)$ means application of a 180°($S$) pulse on the initial state. Then the bilinear rotation pulse $\pi(2I_yS_z)$ is applied, which rotates pure I or S operators about the $y$- and $z$-axis, respectively. Bilinear operators that contain neither $I_y$ nor $S_z$ and the operator $2I_yS_z$ remain unchanged. Bilinear operators containing exclusively either $I_y$ or $S_z$ are inverted. Another implementation of BIRD$_x$ is shown in the frame. (b) BIRD$_y$ can be derived from the BIRD$_x$ inversion pulse by executing a $180_x(I)$ pulse in addition to the $180_x(S)$ pulse on the initial state. Another implementation is shown in the frame. Some of the interesting transformations are $I_z \rightarrow I_z$ in an $IS$ system and $I_z \rightarrow -I_z$ in an $I$ system. Homonuclear decoupling is achieved by BIRD$_y$ because one can invert $I_z$ in an $I$ system and at the same time leave alone (except for the total sign) both transverse components of an I spin in an IS system.

(a)

$$\mathrm{BIRD}_x = \pi_x(S)\ \pi(-2I_yS_z)$$

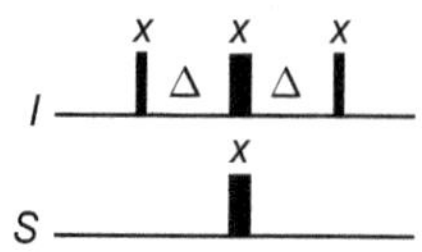


$I_z \rightarrow -I_z$ $\quad$ $I_x \rightarrow -I_x$ $\quad$ $I_y \rightarrow I_y$

I–S

$2I_zS_z \rightarrow 2I_yS_z$ $\quad$ $2I_xS_z \rightarrow 2I_xS_z$ $\quad$ $2I_yS_z \rightarrow -2I_yS_z$

$2I_zS_x \rightarrow 2I_yS_z$ $\quad$ $2I_xS_x \rightarrow 2I_xS_x$ $\quad$ $2I_zS_x \rightarrow -2I_yS_x$

$2I_zS_y \rightarrow -2I_zS_y$ $\quad$ $2I_xS_y \rightarrow -2I_xS_y$ $\quad$ $2I_zS_y \rightarrow 2I_yS_y$

$S_z \rightarrow -S_z$ $\quad$ $S_x \rightarrow -S_x$ $\quad$ $S_y \rightarrow S_y$

I

$I_z \rightarrow I_z$

$I_x \rightarrow I_x$

$I_y \rightarrow I_y$

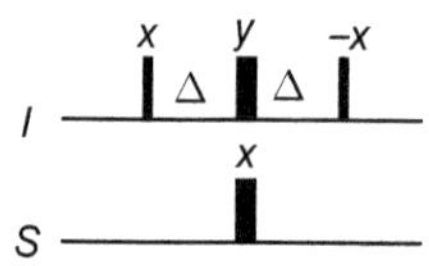


$= \pi_z(I)\ \mathrm{BIRD}_x$

---

(b)

$$\mathrm{BIRD}_y = \pi_x(I,S)\ \pi(-2I_yS_z) = \pi_x(I)\ \mathrm{BIRD}_x$$

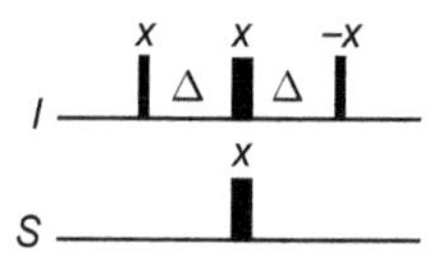


$I_z \rightarrow I_z$ $\quad$ $I_x \rightarrow -I_x$ $\quad$ $I_y \rightarrow -I_y$

I–S

$2I_zS_z \rightarrow -2I_zS_z$ $\quad$ $2I_xS_z \rightarrow 2I_xS_z$ $\quad$ $2I_yS_z \rightarrow 2I_yS_z$

$2I_zS_x \rightarrow -2I_zS_x$ $\quad$ $2I_xS_x \rightarrow 2I_xS_x$ $\quad$ $2I_yS_x \rightarrow 2I_yS_x$

$2I_zS_y \rightarrow 2I_zS_y$ $\quad$ $2I_xS_y \rightarrow -2I_xS_y$ $\quad$ $2I_yS_y \rightarrow -2I_yS_y$

$S_z \rightarrow -S_z$ $\quad$ $S_x \rightarrow -S_x$ $\quad$ $S_y \rightarrow S_y$

I

$I_z \rightarrow -I_z$

$I_x \rightarrow I_x$

$I_y \rightarrow -I_y$

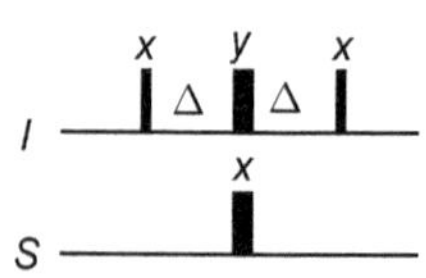


$= \pi_z(I)\mathrm{BIRD}_y$

$= \pi_y(I)\pi_{-x}(I)\mathrm{BIRD}_y$

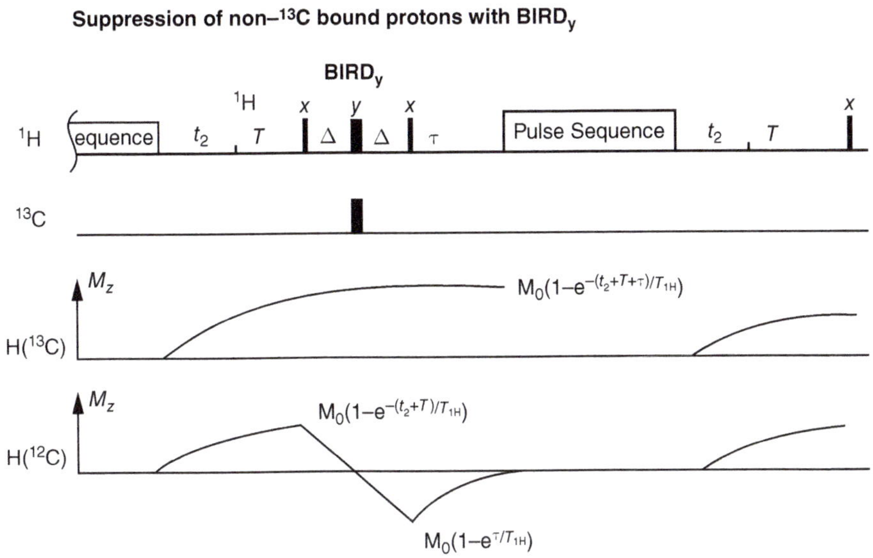


**Figure A.26** Application of BIRD$_y$ for selective inversion of non-$S$-bound $I$ spins at the beginning of a pulse sequence. The delay $\tau$ is tuned such that zero longitudinal magnetization results for non-$S$-bound $I$ spins at the start of the preparation period of the sequence.

### A.3.3.2 BIRD/2 Pulse

The following short section employs product operator formalism alone and therefore is on a more abstract level than the rest of this chapter.

We have discussed the action of the sequence $90(I) - \Delta - 180(I,S) - \Delta - 90(I)$ with $2\Delta = 1/J$. It is essentially an inversion pulse ($\pi$ pulse) around the bilinear axis $-2I_yS_z$. A $\pi/2$ pulse around this axis has other interesting features. This rotation is accomplished by setting $\Delta = 1/(4J)$. Since the rotation angle is only $\pi/2$ we call this bilinear rotation BIRD/2.

The sequence $90_x(I) - \Delta - 180_x(I,S) - \Delta - 90_x(I) = \text{BIRD}_x/2$ can be written $180_x(S)\, \pi/2(-2I_yS_z)$ based on the same reasoning as used for BIRD. The transformations this pulse effects are a bit more involved. It executes a 0° rotation on $I$ spins that are not bound to an $S$ spin. This is immediately clear, because the net rotation for $I$ spins is 360°. $I$ spins in an $IS$ spin system experience a 90° pulse around the $2I_yS_z$ bilinear axis. This means that for $I_x$ or $I_z$ in the product operator before the bilinear pulse, the product operator is rotated by $90_y$ and that an operator $2S_z$ is added or deleted from the product operator if it did not exist or existed already, respectively. The essential transformations are given in Figure A.27.

To account for other phases in the BIRD/2 pulse, for example $90_x(I) - \Delta - 180_x(I,S) - \Delta - 90_y(I)$ we deduce their behavior from the BIRD/2 pulse we have just discussed by appending additional pulses. Application of an additional $90_{-x}(I)90_y(I)$ pulse after the BIRD$_x$/2 pulse: $\text{BIRD}_x/2 - 90_{-x}(I) - 90_y(I) = 90_x(I) - \Delta - 180_x(I,S) - \Delta - 90_x(I) - 90_{-x}(I) - 90_y(I) = 90_x(I) - \Delta - 180_x(I,S) - \Delta - 90_y(I)$ yields the BIRD$_y$/2 pulse. Since we know what the BIRD$_x$/2 pulse does, we in addition need to know what happens after application of a $90_{-x}(I)\ 90_y(I)$ pulse. $90_{-x}(I)\ 90_y(I)$ performs a cyclic permutation of $I_x$, $I_y$ and $I_z$ magnetization. The BIRD$_y$/2 sequence is familiar since it forms the first pulses in INEPT or HSQC where it creates the two-spin order $2I_zS_z$ from $I_z$. Other useful transformations of this sequence are noted in Figure A.27.

Applications of these pulses will be given in the Sections A.3.4 and A.5.2.

#### A.3.3.3 $\beta$-Pulses

A small flip angle pulse $\beta(I)$ may be used to rotate the magnetization of a spin $I_1$ without disturbing the spin levels of a spin $I_2$. We first consider the rotation of the $I_1$ spin under a $\beta_y$ pulse. The following transfer efficiencies are observed:

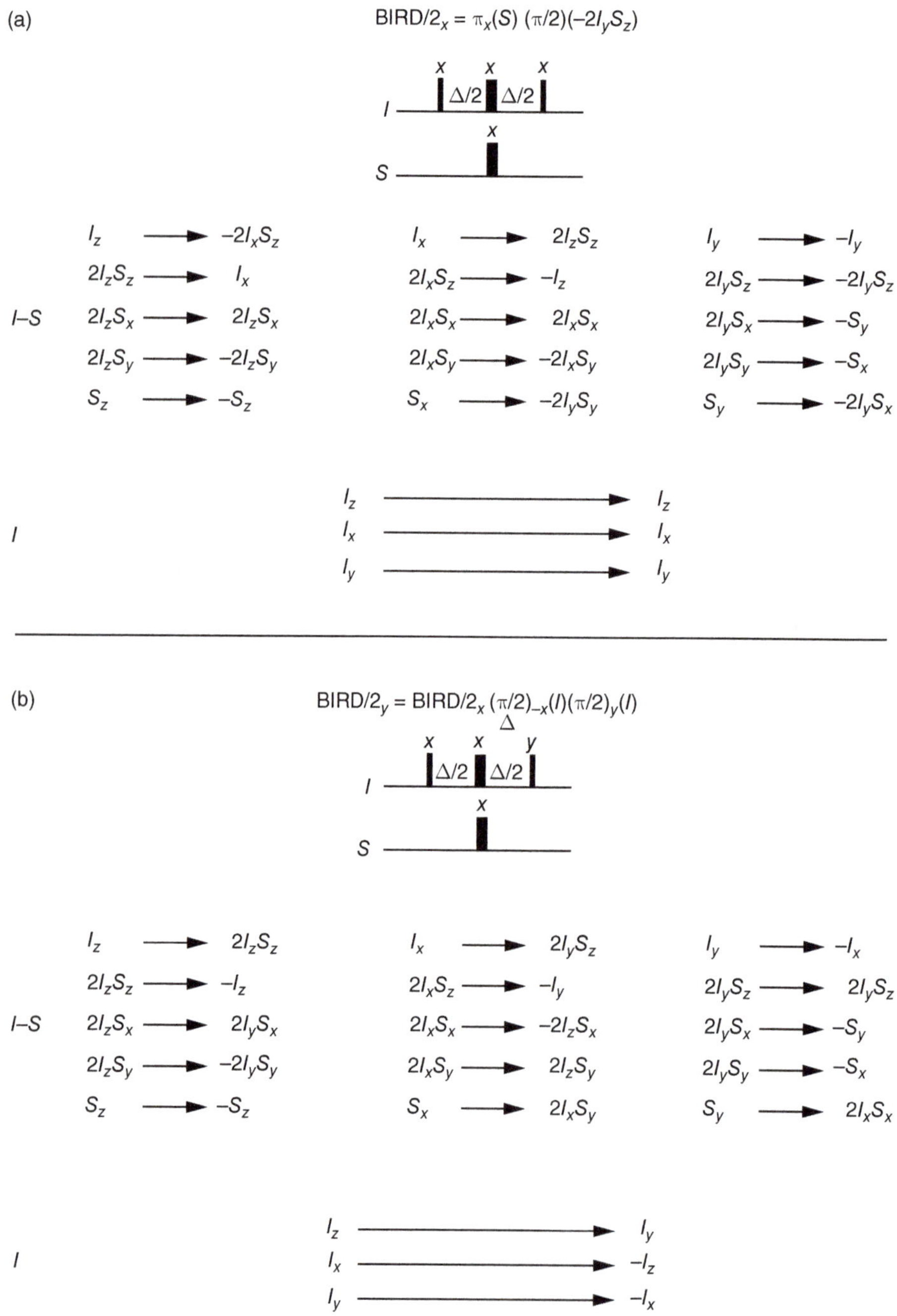


**Figure A.27** BIRD/2 pulses lead to a bilinear $\pi/2$ rotation. The operator transformations effected by BIRD$_x$/2 (a) are denoted in the same way as the ones in Figure A.25. (b) The BIRD$_y$/2 pulse can be constructed from the BIRD$_x$/2 pulse by attaching a sandwich $(\pi/2)_{-x}(I)(\pi/2)_y(I)$.

$$I_{1x} \rightarrow -I_{1z} \sin\beta + I_{1x} \cos\beta$$
$$I_{1z} \rightarrow \quad I_{1x} \sin\beta + I_{1z} \cos\beta$$

In the section about product operators, we have used in addition to $I_x$, $I_y$, and $I_z$ the operators $I_\alpha$ and $I_\beta$ which describe a spin in the $\alpha$ or in the $\beta$ state. We find for these operators the following behavior after the application of a $\beta$ pulse:

$$I_{2\alpha} \rightarrow I_{2\alpha} \cos^2\left(\frac{\beta}{2}\right) + I_{2\beta} \sin^2\left(\frac{\beta}{2}\right) + \text{transverse parts}$$

$$I_{2\beta} \rightarrow I_{2\beta} \cos^2\left(\frac{\beta}{2}\right) + I_{2\alpha} \sin^2\left(\frac{\beta}{2}\right) + \text{transverse parts}$$

Given these different transformation behaviors it is possible to find a flip angle $\beta$ where the spin state of a longitudinal spin is not inverted to a large extent while at the same time the transfer between longitudinal and transverse magnetization is rather efficient. Setting $\beta = 36°$, only one-tenth of the spins change their states from $\alpha$ to $\beta$ and vice versa. On the other hand, transfer of magnetization is effected with 0.59 of the efficiency of a 90° pulse. Therefore $\beta$ pulses can be used to rotate spins by 90° and leave the spin state of a passively coupled spin almost completely untouched. Spin $I_2$ can be regarded as a quasi-heteronuclear spin with respect to $I_1$. This principle will be used in Section A.5.

### A.3.4 Gradients

We have discussed two methods to suppress undesired $^{12}$C-bound protons and solvent signals: phase cycling and the exploitation of $T_1$ relaxation using the BIRD sequence. The subtraction method used with phase cycles requires high stability and uses the dynamic range of the ADC very inefficiently. In addition, for large molecules the BIRD trick does not work: NOE between the inverted $^{12}$C-bound protons and the $^{13}$C-bound protons that are not inverted leads to a smaller initial magnetization for the $^{13}$C-bound protons. Since both $T_1$ and the NOE increase with increasing molecular weight, the BIRD concept is not applicable for macromolecules [77].

$B_0$ or $B_1$ gradients can be used in place of phase cycling to suppress the proton signals of solvent and $^{12}$C-bound protons. Thus $B_0$ and $B_1$ gradients are a third alternative to minimize the signal of non-$^{13}$C-bound protons including protons of the solvent.

We first discuss $B_0$ gradients that can be used with great advantage in homonuclear and heteronuclear sequences. [15, 80–87, 93] Although they have been applied only recently to heteronuclear proton-detected spectroscopy, the results look very promising. $B_0$ gradients are normally implemented as fast,switchable, linearly spatially inhomogeneous $B_0$ fields of several Gauss per centimeter. The time required to switch the gradients on and off must be no more than a few microseconds. Depending on the shielding of eddy currents, homogeneity of the $B_0$ field can be restored a few hundred microseconds after the gradient is switched off. Application of a $B_0$ gradient along the $z$ axis imposes, in addition to the normal evolution of chemical shift, a rotation with a frequency that depends on the $z$-coordinate within the sample. At the end of the gradient pulse, the dephasing results in a complete cancellation of macroscopically observable magnetization. However, the magnetization remains in a highly ordered state. Therefore, macroscopically observable magnetization will result if the net phase imposed on a coherence pathway by a series of $B_0$ gradients during the pulse sequence equals zero. Gradients may be used to select particular coherences and suppress others because the phase acquired depends on the coherence order as well as on the gyromagnetic ratio of the involved nuclei. For example, the phase of a coherence $I_1^+ I_2^+ S^-$ under a gradient $G_z$ is proportional to $(2\gamma_1 - \gamma_s)G_z$. These principles are described in more detail in Chapter 2 of this volume.

The conventional amplitude-modulated HSQC sequence employing gradients for coherence selection illustrates the principle (Figure A.28). Slightly modified sequences have been published. [88–90] For the sake of simplicity we set $\gamma_H/\gamma_C = 4$. The $4G_z$ gradient (two $|\ 2G_z\ |$ gradients) during the duration in which carbon is transverse

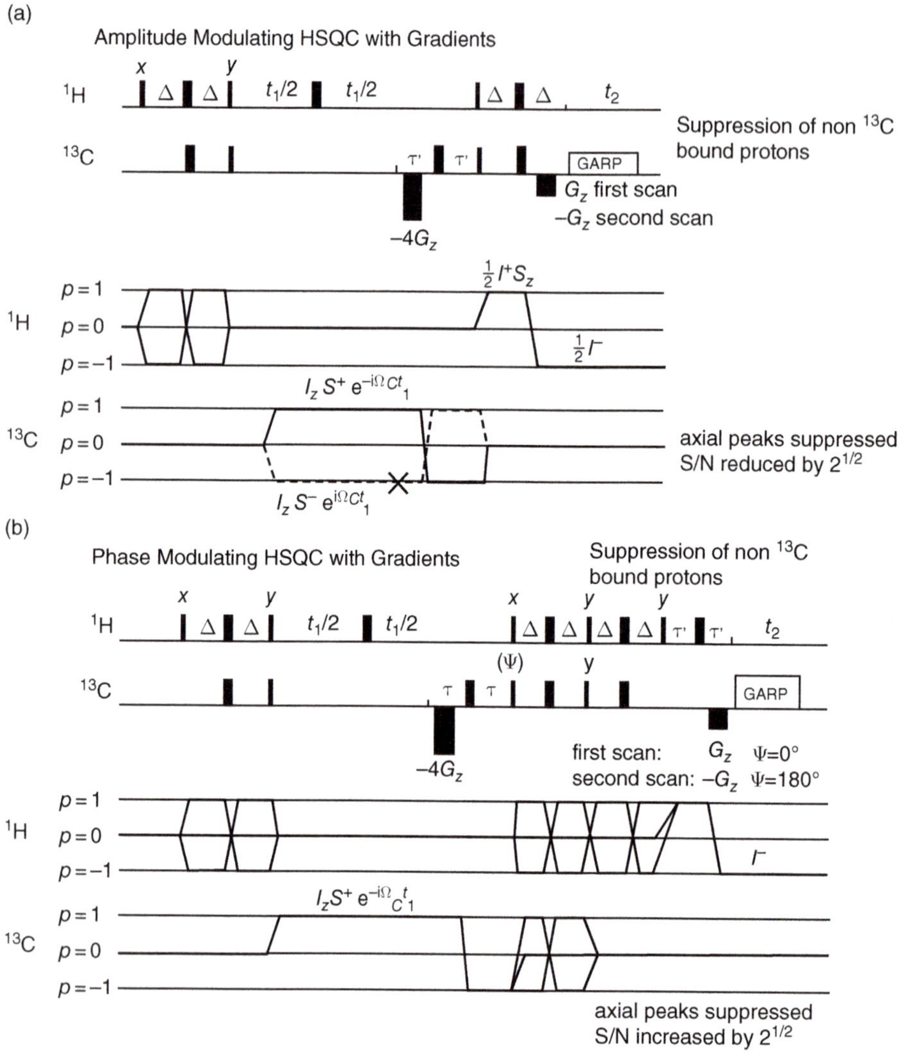


**Figure A.28** (a) Amplitude-modulating HSQC with gradients. The "conventional" amplitude-modulating HSQC sequence (Figure A.13a) is taken. A $4G_z$ gradient is applied during $t_1$. The refocusing gradient is applied in the refocusing delay of the reverse INEPT at the end of the pulse sequence. Only one of the two possible coherence pathways is selected leading to a signal-to-noise loss of $2^{1/2}$ compared with the conventional HSQC sequence (Figure A.13a). (b) Phase-modulating HSQC with gradients. The sensitivity-enhanced HSQC sequence yields a phase-modulated signal based on its property to select the heteronuclear echo and antiecho spectrum depending on the phase of the final $\pi/2(S)$ pulse (Figure A.13b). No signal is lost on incorporation of the gradient pulses because the gradients select echo and antiecho pathways in accordance with the phase $\psi$.

labels the desired coherences $S^+$ (echo) and $S^-$ (antiecho) with $\pm\gamma_C 4G_z$. The gradient before acquisition labels the desired coherence with $-\gamma_H G_z$. Since $4\gamma_C = \gamma_H$, the net phase acquired is zero. Therefore, the desired pathway with carbon magnetization transverse during $t_1$ is selected. Because water- or $^{12}$C-bound protons cannot form such magnetization, their signal is suppressed, since it is dephased by the final gradient. Note that, depending on the sign of the proton gradient, either the $S^+$ or the $S^-$ coherence during $t_1$ is selected. So either the echo or the antiecho spectrum is recorded in one scan.

Because echo and antiecho spectra must be recorded in separate experiments, there is a loss of signal by a factor of $2^{1/2}$ compared with the phase- cycled version of the experiment. This is generally found for amplitude-modulated sequences with gradients in $t_1$ and $t_2$. More specifically, when a gradient echo is formed from two gradients that are applied during $t_1$ and $t_2$, respectively, the loss is $2^{1/2}$. The loss in sensitivity is even larger, namely a factor of 2, when a gradient echo is formed from gradients of which at least one is applied during a fixed delay.

Phase-modulated sequences do not suffer from this sensitivity loss because by definition they select either the echo or the antiecho pathway. Therefore, incorporation of $B_0$ gradients into the sensitivity-enhanced HSQC according to Figure A.28b allows one to record the HSQC without loss in sensitivity, at least for *IS* groups, compared with the sensitivity-enhanced version recorded with conventional phase-cycling procedures. [91] It should be noted that both the relative phase of the two 90°(*S*) pulses in the reverse INEPT and the relative sign of the gradients during $t_1$ and before $t_2$ select echo or antiecho, respectively. If they are not matched, no signal is acquired at all.

For gradient-enhanced sequences, the suppression of undesired signal such as water, including its $t_1$ noise, is tremendous, as can be judged from the spectrum of a 10-mM natural abundance sample of LHRH in $H_2O/D_2O$: 9/1 (Figure A.29a) recorded with the amplitude-modulated HSQC sequence with gradients according to Figure A.28a. The region around the water resonance cannot be interpreted in the conventional experiment using presaturation of water (Figure A.29b). The gradient sequence, however, yields even the $W\text{–}C_\alpha,H_\alpha$ cross peak at 4.7 ppm that is under the water resonance. An inverse probe with actively shielded gradients has been used to acquire this spectrum. Gradients with peak strength of 10 G/cm and duration of 1 ms followed by 200 μs recovery were used. The gradient shape was a half sine wave. The loss of S/N of the amplitude-modulated gradient enhanced

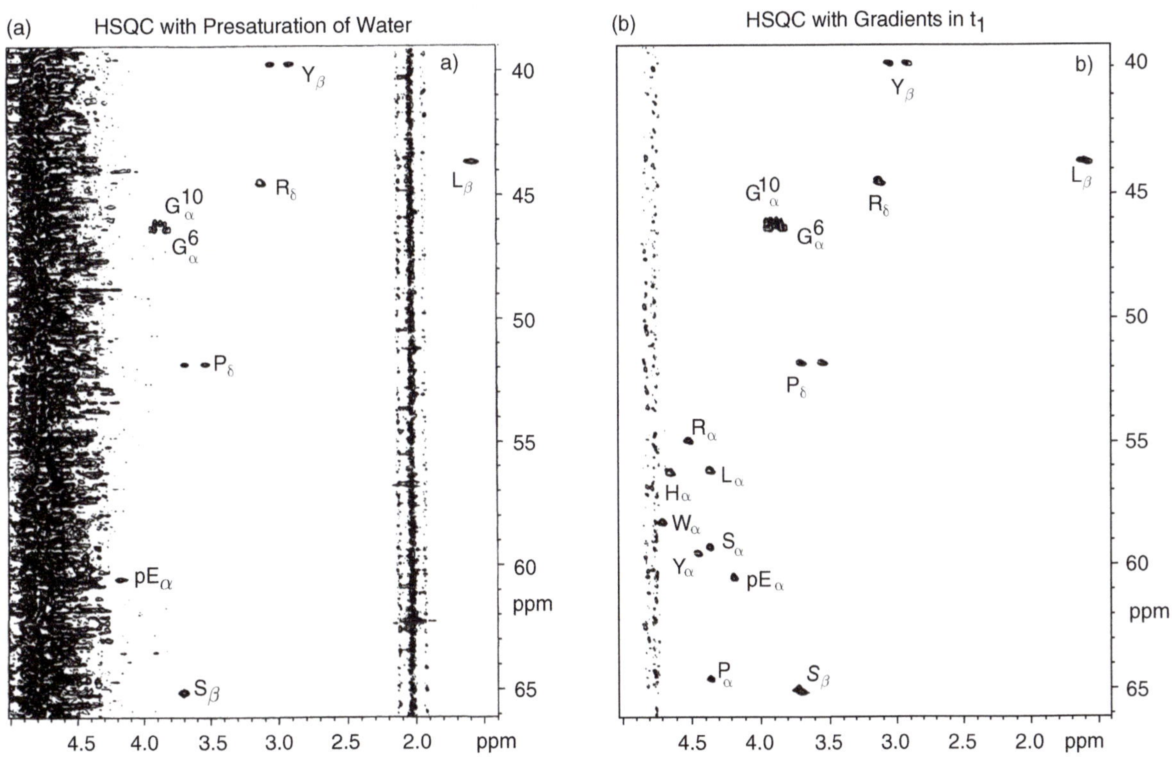


**Figure A.29** (a) Conventional $^{13}C,^{1}H$-HSQC (pulse sequence Figure A.13) of LHRH: PE-H-W-S-Y-G-L-R-P-G-OH at natural abundance in $H_2O/D_2O$: 95/5. The water resonance covers almost all $C_\alpha,H_\alpha$ cross peaks, (b) Amplitude-modulating HSQC with gradients (Figure A.28a) of LHRH. The water suppression is very efficient. Even the $W_\alpha$ resonance at 4.7 ppm directly under the water resonance is observed.

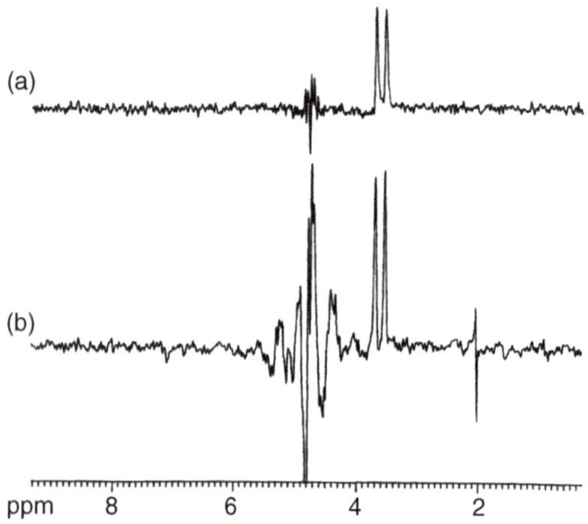


**Figure A.30** $\omega_2$ traces through the $P_\delta$ resonance (a) in the gradient-enhanced HSQC (Figure A.29a) and (b) in the conventional HSQC (Figure A.29b). The water resonance in (a) is very well suppressed. A signal-to-noise ratio that is better by a factor of $2^{1/2}$ is observed in the conventional HSQC (b) compared with the gradient enhanced HSQC (a).

HSQC (Figure A.30a) compared with the conventional HSQC experiment (Figure A.30b) is visible in a slice through the Pro-$\delta$ resonance in the gradient and the conventional HSQC. The loss is of the order of 1.5, which is close to the expected $2^{1/2}$. The almost complete suppression of the water signal, however, favors the gradient experiment.

The combination of sensitivity enhancement with gradient coherence selection is especially valuable for the detection of *IS* spin systems. This applies for most of the triple-resonance experiments in use today for sequential assignment in completely isotopically labeled biomacromolecules that have a reverse INEPT transfer from $^{15}N$ to $^1H$ as the last step (vide infra). For the complex of calmodulin with the peptide C20W, a comparison of traces of a conventional HNCO (Figure A.50; traces in Figure A.32b), an amplitude-modulating HNCO with gradients (Figure A.31; traces in Figure A.32a), and a sensitivity-enhanced HNCO with gradients (Figure A.31; traces in Figure A.32c) is shown. The optimum sensitivity is clearly achieved for the sensitivity enhanced sequence with gradients (traces in Figure A.32c). [92] The extension of sensitivity enhancement for $I_nS$ spin systems in heteronuclear correlation spectroscopy with gradients with $n \geq 1$ has recently been demonstrated [257].

Gradients can be positioned at various places in a sequence. However, there are rules as to which position is optimal:

1. Gradient pulses that cancel each other should be as close together in time as possible to reduce spatial diffusion effects.
2. If gradients are to be applied in evolution or detection periods, they should be applied during these times and not, for example, after a 180° pulse that follows the evolution time or precedes the detection period. Imperfect 180° pulses would otherwise lead to the occurrence of axial peaks and/or quadrature images.
3. For sensitivity reasons gradient pulses should be applied such that a minimum of desired pathways is canceled.

Application of these principles is illustrated in the sequence of Figure A.28. The proton gradient is applied in the second half of the refocusing delay $\Delta$. Half of the $4G_z$ gradient is applied during $t_1/2$. This prevents product operators with transverse proton magnetization during $t_1$ from being detected. It also makes sure that carbon is transverse during $t_1$ and suppresses axial peaks. If the whole "carbon" gradient were applied after the 180°($S$) pulse that follows $t_1$, imperfections of this 180°($S$) pulse would translate into axial peaks [$2I_zS_z$ instead of $2I_zS^+$ (during $t_1 + \Delta'$) $\rightarrow$ $2I_zS_x$ (during $\Delta'$)] or quadrature images [$2I_zS^-$ instead of $2I_zS^+$ (during $t_1 + \Delta'$) $\rightarrow$ $2I_zS^-$ (during $\Delta'$)]. The second $2G_z$ gradient is applied to avoid transverse proton magnetization in the second $t_1/2$ segment; this magnetization leads to a type of artifact at $\omega_1 = \Omega_C \pm \Omega_H/2$.

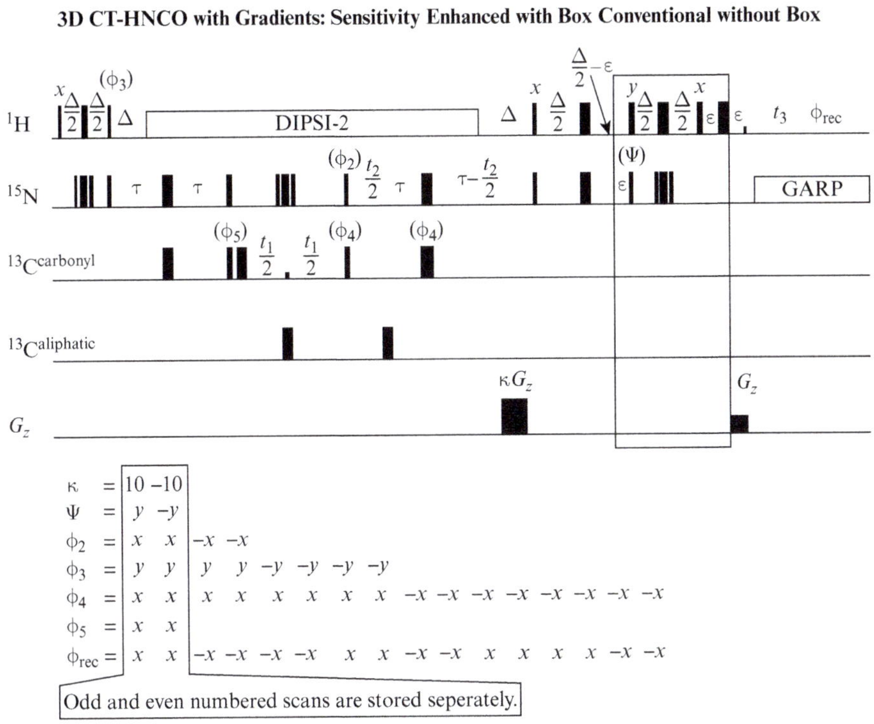


**Figure A.31** Pulse sequence of the sensitivity-enhanced 3D constant-time HNCO with coherence selection by gradients. The amplitude of the gradient applied during $t_2$ on nitrogen coherence and the phase $\psi$ are inverted from scan to scan. The respective FIDs are stored separately. The delays are $\varepsilon = 2.2$ ms, $\Delta = 5$ ms, and $\tau = 14.7$ ms. Omission of the pulses and delays in the box leads to a conventional amplitude-modulating HNCO with reduced sensitivity when coherence selection with gradients is applied. Composite 180° pulses $90_x$ $180_y$ $90_x$ are denoted by thin/thick/thin bars.

In cases where the suppression of undesired non-$^{13}$C-bound protons is not as difficult as with the water resonance, gradients can be employed while at the same time avoiding the reduction in S/N. This can be realized in an HSQC spectrum by inserting a gradient during the polarization transfer pulses. The suppression is achieved exploiting the different phase behavior of the $I_y$ coherence formed by $^{12}$C-H moieties and the $2I_zS_z$ coherence formed by $^{13}$C-H moieties at the end of the initial $\Delta$ delay. After the first three $^1$H pulses in the sequence of Figure A.33, non-$^{13}$C-bound protons form $I_y$ whereas $^{13}$C-bound protons form $2I_zS_z$. The former dephase during a gradient pulse; the latter, however, survive. The same trick is repeated after $t_1$.

An example of this approach implemented on a conventional inverse probe with the 0.6 G/cm "homospoil" pulse on a 400-MHz spectrometer is shown in Figure A.34. The sample is a small organic molecule in $D_2O$. As one can see, the $^{12}$C-bound protons are well suppressed in the gradient HSQC (Figure A.34b). In the conventional HSQC (Figure A.34a) they give rise to $t_1$ noise. The S/N of this sequence is idential compared with the S/N of the conventional HSQC. Application of $B_0$ gradients in 3D HSQC- NOESY has recently been published [93, 94].

HMBC experiments are even more prone to $t_1$ noise due to non-$^{13}$C-bound protons. Gradient coherence selection has proven to be very useful for this type of experiment [83].

The inhomogeneity of the $B_1$ fields ("$B_1$-gradient") of standard probes can be exploited using 2- to 5-ms spin-lock pulses for solvent suppression in heteronuclear correlation spectroscopy. [95, 96] The basic principle here is that after evolution of heteronuclear coupling under refocusing of chemical shift, the phase of $I$ magnetization of

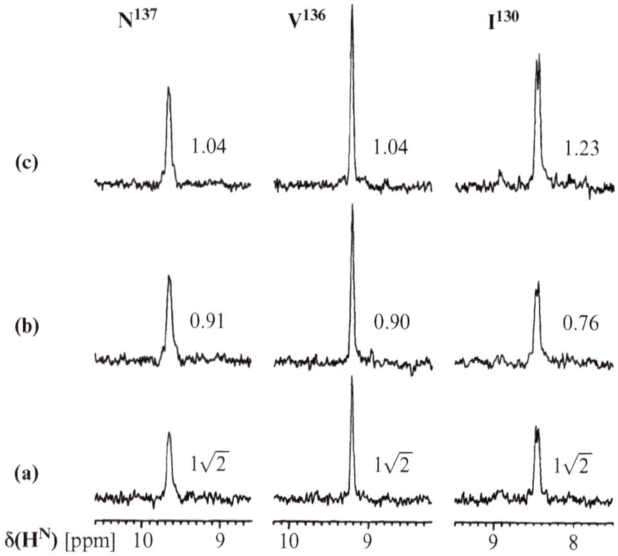


**Figure A.32** 1D slices from the first $\omega_2$, $\omega_3$ spectra ($t_1 = 0$): (a) from the amplitude modulating 3D CT-HNCO with coherence selection by gradients, (b) of the conventional 3D CT-HNCO with presaturation of the water, and (c) of the phase-modulating 3D CT-HNCO with gradient selection. The numbers indicate the relative signal-to- noise ratio with respect to the experiment shown in (a). The measurement time was equal for all experiments.

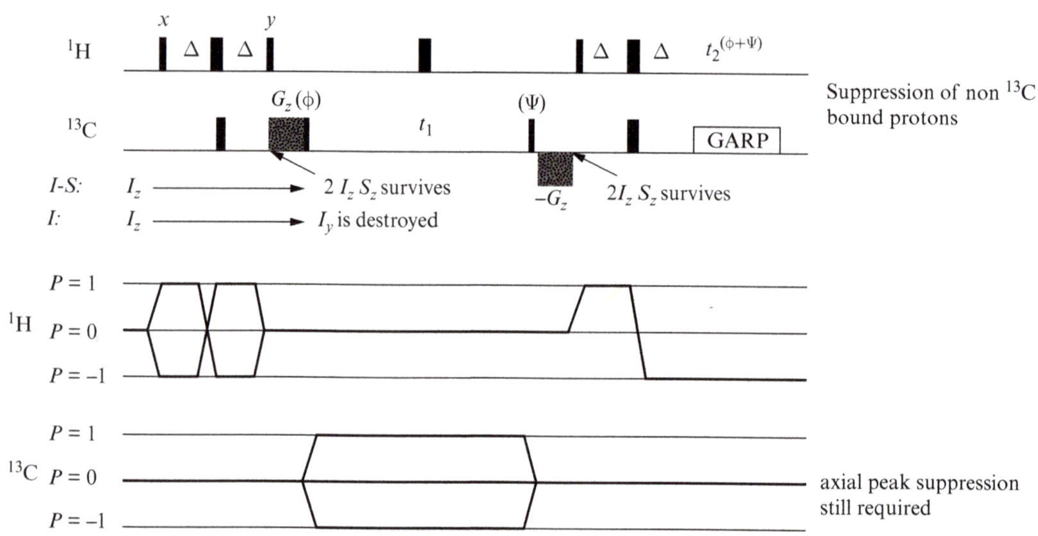


**Figure A.33** Application of $B_0$ gradients during the polarization transfer suppresses magnetization of non-*S*-bound *I* spins. The amplitude of the two gradients should be of opposite sign since the 180°(*I*) pulse inverts the local phases of the undesired signals "engraved" by the first gradient.

*I* spins bound to *S* spins is different from the phase of *I* magnetization of *I* spins that are not bound to *S* spins. Therefore application of such "purge" $B_1$ fields can selectively destroy magnetization of *I* spins not bound to *S* spins. The spin-lock pulse may be inserted before the polarization transfer of an INEPT/HSQC-type experiment or

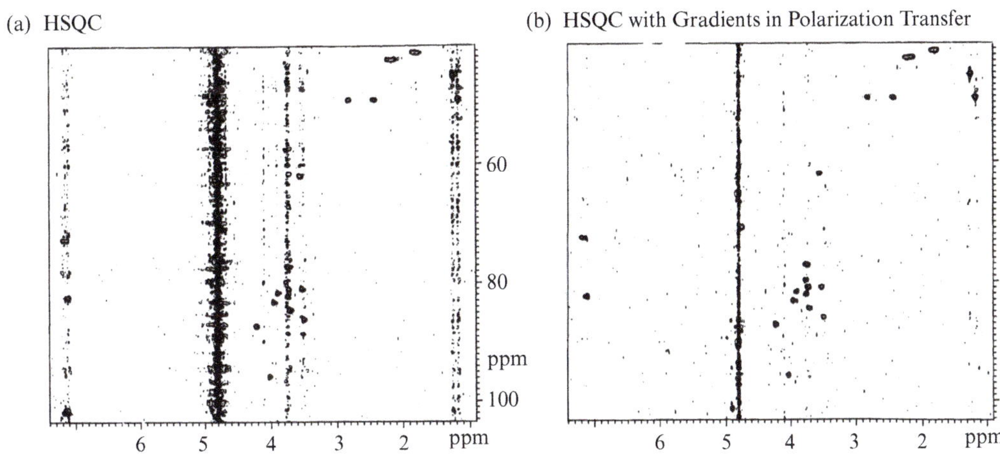


**Figure A.34** (a) Conventional HSQC of a small organic molecule dissolved in $D_2O$. Methyl groups, aromatic protons, and the residual water lead to $t_1$ noise, (b) Application of the $B_0$ gradient during the polarization transfer using "homospoil" pulses on a probe without active shielding leads to a considerably improved spectrum. The $t_1$ noise of the slowly relaxing groups has vanished.

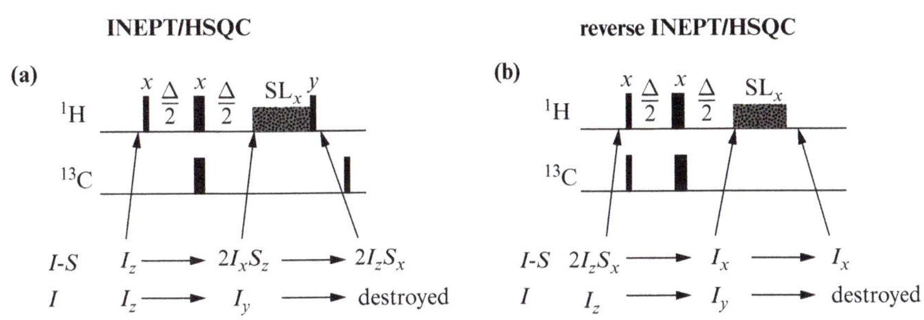


**Figure A.35** Application of $B_1$ gradients (spin-lock pulses). These pulses of 2- to 5- ms duration have an inhomogeneous $B_1$ distribution over the sample. The width of this distribution is normally ± 10% of the nominal $\gamma B_1$. Magnetization that is transverse to the spin-lock direction is therefore suppressed after several milliseconds. This accessory can be applied, for example, after defocusing of the heteronuclear coupling in the initial phase of a HSQC or double INEPT (a). It can equally well be applied before acquisition in the last segment of a HSQC or double INEPT (b).

before detection after reverse INEPT, as indicated in the pulse sequence fragments of Figure A.35. The combination of "$B_1$ gradients" and $B_0$ gradients yields excellent water suppression without the necessity for pre-irradiation of the water.

## A.3.5 Filters

Heteronuclear filters can be used to selectively suppress or retain only protons that share a $^1J$-coupling with a selected heteronucleus. Performing the ID version of the $^{13}$C,H-HMQC (Figure A.8a with $t_1 = 0$), for example, selects only protons bound to carbon. This element can be applied whenever there is transverse magnetization of a proton bound to a heteronucleus to be detected. This pulse sequence element is called an X-filter (Figure A.36a). [97–100] The suppression of the protons not bound to a heteronucleus depends on reproducibility between scans.

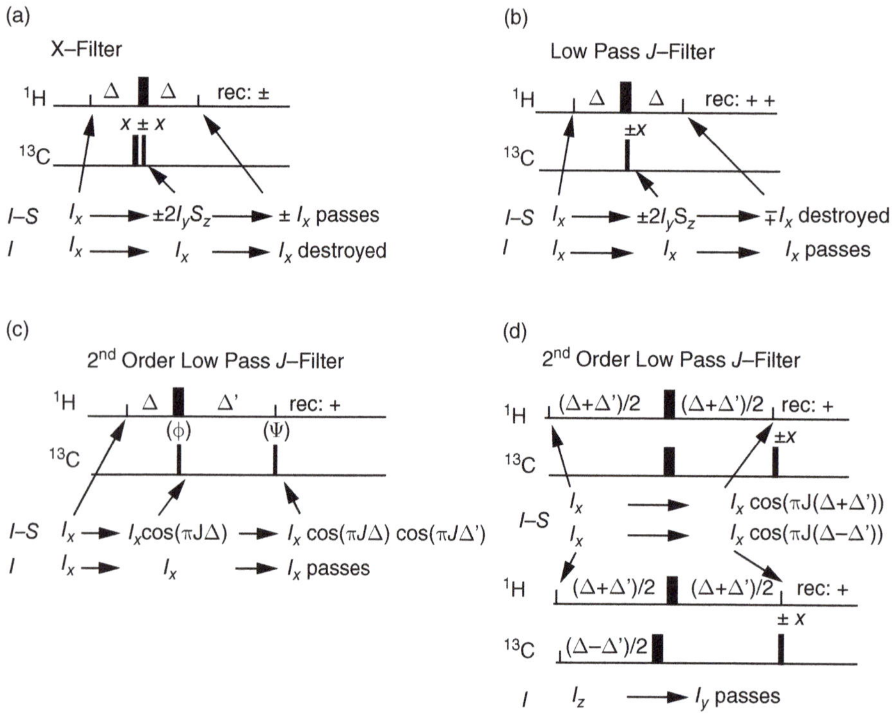


**Figure A.36** Heteronuclear filters for the selection or suppression of $S$-bound $I$ spins, (a) The X-filter selects $S$-bound $I$ spins. The non-$S$-bound $I$ spins are suppressed in consecutive scans by subtraction. Suppression of these $I$ spins would also be possible by the application of $B_0$ gradients (Figure A.28). The efficiency of selection depends on the size of the heteronuclear coupling constant: $\sin^2(\pi J_{IS}\Delta)$. The time for the evolution of chemical shift between the two pulses should be minimized. (b) Low-pass $J$-filter: The $S$-bound $I$ spins are suppressed. For a 90°($S$) pulse the remaining $I$ magnetization of an $S$-bound $I$ spin is $\cos(\pi J_{IS}\Delta)$. Thus the suppression is optimal for just one coupling constant, namely $J_{IS} = 1/(2\Delta)$. (c) Second-order low pass $J$-filter that achieves a suppression factor of $\cos(\pi J_{IS}\Delta)\cos(\pi J_{IS}\Delta')$. This filter function has a broader zero than the filter function of sequence (b). (d) Another implementation of a higher order low-pass $J$-filter employs just one 90°($S$) pulse and a 180°($S$) pulse that is applied after $(\Delta + \Delta')/2$ for the first two scans of the filter and at $\Delta$ for the second two scans.

Low-pass filters [101–104] suppress signal of those protons that are bound to a heteronucleus. Application of a 90°($S$) pulse after a delay $\Delta$ suppresses signals of $S$-bound $I$ spins for $\Delta = (2J_{IS})^{-1}$ (Figure A.36b). However, if the pulse is not exactly 90° but rather $\theta$ or the delay is not exactly matched to the coupling, $I$ magnetization survives with a factor: $\cos(\pi J_{IS}\Delta)\cos\theta$. The dependence on the coupling constant can be removed by design of filters of higher order. A second-order low-pass filter has a filter function: $\cos(\pi J_{IS}\Delta)\cos(\pi J_{IS}\Delta')\cos^2\theta$. An implementation is given in Figure A.36c. [101] Another implementation that makes use of the expansion

$$\cos(\pi J_{IS}\Delta)\cos(\pi J_{IS}\Delta') = 0.5\,(\cos(\pi J_{IS}(\Delta + \Delta')) + \cos(\pi J_{IS}(\Delta - \Delta')))$$

is shown in Figure A.36d [102–104].

Filters can be used to edit 2D proton-proton transfer spectra on the basis of whether the protons in $t_1$ and/or $t_2$ are bound to a heteronucleus. Since the selection can be different in $t_1$ and $t_2$, simplification of spectra is possible with this accessory. Some useful examples follow.

### A.3.5.1 Proton-Proton Correlation Spectra Free of Diagonal Peaks [105]

A transfer experiment (e.g., a COSY or NOESY experiment) with a low- pass $J$-filter in $t_1$ and an X-filter at the beginning of $t_2$ will select transfer between protons in the following moiety: $I_1.\,...\,I_2$-$S$. Since a proton cannot at the

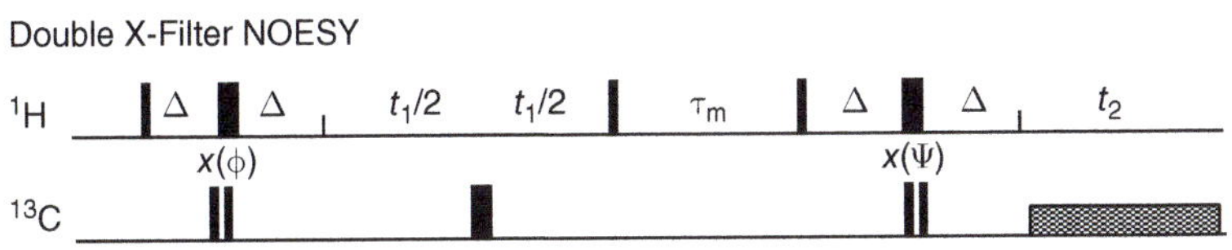


| $t_1$ | $t_2$ | (ϕ) | (Ψ) | rec: |
|---|---|---|---|---|
| $^{13}$C····· $^1$H | $^1$H— $^{13}$C | $x,-x$ | $x,x,-x,-x$ | $x,-x\ -x,x$ |
| $^{12}$C····· $^1$H | $^1$H— $^{13}$C | $x,-x$ | $x,x,-x,-x$ | $x,x,-x,-x$ |
| $^{13}$C····· $^1$H | $^1$H— $^{12}$C | $x,-x$ | $x,x,-x,-x$ | $x,-x$ |
| $^{12}$C····· $^1$H | $^1$H— $^{12}$C | $x,-x$ | $x,x,-x,-x$ | $x$ |

**Figure A.37** Pulse sequence of a double X-filter NOESY experiment. Separate storing of the four indicated $\phi$ and $\psi$ phases allows postacquisition combination to yield the indicated spectra. Isotope editing is thus very easily possible, $\phi$ is the TPPI phase.

same time be bound to a heteronucleus and not be bound to a heteronucleus, this experiment does not contain a diagonal. Cross peaks are retained because magnetization transfer has occurred during the mixing time. Optimum sensitivity is obtained for 50% labeling (where the resultant spectrum has 25% of the sensitivity of a normal experiment). However, the experiment can also be applied in natural abundance samples.

#### A.3.5.2 Different Labeling

If parts of a molecule or a supramolecular complex are labeled with an NMR active heteronuclear spin and others are not, it is possible to selectively observe 2D transfer experiments [106] where

1. the chemical shifts of the protons of the labeled parts only are correlated.
2. the chemical shifts of the protons of the nonlabeled parts only are correlated.
3. the chemical shifts of protons of the labeled parts in $t_1(t_2)$ correlate with the chemical shifts of protons of the nonlabeled part $t_1(t_2)$, and other correlations are suppressed.

For NOESY, for example, it is possible by appropriate choice of phases to record four data sets in one experiment and to combine them according to Figure A.37 to yield the four desired spectra. This technology has been applied successfully to the study of peptide/protein complexes [107–110].

### A.3.6 Decoupling

Decoupling has been discussed already in Chapter 2 of this volume. Decoupling of heteronuclei is performed analogously to the decoupling of protons in heteronuclear-detected spectra. WALTZ-16, [111] GARP-l, [112] and DIPSI [113, 114] can be used. GARP-I [112] has the best ratio of decoupled spectral range to rf power. Since biological samples tend to heat under decoupling, the power should be carefully chosen.

Since carbon spectra, especially of proteins, have nicely separated regions of resonances (much more so than proton spectra) (Figure A.38d), selective decoupling of a certain region of the carbon spectrum is possible. The sequences proposed so far consist of expansions of selective 90° and/or 180° pulses. G3 [21, 22] inversion pulses expanded according to MLEV-16 [115–118] or SEDUCE [117–120] and Hermite pulses expanded according to MLEV-4 [121] are very useful.

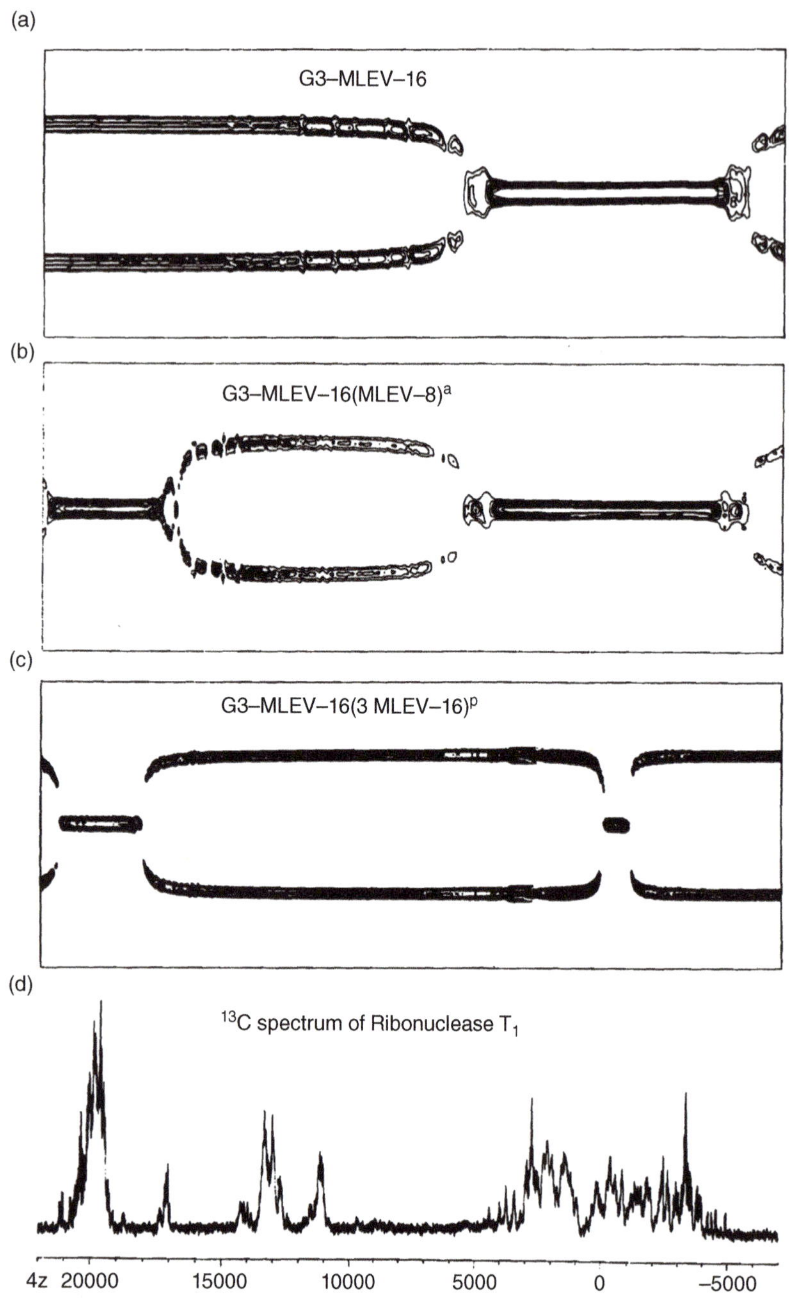


**Figure A.38** Selective decoupling sequences for carbon. The carbon spectrum of proteins contains several well-separated regions (from right to left): aliphatic, aromatic, and carbonyl (d). In Figures a-c the spectrum of the $H_\alpha$ of fully labeled glutamic acid is shown (vertical axis) as a function of the $^{13}C$ offset (horizontal axis) for different decoupling sequences. Selective G3 pulses expanded according to MLEV-16 allow, for example, decoupling of just the aliphatic carbons (a) as indicated by the removal of the doublet splitting. Superposition of expansions of on-resonance G3 pulses and off-resonance G3 pulses allows the simultaneous decoupling of two regions: (b) carbonyl and aliphatic, (c) leucine $C_\beta$ carbons and carbonyls.

Band selectivity as well as the possibility to cover several bands is demonstrated with the G3-MLEV expansion in Figure A.38. The implementation of a multiband decoupling scheme is as follows. If the decoupling pulses are derived from one frequency source, off-resonance pulses can be realized by imposing a phase gradient $\Delta\phi/\Delta t$ on the pulse shape. The frequency shift of such a decoupling sequence is given by $\omega = \Delta\phi/\Delta t$. Several bands can be covered by adding the decoupling schemes of each band. It should be noted that the bands must not overlap in order to avoid interference between the neighboring bands.

### A.3.7 Calibration of Pulses

Heteronuclear pulse sequences normally have many more pulses of different type than homonuclear sequences. In addition, the heteronuclear pulse length must be determined for a circuitry of the spectrometer that normally makes the detection of the heteronucleus impossible. In addition, we find it preferable to determine the pulse durations directly on the sample of interest.

Usually both $90°(S)$ and $180°(S)$ pulses need to be determined. In proton-detected experiments, the $90°(S)$ pulse can be determined by the sequence of Figure A.39a. No decoupling is applied, so that the satellites showing the coupling of the proton to the heteronucleus may be observed in the $^1$H-detected spectrum. The satellites have identical phase with the large central signal. If the central signal must be suppressed, an X-filter must precede the sequence (Figure A.39b).

It is helpful to place the $S$-transmitter on an $S$ resonance that is coupled to a strong and well-resolved proton resonance. Sequences A.39a,b give zero satellite signal for a $\theta(S) = 90°$ pulse, full signal for a $\theta(S) = 0°$ pulse, and inverted signal for $\theta(S) = 180°$ pulse. The $\pi/2$ pulse is intrinsically rather well compensated for off-resonance effects.

To determine the $180°(S)$ pulse, the sequence in Figure A.39c can be applied. The two $S$ pulses should be placed after or in front of the proton 180° pulse to avoid $S$ chemical shift evolution. This sequence can also be used for the calibration of pulses with different rf amplitude. For this purpose, $\theta' = 2\theta$. Since on some instruments the phase of pulses with different rf amplitude may differ, one has to make sure that the phases of the two $S$ pulses are not 90° out of phase. This would lead to no signal irrespective of the flip angle of the pulses. One can utilize the phase dependence of the observed signal from the sequence in Figure A.39c to determine the relative phases of two $S$ pulses with different power levels. In this case we assume $2\theta \sim \theta' \sim \pi/2$. One should first determine the approximate flip angle of each of the two pulses and then adjust the phases of the two pulses to obtain no signal. This is advantageous because the phase determination is most sensitive for zero signal. At zero signal the relative phase of the two $90°(S)$ pulses is 90°. Since chemical shift of $S$ spin evolves during $S$ pulses, for example, during $2/\pi$ of the duration of a rectangular 90° pulse, frequency-dependent phase shifts need also be considered. They can be compensated by introducing a $180°(S)$ pulse and a suitable delay between the 90° pulses. The proton signals should then be zero irrespective of the offsets of the $S$ spins from the $S$ transmitter.

Finally, if selective off-resonance pulses are employed in a sequence, a phase shift of transverse on-resonance magnetization must be considered [117] (see also Chapter 2 in this volume). This is a known effect observed in selective decoupling sequences as the Bloch–Siegert phase shift. A simple example will serve to illustrate the Bloch–Siegert phase shift after a pulse.

If a 180° pulse with a field strength $\gamma B_1$ is applied on resonance, a magnetization at an offset of $\Omega = 3^{1/2}\gamma B_1$ will experience an off-axis field $(\Omega^2 + (\gamma B_1)^2)^{1/2} = (3(\gamma B_1)^2 + (\gamma B_1)^2)^{1/2} = 2\gamma B_1$. The effective field $2\gamma B_1$ is the vector sum of the chemical shift $\Omega = 3^{1/2}\gamma B_1$ which induces a rotation about the $z$ axis, and the rf field $\gamma B_1$ in the $xy$ plane. The effective flip angle for this magnetization is therefore $2 \cdot 180° = 360°$. This means that the off-resonance magnetization is at the same position after the on-resonance 180° pulse as before the pulse. If the 180° on-resonance pulse had not been applied, the phase of the off-resonance spin would have been changed by $\Omega\tau_\pi = 3^{1/2}\gamma B_1 \pi/(\gamma B_1 = 3^{1/2}\pi$ since $\tau_\pi = \pi/(\gamma B_1)$. Thus the application of the on-resonance 180° pulse, which exerts a flip angle of $360° = 0°$ on the off-resonance magnetization, nevertheless changes its phase. Choosing the

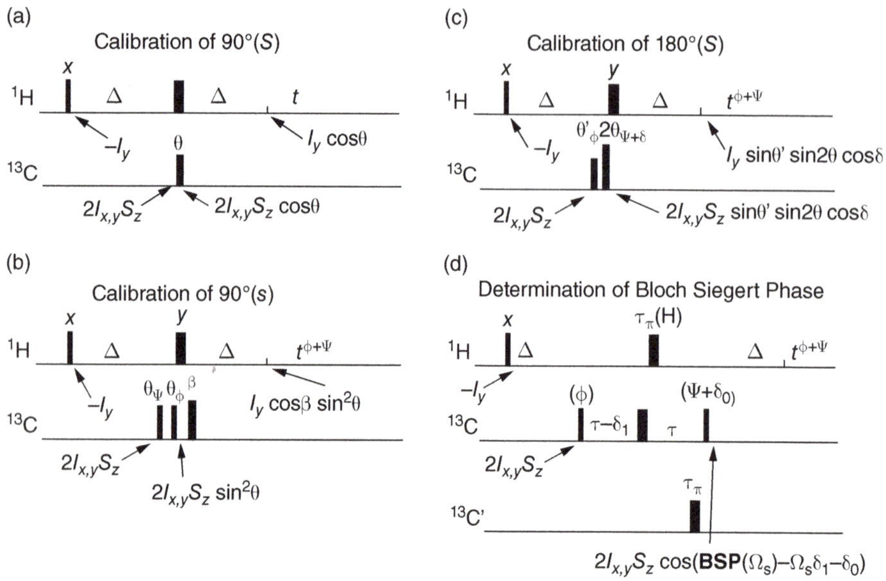


**Figure A.39** Pulse sequences for the determination of heteronuclear pulses with a hardware setup used to detect protons, (a) Determination of a 90°(*S*) pulse. For $\Delta = 1/2J_{IS}$ no satellite signal is left for $\theta = 90°$. (b) Determination of 90°(*S*) for a sample with low abundance of *S* spins when a non-*S*-bound *I*-signal must be suppressed. An X-filter precedes the $\beta(S)$ pulse. The phase cycle is $\phi = \pm x$, $\psi = x, x, -x, -x$. For $\beta = 90°$ no signal results. For optimal sensitivity $\theta$ should be close to 90°. (c) Determination of a 180°(*S*) pulse in an X-Filter-like sequence. For $2\theta = \pi$ no signal is observed, $\theta'$ should be close to 90° for sensitivity reasons. $\phi$ and $\psi$ are cycled as in (b). $\delta$ should be adjusted for maximum signal when $\theta$ and $\theta'$ have different rf power. (d) Determination of Bloch–Siegert phase (BSP) produced by an off-resonance 180°(*S*) pulse acting on $^{13}$C' applied in the presence of transverse resonances in this example on-resonance *S*-magnetization. If no BSP were introduced no signal would be obtained when $\phi$ and $\psi$ are 90° out of phase. By varying the phase $\delta_0$ and the delay $\delta_1$ until no proton signal is observed one can determine the Bloch–Siegert phase produced by the off-resonance pulse.

rotating frame of the off-resonance magnetization as reference frame we can formulate: A 180° pulse applied far off-resonance introduces a Bloch–Siegert phase shift of $-(3^{1/2})\pi$. The effects of Bloch–Siegert phase shifts have to be considered during periods of fixed length, because they change the phases of the desired signals considerably.

We demonstrate this effect using the sequence of Figure A.39d, with selective on-resonance carbon pulses applied on the aliphatic carbons and off- resonance pulses applied on the carbonyls. For the selective pulses we use Gaussian cascades, introduced by Emsley and Bodenhausen (G3 and G4) [21, 22] (see Chapter 2 in this volume).

Application of the sequence in Figure A.39d leads to the spectrum of the protons bound to aliphatic carbons of calmodulin (Figure A.40a). In the absence of Bloch–Siegert phase shift, no signal should result, since the two 90° *S* pulses have been chosen 90° out of phase. Indeed, if the off-resonance carbon pulse applied to the carbonyls is substituted by a delay, the sequence gives zero signal. Setting $\delta_1$ to 25 μs and shifting the phase of the second 90° pulse by $\delta_0 = -120°$ to compensate for the Bloch–Siegert phase leads to the spectrum shown in Figure A.40b. Most of the signal is suppressed. The $^{13}$C transmitter was placed on the methyl groups in the spectra (a) and (b). After moving the $^{13}$C-transmitter to the middle of the region of aliphatic carbons, all signals can be suppressed in the aliphatic part of the spectrum (Figure A.40c), when compensation for the Bloch–Siegert phase is included.

A simulation of the offset-dependent Bloch–Siegert phase is shown in Figure A.41 for a G3 pulse of duration 250 μs and a $\gamma B_1^{max}/2\pi = 14.37$ kHz. As one can see, the dependence of the Bloch–Siegert phase on offset is somewhat nonlinear. This was visible also in the experimental spectrum of Figure A.40b. With a correct transmitter frequency, however, the Bloch–Siegert phase shift can be compensated to a large extent by a phase shift and delay which corresponds to zero and first order phase correction in an indirectly sampled frequency domain.

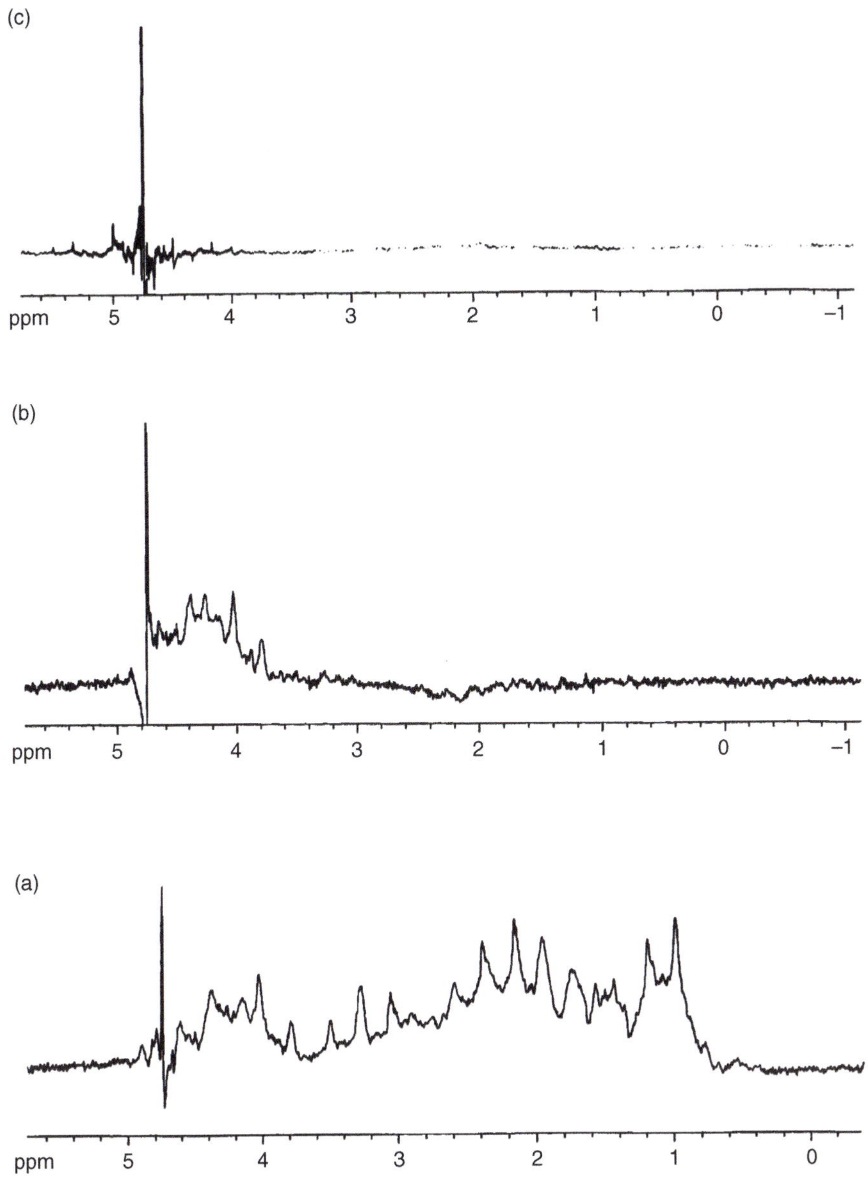


**Figure A.40** (a) The spectrum of $^{13}$C-labeled calmodulin obtained from the pulse sequence of Figure A.39(d) with $\phi = x$, $\psi = y$, $\delta_1 = 0$, and $\delta_0 = 0$. If no Bloch– Siegert phase were produced, the signal should be zero because the two 90°($S$) pulses are 90° out of phase. Changing the phase $\delta_0$ to −120° and the delay $\delta_1 = 25\mu$s yields the spectrum in (b). Only the $H_\alpha$ resonances still produce signal because the carbon transmitter is placed on the methyl groups and the selective carbon pulses therefore do not properly excite the $C_\alpha$ resonances, (c) Setting the transmitter in the middle of the aliphatic region yields a spectrum that is zero.

## A.4 3D Methods

3D spectroscopy is an extension to 2D spectroscopy. [122–125] Introduction of a second evolution time and a second mixing sequence leads to 3D NMR sequences. The overall sequence can be symbolized as in Figure A.42.

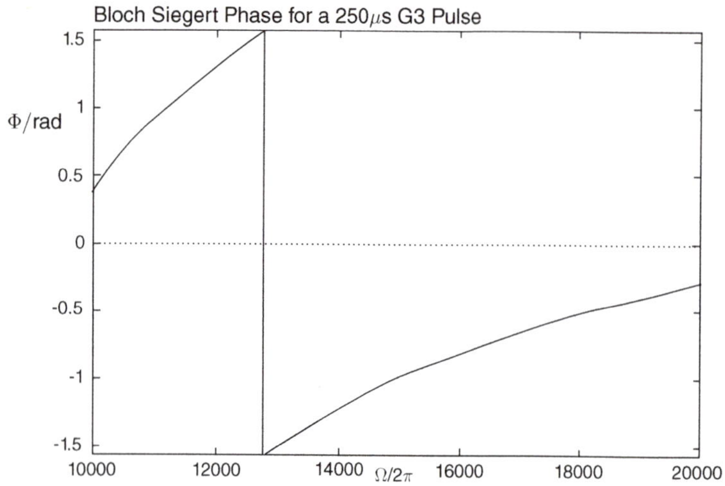


**Figure A.41** Bloch–Siegert phase as a function of offset $\Omega/2\pi$ for a G3 inversion pulse of duration 250 $\mu$s. It is obvious from the plot that the phase is a nonlinear function of the offset. In a small region, however, good compensation can be realized by a linear approximation that can be compensated for by a constant phase and a delay as applied in Figures A.39 and A.40.

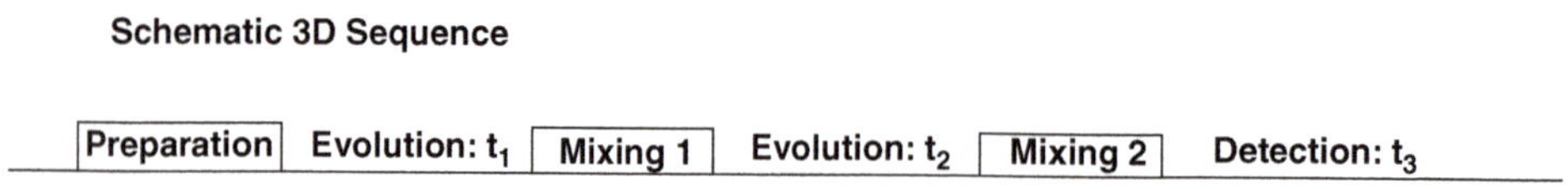


**Figure A.42** Schematic 3D sequence: After the preparation period comes a first evolution period $t_1$, a first mixing, a second evolution period $t_2$, and finally a second mixing and detection during $t_3$. The evolution times have to be incremented separately. TPPI or Ruben-States-Haberkorn has to be applied individually before $t_1$ and before $t_2$. For the 3D and 4D sequences we will use $\phi_1$, $\phi_2$, $\phi_3$ to designate the pulses that should be phase incremented to achieve sign differentiation in $\omega_1$, $\omega_2$, and $\omega_3$.

Fourier transformation along the three time domains yields a 3D spectrum. A thorough and up-to-date review on processing 3D spectra has appeared recently. [126] A schematic spectrum is given in Figure A.43 showing the different types of peaks encountered in 3D NMR spectra as well as the different diagonal planes.

The mixing periods that are applied between $t_1$ and $t_2$ as well as between $t_1$ and $t_3$ are the heart of the method. We will focus the discussion mainly on heteronuclear 3D NMR sequences after discussion of few homonuclear pulse sequences. This is appropriate because the number of heteronuclear sequences has outrun by far the number of homonuclear sequences.

## A.4.1 Homonuclear 3D Experiments

Homonuclear 3D experiments for biomolecules are confined to sequences that transfer in-phase magnetization and have a high efficiency of transfer. Therefore the combinations of TOCSY, NOESY, and ROESY are applied: NOESY-TOCSY, [123, 124, 127–141] TOCSY-TOCSY, [142] and NOESY-NOESY [143–146] (Figure A.44). Homonuclear sequences do not need any special preparation of the sample. Thus biomolecules that cannot be expressed with high yields in organisms growing on isotopically enriched media but must be isolated from higher organisms with low yield can be investigated. However, the sensitivity of these experiments suffers from the fact that two homonuclear transfers must be accomplished, each usually not exceeding an efficiency of 10%. This is in contrast

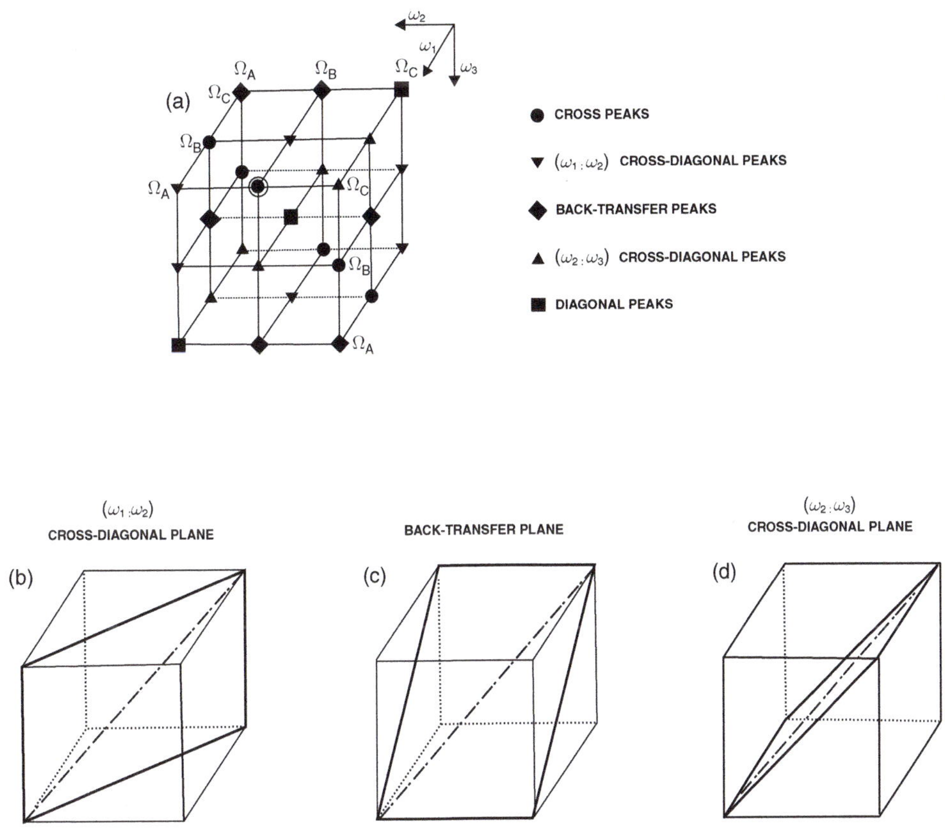


**Figure A.43** Schematic 3D spectrum of a three-spin system. The four different sorts of peaks are cross peaks, cross-diagonal peaks, back-transfer peaks, and diagonal peaks. Only the genuine cross peaks contain new information that is not available in 2D spectroscopy.

to heteronuclear NMR, where the transfer efficiency for the triple-transfer H → C → C → H (Figure A.4) can be close to 1. Therefore, sample concentrations of more than 1 to 2 mM are essential for the investigation. Probe design that allows for sample tube diameters of more than 5 mm (e.g., 8 mm) is helpful in homonuclear applications.

As an example, a NOESY-NOESY spectrum of nonpalindromic DNA 15mer (1 mM concentration, 600 MHz) is shown. Sequential assignment can be obtained from such spectra in cases of overlap. The sequence investigated is

$$5'\text{-C}^{1}\text{-C-G-C-T-G-C-G-A-T-C-C-G-G-C}^{15}\text{-3}'$$

$$3'\text{-G}^{30}\text{-G-C-G-A-C-G-C-T-A-G-G-C-C-G}^{16}\text{-5}'$$

The assignment step between $G^{27}$ and $C^{28}$ (Figure A.45a) is impossible in the 2D NOESY (Figure A.45b) due to overlap in the H(2′, 2″)/base proton region (Figure A.45c) of the 2D NOESY. Since the frequency coordinates of $G^{27}H_8$ and $G^{27}H_{2'2''}$ are known it is possible to search for the $G^{27}H_1/G^{27}H_{2'2''}/C^{28}H_6$ cross peak in the $\omega_3 = G^{27}H_8$ plane of the 3D NOESY-NOESY (Figure A.45d) experiment. One finds the back-transfer peak $G^{27}H_8/H_{2'2''}/G^{27}H_8$ as well as the desired cross peak $G^{27}H_8/G^{27}H_{2'2''}/C^{28}H_6$, which defines the chemical shift of $C^{28}H_6$. Going back to the 2D NOESY, the difficult assignment can be performed (Figure A.45c).

Homonuclear 3D experiments that rely on magnetization transfer via homonuclear couplings suffer from its rather low transfer efficiency in larger molecules. The only experiment that gives increasingly good sensitivity is

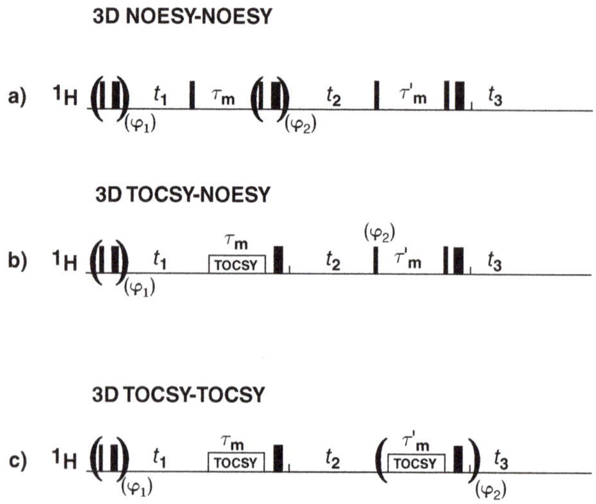


Figure A.44 Homonuclear pulse sequences: NOESY-NOESY, TOCSY-NOESY, and TOCSY-TOCSY. The implementation of these sequences is not different from that of 2D sequences. Water-suppression techniques (like presaturation before the pulse sequence or 1 1 at the end of the pulse sequence) should be applied as for the 2D sequence. The mixing times to be chosen are the same as those for a normal 2D experiment. The 180° pulses before $t_1$ and $t_2$ serve to obtain a vanishing linear phase correction. The pulses in parentheses are cycled simultaneously.

the NOESY. Therefore a great deal of effort has been invested in the assignment of proteins from NOESY-NOESY spectra as well as quantitative evaluation of NOESY-NOESY spectra. [143–148] The latter is especially demanding, since two cross-relaxation rates contribute to each 3D cross peak.

### A.4.2 Assignment-oriented Heteronuclear 3D NMR

The sequences that are employed for correlation in heteronuclear systems are successfully applied to samples at natural abundance, as well as to $^{15}$N- or $^{13}$C-labeled or completely $^{13}$C- and $^{15}$N-labeled proteins. These experiments have been applied, with few exceptions, to proteins. The names used for the designation of the sequences explicitly refer to the spins that are correlated in the sequence rather than to the principles used for the transfer. This is justified from the point of view that the experiments are highly optimized for the moieties to which they are applied. We therefore first discuss assignment strategies for proteins.

Assignment of resonances is achieved by detection of connectivities in the bonding network between nuclei. The nuclei that must be addressed first are the backbone atoms in the peptide chain. A stepwise assignment relies on cross peaks that are resolved and share common chemical shifts. For example, the classical COSY/NOESY [149] (Figure A.46a)-based assignment relies on the fact that the COSY cross peak $H_{\alpha,i}/H_i^N$ and the NOESY cross peak $H_{\alpha,j}/H_{i+1}^N$ share the chemical shift of $H_{\alpha,i}$. So the sequential assignment is a 1D search in the 2D spectra, relying for each assignment step on one resolved resonance, namely here $H_{\alpha,i}$.

3D assignment procedures offer the possibility to correlate three chemical shifts. For each assignment step, changing only one frequency, it is possible to rely on two chemical shifts instead of one. This redundancy is necessary for the assignment of heavily overlapping spectra that are typical for biomolecules.

The assignment strategies can be subdivided in two classes—those that rely on conformationally dependent interactions to achieve sequential assignment and those that rely mainly on conformationally independent NMR parameters such as $^1J$-coupling constants.

The classical NOESY/COSY-based assignment procedure (Figure 3-46a) can be improved by $^{15}N$ labeling (Figure A.46b). Then the chemical shift of an NH can be determined with less ambiguity by measuring not only the $H^N$ chemical shift but also the $^{15}N$ chemical shift. However, the $H_\alpha$ chemical shift cannot be determined in a similar way in molecules labeled only with $^{15}N$. $^{13}C$ labeling is necessary to get access to the $C_\alpha$ chemical shift.

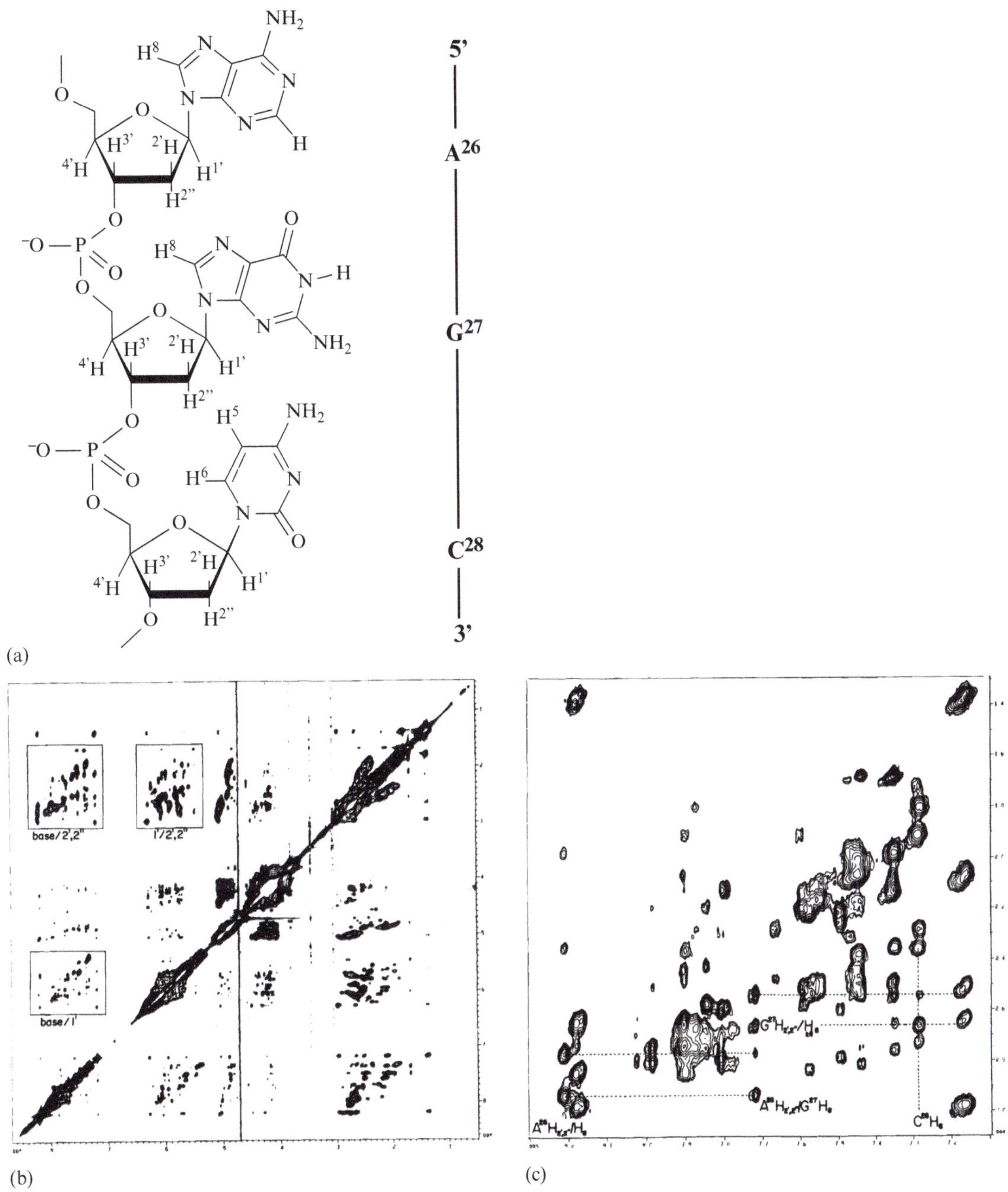


**Figure A.45** (a) Part of a DNA sequence of the nonpalindromic pentadecamer as indicated in the text. (b) 2D NOESY experiment of the substance with the regions of interest for the assignment of the resonances. (c) Sequential assignment of resonances based on the base/H(2′2″) resonances. An ambiguity arises on the $G^{27}H(2', 2'')$ lines, (d) 3D NOESY-NOESY experiment that resolves the overlap present in the 2D NOESY experiment. The resonance frequency of the $G^{27}H(8)/G^{27}H(2', 2'')/C^{28}H(6)$ cross peak identifies unambiguously the base proton $C^{28}H(6)$.

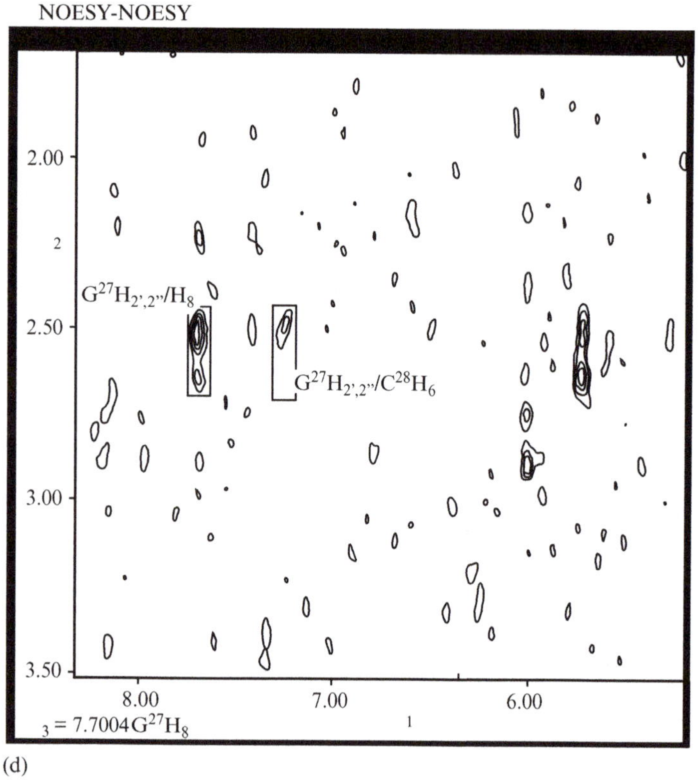


**Figure A.45(d) (Cont'd)**

Then four experiments can be used for an unambiguous assignment (Figure A.46c). These two assignment procedures are implemented by the combination of HSQC or HMQC with homonuclear transfer spectra like NOESY and TOCSY: HSQC-NOESY [150–157] and HSQC-TOCSY. [158–161] A sequential assignment based on the four experiments $^{1}$H,$^{13}$C-NOESY-HMQC; $^{1}$H,$^{13}$C- TOCSY-HMQC; $^{1}$H,$^{15}$N-NOESY-HMQC; and $^{1}$H,$^{15}$N-TOCSY-HMQC has been performed for ribonuclease H (15 kD) $^{13}$C and $^{15}$N labeled at the $C_{\alpha}$ and NH positions, respectively. [162] Resolution is enhanced compared with the 2D homonuclear experiments due to the additional spread of peaks by the $^{15}$N or $^{13}$C chemical shift. At the same time the number of peaks does not increase, which means that the peak density is decreased.

The pulse sequences of the NOESY-$^{15}$N,H-HSQC and TOCSY-$^{15}$N,H-HSQC are given in Figure A.47. It is possible to combine, for example, sensitivity enhanced HSQC with gradients with homonuclear techniques. [91, 163, 164] Suppression of cross relaxation in the TOCSY is recommended in applications to macromolecules [165–168].

### A.4.3 Heteronuclear 3D with $^{15}$N/$^{13}$C-labeled Biomolecules: Backbone Assignment

Assignment of proton resonances in biomolecules has so far been achieved based on the classical method connecting the protons within a spin system by their mutual $^{3}J$-couplings and connecting the spin systems by intraresidual NOEs. The NOEs not only rely on connectivity but also depend strongly on the conformation. Although this is the basis for the eventual determination of the structure of a biomolecule, the conformational dependence of connectivity information is a disadvantage at the stage of assignment, where unbiased connectivity information is

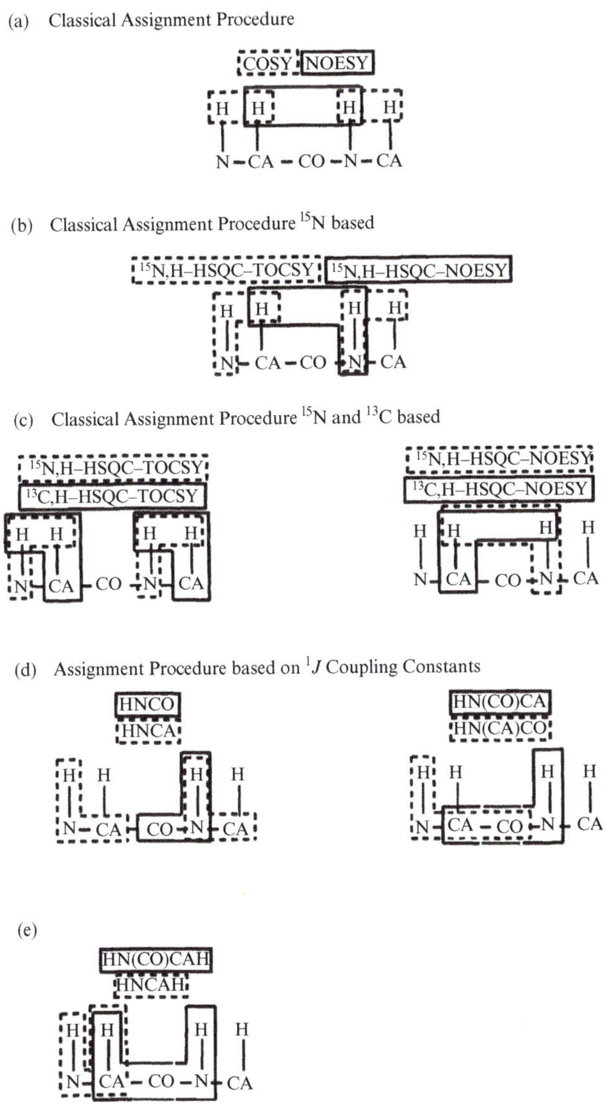


**Figure A.46** Assignment procedures for proteins: (a) The classical assignment procedure relies on the combination of intraresidual $H_\alpha$,$H^N$ and sequential $H_\alpha$,$H^N$ connectivities. In each step, one chemical shift stays constant. The procedure works when the $H_\alpha$ and the $H^N$ chemical shifts are resolved. (b) The classical assignment procedure can be enhanced by correlation of the $H^N$ with $^{15}$N. Experimentally this is achieved by recording HSQC-NOESY and HSQC-TOCSY spectra. The requirement for an unambiguous assignment procedure now is that the $H^N$,N correlation and the $H_\alpha$ resonances are resolved. (c) Recording four HSQC-NOESY and HSQC-TOCSY experiments requires resolved H,C and H,N-HSQC experiments. Due to the fact that in each consecutive assignment step two frequencies remain constant, ambiguities are removed. (d) Assignment based on $^1J$-coupling constants. The four experiments HNCO, HNCA, HN(CO)CA, and HN(CA)CO provide for each consecutive step two constant frequencies because for each amino acid $i$, $C_{\alpha,i}$ and $C'_i$ are correlated with the $H^N_i$, $N_i$ and the $H^N_{i+1}$, $N_{i+1}$. (e) The two 4D experiments, yielding the correlation of each $H_{\alpha,i}$,$C_{\alpha,i}$ with $H^N_i$, $N_i$ in the HNCAH and with the $H^N_{i+1}$, $N_{i+1}$ in the HN(CO)CAH, respectively, can be used for sequential assignment.

desired. This statement is in principle also true for the $^3J_{HH}$-couplings that are the second basis of assignment in the classical assignment strategy. [149] Therefore assignment strategies have been developed earlier [34] that do not depend on the NOE, for example, in peptides using the $^2J_{H^N_i,C'i-i}$ and $^2J_{H_{\alpha,i-1}C'_{i-1}}$-Couplings for the correlation

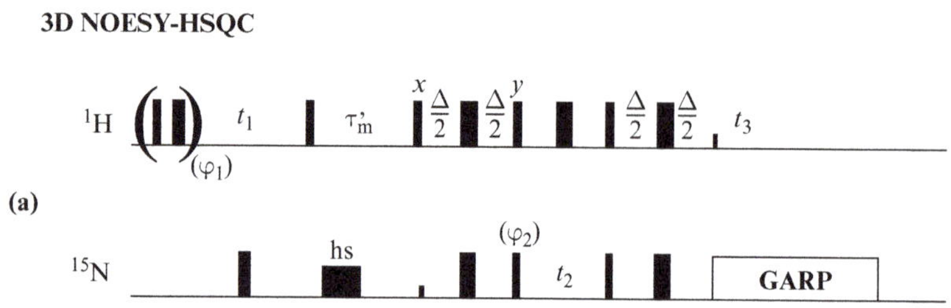


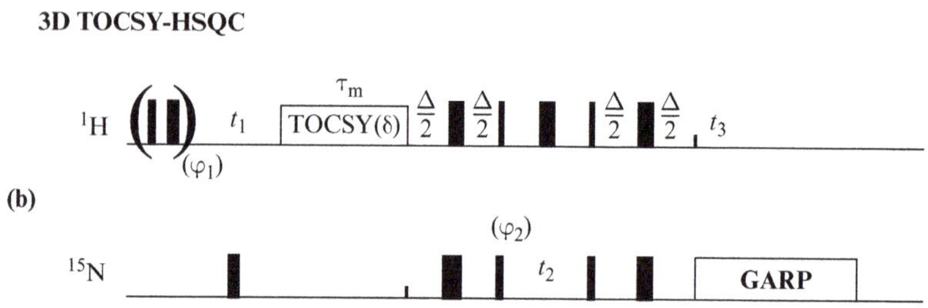


**Figure A.47** Pulse sequence for the 3D NOESY-HSQC (a) and the 3D TOCSY- HSQC (b). The sequences are a concatenation of the respective 2D experiments. For the sequence in (b) the phase difference $\delta$ between the phase of the TOCSY spinlock field in the transverse plane and the $^1$H pulses has to be determined experimentally in order not to loose magnetization. hs is a homospoil pulse to suppress coherence during NOESY mixing.

of $H_i^N$ and $H_{\alpha,i-1}$. Examples were shown in Figures A.10 and A.11 for peptides and oligosaccharides. However, the coupling constants involved are comparable with the linewidth and therefore yield only low transfer efficiency in the application to large biomolecules.

An approach based on the rather large $^1J_{CN}$- and $^1J_{CC}$-coupling constants encountered in the amino acid residues that form the repetitive units of proteins [169–171] was applied for $^{13}C/^{15}N$-labeled proteins. The coupling topology is given in Figure A.48. The combination of this approach with 3D correlation spectroscopy has revolutionized the assignment of proteins. [172] The correlation of the nuclei in the backbone via $^1J$-coupling constants makes it possible to record a number of spectra that give redundant connectivity information devoid of conformational bias and highly resolved in the 3D spectra. The correlation of nuclei that is applied in most cases is presented in Figure A.46d. Correlation of $H_i^N, N_i, C'_{i-1}$ in the HNCO [172, 173] and $H_i^N, N_i, C_{\alpha,i}$ and $H_i^N, N_i, C_{\alpha,i-1}$ in the HNCA [172, 173] experiment is supplemented by the HN(CO)CA [174] and the HN(CA)CO [175] experiments that yield sequential assignment for $H^N$,N,$C_\alpha$, and C′ for each amino acid. Since the first publication of these heteronuclear triple-resonance sequences some improvements have been published, namely the introduction of constant-time evolution periods and of proton decoupling whenever possible to reduce the proton self relaxation influence on heteronuclear transverse magnetization. [175–178] An HNCO using gradients in an improvable manner has recently been published. [179] The sequence for an HNCO with optimum sensitivity employing gradient coherence selection has been shown in Figure A.31 in Section A.3.4.

4D experiments make it possible to reduce the number of different experiments to be recorded for a complete backbone assignment. The first experiment in this context was a HCACON. [180] This experiment is based on detection of $H_\alpha$ chemical shift, which is not optimal for large biomolecules. Alternatively, the combination of a 4D HNCAHA [181, 183–185] with a HN(CO)CAHA [182–184] makes it possible to correlate the $(H^N,N)_i$ with the $(H_{\alpha,i},C_{\alpha,i})$ and with the $(H_{\alpha,i}-1,C_{\alpha,i-1})$. A slightly different implementation of this correlation scheme is

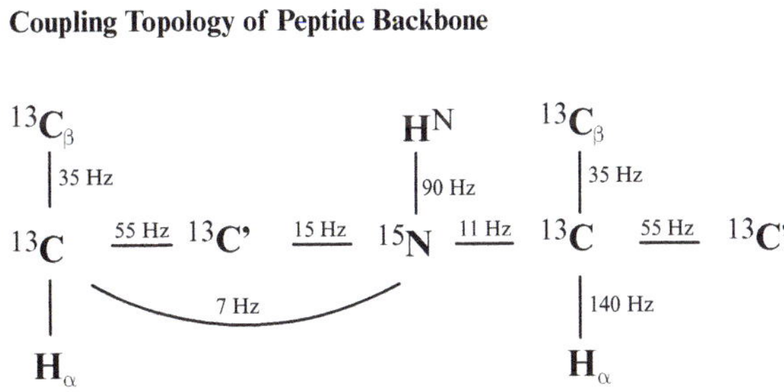


**Figure A.48** Coupling topology of the backbone of a protein. The coupling constants that are employed for coherence transfer in the triple-resonance experiments are indicated.

proposed in the sequences HCANNH and HCA(CO)NNH. [186] Thus again an assignment procedure relying on two chemical shifts that are fixed in each assignment step is implemented. Optimizing the assignment procedures for biomolecules is an objective of current research.

With the sequences discussed in the following the size limitation of roughly 100 amino acids for homonuclear spectroscopy has been shifted to more than 200 amino acids.

The correlation of the backbone and the sidechain nuclei by $^1J$-coupling constants makes it possible to perform experiments that employ rather short delays for the transfer (see Figure A.4) and delivers more redundant information, since the number of nuclei that define a residue is increased compared with homonuclear experiments. The correlation of nuclei employs the building blocks already introduced in this chapter. The sequences are designed to minimize the number of pulses, losses due to relaxation and losses due to evolution of undesired couplings. Since applications of the sequences are described in Chapter 6 of this volume, we will confine the discussion to selected examples of the sequences. Experimental aspects will be discussed rather than applications.

The number of sequences proposed is already large. Most of the sequences that aim at backbone assignment start from the $\mathrm{H^N}$, which is also detected. This is advantageous because the $H_2O$ resonance does not disturb the $\mathrm{H^N}$ spectrum and the $\mathrm{H^N}$ is a narrow resonance due to the low number of homonuclear coupling partners and the weak relaxation by $^{15}\mathrm{N}$. The rest of the sequences start from aliphatic protons. Since the sequences can be implemented in several ways and the name of the sequence does not indicate the way it is implemented, we would like to explain the reasoning behind the sequences with the example of the HNCA, which is a correlation between $\mathrm{H}_i^{\mathrm{N}}$, $^{15}\mathrm{N}_i$, and $\mathrm{C}_{\alpha,i}$. Figure A.48 shows the coupling topology for this experiment. There are two strategies to implement this experiment.

1. Transfer experiment: $\mathrm{H}_\alpha \rightarrow \mathrm{C}_\alpha(t_1) \rightarrow \mathrm{N}(t_2) \rightarrow \mathrm{H^N}(t_3)$
2. "Out and back" experiment: $\mathrm{H^N} \rightarrow \mathrm{N}(t_1) \rightarrow \mathrm{C}_\alpha(t_2) \rightarrow \mathrm{N} \rightarrow \mathrm{H^N}(t_3)$

In the first pathway, magnetization is transferred from the $\mathrm{H}_\alpha$ all the way to $\mathrm{H^N}$. All transfer steps would be of INEPT type. The strategic pulses would be pairs of 90° pulses (Figure A.49a). The transfer amplitude of this experiment is:

$$\sin^2(\pi^1 J_{\mathrm{H}_\alpha\mathrm{C}_\alpha}\Delta')\sin(\pi^1 J_{\mathrm{C}_\alpha\mathrm{N}}(2\tau_1+t_1))\cos(\pi^2 J_{\mathrm{C}_\alpha\mathrm{N}}(2\tau_1+t_1))$$
$$\cos(\pi^1 J_{\mathrm{C}_\alpha\mathrm{C}_\beta}(2\tau_1+t_1))\sin(\pi^1 J_{\mathrm{C}_\alpha\mathrm{N}}(2\tau_2+t_2))\cos(\pi^2 J_{\mathrm{C}_\alpha\mathrm{N}}(2\tau_2+t_2))$$
$$\sin^2(\pi^1 J_{\mathrm{NH}^{\mathrm{N}\Delta}}) \qquad \text{(For Figure A.49a)}$$

The interresidual $^2J_{\mathrm{C\alpha N}}$-coupling of around 7 Hz cannot be ignored since it is only 30% smaller than the $^1J_{\mathrm{C\alpha N}}$-coupling of about 11 Hz. The optimum delays $\tau_1 + t_1$ and $\tau_2 + t_2$ for the transfer via $^1J_{\mathrm{C\alpha N}}$ maximize the products $\sin(\pi^1 J_{\mathrm{C}_\alpha\mathrm{N}}(2\tau_1+t_1))\cos(\pi^2 J_{\mathrm{C}_\alpha\mathrm{N}}(2\tau_1+t_1))\cos(\pi^1 J_{\mathrm{C}_\alpha\mathrm{C}_\beta}(2\tau_1+t_1))$ and $\sin(\pi^1 J_{\mathrm{C}_\alpha\mathrm{N}}(2\tau_2+t_2))\cos(\pi^2 J_{\mathrm{C}_\alpha\mathrm{N}}(2\tau_1+t_1))$ and are therefore 28 ms and 31 ms, respectively. During this rather long time a constant-time evolution of the N chemical shift is advantageous. This yields the transfer amplitude:

$$\sin^2(\pi {}^1J_{H_\alpha C_\alpha}\Delta')\sin(\pi {}^1J_{C_\alpha N}2\tau_1)\cos(\pi {}^2J_{C_\alpha N}2\tau_1)\cos(\pi {}^1J_{C_\alpha C_\beta}2\tau_1)$$

$$\sin(\pi {}^1J_{C_\alpha N}2\tau_2)\cos(\pi {}^2J_{C_\alpha N}2\tau_2)\sin^2(\pi {}^1J_{NH^{N}\Delta}) \quad \text{(For Figure A.49b)}$$

For the "out and back" pathway the $H^N$ magnetization created at the beginning of the sequence is never lost, but is used to obtain sequentially coherence of N and $C_\alpha$. Therefore the sequence is called out and back. The following implementation can be used:

A CT-HSQC or CT-double INEPT from $H^N$ to N sandwiches a HSQC from N to $C_\alpha$. The transfer amplitude for this experiment is given by ($\tau_1 = \tau_2 = \tau$):

$$\sin^2(\pi {}^1J_{NH^{N}\Delta})\sin^2(\pi {}^1J_{C_\alpha N}2\tau)\cos^2(\pi {}^2J_{C_\alpha N}2\tau)\cos(\pi {}^1J_{C_\alpha C_\beta}t_2) \quad \text{(for Figure A.49c)}$$

$$\sin^4(\pi {}^1J_{NH^{N}\Delta})\sin^2(\pi {}^1J_{C_\alpha N}2\tau)\cos^2(\pi {}^2J_{C_\alpha N}2\tau)\cos(\pi {}^1J_{C_\alpha C_\beta}t_2) \quad \text{(for Figure A.49d)}$$

Although the second pathway ("out and back") looks more complicated, it is the advantageous one. In this second experiment the rather small ${}^1J_{NC_\alpha}$- coupling does not need to be defocused while the $C_\alpha$ is transverse whereas this is necessary for the transfer experiments, in Figure A.49a and b. During the long delay the ${}^1J_{C_\alpha C_\beta}$-coupling evolves (Figure A.49a and b), which attenuates the signal. Constant-time evolution is advantageous in these experiments, since one has to wait for the defocusing of the small ${}^1J_{NC_\alpha}$-coupling anyway. The double INEPT version (Figure A.49d) of the experiment is more sensitive because it is less prone to relaxation losses due to short longitudinal proton eigen relaxation. The enhancement trick [29, 30] is easily implemented for the N,H correlation part of the sequence by replacing the $^{15}$N,H back transfer part of the final HSQC by the corresponding part of sensitivity-enhanced HSQC that follows $t_1$ (Figure A.13) [178] including the application of gradient pulses [92].

The carbonyl carbon is refocused by selective pulses applied such that neither the ${}^1J_{NC'}$- nor the ${}^1J_{C_\alpha C'}$-coupling can evolve. Summarizing the reasoning for the implementation of HNCA: Long delays should be used only for resonances that are sharp and have few, small homonuclear couplings. Therefore the $C_\alpha$ should be transverse only during short delays, whereas this restriction does not apply to $^{15}$N because of its much longer $T_2$ and lack of coupling partners. The $H_\alpha$ is normally also broader than the $H^N$ due to the couplings to the $H_\beta$ spins and due to the cross relaxation with the $C_\alpha$ that is as strong as the homonuclear cross relaxation between two geminal protons. In contrast, the $H^N$-$^{15}$N dipolar relaxation is smaller than the $H_\alpha$,$C_\alpha$ dipolar relaxation by a factor of 6.25.

Another key sequence is the HNCO that correlates $H^N$ with $^{15}$N and the preceding $^{13}$C′. The sequence employed is essentially identical with the HNCA (Figure A.50). The correlated nuclei are $^{15}$N, $^{13}$C, and $H^N$ in $t_1$, $t_2$, and $t_3$. The sequence consists first of an INEPT transfer from $H^N$ to N. Refocusing of the ${}^1J_{H^N N}$-coupling is achieved after $\Delta$. Thereafter the $^{15}$N chemical shift evolves during a constant time delay that is also used for the defocusing of the ${}^1J_{NC'}$-coupling. The 180°(C′) and the 180°(N) pulses move in a parallel fashion. To avoid evolution of the ${}^1J_{NC_\alpha}$- and ${}^2J_{NC_\alpha}$-couplings, a 180°($C_\alpha$) pulse is applied after $t_1/2$. Polarization transfer to the carbonyl in an HSQC manner is followed by evolution of the carbonyl chemical shift. The 180°(N) and the 180°($C_\alpha$) pulse in the middle of $t_2$ ensure that no heteronuclear coupling evolves. The protons are still decoupled. After $t_2$ the polarization transfer to the $^{15}$N is performed, and refocusing of the ${}^1J_{NC'}$- and (by switching off the decoupler) defocusing of the ${}^1J_{NH}$-coupling is achieved. Finally, the INEPT transfer to the protons is performed and protons are recorded under $^{15}$N decoupling. The relevant spin states in the course of the sequence are indicated at the most important positions. Sensitivity enhancement by the incorporation of a reverse INEPT according to Figure A.31 has already been discussed.

The transfer amplitude for the sequence is quickly denoted:

$$\sin^4(\pi {}^1J_{HN}\Delta)\sin(\pi {}^1J_{C'N}2\tau_1)\sin(\pi {}^1J_{C'N}2\tau_2)$$

Finally, we would like to discuss two 4D experiments for the assignment of backbone residues: The combination of a HNCAHA [181, 183–185] with a HN(CO)CAHA [182–184] yields correlations of ($H_i^N$, $N_i$) with ($H_{\alpha,i}$,$C_{\alpha,i}$) and

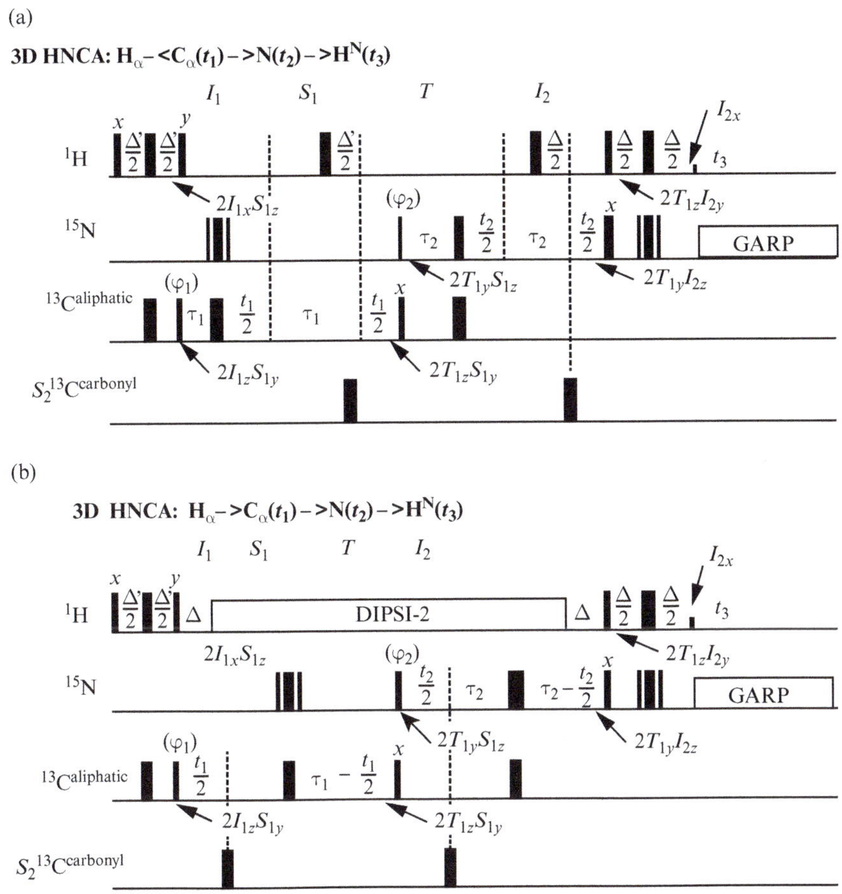


**Figure A.49** Various implementations of the HNCA experiment. The spins are identified in the following way: $H_\alpha = I_1$, $H^N = I_2$, $C_\alpha = S_l$, $C' = S_2$, and $^{15}N = T$. (a) Implementation of the HNCA as a transfer experiment. There are three coherence transfers of the INEPT type with pairs of simultaneous 90° pulses. The delays are set such that the transfer is optimal. $\Delta' = 1/2J_{CH}$, $\Delta = 1/2J_{NH}$, $\tau_i$ is set such that the product $\sin(\pi^1J_{C_\alpha N}2\tau_1)\cos(\pi^2J_{C_\alpha N}2\tau_1)\cos(\pi^1J_{C_\alpha C_\beta}2\tau_1)$ is maximum. The various 180° pulses are set such that all the heteronuclear couplings to the C′ are decoupled. The heteronuclear couplings to the proton evolve neither in $t_1$ nor in $t_2$. For example, the $^1J_{H_\alpha C_\alpha}$-coupling evolves between the first and the second INEPT transfer during: $\tau_1 - t_1/2 - (\tau_1 - \Delta'/2) + (\Delta'/2 + t_1/2) = \Delta'$. This refocuses the coupling before the $C_\alpha \rightarrow N$ transfer. (b) Constant time version of the sequence in (a) with proton decoupling. The strategic three INEPT transfers are the same as in (a). The meaning of the delays is as in (a). The $^{15}$N,$^{13}$C couplings evolve during $\tau_1$ and $\tau_2$. The coupling to the C′ is decoupled between both INEPT transfers. For the $^1J$(C′,N)-coupling in the $2\tau_2$ delay we find an effective evolution during: $t_2/2 - \tau_2 + (\tau_2 - \tau_2/2) = 0$. This sequence contains fewer pulses and decouples the protons during the evolution of heteronuclear chemical shift. (c) Implementation of the HNCA as an "out and back" experiment with constant time $^{15}$N evolution. The sequence can be understood as an inner $^{15}$N,$^{13}$C HSQC embraced by an outer $^1$H,$^{15}$N HSQC. $\tau_1$ and $\tau_2$ are now set so as to optimize the expression $\sin(\pi^1J_{C_\alpha N}2\tau)\cos(\pi^2J_{C_\alpha N}2\tau)$. Again the 180° pulses are set in such a way as to refocus those couplings that are not necessary for the transfers and to avoid first-order phase corrections. For example, the heteronuclear $^1J_{CH}$-and the $^1J_{NH}$-couplings evolve as follows: carbon is transverse only during $t_2$. The 180°($^1$H) pulse in the middle of $t_2$ refocuses the $^1J_{CH}$-coupling during $t_2$. The $^1J_{NH}$-coupling on the other hand evolves between the two $90^\circ_y$(H) pulses during: $t_1/2 - \tau_1 + (\tau_1 - t_1/2) - \tau_2 + \tau_2 = 0$ as it should be for a HSQC. The heteronuclear couplings to the C′ are refocused as in (b). (d) Decoupled version of (c). The two 180°($^1$H) pulses are replaced by a decoupling sequence after refocusing of the $^1J_{NH}$-coupling during $\Delta$. The latter sequence is the optimum sequence of the four shown with respect to signal to noise, because nitrogen is transverse during the defocusing and refocusing delays of the heteronuclear $^1J_{NC_\alpha}$ coupling.

(c)

**3D HNCA: H^N–<N(t1)–>Cα(t2)–>N–>H^N(t3)**

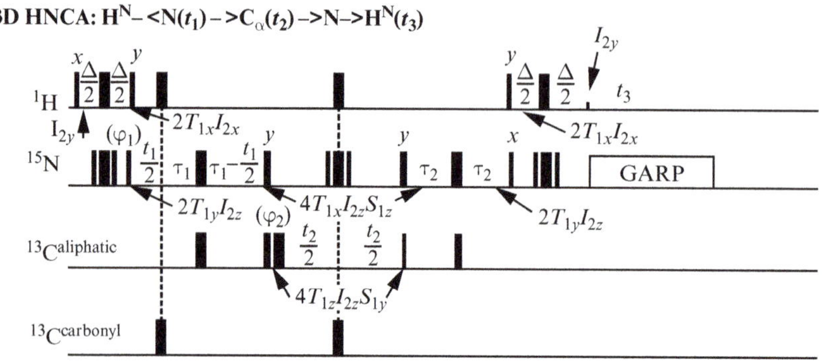


(d)

**3D HNCA: H^N–>N(t1)–>Cα(t2)–>N–>H^N(t3)**

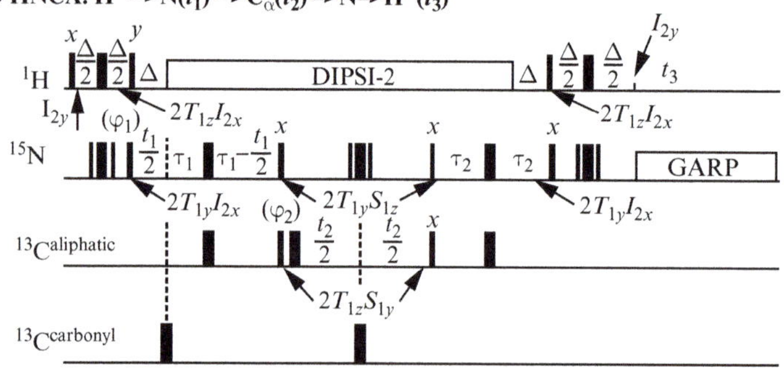


**Figure A.49(c,d) (Cont'd)**

**3D HNCO: H^N–>N(t1)–>C'(t2)–>N >H^N(t3)**

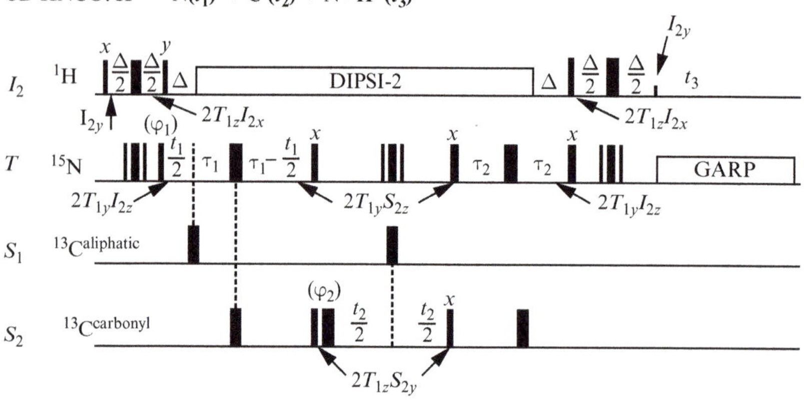


**Figure A.50** HNCO experiment derived from the HNCA experiment in Figure A.49d. The sensitivity-enhanced HNCO with gradients in Figure A.31 is the more advanced implementation of this sequence. If sensitivity enhancement in the $^{15}N \rightarrow {}^{1}H^{N}$ transfer is to be achieved, the constant time evolution of the $^{15}N$ chemical shift is performed during $\tau_2$ instead of during $\tau_l$.

of ($H_i^N$, $N_i$) with ($H_{\alpha,i-1}$,$C_{\alpha,i-1}$). There are a number of pulse sequences used for this combination of experiments. No comparison with respect to S/N ratio has so far been published.

#### A.4.3.1 HNCAHA

The experiment (Figure A.51a) using the transfer $H^N \rightarrow N(t_1) \rightarrow C_\alpha(t_2) \rightarrow H_\alpha(t_3) \rightarrow C_\alpha \rightarrow N \rightarrow H^N(t_4)$ is based on the same rules as the most sensitive HNCA experiment discussed in Figure A.49d. Chemical shift evolution of the $H_\alpha$ can be achieved either in a $^{13}C,^{1}H$-HSQC (with the two $90°(C_\alpha)$ pulses around $t_3$) or $^{13}C,^{1}H$-HMQC (without the two $90°(C_\alpha)$ pulses around $t_3$) fashion. For the HSQC version [181, 182, 184] the $^{1}J_{C_\alpha C_\beta}$-coupling of about 35 Hz does not evolve during $t_3$, as it does for the HMQC [185] version. On the other hand, the $4T_zI_{1y}S_{1z}$ operator present in the HSQC version relaxes faster than the $4T_zI_{1y}S_{1x}$ operator, which is not relaxed by the $J(0)$ (spectral density at frequency zero) term of the $H_\alpha$,$C_\alpha$ dipolar coupling. For molecular weights up to 20 kD the HSQC sequence seems to be slightly more sensitive than the HMQC version. For even larger molecular weights this tendency might be reversed [185].

#### A.4.3.2 HN(CO)CAHA

The HN(CO)CAHA [182–184] supplements the HNCAHA experiment for a complete sequence analysis of the protein backbone. The experiment starts with $H^N$ magnetization that is also detected. During the long delays for the defocusing and refocusing of the $^{1}J_{C'N} = 15$ Hz and the $^{1}J_{C_\alpha C'} = 55$ Hz, N and C′ magnetization are transverse,

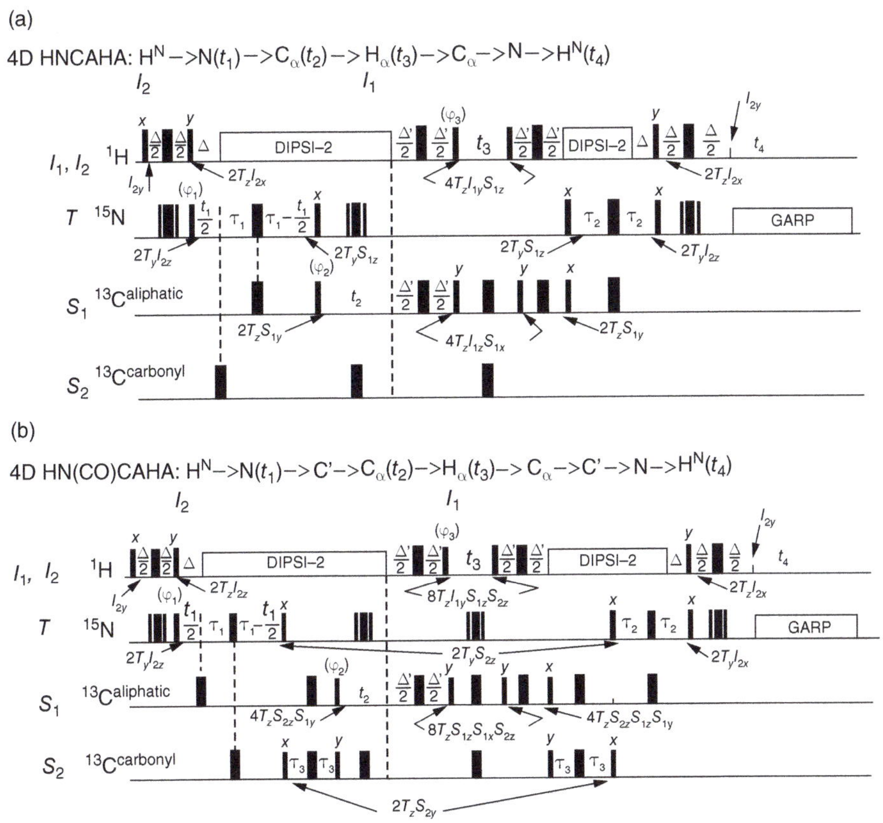


Figure A.51 (a) 4D HNCAHA experiment, "out and back" version with constant-time evolution on the nitrogen. HSQC version for the transfer between $C_\alpha$ and $H_\alpha$. This sequence can be understood as a series of HSQC experiments embracing each other. The innermost HSQC is the $^{13}C,^{1}H$-HSQC, then comes the $^{15}N,^{13}C$-HSQC, and the outermost is the $^{1}H,^{15}N$-HSQC. The meaning of the $\tau$ delays is as in Figure A.52d; the meaning of the $\Delta$ delays as in Figure A.49a. (b) 4D HN(CO)CAHA experiment. There are four HSQC experiments embracing each other. By a consecutive application of pairs of 90° pulses the product operator $8T_zI_{1y}S_{1z}S_{2z}$ is developed. constant-time evolution is applied for the nitrogen chemical shift. $2\tau_3 = 1/2J_{C_\alpha C'}$.

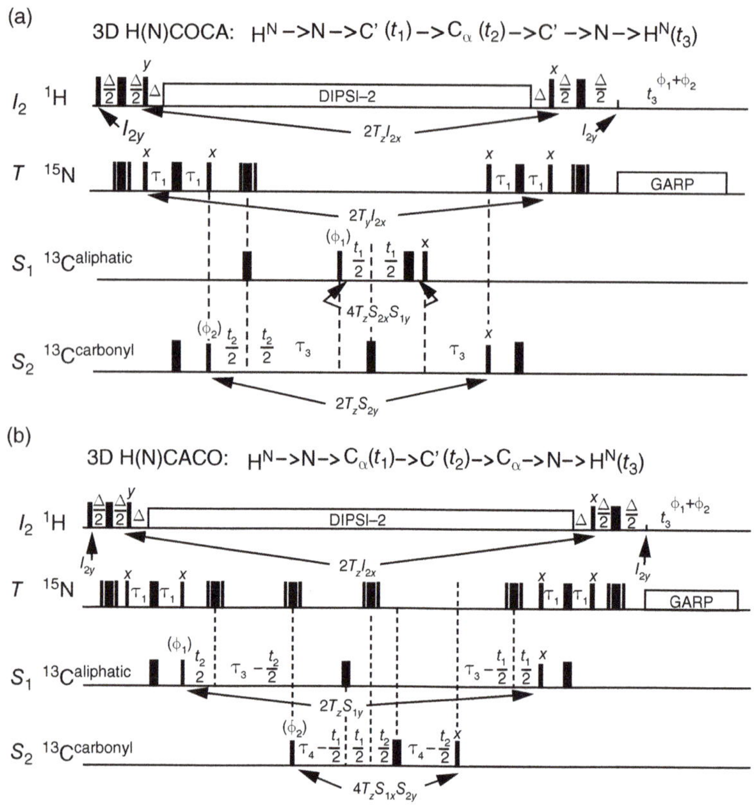


| | $\frac{t_2}{2}$ | $\tau_3-\frac{t_2}{2}$ | $\tau_4-\frac{t_1}{2}$ | $\frac{t_1}{2}$ | $\frac{t_2}{2}$ | $\tau_4-\frac{t_2}{2}$ | $\tau_3-\frac{t_1}{2}$ | $\frac{t_1}{2}$ | |
|---|---|---|---|---|---|---|---|---|---|
| $\Omega(S_1)$ | + | + | + | − | − | − | − | − | $t_1$ |
| $\Omega(S_2)$ | | | + | + | + | − | | | $t_2$ |
| $J(T,S_1)$ | + | − | − | + | − | + | + | − | 0 |
| $J(T,S_2)$ | | | + | + | − | − | | | 0 |

**Figure A.52** (a) 3D H(N)COCA experiment. This is an "out and back" experiment as well. Chemical shift of the C′ and the $C_\alpha$ evolves during a normal evolution time. The delays have the same meaning as in Figure A.51. (b) 3D H(N)CACO in an "out and back" version. The evolution times of both the carbonyl and the $C_\alpha$ are constant time. This has to be done to get rid of the ${}^1J_{C_\alpha C_\beta}$-coupling constant. The total time between the two 90°($S_1$) pulses is $2(\tau_3 + \tau_4) = 1/J_{C_\alpha C_\beta}$. This ensures that the homonuclear ${}^1J_{C_\alpha C_\beta}$-coupling does not deteriorate the sensitivity, $1/(2J_{C_\alpha C'})$. Since the evolution of the various interactions is a bit involved, their evolution is described on the bottom of the sequence.

respectively (Figure A.51b). With complete labeling of the heteronuclei of the backbone as well as the sidechains, essentially every combination of nuclei can be correlated with each other.

A survey of the sequences used until 1991 is given in recent reviews [187–189].

Before leaving backbone assignment with 3D and 4D heteronuclear sequences we would like to present a short example for the application of these sequences. This involves a sequential assignment step in a ${}^{13}C/{}^{15}N$-labeled protein. The experiments used here are a HNCA, HNCO, H(N)COCA, and H(N)CACO. The sequences of the two latter experiments are given in Figure A.52. Since in the H(N)CACO the chemical shift evolution of $C_\alpha$ and C′ is achieved simultaneously in a constant-time HMQC fashion with constant-time delay $1/J_{C_\alpha C_\beta}$ the sign of the Gly residues is opposite to the sign of the amino acids with a $C_\beta$ carbon. So in this sequence, glycines are

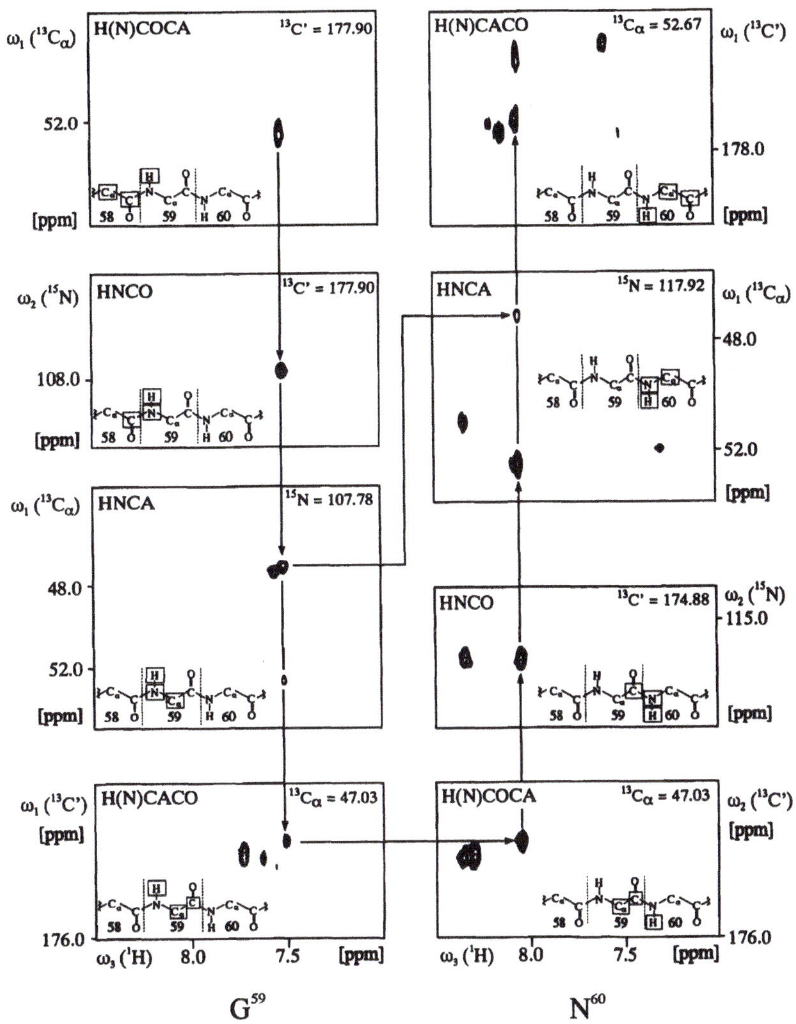


**Figure A.53** Sequential assignment of completely $^{13}$C/$^{15}$N-labeled calmodulin in the complex with C20W based on four experiments: H(N)COCA, HNCO, HNCA, and H(N)CACO. The sequential walk from Gly [59] to Asn [60] is shown with slices through the spectrum. Note that the glycine residues have inverted sign compared with the respective peaks of other amino acids because in the H(N)CACO the constant time delay for the evolution of the $^{1}J_{C_\alpha C_\beta}$-coupling constant is set to $2(\tau_3 + \tau_4) = 1/^{1}J_{C_\alpha C_\beta}$. The cross peaks visible in each of the traces through the 3D spectra are represented in the fragment of the backbone by little squares.

easily identified. A sequential assignment step between G$^{59}$ and N$^{60}$ in $^{15}$N and $^{13}$C-labeled calmodulin is shown. Assignment is done on lines, indicating that two frequencies are fixed in each step, providing optimal reliability of the assignment (Figure A.53).

### A.4.4 Side-Chain Assignments

The protons in the side chains of the amino acids in completely labeled proteins can be assigned using the HCCH-COSY [190–192] experiment or the HCCH-TOCSY [193, 194] experiment. The pulse sequences are given

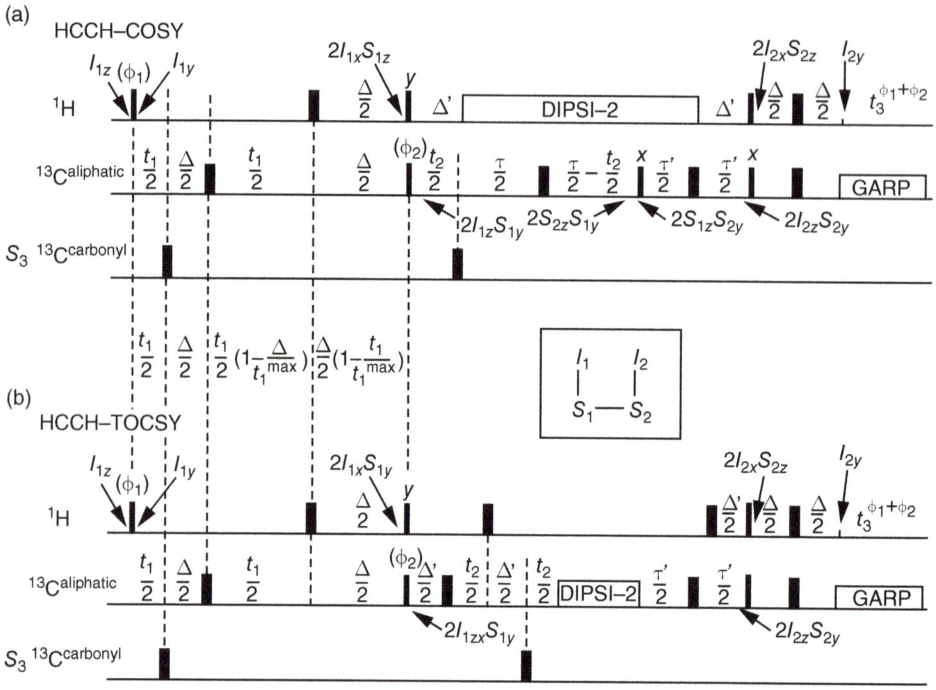


**Figure A.54** (a) HC(C)H-COSY experiment. The coherence transfers are indicated with product operators in a $I_1$-$S_1$-$S_2$-$I_2$ segment. This experiment is a transfer experiment. (b) HC(C)H-TOCSY experiment in a $I_1$-$S_1$-...-$S_2$-$I_2$ segment. $\Delta$ is set to $1/(2J_{CH})$ and $\Delta'$ to $1/(4J_{CH})$ to obtain maximum sensitivity for $CH_2$ groups, $\tau$ and $\tau'$ are set to $1/4J_{CC}$. The mixing time for the C,C-TOCSY should be between 8 ms and 20 ms to enhance direct and long-range cross peaks, respectively. Time shared evolution is demonstrated for $t_1$ as explained in Section A.4.4.

in Figure A.54a and b, respectively. The delays are tuned for the rather large homonuclear C-C coupling constants. Very short TOCSY mixing times can be chosen compared with the usual mixing times in $^1$H homonuclear TOCSY. A study of the mixing time dependence of transfer amplitudes to carbons along amino acid side chains for a mixture of amino acids has been performed. [195] In this sequence the so called "time shared evolution" can be applied for $t_1$ [199, 203]. This makes it possible to use the delay $\Delta$ partly for the evolution of proton chemical shift such that for $t_1^{max}$ the total duration between the 90°(H) pulses is $t_1^{max}$ instead of $t_1^{max} + \Delta$. This minimizes the effect of $T_2$ relaxation of protons. The modified delays in the sequence are given in the middle of the figure.

## A.4.5 Combinations between Backbone and Side-Chain Assignments

Backbone-directed assignment techniques relying also on side-chain resonances have recently been proposed. The correlation of $C_{\beta,i}$ or $C_{\alpha,i}$ with $(NH)_{i+1}$ in the CBCA(CO)NH experiment [196] and $C_{\beta,i}$ or $C_{\alpha,i}$ with $(NH)_i$ in the CBCANH experiment [197] introduces directly information about the type of amino acid into the assignment of the backbone nuclei, due to the characteristic $C_\beta$ and $C_\alpha$ chemical shifts for different amino acids. The CBCA(CO)NH pulse sequence relies on "COSY"-type transfer between carbons whereas the (H)CC(CO)NH sequence employs C,C-TOCSY transfer [198–202]. Correlations with the aliphatic protons instead of carbons can be used [203].

The use of carbonyl resonances for side-chain assignments has been described [204, 205].

### A.4.6 4D HSQC-NOESY-HSQC

Sidechain protons tend to overlap more than backbone protons. The density of resonances is at maximum in this region. Therefore 4D HSQC-NOESY- HSQC [207–209] experiments can be performed to reliably assign the interresidual NOEs. In completely $^{13}C$,$^{15}N$-labeled molecules, it is possible to selectively observe NOEs only in moieties $^{13}CH \cdots H^{13}C$, or only in $^{15}NH \cdots H^{15}N$ moieties or only in $^{15}NH \cdots H^{13}C$ moieties (the broken line indicates spatial proximity necessary for the NOE). It is possible to record the experiments simultaneously as explained in Section A.3.5 by expansion of each of the two X-filters in a full HSQC or HMQC (see Figure A.15) [31, 32].

### A.4.7 Implementation of Triple-Resonance Sequences

The implementation of these sequences requires at least three frequency sources to perform the pulses on the protons, nitrogens, and carbons. Alternatively, four channels are used to generate the pulses for each of the four spin types independently. In the following the implementation of such sequences with three frequency channels is discussed.

The carbon spectrum of a protein is rather well resolved. Aliphatic and carbonyl resonances are well separated (see Figure A.38d).

When all pulses applied to carbons are derived from one frequency source, there are two possibilities to obtain pulses acting on aliphatic and carbonyl carbons:

1. Set the offset frequency of the pulse such as to hit either of the interesting frequency regions. This can be implemented by fast switching of the frequency of the synthesizer. However, the relative phases of two on-resonance pulses will be time dependent if between the pulses the offset of the channel has been switched.
2. Impose a phase gradient on the pulses to shift their frequency. This ensures that the relative phase of on- and off-resonance pulses is known at every stage in the sequence. Using amplitude modulation instead of a phase gradient ensures that there are no zero-order Bloch–Siegert phases. Apart from this the determination of pulses and relative phases is performed as described in Section A.3.7.

## A.5 Determination of Coupling Constants

Heteronuclear as well as homonuclear coupling constants carry useful information. To give just a few examples: $^1J_{XH}$-couplings reflect the electronic structure of the H-X bond, which can also be translated to conformational information in a few cases. [210–213] $^2J_{HX}$-couplings in H-C-C fragments sometimes carry information about the torsional angle about the C-C bond. [209, 210] Finally, $^3J_{HX}$-couplings as well as $^3J_{HH}$-couplings are useful for the determination of local conformations because they strongly depend on the torsional angle about the bond.

The methods for measurement of coupling constants can be subdivided into two classes:

1. The coupling to be measured is not well resolved in the peak but it can be fitted based on a reference experiment.
2. The coupling can be measured from resolved peaks.

### A.5.1 HMBC According to Keeler/Neuhaus

The paradigmatic implementation of the first principle (Figure A.55) is the measurement of heteronuclear long-range couplings, mainly to nonprotonated NMR-active heteronuclei in the HMBC experiment. [214–217] Heteronuclear spins that do not carry a proton can be detected in nonuniformly labeled samples via the long-range heteronuclear coupling in an HMBC experiment.

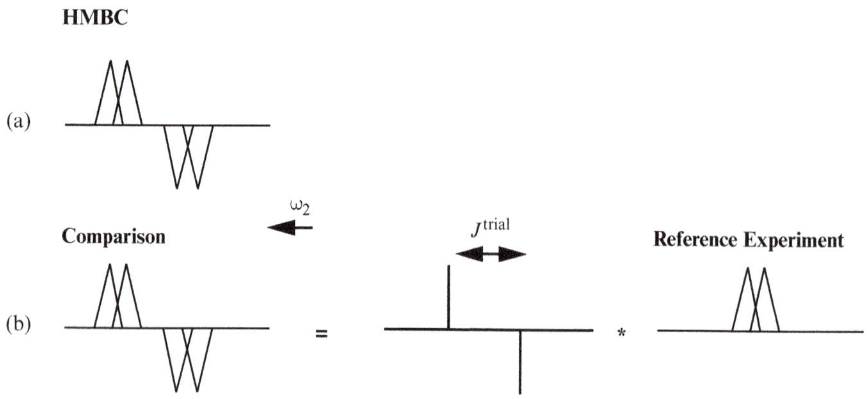


**Figure A.55** Principle of the determination of coupling constants when a spectrum with the coupling of interest and a spectrum in which this coupling is absent are available. This is the case, for instance, for the HMBC experiment (a) where the desired long-range $J_{IS}$-coupling is in antiphase, (b) Recording a reference spectrum that lacks this coupling and convoluting it with a trial antiphase coupling $J^{\text{trial}}$ leads to a "comparison spectrum" that is compared with the coupled spectrum (HMBC). Depending on the nature of the reference experiment and the experiment exhibiting the coupling, both the spectrum amplitude and the coupling may have to be fitted.

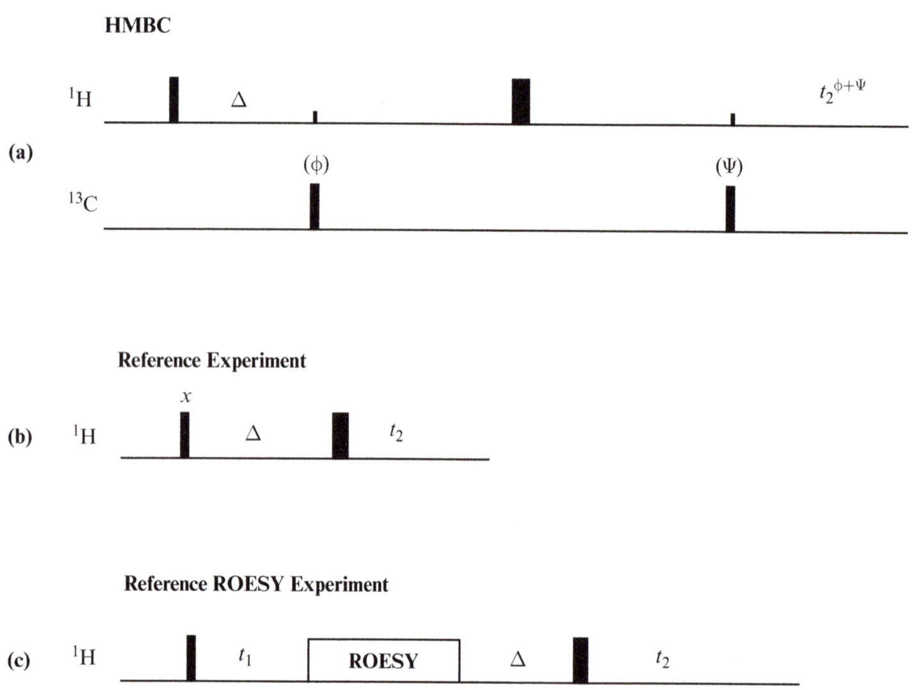


**Figure A.56** Pulse sequence of an HMBC (a) and a 1D reference experiment (b). The reference experiment is equivalent to HMBC for $t_1 = 0$. (c) A 2D experiment that yields in-phase proton multiplets (e.g., a ROESY) can also serve as reference. If $\Delta$ is a multiple of the dwell time of the 2D experiment, then a reference spectrum can be obtained from a standard 2D experiment by throwing away an equivalent number of points acquired in $t_2$.

To apply the Keeler/Neuhaus method, a suitable reference spectrum is needed. Comparison of the HMBC transfer amplitude (Figure A.56a, compare to Section A.2.1) with the transfer amplitude of a reference 1D experiment (Figure A.56b) reveals that there is an additional term $\sin(\pi J_{HC}\Delta)\cos(\Omega_c t_1)\sin(\pi J_{HC} t_2)$ in the HMBC.

The $\sin \pi J \Delta$ term is just an amplitude and the $\cos \Omega_c t_1$ defines the frequency in $\omega_1$. However, the $\sin \pi J_{HC} t_2$ term leads to an additional antiphase, dispersive splitting of the HMBC multiplet in $\omega_2$ compared with the reference multiplet. The relative phase difference between the HMBC and the reference multiplet can be removed by a zero-order phase correction. The $\omega_2$-multiplet pattern of the HMBC can be reconstructed by replicating the 1D experiment with an antiphase displacement by the $J_{HC}$-coupling constant (Figure A.55b). The correct value of $J_{HC}$ is the one that minimizes the difference between the $\omega_2$-multiplet pattern of the HMBC and the reconstructed antiphase multiplet from the reference spectrum. Since two different experiments are combined, for low abundance heteronuclei there are two fitting parameters, namely the intensity of the peaks and the desired coupling constant.

When the resolution in the 1D reference experiment is not sufficient due to the complexity of the spectrum, it can be replaced by rows taken from a 2D experiment that yields in-phase multiplet structures in $\omega_2$, for example, TOCSY, NOESY, or ROESY. The reference 2D experiment can be recorded in the normal way. One should then set the defocusing delay $\Delta$ as a multiple of the dwell time in $t_2$ and throw away the number of points that correspond to the delay $\Delta$. As an example, the measurement of $^3J_{NH}$-coupling, in a nickel complex is shown in Figure A.57a. The traces are obtained from the HMBC spectrum (Figure A.57b).

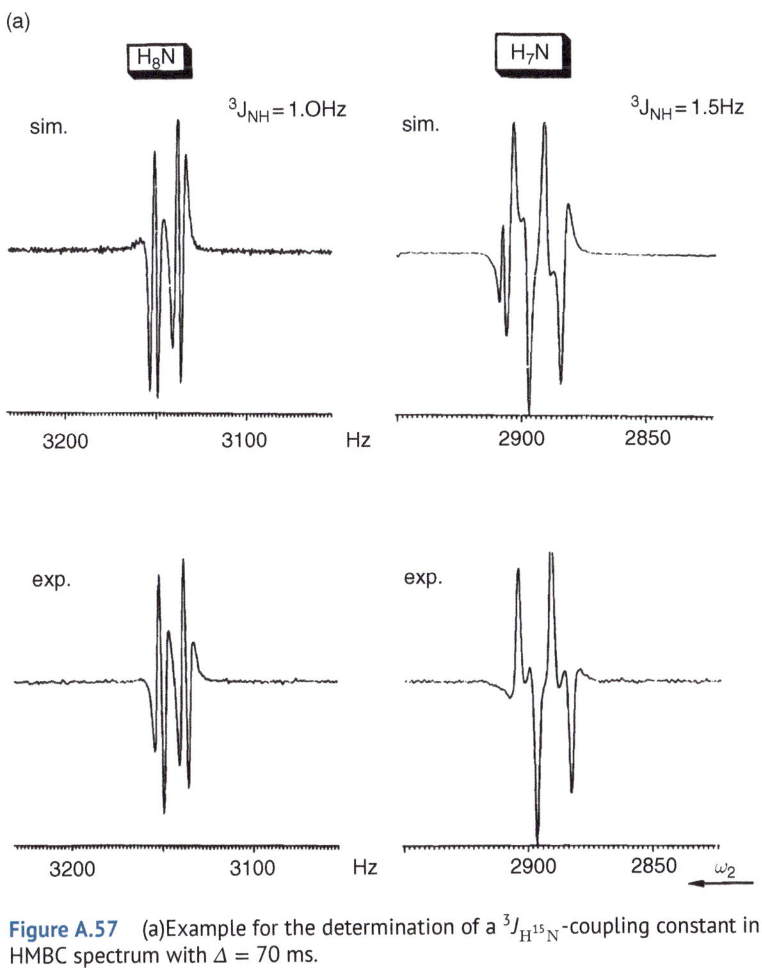


**Figure A.57** (a)Example for the determination of a $^3J_{H^{15}N}$-coupling constant in a nickel complex via HMBC. (b) 2D $^{15}$N, H HMBC spectrum with $\Delta = 70$ ms.

*(Continued)*

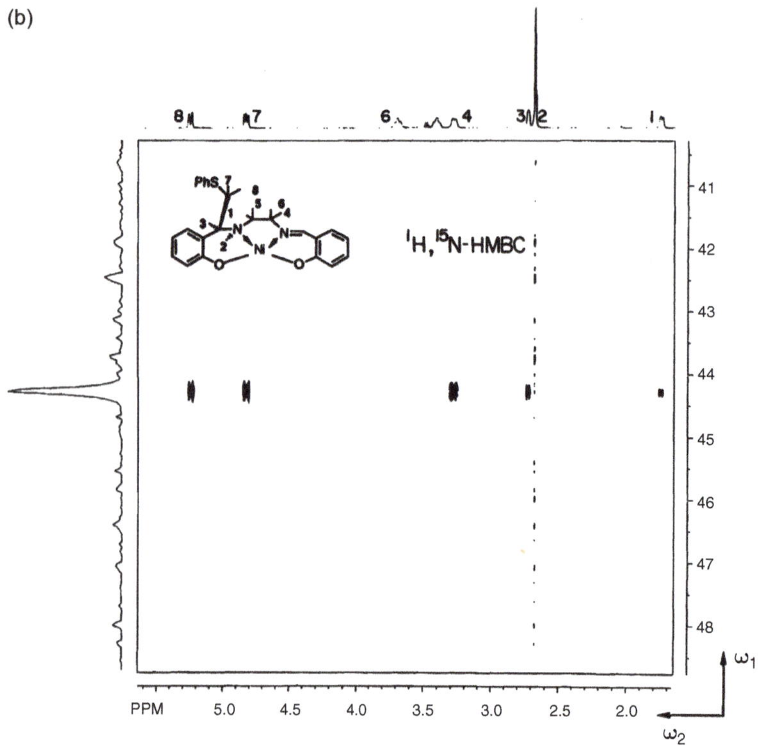


Figure A.57(b) (Cont'd)

#### A.5.1.1 Related Techniques

In applications to a low natural abundance spin the fitting procedure fits both the coupling and the intensity of the two spectra. This is not necessary if couplings to a spin that has a 100% natural abundance are measured, for example, for $^{31}$P that has a natural abundance of 100%. Recording an HSQC with and without $^{31}$P decoupling in $\omega_1$ (Figure A.58a) yields the same spectra except for a splitting due to the $^{31}$P, $^{13}$C coupling in $\omega_1$ [218]. Since this is the only difference between the coupled and decoupled spectrum, only the coupling needs to be fitted and the $J_{^{31}P^{13}C}$-coupling can be determined in an accurate way. The 0°($^{31}$P) pulse ($^{31}$P not decoupled) and the 180°($^{31}$P) pulse ($^{31}$P decoupled) are implemented as $90^\circ_x(^{31}P)\ 90^\circ_{\pm x}(^{31}P)$. This sequence is demonstrated on a dApA dinucleotide in Figure A.59. The coupled and decoupled spectra have the same linewidth. The apparent coupling determined from the splitting is 5.0 Hz. Optimal fitting of the coupling yields $J_{^{13}C(4)^{31}P} = 5.8$ Hz for A$^1$ (Figure A.60).

Finally, a large number of methods based on quantitative comparison of direct and remote cross peak intensities have recently been introduced [219–226].

### A.5.2 E.COSY Type Experiments

If a proton is attached to the heteronucleus that participates in the coupling one wishes to determine, techniques based on the E.COSY principle [227–230] can be used. We first explain the principle:

In the high-field, high-temperature approximation the probability for a nucleus C with spin-1/2 to be in the $\alpha$ or $\beta$ state is equal to 1/2. The lifetime of the states is $T_{1c}$. Provided during a mixing process of a 2D experiment there is no disturbance of this spin $C$, the 2D spectrum that originates from a magnetization transfer $A$ to $B$ looks as shown in Figure A.61. It can be thought of as the combination of the spectra originating from the mixture of two

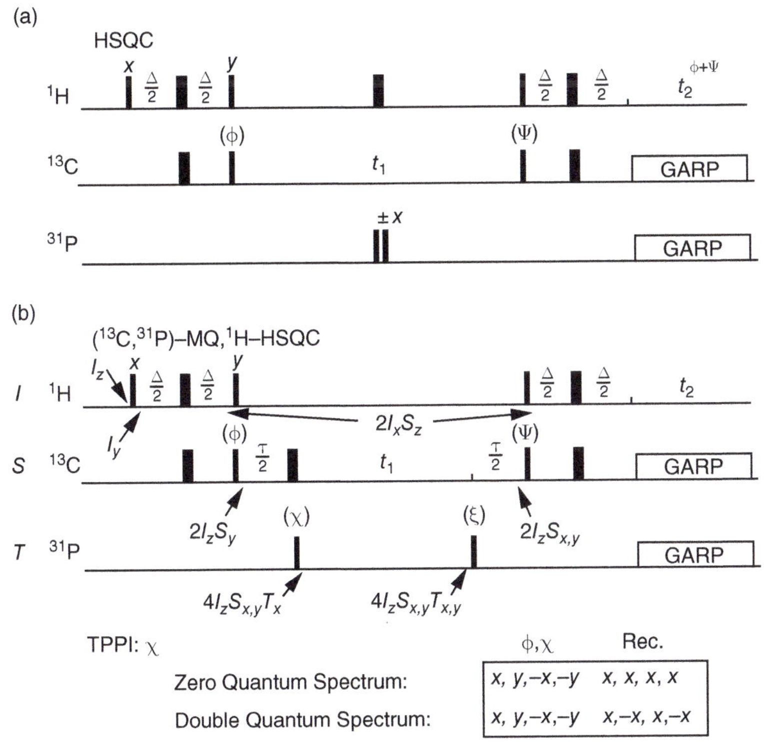


**Figure A.58** (a) $^{13}$C, $^{1}$H HSQC with and without decoupling of $^{31}$P in $t_1$. The $^3J_{^{13}C^{31}P}$ is present in $\omega_1$ when the two 90°($^{31}$P) pulses yield a 0° pulse and it is absent when the two pulses yield a 180°($^{31}$P) pulse, (b) Pulse sequence that correlates $^{31}$P,$^{13}$C double- and zero-quantum coherence with proton magnetization that is detected in $t_2$. Since no proton decoupling is applied during $t_1$, the sum and difference couplings between the proton and the heteronuclear multiquantum coherence evolve. TPPI is applied either on $\phi$ or on $\chi$. $\phi$ and $\chi$ are varied together according to $x$, $y$, $-x$, $-y$. The experiments with $x$ and $-x$ are added. The experiments with $y$ and $-y$ are also added. Additive combination of the two resulting data sets leads to the zero-quantum spectrum. Subtractive combination leads to the double-quantum spectrum.

molecules (spin-isomers). One sort of the molecules has spin $C$ in the $\alpha$ state ($\alpha$-spin isomer, schematic spectrum in Figure A.61a) and the other sort in the $\beta$ state ($\beta$-spin isomer, schematic spectrum in Figure A.61b). The chemical shift of a nucleus $A$ is $\Omega_A - \pi J_{AC}$ for $\alpha$-spin isomers and $\Omega_A + \pi J_{AC}$ for $\beta$-spin isomers. Implementations of 2D sequences that do not disturb the spin $C$ while correlating $A$ and $B$ are shown in Figure A.62. If $C$ is heteronuclear to both spins $A$ and $B$, the sequence simply includes no pulses for $C$ (Figure A.62a). If $C$ is homonuclear to either of the two spins $A$ and $B$, the sequence uses either small flip angle pulses ($\beta$ pulses, Figure A.62b) that conserve the state of $C$ to an extent of $\cos^2(\beta/2)$ or selective pulses that do not touch $C$ and make it again effectively a "heteronuclear" spin (Figure A.62c).

Multiplets of the form in Figure A.61 reflect the couplings of the passive spin $C$ to the two active spins $A$ ($J_{AC}$) and $B$ ($J_{BC}$). The coupling constant $J_{BC}$ can be measured very accurately provided $J_{AC}$ is larger than digitization and resolution in $\omega_1$. The mixing process must be fast compared with $T_{1C}$. This principle can be used for the measurement of coupling constants if three spins are mutually coupled and if at least one of the passive couplings (here $J_{AC}$) is resolved.

### A.5.2.1 Measurement of Heteronuclear Coupling Constants

If the passive spin is a protonated heteronuclear spin X and the X-bound proton $H_1$ is correlated with another proton $H_2$ by an arbitrary mixing process, the coupling $J_{H_2X}$ can be read off as $\omega_2$ shift in the $H_1$,$H_2$ cross

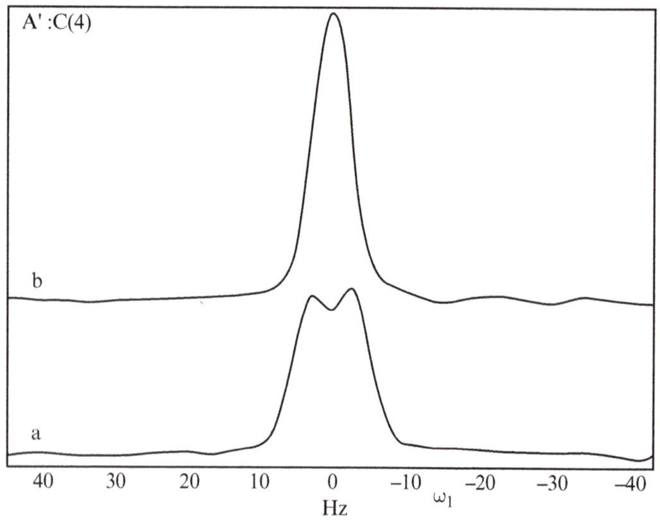


**Figure A.59** $\omega_1$ traces through the HSQC experiment (Figure A.58a) of dApdA: (a) $^{31}P$ coupled and (b) $^{31}P$ decoupled. The two signals have the same linewidth and the intensity of the peaks need not be fitted in determining the coupling constant.

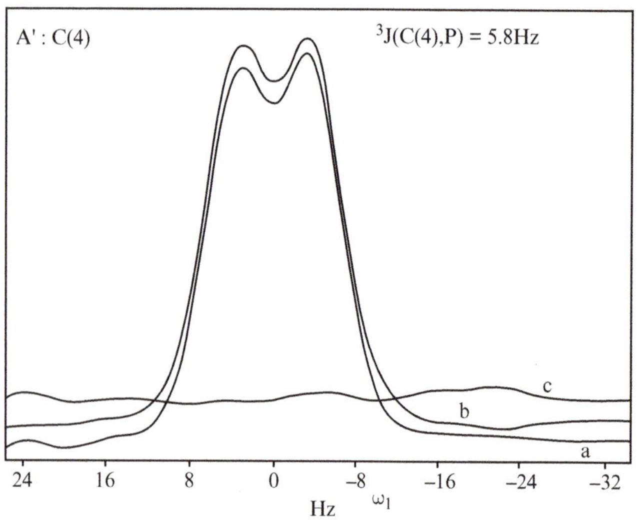


**Figure A.60** Fitting of the spectrum of Figure A.59a with the spectrum of Figure A.59b after convolution with an in-phase splitting of 5.8 Hz. (a) The coupled spectrum is identical to that of Figure A.59a. (b) The spectrum of Figure A.59b after convolution with a 5.8-Hz coupling, (c) The difference between (a) and (b). Zero signal indicates that the coupling is correct.

peak. This technique has been applied to measurement of $^3J_{NH_\beta}$- and $^3J_{N_iH_{\alpha,i-1}}$-coupling constants in $^{15}$N-labeled proteins [231] and peptides [232–237] as well as for the measurement of $J_{HC}$-long-range couplings [229].

In natural abundance samples an X-filter must be included to suppress the signals of the $^{12}$C-bound protons. This is probably the best method to access long-range heteronuclear couplings to protonated heteronuclei.

An application of a pulse sequence of the type shown in Figure A.62a is given in Figure A.64 for a coenzyme essential to methanogenesis in methanogenic archaea. An X-filtered TOCSY including BIRD suppression, also called HETLOC [232] (Figure A.63a), was recorded without heteronuclear decoupling in $\omega_1$ and $\omega_2$.

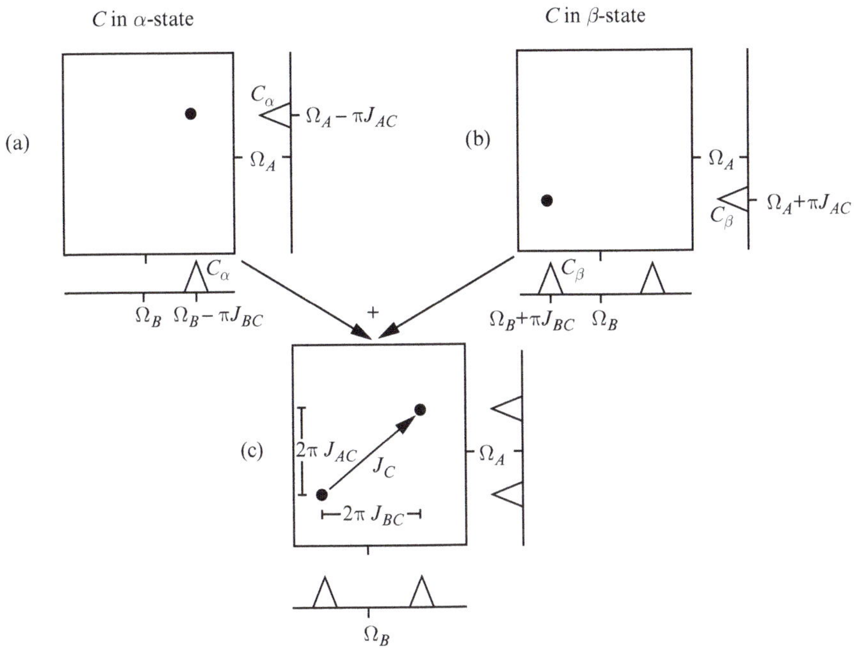


**Figure A.61** Illustration of the E.COSY principle. A mixing process that correlates two spins $A$ and $B$ without touching spin $C$ leads to a multiplet as in (c). $J_c$ is the displacement vector with the components $J_{AC}$ in $\omega_1$ and $J_{BC}$ in $\omega_2$. The spectrum in (c) can be explained by an additive superposition of a spectrum of molecules with $C$ in the $\alpha$ state (a) and in the $\beta$ state (b).

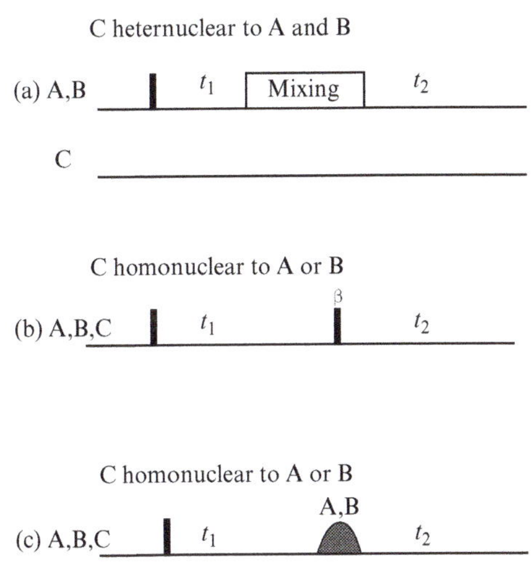


**Figure A.62** Various implementations of the E.COSY principle: (a) C is heteronuclear to A and B. Then a sequence that applies no pulses to C between the beginning of $t_1$ and the end of $t_2$ leads to an E.COSY multiplet, (b) If C is homonuclear to either A or B, a $\beta$ pulse can be used. (c) Selective pulses applied to A and B can be used as a second alternative in the homonuclear case. The selection can be achieved by a frequency selective pulse, or also by topology selective pulses such as BIRD/2.

An example cross peak between C(13a),C(6a)H and C(12a),C(6a)H in the coenzyme $N^5$,$N^{10}$-methenyl-5,6,7,8-tetrahydromethanopterin is given in Figure A.64. A $^3J_{C(13a),C(6a)H}$-coupling constant of 7.2 Hz and a $^3J_{C(12a),C(6a)H}$-coupling constant of 2.4 Hz can be determined from these two cross peaks [238].

Suppression of geminal cross peaks and diagonal peaks along the lines described for diagonal-free proton-proton correlation spectra is achieved by inserting a defocusing period after the spin lock, which is applied along y. During the defocusing period the $S$-bound $I$ spins form an operator of the form $2I_xS_z$ whereas the non-$S$-bound $I$ spins

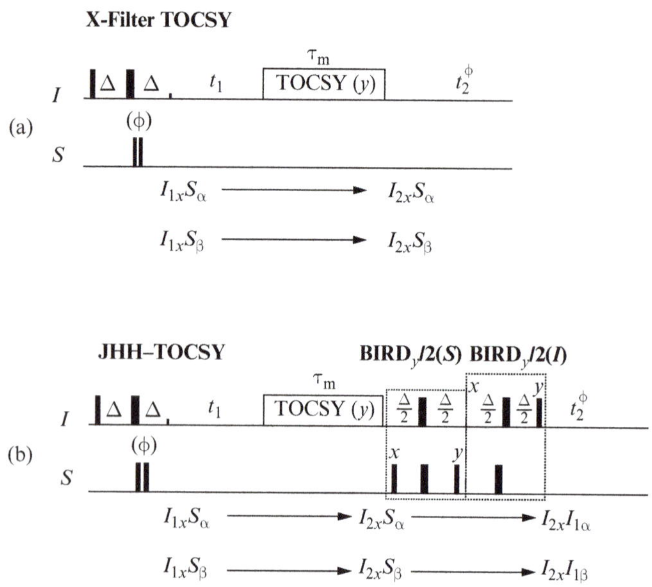


**Figure A.63** (a) Pulse sequence of the X-filter TOCSY that is a combination of the X-filter and a homonuclear TOCSY. No decoupling is applied during $t_1$ in order to retain the splitting due to the $^1J_{CH}$-coupling. (b) Pulse sequence of $J_{HH}$-TOCSY that is applicable for the correlation of $I$ spins when only one of the $I$ spins carries an $S$ spin. During $t_1$ the two multiplet lines of the $I_1$ spin $S_\alpha$ and $S_\beta$ evolve undisturbed. After the mixing process, $S_\alpha$ is transferred exclusively to $I_{1\alpha}$ and $S_\beta$ is exclusively transferred to $I_{1\beta}$. Therefore in $t_2$ the $I_1$ spin leads to the splitting of the $I_2$ multiplet. The displacement vector thus can no longer be assigned to one spin, but it is caused in $t_1$ by the S spin and in $t_2$ by the $I_1$ spin.

remain $I_y$. Application of a $0_y/180_y$ pulse selects the non-$S$-bound $I$ spins, thus suppressing the diagonal and geminal cross peaks in the experiment [239].

When the X-filter TOCSY sequence is applied to completely labeled molecules, selective decoupling of the carbon that is bound to the detected protons is possible and increases the S/N by a factor of 2. The number of coupling constants that can be determined is reduced at the same time, because no couplings to the decoupled nucleus can be measured [236].

Another application of measurement of heteronuclear couplings, this time to nonprotonated heteroatoms, is possible in completely carbon-labeled proteins. Here heteronuclear coupling constants between protons and carbonyl carbons can be accessed due to the large chemical shift difference between the carbonyl carbons and aliphatic carbons. Selective pulses make it possible to selectively excite either aliphatic carbons or carbonyl carbons. The sequences applied here fall in the class of Figure A.62(c). An H,C$^{\text{aliphatic}}$ HSQC yields in the $H_\alpha(\omega_2) \rightarrow C_\alpha(\omega_1)$ cross peaks the $^2J_{H_\alpha C'}$-coupling constants [39] (compare with Figure A.17(b)), whereas the aliphatic selective HCCH-COSY sequence (Figure A.65a) yields for the transfer $H_\alpha(\omega_1) \rightarrow C_\alpha(\omega_2) \rightarrow C_\beta \rightarrow H_\beta(\omega_3)$ the $^3J_{H_\beta C'}$-couplings. [115] The schematic multiplet structure is given in Figure A.66(a). C′ is the passive spin that leads to a splitting of the $C_\alpha$ resonance in $\omega_2$ by $^1J_{C_\alpha C'}$ and of the $H_\beta$ resonance by $^3J_{C'H_\beta}$. Selective decoupling (compare to Section A.3.6) of the aliphatic carbons during $t_3$ is performed to remove all $J_{HC^{\text{aliphatic}}}$-couplings (Figure A.38a).

An example is given for the $^3J_{H_\beta C'}$-coupling of the residue Val [67] in ribonuclease $T_1$ (Figure A.67b).

A second example is provided for the Ala$^{75}$ residue of the same protein in Figure A.68. Since the delay $\tau$ in Figure A.65a is only 14 ms long to obtain optimal transfer via the $C_\alpha$,$C_\beta$ coupling constant, the $^1J_{C_\alpha C'}$-coupling has acquired a phase of only $\pi J_{C_\alpha C'}\tau = 138°$. This is too short to resolve the required 55-Hz splitting (Figure A.68a).

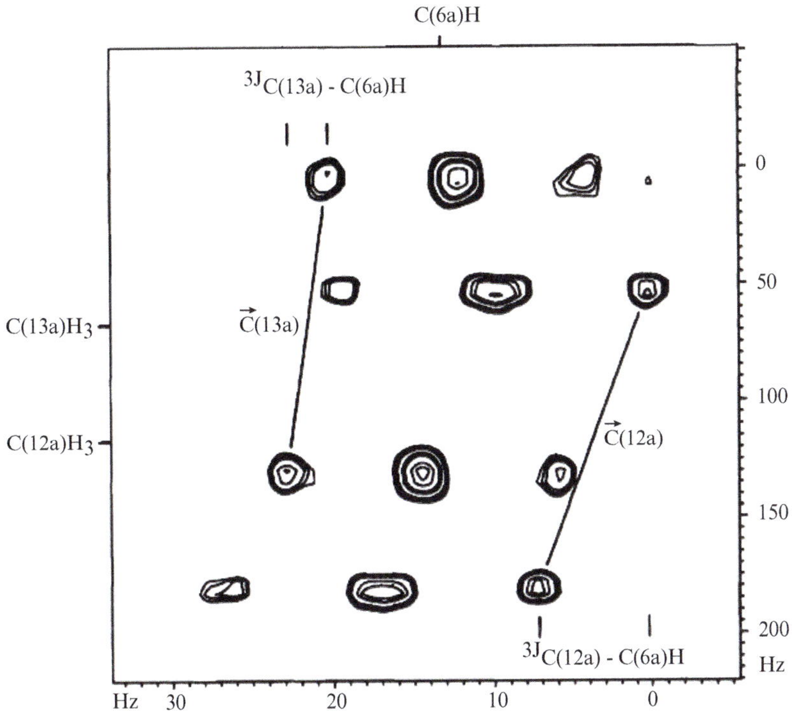


Figure A.64 Contour plot of the $^{13}$C-filtered TOCSY experiment of $N^5$,$N^{10}$-meth- eny1-5,6,7,8-tetrahydromethanopterin showing cross peaks between C(6a)H in $\omega_2$ and C(12a)$H_3$ and C(13a)$H_3$ in $\omega_1$. The large coupling constant $^3J_{\mathrm{C(6a)H.C(12a)}} = 7.2$ Hz and the small coupling constant $^3J_{\mathrm{C(6a)H.C(13a)}} = 2.4$ Hz are clearly visible as displacements in $\omega_2$, while the multiplet component is separated by the large $^1J_{\mathrm{CH}}$-couplings in $\omega_1$.

Mirror image linear prediction, however, leads to an increase of the number of points by a factor of 2. The splitting is then resolved and the desired $^3J_{\mathrm{C'H}_\beta}$-coupling can be extracted (Figure A.68b).

Related sequences can be used to measure C, C coupling constants, for example, $^3J_{\mathrm{C}\gamma\mathrm{C}}$, in valine, threonine, isoleucine, and so forth [240].

The same principle is also applicable, for example in an H,C,P moiety. A $^{13}$C,H-HSQC without decoupling of $^{31}$P allows the measurement of $J^{13}\mathrm{C}^{31}\mathrm{P}$- as well as $J_{\mathrm{H}^{31}\mathrm{P}}$-couplings (Figure A.69).

### A.5.2.2 Measurement of Homonuclear Proton Coupling Constants

The measurement of homonuclear coupling constants can be most effectively accomplished by the E.COSY-derived experiment P.E.COSY. [229, 241, 242] However, the experiment requires a substructure of the form $XH_2$-YH, so that one geminal coupling is available. In addition, the proton linewidths must be smaller than the geminal coupling constants.

For larger proteins with large proton linewidths, these techniques fail. Fortunately, homonuclear couplings can also be measured based on heteronuclear correlation spectroscopy in compounds with heteronuclear labeling at the protons of interest.

#### *Homonuclear Couplings in HCCH Moieties*

Correlation of $C_1$ with $H_2$ in a $H_1$-$C_1$-$C_2$-$H_2$ moiety in a HCCH E.COSY [115, 243, 244] experiment (which features a $\beta$ pulse in the last INEPT transfer in contrast to the HCCH COSY sequence) allows the measurement of the $H_1$,$H_2$ coupling in $\omega_2$. The pulse sequence, which falls in the class of Figure A.62b, is shown in Figure A.65b.

It is a HCCH-COSY experiment that uses $^{13}C^{aliphatic}$ selective pulses to avoid signal loss due to $^{1}J_{C_\alpha C'}$-couplings and a constant-time segment for the transfer from $C_\alpha$ to $C_\beta$. There are no proton pulses (except 180°) applied

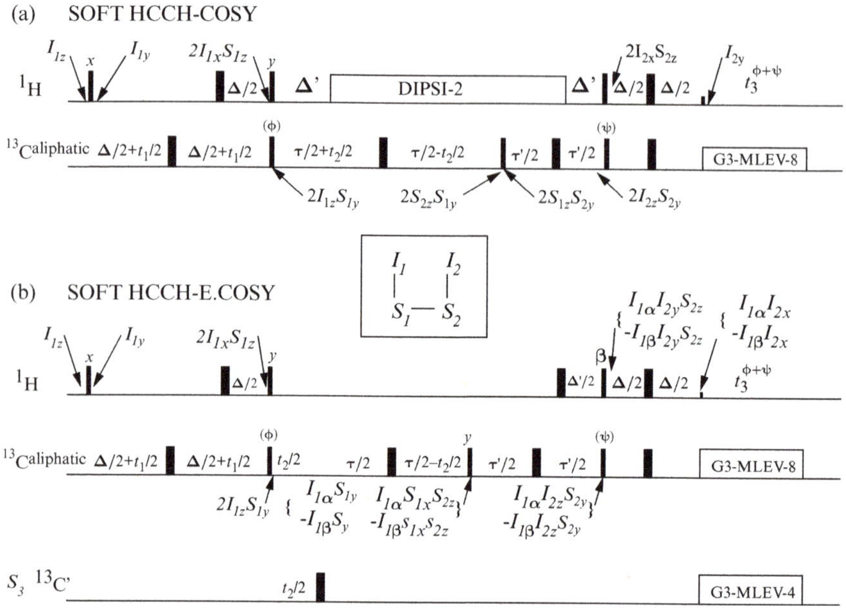


**Figure A.65** (a) Pulse sequence for the HCCH-COSY experiment of $C_\alpha$ and $H_\beta$ with the passive spin C′. The sequence is H,C-COSY-CT-C,C-COSY-C,H-INEPT. Aliphatic-selective decoupling is applied during $t_3$. (b) Pulse sequence of the HCCH-E.COSY experiment. $H_\alpha$ is the passive spin. It is not touched except by 180° pulses and a $\beta$ pulse. Correlation of $C_\alpha$ and $H_\beta$ is achieved, $\tau$ is set to $1/(2J_{C_\alpha C_\beta})$, $\tau'$ to $1/(4J_{C_\alpha C_\beta})$, $\Delta = 1/(2J_{CH})$.

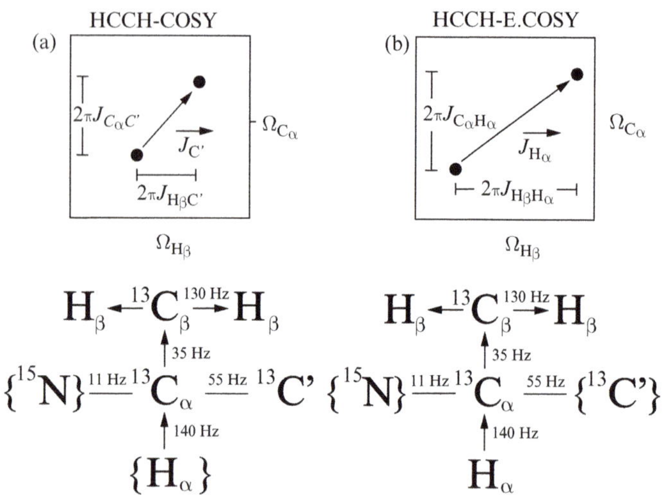


**Figure A.66** (a) Schematic multiplet structure of a $C_\alpha$, $H_\beta$ cross peak in a HCCH-COSY experiment as in Figure A.65a. The C′ leads to a displacement vector with the components $^{1}J_{C_\alpha C'}$ in $\omega_2$ and $^{3}J_{H\beta C'}$ in $\omega_3$. (b) Schematic multiplet structure of a $C_\alpha$,$H_\beta$ cross peak in a HCCH-E.COSY experiment as in Figure A.65b. The $H_\alpha$ leads to a displacement vector with the components $^{1}J_{C_\alpha H\alpha}$ in $\omega_2$ and $^{3}J_{H\beta H_\alpha}$ in $\omega_3$. The coherence transfers are indicated by arrows. Decoupled spins during evolution and detection are enclosed in braces.

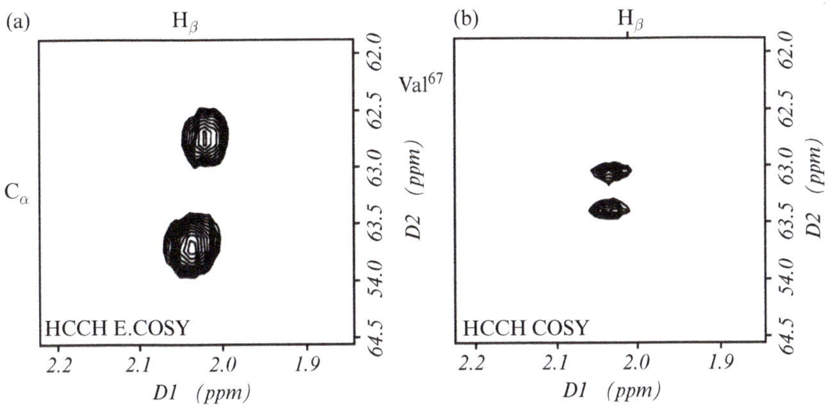


**Figure A.67** (a) Cross peak of Val [67] of ribonuclease $T_1$ in the HCCH-E.COSY experiment described in Figure A.65b. The displacement vector is clearly seen. A $H_\alpha$,H coupling constant of 10 Hz is obtained, (b) Cross peak of Val [67] in the HCCH-COSY experiment described in Figure A.65a. The $^3J_{H\beta C}$-coupling is only 0.8 Hz, which indicates a gauche arrangement of the $H_\beta$ and the C′.

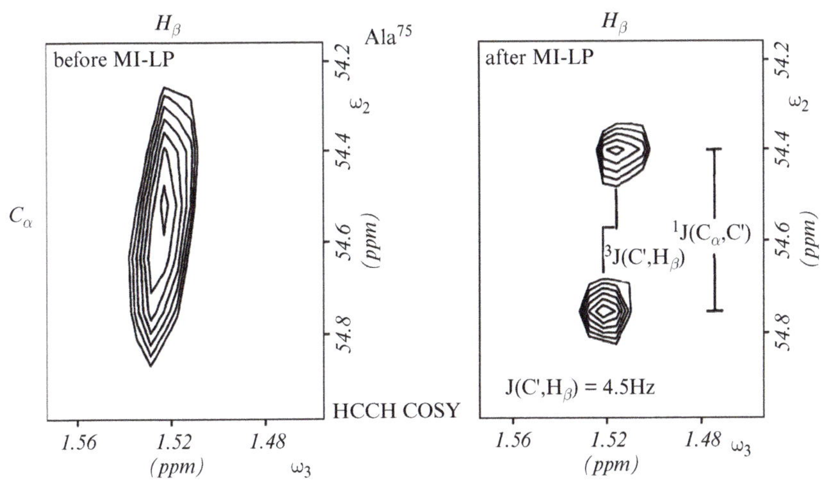


**Figure A.68** Application of mirror image linear prediction on the Ala [75] cross peak from the HCCH-COSY of ribonuclease $T_1$. The resolution within the multiplet can be enhanced such that the $^1J_{C_\alpha C'}$-coupling that is invisible in the left spectrum is resolved in the right spectrum.

from the start of $t_2$ through the end of $t_3$ except for the $\beta$ pulse in the INEPT transfer. This ensures that $H_\alpha$ keeps its spin state during the whole sequence. The schematic multiplet structure is shown in Figure A.66b. The $H_\alpha$ splits the $C_\alpha$ resonance in $\omega_2$ by $^1J_{C_\alpha H_\alpha}$ and the $H_\alpha$ resonance in $\omega_3$ by $^1J_{H_\alpha H_\beta}$. The measurement of H,H coupling constants is demonstrated on ribonuclease $T_1$ (Figure A.67a). $H_\alpha$,$H_\beta$ coupling constants are easily accessible in a quantitative manner. When these coupling constants are used together with the $H_\beta$,C′ coupling constants, stereospecific assignments of diastereotopic protons in amino acids can be achieved and dihedral angles can be measured. [115] An alternative sequence using a selective proton pulse instead of a $\beta$ pulse has been proposed. [245] Sequences based on C,C-TOCSY instead of constant-time C,C-COSY have also been proposed [246–248].

**Proton Homonuclear Couplings in HCNH Moieties**

The measurement of homonuclear couplings in moieties where the protons are connected to two different heteronuclei (here $^{13}C$ and $^{15}N$) is possible by a procedure that is closely related to the measurement of proton-proton couplings in HCCH moieties. Essentially a $C_\alpha$,$H^N$ correlation must be recorded without touching the $H_\alpha$ spin during the polarization transfer from $^{15}N$ to $H^N$. This is easily done by replacing the normal reverse INEPT sequence for the refocusing of the $S_{1z}$ by a BIRD$_x$/2 pulse [243, 249].

The sequence for the measurement is then easily constructed from known elements: INEPT from $H^N$ to $^{15}N$, $^{15}N$,$C_\alpha$ HSQC, and finally back-transfer from N to $H^N$ with BIRD$_x$/2 (Figure A.70). The sequence reproduced here is the latest and probably most sensitive one published for this important coupling constant [250–255].

### A.5.2.3 Homonuclear Coupling Constants in HHX Moieties

The measurement of proton–proton coupling constants in molecules with an isolated heteronucleus in either a natural abundance sample or a selectively labeled sample is possible based on the previously introduced BIRD$_x$/2 and BIRD$_y$/2 pulses. Let us first consider the following transformations (see (Figure A.27):

$$\begin{array}{llll} & & \text{BIRD}_y/2(S) & & \text{BIRD}_y/2(I) & \\ I-S: & S_z & & S_z I_{1z} & & I_{1z} \\ I: & I_{2x} & & I_{2x} & & I_{2y} \end{array}$$

Due to these transformations $S_\alpha = (1 + 2S_z)/2$ and $S_\beta = (1 - 2S_z)/2$ are transformed as follows:

$$\begin{array}{lll} & \text{BIRD}_y/2(S)\ \ \text{BIRD}_y/2(I) & \\ I_1 - S: S_a & & I_{1\alpha} \\ I_1 - S: S_\beta & & I_{1\beta} \end{array}$$

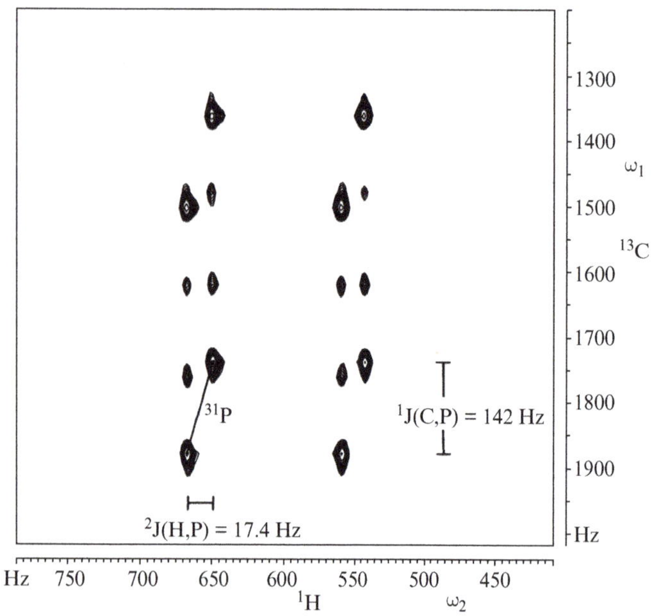


**Figure A.69** Cross peak of the methyl group in a $^{13}C$,$^{1}H$ HSQC in a dinucleotide with a methylphosphonate group. The passive $^{31}P$ leads to a splitting in $\omega_1$ due to the $^1J_{CP}$-coupling. In $\omega_2$ a $^2J_{HP}$-coupling is observed.

Thus the combination of an undecoupled X-filtered proton-proton correlation with the $\mathrm{BIRD}_y/2(\mathrm{S})$-$\mathrm{BIRD}_y/2(I)$ segment makes it possible to measure homonuclear couplings. The sequence is called $J_{\mathrm{HH}}$-TOCSY and is shown in Figure A.63b. The heteronuclear spin $S$ serves as the passive spin during $t_1$. Then, at the end of the sequence, the $S$-spin state is transferred to the directly bound $I$ spin, which serves as the passive spin during $t_2$. The sequence functions only if the proton active in $t_2$ is not bound to an $S$ spin and if only one $I$ spin is bound to the $S$ spin. Thus this method is promising for the determination of $\mathrm{H}_\alpha,\mathrm{H}^\mathrm{N}$ coupling constants in $^{15}$N-labeled proteins [256].

## A.5.3 Measurement of Coupling Constants from Multiquantum Coherence

The E.COSY methods as discussed in the previous section can be used to measure almost all coupling constants. However, two dimensions are needed because the two coupling constants, the associated and the desired coupling, are in orthogonal dimensions. Another principle would be to align the two couplings in one frequency dimension. Then measuring the sum and the difference of the two coupling constants yields the desired coupling provided the sum and the difference of the associated coupling and the desired coupling are both larger than the linewidth. Only one frequency dimension is needed in such experiments. This is especially useful when neither of the two active spins $A$ or $B$ is detected (Figure A.71).

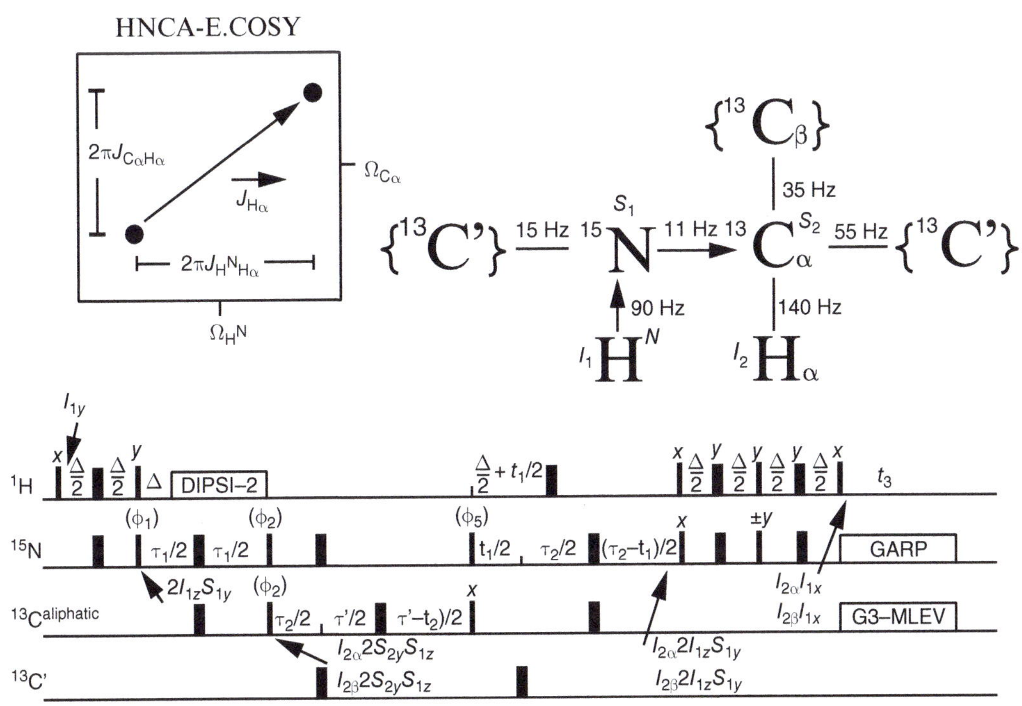


**Figure A.70** HNCA-E.COSY experiment that allows measurement of the $^3J_{\mathrm{H^N C}_\alpha}$-coupling constants in $^{13}$C,$^{15}$N-labeled proteins. The sequence is a HNCA experiment in which the transfer from the nitrogen to the proton at the end of the sequence is implemented in a sensitivity-enhanced way. Compared with the HNCA experiment in Figure A.49c, the experiment lacks the proton $\pi$ refocusing pulse during the evolution of the $\mathrm{C}_\alpha$ to ensure the evolution of the $^1J_{\mathrm{C}_\alpha\mathrm{H}_\alpha}$-coupling constant during $t_1$. The evolution of the nitrogen chemical shift has been moved from the constant-time delay $2\tau_1$ (Figure A.49c) to the delay $2\tau_2$. This allows for the application of the sensitivity- enhanced back-transfer from nitrogen to proton magnetization. The sensitivity enhancement achieved in the sequence is compatible with the requirement that the $\mathrm{H}_\alpha$spins are not touched between the beginning of $t_1$ and the detection. This can be seen in the following way: To derive the action of the proton pulses on the $\mathrm{H}_\alpha$ nuclei, concatenation of the proton pulses after the $2\tau_2$ constant-time delay is allowed, since neither heteronuclear coupling nor chemical shift of the $\mathrm{H}_\alpha$ spins evolves. The total rotation effected by the proton pulses for $z$-magnetization amounts to 180°.

This principle can be implemented by recording during an evolution period double- and zero-quantum coherence between a spin $A$ and $B$. The double-quantum spectrum will reflect the sum of the coupling ($J_{AC}+J_{BC}$), while the zero-quantum spectrum will reflect the difference ($J_{AC}-J_{BC}$). Provided one of the couplings (e.g., $J_{AC}$) is larger than the linewidth, the comparison of the apparent couplings in the two spectra can be used for the measurement of $J_{BC}$. This principle is demonstrated for the measurement of $^3J_{\mathrm{H}^{31}\mathrm{P}}$-coupling constants in DNA. Evolution of the C,P heteronuclear double- and zero-quantum coherence (Figure A.58b) without proton decoupling in the $^{13}$C,$^{31}$P-HMQC segment makes it possible to measure $J_{\mathrm{HP}}$-couplings. The $^{13}$C,$^{31}$P HMQC must be tuned to a $^2J_{\mathrm{CP}}$-coupling; the $^1J_{\mathrm{CH}}$-coupling serves as $J_{AC}$ and is used to measure the $^3J_{\mathrm{PH}}$-Coupling ($J_{BC}$). An example with a P-$CH_3$ group is shown in Figure A.72 where the $J_{\mathrm{PH}}$-coupling is a two-bond and the $J_{\mathrm{CP}}$-coupling is a one-bond coupling. The difference in the splitting amounts to a coupling constant of 17 Hz.

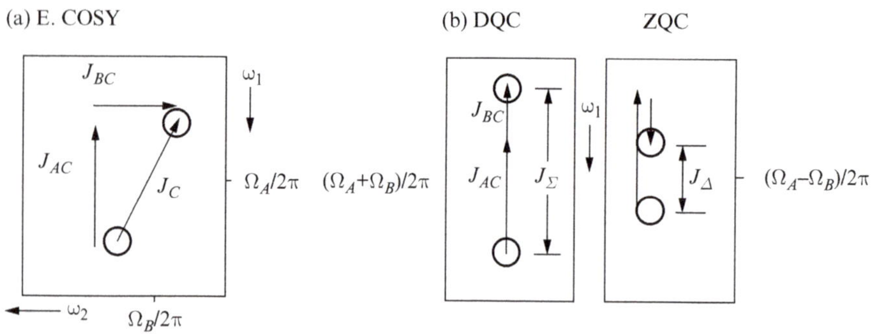


**Figure A.71** Comparison of the E.COSY principle for measuring coupling constants and the measurement from double-quantum and zero-quantum coherence. E.COSY requires two dimensions, since $J_{AC}$ and $J_{BC}$ are in orthogonal dimensions. Measurement of the coupling of the zero- and double-quantum coherenece of spins $A$ and $B$ to spin $C$ allows the determination of the $J_{BC}$ -coupling from one dimension.

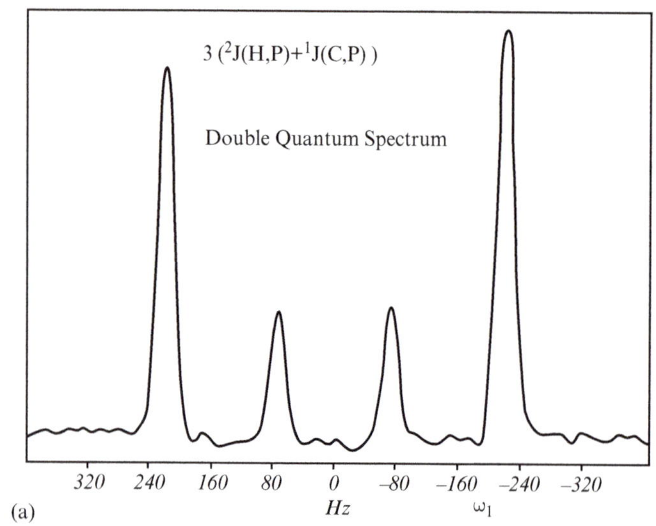


**Figure A.72** Example of the measurement of the $^2J_{\mathrm{HP}}$-coupling from $^{13}$C,$^{31}$P double- and zero-quantum coherence in a $CH_3$-$^{31}$P moiety at natural abundance. The doublequantum trace of the spectrum (a) obtained with the pulse sequence in Figure A.58b reflects the sum of the $^2J_{\mathrm{HP}}$- and the $^1J_{\mathrm{CP}}$-coupling. The zero-quantum coherence (b) reflects the difference. Comparison of the two splittings reproduces the coupling that was already determined in Figure A.69.

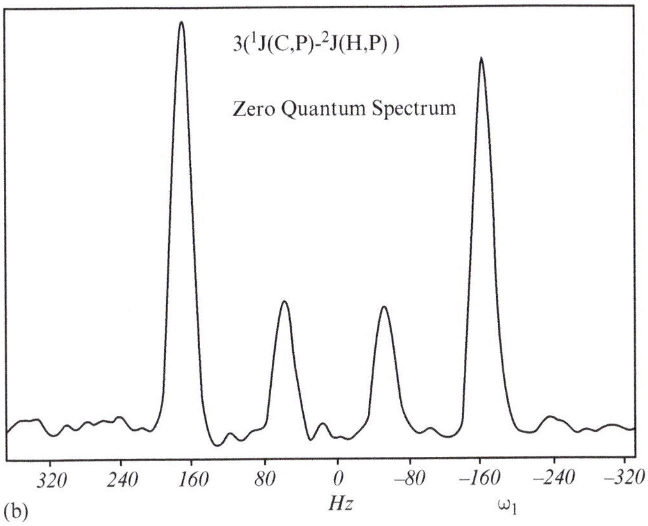


Figure A.72(b) (Cont'd)

## Acknowledgments

We thank the following co-workers for recording the spectra and for preparing some of the figures that are presented in this chapter as well as for careful proofreading of the manuscript: Dr. S. J. Glaser, Dr. M. Köck, J. Quant, A. Ratschinski, P. Schmidt, and Dr. R. Wechselberger. Special thanks go to Dr. M. Schwendiger, Professor Horst Kessler, Dr. W. Croasmun, and Dr. Steffen Glaser who made valuable comments on the manuscript. Collaboration with Professor Ernesto Carafoli and Dr. Joachim Krebs on the calmodulin/C20W in gratefully acknowledged. We are grateful for continuous support by Dr. W. Bermel and Dr. T. Keller. We would like to thank Dr. W. Croasmun for his patience during the preparation of this manuscript.

## References

1 Sørensen, O. W.; Eich, G. W.; Levitt, M. H.; Bodenhausen, G.; Ernst, R. R. *Prog. Nucl. Magn. Reson. Spectros.* 1983, *16*, 163.
2 Sørensen, O. W. *Prog. NucL Magn. Reson. Spectrosc.* 1989, *21*, 503.
3 Ernst, R. R.; Bodenhausen, G.; Wokaun, A. *Principles of Nuclear Magnetic Resonance in One and Two Dimensions*, Clarendon Press, Oxford, 1987.
4 Kessler, H.; Gehrke, M.; Griesinger, C. *Angew. Chem. Int. Ed. Engl.* 1988, *27*, 490.
5 Aue, W. P.; Bartholdi, E.; Ernst, R. R. *J. Chem. Phys.* 1976, *64*, 2229.
6 Bax, A.; Freeman, R. *J. Magn. Reson.* 1981, *44*, 542.
7 Morris, G. A.; Freeman, R. *J. Am. Chem. Soc.* 1979, *101*, 760.
8 Burum, D. P.; Ernst, R. R. *J. Magn. Reson.* 1980, *39*, 163.
9 Bolton, P. H. *J. Magn. Reson.* 1980, *41*, 287.
10 Kessler, H.; Griesinger, C.; Zimmermann, G. *Magn. Reson. Chem.* 1987, *25*, 579.
11 Müller, L. *J. Am. Chem. Soc.* 1979, *101*, 4481.
12 Bendall, M. R.; Pegg, D. T.; Doddrell, D. M. *J. Magn. Reson.* 1983, *52*, 81.
13 Bax, A.; Griffey, R. H.; Hawkins, B. L. *J. Am. Chem. Soc.* 1983, *105*, 7188.

14 Bax, A.; Griffey, R. H.; Hawkins, B. L. *J. Magn. Reson.* 1983, *55*, 301.
15 Aue, W. P.; Karhan, J.; Ernst, R. R. *J. Chem. Phys.* 1976, *64*, 4226.
16 Bax, A.; Summers, M. F. *J. Am. Chem. Soc.* 1986, *108*, 2093.
17 Bermel, W.; Wagner, K.; Griesinger, C. *J. Magn. Reson.* 1989, *83*, 223.
18 Bax, A.; Marion, D. *J. Magn. Reson.* 1988, *78*, 186.
19 Bystrov, V. F. *Prog. NMR Spectrosc.* 1976, *10*, 41.
20 Bauer, C. J.; Freeman, R.; Frenkiel, T.; Keeler, J.; Shaka, A. J. *J. Magn. Reson.* 1984, *58*, 442.
21 Emsley, L.; Bodenhausen, G. *J. Magn. Reson.* 1989, *82*, 211.
22 Emsley, L.; Bodenhausen, G. *Chem. Phys. Lett.* 1990, 165, 469.
23 Bermel, W.; Griesinger, C.; Kessler, H.; Wagner, K. *Magn. Reson. Chem.* 1987, *25*, 325.
24 Bax, A.; Ikura, M.; Kay, L. E.; Torchia, D. A.; Tschudin, R. *J. Magn. Reson.* 1990, *86*, 304.
25 Norwood, T. J.; Boyd, H.; Heritage, H. E.; Soffe, N.; Campbell, I. D. *J. Magn. Re-son.* 1990, *87*, 488.
26 Bodenhausen, G.; Ruben, D. J. *Chem. Phys. Lett.* 1980, *69*, 185.
27 Brühwiler, D.; Wagner, G. *J. Magn. Reson.* 1986, *69*, 546.
28 Davis, D. G. *J. Magn. Reson.* 1990, *90*, 589.
29 Cavanagh, J.; Palmer III, A. G.; Wright, P. E.; Rance, M. *J. Magn. Reson.* 1991, *91*, 429.
30 Palmer III, A. G.; Cavanagh, J.; Wright, P. E.; Rance, M. *J. Magn. Reson.* 1991, *93*, 151–170.
31 Sørensen, O. W. *J. Magn. Reson.* 1990, *89*, 210–216.
32 Farmer II, B. T. *J. Magn. Reson.* 1991, *93*, 635.
33 Bax, A.; Freeman, R. *J. Magn. Reson.* 1981, *44*, 542.
34 Kessler, H. ; Griesinger, C.; Zarbock, J.; Loosli, H. R. *J. Magn. Reson.* 1984, 57, 331.
35 Palmer III, A. G.; Fairbrother, W. J.; Cavanagh, J.; Wright, P. E.; Rance, M. *J. Biomol. NMR* 1992, *2*, 103.
36 Santoro, J.; King, G. C. *J. Magn. Reson.* 1992, *97*, 202.
37 van den Ven, F. J. M.; Philippens, M. E. P. *J. Magn. Reson.* 1992, *97*, 637.
38 Vuister, G. W.; Bax, A. *J. Magn. Reson.* 1992, *98*, 428–35.
39 Vuister, G. W.; Bax, A. *J. Biomol. NMR* 1992, *2*, 401.
40 Kay, L. E.; Bax, A. *J. Magn. Reson.* 1989, *84*, 598.
41 Zhang, X., Wang, C. *J. Magn. Reson.* 1991, *91*, 618.
42 Davis, D. G. *J. Magn. Reson.* 1991, *91*, 665.
43 Kessler, H.; Schmieder, P.; Kurz, M. *J. Magn. Reson.* 1989, *85*, 400.
44 Domke, T.; Leibfritz, D. *J. Magn. Reson.* 1990, *86*, 180.
45 Schmidt, J. M.; Rüterjans, H. *J. Am. Chem. Soc.* 1990, *112*, 1279.
46 Bulsing, J. M.; Brooks, W. M.; Field, J.; Doddrell, D. M. *J. Magn. Reson.* 1984, *56*, 167–173.
47 Kessler, H.; Schmieder, P.; Köck, M.; Reggelin, M. *J. Magn. Reson.* 1991, *91*, 375.
48 Kessler, H.; Schmieder, P.; Oschkinat, H. *J. Am. Chem. Soc.* 1990, *112*, 8598.
49 Schmieder, P.; Kessler, H.; Oschkinat, H. *Angew. Chem.* 1990, *102*, 588.
50 Stelton, D.; Leibfritz, D. *J. Magn. Reson.* 1992, *99*, 170.
51 Freeman, R. *A Handbook on Nuclear Magnetic Resonance,* John Wiley, New York, 1988, p. 171.
52 Marion, D.; Wüthrich, K. *Biochem. Biophys. Res. Comm.* 1983, *113*, 967.
53 Kay, L. E.; Marion, D.; Bax, A. *J. Magn. Reson.* 1989, *84*, 72.
54 Schmieder, P.; Zimmer, S.; Kessler, H. *Magn. Reson. Chem.* 1991, *29*, 375.
55 States, D. J.; Haberkorn, R. A.; Ruben, D. J. J. *Magn. Reson.* 1982, *48*, 286.
56 Bax, A.; Ikura, M.; Kay, L. E.; Zhu, G. *J. Magn. Reson.* 1991, *91*, 174.
57 Marion, D.; Ikura, M.; Tschudin, R.; Bax, A. *J. Magn. Reson.* 1989, *85*, 393.
58 Hore, P. J. In *Maximum Entropy in Action,* Buck, B.; Macaulay, V. A. (eds.), Clarendon Press, Oxford, 1991.
59 Gull, S. F.; Daniell, G. J. *Nature* 1978, *272*, 686.
60 Gull, S. F.; Skilling, J. *Proc. IEEE (F)* 1984, *131*, 646.

**61** Sibisi, S. *Nature* 1983, *301*, 134.
**62** Sibisi, S.; Skilling, J.; Brereton, R. G.; Laue, E. D.; Staunton, J. *Nature* 1984, *311*, 446.
**63** Laue, E. D.; Skilling, J.; Staunton, J.; Sibisi, S.; Brereton, R. G. *J. Magn. Reson.* 1985, *62*, 437.
**64** Makhoul, J. *Proc. IEEE* 1975, *63*, 561.
**65** Markhoul, J. In *Modern Spectral Analysis,* Childers, D. G. (ed.), IEEE Press, John Wiley, New York, 1978.
**66** Kumaresan, R.; Tufts, D. W. *IEEE Trans.* 1982, *ASSP-30*, 833.
**67** Barkhuijsen, H.; de Beer, R.; Bovée, W. M. M. J.; van Ormondt, D. *J. Magn. Reson.* 1985, *67*, 465.
**68** Gesmar, H.; Led, J. J. *J. Magn. Reson.* 1988, *76*, 183.
**69** Gesmar, H.; Led, J. J. *J. Magn. Reson.* 1988, *76*, 575; Led, J. J.; Gesmar, H. *Chem. Rev.* 1991, *91*, 1413–1426.
**70** Hoch, J. C. In *Methods in Enzymology,* Oppenheimer, N.; James, T. L. (eds.), Academic Press, San Diego, 1989, Vol. *176*, p. 216.
**71** Tang, J.; Norris, J. R. *J. Magn. Reson.* 1988, *78*, 23.
**72** Zeng, Y.; Tang, J.; Bush, C. A.; Norris, J. R. *J. Magn. Reson.* 1989, *83*, 473.
**73** Olejniczak, E.; Eaton, H. L. *J. Magn. Reson.* 1990, *87*, 628.
**74** Zhu, G.; Bax, A. *J. Magn. Reson.* 1992, *98*, 192.
**75** Zhu, G.; Bax, A. *J. Magn. Reson.* 1990, *90*, 405.
**76** Garbow, J. R.; Weitekamp, D. P.; Pines, A. *Chem. Phys. Lett.* 1982, *93*, 504.
**77** Bax, A.; Subramanian, S. *J. Magn. Reson.* 1986, *67*, 565.
**78** Bax, A. *J. Magn. Reson.* 1983, *53*, 517–520.
**79** Reynolds, W. F.; Hughes, D. W.; Perpick-Dumont, M.; Enriquez, R. G. *J. Magn. Reson.* 1985, *63*, 413.
**80** Maudsley, A. A.; Wokaun, A.; Ernst, R. R. *Chem. Phys. Lett.* 1978, *55*, 9.
**81** Bax, A.; De Jong, P. G.; Mehlkopf, A. F.; Smidt, J. *Chem. Phys. Lett.* 1980, *69*, 567.
**82** Hurd, R. E. *J. Magn. Reson.* 1990, *87*, 422.
**83** Hurd, R. E.; John, B. K. *J. Magn. Reson.* 1991, *91*, 648.
**84** Hurd, R. E.; John, B. K. *J. Magn. Reson.* 1991, *92*, 658.
**85** Hurd, R. E.; John, B. K.; Plant, H. D. *J. Magn. Reson.* 1991, *93*, 666.
**86** Davis, A. L.; Laue, E. D.; Keeler, J.; Moskau, D.; Lohman, J. *J. Magn. Reson.* 1991, *94*, 637.
**87** von Kienlin, M.; Moonen, C. T. W.; van der Toorn, A.; van Zijl, P. C. M. *J. Magn. Reson.* 1991, *93*, 423.
**88** Davis, A. L.; Keeler, J.; Laue, E. D.; Moskau, D. *J. Magn. Reson.* 1992, *98*, 207.
**89** Tolman, J. R.; Chung, J.; Prestegard, J. H. *J. Magn. Reson.* 1992, *98*, 462.
**90** Boyd, J.; Soffe, N.; John, B.; Plant, D.; Hurd, R. E. *J. Magn. Reson.* 1992, *98*, 660.
**91** Kay, L. E.; Keifer, P.; Saarinen, T. *J. Am. Chem. Soc.* 1992, *114*, 10663.
**92** Schleucher, J.; Sattler, M.; Griesinger, C. *Angew. Chem.* 1993, *105*, 1518; *Angew. Chem. Int. Ed. Engl.* 1993, *32*, 1489.
**93** Vuister, G. W.; Boelens, R.; Kaptein, R.; Hurd, R. E.; John, B. K.; van Zijl, P. C. M. *J. Am. Chem. Soc.* 1991, *113*, 9688.
**94** Vuister, G. W.; Boelens, R.; Kaptein, R.; Burgering, M.; van Zjjl, P. C. M. *J. Biomol. NMR* 1992, *2*, 301.
**95** Otting, G.; Wüthrich, K. *J. Magn. Reson.* 1988, *76*, 569.
**96** Messerle, B. A.; Wider, G.; Otting, G.; Weber, C.; Wüthrich, K. *J. Magn. Reson.* 1989, *85*, 608.
**97** Wörgötter, E.; Wagner, G.; Wüthrich, K. *J. Am. Chem. Soe.* 1987, *108*, 6162.
**98** Otting, G.; Senn, H.; Wagner, G.; Wüthrich, K. *J. Magn. Reson.* 1986, *70*, 500.
**99** Griffey, R. H.; Redfield, A. G. *Q. Rev. Biophys.* 1987, *19*, 51.
**100** Senn, H.; Otting, G.; Wüthrich, K. *J. Am. Chem. Soc.* 1987, *109*, 1090.
**101** Kogler, H.; Sørensen, O. W.; Bodenhausen, G.; Ernst, R. R. *J. Magn. Reson.* 1983, *55*, 157.
**102** Griesinger, C.; Sørensen, O. W.; Ernst, R. R., unpublished.
**103** Griesinger, C. Ph.D. Dissertation, Frankfurt, 1986.
**104** Gemmecker, G.; Olejniczak, E. T.; Fesik, S. W. *J. Magn. Reson.* 1992, *96*, 199.
**105** Otting, G.; Wüthrich, K. *J. Magn. Reson.* 1989, *85*, 586.
**106** Otting, G. ; Wüthrich, K. *Q. Rev. Biophys.* 1990, *23*, 39.

**107** Weber, C.; Wider, G.; von Freyberg, B.; Traber, R.; Braun, W.; Widmer, H.; Wüthrich, K. *Biochemistry* 1991, *30*, 6563.

**108** Fesik, S. W.;Gampe Jr., R. T.; Eaton, H. L.; Gemmecker, G.; Olejniczak, E. T.; Neri, P.; Holzman, T. F.; Egan, D. A.; Edalji, R.; Simmer, R.; Helfrich, R.; Hochlowski, J.; Jackson, M. *Biochemistry* 1991, *30*, 6574.

**109** Fesik, S. W.; Gampe Jr., R. T.; Holzman, T. F.; Egan, D. A.; Edalji, R.; Luly, J. R.; Simmer, R.; Helfrich, R.; Kishore, V.; Rich, D. H. *Science* 1990, *250*, 1406.

**110** Rosen, M. K.; Michnick, S. W.; Karplus, M.; Schreiber, S. L. *Biochemistry* 1991, *30*, 4774.

**111** Shaka, A. J.; Keeler, J.; Freeman, R. *J. Magn. Reson.* 1983, *53*, 313.

**112** Shaka, A. J.; Barker, P. B.; Freeman, R. *J. Magn. Reson.* 1985, *64*, 547.

**113** Shaka, A. J.; Lee, C. J.; Pines, A. *J. Magn. Reson.* 1988, *77*, 274.

**114** Rucker, S. P.; Shaka, A. J. *MoL Phys.* 1988, *68*, 509.

**115** Eggenberger, U.; Karimi Nejad, Y.; Thüring, H.; Rüterjans, J.; Griesinger, C. *J. BiomoL NMR* 1992, *2*, 583.

**116** Eggenberger, U.; Schmidt, P.; Sattler, M.; Glaser, S. J.; Griesinger, C. *J. Magn. Re- son.* 1992, *100*, 604.

**117** McCoy, M. A.; Müller, L. *J. Magn. Reson.* 1992, *99*, 18.

**118** McCoy, M. A.; Müller, L. *J. Magn. Reson.* 1993, *101*, 122.

**119** McCoy, M. A.; Müller, L. *J. Am. Chem. Soc.* 1992, *114*, 2108.

**120** McCoy, M. A.; Müller, L. *J. Magn. Reson.* 1992, *98*, 674.

**121** Zuiderweg, E. R. P.; Fesik, S. W. *J. Magn. Reson.* 1991, *93*, 653.

**122** Griesinger, C.; Sørensen, O. W.; Ernst, R. R. *J. Magn. Reson.* 1987, *73*, 574.

**123** Vuister, G. W.; Boelens, R.; Kaptein, R. *J. Magn. Reson.* 1988, *80*, 176.

**124** Griesinger, C.; Sørensen, O. W.; Ernst, R. R. *J. Magn. Reson.* 1989, *84*, 14.

**125** Ernst, R. R. *Angew. Chem.* 1992, *104*, 817–836.

**126** Pelczer, I.; Szalma, S. *Chem. Rev.* 1991, *91*, 1507.

**127** Oschkinat, H.; Griesinger, C.; Kraulis, P. J.; Sørensen, O. W.; Ernst, R. R.; Gronen- born, A. M.; Clore, G. M. *Nature* 1988, *332*, 374.

**128** Oschkinat, H.; Cieslar, C.; Gronenborn, A. M.; Clore, G. M. *J. Magn. Reson.* 1989, *81*, 212.

**129** Oschkinat, H.; Cieslar, C.; Holak, T. A.; Clore, G. M.; Gronenborn, A. M. *J. Magn. Reson.* 1989, *83*, 450.

**130** Vuister, G. W.; de Waard, P.; Boelens, R.; Vliegenhardt, J. F. G.; Kaptein, R. *J. Am. Chem. Soc.* 1989, *111*, 772.

**131** Oschkinat, H.; Cieslar, C.; Griesinger, C. *J. Magn. Reson.* 1990, *86*, 453.

**132** Padilla, A.; Vuister, G. W.; Boelens, R.; Kleywegt, G. J.; Cavé, A.; Parello, J.; Kaptein, R. *J. Am. Chem. Soc.* 1990, *112*, 540.

**133** Vuister, G. W.; Boelens, R.; Padilla, A.; Kleywegt, G. J.; Kaptein, R. *Biochemistry* 1990, *23*, 1829–1839.

**134** Mooren, M. M. W.; Hilbers, C. W.; van der Marel, G. A.; van Boom, J. H.; Wijmenga, S. S. *J. Magn. Reson.* 1991, *94*, 101.

**135** Simorre, J.-P.; Marion, D. *J. Magn. Reson.* 1991, *94*, 426.

**136** Wijmenga, S. S.; van Mierlo, C. P. *Eur. J. Biochem.* 1991, *195*, 807.

**137** van Mierlo, C. P. M.; van der Sanden, B. P. J.; van Woensel, P.; Muller, F.; Vervoort, J. *Eur. J. Biochem.* 1990, *194*, 199.

**138** Piotto, M. E.; Gorenstein, D. G. *J. Am. Chem. Soc.* 1991, 113, 1438–1440.

**139** Mueller, L.; Campbell-Burk, S.; Domaille, P. *J. Magn. Reson.* 1992, *96*, 408–415.

**140** Radhakrishnan, I.; Patel, D. J.; Gao, X. *Biochemistry* 1992, *31*, 2514–2523.

**141** Bonvin, A. M. J. J.; Boelens, R.; Kaptein, R. *J. Magn. Reson.* 1991, *95*, 626–631.

**142** Cieslar, C.; Holak, T. A.; Oschkinat, H. *J. Magn. Reson.* 1990, *89*, 184–190.

**143** Boelens, R.; Vuister, G. W.; Koning, T. M. G.; Kaptein, R. *J. Am. Chem. Soc.* 1989, *111*, 8526.

**144** Breg, J.; Boelens, R.; Vuister, G. W.; Kaptein, R. *J. Magn. Reson.* 1990, *87*, 646.

**145** Vuister, G. W.; Boelens, R.; Padilla, A.; Kaptein, R. *J. Biomol. NMR* 1991, *1*, 421.

**146** Holak, T. A.; Habazettl, J.; Oschkinat, H.; Otlewski, J. *J. Am. Chem. Soc.* 1991, *113*, 3196.

**147** Ross, A.; Freund, J.; Cieslar, C.; Oschkinat, H.; Schleicher, M.; Holak, T. A. *J. Magn. Reson.* 1991, *95*, 567.

**148** Holak, T. A.; Wiltscheck, R.; Ross, A. *J. Magn. Reson.* 1992, *97*, 632.

**149** Wüthrich, K. *NMR of Proteins and Nucleie Acids,* John Wiley, New York, 1986.
**150** Fesik, S. W.; Zuiderweg, E. R. P. *J. Magn. Reson.* 1988, *78*, 588.
**151** Fesik, S. W.; Luly, J. R.; Erickson, J. W.; Abad-Zapatero, C. *Biochemistry* 1988, *27*, 8297.
**152** Shon, K.; Opella, S. J. *J. Magn. Reson.* 1989, *82*, 193.
**153** Marion, D.; Kay, L. E.; Torchia, D. A.; Bax, A. *J. Am. Chem. Soc.* 1989, *111*, 1515.
**154** Zuiderweg, E. R. P.; Fesik, S. W. *Biochemistry* 1989, *28*, 2387.
**155** Marion, D.; Kay, L. E.; Sparks, S. W.; Torchia, D. A.; Bax, A. *J. Am. Chem. Soc.* *111*, 1571.
**156** Ikura, M.; Kay, L. E.; Tschudin, R.; Bax, A. *J. Magn. Reson.* 1990, *86*, 204.
**157** Zuiderweg, E. R. P.; McIntosh, L. P.; Dahlquist, F. W.; Fesik, S. W. *J. Magn. Reson.* 1990, *86*, 210.
**158** Marion, D.; Driscoll, P. C.; Kay, L. E.; Wingfield, P. T.; Bax, A.; Gronenborn, A. M.; Clore, G. M. *Biochemistry* 1989, *28*, 6150.
**159** Wijmenga, S. S.; Hallenga, K.; Hilbers, C. W. *J. Magn. Reson.* 1989, *84*, 634.
**160** Driscoll, P. C.; Clore, G. M.; Marion, D.; Wingfield, P. T.; Gronenborn, A. *Biochemistry* 1990, *29*, 3542–3556.
**161** Driscoll, P. C.; Gronenborn, A. M.; Wingfield, P. T.; Clore, G. M. *Biochemistry* 1990, *29*, 4668-682.
**162** Nagayama, K.; Yamazaki, T.; Yoshida, M.; Kanaya, S.; Nakamura, H. *J. Biochem.* 1990, *108*, 149.
**163** John, B. K.; Plant, D.; Heald, S. L.; Hurd, R. E. *J. Magn. Reson.* 1991, *94*, 664–669.
**164** Palmer III, A. G.; Cavanagh, J.; Byrd, R. A.; Rance, M. *J. Magn. Reson.* 1991, *96*, 416–424.
**165** Griesinger, C.; Otting, G.; Wüthrich, K.; Ernst, R. R. *J. Am. Chem. Soc.* 1988, *110*, 7870.
**166** Briand, J.; Ernst, R. R. *Chem. Phys. Lett.* 1991, *185*, 276.
**167** Cavanagh, J.; Rance, M. *J. Magn. Reson.* 1992, *96*, 670.
**168** Kerssebaum, U.; Markert, R.; Quant, J.; Glaser, S.; Bermel, W.; Griesinger, C. *J. Magn. Reson.* 1992, *99*, 184–191.
**169** Oh, B. H.; Westler, W. M.; Derba, P.; Markley, J. L. *Science* 1988, *240*, 908.
**170** Westler, W. M.; Srockman, B. J.; Markley, J. L.; Hosoya, Y.; Miyake, Y.; Kainosho, M. *J. Am. Chem. Soc.* 1988, *110*, 6256.
**171** Stockman, B. J.; Reilley, M. D.; Westler, W. M.; Ulrich, E. L.; Markley, J. L. *Biochemistry* 1989, *28*, 230.
**172** Ikura, M.; Kay, L. E.; Bax, A. *Biochemistry* 1990, *29*, 4659.
**173** Kay, L. E.; Ikura, M.; Tschudin, R.; Bax, A. *J. Magn. Reson.* 1990, *89*, 496.
**174** Bax, A.; Ikura, M. *J. Biomol. NMR* 1991, *1*, 99.
**175** Clubb, R. T.; Thanabal, V.; Wagner, G. *J. Magn. Reson.* 1992, *92*, 213–217.
**176** Farmer II, B. T.; Venters, R. A.; Spicer, L. D.; Wittekind, M. G.; Mueller, L. *J. Biomol. NMR* 1992, *2*, 195.
**177** Grzesiek, S.; Bax, A. *J. Magn. Reson.* 1992, *96*, 432–440.
**178** Madsen, J. C.; Sørensen, O. W. *J. Magn. Reson.* 1992, *100*, 431–436.
**179** Davis, A. L.; Boelens, R.; Kaptein, R. *J. Biomol. NMR* 1992, *2*, 395.
**180** Kay, L. E.; Ikura, M.; Zhu, G.; Bax, A. *J. Magn. Reson.* 1991, *91*, 422–428.
**181** Clubb, R. T.; Thanabal, V.; Wagner, G. *J. Biomol. NMR* 1992, *2*, 203–210.
**182** Clubb, R. T.; Wagner, G. *J. Biomol. NMR* 1992, *2*, 389–394.
**183** Kay, L. E.; Wittekind, M.; McCoy, M. A.; Friedrichs, M. S.; Mueller, L. *J. Magn. Reson.* 1992, *98*, 443–450.
**184** Olejniczak, E. T.; Xu, R. X.; Petros, A. M.; Fesik, S. W. *J. Magn. Reson.* 1992, *100*, 444–450.
**185** Seip, S.; Balbach, J.; Kessler, H. *J. Magn. Reson.* 1992, *100*, 406–410.
**186** Boucher, W.; Laue, E. D.; Campbell-Burk, S.; Domaille, P. J. *J. Am. Chem. Soc.* 1992, *114*, 2262.
**187** Clore, G. M.; Gronenborn, A. M. *Prog. Nucl. Magn. Reson. Spectrosc.* 1991, *23*, 43.
**188** Clore, G. M.; Gronenborn, A. M. *Science* 1991, *246*, 411.
**189** Kay, L. E.; Bax, A.; Gronenborn, A. M. *Science* 1991, *246*, 411.
**190** Ikura, M.; Kay, L. E.; Bax, A. *J. Biomol. NMR* 1991, *1*, 299.
**191** Kay, L. E.; Ikura, M.; Bax, A. *J. Am. Chem. Soc.* 1990, *112*, 888–889.
**192** Bax, A.; Clore, G. M.; Driscoll, P. C.; Gronenborn, A. G.; Ikura, M.; Kay, L. E. *J. Magn. Reson.* 1990, *87*, 620–627.

**193** Fesik, S. W.; Eaton, H. L.; Olejniczak, E. T.; Zuiderweg, E. R. P.; McIntosh, L. P.; Dahlquist, F. W. *J. Am. Chem. Soc.* 1990, *112*, 886–888.

**194** Bax, A.; Clore, G. M.; Gronenborn, A. M. *J. Magn. Reson.* 1990, *88*, 425–431.

**195** Eaton, H. L.; Fesik, S. W.; Glaser, S. J.; Drobny, G. P. *J. Magn. Reson.* 1990, *90*, 452.

**196** Grzesiek, S.; Bax, A. *J. Am. Chem. Soc.* 1992, *114*, 6291–6293.

**197** Grzesiek, S.; Bax, A. *J. Magn. Reson.* 1992, *99*, 201–207.

**198** Grzesiek, S.; Anglister, J.; Bax, A. *J. Magn. Reson. Series B* 1993, *101*, 114–119.

**199** Logan, T. M.; Olejniczak, E. T.; Xu, R. X.; Fesik, S. W. *J. Biomol. NMR* 1993, *3*, 225–231.

**200** Clowes, R. T.; Boucher, W.; Hardman, C. H.; Domaille, P. J.; Laue, E. D. *J. Biomol. NMR* 1993, *3*, 349–354.

**201** Lyons, B. A.; Montelione, G. T. *J. Magn. Reson. Series B* 1993, *101*, 206–209.

**202** Richardson, J. M.; Clowes, R. T.; Boucher, W.; Domaille, P. J.; Hardman, C. H.; Keeler, J.; Laue, E. D. *J. Magn. Reson. B* 1993, *101*, 223–227.

**203** Grzesiek, S.; Bax, A. *J. Biomol. NMR* 1993, *3*, 185–204.

**204** Montelione, G. T.; Lyons, B. A.; Emerson, S. D.; Tashiro, M. *J. Am. Chem. Soc.* 1992, *114*, 10974–10975.

**205** Kay, L. E.; Ikura, M.; Grey, A. A.; Muhandiram, D. R. *J. Magn. Reson.* 1992, *99*, 652–659.

**206** Kay, L. E. *J. Magn. Reson. Series B* 1993, *101*, 110–113.

**207** Ikura, M.; Bax, A.; Clore, G. M.; Gronenborn, A. M. *J. Am. Chem. Soc.* 1990, *112*, 9020–9022.

**208** Clore, G. M.; Kay, L. E.; Bax, A.; Gronenborn, A. M. *Bioehemistry* 1991, *30*, 12–18.

**209** Zuiderweg, E. R. P.; Petros, A. M.; Fesik, S. W.; Olejniczak, E. T. *J. Am. Chem. Soc.* 1991, *113*, 370–372.

**210** Kalinowski, H.-O.; Berger, S.; Braun, S. *$^{13}$C NMR Spectroseopy,* G. Thieme Verlag, Stuttgart/, New York, 1984.

**211** Egli, H.; von Phiipsborn, W. *Helv. Chim. Acta* 1981, *64*, 976.

**212** Vuister, G. W.; Delaglio, F.; Bax, A. *J. Biomol. NMR* 1993, *3*, 67.

**213** Mierke, D. F.; Galic-Grdadolnik, S.; Kessler, H. *J. Am. Chem. Soc.* 1992, *114*, 8283.

**214** Keeler, J.; Neuhaus, D.; Titman, J. J. *Chem. Phys. Lett.* 1988, *146*, 545.

**215** Keeler, J.; Neuhaus, D.; Titman, J. J. *J. Magn. Reson.* 1989, *85*, 111.

**216** Keeler, J.; Titman, J. J. *J. Magn. Reson.* 1990, *89*, 640.

**217** Richardson, J. M.; Titman, J. J.; Keeler, J. *J. Magn. Reson.* 1991, *93*, 533–553.

**218** Schwalbe, H.; Samstag, W.; Engels, J. W.; Bermel, W.; Griesinger, C. *J. Biomol. NMR* 1993, *3*, 479–486.

**219** Blake, P. R.; Summers, M. F.; Adams, M. W. W.; Park, J.-B.; Zhou, Z. H.; Bax, A. *J. Biomol. NMR* 1992, *2*, 527–533.

**220** Bax, A.; Max, D.;Zax, D. *J. Am. Chem. Soc.* 1992, *114*, 6924–6925.

**221** Vuister, G. W.; Yamazaki, T.; Torchia, D. A.; Bax, A. *J. Biomol. NMR* 1993, *3*, 297.

**222** Vuister, G. W.; Wang, A. C.; Bax, A. *J. Am. Chem. Soc.* 1993, *115*, 5334.

**223** Grzesiek, S.; Vuister, G. W.; Bax, A. *J. Biomol. NMR* 1993, *3*, 487.

**224** Vuister, G. W.; Bax, A. *J. Am. Chem. Soc.* 1993, *115*, 7772–7777.

**225** Zhu, G.; Bax, A. *J. Magn. Reson.* 1993, *A104*, 353–357.

**226** Vuister, G. W.; Bax, A. *J. Magn. Reson.* 1993, *B102*, 228–231.

**227** Sørensen, O. W. Dissertation ETH Nr. 7658, Zurich, Switzerland, 1984.

**228** Griesinger, C.; Sørensen, O. W.; Ernst, R. R. *J. Am. Chem. Soc.* 1985, *107*, 6394.

**229** Griesinger, C.; Sørensen, O. W.; Ernst, R. R. *J. Chem. Phys.* 1986, *85*, 6837.

**230** Griesinger, C.; Sørensen, O. W.; Ernst, R. R. *J. Magn. Reson.* 1987, *75*, 474.

**231** Montelione, G. T.; Winkler, M. E.; Rauenbühler, P.; Wagner, G. *J. Magn. Reson.* 1989, *82*, 198.

**232** Kurz, M.; Schmieder, P.; Kessler, H. *Angew. Chem.* 1991, *103*, 1341; *Angew. Chem.* (Int. Ed. Engl.) 1991, *30*, 1329.

**233** Schmieder, P.; Kurz, M.; Kessler, H. *J. BiomoI. NMR* 1991, *1*, 403.

**234** Bax, A.; Freeman, R. *J. Magn. Reson.* 1981, *45*, 177.

**235** Kessler, H.; Anders, U.; Gemmecker, G. *J. Magn. Reson.* 1988, *78*, 382.

**236** Zuiderweg, E. R. P.; Fesik, S. W. *J. Magn. Reson.* 1991, *93*, 653.

**237** Sattler, M.; Schwalbe, H.; Griesinger, C. *J. Am. Chem. Soc.* 1992, *114*, 1126.

238 Schleucher, J.; Schwörer, B.; Zirngibl, C.; Koch, U.; Weber, W.; Egert, A. E.; Thauer, R. K.; Griesinger, C. *FEBS LETT* 1992, *314*, 440.

239 Wollborn, U.; Leibfritz, D. *J. Magn. Reson.* 1992, *98*, 142.

240 Schwalbe, H.; Rexroth, A.; Eggenberger, U.; Geppert, T.; Griesinger, C. *J. Am. Chem. Soc.* 1993, *115*, 7878–7879.

241 Müller, L.J. *Magn. Reson.* 1987, *72*, 191.

242 Bax, A.; Marion, D. *J. Magn. Reson.* 1988, *80*, 528.

243 Sørensen, 0. W. *J. Magn. Reson.* 1990, *90*, 433.

244 Griesinger, C.; Eggenberger, U. *J. Magn. Reson.* 1992, *97*, 426.

245 Gemmecker, G.; Fesik, S. W. *J. Magn. Reson.* 1991, *95*, 208.

246 Emerson, D.; Montelione, G. *J. Am. Chem. Soe.* 1992, *114*, 354.

247 Emerson, S. D.; Montelione, G. T. *J. Magn. Reson.* 1992, *99*, 413.

248 Schmidt, P.; Schwalbe, H.; Glaser, S. J.; Griesinger, C. *J. Magn. Reson.* 1993, *B101*, 328–333.

249 Seip, S.; Balbach, J.; Kessler, H. *Angew. Chem.* 1992, *104*, 1656–1658.

250 Weisemann, R.; Rüterjans, H.; Schwalbe, H.; Schleucher, J.; Bermel, W. ; Griesinger, C. *J. Biomol. NMR* 1993 (in press).

251 Montelione, G.; Wagner, G. *J. Am. Chem. Soc.* 1989, *111*, 5474.

252 Montelione, G.; Wagner, G. *J. Magn. Reson.* 1990, *87*, 183.

253 Wagner, G.; Schmieder, P.; Thanabal, V. *J. Magn. Reson.* 1991, *93*, 436.

254 Madsen, J. C.; Sørensen, O. W.; Sørensen, P.; Poulsen, F. M. *J. Biomol. NMR* 1993, *3*, 239–244.

255 Weisemann, R.; Rüterjans, H.; Schwalbe, H.; Schleucher, J.; Bermel, W.; Griesinger, C. *J. Biomol. NMR* 1994, *4*, 231.

256 Willker, W.; Leibfritz, D. *J. Magn. Reson.* 1992, *99*, 421.

257 Schleucher, J.; Schwendinger, M.; Sattler, M.; Schmidt, P.; Schedletzky, O., Glaser, S. J.; Sørensen, O. W.; Griesinger, C. *J. Biomo. NMR,* 1994, *4*, 301.

# Index

Page locators in **bold** indicate tables. Page locators in *italics* indicate figures. This index uses letter-by-letter alphabetization.

## *a*



*Two-Dimensional (2D) NMR Methods*, First Edition. Edited by K. Ivanov, P.K. Madhu and G. Rajalakshmi.

## d

## e

## *i*

## q



## r

## S

Printed in the USA/Agawam, MA
September 16, 2026

910661.196